# 7th ASEAN-FEN International Fisheries Symposium (ASEAN-FEN IFS 2017)



## Projecting ASEA FEN Plus for Sustainable Aquaculture, Fisheries and Aquatic Ecosystem

IOP Conference Series: Earth and Environmental Science
Volume 137

Batu, Indonesia
7 - 9 November 2017

**Editors:**

**Alim Isnansetyo**
**Indun Dewi Puspita**
**Noranizan Mohd. Adzahan**
**Suadi**
**Dini Wahyu Kartika Sari**
**Anes Dwi Jayanti**

ISBN: 978-1-5108-6152-7
ISSN: 1755-1307

**Printed from e-media with permission by:**

Curran Associates, Inc.
57 Morehouse Lane
Red Hook, NY 12571

proceedings.com

**Some format issues inherent in the e-media version may also appear in this print version.**



Printed by Curran Associates, Inc. (2018)

For permission requests, please contact the Institute of Physics
at the address below.

Institute of Physics
Dirac House, Temple Back
Bristol BS1 6BE UK

Phone: 44 1 17 929 7481
Fax: 44 1 17 920 0979

techtracking@iop.org

**Additional copies of this publication are available from:**

Curran Associates, Inc.
57 Morehouse Lane
Red Hook, NY 12571 USA
Phone: 845-758-0400
Fax: 845-758-2633
Email: curran@proceedings.com
Web: www.proceedings.com

# Table of contents

## Volume 137

**Asean-Fen International Fisheries Symposium - 2017**

**7–9 November 2017, Batu City, East Java, Indonesia**

**Accepted papers received: 14 March 2018**
**Published online: 13 April 2018**

### Preface



### Papers

**AQUACULTURE**

**Sustainable fisheries and management**

**AQUATIC RESOURCES, BIODIVERSITY AND ENVIRONMENT**

**FISHERIES ECONOMICS**

IOP Conf. Series: Earth and Environmental Science **137** (2018) 011001 doi:10.1088/1755-1315/137/1/011001

IOP Conf. Series: Earth and Environmental Science **137** (2018) 011001 doi:10.1088/1755-1315/137/1/011001

# Preface

The 7th ASEAN-FEN International Fisheries Symposium was successfully held in Batu, East Java, Indonesia 7 – 9 November 2017. The conference was hosted by Faculty of Fisheries and Marine Science, Brawijaya University Malang Indonesia. The theme of this symposium was "Projecting ASEAN FEN Plus for Supporting Sustainable Aquaculture, Fisheries and Aquatic Ecosystems", with focus on the advanced innovation to address to the newly emerged issues in aquaculture, fisheries and aquatic ecosystems for the synergies between socioeconomic development and protecting natural resources and the environment.

The conference was attended by over 500 researchers from different countries, who presented and discussed the results of their work within the framework of five main areas: 1. Aquaculture, 2. Sustainable fisheries and management, 3. Seafood processing and biotechnology, 4. Aquatic resources, biodiversity and environment, and 5. Fisheries Economic.

ASEAN-FEN IFS 2017 Committee received more than 120 manuscripts from participated universities and research institutes, and 106 manuscripts were accepted for publication. All of the papers were subjected to peer-review by qualified experts in the field selected by the conference committee. The papers selected depended on their quality and their relevancy to the conference.

We would like to thank all the authors who have contributed to this volume and also to the board members, organizing committee, reviewers, speakers, chairpersons, sponsors and all the conference participants for their support to the ASEAN-FEN IFS 2017.

Warm Regards,

**Dr.Sc. Asep Awaludin Prihanto, S.Pi., MP.**
Chairperson of ASEAN FEN, IFS 2017
Faculty of Fisheries and Marine Science,
Brawijaya University, Malang, Indonesia

IOP Conf. Series: Earth and Environmental Science **137** (2018) 011001 doi:10.1088/1755-1315/137/1/011001

# ORGANIZING COMMITTEE

**Chairperson**
Dr. Sc. Asep Awaludin Prihanto., S.Pi., MP, Faculty of Fisheries and Marine Science, Brawijaya University, Indonesia

**International Scientific Board Committee**

**Chair person**
Dr. Sasmito Djati, MS.Vice Rector, Brawijaya University, Indonesia
Assoc. Prof. Dr. Sukree Hajisamae, Dean, PSU, Thailand; Chairman of ASEAN-FEN

**Members**
Dr. Happy Nursyam, Universitas Brawijaya, Indonesia
Prof .Dr.Mazlan bin Ghaffar, Universiti Malaysia Terengganu, Malaysia
Prof. Dr. Siti Azizah Mohd Nor, Universiti Sains Malaysia
Assoc. Prof. Dr. Truong Quoc Phu, Can Tho University, Vietnam
Assoc. Prof. Dr. Tran Ngoc Hai, Can Tho University, Vietnam
Assoc. Prof. Dr. Shettapong Mekrumpun, Kasetsart University, Thailand
Assist.Prof.Dr. Prasert Tongnunui, Rajamangala University of Science and Technology Srivijaya, Thailand
Dr. Mirni Lamid, DVM, M.Sc.Universitas Air Langga, Indonesia
Assoc. Prof. Dr. Nguyen Nhu Tri, Nong Lam University, Vietnam
Dr. Pham Quoc Hung, NhaTrang University, Vietnam
Mr. Seng Samphal, Royal University of Agriculture, Cambodia
Mr.Lam Khannarith, Prek Leap National College of Agriculture, Cambodia
Prof. Myin zu Min, University of Yangon, Myanmar
Assist. Prof. Dr. Ekarut Srisuk, Burapha University, Thailand
Prof. Crispino A. Saclauso, University of Philippines Visayas, Philippines
Prof. Dr. Fatimah Md. Yusoff, Universiti Putra Malaysia, Malaysia

**Secretary**
Rahmi Nurdiani, S.Pi, M.App.Sc, Ph.D

**Treasury**
Citra Satrya Utama Dewi S.Pi., M.Si
Hefti Salis Yufidasari, S.Pi, MP

IOP Conf. Series: Earth and Environmental Science **137** (2018) 011001 doi:10.1088/1755-1315/137/1/011001

**Secretariat**

**Coordinator**

Candra Adi Intyas,S.Pi, MP

Supriyadi, S.Pi

**Web and IT Division**

Dhira Khurniawan S., S.Kel., M.Sc.

Bayu Kusuma, S.Pi, MSc

AnggaWiraperdana, SPi, MP

**Program Division**

Oktiyas Muzaky L, S.T, M.Sc.

Abdul Aziz Jaziri, S.Pi, M.Sc

IOP Conf. Series: Earth and Environmental Science **137** (2018) 011001 doi:10.1088/1755-1315/137/1/011001

# Conference Photographs

IOP Conf. Series: Earth and Environmental Science **137** (2018) 011001 doi:10.1088/1755-1315/137/1/011001

IOP Conf. Series: Earth and Environmental Science 137 (2018) 011001 doi:10.1088/1755-1315/137/1/011001

UB
Ecosystem
HOST
2018

IOP Conf. Series: Earth and Environmental Science **137** (2018) 011002 doi:10.1088/1755-1315/137/1/011002

# Peer review statement

All papers published in this volume of *IOP Conference Series: Earth and Environmental Science* have been peer reviewed through processes administered by the proceedings Editors. Reviews were conducted by expert referees to the professional and scientific standards expected of a proceedings journal published by IOP Publishing.

IOP Conf. Series: Earth and Environmental Science **137** (2018) 012001 doi:10.1088/1755-1315/137/1/012001

# Detection and analysis of hemolysin genes in *Aeromonas hydrophila* isolated from Gouramy (*Osphronemus gouramy*) by polymerase chain reaction (PCR)

**Rozi, K Rahayu, D N Daruti**

Department of Fish Health Management and Aquaculture, Faculty of Fisheries and Marine, Airlangga University, Surabaya 60115, Indonesia

E-mail: rozi@fpk.unair.ac.id

**Abstract.** The goal of this study was to detect of *Aeromonas hydrophila* carrying the *hly*A gene in guramy by PCR assay. A total of 5 *A. hydrophila* strains were isolated from gouramy with different location and furthermore genotypic of all *A. hydrophila* strains havedetected by PCR assay for 16S rRNA gene. The primers used in the PCR targeted a 592-bp fragment of the *hly*A gene coding for the hemolysin gene. Particularly *hly*A genes are responsible for haemolysin toxins production in this genus. After gel electrophoresis, the amplicons from representative strains of the *A. hydrophila* were purified using extraction kit and were subjected to the DNA sequencing analysis. The results showed that: (i) the 592bp amplicon of the *hly*A gene was detected in 5/6 of the *A. hydrophila*; (ii) the nucleotide blast results of hemolysin gene sequences of the strains of *A. hydrophila* revealed a high homology of 90-97 % with published sequences, and;(iii) the protein blast showed 95-98 % homology when compared to the published sequences. The PCR clearly identified the haemolysin-producing strains of *A. hydrophila* by detection in *hly*A genes and may have application as a rapid species-specific virulence test.

## 1. Introduction

*Aeromonas hydrophila* is a bacterium with the following characteristics: gram-negative, short shaped, aerobic and facultatively anaerobic, non-sporing, motile, has one flagel, and lives in the temperature range of 25-30°C [1]. Chronic infection from these bacteria can cause ulceration, inflammation, and dermal lesions with focal haemorrhages Cipriano [2],where the liver and kidneys are the common target organs during acute septicemia [3]. An infection of this bacteria can also cause symptoms of hemorrhagic diseasesepticemia that has a characteristic wound on the surface of the body, gills, ulcers, abscesses, exopthalmia and belly bloating Austin [4], as well as gastroenteritis, diarrhea and extra intestinal symptoms in humans [5]. This bacterium is very influential in freshwater fish farming and often a disease outbreak can cause levels of high mortality (80 – 100 %) in 1 - 2 weeks. The degree of virulence of *A. hydrophila* can cause fish death depending on the poison generated. Genes Aero and *hly*A play an important role in producing aerolysin and hemolysin toxins in the genus Aeromonas [6].The rapid detection of the hemolysin gene from *A. hydrophila* and their characterization is considered very important so appropriate preventive measures can be undertaken to reduce mortality and losses in fish farming. Given the economic importance it has on the health of aquatic animals and

IOP Conf. Series: Earth and Environmental Science **137** (2018) 012001 doi:10.1088/1755-1315/137/1/012001

humans, it is essential that the rapid detection of pathogens and the characterization of these isolates are carried out using molecular techniques. With this in mind, the main purpose of this study is to genetically identify the *A. hydrophila* bacteria and determine the presence of the hemolysin gene in the isolate using PCR assay.

## 2. Methodology

### *2.1. Bacteria isolates*

A collection of five isolates of *A. hydrophila* was taken from the Microbiology Laboratory of the Faculty of Fisheries and Marine at Airlangga University. The identification of bacteria isolates was also carried out by the Faculty of Fisheries and Marine at Airlangga University through PCR amplification with 16S rRNA and was detected for *A. hydrophila* (Rozi, personal communication), while the *A. hydrophila* ATCC® 35654 ™ strain was used as a control strain

### *2.2. Extraction of A. hydrophila isolat DNA at the Faculty of Fisheries and Marine, Airlangga University (extraction using the TIANamp-China kit)*

The *A. hydrophila* isolate was grown on Triple Soya Broth medium (TSB) for 24 hours in about 10 ml bacterial culture suspension in a density of $10^8$ cells / ml. The culture was poured into the microtube and centrifuged at 3000 rpm for three minutes, the supernatant was discarded as much as possible until the pellet remained, then added with an ATL buffer at 200 µl, 20 µl proteinase K, and then incubated at $56^0$C in a water bath for 10 min.Supernatant was vortexed for 15 seconds and then added 200 µl buffer, then divotex again, furthermore 200 µl of absolute ethanolwas added, and vortexed once more. The solution was then placed into the spin column and centrifuged at 8000 rpm for 1 minute. Next, the filter was moved to a new spin column and added with 500 µl of the AW1 buffer and then centrifuged at 8000 rpm for 1 minute. Afterwards, the filter was transferred to a new spin column with the addition of 500 µl of the AW2 buffer, centrifuged at 14.000 rpm for 3 minutes, transferred to a new micro tube and then added with 200 µl of the AE buffer, incubated for 1 minute at room temperature, and centrifuged at 8000 rpm for 1 minute. The filter results are stored at a temperature of - 70°C, while the DNA template supernatant containing bacteria is used directly on specific PCR to detect hemolysin (*hly*) genes.

### *2.3. Amplification of hlyA gene (GoTaq® Green Master Mix, Promega)*

A solution of 6.5 µl dH2O and 12.5 µl Green mix is placed in a microtube, then mixed until homogeneous. The primer used is the oligonucleotide of each 2 µl forward primer (5'-GGCCGGTGGCCCGAAGATACGGG-3') and primary reverse (5'-GGCGGCGCCGGACGAGACGGGG-3') with the final concentration of 0.1 µM and 2 µl DNA templates. The PCR program for DNA amplification is as follows: initial denaturation at 95°C for 5 min followed by 30 cycles at 95°C for 2 min, 55°C for 1 min and 72°C for 1 min. 7 min final extension at 72°C. The PCR results from various treatments were electrophoresed on agarose 1.5 % agar (0.75 g dissolved into TAE buffer 50 ml) and was boiled in the microwave. After warming the agarose and adding 2 µl of EtBr solution with a concentration of 10 mg/ mL, the agarose was then poured into an electrophoresis mold. After the agarose hardened, 0.8 µl of PCR product mixed with 0.2 µl loading dye was inserted to wells 2, 3, 4 and so on. The first well was filled with 0.5 µl of molecular marker. Electrophoresis is run with 100 voltages for 30 minutes, after which the gel is lifted and observed above the UV transilluminator, then documented.

### *2.4. Purification and bioinformatics analysis*

Purification of the PCR sample product was done with microclean. 100 µl of PCR product and 100 µl of microclean (1: 1) was mixed to the homogeneous sample product using a pipette. Incubation was done for 5 minutes at room temperature, and centrifuged at 13.000 rpm for 5 minutes. The supernatant solution is removed, vortexed until the remaining supernatant is completely clean, then the pellet is resuspended with TE of 30 µl andstored at -20°C. The purification and sequencing of *A. hydrophila*

DNA was performed with the automatic sequencer ABI PRISM 377 DNA (Applied Biosystem, USA). Then it was run through the BLAST program (NCBI: http://www.ncbi.nlm.nih.gov/BLASTi/. The philogenetic tree was constructed and similarity analysis was conducted with the program version of MEGA 7.

## 3. Result and Discussion

### *3.1. Detection of hemolysin gene fragments*

The DNA PCR results of the six isolates of *A. hydrophila* on pure cultures using the primer olygonucleotide positively detected hemolysin genes. Overall detection of *hly*A gene fragments showed that 5/6 of the isolates (83.3 %) contained the hemolysin gene in the *A. hydropila* and showed a molecular weight of 592 bp. (figure 1)

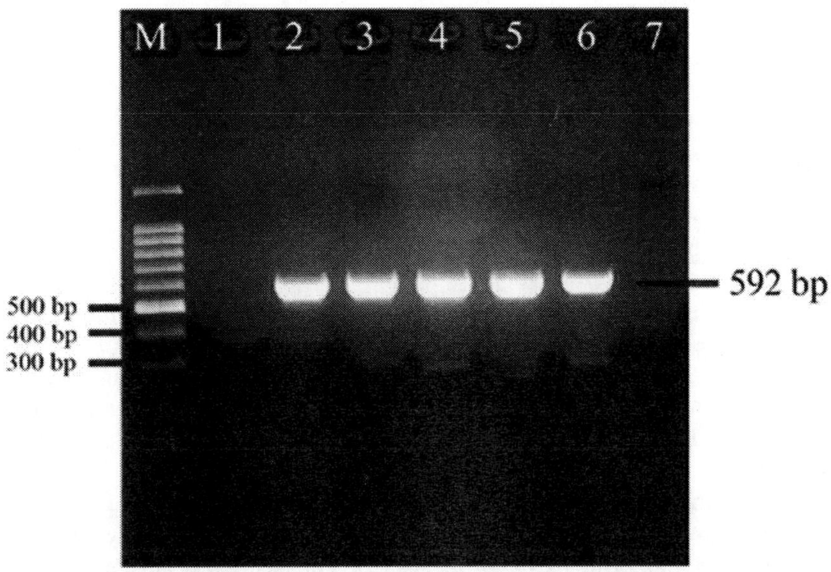


**Figure 1.** Detection and identification of *A. hydrophila* hemolysin gene by amplification of fragments in the PCR assay. M: indicates the 100 bp ladder as a marker, lane 1: the control negative, lane 2: the control ATCC® 35654™, lane 3: isolate AHI1, lane 4: isolate AHI2, lane 5: isolate AHI3, lane 6: isolate AHI4, lane 7: isolate AHI5, respectively.

### *3.2. Homology of hemolysin gene*

The hemolysin gene sequencing results were compared with the GenBank database from the National Center for Biotechnology Information (NCBI) using the BLAST program (http://www.ncbi.nlm.nih.gov/). The homology of the nucleotide sequence of the hemolysin gene sequence of *A. hydrophila* showed a high homology between 90 % - 95 %. The prominent protein concentrations of *A. hydrophila*, showed 95 % - 98 % homology (data not shown) when compared with sequences published in the GenBank database of the National Center for Biotechnology Information (NCBI).

### *3.3. Phylogenetic tree gene hemolysin*

The phylogenetic tree constructed using the neighboring joining method showed five isolates as *A. hydrophila* strains. However, there are several comparative isolates showing different genotypes. Phylogenetically, the relationships of the five isolates tested were divided into three groups. The firstgroup contained was adjacent to the *A. hydrophila*hemolysin gene, while the second group consisting of five isolates have a relationship to one isolate controlled with the *A. hydrophila* ATCC 35654 strain. The third group of the five isolates tested was clustered separately (figure 2)

**Table 1**. Homology levels of hemolysin sequence of *A. Hydrophila*.

| Isolates | Identity closest to the GenBank reference | Accession Number | % Identities |
|---|---|---|---|
| AH1 JA_*hly*A | *A. hydrophila* strain JNC1007 haemolysin (*hly*A) gene, partial cds | JQ003206.1 | 95 % |
| AH2 SA_*hly*A | *A. hydrophila* strain DWCG102 haemolysin (*hly*A) gene, partial cds | JF738031.1 | 95 % |
| AH3 SU_*hly*A | *A. hydrophila* strain SDDLAH/09 hemolysin gene, partial cds | FJ972621.1 | 96 % |
| AHI4 SR_*hly*A | *A. hydrophila* gene for hemolysin, complete cds, strain:AH085 | AB206039.1 | 97 % |
| ATCC _*hly*A | *A. hydrophila* strain CHS-3 haemolysin gene, partial cds | KF483996.1 | 90 % |

Based on the literature, an *A. hydrophila* infection is acute when marked by the very quick death of the fish with the appearance of signs of clinical infection such as: exophthalmia, red patches on the skin, accumulation of fluid in the abdominal pouch, flatulence, bleeding gills and injury to the dermis, and separated scales [20]. According to Kamiso [7] *A. hydrophila* can cause the death of 90-100 % in a short time. Research by Daskalov [8] also shows that *A. hydrophila* is widely distributed through food, drinking water and the environment, and is a pathogen that causes hemorrhagic diseases, zoonotic diseases, and food-borne infections in freshwater fish. Several factors of virulence and pathogenicity are involved with the presence of different exotoxins e.g., haemolysin, enterotoxins and cytotoxins and exoenzymes e.g., proteases and lipases [9]. The two major groups of haemolysins such as extracellular haemolysin and aerolysin [10] are produced by *A. hydrophila*.

Overall, PCR amplification of *hly*A gene fragments showed that 5/6 of the *A. hydrophila* isolates (83.3%) showed a haemolysin gene with a molecular weight of 592 bp. (figure 1). Chacón et al. stated in 2003 [29] a similar observation was in earlier studies on the *A. jandaei* type strain where it presented the aerolysin/hemolysin gene. Pathogenicity of enteric bacteria is produced by cytotoxin [30]. Based on the analysis of phylogenetic trees of the hemolysin gene (figure 2), it shows that the isolates AH1 JA_*hly*A, AH2 SA_*hly*A, AH3 SU_*hly*A, AHI4 SR_*hly*A and ATCC _*hly*ABK1 were the same isolates because they were in one cluster. Both isolates were closely related to refractory strains *A. hydrophila* strains JNC1007 *A. hydrophila* strains DWCG102 haemolysin (*hly*A) gene, partial cds, *A. hydrophila* strains SDDLAH / 09 hemolysin gene, partial cds, *A. hydrophila* gene for hemolysin, complete cds, strains: AH085, and *A. hydrophila* strains CHS-3 haemolysin gene, partial cdshaemolysin (*hly*A) gene, partial cds with a 90-97 % equivalent percentage. In phylogenetic trees, all AH1 JA_*hly*A, AH2 SA_*hly*A, AH3 SU_*hly*A, AHI4 SR_*hly*A and ATCC _*hly*ABK1 isolates were the same isolates. The five isolates are in a branch with the closest species of 26 *A. hydrophila* strains with 97 percent equality percentage with the strain of reference strain of *Vibrio parahaemolyticus, Sterptococcussuis* and *E. coli.*

The virulence rate of *A. hydrophila* which can cause fish mortality depends on the toxicity produced. The Aero and *hly*A genes are responsible for producing aerolysin and hemolysin toxins in the Aeromonas genus [6]. Aerolisin is an extracellular protein produced by some soluble *A.hydrophila*strains, a hydrophilic protein having hemolytic and cytolytic properties. Aerolysin binds

IOP Conf. Series: Earth and Environmental Science **137** (2018) 012001 doi:10.1088/1755-1315/137/1/012001

to specific glycoprotein receptors on the surface of eukaryotic cells before entering into the fat layer and forming a hole. The aerolysin poison that forms a hole passes into the bacterial membrane as a peptide-containing preprotoxine. The toxin can attack epithelia cells and cause gastroenteristis [6]. Haemolysin produced by extracellular products (ECPs) of bacteria become a bacterial defense factor against the defenses of the host blood because it is able to lysis blood cells. The bacteria can survive the blood flow to spread throughout the body of the host cells as well as towards the target organs. Bacteria also have the factor of pathogenicity in the form of enzymes found in ECPs, including caseinase, gelatinase, amylase, lipase, phospolipase, chitinase, collagenase, elastase, hyaluronidase, and proteinases that are able to decipher the complex compounds into simpler compounds, so the bacteria can easily break through the host cell [11].

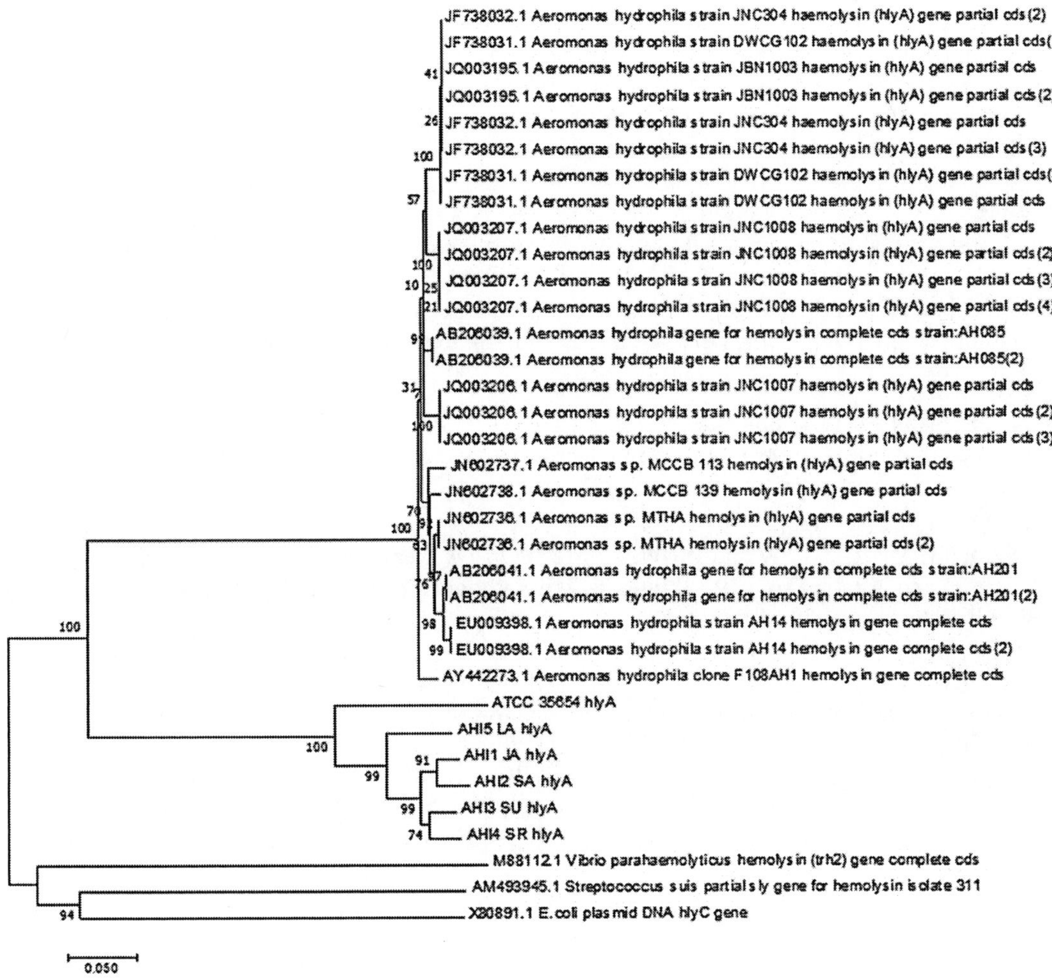


**Figure 2.** Phylogenetic relatedness based on nucleotide sequence hemolysin gene using the Neighbor-Joining method. The percentage of replicate trees in which the associated taxa clustered together in the bootstrap test (500 replicates) are shown next to the branches. The tree is drawn to scale, with branch lengths in the same units as those of the evolutionary distances used to infer the phylogenetic tree. The evolutionary distances were computed using the p-distance method and are in the units of the number of base differences per site. Evolutionary analyses were conducted in MEGA7.

IOP Conf. Series: Earth and Environmental Science **137** (2018) 012001 doi:10.1088/1755-1315/137/1/012001

The presence and frequency of the hemolysin gene (*hly*A) in the Aeromonas strains in the study was consistent with previous PCR surveys [12] that detected *hly*A in all strains including the A. hydrophila species. In this study, the *hly*A gene is widespread. On the contrary, the study of Pollard *et al.* [13] and Lior and Johnson [14] showed only detectable aerA genes in hemolytic, cytotoxic and enterotoxic A. hydrophila but not in A. veronii biotype of sobria and A. caviae. The difference number of strains and the lack of a reliable analysis of the negative strains of PCR, can explain this anomaly. When genotypes of known virulent strains positive of detectable from PCR Wong [15], then all A. hydrophila isolates with *hly*A and aerA genotypes are virulent. These isolates also exhibit hemolytic and cytotoxic activity. Since the aerA gene and hemolysin are found in most isolates from these infection [16].

Other hemolytic/ cytotoxic factors associated with virulence have not been described. It is possible to detect the Aeromonas virulence genes according to characterization into three main groups of virulence markers (ieaerolism, hemolism, and enterotoxin) during a single PCR amplification of clinical, environmental or food isolates, as promoted by Kingombe *et al.* [17]. In this study, the application of this method to detect Aeromonas virulence genes in food and environmental samples has been shown to offer an attractive alternative to the rapid screening of potentially lethal aeromonad in food and the environment.

## 4.Conclussion

*A. hydrophila* is a threat to gourami in Indonesia. Further research needs to be done to determine the pathogenesis of these bacteria and their relationship with host cells. Research on the regional distribution of bacteria and analysis of nucleotide sequences in virulent isolates should also be undertaken to develop DNA vaccines that are strain-specific to these infections.

## 5. References

[1] Post G 1987 *TFH Pub. Inc.* 288
[2] Cipriano R C, Bullock G L, Pyle S W 2001 Aeromonas hydrophila and Motile Aeromonad Septicemias of Fish *Fish disease leaflet 68* (Washington : US Department of the Interior Fish& Wildlife Service)
[3] Huizinga H W, Esch G W and Hazen T C 1979 *J Fish Dis* **2** 263-277
[4] Austin B and Austin DA 1993 Bacterial fish pathogens (New York : Disease In Farmed and Wild Fish Second Edition)
[5] Porteen K, Agarwal R K and Bhilegaonkar K N 2007 *J. Food Technol.* **2** 30-37
[6] Yousr AH, Napis S, Rusul GRA and Son R 2007 *J. Food* **14** 115 – 122
[7] Kamiso HN and Triyanto 1992 *J. Agri. Sci.* **4** 447-464
[8] Daskalov H 2006 *Food Control* **17** 474-483
[9] Allan B J and StevensonR M 1981 *J. Microbiol.* **27** 1114-1122
[10] Samira T L O, Gisele V G Mateus MC 2012 *Pes. Vet. Bras.* **32** 701-706
[11] Pelcza, M J, and Chan E C S 1986 *Elements of microbiology* (Jakarta : UI Press) (In Indonesia)
[12] Husslein V, Chakraborty T, Carnahan A and Joseph S W 1992 *Clin. Infect. Disease* **14** 1061–1068
[13] Pollard D R, Johnson W M, Lior H, Tyler S D and Rozee K R 1990 *J. Clinic Microbiol.* **28** 2477–2481
[14] Lior H, and Johnson W M 1991 *Experientia* **47** 421–424
[15] Wong C Y F, Mayrhofer G, Heuzenroeder M W, Atkinson H M, Quinn D M and Flower R L P 1996 *FEMS Immun. & Med. Microbiol***15**: 233-241
[16] Michelle C, Callegan B D, Jett L E and Michael S G 1999 *Infect. &Immunity* **67** 3357– 3366.
[17] Kingombe C I B, Huys G, Tonolla M, Albert M J, Swings J and Peduzz R 1999 *Appl. & Environ. Microbiol* **65** 5293–5302

IOP Conf. Series: Earth and Environmental Science **137** (2018) 012001 doi:10.1088/1755-1315/137/1/012001

**Acknowledgments**
We are thankful to our team of researchers, both technicians in the laboratory for their valuable helps with fieldwork and sample collection. This study was sponsored of RKAT, Faculty Fisheries and Marine, Univeristas Airlangga. Surabaya.

IOP Conf. Series: Earth and Environmental Science **137** (2018) 012002 doi:10.1088/1755-1315/137/1/012002

# The effect of thallus spreading method on productivity of *Gracilaria* sp. culture

**R Hidayatulbaroroh[1], M Nurhudah[2], M H Edy[2] and Suharyadi[2]**

[1] Student of Aquaculture Study Program, Jakarta Fisheries University, Indonesia
[2] Lecturer at Jakarta Fisheries University, Indonesia

E-mail: mochnurhudah@kkp.go.id

**Abstract.** The aim of this study was to determine growth of (*Gracilaria* sp.) with different spreading time of thallus. The study was conducted from March to April 2017 in pond located in Domas Village, Serang Region, Banten Province. The experiment followed completely randomized design with the treatment of different time on spreading of seaweed thallus during the culture period (45 days). Treatments were without spreading (as control), spreading every 2 weeks, and spreading every 3 weeks. The observed variables were weight of seaweed thallus and several water quality parameters. Analysis of seaweed weight used ANOVA test and Tukey HSD test. The results showed that the spread seaweed thallus had a significant effect on weight gain in 0.05 level. It used 100 gram *Gracilaria* sp. as initial weight, treatment without spreading thallus produced 508 gram, spreading every 2 weeks produced 906 gram and spreading every 3 weeks produced 790 gram. Based on the weight gain of thallus, seaweed culture by spreading thallus every 3 weeks and 2 weeks seem to be able to increase productivity by 56 % and 78 %, respectively.

## 1. Introduction

*Gracilaria* is one of the most in-demand aquaculture commodities in Indonesia. The agar contained in *Gracilaria* is commonly used as an emulsion-making material, thickener, filler and gelling agent. The production value from 2010 to 2014 was 22,474.60 tons; 18,337.00 tons; 77,616.61 tons; 98,475.62 tons and 57,454.00 tons [1].To increase the value, production from cultivation is required.

Despite the high demand of *Gracilaria* sp., its production from cultivation still faces some problems. Although Indonesia has a potential land with potential development up to 2,297.248 ha [1], the productivity is still not optimum. Therefore, some steps are needed to improve the production. One of the solutions to increase the production is to cut thallus periodically or called as spreading.

The purpose of this study is to identify the effect of thallus spreading with different frequencies to increase *Gracilaria* sp. productivity. This study is expected to provide information about the frequency of the spreading of *Gracilaria* sp. to produce maximum weight that can increase the productivity of pond.

## 2. Methodology

The study used such tools as sampling frame (bamboo and net), water level, refractometer, secchi disk, and scales. The materials used were *Gracilaria* sp. seed, pH paper, potassium test kit, phosphate test kit and nitrate test kit. The method used in this study was completely randomized design with three treatments and three replications. The treatment in this study was control without spreading (treatment

IOP Conf. Series: Earth and Environmental Science **137** (2018) 012002 doi:10.1088/1755-1315/137/1/012002

A), spreading every 2 weeks (treatment B), and spreading every 3 weeks (treatment C). The study procedure started from preparation of sampling frame, growth monitoring, until the measurement of water quality.

The preparation of sampling frame was done by installing the bamboo and net with a size of 25 cm x 25 cm x 40 cm. The total sampling frame was 39 frames. The seeds were placed on nine sampling frames. The weight of the seeds used was 100 grams each frame.

Growth monitoring was done by measuring the weight once a week for 45 days of maintenance. At the time of weight measurement, the spreading was done according to the provision of each treatment. In treatment A, spreading was not performed. In treatment B, spreading was performed three times during the experiment, and in treatment C, spreading was done twice during the experiment. The results of weight measurement were analyzed using ANOVA test, followed by Tukey test as advanced test.

Water quality measurements were carried out periodically. Measurements of temperature, salinity and depth were done twice a day, and transparency measurement was done once a day. Parameters of nitrate, phosphate and potassium were measured once a week.

## 3. Results and discussion

### *3.1. Growth monitoring*

The spreading treatment had a beneficial effect. The final weight of the treatment without spreading, spreading every 3 weeks, and spreading every 2 weeks were about 5 times, 8 times, and 9 times respectively, compared to the initial weight. With the initial weight of 100 grams, treatment without spreading produced 508 gram; the treatment of spreading every 2 weeks produced 906 gram, and the treatment of spreading every 3 weeks produced 790 gram.

Spreading provided an opportunity to increase the productivity of pond. Treatment of spreading every 3 weeks increased productivity by 56 % (the difference to control was 282 gram) and treatment of spreading every 2 weeks by 78 % (the difference to control was 398 gram).

The specific growth rate (SGR) of *Gracilaria* sp. fluctuated in every measurement. The average growth rate during observation on treatment without spreading was 3.6 %. With the treatment of spreading every 2 weeks, the growth rate was 5.1 % and the rate from spreading every 3 weeks was 4.7 %. All of the average growth rate was more than 3%. So, the values were profitable because they are more than 3 % [2]. The results of weight measuring are described in table 1, figure 1 and figure 2.

Spreading every 2 weeks produces the highest productivity than other treatments. Spreading every 2 weeks made the *Gracilaria* sp. density in the pond more distributed than spreading every 3 weeks and without spreading. These conditions made the absorption of nutrients more effective to produce maximum growth.

**Table 1.** Weight and growth rate of *Gracilaria* sp.

| Date | Average of weight (g) | | | Growth rate (%) | | |
|---|---|---|---|---|---|---|
| | A | B | C | A | B | C |
| 11/03/2017 | 100 | 100 | 100 | - | - | - |
| 18/03/2017 | 146 | 160 | 137 | 5.5 | 7.0 | 4.6 |
| 25/03/2017 | 152 | 219 | 206 | 0.6 | 4.6 | 6.0 |
| 01/04/2017 | 209 | 285 | 259 | 4.6 | 3.8 | 3.3 |
| 08/04/2017 | 339 | 420 | 378 | 7.2 | 5.7 | 5.5 |
| 15/04/2017 | 423 | 501 | 402 | 3.2 | 2.5 | 0.9 |
| 22/04/2017 | 475 | 785 | 690 | 1.7 | 6.6 | 8.0 |
| 25/04/2017 | 508 | 906 | 790 | 2.2 | 4.9 | 4.6 |

IOP Conf. Series: Earth and Environmental Science **137** (2018) 012002 doi:10.1088/1755-1315/137/1/012002

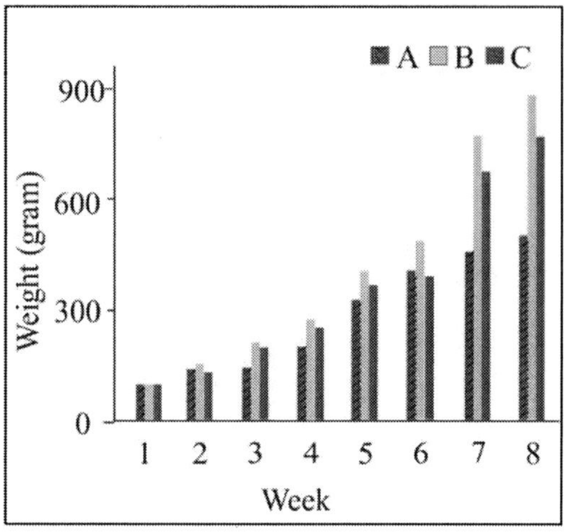


**Figure 1.** Average weight of *Gracilaria* sp.

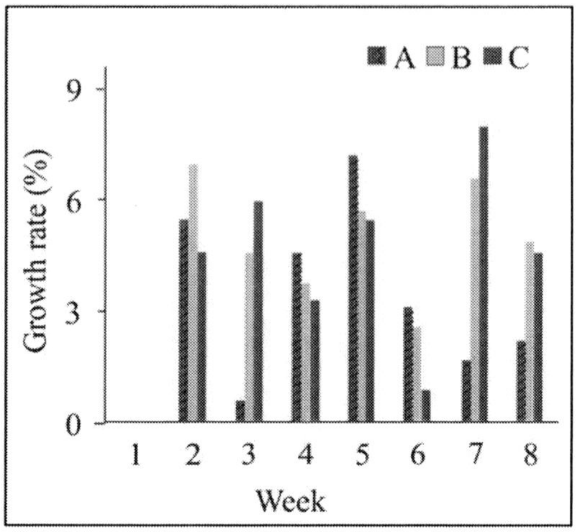


**Figure 2.** Growth rate of *Gracilaria* sp.

Data of *Gracilaria* sp. weight were analyzed with ANOVA test at 0.05 level. After the normality and homogenity of the data had been ensured, the result of F calculate was 24.34 with F standard of 5.14. Because F calculate is more than the F standard, then $H_0$ is rejected. So, spreading thallus strongly affects the growth of *Gracilaria* sp.

Tukey HSD test was used as advance test after the result was obtained from ANOVA test. Two conclusions can be drawn from the Tukey HSD test. First, the treatment without spreading was significantly different from spreading every 2 weeks and spreading every 3 weeks. Second, spreading every 2 weeks was not significantly different from spreading every 3 weeks.

*3.2 Water quality*

The measurement of water quality parameters revealed that the water quality generally did not meet the required standard value of *Gracilaria* sp. The results of water quality parameter measurement during observation are described in table 2.

IOP Conf. Series: Earth and Environmental Science **137** (2018) 012002 doi:10.1088/1755-1315/137/1/012002

**Table 2.** Measurement results of water quality parameters.

| Parameters | Standard value | Reference | Measuring value |
|---|---|---|---|
| Temperature (°C) | 25-28 | SNI 7578:2010 [4] | 27-39 |
| Transparency(cm) | 50-70 | SNI 7578:2010 [4] | 11-22 |
| Salinity ($g.L^{-1}$) | 15-30 | SNI 7578:2010 [4] | 20-31 |
| pH | 6.8-8.2 | SNI 7578:2010 [4] | 6-8 |
| Nitrate ($mg.L^{-1}$) | 0.1 | Effendi, 2003 [6] | 0-12.5 |
| Phosphate | 5-20 $\mu g.L^{-1}$ | Boyd, 1990 [7] | 0.1-0.5 $mg.L^{-1}$ |
| Potassium ($mg.L^{-1}$) | 380 | Effendi, 2003 [6] | 210-380 |
| Water depth (cm) | 50-75 | SNI 7578:2010 [4] | 11-22 |

The measurement showed that the temperature is higher than SNI of 25-28°C. The temperatures directly affected the metabolism rate in thallus. The temperatures which were too high made the thallus wilted because the habitat was not appropriate. This means that the temperatures directly affected plant physiology [3].

Penetration of light or transparency during measurement always reached the bottom because the water level was too low, about 11-22 cm. The depth did not match the standard of 50-70 cm [4]. The transparency and temperature stratification are strongly associated with growth. The higher intensity of the sun and the more optimum temperature conditions result in better growth [5].

The measured nitrate reached 12.5 $mg.L^{-1}$. This value was higher than the value of natural waters of 0.1 $mg.L^{-1}$ [6]. Nitrate was a major component of the nutritional supply for *Gracilaria* sp. However, if the nitrate value reached 0.2 $mg.L^{-1}$ eutrophication will occur [6].

Phosphate measurements showed a high phosphate content of 0.1-0.5 $mg.L^{-1}$. Phosphate was one of essential ingredients for the growth of *Gracilaria* sp.. The phosphate value in natural waters was 5-20 $\mu g.L^{-1}$ [7]. Besides *Gracilaria* sp., in the waters, there were phytoplankton that required orthophosphate of 0.27-5.51 $mg.L^{-1}$ for optimal growth [8]. During the observation, the value obtained was 0.1-2.0 $mg.L^{-1}$. This means that the availability of phosphate in the water was guaranteed. Phosphate has an effect on growth because it is a nutrient needed to grow and develop. It because one form of phosphate was the orthophosphate required in the formation of cells [9].

The minimum potassium value of this area was 210 $mg.L^{-1}$ while its maximum value was 380 $mg.L^{-1}$. The results of potassium measurements reached the standard in natural waters of 380 $mg.L^{-1}$ in only one measurement. The results below the standard values would cause growth problems because potassium was an essential element for plants [6].

The suitable parameters with *Gracilaria* sp. were salinity and pH. Salinity was at optimum value of 15-31 $g.L^{-1}$. The results of pH measurements on all ponds showed the value between 6 and 8. From these results, pH was always close to the standard value of 6.8-8.2 [4]. The pH correlated with oxygen in the water. This stability indicated the fulfillment of oxygen supply from photosynthesis thallus and the need of aquatic organisms because the higher carbon dioxide content caused the pH of water to decrease [10].

## 4. Conclusion

*Gracilaria* need to be spread to increase productivity. Spreading increased productivity by 56% (spreading every 3 weeks) and 78 % (spreading every 2 weeks).

## 5. References

[1] Rahmantya F, Asianto A D, Wibowo D, Wahyuni T and Somad W A 2015 *Maritime and fisheries* (Jakarta: The Center for Data, Statistics and Information) p 59 53 (In Indonesia)

[2] Prabowo G and Farchan M 2008 *Technology of seaweed culture* (Serang: Jakarta Fisheries University) p 70 (In Indonesia)

[3] Parenrengi A, Syah R and Suryati E 2012 *Culture of carragenan producer seaweed (Karaginofit)* (Jakarta: Research and Development Agency of Maritime Affairs and Fisheries of Indonesia) p 23 (In Indonesia)
[4] SNI 7578 : 2010 *Seaweed production Gracilaria (Gracilaria verrucossa) with dispersive methods in pond polyculture* (Jakarta: National Standarization Agency of Indonesia) 2010 p 277-8 (In Indonesia)
[5] Komarawidjaja W 2005 *J. Envi. Technol.* **6** 410-5
[6] Effendi H 2003 *Water quality study* (Yogyakarta: Kanisius) p 134-155 (In Indonesia)
[7] Boyd C E 1990 *Water Quality in Ponds for Aquaculture* (Birmingham: Birmingham Publishing Co.) p 72
[8] Amin, M and Suwoyo H S 2011 Type and composition of plankton on polyculture of tiger shrimp, whiteleg shrimp, milkfish, and seaweed in pond *Proc. Aquac. Technol. Innov. Forum* p 773-8
[9] Komarawidjaja W and Kurniawan D A 2008 *J. Envi. Technol.* **9** p 180-3
[10] Kordi G and Tancung A B 2007 *Water quality management in aquaculture* (Jakarta: Rineka Cipta) p 49 (In Indonesia)

**Acknowledgement**
We thank our institute Jakarta Fisheries University which supported our study. We would also like to show our gratitude to our colleagues Mr. Makmun who provided his pond to become study site, Mr. Yasin as Mr. Makmun's assistant and other colleagues who help us as long as our study.

IOP Conf. Series: Earth and Environmental Science **137** (2018) 012003 doi:10.1088/1755-1315/137/1/012003

# Study on characterization, pathogenicity and histopathology of disease caused by *Aeromonas hydrophila* in gourami (*Osphronemus gouramy*)

**Rozi[1], K Rahayu[1], D N Daruti[1], M S P Stella[2]**

E-mail: rozi@fpk.unair.ac.id

[1] Department of Fish Health Management and Aquaculture, Faculty of Fisheries and Marine, Airlangga University, Surabaya 60115, Indonesia
[2] Department of Aquaculture, Faculty of Fisheries and Marine, Airlangga University, Surabaya 60115, Indonesia

E-mail: rozi@fpk.unair.ac.id

**Abstract.** This study aims to determine the bacterial pathogens that cause disease of the gourami in Blitar (East Java) and Yogyakarta (Central Java), Indonesia. A total of 50 fish samples taken randomly gourami in pond farmers in seventh different locations. There were 18 isolates were isolated and then test Koch's postulates were injected 0.1 ml/fish intraperitoneally to gourami. Characterization is done by using the biochemical tests. Pathogenicity test carried out on 3 isolates of *Aeromonas* spp. with intraperitoneal injection at a dose of $10^4$-$10^8$ CFU/ fish, the value of *Lethal Dosage* 50 (LD50) using the method Dragstedt Behrens. After the treatment, spleen and kidney samples were processed for histopathological analysis. The all of identified bacteria were 5 isolates *Aeromonas hydrophila*. Isolates of *A. hydrophila* in a row AH3 was virulen to gourami with LD50 (4.53 x $10^6$ CFU/fish), while isolate AH4 and AH5 (2.903 x $10^8$, 1.319 x $10^9$ CFU/fish) not be avirulen. Koch's postulates; 3 isolates are pathogenic with mortality of 40-100 % and 2 are non-pathogenic isolates with a mortality of 0 %. Clinically; ulcers, haemorhagic at the base of the fins, body, mouth and exophthalmia. Histopathologically indecated spleen necrosis, piknosis, necrosis and inflammatory cells in kidney.

## 1. Introduction

The giant gourami, *Osphronemus gouramy*, is an Indonesian native fish that has a widespread distribution in Southeast Asia [1,2,3]. It is a commercially important freshwater herbivorous species with a relatively high price [4,5]. Therefore *O.gouramy* is a commodity that is seeded in aquaculture and is an important species in the ornamental as well as edible aquaculture industries [6]. However, production has not been able to meet the market demand.

There are various ways to increase gourami production such as through intensive or extensive cultivation technology, but fish cultivation is prone to the presence of disease as a result of interactions between the host, the pathogen, and the environment [7]. Kamiso [8] stated that the three main barriers

IOP Conf. Series: Earth and Environmental Science **137** (2018) 012003 doi:10.1088/1755-1315/137/1/012003

in fish farming in Indonesia are inferior seed supplies, expensive feed supplies, and disease. Most of the cause of disease in fish is pathogenic bacteria which is more commonly known as bacterial disease, especially gram-negative bacteria such as Aeromonas sp., *Vibrio* sp., *Pseudomonas* SP., *Edwarsiella* sp., and *Flexibacter* sp. [9]. Outbreaks caused by *Aeromonas hydrophila* have resulted in high economical losses and the death of nearly 80-100 % of fish cultivated within 1-2 weeks [10]. Infection from the bacteria can show clinical signs of skin infection in wounds such as ulcers that extend to muscle tissue, fins, rotting gills, protruding eyes (exophthalmia), and swelling of the spleen and the kidneys [11].

Research on disease-causing bacteria found in gourami has been carried out in Bantul, Yogyakarta and Blitar, East Java and very important work has been done by isolating, identifying and inventorying the bacteria. It is hoped that through this research, we can obtain pure isolates and know the symptoms and its pathogenicity so that it can be used in the initial steps of the monitoring, prevention and handling of the onset of the bacterial plague which has widely caused disease in large freshwater fish/gourami.

## 2. Methodology

### *2.1. Sample fish*

About 30 diseased gourami (*O. goramy*) were collected from five farm ponds in Bantul, Yogyakarta, Central Java (7°52'29.34"S, 110°19'31.93"E) and Blitar, East Java (8° 5' 60.0000" S and 112° 9' 0.0072" E.). The fish weighed from 350 to 800 g with the size of 90-10 cm. The fish were anaesthetised with clove oil and then dissected according to Wilson [12] and performed by standard methods [13].

### *2.2. Bacterial isolation*

Lesions from the skin, fins and gills were inoculated onto Trypticase Soy Agar (TSA) and incubated aerobically at 28°C for 24 h. Then internal samples were acquired aseptically from the kidney, liver, spleen and infected muscles. The plates were examined for bacterial growth. Dominant colonies were selected and re-streaked on Trypticase Soy Agar (TSA). Cultures were placed in 20 % glycerol and supplemented in Trypticase Soy Broth (TSB) for storage at -80°C. Dominant colonies were selected for bacterial isolation to establish the optimal number of bacterial cells, and went through a purification procedure until pure colonies were established to be sure that the dominant colonies were not contaminated.

### *2.3. Identification of the isolates*

Bacterial identification was done on the basis of Bergey's Manual of Determinative Bacteriology [14], Biochemical tests for Identification of Medical Bacteria [15], then completed with Bacterial Fish Patogens: Disease in Farmed and Wild Fish [9].

### *2.4. Pathogenicity test*

Patogenicity is determined based on value of LD50 using method Dragstedt Behrens [16]. The calculation of the density of the pathogenicity test is based on the method of counting the bacteria indirectly by the number of colonies that is total plate count (TPC) according to Junoto *et al.* [17].

### *2.5. Histopathological examination*

For histopathological studies, tissue specimens were obtained from the liver and kidney. The tissue specimens were fixed in 10 % neutral buffered formalin. Dehydration and infiltration of tissue were carried out using an automatic tissue processor (Sakura, Japan). Samples were embedded in paraffin, and were sectioned using a rotary microtome (4 to 5 μm) (Sakura, Japan) and stained with Hematoxylin and Eosin (H&E) according to the method described by Carleton *et al.* [18].

## 3. Results and Discussion

### *3.1 Clinical signs*

Gouramis weighing 50-135 grams or (13-20 cm) were obtained from five locations each both in Bantul, Yogyakarta, and Blitar, East Java, Indonesia. Almost the same symptoms were found on the gourami samples taken from each location (table 1).

**Table 1.** The origin of sampling, gross clinical and pathological symptoms.

| Orogins of samples | Fish | Species | | Symptoms |
|---|---|---|---|---|
| | | weighed (gram) | Length (cm) | |
| Bantul, Yogyakarta, Center Java, Indonesia | 30 Goramy | 50-60 | 13-20 | Releasing air bubbles from the operculum, geripis fins, Often appeared on water surface,, Floating on water surface, Red patches, Dark/black body skin |
| Blitar, East Java, Indonesia | 20 Goramy | 80-100 | 16-18 | Eyes were exophthalmia and reddish, torn fins hemorraghic, Releasing air bubbles from the Mouth opens and closes fast and Move slowly and reluctant to feed, swolen kidney |

The symptoms that were generally seen include: the indication of the presence of air bubbles expelled through the operculum, huddled above the water surface, hemorrhagic in some parts of the body especially the base of fin. Some samples of the haemorrhaging were still visibly light, i.e. there were red smudges that were still separate, but other samples like the dorsal, pectoral and tails fins were torn and broken part. The oral cavity was reddish, one or both eyes exhibited exophthalmia and scales were loss in some of the samples; while internal symptoms displayed a change in the color of the kidney and had a tendency to swell. In general, it shows that the more severe the physical (external) symptoms, the more severe the organ(internal) symptoms were. Gourami samples can be seen in the sample (figure 1).

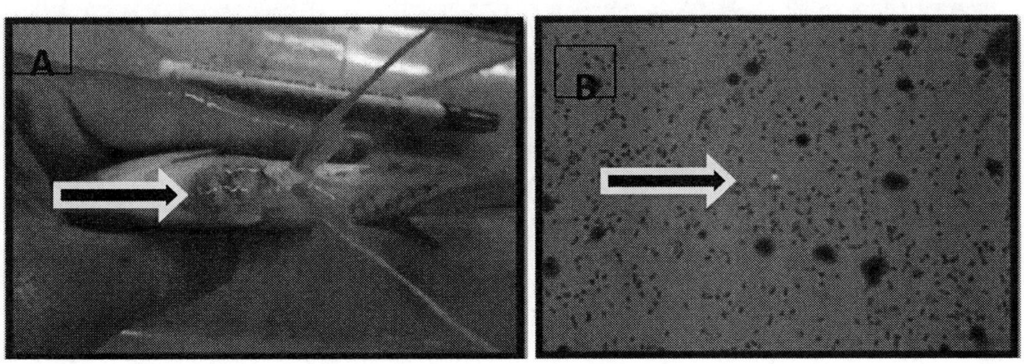


**Figure 1. A.** Hemorrhage of around the abdomen. **B**. Typical morphology of *A. hydrophila.*

### *3.2. Bacterial isolation and identification*

Out of the 50 samples of gourami from the five of ponds of different locations in Bantul and Blitar that were allegedly infected by bacteria, results of the biochemical test showed that 5 out of 18 *A. hydrophila* isolates were successfully obtained. The bacteria that were pathogenic were found in three isolates (40-100 % mortality) and 2 isolates were non-pathogenic (0 % mortality). Biochemical tests were conducted on the identification of 18 isolates of *A. hydrophila*. About five isolates were identified as Gram negative (figure 1B), motile, oxidase-positive, catalase-positive, etc (table 2).

### *3.3. Virulence degree*

The pathogenicity test of *A. hydrophila* in this study can be identified through the $LD_{50}$ test devoted to the death of 50 % of the population due to infection of *A. hydrophila.* After that, results of the observations of $LD_{50}$ of the bacteria dose resulting in death of 50 % of gourami was in the range of $10^5$-$10^7$ CFU/ml. Through the calculation with the method of Bargstedt-Behrens, the $LD_{50}$ value obtained from the AH3 isolates was measured at 4,53 x $10^6$ and was categorized as virulent; while according to Austin & Austin [9] (table 3), the *A. hydrophila* non-virulent isolates were AH4 and AH5 with the LD50 values of 2.90 x $10^8$, 1.31 x $10^9$ respectively.

**Table 2.** LD50 values of the three isolates.

| Isolates | Bacteria | LD50 (CFU/fish) | The degree of Virulence | Reference |
|---|---|---|---|---|
| AH3 | *A. hydrophila* | 4.53 x $10^6$ | Virulen* | 9 |
| AH4 | *A. hydrophila* | 2.903 x $10^8$ | Non Virulen* | 9 |
| AH5 | *A. hydrophila* | 1.319 x $10^9$ | Non Virulen** | 9 |

### *3.4. Clinical signs and gross changes*

Clinical findings observed from the 90 diseased gourami samples for the patogenicity test showed differences among the fish. Some diseased gourami showed clinical symptoms while others showed more than two clinical symptoms. These clinical symptoms include: skin lesions with hemorrhaging, ulcers, inflammation and hyperemia in the fins. There is a "washboard" on the skin, exophthalmia in one or both eyes, a histopathological section observed of liver and kidney; piknosis, necrosis liver and piknosis, necrosis, inflammatory cells kidney (figure 2).

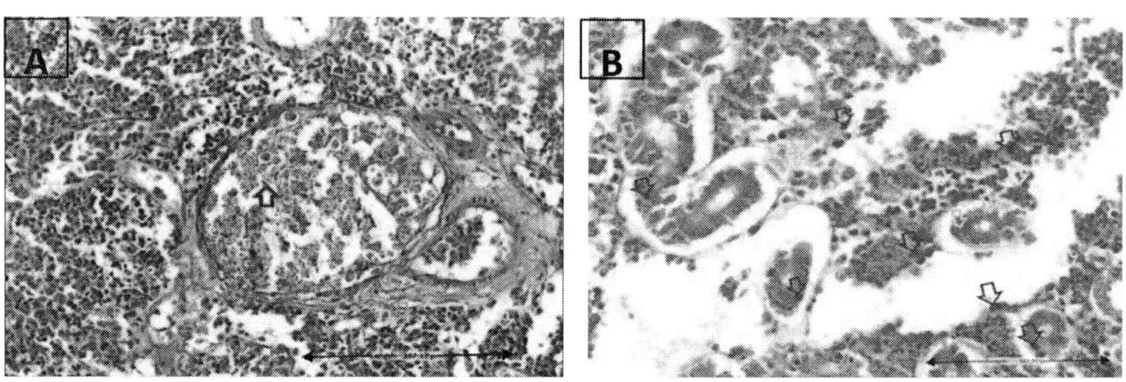


**Figure 2.** Histopathological section of **A:** liver; (yellow row) piknosis, (blue row) necrosis cells (H&E stain 100X). **B:** Kidney; (blue row) piknosis, necrosis (yellow row) inflammatory cells (H&E stain 100X).

IOP Conf. Series: Earth and Environmental Science **137** (2018) 012003 doi:10.1088/1755-1315/137/1/012003

**Table 3.** Characters of bacterial strain AH1.

| Parameter Characters | Characters |
|---|---|
| Colony Morphology - | |
| Colony shape | Circulair |
| - Elevasy shape | Convex |
| - Edge shape | Entire |
| - Oblique agar | Echinulate |
| - Colour | cream |
| - Size | 2 mm |
| - GSP/colour | + (cream)) |
| Cell Morphology | |
| - Gram | - |
| - Shape | rod |
| Phenotype of biochemical | |
| - O/F | F |
| - Motility | + |
| - Produksi: | |
| - Katalase | + |
| - Oksidase | + |
| - H2S | + |
| Lisin dekarboksilase | + |
| Arginine dehydrolase | ND |
| Onitin dekarboksilase | - |
| TSIA | K/A, G |
| LIA | |
| Growth in 30°C | + |
| Growth in NaCl 6.5 % Methyl red | + |
| Voges-Proskaeur | + |
| Sensitivity Novobiocin | + |
| Sensitivity 129.10 | - |
| Simmon citrate | + |
| Gelatin broke | + |
| Acid production | |
| - D - Glucose, acid, D - Galaktose, acid, Maltose | + |
| D - Mannitol, D - Mannose, Sucrose, acid Inosito | - |
| Dulcitol, Raffinose, D - Sorbitol, D – Xylose, Inulin | - |
| Gas production - D - Glucose, gas | + |

| | |
|---|---|
| Prosentase as References | 98 % |

The samples of the 50 gourami fish allegedly infected by bacteria had the size of 13-20 cm. 18 isolates were successfully isolated from the samples from five different locations of both Bantul, Yogyakarta and Blitar, East Java. The samples in the field show symptoms of septicemia. The symptoms generally seen were in the form of hemorrhage, wounds such as sores, reddish wounds around the mouth, and exophthalmia eyes, but overall the samples showed such symptoms such as removed oxygen from the operculum, dark and pale samples, and mass death such as what occured at the Bantul location where the mortality rate reached 100 % of 20.000 stocking densities cultured on the size of consumption. This is consistent with the literature stating that an *A. hydrophila* infection is acute when marked by the very quick death of the fish with the appearance of signs of clinical infection like exophthalmia, red patches on the skin, and accumulation of fluid in the abdominal pouch, flatulence, bleeding gill and injury to the dermis, and separated scales [20]. According to Kamiso, [21] *A. hydrophila* can cause the death of 90-100 % in a short time. Research by Daskalov [22] also described that *A. hydrophila* is widely distributed through food, drinking water and the environment, and is a pathogen that causes hemorrhagic diseases, zoonotic diseases, and food-borne infections in freshwater fish. Several factors are involved in the *A. hydrophila* virulence process. *A. hydrophila* is naturally found in the fish intestinal tract, and is not caused by factors such as: stress, changes in environmental conditions, population density, handling, transport, poor water quality, temperature changes, low dissolved oxygen levels, high $CO_2$ levels, high nitrite levels, and high ammonia levels. This is the most common predisposing factor associated with *A. hydrophila* disease. In addition, the pathogenicity of *A. hydrophila* appears to be related to how stressful the hosts are infected. *A. hydrophila* with high virulence can infect healthy fish; however, stress derived from intensive fish cultivation also contributes to it and can trigger an outbreak [23].

Overall 5 isolates were *A. hydrophila*. The results of biochemical tests are capable of utilizing gelatine as a substrate protein. Bacteria that can take advantage are albumin, casein, fibrinogen and gelatine as a substrate protein, so it can be inferred that strains of this bacteria are proteolytic [24], According to Neito and Ellies [25], there are at least four or perhaps five kinds of protease extracellular of *A. hydrophila*. Based on data from the study of this biochemical test, *A. hydrophila* overall or 100 % produces proteolytic enzymes / enzyme gelatinase. When seen from the relationship symptoms, *A. hydrophila* allegedly caused the death of 40-100 % at 1-2 weeks [21] by utilizing proteolytic enzymes such as protease that functions as the body's defense against the host in the progression of the disease [26]. The internal organs of fish infected by *A. hydrophila* experience hemorrhaging indicated by the occurrence of hemolysis in vitro, so the hemolysin was also considered as a major virulence factor of *A. hydrophila* [27]. Judging from the distribution of the spreading of the bacteria, the one that is most commonly found by percentage (71.42 %) was *A. hydrophila* which was found in almost all sampling sites. One of the factors that cause the extent of the spread of the bacteria is that it is opportunistic in nature and capable of living outside the host in the long term.

The phatogenicity test showed that about 3 out of 5 *A. hydrophila* were virulent to gourami with LD50 (4.53 x $10^6$ CFU/fish), and Koch's postulates; 3 isolates were pathogenic with mortality of 40100 %. There are several factors involved in the *A. hydrophila* virulence process. In this study, the overall virulence factor of *A. hydrophila* isolates includes hemolysin and aerolysin. Aerolysin is the most common virulence factor found in the isolates analyzed [28]. The difference of the level of virulence in each isolate of *A. hydrophila* and *V. cholera* depends on the strain used, the age of the individual, the individual test conditions, and water quality. The onset of the disease is also very important with the factors of bacterial pathogenity, the propagation speed of pathogens, as well as factors of the host defense against pathogens. Activities or leukosidin and haemolysin produced by the Extracellular products (ECPs) of bacteria become a bacterial defense factor against the defence of the host's blood because it is able to lysis blood cells. The bacteria are able to survive and will follow the blood flow to spread throughout the body of the host cells as well as towards the target organs. Bacteria also have pathogenicity factors in the form of the enzymes found in ECPs, including caseinase, gelatinase, amylase, lipase, phospolipase, chitinase, kolagenase, elastase, hyaluronidase, and proteinase. These enzymes are able to decipher the complex compounds into simpler compounds, so the bacteria can easily break through the host cell [29].

The kidneys and liver of fish are internal organs that are sensitive to aeromonas disease attacks. Results of histologic analysis of the kidneys and liver showed a quite significant difference between healthy gourami fish and those exposed to *A. hydrophila* (figure 2). In healthy fish, the slices of liver are brightly colored as well as the hepatocyte cells containing the nucleus and heterochromatin. The condition of liver cells in fish affected by an *A. hydrophila* attack appears damaged because it has an infection, but does not release pus (nonvirulent multifocal hepatitis). Picnosis, necrosis cells and cells blood due to internal bleeding (internal haemoragy) were also found. Death of liver cells (focal necrosis) is a common manifestation of fish affected by *A. hydrophila* [30]. Fish kidneys exposed to *A. hydrophila* show picnosis, inflammatory cells cell death (necrosis) caused by the enzymatic degradation produced by *A. hydrophila,* as stated by Burr et al. [31]. All of the gourami fish strains were stricken under the same pathogenic conditions based on histopathologic analysis of the renal organs and heart. This study showed relatively similar results with Afifi [32], which described that the toxins produced by *A. hydrophila* and extracellular products such as hemolysin, protease, elastase can cause severe necrosis in the liver. The study also observed that the diseased kidneys of the fish were severely damaged and showed degenerative changes in the glomerular epithelium with cytoplasmic vacuol formation, as well as focal lymphocyte infiltration. This is similar to Suprapto's report [33] which states that the kidney is attacked by bacterial toxins that cause kidney cells to lose its structural integrity.

During the past decades, the development of suitable vaccine(s) and vaccination strategies have contributed to controlling infectious diseases in aquaculture in respect to the type of pathogen and cultured species [34]. However, till date only a few vaccines, i.e., against enteric red mouth, furunculosis, vibriosis, pasteurellosis, yersiniosis, etc., are available [34]. These vaccines are mostly polyvalent, containing the whole cell antigen or components of a more pathogen. The major problem associated in immunizing individual fish with antigen is that it lives in an environment where a wide range of other pathogens and secondary invaders are present, and a specific immunization strategy may only be effective in protecting against a specific pathogen or disease.

## 4. Conclusion

These results indicate that the strains in the sample showed symptoms of a disease-causing bacteria which are generally manifested in the form of ulcers, exophthalmia, hemorrhaging at the base of the fin, body and mouth, and air bubbles from the operculum. The isolate of bacterial pathogens that cause disease in gourami have been identified as *A. hydrophila* and five isolates were obtained. The AH3 isolate of *A. hydrophila* is pathogenic to gourami with $LD_{50}$ (4.53 x $10^6$ CFU/fish), while the isolates of AH4 and AH5 (2.903 x $10^8$, 1.319 x $10^9$ CFU/fish) is avirulent.

## 5. References

[1]. Welcomme R L 1988 *J. Fish. & Aqua. Studies* **294** 1-318.
[2]. Setijaningsih L, Arifin O Z and R. Gustiano 2007 *Ind. J. Ichtiologi*; **7** 23-30.
[3]. Nugroho E dan Kusmini I 2007 *J. of Aqua. Research* **2** 51-57.
[4]. Roberts T R 1989 *The freshwater fishes of Western Borneo(Kalimantan Barat Indonesia)*, (San Francisco : California Academy Science)
[5]. Directorate General of Aquaculture 2010 Map Sentra of Aquaculture (Jakarta : Department of Marine and Fisheries of the Republic of Indonesia)
[6]. Amornsakun T, Kullai S, Hassan A 2014 *J. Sci.Technol.* **36** 493-498.

[7]. Anderson D P 1974 *Fish Immunology* (Washington : TFH Publication inc. Ltd)
[8]. Kamiso H N 1986 Fish disease as a problem in intensive aquaculture *Proceeding Seminar Multy Disciplinary Studies on Fisheries and Inshore Coastal Soedarsono (eds)* 416-438.
[9]. Austin B and Austin D A 1987 Bacterial fish pathology (England : Disease in Farmed and Wild Fish, Ellis Horwood Limited)
[10]. Purwaningsih, Uni dan Suwidah 2007 Fish tissue damage due to Kancra (Tor sp.) as a result of artificial Aeromonas hydrophila bacterial infection *Proc. Fish. National Sem.*
[11]. Irianto A 2004 Patologi Ikan Teleoestei. *Gadjah Mada University Press*, Yogyakarta.
[12]. Wilson J M, Bunte R M, Carty A J 2009 *J. The American Association Lab. Anim. Sci.* **48** 785789.
[13]. Buller N B 2004 Bacteria from fish and other aquatic animals: a practical identification manual *CABI Pub.*
[14]. Holt J G, Kreig N R, Sneath P H A, Staley S T, Williams S T 1994 *Bergey's manual of determinative microbiology* (Baltimore : 9th ed. The Williams & Wilkins Co)
[15]. Macfaddin, J F 1980 *Biochemical test for identification of medical bacteria* (USA : 2nd ed Williams & Wilkins, Waverly Press, Inc. Mt. Royal, and Guilford Aves, Baltimore Md. 21202)
[16]. Hubert J J 1980 *Bioassay* (USA : Kendall/Hunt Publishing Company)
[17]. Jutono J, Soedarsono S, Hartadi S, Kabirun S, Suhadi, dan Soesanto 1973 *Microbiology practicum guidelines* (Yogyakarta : University of Gajahmada)
[18]. Carleton H M, Drury R A B, Wallington E A 1967 *Carleton's histological technique.* Fourth Edition pp 20
[19]. Stevenson R M W 1998 *Academic Press* 112-123.
[20]. Cipriano R C 2001 *Aeromonas hydrophila and Motile Aeromonas Septicemias of Fish* (USA : Geological Survey, Leetown Science Center, National Fish Health Research Laboratory)
[21]. Kamiso H N dan Triyanto 1992 *J. Agri. Sci.* **4** 447-464
[22]. Daskalov H 2006 *Food Control* **17** 474-483
[23]. Suomalainen L R, Tiirola M, Valtonen E T 2005 *J Fish Dis* **28** 271-277.
[24]. Shotts E B, Hsu T C and Waltman W D 1985 *Fish. Pathol.* **20** 37-44.
[25]. Nieto T P and Ellis A E 1991 *J. Fish. Dis.* **14** 229-235.
[26]. Angka S L 1990 *Asian Fish. Sci.* **3** 343-351.
[27]. Allan B J and Stevenson R M 1981 *J. Microbiol* **27** 1114-1122.
[28]. Samira T L O, Gisele V G and Mateus M C 2012 *Pesq. Vet. Bras.* **32** 701-706.
[29]. Pelcza M J and Chan E C S 1986 *Elements of microbiology* (Jakarta : UI Press) (In Indonesia)
[30]. Azad I S, Rajendran K V, Rajan J J S, Vijayan K K and Santiag T C 2001 *J. Aqua. Tropics* **16** 265-275
[31]. Burr S E, Pugovkin D, Wahli T, Segner H and Frey J 2005 *Microbiol.* **151** 2111-2118
[32]. Afifi S H, Al-Thobiati S, Hazaa M S 2000 *Assiut. Vet. Med.* **84** 195 205.
[33]. Suprapto H, Sumartiwi L, Prawesthirini S, Handiyatno D, Azmijah A 2005 *Berk. Penel.* **10** 139-141 (In Indonesia)
[34]. Evelyn T P 1997 *Dev. Biol. Stand* **90** 3–12
[35]. Hastein T, Gudding R, Evensen O 2005 *Dev. Biol. (Basel)* **121** 55–74.

**Acknowledgments**
We are thankful to our team of researchers, both technicians in the laboratory for their valuable helps with fieldwork and sample collection. This study was sponsored of RKAT, Faculty Fisheries and Marine, Univeristas Airlangga. Surabaya

IOP Conf. Series: Earth and Environmental Science **137** (2018) 012004 doi:10.1088/1755-1315/137/1/012004

# The growth performance of F1 transgenic mutiara catfish

**Iskandar, I D Buwono, and M U K Agung**

Faculty of Fisheries and Marine Science, Universitas Padjadjaran
Jl. Raya Bandung Sumedang Km 21, Jatinangor 45363, West Java, Indonesia.

E-mail : nday_iskandar@yahoo.co.id

**Abstract.** The growth of catfish (African or Sangkuriang strain) these days is tend to decreased. One of the solutions due to this problem is to improve the genetics of growth using transgenesis technology, toward more profitable. The specific objective of the research is to detect the transmission of exogenous GH (African catfish GH inserts) inside the F1 transgenic Mutiara catfish using PCR (*Polymerase Chain Reaction*) method and to evaluate the growth performance of transgenic Mutiara catfish made using the parameters of feed conversion (*FCR = Feed Conversion Ratio*). Transgenic catfish (strain mutiara) F0 and F1 carried African catfish GH (600 bp) can be produced. Superiority characters of transgenic catfish represented heritability ($h^2$) and heterosis (H), indicating that the offspring of hybrid F1 transgenic mutiara catfish had phenotypes rapid growth ($h^2$ = 17.55% and H = 42.83 %) compared to non-transgenic catfish ($h^2$ = 10.07 % and H = 18.56%). Evaluation of the efficiency of feed use parameters feed conversion ratio, shows that F1 transgenic mutiara catfish (FCR = 0.85) more efficient in converting feed into meat.

## 1. Introduction

Improving the growth of farmed fish can be achieved by applying genetic engineering methods, namely transgenesis or gene introduction which takes a relatively short time compared to selective breeding which requires a relatively long time. Gene transfer techniques (transgenesis technology) has advantages when compared to selective breeding, such as genetic improvement per generation that produces a higher, faster phenotype expression improvement and superior character that can be passed down to the next generation [1]. Research on fish transgenesis has been developing rapidly, especially studies of growth hormone gene transfer (growth hormone / GH) to improve the growth of a species of fish. Results of transgenes have shown that growth in fish such as Atlantic salmon [2] and tilapia [3] increased more than 2-fold compared to non-transgenic fish. While transgenic Pacific salmon has a 10-fold growth and the mud loach fish by 32 times compared to non-transgenic fish [4]. Foreign GH carrier transgenic fish in the second generation has the character of high growth compared to the growth of non-transgenic individuals.

Transgenesis technology is a method of gene transfer using sperm electroporation technique, in principle to modify the permeability of the cell membranes of sperm by using an electric field (electric shock). The electric current short that occurs will open the pores of the cell membrane, allowing the foreign DNA molecules (genes transferred) to penetrate the cell membrane and into the nucleus of sperm cells [5]. The degree of exogenous DNA integration with this sperm genome can reach 50%. The development of mass gene transfer methods is very important because it gives an advantage to the fish gene transfer system, resulting in a high number of fish eggs and external fertilization. Stable transgenic production is necessary in the long run from generation to generation; therefore the inheritance of transgenes in the first offspring of transgenic fish (F1) only reaches 25%. Mass production of transgenic fish was obtained from F3 (the spawning results of F2 transgenic fish and non-transgenic fish) [6].

The B-actin promoter is a promoter capable of driven expression of the transgene in the transgenic fish [7]. In addition to the use of β-actin promoter, another promoter with potential to drive transgene expression is derived from the RSV promoter (*Rous sarcoma virus*) and CMV (*Cytomegalovirus* enhancer) and SV 40 (simian virus 40 enhancer). Promoter CMV has been used in the construct of the transgene pCMV-RGH-IRES2-EGFP with the technique SMGT in the Rohu (*Labeo rohita*) carp in India, with an expression of GH which resulted in an increased 2-3 times growth compared to the control (non-transgenic) [8]. The use of CMV promoters has also been succesfull in driving transgene expression in the embryonic African catfish (*Clarias gariepinus*), zebrafish and *rosy barb* which ranged from 25-50 % [9]. The construction of the recombinant expression vector is pCMV/lac Z containing CMV IE promoter and SV 40 polyadenylation used at a concentration of 50 ug/mL.

The success of gene transfer in the larvae is detected using PCR to confirm the transmission of the transgene in the early generations (F0 = *founder*). Mating between F0 generation produces offspring F1 and the presence of transgenes in individuals F1 were detected to confirm transgene transmission from F0 generation to F1 generation. Research on transgenesis using SMGT in Indonesia have successfully been done on transgenic grouper duck (*Cromileptes altivelis*) containing the construction gene pktBP-ktGH with a concentration of 10 ug/mL on sperm electroporation with 50 $Vcm^{-1}$pulse length, 30 milliseconds shock duration, with a number of 5 pulses and pulse interval of 0.1 seconds [10].

Transgenesis has also been successfully conducted in catfish (*Clarias* sp.) with tilapia GH gene (tiGH) using sperm electroporation, with optimal conditions as follows: electric field strength of 125 $Vcm^{-1}$, pulse length of 30 milliseconds, with a number of 5 pulses and a pulse interval of 0.1 seconds. mBP-tiGH construction transfer efficiency on catfish was around 93 % for a DNA concentration of 80 ug/mL [11]. The results of the study by Dewi *et al.* [12] also successfully inserted stripped catfish GH gene contained in the construction p*Cc*BA-*Ph*GH on stripped catfish (*Pangasionodon hypophthalmus*) with a transfer efficiency of 85.71 % for a plasmid DNA concentration of 90 $ugmL^{-1}$. A shock duration of 30 milliseconds, shock interval of 0.1 seconds, a strong electric field (electric field strength) of 125 $Vcm^{-1}$ were the optimal conditions in the study.

The results of the research in 2015 showed the gene transfer of the African catfish (GH exogenous) growth hormone contained in pTarget-CMV-*Cg*GH that was successfully inserted into mutiara catfish sperm through an electroporation of 125 $Vcm^{-1}$ with a number of 3 pulses . The African catfish GH inserts was detected in 1000 bp size fragment position (96 % identical to the GH *C. gariepinus*) with the bankgen accession number AF416488.1. The body size of F0 transgenic mutiara catfish at the age of four months was around three times larger than the size of the non-transgenic mutiara catfish, which shows the effects of transgenic catfish gigantism (over-expression of exogenous GH) [13].

Based on above description, it is necessary to evaluate the character of the growth of F1 transgenic mutiara catfish (containing GH exogenous) using parameters of heritability ($h^2$) and heterosis (H) as a candidate of superior catfish.

## 2. Methodology

### *2.1. Design of the GH lele dumbo expression vector construction*

The research method of the growth hormone gene transfer of the African catfish (*Cg*GH) to the sperm of mutiara catfish to support the production transgenic mutiara catfish (carrier *Cg*GH) consists of a stage of the *Cg*GH transfer with the sperm electroporation technique and the detection of catfish transgenic F0 and F1 (offspring of the first generation ) by using the PCR (Polymerase Chain Reaction). Based on the results of the research in 2015, the construction of a gene was successfully designed to produce a transgenic mutiara catfish pearl carrier of GH African catfish which was pTarget-CMV-*Cg*GH containing CMV promoter sequences and fragments of African catfish (*Cg*GH) GH as inserts of GH exogenous to the construction of the gene (figure 1). The construction of genes (African catfish growth hormone gene expression vector) has been integrated in the genomic DNA of mutiara catfish sperm as detected in F0 transgenic catfish.

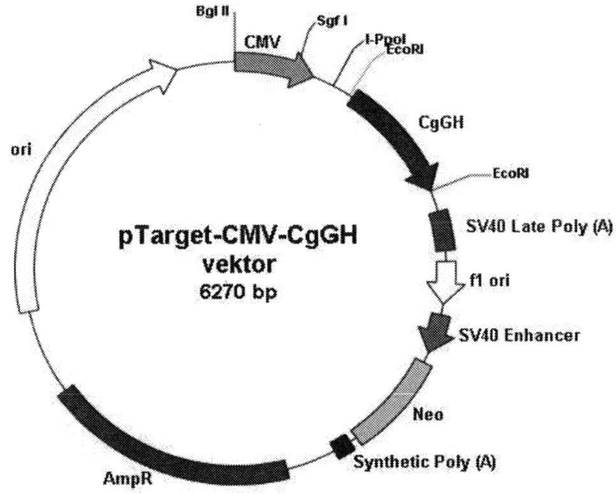


**Figure 1.** Construction pTarget-CMV-*Cg*GH (6270 bp).

Before the transgenes (pTarget-CMV-*Cg*GH) were transferred to the sperm, the sperm of mutiara catfish broodstock (aged 1 year, weight of 1.8 kg) was obtained by injecting ovaprim hormones (at a dose of 0.2 mL per kg weight of the broodstock ) in order to induce the spermiation of the broodstock. As much as 0.5 mL sperm was collected in each petri dish and then diluted with a solution of Na-physiological (isotonic solution) to maintain sperm motility with the ratio of 0.5 mL sperm and 0.5 mL Na-physiological solution (1: 1). The female mutiara catfish broodstock were injected with ovaprim hormones at a dose of 0.5 mLkg$^{-1}$ body weight and stripped 12 hours after injection to ovulate. The eggs were collected in a plastic cup (dry conditions without water).

*2.2. Transfer GH Lele Dumbo With Sperm Electroporation Technique*

The volume of the DNA recombinant vector (pTarget-CMV-*Cg*GH) solution that was used to transfer mutiara catfish sperm (*Clarias* sp.) was at about 60 ng / µl or 60 µg / mL. The electroporation process was done by mixing the sperm and DNA vector into the genepulser II (Biorad) electroporator cuvette machine with the square wave shock type of 125 Vcm$^{-1}$; shock duration of 30 milliseconds; shock interval of 0.1 seconds, and a number of 3 pulses using a 0.4 cm cuvette. The African catfish GH transfer electroporation into mutiara catfish sperm technical flow exercise is presented in figure 2.

The sperm that has been electroporated was then mixed with mutiara catfish eggs and transferred to the aquarium for the fertilization of the egg with a water temperature of about 26-27 $^{0}$C until the eggs hatch into catfish larvae. The detection of the success of gene transfer of the African catfish growth hormone (GH exogenous) that is integrated in the genome of F0 transgenic mutiara catfish and F1 is required to confirm the increased growth of transgenic fish due to over-expression of exogenous GH.

IOP Conf. Series: Earth and Environmental Science **137** (2018) 012004 doi:10.1088/1755-1315/137/1/012004

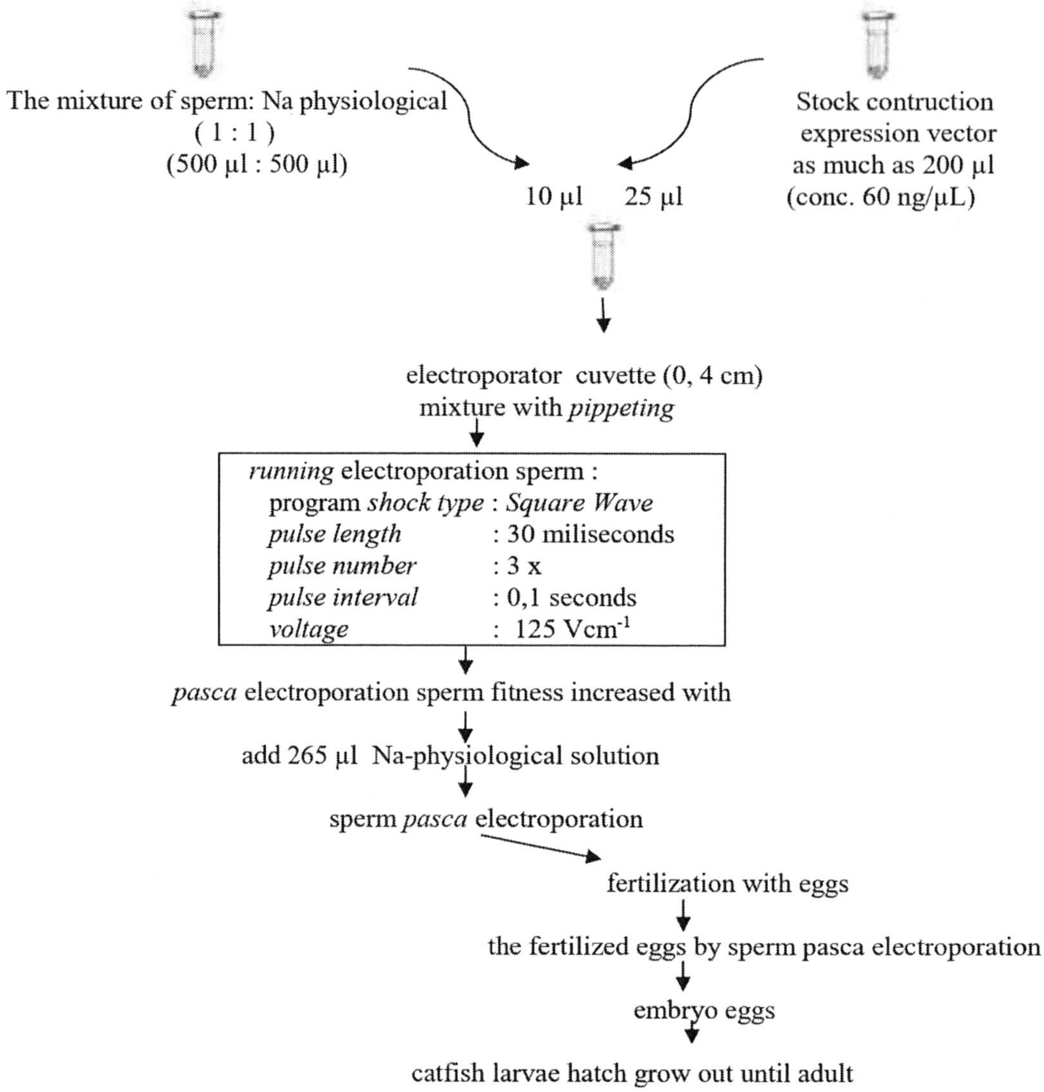


**Figure 2.** The exercise of transfer Target-CMV-CgGH to mutiara catfish sperm.

The existence of the insertions of African catfish growth hormone gene on F0 mutiara catfish broodstock (2015) and mutiara catfish first offspring (F1) (2016) needs to be detected using PCR to confirm the inheritance of the GH exogenous (GH African catfish) from F0 broodstock to offspring F1. Cg-F primer (5'-ATGGCTCGAGTTTGGTGCTGCT-3') and Cg-R(5'-CTACAGAGTGCAGTTGGAATCCAGGG-3') was used to copy sequences of African catfish GH, while CgGH-Fprimer (5'-CAGTACATCAAGTGTATCAT-3') and CgGH-R (5'-AGATAAGTCTCCACTTTGT- 3') was used to copy fragments of CMV-CgGH (transgene). The growth response of fish to the given feed is in correlation with the character of the fish growth which shows an improvement in the growth of fish and was evaluated using the parameters of feed conversion ratio (FCR). The inheritance of superior character (super growth) in the F0 and F1 generation was measured by heritability ($h^2$) which is an indicator of inheritance evaluation of superior character of F0 broodstock to F1 offspring for a quantitative phenotype, while evaluating the performance superiority of fish in F1 offspring was measured using the heterosis (H) test.

IOP Conf. Series: Earth and Environmental Science **137** (2018) 012004 doi:10.1088/1755-1315/137/1/012004

## 3. Results and Discussion

### *3.1. Detection of GH lele dumbo in f0 mutiara catfish sperm*

The results of electroporation in the sperm of mutiara catfish at 125 $Vcm^{-1}$ with a number of 3 pulses shows the success of the gene transfer of African catfish GH (*Cg*GH) in the sperm of these catfish, which is indicated by the detection of 600 bp sized fragments which is a gene of African catfish GH on mutiara catfish sperm (figure 3).

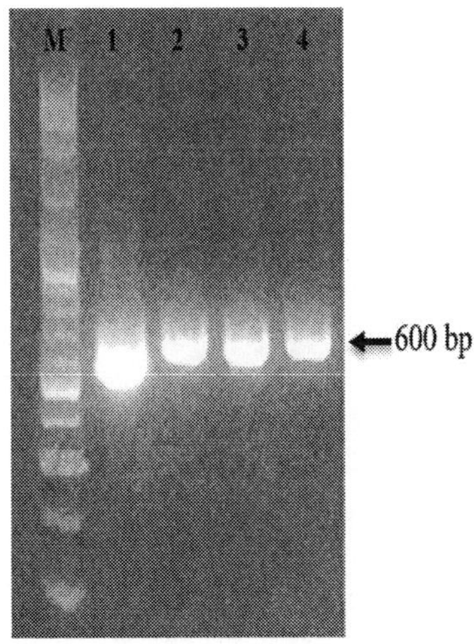


**Figure 3.** Detection *Cg*GH (using primer Cg-F and Cg-R).

The combination treatment of voltage and the number of pulses in sperm electroporation will increase the permeability of the cell, causing exogenous polynucleotide (foreign DNA) into the cell plasma membrane of sperm. More foreign DNA will be transferred by the sperm to the egg at fertilization which is effective in producing transgenic fish carrying a new phenotype (figure 4 and 5). The transgene (African catfish GH gene carrier) can be transmitted back through sexual reproduction of broodstock (F0) to the first offspring (F1) with the expression of a new phenotype.

**Figure 4.** Broodstock mutiara transgenicF0 ♂ (A) and Broodstock mutiara non-transgenic F0 ♂(B) aged 11months.

*3.2. Fragment detection of CMV-CgGH OnF1 transgenic mutiara catfish*

The detection of CMV-fragment transmission *Gg*GH (transgen) in the F1 transgenic catfish was carried out to verify permanent inheritance inserts of African catfish GH in the context of the mass production of transgenic catfish. The design of primers used to copy sequences of CMV-*Cg*GH was designed on-line with Pimer3 software version 4.0 and the size of 1434 bp fragment was obtained, called CgGH-F (5'-CAG TAC AAG TGT ATC ATC AT-3 ') and CgGH-R (5 '-AGA TAA TTG GTC TCC ACT T-3'). Catfish genome amplification on the sample F1 transgenic mutiara with PCR program pre-denaturation at 95 $^0$C (5 minutes), denaturation at 94 $^0$C (45 seconds), annealing at 49.2 $^0$C (45 seconds), extension at 72 $^0$C (2 minutes) and a final extension at 72$^0$C ( 7 minutes) and the number of cycles at 35 times to produce a product size of approximately 1500 bp (Figure 6).

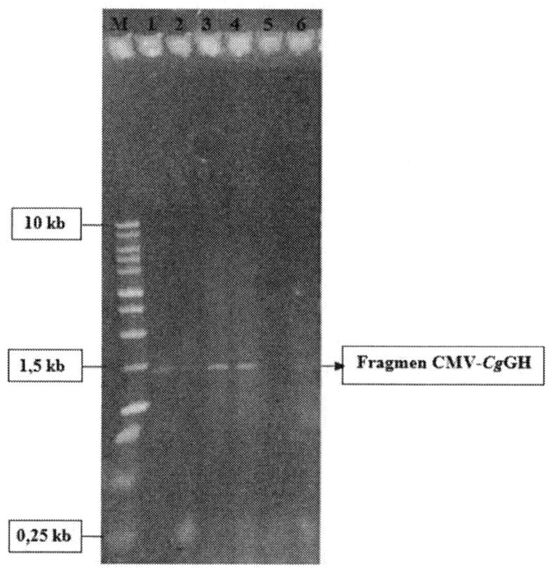


**Figure 5.** Fragments CMV-*Cg*GH in F1 transgenic mutiara catfish. (M = marker DNA ladder 1 kb (Promega), 1 = sample of fin genome of F1 transgenic mutiara catfish (fish number 1), 2 = sample of fin genome of F1 transgenic mutiara catfish (fish number 2), 3 = sample of fin genome of F1 transgenic mutiara catfish (fish number 3), 4 = sample of fin genome of F1 transgenic mutiara catfish (fish number 4), 5= sample of fin genome of F1 transgenic mutiara catfish (fish number 5), 5 = sample of fin genome of F1 transgenic mutiara catfish (fish number 6).)

Transgenic catfish F0 and F1 was detected using this pair of primers, and the insertion transmission of African catfish GH (*Cg*GH) inherited from the broodstock (F0 transgenic mutiara catfish) to the offspring (F1 transgenic mutiara catfish). Based on the designed primer using Pimer3 version 4.0 software on-line, a fragment size of 1434 bp was obtained, hence the primer CgGH-F and CgGH-R is a primer suitable for detecting transgenic catfish containing a portion of the insertion fragment promoter CMV connected with partial fragment inserts of African catfish GH (CMV-*Cg*GH fragment size of about 1500 bp) (figure 6). Similar results were obtained from the studies of Dewi *et al.* [14], that the size of the transgene (fragment p*Cc*BA-*Ph*GH) was detected in F3 transgenic African catfish of 1500 bp. Transgene expression (CMV-*Cg*GH) in F1 transgenic mutiara catfish is driven by the CMV promoter as a transgene expression regulator and demonstrated that the transgene was stably transmitted from parent F0 to F1 generation.

*3.3. Performance growth F1 transgenic mutiara catfish*

To prove the inheritance of the growth rate characterization of F0 transgenic mutiara catfish which had properties that grew threefold (474 g/133.9 g) compared to non-transgenic catfish (normal mutiara catfish), the calculation of the value of heritability in the first offspring (F1) transgenic mutiara catfish was required. Tsai [5] reported that the growth of F2 transgenic fish loach (*Misgurnus anguilicaudatus*) was twice the growth of the non-transgenic. Over expression of growth also occurred in F2 transgenic salmon (*Salmo salar*) of 2.62 to 2.85 times the growth of non-transgenic salmon. Results with a cross of F0 male transgenic mutiara catfish with a female Sangkuriang catfish produced a hybrid offspring F1 mutiara transgenic showing rapid growth as the broodstock when the fish was aged 4.5 months. The average growth of F1 transgenic mutiara catfish hybrid (age 4.5 months) at 460.74 g is close to the average growth of F0 transgenic mutiara catfish (age 4 months) which amount to 474 g (figure 7).

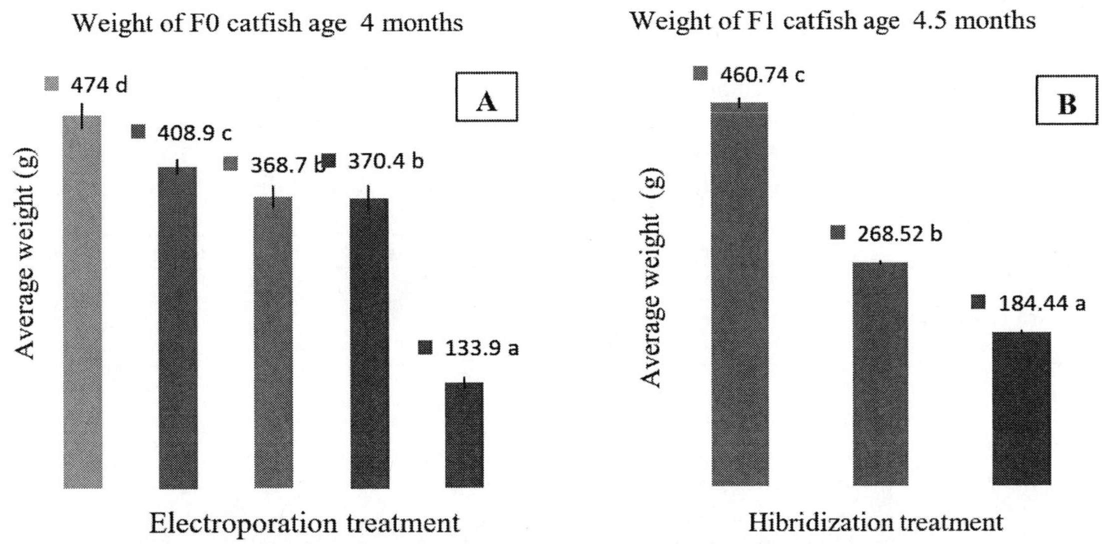


**Figure 6.** Inheritance character catfish growth of transgenic. F0 (A) to F1 (B).

Rapid growth properties that is possessed by the offspring of the hybrid of F1 transgenic mutiara catfish in particular is the average weight of fish F1 97.19% (460.74 g / 474 g) was identical to the average weight of the parent (F0 transgenic mutiara catfish) when the fish was aged 4.5 months. This is an indication that the additive gene effects on the parent is fully inheritable on to the offspring (F1) [15]. The level of inheritance contained in the F1 generation (offspring of hybrid transgenic mutiara catfish) could explain that the additive effect is inheritable from the parent (F0 transgenic mutiara catfish), especially the character of high growth (super growth) that is ia trait inherited by its broodstock. The heritability weight growth of transgenic mutiara catfish offspring F1 hybrid aged 4.5 months is presented in table 1 below.

**Table 1.** Heritability ($h^2$) of transgenic mutiara catfish offspring F1 hybrid age 4.5 months.

| Grwoth character | Purebreed of transgenic hybrid | | Control |
|---|---|---|---|
| | F1 (♂ MT x ♀ S) | F1 (♂ MNT x ♀ S) | F1 (♂ S x ♀ S) |
| Average weight (g) | 460.74 | 268.52 | 184.44 |
| Deviation Standart | 6.32 | 2.31 | 2.22 |
| Varians | 39.92 | 5.35 | 4.94 |
| Heritability ($h^2$) | 17.55 % | 10.07 % | - |

Note:

♂ MT = Male of Mutiara Transgenic
♀ S = Female Sangkuriang
♂ MNT = Male of Mutiara Non-Transgenic

The results presented in table 1 above show that the transgenic mutiara catfish offspring F1, especially transgenic mutiara catfish hybrid (♂ MT x ♀ S) has a value of $h^2$ of 17.55%, which means that the growth rate of phenotype inheritance (the size of the individual weight) of parent (F0) was inherited in the offspring (F1), with a potential inheritance that is quite high, therefore the value of $h^2$ was above 15% [16]. Heritability measurement results of the other catfish group, namely the stripped catfish (*Pangasionodon hypophthalmus*) from the cross three strains by diallel crossing 3 x 3, yielding a value $h^2$ by 17% in the offspring hybrid third generation (G3) [17]. The heritability value of the hybrid offspring of F1 transgenic mutiara catfish ($h^2$ = 17.55%) is slightly higher than stripped catfish of the third-generation hybrid offspring with a heritability amount of 17%. The calculation of the value of heterosis (H) weight of fish obtained from F1 transgenic mutiara catfish offspring hybrid (♂ MT x ♀ S) amount to 42.83% higher compared to F1 non-transgenic mutiara catfish hybrid (♂MNT x ♀ S) which is only 18.56% of the average weight of pure strains of catfish (♂S x ♀ S) (table 2).

**Table 2.**Heterosis (H) value of hybrid F1 transgenic mutiara catfish.

| Growth fenotipe | Purebreed of transgenic hybrid | | Purebreed |
|---|---|---|---|
| | F1 (♂ MT x ♀ S) | F1 (♂ MNT x ♀ S) | F1 (♂ S x ♀ S) |
| Average weight (g) | 460,74 | 268.52 | 184.44 |
| Heterosis (H) | 42,83 % | 18.56 % | - |

Note:

♂ MT = Male of Mutiara Transgenic
♀ S = Female of Sangkuriang
♂ MNT = Male of Mutiara Non-Transgenic

This evidence shows that the growth of the highest weight obtained from the cross male transgenic mutiara catfish (♂ MT) and female Sangkuriang (♀ S) than the cross pure strains (♂S x ♀ S) or weight growth of hybrid transgenic mutiara is 42.83% more than the weight of the pure strain. The heterosis value (42.83%) was obtained from the results of this study (cross ♂ MT x ♀ S) by 2-fold (42.83%: 18.56%) of heterosis in hybrid cross ♂ MNT x ♀ S ) (table 2).

Feed conversion ratios for growth show that F1 transgenic mutiara catfish are more feed efficient (FCR = 0.85) compared to non-transgenic mutiara catfish (FCR = 1.38) and sangkuriang catfish (FCR = 1,5) (table 3).

**Table 3.** Feed convertion ratio of F1 mutiara catfish hybrid offspring.

| F1 offspring hybridzation | ♂ S x ♀ S (pure strain) | ♂ MNT x ♀ (non-transgenic) | ♂ MT x ♀ S (transgenic) |
|---|---|---|---|
| Weight fenotipe : | | | |
| -$Wt_1$ | 3500 g | 4860 g | 5400 g |
| .$Wt_{30}$ | 4980 g | 7250 g | 12420 g |
| Feed amount 1 months | 2178 | 3307 g | 6000 g |
| *FCR* | 1.5 | 1.38 | 0.85 |

Note :
$Wt_1$ = biomass weight of early treatment (1th days))
$Wt_{30}$ = biomass weight of end treatment (30th days)

MT = mutiara transgenic
MNT = mutiara non transgenic

These results confirm the fact that the offspring of transgenic mutiara catfish F1 by consuming feed (feed protein content hi-provite 30%) of 850 g can grow up to 1000 g (1 kg), thus economizing the use of 150 g of feed. This was considered efficient for the fish farming business where 60% of production costs are for the fish farming feed requirements [18]. This efficiency is associated with the over-expression of the two genes contained in the GH transgenic fish so that the synthesis of growth hormone levels causes an excessive and rapid growth rate of fish. The research of Marnis *et al.* [19] showed that the F2 transgenic fish of African catfish had a high feed efficiency (FCR 0.86) compared to non-transgenic catfish (FCR 1.79). Kobayashi *et al.* [3] also showed that transgenic tilapia had a 35% higher fish feed efficiency than non-transgenic tilapia which led to increased growth. The high feed efficiency of transgenic fish was due to the exogenous GH (GH inserted) which stimulated the production of IGF-I (Insulin-Like Growth Factors) in the liver in response to GH accelerating their metabolic rate to transform feed protein into the growth of fish. Plasma GH and transgenic IGF-I catfish increases higher than the non-transgenic fish, showing that the synthesis of GH transgenic fish was much higher than non-transgenic fish, causing the fish to be very responsive to feed due to increased levels of GH plasma and tend to be satiated, thereby reducing the consumption of feed [18, 20].

## 4. Conclusion

- The transgenic catfish (mutiara strain) F0 and F1 carrier inserts African catfish GH (600 bp) can be produced.
- CMV fragment connected with *Cg*GH fragment (1500 bp) was detected using the primers CgGH-F and CgGH-R in the F1 transgenic mutiara catfish.
- The value of heritability ($h^2$) of F1 transgenic mutiara catfish offspring was 17.55%.
- The heterosis (H) growth of F1 transgenic mutiara catfish hybrid offspring was 42.83%.
- F1 transgenic mutiara catfish hybrids provide the best growth performance (FCR 0.85).

## 5. References

[1] Yaskowiak E S, Shears M A, Agarwal-Mawal A, Fletcher G L 2006 *Trans. Res.* **15** 465-480
[2] Du S J ,Z Gong, G L Fletcher and C L Hew 1993 *J. Mar.Biotech.* **3** 216-219
[3] Kobayashi S-I A, T Morita, M Miwa, J Lu, M Endo, T Takeuchi and G Yoshizaki 2007 *Aquaculture* **270** 427-435
[4] Nam Y K, J K Noh, Y S Cho, H J Cho, K N Cho, C G Kim and D S Kim 2001 *Trans. Res.***10** 353 – 362
[5] Tsai H J 2000 *Mol. Reprod.Dev.* **56** 281- 284
[6] Hostetler H A, L P Stephanie and W M Muir 2003 *Trans. Res.* **12** 413-424.
[7] Noh J K, K N Cho, E H Han, A Kim, J S Lee, D S Kim and C B Kim 2003 *Mar. Biotechnol.* **5** 244-252
[8] Pandian T J and T Venogupal 2004 Contribution to transgenesis in Indian major carps *Labaeo rohita* In: *Fish Genetics and Aquaculture Biotechnology* (USA: Science Publishers, Inc.) p1-20 (edited by T J Pandian, C A Strussman and M P Marian)
[9] Muller F, L Zsolt, V Laszio, M Lazsio and O Laszio 1993 *Federation of European Biochemical Societies* **324** 27-32
[10] Subyakto S, Alimuddin, Rustidja, M S Jati, I Faizal, R S Aliah, G Astutik and K Sumantadinata 2011 *Biotropia* **18** 13 – 23
[11] Gusrina A, K Sumantadinata and U Widyastuti 2009 *J. Res. Aquac.* **4** 333-340
[12] Dewi R R S P S, Alimuddin, A O Sudradjat and K Sumantadinata 2010b Effectiveness of transfer and gene expression *Ph*GH In stripped catfish (*Pangasionodon hypophthalamus*). 2010. Final Report of Research Results LRPTBPAT Sukamandi

IOP Conf. Series: Earth and Environmental Science **137** (2018) 012004 doi:10.1088/1755-1315/137/1/012004

[13] Buwono I D, Iskandar, M U K Agung and U Subhan 2015 Production larva local catfish (*Clarias batrachus*) transgenic with sperm electroporation techniques for the introduction of aquaculture in Cirata. The final report leading research universities (PUPT). LPPM Unpad. 76 p
[14] Dewi R R S P S, J Darmawan and I Nurlaela 2014 *J. Res. Aquac.* **9** 31-37
[15] Stuart G W, Mcmurray J V and Westerfield M 1988 *Development* **103** 403–412
[16] Gjedrem T 2005 *Selection and breeding programs in aquaculture* (New York: Published by Springer, Berlin, Netherlands) 364 p
[17] Phuc T H 2016 Analysis of crosses between three strains of strin ipped catfish (*Pangasionodon hypophthalmus*) in Viet Nam (Thesis: Department of Animal and Aquacultural Sciences Norwegian University of Life Sciences) 27 p
[18] Cook J T, M A McNiven, G F Richardson and A M Sutterlin 2000 *Aquaculture* **188** 15-32
[19] Marnis H B, Iswanto, R Suprapto, Imron and R R S P S Dewi 2015 *J. Res. Aquac.* **10** 161 – 168
[20] Raven P A, M Uh, D Sakhrani, B R Beckman, K Cooper, J Pinter, E H Leder, J Silverstein, R H Devlin 2008 *Gen Comp Endocrinol* **159** 26-37

**Acknowledgment**
The authors would like to thank the Director General of Strengthening Research and Development Research and Community Service of the Ministry of Research, Technology and Higher Education that has funded this study with the DP2M 2015 Fund. Similarly, we would also like to thank the team who help in completing this research.

IOP Conf. Series: Earth and Environmental Science **137** (2018) 012005 doi:10.1088/1755-1315/137/1/012005

# Nitrite oxidizing bacteria for water treatment in coastal aquaculture system

**S Noorak[1], S Rakkhiaw[1], K Limjirakhajornt[1], A Uppabullung[1,2], T Keawtawee[1,2] and Y Sangnoi[1,2*]**

[1]Department of Aquatic Science, Faculty of Natural Resources, Prince of Songkla University, Hat Yai, Songkhla 90110 Thailand
[2]Discipline of Excellence for Sustainable Aquaculture, Prince of Songkla University, Hat Yai, Songkhla 90110 Thailand

E-mail: yutthapong.s@psu.ac.th

**Abstract.** This research aimed to isolate and characterize nitrite oxidizing bacteria and to study their capability for water quality improvement. Fourteen strains of bacteria with nitrite-oxidizing character were isolated after 21 days of enrichment in Pep-Beef-NOB medium contained $NaNO_2$. Two strains, SF-1 and SF-5, showed highest nitrite removal rate for 42.42% and 37.2%, respectively. These strains were determined an efficiency of open-system wastewater treatment for 14 days. The results showed that control, SF-1 and SF-5 had remove ammonia from day 1 to day 6. At the end of the study, ammonia was removed by the control, SF-1 and SF-5 for 81.27%, 70.1% and 69.82%, respectively. Nitrite concentration was lowest at day 8 with removal rate of 98.73%, 98.3% and 97.24% from control, SF-1 and SF-5, respectively. However, nitrite concentration in control experiment was increased again at day 11 whereas in SF-1 and SF-5 were increased at day 13. Chemical Oxygen Demand (COD) was decreased by 77.78%, 73.50% and 78.63% in the control, SF-1 and SF-5, respectively. Biological Oxygen Demand (BOD) in the control, SF-1 and SF-5 were reduced by 85.92%, 79.53% and 82.09%, respectively. Based on 16S rRNA gene, SF-1 and SF-5 were identified as *Bacillus vietnamensis* and *B. firmus*, respectively.

## 1. Introduction

Coastal aquacultures have an important economy of Thailand [1]. These are including of marine shrimp and seabass. However, the accumulation of inorganic nitrogen compounds such as ammonia, nitrite and nitrate in coastal aquaculture system is serious toxic to aquatic animals. Biological process of nitrification including ammonium oxidation and nitrite oxidation has transformed ammonia and then nitrite to nitrate. These two process are involved with nitrifying bacteria; ammonium oxidizing bacteria (AOB) and nitrite oxidizing bacteria (NOB). Our previous study suggested that heterotrophic nitrifying bacteria have ability to oxidize ammonia to nitrite but they cannot oxidize nitrite to nitrate. Therefore, this study aimed to isolate heterotrophic nitrite oxidizing bacteria and determine the water treatment efficiency.

## 2. Methodology

### *2.1. Isolation and screening of NOB*

Water and sediment samples were collected from seabass farms located in Koh Yor Island, Songkhla province. Each 1 g or 1 ml of sample was enriched into 100 ml of Pep-Beef-NOB medium (peptone 5 g, beef extract 3 g, $NaNO_2$ 0.2 g, $Na_2CO_3$ 1 g, $K_2HPO_4$ 0.5 g, $MgSO_4$ 0.5 g, $FeSO_4$ 0.4 g, sea salt 7 g, $H_2O$ 1000 ml) at 28 °C and shaken at 150 rpm for 21 days [2]. At days 7 and 14, ten milliliters of suspension were transferred to fresh Pep-Beef-NOB medium with 0.3 g and then 0.4 g of $NaNO_2$, respectively. Nitrite oxidizing reaction was tested by Griess-Ilosvay method [3, 4, 5]. Positive tested suspensions were further isolated by using serial dilution technique spreading on Pep-Beef-NOB agar medium.

### *2.2. Morphological characteristic and optimal salinity of NOB*

The NOB isolates were Gram stained and observed morphology under a light microscope. Salinity tolerance ranging of 0, 5, 10, 15, 20, 25, 30, 35 and 40 ppt were evaluated.

### *2.3. 16S rRNA gene sequencing and phylogenetic analysis*

DNA of NOB isolates were extracted by using Phenol/Chloroform method. The 16S rRNA genes were amplified by PCR using 27F and 1492R primers. The DNA sequences were analyzed and compared within the GenBank/EMBL/DDBJ database. Phylogenetic tree was constructed by using MEGA6 program [6].

### *2.4. Nitrogen removal efficiency of NOB in waste water treatment*

Seabass-cultured water was mixed and fermented with fish feed (200 g/40 L) for 3 days before used. NOB was prepared to $10^7$ CFU/ml and then 50 ml (1%) were inoculated into open container of 5 L fermented wastewater. The experiment was studied for 14 days. Every day, wastewater samples were collected in order to examine the quantities of ammonia, nitrite and nitrate. While, the amounts of BOD and COD were determined at days 7 and 14.

## 3. Results and Discussion

### *3.1. Isolation and characterization of NOB*

Fourteen bacterial strains including 10 of Gram negative and 4 of Gram positive were isolated. However, 2 from 4 Gram positive strains, SF-1 and SF-5, showed a highest nitrite removal rate in preliminary study. Strains SF-1 and SF-5 were rod shape (figure 1) and isolated from sediment samples. Optimal salinity for growth of strains SF-1 and SF-5 were 20-40 ppt and 20-30 ppt, respectively.

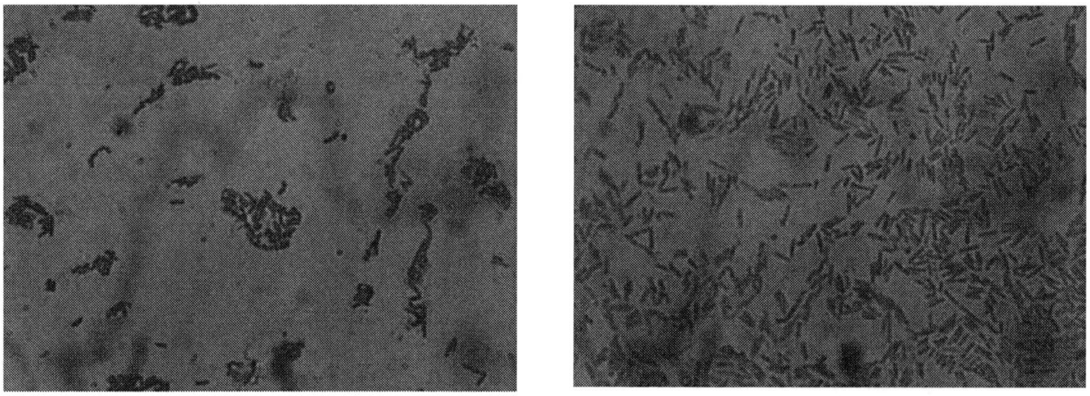

**Figure 1.** Vegetative cells of isolated SF-1 (left) and SF-5 (right). Bar = 5 μm.

### *3.2. 16S rRNA gene sequence and phylogenetic analyses*

Comparison result of partial 16S rRNA gene sequence showed that strain SF-1 has relatedness with *Bacillus vietnamensis* for 92 % similarity. Whereas, strain SF-5 has relatedness with *Bacillus firmus* for 99% similarity. These result were relevance with phylogenetic tree analysis. It demonstrated strains SF-1 and SF-5 were located into the group of *Bacillus* and showed evolutionary linear distance from relatedness neighbor species (figure 2).

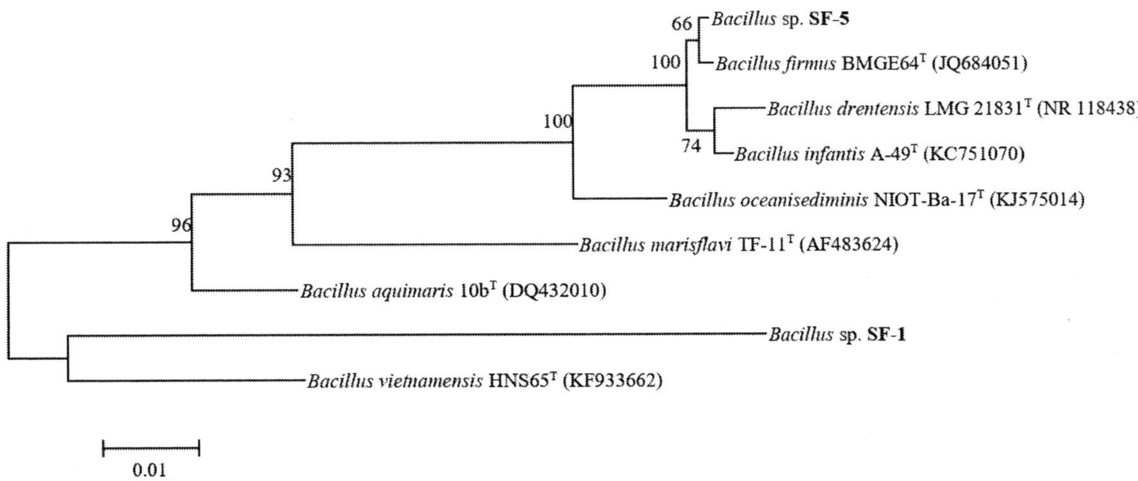


**Figure 2.** Phylogenetic tree of 16S rRNA gene of *Bacillus* strains SF-1, SF-5 and neighbor species. Bar = 0.01.

### *3.3. Efficiency of waste water treatment*

The results showed that experiments of control (no bacteria added), SF-1 and SF-5 have successively remove ammonia from day 1 to day 6. Then ammonia concentration were continuously decreased until day 14. At the end of the study (day 14), ammonia was totally removed by the control, SF-1 and SF-5 for 81.27%, 70.1% and 69.82%, respectively (figure 3).

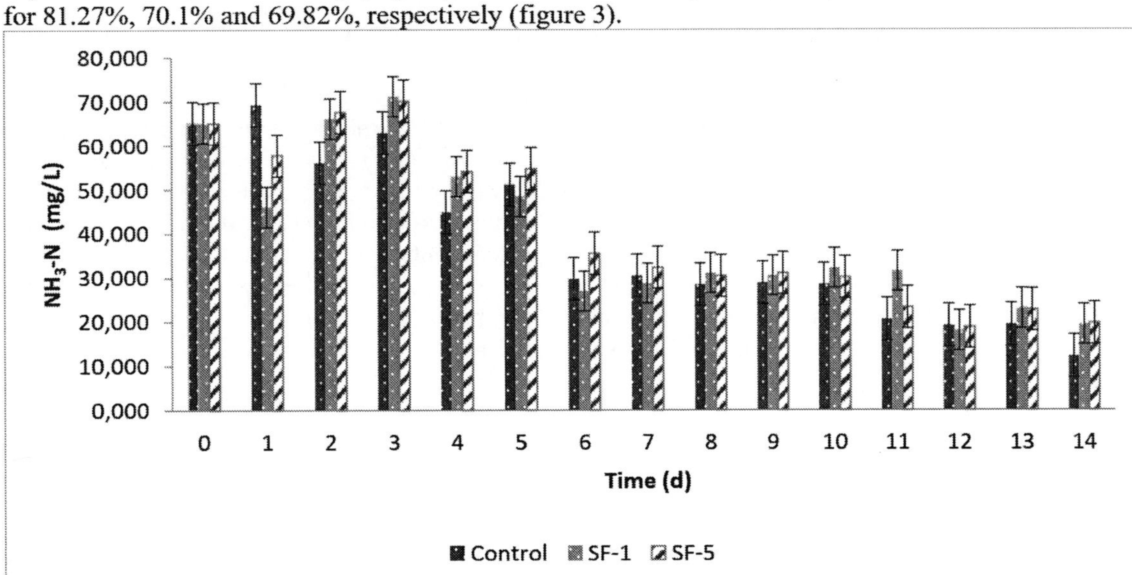


**Figure 3.** Ammonium removal efficiency of strains SF-1 and SF-5.

The initial nitrite concentration of fermented wastewater was very low (<1 mg-N/L) when compared with Pep-Beef-NOB medium. This resulted to no more nitrite for assimilation of *Bacillus* strains. However, the nitrite concentration was lowest at day 8 with removal rate 98.73%, 98.3% and 97.24% from control, SF-1 and SF-5, respectively. Nitrite concentration in control experiment was increased again at day 11 and has very high (3.09 mg-N/L) at day 14. Whereas the experiments of SF-1 and SF-5, nitrite was increased at day 13. This suggested that these two NOB isolates could remove and control nitrite better than the control experiment (figure 4).

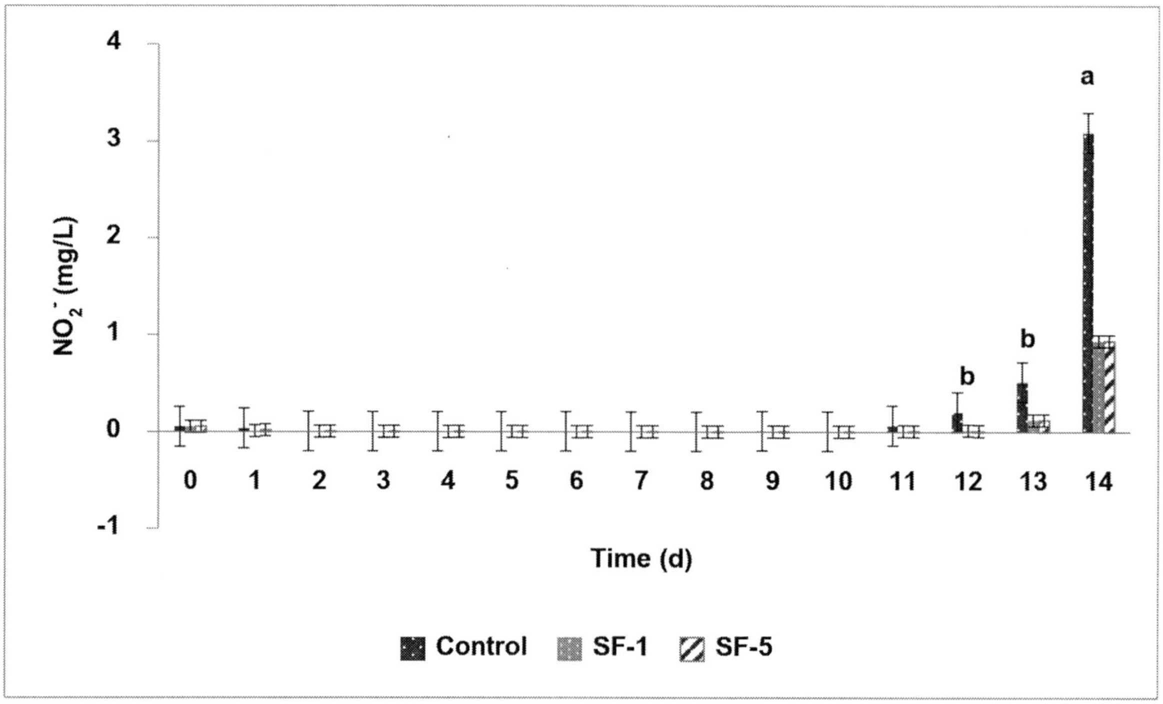


**Figure 4.** Nitrite removal efficiency of strains SF-1 and SF-5.

In case of biological oxygen demand (BOD), the control, SF-1 and SF-5 experiments showed the capable of decreasing for 85.92%, 79.53% and 82.09%, respectively. Likewise, chemical oxygen demand (COD) was reduced by 77.78%, 73.50% and 78.63% in the control, SF-1 and SF-5 experiments, respectively (figure 5-6). Although overall efficiency of wastewater treatment by NOB *Bacillus* SF-1 and SF-5 was no significance with the control experiment. Also, they have eliminated nitrite better than control experiment. This study, we added only 1 time and 1% of NOB inoculum which it may not enough quantity of cells. This resulted to reduce an efficiency of NOB. Therefore, the increasing of initial amount inoculum and frequency of inoculum adding will significant increase an efficiency of wastewater treatment of NOB *Bacillus* isolates.

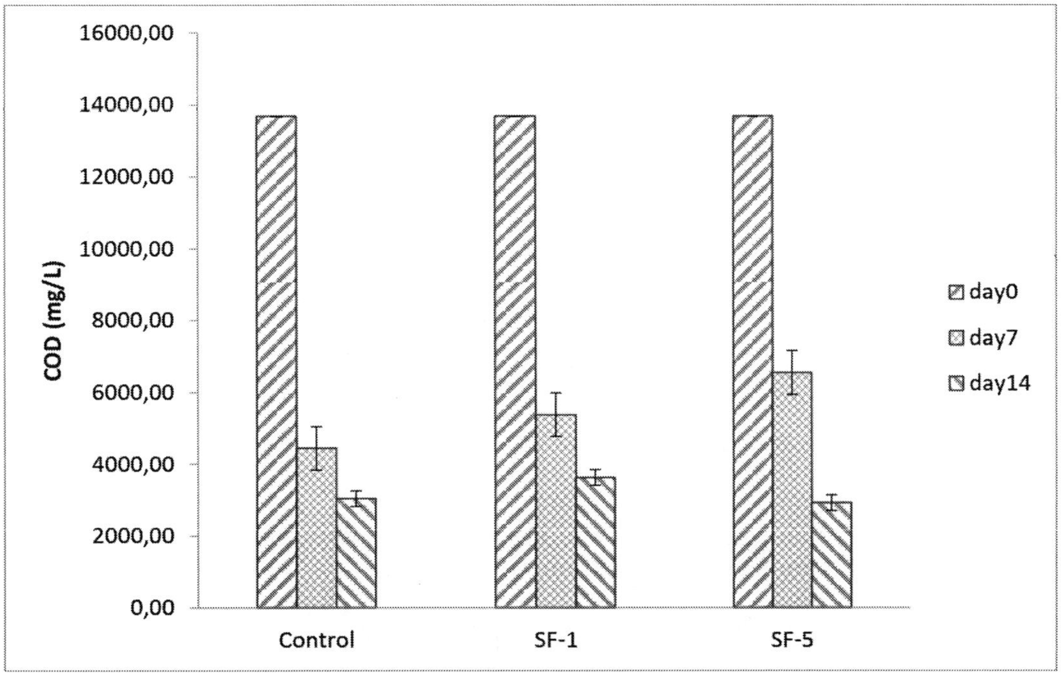


**Figure 5.** Efficiency of COD reduction by strains SF-1 and SF-5.

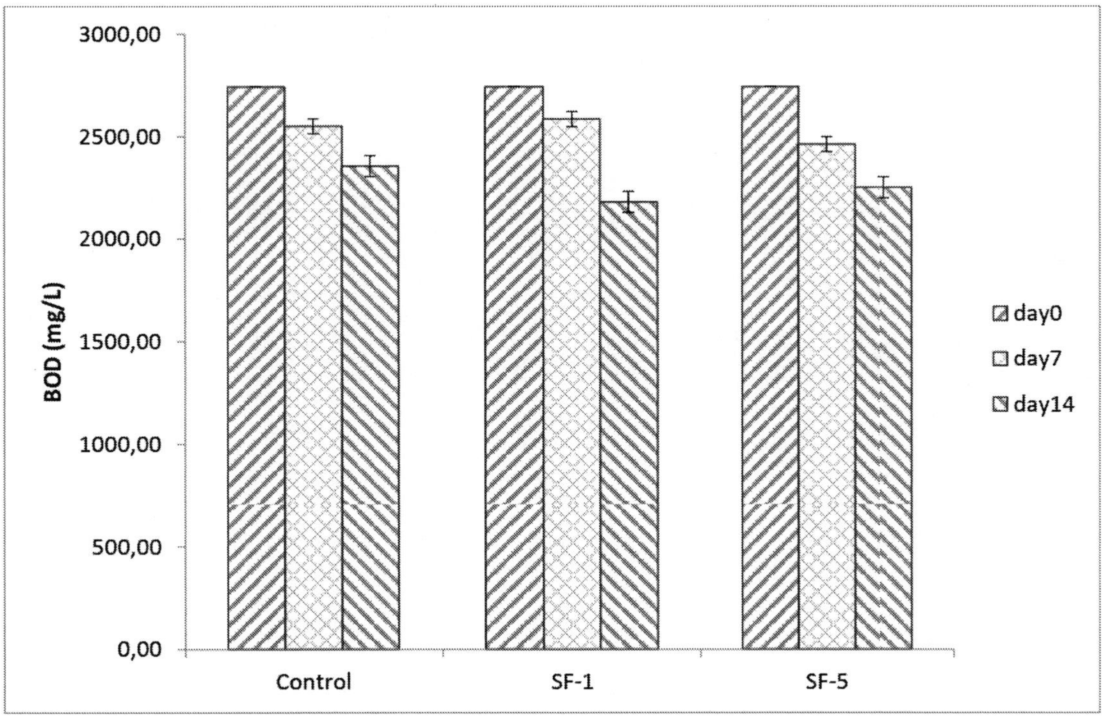


**Figure 6.** Efficiency of BOD reduction by strains SF-1 and SF-5.

## 4. Conclusion

Two NOB *Baciilus* SF-1 and SF-5 were isolated and identified as *B. vietnamensis* and *B. firmus*, respectively. Efficiency of these isolates for wastewater treatment including ammonia, BOD and COD removal was no significance when compared with control experiment. However, these NOB isolates have ability to remove nitrite better than the control experiment.

**Acknowledgements**
The authors are grateful Department of Aquatic Science, Faculty of Natural Resources, Prince of Songkla University and Discipline of Excellence for Sustainable Aquaculture, Prince of Songkla University.

## 5. References

[1] Paungfoo, C, Prasertsan, P, Burrell, P C, Intrasungkha, N and Blackall, L L 2007 Nitrifying bacterial communities in an aquaculture wastewater treatment system using fluorescence in situ hybridization (FISH), 16S rRNA gene cloning, and phylogenetic analysis *Biotechnology and Bioengineering* **97**: 985–990

[2] Ma S, Zhang D, Zhang W and Wang, Y 2014 Ammonia stimulates growth and nitrite oxidizing activity of *Nitrobacter winogradskyi Biotechnology and Biotechnological Equipment* **28**: 27–32

[3] Lu Y, Wang X, Liu B, Liu Y and Yang X 2012 Isolation and characterization of heterotrophic nitrifying strain W1 *Chinese Journal of Chemical Engineering* **20**: 995–1002

[4] Sangnoi Y, Chankaew S and O-Thong S 2017 Indigenous *Halomonas* spp., the potential nitrifying bacteria for saline ammonium waste water treatment *Pakistan J. Bio. Sci.* **20**: 52–58

[5] Yang X P, Wang S M, Zhang D W and Zhou L X 2011 Isolation and nitrogen removal characteristics of an aerobic heterotrophic nitrifying–denitrifying bacterium, *Bacillus subtilis* A1 *Bioresource Technology* **102**: 854–862

[6] Tamura K, Stecher G, Peterson D, Filipski A and Kumar S 2013 MEGA6: molecular evolutionary genetics analysis version 6.0. *Molecular Biology and Evolution* **30**: 2725–2729

IOP Conf. Series: Earth and Environmental Science **137** (2018) 012006 doi:10.1088/1755-1315/137/1/012006

# Effect of alkaloids derived from jellyfish *(Aeginura* sp.) on the intestinal histopathology and relative percentage survival (RPS) of tiger grouper *(Epinephelus fuscoguttatus)* infected by *Vibrio harveyi*

**S Andayani[1], M Fajar[1] and M F Rahman[2]**

[1]Department of Aquaculture, Faculty of Fisheries and Marine Science, Brawijaya University, Malang East Java, Indonesia
[2]Laboratory Organic Chemistry, Faculty of Mathematic and Life Sciences, Brawijaya University, Malang East Java, Indonesia

Email: yanik@ub.ac.id ; srianday_08@yahoo.com

**Abstract.** The purposes of this research were to determine the effect of alkaloid jellyfish compounds on intestinal histopathology of tiger grouper and to determine the best doses to the relative percent survival (RPS) of tiger grouper. The method of this research was descriptive with completely randomized design. The treatment of active alkaloid compound on feed was investigated for 28 days. The fish were then challenged with *Vibrio harveyi* at $10^5$ CFU/cell for 7 days. Alkaloids were added to the feed with the doses (g alkaloid/kg feed) of 0 (control); A = 0.5; B = 0.75; C = 1.0; and D = 1.25. The intestinal histopathology and RPS were observed. The best RPS was found at a treatment of C with the value of 100 %.

## 1. Introduction

The intensification of grouper culture has led to a number of disease outbreaks with an increasing range of pathogens causing them. *Vibriosis* is a common disease caused by *Vibrio carchariae, Vibrio alginolyticus, Vibrio parahaemolyticus and Vibrio harveyi* and is one of the most serious problems in various stages of grouper culture. The mortality rate of the seeds in farming can reach up to 99 %, and this is mainly caused by infections by pathogenic bacteria [1].

The rate of *Vibrio harveyi* outbreaks in the hatchery of tiger grouper can be calculated within a few hours. *Vibrio* sp. attack can lead to the destruction of organs in fish and wounds on the skin [2]. To control this disease, particularly a bacterial disease, various types of antibiotics such as chloramphenicol, erythromycin and oxytracycline have been used. But apparently, a lot of antibiotics raise the resistance of new bacterial strains in response to the disease [3, 4]. Thus, it is necessary to control the disease using natural materials which are still limited to saponin and rotenon. Consequently, a breakthrough in the utilization of jellyfish that are environmentally friendly as immunostimulants should be made [1, 5].

IOP Conf. Series: Earth and Environmental Science **137** (2018) 012006 doi:10.1088/1755-1315/137/1/012006

Jellyfish, from phylum *Coelenterata*, is both poisonous and edible and has bioactive compounds that prevent diseases in fish and shrimp. Bioactive compounds derived from jellyfish such as alkaloids, phenolic, steroids and terpenoid show a strong activity against bacteria in fish with a bigger diameter of resistance compared to chemicals and antibiotics [6]. Research on jellyfish is uncommon in spite of its potential utilization. Yahya *et al.* [7] states that *Pysalia* sp can inhibit the growth of bacteria in shrimp larvae. In addition to *Pysalia* sp., there are other jellyfish species with potential as immunostimulants such as *Gonionemus* sp., *Obelia* sp., *Hydra* sp., *Pennaria, Aurelia* sp., *Aeginura* sp., *Bougainvillia* sp. etc. [8].

Based on previous findings, it is necessary to explore unconventional natural resources in order to get the benefits of bioactive compounds from jellyfish as immunostimulants, particularly for tiger grouper seeds. Our study evaluated the effect of different fish feeds in the alkaloid substances of jellyfish on intestinal changes pathologies and its impact on the increase in the Relative Percetage of Survival.

## 2. Methodology

The extraction and isolation of the alkaloids in jellyfish (*Aeginura* sp.) was done using the modified method by Maldoni [9]. The filtered jellyfish (250 grams) was extracted with petroleum benzene. This process was repeated four times. Four days later, the extract was evaporated in vacuum at a temperature of 40°C. The resulting extract was filtered and evaporated into 10 mL of chloroform extract. After settling, the sediment of chloroform extract that contained alkaloids was separated using Thin Layer Chromatography (TLC), UV spectroscopy, infrared (IR) spectroscopy and Nuclear Magnetic Resonance Hydrogen ('H-NMR).

The methods used in this research were descriptive and experimental methods with Completely Randomized Design (CRD). The alkaloid substances of *Aeginura* sp. were used in the feed given at certain doses as follows: control = 0 g alkaloid/kg feed; A = 0.5 g alkaloid/kg feed; B = 0.75 g alkaloid/kg feed; C = 1.0 g alkaloid/kg feed; and D = 1.25 g alkaloid/kg feed. They were tested with *Vibrio harveyi* in $10^5$cfu/mL for seven days. The fish tested was tiger grouper (*Epinephelus fuscoguttatus*), 7-8 cm in size, at the age of D 90. The fish were kept in a glass tank with a volume of 15 liters (five levels of treatment) at a density of 10 fish per tank. The intestinal sample for the histopathological analysis was taken after 28 days of the administration of immune-stimulant and after the bacterial infection on Day 35. The preparation of the histological analysis of intestine used method by Lightner [10]. The RPS of grouper fish was recorded at the beginning (day 1) and the end (day 38) of the study.

## 3. Results and discussion

### *3.1. Characterization of alkaloids Aeginura* sp.

#### *3.1.1 Characterization of* spectroscopy *UV*

For preparative TLC analysis, the ratio of chloroform to methanol was 2:8. This dilution was used for the preparative TLC analysis, UV spectroscopy, infrared spectroscopy and H'NMR.

The characterization of molecule by UV yielded the results of spectra, absorbance data and maximum wavelength of jellyfish, which are presented in table 1 and figure 1.

According to the qualitative results, the absorption of chloroform extract showed strong wavelengths of about 239.4 nm and 284.8 nm.

**Table 1.** UV Maximum Wavelength.

| Species | $\lambda_{maks}$ (nm) | $A_{maks}$ |
|---|---|---|
| *Aeginura sp* | λ1 = 239.4 | A1 = 0.2002 |
| | λ2 = 284.8 | A2 = 0.0406 |

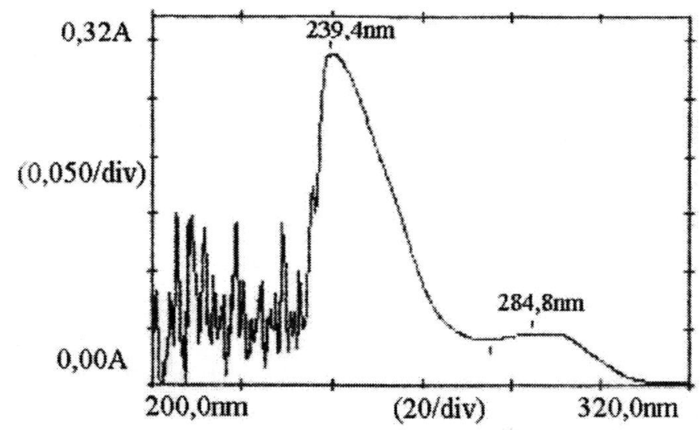


**Figure 1.** Spectra UV of *Aeginura* sp.

The transition could relate to carbonyl bond (C=O) or C=N and N-C=C with alkaloid characteristics. Based on the molecular analysis of the chloroform extract, there were unsaturated molecules containing aromatic nucleus and heteroatom, carbonyl bond C=N and characteristic alkaloid.

*3.1.2 Characterization with infrared spectroscopy*

Infrared spectroscopy was used to support the H'-NMR spectra data in determining the chemical structure of alkaloid extract. The results of the identification using infrared spectroscopy indicated that the molecules analyzed contained some functional chains, each of which had an absorption band at specific wavenumber as shown in figure 2.

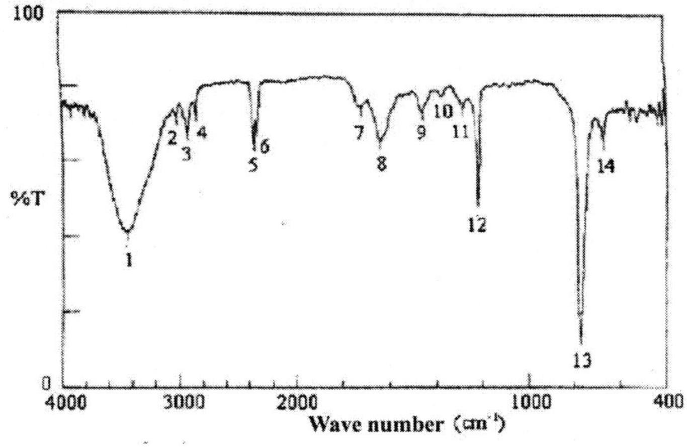


**Figure 2.** Spectra IR molecular of *Aeginura* sp. isolate.

The occurrance of absorption band at a wavenumber of 3.441.32 $cm^{-1}$ shows the existence of stretching vibrations in the functional bond O-H. The N-H stretching vibrations of primary and secondary amine showed an absorption band at a wavenumber of 3.750—3.000 $cm^{-1}$. There is a possibility that the molecular functional chain of alkena in aromatic system supported by the occurrence of absorption band at a wavenumber of 3.022.73 $cm^{-1}$ arose out from C-H stretching vibrations. The aromatic system in the molecular analysis was also supported at a wavenumber 1.714.87 $cm^{-1}$. According to Silverstein *et al.* [11], aromatic hydrocarbon is at a wavenumber of 2.000—1.650 $cm^{-1}$.

The pyrimidine ring is a molecule supported by absorbance at a wavenumber of 1.643.50 $cm^{-1}$. C-N chain is at a wavenumber of 1.280.85 $cm^{-1}$ and 1.360.10 $cm^{-1}$. The C=N stretching vibrations will show an absorption band at a wavenumber of 1.689—1.471 $cm^{-1}$.

### *3.2 Intestine histopathology*

The results of the study on the condition of the tiger grouper's intestine after being given an alkaloid immune stimulant showed a normal histological form with the length of villi, width of lamina propria, width of villi, thickness of muscle layers and number of goblet cells per segment being measured (figure 3). The samples were taken from all three parts of the intestine (proximal, medial and distal) [12].

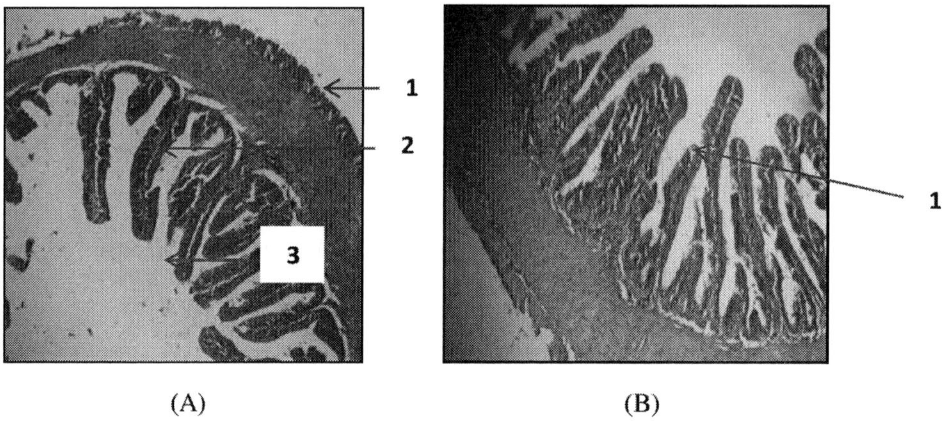


**Figure 3.** (A) Normal cross section of the intestine after an alkaloid immune stimulant consisting of (1) muscularis external (2) Villi (3) goblet cells, 400x HE (bar = 100 um). (B) After tested challenge with *Vibrio* liver damage such as (1) fusion 400 x HE, (bar = 100 μm).

Figure 3 shows the intestinal bacterial villi controlled with fusion-attached bacteria, and there was no tissue damage in the fish in treatments A, B, C and D. As a result, the bacterial attack of *V. harveyi* in the gut does not necessarily cause anatomical changes.

*V. harveyi* causes death from decreased appetite or abnormal movement of the fish. As suggested by Karthigayani *et al.* [13], some illnesses appear to involve functional impairment (e.g.: decreased appetite or abnormal movement) which actually reaches clinical symptoms even though at that time no anatomical abnormality was found.

IOP Conf. Series: Earth and Environmental Science **137** (2018) 012006 doi:10.1088/1755-1315/137/1/012006

*3.3 Relative percentage of survival*

The results showed that after the immune stimulation was given, the RPS value was 100 % for all treatments. But after the administration of immune stimulation, the best RPS was gained by the tiger grouper in treatment C with 1 gr alkaloids/kg of feed (100 %), followed by treatment B (70.59 %) and treatment D (35.29 %), and the smallest RPS value was gained by the fish in treatment A (23.53 %). The average RPS obtained at the end of the study of each treatment with different replicates can be seen in table 2.

**Table 2.** Relative Percentage of Survival.

| Treatment | Replicates | | Total | SR (%) | Death (%) | RPS (%) |
|---|---|---|---|---|---|---|
| | 1 | 2 | | 0 ± SD | 0 ± SD | 0 ± SD |
| 05 gr alkaloid | 30 | 40 | 70 | 35 ±0.7070 | 65±0.7071 | 23.53±8.3191 |
| 0.75 gr alkaloid | 80 | 70 | 150 | 75± 0.7070 | 25±0.7071 | 70.59±8.3198 |
| 1 gr alkaloid | 100 | 100 | 200 | 100 ± 0 | 0 | 100±0 |
| 1.25gr alkaloid | 50 | 50 | 100 | 55 ±0.7071 | 45±0.701 | 35.29±8.3184 |
| Control | 10 | 20 | 30 | 15 ±0.7071 | 85±0.7071 | |

The lowest RPS value was 23.53 %. According to Sakai [5], the main factors determining the effect of a compound are the dose and the concentration of the compound. The concentration of alkaloids *of Aeginura* sp. as an immune stimulant ingredient fed to the tiger grouper did not provide any immune response. The dose was too high that the immune stimulant effect could not increase the immunity. Because the fish body was unable to respond to the mechanisms of cellular and humoral response, antibodies were not formed and toxic effects may be resulted.

*Aeginura* sp. has cytotoxic bioactive compounds. These toxins can disrupt or destroy cell membranes, which can disrupt the transport of compounds in and out of cells. However, the results of this study were better than those of Burrels *et al.* [14], which used 2 % β-glucan + 2% immune stimulant nucleolide for 21 days with RPS maintained at 37 % in Rainbow Trout after *Vibrio anguillarum* infection.

## 4. Conclusion

The administration of immune stimulant did not cause damage to tiger grouper intestine after being tested with *Vibrio harveyi.*

The highest Relative Percentage of Survival value was gained in treatment C with 1 gr alkaloid / kg of feed (100 %), followed by treatment B (70.59 %) and treatment D (35.29 %), and the smallest RPS value was gained in treatment A (23.53 %).

## 5. References

[1]. Harikrishnan R, Balasundaram C and Heo M S 2010 *J. Aquac.* **30** 1-14
[2]. Moriarty D J W 1997 *J. Aquac.* **151** 333-349
[3]. Harikrishnan R, Balasundaram C, and Heo M S 2011 *J.Fish Shellfish Immunol.* **30** 128-134
[4]. Hamed A S, Rahaman K H, Alagan A and Yoganandhan K. 2003 *J. Aquac.* **217** 39-48
[5]. Sakai M 1999 *J. Aquac.* **172** 63-92
[6]. Andayani S 2007 Effect of bioactive alkaloids jellyfish (*Bougainvillia* sp) as immunostimulant toward bactericidal activity, non-specific immune responses and survival (RPS) of tiger grouper

(*Ephinephelusfuscoguttatus*) (Malang : Dissertation.Graduate Program in University of Brawijaya) 155 p (In Indonesia)
[7]. Yahya, Sukoso, Aulanni'am, Bagus S B U, and Basmal J 2002 *J. Life Sci.* **14** 106-119.
[8]. Barnes R D 1974 *Invertebrata zoology* (Toronto : W.B. Saunders Company. Toppan Company, LTD. Piladelpia. London) 870 p
[9]. Maldoni B 1991 *J.Chem.Edu.* **68** 700-707.
[10]. Lightner D V 1996 *Handbook of shrimp pathology and diagnostic procedures for diaseases of cultured penaeid shrimp* (Lousiana : Word Aquaculture society) 157 p.
[11]. Silverstein R M, Bassler G C and Morrill T C 1981 *Spectrometric Identification of Organic Compounds* 4$^{th}$ Ed (Singapore : John Wiley and sons, Inc.)
[12]. Silva M R, Natali M R M and Hahn N S 2012 *Bio. Sci.* **34** 319-326 p.
[13]. Karthigayani T, Denis M, Rexlin A, Remy A and Shettu A. 2014 *J. Mod. Biotech.* **3** 48-53
[14]. Burrells C, William P D and Forno P F 2001 *J. Aquac.* **199** 159-169

**Acknowledgments**

This research was supported by DP2M PHB. The authors would like to thank Afifah, Ninis, Indri, Nafis, Dina, Anggi, Merry, Firdi, Esti kurniati and Afriandini from the Faculty of Fisheries and Marine Sciences, Brawijaya University, Malang, East Java, Indonesia.

# The influence of supplemented *Curcuma* in feed formulation to improve growth rate and feed efficiency of catfish (*Clarias* sp.)

**M M Ulum [1], M Zubaidah[1], M Arief [2] and Prayogo [2]**

[1] Department Aquaculture, Faculty of Fisheries and Marine Airlangga University, Surabaya, Indonesia 60115
[2] Departement of Fish Health Management& Aquaculture, Faculty of Fisheries and Marine Airlangga University, Surabaya, Indonesia 60115

E-mail: misbachul.ulum15@gmail.com

**Abstract.** Catfish (*Clarias* sp.) is very potential to be developed as a food fish. The use of the feed additive in feed intended to improve the health, productivity and compliance with animal nutrition. Feed additive is to use ginger are derived from nature.The purpose of this study was to determine the potential increase in feed formulation. The *Curcuma* effect on growth rate and feed efficiency of catfish (*Clarias* sp.). The method used experimental methods and design complete random with 4 treatments and 5 replications. The parameters examined in this study aregrowth rate and efficiency feed.The research showed that the ANOVA markedly dissimilar ( $P< 0.05$ ) against growth rate and efficiency feed catfishes . The average is 0.275 - 0.33 growth and the results efficiency feed catfishes of 60.70 % - 73.10 % and the results obtained temperature water quality around 27-30°C , pH 7-8 , and DO 3.09-6.4 . Based on this research cultivator of the suggested use flour curcuma as *feed additive* on formulations feed .

## 1. Introduction

The catfish(*Clarias* sp.) is a species with the potential to be developed as a food fish.These fish have several advantages, such as they arerelatively resistant to disease,have rapid growth and very responsive to the feed given. Catfish production in Indonesia has increased from year to year.In 2011, national catfish production reached 337.577 tons,while it reached 722.623 tons in 2015 [1]. Feed is an important element of the cultivation process that can support the growth and survival of the fish.Feed costs take up about 60-70 % of production costs incurred by the cultivator. Enhancing the growth rate of catfish (*Clarias* sp.) has been done through efficient feed given that the use of more efficient feed would eventually reduce production costs [2].

A good availability of feed, timely feeding and good nutrition are very important in aquaculture. Providing feed which is not in accordance with the quantity of fish maintained as well as low quality feed will not meet the nutritional needs of the fish, thus causing slow growth rates and unable to attain the expected fish production. One effort to improve the efficiency of feed is by using feed additives [3]. The use of additives in feed aims to improve the health and productivity of cattle as well as nutrition [4].

An alternative that can be used to solve this problem is by using natural materials such as curcuma.

Natural materials like curcuma have several advantages, namely: it is more feasible and cheaper than chemical remedies; contains fiber, starch and potassium oxalic, volatile oil and flavonoids; has antibacterial and anti-inflammation properties; prevents the collection of blood; launches the metabolism and maintaining the function of organs [5].

The curcumin in *Curcuma rhizomes* ranges between 1.6 – 2.22 percent calculated based on dry weight. The curcumin and the substances inits volatile oil allegedly causes curcuma to beefficacious. *Curcuma* can affect the pancreas, increases appetite and can speed the emptying of the stomach. Thus it will cause hunger and stimulate the appetite [6]. Along with curcumin, the *Curcuma* and volatile oil content serves as an antiseptic and antibacterial so as to induce growth effectively and can improve endurance [7]. *Curcuma rhizomes* containing curcuminoidis believed to be able to increase employment of the kidney and increase the appetite, as well as having anti-inflammation, anti-cholesterol, anti-oxidant, anti-cancer and anti-microbial properties [8]. This report aims to understand the effects of additional curcuma in feed formulationstowards the growth rate and feed efficiency of catfish.

## 2. Methodology

### *2.1. Preparation of tool and materials*

Research was carried out in the laboratory at the Faculty of Fisheries and Marine, while proximate analysis was conducted at the Faculty of Veterinary Medicine, both at Airlangga University, Surabaya. The tools and materials used were 20 of aquarium sized 50 x 30 x 30 $cm^3$, aeration, hose aeration, stone aeration, trays, printer pellets, loaves, digital scale, fish nets, spoons, siphons, filters, thermometers, pH papers, DO meters and ammonia test-kits.

The material used in the research are catfish (*Clarias* sp.) measuring 9-12 cm in lengthand 4-8 grams in weight which were obtained from the Gunungsari Fish Market. There were 4 treatments that were tested 5 times.Each aquarium was filled with 10 fish, so 200 fish were needed in total. The materials used for research were fish flour, soybeanflour, ricebran, *Curcuma* powder and tapiocaflour.

- Experimental procedures: 20 fish were placed inside a 50 x 30 x 30 $cm^3$ aquarium. The aquarium was cleaned and sterilized before use to avoid disease. The equipment was washed with soap and then flushed with chlorine before being dried.The dried aquarium was then filled with fresh water. The aquariums were aerated for 24 hours to improve the level of dissolved oxygen and to remove undesirable chemicals that may still bein the aquarium. The catfishwere acclimatized for one day.
- The fish were not given food for one day to remove the effects of previous feed given . The fish used in this research were dumbo catfishes (*Clarias* sp.) with the size of 9-12 cm and weight of 4-8 grams which were free from disease. The fish seeds were stocked during low temperatures,namely in the morning or evening. The catfishes were acclimatized for one day.
- The formulations of feed material was performed at the Feed Laboratory at the Faculty of Fisheries and Maritime,Airlangga University. The coarse materials were sifted beforehand using a sieve to obtain finer material. The fine feed materials were weighed using analytical scales to measure out the appropriate feed formulations.The weighed ingredients were then blended until homogeneous. Materials were added according to the composition amount of the ingredient, from the smallest size to the largest size. Feed material that were evenly mixed were put into loaves and then was run through a pellet-making machine to obtain the feed pellets. The pellets were then dried by the sun and were ready to use.
- Catfish were maintained in a 50x30x30 $cm^3$aquarium.Each aquarium was filled with fish and kept for 21 days. The formulated feed pellets were shaped according to the size of the mouths of the fishes.
- The fish were fed three times a day with a dose of 5 % of the total weight of fish in each meal.For maintenance,aspiration was performed in the morning to clean the uneaten feed and fish waste. The aspiration was done to maintain good water quality so as to not create a poisonous environment for the fish.

IOP Conf. Series: Earth and Environmental Science **137** (2018) 012007 doi:10.1088/1755-1315/137/1/012007

The main parameters of the research were growth rates and feed efficiency, while the supporting parameters werepH , temperature, DO and ammonia levels.Statistical analysis:Analysis of data was carried out through variant analysis (ANOVA) with the study design of Completely Randomized Design (CRD) to determine the differences in treatment. If there is a difference in treatment,the Duncan range test was conducted with a confidence level of 0.05 to determine the differences among all treatmentsas well as the use of SPSS 22 software.

## 3. Results

The results of the growth rate of the catfish for 21 days can be seen in table 1. ANOVA test results show that the addition of *Curcuma* powder (*Curcuma xanthorrizha*) in feedformulation show markedly dissimilar ($P < 0.05$) results of growth rate in dumbo catfishes, so that it can be followed by the Duncan multiple range test.

The results of the Duncan multiple range test show that that the lowest growth rate was seen in treatment pd with an average of 0.275 grams per day. The highest growth rate was seen in treatment pc by an average of 0.33 grams per day.The average growth rate of catfish is presented infigure 1.

The average growth rate of the catfish for 21 days can be seen in treatment $P_A$ with the average growth of 0.2756 g/day, $P_B$ treatment with the average of 0.306 g/day, $P_C$ treatment with the average growth of 0.33 grams/day and the treatment of $P_D$ with the average growth of 0.275 grams/day. The highest average growth rate was seen in treatment $P_C$, while the lowest average growth rate was seen in treatment $P_D$.

Feed efficiency: the research results for catfish feed efficiency for 21 days can be seen in table 2. The calculation of the value of feed efficiency can be found in Annex 9. Statistical analysis of feed efficiency of can be found in Annex 10. ANOVA test results shows that the *Curcuma* powder (*Curcuma xanthorrizha*) in feed formulations show markedly dissimilar ($P < 0.05$) results to catfish feed efficiency, so that it can be followed by the Duncan multiple range test. The results of the Duncan multiple range test show that the highest feed efficiency is seen in treatment pc with the average of 73.103 % and the lowest feed efficiency can be found in treatmentpd with an average of 60.076 % (table 2)

**Table 1.** The growth rate (gram/day) of catfish during the 21-day study information as follows: $P_A$ (0 % flour) curcuma, $P_B$ (2 % flour) curcuma, $P_C$ ( 4 % flour) curcuma, $P_D$ (6 % flour) curcuma. Superscript different show there are significant differences ( $P < 0.05$).

| Treatment | Grouth Rate± SD |
|---|---|
| $P_A$ | $0{,}2756^b \pm 0{,}02$ |
| $P_B$ | $0{,}306^{ab} \pm 0{,}05$ |
| $P_C$ | $0{,}33^a \pm 0{,}02$ |
| $P_D$ | $0{,}275^b \pm 0{,}04$ |

IOP Conf. Series: Earth and Environmental Science 137 (2018) 012007 doi:10.1088/1755-1315/137/1/012007

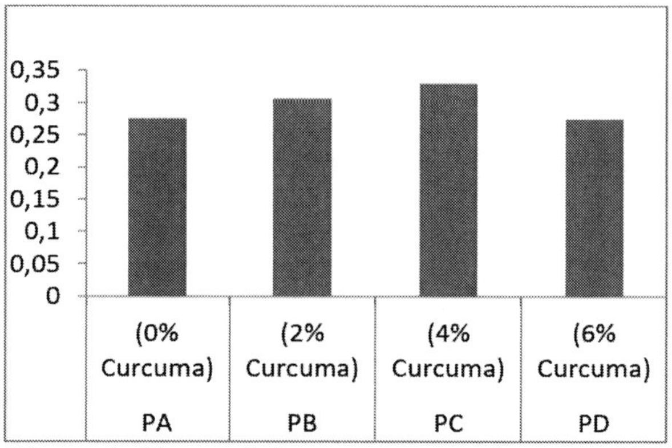


**Figure 1.** Charts of the average growth rate $P_A$ (0 % flour) curcuma, $P_B$ (2 % flour) curcuma, $P_C$ (4 % flour) curcuma, $P_D$ (6 % flour) curcuma.

The average yield of catfish feed efficiency for 21 days can be seen in treatment $P_A$ with an average of 62.897 %, $P_B$ treatment with an average of 63.29 %, $P_C$ treatment with an average of 73.103 % and treatment $P_D$ with an average of 60.076 %. The highestaverage growth rate was seen in treatment $P_C$, while the lowest average was seen in treatment $P_D$. Water quality data during the maintenance of the catfish (*Clarias* sp.) can befound in table 3. The water quality parameters observed during the survey were temperature, pH, dissolved oxygen and ammonia levels.

**Table 2.** Charts the average efficiency feed on fish catfish.

| Treatment | Feed Efficiency ± SD (%) |
|---|---|
| $P_A$ | $62{,}897^b$ ± 5,44 |
| $P_B$ | $63{,}29^b$ ± 2,84 |
| $P_C$ | $73{,}103^a$ ± 8,98 |
| $P_D$ | $60{,}076^b$ ± 2,82 |

**Table 3**. Data the water quality maintenance catfishes for 21 days.

| Treatment parameters | parameters | | | |
|---|---|---|---|---|
| | temperature (°C) | pH | DO (mg/L) | Amoniak (mg/L) |
| PA | 27-30 | 7-8 | 3,77 – 6,01 | 0,5 – 1 |
| PB | 27-30 | 7-8 | 3,09 – 5,9 | 0,5 – 1 |
| PC | 27-30 | 7-8 | 3,42 – 5,49 | 0,5 – 1 |
| PD | 27-30 | 7-8 | 3,18 – 5,22 | 0,5 – 1 |

## 4. Discussion

Growth is measured through size, length or weight in a period of time. Growth occurs due to extra tissue from cell division in mitosis, excess input energy, and a protein derived from feed [9]. Growth is very closely related to the feed given. Feed that fulfils nutritional requirements highly influences the rapidity of growth. Feed that meets nutritional requirements will also guarantee life and expedite the growth of [10].

Based on the calculation of statistics in table 1, we can see that the growth rate for all treatments of catfish ranges from 0.275-0.33 g/day. The highest growth rate was found in treatment pc which was added with about 4 % curcuma powder with an average growth of 0.33 g/day. The lowest growth rate was found in the treatment $P_D$ which was added with 6 % *Curcuma* powder with an average growth of 0.275 g/day.

Results of the ANOVA test show that the growth rates of the catfish were markedly dissimilar ($P < 0.05$). Next, a Duncan multiple range test was performed and it can be seen that treatment $P_B$ did not show much difference to treatment $P_C$, althougt treatment pb was significantly different with treatment pa and pd. The lowest growth rate was seen in treatment pd with average value of 0.275 g/day.

The highest growth rate was seen in treatment $P_C$ by an average of 0.33 grams per day. The best feed formulation was obtained in treatment pc by the addition of a 4 % concentration of *Curcuma* powder in the feed formulation. The addition of 5 percent curcuma powder gives the highest growth rate. It is suspected that the nutrition from curcuma powder increase the growth rate [11]. The treatment pd is the lowest result of the addition of curcuma in feed. It is suspected that it is due to the excess volatile oil content and curcumin in the feed, so that it may cause it to taste bitter and be less tasty for the fish.

The increase in weight may be due to all the feed given to the fish are likely responded to by the fish and applied to the process of metabolism and growth. Growth can be influenced by the balance of the nutrients that are in the feed. Growth occurs when there is excess energy due to metabolism after it is used to maintain the body and activity [9].

The curcumin content as well as the volatile oil in curcuma serves as an anti-biotic, and can also neutralize venom and increase the secretion bile, so that it can increase the appetite in fish. This is because curcumin and volatile oil can improve the digestive system and be used as a pacemaker of growth and increase digestion [12].

Positive growth is supported by the good health of fish and would improve the efficiency of food absorption to meet their life requirements which can be seen by its weight [13]. The rate of growth in treatment d was lower than the treatment c which was caused by the higher dose of curcuma. The high dose had an effect on the bitter taste of the feed produced. The bitter taste causes a lack of response from the fish thus reducing the consumption of feed which is influential to their daily weight.

The *Curcuma* given has a variety of benefits for the fish especially for health and growth. Apart from containing antibiotics, curcuma also contains volatile oil and curcumin. Curcumin serves to increase the appetite and play a role in increasing the employment of digestive organs, stimulating the issuing of wall bile liquid and stimulate the pancreas that contains the enzymes of amylase, lipase and protease to increase the digestion of feed material such as carbohydrates, fat and protein [14]. Its antibacterial properties are also beneficial to lysis the poison attached to the walls of the intestines, so that there is better absorption of substance nutrients and can trigger growth [15].

### *4.1. Feed efficiency*

Feed efficiency is the value of ratio between the weight of organism and the feed consumed in percetage [16]. Based on the known statistics, the feed efficiency for all treatments of catfish ranged from 73.103 % to 62.897 %. The highest feed efficiency was obtained on treatment $P_C$ by the concentration of 4 % curcuma powder with the average value of 73.103 %. The lowest feed efficiency was obtained on treatment $P_D$ by the concentration of 6 percent curcuma powder with the average value of 62.897 %.

Based on the results of the ANOVA test, the feed efficiency of catfish is markedly dissimilar ($P < 0.05$). Following a Duncan multiple range test,it can be seen that treatment $P_A$, $P_B$ and pd are significantly different with treatment $P_C$.The highest feed efficiency was seen in treatment pc with the average value of 73.103 %, while the lowest feed efficiency was seen in treatment $P_D$ with the average value of 62.897 % .

IOP Conf. Series: Earth and Environmental Science **137** (2018) 012007 doi:10.1088/1755-1315/137/1/012007

Factors that influence feed efficiency are the type of nutrition source and the amount of each nutrient component in the feed. The quantity and quality of feed given to fish is influential on the growth of fish .The higher the efficiency of the feed, the better the response of the fish to the feed [17].

Water quality is vital in the success of fish cultivation. Water quality is defined as water conformity for the continuity and growth of fishes which are generally determined by several parameters. The maintenance media should be done in accordance to the natural habitat of the fish during the maintenance process, so that the physiology and morphology of the fish are not disturbed. The parameters observed were temperature, pH and ammonia. The range of water quality during the study saturation is 4 mg/L. Saturation in this study was not a problem for fish maintenance because physiologically the fish are still able to swim normally. The temperature of media during the study maintenance period was between 27-30°C. The state of temperature between 26-30°C is good for the maintenance of catfish. The pH values were in the range of 7.0-8.0, where the optimal range of the waters is 7.5-8.5. The ammonia concentration was optimal for cultivation which was no more than 1 ppm.

## 5. Conclusion

It can be concluded that curcuma powder added at a dose of 4 % can be used as a feed additive in feed formulation. Further research needs to be done regarding the increase of curcuma powder in feed formulations with different percentages of protein.

## 6. References

[1]. Ministry of Marine Affairs and Fisheries Republic of Indonesia 2016 Report of the ministry affairs and fisheries, aquaculture production year (Jakarta: KKP) p 67 (In Indonesia)

[2]. Kusuma D M, Afrianto E and Priyadi A 2012 *J. Fish. Mar.* **3** 40 (In Indonesia)

[3]. Widianto B, Prayogi, H S and Nuryadi, N 2013 *J. Vet. Sci.* **25** 28-35 (In Indonesia)

[4]. Adams C A 2000 *Proc. Aust. Poult. Sci.* **12** 17-24

[5]. Indonesia D K R 2000 *Common standard parameters of medicinal plant extracts* (Jakarta: Departemen Kesehatan Republik Indonesia) p 9-12(In Indonesia)

[6]. Wijayakusuma H 2003 *Healing with temulawak* (Jakarta: Milenia Populer) (In Indonesia)

[7]. Ahira 2011 Temulawak function http://temulawak.co.id accessed at 20 January 2017 (In Indonesia)

[8]. Purnomowati S 2008 Benefits of temulawak: review of literature year 1980-1997 htttp://www.indofarma.co.id/index.php?option=com_content&task=view&id=21&Itemid=125 . accessed at 20 January 2017 (In Indonesia)

[9]. Effendie M I 1997 *Fisheries biology* (Yogyakarta: Yayasan Pustaka Nusatama) p 163 (In Indonesia)

[10]. Arief M, Dwi K P and Yudi C 2011 *J. Fish. Mar. Sci.* **3** (In Indonesia)

[11]. Insana N and Wahyu F 2015 *J. Mar. Sci.* **4** 381-391(In Indonesia)

[12]. Setianingrum 1999 Effect of ginger (Curcuma xanthorrhiza Roxb) to increase appetite in primary anorexia patients (Semarang : FK UNDIP) p 57 (In Indonesia)

[13]. Koesdarto S, Subekti S and Studiawan H 2001 *J. PME.* **2** 114-122 (In Indonesia)

[14]. Sastroamidjojo S A 2001*Original medication from Indonesia* (Jakarta: PT. Dian Rakyat) p. 57-63(In Indonesia)

[15]. Samsundari S 2006 *J. Gamma.* **2** (In Indonesia)

[16]. Agustono, Yudiarto S andArief M 2012 *J. Fish. Mar. Sci.* **4** 135-140 (In Indonesia)

[17]. Haryadi B, Haryono A and Susilo U 2005 Evaluation of feed and protein efficiency in grass carp fish (*Ctenopharyngodon idella* Val.) fed with different carbohydrate and energy level (Purwokerto : Faculty of Biology) **4** pp 87-92(In Indonesia)

IOP Conf. Series: Earth and Environmental Science **137** (2018) 012008 doi:10.1088/1755-1315/137/1/012008

# The effect of differences in altitude location of an aquaculture on fish's hematocrit and fish's haemoglobin of Carp fish and resistance to bacterial attack

**Rosidah, A Rizal, I Rustikawati and F Octavia**

Faculty of Fishery and Marine Science, Universitas Padjadjaran
Bandung-Sumedang, Km 21, Jatinangor Campus 40600 Sumedang 453 63 West Java

E-mail: ros_ahdi@yahoo.com

**Abstract.** The aim for this research is to analyze the relation between the height place of aquaculture and life directness towards Aeromonas hydrophila attack for hematocrit and hemoglobin degrees on Cyprinus carpio L. This research have been doing on February - March 2017 at UPT PTBP Kepanjen (as Group 1), Malang (332 amsl) and UPT PBAT Umbulan IBAT as Group 2 (955 amsl). The treatment that be used is *Cyprinus carpio* infected by *Aeromonas hydrophila*. Parameters was observed, are hematocrit, hemoglobin degrees, clinical symptom of fish and the quality of water. The result of this investigation, hematocrit and hemoglobin degrees of *Cyprinus carpio* in level low land (Group 1, 332 amsl) around 12.8 % and 9.58 g/dL, meanwhile hematocrit and hemoglobin degrees in level up land (Group 2, 955 amsl) around 24.19 % and 12.64 g/dL. *Cyprinus carpio* L. after infected by *Aeromonas hydrophila*, it shows the clinical symptoms : its mucus be left, bleeding, its scale get peeled off and dropsy. The life directness of that *Cyprinus carpio* L. maintaining in 15 days 0 %. The differences between the height level of aquaculture for hematocrit and hemoglobin degrees also life directness towards Aeromonas hydrophila attack on *Cyprinus carpio* L.

## 1. Introduction

Goldfish or Carp fish (*Cyprinus carpio* L.) is an important commodity in freshwater aquaculture that is relatively easy to cultivate, so it is popular in almost every region of Indonesia [1]. Cultivation of carp can be done in various altitude places, both in the lowlands and in the highlands. Differences in altitude of a place result in differences in local environmental conditions.

Increased or decreased air pressure and oxygen content in the blood will affect the content of hematocrit and haemoglobin in fish. Examinations of hematocrit and haemoglobin can be used as one of the benchmarks to determine the state of fish health or as an early diagnosis in determining fish health. The hematocrit value in fish of less than 22 % indicates anaemia [2].

According to Amanda [3], most fish that are in stress condition will get sick and die. One of the causes of stress in fish is inappropriate oxygen content of the water. Stress in fish will affect the resistance of the fish to disease. Disease that can easily attack the fish is haemorrhagic septicaemia caused by the bacteria *Aeromonas hydrophila* [4].

According to Kabata [5], *Aeromonas hydrophila* bacteria cause a high mortality rate (80-100 %) within 1-2 weeks. Fish infected by the bacteria will experience bleeding on several body parts such as the chest, the abdomen, and the base of the fins [6]. So far, there has been no research on the effect of differences in height of the cultivation area on hematocrit and haemoglobin of carp fish associated with the resistance to bacterial attack from *Aeromonas hydrophila*; therefore, it needs to be researched

## 2. Methodology

This research was conducted in February until March 2017, located at UPT PTBP Kepanjen and UPT PBAT Umbulan IBAT Punten, Malang. This research was a descriptive study conducted to determine the relationship between the altitude of fish pond with hematocrit and haemoglobin levels in carp fish and endurance against bacteria attack from *Aeromonas hydrophila.* At each location, fish were taken as many as 65 heads weighting of 200 g/tail, 5 tails were used for blood sampling and 60 tails were for test challenge with the infecting bacteria *Aeromonas hydrophila.*

## 3. Results and Discussion

### *3.1. Hematocrit*

Hematocrit is the volume of blood cells compared with blood plasma that is expressed as a percentage. Based on observations of the hematocrit, carp fish that have been maintained at different altitudes had different hematocrit levels (table 1).

**Table 1.** Hematocrit level of carp fish.

| Altitude (amsl) | Average Hematocrit Level (%) |
|---|---|
| 332 (Group 1) | 12.8 |
| 955 (Group 2) | 24.19 |

Table 1 shows that carp fish that was maintained at 332 amsl (Group 1) had an average hematocrit level of 12.8 %; while in carp fish cultivated at 955 amsl (Group 2), the average hematocrit level was 24.19 %. The hematocrit of carp fish cultured at 332 amsl (Group 1) was below normal; while the hematocrit of carp fish cultured at 955 amsl (Group 2) was within normal limits. From the results, it can be concluded that carp fish from higher than 332 amsl (Group 1) had anaemia. This is in accordance with the statement of Gallaugher [2] that a hematocrit value lower than 22 % indicated the fish is anaemic. According to Handayani [7], anaemia is a circumstance in which circulating erythrocytes do not fulfil their function to provide $O_2$ for body tissues. Anaemia can be affected by several factors including age, sex, and altitude.

Higher hematocrit values of carp were found in Group 2 (955 amsl) compared with the hematocrit of carp fish at 332 amsl (Group 1) that was due to dissolved oxygen content (DO) in setting media below normal range of 2-3 mg/L (Table 2), thus causing a lack of oxygen ($O_2$) supply in the blood. Blood is divided into three parts: red blood cells (erythrocytes), white blood cells (leukocytes), and platelets (thrombocytes).

White blood cells (leukocytes) are one component of the blood that serves as a non-specific defence that inhibits the attack of pathogenic bacteria through phagocytosis. Leukocytes can also be used as an indicator of infection in the body. The body will produce more leucocytes when there is a foreign matter that enters the body [8]. Based on the observation, the fish leucocytes were different between those cultured at the altitude of 322 amsl (Group 1) and at 955 amsl (Group 2) (table 2).

**Table 2.** Leucocytes of carp fish.

| altitude (amsl) | Average Leukocytes Level (cell/mm$^3$) |
|---|---|
| 332 (Group 1) | 26,896 |
| 955 (Group 2) | 28,832 |

In Table 2, it can be observed that carp fish kept at 332 amsl (Group 1) had a lower average leukocyte value (26,896 cells/mm$^3$) compared with carp fish cultured at 955 amsl (Group 2) (28,832 cells/$mm^3$). The levels of carp fish leucocytes from both locations were within the normal range. According to Lagler [9], the number of white blood cells in fish ranged from 20,000 to 150,000 $mm^3$.

Increasing number of leucocytes can be used as a sign that the fish is in stress condition. The stress can be caused by deterioration in water quality [10].

### *3.2. Hemoglobin*

Haemoglobin (Hb) plays an important role in the transport of gases, especially oxygen ($O_2$) from the gills pumped by the heart to all cells and organs, transporting nutrients into cells, disposing metabolic waste, and so on. Based on the observed haemoglobin concentration of carp fish cultured at 332 amsl (Group 1) and 955 amsl (Group 2) had different haemoglobin levels (table 3).

**Table 3.** Hemoglobin level of carp fish.

| Altitude (amsl) | Average Hemoglobin Level (g/dL) |
|---|---|
| 332 (Group 1) | 9.58 |
| 955 (Group 2) | 12.6 |

Table 3 shows that carp fish cultivated at 332 amsl (Group 1) had lower average haemoglobin level (9.58 g/dL) compared to carp fish cultivated at 955 amsl (Group 2) which had an average haemoglobin level of 12.64 g/dL. Both groups were classified as above normal, since normal haemoglobin level for carp is $6.22 \pm 0.91$ g/dl [11]. The value of haemoglobin may be affected by the age of the animal, species, environment, feed, presence or absence of erythrocyte damage, and blood handling at the time of examination [12].

The average value of haemoglobin level of carp fish from low (Group 1, 332 amsl) and high (Group 2, 955 amsl) altitudes had considerable differences compared to normal carp fish.

The higher the haemoglobin level, the oxygen ($O_2$) transport process to all parts of the fish body tissue will be more optimal, thus affecting the maximum consumption of $O_2$; whereas the lower the haemoglobin level or the more severe the anaemia, the decrease in maximum $O_2$ consumption is also greater [13,14].

From the description above, it shows that hematocrit, leukocyte and haemoglobin levels in carp fish are affected by the difference in altitude, as seen from the examination of hematocrit (table 1), leukocytes (table 2) and haemoglobin levels (table 3) of fish culture at the altitudes 332 amsl (Group 1) and 955 amsl (Group 2).

### *3.3. Clinical symptoms of fish infected with Aeromonas hydrophila*

Observation of clinical symptoms in fish from Group 1 (332 amsl) and Group 2 (955 amsl) was performed for 15 days after the fish got bacterial infection of *Aeromonas hydrophila* by injection. The Observation included changes in morphology and behaviour. Morphological changes were seen with damage to the surface of the body, such as excess mucus, bleeding, peeling of scales, prominent eyes (exophthalmia), and dropsy (table 4), whereas behavioural changes were seen from reduced response to feed (table 5) and to the given shock (table 6).

IOP Conf. Series: Earth and Environmental Science **137** (2018) 012008 doi:10.1088/1755-1315/137/1/012008

**Table 4.** Clinical symptoms of carp fish during the challenge test.

| Day of Observation | Altitude (amsl) | |
|---|---|---|
| | 332 (Group 1) | 955 (Group 2) |
| 1 | - | - |
| 2 | - | - |
| 3 | A | - |
| 4 | A | A |
| 5 | a and b | A |
| 6 | a and b | A |
| 7 | a, b and c | a and b |
| 8 | a, b, c and e | a and b |
| 9 | a, b,c and e | a and b |
| 10 | a, b, c and e | a, b, c and e |
| 11 | 0 | a, b, c and e |
| 12 | 0 | a, b, c and e |
| 13 | 0 | a, b, c and e |
| 14 | 0 | 0 |
| 15 | 0 | 0 |

Information of Clinical Symptoms :
- = There are no Clinical Symptoms
0 = Death
a = Excessive Mucus
b = Bleeding
c = scales peel off
d = The fish eyes stand out (*eksopthalmia*)
e = abdominal bloating (*dropsy*)

Table 4 shows that on Day 1 and 2 after bacterial infection of *Aeromonas hydrophila*, carp fish from the altitude 332 amsl (Group 1) and 955 amsl (Group 2) did not show any damage to the body and the fish were still in healthy condition (figure 1). Bacteria *Aeromonas hydrophila* has an optimal growth in the exponential phase occurring at the 4$^{th}$ hour to the 12$^{th}$ hour [15]. But on the basis of observation, the clinical symptoms began to appear on the 3$^{rd}$ day. It is suspected that the bacteria *Aeromonas hydrophila* that entered the body of carp did not directly become virulent, because they were still in sticking stage that takes time to enter the stage of infection or virulence [16].

**Figure 1.** Healthy carp fish.

The third day after the infection, the fish cultured at 332 amsl (Group 1) began to show symptoms of excess mucus on the body surface (figure 2), while those at 955 amsl (Group 2) started to show similar symptom on Day- 4 (figure 2). The mucus produced by the carp fish from Group 1 was more than the mucus produced by the carp in Group 2 (figure 2). Once *Aeromonas hydrophila* starts to infect a fish, the fish will release more mucus as a defence.

**Figure 2.** Fish with excessive mucus.

Another clinical symptoms began to appear on the 5$^{th}$ day In carp fish cultured at 332 amsl (Group 1), which was bleeding (figure 3), followed by peeling of the scales and inflation of the abdomen on the next day (figure 4). For Carp fish cultured at 955 amsl (Group 2) similar symptoms started to appear on Day 7 and Day 10 respectively (figure 4).

**Figure 3.** Bleeding.

**Figure 4.** Abdominal bloating.

The carp fish from Group 1 suffered severe bleeding causing the cultivating water media to turn red, while the bleeding in carp fish from Group 2 was not too severe so that it did not change the colour of the water. This condition is associated with the attack of bacteria, so that fish experience disruption or physiological changes. Fish infected with *Aeromonas hydrophila* bacteria will show clinical symptoms of haemorrhage and abdominal bloating due to fluid accumulation in the abdominal cavity [15]. The physiological changes in carp that were infected by *Aeromonas hydrophila* bacteria were a defence mechanism of fish itself. According Syawal [17] in Nurjannah[18], physiological changes in fish, such as inflammation due to infection, is a nonspecific immune response characteristic.

In addition to morphological changes, carp infected by *Aeromonas hydrophila* bacteria also experienced behavioural changes such as decreased appetite (table 5) and reduced reflexes to shock (table 6).

**Table 5.** Feeding response of carp fish during the challenge test.

| Day of Observation | Altitude (amsl) | |
|---|---|---|
| | 322 (Group 1) | 955 (Group 2) |
| 1 | + | + |
| 2 | + | + |
| 3 | - | + |
| 4 | - | + |
| 5 | - | - |
| 6 | - | - |
| 7 | - | - |
| 8 | - | - |
| 9 | - | - |
| 10 | 0 | - |
| 11 | 0 | - |
| 12 | 0 | - |
| 13 | 0 | 0 |
| 14 | 0 | 0 |
| 15 | 0 | 0 |

Information of Feeding Response:
( + ) = Normal Feeding Response
( - ) = Less Feeding Response
( 0 ) = No Feeding Response

Table 5 shows that the carp fish response to feeding at different altitudes was still normal until Day 2. This shows that the fish was still in good condition, so they could respond to the feeding properly. Fish was still in good health because *Aeromonas hydrophila* that infected the fish did not directly become virulent, but it still in the adhesive stage on the fish body and it took time to enter the infection stage [16].

The less of feeding response (-) appetite was seen on Day 3 to Day 9 in carp fish cultured at 332 amsl (Group 1) while the fish from e 955 amsl (Group 2) showed similar symptom on Day 5 to Day 12. Less feeding response appetite was seen with the rest of the feed into a precipitate to the bottom of aquarium. Decreased feeding response was due to the metabolic process that is disrupted. Decrease in response to feeding in fish is caused by metabolic disorders in the body post bacterial infection of *Aeromonas hydrophila* [19]. The absence of feeding response or no feeding response (0) in goldfish from the altitude of 332 amsl (Group 1) occurred on the 10th to 15th day of observation, whereas in carp fish from the altitude of 955 amsl (Group 2) occurred on the 13th day of observation until to-15.

In addition to decreasing response to feeding, carp fish that was infected by bacterium *Aeromonas hydrophila* also showed decreased response to shock. Reflex tests were performed by tapping the aquarium walls at each treatment and fish infected with *Aeromonas hydrophila* tended not to respond (table 6).

**Table 6.** Fish response to shock during the challenge test.

| Day of Observation | Altitude (amsl) | |
|---|---|---|
| | 332 (Group 1) | 955 (Group 2) |
| 1 | + | + |
| 2 | + | + |
| 3 | + | + |
| 4 | - | + |
| 5 | - | - |
| 6 | - | - |
| 7 | - | - |
| 8 | - | - |
| 9 | 0 | - |
| 10 | 0 | - |
| 11 | 0 | - |
| 12 | 0 | 0 |
| 13 | 0 | 0 |
| 14 | 0 | 0 |
| 15 | 0 | 0 |

Information of Feeding Response:
( + ) = Normal Feeding Response
( - ) = Less Feeding Response
( 0 ) = No Feeding Response

Table 6 shows that response to the shock in carp fish cultured at 332 amsl (Group 1) was still normal until the 3rd day, while those kept at 955 amsl (Group 2) still showed normal response until the 4th day of observation. This indicated that the fish were still fairly healthy and the immune system was good, so that bacteria did not cause damage yet.

In carp fish cultured 332 amsl (Group 1); there was a lack of response to the shock on the 4th day until the 8th day, while in those from 955 amsl (Group 2) similar symptom was seen on Day 5 to Day 11. This shows that the immunity of carp fish began to decline due to bacterial attack of *Aeromonas hydrophila*.

In carp fish kept at 332 amsl (Group 1), the disappearance of response to shock occurred on the 9th day until the 15th day, whereas in carp fish from 955 amsl (Group 2) similar symptom occurred on the 12th day until the 15th day. The fish immune system had been greatly compromised so they were not able to resist attacks against the bacteria *Aeromonas hydrophila*.

### *3.4. Survival rate of carp fish*

Observation of the survival rate of carp fish was conducted after the challenge test by infecting *Aeromonas hydrophila* bacteria for 15 (fifteen) days. Fish culture at different altitudes produced different survival rate (figure 5).

IOP Conf. Series: Earth and Environmental Science **137** (2018) 012008 doi:10.1088/1755-1315/137/1/012008

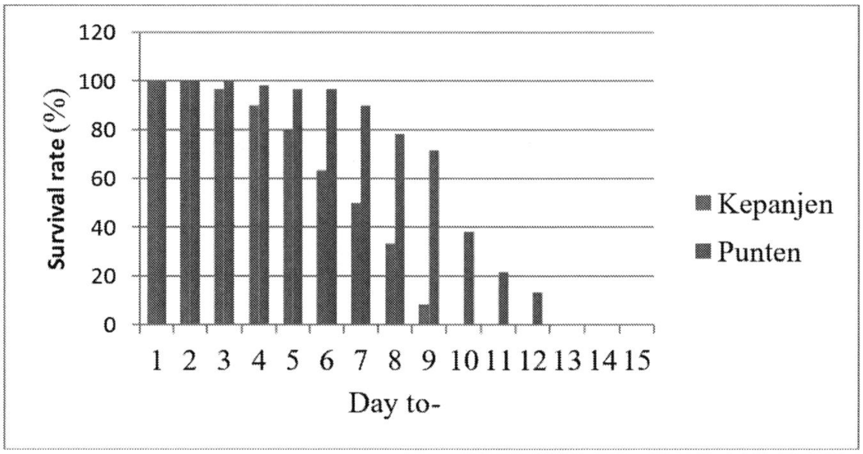


**Figure 5.** Survival rate of carp fish.

Based on figure 5, carp fish cultured at 332 amsl (Group 1) had a lower survival rate compared to those kept at 955 amsl (Group 2). Carp fish originated from Group 1 only lasted for 10 (ten) days from the total cultivation period of 15 (fifteen) days after challenge test. The death of carp fish originating from Group 1 started from the 3$^{rd}$ day with 3.34 % of the population died after the injection of *Aeromonas hydrophila* and each day the death rate was different. In carp fish originating from Group 2, the mortality started from Day 4 with 1.67 % and the number of fish that died each day was different.

Deaths in carp fish from both altitudes were due to hematocrit, leukocyte, haemoglobin, and dissolved oxygen in water (DO). In carp fish cultured at 322 amsl (Group 1) the death rate was more rapid due to lower hematocrit, leukocyte and haemoglobin levels. Reduction of all those factors can affect the immune system. Moreover, fish resistance decreased due to lack of drug delivery as prevention or treatment.

### *3.5. Water quality*

Water quality analysis was performed three times during the challenge test. Water quality observations during the study are shown in table 7.

**Table 7.** Water quality of cultivate media during the challenge test.

| Altitude (amsl) | Temperature ($^0$C) | pH | DO (mg/L) |
|---|---|---|---|
| 332 (Group 1) | 25 – 26 | 7 – 7.5 | 4 – 6 |
| 955 (Group 2) | 20 – 23 | 6 – 7 | 2 – 3 |
| Normal | 25 – 30 | 6.5 – 9 | >3 |

In fish culture at 332 amsl (Group 1), the temperature range, pH, and DO lower normal conditions were 25-26 °C, pH 7-7.5 and DO 4-6 mg/L, whereas at 955 amsl (Group 2), the temperature range and DO were under normal conditions and the pH was within the normal range. The temperature values at the time of observation were in the range of 20 - 23°C, pH 6 - 7 and DO 2 - 3 mg/L. According to Supriyatna [1], the ideal temperature for fish life is 25 - 30 °C and according to Paulo [20] the normal pH value for fish life ranges from 6.5 to 9, while the normal dissolved oxygen (DO) value for fish life should be higher than 3 mg/L.

IOP Conf. Series: Earth and Environmental Science **137** (2018) 012008 doi:10.1088/1755-1315/137/1/012008

If the condition of water quality is abnormal then it affects the survival rate of the fish. Water temperature affects growth rate, metabolism rate, and appetite [21]. Abnormal acidity levels (pH) can decrease reproductive capacity and fish growth [20]. The DO concentration is influential in the metabolic process, which can affect growth rate and feed conversion [21].

## 4. Conclusion

1. Differences in altitude affect the differences in hematocrit and haemoglobin levels of carp fish.
2. Carp fish cultured at 332 amsl (Group 1) had hematocrit and haemoglobin values lower than those kept at 955 amsl (Group 2).

## 5. Reference


[1] Supriatna Y 2013 *Fish cultivation in water-saving pool* (Jakarta : AgroMedia Pustaka) 78 (In Indonesia)

[2] Gallaugher P H, H Thorarensen and A P Ferrel 1995 *Resp. Phys.* **102** 279-292

[3] Amanda 2012 Observation of hematocrit value of tombro fish blood (*Cyprinus carpio l.)* preserved at elevation place (Yogyakarta) 76

[4] Mandiri T K T 2009 *Guidelines for farm and cultivation of carp fish* (Bandung : CV. Nuansa Aulia) p 194

[5] Kabata Z 1985 *Parasites and diseases of fish cultured in the tropics* (Taylor and Francis) 318

[6] Hidayat R, Patana P and Lesmana I 2015 *J Agrobiol.*

[7] Handayani W and Haribowo A S 2008 *Salemba Medika* 150 (In Indonesia)

[8] Dianti L, Prayitno S B, Ariyati R W 2013 *J.Aquac. Man. Tech.* **2** 63-71.

[9] Lagler K F, Bardach J E, Miller R R, and Passino D R M 1977 *Ichthiology* p 506

[10] Paulo C F C, Pedro H S K, Elaine A, Correia S, and Bernardo B 2009 *Neotrop. Ichthyol.* **7** 283-288

[11] Yanto H, Hasan H and Sunarto 2015 *J. Aquac.* ISSN 0853-2532 11-20 (In Indonesia)

[12] Wardhana A H, Kencanawati E, Nurmawati, Rahmaweni, and Jatmiko C B 2000 Influence of provision of buffal stock (*Eurhorbia hirta l*) on erotrocytes, hemoglobin and hematocrit value in infected chickens with eimeria tenella (Surabaya : University of Airlangga)

[13] Anderson D P and Siwicki A K 1993 Injection or immersion delivery of selected immunostimulant to trout demonstrate enhancement of non spesific defence mechannism and protective immunity pp 185-426

[14] Huldani 2010 *J. Vet. Sci. Med.*

[15] Wahjuningrum D 2013 *J. Aquac. Ind.* **12** 94-104 (In Indonesia)

[16] Lestari F N 2015 Effectiveness of addition of cambodian leaf extract on feed to prevent infection of aeromonas hydrophila in carp fish juvenile (Cyprinus carpio) (Bandung : University of Padjadjaran)

[17] Syawal H and Siregar Y I 2010 Immunization of siam jambal fish with ichthyophthirius multifilis vaccine (Pekanbaru : University of Riau) 163-167

[18] Nurjannah R D D, Prayitno S B, Sarjitro and Lusiastuti A M 2013 *J. Aquac.Management Tech.* **2** 72-83

[19] Muslim M P, Hotly and Widjajanti H 2009 Use of garlic extract (*Allium sativum*) to treat seeds of patin siam (*Pangasius hypopthalmus*) infected bacteria aeromonas hydrophila **8** 91-100

[20] Boyd C E 1982 Water quality in warm fish ponds for aquaculture (Auburn University) 359

[21] Agustiningtyas N 2014 Utilization of heterotrophic bacteria on dumbo catfish cultivation (*Clarias* sp.) with water convertion system against FCR (Food Convertion Rsate) and protein retention (Surabaya : Universitas Airlangga)

IOP Conf. Series: Earth and Environmental Science **137** (2018) 012009 doi:10.1088/1755-1315/137/1/012009

# Characterization of phytase enzymes as feed additive for poultry and feed

**M Lamid[1,2,3], A Al-Arif[1], O Asmarani[3] and S H Warsito[1,4]**

[1]Faculty of Veterinary Medicine, Airlangga University, Surabaya 60115 Indonesia
[2]Faculty of Fisheries and Marine, Airlangga University, Surabaya 60115 Indonesia
[3]Proteomics Laboratory, Institute of Tropical Disease, Airlangga University, Surabaya 60115 Indonesia
[4]Faculty of Vocational, Airlangga University, Surabaya 60115 Indonesia

E-mail: mirnylamid@yahoo.com

**Abstract.** One of the obstacles to utilizing rice bran as feed is the presence of antinutrition in the form of phytic acid which binds in minerals to form complex compounds with P, Mg, Mn, Fe, Zn, Ca. Phytic acid and its salts are the main forms of P, Mg, Mn, Fe, Zn, Ca deposits contained in cereals, legume and grains, about 60-90% of total minerals P, Mg, Mn, Fe, Zn, Ca in the form of phytic acid or phytate salts. Phytate is one of the enzymes belonging to the phosphatase group capable of hydrolyzing phytate compounds of myo-inositol (1,2,3,4,5,6) hexsa phosphatase into myo-inositol and organic phosphat. The aim of this study was to obtain characterization of phytase enzymes from isolate *Actinobacillus* sp., *Bacillus pumilus*, *Bacillus vallimortis* and IBR-1. Determination of phytase activity and the absorbance was measured using a UV-Vis spectrophotometer at a wavelength of 392 nm. The result of *Actinobacillus* sp, *Bacillus pumilus*, *Bacillus vallimortis*, IBR-1 each having optimum temperature were 50 °C, 40 °C, 45 °C, 45 °C, and optimum pH were 4, 4, 5.5. Bacteria especially *Actinobacillus* sp, *Bacillus pumilus*, *Bacillus vallimortis*, IBR-1 are proven capable of producing the high enough phytase enzymes required for mineral availability for livestock and fish.

## 1. Introduction

Phosphor has an important role in the broilers metabolic and growth processses in the chicken's body. This is due to the fact that phosphorus is a nutrient that has the third highest economic value in broiler diet formulation after energy and amino acids, and its use should be optimized [1]. The use of phytic acid as a source of phosphorus in chicken's diet through phytate-bound phosphate hydrolysis process is expected to improve the efficiency of phosporus use.

The drowback of the use of rice bran as poultry feed in high phytic acid content is phytic acid will form insoluble salts when the phytic acid binds with phosphorus and other minerals. Thus these minerals cannot be absorbed by the digestive tract of poultry. Phytic acid has a negative charge at low, neutral and high pH. This enables phytic acid to bind with metal ions such as P, Ca, Mg and Zn as well as positive proteins as a terminal amino group at pH below the isoelectric point. The formation of insoluble phytate-mineral or phytate-protein compounds can cause a decrease in the availability of minerals and the nutritional value of proteins [2]. Phytate is the primary form of phosphorus and is found in cereal grains, legumes and oilseed meal used in the diet of monogastric animals such as poultry which be used due to their lack of endogenous phytase enzymes. To meet poultry's need for phosphor, inorganic phosphate is added to poultry's feed. The phosphorus excreted in feces becomes the primary cause of the problems of environmental pollution. As an alternative to this, microbial phytase can be used. Microbial phytase has a beneficial effect on the growth performance, feed efficiency, protein digestibility, energy utilization, mineral retention and bone growth of broilers as a direct hydrolytic effect on phytate [3,4]. Phytase is one of the enzymes belonging to phosphatase group that is capable of hydrolyzing phytic compounds such as myo-inositol 1,2,3,4,5,6-hexa into myo-inositol phosphatase and organic phosphate.

The abovementioned problem can be overcome by using phytase as a feed additive in animal feed. Degradation of phytic acid is the disconnection of bond between inositol group and myo¬phosphoric acid group of phytase produced by rumen microbes [5]. The phosphate released will be used as a source of phosphorus minerals for livestock [6], and the beneficial effect of exogenous phytase in poultry diet is the direct hydrolytic effect on the increase in the availability of phytate and minerals, amino acids, and energy [7]. There are two types of phytase, namely 6-phytase, which is found in plants, and 3-phytase, which is produced by fungi [5]. Phytic acid can be classified as an anti-nutrient component in feed, thus a sitrate-producing enzyme capable of hydrolyzing phytic acid is needed. Rumen bacteria from ruminant animals (*Actinobacillus* sp. and *Bacillus pumilus*) are expected to produce phytase as a feed additive to produce good feed quality in terms of high availability of proteins and minerals P, Mg, Mn, Fe, Zn and Ca. Local feed ingredients that can potentially be used as poultry feed are rice bran, palm kernel cake, sludge palm oil, coconut cake and other agricultural and industrial waste. Rice bran has been widely used as an animal feed ingredient for poultry. The use of higher portion of rice bran in diet will lower the production costs as its price is relatively cheap. The aim of this study was to find out the characterization of phytase originating from isolates *Actinobacillus* sp., *Bacillus pumilus*, *Bacillus vallimortis* and IBR-1.

## 2. Methodology

### 2.1 *Phytase producing bacteria*

The bacteria, *Actinobacillus* sp., *Bacillus pumilus* and *Bacillus vallimortis*, are the pure isolates from proteomic, Laboratory Institute of Tropical Disease. The bacteria were from the rumen of cow.

### 2.2 *Production of crude extracts of phytase*

The single rumen selected was rumen isolate that had the highest phytase activity, grown in 20 mL of liquid LB medium at a temperature of 52°C, shuffled using a 200 rpm incubator shaker for 24 hours. Furthermore, 10 % of the liquid culture was inoculated in 100 mL of fittings filtering medium at a temperature of 50°C, shuffled using a 200 rpm incubator shaker for 16-18 hours. The suspension was centrifuged at 4.000 rpm at a temperature of 4°C for 20 minutes. The supernatant obtained was the crude extract of phytase.

### 2.3 *Determination of phytase activity*

The Liquid LB medium was composed of: 1 % (w/v) bacto-tryptone, 0.5 % (w/v) yeast extract and 1 % (w/v) NaCl. The composition of the solid LB medium was the same as that of the liquid LB medium, but in the solid LB medium 2 % (w/v) bacto-agar was added. The pharmaceutical screening medium was composed of: 2 % (w/v) glucose, 0.4 % (w/v) Na-phytate, 0.2 % (w/v) $CaCl_2$, 0.5 % (w/v) $NH_4NO_3$, 0.05 % (w/v) KCl, 0.05 %(w/v) $MgSO_4.7H_2O$, 1.5 % bacto-tryptone, 0.001 % (w/v) $FeSO_2.7H_2O$ and 0.001 % (w/v) $MnSO_4.H_2O$ [8]. The method used in the determination of physiological activity was based on the method of Wang [9] with a slight modification. One mL of supernatant from each liquid culture on the

IOP Conf. Series: Earth and Environmental Science **137** (2018) 012009 doi:10.1088/1755-1315/137/1/012009

screening medium was incubated for 5 minutes at a temperature of 55°C. Also incubated was 2 mL of 0.1 M phosphate buffer solution at pH 6 containing 2 mM of Na-phytate and 1 mM of $CaCl_2$ at a temperature of 55°C for 5 minutes. Furthermore, this solution was added to 1 mL of supernatant from liquid culture, mixed and incubated again for the next 15 minutes. Afterwards, 3 mL of 5 % TCA was added, and the mixture of the substrate was kept at a temperature of 4°C for 15 minutes and then centrifuged at 4.000 rpm for 20 minutes. In addition to the supernatant, vanadate-molybdate reagent was added at the same volume ratio, then they were mixed homogeneously. Finally, the absorbance was measured using a UV-Vis spectrophotometer at a wavelength of 392 nm.

## 3. Results and Discussion

### *3.1 Characterization of phytase*

*Actinobacillus* sp., *Bacillus pumilus*, *Bacillus vallimortis*, IBR-1 produce phytase enzymes. The production capabilities of phytase enzymes vary from one another (table 1 and table 2)

**Table 1.** Optimum temperature of enzymes phytase.

| Temperature $^0$C | Phytase Enzymes Activities (Unit/mL) | | | |
|---|---|---|---|---|
| | *Actinobacillus* sp | *Bacillus pumilus* | *Bacillus vallimortis* | IBR-1 |
| 40 | 0.246 | 0.267 | 0.202 | 0.214 |
| 45 | 0.257 | 0.237 | 0.254 | 0.229 |
| 50 | 0.278 | 0.144 | 0.212 | 0.203 |
| 55 | 0.175 | 0.126 | 0.180 | 0.139 |
| 60 | 0.147 | 0.120 | 0.209 | 0.154 |

**Table 2.** Optimum pH of enzymes phytase.

| pH | Phytase Enzymes Activities (Unit/mL) | | | | |
|---|---|---|---|---|---|
| | Buffer | *Actinobacillus* sp | *Bacillus pumilus* | *Bacillus vallimortis* | IBR-1 |
| 4 | Acetate | 0.237 | 0.257 | 0.216 | 0.265 |
| 5 | Acetate | 0.230 | 0.238 | 0.228 | 0.272 |
| 6 | Tris-maleat | 0.168 | 0.175 | 0.199 | 0.196 |
| 7 | Tris-maleat | 0.149 | 0.126 | 0.188 | 0.185 |
| 8 | Tris-maleat | 0.125 | 0.129 | 0.171 | 0.178 |

Temperature affects the production of phytase. *Actinobacillus* sp. has an optimum activity at a temperature of 45-50°C, *Bacillus pumilus* 40-45°C, *Bacillus vallimortis* 45-50°C and IBR-1 40-45°C (table 1). At higher temperatures, the production of phytase by each bacterium will decrease. The fermentation temperature will affect the growth rate of biomass and enzyme production. The effective temperatures for growth depend on the type of microorganisms, be it bacterium or fungi. *Rhizopus oligosporus* grown on coconut husk medium at pH 5.3 resulted in the highest activity of phytase in 4 days of fermentation. *A. niger* produced the highest activity of phytase after 3 days of incubation on a medium containing 1 g/L of gum at a temperature of 30°C [10].

The production of phytase is also affected by pH. *Actinobacillus* sp., *Bacillus pumilus, Bacillus vallimortis* and IBR-1 had optimum activity at pH 4-5 (table 2). At higher pH the production of phytase by each bacterium will decrease. The pH factor affects the growth rate of biomass and enzyme production. The highest amount of phytase obtained from *Actinobacillus* sp. was 0.230 unit/L. The pH factor affects the growth of biomass and the production of phytase. *Aspergillus ficuum* produces phytase optimally at pH 6.8 -7.0 [11,12]. Research on the production of phytase using tofu pulp is still not common. In general,

IOP Conf. Series: Earth and Environmental Science **137** (2018) 012009 doi:10.1088/1755-1315/137/1/012009

research on the production of phytase uses by-products of coconut oil, olive oil, palm oil and cottonseed [12]. The addition of phytase to soybean curd residue increases the nutritional value and promotes the growth of fish [13]. This research succeeded in using tofu pulp as a medium for producing phytase using *A. niger*, *R. oryzae* and *N. sithophila*.

Phytic acid is an anti-nutrient substance that can interfere with the nutritional function and components of food. Phytic acid is a very powerful chelating agent that can bind to minerals, proteins and some amino acids [14]. Livestock and fish feed generally contain phytic acid. The phytic acid in livestock and fish feed can decrease mineral availability for livestock and fish. Phytic acid in livestock and fish feed can be hydrolyzed by phytase by breaking phosphate bond from phytic acid into inositol and inorganic phosphate [15]. The presence of phytase enzymes in livestock and fish feeds can improve the nutrients for livestock, especially monogastric livestock and fish.

The presence of phytic acid that can bind to metal ions, proteins and amino acids can interfere with the absorption of these compounds by livestock and human gut. So it is absolutely necessary to have an enzyme that can break phytic acid bonds with proteins, amino acids and micro elements. The addition of the phytase is aimed to break the ester bonds that bind amino acids, fats and microelements to phytic acid. Bacteria, especially *Actinobacillus* sp., *Bacillus pumilus*, *Bacillus vallimortis* and IBR-1, are proven to be capable of producing high enough phytase required for the availability of minerals for livestock and fish.

## 4. Conclusion

The optimum temperatures of *Actinobacillus* sp., *Bacillus pumilus*, *Bacillus vallimortis* and IBR-1 were 50°C, 40°C, 45°C, 45°C, respectively, and the optimum pH were 4, 4, 5.5. Bacteria, especially *Actinobacillus* sp, *Bacillus pumilus*, *Bacillus vallimortis* and IBR-1 have been proven to be capable of producing high enough phytase required for the need of minerals of livestock and fish.

## 5. References

[1]. Woyengo T A and Nyachoti M 2013 *Canadian J. Am Sci.* **93** 9-21.
[2]. Kornegay E T 2001 Digestion of phosporus and other nutrients: the role of phytates and factors influencing their activity (Virginia : Virginia Polytechnic Institute and State University Blacksburg)
[3]. Yu B, Wu S T and Liu C C 2007 *Am. Feed Sci. Technol.* **134** 283–294.
[4]. Khan S A, Haroon R C, Yasser, Saleem M and Tariq J 2013 *Biologia (Pakistan)* **59** 99-106
[5]. Bedford M R and Patridge G G 2001 *Enzymes in farm animal nutrition* (CAB International)
[6]. Morse D H H and Wilcox D J 1992 *J. Dairy Sci.* **75** 1979 – 1986
[7]. Selle P H and Ravindran V 2007 *Anim. Feed Sci. Tech.* **135** 1-41.
[8]. Kerovuo J, Kerovuo J, Marko L, Päivi N, Nisse K, and Juha A 1998 *J. Appl Environ Microbiol* **64**
[9]. Wang X Sun Q, Ding Z, Ji J, Wang J, Kong X, Yang J, Cai G 2004 *Cell. Res.* **24** 796-808.
[10]. Papagianni M, Nokes S E and Filer K 2001 *Food Technol. Biotech.* **39** 319-326.
[11]. Han Y W, Gallagher D J and Wilfred A G 1987 *J. Industrial Microbiol.* **2** 195-200
[12]. Ramachandran S, Roopesh K and Nampoothiri K M. 2005 *Process Biochem.* **40** 1749-1754
[13]. Selle P H and Ravindran V 2007 *Anim. Feed Sci. Technol.* **135** 1-41
[14]. Ahmad T, Rasool S, Sarwar M, Haq A U and Hasan Z U 2000 *Anim. Feed, Sci. Technol.* **83** 103-114
[15]. Guo M J, Zhuang Y P, Chu J, Zhang S L, Xiong A S, Peng R H and Yao Q H 2007 *Process Biochem.* **42** 1660-1665

# The effect of the addition of cow brain powder in commercial feed on the gonadal maturity of comet goldfish (*Carassius auratus auratus*)

**Y Andriani[1], U Subhan[1], Rosidah[1], Iskandar[1], I Zidni[1], A M Abdillah[2]**

[1]Lecturer of Faculty of Fisheries and Marine Science, Universitas Padjadjaran
[2]Alumny of Faculty of Fisheries and Marine Science, Universitas Padjadjaran.

E-mail: yuliyusep@yahoo.com

**Abstract.** The aim of this research was to analysis the effect of addition bovine's brain meal in artificial feed on gonad maturity and to find out the best time of gonad maturity in comet fish.This research was conducted at Fourth Building Hatchery Faculty of Fisheries and Marine Sciences Padjadjaran University on November 2014 until Januari 2015. Freeze drying of bovine brain was conducted at Research Center Inter University Bandung Institute of Technology. The research was using Completely Randomized Design (CRD) with four treatments and three replications.The treatment were 20 mg/kg, 35 mg/kg, 50mg/kg and control. The parameters of this research are Gonado Somatic Index (GSI) and egg maturity level. Addition of bovine brain meal in feed with the dose of 50 mg/kg are giving the best result until 45 days of the care time against gonad maturity of comet fish with GSI result 12.93 %, egg maturity level ripe phase 21.115 and fecundity 1520 grain/g.

## 1. Introduction

Gonadal maturity and egg quality can be improved by good broodstock management. The external factors that need to be considered in broodstock management are, among others, cultivation container, water quality, temperature, broodstock density, and feed. Internally, the use of natural or synthetic hormones can accelerate the maturity of the gonads.

Cow brain is one of the natural ingredients that can be used to accelerate gonadal maturity. The brain is an organ that is extremely important in the reproductive system, because Gonadotropin-Releasing Hormone (GnRH), which consists of 10 amino acids (decapeptide) that stimulate the release of gonadotropin (GTH) in the pituitary gland, is found in every part of it [1]. In a cultivation container, environmental signals are often weak or nonexistent that the reproduction process cannot run perfectly [2]. Stimuli from the environment, such as temperature, are processed through sensory receptors. The resulting neural signals, when reaching the hypothalamus in the brain, affect the pituitary gland through a chemical messenger which is referred to as releasing hormone. The releasing hormone stimulates the pituitary to produce a hormone with the gonads being the target organs. This hormone is called gonadotropin. This hormone affects the production of sex steroids in the gonads that eventually play a role in the maturation of gametes [3]. Therefore, hormonal preparations such as ovaprim, hCG (Human Chorionic Gonadotropin), and LHRH (Lutenizing Hormone Releasing Hormone) are used.

Cow brain that is processed into powder and mixed in feed can be used as an alternative ingredient for stimulating the gonadal development in comet goldfish. Comet goldfish (*Carassius auratus auratus*) is one of goldfish strains with physical beauty represented by a longer and more beautiful tail, better endurance, and relatively lower price compared to the price of goldfish. There has been high demand for comet goldfish in the market that it is necessary to have a strategy to increase its production. One of the strategies that can be applied is having appropriate and efficient broodstock management. This study examined the extent to which cow brain powder mixed in feed affects the maturation time of gonads in comet goldfish.

## 2. Methodology

### *2.1. Tools and materials*

This study was conducted in the Hatchery of the Faculty of Fisheries and Marine Sciences, Universitas Padjadjaran. The freeze-drying of cow brain was conducted at the Inter-University Research Center of Institut Teknologi Bandung.

The materials used in the study were: female comet goldfish broodstocks weighing 9.67-12.22 g; cow brains from Caringin market, Bandung; fish pellets with 28 % protein content; carboxymethyl cellulose (CMC); sera solution made of 99 % alcohol solution; 40 % formaldehyde solution; and 100 % acetic acid solution at a ratio of 6:3:1 [4]. The tools used were: twelve 60 x 30 x 30 cm aquariums; aeration installation; DO meter from Lutron with a precision of 0.01 mg/L; pH Meter from Lutron; digital thermometer; blender; dissecting kit; thermostats for stabilizing the temperature of the aquarium water; analytical balance; microscope; and digital camera.

### *2.2. Method*

This study used an experimental method with Completely Randomized Design (CRD), four treatments, and three replications. The treatments in this study were distinguished based on the administration of different doses of cow brain powder mixed with commercial feed as follows:

- Treatment A = Feed without cow brain powder (Control)
- Treatment B = Feed with cow brain powder, 20 mg/kg broodfish
- Treatment C = Feed with cow brain powder, 35 mg/kg broodfish
- Treatment D = Feed with cow brain powder, 50 mg/kg broodfish

### *2.3. Procedure*

#### *2.3.1. Test feed preparation*

Fresh cow brains were cleaned and then cut/chopped into small pieces and inserted into a freeze-drying instrument. The freeze-dried brains were ground in a blender and the resulting powder was placed into a plastic container. This cow brain powder was then mixed with commercial feed according to the treatments. Adhesive (CMC binder) and 2 % of the commercial feed were added into the mixture. The mixed feed was than pelettized using a pellet mill. The pellets produced were sun dried and then stored in jars to be fed to the comet goldfish broodstocks until the gonads matured.

#### *2.3.2. Comet goldfish broodstocks preparation*

At the time of the study, the comet goldfish were in the initial maturation phase of gonads. The broodstocks used in the study were spawned first. Afterwards, stripping was performed on the comet goldfish broodstocks to remove eggs from the fish bodies.

#### *2.3.3. Conduct of study*

A total of 12 comet goldfish broodstocks were placed in individual aquariums. Feed was administered twice a day at 09.00 pm and 15.00 pm for 2 months of rearing. The amount of feed given was 3 % of the fish biomass weight. Container maintenance cleaning was performed every other day using siphon. The water quality parameters observed included temperature, pH, and dissolved oxygen content (DO). Observations were performed three times i.e., at the beginning, middle, and end of the study.

The comet goldfish gonadal maturity checks were conducted at 3 different times. The first check was performed after 1 month of rearing by examining 1 fish from every treatment to check the

IOP Conf. Series: Earth and Environmental Science **137** (2018) 012010 doi:10.1088/1755-1315/137/1/012010

development of the gonads. The second check was performed after 1 ½ months of rearing, while the third check was performed after 2 months of rearing using the same method i.e., examining 1 fish from every treatment to check the development of the gonads. The Gonad Maturity Index (GMI) was measured by deducting the weight of the dissected gonad from the weight of a comet goldfish broodstock. After the gonadal maturity examination was completed, the data obtained were further analyzed.

*2.4. Observation Parameters*

*2.4.1. Gonad Maturity Index (GMI)*

The GMI is calculated using the following formula of Effendi [5] :

$$GMI = GW/BW \text{ X } 100\ \% \quad (1)$$

Notes:

GMI = Gonad Maturity Index

GW = Fish gonad weight in gram

BW = Body weight in gram

*2.4.2. Percentage of Fish Egg Maturity Rate*

The Egg Maturity Rate (EMR) was calculated based on the following formula [4]:

$$\text{EMR of vitellogenic phase} = \frac{\text{Number of eggs centered nucleus x 100 \%}}{\text{Number of eggs observed}} \quad (2)$$

$$\text{EMR of early maturity phase} = \frac{\text{Number of eggs with non-centered nucleus x 100 \%}}{\text{Number of eggs observed}} \quad (3)$$

$$\text{EMR of mature phase} = \frac{\text{Number of eggs with fused nucleus x 100\%}}{\text{Number of eggs observed}} \quad (4)$$

*2.5. Data analysis*

The data were analyzed using variance analysis with F-test to determine the effect of each treatment. If there were differences among the treatments, further analysis using Duncan's multiple range test with a significance level of 5 % was performed [6].

## 3. Results and Discussion

*3.1. Gonad maturity index*

Gonad maturation index (GMI) is an index that is used to determine the percentage of gonad weight in comparison to broodstock body weight. One of the factors influencing the maturity level of the gonads is the feed. Good quality feed contains compositions that support successful gonadal and spawning processes i.e., the compositions that include vitamin, 28-40 % protein content, and hormones, both synthetic and natural.

**Table 1**. Average of gonad maturity index.

IOP Conf. Series: Earth and Environmental Science **137** (2018) 012010 doi:10.1088/1755-1315/137/1/012010

| Treatment | Gonad Maturity Index (%) | | |
|---|---|---|---|
| | Day 30 | Day 45 | Day 60 |
| A (control) | 10.54 | 9.66 | 12.13 |
| B (20 mg/kg broodfish) | 10.44 | 6.93 | 10.90 |
| C (35 mg/kg broodfish) | 7.00 | 10.94 | 11.81 |
| D (50 mg/kg broodfish) | 12.44 | 12.93 | 11.54 |

The results of the study showed that the average GMI on the 30$^{th}$ day of rearing ranged from 7 to 12.44 % (table 1). The highest GMI (12.44 %) after 30 days of rearing was found in Treatment D, and the lowest (7 %) was found in Treatment C. The average GMI on the 45$^{th}$ day of rearing ranged from 6.93 to 12.93 %. The highest GMI (12.93 %) at this point of time was seen in Treatment D, and the lowest (6.93 %) was found in Treatment B. The average GMI on the 60$^{th}$ day of rearing ranged from 10.90 to 12.13 %. Based on the analysis results that showed different GMIs, it was revealed that the addition of cow brain powder into the feed did not significantly affect the GMI of comet goldfish broodstocks. Overall, the GMIs in this study were still within the normal GMI range for comet goldifish, i.e., 7.28 to 19.11 % [7].

The GMI in Treatment D was the largest, and an increase was seen between the 30$^{th}$ day and 45$^{th}$ day of rearing (table 1). This is assumed to be the effect of GnRH in cow brain powder mixed with the feed that increased the weight of the test fish gonads. The GMI in Treatment C was the lowest on the 30$^{th}$ day of rearing, and the GMI in Treatment B was the lowest on the 45$^{th}$ day of rearing. This may have been caused by lower GnRH content in these treatment compared to that in Treatment D. According to Hartika [8], ovaprim induction (GnRH) needs a long pathway to be able to affect gonadal maturation. GnRH will cause hypophysis to secrete Gonadotropin (GtH), which is then passed into the blood and only at a certain level stimulates the final gonadal maturity through stimuli to synthesize the steroid hormone maturation by follicles in the ovaries. Based on this statement, it can be concluded that higher doses of GnRH will be more effective in stimulating gonadal maturation.

The feed that is mixed with GnRH-containing cow brain powder is consumed by the fish and then enters the intestine to be digested. GnRH in the feed is absorbed by the intestine and bound by a receptor through the blood and then brought to the target organ, which is the hypophisis. GnRH stimulates the pituitary gland to release gonadotropin, i.e. the follicle-stimulating hormone (FSH). This gonadotropin hormone is then carried by the receptor through the blood into the gonadal theca cells to stimulate the formation of testosterone (figure 1). The testosterone formed then enters the granulosa cell to be converted by the aromatase enzyme intone estradiol-17$\beta$ hormone. Estradiol-17$\beta$ is a steroid hormone derivative of cholesterol that plays an important role in the vitellogenesis [9]. Some of the estradiol-17$\beta$ will go to the liver to form vitellogenin, which is the main component of egg yolks. The resulting vitellogenin is then carried by the blood into the gonads to be absorbed by the oocytes, leading to egg diameter expansion. The process of vitellogenin will continue in the fish body. According to Effendie [5], changes in GMI are closely related to the stages of egg development.

IOP Conf. Series: Earth and Environmental Science 137 (2018) 012010 doi:10.1088/1755-1315/137/1/012010

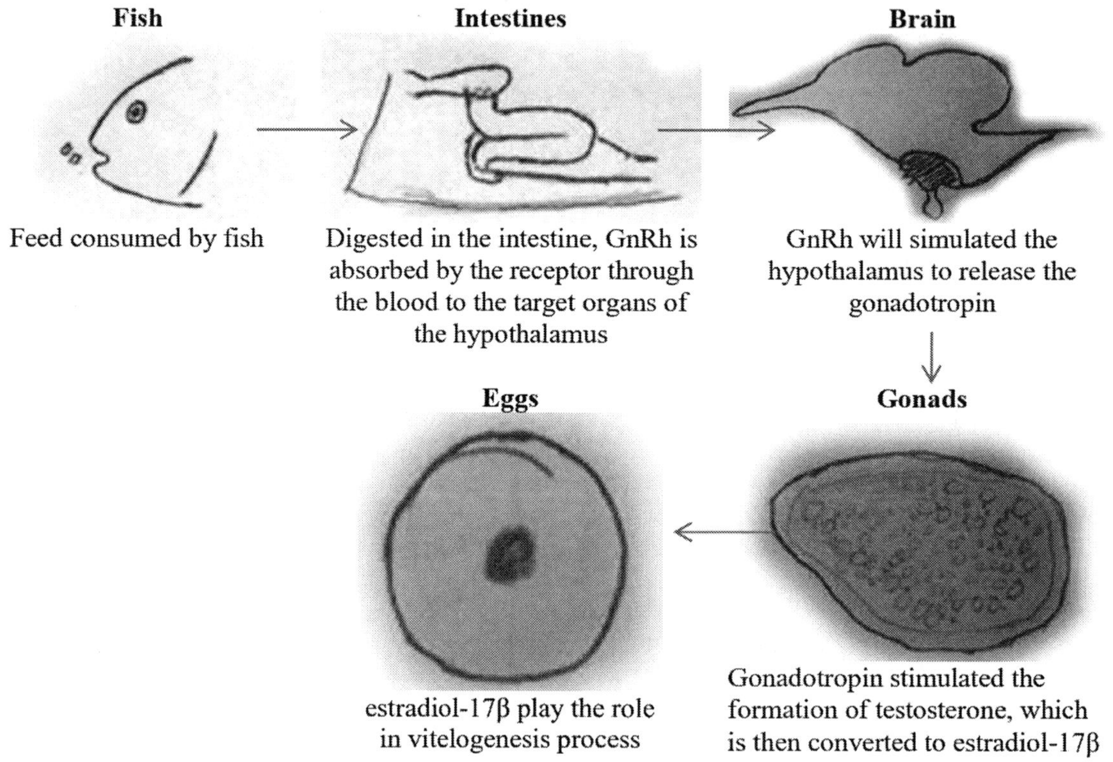


**Figure 1.** Flowchart of GnRH effect in feed.

The GMI value in Treatment D on the 60th day of rearing decreased when compared with the values of other treatments which increased. This is allegedly due to the process of atresia, which is the absorption of eggs that have reached the maximum development but not ovulated [10]. In figure 2, it is depicted that after going through the dormant phase (the nucleus of the cell is located in the centre), if the level of gonadotropin is adequate, the egg will progress to the next stage which is marked by the shifting of the egg nucleus to the edge. Atresia will occur if the eggs that are ready to ovulate are ignored for too long or not fertilized.

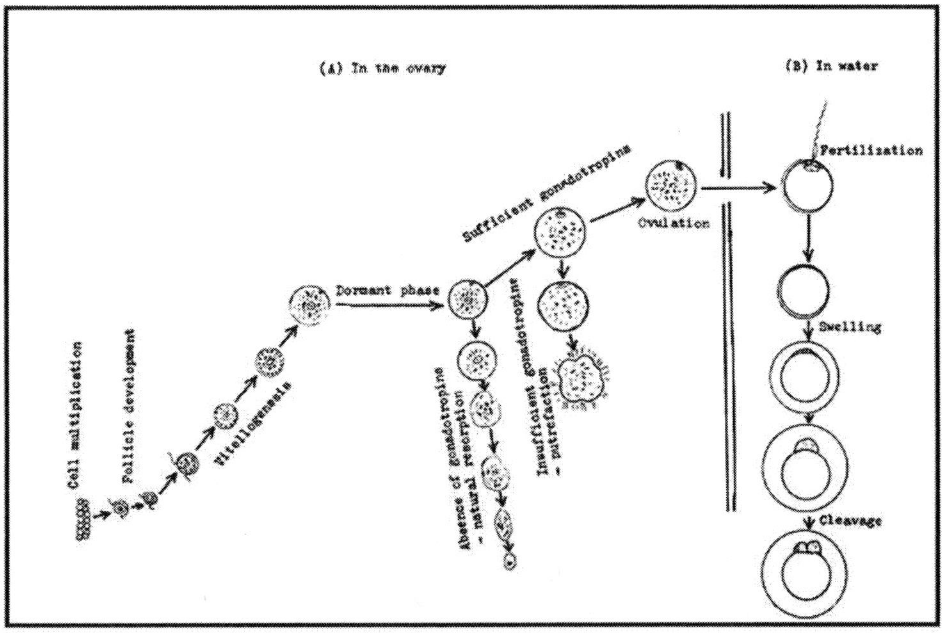


**Figure 2**. Egg development process.

### *3.2. Egg maturity level*

Based on the nucleus positions, the egg maturity rate (EMR) is divided into three according to different phases of development, namely before the egg is ovulated—which is called vitellogenic phase, initial maturition phase, and mature phase. The vitellogenic phase is characterized by the position of the egg nucleus at the center. The nucleus subsequently migrates from the centre to the edge during the initial maturation phase, and then fuses in the mature phase (figure 3). Gonadotropin and steroid hormonal actions cause the nucleus that is originally centered to move to the edge of the egg, near the microphyllid, and to fuse just before the ovulation [11]. This process is known as GVBD (Germinal Vesicle Breakdown) which indicates that the egg has entered the final maturation process and it may ovulate.

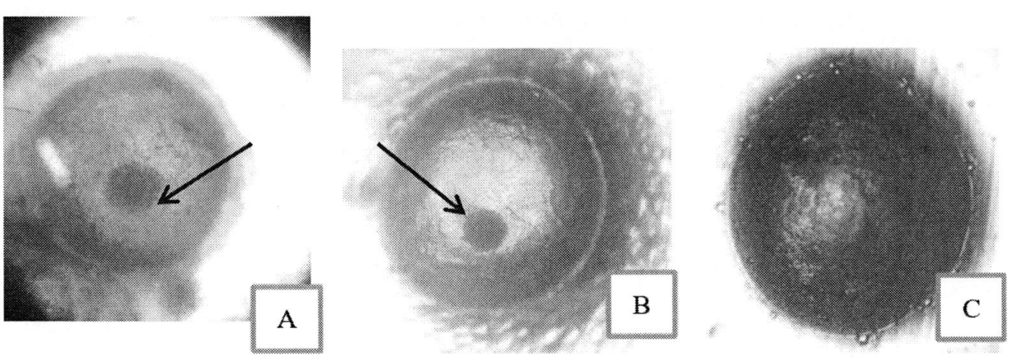


**Figure 3.** Nucleus movement phases of comet fish eggs (A) Edged nucleus; (B) Centered nucleus; (C) Fused nucleus (GVBD).

According the results of the observation on the 30$^{th}$ day, overall increases consistent with the increased doses of cow brain powder in feed are presented in figure 4. The average EMRs in

Treatment A were 90.56 %, 9.44 %, and 0 % for the vitellogenic phase, initial maturation phase, and mature phase, respectively. The average EMRs in Treatment B were 59.44 % for the vitellogenic phase, 33.89 % for initial maturation phase, and 6.67 % for the mature phase. The average EMRs in Treatment C were 82.22 %, 13.33 %, and 4.44 % in the vitellogenic phase, initial maturation phase, and mature phases, respectively. The average EMRs in Treatment D were 38.89 % for the vitellogenic phase, 38.89 % for the initial maturation phase, and 22.22 % for mature phase. Based on the analysis of variance on the maturity level of the eggs on the 30th day of rearing, it is apparent that there was a significant difference in the mature phase between treatments.

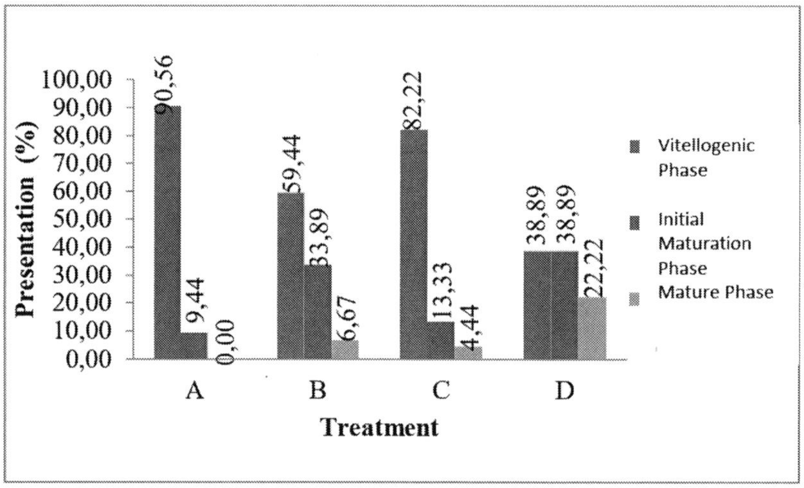


**Figure 4.** Egg maturity rate on the 30th day of maintenance.

Based on the results of the Duncan's follow-up test, Treatment A was not significantly different from Treatment B and Treatment C but significantly different from Treatment D (table 2). Treatment D had higher values than those of other treatment groups, which was caused by the high level of GnRH contained in the feed that put more trigger on the mature eggs. Murtejo [11] stated that gonadotropin and steroid hormone actions cause the nucleus that is initially located in the centre to move to the edge, near the microphyllid, and then it fuses just before ovulation. This process is known as GVBD (Germinal Vesicle Breakdown) which indicates that the egg has entered the final maturation process and can ovulate. When eggs are in the GVBD, the process of vitellogenesis or absorption of the vitellogenin in oocyte will be stopped and the egg diameter reaches the maximum size.

**Table 2.** Average Movement of Fish Comet Egg Nucleus on the 30th Day of Maintenance.

| Cow Brain Meal (mg per kg broodfish) | Vitellogenic Phase (%) | Initial Maturation Phase (%) | Mature Phase (%) |
|---|---|---|---|
| A (Control) | 9 0.56 ± 3.47 | 9, 44 ± 3.47 | 0 ± 0 [a] |
| B (20 mg/kg broodfish) | 59, 44 ± 37.28 | 33.89 ± 34.17 | 6.67 ± 7.64 [a] |
| C (35 mg/kg broodfish) | 82, 22 ± 10.05 | 13, 33 ± 8.82 | 4.45 ± 2.55 [a] |
| D (50 mg/kg broodfish) | 38.89 ± 16.68 | 38.89 ± 3.47 | 22.23 ± 13.47 [b] |

Notes: Figures followed by the same lowercase letters mean no significant difference at the significance level of 95%

The results of the observation on the 45th day still showed that Treatment D presented better results than those of other treatments in terms of egg maturity rate (figure 5). The average EMRs in Treatment A were 73.33 %, 19.44 %, and 7.22 % for the vitellogenic, initial maturation, and mature phases, respectively. Meanwhile, for Treatment B, the average EMRs were 78.33 % for the vitellogenic phase,

1.11 % for the initial maturation phase, and 0.56 % for the mature phase. The average EMRs for Treatment C were 71.67 %, 27.78 %, and 0.56 % for the vitellogenic, initial maturation, and mature phases, respectively. For Treatment D, the values were 38.33 %, 40.56 %, and 21.11 % for the vitellogenic, initial maturation, and mature phases, respectively.

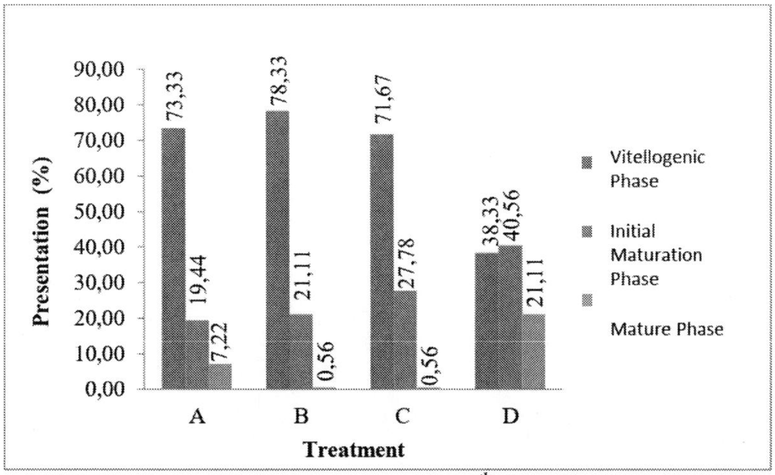


**Figure 5.** Egg maturity rate on the 45th day of maintenance.

The results of Duncan's follow-up test showed that there was a significant difference in the vitellogenic phase on the 45th day of rearing (table 3). Treatment A was significantly different from Treatment C and Treatment D, but it was not significantly different from Treatment B. Based on the analysis of variance, there was no significant difference between treatments during early maturation phase. In the mature phase, there were significant differences between treatments. Treatment B was not significantly different from Treatment A and Treatment C, but it was significantly different from Treatment D. Treatment A was not significantly different from Treatment D. Treatment D in the mature phase has higher value compared to that of other treatments, which was proportional to the increasing amount of cow brain powder. Based on the egg maturity rates on the 45th day of rearing, Treatment B and Treatment C had the smallest EMRs in the mature phase which probably due to less GnRH content in the feed than the GnRH content of Treatment D. However, an increase was still seen in the early maturation phase.

**Table 3.** Average egg nucleus movement of comet fish eggs on the 45th Day.

| Cow Brain Meal (mg per kg broodfish) | Vitellogenic Phase (%) | Initial Phase (%) | Mature Phase (%) |
|---|---|---|---|
| A (Control) | 73.33 ± 3.33 $^a$ | 19.44 ± 8.22 | 7.22 ± 5.36 $^a$ |
| B (20 mg/kg broodfish) | 78.33 ± 5.77 $^a$ | 21.11 ± 4.81 | 0, 56 ± 0.96 $^a$ |
| C (35 mg/kg broodfish) | 71.67 ± 1.67 $^b$ | 27.78 ± 0.96 | 0, 56 ± 0.96 $^a$ |
| D (50 mg/kg broodfish) | 38.33 ± 12.58 $^b$ | 40.56 ± 20.02 | 21.11 ± 16.69 $^b$ |

Notes: Figures followed by the same lowercase letters mean no significant difference at the confidence level of 95%

The results from the observation on the 60th day of rearing were different from the previous days of rearing (figure 6). Treatment C had the highest EMR in the mature phase compared to other treatments. The average EMRs for Treatment A were 52.22 %, 31.67 %, and 16.11 % in the vitellogenic, initial maturation, and mature phases, respectively. In Treatment B, the EMRs for the vitellogenic phase were 62.22 %, while the values for the initial maturation phase and mature phase were 29.44 % and 13.89 %, respectively. In Treatment C, the EMRs were 22.78 % for vitellogenic phase, 40.56 % for initial maturation phase, and 36.67 % for mature phase. The average EMRs in Treatment D were 37.22 %, 50 %, and 23.78 % for vitellogenic, initial maturation, and mature phases, respectively.

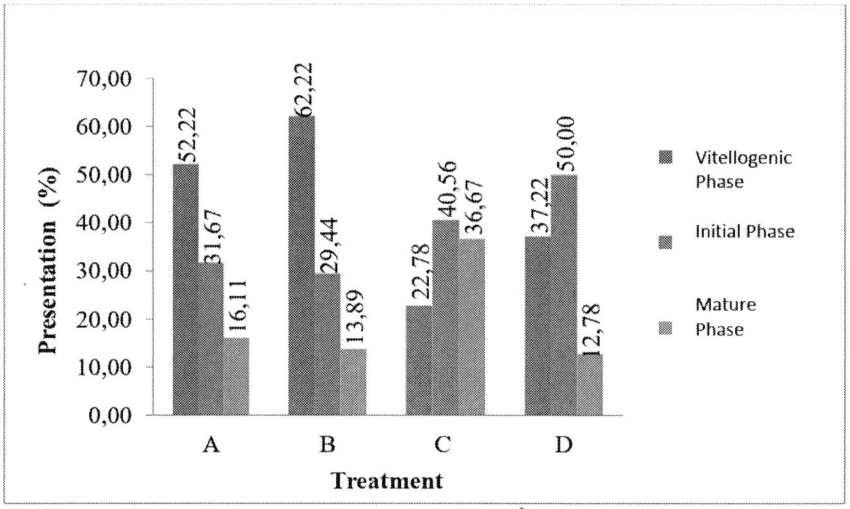


**Figure 6.** Egg maturity rate on the 60th day of maintenance

The results of the analysis of variance revealed that the egg maturity levels on the 60th day in the initial maturation and maturation phases were not significantly different, but they were significantly different from that of the vitellogenic phase. Based on the results of Duncan's follow-up test in vitellogenic phase, Treatment A was not significantly different from Treatment B, but it was significantly different from Treatment C and Treatment D (table 4).

**Table 4.** Average nucleus movement in comet fish on the 60th day of maintenance.

| Cow Brain Meal (mg per kg broodfish) | Vitellogenic Phase (%) | Initial Phase (%) | Mature Phase (%) |
|---|---|---|---|
| A (Control) | 52.22±11.71[a] | 31.67± 16.07 | 16.11± 11.10 |
| B( 20 mg/kg broodfish) | 62.22±13.37[a] | 29.44± 8.39 | 13.89± 14.17 |
| C (35 mg/kg broodfish) | 22.78±13.47[b] | 40.56± 14.17 | 36.67± 25.22 |
| D( 50 mg/kg broodfish) | 37.22±19.32[b] | 50.00± 16.41 | 12.78± 7.52 |

Overall, the EMR of comet goldfish on the 60th day of rearing showed an increased egg maturity level in the vitellogenic phase compared to the EMR on the 45th day, which experienced a decrease. Decreased EMR in the mature phase of Treatment D was caused by atresia, which is the re-absorption of eggs that have reached the maximum development but not ovulated [10].

## 4. Conclusions
The addition of 50 mg/kg broodfish cow brain powder to the feed leads to the best effect on comet goldfish gonadal maturity on the 45$^{th}$ day of rearing with EMRs of 12.93 % and 21.11 % in the initial maturation phase and mature phase, respectively.

## 5. References

[1]. Subhan 2011 Evaluate the effectiveness of catfish fish extract in inducing fish spawning lele sangkuriang, *Clarias sp.* (Bogor : Bogor Agricultural Institute) (In Indonesia)

[2]. Hunter GA Donaldson EM 1983 *Fish Physiology* **11** *Academic Press*

[3]. Harvey B Dan Carolsfeld J 1993 *Induced breeding in tropical fish culture* (Canada : International Development Research Centre)

[4]. Nurmadi T 2005 The effect of hormone HCG use with different doses on the development of gonad baung fish (Hemibagrus nemurus) (Bogor : Agricultural Institute) (In Indonesia)

[5]. Effendie M I 1997 *Fisheries biology* (Jakarta : Yayasan Pustaka Nusantara) (In Indonesia)

[6]. Gasperz V 1991 *Experimental design method* (Bandung : CV Amrico) 472 p (In Indonesia)

[7]. Najim S Raja A Furat K 2012 *J. Aquac.* 9:83-94

[8]. Hartika R 2013 The effect of ovaprim injections with different doses of ovulation and quality of pearl fish eggs (Pekanbaru : University of Riau) (In Indonesia)

[9]. Sinjal HJ 2007 Study of reproductive performance of catfish (*Clarias gariepinus*) through females addition of ascorbyl phosphate magnesium as a source of vitamin c and implantation with estradiol-17 $\beta$ (Bogor : Agricultural Institute) (In Indonesia)

[10]. Yudhistira T.E 2013 Effect of cow testis feeding on maturity level of gonad and fecundity of nilem fish (Osteochilus hasselti) (Jatinangor : University of Padjadjaran) (In Indonesia)

[11]. Murtejo H E 2008 Effectiveness of egg stimulant in feed against gonad ripening and productivity of red fin shark fish (*Epalzeorhynchos frenatum*) (Bogor : Agricultural Institute) (In Indonesia)

IOP Conf. Series: Earth and Environmental Science **137** (2018) 012011 doi:10.1088/1755-1315/137/1/012011

# The effect of colchicine on the size and bioactive compound of microalgae *Spirulina platensis*

**A Mahardika**[1], **A T Mukti**[2] **and M Arief**[2]

[1] Graduate, Study Program of Aquaculture, Faculty of Fisheries and Marine, University of Airlangga, Campus C Unair Jl. Mulyorejo Surabaya 60115, Indonesia
[2] Department of Fish Health Management and Aquaculture, Faculty of Fisheries and Marine, University of Airlangga, CampusC Unair Jl. Mulyorejo Surabaya 60115, Indonesia

Email: atm_mlg@yahoo.com

**Abstract.** Polyploidy is one of the techniques used to increase the genetic variant and once used as a breeding method of plant. Colchicine is one of the chemical which apply to produce polyploid organisms, such as plant. This study aimed to determine the effect of colchicine on the size and phycocyanin content of *Spirulinaplatensis*.Research was used six treatments of colchicine concentration with three replications. *S. platensis* were immersed in the colchicine solution for 12 hours and were observed for 5 days culture. This research was showed that colchicine concentration of 0.1 % were resulted highest diameter of *S. platensis*(12.57 μm) while highphycocyanin content obtained by treatment of 0.025 % (0.091 mg/ml).

## 1. Introduction

Microalgae have been developed for the purpose of research and technology. The development of microalgae has advantages in terms of fast growth and high fat and protein content [1]. *S. platensis*are microalgae with complete and high protein content [2]. *S. platensis* are small in size, and their threads consist of a series of cylindrical cells of trichomes with thin cell wall diameter of 1-12 μm [3]. The high phycocyanin content in these microalgae causes blue-green color. *S. platensis*'strichomes have a spiral structure with filaments, but they have no heterocyst [4]. High demand for *S. plantesis*greatly impacts on the need for the improvement of their quality. One of the alternative to improve *S. plantesis*'squality is changing them into polyploid. Polyploidy has been used to increase genetic variance [5]. Polyploidy can be obtained by using colchicine.

Colchicine, chemicals when administered to plants, can resulted in polyploid individuals. Common traits displayed by polyploid plants are bigger size and larger parts such as roots, stems, leaves, flower and fruit [6]. High colchicine concentration and soaking time are not sufficient for producing polyploid individuals [7]. Most studies on polyploidy have been conducted on plants, such as onion [8], garden balsam and soybean, while on microalgae, the effects of colchicine treatment are yet to be known. This study was aimed to investigate the effect of colchicine treatment on the diameter and phycocyanin content of *S. platensis*.

## 2. Methodology

This research was conducted at Laboratory of Fisheries Education, Faculty of Fisheries and Marine, University of Airlangga Surabaya.

### *2.1. Preparation of S. platensis culture*
Some instruments and media were used for *S. platensis* culture prior to sterilization using autoclave [9,10]. The *S. platensis* used for culture were collected from Brackishwater Aquaculture Development Center (BADC) Situbondo. Walne fertilizer with a concentration of 0.1 % (v/v) was used to improve the growth of *S. platensis* population. The initial stock density of *S. platensis* was $1 \times 10^4$ cells/mL.Calculation of the cell according to Satyantini [11].

$$V1 = \frac{N2 \times V2}{N1} \quad (1)$$

Note:
V1 = volume of seed for initial stocking (ml)
N1 = density of plankton seeds (cells / ml)
V2 = volume of desired culture media (ml)
N2 = density of desired plankton seed (cells / ml)

### *2.2. Calculation of S. platensis Density*
The population growth of *S. platensis* was observed every day in five days of culture by calculating the density according to Octhreeani [12].

$$\text{Phytoplankton density (cells / mL)} = \frac{na+nb+nc+nd+ne}{5 \times 4 \times 10^{-6}} \quad (2)$$

Note:

| | |
|---|---|
| na, nb, nc, nd, ne | = number of cells *of S. platensis* in box a, b, c, d, e |
| 5 | = number of boxes counted |
| $4x10^{-6}$ | = area of small box ( a, b, c, d or e) |

### *2.3. Treatment of Colchicine Solution*
The doses of colchicine used were 0.01, 0.025, 0.05, 0.075 and 0.1% (w/v), and one treatment without colchicine solution served as control. Each treatment was triplicates. The colchicine solution treatments were administered 90 minutes after the initial culture of *S. platensis*. The immersion of *S. platensis* in the colchicine solution was conducted for 12 hours.

### *2.4. Measurement of S. platensis's Size*
The diameter of *S. platensis*was measured every day using a microscope camera completed with OpticLab and ImageRastersoftware on the computer.

### *2.5. Measurement of S. platensis's Phycocyanin Content*
The phycocyanin extract of *S. platensis* was modified according to Lorenz [13] method. An acetic acid solution (pH 7) was added to the *S. platensis* sample at aratio of 1:5 (v/w). Then, the mixture of acetic acid solution and *S. platensis* was shaken using a vortex. The sample was stored for 24 hours. The mixture was shaken and centrifuged to separate phycocyanin from biomass. The centrifugation was conducted at a minimum speed of 3.500 rpm for 5 minutes. Afterwards, the extraction was tested on a spectrophotometer at wavelengths of 652 and 620 using equation 1 [14]:

$$CPC = \frac{(OD_{620} - 0.474 OD_{652})}{5.34} \quad (3)$$

## 3. Result and Discussion
### *3.1. Diameter Spirulina platensis*
Based on the observations of the diameter of *S. platensis* from day 1 to day 5 showed that the dose of colchicine had a significant effect on the diameter of *S.platensis*, which is presented in table 1. The

IOP Conf. Series: Earth and Environmental Science **137** (2018) 012011 doi:10.1088/1755-1315/137/1/012011

data was analysed using analysis of varians and Tuckey test. The Tuckey test results showed that the largest diameter observed at a concentration of 0.1. The smallest diameters of control observed on day 1 and 2 were 7.91 and 8.23 μm, respectively. In addition, the smallest diameters observed at a concentration of 0.01 on day 1 and 2 were 8.45 and 9.34 μm, respectively. This happened because the doses administered were easily absorbed by *S. platensis*. The colchicine concentration and the dipping duration were not appropriate, thus polyploid individuals would not be produced by nature [7].

**Table 1.** Diameter of colchicine-immersed *S. platensis*in different concentrations.

| Day | Concentration of Colchicine (%) | | | | | |
|---|---|---|---|---|---|---|
| | 0 (Control) | 0.01 | 0.025 | 0.05 | 0.075 | 0.1 |
| 1 | $7.91^{a}$±0.02 | $8.44^{b}$±0.02 | $8.93^{c}$±0.01 | $9.34^{d}$±0.01 | $9.59^{e}$±0.01 | $11.25^{f}$±0.01 |
| 2 | $8.23^{a}$±0.21 | $9.33^{b}$±0.02 | $9.83^{c}$±0.01 | $10.02^{d}$±0.03 | $11.18^{e}$±0.01 | $11.43^{f}$±0.06 |
| 3 | $9.10^{a}$±0.02 | $9.37^{b}$±0.01 | $11.28^{c}$±0.02 | $11.67^{d}$±0.01 | $11.72^{d}$±0.01 | $12.20^{e}$±0.01 |
| 4 | $9.59^{a}$±0.02 | $10.50^{b}$±0.02 | $11.41^{c}$±0.02 | $11.98^{d}$±0.01 | $12.09^{e}$±0.01 | $12.19^{f}$±0.01 |
| 5 | $9.70^{a}$±0.07 | $10.54^{b}$±0.01 | $11.65^{c}$±0.01 | $12.17^{d}$±0.02 | $12.46^{e}$±0.01 | $12.47^{e}$±0.02 |

Note: Data represent as means±SD. Different superscript in the same row indicates significant differences ($P<0.05$).

*S. platensis* undergoes four phases, namely adaptation phase, exponential phase, stationary phase and death phase. Adaptation phase, was shown on day 1 in all treatments. The second phase was the exponential phase, which began on day 2 in all treatments. In this phase cell division started. The third phase is the exponential phase, which was characterized by increasing cell density.

The fourth phase was the stationary phase, which was characterized by slow increase of cells although the number of living cells remains. Every treatment in this research demonstrated stationary phase, which began on the third day. The last phase was marked with increased number of cell deaths and decreased density of *S. platensis*.

*3.2. S. platensis'sphycocyanin content*

Phycocyanin is a protein compound that belongs to the phycobilliprotein group like allophycocyanin and phycoeritrin. The whole phycobilliprotein group is insoluble in water and forms a compound attached to the phycobilisometilacoid membrane. Phycocyanin functions as the main photosynthetic pigment in *S. platensis*and as a store of reserves of nitrogen and amino acids.

Based on the results of this study, *S. Platensis*'s highest phycocyanin content was found after the administration of colchicine at a dose of 0.025 % and concentration of 0.1 %, which showed no significant difference from the result demonstrating the lowest phycocyanin content at a concentration of 0.1 % and colchicine highest dose of 0.1 %. This is consistent with the statement of Sofia [7] that administration of colchicine at improper concentration can cause a failure in plant breeding. The phycocyanin content in *S. platensis*can be seen in table 2.

**Table 2**. Average content of phycocyanin *S.platensis*.

| Concentration of Colchicine (%) | Content of Phycocyanin |
|---|---|
| 0 (Control) | $0.034^{a}$ ± 0.00 |
| 0.01 | $0.081^{d}$ ± 0.00 |
| 0.025 | $0.091^{e}$ ± 0.00 |
| 0.05 | $0.071^{c}$ ± 0.00 |
| 0.075 | $0.058^{b}$ ± 0.00 |
| 0.1 | $0.033^{a}$ ± 0.00 |

Note: Data represent as means±SD. Different superscript in the same row indicates significant differences ($P<0.05$).

## 4. Conclusion

The research results showed that the administration of colchicine at different doses affected the diameter and phycocyanin content of *S. platensis*.

## 5. References

[1]. Sheehan et al 1998 A Look Back at the US Department of Energy's Aquatic Species Program Biodiesel From Algae (Colorado: National Renewable Energy Laboratory)
[2]. Amanatin D R danNurhidayati T 2013 *Sci. Art Pomits* **2** 2337-3520 (In Indonesia)
[3]. Hariyati R 2008 *J. Bioma Res.* **10** 19-22 (In Indonesia)
[4]. Kabede E and Ahlgren G 1996 *Hydrobiology* 332: 99-109
[5]. Dinarti D; Yudiwanti and Rahayuningsih S 2006 *J. Sci. Art Pomits* **2** 88-91(In Indonesia)
[6]. Chahal G S and Gosal S S 2002 Principles and procedures of plant breeding biotechnological and conventional approaches (United Kingdom: Alpha Science International Ltd. Harrow)
[7]. Sofia D 2007 Effect of concentration and duration of colchicine administration on growth and polyploidy in vitro-cultured young soybean (North Sumatera: Faculty of Agriculture University of North Sumatera) (In Indonesia)
[8]. Suminah, Sutarno and Setyawan A D 2002 *Biodiver.* **6** 174-180 (In Indonesia)
[9]. Ekawati A W 2005 *Module of Live Feed Culture* (Malang: Faculty of Fisheries Brawijaya University) p 91(In Indonesia)
[10]. Purnamawati F S, Soeprobowati T R and Izzati M 2013 Growth of *Chlorella vulgaris* Beijerinck in medium with Cd and Pb content at laboratory scale (Semarang : National Seminar on Biology) pp 104-116 (In Indonesia)
[11]. Satyantini W H and Masithah E D 2008 Practical modul of live feed culture (Surabaya: Faculty of Fisheries and Marine University of Airlangga) p 91 (In Indonesia)
[12]. Octhreeani A M, Supriharyono and Soedarsono P 2014 *J. Maquares* **3** 102-108
[13]. Lorenz R T 1998 *J Cell Biol.* **58** 419
[14]. Arlyza I 2005 *J. Oceano.* **3** 27-36 (In Indonesia)

IOP Conf. Series: Earth and Environmental Science **137** (2018) 012012 doi:10.1088/1755-1315/137/1/012012

# Quality characteristics of Bali sardinella (*Sardinella lemuru*) oil purified with bentonite as an adsorbent

**U Nadhiro[1], S Subekti[2], W Tjahjaningsih[2]and Patmawati[2]**

[1] Aquaculture Study Program, Faculty of Fisheries and Marine, Universitas Airlangga.
[2] Department of Marine, Faculty of Fisheries and Marine, Universitas Airlangga.

E-mail: patmawati.wahyudin@gmail.com

**Abstract.** Crude fish oil extracted from fish canning industry a low quality, therefore refining process is required to obtain feasible fish oil for food purposes. Purification of fish oil can through steps of degumming, neutralization, and bleaching by using bentonite as the adsorbent. This study aims to analyze the results of the purification process of crude fish oil by-product of canning industry of lemuru fish by using bentonite adsorbent with different concentrations. The method used was an experimental method by descriptive data analysis. The results showed that the highest yield (33.418 %) obtained from oil purification of lemuru with bentonite concentration of 6 % are classified as follows: free fatty acid content of 0.265 %, peroxide value of 6.343 mEq / kg, produce clarity 60.275 % T, 88.075 % T, 87.5 % T, 87.425 % T, 87.975 % T at a wavelength (λ) of 450 nm, 550 nm, 620 nm, 665 nm, 700 nm, para-anisidine value of 3.725 mEq / kg; and value of oxidation total of 16.41 meq / kg.

## 1. Introduction

Fish canning industry produces a low quality liquid by-product, which is a mixture of oil fraction, water fraction and suspended solids obtained in the steaming stage (pre-cooking) [1,2]. To increase the value of Indonesia's fish oil exports, purification of fish oil should be performed [3].

Purification of fish oil consists of the stages of degumming, neutralization and bleaching [1]. According to Rubio-Rodriguez *et al.* [4], degumming aims to eliminate phospholipids by adding phosphoric or citric acid, and neutralization aims to eliminate free fatty acid with sodium hydroxide, while bleaching aims to absorb oxidation products and color with active clay (activated clay). In this study, bentonite was used in the fish oil bleaching stage. According to Faisal [5], bentonite (aluminosilicate mineral is widely used as a catalyst, buffer, bleaching agent and adsorbent.

The use of bentonite as an adsorbent has an advantage because bentonite has an interlayer structure that can be modified easily, which will improve its absorption properties. This study aimed to analyze the results of the purification process of crude Bali sardinella oil in the fish canning industry using bentonite as an adsorbent at different concentrations.

## 2. Methodology

The research materials consisted of oil of Bali sardinella (*S. lemuru*) from the Canary of Bali; bentonite, $H_2SO_4$ 5N, distilled water, 3 % citric acid, 9.5 % NaOH, 95 % ethanol, KOH 0.1N, indicator of phenolphthalein, acetic acid, chloroform, potassium iodide, sodium thiosulfate ($Na_2S_2O_3$) 0.1N, 1 % starch solution, isooctane (2,2,4-trimethylpentanes), and p-anisidine.

IOP Conf. Series: Earth and Environmental Science **137** (2018) 012012 doi:10.1088/1755-1315/137/1/012012

This research used experimental method with completely randomized design (CRD), and the data were analyzed descriptively. The data obtained included free fatty acid value, peroxide value, para-anisidine value, oxidation number and clarity, and the results were compared with the fish oil standard Codex Alimentarius Commission (CAC) [6] or International Fish Oil Standard (IFOS).

The study began with the activation of bentonite chemically by mixing bentonite and $H_2SO_4$ 5N (1 gram of bentonite:10 mL of acid) in a Beaker glass. The activation was done in water bath for two hours at 70°C. The following step was filtering and washing with water at a temperature of 75 °C until the pH of the wash water became neutral. The bentonite was subsequently dried using an oven at 105°C until it reached a constant weight [7].

Before the purification was done, initial characterization of Bali sardinella oil was carried out, and the characteristics consisted of free fatty acid value, peroxide value, para-anisidine value, oxidation number and clarity. The procedure of crude fish oil purification was based on the study of Sari *et al.* [2] that was modified.

In the degumming stage, 100 grams of fish oil was heated at 70°C for one minute and then 3 mL of 3 % citric acid solution was added. This was reheated to a temperature of 70°C for one minute while being stirred. Next, the oil was kept at a room temperature and centrifuged at a speed of 2.600 rpm for 10 minutes.

In the neutralization stage, the degummed oil was added with 9.5 % NaOH solution as much as 50.3 % (w/w) and heated at 65°C for 20 minutes while being stirred. The next step was keeping it at a room temperature and centrigufing it at a speed of 2.600 rpm for 10 minutes. The result of centrifugation of fish oil was rinsed with distilled water three times to obtain oil and soap.

In the bleaching stage, some of the result of oil neutralization was added with bentonite at concentrations of 2 %; 4 %; 6 %; 8 % of the oil weight in different treatments, and some other was not added with bentonite (control). After the addition of bentonite, the fish oil was heated at 80°C for 20 minutes while being stirred. The fish oil that was heated and centrifuged at a speed of 6.500 rpm for 10 minutes became pure fish oil. The result of the purification process of Bali sardinella oil was analyzed in terms of: free fatty acid value, peroxide value, para-anisidine value, oxidation number, clarity, yield [8, 9, 10].

## 3. Results and Discussion

The Bali sardinella oil used was the rough, dark brown by-product of the canning process in pre-cooking stage. The results of the baseline characteristics of crude Bali sardinella oil in the form of free fatty acid value, peroxide value, para-anisidine (p-anisidine) value, and oxidation number did not meet the IFOS or CAC standard, where the free fatty acid value was < 2 %, the peroxide value was ≤ 5 mEq/kg, the para-anisidine (p-anisidine) value was ≤ 20 mEq/kg and the oxidation number was ≤ 26 mEq/kg (table 1). The results of the baseline characteristic of clarity of crude Bali sardinella oil did not meet the commercial fish oil standard due to low percent trasmittance (%T) (table 2). The results of the final characteristics of Bali sardinella oil (*S. lemuru*) after purification are listed in table 5.3 and table 5.4. The results of the final characteristics of Bali sardinella oil (*S. lemuru*) after purification are listed in table 3 and table 4.

**Table 1.** The results of the baseline characteristics of lemuru fish oil.

| Characteristic | Value |
|---|---|
| Free fatty acid content (%) | 32.25 |
| Peroxide value (meq/kg) | 10.95 |
| Paranisidine (p-anisidin) value (meq/kg) | 41.83 |
| Value of oxidation total (meq/kg) | 63.72 |

**Table 2.** The result of the baseline characteristics of the clarity of lemuru fish oil.

| Fish Oil | λ450 nm | λ550 nm | λ620 nm | λ665 nm | λ700 nm |
|---|---|---|---|---|---|
| Commercial fish oil (standard) | 89.5%T ± 0.71 | 96.05%T ± 0.91 | 97.1%T ± 0.99 | 97.45%T ± 0.92 | 97.95 %T ±1.34 |
| Baseline of fish oil | 12.5 %T ±0.90 | 57.2 %T ± 9.76 | 71.05%T ± 4.03 | 70.05%T ± 6.15 | 78.5 %T ± 1.70 |

**Table 3.** The results of the final characteristics of lemuru fish oil.

| Characterisics | Treatments | | | | | |
|---|---|---|---|---|---|---|
| | $P_0$ | $P_2$ | $P_4$ | $P_6$ | $P_8$ | IFOS/CAC |
| Free fatty acid content(%) | 0.38 ± 0.13 | 0.33 ± 0.06 | 0.365 ±0.16 | 0.265 ± 0.10 | 0.245 ±0.10 | < 2* |
| Peroxide value(meq/kg) | 4.213 ± 2.24 | 3.815 ±1.60 | 8.553 ±4.40 | 6.343 ± 3.00 | 10.16 ±5.54 | < 5** |
| Paranisidine (p-anisidine) value(meq/kg) | 3.545 ±2.64 | 5.32 ± 6.47 | 7.29 ± 2.32 | 3.725 ± 3.72 | 7.84 ± 4.27 | < 20** |
| Value of oxidation total (meq/kg) | 11.97 ±6.43 | 12.95 ± 5.81 | 24.395 ±7.97 | 16.41 ± 7.07 | 28.17±11.79 | < 26** |
| Rendement (%) | 28.763±6.00 | 29.228 ±7.78 | 27.87 ±7.46 | 33.418 ± 1.22 | 27.305±5.50 | |

Description: P0 = 0% bentonite, bentonite P2 = 2%, 4% bentonite = P4, P6 = 6% bentonite, P8 = 8% bentonite. * IFOS; ** CAC standard.

**Table 4.** The results of test of clarity of lemuru fish oil after purified.

| Fish Oil | λ 450 nm | λ 550 nm | λ 620 nm | λ 665 nm | λ 700 nm |
|---|---|---|---|---|---|
| Commercial fish oil (standard) | 89.5 %T ± 0.70 | 96.0%T ± 0.92 | 97.1%T ± 0.99 | 97.45 %T ± 0.92 | 97.95 %T ± 1.34 |
| $P_0$ | 49.4%T± 2.13 | 87.2%T± 4.76 | 89.2%T ± 4.51 | 82.5%T ± 5.09 | 90.12%T ± 6.00 |
| $P_2$ | 40.7%T± 24.22 | 89.8%T± 8.12 | 89.7%T±6.27 | 86.9%T ± 5.43 | 90.35%T ± 6.74 |
| $P_4$ | 58.9%T±2.12 | 88.0%T ± 4.00 | 88.4%T±3.74 | 85.6%T ± 3.15 | 90.28%T ± 3.28 |
| $P_6$ | 60.3%T± 3.77 | 88.1%T± 2.45 | 87.5%T ± 2.43 | 87.4%T ± 5.07 | 87.98%T ± 2.69 |
| $P_8$ | 64.6%T± 5.27 | 90.7%T ± 0.93 | 89.5%T± 4.50 | 89.1%T ± 3.49 | 90.25%T ± 2.98 |

Description: P0 = 0% bentonite, bentonite P2 = 2%, 4% bentonite = P4, P6 = 6% bentonite, P8 = bentonite 8%

After the purification of Bali sardinella oil with bentonite at concentrations corresponding to different treatments, the results showed that the free fatty acid values in all treatments met the International Fish Oil Standards (IFOS), that were <2 %. The lowest free fatty acid value (0.245%) was obtained at a bentonite concentration of 8 %. The free fatty acid value at maximal decreased from the baseline by 32.245 %. According to Bahri [11], the higher the mass of bentonite in the bleaching stage, the lower the free fatty acid value. It was also stated by Okolo and Adejumo [12] that the free fatty acid value, iodine value, peroxide value, refractive index and viscosity decrease during bleaching. According to Arita *et al.* [13], Additional NaOH can attach free fatty acids, which leads to decreased levels of free fatty acids (FFA). Estiasih [1] also argued that free fatty acids will react to form colored compounds that affect the color of the fish oil. According to Aditia *et al.* [14], the carbon chain with a double bond in unsaturated fatty acids will react to heat and form free fatty acids which can affect the quality of fish oil.

The lowest value of peroxide of Bali sardinella oil after being purified with bentonite at a concentration of 2 % is 3.815 mEq/kg. This value has yet to meet the CAC standard, i.e. <5 mEq/kg, but this value has the greatest decrease from the baseline characteristics of peroxide of Bali sardinella oil, amounting 10.945 mEq/kg. According to Dewi and Hidayati [15], bentonite contains alumina and silicate that are effective in attracting adsorbates. Palanisamy *et al.* [16] also stated that the effectivity of an adsorbent to reduce peroxide depends on the type of fish oil being purified, time, temperature and the concentration of adsorbent being used. According to Estiasih [1], purification using alkali can also reduce fat oxidation products such as peroxides. According to Andarwulan *et al.* [9], peroxide is a product that is formed at the beginning of fat oxidation process. The decomposition of peroxide occurs in several steps. The first step is the dissolution of the oxygen bond in peroxide group, which will produce alkoxyl and hydroxyl radicals, while the second step is the breaking of carbon bond.

The p-anisidine values in all treatments are below the CAC standard, i.e. <20 mEq/kg. The p-anisidine values of Bali sardinella oil after purification has decreased as the activated bentonite absorbed the oxidation products. The lowest p-anisidine value of Bali Sardinella oil (3.545 mEq/kg) was obtained from the treatment without bentonite. According to Guillen and Cabo [17], the para-anisidine (p-anisidine) value is not always in line with high peroxide value.

Total oxidation is the relationship between peroxide value and anisidine value that indicates the level of oxidation of fats/oils [1]. The lowest total oxidation value was obtained with bentonite concentration of 0 %, whereas the bentonine concentration of 8 % yielded the highest total oxidation value (28.17 mEq/kg). This value did not meet the CAC standard i.e. <26 mEq/kg. This is due to the fact that the activated bentonite adsorbtion was not perfect in the bleaching stage. According to O'Brien [18], the factors that affect the bleaching process are the procedure, the dosage of adsorbent, temperature, time, humidity and filtration. Similarly, the pore size plays an important role in the process of adsorption [19].

As a result, the quality and quantity of the result of fish oil purification is directly influenced by the condition of the purification process. The resulted Bali sardinella oil was the highest in the treatment with a bentonite concentration of 6 % (33.418 %). This value indicates that the higher the yield obtained, the higher the economic value of the product. The research of Kurniati and Susanto [20] that used 9.5 % NaOH in the neutralization process could produce a high yield. The research of Suseno *et al.* [21] showed that the centrifugation at a speed of 6,500 rpm produced the lowest yield. In this study, the centrifugation speed of 6.500 rpm for 10 minutes became a controlled variable, thus it did not affect the yields of all treatments.

According to Hastarini *et al.* [22], the decrease in the oil weight from the baseline weight is the result of the purification process that can eliminate the impurities found in crude Bali sardinella oil before the purification process. Estiasih [1] also stated that the decrease of yield of oil purification occurs because the amount of alkali given is excessive, which causes hydrolysis reaction of triglycerides and forms soap in an excessive amount. The type of alkali, concentration, the amount of dye, and the technique used (e.g. agitation and separation between saponified and unsaponified fractions) are the important factors that must be considered as they affect the purification process.

The highest clarity of Bali Sardinella oil from the purification process was obtained from the treatment using a concentration of of 8 %, with the values at wavelengths ($\lambda$) of 450 nm, 550 nm and 665 nm being 64.625 %T, 90.7 %T and 89.1 %T, respectively (table 4). According to Hasibuan *et al.* [23], the percent transmittance of oil is used as an indicator of change in the levels of dye contained in the oil. The correlation between the percent transmittance and oil clarity level is that the higher the percent transmittance, the lower the level of dye contained in the oil.

The results showed that the clarity of the Bali sardinella oil after the purification was higher than that before purification because the activated bentonite was able to improve the color of the oil. According to Dewi and Hidajati [15], the activation of bentonite using mineral acids ($H_2SO_4$/HCl) produces clay with greater active site because the mineral acids dissolve and react with components in

IOP Conf. Series: Earth and Environmental Science **137** (2018) 012012 doi:10.1088/1755-1315/137/1/012012

the forms of tar, Ca salt and Mg covering the pores of the adsorbent, thus bentonite has a surface with high acidity that allows for higher adsorption capacity than its adsorption capacity prior to activation.

## 4. Conclusion

The purification of Bali sardinella oil with bentonite adsorbent at a concentration of 6 % produced the highest yield (33.42 %) with quality characteristics like free fatty acid value of 0.27 %, peroxide value of 6.34 mEq/kg, para-anisidine value of 3.725 mEq/kg, oxidation number of 16.41 mEq/kg, and clarity of 60.28 %T, 88.08 %T, 87.50 %T, 87.43 %T, 87.98 %T at wavelengths (λ) of 450 nm, 550 nm, 620 nm, 665 nm, 700 nm, respectively. This research revealed that the addition of activated bentonite can increase the quality of fish oil by-product from the Bali sardinella canning industry.

## 5. References


[1] Estiasih T 2009 Fish oil technology and its application for food and health (Yogyakarta: Graha Ilmu) pp: 1-100. (In Indonesia)

[2] Sari R N, Utomo B S B, Basmal J, Kusumawati R 2015 JPHPI **18** 276-86. (In Indonesia)

[3] Indonesian Ministry of Marine Affairs and Fisheries 2012 Export statistics of fishery products by commodity, province, and port of origin export Data Center, statistics and information secretary general, ministry of marine affairs and fisheries Jakarta pp 35-41. (In Indonesia)

[4] Rubio-Rodriguez N, Beltran S, Jaime I, de Diego S M, Sanz M T, Carballido J R 2010 *Innov. Food Sci. Emerg. Technol.* **11** 1-12.

[5] Faisal M 2015 *J. Tek. Kim* **4** 20-4. (In Indonesia)

[6] [CAC] Codex Alimentarius Commission 2017 Standard for fish Oils CODEX STAN 329-2017. Rome, Italy.

[7] Handayani K dan Yusnimar 2013 *J.Teknobiologi* **4** 117-21. (In Indonesia)

[8] Association of Official Analytical Chemists (AOCC) 1995 Official Methods of Analysis (USA: Marylan) pp 2-53.

[9] Andarwulan N, Feri K, Dian K 2011 *Food analysis* (Jakarta : Dian Rakyat) pp 198- 219. (In Indonesia)

[10] Suseno S H, Tajul A Y, Wan Nadiah W A 2011 *Int. Res. Biochem. Bioinform.* **1** 103-13.

[11] Bahri S 2014 *JDPI* **25** 63-9. (In Indonesia)

[12] Okolo J C and Adejumo B A 2014 *J. Eng.* **4** 25-8.

[13] Arita S, Anindya S A, Wildayani H 2009 *J. Chem. Tech.* **16** 58-65. (In Indonesia)

[14] Aditia R P, Darmanto Y S, Romadhon 2014 *J. Fish. Biotech.* **3** 55-60. (In Indonesia)

[15] Dewi M T, I Hidajati N 2012 *J. Chem.* **1** 47-53.

[16] Palanisamy U D, Sivanathan M, Radhakrishnan A K, Haleagrahara N, Subramaniam T, and Chiew G S 2011 *Molecules* **16** 5709-19.

[17] Guillen M D and Cabo N 2002 *Food Chem.* **77**: 503-10.

[18] O'Brien R D 2004 *Fats and oils formulating and processing for applications* Second Edition (Washington DC: CRC Press) pp: 79-107.

[19] Batafor Y M J 2014 Improved quality of sardinian fish oil (Sardinella sp.) by centrifugation and adsorbent Thesis (Bogor: Institut Pertanian Bogor) pp 1-75. (In Indonesia)

[20] Kurniati Y and Susanto W H 2015 *J. Food Agro Industry* **3** 193-202. (In Indonesia)

[21] Suseno S H, Nurjanah, Jacoeb A M, and Saraswati 2014 *J. Food Sci. Technol.* **6** 60-7

[22] Hastarini E, Fardiaz D, Irianto H E, and Budijanto S 2012 *Agritech.* **32** 403-10.

[23] Hasibuan S, Sahirman and Ni Made A Y 2013 *Agritech.* **33** 311-19.


## Acknowledgments

This research is supported by Patmawati, Lecturer of the Faculty of Fisheries and Marine Universitas Airlangga who received The Research Grant from the Faculty of 2016 based on the Decree of Dean of Faculty of Fisheries and Marine Airlangga University No. 01/UN3.1.12/KP/2016

# Effect of feeding silkworm on growth performance and feed efficiency of snakehead (*Channa striata*)

**U Firmani and Lono**

Department of Aquaculture, Aquaculture Laboratory, Faculty of Agriculture
Muhammadiyah University

E-mail: ummul.buper@yahoo.com

**Abstract.** The snakehead, *Chana striata* is a carnivorous freshwater fish and widely distributed in Asia. High demand of this fish has been triggering many aquaculturist to culture *C. stiata*. Feed was the important factor for fish growth. Silkworm has high protein content, low fat and can be used as natural feed for finfish. This study investigate the silkworm feed in *C. striata*. The treatment of this research were A (100 % pellet); B (100 % silkworm); C (combination of 75 % pellet and 25 % silkworm); D (combination of 50 % pellet and 50 % silkworm); and E (combination of 25 % pellet and 75 % silkworm). The variables measured in this study were relatif growth, specific growth rate, feed efficiency, feed conversion ratio, and survival rate. The result show that silkworm gave the high growth performance, feed efficiency and survival rate of the snakehead (*Channa striata)* compared with the control.

## 1. Introduction

The snakehead, *Channa striata* is a carnivorous freshwater fish, which distributed in Asia. It has high protein content, rich of albumin, and low lipid content [1]. For decades, the minimum nutrient requirement and lower cost of feed for normal fish growth is a focus of fish nutrition of aquaculture [2]. Feed for carnivorous fish is generally contain high protein. Protein from the feed is required for the fish to utilize amino acids either to build new proteins during growth and reproduction or to replace existing ones during maintenance [3].

Silkworm has nutritional content of crude protein 64.47 %, crude fat 17.63 %, crude fiber 0 %, ash 7.8 % and BETN 0.06 % [3]. Result of research by Kusumaningrum [4] showed that feed with protein content 32 % affect the highest growth rate of Snakehead about 0.269 g/day with 1.73 % albumin level. Furthermore, Amanta [5] study on the effect of feeding combination pellet and silk worms are 75 % and 25% improve the length and weight of catfish that 3.32 to 3.40 cm and 0.37 to 0.38 g for 40 days maintaining resulted in a final length growth of 3.51 cm and an absolute weight of 1.66 g. Considering the amount of benefits of albumin, especially for health and industry and the high nutritional value of snakehead. Formulated diets for the snakehead is not already axist especially in Indonesia.

Fish meal has been used in aquaculture and industrial as fish feed because of its high protein content. Due to the high price of the fish meal, its need to find other source as fish feed that has high nutritional content, lower of cost and easy to be produced. Silkworm has high nutritional content especially protein and can be cultured easily. Cost production of Silkworm is cheap in Indonesia because the needed of its food can be from. This study wanted to evaluate the effect of feeding combination pellet and silkworm on the growth performance and feed efficiency of snakehead.

## 2. Experimental

### *2.1. Time, location and fish husbandry*

This research was conducted from May to July 2017 at the Laboratory of Aquaculture, Departement of Aquaculture, Faculty of Agriculture, Muhammadiyah University of Gresik. A total of 450 snakehead in length about 5-7 cm were cultured in 30 L concrete tank. The fish density was 1 fish/L. Fish was feed following the treatments in three times a day at 08.00 a.m, 12.00 a.m and 05.00 p.m.

### *2.2. Experimental design*

Experimental design was Completely Randomized Design (CRD). The treatments were A (100 % Pellet); B (100 % Silkworm); C (combination of 75 % Pellet and 25 % Silkworm); D (combination of 50 % Pellet and 50 % Silkworm); E (combination of 25 % Pellet and 75 % Silkworm); which was triple replicated.

### *2.3. Experimental variable*

The snakehead was cultured for 28 days. Every 7 days, fish was wighed and analyzed. Water temperature and pH were recorded twice daily in the morning and evening. The variables measured in this research were relative growth (RG), specific growth rate (SGR), feed efficiency (FE), feed Convertion Ratio (FCR), and survival rate (SR).

### *2.4. Data analyzed*

Data were analyzed using one way analysis of variance (ANOVA). The difference between treatmens means to be significant at 95 % were analyzed with Duncan's Duncan's Multiple Range Test (DMRT). SPSS software were applied for the data analysis.

## 3. Result and Discussion

### *3.1. Absolute growth rate*

Absolute growth rate of the snakehead in 28 days culture is shown in figure 1.

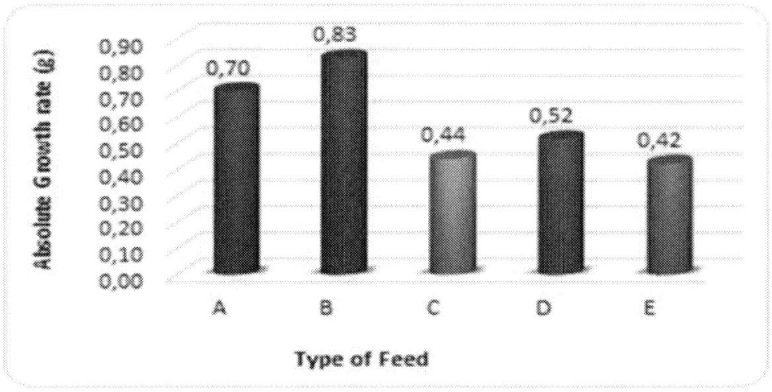


**Figure 1.** Absolute growth rate of the snakehead in 28 days culture.

The result shows that silkworm has higher of absolute growth rate 0.83 g ($P<0,05$). Its follows by treatment of A 0.70 g, D 0.52 g, C 0.44 g and E 0.42 g, respectively.

*3.2. Specific growth rate*
The specific growth rate of the snakehead in 28 days culture is shown in figure 2.

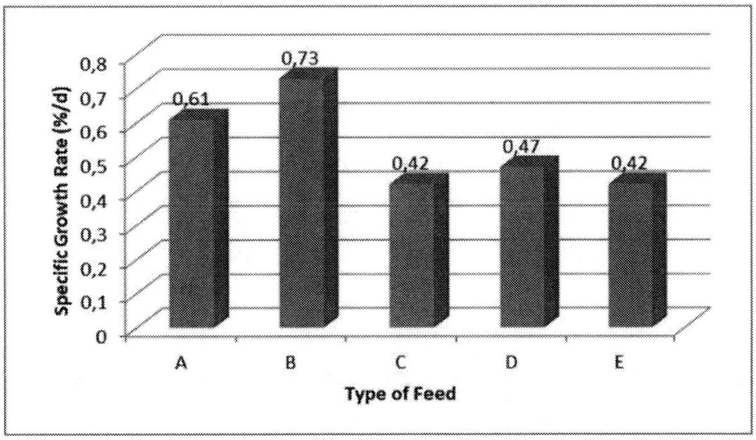


**Figure 2.** The specific growth rate of snakehead in 28 days culture.

The highest spesific growth rate is snakehead that given feed 100% silkworm, and follows by another treatment same as the absolute growth rate.

*3.3. Absolute length rate*
The highest absolute lenght rate shown in the treatment C, A, D and E that significantly different with the treatment B 100 % silkworm. The absolute lenght rate of snakehead feed by 100 % silkworm has the lowest lengh rate.

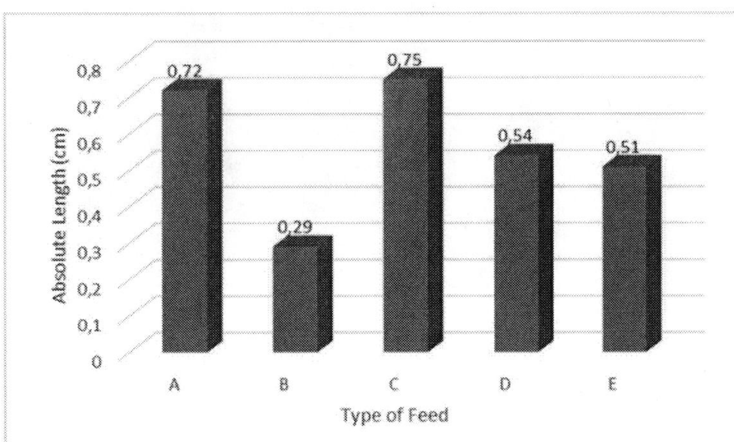


**Figure 3.** Absolute length rate of the snakehead in 28 days culture.

*3.4. Feed efficiency*

Figure 4 shows that the best feed efficiency of the treatment is the snakehead feeding by 100 % silkworm.

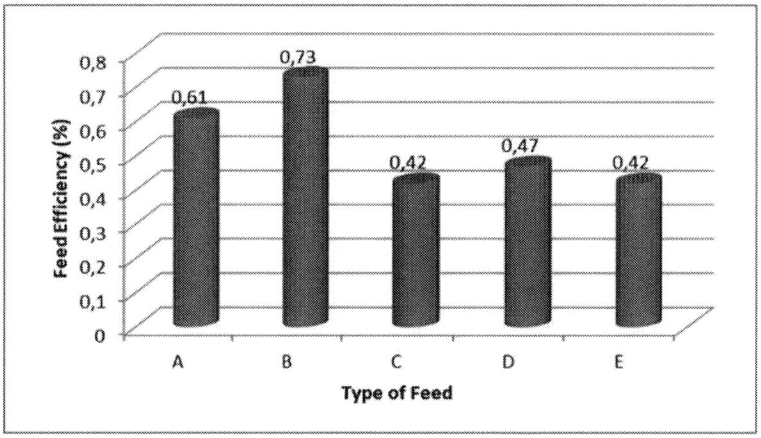


**Figure 4.** Feed efficiency of the snakehead in 28 days culture.

*3.5. Feed convertion ratio*
Feed convertion ratio was shown in figure 5. Feed convertion ratio of the snakehead feeding by 100 % silkworm is the best result (0.98), followed by other treatment that significantly different (P>0.05), there were A (1.15), D (1.41), C (1.60) and E (1.78).

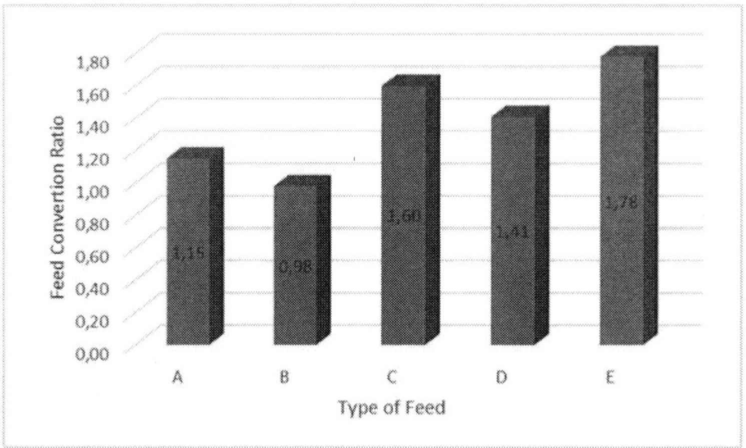


**Figure 5.** Feed convertion ratio (FCR) of the snakehead in 28 days culture.

*3.6. Survival rate*
The result in figure 6 showed that survival rate for all the treatment is 100 %. All fish from all treatments were survive until the end of treatment.

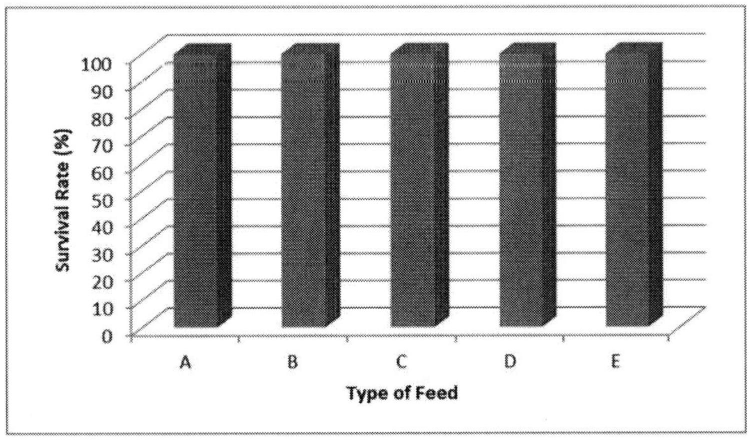


**Figure 6.** Survival rate of the snakehead in 28 days culture.

### *3.7. Water quality*

Water temperature and pH of the snakehead culture in 28 days is shown in table 1. Water temperature of all treatments is in the normal range.

**Table 1.** The average of water temperature and pH of the snakehead culture.

| Treatments | Time a day | | | |
|---|---|---|---|---|
| | Morning | | Evening | |
| | Temp (°C) | pH | Temp (°C) | pH |
| A | 28.9 | 9.06 | 26.7 | 8.17 |
| B | 29.1 | 8.81 | 26.7 | 8.53 |
| C | 27.7 | 8.82 | 26.1 | 8.28 |
| D | 27.1 | 8.86 | 26.7 | 8.48 |
| E | 27.9 | 8.82 | 26.8 | 8.31 |

## 4. Discussion

Growth involves several processes. It is highly influenced by the digestive rate and absorptive physiology of an organism, which affect the rudimentary dynamic of utilization of ingested nutrient and control the extent of a fish's response. The activity of digestive enzyme is influence by the composition of diet and ability of fish to digest and absorb diet effectively [6].

In this research, 41% dietary protein level in pellet produced low absolute wight, specific growth rate, and feed efficiency. Protein level of silkworm is about 60 %. Fish fed diet pellet or all the combination pellet and silkworm had the lower growth performance. It suggested that the combination failed to meet nutritional requirements of *C. striata* under the conditions of this research.

Juvenile snakehead in this research need higher protein content in the diet more than 41 %. However, this field of research seems largely unexplored the albumin level, digestive enzyme and the relationship between digestive enzyme activity and growth performance.

Silkworm (*Tubifex* sp.) is good for the growth of juvenile of snakehead [7]. Then, the silk worm emits a distinctive aroma that might attract the attention of cork fish to eat it. Silk worm contains vitamin B12, calcium, pantothenic, nicotinic acid and vitamin B2. In addition, silk worms are easily digested by carnivorous fish because they do not contain crude fiber.

The growth rate is closely related to the composition of the material and nutrient content of the feed. Based on the results of statistical analysis, indicated that the absolute growth and the highest specific growth rate was found in cork fish fed 100 % silk worm (treatment B) which was significantly different from other treatments.

Snakehead fish in the B treatment is able to digest the feed better than other treatments. B well-digested feed will produce an energy supply. Energy derived from this feed is used to improve body structures. Excess energy is used for growth. Growth occurs when there is excess energy after energy used for body maintenance, basal metabolism, and activity [8,9].

## 5. Conclusion

Silkworm can be used as an alternatif source of fish fed especially for juvenile snakehead because of high protein content. Snakehead which is feed by 100 % silkworm give the best growth performance, feed efficiency and survival rate significantly different from the other treatment.

## 6. References

[1] Amalia R, Subandiyono and Arini E 2013 *J. Aqua. Man.&Tech.* **2** 136 – 143
[2] Amanta R, Usman S and Lubis M R K 2014 The effect of natural food combination with artificial feed on growth of fry catfish (*Clarias gariepinus*) (Medan : University of North Sumatra)
[3] Annasari M, Aris M and Yohanes W 2002 *IJSTE.* **2** 1-8
[4] Hariati E 2010 The potential of silk worm flour (Tubifex sp) and tapioca starch for substitution of commercial feed Pangasius (*Pangasius hypophtalmus*) (Yogyakarta : Atmajaya University)
[5] Kusumaningrum G A, Alamsjah M A and Masithah E D 2014 *J Fish Sci&Mar* **6** 25-29
[6] Wijayanti K 2010 The influence of different natural feeding against the validity and growth of palmas fish seeds (*Polypterus senegalus*) (Depok : Faculty of Math and Science University of Indonesia)
[7] National Research Council (NRC) 2011 Nutrient requirements of fish and shrimp (USA : National Academy Press)
[8] Perez-Jimenez A, Cardenete G, Morales A E, Garcia-Alcazar A, Abellan E and Hidalgo M C 2009 *Comp Biochem Phys.* **154A** 157-164
[9] Subandiyah S, Satyani D, and Aliyah 2003 *J Ichtyol. Ind.* **3** 67–72 (In Indonesia)

**Acknowledgement**

This reseacrh was supported by Muhammadiyah University of Gresik, East Java Indonesia. The authors thanks to Andi Rahmad Rahim for supported data analyzed.

# The identification of plankton, water quality, blood cell, and histology in culture pond of tilapia *Oreochromis niloticus* which infected by viral nervous necrosis (VNN)

**U Yanuhar, D T Rahayu, M Musa and D Arfiati**

Laboratory of Environment and Aquatic Biotechnology, Management of Waters Resources, Faculty of Fisheries and Marine Sciences
Jl. Veteran, 65144. Malang, Jawa Timur , Indonesia

E-mail: uunyanuhar@yahoo.com; doktoruun@ub.ac.id

**Abstract.** Currently, Viral Nervous Necrotic (VNN) is not only attacking the marine fish but also the freshwater fish like tilapia (*Oreochromis niloticus*). The aims of study to identify the type of plankton, water quality status, blood cell status, also histology of VNN infected tilapia obtained in culture ponds. The methods included plankton identification and water quality analysis from the infected fish pond in the Krakal, Blitar. The quality of blood cells and the histology of tilapia infected by VNN observed using a microscope with Hematoxylin-Eosin staining. The result show plankton in a fish pond of infected tilapia includes 3 divisions: Chlorophyta, Cyanophyta, and Bacillariophyta and 2 phyla: Arthropoda, and Rotifera. The values of erythrocyte, hematocrit, and hemoglobin were smaller than normal tilapia, however, the leukocyte and macronucleus values of VNN-infected fish were higher than normal fish. The fish histology shows the vacuolation in the brain and eyes tissue. The water quality of the culture pond have the temperature, pH, turbidity, DO, $CO_2$, $NO_3$, $PO_4$, TOM in the range of 30-32°C; 7.0-9.0; 25cm; 6.082–7.44mg/L; 3.98–9.08mg/L; 1.039–1.139 mg/L; 0.051-0.054mg/L; and 11.377-13.905mg/L, respectively. VNN causing high leukocyte and macronuclei and the damaging in brain and eyes tissue in infected tilapia.

## 1. Introduction

Viral Nervous Necrosis (VNN) is an RNA virus that commonly attacks marine fish commodities. However, the virus has been found to attack not only marine fish such as groupers but also tilapia cultivated in many cultivation centers in East Java, Indonesia. The VNN outbreak caused tilapia to die in January-March 2016 and 2017 in the village of Krakal, district of Wlingi, Blitar. Tilapia is one of the main commodities of the fisheries sector in Indonesia because it is easy to breed [1]. Tilapia can be found in various regions in Indonesia such as in Blitar, East Java.

The environment is one of the factors that should be considered in cultivation activities. One of the processes of the aquatic environment management can be seen from the quantity and quality of water [2]. The contaminated environment and poor water quality will trigger the increase of VNN infection [3]. In addition, VNN will infect stressful fish due to high stocking density and high water temperatures in cultivation systems [4].

VNN is a virus that has genetic materials in the form of single-stranded RNA. This virus attacks the grouper fish in the larvae and juvenile stadia and can cause a high mortality rate, up to 100% [5]. One of the symptoms is spinning fish. Some fish are seen sleeping or dying at the bottom. Also, some fish become abnormal [6]. The clinical symptoms of Blitar tilapia are similar to the clinical symptoms of VNN infected snapper. The study aims are to determine the types of plankton, water quality status, blood cell status, and histology of VNN infected tilapia obtained in culture ponds.

## 2. Methodology

Fish samples were taken at the time of the disease outbreak, in January-March 2016-2017 in Krakal, Wlingi district, Blitar. The research methods used were plankton identification using microscopes and quality analysis of water from the infected fish pond. Plankton samples were taken by filtering 25 L water samples using plankton net no. 25. The samples were put into the bottle film and preserved with Lugol's (3 drops). The water quality parameters were observed including turbidity, temperature, pH, DO, $CO_2$, $NO_3$, $PO_4$, and TOM. The observation of water quality was done 3 times and the interval was one week. Water sampling used Wrinkle bottle and a 600ml bottle.

PCR assay with VNN target for DNA length band of 294 base pairs (bp) was performed to confirm the infected fish. The quality of blood cells and the histology of digestive systems infected by VNN were observed using microscopes with Hematoxylin and Eosin (HE) staining. Blood taking was done 3 times, on day-5, day-9, and day-14. The fish were placed with head on the left before the syringe was rinsed with Na-citrate 3.8% as an anticoagulant. Blood samples were taken with a syringe needle injected in the musculus at the midline of the body of the anal fin. Blood samples were taken, inserted into 1.5 ml tubes to observe blood image for the measurement of Erythrocyte, Leucocytes, Hemoglobin, Hematocrit, Macronucleus, and Micronucleus.

## 3. Results and Discussion

### *3.1. Tilapia disease analysis*

The results of morphological observations and the behavior of tilapia indicate the presence of symptoms. They showed that the fish were infected by VNN. With regard to the visible symptoms of fish swimming on the surface, the movement was weak and skewed. The eyes bulged and the skin contained mucus. The body looks pale and some fish were found to have body color darkening.

There was a change in swimming habits of VNN infected fishes such as swirling and whirling or abdominal position above because of swelling of bubbles [7]. Sometimes, VNN infected tilapia showed vertical swimming habits (in the bottom of the pond or spinning swiftly or stomping to the surface of the water). Detection of VNN on tilapia by PCR assay on primer 294 bp revealed that the brain, kidney, liver, intestine, and flesh were positively infected by VNN.

### *3.2. Water Quality Parameters on Tilapia Maintenance Pond*

Table 1 shows that the water quality in the tilapia pond is still good for the growth of aquatic organisms, such as plankton and tilapia. The results of the measurement of the water quality parameters for 3 weeks are still in the range of quality standard for aquaculture. However, the pH and temperature measurements show that the ideal water quality condition for the retention of VNN infection in tilapia is in the pH range of 2-9 and temperature of 30-32$^{\circ}$C.

### *3.3. Plankton Analysis*

The identification of plankton in tilapia ponds from week 1 to 3 obtained phytoplankton and zooplankton. Phytoplankton consists of 3 divisions of Chlorophyta (2 genus), Cyanophyta (1 genus) and Bacillariophyta (7 genus). Zooplankton consists of 3 phylum including Arthropod (2 genus) and Rotifera (1 genus). Cyanophyceae is able to survive in non-light condition and indicates the presence of high organic matter content. This alga is able to take $CO_2$ and phosphorus in low concentration conditions in waters [8]. The abundance of zooplankton has increased and decreased due to the factors, including growth, death, vertical distribution and different migration as well as changes in water quality [9].

The calculation results of phytoplankton and zooplankton diversity index show that the ponds of tilapia belong to the low moderate H' criteria. Based on the calculation of phytoplankton dominance index and zooplankton at week 1 and 2, there is dominant phytoplankton from the division of Cyanophyta (genus Merismopedia), while there is no dominant zooplankton in the waters. Each type of phytoplankton has a different response to the nutrient types dissolved in water bodies. This condition can cause the phytoplankton community in a water body to have structure and dominance of different types with other water bodies [13].

**Table 1.** Results Measurement of Water Quality Parameters on Tilapia Maintenance Pond.

| No | Parameter | Unit | Average | Quality standard |
|---|---|---|---|---|
| 1 | Temperature | °C | 31 | 25 – 32 (*) |
| 2 | Turbidity | cm | 25 | 30 – 40 (*) |
| 3 | pH | - | 7 | 6.5 – 8.5 (*) |
| 4 | DO | mg/L | 6.757 | ≥ 3 (*) |
| 5 | $CO_2$ | mg/L | 6.32 | 1 – 10 (**) |
| 6 | $NO_3$ | mg/L | 1.119 | 10 (***) |
| 7 | $PO_4$ | mg/L | 0.052 | ≤0.2 (***) |
| 8 | TOM | mg/L | 12.640 | ≤50 (***) |

Note:(*) [10]; (**) [11]; (***) [12]

### *3.4. The Presence of Plankton and VNN in Tilapia Maintenance Pond*

The phytoplankton identified in the digestive tract of tilapia are genus Nitzchia, navicula, Achnantes, Tabellaria, Surirella and Diatoma and genus Netrium. The zooplanktons identified in the digestive tract of tilapia are from genus Calanus and Karatella. The plankton contains several amino acids. The genus Nitzhia, Navicula and Achnantes contain amino acids, including L-aspartic acid and l-lysine [14]. Genus Calanus contains amino acids: lysine and arginine [15]. The genus Keratella has amino acid contents of L-aspartic acid, arginine, and L-lysine [16]. Protein is a virus capsid compiler. Betanodavirus utilizes the amino acid cysteine as a capsid; L-aspartic acid plays an important role in the cleavage of proteins in the Betanodavirus capsid. The amino acid arginine and lysine also play a role to bind the RNA genome into the inner layer of the capsid wall [17]. This suggests that amino acids contained in the host body (plankton) have corresponding amino acids the virus needed for reproductions. It can be indicated that genus Nitzshia, Navicula, Achnates, Keratella and Calanus can be vectors against the spread of VNN on tilapia bodies by predation process, which can also be referred to as horizontal transmission.

### *3.5. Observation of Erythrocytes, Leucocytes, Hematocrit, Hemoglobin*

The results of erythrocyte analysis can be seen in figure 1.

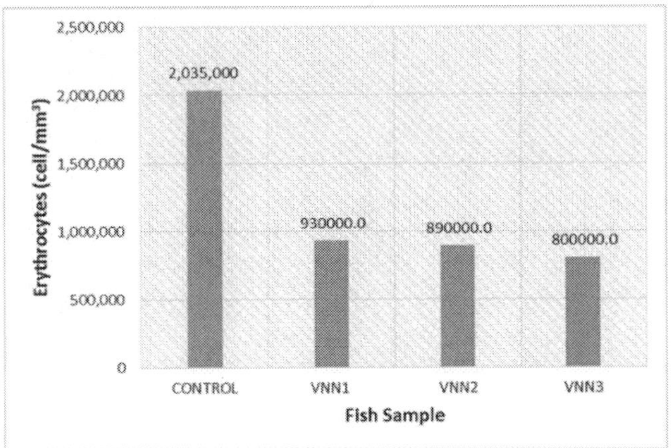


**Figure 1.** Amount of erythrocyte in tilapia.

The observation results on the first, second and third samples show a number of erythrocyte on VNN infected tilapia, which were 930.000 cell/mm$^3$, 890.000 cell/mm$^3$, dan 800.000 cell/mm$^3$. However, the total of erythrocyte on healthy tilapia fish was 2.035.000 cell/mm$^3$. The low number of erythrocyte is due to poor water quality in Krakal that affects the stress of tilapia. Erythrocytes have a nucleus and the number of variations depends on the species, stress condition, and temperature of the environment. The erythrocytes are in the range of 1.05 – 3.0 $10^6$ cell/mm$^3$ [18]. Decreasing erythrocyte is caused by water quality during the maintenance of the pond which is not in accordance with the quality standard. High levels of ammonia can cause fish stress, leading to diseases such as ectoparasite and VNN.

From the observations on first, second and third samples, it was found that the Hb levels on tilapia infected by VNN were 5 gr/100 ml, 4.5 gr/100 ml, dan 4 gr/100 ml. In a healthy fish, the total Hb is 6.7 gr/100ml. The results of hemoglobin analysis (Hb) can be seen in figure 2.

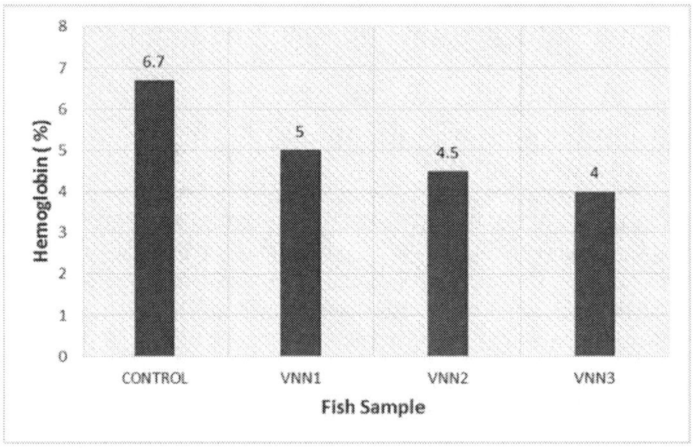


**Figure 2.** Amount of Hemoglobin (Hb) in tilapia.

The hemoglobin of the fish decreased due to the unbalanced ability to bind oxygen although the metabolism to produce energy is inhibited. Decreased hemoglobin is also seen in the decreased amounts of erythrocytes. Red blood cells decline is due to the lack of required oxygen levels. The low Hb causes the metabolism to decrease resulting in low energy production. This makes the fish weak and has no appetite. The fish were seen at the bottom or hanging below the water surface [19]. The results of hematocrit analysis can be seen in figure 3.

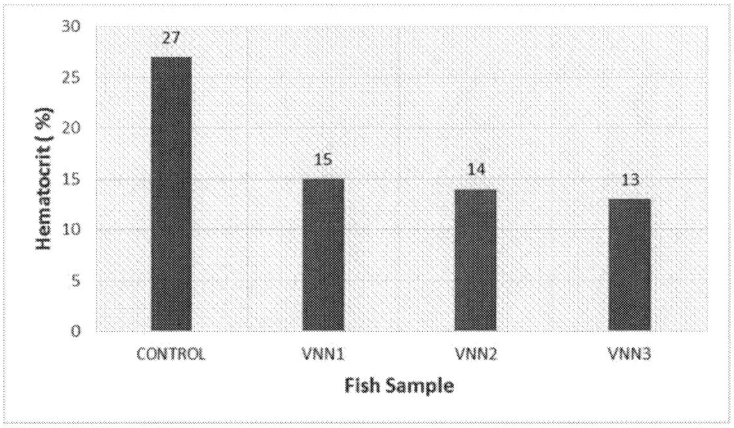


**Figure 3.** Amount of hematocrit in tilapia.

The results of the first, second and third fish samples indicated that the hematocrit in VNN infected tilapia were 15%, 14%, dan 13%. However, the total of hematocrit in healthy fish is 27%.

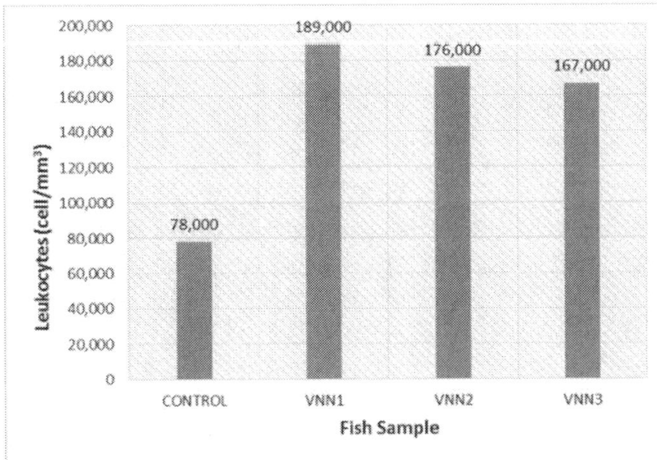


**Figure 4.** Amount of leukocytes in tilapia.

The observation results show that the total hematocrit from the first week to the third week decreased. The low hematocrit is due to the number of red blood cells and low hemoglobin. Hematocrit has a strong correlation with the amount of hemoglobin. The lower red blood cells result in lower hemoglobin and hematocrit. Hematocrit will decrease in cases of anemia. It can be used as guidance for the low levels of protein, vitamin deficiency or infected fish [20].

Observations on leukocytes are shown in figure 4. Total leukocyte in tilapia infected by VNN on first, second and third samples were 189,000 cell/mm$^3$, 176,000 cell/mm$^3$ and 167,000 cell/mm$^3$. The total of leukocytes of healthy fish was around 78,000 cell/mm$^3$. The total of leukocytes was lower than the total red blood cells. The total leukocytes of every mm$^3$ of teleost fish blood cells were around 20,000 cell/mm$^3$ – 150,000 cell/mm$^3$ [21].

### *3.6. Micronucleus (Mn)*

The observations results on micronucleus were shown in figure 5.

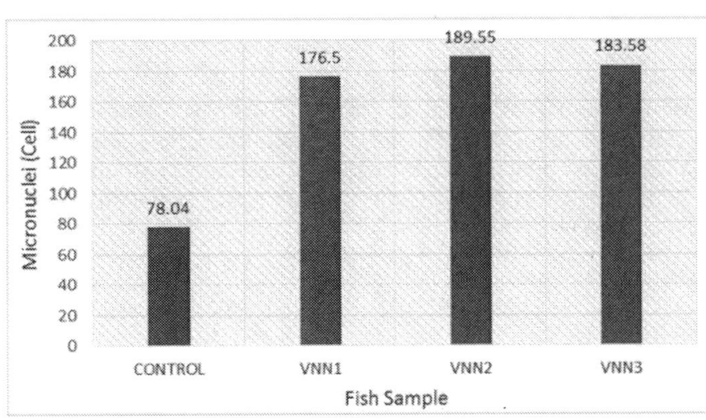


**Figure 5.** Amount of micronucleus in tilapia.

Based on figure 5, it is known that the number of micronucleus in tilapia affected by VNN on first, second and third samples were 176.5 cells, 189.55 cells, dan 183.58 cells. However, the observation results of healthy tilapia fish show that the total average of micronucleus was 78.04 cells.

The increase of micronucleus is due to erythrocytes of chromosomal damage and disruption in DNA duplication. Micronucleus observations were performed to determine the chromosomal damage seen from erythrocyte and to detect genotoxicity in the waters (figure 6).

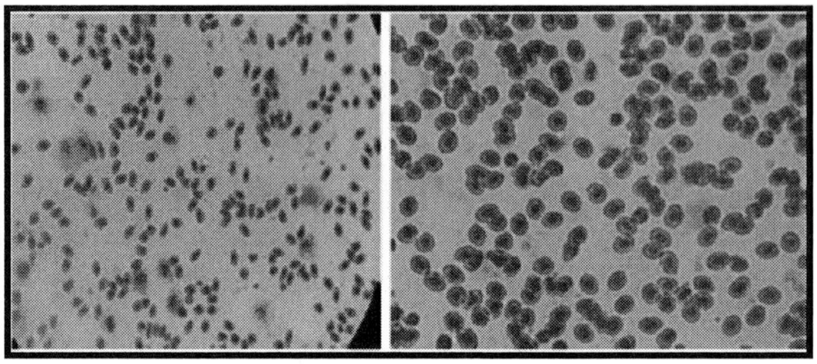

**Figure 6.** Observation results of Micronucleus on Interphase, Prophase, Metaphase, Anaphase, Telophase, and MN.

Micronucleus were conducted from acentric fragments that fail to join the cells during the process of cell divisions. They can also be formed from the chromosome left behind or not carried away in the process of mitosis or occurs due to complex chromosome configuration at the time of the anaphase process. Micronucleus criteria include a diameter of less than one-fifth of the nucleus (10μm) located in the cytoplasm and out of the nucleus, no contact with a nucleus [22].

*3.7. Histopathological Analysis of Brains Tilapia Infected by VNN*

VNN infects the tilapia brain organs, directly attacking the fish receptors because VNN is a virus that does not have envelopes. Then, VNN is spread to the brain tissue through the blood circulation. The VNN is also stable in extreme environments, and the difference is that if the environment is appropriate, then this VNN will be endemic. VNN attacks the brain through blood circulation [23]. VNN is directly attached to the host's receptor, then introducing genetic material into the host cell or intracellular infection by leaving the protein coat outside the cell [24]. In the histopathological observation of infected tilapia, VNN used pieces of brain tissue of tilapia that have been prepared and colored by using HE staining. They were then observed using a microscope as shown in figure 7.

Figure 7(A) shows the healthy tilapia brain tissue, the structure of brain tissue that is still good, regular and does not show any damage. The existence of healthy fish here was used as a comparison in the observation of normal brain tissue and that which has been damaged. The histopathology of healthy fish brains indicates a normal state of the histological structure without any indication of damages and defects. In the molecular layer, the granular layer and pyramid layer appear normal and there is no sign of necrosis and no apparent hemorrhage occurs [25]. For the first observation, the tilapia showed disruption to He, Va and Ko (figure 7B). The second observation showed the disruption of He, Va and Ne (figure 7C). The third observation showed disruption to He, Ne and Va (figure 7D). He of VNN infected tilapia brain tissue was observed in all observations (figure 7B, 7C, and 7D) which occurred in neuronal cells/nerve. Ko was found in the first observation (figure 7B) occurring on brain nerve cells of tilapia.

IOP Conf. Series: Earth and Environmental Science **137** (2018) 012014 doi:10.1088/1755-1315/137/1/012014

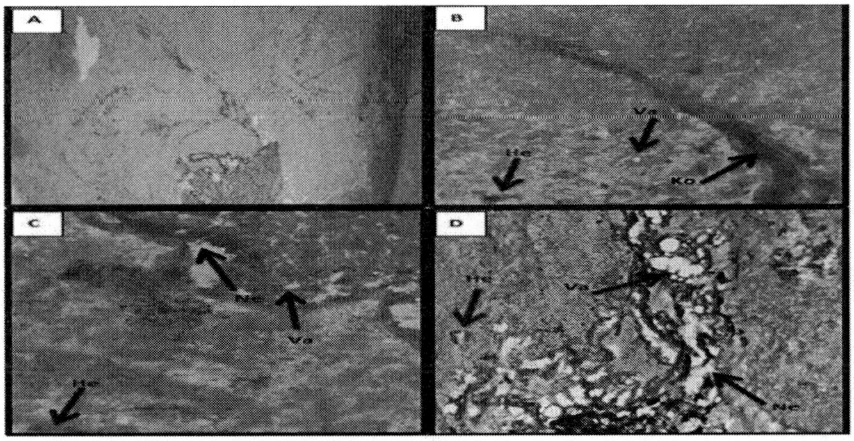


**Figure 7.** Histopathology of Tilapia Brain Tissue. (A) Healthy brain tissue. (B) The first observation shows the Kongesti(Ko), Vacuolization (Va) and Hemorrhage (He). (C) The second observation shows the damage of Va, He and Necrosis(Ne). (D) The third observation shows He., Ne., and Va.

Congestion can be seen microscopically in the erythrocytes collected in blood vessels. The fish brains exhibited neurological signs, capillary congestion, and venoms clearly visible in the cerebellar valvula and medulla oblongata, which are associated with dissociation of fiber edema in nerves [26]. Damage to Ne was found in the second and third observations (figure 7C and 7D)). Cells experienced necrosis that caused the cells to be destroyed so that only empty space was left in the brain tissue or vacuolization that occurred due to cell damage. This is due to VNN infection through the bloodstream and into the brain, causing damage to the tissues. Damage to vacuolization (Va) was found in the 1st observation (figure 7B), the second observation (figure7C), and the third observation (figure 7D) characterized by the appearance of a nucleus that appeared enlarged and bubbly and not eosinophils (redness) as a result of changes in fluid balance in the cell due to increased amount of fluid. Vacuolization occurred due to cell damage, then remained empty in the brain tissue.

## 4. Conclusion

The identified plankton consist of Chlorophyta (2 genus), Cyanophyta (1 genus) and Bacillariophyta or Chrysophyta (7 genus) and Arthropoda (2 genus) and Rotifera (1 genus) phyla. Parameters of measured water quality support the planktons growth and aquaculture activities as well as the presence of VNN in tilapia ponds. The hematology and micronucleus status of VNN-induced tilapia show that the deterioration of the condition became worse, which could cause fish death.

## 5. References

[1] Setijaningsih L 2011 Plankton Diversity On The Best Tilapia Fish Cultivation (Oreochromis niloticus) with Addition of C-Organic *Proc. of the Aquaculture Techn. Innov. Forum.* 1019-1030. (in Indonesian).

[2] Ciptanto S 2010 *Top 10 Fresh water fish* (Jakarta : Lily Publisher) (in Indonesia).

[3] Novriadi R, Sri A, and Tanjung D O N 2015 *J. Omni-Akuatika* **14** 54-62. (In Indonesia).

[4] Putri R R, Uun Y and Asus M S 2013 *MSPi Student J* **1**(1) :1-10. (in Indonesian).

[5] Yuwanita R, Yanuhar U and Hardoko 2013 *Adv. Environ. Biol.* **7** 1074–1081.

[6] Yanuhar U, Gusman E and Arfiati D 2012*Adv. Environ. Biol.* **6** 388–396.

[7] OIE 2013Viral *Encephalopathy and retinopathy. in: manual of diagnostic tests for aquatic animals* 4th Edition (Paris : Office International des Epizooties)

[8] Erdina D, Yuliati and Efawani 2013 The type and abundance of phytoplankton in the village pond sialang buah, teluk mengkudu sub-regency, serdang bedagai regency, sumatera utara province (Pekanbaru : FPIK Riau University)

[9] Pranoto B A, Ambariyanto and Zainuri M 2005 *J. Mar. Sci.* **10** 90 – 97. (in Indonesian).

[10] SNI 2009 Tilapia Production (*Oreochromis niloticus Bleeker*) Enlargement Class in calm water cultivation pond (Jakarta) (in Indonesian).
[11] Marion J E 1998 Water quality of pond aquaculture. international center for aquaculture and aquatic environment (Alabama : Auburn University)
[12] PP No. 82 2001. Water Quality Management and Water Pollution Control. Online. http://jdih.menlh.go.id/pdf/ind/IND-PUU-3-2001-lLampiran.pdf. Accessed by 1 March 2016 at 07.06 WIB. (in Indonesian).
[13] Pamukas N A 2011 *J. Coral Reefs Fish.* **39** 79 – 90. (in Indonesian).
[14] Scholz B and Liebezeit G 2012 *Eur J. of Phycol* **47** 393 – 407.
[15] Cowey B 1963 *J. Mar. Biol. Ass. U. K* **43** 485 – 493.
[16] Guisande C, Loreinco C G and Leon D 2008 *Limnetica* **27** 273 – 284.
[17] Costa J Z and Thompson K D 2016 *Fish. Shellfish Immunol.* **53** 35 – 49.
[18] Irianto A 2005 *Gadjah Mada University press* (in Indonesian).
[19] Alamanda E I., Handajani H N and Budiharjo A 2007 *Diversity* **8** 4-38. (in Indonesian).
[20] Mudjiutami E, Ciptoroso and Zainun Z 2007 *J. Freshwater Aquac.* **4** 37 – 41. (in Indonesian).
[21] Moyle P B, Cech Jr JJ 1988 *Fishes: An Introduction to Icthyology* (USA : Prentice Hall, Inc)
[22] Lusiyanti Y and Alatas Z 2011 The Utilization of Premature Chromosome Condensation Technique and Micronuclei Assay In Biological Dosimetry *Proc. of National. Sem. on Occup. Health and Safety* **7** 57 – 71. (in Indonesian).
[23] Chi S C, Lo B J and Lin S C 2001 *J. Fish Dis* **24**: 3-13.
[24] Yanuhar U 2011 *Int. J. Biosci. Biochem. Bioinforma* **1** 119-124.
[25] Berilis P, Papadimitriou T, Petridou E, Konstantinos K and Ilfigenia K 2014 *Turk. J. Fish.Aquat. Sci.* **14** 213-219
[26] Michel B, Fournier G, Lieffrig F, Costes B and Vanderplasschan A 2010 *Emerg. Infect. Dis.* **16** 1835-1843.

**Acknowledgements**
This work was funded by the Ministry of Research, Technology and Higher Education of the Republic of Indonesia through a scheme of the PUPT Program of Brawijaya University with contract no: 460.137/UN10.C10/PN/2017.

IOP Conf. Series: Earth and Environmental Science **137** (2018) 012015 doi:10.1088/1755-1315/137/1/012015

# Effect of mercury chloride to number of melano-macrophage centers on the kidney of carp fish (*Cyprinus carpio*)

**L Mubarokah[1], W Tjahjaningsih[2] and L Sulmartiwi[2]**

[1] Aquaculture Program, Faculty of Fisheries and Marine, Airlangga University, East Java, Surabaya 60115
[2] Department of Marine, Faculty of Fisheries and Marine, Airlangga University, East Java, Surabaya 60115

E-mail: wahju_fpk@yahoo.com

**Abstract**. Mercury chloride can cause immunotoxic effects on fish. The accumulation or aggregate of melano-macrophages centers (MMCs) in the kidney is a feature of cellular immune response, so it can be used as a bioindicator of heavy metal toxicity in waters. This research aims to determine an effect of heavy metal exposure concentration of mercury chloride ($HgCl_2$) on a number of MMCs from common carp kidney. This research using four treatments of mercury chloride: 0 ppm, 0.01 ppm, 0.05 ppm and 0.1 ppm. The main parameters were the number of MMCs in kidney common carp. Supporting parameter was behavior change, water quality and concentration of mercury in the water and kidney. The results of the research showed that the concentration the heavy metal exposure concentration of mercury chloride ($HgCl_2$) affected the number of MMCs from common carp (*Cyprinus carpio*) kidney. The number of MMCs in the kidney of common carp exposed to mercury chloride 0.01 ppm, 0.05 ppm and 0.1 ppm increased compared to carp was not exposed to mercury chloride (0 ppm).

## 1. Introduction

Heavy metals are part of industrial waste as a result of the pace of development in the industrial sector [1, 2]. Heavy metals that are toxic include mercury (Hg), cadmium (Cd) and lead (Pb) [3]. The heavy metal mercury can cause damage to the liver and kidneys of exposed organisms, especially the type of inorganic mercury that is mercury chloride ($HgCl_2$) which is also corrosive to the intestine [4]. Mercury chloride ($HgCl_2$) can cause changes in fish behavior and can give an immunotoxic effect [5]. The accumulation of metals in fish begins with the process of uptake through the gills and then is absorbed into the entire body tissue and stored in the organs [6]. The accumulation of metals is present in fish tissues, therefore the monitoring of heavy metals in fish tissues can help to assess the health of aquatic ecosystems [7]. Al-Madani *et al.* [8] state that one of the organs that can accumulate mercury chloride ($HgCl_2$) is the kidney, which functions for excreting waste products and plays an important role in maintaining homeostasis by regulating body fluids.

The research results of El-Boshy and Taha [9] showed that mercury chloride causes impaired cellular or humoral immune response in tilapia. The cellular immune response can be an indicator of exposure to heavy metals through the changes and appearance of pigments in MMCs in the liver and kidneys of the anterior part. [10]. Visible changes in MMCs resulting from pollution response include the quantities, sizes, and pigments seen. [11]. Melano-macrophages centers are the accumulation or aggregate of macrophages [12]. Observation of the structure of MMCs can be seen from changes in

IOP Conf. Series: Earth and Environmental Science **137** (2018) 012015 doi:10.1088/1755-1315/137/1/012015

the number and appearance of pigments [13]. The MMC structure is easily visualized histologically by the presence of three pigments namely hemosiderin, melanin and lipofuscin. [11, 14].Hemosiderin is a storage form of iron that is derived from food sources and from the red blood cell hemolysis [13].

The results of the study from Reddy [10] concluded that an increase in the number and size of MMCs in the liver and kidney of *Cyprinus carpio* exposed to cadmium for 30 days is considered a histopathologic biomarker due to stress caused by the metabolic toxicity of the aquatic environment. Based on this background, a study was conducted to determine the change in the number of MMCs in the goldfish kidney exposed to mercury chloride. The anterior part of the kidney organ of where most of the complex hematopoietic tissue is located. Histopathological features of kidneys' anterior showed the cellular structure melano-macrophage centers [15,16].

The use of carp as a test fish in this study was based on the Environmental Protection Agency [17] which mentions that carp can be used as bioindicators. Coban [18] state that carp (*Cyprinus carpio*) is a water biota that has met the requirements to be used as bioindicators because it is easily adapts to contaminated water conditions. According to Osswald [18], the extensive distribution of carp makes these fish as ecologically compatible aquatic organisms for the study of the impact of water toxicity.

## 2. Methodology

### 2.1. *Materials*

The test fish used were carp fish obtained from the fish market of Gunung Sari with lengths of 18-20 cm and weight of 75-80 gr. Other materials used in the research were mercury chloride ($HgCl_2$), pellet feed, aquadest, formalin buffer, liquid paraffin, 70 %, 80 %, 90 %, 96 % alcohol, absolute alcohols I, absolute alcohols II, xylol I, xylol II and substances color hematoxylin-eosin (HE).

### 2.2. *Methods*

#### 2.2.1. *Preparation*

An aquarium measuring 50 cm x 35 cm x 30 cm was first washed with liquid soap, followed by disinfecting chlorine with a dose of 100 ppm, according to Wahjuningrum [19] and dried in the sun. The aquarium was filled with 25 liters of water and given the aerator. Each treatment used five fish with five replications. Before being treated, the carp fish were acclimated beforehand.

Measurements of mercury in water were carried out before the water medium was treated with mercury chloride, while the mercury content in the kidney organ was measured after the carp was exposed to mercury chloride for 21 days. The measurement of mercury content was carried out using the Atomic Absorption Spectrophotometer (AAS) at the Environmental Health Engineering and Disease Control (Balai Besar Teknik Kesehatan Lingkungan dan Pengendalian Penyakit), Surabaya.

#### 2.2.2. *Procedures*

The study was conducted by exposing the carp to mercury chloride concentrations of 0.01 ppm, 0.05 ppm. and 0.1 ppm for 21 days [9]. Pellet feed was given ad libitum in the morning and evening. Substitution of water was done every three days by adding water with the same mercury chloride ($HgCl_2$) concentration. Water quality measurements were performed in the morning at the beginning and end of the study.

After 21 days of exposure to mercury chloride, the carp were autopsied to retrieve the kidney organs as histology preparation material. The steps for making histologic preparations consist of fixation, dehydration and clearing, infiltration, paraffin blocking, microtome cutting and staining. Organ fixation is done by soaking the organs with formalin buffer solution for at least 24 hours. The stages of dehydration were done using 70 % alcohol for one hour, 80 % alcohol for two hours, 90 % alcohol for two hours, 96 % alcohol for two hours, absolute alcohol for two hours. The clearing process aimed to replace the alcohol solution from the tissue by inserting a piece of kidney organ into the xylol for two hours. The next stage was infiltration which was entering pieces of kidney organ into liquid paraffin with a temperature of 56 - 60°C for two hours. To facilitate the cutting of the tissue to 4-7 microns thick, paraffin blocks with iron molds were made. In order for the tissue to expand properly, it was

IOP Conf. Series: Earth and Environmental Science **137** (2018) 012015 doi:10.1088/1755-1315/137/1/012015

dipped in a warm water temperature of 42-45°C [20]. Next it was put on an object glass and dried over the hot plate. According to Hibiya [21] and Suresh [14], to facilitate the observation of the hemosiderin pigment of the anterior renal tissue, the hematoxylin-eosin staining method was used to show a brownish yellow color.

*2.2.3. Parameters*

The main parameter in this study is the number of the MMCs of kidney organs exposed to mercury chloride for 21 days. The observation of the number of MMCs in the anterior kidney from histopathology was conducted using a 400x microscope connected to the computer to determine the bar size of the drawn image.

Supporting parameters include the change of carp behavior during the study, the mercury content in carp kidney organs and water media prior to treatment. Water quality parameters measured include temperature, pH and dissolved oxygen (DO) at the beginning and end of the study.

*2.2.4. Analysis of data*

Data on the number of MMCs were analyzed using Analysis of Varians (ANOVA) with SPSS version 17, followed by the Duncan's test to determine the differences between treatments.

## 3. Result

The results of the measurements of mercury in water before treatment were 0 ppm (treatment A), 0.0164 ppm (treatment B), 0.0546 ppm (treatment C) and 0.0700 ppm (treatment D). The results of the measurements of mercury content in the renal carp organ after exposure to mercury chloride for 21 days were 0 ppm (treatment A), 3.4032 ppm (treat B), 30.422 ppm (treatment C) and 26.7605 ppm (treatment D).

The histopathological observation data is the number and size of the visible melano-macrophage centers. The observations showed differences in the number and size of melano-macrophages in the carp kidney exposed to mercury chloride ($HgCl_2$) characterized by the hemosiderin pigment (Figure 1). The average data on the number of MMCs of carp kidney exposed to mercury chloride ($HgCl_2$) concentrations of 0 ppm, 0.01 ppm, 0.05 ppm, and 0.1 ppm are presented in table 1. The result of Analysis of Varians (ANOVA) with 95 % confidence interval showed that mercury chloride concentration ($HgCl_2$) had a significant effect ($p < 0.05$) on the number of MMCs of the carp kidney. Duncan Multiple Range Test results showed significant differences ($p < 0.05$) of MMCs at 0.05 ppm treatment with 0 ppm, 0.01 ppm, and 0.1 ppm.

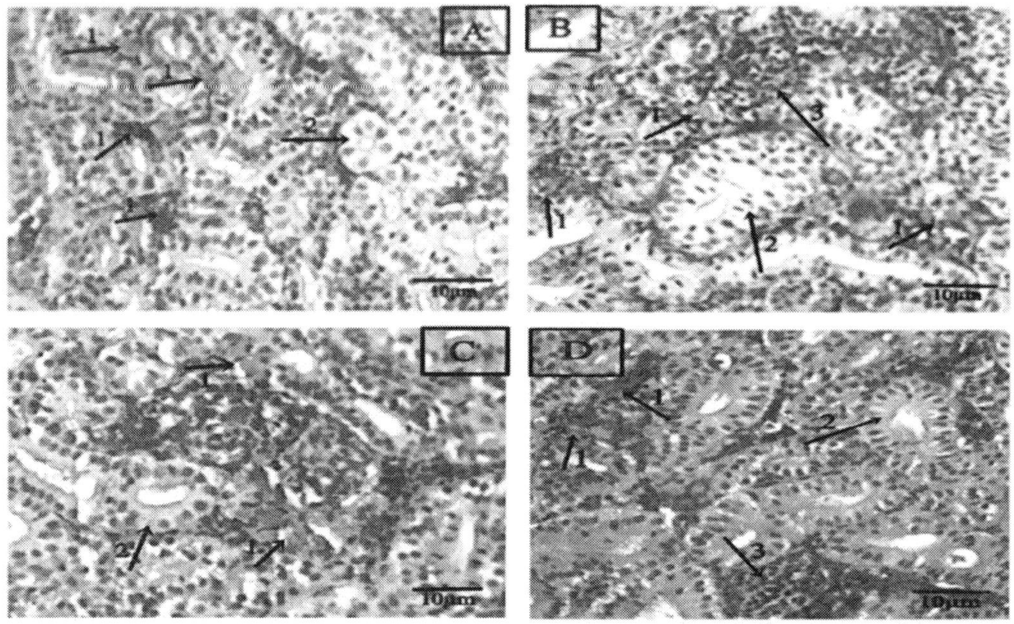


**Figure 1**. Kidney histopathology of carp (400x magnification). A: $HgCl_2$ concentration 0 ppm, B : $HgCl_2$ concentration 0.01 ppm, C : $HgCl_2$ concentration 0.05 ppm and D: $HgCl_2$ concentration 0.1 ppm.Melano-macrophage centers (1), tubules (2), glomeruli(3).

**Table 1.** The average number of melano-macrophage in the carp kidney (*Cyprinus carpio*).

| Treatment | Number ($mm^{-2}$) |
|---|---|
| A (0 ppm) | 21.45 $^{d}$ ±2.39 |
| B (0.01 ppm) | 50.35 $^{c}$ ±3.88 |
| C (0.05 ppm) | 78.95$^{a}$ ±5.47 |
| D (0.1 ppm) | 65.25 $^{b}$ ±2.61 |

Description: Different superscripts in one column show no significant difference at significance level (α = <0.05)

Data on the behavior of carp fish exposed to mercury chloride are presented in table 2.

**Table 2.** Data changes in behavior of carp.

| Treatment | Change of Behavior |
|---|---|
| A (0 ppm) | No change in behavior, normal swimming fish |
| B (0.01 ppm) | Fish tend to swim faster, the accumulation of mucus |
| C (0.05 ppm) | Fish tend to be hyperactive, lots of mucus accumulation and fish try to jump out of the aquarium |
| D (0.1 ppm) | Fish tend to be hyperactive, very much mucus accumulation, loss of balance and try the fish jump out of the aquarium |

Water quality measurements aim to monitor water quality conditions for carp during maintenance. Water quality measurements during the study ranged from 25.8 - 29.04°C, pH 7.1 – 7.8 and DO 3.95 – 4.14 mg / L. The value was still in the optimum range for the life of the carp, namely the temperature of 23 – 30°C, pH 6.5 - 9 [22, 23] and dissolved oxygen 3-5 mg / L [24]. This suggests that the change

IOP Conf. Series: Earth and Environmental Science **137** (2018) 012015 doi:10.1088/1755-1315/137/1/012015

in the number of melano-macrophages of the carp of each treatment group is caused by exposure to mercury chloride, rather than influenced by water quality conditions.

## 4. Discussion

Mercury can enter through food and then into the digestive tract after experiencing absorption in the intestine, mercury compounds will be brought into the liver. Mercury entering into the heart will be divided into two; some will accumulate in the liver and some will be sent to the bile. Mercury in the bile will be carried to the excretory organs of the kidney, where some will accumulate in the kidney and some will be excreted through urine [25]. Subanri [25] states that mercury is a systemic poison and usually accumulates in the liver, kidneys, and spleen. The result of the highest measurement of mercury content in the renal organ is at treatment with a mercury concentration 0.05 ppm that was 30.422 mg/kg, but at treatment with a mercury concentration of 0.1 ppm it is seen that there is a decrease of mercury content in the kidney which was 26.7605 mg/kg.

The result of the Javed and Usmani [26] study proved that the concentration of heavy metals in the kidney organs are higher than in the gills, liver, muscles, and integument. Thophon [27] also proved that renal tubules and gills lamellae were the primary target organs for the effects of acute toxicity from cadmium while in subchronic exposure, but toxic effects on the gills were less than those of the kidneys and liver. Sadeghi [28] also proved that there was a tendency of chromium accumulation in kidney organs of Epaulet grouper (*Epinephelus stalickae*) exposed to chromium pollution.

Mercury chloride ($HgCl_2$) can cause cellular or humoral immune response [9]. Cellular immune responses are indicators of changes in the number of melano-macrophages in the liver and kidneys, especially the anterior kidney [10]. Melano-macrophages plays an important role in the response of fish to foreign materials, including infectious agents [29,30]. Melano-macrophages are an aggregation of macrophages [15]. Macrophages function as the main phagocytic cells that play a role in the process of phagocytosis [11, 31, 32, 33].

The highest number of MMCs in the kidney was found in the 0.05 ppm treatment which significantly differed ($p < 0.05$) to the treatment of 0 ppm, 0.01 ppm, and 0.1 ppm. The presence of heavy metals in organs will cause cell or tissue damage [34]. The damaged cells or tissue will be digested by phagocyte cells and will be concentrated in macrophages [15]. This mechanism will stimulate the formation and increase the number of MMCs [30]. According to Bols [11], the visible changes in melano-macrophages are resulted from pollutant responses include the amount, size and pigment seen. The results of Sayed and Younes [34] prove the immune role of MMCs as indicated by an increase in the number and size of MMCs in the spleen, kidney and liver organs of *Clarias gapriepinus* exposed to silver nanoparticles. Tjahjaningsih [35] also proved that there is accumulated macrophage activity in the kidney, spleen, and liver of carp due to the exposure to mercury chloride. The macrophage activity illustrates the role of melano-macrophage centers (MMCs) in capturing non-antigenic materials that were damaged cells by exposure to mercury chloride. Reddy [24] also proves that macrophage activity in the kidneys and liver of carp can play a role in the clearance of the macromolecular circulation and foreign particles due to cadmium exposure.

In the treatment of 0.1 ppm concentration of mercury chloride ($HgCl_2$), a visible decrease in the number of MMCs was observed. Excessive mercury chloride exposure can be very toxic, thus causing a decreased immune response expression due to an immunosuppression mechanism. According to Middleton [36], a substance that toxic can inhibit macrophage phagocytosis activity at a certain concentration limit. This is also evident from the changes in the behavior of carp fish in the treatment of 0.1 ppm mercury concentration that appeared hyperactive, tried to get out of the aquarium, irregular swimming and high mucus accumulation. Nirmala [37] state that mercury can disrupt the taste sensors and receptors of the sight of fish so that there are changes in behavior. Changes in the behavior of carp in the mercury concentration treatment of 0.1 ppm are very visible i.e hyperactive fish, trying to get out of the aquarium, irregular swimming, and accumulation of high mucus. This is in accordance with the statement of Guedenon [38] which stated that changes in fish behavior in response to mercury chloride toxicity ($HgCl_2$) include fish becoming hyperactive by attempting to jump out due to skin irritation, respiratory distress, loss of balance, swimming backwards, swirling, excessive mucus accumulation and ending with death.

## 5. Conclusion

Based on this research, it can be concluded that the levels of exposure to the heavy metal mercury chloride ($HgCl_2$) has an effect on the change of the number of melano-macrophage of the kidney of carp fish (*Cyprinus carpio*). The melano-macrophage centers of carp that were exposed to 0.01 ppm, 0.05 ppm and 0.1 ppm mercury chloride had increased MMCs compared to carp fish that were not exposed it (0 ppm).

## 7. References

[1]. Kurniawan Y B 2003 Mercury analysis (hg) in soil agriculture using neutron activation analysis method (Sebelas Maret University : Undergraduate Thesis Faculty of Mathematics and Natural Sciences) 73p.
[2]. Rochyatun E, Kaisupy M T, and Rozak A 2006 *Makara Sains* **10** 35-40
[3]. Darmono 1995 Metals in biological systems of living beings (Jakarta: Universitas Indonesia Press) 140p
[4]. Supriharyono 2002 *Conservation and management of natural resources in tropical coastal areas* (Jakarta: PT Gramedia Pustaka Utama) 156p.
[5]. Hedayati A, Safahieh A, Savari A, and Marammazi J G 2010 *Iranica J. Energy Environ.* **1** 228-33
[6]. Lasut M T 2009 *J. Math. Sci.* **4** 89-95 (In Indonesia)
[7]. Adam S M 2002 *Biological indicators of aquatic ecosystem stress* (Bethesda: American Fisheries Society) 565 p
[8]. Al-Madani W.A, Siddiqi N J, and Alhomida A S 2009 *Biochem. Insights* **2** 37-45
[9]. El-Boshy M E and Taha R 2011 *Nat. Sci.* **9** 7-15
[10]. Reddy S J 2012 *World J. Fish Mar. Sci.* **4** 179-84
[11]. Bols N C, Brubacher J L, Ganassin R C, and Lee L E J 2001 *Dev. Comp.Immunol.* **25** 853-73.
[12]. Fournie J W, Summers J K, Courtney L A, and Engle V D 2001 *J. Aquat. Anim. Health* **13** 105-16.
[13]. Wolke R E, Murchelano R A, Dickstein C D, and George C J 1985 *Bull. Environ. Contam. Toxicol.* **35** 222-7.
[14]. Suresh N 2009 *J. Environ. Biol.* **30** 505-8.
[15]. Genten F, Terwinghe E, and Danguy A 2009 *Atlas of Fish Histology* (United States of America: Science Publishers) 57p
[16]. https://www.epa.gov/sites/production/files/2015-07/documents/850-1075.pdf
[17]. Coban M Z, Eroglu M, Canpolat O, Calta M, and Sen D 2013 *J. Anim. Plant. Sci.* **23** 1455-59
[18]. Osswald J S, Rellan A, Carvalho P, Gago A, and Vasconcelos V 2007 *Toxicon* **49** 693-8
[19]. Wahjuningrum D, Solikhah E H, Budiardi T dan Setiawati M. 2010. *J. Aquac. Ind.* **9** 93-103 (In Indonesia)
[20]. Brunt E M 2000 *J. Hepatol.* **31** 241-246
[21]. Hibiya T 1982 *An atlas of fish histology normal and pathological features* (Tokyo: Kodansha Scientific Books) 56 p
[22]. http://www.fao.org/fishery/culturedspecies/Cyprinus_carpio/en
[23]. Rudiyanti S and Ekasari A D 2009 *J. Sci. Tech. Fish.* **5** 39-47
[24]. Destiany M 2007 The effect of chloride mercury on the structure of liver microanatomy goldfish (Semarang State University : Undergraduate Thesis) 63 p
[25]. Subanri 2008 The assessment of mercury contamination in Menyuke river and it`s health to mine workers from the impact of illegal gold mining in menyuke distrik landak regency West Kalimantan (Diponegoro University : Thesis Master of Environmental Health) 102 p
[26]. Javed M and Usmani N 2011 *Int. J. Environ. Sci.* **12** 671-82

[27]. Thophon S, Kruatrachue M, Upatham E S, Pokethitiyook P, Sahaphong S, and Jaritkhuan S 2003 *Environ. Pollut.* **121** 307-20.
[28]. Sadeghi P, Kazerouni F, Savari A, Movahedinia A, Safahieh A, Ajdari D 2015 *Sci. Total Environ.* **518-519** 554-61
[29]. Agius C and Roberts R J 2003 *J. Fish Dis.* **26** 499-509
[30]. Herraez M P and Zapata A G 1986 *Vet. Immunol. Immunopathol.* **12** 117-26
[31]. Wolke R E 1992 *Ann. Rev. Fish Dis.* **2** 91-108
[32]. Thomas A W, Chawla A, and Pollard J W 2013 *Nature* **24** 496: 445
[33]. Wikiandy N, Rosidah, and Herawati T 2013 *J. Mar. Sci.* **4** 215-25 (In Indonesia)
[34]. Sayed A E H and Younes H A F M 2017 *J. Microsc. Ultrastruct.* **5** 97-104.
[35]. Tjahjaningsih, W, Pursetyo K T, and Sulmartiwi L 2017. *AIP Conf. Proc.* **1813** 20012
[36]. Middleton E, Kandaswami C, and Theoharides T C 2000 *Pharmacol. Rev.* **52** 673-751
[37]. Nirmala K, Hastuti Y P, and Yuniar V 2012 *J. Aquac. Ind.* **11** 38-48 (In Indonesia)
[38]. Guedenon P, Edorh A P, Hounkpatin A S Y, Alimba C G, Ogunkanmi A, Nwokejiegbe E G, and Boko M 2012 *Res. J. Chem. Sci.* **2** 41-5.

**Acknowledgments**
This research is supported by Wahju Tjahjaningsih, Lecturer of the Faculty of Fisheries and Marine Universitas Airlangga who received The Research Grant from the Faculty of 2014 based on the Decree of Dean of Faculty of Fisheries and Marine Airlangga University No. 1271/UN3/2014

IOP Conf. Series: Earth and Environmental Science **137** (2018) 012016 doi:10.1088/1755-1315/137/1/012016

# The effects of season, aeration and light intensity on the performance of pacific whiteleg shrimp (*Litopenaeus vannamei*) polycultured with seaweed (*Gracilaria verrucosa*)

**T Susilowati, Desrina, J Hutabarat, S Anggoro, M Zainuri, Sarjito, F Basuki and T Yuniarti**

Aquaculture Department. Fisheries and Marine Science Faculty, Diponegoro University. Jl. Prof. Soedharto SH, Lingkar Tembalang, Semarang, Central Java, Indonesia.

E-mail :susilowatibdp@gmail.com

**Abstract.** This study was aimed to determine impact of stocking season, additional oxygen supply and light intensity on performance of pacific white leg shrimp (*Litopenaeus vannamei*) polycultured with seaweed *Gracilaria verrucosa*. Three sets of experiments were used and each experiment was conducted separately in 3 different season (factor W). Three factors and the interaction, that is, stocking seasons (W1: March-June;. W2 : July-October and W3 : November-February), Oxygen supply (O, with or without supply aeration of 6.5 ppm) and light intensity (with or without addition of light 640 lux) was observed. The experiment was conducted in 16 polyethilen net place in concrete tanks (1.2 $m^3$). Shrimp (average weight 0.1 g and lenght 1.2 cm) with density 94 shrimp/$m^3$ and seaweed 2.188 g/$m^3$, cultured for 94 days. Data collected included absolute growth, specific growth rate (SGR), survival rate (SR) and biomass production of shrimp and seaweed. The result showed that culture period March to June, additonal light and suply DO gave the best result with shrimp absolute growth 13.23 g, SGR 5.09 %/day, SR 99.64 % and biomass production 1.256.36 g/$m^3$. Absolute growth of *G. verrucosa* was 5.223.75g, SGR 268 %/day and biomass production 12.608.55 g/$m^3$.

## 1. Introduction

Polyculture of pacific whiteleg shrimp (*Litopenaeus vannamei*) and seaweed (*Gracilaria verrucosa*) is a common practice in Southeast Asia. Pacific whiteleg shrimp (*L. vannamei*) are suitable for this type of polyculture because they are highly adaptive to different environmental conditions. For example, *L. vannamei* are cultured in "ponds", formerly excavated sand, from which 4 tons of *L. vannamei* are produced after 90 days of culture [1]. The use of *G. verrucosa* in the polyculture system benefits the environment by trapping excess nutrients normally occurring in shrimp ponds but possibly leaving negative effects on shrimp's health. *G.verrucosa* also absorbs toxic substances from organic waste while producing oxygen supply during the day. Hence, it reduces waste contamination and converts the waste contamination into nutrients for its growth. Most of the time, polyculture is usually applied in extensive ponds. However, it is also applicable in semi-intensive shrimp ponds with additional oxygen supply. In addition, the season for farming can also affect shrimp's survival, growth and, in turn, production. Season affects the pond environment, including the salinity, dissolved oxygen, temperature and consentration of N (nitrate, nitrite and ammonia) in the water, which in turn, contributes to the shrimp's well-being and seaweed growth. Selecting the right period for stocking and culturing shrimp is important for the success of the polyculture of shrimp and seaweed.

## 2. Methodology

The study was conducted in the Laboratory of Coastal Development, Faculty of Fisheries and Marine Sciences, Universitas Diponegoro. The experiment was done in different seasons in three periods, namely March-June (post-wet season), July-October (dry season) and November-February (wet season). The experiment used 16 polyethilene nets (1x1x1.2 m in size) placed in four concrete tanks (each tank was 6x5x1.2 m in size). For each period the shrimp (average weight of 0.1 g/individual average length of 1.2 cm) were stocked at a density of 94 shrimp/m$^3$ along with sea weed at a density of 2.188 g/m$^3$. The experiment was carried out in split plot in time design with quadreplicate. There were three factors used in this study, namely factor W (season), O (oxygen supply) and I (light intensity). Factor W was the season in which the experiment was started and the stocking was conducted (W1: March-June; W2: July-October; and W3: November-February). Factor O consisted of oxygen supply (2 levels, with or without aeration) and light intensity (with or without additional light). The experiment was designated as follows: O1 (an addition of light intensity of 640 lux without oxygen supply); O2 (an addition of oxygen supply of 1.5 m/second, equal to 6.5 ppm and without additional light); O3 (an addition of oxygen supply of 1.5 m/second and a light intensity of 640 lux) and O4 (without supply aeration and light (control)). A TL lamp was used as the source of illumination in this study because it emits light similar to the sunlight [1].

The shrimp and the seaweed were weighted once every 12 days for 92 days. The growth parameters measured were the absolute growth, specific growth, survival rate and production. The data were analyzed statistically using Analysis of Variance and Duncan's Multiple Range Test (DMRT) [2].

## 3. Results and Discussions

### *3.1 Results*

#### *3.1.1 Growth, survival rate and production of pacific white leg shrimp (L.vannamei)*

The average absolute weight growth of shrimp (*L. vannamei*) ranged from 8.93 ± 0.0 g (the lowest, treatment W3O4) to 13.13 ± 0.1 g (the highest, treatment W1O3). The specific growth rate (%/day) ranged from 4.68 ± 0.00 (treatment W3O4) to 5.09 ± 0.00 (treatment W1O3). The survival rate (%) ranged from 40.36 (W3O4) to 99.64 (W1O3). The production (g) ranged from 376.87 ± 0.34 (W3O4) to 1,005.08 ± 2.47 (table 1).

**Table 1**. Average absolute growth, specific growth rate, survival rate and production of pacific white shrimp *(L.vannamei)* measured in this study.

| Treatment | Average absolute growth (g) | Specific growth rate (%/day) | Average survival rate(%/) | Average production(g) |
|---|---|---|---|---|
| W1O1 | 11.42±0.020 | 4.94±0.002 | 80.00±0.290 | 865.85±0.760 |
| W1O2 | 12.71±0.040 | 5.05±0.003 | 93.22±0.120 | 964.71±0.220 |
| W1O3 | 13.23±0.000 | 5.09±0.000 | 99.64±0.180 | 1005.08±2.470 |
| W1O4 | 9.95±0.0100 | 4.79±0.001 | 50.36±0.180 | 467.30±0.380 |
| W2O1 | 11.28±0.000 | 4.92±0.000 | 85.00±0.200 | 851.53±0.680 |
| W2O2 | 12.63±0.050 | 5.04±0.004 | 73.22±0.180 | 944.37±1.370 |
| W2O3 | 13.13±0.000 | 5.08±0.000 | 93.93±0.180 | 972.46±0.080 |
| W2O4 | 9.85±0.0100 | 4.78±0.001 | 45.00±0.200 | 394.98±2.610 |
| W3O1 | 10.75±0.000 | 4.90±0.013 | 67.50±0.180 | 509.90±0.620 |
| W3O2 | 11.83±0.010 | 4.97±0.001 | 81.25±0.630 | 592.97±0.320 |
| W3O3 | 12.57±0.000 | 5.04±0.000 | 90.36±0.340 | 646.80±0.220 |
| W3O4 | 8.93±0.000 | 4.68±0.000 | 40.36±0.030 | 376.87±0.340 |

*3.1.2. Effect of season-oxygen supply interaction on the growth of Pacific whiteleg shrimp (L. vannamei)*

The results of the Analysis of Variance showed that there was no effect of season and oxygen supply on the absolute growth and specific growth rate of the experimental shrimp in this study, but there was an effect only on the survival rate (figure 1).

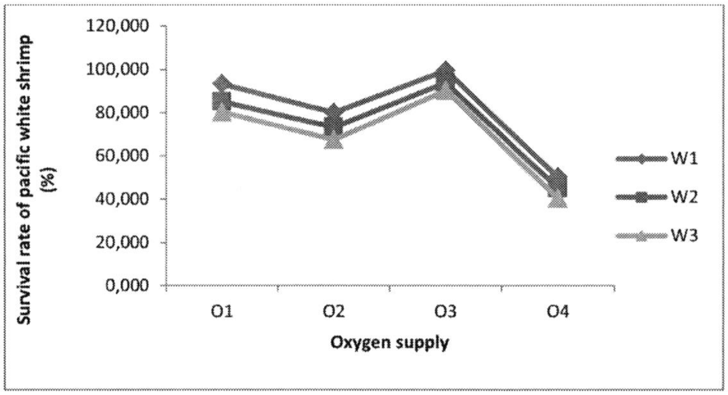


**Figure 1**.Effect of intreaction of season and oxygen supply on the survival of the pacific white shrimp (*L. vannamei*).

*3.1.3. Absolute growth, specific growth rate and production of G. verrucosa*

The average absolute growth of seaweed ranged from 2,496.5 ± 0.32 g/m$^3$ (the lowest, treatment W3O4) to 5,223.75 ± 0.31 g/m$^3$ (the highest, treatment W1O3). The specific growth rate of seaweed obtained in this study ranged from 1.750 ± 0.00 %/day (treatment W3O4) to 2.68 ± 0.00 %/day (tretament W1O3). The average production of seaweed ranged from 3,916.39 ± 0.64 g (W3O4) to 10,086.85 ± 0.79 g (treatment W1O3) (table 2).

**Table 2** .Average absolute growth, specific growth rate, and production of sea weed *G.verrucosa*) in present study.

| Treatment | Average absolute growth (g) | Specific growth rate (%/day) | Production (g) |
|---|---|---|---|
| W1O1 | 3024±0.200 | 1.97±0.000 | 4994.19±0.430 |
| W1O2 | 4895±0.540 | 2.59±0.000 | 9273.08±1.320 |
| W1O3 | 5223.75±0.310 | 2.68±0.000 | 10086.85±0.790 |
| W1O4 | 2663.25±0.130 | 1.82±0.000 | 4249.98±0.250 |
| W2O1 | 2823.25±0.430 | 1.89±0.000 | 4576.35±0.880 |
| W2O2 | 4721.75±0.380 | 2.54±0.000 | 8850.98±0.910 |
| W2O3 | 5064±7.170 | 2.64±0.000 | 9724.65±0.590 |
| W2O4 | 2518±2.190 | 1.76±0.000 | 3950.08±0.640 |
| W3O1 | 2638.25±3.880 | 1.82±0.000 | 4215.66±0.480 |
| W3O2 | 4611±0.200 | 2.51±0.000 | 8583.67±0.490 |
| W3O3 | 4887.5±0.140 | 2.59±0.000 | 9372.50±8.690 |
| W3O4 | 2496.5±0.320 | 1.75±0.000 | 3916.39±0.640 |

*3.1.4. Effect of season-oxygen supply interaction on the specific growth rate of the seaweed G. verrucosa*
The interaction between season and oxygen supply had a significant effect on the absolute growth of Pacific whiteleg shrimp (*L.vannamei*), but no effect on the specific growth rate of seaweed (*G. verrucosa*) (figure 2).

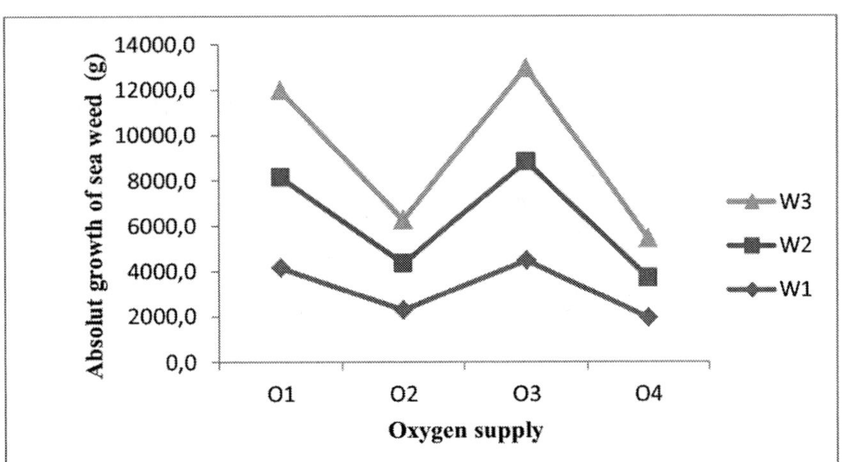


**Figure 2.** Interaction of season of stocking and oxygen supply on the absolute growth of sea weed (*G. verrucosa*).

IOP Conf. Series: Earth and Environmental Science **137** (2018) 012016 doi:10.1088/1755-1315/137/1/012016

### *3.1.5. Water quality during the study*

The water quality parameters measured were the dissolved oxygen (DO), temperature, salinity, pH, transparancy, nitrate, nitrite, phosphat, $CO_2$ and total ammonia (table 3).

**Table 3.**Water quality measurement during the experiment.

| | Treatments | | | | | | | | | | | |
|---|---|---|---|---|---|---|---|---|---|---|---|---|
| Parameters | W1O1 | W1O2 | W1O3 | W1O4 | W2O1 | W2O2 | W2O3 | W2O4 | W3O1 | W3O2 | W3O3 | W3O4 |
| DO (ppm) (night) | 5.7 | 6.6 | 6.5 | 3.3 | 5.7 | 6.7 | 6.7 | 3.3 | 5.7 | 6.4 | 6.7 | 2.3 |
| DO(ppm) Day | 6.7 | 6.7 | 6.7 | 6.6 | 6.9 | 6.9 | 6.9 | 6.9 | 6.7 | 6.7 | 6.7 | 6.7 |
| pH | 7.5-8.2 | 7.3-8.1 | 7.5-8.2 | 7.2-7.8 | 7.3-7.9 | 7.2-7.9 | 7.2-7.9 | 7.1-7.8 | 7.2-7.8 | 7.1-7.8 | 7.2-7.8 | 7.7-7.8 |
| Temperature (°C) | 27.5-31.1 | 27.1-31.2 | 27.5-31.2 | 27.4-31.1 | 28.7-33.0 | 27.7-33.1 | 28.7-33.2 | 27.2-33.3 | 27.1-30.2 | 26.8-30.1 | 27.5-.30.1 | 25.1-30.1 |
| Salinity (ppt) | 30 | 30 | 30 | 30 | 30 | 30 | 30 | 30 | 30 | 30 | 30 | 30 |
| NO3 (ppm) | tt-0.005 | tt-0.006 | tt-0.004 | 0.025 | tt-0.002 | tt | tt | 0.001-0.003 | tt-0.013 | tt-0.003 | tt-0.025 | 0.045 |
| Total NH3 (ppm) | 0.011-0.01 | 0.01-0.02 | 0.01-0.01 | 0.01-0.02 | 0.01-0.02 | 0.01-0.03 | 0.01-0.02 | 0.01-0.03 | 0.011-0.028 | 0.011-0.047 | 0.011-0.031 | 0.011-0.054 |
| Fosfat (ppm) | 0.21-0.32 | 0.32-0.43 | 0.32-0.35 | 0.28-0.33 | 0.22-0.31 | 0.33-0.43 | 0.33-0.44 | 0.29-0.35 | 0.26-0.327 | 0.285-0.328 | 0.326-0.358 | 0.224-0.327 |
| CO2 (ppm) | tt-0.48 | tt | tt | 0.44-3.98 | tt-0.45 | tt | tt | 0.44-3.98 | tt-0.44 | tt | tt | 0.42-3.98 |
| Transparancy (cm) | 80 | 80 | 80 | 80 | 80 | 80 | 80 | 80 | 80 | 80 | 80 | 80 |
| Nitrite mg/l | 0.025 | 0.023 | 0.015 | 0.084 | 0.036 | 0.027 | 0.027 | 0.096 | 0.041 | 0.029 | 0.019 | 0.098 |

### *3.2. Discussion*

Overall, the results showed that stocking and rearing shrimp and seaweed in the post-wet season (March–June) was advantageous for the cultured organisms. Stocking and culturing shrimp and seaweed in the period March-June with oxygen supply and light (W1O3) resulted in the highest absolute growth. This result was higher than that of the other seasons with and without oxygen supply and light. The lowest absolute growth was obtained in the period November-February without oxygen supply and light (W3O4). This may be related to water quality, especially the DO and $CO_2$ at night. The DO at nighttime in the treatment with the highest absolute growth was much higher (6.7 ppm), and the $CO_2$ was undetected. Quite the contrary, the treatment with the lowest absolute growth had low DO (2.3 ppm) and $CO_2$ of 0.42-3.98 ppm. Low concentration of DO at night usually causes a problem in shrimp farming [3]. Low DO in treatment W3O4 may have caused the shrimp to spend extra energy to compensate for the dissolved oxygen condition to meet physiological needs. For example, this may be done by pumping more water through the gills, lowering the oxygen partial pressure (pO2), regulating the amount of gill surface used for respiration, which may affect the osmoregulation [4]. During the experiment, the salinity was maintained at 30 ppt, which was isosmotic for shrimp. Therefore, the energy spent for osmoregulation was minimal. This may contribute to better growth because at isosmotic condition, the Na+/K+-ATPase activity of shrimp is low, and the food utilization will be efficient [5]. It is possible for oxygen supply and additional light to increase water temperature, which increases the metabolic activities and moulting frequency and results in higher absolute growth of the shrimp [6, 7].

Water temperature is naturally the lowest in the period November-February, averaging 25.1–30.1 °C. This is lower than the optimal temperature for the survival rate (SR) of Pacific whiteleg shrimp, showing a paralel result with that of growth of shrimp. Previous study showed that the SR of shrimp is affected by stocking time and oxygen supply [7,8,9]. The mortality of shrimp in this study may be related to season, which in turn, will affect the water temperature and dissolved oxygen concentration that is less than optimal.

During wet season (November to February) water temperature ranges between 25.1 °C and 27.1 °C. On the other hand, high day temperature in the dry season can also cause high mortality shrimp production is population weight gain over certain period of time [10] determined by growth rate, mortality rate and biomass in a certain period of time.[11] One of the natural factors determining production is stocking season. The addition of oxygen supply and light in this study aimed to alleviate the low DO at night. Light is necessary for facilitating photosynthesis by seaweed for it to provide extra DO for shrimp. The gain in treatment W3O4 (wet season, no oxygen supply and no additional light) showed how all of the three factors could impact shrimp production negatively [12]. Seaweed growth is affected by environmental conditions such as light, temperature, pH and nutrient. Light intensity and the length of lighting period are important factors for the culture of algae, including seaweed. The results of this study showed that oxygen supply and additional light had no significant effect on seaweed growth despite longer period of photosynthesis (W1O3). However, the light intensity of 3.500 lux was optimum for the production of good absolute growth of seaweed. In addition, salinity of 30 ppt is suitable for seaweed as it is the salinity of the coast from which the seaweed used in this study originated. The lack of light for the seaweed cultured in November-February, and the lack of additional light resulted in low weight gain. Light intensity influences spore production and the growth of ectocarpus because tetraspora and karpospora production occurs only in the summer [13, 14].

## 4. Conclusion

This study showed that the most suitable period for the polyculture of the shrimp *L. vannamei* and the seaweed *G. verrucosa* is March to June. An additional light intensity of 640 lux and oxygen supply of 6.5 ppm can produce high growth rate, survival rate and production of shrimp and seaweed.

IOP Conf. Series: Earth and Environmental Science **137** (2018) 012016 doi:10.1088/1755-1315/137/1/012016

## 5. References

[1]. Trobos 2013 *The Magazine Agribisnis of Marine and Fisheries* **12** (1) 24-26
[2]. Srigandono B 1990 Experimental Design Faculty of Husbandry UNDIP Semarang 178 p
[3]. Zonneveld N, Huisman E A and J H Boon 1991 Principles of fish culture (Jakarta: Gramedia Pustaka Utama) 54 p (In Indonesia)
[4]. Fujaya Y 2004 Basic of Fish Fisiology for Developing Fisheries (Jakarta: Technology Rineka Cipta) 179 p (In Indonesia)
[5]. Anggoro S 2000 Osmotic regulatory patterns and enzyme work Na-K-ATPase tiger shrimp (*Penaeus monodon*) of various molting phases *J. Aquac.Ind.* **1** (2) : 15-20. (In Indonesia)
[6]. Vinatea L W, Muedas, R Arantes 2011 The Impact of Oxygen Consumption by Shrimp *Litopenaeus vannamei* According to Body Weight.temperature. Salinity and Stocking Density on Pond Aeration *a Simulation Maringi* **33**(2) 125-132
[7]. Spanopoulos-Hernandez M, C A Martinez-Palacios, R C Vanegas-Perez, C Rosas and L G Ross 2005 The Combined Effects of Salinity and Temperature on The Oxygen Consumption of Juvenile Shrimps Litopenaeus stylirostris (Stimpson 1874) *J. Aquac* **244**:341-348
[8]. Anongponyoskun M A Choksuchart, J Salaeno and P Aranyakananda 2012 *J. Nat. Sci.* **46** (5) : 751-758
[9]. Ruiz-Velazco J M J, A Hernandez-Llamas .and V M Gomez-Munoz 2010 *J. Aquac. Engineering* **43** : 114-119
[10]. Edmonson W T and G G Winberg 1971 *A manual and methods for assesment of secondary productivity in freshwater* (London : IBP Blackwell Sc 1 Pub) 358 p
[11]. Taqwa F H 2008 The effect of potassium addition on adaptation time of salinity decrease and time of natural feed replacement by artificial feed to performance pasca larva vannamei Shrimp (*Litopenaeus vannamei*) (Bogor : Bogor Agricultural Institute) 83 p (In Indonesia)
[12]. Salisbury F B dan C W Ross 1995 *Plant physiology* Volume 2 (Bandung : Institute of Technology Bandung) 173 p (In Indonesia)
[13]. Masyahoro and Mappiratu 2010 Respons from a Variety of Different Dept Againts Seeds of Seaweeds and Harvest Time *Eucheuma cottonii*) in Palu Bay. *The Magazine of Litbang Sulteng* **3** (2) 104-111. (In Indonesia)
[14]. Alamsjah M A, N O Ayuningtias and S Subekti 2010 *J. Fish. Mar. Sci.* **2** 21-30

IOP Conf. Series: Earth and Environmental Science **137** (2018) 012017 doi:10.1088/1755-1315/137/1/012017

# The Effect of maceration period on contents and color brightness of phycoerythrin from *Gracilaria* sp.

**H Lidiana[1], L Sulmartiwi[2] and S Andriyono[2]**

[1]Departement of Aquaculture, Faculty of Fisheries and Marine, Universitas Airlangga, Surabaya 60115 Indonesia.
[2]Department of Marine, Faculty of Fisheries and Marine, Universitas Airlangga, Surabaya 60115 Indonesia.

E-mail: laksmi-s@fpk.unair.ac.id

**Abstract.** Natural pigment from seaweed, is currently required by humans as one of active compounds that are usefull in the field of health, cosmetics and food. One of the pigments which potentially can be developed is red phycobilin pigment. The pigment can be obtained from *Gracilaria* sp. throught maceration method. This research aims to find out wheter there is influence from the long maceration time toward to the content and the phycoerytrin color brightness of *Gracilaria* sp. The reseach method used experimental experiment with complete randomized design (CRD) which was consisted of nine treatments of long time maceration period three replications. Data analysis was used the Analysis of Variance (ANOVA) and continued to Duncan's Multiple Distance Test to. The result showed that treatments with the time of maceration for 30 hours was the best treatment on phycoerytrin content about 0.98 mg/g with the brightness value was 19.73 L*, reddish value was 7.23 a*, and yellowish value was 2.87 b* of *Gracilaria* sp.

## 1. Introduction

Pigment in seaweed consists of the main pigments which are chlorophyll and two additional pigments (accessories) in the form of carotenoids and phycobylin. The Main pigment has function for photosynthesis while additional pigments are used to absorb sunlight during photosynthesis process [1]. Other than used by seaweed for photosynthesis, the pigment is currently used by human as one of the active compounds useful in the field of health, cosmetics, and food [2]. One of the pigments that are potential to be developed is phycoerythrin, which is a group of phycobylin protein pigment with red color [3]. Phycoerythrin has potential as natural red food coloring, cosmetics, and also as a source of antioxidants in health supplements [4]. This pigment can be obtained from the macroalgae Rhodophyta, and one of them is *Gracilaria* sp. [5].

Phycoerytrin can be obtained from *Gracilaria* sp. through extraction with maceration method. This method is selected because phycoerythrin has characteristics that are easily degraded by high intensity of direct light and heat [6]. Maceration is extraction method with sample immersion in suitable solvent where active compounds can be obtained without heating process, so it will guarantee the active compound that

IOP Conf. Series: Earth and Environmental Science **137** (2018) 012017 doi:10.1088/1755-1315/137/1/012017

are extracted will not be damaged [7]. Maceration of *Gracilaria* sp. needs an effective maceration time to separate the pigment algae cell so it will affect to the obtained phycoerythrin [8]. This is because the longer time of solvent agent contact with sample than the more phycoerytrin will be dissolved [9]. Research on extraction of phycoerytrin from *Gracilaria* sp. with 24 h maceration has been done and it could obtain phycoerytrin as much as 0.24 mg from every gram of sample [10]. Therefore, in this research, maceration method was used but with longer immersion time to know the maceration time that yield the highest phycoerytrin content and the lowest color brightness from *Gracilaria* sp.

## 2. Methodology

### *2.1. Time and place of research*

This research was held at Educational Laboratory of Faculty of Fisheries and Marines, Airlangga University, Surabaya included maceration of *Gracilaria* sp., determination of phycoerytrin contents, and observation of phycoerytrin color brightness.

### *2.2. Tools and ingredients*

Tools used in this research were dark colored bottle, scissors, mortar and crusher, digital scales, measuring cup, pH meter, refrigerator, UV-Vis spectrophotometer, centrifuge, and color reader. Materials used in this research were red algae *Gracilaria* sp., distilled water, acetone, and phosphate buffer solution with pH 6.8 ($NaH_2PO_4.H_2O$ and KOH).

### *2.3. Working procedures*

Seaweed was obtained from traditional farmers at Jabon Village, Sidoarjo. Obtained seaweed was then inserted into polybags and stored in cool box filled with ice. Seaweed was then washed from dirt, cut into small pieces and then crushed with mortar and crusher. To make solution of buffer phosphate with pH 6.8, it was started by dissolving 13.79 g$NaH_2PO_4.H_2O$ in distilled water till the border line of 1 L measure cup. Then, $NaH_2PO_4.H_2O$ solution was added by KOH gradually till pH 6.8 [11]. Two grams of *Gracilaria* sp. wet sample that was already crushed with mortar and crusher were inserted into the test tube. After that, 5 mL of phosphate buffer solution was given to the extracts the phycoerytrin, while for chlorophyll extraction 5 mL of acetone were given [10]. Next, samples were being macerated according to the treatments (6, 12, 18, 24, 30, 36, 42, and 48 h) and stored in the refrigerator at 4 °C [12]. Sample without maceration (0 h) was directly calculated for the content of phycoerytrin and chlorophyll.

### *2.4. Test procedures*

a. Calculation of *Gracilaria* sp. Phycoerytrin Content

Phycoerytrin content was calculated from every solvent used, based on the following formula:

$$ {}^{mg}/_{ml}\ Phycoerytin = 0.1247[(A_{564} - A_{730}) - 0.4583(A_{618} - A_{730})] \quad (1) $$

Description:
A = Absorbance value at the set wavelength

b. Calculation of *Gracilaria* sp. Chlorophyll Content

Chlorophyll content was calculated following the formula from [13] to know the concentration of the chlorophyll in every solvent:

IOP Conf. Series: Earth and Environmental Science **137** (2018) 012017 doi:10.1088/1755-1315/137/1/012017

$$mg/ml\ Chlorophyll\ a = (11.75\ x\ A662) - (2.350\ x\ A645) \quad (2)$$

Description:
A = Absorbance value at the set wavelength

c. Calculation of Phycoerytrin Production and Chlorophyll of *Gracilaria* sp.
Calculation of Phycoerytrin Production and Chlorophyll were obtained to know the phycoerytrin and chlorophyll content in every sample:

$$Pigment\ Production\left(\frac{mg}{g}\right) = \frac{CxV}{M} \quad (3)$$

Description:
C = Phycoerytrin Content (PE) or Chlorophyll (Chl) (mg/mL)
V = Extract volume of phycoerytrin or chlorophyll (mL)
M = Sample weight (mg)

*2.5. Data Analysis*
This research used Completely Randomized Design (CRD) consisted of nine treatments and three replications. Treatment in this research was maceration time for 0, 6, 12, 18, 24, 30, 36, 42, and 48 h to obtain phycoerytrin from *Gracilaria* sp. Data analysis in this research used Analysis of Variance and continued with Duncan's Multiple Range Test to determine which treatments gave different effects[14].

## 3. Results and Discussion
Table 1 shows that every treatment of maceration time gave significantly different effect to phycoerytrin content ($p<0.05$). Statistical analysis was then continued to Duncan test and it showed that the highest phycoerytrin content on *Gracilaria* sp. was obtained from treatment I (48 h maceration) with 1.56 mg/g, while the lowest phycoerytrin content was obtained from treatment A (0 h) with 0.41 mg/g. According to [7], longer time of maceration will increase the extraction volume. This is because longer maceration time allows longer contact of solvent with sample, so it will balance concentration of the solution inside and outside of the extracted ingredients [15]. Table 2 shows that every treatment of maceration time gave significantly different effect to phycoerytrin content ($p<0.05$). Analysis was then continued with Duncan test and the results showed that the maceration time that gave bright color phycoerytrin was treatment F (30 h) with high value of brightness and redness. High value of redness indicated high phycoerytrin got dissolved during maceration. Immersion for 30 enough provided sufficient time to dissolve phycoerytrin maximally. Limit time of contact happened after the creation of equilibrium states in the cell to dissolve phycoerytrin which is increasing the maceration time to dissolve other pigment. Improved maceration time dissolved other pigments so it affected the increase of dissolved pigment content and the decrease of obtained color brightness.

**Table 1**. Phycoerytrin content of *Gracilaria* sp. (mg/g) at different maceration time.

| Treatment (hour) | Average of phycoerytrin content (mg/g) |
|---|---|
| A (0) | $0.41^{a} \pm 0.02$ |
| B (6) | $0.58^{ab} \pm 0.03$ |
| C (12) | $0.62^{ab} \pm 0.02$ |
| D (18) | $0.74^{bc} \pm 0.28$ |
| E (24) | $0.95^{cd} \pm 0.05$ |
| F (30) | $0.98^{cd} \pm 0.26$ |
| G (36) | $0.99^{cd} \pm 0.16$ |
| H (42) | $1.04^{d} \pm 0.08$ |
| I (48) | $1.56^{e} \pm 0.01$ |

Note: different alphabet notation on the graphic shows significant differences among treatments (p<0.05).

There are other factor that can affect the increase of phycoerytrin content during maceration other than contact time and types of solvent; they are light intensity, solvent, pH and temperature during maceration. chromoprotein (polypeptide α and β) as pigment composer is very sensitive to temperature, light intensity, and pH so damaged done on the cell is not followed by pigment denaturation during extraction [16]. Those factors can affect the content of phycoerytrin and color brightness.

**Table 2** Phycoerytrin Color Brightness of *Gracilaria* sp. (L*) in different maceration time.

| Treatments (hour) | Average value of brightness (L*) | Average value of reddish (a*) | Average value of yellownish (b*) |
|---|---|---|---|
| A (0) | $22.57^{i} \pm 0.06$ | $5.10^{b} \pm 0.17$ | $2.43^{a} \pm 0.06$ |
| B (6) | $19.20^{d} \pm 0.00$ | $6.00^{de} \pm 0.10$ | $2.77^{b} \pm 0.15$ |
| C (12) | $19.03^{c} \pm 0.06$ | $5.87^{cd} \pm 0.15$ | $2.77^{b} \pm 0.11$ |
| D (18) | $19.97^{g} \pm 0.15$ | $5.67^{cd} \pm 0.25$ | $3.33^{cd} \pm 0.15$ |
| E (24) | $18.57^{b} \pm 0.06$ | $5.50^{c} \pm 0.10$ | $3.30^{cd} \pm 0.10$ |
| F (30) | $19.73^{f} \pm 0.06$ | $7.23^{f} \pm 0.06$ | $2.87^{b} \pm 0.06$ |
| G (36) | $20.47^{h} \pm 0.12$ | $5.97^{de} \pm 0.11$ | $4.23^{e} \pm 0.15$ |
| H (42) | $19.43^{e} \pm 0.06$ | $6.47^{e} \pm 0.06$ | $3.53^{d} \pm 0.06$ |
| I (48) | $15.53^{a} \pm 0.15$ | $2.97^{a} \pm 0.25$ | $3.17^{bc} \pm 0.23$ |

Note: different alphabet notation on the graphic shows significant differences among treatments (p<0.05).

Ttreatment I (48 h) had the highest phycoerytrin content and the lowest brightness value (L*) , but it was not followed by increase in high reddish value (a*). Obtained a* value was low and close to greenish scale value (0 to -80), so it could be stated that the dissolved pigments was not only phycoerytrin. Dissolved pigments could be chlorophyll, carotenoids, as well as other phycobylin. Other dissolved pigments other than phycoerytrin can affect the decrease of brightness value because there were other color additions. *Gracilaria* sp. has red colored thallus that is the main characteristic of phycoerytrin pigment, but *Gracilaria* sp. also has chlorophyll pigment as its main pigment [4]. Chlorophyll is the main pigment found in every organism that performs photosynthesis [17]. Rhodophyta, a phycoerytrin pigment, will absorb sunlight and forward it to chlorophyll α to maximize photosynthesis process [1]. This is because the ability to chromatic adaptation of Rhodophyta to change characteristic of light absorption

IOP Conf. Series: Earth and Environmental Science **137** (2018) 012017 doi:10.1088/1755-1315/137/1/012017

during photosynthesis according to light source in the environment. After that, the pigment will absorb the strongest wavelength to become the dominant pigment.

**Table 3** Chlorophyll content of *Gracilaria* sp. (mg/g) on different maceration time.

| Treatments (hours) | Average content of chlorophyll (mg/g) |
|---|---|
| A (0) | $0.16^{a} \pm 0.05$ |
| B (6) | $0.19^{a} \pm 0.05$ |
| C (12) | $0.29^{ab} \pm 0.07$ |
| D (18) | $0.33^{ab} \pm 0.07$ |
| E (24) | $0.39^{bc} \pm 0.08$ |
| F (30) | $0.44^{bcd} \pm 0.06$ |
| G (36) | $0.54^{cde} \pm 0.18$ |
| H (42) | $0.61^{de} \pm 0.13$ |
| I (48) | $0.69^{e} \pm 0.69$ |

Description: different alphabet notation on the graphic shows significant differences among treatments ($p<0.05$).

Table 3 shows that longer maceration time also gave effect to chlorophyll content of *Gracilaria* sp. The highest chlorophyll content was obtained from treatment without maceration for 48 h, while the lowest chlorophyll content was obtained in treatment without maceration. Chlorophyll content in *Gracilaria* sp. is lower than phycoerytrin content. According to Veronica [18], chlorophyll content is relatively lower than phycoerytrin in Rhodophyta.

## 4. Conclusion

Different maceration time gave significantly different effect on phycoerytrin content and color brightness from *Gracilaria* sp. Treatment with 30 h of maceration time was the best treatment that obtained phycoerytrin as much as 0.98 mg/g and brightness value of 19.73 L*. reddish value of 7.23 , and yellowish value of 2.87 from *Gracilaria* sp.

## 5. References

[1] Mustofa 2013 *Light spectrum effect on Gracilaria verrucosa growth* (Jember: University of Jember) p79
[2] Karseno H I, Setyawati R 2013 *Agritech.* **33** 371–375
[3] Abfa I K, Prasetyo B, Susanto A B 2013 *Characteristics of ficoeritrin as pigment accessories in red seaweed and its benefits* (Semarang: Semarang State University) p 7
[4] Ichwan, Zakki, Winarti, Yayuk 2014 *Thallophyta (Alga)* (Semarang: Semarang State University) p 16
[5] Merdekawati, Windu, Susanto A B 2009 Content and composition of seaweed pigments and its potential for health Squalen **4** 41 - 45
[6] Agustini, Ni Wayan S 2013 *Antioxidant activity and biometric toxicity test of phycobiliprotein pigment from spirulina platensis extract* (Semarang: Semarang State University) p 7
[7] Nurdiansyah, Abdi R 2011 *J. Belian* **10** 218-224
[8] Anggiary R D 2012 *Comparison of cell density and synthesis of chlorophyll Synechococcus sp. RDB001 grown at 30 ± 5 °Cand 50 ± 5 °C* (Jakarta: University of Indonesia) p 93
[9] Sudarmi S, Subagyo p, Susanti A, Wahyuningsih A S 2015 *Simple anthocyanin extraction from dragon fruit skin (hylocereus polyrhizus) as natural dye* Eskergi **12** 5 - 7

[10] Sudhakar M P, Saraswathi M, Nair B B 2014 *J. Agardh var. corticata. Indian J. Nat. Prod. Resour.* **5** 371-374
[11] Wiley, Priya S, Christopher D, Neefus 2007 *J. Appl. Phycol.* **19** 123-129.
[12] Jos B, Setyawan P E, Satria Y 2011 *J. Tek.* **32** 1-7.
[13] Lichtenthaler H K, Wellburn A R 1983 *Biochem. Soc. Trans.* **11** 591-592.
[14] Kusriningrum R S 2012 *Experimental design* (Surabaya: Airlangga University) p 43-63 (Text in Indonesia)
[15] Wahyuni D T, Widjanarko S B 2015 *J. Pangan dan Agroindustri* **3** 390-401
[16] Farihah S, Yulianto B, Yudiati E 2014 *J. Mar. Res.* 140-146
[17] Suparmi, A Sahri 2009 *Sultan Agung XLIV* 105 – 10
[18] Veronica, Izzati M 2009 *J.Anat. Fisiol.*

IOP Conf. Series: Earth and Environmental Science **137** (2018) 012018 doi:10.1088/1755-1315/137/1/012018

# Culture of *Daphnia* sp. (crustacean – cladocera): the effect of manure variation on the growth, natality, and mortality

**H Herman, Y Andriani, A Sahidin, T Hidayat and T Herawati**

[1]Fisheries Study Program, Faculty of Fisheries and Marine Science, Universitas Padjadjaran, 45363 UNPAD Jatinangor Sumedang, Indonesia

E-mail: hamdani_herman@yahoo.com

**Abstract.** The objective of this research was to analyze the growth rate, reproduction rate, and mortality rate of *Daphnia* sp. which cultured in variant organic manure. This research used experimental method Randomized Completely Block Design (RCBD) with four treatment and three replications. The treatments in this research were the variant manures from chicken, quail, goat, and cow with same of growth (2,4 g/l). *Daphnia* cultured with using 100 breeders since from neonets (0 day) until growth up and died in one life cycle. At the 3-days, culture of *Daphnia* sp was give peak population with maximum age of culture using quail manure is 7 days, and other treatments are 8 days . The growth rate and the reproduction rate of using quail manure was higher than using chicken manure, goat manure, and cow manure (mean GR = 3.68 : 2.32 : 2.74 : 2.97; mean RR = 3.87 : 2.59 : 3.00 : 3.31; $p < 0,05$). Although all the breeders of *Daphnia* sp. died at 8$^{th}$ day of culture, quail manure give the lowest of mortality rate than using chicken manure, goat manure, and cow manure (mean MR = 0.19 : 0.28 : 0.26 : 0.34).

## 1. Introduction

*Daphnia* sp. is the best natural feed for tropical fry fish and it has become an alternative for natural fish feed [1]. The advantages of using *Daphnia* for aquaculture are its low cost and convenient to culture [2]. *Daphnia* is one of micro crustaceans from *Brachiopoda* family and it lives in fresh water [3]. All individual of *Daphnia* are females, but when the environmental quality is poor, *Daphnia* will produce male for sexual reproduction [4].

Growth and development of *Daphnia* sp. s affected by nutrient, age, temperature, respiration, and filtered particle shape [5]. *Daphnia* sp. has two reproduction methods, parthenogenesis and sexual. Parthenogenesis occurs when the environment is suitable for *Daphniai*'s life [1]. *Daphnia* is non-selective filter feeders and it is able to eat bacteria, phytoplankton, ciliate, and detritus [6]. The food is collected by filtering apparatus, while the legs make some water to flow from anterior to posterior and move the particle to the gut by setae [4].

Natural feed brings risks for of diseases and heavy metal accumulation [7]. To minimize the risk, culturing the natural feed is the safest way. Other advantages of using natural feed culture are its high quality, parasite-free or pollution-free characteristic, and sustainability. More importantly, *Daphnia* sp. lives in aquaculture media without affecting water quality [6]. There are many food sources for cultured *Daphnia,* such as organic manure, particularly chicken manure [8], cow manure, and goat manure. Nutrient from poultry waste contains nitrogen (N) 1.48 % and phosphorus (P) 0.4 %, cow manure contains nitrogen (N) 2.03 %, and phosphorus (P) 0.21 %, and goat manure contains nitrogen

IOP Conf. Series: Earth and Environmental Science **137** (2018) 012018 doi:10.1088/1755-1315/137/1/012018

(N) 2.10 %, and phosphorus (P) 0.11 % [9]. Manure in the water will produce bacteria, detritus, and other materials, and become food for *Daphnia* sp. [4]. The abundance of food source in culture media will increase the growth of *Daphnia*. In good water condition with lack of food, *Daphnia* sp. use parthenogenesis as the main reproduction pathway. Parthenogenesis produces many of neonates and short of performance [10]. Each individual of *Daphnia* can produce 17 – 20 neonates [5]

Ammonia is a key for *Daphnia* culture, which is essentially the main component in animal waste. Ammonia comes in the form of uric acid in poultry manure and urea in mammals' manure [11]. Ammonia is very toxic for all organisms, but in acidic condition, the toxicity is decreased, therefore pH monitoring is required for *Daphnia* culture [12].

The best method to culture a high population of *Daphnia carinata* is within 9 days with 500 ppt of chicken manure [8]. Another research from Sulasingkin in Izza *et al.* [13] reported the culture of Indonesian *Daphnia* using chicken manure with the dose of 12 g/5 L or 2.4 g/L [13]. With those theories, the hypothesis of this study was:

$H_0$: The best manure for culture *Daphnia* sp. is chicken manure

$H_1$: The best manure for culture *Daphnia* sp. is not chicken manure

Using organic manure for *Daphnia* culture gives good performance in terms of growth rate, natality rate, and mortality rate [14]. The biological factor of organic manure to culture *Daphnia* sp. is now known comprehensively yet. The objective of this research was to analyze the growth rate of *Daphnia* sp. cultured in different nutrient from variety of organic manure.

## 2. Methodology

### *2.1. Research design and formula.*

This research used Randomized Complete Block Design (RCBD) with four treatments and three replications. The treatments were the types of manure, which were from chicken, quail, goat, and cow with same growth (2.4 g/l). If the result was significant, data would be analyzed with Honestly Significant Difference test. Growth rate of *Daphnia* sp. was calculated using a formula taken from Birch [15].

$$\sum e^{7-rx} l_x m_x = 1097 \tag{1}$$

| | |
|---|---|
| e | = natural logarithmic value |
| r | = growth rate |
| x | = maintenance time |
| lx | = life individual in the time (x) |
| mx | = neonates F1 in the time (x) |

Natality rate of *Daphnia* sp. was calculated using a formula:

$$b = \frac{r\beta}{e^r - 1} \tag{2}$$

$$1/\beta = \sum L_x e^{-r(x+1)} \tag{3}$$

| | |
|---|---|
| B | = natality rate |
| Lx | = (lx + lx+1)/2: average of individual in KU-x and KU-x+1 |
| e | = natural logarithm |
| r | = growth rate |

IOP Conf. Series: Earth and Environmental Science **137** (2018) 012018 doi:10.1088/1755-1315/137/1/012018

Natality rate of *Daphnia* sp. was calculated using the follow formula:

$$d = b - r \tag{4}$$

| | |
|---|---|
| d | = mortality rate |
| b | = natality rate |
| r | = growth rate |

*2.2. Research procedure.*
The research consisted of preparation, cohort formation, and main research. We used *Daphnia* with similar age for this research. There were two methods of sample preparations. First, manure was be dried and weighed 2.4 gram. Next, the manure was wrapped and tied. The second preparation was media preparation. One L of water was put in a container and installed on the aeration set. after the manure was added, the media was aerated for 3 days.

To make cohort formation, mature *Daphnia* sp. (F0) was cultured in media for 1 day. The new neonates (F1) were taken to another culture media. The neonates (F1) were cultured to adulthood (F2). The new generation (F2) was considered of the same age and used for the main research.

The second generation (F2) of *Daphnia* sp. was added in every treatment at a dose of 100 individual/L. *Daphnia* (F2) was counted everyday and moved into new media. If neonates (F3) appeared, they were (F3) separated from the previous media and counted in different place to avoid the neonates (F3) from growing up and getting difficult to be distinguished. The main research was carried out when all of the second generation of *Daphnia* (F2) died.

*2.3. Water quality.*
Water quality is an important key of *Daphnia* sp. culture. Water quality was checked every day (Table 1); involving dissolved oxygen (DO), pH, and temperature. Ammonia was checked three times: the first sample was taken from the early culture, the second sample was taken at the first spawning (4$^{th}$ day), and the third one was taken from the last day.

**Table 1.** Water Quality Control Method.

| Parameter | Unit | Method | Tools |
|---|---|---|---|
| Temperature | °C | Reading of scale | Termometer ($\pm$ 0.01 carefulness) |
| DO | mg/l | Reading of scale | DO-meter ($\pm$ 0.01 carefulness) |
| pH | - | Reading of scale | pH-meter ($\pm$ 0.1 carefulness) |
| $NH_3$ | mg/l | Nessler method | Spectofotometer ($\pm$ 0.001 carefulness) |

## 3. Results and Discussion

*3.1. The number of Daphnia sp. (F2)*
Parent of *Daphnia* sp. (F2) decreased since 1$^{st}$ day of culture. The first day was the adaptation stage for *Daphnia* sp. to grow. Adaptation phase in culture media started from Day 0 to 3 [16]. Declamation of population occurred linearly until all of the population (F2) died. All *Daphnia* sp. from quail treatment died on 8$^{th}$ day, while those of the other treatment died on the 10$^{th}$ or 11$^{th}$ day. This condition was in accordance with Deckaestacker *et al.* [10] who reported 9 – 11 days as the generation time of *Daphnia* sp.. Life span of *Daphnia* sp. from being in brood chamber to adulthood and death depends on the species and environmental condition [12].

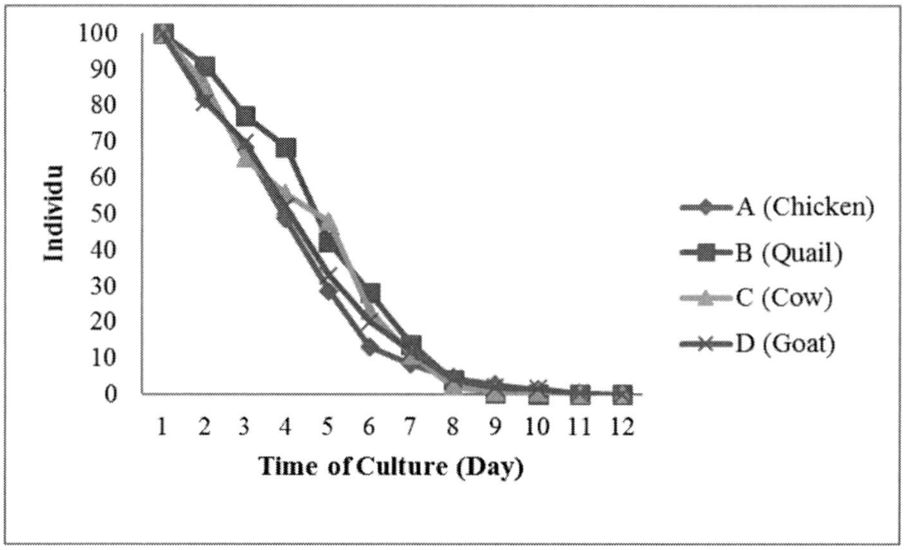


**Figure 1.** Parent of *Daphnia* sp. per-day.

The difference in culture time caused *Daphnia* sp (F2) from quail manure to produce more neonates than other treatment. Continuous reproduction can reduce the capacity of brood chamber or carapax to recuperate and consequently *Daphnia* parent dies faster. Reproduction of *Daphnia* sp. from other manure was lower based on the number of neonates (F3). The birth of F3 in every treatment created competition for food. Deficiency of food in medium affects the life of *Daphnia* [17]. The high density of *Daphnia* sp. in media creates a competition for survival [6].

Using quail manure shortened the culture time (x = 8 days) compared to using other manures (figure 2). Quail manure gave more food for *Daphnia*, thus they bred more than normal situation, resulting in large number of neonates in every reproduction and high frequency of breeding. The high reproduction rate and total neonates in every reproduction decrease the energy needed for *Daphnia* sp. to maintain their life, metabolism and also to recover the brood chamber; thus, the situation shortened the age of *Daphnia* sp..

### *3.2. Neonate (F3) of Daphnia*

The neonates (F3) of *Daphnia* were related with reproduction of F2. The first and the second day, parent (F2) did not produce neonates yet because F2 was too young. The first reproduction of *Daphnia* sp. (F2) occurred on the third day. *Daphnia* sp. produce eggs every 3-4 day until they die [4]. The highest neonates in the culture occurred on the third day with 104 individuals for chicken manure, 651 individuals for quail manure, 262 individuals for cow manure, and 250 individuals for goat manure. The high population indicated that the condition of the culture media was well accepted by *Daphnia* sp. number of neonates declined every day after population peak. Reproduction stopped on the 9$^{th}$ until 11$^{th}$ day. This is in accordance with Deckaestacker *et al.* who claimed the generation time of *Daphnia* sp. is between 9- 11 days in 20°C [10].

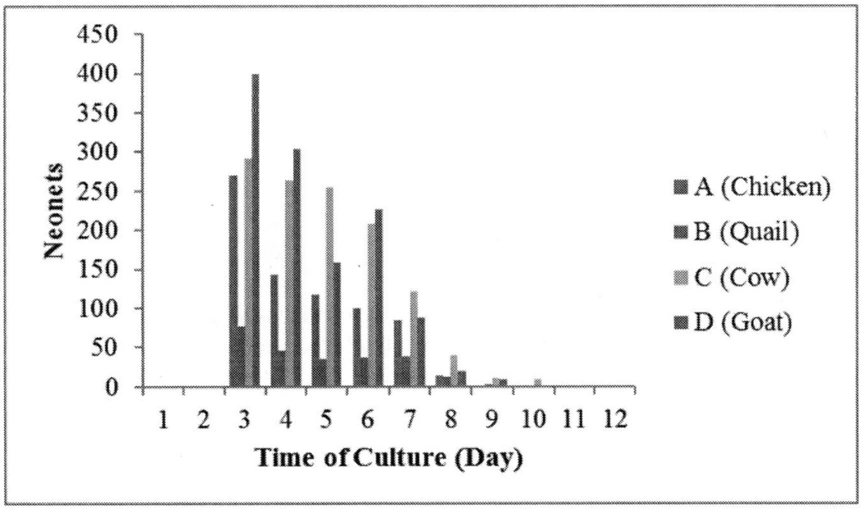


**Figure 2.** Neonates of *Daphnia* sp. per-day.

The difference of manure affecting the quantity of neonates. Quail manure gives the highest of neonates between other treatments. Quail media give many of food sources for *Daphnia* sp, so that the growth of *Daphnia* and the reproduction will be faster and higher than a media with poor of food source. Quail manure give chance for *Daphnia* to make more reproduction frequency and neonate's quantity than organic manure. The abundant of food source give a good chance for *Daphnia* to make a reproduction with parthenogenesis, but if the food source is low, *Daphnia* sp. will produce male *Daphnia* for sexual reproduction. Parthenogenesis gives more of neonates then sexual reproduction in *Daphnia* [10]. Neonate's density from parthenogenesis is different according to the food.

*3.3. Life table of Daphnia*

Growth rate counted by Birch [15], formula with counting the entire individual of *Daphnia* sp. in culture media. Growth rate of *Daphnia* sp. in the treatment give significantly ($p < 0.05$) with correlation coefficient amount 0.02. Tukey test method ($p < 0.05$) gives significant level for B treatment (using quail manure) for other treatment. Natality rate and mortality rate were calculated also with insignificant value ($p > 0.05$). The result showed in table 2.

**Table 2.** Life table of *Daphnia* sp..

| n | Growth Rate | Natality Rate | Mortality Rate | Culture Time (x) |
|---|---|---|---|---|
| A | 2.32 ± 0.14 | 2.59 ± 0.08 | 0.28 ± 0.08 | 10,0 ± 0,57 |
| B | 3.68 ± 0.16 | 3.87 ± 0.08 | 0.19 ± 0.24 | 8,0 ± 0,57 |
| C | 2.74 ± 0.79 | 3.00 ± 0.78 | 0.26 ± 0.10 | 9,0 ± 1,15 |
| D | 2.97 ± 0.27 | 3.31 ± 0.15 | 0.34 ± 0.13 | 10,0 ± 1,00 |

The highest growth rate was shown by treatment with quail manure, with mean value of 2.23 ± 0.07, followed by goat manure 1.8 ± 0.06, cow manure 1.76 ± 0.08, and chicken manure 1.48 ± 0.05. Those numbers mean how much *Daphnia* could survive in the environmental condition. For example, quail manure from B2 gave growth rate of 2.31, which means that the population would increase by amount 2.31 times. Treatment B also gave significant results compared to other treatments. The difference was impacted by some factor, including food availability, reproduction frequency, and water quality of culture media.

Growth rate and natality rate gave similar results. treatment with Quail manure showed the highest natality rate at 2.37 ± 0,07, followed by goat 1.99 ± 0.44, cow 1.98 ± 0.28, and chicken 1.10 ± 0.5. Natality rate gives probability of recruitment for the new generation of the parent. Quail treatment gave more food for *Daphnia* sp, so that it affected the growth of *Daphnia* and provided impulse to breed. According to nitrogen test, quail manure contained N-total of 2.86 %, while the N-total from chicken manure was 2.8 %. Meanwhile, cow manure and goat manure had similar N-total of 1.97%. Nitrogen functions as one of nutrients needed for the growth of microbes. Decomposition of organic material from manure helps the growth of bacteria that later becomes food for *Daphnia* sp. [13].

Food also affects mortality rate. The lowest mortality rate was from quail manure with 8.0 ± 0.57. Although the age of the parent was shorter than in other treatments, quail manure gave more neonates and affected mortality rate. The high amount of food can extend the life of *Daphnia* [17]. If the environment is not hospitable, growth, development, and metabolism of *Daphnia* sp. decrease [18].

Chicken, cow, and goat manures gave longer culture time than quail manure, but it had lower growth and natality rate and higher mortality rate. Decomposing Microbes in other manures were lower than in quail manure, so that the growth of *Daphnia* was not optimum. Declined filtration rate at 54% after 70 hours was due to low concentration of feed (0.025 mgC/L) [17]. Declamation of filtration rate decreases the population growth and reproduction rate. Nevertheless, *Daphnia* sp. can still survive and adapt, but more death occurs [18].

With these results, we ignored the hypothesis ($H_0$) that claims the best of organic manure to culture *Daphnia* was chicken manure. We found a new and better manure to culture *Daphnia* that was quail manure, with high growth rate and natality rate, and low mortality rate.

### *3.4. Water quality*

Growth rate of *Daphnia* sp. was affected by some factors, such as food source, species, and environment. Environment control provides opportunity to make the media anoxic. Temperature, DO, and pH in every treatments were controlled with temperature ranged from 24.1 to 25.7 °C, DO ranged from 3.6 to 4.2 mg/L, and pH ranged from 7.5 to 8.2.

**Table 3.** Water quality of media culture.

| n | Temperature (°C) | DO (mg/l) | Ph | Ammonia (mg/l) |
|---|---|---|---|---|
| A | 24.1 – 25.6 | 3.7 – 4.2 | 7.7 – 8.1 | 0.08 – 0.15 |
| B | 24.3 – 25.6 | 3.6 – 4.1 | 7.6 – 8.3 | 0.10 – 0.38 |
| C | 24.1 – 25.7 | 3.7 – 4.2 | 7.5 – 8.0 | 0.01 – 0.13 |
| D | 24.2 – 25.7 | 3.7 – 4.2 | 7.6 – 8.2 | 0.01 – 0.13 |

Temperature of media = affects metabolism including for *Daphnia* as an aquatic organism. Cultured *Daphnia* sp. has good performance at temperature of 20 °C [19]. Khan and M Khan [20] stated that *Daphnia* has optimum life span at 25 °C. Ocampo *et al.* [1] suggested 20- 25 °C as the temperature for *Daphnia* culture to have normal life span. Oxygen in culture media supports respiration of *Daphnia* sp. without creating competition between *Daphnia* and decomposing microbes. DO in the culture was controlled by aeration to be within the tolerance range for *Daphnia*. *Daphnia* sp. can survive in DO > 3 mg/L, but they grow better at 6 mg/L [1]. Oxygen also gives chance for microbes to perform their metabolism and decompose organic matter in media, so that the water quality is good.

The power of hydrogen (pH) affects eggs life span zooplankton *microcrustacean* [6]. According to Ocampo *et al.* [1], optimum pH for *Daphnia* culture is between 7- 8.6. Darmawan [6].stated that neutral or base (pH 7.0- 8.2) is good for *Daphnia magna* growth. pH determines the toxicity of medium from ammonia.

Ammonia is the first substance from organic matter decomposition. Ammonia is divided into un-ionized ammonia ($NH_3$) and ionized ammonia ($NH_4^+$) [21]. Ammonia in water gets more toxic with increased pH [11]. According to Delzer and McKenzie [22], nitrogen decomposition from ammonia: nitrite, and nitrate, were formed by microbes starting from the 6$^{th}$ day at temperature of 20 °C. Normally, decomposition is divided into two stages: carbonaceous (carbon decomposition that starts from day 0) and nitrogenous (nitrogen decomposition). Existence of Ammonia in media is good for *Daphnia* culture (table 3). Although there was fluctuation, but *Daphnia* still survived in every treatment. in this research, Ammonia was not the prime factor that affected mortality rate. *Daphnia* can maintain life by adapting even at the worst condition [18]. The main factor that affects the mortality of *Daphnia* culture is frequency of reproduction, even though the water condition has contribution too.

This current study used manure that was soaked for three days. the soaking time was not sufficient for ammonia decomposition, because decomposition of ammonia starts from so that the concentration of ammonia in the treatments was high, particularly in quail manure. The lowest concentration for un-ionized ammonia that gives effect for *Daphnia* sp. is 1.3 mg/L [23]. Concentration of ammonia in treatment with quail manure was between 0.10- 0.38 mg/L. The high ammonia provided opportunity for decomposing microbes to perform their function and then they became food for *Daphnia*. Cow manure, goat manure, and chicken manure gave lower concentration of ammonia (table 3) than quail manure, so that lower bacteria was available as food for *Daphnia*.

## 4. Conclusion

Growth rate and mortality rate of *Daphnia* sp. cultured using quail manure was higher than chicken, cow, and goat manures. Although all the breeders of *Daphnia* sp. died on the 8$^{th}$ day of culture, quail manure gave the lowest of mortality rate. thus, null hypothesis ($H_0$) was ignored ($H_a$) was accepted.

## 6. References

[1] Ocampo L, M Botero, and L F Restrepo 2012. Measurements population growth and fecundity of *Daphnia magna* to different levels of nutrients under stress condition *Intech Open; Aquaculture*, Dr. Zainal Muchilisin (Ed). ISBN 978-953-307-974-5.
[2] Ernwati and Rekna 2007 *Jurnal Yudharta* **10**
[3] Pangkey H 2009 *J. Fish. Mar.* **5** 33-37
[4] Ebert D 2005 *Ecology, Epidemiology, and Evolution of Parasitism in Daphnia* (USA: National Center for Biotechnology Informatio)
[5] Lawrence S G 1981 *Manual for Culture of Selected Freshwater Invertebrates* (Canada: Canadian Special Publication of Fisheries and Aquatic Sciences) **54** 169 p.
[6] Darmawan J 2014 Population growth of *Daphnia* sp. in aquaculture media with catfish wastewater addition (*Clarias gariepinus* Burchel 1822) *Berita Biologi* **13** 57
[7] Celia *et al* 2000 *Limnol. Oceanogr.* **45** 1525-1536
[8] Paray B and M K Sadoon 2016 *Indian J. Geomarine. Sci.* **45** 399-404
[9] Damle D K and M S Chari 2011 *J. Fish. Aquat. Sci.* **6** 57 – 6
[10] Deckaestacker *et al.* 2009 Cyclical parthenogenesis in *daphnia*: sexual versus asexual reproduction *Researchgate* Chapter September 2009
[11] Gay S W and K F Knowlton 2009 *Ammonia emission and animal agriculture* (Virginia: Cooperative Extension) Publication 442 – 110
[12] Clare J 2009 *Daphnia: An Aquarist's Guide*. Caudata.Org.
[13] Izzah N 2014 *J. Aquat. Manag. Tech.* **3** 44 – 52
[14] Utarini D, Carmudi and Kusbiyanto 2012 Population growth of *daphnia* sp. in combination media between quail manure and chicken manure with different density *Proc. Conf. Nat. Rural Resources Development and Sustainable Local Wisdom II.* ISBN: 978-979-9204-79-0, November: p 57 [Text in Indonesia)
[15] Birch L C 1948 *J. Anim. Ecol.* **17** 15-26

IOP Conf. Series: Earth and Environmental Science **137** (2018) 012018 doi:10.1088/1755-1315/137/1/012018

[16] Zahidah, W Gunawan and U Subhan 2012 *J. Aquat.* **3** 84-94
[17] Plath K 1998 *Limnol. Oceanogr.* **43** 593 – 599
[18] Maier R M 2009 *Bacteria Growth* Environmental Microbiology Academic Press Inc.
[19] Ranta E *et al* 1993 *Ann. Zool. Fennici.* **30** 299 – 311
[20] Khan Q and M Khan 2008 Effect of temperature on waterflea *Daphnia magna* (Crustacea: Cladocera) *Respository Nature Precidings* University of Illinois Chicago.
[21] Sriyasak P *et al.* 2015 *Int. Aqua. Res.* **7** 287 – 299.
[22] Delzer G C & S W McKenzie 2003 *Five-Day biochemical oxygen demand* USGS TWRI Book 9 – A7 3rd ed.
[23] Reinbold K A and S M Pescitelli 1990 *Effect of Exposure to Ammonia on Sensitive Life Stages of Aquatic Organism: Center For Aquatic Ecology* (Chicago: University of Illinois)

**Acknowledgement**
sincere appreciation goes to Rector Universitas Padjadjaran for Internal Scheme Grant Funds for Fundamental Research in Universitas Padjadjaran.

IOP Conf. Series: Earth and Environmental Science **137** (2018) 012019 doi:10.1088/1755-1315/137/1/012019

# The effects of salinity and temperature shock on *Kappaphycus alvarezii* seaweed spores release

**F K Harwinda[1], W H Satyantini[2], E W Masithah[3]**

[1] Aquaculture,Faculty of Fisheries and Marine Airlangga University,Surabaya60115
[2] Departement of Fish Health Management and Aquaculture, Faculty of Fisheries and Marine Airlangga University,Surabaya60115
[3] Departement of Marine,Faculty of Fisheries and Marine Airlangga University,Surabaya60115

E-mail: farizkukuh@hotmail.com

**Abstract.** One of the reproductive aspects of development step that is considered as the solution of this issue is seaweed sporulation technique through which is induced through salinity and temperature shock. This study aims to determine the effect of combination and interaction of salinity and temperature shock on the release of K. alvarezii spores in order to produce superior seeds. This research was conducted using Complete Randomized Design Factorial which consists of nine combinations of treatments and three replications. The used treatment in this study is the combination of different environmental factors such as salinity shock and temperature shock. The data were analyzed using ANOVA (Analysis of Variance) followed by Duncan Multiple Range Test. The results showed that salinity (31 ppt, 33 ppt, and 35 ppt) and temperature (30°C, 32°C, and 34°C). shock affected the osmoregulation system and the release of K. alvarezii spores. The salinity shock and temperature shock had interaction with K. alvarezii spore release on the sixth and seventh day with the best treatment at 32°C temperature and 31 ppt salinity and released 5413 cells/ml spores on the seventh day.

## 1. Introduction

Seaweed as a fishery commodity that produces carrageenan has an important role in the industrial sector. It is much utilized in the food industry, cosmetics, pharmaceutical, and other industries such as the paper industry, textiles, photography, pasta, and fish canning [1]. The current market demand for Kappaphycus alvarezii has increased each year in line with the development of seaweed processing industry in Indonesia. The Ministry of Marine and Fisheries expects that the cultivation of seaweed and the processing industry sector can be developed to compete in the world market. [2] stated the efforts to increase the production of *K. alvarezii*require the availability of superior seaweed seed continuously.

Since 1970, the seed of *K. alvarezii* has always been obtained from vegetative reproduction. The repetitive clonal of vegetative development will result in a reduction in genetic variability, leading to the fall in the growth rate, levels of carrageenan, and gel strength [2]. Based on that reason, the development of reproductive aspects needs to be considered in order to increase productivity. One of the aspects of reproductive development step is the growing of seed through seaweed sporulation. In addition, [3] stated that the reproduction using spores is considered to be more efficient, fast-growing,

IOP Conf. Series: Earth and Environmental Science **137** (2018) 012019 doi:10.1088/1755-1315/137/1/012019

and capable of producing superior seeds. It is also explained that there is a possibility of genetic engineering that could be done through spores reproduction [4]

However, reproduction by using spores depends on the season in which the release of spores is very rare. The reproductive cycle of *K. alvarezii* happens only once a year with the release of spore in August to September [5]. Therefore,research on the implementation of environmental manipulation in the provision of sustainable seaweed seeds throughout the year needs to be conducted.

It is explained that the release and germination of *K. alvarezii* seaweed spores are influenced by biotic and abiotic components [6]. Factors of the abiotic component are one of the factors that needs to be developed in the application. Some influential abiotic components in the reproduction of the seaweed are salinity, temperature, light, pH, and waves [7]. Temperature and salinity are important factors that determine the growth and reproduction of seaweed [2]. The rise or decline in the level of salinity could influence the turgor pressure that will induce seaweed to do an osmotic adjustment [8]. Meanwhile, temperatures have a fundamental effect on chemical reactions and metabolism level of seaweed. Both of these factors greatly affect the release of seaweed spores.

Thus, research on the effect of the application of salinity and temperature shock on spores release and germination is needed so that the optimal combination of temperature and salinity shock to produce *K. alvarezii* superior and sustainable seeds could be identified and understood.

## 2. Methodology

### *2.1. Research time and place*

This research was conducted from February to March 2017 at Seaweed Tissue Culture Laboratory, Marine Aquaculture Development Center Lombok.

### *2.2. Materials*

The equipments used in this study were petridish, stopwatch, thermostat heater, aerator set, bottles with a diameter of 5 cm and height of 12 cm, an aquarium with a length of 60 cm, width of 30 cm and height of 30 cm, stereo microscopes, pipette, a beaker glass, hand counters, Sedgwick Raffter counting cell and thermometer. The research materials used were fresh water, sea water, K. alvarezii seaweed obtained from Lombok island.

### *2.3. Procedures*

K. alvarezii seaweed was obtained from Gerupuk, Lombok. The first stage before doing the temperature and salinity shock towards K. alvarezii seaweed spores wasthe selection of seeds. Seaweed seedling selection was done to obtain eligible explant to be reproduced, cultivated and tissue cultured. K. alvarezii seaweed from nature were acclimatized in the aquarium for a week. Acclimatization in greenhouses was done by moving the seaweed on media containing sea water in aquarium.

The minimum size of the aquarium used was 90 cm x 30 cm x 35cm with a recirculating water system. A recirculation system was implemented using a multilevel aquarium. The arrangement of the aquarium water temperature allows the sea-floor to remain stable because the heat of the sun was covered by the aquariums placed above.

An increase of temperature of water in a beaker glass was done using a commercial thermostat to regulate the temperature [9]. Salinity shock treatment was done using three different salinity levels at each beaker glass. Certain salinity obtained by giving a certain appropriate weight of salts with the desired salinity. It is in accordance with Ask [10] that used this comparison:

$$\text{Salinity (ppt)}=\frac{\text{Salt Concentration (gr)}}{\text{Volume of Water (1 L)}} \tag{1}$$

Before the treatment was done in mini media (petridish), seaweed was washed with sea water, cleaned from other algae and attached dirt. Seaweed used as a sample in the shock process was the one

IOP Conf. Series: Earth and Environmental Science **137** (2018) 012019 doi:10.1088/1755-1315/137/1/012019

having sporophyte. The making of sterile seaweed explant was done by selecting the thallus and sterilizing the thallus. The sterilization process wasconducted according to Harley [2] with some modifications. The steps are: (1) the young thallus was cut with the length of 5 cm using scalpel and tweezers on sterilized petridish, then rinsed with sterile sea water twice; (2) the thallus or explant were soaked for 8-10 minutes in a solution of liquid soap with a concentration of 2 drops per 200 ml, and then rinsed with sterile sea water 2 times; (3) explant were soaked for 2-3 minutes in 100 ml of seawater containing Povidone Iodine 1 % (10 ml of Betadine and 90 ml of sterile sea water), then rinsed two times with sterile seawater; (4) the explant were dried with paper towels and put into sterile media shock treatment temperature and salinity.

*2.4. Environmental shock*

The media used for the temperature and salinity shock treatment was sea water with different salinity levels and a certain temperature in a beaker glass. The aalinity and temperature shock treatment were given with an increase in salinity i.e. 31 ppt, ppt, 33 and 35 ppt at different temperatures i.e 30°C, 32°C, and 34°C.

The temperature and salinity shock treatment were done by putting the selected and cleaned thallus explant to beaker glasses, 6 explants for each beaker glass. Temperature and salinity shock was done for 30 minutes. After that, the seaweed was restored into a culture bottle of 100 ml with original water conditions on a rotary shaker for 7 days with a room temperature of 22-25°C and 1500 lux fluorescent light for 12 hours of light and 12 hours of dark.

Seaweed spores were counted using Sedgwick Rafter counting chamber [11]. The calculations were performed every day during the maintenance period after treatment to observe the number of spores released per day. The calculation was done based on the formula used by Bindu [12]:

$$\text{Cells/ml} = \frac{\mathit{Counted\ cells\ x\ 1000\ mm^3}}{\text{A x D x F}} \qquad (2)$$

Description:
A = Area of field
D = Depth of Sedgwick Counting Cell
F= Counted fields

*2.5. Data analysis*

The data of released spores yield obtained from the results of this research were analyzed using ANOVA (Analysis of Variance) to identify whether there was a difference of each treatment, according to the research experimental design used. The analysis of multiple distance trials was followed by Duncan with the SPSS program. Further tests using Test Duncan were needed to compare the treatments.

## 3. Result and Discussion

The results of the variant analysis show that there is no significant difference ($p > 0.05$) on the interaction of temperature and salinity shock treatment on the first day until the second day, while the third to seven days shows a significant difference ($p < 0.05$). The result of the *K. alvarezii* spores release is shown in table 1.

*K. alvarezii* spores had been released gradually every day. Seaweed thallus used as samples has different spore release responses. A3B2 treatment showed exponential phase on the second day, while other treatments experienced exponential phase on the fourth day. A1B2, A3B1, and A3B2 treatment showed declination phase on the sixth day, while the other treatments had not showed declination phase until the seventh day of observation.

The temperature and salinity shock treatment influenced spore releases significantly. Seaweed spores had also been released gradually during the 7 days of observation. According to Clesceri [6],

IOP Conf. Series: Earth and Environmental Science **137** (2018) 012019 doi:10.1088/1755-1315/137/1/012019

some abiotic components such as salinity, temperature, light, pH, current and waves are able to affect the reproductive system of seaweed. The results of the variant analysis showed that the given treatment had no significant difference ($p > 0.05$) on the interaction of temperature and salinity shock treatment on the first day to the fourth day, while on the fifth day to seven shows a significant difference ($p < 0.05$). This indicates that the spores of seaweed have a cell maturation phase before being released into the environment. The discharge of spores has a gradual releasing system which is divided into several processes such as spores formation, spores maturation, and spores release. These processes are strongly influenced by factors such as changes in temperature, salinity, and other abiotic components [2].

Spore release yield, shown in figure 1. and table 1., indicates that there are fluctuations and the phase of spores beginning to release. It also shows significant differences, indicating that the release of spores has a certain pattern, and does not directly occur after experiencing the stress on the environment. In addition, the data showed that the *K. alvarezii* seaweed needs a process for certain days to be released. It is in accordance with Ask [10] who mentioned that the release, growth and germination process of seaweed spores can be observed after five days. *K. alvarezii* seaweed spores experienced a maturation process before they were released. The transition process of spore filaments formation consists of thin layer formation, filament elongation, and formation of vegetative conchospore (spore) which is strongly influenced by the extreme change in temperature [13].

**Table 1**. *Kappaphycus alvarezii* seawed spores release during 7 days

| | Amount of Released *K. alvarezii*Spores(cell/ml) ± SD | | | | | |
|---|---|---|---|---|---|---|
| | Day 2 | Day 3 | Day 4 | Day 5 | Day 6 | Day 7 |
| A1B1 | $132{,}67^{a}$ ± 37,52 | $123{,}83^{a}$ ± 30,63 | $123{,}83^{a}$ ± 30,63 | $3078{,}03^{c}$ ± 1518,71 | $4484{,}39^{abc}$ ± 37,49 | $4953{,}15^{de}$ ± 457,56 |
| A1B2 | $141{,}52^{ab}$ ± 30,69 | $159{,}21^{a}$ ± 75,05 | $159{,}21^{a}$ ± 53,07 | $1362{,}06^{abc}$ ± 220,96 | $4970{,}89^{c}$ ± 353,35 | $1804{,}34^{abc}$ ± 1276,94 |
| A1B3 | $238{,}81^{ab}$ ± 37,52 | $159{,}21^{a}$ ± 51,05 | $106{,}14^{a}$ ± 75,05 | $424{,}56^{a}$ ± 231,32 | $2998{,}45^{ab}$ ± 37,52 | $3856{,}33^{cde}$ ± 1649,73 |
| A2B1 | $159{,}21^{a}$ ± 75,05 | $212{,}28^{a}$ ± 53,07 | $106{,}14^{a}$ ± 75,05 | $2618{,}06^{bc}$ ± 81,05 | $4033{,}29^{abc}$ ± 675,43 | $5413{,}06^{e}$ ± 414,49 |
| A2B2 | $238{,}81^{ab}$ ± 37,52 | $300{,}73^{a}$ ± 30,63 | $132{,}67^{a}$ ± 37,52 | $618{,}98^{ab}$ ± 110,67 | $371{,}49^{a}$ ± 243,19 | $1450{,}55^{ab}$ ± 615,06 |
| A2B3 | $159{,}21^{ab}$ ± 53,07 | $159{,}21^{a}$ ± 53,07 | $212{,}28^{a}$ ± 140,41 | $902{,}17^{ab}$ ± 424,53 | $689{,}91^{ab}$ ± 53,07 | $2794{,}96^{abcd}$ ± 425,65 |
| A3B1 | $106{,}14^{a}$ ± 51,79 | $123{,}83^{a}$ ± 30,63 | $371{,}49^{a}$ ± 231,32 | $1892{,}77^{abc}$ ± 429,00 | $2282{,}01^{abc}$ ± 265,35 | $919{,}86^{a}$ ± 376,48 |
| A3B2 | $283{,}04^{b}$ ± 82,39 | $972{,}9^{b}$ ± 391,14 | $1406{,}34^{b}$ ± 712,98 | $1432{,}83^{abc}$ ± 486,36 | $3060{,}37^{abc}$ ± 1648,87 | $1468{,}23^{ab}$ ± 1327,09 |
| A3B3 | $212{,}28^{ab}$ ± 54,93 | $247{,}66^{a}$ ± 61,27 | $636{,}84^{ab}$ ± 231,32 | $3378{,}75^{c}$ ± 2869,85 | $3414{,}12^{bc}$ ± 963,07 | $3325{,}69^{bcde}$ ± 2575,01 |

A1B1 = Temperature30ºC (A1) dan Salinity 31 ppt (B1)
A1B2 = Temperature30ºC (A1) dan Salinity 33 ppt (B2)
A1B3 = Temperature30ºC (A1) dan Salinity 35 ppt (B3)
A2B1 = Temperature32ºC (A2) dan Salinity 31 ppt (B1)
A2B2 = Temperature32ºC (A2) dan Salinity 33 ppt (B2)
A2B3=Temperature32ºC (A2) dan Salinity 34 ppt (B3)
A3B1 = Temperature34ºC (A3) dan Salinity 31 ppt (B1)
A3B2 = Temperature34ºC (A3) dan Salinity 33 ppt (B2)
A3B3 = Temperature34ºC (A3) dan Salinity 34 ppt (B3)

Temperature and salinity shock treatment simultaneously showed a significant difference on day three to seven. The results of Duncan Multiple Distance Test on day three to five showed that an increased temperature shock to 34°C could give a significant influence on the release of spores. In addition, the data of thallus weight decrease indicate that the higher the temperature, the higher the thallus lose its water from within. Seaweed thallus weight-loss can be seen in table 2. Temperature has a fundamental effect on the degree of chemical reactions, reproduction, and metabolism of seaweed [8]. The effect of high-temperature shock will gradually affect photosynthesis, respiration, membrane stability and physiological changes [14]

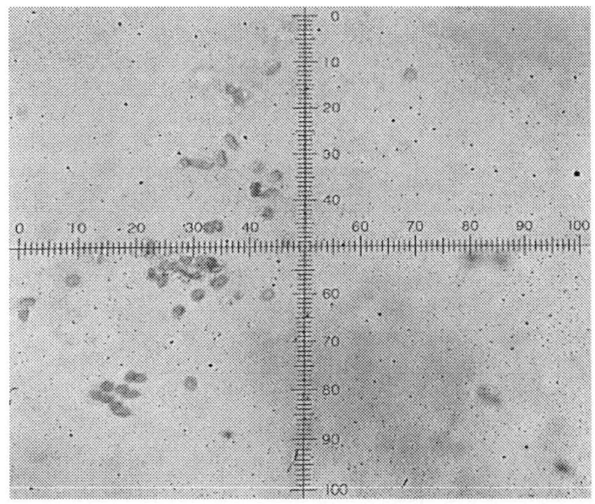

**Figure 1.** Released spores from seaweed *K. alvarezii* Thallus with 200x magnification.

Rising temperatures in the form of temperature shock treatment also elevated the cell membrane activity and water flow in plants. The rise in temperatures will increase water flow in plants that will further enhance the turgor pressure within the plant cell [15]. It urged the cell to increased pressure from within. Then, the fluid within the cell is pushed to out. This term is called osmosis. In addition, a decrease in the weight of the thallus of *K. alvarezii* also caused the cell membrane to loose which will simplify the osmosis process. Meanwhile, the higher temperature could loose the pores on the surface of the plants [16. When the pores which are considered as semipermeable membrane open up wider, the water-loss level will be higher [17]. This condition causes the plants to lose more fluids and pushes the spores to be discharged. The influence of force, pressure, and the weakening of the liquid on the layers of the skin pore could trigger algae cystocarp to force the seaweed spores to discharge through the venturi tube [18].

Salinity shock treatment has a significant difference and the best result was shown by 31 ppt salinity. The Duncan test showed that during the fifth day to the seventh day, 31 ppt salinity was very influential on the release of spores with the largest results. Seaweed spores can be released optimally on the salinity range of 30 ppt [19]. The development of good seaweed spores occurs in the range of salinity of 30 ppt [20]. The release of seaweed spores also has a narrow salinity range. It is also stated that the most suitable salinity for *K. alvarezii* spores to release is in the range of 29-34ppt. In addition, the thallus weight decrease data indicate that the higher the salinity, the higher thallus loses its water from within. The result of weight-loss of the thallus can be seen in table 2. This indicates that there is a presence of osmoregulation activity within seaweed cells. It is also considered as the most effective treatment in the weight-loss of *K. alvarezii* seaweed. Salinity shock directly affects the osmoregulation system on seaweed. The increase in salinity will directly increase the concentration of seawater correspondingly. So, the osmotic pressure will change differently. Increased salinity of sea water will result in the loss of water and ion absorption of the thallus. The entire process caused the change in osmotic pressure so that the seaweed will suffer osmosis. If the concentration of the environment outside the cell is higher than the concentration within the cell, this difference will cause the water to shrink out [8]. This progressive water-loss will be correspondingly loosening the space among the cells.

**Table 2.** Weight loss of *Kappaphycus alvarezii* thallus.

| | Given Treatment | | Thallus Weight Reduction (gr) ± SD |
|---|---|---|---|
| | Temperature Shock (°C) | Salinity Shock (ppt) | |
| A1B1 | | 31 | $0^{a} \pm 0$ |
| A1B2 | 30 | 33 | $0{,}016^{b} \pm 0{,}005$ |
| A1B3 | | 35 | $0{,}026^{c} \pm 0{,}005$ |
| A2B1 | | 31 | $0{,}013^{b} \pm 0{,}005$ |
| A2B2 | 32 | 33 | $0{,}026^{c} \pm 0{,}005$ |
| A2B3 | | 35 | $0{,}036^{d} \pm 0{,}005$ |
| A3B1 | | 31 | $0{,}013^{b} \pm 0{,}005$ |
| A3B2 | 34 | 33 | $0{,}026^{c} \pm 0{,}005$ |
| A3B3 | | 35 | $0{,}070^{e} \pm 0{,}010$ |

A1B1 = Temperature 30°C (A1) dan Salinity 31 ppt (B1)
A1B2 = Temperature 30°C (A1) dan Salinity 33 ppt (B2)
A1B3 = Temperature 30°C (A1) dan Salinity 35 ppt (B3)
A2B1 = Temperature 32°C (A2) dan Salinity 31 ppt (B1)
A2B2 = Temperature 32°C (A2) dan Salinity 33 ppt (B2)
A2B3=Temperature 32°C (A2) dan Salinity 34 ppt (B3)
A3B1 = Temperature 34°C (A3) dan Salinity 31 ppt (B1)
A3B2 = Temperature 34°C (A3) dan Salinity 33 ppt (B2)
A3B3 = Temperature 34°C (A3) dan Salinity 34 ppt (B3)

Disruption of the function of the plant cell wall is affected by the role of Golgi apparatus that are not optimal so that the cells will undergo osmosis. According to Davis [21], a different concentration of the liquid inside and outside the cell will push the Golgi apparatus to continue to strive balancing the nature until the environment condition gets isotonic. The cell will become increasingly dense, but the size will stay or become relatively smaller than the size of the previous condition [22].

The results of this research show that temperature and salinity affect the osmoregulation and release of spores. The results of the variant analysis showed that the interaction of temperature and salinity could obtain the biggest amount of spore cells on the seventh day with the A2B1 treatment. This treatment, with a temperature of 32°C and 31 ppt salinity, could obtain 5413.06 cells/ml. It shows the positive interaction between the two factors. Temperature and salinity are the factors that interact in giving the influence of seaweed physiology and reproduction. The temperature and salinity factors correspondingly interact with significant spores release. Seaweed reproductive system chain is affected by some combination of biotic and abiotic factors [6]. Temperature and salinity are the most important abiotic component factors which determine the growth and reproduction of seaweed [7].

*K. alvarezii* spores release is closely related to seaweed physiological system and environmental factors. The osmosis from seaweed cell, which is caused by the different concentration of the environment and spaces between cells that loosen, will carry spores out. The opening of the cystocarp pores will force seaweed spores to be released through the venturi tube with the influence of force, pressure, and the weakening of the adhesive liquid on the skin layer of algae [18]. Environmental factors such as temperature and salinity affect the connection between cells (the pit connection) to detach and release the reproduction cell [22].

## 4. Conclusion

Based on the discussion, it can be concluded that environmental manipulation using salinity and temperature shock have several effects on osmoregulation system and *K. alvarezii* spores release with 31 ppt salinity on the fifth day to the seventh day and 30°C temperature on the sixth day and seventh day. The addition of salinity and temperature shock also have an interaction with the release of *K. alvarezii* spores on the third day to the seventh day with A2B1 as the best treatment. A2B1 is the combination of 32°C temperature shock and 31 ppt salinity shock that could produce 5413.06 cells/ml on the seventh day. The temperature and salinity shock can be used as the candidate of *K. alvarezii* seaweed reproductive method to obtain superior seeds which can be used for fishery development.

## 5. References

[1] Jaya D, Hadi F, Kusumasari D and Riswardani E 2012 *Drying carrot (Daucus carota) by dehydration osmosis* (Jakarta : National Seminar on Chemical Engineering Soebardjo Brotohardjono) **11** 1-6
[2] Harley C D G, Anderson K M, Demes K W, Jorve C P, Kordas R L and Coyle T A 2012 *J Phycol.* 1-15
[3] Bulboa C, De Paula E J and Chow F 2008 *Phycol. Res* **56** 39-45
[4] Ask E I, Batibasaga A, Zertuche-González J A and de San M 2003 *Kappaphycus alvarezii Introduction* pp 49-57
[5] Agrawal S C 2009 *Folia Microbiol* **54** 273-302
[6] Clesceri L S, Greenberg A E and Eaton A D 1999 *Standard methods for the examination of water and wastewater* (America : American Public Health Association, American Waterworks Association, and Water Environment Federation ) 2671 p.
[7] Giplin M 1991 *Biol. J. Linn. Soc.* **42** 165-175.
[8] Azanza R V and Ask E I 2001 *Int. Seaweed Sym.* 1-10
[9] Arisandi A, Marsoedi, Nursyam H and Sartimbul A 2011 *Ilm. Kel* **16** 143-150 (In Indonesia)
[10] Ask E I and Azanza R V 2002 *J. Aquac.* **206** 257-277.
[11] Azanza-Corrales R, Mamauag S S, Alfiler E and Orolfo M J 1993 *J. Aquac.* **103** 29-34.
[12] Bindu M S and Levine I A 2011 *J Appl. Phycol.* **23** 789-796.
[13] Jensen R D and Taylor S A 1961 *Plant Phsy.* **171** 639-642
[14] Dickson L G and Waaland J R 1985 *Planta.* **165** 548-553.
[15] Guiry M D and Cunningham M H 1984 *Phycologia* **23** 357-367.
[16] Davis T W, Berry D L, Boyer G L and Gobler C J 2009 *Harm. Algae* **8** 715-725.
[17] Guzmán-Urióstegui A and Robledo D 1999 *J. Hydrobiol.* **389/399** 285-290.
[18] Hurtado A Q and Cheney D P 2003 *Bot. Mar.* **46** 338-341.
[19] Hay M E 1991 *Academic Press Inc. North Carolina* 96-108.
[20] Anggraeni S R, Sudarsono and Soedharma D 2008 *Bionatura* **10** 196-208. (In Indonesia)
[21] Forslund H 2009 *Grazing and the geographical range of seaweeds : the introduced fucus evanescens and the newly described fucus radicans* (Sweden : Licentiate Thesis. Stockholm University) 30 p.
[22] Dhargalkar V K and Kavlekar D 2004 *Seaweeds – A Field Manual* (Dona Paula : National Institute of Oceanography) 42 p.

## Acknowledgement

This research is dedicated to Indonesia future seaweed aquaculture. I would also like to thank all of the supervisors and Faculty of Fisheries and Marine, Universitas Airlangga.

IOP Conf. Series: Earth and Environmental Science **137** (2018) 012020 doi:10.1088/1755-1315/137/1/012020

# Effectivity of immunostimulant from *Zoothamnium penaei* protein membrane for decreasing the mortality rate of white shrimp (*Litopenaeus vannamei*) in traditional plus pond

**G Mahasri[1*], R Kusdarwati[1], Kismiyati[1], Rozi[1], H Gustrifandi[2]**

[1]Department of Fish Health Management and Aquaculture, Faculty of Fisheries and Marine, Universitas Airlangga, Campus C Mulyorejo Surabaya, Indonesia
[2]Fish Quarantine Center and Quality Control of Fishery Products, Class I, Surabaya, Indonesia

E-mail: mahasritot@gmail.com

**Abstract.** The purpose of this research was to analys immunogenic membrane protein as immunostimulant development material to control the mortality of white shrimp in traditional plus pond. This research was designed to use explorative experiment and experimental laboratory methods which used completed random sampling design. Collected data was analyzed with analysis of variance for examination of survival rate (SR), total haemocyte count (THC) and differensial haemocyte Count (DHC). The research divided into 2 part of riset: (1) Identification, cultivation *Zoothamnium penaei*, analysed of membrane protein by SDS-PAGE, (2) Field test protein membran on Survival Rate level, immune response (THC and/or DHC level) and infestation of *Zoothamnium penaei* in traditional plus pond. The result showed that there were seven bands membrane protein of *Zoothamnium penaei* with molecular weight 38 kDa, 48 kDa, 67 kDa, 71 kDa, 77 kDa, 98 kDa dan 104 kDa by using SDS-PAGE. Immunogenicity tested decrease by using ELISA and western blotting there are only found three bands with molecular weight 38 kDa, 48 kDa dan 67 kDa. The membrane protein could increase the immun respons and decrease the mortality, by subsequenly, it could increase the survival rate from 17% until 68% and pressured the parasite infestation of white shrimp.

## 1. Introduction

The vaname shrimp (*Litopenaeus vannamei*) is one of imported species from America entering Indonesia in 1998. After various technological applications and trials, in 2002 the Indonesian Government legalised that vannamei shrimp could be developed in Indonesia. The Government expected that the emergence of shrimp vannamei would replace the position of tiger prawns, which were still one of the main non-oil export commodities from the fishery sector until 1993. The value of vannamei shrimp exports in 2006 reached 6000 billion US Dollars. Vannamei shrimp also has a big role in the fulfillment of animal protein from fish, because shrimp has a high nutritional value [1].

The production of vaname shrimp has been increasing, meeting the needs of the market especially for export. One of the main constraints that is difficult to control is the presence of disease, which causes the death of shrimp up to 100 % [2]. Furthermore, it is also stated that in general the losses caused by the disease reach billions of rupiah. This is caused by a very sudden onset of illness and the total death will occur in just 2 to 3 days of the post infection. The attack of this disease had caused many shrimp-culture farmers to close their business and 80 % of shrimp farmings and shrimp hatcheries not to operate.

One of the diseases that can cause the death of shrimp both in pond and hatchery is Zoothamniosis. This disease is one of the parasitic diseases found in vannamei shrimp caused by *Zoothamnium penaei*. This disease causes the shrimp to breathe hard, difficult to move and cannot find food [3, 4], difficult to moult, inhibit growth, reduced economic value and cause death to 91 % [5]. In addition, this disease is a predisposing factor of secondary infection by bacteria and viruses.

Prevention efforts have been done, such as using the circulatory system, formaline treatment, and antibiotics. According to Chamratchakool [6], the treatment of Zoothamniosis with 30 ppm of formaline could suppress the incidence of this disease, 2 ppt effective acetic acid for larvae [5]. Many also used chemicals such as malacyt green, and methylene blue. Mahasri [7] emphasized *Zoothamnium penaei* infestation using aeration and stocking solid arrangement, which showed that Zoozid of *Zoothamnium penaei* infestation decreased at high aeration level and low to medium density of stocking. The use of biological filters of milkfish could also suppress Zoothamniosis attacks from 89 % to 14 % [8].

Rukyani [9], argued that the resilience improvement of shrimp bodies both in hatcheries and ponds could be done by using the immunostimulant. Furthermore, Mahasri [8] added that the immunization with immunogenic membrane protein isolated from *Zoothamnium penaei* could increase the shrimp life rate up to 93%. It is also stated that the isolation of immunogenic membrane proteins could be done with SDS-PAGE, ELISA and Western Blotting. The results showed that found 3 proteins found were immunogenic, i.e. MP38, MP48 and MP67 membrane proteins.

The isolation of protein has been performed by Amos *et al.* [10] and Routledge [11], who characterized the calcium bond with proteins using the SDS-PAGE method. The synthesis of Telotroch proteins (taken from the part of *Zoothamnium arbuscula* during free swim stadium) Pylawka *et al.* [12] and characterization of *Zoothamnium arbuscula* spore protein with SDS-PAGE [13]. Itabashi *et al* [13] also stated that the spore had spasmin-1 protein in its spasmonema. The analysis of immunoblotting results showed that the antigen protein had a molecular weight of 68 kDa, 55 kDa and 71 kDa.

Spasmine protein is a *Zoothamnium arbuscula* trophozooid protein presented in spasmonema and contractile stalk. The genome length of this spasmin protein is 531 bp, which is predicted to consist of 177 amino acids with a molecular weight of 19.659 kDa [13]. The characterization of spasmine protein from *Zoothamnium arbuscula* (Familia Vorticellidae) using SDS-PAGE indicated that 60% protein bands were spread over a molecular weight of about 20,000 kDa. The rich amino acid contents found were glysine and serine, but were poor in aromatic amino acids and had no cysteine and methionine [10]. Routledge [11] has isolated proteins from contractile spasmony from *Vorticella convollaria*, *Carcheslium polypinum*, and *Zoothamnium geniculatum* extracted in detergent, Sodium Dedocyl Sulfate (SDS), such as urea and Guanidine Hydrochloride (GuCl). After extracting with SDS, the molecular weight distribution of proteins was identified with SDS-PAGE. Clearly, there were no actine and tubuline proteins. The molecular weight of the contractile organ protein was closed to 20,000 kDal. Moriyama *et al.* [14], had isolated proteins from *Zoothamnium* sp. spasmonema and stated that the molecular weight of the protein chain was estimated to be closed to 50 kDal and 20 kDal synthesized from sponical protein binding [15]. The results show that 1 spasmin gene has no intron and 531 bp genome length. It is predicted to produce 177 amino acids with molecular weight approaching 19659 Da (19,659 kDa) . The acid sequence has two calcium bonds.

The body's defense system in invertebrates (including shrimp) plays a role in the body's defense mechanism by haemocytes, where the spread and increased number of haemocytes is assumed to be a form of cellular immune response in the shrimp body [16, 17]. To perform phagocytic activity, encapsulation, nodulation, activation of prophenoloxidase system, anti microbial or toxic compound, it is necessary to release some proteins to overcome the foreign object or the incoming agent [18]. Johanson *et al.* [19] mentioned that hyalin cells also play a role in the shrimp body's defense system. This hyalin cell is activated by Opsonin Factor that results from proPO activation into PO in granular cells. So, it can phagocytosis both bacteria and virus, but the most important role in shrimp body defense system is granulocyte.

Increased body resistance can be known from the increased activity of phagocyte cells from hemosite. These phagocytic cells function to perform phagocytosis of foreign objects entering the host's body. Phagocytosis is a non-specific defense mechanism that generally protects against disease. Hemocytes are known to be very important factors in non-specific cellular defense systems. To know that the hemoglobin is a cellular defensive body, it can be seen from its ability in phagocytosis activity that can be increased in the incidence of infection. The presence of three forms of mechanisms of observable hemosite activity, namely (1) the mechanism of entrapment (encapsulation) of a foreign material, (2) the joint phagocytic mechanism of some hemorrhoid which form larger aggregates, and (3) aggregates of many hemoglyms forming a pigmented layer [20].

Kondo *et al.* [21] found 5 types of phagocytes in kuruma shrimp (Penaeus japonicus); two of which are phagocyte cells and the rest are hemocytes. The first two types of phagocytes are found in the heart of the basal lamina covering the sarcolemma of the heart muscle and its cells have lysosomal granules of 0.1 μm diameter, the second type of phagocytic cell found in lymphoid organs. In the five types of phagocytes were found lysosomal enzymes such as acid phosphate, β-glukronidase and non-specific esterase. Very high estersae activity was found in granular cells. Phenoloxidase (PO) activity was also found in all phagocyte cells but it was very weak, whereas prophenoloxidase (proPO) activity was found only in granular cells and semigranular cells.

Humoral body defense systems include phenoloxidase (PO), prophenoloxidase (proPO), lectin and aglutinin. Both of these defense systems work together to provide body protection against infection of pathogenic organisms from the environment [16]. ProPO is activated by enzyme prophenoloxidase actuating enzyme (PPA). Meanwhile, the prophenoloxidase activating enzym can be activated by lipopolysaccharide. ProPO and PPA are proteins located in granular hemosite. As a result of the activation of proPO, into Protein was produced as Opsonin Factor that stimulates phagocytosis of hyalin cells (agranular cell) [19].

Van de Braak [17] proposed that the immune response of shrimp among others can be seen with the increase and change of hemocytes, namely Total Haemocite Count (THC) and Differential Haemocite Count (DHC). He further explained that THC and DHC will increase in shrimp in the affected area, resulting in a decrease in the flow of blood circulation. This increase of THC and DHC signifies an increase in the body's defense of shrimp, thus reducing the pathogen infection and consequently increasing the survival rate of shrimp.

McKay *et al.* [22] said that immunization in crayfish (Parachaeraps bicarinatus) 4 times with endotoxin 2.5 μg and 2.5 X 107 vaccine from bacteria Pseudomonas sp. with 4 injections resulted in very significant differences between the amount of hemocytes in the control group and the immunized ones. Proteins released during vaccination are used for immune responses such as phagocytosis, encapsulation, melanization, coagolation, peroxidase activity, opsonization, anti-microbial activity and other humoral and cellular activity processes [17]

Giving immunostimulants to crustaceans, which is also called vaccination, has no side effects and is very good to be applied to organisms that do not have memory cells in the immune system, so as to stimulate and or maximize non-specific immune responses [23]. The provision of Lipopolysaccharide (LPS) from the Vibrio harveyi bacterial cell wall orally to the vannamei shrimp (*Litopenaeus vannamei* Fab.) within 42 days can increase the total amount of hemocytes and the granulocyte cell type, while the phagostic activity of the hemosite cells has shown an increase on day 28 and higher on day 42. The occurrence of increased phagocytic activity of hemositic cells along with the increasing total hemocytes and granulocyte cell types is due to Lipopolisaccharide.

Based on the above background, it is necessary to examine the immune response (THC and DHC) and the survival and parasitic infestations of vannamei shrimps given immunogenic membrane proteins MP38, MP48 and MP67 orally and infested with *Zoothamnium penaei.*

IOP Conf. Series: Earth and Environmental Science **137** (2018) 012020 doi:10.1088/1755-1315/137/1/012020

## 2. Methodology

### *2.1. Materials Research*

The materials used in this research were *Zoothamnium penaei* from Gresik and Lamongan, East Java. The species sample then was cultivated by means of cohabitation at the Aquarium in order to obtain the amount of zooid in large quantities planted in the vaname shrimp seed aged PL30-PL40 as much as 10.000 shrimps.

The materials used for the isolation of whole protein using the Sodium Dodecyl Sulphate Polyacrylamide Gel Electrophoresis (SDS-PAGE) were Acrylamide (Sigma Chemical Company 3050, PO BOX Spruce Street 14508, St Louise, MO 63178, USA), bis- (N, N '-Methylene-bis-Acrylamide) (Sigma Chemical Company 3050, PO BOX Spruce Street 14508, St. Louis, MO 63178, USA), Sodium Dodecylsulfate or SDS (Merck, Postfach 4119, D-6100, 64271 Dannstadt 1 Germany), PMSF, TPCK, TLCK and EDTA.

### *2.2. Research instruments*

The instruments used in this research were micro pipettes (Gilson), J-6B centrifuge (Beckman), microfugeTM11 (Beckman), stereo type 102 (Nixon) microscope. The immunostimulatory apparatus equipments were 3 plot ponds, 2 sets of saprodi, and a genset for water pump.

### *2.3. Research procedure*

#### *2.3.1. Zoothamnium penaei protein membrane isolation*

6 x 108 Zooid/mL result of in vitro cultivation was resused with 10 mM PBS to obtain final concentration of 6 x 108 Zooid/mL, then added with protease inhibitor (for 100 mL PBS plus 100 □l 40 mM PMSF, 7.3 mg TLCK, 7 mg TPCK and 7.5 mg EDTA). The suspension was synchronized with 4 x 0.5 minutes in ice. The suspension was then centrifuged at 4000 rpm for 1 hour. The supernatant was taken. The cytoplasmic protein (soluble protein) and pellets contain whole protein. The pellet was further resolved with 0.5 % Nonidet P40 in a protease inhibitor solution with the same composition as in the previous composition. Sonication was performed with 4 × 0.5 min in the subsequent ice incubated overnight at $4^0$C while stirring with stirrer at medium speed. The suspension was then centrifuged at 4,000 rpm for 30 min at 4 ° C and the supernatant was taken for measurement of its protein concentration.

#### *2.3.2. Protein concentrations determine*

To determine the concentration of homogeneous protein, the Bio-Rad Protein Assay Method was used. It was read using UV-Visible Spectrophotometer with 595 nm wavelength.

#### *2.3.3. The characterization and purification of protein membranes with SDS-PAGE*

The purpose of this activity is to look at the molecular weight pattern of the whole protein fraction. Running gel was made and inserted into the glass plate, after hardening the prepared stacking gel was inserted at the top. The composition of running gel and stacking gel was made by mixing Acrylamid, Tris, SDS 0.8 %, Temed, APS and aquades in a glass beaker. A total of 10 µg samples added with laemly buffers with a 2: 1 ratio were boiled at 1000C for 5 min, then put into a molding column located on the stacking gel and performed running on a chamber that had been filled Electrode Buffers 1X with 100 volt, 40 mA. After running, the gel was introduced into a washing solution consisting of 25 mL methanol, acetic acid 3.7 mL, and Aquades ad 100 mL. It was then stirred over sackher for 30 minutes. Repeat washing was done with the same solution with a reduction of the ethanol composition and the addition of acetic acid from the previous half for 30 minutes. Next, it was washed with 10 % glutaraldehyde and Aquades for 30 minutes. After washing, the gel was stained with AgNO3 for 15 minutes, then washed with aquades twice, each for 2 minutes. A color development solution was given comprising formaldehyde 3.7 %, zitronsauce 5 % and aquades. After the protein bands were seen, then the reaction was stopped by adding 10 % acetic acid. The resultant gel of the protein bands

IOP Conf. Series: Earth and Environmental Science **137** (2018) 012020 doi:10.1088/1755-1315/137/1/012020

was stored in a 10% glycerol solution, ready for documentation. Calculation of molecular weight was done by comparing the standard marker.

*2.3.4. Inspection of zoothamnium penaei infestation in shrimp*
Observation of ectoparasite infestation was done natively using Jhonson's method (1986) that is by doing scrapping on the entire surface of the shrimp body. The result of the scrap was placed on top of the glass of the object, given water and observed with a microscope with 100X magnification. Parasitic infestation was calculated with a positive shrimp percentage on the number of shrimp examined.

*2.3.5. Total haemocyte count (THC)*
Haemocytes were taken in the ventral portion of the second abdominal segment by using a 25 G needle and a mL syringe which had been inserted 0.2 milliliters of cold Alsever modified solution (AS 19.3 mM, 239.8 mM Na citrate, NaCl 182.5 glucose and 6.2 mM EDTA, pH 7.2) as an anticoagulant (Van de Braak, et al., 2000). The calculation of the number of haemocytes was done by the method of Van de Braak, et al. [17] using a bright microscope (LM) with 1000 magnification, then calculated by ZM model counter (Counter Electronic Ltd). The size of the hemosite particle ranged from 0.4 to 800 µm. As the supporting data, it can also be observed with an electron microscope (EM) with first centrifuge 700 X gravity at 40 C for five minutes.

*2.3.6. Differential Haemocyte Count (DHC)*
Blood was dripped on a glass of the object and a blood pillow with giemsa staining was made (Nash et al., 1993). Then, the cell type was identified. This hemositic differential was aimed at identifying the number, type and percentage of hemosite cells. The analyzed hemocytes were classified according to the instructions of Owens and O'Neill [25]. The amount of hemosite was calculated to 100 cells. The percentages of each species was identified, performed at the beginning of the study.

*2.3.7. Vaname shrimp survival rate (SR) calculation*
The survival rate (SR) was calculated by the percentage of shrimp that lived during the maintenance of the total shrimp kept, which was done at the end of maintenance, ie at harvest time of 90 days in pond. The formula used to measure according to Effendi [26] was:

$$SR = \frac{Total\ shrimp\ were\ alive}{Total\ shrimp\ stocked}\ x\ 100\ \% \quad (1)$$

Information: SR = *Survival rate* / Kelangsungan hidup (%)
Nt = The number of shrimp lives at the end of the study
No = The number of shrimp lived at the beginning of the study

**2.4. Data Collection and Analysis**
The data obtained were divided into two kinds, namely qualitative and quantitative data. The qualitative data were the data of isolation and molecular characterization, namely Spasmin Performin with SDS-PAGE. On the other hand, the quantitative data were protein data of vannamei shrimp on *Zoothamnium penaei* at a certain time, including: the amount of hemosite (THC), differential hemosit (DHC), parasitic infestation and number of survival of vaname shrimp used by ANOVA test if there is difference followed by Duncan [27].

IOP Conf. Series: Earth and Environmental Science **137** (2018) 012020 doi:10.1088/1755-1315/137/1/012020

## 3. Results And Discussion

### *3.1. Infestation in Vaname shrimp*

The results of the parasite infestation examination on vaname shrimp were presented in table 1. Meanwhile, the prawn image infested with *Zoothamnium penaei* was presented in figure 1.

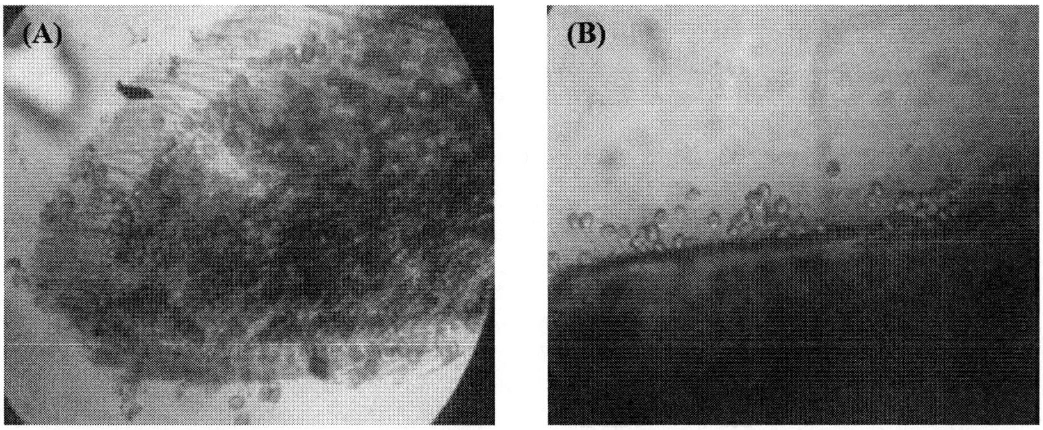


**Figure 1.** Description of the *Zoothamnium Penaei* Infestation on *Some* Organs. A: Tail and B: Back (100x Magnified Microscope Lens).

The results of examination of samples in the form of immunized and un-immunized vaname shrimp showed positively infested by *Zoothamnium penaei*. Clinical symptoms of shrimp infested with this parasite are similar to the shrimp attacked by zoothamniosis; the shrimp tail does not look like a fan if swimming, swimming on the surface and clustered. On the entire body surface and gills, there are parasites stick to the white brownish, empty digestive tract, the surface of the body and gills are muddy like moss. Moderate healthy shrimp (not attacked by zoothamniosis) looks clear, transparent, clean and there is no color change on the entire surface of the body and gills, active swimming prawns and the tail opening like a fan. *Zoothamnium* identification results found align with the results of identification [28], which has a circular peristome surrounded by cilia. In the inside, there were a contractive vacuola and some feed vacuola, nucleus. It has a stem on the posterior part of the spasmonema. Vaname shrimp had already started to infest at the age of 30 days, both immunized and unimmunized, but the prevalence value indicated thr value in unimmunized shrimp. The highest prevalence was found in non-immunized shrimp at the age of 90 days, on average 67.22 % and the lowest at the age of 90 days. The lowest occurred in 30-day immunized shrimp ie 5.55 %. The result of zoothamniosis prevalence on vaname shrimp can be seen in table 1.

**Table 1.** *Zoothamnium Penaei* infestation determination in vaname shrimp.

| Age of Maintenance (Day) | Infestation *Zoothamnium penaei* Shrimp Vaname (%) | |
|---|---|---|
| | Not Immunization | Immunization |
| 30 | $19{,}99^{a} \pm 2{,}9829$ | $5{,}55^{f} \pm 1{,}7196$ |
| 60 | $38{,}20^{b} \pm 1{,}9839$ | $14{,}44^{g} \pm 2{,}7189$ |
| 90 | $67{,}22^{c} \pm 5{,}3395$ | $15{,}18^{h} \pm 8{,}6482$ |

Different superscripts on different columns and rows shows a very real difference (p<0,01)

Anova results showed that the prevalence of zoothamniosis in vaname shrimp immunized before being stocked was very significantly different (P <0.01), when compared to the prevalence of unimmunized shrimp (table 1).

IOP Conf. Series: Earth and Environmental Science **137** (2018) 012020 doi:10.1088/1755-1315/137/1/012020

### *3.2. Immune response (THC and DHC) of vaname shrimp seeds*

The immune response of seed (PL 30) was not measured, because the seed was still too small and had difficulties in taking the blood sample. Table 2 shows the calculated total haemocyte count (THC) of vaname shrimp seeds that are not immunized and immunized at 30, 60 and 90 days on the pond maintenance, while the Differential Haemocyte Count (DHC) count can be seen in table 3. The results of anava showed that there was a significant difference between THC in immunized and unimmunized vaname shrimp (p <0.01). The highest THC was found in 60-day immunized shrimp, 54.08 x 106 cells/mL, while the lowest was in 30-day shrimp, 25.27 x 106 cells/mL.

**Table 2.** Total haemocyte count (THC) in vaname shrimp.

| Age of maintenance (Day) | Total Haemocyte Count (THC) Shrimp ($10^6$ Sel/mL) | |
|---|---|---|
| | Not Immunization | Immunization |
| 30 | 25,27$^a$ ± 0,4001 | 39,14$^f$ ± 0,2984 |
| 60 | 26,95$^b$ ±0,3345 | 54,08$^g$ ± 1,1259 |
| 90 | 27,08$^c$ ± 0,4334 | 41,55 ± 3,7776 |

Different superscripts on different columns and rows shows a very real difference (p<0,01)

Immunization of immunogenic membrane protein *Zoothamnium penaei* gave different effects on vaname shrimp immune response (DHC). There was a very significant effect difference between treatments (p<0.01).

**Table 3.** Differential determination of haemocyte count (DHC) on vaname shrimp.

| Age of Maintenance (Day) | Diffrential Haemocyte Count (DHC) Shrimp (%) | |
|---|---|---|
| | Not Immunization | Immunization |
| 30 | 13,09$^f$ ± 0,2256 | 15,73$^f$ ± 0,4658 |
| 60 | 14,20$^g$ ± 0,1989 | 17,20$^g$ ± 0,1150 |
| 90 | 12,99$^h$ ± 0,2267 | 24,03$^h$ ± 0,3135 |

Different superscripts on different columns and rows shows a very real difference (p<0,01).

Table 3 also shows that the highest DHC occurred in 90-day immunized vaname shrimp, while the lowest was found in unimmunized shrimp at 90 days, i.e. 24.03 % and 12,99 %.

### *3.3. Survival rate (SR) of vaname shrimp*

The result of determining the survival rate of vaname shrimp can be seen in table 4.

**Table 4.** Vaname shrimp survival rate.

| Treatment | SR (%) |
|---|---|
| Not Immunization | 17 |
| Immunization | 68 |

The survived vaname shrimp immunized before being stocked and maintained in ponds reached 68%, higher than the unimmunized which was only 17% at the end of the maintenance period of 90 days.

IOP Conf. Series: Earth and Environmental Science **137** (2018) 012020 doi:10.1088/1755-1315/137/1/012020

*3.4. Water quality*
The Water quality check results for 90 days of maintenance can be seen in table 5.

**Table 5.** Pond Water Quality Inspection for 90 Days Period of Maintenance.

| Parammeter | Average Parameters / Water Quality During Shrimp Maintenance | Range Normal Value |
|---|---|---|
| Suhu ( °C ) | 27 – 29 | 27 – 32 |
| Salinitas ( °/oo ) | 21 – 24 | 16 – 30 |
| pH | 7,6 – 8,8 | 7,5 – 8,5 |
| Disolvet Oxygen ( ppm ) | 3,4 – 6,2 | >3 – 7 |
| Amoniak ( ppm ) | 11 -13 | <15 |

Table 5 demonstrate that pond water quality during 90 days of vaname shrimp using average immunostimulant was still within the normal range. So, it is in accordance with vaname shrimp maintenance requirements.

The results showed that immunization with immunogenic membrane proteins *Zoothamnium penaei* on shrimp before stocking could decrease the prevalence of zoothamniosis and increase shrimp survival rate as well as increase immune response of THC and DHC. Table 1 shows that *Zoothamnium penaei* infestation had begun to be found in 30-day-old shrimp whether immunized or not immunized. The prevalence of non-immunized vanished zoothamniosis of vaname shrimp always increases with the increasing maintenance period, where the lowest prevalence occured in shrimp with a 30-day maintenance period of 19.99 % followed by maintenance period of 60 and 90 days, with prevalence values of 38.20 % and 67, 22 %. In vaname shrimp immunized with immunogenic membrane protein *Zoothamnium penaei*, which showed that prevalence increased during the 60 days maintenance period from 5.55 % to 14.44 % and 15.18 % on day 90 (end of maintenance).

Immune responses (THC and DHC) of vaname shrimp immunized with immunogenic membrane proteins *Zoothamnium penaei* were increased and from day 30 maintenance from 25.27 x 106 se / mL to 39.14 x 106 se / mL. While at shrimp aged 60 days THC shrimp increased from 26.95 x 106 se / mL to 54.08 x 106 se / mL and at age 90 day increase from 27.08 x 106 se / mL become 41.55 x 106 se / mL. When viewed as a whole than the immature vaname shrimp always increased during the 90 days maintenance (harvest). Meanwhile, in the non-immunized shrimp THC increased up to 60 days maintenance, however there was a decrease on day 90 with not much different value, although the result of anava was different (table 2). Table 2 also shows that immunized shrimp THC increased up to day 60 and decreased on day 90.

The immunized DHC shrimp vaccine (table 3) also increased from 13.09 % to 15.73 % in 30-day shrimp, and from 14.20 % to 17.20 % in shrimp aged 60 days and from 13.99 % to 24.03 % at age 90 days. Table 3 also shows that immunized vaname DHC shrimp also increased during maintenance ie 15.73 %, 17.20 % and 24.03 % at 30, 60 and 90 days day, while those not immunized ranged from 13 %.

The ability of immunogenic membrane proteins *Zoothamnium penaei* as immunostimulant could also be seen from the survival rate of vaname shrimp immunized before being dispersed higher when compared with that not immunized, i.e. between 68 % and 17 %. This could mean that the immunostilant was able to provide protection on shrimp that are maintained in the pond, especially against zoothamniosis. The immunostimulants that entered the body of the shrimp would stimulate the haemocytes activity cells in shrimp, attempting to fight the pathogens entered the body of the shrimp during maintenance. This is in accordance with Van de Braak [17], which stated that hemocyte-activated immunostimulant cells would perform phagocytic activity on shrimp by hyalin (granular) and semi granular cells.

According to Soderhall and Cerenius [18], the immune system in shrimp was still primitive and unlike that in fish and mammals that has immunoglobulins. So, immunoglobulin in shrimp was replaced by Prophenoloxidase Activating Enzyme (PPA). The PPA is a protein located in granular haemocytes. This PPA can be activated by lipopolysaccharides and β 1.3-Glucan, which will stimulate prophenoloxidase into phenoloxidase. As the result of the activation, it will then produce a Opsonin Factor protein that can induce hyalin cells to perform the process of phagocytosis. Van de Braak [17] and Rukyani [9], also supported the statement above that the haemolytic cells will degranulate, and some proteins would be released for immune response, such as increased haemocytes, increased trapping activity, and phagocytosis. In addition, the immunogenic membrane proteins would stimulate haemocytes to release proPO and protein-binding PPA, thereby causing hemocyte cells to increase their activity for trapping and phagocytosis against disease agents, which related in this case was the *Zoothamnium penaei*. It is evident that the prevalence of zoothamniosis in immunized and different immunized shrimp was very marked with the prevalence of unimmunized shrimp. The presence of infested shrimp *Zoothamnium penaei*, because this parasite was opportunistic, so that in normal water condition, this parasite still grows but it develops long, which does not cause disease in the shrimp.

If the immunostimulant goes into the body of the shrimp then it will cause an increase in the number of haemocytes (THC) and haemocyte cell differential (DHC). This is an indication of increased shrimp body defense against zoothamniosis attacks [8]. Furthermore Itami *et al.* [16] supports the basic theory, which says that vaccine administration can prevent infectious diseases in the host body and lead to increased phagocyte activity of haemocytes and proPO enzymes. This statement is reinforced by Soderhall *et al.* [30] and Van de Braak [17], that membrane proteins entering the shrimp body will induce antibodies capable of neutralizing *Zoothamnium penaei* infestation, thus unable to infest the shrimp.

According to Mahasri [8], the immunogenic membrane proteins which entered the body could increase the tiger shrimp survival rate from 17 % to 68 % in 90 days old shrimp (the end of maintenance). Furthermore, it was also said that the immune response of tiger shrimp also increased as indicated by the increase of THC and DHC, because immunogenic membrane protein *Zoothamnium penaei* had a large molecular weight that was greater than 1000 Dalton, thus making it immunogenic. Proteins that had high molecular weight and had a high level of immunogenity, the protein must have a complex structure. In the opinion of Tizard [31] and Baratawidjaja [32], proteins that were immunogenic had a large molecular weight of more than 1000 Dalton and had a complex structure.

Increased total haemocyte (THC) and DHC cells could be used as the indicators or signs of pathogen infection in the host body. This infection would cause inflammation, which was a non-specific body defense characteristic due to influencing factors such as parasites, bacteria, fungi, viruses and nonliving agents [11, 18]. The results showed that there was an increase in THC and DHC in shrimp immunized with membrane proteins and different from those not immunized. THC in immunized shrimp increases in shrimp from 30 days to 60 days, but decreases at age 90 days. This was because the shrimp immune system would increase along with the increasing shrimp life and but at certain age would decrease again. The high THC indicated that immunogenic membrane proteins could improve the shrimp immune response, because high THC in shrimp is one indicator of increased shrimp body resistance. This is in accordance with the opinion of Soderhall *et al.* [30], who suggested that increased immune responses to invertebrates were indicated by an increase in THC. Increased THC and DHC could be used as the indication of shrimp body defense reaction with Z. *Penaei* infestation. The differences between THC and DHC among shrimp treatment groups in this protein-protective testing test, probably due to the immunogenicity of different membrane proteins. According to Baratawidjaja [32], protein immunogenity was determined by large molecular weight.

The results also showed that there was a significant difference ($p < 0.05$) between THC and DHC shrimp immunized with immunogenic membrane proteins at different doses and infested with the same zooid *Zoothamnium penaei*. The main factor causing this difference is the immunogenity of the membrane proteins as immunostimulants. There is a significant increase in DHC (granular haemocytes), presumably because shrimps did not have memory cells in the immune system, which were unable to detect pathogens that had been exposed. Thus, it can be argued that immunostimulant membrane proteins could induce the defense mechanism of the shrimp body. However, it took time to stimulate the haematopoietic organ to produce granulocytes to counteract zoothamniosis attacks [8]. These granulocytes would then destroy pathogens by swallowing the pathogens, so these granulocyte cells would migrate to parasitic infestation areas. On the other hand, the water quality maintenance also affected the infestation and immune response in the shrimp, but based on the results of research, it showed that the water quality was in the optimal conditions for shrimp life during maintenance.

## 4. Conclusion

The conclusions that can be drawn from the results of this study: 1) Immunization with immunostimulant from membrane *Z. penaei* protein could decrease *Zoothamnium penaei* infestation on vaname shrimp maintained in pond from 68 % to 17 % during 90 days of maintenance period; 2) Immunization with immunostimulant of immunogenic membrane proteins *Z. penaei* was able to increase the immune responses (increase THC and DHC) in the vaname shrimp, from 27.08 x $10^6$ cells/mL to 41.55 x $10^6$ cells/mL for THC and 12.99 % to 24.03 % for DHC; and 3) Immunization with immunostimulant from immunogenic membrane protein *Z. penaei* was able to increase the survival rate of vaname shrimp from 17 % to 68 %.

## 5. References

[1] Taslihan A 2006 Impelentation of shrimp cultivation techniques (bmp) by good practice to anticipate viral white patch disease in semiintensive ponds *Workshop on Implementation of BMP Technology to Support Shrimp Culture Revitalization* Surabaya 6-8 Maret 2006
[2] Buchholz F 2003 *Mar. Freshw Behav. Physiol.* **36** 229–247
[3] Foster C A, Sharpie T G and Hawkins W E 1998 *Fish. Dis.* **1** 321 – 335.
[4] Sindermann C J 1997 *Trans. Am. Microse. Soc.* **98** 136 –138.
[5] Tonguthai K 1991 Diseases of the Freshwater Prawn, *Machrobrachium rosenbergii*, AAHRI Newsletter Article Bangkok Thailand **4**
[6] Chamratchakool P 1996 Health Management in Shrimp Ponds, Health Research Institute, Bangkok Thailand pp 50 – 53.
[7] Mahasri G 1996 Effect of manipulation of aerated and dense stocking level against parasite infestation of ciliata protozoa class on shrimp fries, Thesis, IPB, Bogor : 67 hlm.
[8] Mahasri G 2007 Immunogenic membrane protein zoothamnium penaei as an immunostimulant development material on windu shrimp (*Penaeus monodon* Fabricus) *Against Zoothamniosis. Paper Presented at Int. Conf.* FKH – Unair. Surabaya.
[9] Rukyani A 1996 Types of Shrimp disease in ponds and how to control it, application of agricultural technology tool application paper date: 9 – 11 Januari di BIP Lampung.
[10] Amos W B, Routledge L M, and Yew F F 1975 *J. Cell. Sci.* **19** 203-213.
[11] Routledge L M 1978 *J. Cell. Biol.* **77** 358-370
[12] Pylawka, S and Buhse Jr 2003 *J. Eukaryot Microbiol* **198** 45 - 52
[13] Itabashi T ,Terasaki T and Asai H 2004 *J. Biochem.* **136** 651- 657.
[14] Moriyama Y, Okamoto H and Asai H 1999 *J. Biophys.* **76** 993–1000
[15] McCutcheon S M, McLaughlin N B, and Buhse H E 2002 *Mol. Biol. Cell.* **13** 331a.
[16] Itami T 1994 Body Defense System of *Penaei*d Shrimp, Seminar on Fish Physiology and Prevention of Epizootics Department of Aquaculture and Biology Shimonoseki University of Fisheries Japan **7** : 59-65.
[17] Van de Braak K 2002 Haemocytic defence in black tiger shrimp (*Penaeus monodon*), Disertation, van Wareningen Universiteit, Germany.
[18] Sőderhąll K and Cerenius L 1978 *Annu. Rev. Fish. Dis.* pp 3 – 23.
[19] Johansson M W, Keyser P, Sritunyalucksana K, Söderhäll K 2000 *Aqucat* **191** 45–52
[20] Fontaine C T and Lightner D V 1974 *J. Invertebr. Pathol.* **24** 141-148

[21] Kondo T, Sugiura M Kamada, H Imagawa 1996 *J. Equine. Sci.* **7** 63-66
[22] McKay D, Jenkin C R, Rowley D 1969 *Aust. J. Exp. Biol. Med. Sci.* **47** 125–134 [PubMed]
[23] He F, Fenner B J, Godwin A K and Kwang J 2006 *J. Virol.* **80** 3884-92
[24] Johnson R T and Johnson D W 1986 *Science and Children* **24** 31-32.
[25] Owens L and A O'Neill 1997 *Dis. Aquat. Organ.* **31** 147–153.
[26] Effendie M I 1979 *Method of fisheries biology* (Bogor: Yayasan Dewi Sri) 111 p
[27] Steel R G and Torrie J H 1993 *Principles of statistical procedures* (Jakarta: Gramedia) Translation by Bambang Sumantri p 425-478
[28] Debaufer P J and Buhse H E 1998 A Possible Mechanism for Initiating Stalked Zooid to Telotroch Transformation in Vorticella *Proc. of the 51st Annual Meeting of The Society Of Protozoologists*
[29] Smith R, Chaffin M K, Cohen N D and Martens R J 2002 *Am. J. Vet. Res.* **63** 531–537
[30] Soderhall *et al.* 1996 *New directions in invertebrate immunology* (SOS Publications: Fair Haven, NJ) pp 131-154
[31] Tizard I R 1988 *Introduction to veterinary immunology* (Translation) (Surabaya: Airlangga University Press)
[32] Baratawidjaja C 2004 basic immunology (Jakarta: Balai Penerbit Fakultas Kedokteran Universitas Indonesia) 4th ed.

**Acknowledgment**
We are thankful to the staff of Balai Karantina Ikan dan Pengendalian Mutu Hasil Perikanan, Kelas I, Surabaya, Indonesia for their valuable helps with fieldwork and sample collection. This study was sponsored of DRPM, Ministry of Research Technology and Higher Education Republic of Indonesia.

IOP Conf. Series: Earth and Environmental Science **137** (2018) 012021 doi:10.1088/1755-1315/137/1/012021

# The fecundity of fork-tailed threadfin bream *(Nemipterus furcosus)* in Bangka, Bangka Belitung

**E Utami[1], E Safitriyani[2] and Leo Gatra Persada[2]**

[1]Department of Marine Science, Bangka Belitung University, Bangka Belitung, Indonesia.
[2]Department of Aquatic Management Resources, Belitung University, Bangka Belitung, Indonesia.

E-mail: evautami.eva@gmail.com

**Abstract.** Fork-tailed threadfin bream *(Nemipterus furcosus)* is one of important economic fishes in Bangka. The sustainability of fork-tailed threadfin bream is threatened by degradation of natural habitat. Information of reproductive is needed for further management. The objective of this study was to examine fecundity of fork-tailed threadfin bream. The mean values of temperature was 28.83 ± 0,37°C, respectively. Sex ratio during sampling showed that female fork-tailed threadfin bream greater than male population. Berried female fork-tailed threadfin bream found from March until November. The greatest number of berried female fork-tailed threadfin bream showed in July with berried female value of 25. Fork-tailed threadfin bream fecundity was 19951 and 66628, respectively. The fecundity data can be used to access the reproductive potential of fish stock and also as an assessment on stock size of their natural population.

## 1. Introduction

Fork-tailed threadfin breams (*Nemipterus furcosus*), are indigenous from Southern Japan to north-eastern Australia and the Indian Ocean.This species is a valued food source and has beenfished commercially by hook and line. Fork-tailed threadfin breams areone of the important economic fishes in Bangka, Indonesia. The production volume of Fork-tailed threadfin breamswas approximately 269.289 kg in 2010 and 293.618 kg in 2011 with the percentage of production at around 9. 03 %. In 2012, catch increased to 311.502 kg or 6.09 %, while it reached a peak in 2013at 434.774 kg (39.57 %) [1]. The catch may obtain high yields in a given year, but if fishing efforts continued it would decline in the years that followed. This is due to its resources been caught in the previous years. In line with the increase of production volume of the catch, there needs to be measure to manage and maintain the availability of Fork-tailed threadfin breams. However, studies about Fork-tailed threadfin bream fecundity and aspects of reproduction in Bangka is still lacking. Fecundity study is important in estimating the stock size of their natural population. Due to the importance of *Nemipterus furcosus* for commercial fisheries, much is known about their ecology, reproductivebiology and behavior. The objective of this study was to provide information on the reproductive biology including sex ratio, fecundity, spawning season,and fishing ground of *Nemipterus furcosus* in Bangka.

## 2. Methodology

IOP Conf. Series: Earth and Environmental Science **137** (2018) 012021 doi:10.1088/1755-1315/137/1/012021

A total of 456 specimens were randomly collected each month from January until December 2015 at Nusantara Fishery Port, Sungailiat, Bangka [2]. *Nemipterus furcosus*was caught using handline fishing with the fishing ground located onthe north-eastern parts of Bangka Island.Total length (TL) was measured to the nearest 0.1 mm. Ovaries were macroscopically examined to determine the condition of maturity. Each ovary was divided into 3 portions (anterior, middle and posterior) and one third wastaken to estimate fecundity. Fecundity was analyzed using Efendie [3]. In total,108 ovaries were used to estimate fecundity. The relationship of fecundity (F) compared to body weight (BW) was described by the algometric equation Ln F=b Ln BW + a, while therelation of fecundity (F) comparedtototal length (TL) was described by the algometric equation Ln F=b Ln TL + a.

## 3. Result and Discussion

### *3.1. Sex Ratio*

Overall, sampling revealed that the sex ratio of Fork-tailed threadfin breams were 263 female, 193 male and 108 berried female. The sex ratio showed that during 2015 the number of female *Nemipterus furcosus*was significantly greater than the male population, with a ratio of Female: Male = 1.4: 1. This ratio is not harmful to fish stocks. Females outnumbered males throughout the yearexcept in January and March. The proportion of Female: Male in January and March were 1: 1.2; 1:1.7, respectively. Meanwhile in May, the ratio between female and male remained equal at1:1.Furthermore, the highest gap in the number between female and male of the fish was found in July with the sex ratio of 6:1. Comparisons between both sexes of female and male consecutively from June until December,show that the number of females wasgreater than males.The sex ratio for female fish rangedfrom the lowestof1 in January, March and May to the highest of6 in July; whilst the proportion for males reached the highest at 1.7 in March.

### *3.2. Fecundity*

The monthly mean length and weight of berried Fork-tailed threadfin breams ranged between 19.10 ± 0.71cm – 22.12 ± 2.50 cm and 76.04 ± 18.39 gr – 114 ± 14.26 gr. Fecundity (F) compared to total length (TL) was Ln F= 5.4275 Ln TL – 5.7945. Fecundity (F) compared to body weight (BW) was Ln F=2.0188LnBW +1.3817.The development of the five stagesof maturityfor Fork-tailed threadfin breams gonad was determined by visual examination.Stage I: testes and ovaries are flat transparent. Stage II: testes are white flat and ovaries are white tubular with blood vessels surrounding the gonad. Stage III: testes are white tubular with 2/3 – 3/4 body cavity in length; ovaries are light yellow tubular. Stage IV: testes are white tubular, fulfilling the body cavity and ovaries are yellow; eggs can be seen by visual examination. Stages V: gonads are red creased. Eggs will release if female fish are in stage IV of maturity gonad.

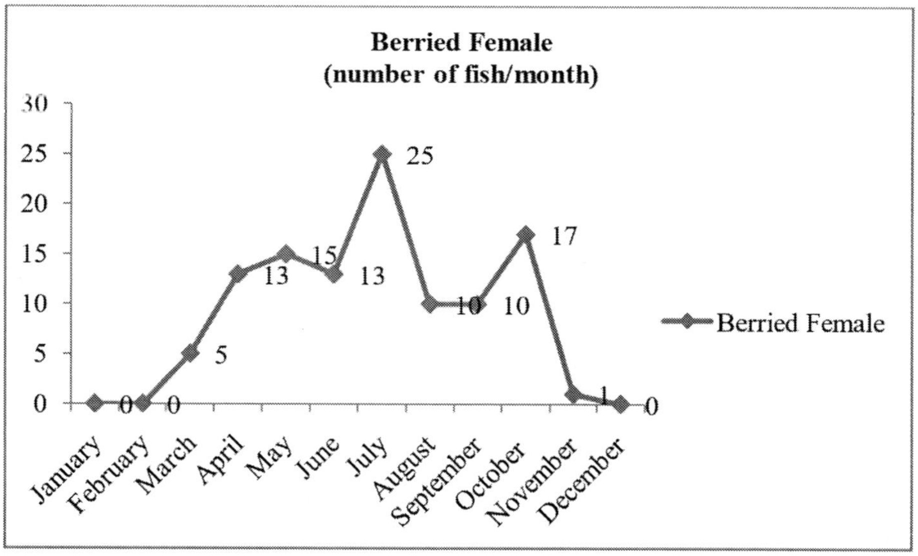


**Figure 1**.Monthly number of berried female Fork-tailed threadfin bream (number of fish/month).

The amount of berried female Fork-tailed threadfin bream each monthduring 2015fluctuated considerably. The greatest number of berried female Fork-tailed threadfin bream occurred in July with 25 berried female fish. Beginning in January and February, none of the female ovaries were indicated to be in the mature stage. Nevertheless, there was a gradual rise of the amount of berried females in March, April and May, at 5, 13 and 15 fish, respectively, followed by a marginal dip until 13 berried female fish was caught in June.Afterreaching a peak of approximately 25 fish in July, it decreased sharply to a mere 10 fish. The amount of berried females remained the same for the following month, while, it increased negligibly in October to roughly 17 fish. It dropped sharply to only 1 berried female in November and leveled off thereafter. The macroscopic appearance of ovaries indicated that most immature ovaries were presented from November – March. Spawning season appeared from April until October with a peak in July. The protection of adults during spawning season is needed to ensure the sustainability of fish stock [4].

### *3.3. Fishing Ground*

The fishing ground of the Fork-tailed threadfin bream is in the north-eastern part of Bangka Island withthe range of coordinatesbetween 105°50'-106°50'E;0°50'-2°1'S. In Bangka Belitung Province, Fork-tailed threadfin bream are caught by handline. Handline is a fishing technique where a single fishing line is held in the hands with one or more hooks and fishing lures attached to the line.

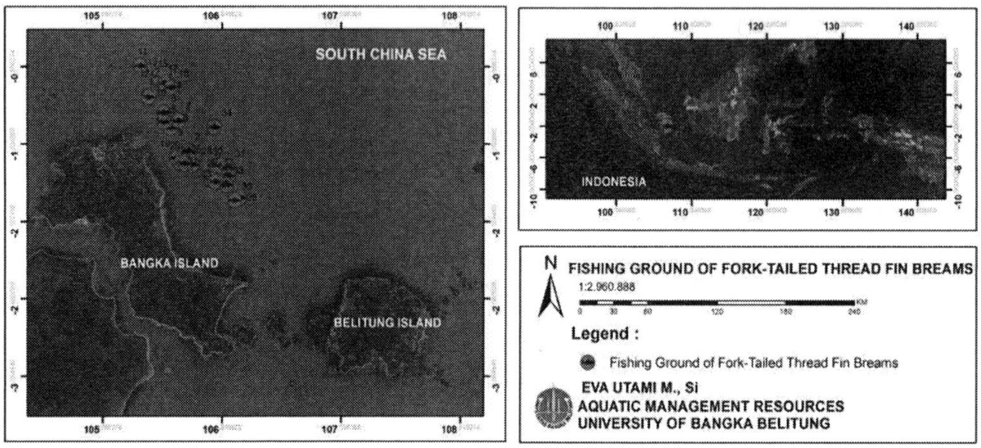


**Figure 2.** Map showing the fishing ground of Fork-tailed threadfin bream.

Fork-tailed threadfin bream are demersal fish. It is found in 15 meters to 40 meters under the sea surface with a sandy seabed and coral reefs. The temperature of sea water was measured between 27°C at the lowest to 31°C for the highest.The average temperature of the Bangka sea water during the year 2015 was 28.8±0.39 °C.

## 4. Conclusions:

The fecundityof Fork-tailed threadfin bream in Bangka appeared from April until October, with a peak in July. The sex ratio showed that the female *Nemipterus furcosus*population was significantly greater than its male population, with a ratio of Female: Male = 1.4 : 1. The fishing ground of Fork-tailed threadfin bream is in the north-eastern part of Bangka Island.

## 5. References

[1]. Nusantara Fishery Port Sungailiat 2013 Data series of Fork-tailed threadfin bream (*Nemipteridae*) production and fishing gear in Bangka
[2]. Fauzi A 2001 An Economic Analysis of The Surplus Production: An Application for Indonesian Small Pelagic Fisher
[3]. Effendie MI 1979 *Methode of Fisheris Biology.* (Bogor: Yayasan Pustaka Nusantara) p155
[4]. Wu C C, Weng J S, Liu K M and Su W C 2008 *J. Zool Stud.* **47**

IOP Conf. Series: Earth and Environmental Science **137** (2018) 012022 doi:10.1088/1755-1315/137/1/012022

# The effect of *Chaetoceros calcitrans* extract on hematology common carp (*Cyprinus carpio*) infected by *Aeromonas salmonicida*

**Maftuch[1,4], N D A Wulan[1], H Suprastyani[1], E Wijayanto[1], M Noercholis[1], A A Prihanto[2,4] and A Kurniawan[3,4]**

[1]Dept. Aquaculture, Faculty of Fisheries and Marine Science, Brawijaya University, Malang, Jl. Veteran, Malang 65145, East Java, Indonesia
[2]Dept of Fishery Product Technology, Faculty of Fisheries and Marine Science, Brawijaya University, Malang, Jl. Veteran, Malang 65145, East Java, Indonesia
[3]Dept Aquatic Resources Management, Faculty of Fisheries and Marine Science, Brawijaya University, Malang, East Java, Indonesia
[4]Coastal and Marine Research Center, Brawijaya University, Malang, East Java, Indonesia
E-mail : maftuch@ub.ac.id

**Abstract.** The application of *C. calcitrans* extract in carp (*C. carpio*) is expected to inhibit the growth of *A. salmonicida. A. salmonicida*-infected common carp (*C. carpio*) were treated with the extract of *C. calcitrans*. Hematology, erythrocyte, leukocyte, hematocrit and hemoglobin test.analysis was observed. The result indicated that the extract can be used to treat the infected fish. The best dose was treatment of D with 45.3 ppm.

## 1. Introduction

The fishery and marine sectors is leading sectors for promoting economic growth in Indonesia [1]. Aquaculture is an endeavor to increase the fisheries production. According to Suseno [2], carp (*Cyprinus carpio*) is an example of freshwater fishery which has an economical important value. Hence, thi species is one of important species to be intensively cultured.

One dracbacj for the aquaculture is bacterial infection, such as *A. salmonicida. A. salmonicida* is a gram-negative bacterium categorized as must Quarantine Disease and Pest of Fish. This bacterium is designated as a quarantine pest of fish-bacteria class II based on Ministerial Decree No.03/MEN/2010 [3]. The attack of this bacterium is extremely severe.

An alternative agent for combating this infection is by using a natural product. Anti-bacterial agent which are natural and effective to kill and inhibit the growth of bacteria. The extract of *Chaetoceros calcitrans* contains many metabolites and antibacterial compounds composed of fatty acids, alkaloids, and terpenoid [4]. Hematology analysis is a very useful diagnostic method to determine the health status of fish. Blood tests can be used as an indicator of the severity of a particular disease. The

IOP Conf. Series: Earth and Environmental Science **137** (2018) 012022 doi:10.1088/1755-1315/137/1/012022

parameters that can show pathological changes in blood are the number of leukocytes, leukocyte differential, and erythrocytes [5].

This study was aimed to determine the effect of *C. calcitrans* extract on the hematology of common carps (*C. carpio*) and to prescribe the best dose of *C. calcitrans* extract for common carps (*C. carpio*) infected by *A salmonicida*.

## 2. Methodology

### 2.1. Materials

The research was conducted from February 20, 2017–April 20, 2017 at Fish Cultivation Laboratory, Division of Disease and Health of Fish, Faculty of Fisheries and Marine Sciences and Central Laboratory of Life Sciences of Brawijaya University.

The following materials were used in this research: *C. calcitrans* extract and *A. salmonicida* were obtained from BPBAP (Center of Brackish Water Aquaculture) Situbondo. Common carps (*C. carpio*) in the size of 5–7 cm with the weight of 6–10 grams obtained from local breeder. The research design used was complete randomized design (CRD). This research consisted of four treatments and control with three replications.

### 2.2. Preparation and infection of animal test

The fish used in this research were common carps (*C. carpio*) that were obtained from fish farmers in Malang. As many as 210 fish in the size of 5–7 cm were required in this research. According to [10], during the study, the carps were given pellet feed as much as 4% of their biomass weight 2 times (morning and evening) in a day. After being acclimatized, the fish were ready to use.

### 2.3. Infecting common carp (C. carpio) with A. salmonicida

Bacteria *A. salmonicida* were immersed in an aquarium containing the carps (*C. carpio*). The infection process followed [11]. The bacterial density of $10^7$ cells/mL was applied for the infection. Fish was soaked for 2 hours.

### 2.4. Treatment of C. Calcitrans extract in carps (C. carpio)

The administration of *C. calcitrans* was done by immersion method. The *C. calcitrans* extract was administered at doses of 0.3, 15.3, 30.3, and 45.3 ppm (consistent with in vitro dosage). Then, the carps were kept for 7 days to allow the extract to function as an antibacterial agent. Then, blood sample was taken as a negative control.

### 2.5. Hematology analysis

Fish blood sampling was performed after 12 hours of infection. Blood was taken from caudal peduncle of carps (*C. carpio*) using a syringe needle. 3.8% citrate was added to the blood as anti-coagulant.

a. Erythrocytes

The number of erythrocytes was counted according to Blaxhall and Daisley [14]. Blood was dropped onto the hemocytometer (Neubauer type) and covered with a glass cover. The total erythrocyte was counted in five small boxes and the amount was calculated using the following formula:

IOP Conf. Series: Earth and Environmental Science **137** (2018) 012022 doi:10.1088/1755-1315/137/1/012022

$$Total\ erythrocytes = \sum N\ x\ 10^4\ {}^{cell}/_{mm^3} \qquad (1)$$

b. Leukocytes

The number of leukocytes was measured according to Blaxhall dan daisley [14]. Then, turk solution was added up to scale 11. After that, the first two drops of blood solution from the pipette were removed, and the next drop was dropped into the solution on the hemocytometer. The total leukocyte count was calculated following the formula:

$$Total\ leukocytes = \sum n\ x\ 50\ {}^{cell}/_{mm^3} \qquad (2)$$

Note:

$\sum$ n : Total number of leucocytes in 4 field of view
50 : Dilution factor

c. Hematocrit

The level of hematocrit was measured according to Anderson and Siwicki [15]. The end of the bloodfilled tube was covered with crytoceal by placing the tip of the tube into crytoceal approximately 1 mm deep to form a crytoceal plug. The microhematocrit tube was centrifuged for 5 minutes at a speed of 12,000 rpm with the position of the tube having the same volume to face for a balanced centrifuge rotation. The length of the deposited blood portion (a) and the total volume of blood contained in the tube (b) were measured using a microhematocrit tube. Measurement of hematocrit grade values was done by comparing the volume of red blood cell solids with total blood volume using hematocrit scale.

d. Hemoglobin

The calculation of hemoglobin followed Wedemeyer and Yasutake method [16]. Sahlinometer tube was filled with HCl 0.1 N solution until number 10 showed (the lowest scale line on the sahlinometer tube). Then, the tube was placed between 2 tubes in standard color. Next, 0.02 mL fish blood was taken from the tube with sahli pipette. The next steps were cleaning the tip of the pipette, putting the blood into the Sahli tube, and waiting for 3 minutes. Lastly, distilled water was added gradually by using a dropper, and it was stirred by using a stirrer glass until the color turned exactly the same as the standard color.

## 3. Results and Discussion

### *3.1. Erythrocytes*

Diagram of erythrocyte observation results during observation and in figure 1. From the result it can be seen that the highest erythrocytes (2.4 cell/ml) was obtained from the extract of 45.3 ppm. The lowest cell was from 12 hours and the dose of 0.3 ppm extract.

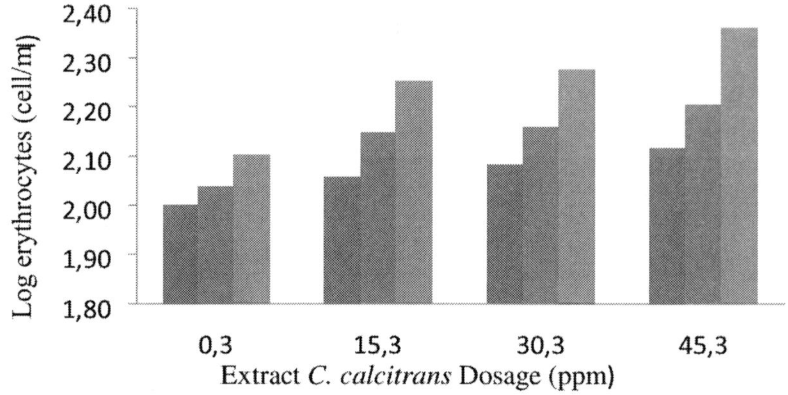


**Figure 1.** Number of Erythrocyte of infected-Common Carp (*C. carpio*). (Blue is 12 hours, Orange is 24 hours, Grey is 36 hours).

Regression analysis suggested that the erythrocytes was followed of y = 2.12 + 0.005x function, which had a coefficient determination of 0.8266. This suggested that the higher the dose given, the higher the number of erythrocytes. Erythrocytes in the body play a role in the transfer of oxygen throughout the body. The decrease of the number of erythrocytes leads to lower oxygen supply to all organs and reduces food supply to the cells.

### *3.2. Leukocytes*

The average results obtained from leukocyte observation can be seen in figure 2.

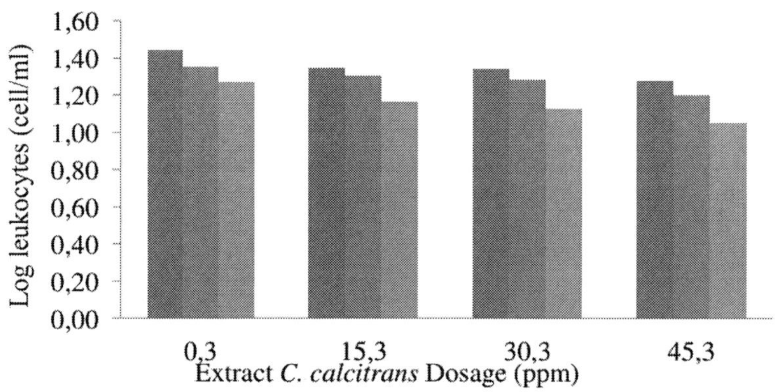


**Figure 2.** Number of Leukocytes in infected-Common Carp (*C. carpio*). (Blue is 12 hours, Orange is 24 hours, Grey is 36 hours).

Orthogonal polynomial analysis showed that the result followed the function ofy = 1.256-0.004x, which has a coefficient determination of (R2) of 0.740. The higher the dose of *C. calcitrans* extract given, the lower the number of leukocytes in the blood of carp. Leukocyte is one of blood components

IOP Conf. Series: Earth and Environmental Science **137** (2018) 012022 doi:10.1088/1755-1315/137/1/012022

that has a function in the non-specific body defense system that will localize and minimize pathogens through phagocytosis [17].

### *3.3. Hematocrit*

The average results obtained from hematocrit observations during the observation can be seen in figure 3.

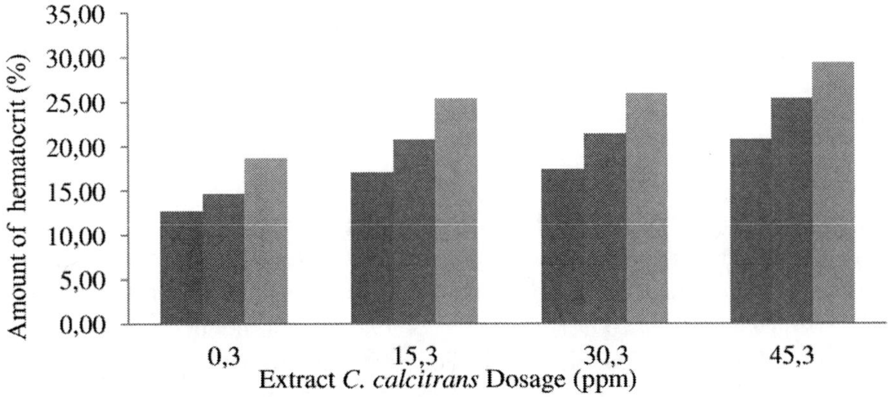


**Figure 3.** Number of Hematocrits in infected-Common Carp (*C. carpio*). (Blue is 12 hours, Orange is 24 hours, Grey is 36 hours).

Regression analysis revealed that the hematocrit followed the function of y = 19.86 + 0.217x, which had a coefficient determination of 0.801. The extract affected the hematocrit levels of the fish (*C. carpio*). This is shown by the fact that the higher dose of extract *C. calcitrans* given, the higher the hematocrit level in carp blood (*C. carpio*). This is in line with Wahjuningrum *et al.* [18], which revealed that a hematocrit level of less than 30 % indicates that the fish has vitamin deficiency, low protein in the feed, and infection. If the hematocrit is lower than normal fish, the fish is in anaemia condition. Contrastingly, if the hematocrit level is above normal, the fish is stress or infected.

### *3.4. Hemoglobin*

The average results obtained from hemoglobin observation can be seen in figure 4.

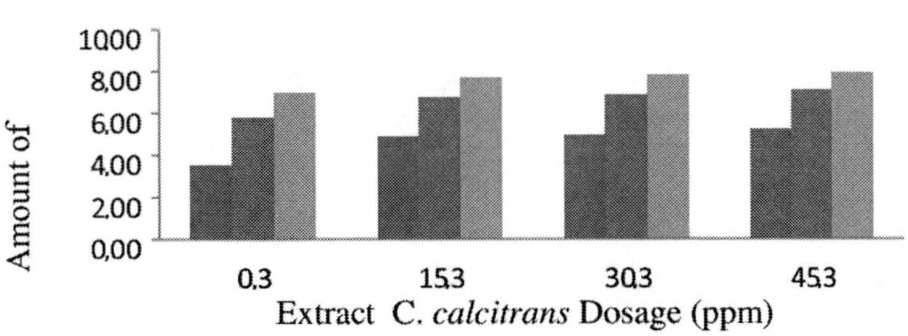


**Figure 4.** Number of Hemoglobin in infected-Common Carp (*C. carpio*). (Blue is 12 hours, Orange is 24 hours, Grey is 36 hours).

IOP Conf. Series: Earth and Environmental Science **137** (2018) 012022 doi:10.1088/1755-1315/137/1/012022

The hemoglobin resulted in the fuction of y = 7.107 + 0.020x, which had a coefficient determination of 0.730. According to Sukenda *et al.* [17], the hemoglobin contained in red blood cells has the function in the transport of oxygen from the gills to the tissues. Research from Serapse *et al.* [19] suggested that the extract of *Chaetoceros sp.* which is dissolved with ethanol and various organic materials, produces terpenoid compounds. The results of the study also showed that *Chaetoceros sp.* extract had a wide spectrum of activity. Hence, *C. calcitrans* extract could be used to inhibit bacterial growth.

## 4. Conclusion

Treatment of *C. calcitrans* extract affected the hematology of infected common carp (*C. carpio*). Based on the result *C. calcitrans* extract has a potency to be used as the treatment agent for *A.salmonicida*

## 5. References

[1]. Nadeak A 2009 *J. Plan. Reg. Dev.* **4** 102-110 (In Indonesia)
[2]. Suseno 2000 Carp breeding bussiness management (Jakarta : Penebar Swadaya) 74 p
[3]. Retnoningsih S, Nitimulyo K H, Lanadimulya K, Suprayogi, Supardi, Darmantani D, Panca I P, Hasnah, Soefaad andMilis. 2009 *J. Fish. Sci.* **11** 192-200 (In Indonesia)
[4]. Setyaningsih I, Linawati H, Monintja D R, Sondta M F A, Bintang M, Lailati N and Panggabean L 2010 Cultivation and characterization of active components and nutrients of the marine microchannel *Chaetoceros gracilis* (Bogor : Bogor Agricultural Institute Disertation) 134 p
[5]. Maftuch, Nursyam H and Sukarni 2012 *J.Exp. Life Sci.* **2** 65-69.
[6]. Kwangdinata R., Raya I and Zakir M 2014 *The Sci. World J.* 1-5
[7]. Sari I P and Manan A 2012 *J. Fish. Mar. Sci.* **4** 1-5 (In Indonesia)
[8]. Rustanti, E, Jannah A and Fasya A G 2013 *J. Alchemy.* **2** 138-149
[9]. Pastra, D A, Melki and Surbakti H 2012 *Maspari J.* **4** 77-82 (In Indonesia)
[10]. Widiastuti I M 2009 *Med. Lit. Sulteng.* **2** 126–130
[11]. Mariyono and A. Sundana. 2002 *Bull. Tech. Agri.* **7** 33-36 (In Indonesia)
[12]. Kusriani, Widjanarko P and Rohmawati N 2012 *J. Fish. Res.* **1** 36-42 (In Indonesia)
[13]. Lestari A T and Sudaryatma P E 2014 *J. Vet. Sci.* **32** 85-92 (In Indonesia)
[14]. Blaxhall P C and Daisley K W 1973 *J. Fish Biology.* **5** 577-581.
[15]. Anderson, D. P and Siwicki A. K. 1993. Basic hematology and serology for fish health programs (Thailand : Symp. on diseases in Asian Aquaculture "Aquatic Animal Health and the Environment") 202 p
[16]. Wedemeyer G A and Yasutake W T 1977 *Clinical methods for the assessment of the effect environment stress on the fish health* (US : Depart of the Interior Fish and Wildlife Service) **89** 1-17
[17]. Sukenda L J, Wahyuningrum D and Hasan A 2008 *J. Aquac. Ind.* **7** 159-169 (In Indonesia)
[18]. Wahjuningrum D, Angka S L, Lesmanawati W, Sa'diyah and Yuhana M 2007 *J. Aquac. Ind.* **6** 109-117
[19]. Serapse E B, Ticar B F, Formacion M J, Pahila, Milagros R and Amar E C 2012 *Asian Fish. Sci.* **25** 343-356

IOP Conf. Series: Earth and Environmental Science **137** (2018) 012023 doi:10.1088/1755-1315/137/1/012023

# The motility and motion duration of jatimbulan tilapia (*Oreochromis niloticus*) spermatozoa in different salinity

**J Triastuti[1], D Kintani[2], E M Luqman[3] and D Y Pujiastuti[1]**

[1]Marine Departement, Faculty oof Fisheries and Marine, Universitas Airlangga
[2]Aquaculture Study Programme, Faculty oof Fisheries and Marine, Universitas Airlangga
[3]Anatomy Veterinary Department, Faculty of Veterinary Medicine, Universitas Airlangga Jl. Mulyorejo Kampus C Unair, Surabaya

E-mail: jt.astuti@gmail.com

**Abstract**. Tilapia hatchery is still conducted in freshwater and seeds are death simultaneousy when cultivated in high salinity due to the acclimatization process. An alternative method to implement hatchery at high salinity is required. This study aims to determine the salinity of activation medium that provides the best Jatimbulan Tilapia sperm motility and motion duration at high salinity. The study applies completely randomized design (CRD), which consists of 5 treatments (0 ppt, 4 ppt, 9 ppt, 14 ppt and 19 ppt) and 4 repetitions. The parameters consists of sperm motility, motion duration, fresh sperm data (volume, color, odor, pH, consistency, and the concentration of sperm) and sperm abnormalities. The results exhibited that salinity significantly ($p < 0.05$). Influeneed the sperm motility and motion duration. Motility reaches its best at 0 ppt and 4 ppt (93.4 % and 87.8 %). For motion duration, best condition was in 0 ppt and 4 ppt treatments, totaling 2128 seconds and 1961.5 seconds. Meanwhile, sperm did not move when treated in waters with 9 ppt, 14 ppt and 19 ppt salinities.

## 1. Introduction

Tilapia is a species of fish with a high economic value [1] and *euryhaline*, which means that they are able to live and reproduce at a salinity above 30 ppt. The Freshwater Aquaculture Development Center (BPBAT) in Umbulan, East Java [2] has developed a superior tilapia seed through individual selection and has named as Jatimbulan tilapia (Jatimbulan stands for Jawa Timur (East Java) Umbulan). Similar to the case of milkfish *(Chanos chanos),* fish farmers have began to consider tilapia as a main commodity. Tilapia breeding has been cultivated in freshwater so when farmers cultivate seeds at high salinity, the seeds often die because of the abrupt change to salinity [3]. An alternative method that can be done to cope with death in high salinity is by hatching Jatimbulan tilapia in high salinity.

Research has been conducted as an efforts to hatch Jatimbulan tilapia at high salinity. Jatimbulan tilapia can tolerate 0 to 15 ppt salinity, but 10 ppt salinity is more recommended for to obtain optimum results. However, the results were only based on experiments performed in the laboratory with artificial fertilization. Hence further research on natural fertilization is required to prove the success of Jatimbulan tilapia hatchery at high salinity. The inhibiting factor in the natural fertilization of Jatimbulan tilapia is the suitable salinity for the motility and motion duration of the sperm remains unknown.

Fertilization in tilapia occurs externally [4]. According to Morita *et al.* [5] sperm motility is regulated by osmolality and calcium ion [Ca $2^+$]. When the sperm comes in contact with a hypotonic environment, it turns hypertonic to the environment. This causes the Ca $2^+$ ion to increase and protein phosphorylation to take place so that the sperm becomes motile. According to Takai and Morisawa [6] argues that isotonic solution makes the sperm become nonmotile. When tilapia sperm is in contact with a hypertonic environment, its sleeve shrinks and its intracellular ion [$Ca^{2+}$] is not increased so the sperm becomes nonmotile and dies [5]. The increase in salinity causes a decrease in the sperm motion duration of carp fish. If sperm motion duration and motility decreases, the ability of sperm to penetrate the micropyle is weak. Therefore, further research is required on the motility and motion duration level of Jatimbulan tilapia sperm at different salinities.

## 2. Methodology

The research was conducted using sperm derived from a Jatimbulan strain tilapia male parent, sperm coloring in form of eosin 2 %, NaCl 3 %, physiological NaCl, saltwater, and freshwater. This research employed experimental methods with a completely randomized design (CRD), which consists of five salinity treatments, i.e: 0 (control), 4, 9, 14 and 19 ppt with four repetitions for each treatment.

Sperm retrieval was performed with the stripping technique, which provides subtle pressure on the belly of the fish starting from the linea lateral to the genital orifice. Sperm was collected in a container for different salinity treatments as stipulated in this study. Sperm was stored in a physiological NaCl with a 1: 9 ratio of sperm to diluent. The sperm was stored in a refrigerator [7].

The main parameters of this study were sperm motility and motion duration. Observation of sperm motility was conducted through a light microscope with 400x magnification, while a handtally counter was used to calculate the progressive motile sperm. Observation of motion duration was done through a 100x-magnification microscope, while a stopwatch was used to calculate its duration. Support parameter observations in this research includes the quality of fresh sperm (volume, color, odor, pH, and consistency) which was observed at baseline and sperm abnormalities that are observed at the end of the study. Support parameter observations include the examination of fresh sperm quality and sperm abnormalities. Macroscopic examinations covers volume, color, odor, pH, and consistency. Microscopic examination include sperm concentration [8]. Abnormality observations are carried out by examining the preparation of sperm smear.

Data of sperm motility and motion duration were analyzed using ANOVA. The mean were also compered using the Duncan's Multiple Range Test with the preset level of significant at $p < 0.05$.

## 3. Results and discussion

Fresh sperm examination was conducted in macroscopic and microscopic manners. The results of the examination of Jatimbulan tilapia fresh sperm is illustrated in table 1. Table 1 exhibits the condition of fresh sperm from a Jatimbulan tilapia parent aged seven months weighing 502 grams and has undergone two spawnings. It is generally seen that the sperm volume was higher than the literature. The color, odor, and pH indicated the same condition with several results in the literature. Meanwhile, the concentration of sperm was higher than the literature.

The motility rate of the fish was observed by calculating the percentage of progressive motile sperm. Observations of motility rate are displayed in table 2. ANOVA test results indicated that salinity significantly ($p < 0.05$) affected the sperm motility. The Duncan's multiple range test indicates that the highest average of motility is present in 0 ppt treatment (controls) that significantly differs ($p < 0.05$) from 4 ppt, 9 ppt, 14 ppt, and 19 ppt treatments.

**Table 1.** Jatimbulan tilapia fresh sperm data.

| Parameter | Research result | Reference |
|---|---|---|
| Macroscopic | | |
| Volume | 2.3 ml | 0.57 ml [9] |
| Color | White milk | Creamy-white milk [10] |
| Odor | Fishy | Distinctive smell sperm / amis [11] |
| Ph | 7 | 7 [7] |
| Consistency | Viscous / Pekat | Medium-thick [10] |
| Microscopic | | |
| Concentration (cells / ml) | $9.00 \times 10^{8}$ | $6.83 \times 10^{8}$ cells / ml (Moses, 2010) |

The average of motility in the Jatimbulan sperm motion duration was measured using a stopwatch and the sperm was observed in a single field of view. Table 2 below illustrates the motion duration.

**Table 2.** Jatimbulan tilapia sperm motility rate.

| Treatment | Motility Average (%) $\pm$ SD |
|---|---|
| 0 ppt | $93.4 \pm 1.5^{c}$ |
| 4 ppt | $87.8 \pm 2.6^{b}$ |
| 9 ppt | $0.0 \pm 0.0^{a}$ |
| 14 ppt | $0.0 \pm 0.0^{a}$ |
| 19 ppt | $0.0 \pm 0.0^{a}$ |

Note: Different superscript in one column indicates significant difference ($p < 0.05$)

Jatimbulan sperm motion duration was measured using a stopwatch and the sperm was observed in a single field of view. Table 3 below illustrates the motion duration.

**Table 3.** Motion duration of Jatimbulan tilapia sperm.

| Treatment | Average of motion duration (second) $\pm$ SD |
|---|---|
| 0 ppt | $2128 \pm 50.2^{c}$ |
| 4 ppt | $1961.5 \pm 129.7^{b}$ |
| 9 ppt | $0.0 \pm 0.0^{a}$ |
| 14 ppt | $0.0 \pm 0.0^{a}$ |
| 19 ppt | $0.0 \pm 0.0^{a}$ |

Note: Different superscript in one column indicates significant difference ($p < 0.05$)

ANOVA test results indicated that salinity had a significant impact ($p < 0.05$) on the motion duration. The Duncan's Multiple Range Test results show that the highest average of motion duration is in 0 ppt (control) treatment and was significantly different ($p < 0.05$) to 4 ppt, 9 ppt, 14 ppt, and 19 ppt treatments. The second highest average of motion duration was found in 4 ppt treatment which was significantly different ($p < 0.05$) from the lowest average low of motion duration found ini 9 ppt, 14 ppt, and 19 ppt treatments.

Observation of sperm abnormalities was conducted by observing the smear and calculating the percentage of sperm abnormality. The observation of sperm abnormality is exhibited in Table 4.

Abnormal sperm is found was 14 ppt treatment (tail was broken off, tailless head, curled tail) and 19 ppt treatment (curled tail and broken tail). Normal and abnormal sperm can be viewed in figure 6 and figure 7.

**Table 4.** Jatimbulan tilapia sperm abnormality.

| Treatment | Abnormality (%) |
|---|---|
| 0 ppt | 0 |
| 4 ppt | 0 |
| 9 ppt | 0 |
| 14 ppt | 34 |
| 19 ppt | 61 |

## 4. Discussion

The fresh sperm volume of 2.3 ml signifies that Jatimbulan tilapia sperm volume is higher than that of common tilapia. According to Herdiana [9] suggested that red tilapia sperm volume is just 1 ml. The sperm concentration of 9.00 x $10^8$ cell/ml is still above the lowest range when compared to that of other tilapia. According to Musa [12], the highest concentration of tilapia sperm was 7.75x $10^9$ cell/ml and the lowest one was 6.83x$10^8$ cell/ml.

The characteristics of Jatimbulan tilapia sperm are milky white, quite thick, and fishy-smelling. This was similar to the physical characteristics of GIFT tilapia sperm: thick and milky white [13]. According to Mar'ati [11], one character of good sperm is the fishy odor (typical sperm odor). The result also indicates that the sperm has a pH of 7 (neutral). According to Condro *et al.* [7], good sperm has a pH of 7.

The highest level of sperm motility was found at 0 ppt (control) amounting to 93.4%, which was significantly different from that of 4 ppt treatment (brackish water), which amounted to 87.8 %. Both were still in the normal range and were even far above the bottom threshold. According to Mar'ati [11] contends that normal motile sperm has motility of 70 % to 90 %.

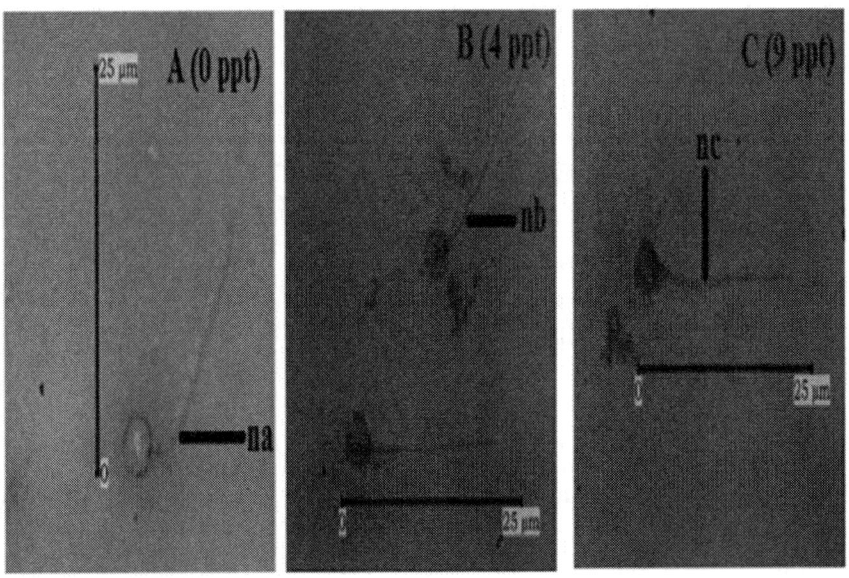


**Figure 1.** Jatimbulan tilapia normal sperm (1000x magnification).
Note: na: normal (at 0 ppt); nb: normal (at 0 ppt); nc: normal (at 0 ppt).

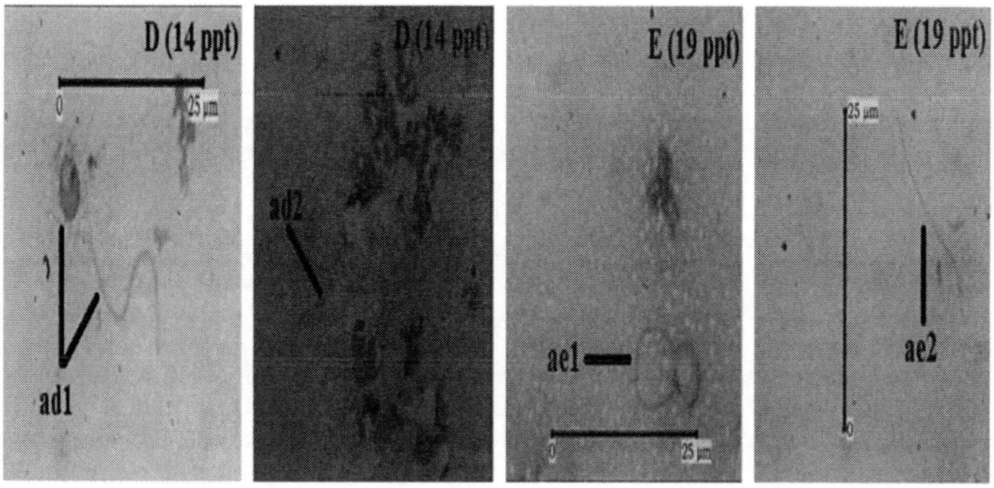


**Figure 2**. Jatimbulan tilapia normal sperm (1000x magnification).
Note: ad1: broken tail and tailless sperm;ad2: curled tail. ae1: rolled tail; ae2: broken tail

Fish sperm turns motile when it is in contact with a hypotonic environment. This causes intracellular ion [$Ca^{2+}$] to increase and the midpiece part of the sperm containing the mitochondria (sleeve) to swell, so that the mitochondria is able to form ATP to become motile [5]. The motility rate at 9 ppt treatment reaches 0 % because the sperm does not move at all when treated with 9 ppt water. Takai and Morizawa[6] suggests that isotonic solution causes the sperm to be nonmotile.

The sperm motility rate in 14 ppt and 19 ppt treatments that reaches 0% is supported by Morita *et al.* [5] who suggested that when fish spermatozoa comes in contact with a hypertonic environment, it causes its sleeve containing mitochondria to contract and its intracellular ion [$Ca^{2+}$] not to increase. This makes ATP to become unformed, the flagellar protein phosphorylation not happen, and eventually the sperm become nonmotile. According to Boryshpolets *et al.* [14] added that hypertonic solution can cause sperm cells to shrink and cause damage to the plasma membrane so that the sperm becomes nonmotile. The motility rate at 0 ppt treatment was significantly different ($p < 0.05$) from that at 4 ppt treatment it was consistent with the statement of Guthrie *et al.* [15]: the increase in salinity causes a decrease in the percentage of sperm motility. According to Janani *et al.* [16], the higher the osmolality of the activation media, the lower the amount of ATP will be. Moreover, the lower the amount of ATP, the lower the motility and the shorter the motion duration will be.

The longest sperm motion duration was found at 0 ppt treatment which endured 2128 seconds (35 minutes 46 seconds). Yet the motion duration found at 4 ppt treatment which lasted 1961.5 seconds (32 minutes 69 seconds) was still higher than that of other *Oreochromis* fish which are only able to move for 1800 seconds or about 30 minutes [5]

The sperm does not move when activated with 9, 14 and 19 ppt treatment waters. This is why 0 ppt and 4 ppt treatments were significantly different ($p <0.05$) from 9 ppt, 14 ppt and 19 ppt treatments. At the interim, 0 ppt and 4 ppt treatments cause sperm to be motile. According to Morita *et al.* [5] asserted that freshwater fish sperm turns motile when it is in contact with a hypotonic environment. The sperm did not move when activated with 9 ppt treatment. According to Hannover [17], isotonic solution has a concentration of 9 ppt. According to Ridwan[18], argued that isotonic solution causes spern to turn nonmotile. The sperm motion duration in 14 ppt and 19 ppt treatments which only lasted 0 seconds is in accordance with the opinion of Morita *et al.* [5] which argues that when fish sperm comes in contact with a hypertonic environment, it turns nonmotile.

The treatments of 0, 4 and 9 ppt show an abnormality percentage of 0 %. This is consistent with the statement Sousa *et al.* [19] that the sperm plasma membrane is intact when in hypotonic condition (<180 mOsmol/kg). According to Gao *et al.* [20], isotonic solution has the same osmotic pressure with

IOP Conf. Series: Earth and Environmental Science **137** (2018) 012023 doi:10.1088/1755-1315/137/1/012023

that of sperm, thereby maintaining between the balance of osmotic pressure inside sperm cells and the one outside environment and making the cell membrane unbroken and intact. The treatments of 14 and 19 ppt show abnormality percentages of 34% and 61% respectively. According to Gao *et al.* [20], hypertonic solution may cause abnormality in the form of curled tail sperm. Hypertonic solution will make cytoplasmic vacuoles open and tail membranes become more permeable; hence the curled tail. The 19 ppt treatment caused sperm tail to break off. According to Sousa *et al.* [19] explains that the damage to the sperm tail can be caused by the difference in osmotic pressure between the sperm and the external environment. This ruptures the cell membrane. Tailless sperm may exist because hypertonic solution causes the cell membranes to contract so that they lose membrane integrity [20].

## 5. Conclusion

The higher salinity of the water and the activation medium lead to lower motility and shorter motion duration. The best motility rate and motion duration was found at 0 ppt and 4 ppt of salinity treatments. This study suggests that 0 and 4 ppt salinities can be performed as a benchmark to generate good fertilization because the motility rate and motion duration of Jatimbulan tilapia sperm are still in the normal range. Accordingly, Jatimbulan tilapia hatchery at high salinity can use 4 ppt salinity media with a natural fertilization method.

## 6. References

[1] Darwisito S, Sjafei M Z Jr D S, Manalu W and Sudrajat A O 2008 *JAI.* **7** 1-10
[2] Office of Fisheries and Marine Affairs East Java Provincial Goverment 2008 Jatimbulan tilapia superior parent umbulan freshwater aquaculture development center Pasuruan pp 5-6
[3] Lutfiyah L 2010 Development of organ and survival rate of larvae of tilapia (Oreochromis niloticus) at different salinity hatchery media Thesis Magister Program of Fisheries and Marine Biotechnology Airlangga University Surabaya p 47
[4] Bucur C, Costache M, Daniel O and Nino M 2012 *Anim.Sci. Biotech.* **45**1-6
[5] Morita M, Takemura A and Okuno M 2003 *J.Exp. Biol.* **206** 913-921.
[6] Takai H and Morisawa M 1995 *J. Cell Sci.* **108** 1175-1181
[7] Condro H S, Mubarak A S and Sulmartiwi L 2012 *JMCS.* **1** 1-12
[8] Rahardja B S, Mubarak A S and Rini P S 2010 *Sci. J. Fish. Mar. Res.* **2** 185-191
[9] Herdiana L 2013 Study of reproduction aspect of male red tilapia (Oreochromis sp.) maintained at different environmental conditions (Thesis School of Fisheries and Marine Sciences Bogor Agricultural Institute Bogor) p 42
[10] Japet N 2011 Semen characteristics of commercially cultivated fish: carp (Cyprinus carpio) and catfish (Pangasius hypophthalmus) Thesis Department of water resource management Bogor Agricultural Institute Bogor p 46
[11] Mar'ati K 2007 Effect of dose and duration storage of egg yolk skim milk diluent on semen quality of Carp (Cyprinus carpio L.) (Thesis School of Science and Technology State Islamic University Malang) p 54
[12] Musa N 2010 Sperm activation in Nile Tilapia Oreochromis niloticus and the effect of environmentally relevant pollutants on sperm fitness (Thesis Institute of AquacultureUniversity of Stirling Scotland) p 182
[13] Nainggolan R, Monijung R D and Mingkid W 2015 *J. Aquatic.***3** 131-140
[14] Boryshpolets S, Dzyuba B, Stejskal V and Linhart O 2009 *Theriogenol.* **72** 851-859
[15] Guthrie H D, Woods L C, Long J A and Welch G R 2008 *Theriogenol.* **69** 1007-1012
[16] Janani D, Prathibanandhi J, Vidya P M and Sujatha K S 2014 *Proc. Nat. Conf. on VLSI Embedded and Comm. and Networkson.* pp 1-4
[17] Hannover T H 2012 Effects of cryopreservation on osmotic membrane properties intracellular ion distribution and ion channels of bovinesperm (Dissertation Uniersitats kliniku Hamburg) p 94
[18] Ridwan 2009 *Agroland Journal.* **16** 187-192

[19] Sousa C P, Santos E A A, Souza A L P, Lima G L, Barros F F P C, Oliveira M F and Silva A R 2013 *Pesqui. Veterin. Bras.* **33** 924-930
[20] Gao D Y, Ashworth E, Watson P F Q, Kleinhans F W, Mazur P and Critser J K 1993 *Biol. Reprod.* **49** 112-123

IOP Conf. Series: Earth and Environmental Science **137** (2018) 012024 doi:10.1088/1755-1315/137/1/012024

# Immune response and parasitic infestation on Pacific white shrimp (*Lithopenaeus vannamei*) in immuno-probio circulation system (SI-PBR) in ponds

**G Mahasri[1,2], P D W Sari[1] and Prayogo[1]**

[1]Department of Fish Health Management and Aquaculture, Faculty of Fisheries and Marine, Universitas Airlangga, Campus C Mulyorejo 60115, Indonesia

E-mail: mahasritot@gmail.com

**Abstract.** The main causes of death of pacific white shrimp in aquaculture are diseases. One effort to control deseases by improving the defense ability of shrimp body against disesases and optimizing water quality during farming through the application of a new aquaculture technology called Immuno-Probiocirculation System (SI-PBR). This research aimed to analyze immune response on Total Haemocyte Count (THC) and Differential Haemocyte Count (DHC), parasitic infestation on pacific white shrimp in many ages, survival rate of pacific white shrimp during farming period for 90 days in SI-PBR. The results of this research showed that the lowest parasitic infestation (*Zoothamnium penaei*) is 12.46 % that happened on 90-days-old shrimp in SI-PBR pond, while the highest infestasion is on the shrimp not given SI-PBR, reaching 54.65 %. In addition, the immune response (THC and DHC) also increased. The highest survival rate discovered in 90 days shrimp farming is 80% using SI-PBR. This is higher than the pond without SI-PBR, which is 22 %. Therefore, SI-PBR in shrimp farming in tradisional ponds is able to increase immune response, survival rate, and is also able to decrease parasitic infestation during 90 days of farming.

## 1. Introduction

Since late 1993, production of tiger shrimp has been declining drastically; only around 30 % of ponds are still operating till now, while the rest have become unproductive and idle [1]. The reality shows that increased production of vannamei shrimp is inhibited due to diseases caused by viruses, bacteria, fungi and parasites [2].

The main problems hampering revitalization efforts of shrimp ponds in Indonesia, particularly in East Java, are low level of growth and high number of shrimp death due to declining water quality and diseases/pathogens. White spot virus disease, also known as "white spot" (WSSV), is the main culprit and it has become a serious problem in shrimp farming activities. Moreover, traditional shrimp ponds farming activities are highly dependent on tidal conditions.

IMNV Infection on vannamei shrimp starts with the weakening of the shrimp body, decreased appetite, unbalanced swimming, and movement towards the surface. After that, around two segments on the base of the tail, and the base itself, become opaque white to reddish. Histopathologically, meat or muscle tissue, especially in the last part of the IMNV-infected segments, exhibits necrosis [3]. In addition, histopathology testing shows that IMNV-infected shrimp can be identified through lesions on skeletal muscles, including multifocal necrosis, haemolytic blockage, fibrocystic inflammation, phagocytosis, and appearance of cytoplasmic body inclusion. IMNV appears to replicate in shrimp cytoplasmic muscle cells and exhibits a lack of conclusive evidence for protein incorporation.

Bacterial diseases that often emerge to infect shrimp in ponds and hatcheries include diseases caused by *Vibrio parahaemoliticus*, *Vibrio harveyii,* and others. Itami *et al.* [4], stated that shrimp infected by these vibrio bacteria will swim on the surface; their gills are dirty and their tails, fins and the anterior part of the body suffer many reddish wounds. This disease can cause death up to 86 %.

One of the parasitic diseases that can cause vannamei shrimp death both in ponds and hatcheries is Zoothamniosis. This disease is caused by *Zoothamnium penaei*. It causes shrimp to suffer difficulty in breathing, moving, searching for food Sindermann [5] and Foster *et al.* [6], and moulting. It also impedes growth, reduces economic values, and causes death up to 91 % [7]. In addition, this disease is a predisposing factor of secondary infection by bacteria and viruses.

Smith *et al.* [8], wrote that the efforts to improve the immunity of shrimp body in both hatcheries and ponds can be performed using immunostimulants. Furthermore, Mahasri [9] reported that immunization using immunogenic membrane protein from *Zoothamnium penaei* could increase survival rate of shrimp up to 93 %. In addition, isolation of immunogenic membrane proteins can be conducted using SDS-PAGE, ELISA, and Western Blotting methods. The results showed that 7 proteins were discovered and 3 of them were immunogenic, namely membrane proteins MP38, MP48 and MP67.

Generally, crustacean group has non-specific immune system because it lacks the ability to memorize antigens. In the event of bacterial, viral, or fungal infection, the shrimp's hard cuticle is the first physical defence inhibiting the entry of pathogens. If the pathogens pass through this external defence, the internal defence of the shrimp body becomes the second defence performed by its cellular and humeral responses [10].

Immunity system in shrimp is different from that of vertebrates. What plays a role in the immunity system of shrimp is the defence mechanism through haemocytes and plasma proteins [11]. Haemocytes are cells that possess functional devices such as macrophages, granulocytes, and natural killer cells (NK cells) in vertebrate animals [12]. Haemocytes play an important role in the immune system of crustacean body by expelling foreign particles through phagocytosis, encapsulation, and nodular aggregation [13]. The marker parameters for the evaluation of shrimp defence system involve humeral and cellular immunity, which include haemocyte counts [14], prophenoloxidase (proPO) enzyme and phenoloxidase (PO) enzyme activities [14,15]. The immune response is characterized cellularly by increased haemocytes in shrimp and marked molecularly by increased plasma proteins, which can be detected and characterized by SDS PAGE electrophorus, as well as increased phenoloxidase activity as the parameters of increasing immune system in shrimp. Haemocytes have an important role in the internal immune system of shrimp. Shrimp has 3 types of haemocytes, namely hyalin that plays a role in the process of phagocytosis and granular and semi granular cells that are important in the process of activation of phenoloxidase.

Shrimp farming using Imuno-Biocirculation System (SI-PBR) is an effort to increase the harvest yield by applying immunization and biological filter with milkfish. This system is a combination of immunostimulant use in shrimp seeds before the stocking to increase the resistance of shrimp. The immunostimulant used is an immunostimulant obtained from immunogenic membrane protein of *Zoothamnium penaei* in order to improve the immunity of shrimp and to prevent the ectoparasitic diseases [9]. In addition, this system also uses probiotics to maintain water quality. Gunarto *et al.* [16], reported that utilization of probiotics, using various types of bacteria that decompose organic materials, on tiger shrimp farming has been widely implemented to increase shrimp life up to 86 - 93%.

The purpose of the use of filter is to balance the biomass of plankton and reduce nitrite and ammonia compounds, because milkfish is able to accumulate compounds in shrimp, so there is no decay inside the ponds. According to Mahasri [9], immuno-probiorecirculation system (SI-PBR) was able to laboratorically increase tiger shrimp life from 40 % to 82 %.

IOP Conf. Series: Earth and Environmental Science **137** (2018) 012024 doi:10.1088/1755-1315/137/1/012024

Based on the background of the problem, the application of SI-PBR method needs to be studied to increase shrimp production in traditional ponds and to analyze immune response (THC and DHC), ectoparasitic infestation, and survival rate (SR) of vannamei shrimp (*Litopenaeus vannamei*) nurtured in ponds using immuno-probiocirculation system (SI-PBR).

## 2. Materials and method

### *2.1 Research methodology and design*

The research method used in this research was field experimental method (in ponds) to discover the values of independent variables, either one or more (independent) variables, without making comparison or links with other variables. It was expected that this research would be able to achieve the objective of analyzing immune response and survival rate of vannamei shrimp in ponds with immuno-probiocirculation system, using the following treatments:

K1 → Reservoir plot and biological filter of milkfish and Recirculation

K2 → Shrimp control group nurtured in the pond without Immuno-Probio-circulation system (SI-PBR)

K3 → Shrimp group kept in the pond using Immuno-probio-circulation system (SI-PBR)

### *2.2 Materials and equipments*

The main materials in this study were 1000 healthy juvenile (40-days-old) vannamei shrimp and membrane protein of *Zoothamnium penaei*, as a substance for developing immunostimulant, that had been laboratorically tested by Mahasri [18]. The main materials for protein isolation and characterization include saline, ethanol, gradient Percoll solvent, pepsin, HCl, EDTA, KCl, $KH_2PO_4$, $Na_2HPO_4$, trypsin, sodium citrate, $NaHPO_4$, $H_2NO_3$, $NaHCO_3$, glucose, phenol red 0.5 %, NaOH, 0.22 μm filter, bovine serum albumin, dextrose, EtOH, proteinase, Forward ERIB1 5' primary pair – ACCTGGTTGATCCTGCCAG-3' (2-20) and Reverse ERIB10 5 '-CCTCCGCAGGTTCACCTACGG-3' 2079- 2059), 400 μM DNTP, 3 μm of MgCl2, yellow and blue dye, agarose, TAE buffer, sybrsafe, 100 bp and 1 bp of DNA ladder, loading dye, tris-HCL, 2-mercaptoethanol, sodium dodecyl sulphate (SDS) bromophenol blue, glycerol, SDS loading buffer, polyacrylamide, stacking gel, ammonium persulfate (APS), TEMED, and glycine.

The material for biological filter was 3000 milkfish with a length of 10-15 cm. The materials for ponds preparation were: 1,500 kg of dolomite lime, 150 kg of urea fertilizer, 75 kg of TSP fertilizer, and 2 ampoules of vitamin.

The equipments used were 3 ponds (each was 500 $m^2$), 2 sets of production facilities, 1 8-DHIM generator. To isolate protein, the tools used were haemocytometer, micrometer microscope, autoclave, centrifuge tube, swinging rotor, water bath sonicator and 1 set of electrophoresis equipment for SDS-PAGE [17].

### *2.3 Research procedure*

#### *2.3.1 Preparation of ponds*

The ponds for the application of SI-PBR consisted of a reservoir plot, which was also used as recirculation plot with milkfish as biological filter, and shrimp nurturing plots. Each of these plots was connected to two 8 cm pipes for recirculation. Recirculation was run by a water pump with a size of 20 hp and 8 cm. The design of SI-PBR-Plus ponds can be seen in Appendix 1.

#### *2.3.2 The making of immunostimulant materials from Zoothamnium penaei protein*

The immunostimulant development material used in this study was the immunostimulant material produced by Mahasri [18]; The stages in the making of the material were: 1) in vitro farming and isolation of Zooid immunogenic proteins from *Zoothamnium penaei*, 2) characterization and purification of immunogenic proteins using SDS-PAGE, 3) determination of protein concentrations, and 4) testing of immunostimulant material on vannamei shrimp (*Litopenaeus vannamei*) in traditional-plus pond.

*2.3.3 Application of SIBR-Plus*

This stage was the stage of direct implementation of vannamei shrimp farming using Imuno-Probiocirculation System (SIBR-Plus) in traditional-plus pattern on ponds. This stage was started by re-checking biorecirculation and immunostimulant plots. The biofilter used was milkfish. in one cycle (90 days) of shrimp nurturing period, the activities conducted included data collection of the circulation plot, water quality, and shrimp health and growth as supporting data on the success rate of SI-PBR implementation. The area of shrimp nurturing plot was 500 $m^2$ with a dense stocking of 50 shrimp per $m^2$ and 3000 milkfish. Circulation was done 2 times every 24 hours, namely at 06.00 to 08.00 for 2 hours, and 24.00 - 02.00 for 2 hours.

*2.3.4 Examination of ectoparasitic infestation on Vannamei Shrimp*

Observation of ectoparasitic infestation was conducted natively by scrapping the entire surface of the shrimp body [19]. The result was placed on top of an object glass, given a drop of water, and examined using a microscope with 100× magnification. Parasitic infestation was calculated with a percentage of positive shrimp on the number of shrimp examined.

*2.3.5 Total haemocyte count (THC)*

Haemocytes were taken at the ventral portion of the second abdominal segment using 1 ml syringe containing 0.2 ml of cold Alsever modified solution (AS 19.3 mM; 239.8 mM Na citrate, 182.5 glucose NaCl, and 6.2 mM EDTA; pH 7.2) as an anticoagulant [20]. Furthermore, calculation of the number of haemocytes was performed using Haemocytometer seen under the microscope with 1000× magnification and was calculated with a hand counter. Total haemocyte count was conducted at the age of 30, 60, and 90 days in the ponds using the following formula [21]:

$$\text{PTH}/\text{mm}^2 = \frac{(C)(D)(100)(10)}{(S)(4)} \quad (1)$$

Notes:

C = Number of countable cells
D = Dilution factor
S = Number of countable 1 mm boxes
PTH = Total haemocyte count /$mm^2$

*2.3.6 Differential Haemocyte Count (DHC)*

Blood was dripped on the object glass and blood smear was made by giemsa staining method [22]; therefore the cell type could be identified. Differential haemocyte count was aimed to discover the number, type, and percentage of haemocyte cells. The haemocytes analyzed were classified according to the method by Owens and O'Neill [23]. The number of granular haemocytes cells (DHC) was calculated up to 100 cells and its percentage was searched for [9]. Calculation of DHC was conducted at the age of 30, 60, and 90 days in the ponds.

*2.3.7 Determination of survival*

Based on a research conducted by Nuhman [24], observation on survival was performed through direct observation by observing live shrimp at the beginning until the end of the study. The formula used to measure it is as follows:

$$SR = \frac{Nt}{No} \times 100\% \quad (2)$$

Notes: SR = Survival rate (%)
Nt = Live shrimp at the end of the research
No = Live shrimp at the beginning of the research

IOP Conf. Series: Earth and Environmental Science **137** (2018) 012024 doi:10.1088/1755-1315/137/1/012024

## 3. Results

### *3. 1 Analysis of Zoothamnium penaei infestation on Vannamei Shrimp*

Examination result of parasitic infestation on vannamei shrimp showed that there were shrimps positively infested with *Zoothamnium penaei*; both in the pond with and without SI-PBR. Parasitic infestation on the shrimp nurtured using SI-PBR was 17.34 % on 90-days-old shrimp, while on the shrimp pond nurtured without SI-PBR, the highest level of infestation was 54.65 %. The lowest parasitic infestation on the shrimp was found on the shrimp nurtured using SI-PBR, which was 7.68%, after 30 days in the pond. The shrimp infected by parasites showed several clinical symptoms: there were objects looking like fibres attached to the tails, their swimming and walking legs became brownish white, their tail did not look like fans when swimming, they swam clustered on the surface, their digestive tracts were empty, and their body surface and gills seemed turbid like containing moss. Healthy shrimp (not infested by zoothamniosis) looked clear, transparent, and clean and there was no discoloration on the entire surface of the body and gills, they actively swam, and their tail opened like fans. The examination result of *Zoothamnium penaei* infestation on vannamei shrimp can be seen in table 1.

**Table 1.** Examination Result of *Zoothamnium penaei* on Vannamei Shrimp.

| Age of Shrimp Nurturing (Day) | Infestation of *Zoothamnium penaei* on Vannamei Shrimp (%) | |
|---|---|---|
| | Shrimp nurtured using SI-PBR | Shrimp nurtured without SI-PBR |
| 30 | 7.68 | 14.76 |
| 60 | 9.24 | 24.74 |
| 90 | 17.34 | 54.65 |

### *3.2 Analysis of immune response (THC and DHC) of vannamei shrimp seeds*

Examination of immune response of the seeds (logs/PL 30) stock could not be conducted because the seeds were too small, so there was difficulty in taking blood sample. The calculation result of total haemocyte count (THC) on the vannamei shrimp nurtured in SI-PBR pond at the age of 30, 60, and 90 days is presented in table 2. The highest THC was found on 60-days-old shrimp nurtured in the pond using SI-PBR, reaching 61.23 x $10^6$ cells/mL, while the lowest was found on 30-days-old shrimp in the pond without SI-PBR, which was 29.34 x $10^6$ cells/mL.

**Table 2.** Determination of Total Haemocyte Count (THC) of Vannamei Shrimp.

| Age of Nurturing (Day) | Total Haemocyte Count (THC) of Vannamei Shrimp ($10^6$ cells/ml) | |
|---|---|---|
| | Shrimp Nurtured using SI-PBR | Shrimp Nurtured without SI-PBR |
| 30 | 29.34 | 21.56 |
| 60 | 61.23 | 36.35 |
| 90 | 48.96 | 32.54 |

Table 3 shows the differential haemocyte count (DHC) of vannamei shrimp nurtured in the ponds with and without SI-PBR.

**Table 3.** Differential Haemocyte Count (DHC) of Vannamei Shrimp.

| Age of Nurturing (Day) | Differential Haemocyte Count (DHC) of Vannamei Shrimp (%) | |
|---|---|---|
| | Shrimp Nurtured Using SI-PBR | Shrimp Nurtured without SI-PBR |
| 30 | 25.32 | 11.23 |
| 60 | 19.73 | 15.84 |
| 90 | 15.54 | 10.79 |

The results showed that the highest DHC was found in 90-days-old vannamei shrimp nurtured in the pond using SI-PBR, which reached 25.32 %, while the lowest was found in 90-days-old shrimp nurtured in the pond without SI-PBR, which was 10.79 %

*3.3 Determination of Survival Rate (SR) of Vannamei Shrimp*
The survival rates of vannamei shrimp are shown in table 4.

**Table 4.** Survival rates of vannamei shrimp.

| Treatment | Survival Rate (%) |
|---|---|
| Shrimp Nurtured in the pond using SI-PBR | 80 |
| Shrimp Nurtured in the pond without SI-PBR | 22 |

Table 4 shows that the survival rate of 90-days-old vannamei shrimp in the pond that used SI-PBR reached 80 % at harvest, higher than the shrimp nurtured in the pond without SI-PBR, which reached only 22 % at the end of nurturing period.

*3.4 Examination of water quality*
Table 5 shows the water quality during 90 days of nurturing period in the ponds.

**Table 5.** Water quality in the Ponds during 90 Days of Nurturing Period.

| Parameter | Average Parameter/Quality of Water during Shrimp Nurturing | Normal Range |
|---|---|---|
| Temperature (°C) | 27 – 29 | 27 – 32 |
| Salinity (°/oo) | 26 – 26 | 16 – 30 |
| pH | 7.7 – 8.3 | 7.5 – 8.5 |
| Dissolved Oxygen (ppm) | 4.1 – 6.2 | >3 – 7 |
| Ammonia (ppm) | 0.8 - 0.96 | <1 |

Table 5 shows that the water quality in the ponds for 90 days of shrimp nurturing was generally within the normal range, thus it was in accordance with the nurturing requirements of vannamei shrimp.

## 4. Discussion

Results showed that shrimp farming using SI-PBR system was able to reduce parasitic infestation (particularly *Zoothamnium penaei*), improve the immune response (based on the increase in THC and DHC), and increase the survival of vannamei shrimp. Parasitic infestation occurred during nurturing both with and without SI-PBR, but infestation on the shrimp with SI-PBR was lower and showed a declining trend at the age of 30, 60, and 90 days. Table 1 also shows that parasitic infestation always increased with increased nurturing period, starting from 7.68 %, 9.24 %, and 17.34 % on the shrimp nurtured using SI-PBR system. It also happened on the shrimps without SI-PBR in its nurturing, which were 14.76 %, 24.74 % and the highest was at the age of 90 days during the nurturing that reached 54.65 %.

Table 1 and 2 show that the immune response (THC and DHC) of vannamei shrimp nurtured in the pond using SI-PBR increased, starting from the 30$^{th}$ day of the nurturing, from 21.56 x 10$^6$ cells/mL to 29.34 x 10$^6$ cells/mL. While in 60-days-old shrimp, there was an increase in THC from 36.35 x 10$^6$ cells/mL to 61.23 x 10$^6$ cells/mL and at the age of 90 days, it increased from 32.54 x 10$^6$ cells/mL to 48.96 x 10$^6$ cells/mL. Table 1 shows that vannamei shrimp nurtured with and without SI-PBR experienced an increase during the nurturing period until the age of 60 days and then it decreased at the age of 90 days (harvest).

DHC of the vannamei shrimp (table 3) nurtured using SI-PBR also increased from 11.23 % to 25.32 % in shrimp aged 30 days, from 15.84 % to 19.73 % in shrimp aged 60 days, and from 10.79 % to 15.54 % in shrimp aged 90 days.

The increased immune response in both THC and DHC, along with the increase in survival rate of vannamei shrimp (between 22 % and 81 %), were indicators that nurturing using SI-PBR caused the shrimp to have higher immunity, lower parasitic infestation, and higher survival rate. Existence of immunostimulant was able to provide protection to the shrimp nurtured in the ponds, especially against parasitic infestation particularly *Zoothamnium penaei* since the immunostimulant entering the body of the shrimp would stimulate the activities of haemocyte cells in the shrimp as an effort to fight the pathogens. This is in accordance with Van de Braak in Mahasri [9], who stated that immunostimulant-activated haemocyte cells will perform phagocytic activities on the shrimp by hyalin (granular) and semi-granular cells.

Soderhall and Cerenius [25], said that the immune system in shrimp is still primitive, unlike the systems in fish and mammals that have immunoglobulin, so immunoglobulin in shrimp is replaced by prophenoloxidase activating enzyme (PPA). PPA is a protein located in granular haemocyte cells. it can be activated by lipopolysaccharides and β 1,3-Glucan, which will stimulate prophenoloxidase to change into phenoloxidase. This transformation produces a compound similar to Opsonin Factor protein that can induce hyalin cells to perform the process of phagocytosis. Van de Braak [26], also supported this statement and added that haemocyte cells will degranulate and some proteins would be released for the immune responses, such as: increased haemocytes, trapping activity, and phagocytosis. In addition, immunogenic membrane proteins will stimulate haemocytes to release proPO and protein-binding PPA, thus causing haemocyte cells to increase their activities to perform trapping and phagocytosis from the agents of diseases, which in this case is *Zoothamnium penaei*. Table 1 shows that parasitic infestation on the shrimp nurtured using SI-PBR exhibited a low tendency in the shrimp aged 30, 60, and 90 days in the pond, or equal to 7.68%, 9.24%, and 17.34%. Meanwhile, parasitic infestation on the shrimp nurtured in the pond without SI-PBR showed higher tendency, which were 14.76%, 24.74%, and 54.65%.

If immunostimulant enters the body of the shrimp, it will increase the number of haemocytes (THC) and differential haemocyte cells (DHC). This is an indication of increased immunity of vannamei shrimp [9]. Furthermore, Itami [27] supported the theory by stating that providing immunostimulant may prevent infection of diseases within the host body and increase phagocyte activities of haemocytes and proPO enzymes. This statement was also supported by Van de Braak [26] who wrote that immunostimulant substances entering the shrimp body would produce antibodies capable of neutralizing pathogens, so pathogens cannot infect the shrimp, resulting in lower infestation (infection) by pathogens.

Mahasri [9], stated that the immunogenic membrane protein of *Zoothamnium penaei* can increase the life of vannamei shrimp from 22 % to 81 % in the shrimp aged 90 days old (end of nurturing). In addition, it was also reported that the immune response of tiger shrimp also increased, marked by the increase of THC and DHC since the immunogenic membrane protein of *Zoothamnium penaei* had a high molecular weight, greater than 1000 Da. Proteins with high molecular weight and high level of immunogenity must have a complex structure.

The increase in THC and DHC can be used as an indicator for pathogenic infection in the host body. Infection will cause inflammation, which is a non-specific body defence mechanism caused by several factors such as parasites, bacteria, fungi, viruses and nonliving agents Soderhall and Cerenius [25] THC in the shrimp given immunostimulant increased starting from the age of 30 days until 60 days, but decreased at the age of 90 days. shrimp immune system increases along with the increased age, but at certain age it decreases back. The high level of THC indicated the added crude protein could improve immune response in shrimp since high THC is an indicator of increased body resistance. This is in accordance with the statement by Soderhall and Cerenius [25] who asserted that increase in immune response on invertebrates is indicated by an increase in THC. Increased THC and DHC can be used as indicators of the immune system reaction in shrimp in the presence of pathogens. The increase in DHC (granular haemocyte cells) was presumed to be due to the lack of memory cells in shrimp immune system, so they were not able to detect pathogenic substances they had been exposed to. Therefore, it can be concluded that the immunostimulant was able induce the immunity mechanism in shrimp body. However stimulating the hematopoietic organ to produce granulocytes to fight pathogens needs time [9]. Granulocytes destroy pathogens by swallowing them, so they have to migrate to parasite-infested areas first.

Water quality during nurturing also affects infestation and immune response of shrimp. Based on the results, the water quality was in optimum condition for shrimp. good water condition of made opportunistic pathogens unable infect shrimp, thus the shrimp stayed in good health and immune system. Maintenance of water quality causes shrimp to live well and prevent diseases from emerging so the shrimp can grow well.

Mahasri [28], stated that probiotic is a material containing various types of bacteria that can play a positive role in decomposing organic materials in ponds and helping metabolism when mixed in shrimp feed; while biofilter, a role that can be played by milkfish (*Chanos chanos Forsk*) or seaweed, can be used to balance biomass (plankton) in the pond water. Furthermore, water circulation is necessary to be implemented in shrimp farming activities to maintain water quality.

application of probiofilter technology can result in high quality and disease-free shrimp harvest. the ponds Water remain in good condition, so diseases do not emerge during farming and the shrimp can grow well and healthy.

it appears that shrimp nurtured using SI-PBR reached 80 % of survival rate, while those without the addition of crude protein only reached 22 %. it indicated the crude protein given was able to improve defence system in the shrimp, and good quality of pond water during the research caused the shrimp to stay in good condition, resulting in increased growth rate of the shrimp.

## 5. Conclusions

There was an increase in immune response caused by provision of immunostimulant, probiotics, biological filter, and circulation in SI-PBR, resulting in a decrease in *Zoothamnium penaei* infestation on vannamei shrimp. Provision of immunostimulant improved immune response (increased THC and DHC) on vannamei shrimp nurtured in the ponds. The immune response of vannamei shrimp increased for 60 days and started to decrease after 90 days in the ponds. Provision of immunostimulant, probiotics, biological filter, and circulation of SI-PBR were able to improve the survival rate of vannamei shrimp from 22 % to 80 % in 90 days of nurture in the ponds.

## 6. References

[1] Department of Fisheries and Marine 2001 Type of shrimp disease on brackish water cultivation artikel 02/09/04 Mina diklat BPPP Belawan Medan

[2] Lightner D V 2002 *Prevalence and geographic distribution of mbv and other diseases in cultured Giant Tiger Prawns (Penaeus monodon) in the Philippines* (Hawaii : The Oceanic Institute, Honolulu)

[3] Lightner D V 2003 *Exclusion of specific pathogens for disease prevention in a penaeid shrimp biosecurity program* In: Lee, CS, O'Bryen P J (Eds) Biosecurity in Aquaculture Production Systems: Exclusion of Pathogens and Other Undesirables. The World Aquaculture Society (Baton Rouge, Louisiana) p 81–116.

[4] Itami T, Kondo M and Takahasi Y 1996 Enchancement of disease resistance of kuruma prawn (*Penaeus japonicas)* After Oral Administration of Peptidoglucan (Japan : National Fisheries University) p 59-65.

[5] Sindermann C J 1997 *Trans. Am. Microsc. Soc.* **98** 136 –138.

[6] Foster C A, Sharpie T G and Hawkins W E 1998 *J. Fish. Dis.* **1** 321-335

[7] Tonguthai K 1991 Diseases of the freshwater prawn *Machrobrachium rosenbergii*, AAHRI Newsletter Bangkok Thailand Article **4**(2)

[8] Smith V J, Brown J H and Hauton C 2003 *Fish Shellfish Immun* **15** 71 – 90.

[9] Mahasri G 2007 *Immunogenic membrane protein zoothamnium penaei as an immunostimulant development in Shrimp Windu (Penaeus monodon Fab.) To Prevent Zoothamniosis* (Surabaya : Universitas Airlangga).

[10] Van de Braak K 2002 *Haemocytic defence in black tiger shrimp (Penaeus monodon)* Disertation (Netherland: Wageningen University).

[11] Vargas-Albores 1996 *Dev Comp Immunol* **20** 299-306

[12] Raa J 2000 *The use of immunostimulant in fish and shellfish feeds*, In : LE Cruz Suarez, D Richie-Marie, M Tapia-Salazar, MA Olver-Novoa, R Civera-Cerecedo (Eds) Avences en Nutricion Acuicola, Merid, Yucatan, Mexico : 47-54.

[13] Rodríguez J, Le Moullac G 2000 *Aquacult* **19** 109-119.

[14] Le Moullac, G Soyez, C Saulnier, D Ansquer, D Avarre, J C Levy P 1998 *Fish Shellfish Immunol* **8** 621–629.

[15] Hernández-López J, Gollas-Galván T and Vargas-Albores F 1996 Activation of the prophenoloxidase system of the brown shrimp Penaeus californiensis Holmes. Comparative Biochemistry and Physiology **113C**: 61-66.

[16] Gunarto, Mansyur A dan Muliani 2009 Application of different probiotic fermentation doses on the cultivation of vaname shrimp (*Litopenaeus vannamei*) with intensif pattern. Center for Brackishwater Aquaculture Research, Maros, South Sulawesi.

[17] Rantam FA 2003 Metode Imunologi (Surabaya-Airlangga University Press).

[18] Mahasri 2015 Respon Imun Dan Kelulushidupan Udang Vannamei (*Lithopenaeusvannamei* Yang Dimunisasi Dengan Protein/Membran Imunogenik *Zoothamnium penaei. Article of Konggres and International Seminar IFS, Can Tho University, Vietnam.*

[19] Johnson 1986 *J. Parasitol* **72** 517–520.

[20] Van de Braak CBT, N Taverne, MHA Botterblom, WPW Van der Knaap and JHWM Rombout 2000 *Fish Shellfish Immuno* **10** 515-530.

[21] Blaxhall P and K Daisley 1993 *J. Fish. Biol.* **5** 1-8.

[22] Nash G L, Anutara and W Boonsirm 1993 Rapid Diagnosis of yellow head diseases in black tiger shrimp culture reesearch centre. AHHRI NEWSLETTER (Choroen Phokphand Feedmil Co. Ltd, Smuth Sakorn Thailand) p 87.

[23] Owens L and O'Neill A 1997 *Dis. Aquat. Org.* **31** 147-153.

[24] Nuhman 2008 *Scientific J. Fish. Sci.* **1**

[25] Sŏderhậll K and Cerenius L 1992 *Ann. Rev. Fish. Dis.* 3 – 23.

[26] Van de Braak CBT, MHA Botterblom, N Taverne, JHWM Rombout and WPW Van der Knaap 2002 The roles of haemocytes and the lymphoid organ in the clearance of injected Vibrio bacteria in Penaeus monodon shrimp *Fish Shellfish Immunol* **13**: 239-309.

[27] Itami T 1996 *body defense system of penaeid shrimp, seminar on fish physiology and prevention of epizootics* (Japan : Department of Aquaculture and Biology, Shimonoseki University of Fisheries) **7** : 59-65.

[28] Mahasri 2004 Effectiveness of milkfish as biological filters suppress zoothamniosis on intensive cultivation national seminar review of the development of science and technology of fisheries, on 30 November, in Semarang.

**Acknowledgments**
The authors would like to express gratitude to the Dean of the Faculty of Fisheries and Marine in Airlangga University who had provided funds through RKAT for conducting this research.

IOP Conf. Series: Earth and Environmental Science **137** (2018) 012025 doi:10.1088/1755-1315/137/1/012025

# The effects of different concentrations of ccBA-GFP promoter with electroporation methods on the quality of koi sperm (*Cyprinus carpio* var. koi)

**A Soeprijanto [1] and D Aisyah [2]**

[1]Faculty of Fisheries and Marine Science, Brawijaya University, Malang-Indonesia 65145
[2] Master Programme. Faculty of Fisheries and Marine Science, Brawijaya University, Malang-Indonesia 65145

E-mail : goes_pri@ub.ac.id

**Abstract.** The effectiveness of the use of promoter concentration which will be inserted into the Koi sperm as the medium of gene transfer is important. The objective of this research is to find out the influence of the adding of different concentrations of the ccBA-GFP promoter with electroporation methods to the motility, viability and the fertilization rate of the Koi sperm . This study was conducted at Central Lab of Life Sciences Brawijaya University in April 2017. Electroporation methods were conducted by using 30-volt voltage, 4 times shocks with 0.5 seconds per shock. The treatment of different concentration was done through 3 types of ccBA-GFP promoter concentration, namely: 10 ng/μl, 30 ng/μl, and 50 ng/μl. The best motility percentage with the score of 4 is at the treatment A (10ng/μl concentration), the best viability percentage is 77.83 % at the treatment A (10 ng/μl concentration) and the best fertilization rate is 73.09 % at the treatment A (10 ng/μl concentration). The result shows that there is a relationship between the treatment given to the motility and viability of the Koi sperm, at which, the higher the shocks, the lower the percentage of the motility and viability of the Koi sperm.

## 1. Introduction

Koi fish (*Cyprinus carpio*) is one of the leading tropical fish commodities that has a big potential in Indonesia, especially in West Jakarta, Blitar, Cianjur, Sukabumi, and Makassar. Japan is known as the biggest Koi cultivator country in the world and it is facing an obstacle on the cultivation area, hence Indonesia has a potential to utilize this big opportunity [1]. For that reason, the genetic quality of the local Koi has to be improved to make them able to compete with the imported ones. One of the genetic engineering techniques the can be applied is transgenesis.

Electroporation is one of effective and efficient transfer methods to be applied. The basic principle of this method is creating reparable-holes in the cell membrane with the help of power supply and then the cell is suspended in the DNA solution that later on will come into the cell through the whole shaped [2]. Electroporation methods conducted to the sperm functions as the vector of foreign gene holder that is going to be introduced (Sperm Mediated Gene Transfer). Several previous studies have shown that sperm has a unique ability to be natural foreign gene holder vector [3].

The success of gene transfer with electroporation methods depends on the level of voltage and the DNA concentration used [4]. In addition to the effective electroporation methods, it is also important to know the use of promoter concentration or foreign DNA that will be inserted. For that matter, proper concentration of promoter is essential and so is the effect of the sperm quality on the

IOP Conf. Series: Earth and Environmental Science **137** (2018) 012025 doi:10.1088/1755-1315/137/1/012025

success of transgenesis by using the sperm as the gene transfer medium. During the transgenesis, the promoter has an important role to decide if a character coded by the gene transferred meets the expectation. The quality of the sperm is the main factor in succeeding the transgenesis process during fertilization. The quality of koi sperm includes sperm motility and viability.

The objective of this research was to find out the effects of the use of different concentrations of ccBA-GFP promoter inserted in sperm through electroporation process on the quality of the Koi sperm, which includes sperm motility, sperm viability, and fertilization rate of the sperm to the Koi fish eggs (*Cyprinus carpio*).

## 2. Methodology

This study was conducted in April 2017 at Central Lab of Life Sciences, University of Brawijaya (LSIH UB). The materials needed to conduct this study were Koi sperm, saline, eosin-nigrosin, ccBA-GFP DNA promoter, aluminum foil, and distilled water. The equipments used were a set of Gene Pulser Xcell Tm of BIO-RAD consisted of electroporator machine, shock pod electroporator, and a 0.2 mm cuvette, Olympus BX 51 inverted microscope, Olympus CX 41 binocular microscope, micropipette, microtube, tip, digital camera, microscope slide, Petri dish, and hand tally counter.

The research followed a Completely Randomized Design (CRD) with three treatments and one control with each of them was repeated three times. The treatments applied are as follows:
Treatment A = 10 ng/µl ccBA-GFP promoter concentration.
Treatment B = 30 ng/µl ccBA-GFP promoter concentration.
Treatment C = 50 ng/µl ccBA-GFP promoter concentration .
Treatment K = no treatment

### *2.1. Sperm Collection*

The sperm was collected from the male parent Koi gonad through hand-stripping. It was then put into a 1.5 ml syringe and added with saline to dilute it with the ratio of 1:1, the next step was moving it into a microtube and keeping it in a room temperature [5].

### *2.2. Koi Fish Eggs Collection*

Koi fish eggs were collected from the female parent gonad through hand-stripping. The female parent Koi was previously given no food for 24 hours to empty the stomach and intestine and then injured using Ovaprim with a dose of 0.5 ml/weight and the last step was conducting an artificial spawning.

### *2.3. Electroporation*

The koi sperm, which had been prepared in a microtube, was then electroporated using 1 set of Gene Pulser Xcell TM Electroporation System Bio-Rad. The mechanism was started with putting 25 µl of sperm into an electroporator cuvette and adding 10 µl of ccBA-GFP promoter according to the designated treatment [6]. Next, the electroporator was then pulsed by 30-volt electric field strength with the period of shocks was 0.5 m/s, and the process was repeated four times [7, 8]. The sperm was then added with 275 saline (to make total liquid of 300 µl) and t the sperm motility and viability were observed. Electroporation methods used on the koi sperm referred to the previous study conducted on goldfish sperm (*Cyprinus carpio*) [9].

### *2.4. Sperm Motility*

Post-electroporation sperm motility was observed by collecting 1 µl Koi sperm and dropping it on a microscope slide. It was then dripped on with water to activate the sperm. Next, the slide was covered by glass cover and the motility was observed for 7 minutes under the inverted microscope that had been connected to the computer. Table 1 shows the percentage of sperm assessed by scoring methods according to Dewi [10].

**Table 1.** Motility scoring.

| Percentage | Criteria | Score |
|---|---|---|
| >70 % | Spermatozoa moves forward and quickly with the variation of tail movement | 5.0 |
| 55 – 70 % | Spermatozoa moves forward and shows a quick movement | 4.0 |
| 40 – 55 % | Spermatozoa moves forward and several of them show a quick movement | 3.0 |
| 25 – 40 % | Spermatozoa moves forward | 2.0 |
| 10 – 25 % | Spermatozoa moves | 1.0 |
| 1 – 10 % | Mostly spermatozoa are not moving | 0.5 |
| <10 % | Spermatozoa are not changing | 0.0 |

*2.5. Sperm Viability*
An observation of the sperm viability or sperm survival was performed by coloring them using eosin-negrosin staining method. After being electroporated, 5 μl sperm was taken and 1 μl was dripped on a microscope slide. Next it was added with eosin-negrosin and painted using a cover glass. After dry, it was then observed under the CX41 Olympus microscope. Living sperm was marked by clear color and red color indicated dead sperm. According to Susilowati [11], the formulation used to assess sperm viability is:

$$\text{Viability} = \frac{\sum \text{Viable Sperm}}{\sum \text{Non Viable sperm}} \text{ x } 100\ \% \qquad (1)$$

*2.6. Fertilization Rate*
The number of Koi fish eggs that had been fertilized by the sperm of electroporation was counted after 3 hour. Fertilization rate was calculated following the following formula [12]:

$$\text{Fertilization rate} = \frac{\sum \text{Fertilized egg}}{\sum \text{Egg}} \text{ x } 100\ \% \qquad (2)$$

*2.7. Data Analysis*
Data was analyzed using Analysis of Variance (ANOVA). If there was significant different, LSD test was performed. In order to find out the relationship between treatment and the result, regression analysis was applied to decide the character and function of regression, giving a decision for the best treatment.

## 3. Result and Discussion

*3.1. Pre-Treatment Sperm Quality*
The success of electroporation includes motility and viability of sperm and also the sperm characteristics like color, smell, and pH.

**Table 2.** Sperm quality.

| Parameter | Result |
|---|---|
| Concentration | 11.7 x 109 cell.mL-1 |
| Volume | 2.3 mL |
| pH | 7 |
| Colour | White Milk |
| Head Diameter | ± 3 μm |
| Length of Tail | ± 8 μm |

| | |
|---|---|
| Motility | Score of 5 |
| Movement Intensity | Progressive |
| Viability | 90 % |

Table 2 shows that the quality of the Koi sperm pre-electroporation was good and could be used to conduct the research. The condition or the quality of koi sperm was essential since it would be a vector of the foreign genetic carrier to maximize the method of SMGT (Sperm Mediated Transfer Gen) and the process of foreign gene binding. It would affect the success of the sperm in the insemination process with the egg cell after electroporation process.

According to Dewi [13], good quality sperm can fertilize egg cell well and changes occur to the sperm will affect its performance. The changes refer to motility and survival. According to Havez [14], the cement, which meets the requirement of fertilization, has to contain a living and progressive sperm and the percentage of motile sperm is not more than 75 %.

### 3.2. *Post-Electroporation Sperm Motility*

Post-electroporation, each treatment showed a different result on the percentage of Koi sperm motility. The percentages of motility are shown in table 3.

**Table 3.** Sperm motility (%).

| Treatment | Repetition | | | Total | Average |
|---|---|---|---|---|---|
| | 1 | 2 | 3 | | |
| A | 79 | 79.5 | 75 | 233.5 | 77.83 |
| B | 74 | 77 | 75.5 | 226.5 | 75.50 |
| C | 75.5 | 73.5 | 75.5 | 224.5 | 74.83 |
| Total | | | | 684.7 | |
| Control | 89 | 93 | 90 | 272 | 90.66 |

From table 3, it can be seen that the highest average score of Koi sperm motility percentage gained by treatment A (10/μl concentration) with the score of 4, which means 55-70 % spermatozoa moved forward and showed a quick movement, while the lowest score was gained by treatment C (50 ng/μl concentration) with the score of 3.33, which means 40-55 % spermatozoa moved forward and several of them showed a quick movement. The analysis of the variety of Koi sperm motility can be seen in table 4.

**Table 4.** Analysis of the variety of koi sperm motility after electroporation treatment.

| Source Diversity | Db | JK | KT | F Test | | |
|---|---|---|---|---|---|---|
| | | | | F | 5 % | 1% |
| Total | 2 | 0.666 | 0.333 | | | |
| Random | 6 | 3.334 | 0.555 | 0.6 (ns) | 5.14 | 10.92 |
| Treatment | 8 | | | | | |

Table 4 shows that F-table 1 % > F-count (6.21) < F-table 5 %, which means different concentration of ccβA-GFP promoter with electroporation method did not give significant influence on Koi motility.

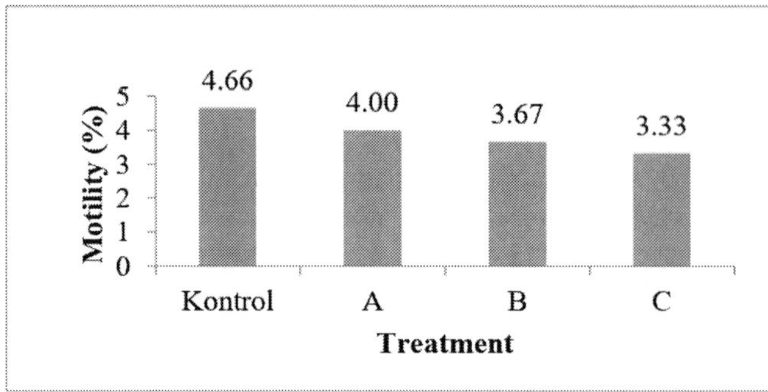


**Figure 1.** Sperm motility graph.

Figure 1 shows comparison between motility value of control (untreated) and treated sperm. The control sperm showed the highest result of motility that means the koi sperm used had a good quality. The graph indicates that the higher the concentration of the ccBA-GFP promoter given, the lower the percentage of the koi sperm motility.

The different concentration given had an influence on dilution ratio used in producing ccBA-GFP promoter. Different concentrations were collected from dilution using TE buffer according to the dilution formula. At lower concentration, more dilution solution was added compared to the amount added at higher concentration, causing differences in osmolarity value. Different dilution ratio where thicker and thinner solutions were separated by semipermeable membrane, made the diluted molecules flow from the thinner solution to the thicker one, similar to osmosis process. Therefore, the use of lower concentration caused easier diffusion into the sperm cell and resulted in a higher intensity of luminescence if compared to higher concentration with high numbers of DNA. In addition, higher concentration caused an incompatibility as a living medium for sperm; hence it decreased the rate of sperm motility and viability. There are at least two important parameters that have to be optimized in applying SMGT to reach an efficiency, namely cement quality and the entry process of sperm. Entry process of the sperm highly depends on the sperm viability and motility [15]. DNA concentration in the dilution also affects the success of gene transfer [16]. Spermatozoa membrane permeability tightly relates to the spermatozoa motility since membrane permeability is related to the nutrition transportation, which is important for cell metabolism. The sperm cell tends to be smaller after performing electroporation causing a decrease in sperm motility percentage [17, 18].

### *3.3. Post-Electroporation Sperm Viability*

Sperm viability is the percent comparison of living and dead sperm. The average values of viability percentage can be seen in table 5.

**Table 5.** Sperm viability (0%).

| Treatment | Repetition | | | Total | Average |
|---|---|---|---|---|---|
| | 1 | 2 | 3 | | |
| A | 79 | 79.5 | 75 | 233.5 | 77.83 |
| B | 74 | 77 | 75.5 | 226.5 | 75.50 |
| C | 75.5 | 73.5 | 75.5 | 224.5 | 74.83 |
| Total | | | | 684.7 | |
| Control | 89 | 93 | 90 | 272 | 90.66 |

According to table 5, the highest result was given by treatment A (10 ng/μl concentration), which was 77.83 %, while the lowest result was shown by treatment C (50 ng/μl concentration) with 74.83

%. The control (untreated sample) was higher than treatment A by 90.66 %, which showed the higher the ccBA-GFP promoter concentration given, the lower the viability percentage resulted. Improper living medium caused sperm viability decreased.

According to Table 6, F table 1 % > F count (2.279) < F table 5 %, which showed different concentration of ccβA-GFP promoter given with electroporation methods did not give significant influence over the Koi viability. The analysis of the variety of Post-electroporation Koi Sperm Viability is shown in table 6.

**Table 6.** Analysis of the variety of post-electroporation koi sperm viability.

| Source Diversity | Db | JK | KT | F Test | | |
|---|---|---|---|---|---|---|
| | | | | F | 5% | 1% |
| Total | 2 | 14.68 | 7.34 | | | |
| Random | 6 | 19.36 | 3.22 | 2.279 | 5.14 | 10.92 |
| Treatment | 8 | | | | | |

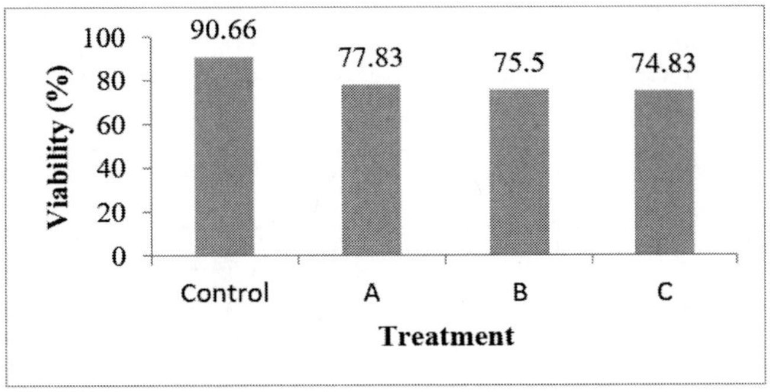


**Figure 2.** Viability of control and treated sperm.

Sperm viability normally ranges from 1-2 minutes after coming out of the testes. Change of plasma membrane structure and the loss or decrease of some of the mitochondrial matrix can result in loss of spermatozoa viability [19].

Electroporation treatment with high electric field can stretch the sperm cell and temporarily destabilize the membrane cell. During that period, the membrane is highly permeable to exogen molecules (DNA or RNA) around the cell medium. DNA can then move into the cell (internalization process). When the electric field is turned off, the membrane's deep hole closes and the exogen DNA comes into the cell [20, 21].

Therefore, the best result was gained when low concentration of promoter was used to produce transgenic Koi fish. During the transgenesis using sperm as the genetic holder vector, the motility and viability of sperm were important.

### *3.4. Sperm Fertilization Rate at the Cell Eggs*

Fertilization rate is the number of eggs cell that can be fertilized, i.e., the percentage of ratio between number of eggs cell spread and fertilized by sperm. A total of 0.25 gram of eggs (±114 eggs) was fertilized The egg cell was fertilized with the electroporated sperm . Fertilized eggs cells were then observed after the fertilization process and calculated according to the fertilization rate formulation. The percentage of successfully fertilized eggs is presented in table 7.

**Table 7.** Fertilization rate (%).

| Treatment | Fertilization Rate | | | Total | Average |
|---|---|---|---|---|---|
| | 1 | 2 | 3 | | |
| A (10 ng/µl) | 76.3 | 74.5 | 68.4 | 219.2 | 73.09 |
| B (30 ng/µl) | 73.6 | 52.6 | 66.6 | 192.9 | 64.32 |
| C (50 ng/µl) | 39.4 | 58.7 | 68.4 | 166.6 | 55.55 |
| Total | | | | 578.92 | |
| Control | 85,09 | 83,33 | 71,92 | 240,34 | 80,11 |

The highest average of fertilization rate was 73.09 % given by treatment A, followed by treatment B with 64.32% and the lowest was treatment C with 55.55 % (table 7). The control sample had higher fertilization rate than treated samples. Fertilization rate was equal to sperm quality used in fertilization including sperm motility and viability. The best quality sperm came from treatment A, thus the best fertilization occurred at treatment A since sperm quality of highly affects fertilization rate. Sperm with good motility can move maximally during fertilization.

Table 8 shows, F table 1 % > F count (1,982428) < F table 5 %, which showed different concentration of ccβA-GFP promoter given with electroporation methods did not give significant influence over the Koi Fertlization Rate. The analysis of the variety of Post-electroporation Koi Fertlization Rate is shown in table 8.

**Table 8**. Analysis of the varians of post-electroporation koi fertilization rate.

| Source Diversity | db | JK | KT | F.Hit | | |
|---|---|---|---|---|---|---|
| | | | | F | F 5% | F 1% |
| Total | 2 | 461.6528 | 230.8264 | 1.982428 | 5.14 | 10.92 |
| Random | 6 | 698.6172 | 116.4362 | ns | | |
| Treatment | 8 | | | | | |

Figure 3 shows the fertilizaton rate control showed the highest result of fertilizaton rate that means the koi sperm used had a good quality. The graph indicates that the higher the concentration of the ccBA-GFP promoter given, the lower the percentage of the koi fertilizaton rate.

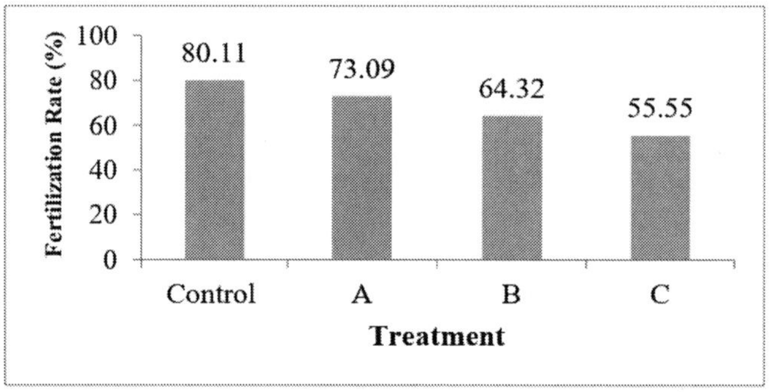


**Figure 3.** Fertilization Rate of control and treated sperm.

The difference between control sample and treated samples was caused by the difference in sperm quality was used to fertilization. The control sperm had better quality since it was not given electroporation so it did not experience environment stress and it still had energy reserve coming from the medium used to protect the sperm immotility. In addition, the control sperm was able to

IOP Conf. Series: Earth and Environmental Science **137** (2018) 012025 doi:10.1088/1755-1315/137/1/012025

maintain the sperm quality so that during the artificial fertilization, the sperm went into cell micropyle and had adequate energy to go through embryogenesis process. Moreover, unlike the electroporated sperms, the control sperm did not cause the plasma membrane to get broken and it did not lose the mitochondria contained in its tail. The loss of mitochondrial matrix can decrease viability and percentage of the sperm can influence the rate of the hatching eggs.

The electroporated sperm decreased the viability and the fertilization rate due to its broken tail [22]. The low rate of spermatozoa fertilization is also influenced by the short time of motility and the sperm viability, while the success of egg fertilization is highly influenced by the sperm motility [23, 24]. Long viability does not guarantee high fertilization rate, thus spermatozoa needs high energy to fertilize cell. The success of post-electroporation sperm fertilization is highly influenced by the sperm motility. There is a positive correlation between the capability of foreign DNA to come and integrate into the main sperm with the quality of cement and sperm motility rate. Therefore, it needs the sperm mobility of at least 80% and not lowers than 65%. In addition, a motile sperm is able to tie foreign DNA transferred up to 30% from the total [25].

## 4. Conclusion

Different concentration of ccBA-GFP promoter given with electroporation method had effect on sperm motility and viability. The higher the concentration promoter given, the lower the quality of Koi sperm. The best result was shown by treatment A that had the lowest concentration of promoter (10 ng/μl).

## 5. References


[1] Kusrini E 2015 *J. of Media Aquat.* 259 – 64.

[2] Alimuddin, Yhosizaki, Carman and Sumantadinata 2003 *Ind. J. Aquat.* **2** 41-50.

[3] Lavitrano M, Busnelli, Cerrito, Giovannoni, Manzini and Vargiolu 2006 *Reprod. Fertil. Dev.* **18** 19-23.

[4] Tsai H J 2000 *Mol. Reprod. Dev.* **56** 281-284.

[5] Gusrina 2011 Application of gene transfer in introduction and expression of the growth hormone gene of nila fish (*Oreochromis niloticus*) on Catfish (*Clarias* sp.) (Disertasion: Bogor Agricultural University)

[6] Faqih A 2011 *J. Exp. Life. Sci.* **1** 56 – 110.

[7] Anitasari S, A Soeprijanto, A R Faqih 2015 *J. Exp. Life. Sci.* **5** 82- 88.

[8] Buwono I D, Iskandar M U K, Agung U, Bubhan 2016 *J. Biol.* **20** 17-28.

[9] Anitasari S, A Soeprijanto, A R Faqih 2015 *J. Exp. Life. Sci.* **5**(2) 82- 88.

[10] Dewi R R S P S 2010 Study of over expression of growth hormone gene through sperm electrophoresis to make transgenic siamese catfish grow fast (Disertasion: Bogor Agricultural University)

[11] Susilowati T 2011 *Spermatology* (Malang: UB Press Brawijaya University)

[12] Rustidja 2000 Prospect of fish sperm freezing Faculty of Fisheries and Marine Science . Brawijaya University Malang

[13] Dewi R R S P S 2010 Study of over expression of growth hormone gene through sperm electrophoresis to make transgenic siamese catfish grow fast (Disertasion: Bogor Agricultural University)

[14] Hafez E S E 1987 *Reproduction in Farm Animal* Philadelphia 5th ed.

[15] Lavitrano M, M Busnelli, M G Cerrito, R Giovannoni, S Manzini, A Vargiolu 2006 *Reprod. Fertil. Dev.* **18** 19-23

[16] Walker S P, Symonds J E, Sin I L, Sin F Y T 1995 *J. Mar. Biotechnol.* **3** 232-234

[17] Sin F Y T, Walker S P, Symonds J E, Mukherjee U K, Khoo J G I, Sin I L 2000 *Mol. Reprod. Dev.* **56** 285-288

[18] Effendi M I 1997 *Fisheries biology* (Bogor: Nusatama Foundation)

[19] Woynarovich E, L Horvarth 1980 *The artificial propagation of warm-water fin fish. a manual for extention FAO Fish. Tech.* No. 201 183 p.

[20] Toelihere, M R 1985 Artificial insemination in livestock (Bandung: Angkasa)
[21] Spadafora C 2008 *J. Hum. Reprod.* 735-740.
[22] Sin F Y T, Walker S P, Symonds J E, Mukherjee U K, Khoo J G I, Sin I L 2000 *Mol. Reprod. Dev.* **56** 285-288.
[23] Nurman 1998 *J. Fish.* **7** 3-42
[24] Hidayaturrahmah 2007 Time motility and viability of spermatozoa fish mas (*Cyprinus carpio* L.) on different concentrations of fructose solution (South Kalimantan: Biology Study Program Faculty of Mathematics and Natural Sciences Lambung Mangkurat University) p 9 – 18
[25] Lavitrano M, M Busnelli, M G Cerrito, R Giovannoni, S Manzini, A Vargiolu 2006 *Reprod. Fertil. Dev.* **18** 19-23

**Acknowledgement**
Dr. Alimuddin, S.Pi., M.Sc (IPB Bogor), for providing the material (pmβ-Actin-GFP) for this research.

IOP Conf. Series: Earth and Environmental Science **137** (2018) 012026 doi:10.1088/1755-1315/137/1/012026

# Analysis of growth performance and benefits of a high density catfish *Clarias gariepinus* Burchell culture in biofloc system

**F Basuki, T Yuniarti, D Harwanto and T Susilowati.**

Aquaculture Department, Faculty of Fisheries and Marine Science Diponegoro University. Jl. Prof. H. Soedharto SH, Tembalang Semarang, 50275. Indonesia

E-mail : fajar.basuki@live.undip.ac.id

**Abstract**. The aim of the study was to determine the growth performance and calculation of business profit of catfish in various densities using biofloc system. Catfish fingerlings, 8±0.2 g each, were used in this research. There were three treatments in various densities, i.e. $P_1$: 1,100; $P_2$: 1,200 and $P_3$: 1,300 fingerlings/$m^3$. After 4 months observation, the results showed that growth performance regression from $P_1$, $P_2$, and $P_3$ was Y ($P_1$) = 7.66 + 30.7 X (r2 = 0.971); Y ($P_2$) = 7.45 + 23.6 X (r2 = 0.996) and Y ($P_3$) = 6.5 + 23.63 X (r2 = 0.996), respectively. The $P_1$ harvest time was achieved within 3 months; while $P_2$ and $P_3$ were 4 months. Total yield/tank in $P_1$, $P_2$ and $P_3$ were 297.00, 295.29 and 309.52 kg, respectively. The calculation of business profit per tank per month of $P_1$, $P_2$ and $P_3$ were 626,784.17; 399,058.77 and 351,793.65 IDR (Indonesian Rupiah), respectively. Meanwhile, result of R/C ratio of $P_1$, $P_2$ and $P_3$ were 1.60, 1.5 and 1.4, respectively. Since the calculation of R/C ratios were $\geq 1$, this culture method is considered feasible as a business.

## 1. Introduction

Based on statistical data, the production of catfish *Clariasgariepinus* (Burchell 1822) in Central Java has been increasing rapidly. In 2013, the production was recorded at 62.686 tons and reached 75,236 tons in 2013 or an increase of 20.02% within a year [1].The increase in catfish culture has been supported by several reasons such as: biological reasons (fast growing, easy cultivation, readily consumes fabricated feed, and tends to be resistant to disease), social reasons (good market value), and physical reasons (resistant to environmental changes) [2].

Fish farmers and aquaculturists have tried to implement high density catfish farming. The catfish is labyrinth fish, so it is possible to cultivate it using high density technology such as the biofloc technique. According to a previous study [3], high density catfish culture using biofloc has reached 50 catfish/100L of water or 500 catfish/1000 L of water with the weight of the catfish at 2.3±0.12 g/fish. High density catfish culture using biofloc means that catfish were stocked into concrete tanks (vol. 1 $m^3$), where each tank contained 1,000 juveniles (mean bodyweight: 2.5 g) [4].

Catfish culture with the biofloc system in closed recirculation containers is one technique that can be implemented to increase catfish production. The biofloc system allows high density catfish culture so that production increases [5]. Biofloc is a collection of various organisms including bacteria, fungi, protozoa or algae incorporated in a floc. High density catfish culture produces more waste so it needs regular stirring that allows waste to be suspended which is then processed aerobically by heterotrophic bacteria into inorganic compounds. The remaining organic material derived from feed or dirt will cause a decrease in water quality. Biofloc technology can add organic carbohydrates through an increase in C/N ratio and stimulate the growth of heterotrophic bacteria. Probiotics also serve as immunostimulants to decrease FCR, inhibit pathogenic bacteria and improve water quality [3].

The aim of this research was to observe the performance of growth and to predict the harvest weight and analyze the profit per pool with various densities and R/C ratio.

## 2. Methodology

**Tested Fish :** Catfish fingerlings weighing 8 ± 0.2 grams/fingerling or 8 ± 0.2 g/fingerling in various densities as treatment: $P_1$ (1.100 fingerlings/m$^3$ or 3.300 fingerlings/pool); $P_2$ (1,200 fingerlings/m$^3$ or 3,600 fingerlings/pool) and $P_3$ (1,300 fingerlings/m$^3$ or 3,900 fingerlings/pool).

**Pool:** The pool was made from plastic with a diameter of 3 m and equipped with a 6 mm iron frame. The total volume of water in each pool was 3000 L. The fingerlings were adapted and stocked into the pool randomly.

**Feed Preparation:** The feed was a commercial pellet containing 31-33% protein; 3-5% fat; 4-6% fiber, 10-13% ash, and 11-13% water. Each kilogram of pellet was added with 5cc EM4 bacteria in 150 cc fresh water. Each ml of EM4 contains *Lactobacillus casei* (2.0x10$^6$ cell/ml) and *Sacharomyces cerevisiae* (3.5x10$^5$ cell/ml).

**Pool Preparation:** Before stocking the fingerlings, pools were filled with fresh water until reaching the height of 43 cm, so that the volume of fresh water would be ± 3 m$^3$, salt was added at 600 grams/pool or 200 grams/m3, molasses as much as 300 cc/pool or 100 cc/m$^3$ and 5 cc of EM4 [4]. The water was replaced periodically if the water quality variable was exceeding the allowed threshold.

**Pool Recirculation:** For pool recirculation, three water pumps powered by 33 watt electricity were used in each pond.

**Experiment Design:** Adapted catfish fingerlings were adapted and stocked into pools randomly, according to the treatments of $P_1$ (1,100 fingerlings/m$^3$ or 3,300 fingerlings/pool); $P_2$ (1.200 fingerlings/m$^3$ or 3.600 fingerlings/pool) and $P_3$ (1.300 fingerlings/m$^3$ or 3.900 fingerlings/pool).

**Fish Growth Measurement:** Individual growth rates were measured every two weeks (14 days) after the catfish were cultured. Duration of the culture was 16 weeks or 4 months. Data was then analyzed based on regression analysis using SPSS 20.

**Survival Rate (SR)**: Measured from the final number of fish divided with the initial number of fish then multiplied by 100% [3].

**Food Convertion Rasio (FCR)**: Measured by comparing the total feed (g) with the total fish weight (g) [6].

**Benefit-Cost Ratio:** Calculating the ratio of revenue (R) and cost (C).

**Criteria:** R/C >1: feasible; and R is revenue or total revenue, which is the total of the catfish selling price, while C is cost or total production cost [7].

## 3. Results and Discussion

### *3.1. Growth Performance*

Weight observation is important since the selling of harvest is based on the total weight of catfish. Based on the observation, high-density catfish culture in biofloc shows that in week 4 (month 1) all treatments show a considerable weight increase. It shows that the quality of feeding and water quality have positive effects on the fish but the levels of response was different in each treatment. P1 shows a weight increase of 37.92 ± 0.74 gr; P2 shows a weight increase of 30.25 ± 0.99 gr and P3 shows a weight increase of 27.65 ± 1.78 gr.

The observation on growth was continued. In the 3$^{rd}$ month (week 12), P1 showed the best results, reaching the weight of 100 ± 3.54 gr or ready for harvest. P2 reached the weight of 78.54 ± 0.63 gr, which was higher than P3 at 77.9± 0.67 gr. The weight of catfish in P2 and P3 do not meet the criteria for harvesting i. e. 100 gram. The dynamics of growth can be seen in figure 1. The regression analysis on P1 shows the equation YP1 = 6.46 +7.81 X which means that after 12 weeks or 3 months it can be predicted that the catfish have reached 100g or the ideal harvest weight. While for treatment P2, the equation is YP2 = 6.53 + 6.0 X meaning that after 12 weeks or 3 months culture, the weight of catfish is estimated at 78.54 ± 0.63 gr or not ready for harvest and requires 4 more weeks to reach ideal weight. The treatment P3 shows the regression equation YP3 = 6.71 + 5.94 X meaning that after culturing for 12 weeks or 3 months, the weight of catfish is estimated as 77.9± 0.67 gr or not ready for harvest and needs 4 more weeks to reach ideal weight.

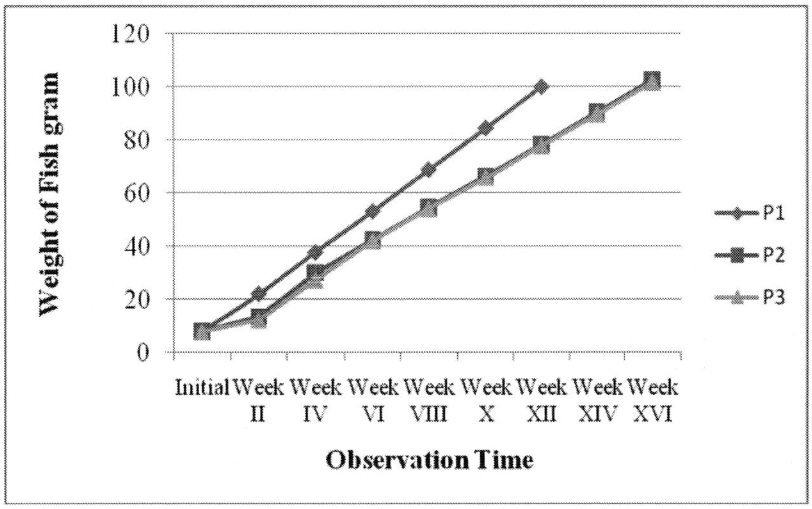


**Figure 1**.The dynamics of Catfish growth rate based on the weight in various densities which are maintained based on biofloc.

*3.2. Survival rate (SR)*
As soon as the fingerlings were stocked into the pools, there was a change to the water quality. The fingerlings have a good chance of survival as long as the water quality can be well maintained. However, sometimes the quality of water decreases suddenly, leading to the death of fish. The quality of water used in the research was closely monitored, yet the survival rate changes every four weeks or one month (table 1).

**Table 1.** Survival rate (SR) per month.

| Treatment | Initial (%) | Weeks IV (%) | Weeks VIII (%) | Weeks XII (%) | Weeks XVI (%) |
|---|---|---|---|---|---|
| P1 | 100 | 98 | 96 | 95 | ---- |
| P2 | 100 | 96 | 92 | 87 | 85 |
| P3 | 100 | 95 | 88 | 83 | 78 |

*3.3. Production per pool and Total Revenue (IDR)*
After culturing for 3 months, the individual weight of the fish in $P_1$ reached 100 grams with a survival rate (SR 95 %) hence the total production of P1 amounted to 313.5 kg. At the price of 16,000 IDR/kg, the total revenue was estimated at 5,016,000 IDR. Treatment $P_2$ and $P_3$ weighed under 100 g/fish so the culture continued to 16 weeks or 4 months. After 4 months, P2 reached the individual weight of 102.53 grams and with a Survival Rate (SR) of 85 %, so total production of P2 amounted to 313.74 kg. At the price of 16,000 IDR/kg, the total revenue was estimated at 5,019,869

IOP Conf. Series: Earth and Environmental Science **137** (2018) 012026 doi:10.1088/1755-1315/137/1/012026

IDR. P3 reached the individual weight of 101.75 grams with the Survival Rate (SR) at 78 %, so total production of P3 amounted to 309.52 kg. Selling at the price of 16,000 IDR/kg, the total revenue was estimated at 4,952,376 IDR. The dynamics of production and total revenue can be seen on table 2.

**Table 2.** Production and total revenue (IDR).

| Treatment | Total Stocking (fish) | Weigh (gr) | SR (SR) | Total Production (kg) | Price | Total Revenue (IDR) |
|---|---|---|---|---|---|---|
| P1 | 3,300 | 100 | 95 | 313.5 | 16,000 | 5,016,000 |
| P2 | 3,600 | 102.53 | 85 | 313.7418 | 16,000 | 5,019,869 |
| P3 | 3,900 | 101.75 | 78 | 309.5235 | 16,000 | 4,952,376 |

*3.4. Cost*

In fish cultivation, cost estimation is the determining factor of continuation. This research provides cost estimation for high-density catfish culture using biofloc. The costs estimated in these activities are the cost of feed, the depreciation cost of tarp, depreciation cost of pumps, electricity, labor, and cost of fingerlings.

The research shows the lowest cost is $P_1$, which amounts to 2,980,213 IDR, followed by $P_2$ at 3,325,163 IDR, while the highest is $P_3$ with the amount of 3,360,163 IDR. The dynamics of cost is presented in table 3.

**Table 3.** Total cost (IDR).

| Treatment | Feeding cost | Depreciation Cost of Tarp | Depreciation Cost of Pump | Electricity | Labor | Fingerling Cost | Total Cost |
|---|---|---|---|---|---|---|---|
| P1 | 166,6000 | 62,500 | 113 | 291,600 | 300,000 | 660,000 | 2,980,213 |
| P2 | 169,4000 | 83,333 | 150 | 427,680 | 400,000 | 720,000 | 3,325,163 |
| P3 | 172,9000 | 83,333 | 150 | 427,680 | 400,000 | 780,000 | 3,360,163 |

*3.5. Benefit cost ratio*

The estimations of business profit per pool per month are $P_1$: 678,595.83 IDR; $P_2$: 423,676.37 IDR and $P_3$: 398,053.17 IDR. While the result of R / C ratio shows $P_1$: 1.68; $P_2$: 1.51 and $P_3$: 1.47. The dynamics of benefit-cost ratio can be seen in table 4.

**Table 4.** Benefit-Cost Ratio (IDR).

| Treatment | Total Revenue | Total Cost | Profit (3-4 months) | Profit per Month | R/C |
|---|---|---|---|---|---|
| P1 | 5,016,000 | 2,980,213 | 2,035,788 | 678,596 | 1.68 |
| P2 | 5,019,869 | 3,325,163 | 1,694,706 | 423,676 | 1.51 |
| P3 | 4,952,376 | 3,360,163 | 1,592,213 | 398,053 | 1.47 |

## 4. Discussion

Initial growth is very important in the maintenance of fish, especially catfish. The result shows that water quality at the beginning of the experiment was in accordance with the standard of maintenance [8] as well as good feed management, which can be seen from the growth of the fish. The research shows the daily growth rate ranged from 0.87 – 1,02 grams/day. According to a previous study [3], high-density cultured catfish in biofloc with an initial weight of 2.3 gram fish$^{-1}$ show a specific growth rate (SGR) as much as 6.35 % after cultured for 42 days. Meanwhile, another study [4] showed that high-density cultured catfish in biofloc with an initial weight of 2.5 gram fish$^{-1}$ result in a final weight of around 4.9 – 8.7 grams after cultured for 35 days. The research conducted by Channel Catfish using biofloc technology [8] showed that fish having an initial weight of 217 g/fish, stocking density 1.4, 2.1, or 2.8 kg/m$^3$, resulted in an individual harvest weight ranging from 658-829 g/fish. The research of the channel catfish stocker [9] showed that catfish (0.26 kg/fish) stocked into earthen ponds at approximately 0.22 kg/m3 (11.115/ha) grew at 4.0 g/d, achieved a mean final weight of 0.91 kg/fish in 164 days, and consumed 1.17 kg feed/fish.

Observations on survival rates (SR/%) revealed the result between 78-95 %, while previous research Yusuf *et al.* [3], showed that high-density cultured catfish in biofloc with an initial of weight 2.3 gram/fish resulted in 86.67-89.33% of SR after cultured for 42 days. Additionally, another research Hapsari [4], founded that high-density cultured catfish in biofloc having an initial weight of 2.5 gram/fish resulted in 85.8-91.4 % of SR after cultured for 35 days. Furthermore, the Tilapia survival rate was observed at 98-99 % using biofloc technology [10].

Fish cultured for 3-4 months showed an individual growth between 100-102.53 and a SR between 78-95 %. Therefore, the production of each 3m$^3$ pool ranged between 309.52-313.5 kg or 103.57 - 104.5 kg/m$^3$. The growth of catfish culture based on biofloc shows a higher rate compared to tilapia [10]. Research shows that daily tilapia growth ranges between 0.1 g/fish/day, with the initial fingerling weight at 3.37-3.44 gram/fish. Moreover, tilapia with growth rates of up to 0.3 g/day yielding up to 300 mt/ha have been reported in well managed bioflocs ponds, so that production reaches 30kg/m$^3$ water [11].

High-density catfish culture using biofloc is a profitable business; it can even be an option for home industries, or small and large scale industries. The research shows that the growth of fish, survival rate, food conversion ratio, total expenditure, total income, and profit can be well calculated. The research also shows that R/C ratio of the business ranges from 1.47 – 1.68 meaning that the business is profitable [7,11].

## 5. Conclusion

The growth performance of catfish cultured with biofloc shows a simple regression line for treatment P1, showing the equation YP1 = 6.46 +7.81 X which means that after 12 weeks or 3 months it can be predicted that the catfish will have reached 100g or the ideal harvest weight. While for treatment P2, the equation is YP2 = 6.53 + 6.0 X meaning that after the culture of 12 weeks or 3 months, the weight of catfish is estimated as 78.54 ± 0.63 gr or not ready for harvest and needs 4 more weeks to reach ideal weight. The treatment P3 shows the regression equation YP3 = 6.71 + 5.94 X meaning that after culturing for 12 weeks or 3 months, the weight of catfish is estimated as 77.9± 0.67 gr or not ready for harvest and needs another 4 weeks to reach ideal weight. The estimation of business profit per pool per month of $P_1$, $P_2$, and $P_3$ were 678,596; 423,676 and 398,053 IDR, respectively. While the result of the R / C ratio showed $P_1$ was 1.68; $P_2$ was 1.51 and $P_3$ was 1.47. Since the calculation of R/C ratios were $\geq 1$, high-density catfish culture using the biofloc technique is considered feasible as a business.

Based on the research, it is advised that high density catfish culture based on biofloc is proven feasible. The ideal density should be 1,100 fingerlings/m$^3$ or 3,300 fingerlings/pool to obtain the monthly profit of 678,596 IDR. To reach a minimum monthly income of 2,000,000 IDR, three ponds should be maintained.

## 6. References

[1]. Director General of Aquaculture 2013 Statistical Data Series Cultivation Fishery Production-Indonesia, Production of Catfish, Gurame, Other (Dirjen Perikanan Budidaya 2013 Data

Statistik Series Produksi Perikanan Budidaya-Indonesia, 2009-2013 Produksi Lele, Gurame, Lainnya)

[2]. Jamabo, N A and R I Keremah 2009 Effect of Stocking Density on Growth and Survival Rate of the Fingerling on Clarias gariepinus (Burchell, 1822) *Int. J. Fish.* **4** 55-57

[3]. Yusuf, M W, N B P Utomo, M Yuhana and Widanarni, 2015 *J .Fish. Aquat. Sci* **10** 523-532

[4]. Hapsari, F, 2016 *AACL Bioflux,* **9**

[5]. Kareem, O K and A N Olanrewaju 2015 *J. Fish. Aquat. Scie* **10** 400-404.

[6]. Islam M Z , Md R I Sarder and Md R I Akhand 2015 *IJFAS* **2** 186-191.

[7]. Hidayah, I , 2010 *Jurnal Budidaya Pertanian* **6**

[8]. Green B W and K Schrader 2015 *J. Aquac. Res. Dev.* **6** 314

[9]. Green B W and C R Engle 2004 Growth of stocker channel catfish to large market size in single-batch culture *J. World Aquac. Soc.* **35** 25-32

[10]. Abduljabbar A A , A M Nour. T Srour, N El-bermawy, W A Fayed and A T Mansou 2015 *Global Journal of Fisheries and Aquaculture Researches* **2** 64-80

[11]. Ogello, E O, S M Musa, C M Aura, J O Abwao and J M Munguti 2014 *Int. J. Aquac. Fish. Sci.* **5** 21-39

**Acknowledgments**

This research is supported by the Rector of Diponegoro University and the Dean of the Faculty of Fisheries and Marine Sciences under contract number 2270 / UN7.3.10 / LT / 2017 fiscal year 2017

IOP Conf. Series: Earth and Environmental Science **137** (2018) 012027 doi:10.1088/1755-1315/137/1/012027

# Performance efficiency of feed utilization, relative growth rate, and survival rate of common carp (*Cyprinus carpio*) through the addition of phytase in the feed

**D Rachmawati and I Samidjan**

Aquaculture Department of Fisheries, Faculty of Fisheries and Marine Sciences, Diponegoro University, Jl. Prof. Soedarto, SH, Tembalang, Semarang, 50275 Indonesia

Email: dianarachmawati1964@gmail.com

**Abstract**. The purpose of this study was to determine the effect of adding phytase enzyme in the feed on digestibility of feed, efficiency of feed utilization, relative growth rate and survival rate of Common carp (*Cyprinus carpio*). Fish samples in this research were Common carp with an average - weight of 3.34 ± 0,16 g/fish. The treatments were adding the phytase enzyme in the feed with the different level of doses. Those were A (0 U $kg^{-1}$ feed), B (500 U $kg^{-1}$ feed), C (1.000 U $kg^{-1}$ feed g) and D (1.500 U $kg^{-1}$ feed). Observation was conducted on digestibility of protein ($ADC_P$), digestibility of phosphor ($ADC_F$), efficiency of feed utilization (EFU), relative growth rate (RGR), protein efficiency ratio (PER), feed conversion ratio (FCR), survival rate (SR) and water quality parameters. The results show that the addition of phytase enzyme significantly ($P<0.01$) affected on $ADC_P$, $ADC_F$, EFU, RGR, FCR, and PER, on the other hand it insignificantly ($P>0.05$) affected on SR of common carp. Based on results, it was concluded that optimum doses of phytase enzyme feed in terms of digestibility of feed, efficiency utilization of Feed and growth rate of Common carp ranges from 943 to 1100 U $kg^{-1}$ feed

## 1. Introduction

Intensive common carp (*Cyprinus carpio*) aquaculture needs feed that contains complete nutrients. It should also be efficient and economical. Common carp (*C. carpio*) highly needs complete feed, which contains fat, protein, carbohydrate, vitamins, and minerals. Protein is the most important element in the feed to support growth [1]. Protein in the feed can be from plant-based protein or animal-based protein. However, the use of such plant ingredients in aqua feed is limited because of the presence of wide variety of anti-nutrient compounds [2], which have deleterious effects on the physiology and morphology of the digestive tract [3, 4], thereby affecting the overall fish growth [5, 6]. Plant-based feed from soybean meal has a drawback because soybean meal contains the anti-nutrient phytic acid. Phytic acid can hinder the absorption of plant-based protein by fish. Around 80% of the available phosphor (P) is in phytic acid [7]. Phramkunthong *et al.* [9] mentioned that the content of phytic acid in soybean meal was 3.88%, with as much as 59.9% of it was phosphor [8]. Gatlin *et al.* [10] stated that most plant-based feed contained high fiber that was difficult to digest; therefore, fish cannot utilize the feed well [9]. Moreover, Masumato *et al.* [11] reported that phytic acid can reduce digestibility of protein and minerals such as K, Mg, Ca, Zn, Fe, and Cu [10]. To solve these problems the enzyme phytase can be added in the feed.

An increase in the use of plant-based feed creates another problem. It produces phosphoric pollutant in the water. According to Masumoto *et al.* [11], Cheng *et al.* [12], Depnath *et al.* [13], Rachmawati *et al.* [14], phosphor in plant-based feed cannot be fully utilized by the fish due to lack of the enzyme phytase that decomposes phytic acid. Moreover, Kumar *et al.* [15] explained that phytic acid bound 80% of the total phosphor available in plant-based feed. Phytic acid in the feed is excreted along with feces into the environment, then it is degraded by microbes that produce phytase and the phosphor is then released into the water. High phosphor concentration in the water will trigger eutrofication that hinders cultivation of fish [7]. Jagannathan *et al.* [16] reported that phosphor is a macro nutrient that is needed by animals including fish.

Chung [17] suggested that to add phytase to overcome the problem of phytic acid. This enzyme can hydrolyze phytic acid into inositol and phosphoric acid; in turn, it can eliminate metal-chelating character [8]. Moreover, Chung [17] mentioned that phytase increased nutrient absorption and regulated nutrient excretion, such as phosphor, nitrogen, and minerals. Some researchers studied on addition of phytase in artificial feed, such as, by [14, 18, 19, 20, 21, 22]. Husain *et al.* [18] and they found that the optimal dose of phytase on the fingerlings of *Labeo rohita* was 750 FTU $kg^{-1}$ feed. Shapawi *et al.* [19] suggested that 30 % of soy bean meal with the addition of phytase (2000 FTU $kg^{-1}$ feed) formed an optimal dose for *Ephinephelus fuscoguttatus* fish. Moreover, Bulbul *et al.* [20] reported that supplementing phytase in feed can increase nutrient digestibility and growth of shrimp (*Marsupenaeus japonicas*). Danwitz *et al.* [21], also mentioned that the dose of phytase at 2000 $Ukg^{-1}$ feed could increase growth, protein digestibility, and phosphor of turbot fish (*Psetta maxima* L.). Rachmawati and Samidjan [22], also reported that the dose of phytase at 1000 FTU $kg^{-1}$ feed increased growth of shrimp (*Penaeus monodon*). Moreover, Rachmawati *et al.* [14], stated that 1000 FTU $kg^{-1}$ feed of phytase increased digestibility and growth of fish (*Channos channos*). Based on the above discussion, this research was aimed to study effects of phytase on digestibility of feed, efficiency of feed utilization, relative growth rate, and survival rate of common carp (*C. carpio*) through the addition of phytase in the feed and also to determine the optimal dose of phytase.

## 2. Methodology

Fish used in this study was fingerlings of common carp (*C. carpio*) with the average weight of 3.34 ± 0.16 g/fish. The fish was obtained from Magelang, Central Java, Indonesia. Density of fingerlings was 25 fish per $m^3$ [23]. The fish was selected on the base of size uniformity, health, completeness of organs, and lack of potential diseases. The selected fish was placed in to the cultivation container for 7 (seven) days in order to adapt to the environment and fed with feed without phytase. After 7 (seven) days, the fish was left to fast for one day to empty waste of its metabolism; therefore, at the beginning of the study, the fish could be weighed without any waste of metabolism in the body [14]. The sampled fish was raised for 42 days and fed at satiation 3 (three) times a day at 08:00, 12:00, and 16:00 o clock. The sampled fish was weighed every 7 (seven) days.

Containers used in the study were happas with the dimensions of 1 m x 1 m x 0.6 m. The experiment was conducted with 4 (four) treatments and 3 (three) repetitions. happas were Then put in the pond with the dimensions of 13 m x 12 m x 1 m equipped with inlet and outlet to circulate the water [23].

The study used artificial feed in the form of pellet containing 30% protein and added phytase. The doses of enzyme were A (0 U $kg^{-1}$ feed), B (500 U $kg^{-1}$ feed), C (1000 U $kg^{-1}$ feed), and D (1500 U $kg^{-1}$ feed). The enzyme doses used in this study were modified from the results of a study by Baruah *et al.* [24] that suggested phytase, as much as 750 U $kg^{-1}$ feed was the optimum level for the growth of fingerlings *Labeo rohita*. Each feed was fed thrice daily (07.00, 12.00, and 17.00 h) to approximate satiation for 42 d. Initially, feeds were fed at 3% of the body weight and subsequently adjusted based on daily intake. Round-the-clock aeration was provided to all the containers from a compressed air pump, and manual water exchange was carried out daily. Water quality parameters were checked every week using the methods as described by APHA [25].

The feed was made of fish meal, soybean meal, corn meal, rice bran, wheat flour, fish and corn oils, vitamins and minerals, CMC, phytase, and $Cr_2O_3$, where fish meal functioned as a source of animal protein, soybean meal as a source of plant-based protein, corn meal, rice bran, and wheat flour as sources of carbohydrate, fish oil and corn oil as sources of fat, vitamin mix and mineral mix as sources of vitamins and minerals, CMC as a binder, phytase to unbind phytic acid; and $Cr_2O_3$ as much as 0.5 % [1]. Proximate analysis was conducted to determine the nutrient content of feed and feed composition used in this study [26]. Phytase used was purchased commercially with the brand name Natuphos 5000G, produced by PT. BASF Indonesia. One unit of phytase activity (*Phytase Unit*) was defined as the amount of enzyme that releases 1 micro molecule of non organic per minute from 0.0051 mol/l of phytic acid at pH 5.5 and 37 °C [24].

The procedure to prepare feed was started with dissolving an appropriate dose of phytase into warm water (45° C) and then mixing with soybean meal evenly. The mixture was stored in the air-sealed container for 24 h [27]. According to National Research Council (NRC) [1], feed is made by mixing the least amount of the ingredients first and gradually adding and mixing other ingredients with larger amounts except fat sources that are added after all the ingredients have been mixed. The mixture was then formed into granules with diameter of 1-2 mm. Then, the feed was dried in the oven at 40 °C. The composition and results of proximate analysis of the feed can be seen in table 1.

**Table 1.** Composition and proximate analysis in the artificial feed.

| Ingredients | Composition | | | |
|---|---|---|---|---|
| | A | B | C | D |
| Phyetase enzyme | 0 | 0.5 | 1 | 1.5 |
| Fish meal | 26 | 26 | 26 | 26 |
| Soybean meal | 24 | 24 | 24 | 24 |
| Corn meal | 16.5 | 16.5 | 16.5 | 16.5 |
| Rice bran | 14 | 14 | 14 | 14 |
| Wheat flour | 12.0 | 11.5 | 11.0 | 10.5 |
| Fish oil | 1.00 | 1.00 | 1.00 | 1.00 |
| Corn fish | 1.00 | 1.00 | 1.00 | 1.00 |
| Vit Min Mix | 3.00 | 3.00 | 3.00 | 3.00 |
| $Cr_2O_3$ | 0.5 | 0.5 | 0.5 | 0.5 |
| CMC | 2.00 | 2.00 | 2.00 | 2.00 |
| Total (g) | 100 | 100 | 100 | 100 |
| Proximate analysis results | | | | |
| Protein (%)* | 30.18 | 30.17 | 30.14 | 30.16 |
| Fat(%) * | 7.39 | 7.38 | 7.34 | 7.37 |
| Extracted materials without nitrogen (%)* | 42.52 | 42.49 | 42.48 | 42.41 |
| Energy (kkal) | 271.79 | 271.62 | 271.13 | 271.29 |
| Ratio E/P | 9.01 | 9.00 | 8.98 | 9.00 |

Notes:
a.The values were calculated based Digestible Energy for 1 g protein equals 3.5 kcal, 1 g fat equals 8.1 kcal, and 1 g carbohydrate equals 2.5 kcal [19].
b. According to Bulbul *et al.* [20], the optimal E/P ratio for growth ranges from 8 kcal/g to 12kcal/g.
*Animal Nutrient Laboratory, Faculty of Husbandry and Agriculture, Diponegoro University (2017)

Data collected were $ADC_P$ and $ADC_F$ according to Fenucci [30], EFU, RGR, PER, FCR according to Tacon [31], and SR according to National Research Council [1]. The chromic oxide level in feeds and feces were analyzed using a modified colorimetric method Tacon [31] using a spectrophotometer (at 540 nm) (Shimadzu UV-2102 PC, UV-visible Scanning Spectrophotometer) after perchloric acid oxidation and formation of a colored complex with diphenylcarbazide (DPC). Samples were analyzed to determine concentration of phosphor (P) by flame atomic absorption spectrophotometer Shimadzu

AA6800 (Shimadzu, Japan). Variables of water quality that were tested were pH, DO, temperature, and Ammoniac (HANNA: HI. 8633). Aerator to recirculate the water was placed in every container.

$$ADC_P : 100\% \left\{ \frac{Cr_2O_3 \text{ in the feed}}{\% \ Cr_2O_3 \text{ in the feces}} \times \frac{\% \text{ protein in the feces}}{\% \text{ protein in the feed}} \right\} \quad (1)$$

$$ADC_F : 100\% \left\{ \frac{Cr_2O_3 \text{ in the feed}}{\% \ Cr_2O_3 \text{ in the feces}} \times \frac{\% \text{ phosphor in the feces}}{\% \text{ phosphor in the feed}} \right\} \quad (2)$$

$$EFU : \frac{\text{Final weight} - \text{Initial weight}}{\text{The amount of feed consumed}} \times 100\% \quad (3)$$

$$RGR : \frac{\text{Final weight} - \text{Initial weight}}{\text{Initial weight x Time experiment}} \times 100\% \quad (4)$$

$$PER : \frac{\text{Final weight} - \text{Initial weight}}{\text{The amount of feed consumed x Protein content of feed}} \times 100\% \quad (5)$$

$$FCR : \frac{\text{The amount of feed consumed}}{(\text{Final weight} + \text{Total weight fish death}) - \text{Initial weight})} \times 100\ \% \quad (6)$$

$$SR : \frac{(\text{Final count})}{\text{Initial count}} \times 100\%$$

Analysis of variance (ANOVA) was used to analyze the data. Before analysis, the data were first tested for normality, additivity, and homogeneity. The tests was to make sure that the data were normal, homogenous, and had additive property. If the analysis of variance was significant ($p<0.05$) or highly significant ($p<0.01$), Duncan test was conducted to find out the differences among the treatments, while water quality data were descriptively analyzed. To determine optimal dose of phytase, polynomial orthogonal test was conducted using SAS9 and Maple12 [23].

## 3. Results and Discussion

The results of study on common carp (*C. carpio*) for $ADC_P$, $ADC_F$, EFU, RGR, PER, FCR, and SR are shown in table 2.

**Table 2.** The values of $ADC_P$, $ADC_F$, EFU, RGR, PER, FCR, and SR common carp.

| Treatment | Data | | | | | | |
|---|---|---|---|---|---|---|---|
| | $ADC_P$ (%) | $ADC_F$(%) | EFU (%) | RGR (%/hari) | PER | FCR | SR (%) |
| A | 63.68± 0.07$^c$ | 54.65± 0.06$^c$ | 50.15±0. 81$^c$ | 2.17 ± 0.06$^c$ | 1.68± 0.03$^c$ | 1.70± 0.07$^c$ | 89.00±4.0 0$^a$ |
| B | 78.91± 0.07$^b$ | 74.41± 0.06$^b$ | 57.23±3. 46$^b$ | 2.63.± 0.26$^b$ | 1.98±0.12 $^b$ | 1.48±0.04 $^b$ | 87.78±6.1 1$^a$ |
| C | 85.13± 0.07$^a$ | 81.23± 0.06$^a$ | 67.34±2. 09$^a$ | 3.16 ± 0.23$^a$ | 2.18±0.07 $^a$ | 1.39±0.03 $^a$ | 89.00±4.0 0$^a$ |
| D | 77.52± 0.07$^b$ | 73.52± 0.06$^b$ | 54.23±4. 15$^b$ | 2.43 ± 0.25$^b$ | 1.78±0.14 $^b$ | 1.59±0.12 $^b$ | 90.37±2.3 1$^a$ |

*Note: The Values with the same superscripts in the column show that there was no difference*

The addition of phytase significantly ($P<0.01$) affected protein and phosphor digestibility of common carp (*Cyprinus carpio*). Table 2 shows that addition of phytase at 500-1000 U $kg^{-1}$ feed could increase the digestibility of protein and phosphor. Similar finding was also reported by Storebakken *et al.* [33] where addition of enzyme in the feed increased digestibility and retention of protein. Rachmawati and Samidjan [22] also suggested that the raw protein and total protein digestibility depended on the ability of the fish to absorb the nutrients. The increased feed digestibility was followed by the increase of PER (2.18) and EFU (67.34 %); therefore, it positively affected the RGR (3.16 %/day). Debnath *et al.* [13] also found that Atlantic salmon increased its protein digestibility and retention when phytase was added to its feed; meanwhile, without addition of phytase, it resulted in low digestibility and retention of protein. Hunter [34] found that the addition of enzyme increased protein digestibility from 84.5 % to 87.7 %. Similar results were found in the species of carp [35], rainbow trout [36, 37], and *Labeo rohita* [18]. Other researchers have observed that addition of phytase in feed made of plant ingredients increased protein digestibility due to the breakdown of phytin-protein complex [14, 22, 24]. According to Husain *et al.* [8], also reported that phytase can unbind anti-nutrient in feed, such as phytic acid, non-starch polysaccharide, and trypsin inhibitor, and it also improved nutrient digestibility. The equation resulted from orthogonal test was a cubical equation as $Y = -0.0000000006x^3 - 0.0000008x^2 + 0.0363x + 63.683$, $R^2 = 0.99$ figure 1). The optimum dose of the phytase in feed was 1040 U $kg^{-1}$ feed, with the maximum value of $ADC_P$ 86.03 %.

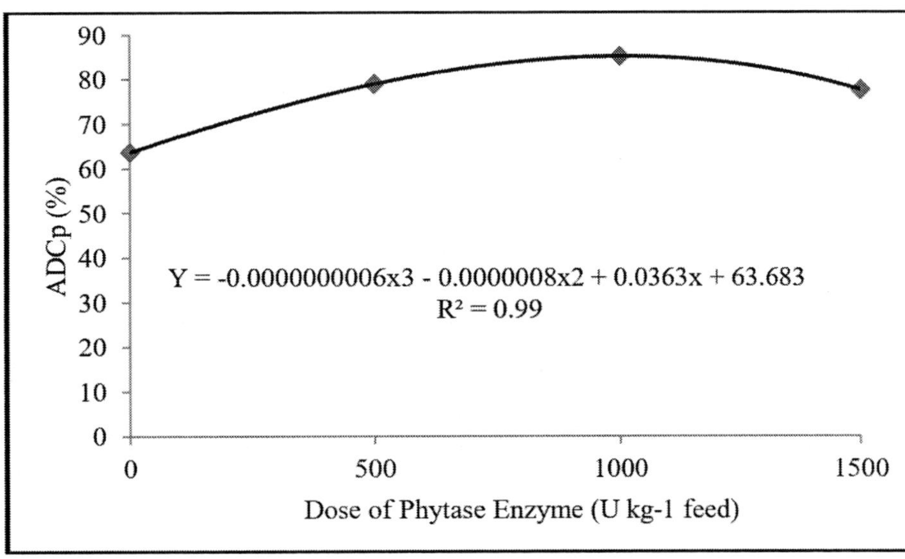


**Figure 1.** $ADC_P$ polynomial orthogonal (%) of common carp (*C. carpio*).

Digestibility of phosphor is one of the most sensitive criteria for assessing the influence of phytase on phosphor utilization in fish [38]. In the present study, digestibility of phosphor was increased with phytase supplementation, confirming the established properties of phytate with respect to nutrient availabilities in feed. phosphor digestibility increased from 54.65 to 81.23%, but the relationship was not linear. The equation resulted from orthogonal test was a cubical equation as $Y = -0.0000000002x^3 - 0.000002x^2 + 0.0514x + 54.653$, $R^2 = 1$ (figure 2). The optimum dose of phytase in the feed was 1100 U $kg^{-1}$ feed, with the maximum value of $ADC_F$ 84.24 %.

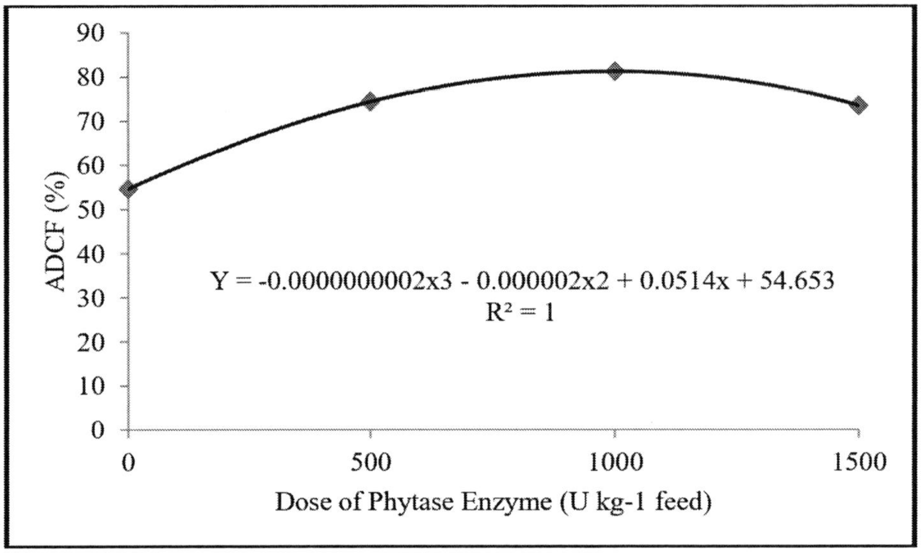


**Figure 2.** $ADC_F$ Polynomial Orthogonal (%) of Common carp (*C. carpio*)

The highest efficiency of feed utilization was in treatment C with (1000 U $kg^{-1}$ feed phytase) with 67.34 %, while the lowest was treatment A (0 mg $kg^{-1}$ feed phytase) with 50.15 %. treatment C (1000 U $kg^{-1}$ feed) had the highest growth as it was expected, due to the suitable dose of phytase to effectively hydrolyze phytic acid in soybean meal. In turn, it could help the fish to digest nutrient and mineral very well. The same results were also reported by several author [14, 20, 21, 22]. The high efficiency of feed utilization shows that the fish can efficiently absorb protein to support their growth instead of using it for metabolism, osmoregulation, and reproduction. growth happens when there is an excess energy from activities for survival, such as breathing, swimming, metabolism, and maintenance. low efficiency of feed utilization indicated that the fish need more feed to grow. Density and protein content in the feed also influences feed utilization efficiency [39]. Addition of phytase can break down complex compound, therefore they were able to be absorbed by the digestion system and distributed to the whole body optimally, as suggested by Bulbul *et al.* [20] who stated that addition of phytase in soybean meal could increase feed utilization efficiency better than that of fish meal.

The lowest efficiency of feed utilization was in treatment A (0 U $kg^{-1}$ feed phytase), since the fish were not able to absorb the nutrient efficiently due to the existence of phytic acid. Low efficiency of feed utilization also indicated the decrease of feed quality, as suggested by Qinghui dan Xiaojun [40] who mentioned that the absence of phytase in the digestion system of fish caused the nutrient in the soybean meal difficult to be absorbed, thus the nutrients would be excreted with feces. The equation resulted from orthogonal test was a cubical equation as $Y = -0.000000004x^3 + 0.000006x^2 - 0.0064x + 50.51$, $R^2 = 0.88$ (figure 3). The optimum dose of phytase in the feed was 943 U $kg^{-1}$ feed with the maximum value of EFU at 63.93%.

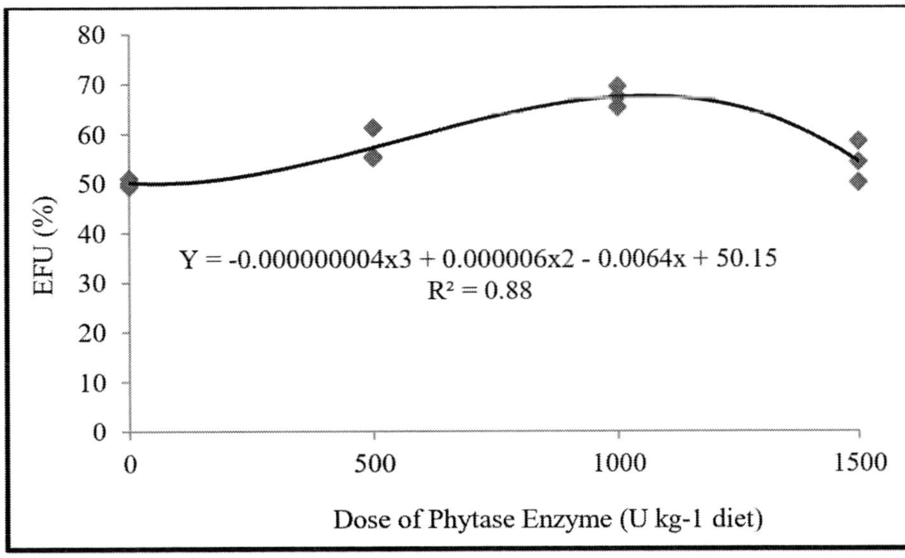


**Figure 3.** EFU polynomial orthogonal (%) of common carp (*C. carpio*).

The Analysis of Variance showed that different doses of phytase in the artificial feed had significant effects (P<0,01) on RGR of common carp (*C. carpio*). growth rate is related to addition of body weight from utilization of protein in the feed. protein could be hardly absorbed due to the existence of phytic acid that hinders the availability of minerals and suppresses growth of the cultivant. phytase is able to hydrolyze protein bound by phytic acid into amino acids that are readily digested by the fish. The more protein digested, the higher the energy the fish get to grow [41, 42]. palatability is also increased with addition of phytase since it breaks down phytate-protein and phytate-minerals bounds, thus It can increase the fish growth . This phenomenon was proven by Rachmawati and Istiyanto [22] in their study that showed phytase broke down phytic acid into inositol and phosphate acid. Inositol is a compound that is needed to support normal growth and maintain reproduction. Baruah *et al.* [24] and Fox *et al.* [27] also reported that there was an increase in the growth, raw protein coefficiency, sulfur, total phosphor, and phytate-phosphor of rainbow trout, after they were fed with plant-based feed supplemented with phytase.

A significant increase in growth was concurrent with increase in phytase concentration to a level of 1000 U $kg^{-1}$ feed, after which it decreased. it indicated that concentration of 1000 U $kg^{-1}$ feed was effective in enhancing the bioavailability of nutrients for common carp (*C. carpio*). The positive effect of phytase on growth performance of the fingerlings in the present study was consistent with the results obtained by various authors [12, 13, 20, 24, 35, 37, 43, 44].

The highest growth was obtained treatment C (1000 U $kg^{-1}$ feed) with 3.16 %/day, and the lowest growth was treatment A (0 U $kg^{-1}$ feed) with 2.18 %/day. treatment C had the highest growth as expected due to suitable dose of phytase; therefore it could effectively hydrolyze phytic acid. In turn, it could make the fish to digest nutrient and minerals very well. This results was also supported by Nwanna *et al.* [45], in their study that showed addition of phytase could release phosphor and minerals from phytic acid, so the fish could absorb the nutrient effectively. Moreover, Carter and Sajjadi [6] also proved that addition of phytase could increase digestibility of phosphor and reduce the effect of anti-nutrient compounds. treatment A (0 U $kg^{-1}$ feed) had the lowest RGR. It was suggested that phytic acid in the feed was still intact and it caused the fish not to be able to digest nutrients and minerals. similar results were also obtained in a study on carp [47], African catfish [48], striped bass [49], rainbow trout [35], and Atlantic salmon [38], Korean Sebastes schlegeli [44], *Penaeus monodon* [22], and *Channos channos* [14]. The equation resulted from the orthogonal test was cubical equation as $Y = -0.0000000002x^3 +$

$0.0000003x^2 - 0.000007x + 2.17$, $R^2 = 0.81$ (figure 4). The optimum dose of phytase in the feed was 988 U/kg feed, with the maximum value of RGR 3.10 %/day.

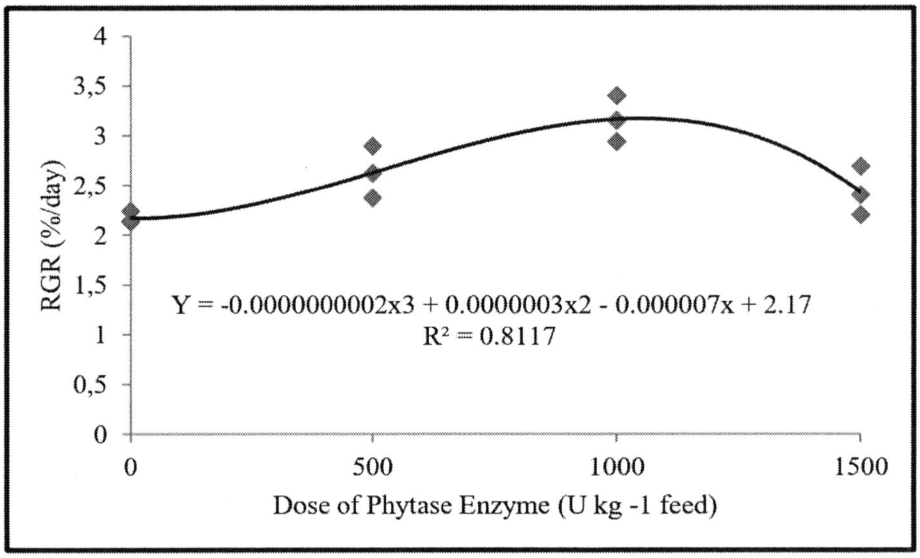


**Figure 4.** RGR polynomial orthogonal (%/day) of common carp (*C. carpio*).

The results of variance analysis showed that addition of phytase significantly (P<0,01) influenced PER of common carp (C. *carpio*). The highest protein conversion was showed by treatment C (1000 U phytase $kg^{-1}$feed). It was suggested that the protein broke down into amino acids, so the fish could easily digest it. In turn, it increased protein efficiency ratio, as reported by Rachmawati and Istiyanto [22]. Moreover, Danwitz *et al.* [21] found addition of phytase in feed increased nutrient digestibility, feed utilization, and growth of turbot fish. He also suggested that the enzyme addition could increase protein utilization. These findings were also supported by Trichet *et al.* [50], who reported that addition of phytase could hydrolyze phytic acid, so the fish could easily digest raw protein.

The lowest protein efficiency ratio was 1.68 in treatment A (0 U phytase $k^{-1}$g feed) . It was expected that phytic acid in the feed had not been broken down yet into amino acids. This finding was supported by Gatlin *et al.* [10] who found phytic acid could create complex compound by capturing calcium, magnesium, cuprum, zinc, carbohydrate, and protein. The equation resulted from orthogonal test was cubical equation as $Y = - 0.00000000007x^3 + 0.00000009x^2 + 0.0003x + 1.68$, $R^2 = 0.85$ (figure 5). The optimum dose of phytase in feed was 1000 U $kg^{-1}$ feed with the maximum value of PER 2.18.

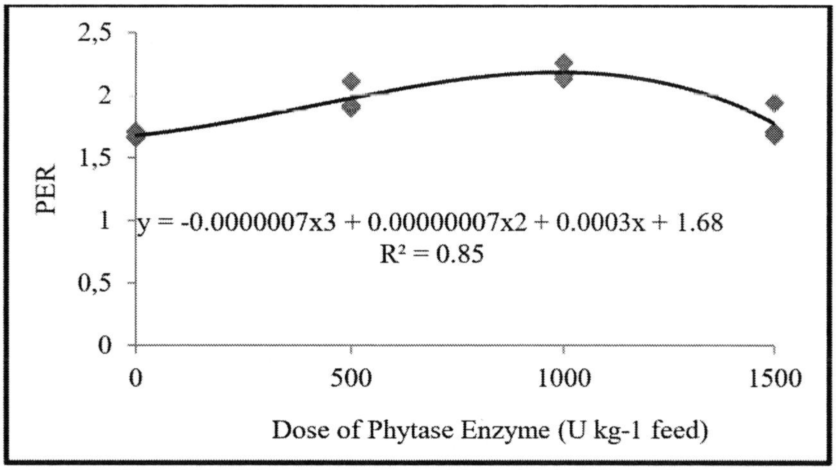


**Figure 5**. PER polynomial orthogonal of common carp (*C. carpio*).

The enzyme addition can increase palatability and FCR since phytase breaks down the bound between phytic acid and protein and minerals [51]. The results indicated that treatment C (1000 U phytase $kg^{-1}$ feed) had the lowest FCR, i.e. 1.39. It was suggested that the dose had suitable amount of phytase, so it resulted in the lowest FCR. Feed conversion is highly related to the efficiency of feed utilization. optimum feed utilization would generate abundant energy. Energy was produced from the breakdown of protein into amino acids. Due to the availability of simple compounds so the fish were able to digest nutrients very well [39]. Feed conversion was an indicator of feed utilization efficiency. The lower the feed conversion, the more efficient the use of feed by the fish.

high growth and feed efficiency make feed conversion low. treatment C (1000 U phytase $kg^{-1}$ feed) increased growth and feed efficiency; therefore, they made feed conversion low. These findings were supported by Storebakken *et al.* [33], who suggested that Atlantic salmon fed with protein made of soybean meal with addition of phytase resulted in better feed conversion. Moreover, Hassan *et al.* [51], also reported that addition of phytase in the feed increased the palatability and feed conversion because it could break down the bound between phytic acid and protein and minerals. According to Debnath *et al.* [13], also suggested that addition of phytase in feed can generate feed conversion better in catfish (*Pangasius pangasius).* The equation resulted from orthogonal test was cubical equation as $Y = -0.00000000002x^3 - 0.000000006x^2 – 0.0005x + 1.7033$, $R^2 = 0.80$ (figure 6). The optimum dose of the phytase in the feed was 1000 U $kg^{-1}$ feed with the maximum value of FCR 1.39.

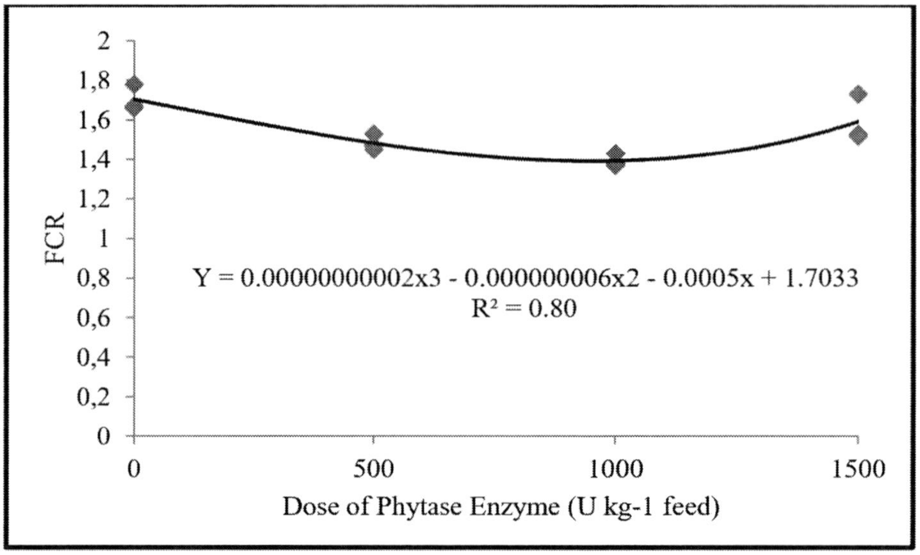


**Figure 6.** FCR polynomial orthogonal of common carp (*C. carpio*)

The results of variance analysis showed that addition of phytase in feed insignificantly (P>0.05) affected SR of common carp (*C. carpio*). According to Yakuputiyage [52], feed is not a factor that influences survival rate, because survival rate is affected by the initial treatment of the fish and quality of the cultivation media. The same results were also reported by Rachmawati *et al.* [14], Rachmawati and Istiyanto [22], Viela *et al.* [35], Sajjadi and Carter 38], Yoo *et al.* [44], Schaefer *et al.* [47], Weerd *et al.* [48], Papatryphon *et al.* [49], who reported that survival rate was affected by fish gender, heredity, age, reproduction, disease resistance, and external factors such as water quality, density, amount, and composition of amino acids in the feed [53]. quality of Water during the research was on viable condition for common carp (*C. carpio*). Measurements of water parameter during cultivation of common carp (*C. carpio*) are shown in table 3.

**Table 3**. Parameters of water quality for common carp (*C. carpio*) cultivation.

| Treatment | Water Quality | | | |
|---|---|---|---|---|
| | Temperature (°C) | pH | DO (mg/l) | $NH_3$ (%) |
| A | 26 – 33 | 7.50 – 7.85 | 3.30 – 3.55 | 0.0072 – 0.0074 |
| B | 26 – 33 | 7.50 – 7.82 | 3.24 – 3.48 | 0.0072 – 0.0074 |
| C | 26 – 33 | 7.50 – 7.81 | 3.28 – 3.58 | 0.0072 – 0.0074 |
| D | 26 – 33 | 7.50 – 7.81 | 3.32 – 3.53 | 0.0072 – 0.0074 |
| Feasibility | 14-38* | 6.50 – 8.5* | >2* | <0.1* |

Note : * [54]

## 4. Conclusion

Addition of phytase in feed significantly increased digestibility of feed, efficiency of feed utilization, and relative growth rate of common carp (*Cyprinus carpio*). The optimal doses of phytase based on $ACD_p$, $ACD_f$, EFU,RGR,PRR and FCR in common carp (*C. carpio*) were 1040, 1100, 943, 988, 1000 and 1000 U $kg^{-1}$ feed respectively.

IOP Conf. Series: Earth and Environmental Science **137** (2018) 012027 doi:10.1088/1755-1315/137/1/012027

## 5. References

[1] National Research Council (NRC) 1993 *Nutrient requirements of fish* (Washington D. C.: National Academy Press)
[2] De Silva S S and T A Andersons 1995 *Fish nutrition in aquaculture* (London: Chapman and Hall aquaculture series 1 UK)
[3] Grant G 1989 *J. Food Nutritional Science* **13** 317–348
[4] Van den Ingh T S G A M, A Krogdahl, J J Olli, H G C J M Hendricks and J G J F Koninkx 1991 *J. Aquaculture* **94** 297-305
[5] Hendricks J O and G S Bailey 1989 *Fish nutrition ed J E Halver* (USA: Academic Press Inc., San Diego California) pp 606-644
[6] Usmani N and A K Jafri 2002 *J. World Aquaculture Society* **33** 199-204
[7] Baruah K, Sahu N P, Pal A K, Debnath D 2004 *J NAGA World Fish Center Quart* **27** 15-29
[8] Cao L, W Wang, C Yang, Y Yang, J Diana, Yakupitiyage A, Luo dan D Li 2007 *J. Enzym and Microbial Technology* **40** 497-507
[9] Phromkunthong W, N Nuntapong and J Gabaudan 2010 *J. Sci. Technol.* **32** 547-554
[10] Gatlin III D M, Barrows F T, Brown P, Dabrowski K, Gaylord T G, Hardy R W, Herman E, Hu G, Krogdad S, Nelson R, Overturf K, Rust M, Sealay W, Wurtele E 2007 *J. Aquacult. Res.* **38** 551-579
[11] Masumoto T, Tamura B and Shimeno S 2001 *J. Fish. Sci.* **67** 1075-1080
[12] Cheng Z J and Hardy R W 2003 *Aquaculture* **218** 501-514
[13] Debnath D, Pal A K, and Sahu N P 2005 *J. Aquacult. Res.* **36** 180-187
[14] Rachmawati D, Istiyanto S and Maizirwan M 2017 *Philipp. J. Sci.* **146** 237-245
[15] Kumar V, Sinha A K, Makkar H P S, De Boeck G and Becker K 2011 *J. Anim. Physiol. Anim. Nutr.* **96** 335-364
[16] Jagannathan K R and Nielsen P H 2013 *J. Clean. Prod.* **42** 228-240
[17] Chung T K 2001 *Sustaining livestock production and environment* (Singapore: Food and Agriculture Asia Pacific Development) pp 52-54
[18] Husain S M, T Hameed, M Afzal, M S Mubarik, M Asrar, S Z H Shah, S Ahmad, M Z H Arzalan, D Riaz, N Tahir, F Amber, M M Shahzad and Tanwir Ahmad Abbas Khichi 2011 *Int. J. Biosci.* **5** 173-181
[19] Shapawi R, I Ebi, A Yong, M Chong, LK Chee, A Sade 2013 *J. Agric. Sci.* **4** 19-24
[20] Bulbul M, Md A Kader, M A Ambak, Md S Hossain, M Ishikawa dan S Koshio 2015 *J. Aquaculture Elsevier* **438** 98-104
[21] Danwitz A, Von, C G J van Bussel, S F Klatt and C Schulz 2016 *J. Aquaculture Elsevier* **450** 405-411
[22] Rachmawati D and Istiyanto S 2016 *J. Teknologi* **78** 39-43
[23] Dasuki A, Auta J and Oniye S J 2013 *J. Pure. Appl. Sci.* **6** 112-117
[24] Baruah K, Sahu N P, Pal A K, Debnath D and Mukherjee S C 2007 *J. Aquaculture Study* **38** 109-120
[25] APHA 1998 *Standard methods for the examination of water and wastewate* 20th edition ed LS Clesceri, A E Greenberg and A D Eaton (American Public Health Association, American Water Works Association, Water Environment Federation, Washington, DC, USA)
[26] AOAC 1990 *Official methods of analysis* (Washington D C: Association of Official Analytical Chemists) 50th ed. 1298 pp
[27] Fox J M, Addison L L, Anthony J S, D Allen D, Denis R M, Elizabeth C D, Tzachi M S 2006 *Phytase supplementation in aquaculture feeds improves fish, shrimp growth performance* (Global Aquaculture Alliance) 66 p
[28] Wilson RP 1982 *Nutrition and feeding of channel catfish* ed RR Stickney and RT Lovell (Southern Cooperative Series) 193-201 p
[29] De Silva SS 1987 *finfish nutrition study in asia* (Proceeding of The Second Asian Fish Nutrition Network Meeting Heinemann Singapore) 128 p

[30] Fenucci J L 1981 *Studies on the nutrition of marine shrimp of the penaeus* Faculty of Department of Biology, University of Houston, Houston, Texas, USA (Ph D Thesis)
[31] Tacon A G 1995 *The Nutrition and feeding of farmed fish and shrimp-a traning mannual* (FAO of The United Nations Brazil) 106-109 p
[32] Steel R G D, Torrie J H and Dickey DA 1996 *Principles and Procedures of Statistics* (New York: McGraw Hill International Book Company, Inc) 3rd ed.
[33] Storebakken T, Shearer K D and Roem A J 1998 *J. Aquaculture* **161** 365-379
[34] Hunter B 2002 *Phytase Aplications in Aquaculture* (Bangkok: Roche Aquaculture Center Asia Pasific) 425 p
[35] Vielma J, Ruohonen K, Gabaudan J dan Vogel K 2004 *J. Aquaculture Res.* **35** 955-964
[36] Sugiura SH, Gabaudan J, Dong FM, Hardy RW 2001 *J. Aquacult. Res.* **32** 583-592
[37] Forster I, Higgs D A, Dosanjh B S, Rowshandeli M 1999 *J. Aquaculture* **179** 109-125
[38] Sajjadi M and Carter C G 2004 *J. Aquacult Nutr.* **10** 135-142
[39] Tawwab M A 2012 *J. Int. Aquatic. Study* **4** 1-13
[40] Qinghui Ai and Xiaojun Xie 2005 *J. Comparative Biochemistry and Physiology* **14** 461-469
[41] Amoah Y T, Thorarensen H and O Sigurgeirrson 2011 *Effect of Feedary Protein Levels on Growth and Protein Utilization in Juvenile Arctic Char (Salvelinus alpinus)* (Fisheries Training Programme, United Nations University) 26 pp
[42] Haghbayan S and M S Mehrgan 2015 *J. of Molecules* **20** 258-266
[43] Li M H and Robinson EH 1997 *J. World Aquacult. Soc.* **28** 402-416
[44] Yoo GY, X Wang, S Choi, K Han, J C Kang and S C Bai 2005 *J. Aquacul.* **243** 315-322
[45] Nwanna L C, R Eisenreich and F J Schwarz 2007 *J. Aquaculture* **271** 461-468.
[46] Carter C G And Sajjadi M 2011 *J. Aquacult. Int.* **19** 431-444
[47] Schaefer A, W M Koppe, K H Meyer-Burgdorff and K D Guenther 1995 *J. Water Sci. Tech.* **31** 149-155
[48] Weerd V J H, Khalaf K H, Artsen EJ and Tijssen P A 1999 *J. Aquacult. Nutr.* **5** 135-142
[49] Papatryphon E, Howell R A and Soares J H 1999 *J. World Aquacult. Soc.* **30** 161-173
[50] Trichet V V, J Vielma, J Dia, P Rema, E Santigosa, T Wahli And K Vogel 2014 *J. of the World Aquaculture Society* **45** 267-279
[51] Hassan M S, MA Soltan, H M Agouz and A M Badr 2013 *J. Egyptian J. of Aquatic. Res.* **39** 205-213
[52] Yakuputiyage A 2013 *On-Farm Feeding and Feed Management Strategies in Tropical Aquaculture* ed MR Hasan and MB New (FAO Fisheries and Aquaculture Technical Rome, FAO) p 361-376
[53] Hepher B 1988 *Nutrition on Pond Fisheries* (Cambridge University Press Cambridge USA) p 388
[54] Boyd C E 2003 *J. Aquaculture* **226** 101–112

**Acknowledgments**
Appreciation are expressed to those who helped this study, especially for the Head of the Hatchery Center and Freshwater Aquaculture, Muntilan, Central Java, Indonesia who has provided its laboratory to do the study

IOP Conf. Series: Earth and Environmental Science **137** (2018) 012028 doi:10.1088/1755-1315/137/1/012028

# The effect of hydrogen peroxide on N/P ratio and phytoplankton diversity in Vannamei shrimp (*litopeneus vanname*) ponds in Banyuwangi, East Java

**D N Daruti*, Rozi, and K Rahayu**

[1]Department of Fish Health Management and Aquaculture, Faculty of Fisheries and Marine, Universitas Airlangga, Surabaya, Indonesia

E-mail: darutinindarwi@ymail.com

**Abstract**. Abundance of plankton in a waters is influenced by environmental parameters and physiological characteristics. Complex factors support phytoplankton growth. They are light intensity, dissolved oxygen, temperature stratification, salinity and availability of nitrogen and phosphorus. Nitrogen is an important element in the formation of chlorophyll, protoplasm, protein and nucleic acids. Observation of the three ponds in North Banyuwangi (Wongsorejo, Bangsring and Bulusan) and three ponds area of South Banyuwangi (Bomo, Jatisari and Badean). The observed parameters were $NH_4$, $NO_2$, $NO_3$, $PO_4$. Hidrogen peroxide effectively improve water quality, however it has a different effect in each different area.

## 1. Introduction

*Litopeneus vannamei* is one of the main commodities of the Indonesia. Several technology have been developed for improving the productivity. One of the effort is intensive culture method. Interaction biotic and abiotic factors affects the water quality. Hence, it will also affect the plankton in the water.

Plankton is a major component of the food chain. It is also an indicator of water quality. Several environmental parameters and physiological characteristics affect the growth and abundant of the plankton. The composition of plankton was affected by environmental conditions. Nitrogen and phosphorus (N/P Ratio) are two parameters that affect the waters [1]. Although there are nutrients in the water system, only a few can be utilized by algae and aquatic plants.

The nitrogen elements can be utilized as nitrite and nitrate, while for phosphorus are ortho phosphate compounds. The presence of nitrogen and phosphorus in aquatic environments have a positive impact. Excess N/P will negatively affects the water quality. The positive impact is the increase in plankton production and total fish production. The negative impact are decreasing the oxygen in the waters, and raising Harmful Alga Blooms (HABs) incident . The administration of hydrogen peroxide in the water will affect density, abundance and dominance of plankton [2]. $H_2O_2$, an anorganic chemical agent is weak acid and strong oxidizer. The $H_2O_2$ influence the dynamics of the NP ratio. Hence, it will affect the dynamics of plankton in the aquaculture waters [2]. This research was to investigate the effect of $H_2O_2$ at six different shrimp ponds in Banyuwangi, East Java, Indonesia.

## 2. Methodology

### *2.1. Plankton Samples*

The areas of the research was North Banyuwangi (Wongsorejo, Bangsring and Bulusan village) and South Banyuwangi (Bomo, Jatisari and Badean Vilage). Each village represented one sample of the pond.

### *2.2. Water quality parameters*

The observed parameters were $NH_4$, $NO_2$, $NO_3$, $PO_4$ which were observed once every three days, while the plankton diversity were daily investigated. Supporting parameters were observed were DO (Dissolved Oxygen), pH, temperature, salinity, brightness, and the water color [3].

## 3. Result and Discussion

Based on the result of the research. The application of hydrogen peroxide affected the chemical and biological processes in the water. NP ratio is one of the many parameters in water. The role of the hydrogen peroxide in the water can help to improve nitrification.

### *3.1. Wongsorejo village*

The amount of Ammonium ($NH_4$) fluctuated during the middle study between weeks 8 and 9; during the interval of the following week ie. week 10, $NH_4$ decreased and Nitrite (NO2) increased which followed by an increase on nitrate ($NO_3$) in the next following week. Phosphate content was relatively stable (figure 1A). The plankton observation revealed that chrysophyta, cyanophyta and chlophyta was differently fluctuated. On the first week, it appears that chrysophyta dominated the waters. It will be decrease and replaced by cyanophyta. On the contrary, chlorophyta species was relatively stable in number (figure 1B).

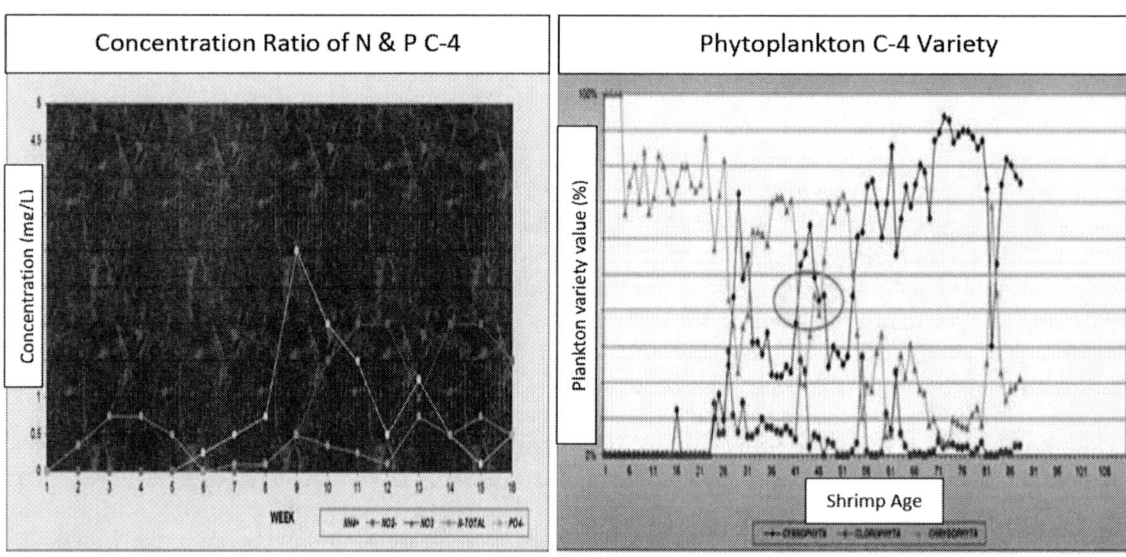


**Figure 1.** The N/P and phytoplankton dynamic in Wongsorejo shrimp pond. A= N/P value, B= phytoplankton diversity.

### *3.2. Bangsring village*

Phosphate increased at week 4$^{th}$, then decreased at week 5$^{th}$. The value of phosphate stable after the following week. $NH_4$ increased at the end of week 8$^{th}$ to week 9$^{th}$, then decreased at week of 10$^{th}$ which was accompanied by the increase of $NO_2$ at week 10$^{th}$ (figure 2A). The observation data of phytoplankton showed that phytoplankton diversity of the cyanophyta and chlorphyata groups was high at the beginning of cultivation. This was also followed by high phosphate values at the beginning of the vanname shrimp cultivation period. Both types of phytoplankton were decreased in DOC 42 and 43. The dynamics of the phytoplankton species were fluctuative (figure 2B.) in this pond. This phenomena was due to the high nutrient in the water [4].

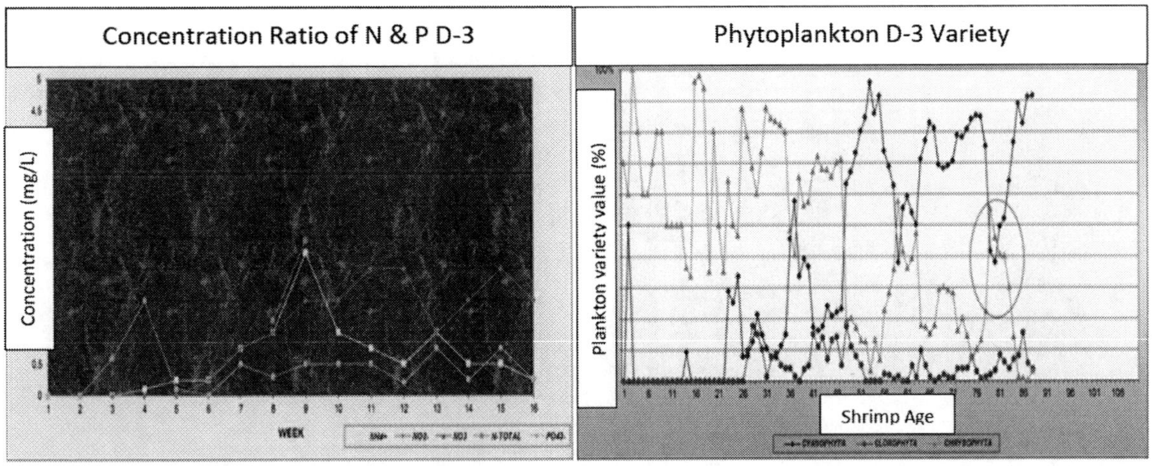


**Figure 2.** The N/P and phytoplankton dynamic in Bangsring shrimp pond. A= N/P value, B= phytoplankton diversity.

*3.3. Bulusan village*

The results showed that the phosphate content at the beginning of cultivation increased on week 3th and decreased on week 4th. For $NH_4$, there was a high increase on week of 8th to week 9th. it decreased again at week of 10 th. It started increase again on week of 11st and decreased on week 12nd. For $NO_2$, it began dramatically increaded on the week of 11th (figure 3A). The results indicated that chrysophyta, cyanophyta and chlophyta fluctuated. At the beginning of cultivation, it appears that chrysophyta dominate the waters, then along with longer cultivation, the population of chrysophyta decreased. It was replaced by the cyanophyta. The chlorophyta species was relatively stable in number from the beginning to end of cultivation (figure 3B).

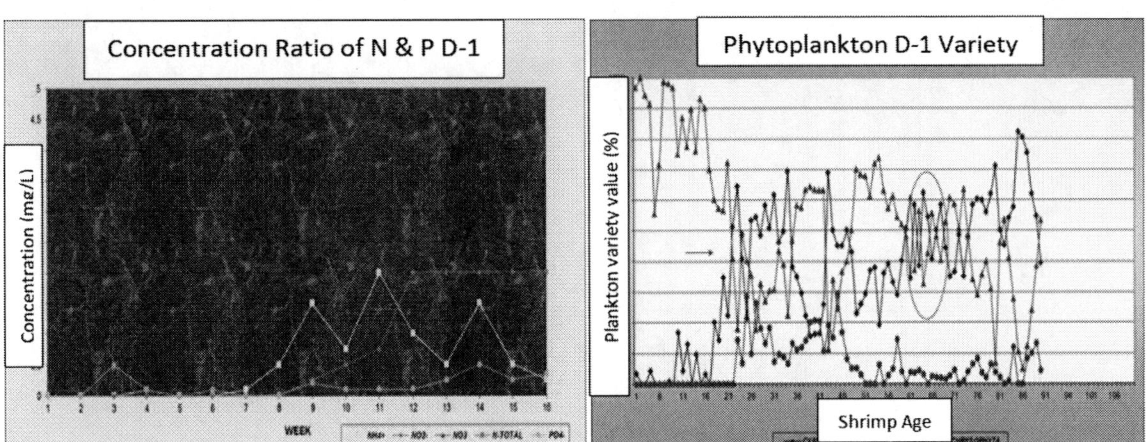


**Figure 3.** The N/P and phytoplankton dynamic in Bulusan shrimp pond. A= N/P value, B= phytoplankton diversity.

### *3.4. Bomo village*

All parameters, $NH_4$, $NO_2$ and $NO_3$ experienced steady movement until week of 11st. All started to increase on week 12nd. (figure 4A.). Results of the dynamics of phytoplankton diversity showed that fluctuations in the dynamics of phytoplankton diversity were stable irrelatively stable. Chrysophyta, cyanophyta and chlorophyta grew stable until the 90th days (figure 4B).

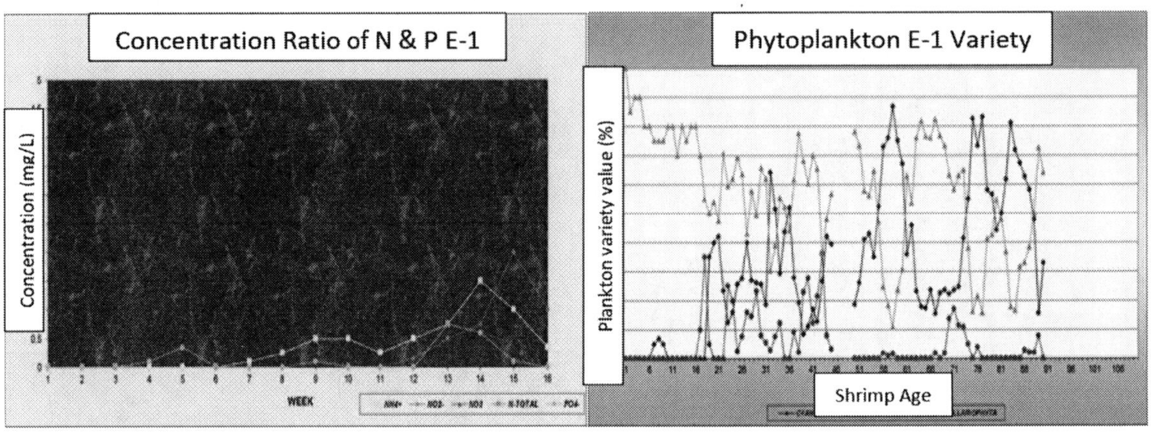


**Figure 4.** The N/P and phytoplankton dynamic in Bomo shrimp pond. A= N/P value, B= phytoplankton diversity.

### *3.5. Jatisari village*

Ammonium ($NH_4$) showed dramatically fluctuation in the week of 8th. It started decrease after week 10th, The decreased of $NH_4$ and increased of Nitrite of ($NO_2$) followed by an increase of nitrate ($NO_3$) on the next following weeks. Phosphate content was relatively stable (figure 5A.). Ddifferent fluctuations on chrysophyta, cyanophyta and chlophyta occurred. At the beginning of cultivation ie. the first week, it appears that chrysophyta dominated the waters, then along with longer cultivation and accompanied by the increase in $NH_4$, the population of chrysophyta decreased and replaced by cyanophyta. The chlorophyta species were relatively stable during cultivation (figure 5B).

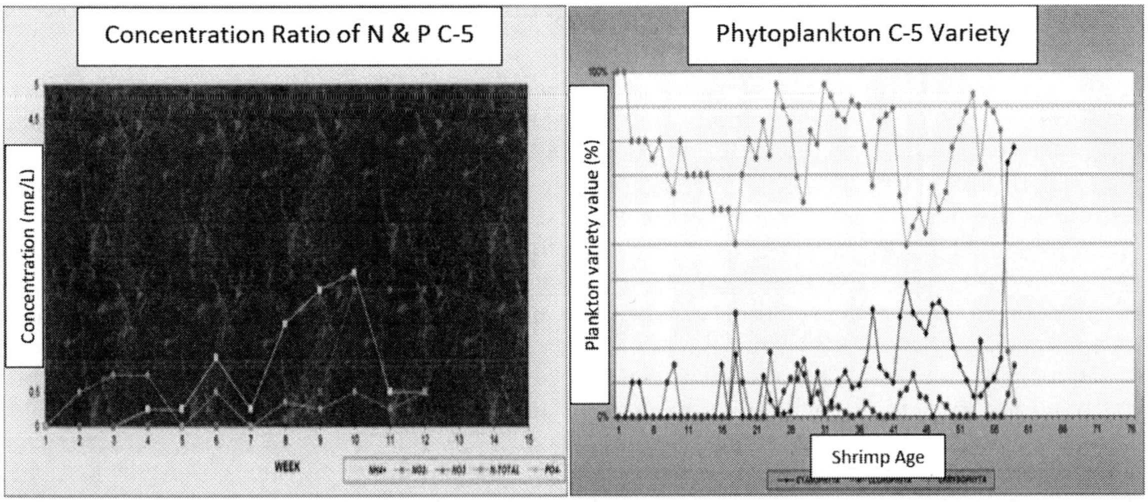


**Figure 5.** The N/P and phytoplankton dynamic in Jatisari shrimp pond. A= N/P value, B= phytoplankton diversity.

*3.6. Badean village*

The observations of research results in the ponds of Badean village were relatively stable. The content of phosphate, $NH_4$, $NO_2$ and $NO_3$ experienced a steady movement from week 11st to week 15th. Then all parameters decreased at the end of week 15th (figure 6A.). Phytoplankton analysis revealed that fluctuations in the dynamics of phytoplankton diversity were also stable. Chrysophyta, cyanophyta and chlorophyta growth were dynamic until the 90th day or DOC 90 (figure 6B.).

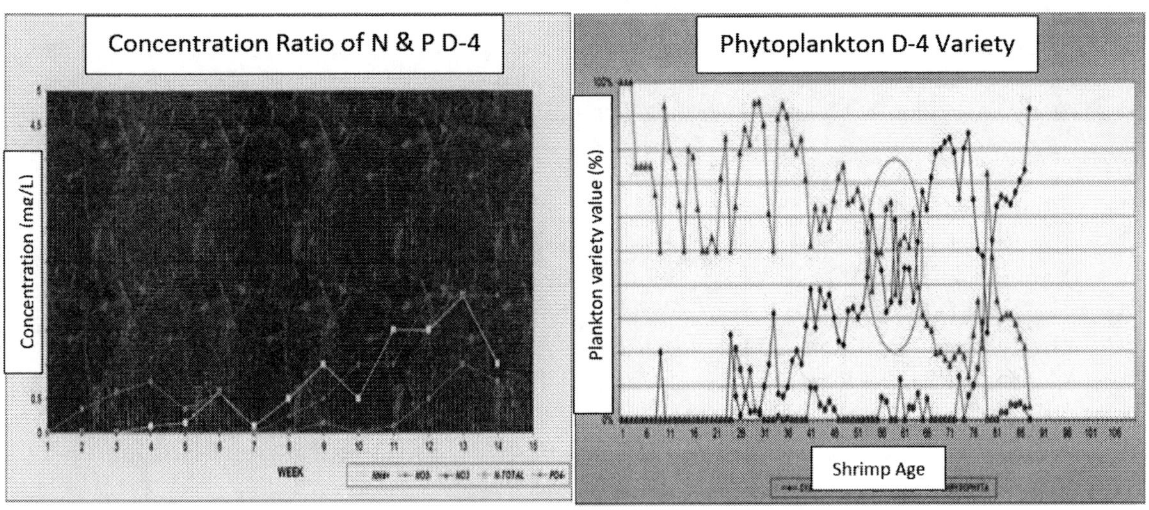


**Figure 6.** The N/P and phytoplankton dynamic in Jatisari shrimp pond. A= N/P value, B= phytoplankton diversity.

The existence of plankton has a very diverse role in the waters, such as aquatic feed, oxygen provider, producer of carbondioxide, water coloring, and as an indicator of water fertility. Plankton abundance is strongly influenced by inorganic compounds in the waters, namely Nitrogen and Phosphate in the ratio of total ppm of Nitrogen and Phosphate, often called the N/P ratio. It is understood that the presence of the N/P ratio hugely determines the type and abundance of plankton. The dynamics of plankton and movement of N/P ratio is strongly influenced by the state of water quality[7,8].

Oxygen solubility greatly affects the chemical balance of water and plankton life, but it also affects the amount of organic matter in the waters. The lower the oxygen content, the higher the organic matter. Planktons are the largest oxygen provider in the waters. The ratio of N and P is largely the ratio of the total N derived from ammonium and nitrate, and P from phosphate. According to Cardona *et al.*, [2], the distribution of planktons such as cyanophyta and dinoflagellate groups are determined by the value of the proportion of nitrogen to phosphorus in the eutrophic environment with the ratio of total ppm N and ppm P. If under 10 it will be dominated by cyanophyta because it is able to absorb free nitrogen from the air, so the growth is not limited. Therefore, this does not support the required water balance, nitrogen and phosphate levels for the plankton [9].

Hydrogen peroxide can affect the dynamics of the N: P Ratio and cause the decrease of phosphorus so that it can control the dynamics of plankton abundance. This process occurs because it is influenced by the pH; if the pH decreases it contains many hydroxyl ions that allow the process of Ligand Exchange and increase the ability of phosphate ion absorption. The number of hydroxyl ions decreases further in increasing pH conditions, leading to a decrease in Ligand Exchange capability and decreased ability to absorb phosphate ions [1]. The administration of Hydrogen peroxide in waters can affect the presence of Hydroxide (OH) ions in the waters [10]. The OH ion has been known to be a very strong oxidizer produced from the reaction between the holes with $H_2O$ / OH-adsorbed on the surface of the semiconductor [11, 12]. The electrons will adsorp molecules of O2 / H2O to form superoxide anion radicals (• O2-) which is a reducing species. The dissolved contaminant in the system is attacked by the oxidizer and reducing agent and degrades it into a harmless compound, and converts the organic compound to inorganic [13]. An increase in OH ions will also increase the salinity in aquatic areas [14]. The administration of Ca (OH) 2 (calcium hydroxide) and $NaHCO_3$ (sodium bicarbonate) in the water can affect the dynamics of N / P Ratio and the abundance of plankton [15].

## 4. Conclusion

- Hydrogen peroxide is effective in improving water quality when administered regularly and at appropriate dosage.
- The administration of Hydrogen peroxide has a different effect in each area.

## 6. References

[1] Rahman M, Herliwati, Prihanto A A. AACL Bioflux. **10** 1091-1097
[2] Cardona E, Gueguen Y, Magré K, Lorgeoux B, Piquemal D, Pierrat F, Noguier F, D Saulnier 2016 BMC Microbiol. **16** 157-166
[3] Fu Y , D D I Chung 2011 *Appl. Clay Sci.* 634-641
[4] Brady M and S C Weil 2002 *Global Biogeochemical Cycles* **21** 1018-1029
[5] Gardner F P, R B Pearce and R L Mitchell 1991 *Physiology of Crop Plants* (Jakarta: Ul-Press) 18-21 (Text in Indonesia)
[6] Horne, A J and C R Goldman 1994 *Limnology.* (New York: McGraw-Hill Inc.) p 9-11 2nd ed.
[7] Legendre P and L Legendre 1983 Numerical Ecology.. (Elsevier) p 852 2nd ed.
[8] Jonson C B and G Gage 1997 *The effect of nitrogen application timing on plant available phosphorus* (USA: Thesis, Graduate School of the Ohio State University) p 41-43
[9] Lagus A 2009 Role of Nutrients in Regulation of the Phytoplankton Community in the Archipelago Sea, Northern Baltic Sea, Turun Yliopiston Julkaisuja Annales Universitatis Turkuensis p 5-43
[10] Lambers HF S Chapin and T L Pon 2008 *Plant physiological ecology* (Springer) p 108-111

[11] Mason C F 1980 *Biology of freshwater pollution. longman scientific and technical* (Singapore: Longman Singapore Publisher Pte. Ltd.)
[12] Muchtar M 2012 *Distribusi zat hara fosfat, nitrates and silicates in the waters of the Natuna Islands J. Sci. Techno. Tropical Mar.* 304-317
[13] Odum E D P 1971 *Fundamental of ecology* (Philadelphia WB Saunders) 3rd ed.
[14] Reynolds W A, Pratt and Robertson 2001 *Americ. J. Botany* **20** 117-132
[15] Boyd C E 1979 *Water quality in warmwater fish ponds* (Alabama: Auburn University)

**Acknowledgments**
We are thankful to our team of researchers, both technicians in the laboratory for their valuable helps with fieldwork and sample collection. This study was sponsored of RKAT, Faculty Fisheries and Marine, Univeristas Airlangga. Surabaya

IOP Conf. Series: Earth and Environmental Science **137** (2018) 012029 doi:10.1088/1755-1315/137/1/012029

# The identification of plankton tropical status in the Wonokromo, Dadapan and Juanda extreme water estuary

**L A Sari[1], W H Satyantini[1], A Manan[1], K T Pursetyo[1] and N N Dewi[1]**

[1] Department of Fish Health Management and Aquaculture,
Faculty of Fisheries and Marine, Universitas Airlangga, Jl. Airlangga, Surabaya 60115, East Java, Indonesia

E-mail: luthfianaaprilianitas@gmail.com

**Abstract**. Wonokromo, Dadapan and Juanda estuaries are extreme waters located around Surabaya environment. This is because of a lot of organic material intake, which provided nutrients for plankton growth. In addition, the waters is also dynamic in reason of physico-chemical, geological and biological processes controlled by the tides and freshwater run-off from the river that empties into it. The objective of this study was to identify the presentation of plankton in extreme waters based on brightness and ammonia level. The study was conducted in January 2017. Three sampling locations were Wonokromo, Dadapan and Juanda estuaries. Each station consists of three points based on distances, which were 400, 700, and 1000 meters from the coastline. The brightness in Wonokromo, Dadapan, and Juanda environment was 60, 40, and 100 cm, respectively. The result of ammonia in Wonokromo, Dadapan, and Juanda estuary was 0.837, 0.626, and 0.396 mg/L, correspondingly. Nine classes of phytoplankton's were found in three locations (bacillariophyceae, dynophyceae, chlorophyceae, cyanophyceae, crysophyceae, euglenoidea, trebouxlophyceae, mediophyceae, and nitachiaceae) and five classes of zooplanktons (maxillopoda, hexanuplia, copepoda, malacostraca, and oligotrichea). The density of plankton in Wonokromo, Dadapan and Juanda environments, was 37.64, 63.80, and 352.85 cells/L, respectively.

## 1. Introduction

The fertility of waters can be seen from the existence of plankton. According to Sofarini [1], stated that the presence and density of phytoplankton is one indicator of the fertility of the aquatic environment. According to Hemraj *et al.* [2], also stated that planktons are a bio-indicator of the water conditions of coastal lagoons. This indicator is performed by measuring the concentration of the chlorophyll produced by plankton.

The plankton density in waters is influenced by a number of environmental parameters and physiological characteristics. The composition and plankton density change at various levels in response to changes in physical, chemical, and biological conditions. These include physical and chemical changes such as the intensity of light, dissolved oxygen, temperature stratification, and the availability of nutrients, nitrogen and phosphorus; while the biological aspect is the existence of animal predation activities, death, and decomposition [3].

IOP Conf. Series: Earth and Environmental Science **137** (2018) 012029 doi:10.1088/1755-1315/137/1/012029

The east coast of Surabaya is located in the northwestern part of Madura Strait, covering the area between Tambak Wedi village and the Dadapan river estuary (border area with Sidoarjo). The waters of the east coast of Surabaya can be regarded as extreme waters due to several factors. One of the factors is that there are approximately 10 rivers that empties into it [4]. These waters make the coastal waters of Surabaya's east coast an estuary area with an input of numerous organic material intake as nutrients for the growth of plankton. Moreover, the coastal waters of Surabaya's east coast can be considered as very dynamic waters because there are various physical-chemical, geological and biological processes controlled by the sea water tides and freshwater run-off from the rivers that empties into it.

According to Winarni *et al.* [5], have stated that the structure of communities in the east coast of Surabaya have led to the discovery of seven types of sea cucumbers with high to low distribution indexes, none very high. According to Affandi *et al.* [4], mentioned that there were four species of *kupang* (local name) and one *lorjuk* (local name) species in the east coast waters of Surabaya with a distribution that did not occur spontaneously or randomly but were closely related to preferences or selection of suitable habitats.

This research is a preliminary study to provide basic data and will be continued in several parts. The research aims to assess the extreme coastal waters of Surabaya with extreme water condition and high organic material which have dominance, diversity and plankton density. Extreme waters with good fertility will have a variety of species and good plankton density as well. Therefore, a study on the identification and fertility levels of plankton in the extreme waters in the east coast of Surabaya was carried out.

Seawater as public waters may at any time experience changes in conditions and quality caused by surrounding environmental factors, weather, climate as well as human activities conducted on these waters. Changes that occur in sea water will cause fluctuations in water quality such as temperature, salinity, water brightness (turbidity), degree of acidity (pH) and dissolved oxygen (DO). Changes that occur to water quality will influence the sustainability of aquatic organisms. Planktons are one of the most influential organisms in extreme water conditions [6].

Some types of plankton are able to survive and grow in extreme waters. Plankton is able to adapt to extreme environments. According to Seckback [6], explained that some algae are able to survive under conditions of low temperature, high temperature and high salinity waters. Tolerant organisms that can adapt to extreme waters will easier adapt to within the culture.

Research from Hemraj *et al.* [2], stated that coastal lagoons are characterized by extreme environments especially high salinity. This is led to community changing. Studies have been performed which have used plankton as an indicator of extreme environmental fluctuations. Two different communities of phytoplankton and zooplankton were identified, with salinity and nutrients being the main factors affecting species distribution. Polychaete and gastropod larvae showed positive indicators of high salinity fluctuations.

## 2. Methodology

The study was conducted in January 2017. The research included field observations and laboratory analysis. Sampling was carried out in the coastal waters of Surabaya on three stations, where each station consisted of three points. The identification of plankton samples were performed at the Dry Laboratory of the Faculty of Fisheries and Marine at Airlangga University, Surabaya and the water sample was measured there.

The research material used was plankton found in the Wonokromo, Juanda and Dadapan waters. The research materials used were 4% formalin, lugol, walnut nutrient ($NaNO_3$, $Na_2EDTA$, $H_3BO_3$, $NaH_2PO_4$, $2H_2O$, $FeCl_3$, $6H_2O$, $MnCl_2$, $4H_2O$), vitamin solution stock, stock of micro metal solution (BBPAP Jepara), sea water chlorine (Brataco), Na Thiosulphate (Brataco), alcohol (Brataco) paper, gloves and masks.

The equipment used in the research were: sample bottles, plastic clips, bottle insulation, gel ice, GPS, styrofoam, aerator set, aerator hose, cork, measuring cup, Erlenmeyer, dropper, volume pipette,

IOP Conf. Series: Earth and Environmental Science **137** (2018) 012029 doi:10.1088/1755-1315/137/1/012029

microscope, homogenizer, hand tally counter, autoclave American 25H, Haemocytometer, Sedgwick-Rafter, Spectrophotometer, Hettich EBA-20 centrifuge, hot plate stirrer, refractometer, pH paper, thermometer, ammoniac test kit, DO test kit, Ohaus PA 2102 digital scales, OHAUS digital scales, Analytical Balance PA413, TL 40 watts, cotton, water funnel, gauze and aluminum foil.

The plankton identification books used were: Freshwater Algae Identification, Enumeration and Use as Bioindicators [10]; Methods for the Study of Marine Benthos (Athlete of March) Ostracods, and Chaetognaths [16]; Algae [9]; Identifying Marine Phytoplankton [22]; and Plankton: A Guide to their Ecology and Monitoring for Water Quality [3].

Plankton and water samples were obtained from three stations in Wonokromo, Juanda and Dadapan waters. Each station consisted of 3 water sampling points at a distance of 200, 500, 800 meters from the coastal waters.

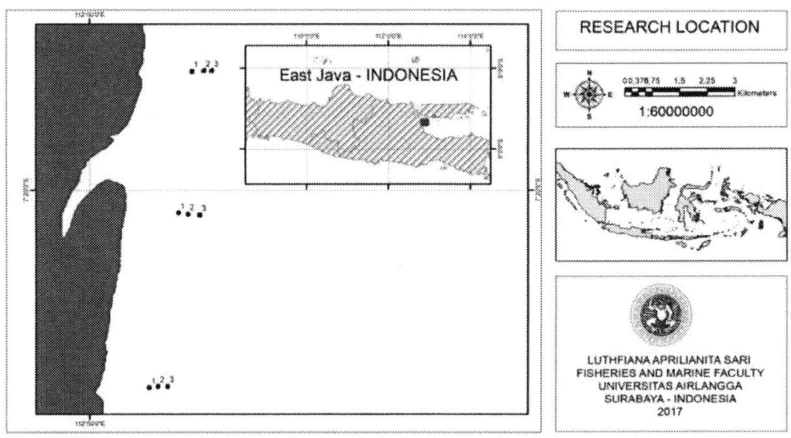


**Figure 1.** Research location and sampling site at Wonokromo, Dadapan and Juanda waters.

*2.1. Sample collection*

The sampling of plankton and water was done at each sampling location. Sampling of microalgae (phytoplankton) was carried out using a 10 μm diameter plankton net. Zooplankton sampling was done using a 80 μm diameter plankton net. First, 50 liters surface water was filtered by a plankton net in each station. Then the filtered water was put into a sample container. The ordinate sampling station was recorded. Bottles were labeled with information such as times, dates, points and sampling stations. Two sample bottles were collected at each sample site location, one bottle of sample was preserved by using lugol (sample for density calculation and plankton diversity), while the other contained a sample without additional preservatives but stored in a gelled ice coolbox (sample for plankton insulation) [3].

The identification of samples was perfomed at the laboratory of the Faculty of Fisheries and Marine at Airlangga University. Plankton analyses was determined by dripping 1 ml of sample water on the haemacytometer, then observed under the microscope with 100 enlargement. The checking of parameters of water samples such as Dissolved Oxygen (DO), temperature, brightness, current and pH were done directly at the sampling site.

*2.2. Identification and Calculation of Plankton Density*

The identification of the plankton is based on its morphological form observed by using a microscope [23].The identification of zooplankton is based on [16] the calculation of plankton density using the formula from [20], while the Calculation of Dominance Index and Diversity Index was calculated using the formula from [19].

IOP Conf. Series: Earth and Environmental Science **137** (2018) 012029 doi:10.1088/1755-1315/137/1/012029

## 3. Result and Discussion

The research data stated that the extreme waters at the locations of Wonokromo, Dadapan and Juanda was caused by the condition of water brightness. The brightness at 3 stations were 60 cm at Wonokromo, 40 cm at Dadapan and 100 cm at Juanda waters. Ammonia at the three stations were measured at 0.837 mg/L in the waters of Wonokromo, 0.626 mg/L in Dadapan and 0.396 mg/L in Juanda.

**Figure 2.** Location of sample waters.

The development of phytoplankton is determined by the intensity of sunlight, temperature and nutrients. The high growth of phytoplankton may not always be beneficial for aquatic conditions; it can also cause a population explosion (blooming), which can produce harmful toxic substances.

The structure of the phytoplankton community is a collection of populations that live in a particular region or habitat that interconnect and interact or have a reciprocal relationship of a particular zone [19]; this includes the index of diversity, index dominance, uniformity index and species wealth index (Kamali, 2004). The Shannon-Wiener species diversity index is a mathematical calculation that describes the analysis of information about the number of individuals in each species, the number of species and the total of individuals within a community (Masson, 1981). The uniformity index (ETIsability) is a description of the uniformity of the individual distribution of the phytoplankton species in a community [19]. Simpson's dominance index represents the presence or absence of a dominant species in a community. The disappearance of the dominant species causes a change in the biotic community and its physical environment [19]. The richness index is used to find out the least amount of taxa and the concentration of biota in a community (Margalef, 1951 in Romimohtarto, 2001).

From the sampling sites, we found 10 classes of phytoplankton (bacillariophyceae, dynophyceae, chlorophyceae, cyanophyceae, crysophyceae, euglenoidea, trebouxlophyceae, mediophyceae, nitachiaceae) and 5 classes of zooplankton (maxillopoda, hexanuplia, copepoda, malacostraca, oligotrichea). The density of plankton were 37.64 cells/liter in Wonokromo, 63.80 cells/liter in Dadapan, and 352,85 cells/liter in Juanda waters.

Phytoplankton have chlorophyll that the function in photosynthesis to produce organic and oxygen in the water, However, certain phytoplankton may degrade the quality of the waters if the amount is excessive (blooming). The high populations of toxic phytoplankton in waters can cause negative consequences for aquatic ecosystems, such as reduced oxygen in the water that can cause the death of various aquatic creatures.

**Table 1**. Identification of Plankton Density at the Wonokromo, Juanda, and Dadapan locations.

| Species of | Wonokromo | | | Juanda | | | Dadapan | | |
|---|---|---|---|---|---|---|---|---|---|
| Plankton | 400 m | 700 m | 1000 m | 400 m | 700 m | 1000 m | 400 m | 700 m | 1000 m |
| *Amphora* | | x | x | | | | | x | x |
| *Amphidinium* | x | | | | x | | | x | |
| *Bidulphia* | | | | | | x | x | x | x |
| *Ceratium* | | | x | | | | | | |
| *Chaetoceros* | | x | x | x | x | x | | | |
| *Chromulina* | | | x | x | x | x | | | |
| *Coscinodiscus* | | x | x | x | x | x | x | | |
| *Cyclotella* | x | | x | x | | | x | x | x |
| *Dynophisis* | | | | | x | | | | x |
| *Dythilum* | | | x | | x | | | | x |
| *Euglena* | | | | | | | | | x |
| *Gymnodinium* | | | | | | | | | |
| *Lyngbia* | x | | | | x | | | x | |
| *Naupli* | | | x | x | | | | | |
| *Navicula* | x | | x | x | x | x | | | X |
| *Nitzchia* | | x | x | | x | x | | | |
| *Oocystis* | x | x | | | | | | | |
| *Oscillatoria* | x | x | | | x | | x | x | X |
| *Parafavella* | | | | x | x | x | | | |
| *Peridinium* | | | | x | | | | x | |
| *Planktoniella* | | | x | x | x | x | | | |
| *Pleurosigma* | | | | x | | | | x | |
| *Rhizosolenia* | x | x | x | x | x | | | x | |
| *Siola* | | x | | | | | | | X |
| *Tabellaria* | | x | | | x | | | | |
| *Tintinopsis* | | | | | | x | | | |
| *Thalassiothrix* | | | | | | | | | |
| *Skeletonema* | | | x | x | x | x | | | |

Phytoplankton can be found throughout the mass of water from the surface to a depth where it is still possible for the intensity of sunlight to be used in the process of photosynthesis. This phytoplankton is the largest flora component of its role as a primary producer in a waters [19]

## 4. Conclusion

Our gratitude goes out to the Faculty of Fisheries and Marine who have funded this research. We would also like to thank our team of researchers: both technicians in the laboratory and in the field. We hope that this research can be beneficial for future studies on this topic. The brightness in Wonokromo, Dadapan, and Juanda environment was 60, 40, and 100 cm, respectively. The result of ammonia in Wonokromo, Dadapan, and Juanda estuary was 0.837, 0.626, and 0.396 mg/L, correspondingly. Nine classes of phytoplankton's were found in three locations (bacillariophyceae, dynophyceae, chlorophyceae, cyanophyceae, crysophyceae, euglenoidea, trebouxlophyceae, mediophyceae, and nitachiaceae) and five classes of zooplanktons (maxillopoda, hexanuplia, copepoda, malacostraca, and oligotrichea). The density of plankton in Wonokromo, Dadapan and Juanda environments, was 37.64, 63.80, and 352, 85 cells/L, respectively.

## 5. Refference

[1] Sofarini D 2012 *Enviro Scienteae* **8** 30-34

IOP Conf. Series: Earth and Environmental Science **137** (2018) 012029 doi:10.1088/1755-1315/137/1/012029

[2] Hemraj DA Md A Hossain Q Ye J G Qin and SC Leterme 2017 *ECSA* **184** 102-114
[3] Suthers IM and D Rissik 2008 Plankton: a guide to their ecology and monitoring for water quality *National Library of Australia Cataloguing-in-Publication* entry CSIRO Publishing p 273
[4] Affandi MB Irawan A Soegianto and Sosmanida 2009 *J. Biol. Researches* **3B** 37–41
[5] Winarni D M Affandi ED Masithah and AN Kristanti 2014 *JPMIPA* **117**
[6] Seckbach J 2007 Algae and Cyanobacteria in extreme environments (cellular origin life in extreme habitats and astrobiology) *Dordrecht Springer* p 786
[7] Ahuja S 2013 Monitoring water quality pollution assessment analysis and remediation *Elsevier* p 374
[8] Barnes RSK and KH Mann 1992 Fundamental of aquatic ecology second edition Blackwell Science p 271
[9] Barsanti L and P Gualtieri 2014 Algae CRC Press Taylor and Francis p 344
[10] Bellinger EG and DC Sigee 2015 Freshwater algae : identification enumeration and use as bioindicators *Wileyblackwell* p 275
[11] Boyd CE 2015. Water quality an introduction second edition *Springer International Publishing Switzerland* pp 14-16
[12] Darley WM 1982 Algal biology: a physiological approach blackwell scientific publications. Pp 97-98
[13] Dolan JR DJS Montagnes S Agatha DW Coats and DK Stoecker 2013 The biology and ecology of tintinnid ciliates: models for marine Wiley-Blackwell p 319
[14] Effendi H 2003 Water quality study for resource management and aquatic environments Kanisius Yogyakarta p 250
[15] Eleftheriou A 2013 Methods for the study of marine benthos Wileyblackwell p 502
[16] Guglielrno L and A Ianora 1997 Atlas of marine zooplankton 1straits of magellan amphipods, euphausiids mysids ostracods and chaetognaths Spinger p 263
[17] Li Y and K Migliaccio 2011 Water quality concepts sampling and analyses CRC Press p 335
[18] McVey JP 1983 CRC Handbook of marine culture volume I CRC Press Inc Boca Rton Florida pp 33- 36
[19] Odum EP 1993 The third edition of the ecology base (*Yogyakarta: Gadjah Mada University Press*) p 697
[20] Rice EW RB Baird AD Eaton and LS Clesceri 2012 Sandard methods for the examination of water and wastewater 22nd edition *American Public Health Association American Water Works Association Water Environment Federation* 1496p
[21] Soccodato A F d'Ovidio M Lévy O Jahn MJ Follows and SD Monte 2016 **29** 9–17
[22] Tomas CR 1997 Identifying marine phytoplankton Printed in The United States of America p 875
[23] Wehr JD RG Sheath and JP Kociolek 2015 Freshwater algae of North America ecology and classification Academic Press is an imprint of Elsevier p 1045

IOP Conf. Series: Earth and Environmental Science **137** (2018) 012030 doi:10.1088/1755-1315/137/1/012030

# Effect of maggot (*Hermetia illucens*) flour in commercial feed on protein retention, energy retention, protein content, and fat content in tilapia (*Oreochromis niloticus*)

**D R Kurniawan[1], M Arief[2], Agustono[2] and M Lamid[2,3]***

[1]Aquaculture, Faculty of Fisheries and Marine, Airlangga University, Surabaya 60115
[2]Department of Fish Health Management and Aquaculture, Faculty of Fisheries and Marine, Airlangga University, Surabaya 60115
[3]Department of Animal Husbandary, Faculty of Veterinary Medicine, Airlangga Univeristy, Surabaya 60115

E-mail:mirnylamid@yahoo.com

**Abstract.** Nile tilapia is a fish with a nature of breeding throughout the year and has a high adaptation. Feed is one of the most important factors that will affect the growth and the sustainability of fish. Maggot is one of the alternative feeds which has function as source of protein.The aim of this research is to ascertain the influence of maggot (*Hermetia illucens*) flour substitution on commercial feed which affect the retention of energy, retention of protein, and also rough fat and protein of Nile tilapia flesh. The treatment was to substitute the commercial feed with different doses of maggot flour. In this study, the amount of maggot flour substituted on commercial feed is as follows: treatment P0 (0 %), P1 (12 %), P2 (14 %), P3 (16 %) and P4 (18 %). The parameter observed in this study is an effect on to protein retention, energy retention, and also rough fat and protein in Nile tilapia flesh. The result showed the highest protein content was gained from the treatment P4 with 17.267 %. The lowest protein content was 16.344 % obtained in the treatment P0. There was an increasing protein content in Nile tilapia along with the increase of the maggot flour addition in each treatment.

## 1. Introduction

One of the freshwater fish commodities favored by consumers considerably is Nile tilapia (*Oreochromis niloticus*). Nile tilapia (*Oreochromis niloticus*) is a fish with a nature of breeding throughout the year and has a high adaptation [1]. The production of Nile tilapia fishery has enhanced significantly from year to year, from 567.078 tons in 2011 to 695.063 tons in 2012 and 909.016 tons in 2013 [2]. Nile tilapia is a freshwater fish that contains high protein and has a preeminence of breeding rapidly. Nutritional content of Nile tilapia are 16-24 % of protein, 0.2-2.2 % of fat, carbohydrate, minerals, and vitamins [3]. According to Venugopal [4], fish is a good source of nutrients to fulfill nutritional needs of humans since fish contains protein, vitamins, minerals, calcium, taurine, and it is a source of good fats such as omega 3 (EPA and DHA) and omega 6.

IOP Conf. Series: Earth and Environmental Science **137** (2018) 012030 doi:10.1088/1755-1315/137/1/012030

Feed is one of the most important factors that will affect growth and sustainability of fish life [1]. Maggot is one of the alternative feed which has a function as a source of protein, maggot contains about 45.01 % protein [5]. Maggot has amino acids, 44.26 % protein, 29.65 % fat, and 0.7 % linoleic acid and 2.24 % linolenic acid as the source of EPA and DHA [6]. Maggot can be given in fresh form to the fish and also in a pellet form. It can be processed as maggot flour before it is formed as a maggot pellet. The maggot which is made into a pellet must be steamed with a certain time where the time of maggot steaming process could affect the nutritional content of the maggot pellet [5].

To improve the quality of the fish meat, one thing to do is the application of feed with a balanced nutrition. Protein is one of the important nutrients for fish and it needs to be fulfilled to achieve optimal growth. On the other side, according to Diana [7], protein inside the body is the main source of energy other than carbohydrates and fats, it also functions as building substance, and as regulatory substance. Proteins regulate metabolic processes in the form of enzymes and hormones, as well as the defense mechanism of the body against various microbes and other toxic coming from outside the body, along with retaining cells and body tissue.

The inclusion of maggot flour in commercial feed is expected to be an alternative feed for Nile tilapia farmers to yield efficient feed cost, and high retention of energy, retention of protein, protein content, and fat content of Nile tilapia flesh.

## 2. Methodology

### *2.1. Maggot flour making process*

Maggot (*Hermetia illucens*), which was cultivated for 17 days, was obtained from Puspa Agro Sidoarjo. The maggot flour making process was as follows: (1) 17 days old maggots were harvested, then it was cleaned to remove dirt, (2) clean maggots were boiled for 5 minutes, and subsequently squeezed using muslin cloth to reduce water and fat content, (3) afterward, maggots were steamed for 20 minutes at 80 °C, (4) after steaming, the maggots were put into the oven at 50v°C for 2-3 days, and (5) dried maggots were mashed with a blender and sieved to obtain maggot flour which was ready to use as feed ingredients.

### *2.2. Feed making preparation*

The first step of feed making process is to calculate the composition of each feed ingredient in every treatment. The Second step was mixing the ingredients gradually starting from the lightest weight to the heaviest weight into a tub and stirred until it mixed well before added with water. The third step was, after the feed ingredients were evenly mixed, they were put into the pellet maker machine. After the pellet feed had been shaped with the diameter of 1-2 mm, they were either chilled or sun-dried. Lastly, the pellet feed was ready to be given to Nile tilapia.

### *2.3. Data analysis*

Data was analyzed using Analysis of Variance with Completely Randomized Design. When there were any differences, then Duncan's Multiple Range Test was used with a degree of confidence 0.05 to know the differences among treatments [8].

## 3. Results and Discussion

### *3.1. Protein retention*

The results showed that protein retention of Nile tilapia (*Oreochromis niloticus*) flesh ranged between 58.217 % - 91.505 %. The average data of protein retention of Nile tilapia fish flesh can be seen in table 1.

IOP Conf. Series: Earth and Environmental Science **137** (2018) 012030 doi:10.1088/1755-1315/137/1/012030

**Table 1**. The average of nile tilapia fish flesh protein retention.

| Treatments | Protein Retention ± SD | Transformation($\sqrt{y + 0.50}$)±SD |
|---|---|---|
| P0 (0 %) | $89.322^{b}$± 13.439 | $9.456^{b}$± 0.727 |
| P1 (12 %) | $73.840^{ab}$ ± 14.978 | $8.587^{ab}$± 0.897 |
| P2 (14 %) | $91.505^{b}$± 22.274 | $9.538^{b}$± 1.165 |
| P3 (16 %) | $61.322^{a}$± 11.295 | $7.837^{a}$± 0.722 |
| P4 (18 %) | $58.217^{a}$± 14.210 | $7.621^{a}$± 0.920 |

Note: Different superscripts in the same column are indicated (p <0.05)differences

The highest protein retention (91.505 %) was obtained from the Treatment P2 with 14 % of maggot flour substitution. The lowest protein retention was 58.217 % that was found in Treatment P4 with 18 % of maggot flour substitution. Treatments P1 and P3 showed an enhancement in protein retention of Nile tilapia fish flesh, yet the protein retention in P0 (89.322 %) was still greater than P1 and P3. The treatment P2 showed the highest protein retention (91.505 %) compared to other treatments (P0, P1, P3, and P4). results of the Analysis of Variance (ANOVA) showed a significant difference (P<0.05) between each treatment.

Protein retention indirectly describes the amount of feed protein consumed and used to build the body's protein tissue (beneficial for growth) [9]. Based on the results of statistical analysis, it indicated that maggot substitution on commercial feed significantly affected (P<0.05) protein retention in Nile tilapia. This occurred because the proteins in the feed with high biological values will lead the protein in the body to have good biological value as well [10].

The highest protein retention rate was found in treatment P2 with 14 % maggot flour that contained 9.538% feed protein. This exemplifies that feeding Nile tilapia fish with 29.8677 % protein in resulted in 9.538% protein retention.

### *3.2. Energy retention*

Energy retention in Nile tilapia fish flesh (*Oreochromis niloticus*) ranged between 26.312 kcal/kg - 47.672kcal/kg. The average data of energy retention can be seen in table 2.

**Table 2**. The average energy retention of Nile tilapia fish flesh.

| Treatments | Energy Retention ± SD | Transformation($\sqrt{y + 0.50}$)±SD |
|---|---|---|
| P0 (0 %) | $47.672^{c}$± 14.926 | $6.876^{c}$± 1.084 |
| P1 (12 %) | $30.112^{ab}$± 4.274 | $5.522^{ab}$± 0.397 |
| P2 (14 %) | $42.485^{bc}$± 10.692 | $6.517^{bc}$± 0.826 |
| P3 (16 %) | $27.217^{a}$± 4.567 | $5.251^{a}$± 0.433 |
| P4 (18 %) | $26.312^{a}$± 6.163 | $5.152^{a}$± 0.590 |

Note: Different superscripts in the same column are indicated (p<0.05) differences

IOP Conf. Series: Earth and Environmental Science **137** (2018) 012030 doi:10.1088/1755-1315/137/1/012030

The highest energy retention was produced from the control treatment (P0), which was 47.672 %. The lowest energy retention was 26.312 % found in treatment P4 with 18% maggot flour substitution. Treatments of P1 and P3 showed an enhancement of energy retention, yet the energy retention in P2 (42.485 %) was still higher than P1 and P3. Treatment P0 showed the highest energy retention by 47.672% compared to other treatments. Analysis of Variance (ANOVA) showed significant difference (P<0.05) among treatments.

The use of energy in Nile tilapia is influenced by the amount of feed consumed. Energy is obtained from reshuffling the chemical ligament through the oxidation reaction process towards feed components, specifically on proteins, fats, and carbohydrates to produce simpler compounds (amino acids, fatty acids, and glucose), so that it can be absorbed by the body to be either used or saved. Energy retention is a depiction of the amount of energy stored in tissues in the fish's body divided by the amount of energy in the feed consumed [11].

Average energy retention ranged from 26.312 kcal/kg–47.672 kkal/kg, with the highest energy retention found in control treatment (P0). Analysis of Variance (ANOVA) showed significant differences (p<0.05) among treatments. This signifies that the energy content in feed comes from proteins, causing it to be used to synthesize proteins in the body instead of as energy storage.

### *3.3. Crude fat*

Crude fat of Nile tilapia fish flesh ranged from 3.572% - 4.779%. Analysis of Variance (ANOVA) showed a significant difference (P<0.05) among treatments. Average data of fat content can be seen in table 3.

**Table 3.** The average of crude fat content of Nile tilapia (*Oreochromis niloticus*) flesh.

| Treatments | Rough Fat Content ±SD | Transformation $\sqrt{y+0.5}$±SD |
|---|---|---|
| P0(0 %) | $3.574^{a}$± 0.288 | $2.017^{a}$± 0.071 |
| P1(12 %) | $3.572^{a}$± 0.305 | $2.016^{a}$± 0.075 |
| P2(14 %) | $3.898^{a}$± 0.301 | $2.096^{a}$± 0.071 |
| P3(16 %) | $4.772^{b}$± 0.137 | $2.295^{b}$± 0.029 |
| P4(18 %) | $4.779^{b}$± 0.321 | $2.296^{b}$± 0.069 |

Note: Different superscripts in the same column are indicated (p<0.05) differences

The highest crude fat content was obtained from treatment P4 with 4.779 % with 18 % maggot flour substitution. The lowest crude fat content was 3.572 % that was found in treatment P2 with 12% maggot flour substitution. The Duncan test results showed that substitution of maggot flour in commercial feed had increased crude fat content in Nile tilapia, but the substitution of 14 % maggot (P1) had decreased the crude fat content which was not significantly different with P0 and P2. it can be seen that the substitution of maggot flour in commercial feed can increase the crude fat content of Nile tilapia fish flesh between 3.572 % - 4.779 %. Nile Tilapia is a low fat type of fish with 2.54 % fat content [12]. The increase in crude fat of Nile tilapia was caused by high content of fat in maggot flour that was 13.945 %. The commercial feed ration substituted with maggot flour yield a crude fat content of 11.242 % - 11.709 %. A proper growth will be generated when the fat content in the fish ration is between 4 – 8 %, but the content of 10 – 11 % of crude fat in the feed is still tolerable by the fish in general [13]. The decrease of the crude fat content in treatment P1 was caused by the low temperatures during the maintenance. Fat content decreases at low temperatures and increases at high temperatures. Water temperature greatly affects the activity of the digestive tract of Nile tilapia [12]. Nile tilapia is a type of thermopile animal whose feeding ability will rise according to the increasing of the water temperature [14].

### *3.4. Protein*

Protein content of Nile tilapia fish ranged between 16.344 % - 17.267 %. Analysis of Variance (ANOVA) showed a significant difference of ($P<0.05$) among treatments. Protein content of tilapia fish can be seen in table 4.

**Table 4**. The Average protein content of nile tilapia fish flesh.

| Treatments | Protein Content ± SD | Transformation $\sqrt{y+0.5}$ ± SD |
|---|---|---|
| P0(0 %) | $16.344^{a} \pm 0.252$ | $4.104^{a} \pm 0.030$ |
| P1(12 %) | $16.876^{a} \pm 0.518$ | $4.168^{a} \pm 0.061$ |
| P2(14 %) | $17.061^{ab} \pm 0.270$ | $4.190^{ab} \pm 0.032$ |
| P3(16 %) | $17.094^{b} \pm 0.647$ | $4.194^{b} \pm 0.077$ |
| P4(18 %) | $17.267^{b} \pm 0.093$ | $4.215^{b} \pm 0.011$ |

Note: Different superscripts in the same column are indicated ($p<0.05$) differences

The highest protein content was gained from the treatment P4 with 17.267 %. The lowest protein content was 16.344% obtained in the treatment P0. There was an increasing protein content in Nile tilapia along with the increase of the maggot flour addition in each treatment. maggot flour substitution in commercial feed can increase the protein content of Nile tilapia fish with the range between 16.344 % - 17.267 %.

Nile tilapia is the type of fish which has a low protein (16 %) [3]. The increase of protein content in Nile tilapia found in this study was caused by the high content of protein in the maggot flour (29.8 %). The commercial feed that was substituted with maggot flour gained 29.22 % - 29.29 % of protein content. Protein in the Nile tilapia flesh indirectly represented the amount of protein in the feed consumed and used to build tissues in the body and for growth [9]. Fish proteins is influenced by the feeding quality and their energy rate. the right balance of protein energy with the right amount of feeding will produce the best growth, feed conversion, and feeding efficiency. The amount of protein in the feed will affect the feed consumption leading to the level of protein absorption in the fish flesh [15].

## 4. Conclusion

The current study has demonstrated that the addition of maggot flour can improve protein retention, energy retention, protein content, and fat content in Nile tilapia. The treatment P4 with the addition of maggot flour 18% has effect on increasing the fat and protein contents of 4.779% and 17.267%, respectively.

## 5. References

[1] Anggriani I and Ankiq T 2012 *J. Per. Kel.* **3**
[2] Directorate General of Aquaculture-Ministry of Marine Affairs and Fisheries 2014 Aquaculture Production Supports National Economy and Increases Welfare
[3] Widaryati R 2016 *J. Ilm. Hewani Trop* **5**
[4] Venugopal 2010 *J. Apl. Tek. Pang.* **1**

[5] Haryati 2010 *Effect of substitution of fish starch with maggot flour on retention and efficiency of nutrition utilization in the milkfish body (Chanos chanos)* (Makasar : Universitas Hasanudin)
[6] Fahmi M R, Saurin, H, and I Wayan S 2009 Maggot potential as one of protein source of animal feed loka research on freshwater ornamental fish national *Seminar on Technology Support To Improve Animal Food Products In Order To Fulfill People's Nutrition*
[7] Diana F M 2010 *J. Kes. Masy.* **4**
[8] Kusriningrum R S 2012 *Experiment Design* (Surabaya : Airlangga University Press)
[9] Halver J E 1988 *Fish Nutrition* (London Academis Press, INC) p 798
[10] Johnson E G, W O Watanabe and S C Ellis 2002 *North American J. Aquacult.* **64** 47-54
[11] Haryati E, Saadedan A, Pranata 2011 *The effect of substitution rate of fish flour with maggot flour on the retention and efficiency of nutrient utilization in milkfish body* (Makassar : Universitas Hassanudin) p 31
[12] Isa M, Rinidar, Zalia T, Harris A, Sugito dan Herrialfian 2015 *J. Med. Vet.* **9** 60-63
[13] Benyamin B 2002 *The effect of using kaliandra leaf flour in an artificial feed against growth of freshwater pomfret seeds (Collosuma macropomum)* (Bandung : Unversitas Padjajaran)
[14] Turker C S 2004 *Biology and culture of channel catfish* (Elsevier) **34** p 686
[15] Fran S and Akbar J 2013 *Effect of Protein Level Difference and Feed Protein Ratio on Growth of Trichogaster pectoralis* (Banjarmasin : Universitas Lambung Mangkurat)

IOP Conf. Series: Earth and Environmental Science **137** (2018) 012031 doi:10.1088/1755-1315/137/1/012031

# Anti-leech activity of *Scutellaria baicalensis* and *Morinda citrifolia* extracts against *Piscicola geometra*

**P N Rizky[1,2], T C Cheng[1] and H Nursyam[2]**

[1]Department of Tropical Agriculture and International Cooperation, National Pingtung University of Science and Technology, Pingtung 91201, Taiwan
[2] Faculty of Fisheries and Marine Science, University of Brawijaya, Malang 65145, Indonesia

E-mail: putrimarine92@gmail.com

**Abstract.** *Piscicola geometra* leeches are naturally infecting cobia juvenile. The leeches attach to cobia by sucking and biting its surface and provide the gate of second infection. Water extracts of *Scutellaria baicalensis* root and *Morinda citrifolia* leaves were used to be tested through *In Vitro* method to look for the anti-leeches activity against *Piscicola geometra*. In this study, a total number of 800 leeches from infected cobia were prepared. The anti-leech activity from water extract of *S. baicalensis* root and *M. citrifolia* leaves were compared in different dilutions of plant extracts for 96 hours. Significant anti-leech activity was observed with *M. citrifolia* leaves with 80% mortality of leeches. *S. baicalensis* root showed higher anti-leech activity with 100% mortality of leeches. The average time was needed for *S.baicalensis* root to paralyzing and kill the leeches were 8h, 40h, 48h, 72h, and 96h in various dilutions of *S. baicalensis* root. This study indicated that *S. baicalensis* water extract had a potent for new anti-leeches agent.

## 1. Introduction

Cobia (*Rachycentron canadum*) culture has become popular in more than 23 countries in the Asia-Pacific region [1]. However, rapid expansion and intensification of cobia culture has resulted in increasing incidences of disease outbreaks [2]. Parasitic diseases are particularly common to infect the gastrointestinal tract, gills, and skin of cobia, which have caused economic losses [3, 4]. Moreover, leeches infestation are considered as the portal for other pathogens that develop secondary infection and lead to death [5].

It has been demonstrated that the impact of leeches are correlated with the level of infestation [6]. *P. geometra* is a member of Piscicolidae family which is considered as blood-sucking marine leeches [7]. It has a cylindrical body shape and an anterior bell-shaped sucker -[8]. Moreover, it has much higher rate of oxygen consumption, it prefers running water and it is oligotrophic with pH range varies from 7.0 to 9.0 [9]. Since the leeches are found in wide variety of habitats, it is conceivable that the leeches can tolerate much broader ranges of conditions [10]. *P. geometra* was previously reported to be found in *Pleuronectes platessa* and *Myoxocephalus* [11], *Rutilus rutilus, Scardinius erytropythalmus, Blica bjoerkna, Tinca tinca [12], Esox lucius, Barbus rajanorum, Salmo salar, Salmo girdneri, Rutilus rutilus* [13], *Gasterosteus aculeatus*, *Cottus scorpius* and *Platichtys flesus* [14].

Medicinal plants contain bioactive compounds and shows impact in controlling fish diseases as a low-cost and eco-friendly method and it has a broad spectrum of activity [15, 16]. *Scutellaria baicalensis* and *Morinda citrifolia* have been widely used as medicinal plant [17, 18]. unique compounds contained in those plants are for example baicalein, wogonin, and oroxylin that are highly present in the root of *S. Baicalensis* and they can directly bind with the active site of enzymes to increase the stimulation of protein synthesis [19]. *S. baicalensis* root has been reported to be useful to treat liver problems [20], diarrhea, hypertension, hemorrhaging, inflammations, and respiratory infections, as well as to act as antimicrobial, , anti toxoplasm [21] and anti trypanosomal activities [22]. The leaf of *M. citrifolia* has been studied for its antimicrobial, anticancer [23], antioxidant, and larvicidal activities [24, 25]. Moreover, the plant extract has also been used as anti-helminthic and anti-parasitic agent [26-29]. The aim of this study was to determine the anti-leech activity of water extracts of *S. baicalensis* root and *M. citrifolia* leaves against *Piscicola geometra*.

## 2. Methodology

### 2.1. Sample collection

Eight hundred of leeches (1.27±0.44 cm in length) were collected from infected cobia in Tungkang Marine Biotechnology Research Station, Taiwan. The bell shape of anterior sucker, cylindrical body shape, and clear segments with longitudinal white spots on the body were the main signs for detection of *Piscicola geometra* species (figure 1).

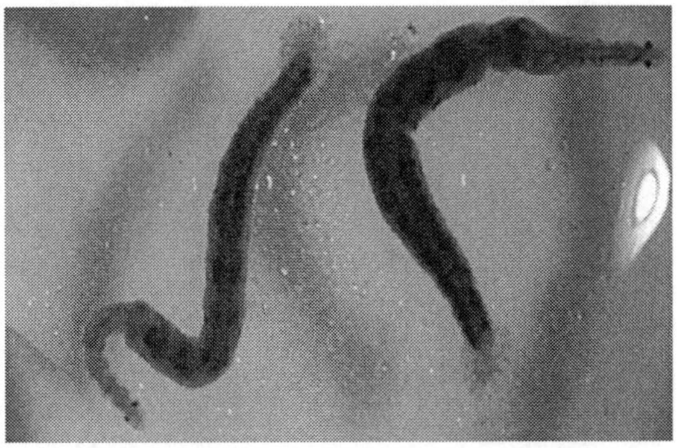

**Figure 1.** *Piscicola geometra.*

### 2.2. Plant extract preparation

The dried roots of *S. Baicalensis* were purchased in the local market of Neipu, Pingtung, Taiwan. Fresh *M. citrifolia* leaves were collected from noni farm, Pingtung, Taiwan. The leaves were chopped into small pieces and rinsed with distilled water before put into the oven for 12 h. The dried plant was grounded separately in a mechanical grinder and stored in an airtight container. Water extract was prepared by boiling 150 mL distilled water with 10 g of plant powder at 95 °C. After extraction, the supernatant was centrifuged at 4 °C, 1000 g for 10 min. The clear supernatant was filtered and stored at -20 °C.

### 2.3. Anti-leech assay

The clear supernatant of *S. baicalensis* roots and *M. citrifolia* leaves was diluted with sea water to obtain concentrations of 50x, 100x, 200x, 400x, 500x, 700 x, and 1000x dilutions.

Five leeches per petri dish were used in three replicates involving eight petri dish of seven different concentrations and a control (sea water). Their effects were screened for 96 h with 8 h interval for observation of mortality. The behavior of the leeches and the time used to paralyze and kill them were recorded. The dilutions were renewed every 12 h. The efficacy of the plant extract that were able to kill the leeches within 1–96 h reflects the anti-leech activities and would be used to treat infected cobia

IOP Conf. Series: Earth and Environmental Science **137** (2018) 012031 doi:10.1088/1755-1315/137/1/012031

in the future. The differences between control and treated groups were analyzed by one-way ANOVA.

## 3. Results

The anti-leech activities of *S. baicalensis* root and *M. citrifolia* leave water extracts against *P. geometra* infestation are shown in table 1 and table 2, respectively. Introduction of plant extract caused the reduction of swimming activity of *P. geometra*. Total mortality (100 %) of the leeches was observed in *S. baicalensis* root water extracts (table 1). During the observation, leeches gradually became weak and their bodies turned to be wrinkled. They moved weakly and finally died. On the other hand, *M. citrifolia* leaves extract was able to cause about 60 % of mortality in leeches. According to the statistical analysis, 50x and 100x dilutions exhibited the strongest effect of both plant extract, while 700x and 1000x dilutions showed the lowest. At 500x, 700x and 1000x dilutions, all the leeches immersed in *S. baicalensis* root died while few leeches could survive in *M. citrifolia* extract (table 2). Control sample showed no anti-leech effect in both studies. It was obvious that the water extract of *S. baicalensis* root showed higher anti-leech activity than *M. citrifolia* leaves extract, which was characterized by the paralysis and death of leeches in a short time.

**Table 1.** Mortality of leeches for 96 h immersion in different serial dilution of *S. baicalensis* root extract (N=5 leeches per treatment in 3 replicate). a, b, c, d, e character show that there are no significant differences among treatments.

| Time observed (hour) | Serial dilution | | | | | | | |
|---|---|---|---|---|---|---|---|---|
| | Control[e] | 1/50x[a] | 1/100x[a] | 1/200x[b] | 1/400x[bc] | 1/500x[c] | 1/700x[cd] | 1/1000x[d] |
| 1 | 0 | 5 | 5 | 0 | 0 | 0 | 0 | 0 |
| 8 | 0 | 5 | 5 | 0 | 0 | 0 | 0 | 0 |
| 16 | 0 | - | - | 2 | 0 | 0 | 0 | 0 |
| 24 | 0 | - | - | 1 | 1 | 1 | 1 | 0 |
| 32 | 0 | - | - | 0 | 0 | 0 | 0 | 0 |
| 40 | 0 | - | - | 2 | 0 | 1 | 1 | 0 |
| 48 | 0 | - | - | - | 4 | 0 | 0 | 0 |
| 56 | 0 | - | - | - | - | 1 | 1 | 1 |
| 64 | 0 | - | - | - | - | 1 | 1 | 1 |
| 72 | 0 | - | - | - | - | 1 | 1 | 0 |
| 80 | 0 | - | - | - | - | - | - | 1 |
| 88 | 0 | - | - | - | - | - | - | 0 |
| 96 | 0 | - | - | - | - | - | - | 2 |
| **Total** | 0 | 5 | 5 | 5 | 5 | 5 | 5 | 5 |

**Table 2.** Mortality of leeches for 96 h immersion in different serial dilution of *M.citrifolia* leaves extract (N=5 leeches per treatment in 3 replicate). a, b, c, d, e character show that there are no significant differences among treatments.

| Time observed (hour) | Serial dilution | | | | | | | |
|---|---|---|---|---|---|---|---|---|
| | Control[e] | 1/50x[a] | 1/100x[ab] | 1/200x[b] | 1/400x[c] | 1/500x[d] | 1/700x[d] | 1/1000x[de] |
| 1 | 0 | 0 | 0 | 0 | 0 | 0 | 0 | 0 |
| 8 | 0 | 2 | 0 | 0 | 0 | 0 | 0 | 0 |
| 16 | 0 | 0 | 2 | 0 | 0 | 0 | 0 | 0 |
| 24 | 0 | 2 | 0 | 1 | 0 | 1 | 0 | 0 |
| 32 | 0 | 1 | 2 | 0 | 0 | 0 | 0 | 0 |
| 40 | 0 | - | 0 | 2 | 1 | 2 | 1 | 0 |
| 48 | 0 | - | 1 | 1 | 1 | 0 | 0 | 0 |
| 56 | 0 | - | - | 1 | 0 | 0 | 0 | 0 |
| 64 | 0 | - | - | - | 0 | 0 | 1 | 0 |
| 72 | 0 | - | - | - | 0 | 1 | 1 | 0 |
| 80 | 0 | - | - | - | 1 | 0 | 1 | 1 |
| 88 | 0 | - | - | - | 1 | 0 | 0 | 0 |
| 96 | 0 | - | - | - | 1 | 0 | 0 | 2 |
| **Total** | 0 | 5 | 5 | 5 | 5 | 4 | 4 | 3 |

### *3.1. Behavior of leeches*

Leeches (*Piscicola geometra*) use their sucker for locomotion; after the posterior sucker attached, they extend their bodies and the oral sucker will be attached (figure 2A). The posterior sucker is drawn forward, it reattaches near the anterior sucker and it will be released (Figure 2B). This movement is rapid and efficient. After detachment, the posterior sucker is applied high up on the ventral surface,

IOP Conf. Series: Earth and Environmental Science **137** (2018) 012031 doi:10.1088/1755-1315/137/1/012031

producing an aggregate loop (figure 2C-D) before the body reaches its usual position (figure 2E). When the leeches were immersed in several dilutions of extract, they swayed in random movement until the posterior sucker reached the bottom of the petri dish. The leeches gradually became weak and their bodies turned to be wrinkled over the increasing time of immersion.

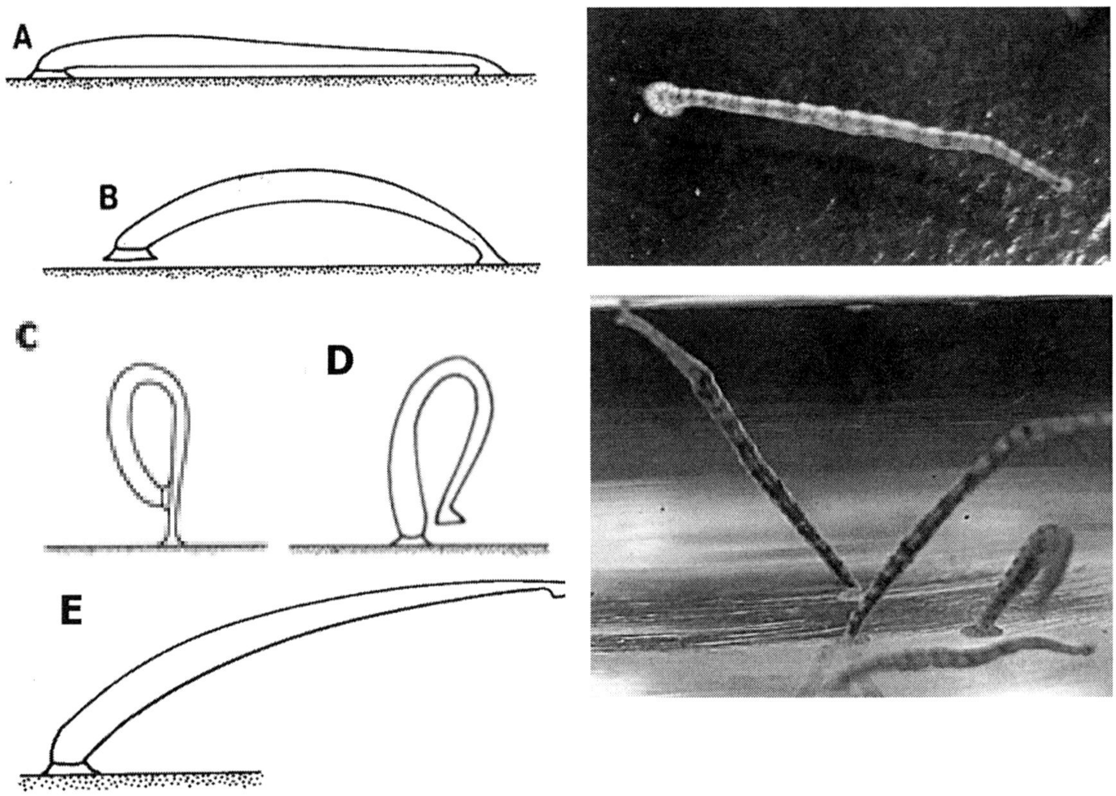


**Figure 2.** Behavior of *Piscicola geometra*

## 4. Discussion

Infestation of *P. geometra* has caused extensive damage characterized by small circular lesions with the presence of blood on the skin, eyes, and gill of cobia. The damaged scale is considered as the portal of bacteria and fungi infections that can lead to second infections [30]. Leeches are rarely pathogens to their host and only become a serious problem when the infestations are high [31].

No standard medicine has been registered to treat leeches infestation. Anti-histamine, calcium hydroxide, lidocaine, hypertonic saline, bicarbonate, and citrus extract are traditional treatments to treat leeches infection and have been noted as ineffective treatments [32]. Pharmaceutical plants can be used as a source for a wide variety of natural antioxidants and they have been proven as safe therapeutic effects to treat fish disease [33, 34]. Flowering plant *Scutellaria baicalensis* and *Morinda citrifolia* have been listed in a traditional Chinese medicine (TCM) to treat many diseases [35]. The dried root of *S. Baicalensis* and dried leaves of *M. citrifolia* produce large amounts of bioactive flavones and provide a variety of specific health benefits [36, 37].

*S. baicalensis* and *M. citrifolia* have been reported to support immunostimulatory activity, antitumor, antibacterial, antiseptic, anti-fungal, anti-inflammatory, and anti-helminthic activity with no

IOP Conf. Series: Earth and Environmental Science **137** (2018) 012031 doi:10.1088/1755-1315/137/1/012031

side effects [38, 39]. As an effective drug with no side effects, this study has offered the use of *S. baicalensis* root and *M. citrifolia* leaves against leeches' infestation. The result of this study shows that high rate of mortality was observed within 96 h of immersion of leeches with *S. baicalensis* root water extract indicating that the herb was effective to control leeches in cobia culture. *S. baicalensis* root was able to kill the leeches in 8 h, 40 h, 48 h, 72 h, and 96 h, respectively. The average time for leeches to die with *M. citrifolia* were 32 h, 48 h, 56 h, and 96 h with 50x, 100x, 200 x, and 400x dilutions, respectively. Meanwhile, few leeches could still survived at 500x, 700x, and 1000x dilutions of *M. citrifolia* leaf after 96 h of immersion.

A preliminary investigation on the toxicity of herbal extract on *Piscicola geometra* was reported by. It was shown to reduce swimming activity of leeches in the 42 h, 30 h, 18 h, and 12 h for 5 %, 10 %, 15 %, and 20 % concentration of *Tetracer alnifolia*. Toxiceffect of *Raphia vinifera* on *Piscicola geometra* has been demonstrated previously in a study by [40]. Terrestrial blood sucking leeches (Haemadipsidae) population can be reduced by an aqueous extract of *C. spathulifolia* (62.6%), *V.elaeagnifolia* (63.0%), and *S.rarak* (82.6%) [6]. The root of *S baicalensis* has been effective as anti-parasitic agent against *Toxoplasma gondii* and *Trypanosoma cruzi* [41]. However, there is no report of the anti-parasitic or anti-leech activity of *S. baicalensis* and *M. citrifolia* on *P. geometra*.

## 5. Conclusions

This study has demonstrated that water extract of *S.baicalensis* dried root has stronger anti-leech activity than the water extract of *M. citrifolia* dried leaves and it has ability to inhibit the swimming activity of *P. geometra*. *S. baicalensis* water extract demonstrated an effective anti-leech activity against *P. geometra* infection. Further study on bath treatment of infected cobia and analysis of active compounds of *S. baicalensis* root extract will be required in the future.

## 6. References

[1] Lee D-H, Lim S-R, Han J-J, Lee S-W, Ra C-S, *et al* 2014 *J. Anim. Sci.* **27** 1303-1310
[2] Chu KB, Abdulah A, Abdullah SZ and Bakar RA 2013 *Trop. Life Sci. Res.* **24** 77-84
[3] Shaffer RV and Nakamura EL 1989 Synopsis of biological data on the cobia rachycentron canadum (pisces: Rachycentridae)
[4] Megharajan S, Ranjan R, Xavier B, Muktha M, Edward L, et al 2017 *J. Asia Pac.* **13** 12-16.
[5] Vongsombath C, De Boer HJ and Pålsson K 2011 *Acta trop.* **119** 178-182
[6] McLean E, Salze GandCraig SR 2008 *Ribarstvo* **66** 1-16
[7] Mann KH 1962 *Chapter 1 - introduction* Book title: Pergamon; p 1-4
[8] Kearn GC 2004 *Leeches, Lice and Lampreys a Natural History of Skin and Gill Rarasites of Fishes* (Netherland : Dordrecht-Springer)
[9] Mann KH 1962 *Chapter 2 - The Medicinal Leech, Hirudo Medicinalis* Book title: Pergamon; p. 5-21
[10] Landsberg JH 1982 *Infection Strategies of Fish Ectoparasites* [Thesis type]
[11] Mann KH 1962 *Chapter 3 - A Survey of The Group: A More Detailed Systematic Survey is Given in the Appendixes* Book title: Pergamon p 22-35
[12] Demirtas M 2011 *Turkiye Parazitol Derg* **35** 159-163
[13] Ceylan M, Boyaci YO, Meke T, Inceoglu H and Kara A 2011 *Turkiye Parazitol Derg* **35** 207-209
[14] Kendzierska H, Dąbrowska A, Cichocka J, Janas U and Bielecki A 2014 The first record of piscicola pojmanskae bielecki, 1994 in the gulf of gdańsk (the southern baltic sea) with species characteristics distinguishing it from Piscicola geometra (linnaeus, 1758) ***Secondary title*** p 324
[15] Madhuri S, Mandloi A, Govind P and Sahni Y 2012 *IOSR J Pharm* **3** 28-30
[16] Mehana EE, Rahmani AH and Aly SM 2015 *Annu Res Rev Biol* **5** 477
[17] Abou Assi R, Darwis Y, Abdulbaqi IM, khan AA, Vuanghao L, et al 2017 *Arab. J. Chem.* **10** (5): 691-707

[18] Zhao Q, Chen X-Y and Martin C 2016 *Sci. Bull.* **61** (18): 1391-1398
[19] Mahmoud M, El-Lamie M, Dessouki A and Yusuf M 2014 *Glob. Res J. Fis. Sci.* **1** 26-33
[20] Liu A, Huang L, Fan H, Fang H, Yang Y, et al 2015 *Int. Immunopharmacol* **24** 72-79
[21] Yang X, Huang B, Chen J, Huang S, Zheng H, et al 2012 *J. Parasitol. Res.* **110** 2221-2227
[22] Mamadalieva NZ, Herrmann F, El-Readi MZ, Tahrani A, Hamoud R, *et al* 2011 *Afr. J. Pharm. Pharmacol.* **63** 1346-1357
[23] Sang S, Wang M, He K, Liu G, Dong Z, et al 2002 Chemical components in noni fruits and leaves (morinda citrifolia l.) Book title: *ACS Publications* p 33
[24] Kovendan K, Murugan K, Shanthakumar SP and Vincent S 2012 *Asian Pac. J. Trop. Dis.* **2** S362-S369
[25] Kovendan K, Shanthakumar SP, Praseeja C, Kumar PM, Murugan K, et al 2014 *Asian Pac. J. Trop. Dis.* **4** S173-S180
[26] Militz TA, Southgate PC, Carton AG and Hutson KS 2014 *J. Fish. Dis.* **37** 451-461
[27] Chitmanat C, Tongdonmuan K, Khanom P, Pachontis P and Nunsong W 2003 *Antiparasitic, antibacterial, and antifungal activities derived from a terminalia catappa solution against some tilapia (oreochromis niloticus) pathogens*. Secondary title p 179-182.
[28] Huang A-G, Yi Y-L, Ling F, Lu L, Zhang Q-Z, et al 2013 *J. Parasitol. Res.* **112** 4065-4072
[29] Ling F, Wang J-G, Lu C, Wang G-X, Lui Y-H, et al 2012 *J. Parasitol. Res.* **111** 841-848
[30] Nwabueze AA 2013 *J. Agric. Sci.* **1 1-9**
[31] Bielecki A, Cichocka J, Terlecki J and Witkowski A 2011 *Biologia* **66** 294-298
[32] Bahmani Ma nd Rafieian-Kopaei M 2014 *Asian. Pac. J. Trop. Dis.* **4** 315-316
[33] Rafieian-Kopaei M 2013 *J. Herb. Med.* **1** 1-2
[34] Shirzad H and Nasri H 2014 *J. Herb. Med.* **2** 21-22
[35] Calabrò S 2015 *Plant secondary metabolites* Book title: Springer p 153-159
[36] Zhao Q, Zhang Y, Wang G, Hill L, Weng J-K, et al A specialized flavone biosynthetic pathway has evolved in the medicinal plant, scutellaria baicalensis *Sci. Adv.* **2** e1501780
[37] Jeyabalan S, Subramanian K, Cheekala UMR and Krishnan C 2017 *J. Appl. Pharm. Sci.* **7** 089-095
[38] Torres MAO, Fátima Braga Magalhães I, Mondêgo-Oliveira R, Sá JC, Rocha AL, *et al* 2017 *Phytother. Res.* **7** 5817 - 5821
[39] Muluye RA, Bian Y and Alemu PN 2014 *J. Tradit. Complement Md.* **4** 93-98
[40] Fafioye O and Adebisi A 2001 Toxic effect of raphia vinifera on fish leech (piscicola geometra)
[41] Schinella GR, Tournier HA, Prieto JM, Ríos JL, Buschiazzo H, et al 2002 *Fitoterapia* **73** 569-575

**Acknowledgements**
This work was supported by Aqua Molecular Genetic Laboratory at Department of Tropical Agriculture and International Cooperation, NPUST, Taiwan and Tungkang Biotechnology Research Center, Pingtung, Taiwan.

IOP Conf. Series: Earth and Environmental Science **137** (2018) 012032 doi:10.1088/1755-1315/137/1/012032

# Effect of earthworm (*Lumbricus rubellus*) in feed formulation to improve fatty acids profile in eel (*Anguilla bicolor*) meat

**K Farah[1], I R Gunawan[1], G B Putra[1], Agustono[2], W P Lokapirnasari[4], M Lamid[2], E D Masithah[3], T Nurhajati[4] and Rozi[2]**

[1] Faculty of Fisheries and Marine, Airlangga University, Jl. Airlangga 60115, Surabaya, East Java, Indonesia.
[2] Department of Fish Health Management and Aquaculture, Faculty of Fisheries and Marine Airlangga University, Jl. Airlangga 60115, Surabaya, East Java, Indonesia.
[3] Departement of Marine, Faculty of Fisheries and Marine, Airlangga University, Jl. Airlangga 60115, Surabaya, East Java, Indonesia.
[4] Department of Animal Science, Faculty of Veterinary Medicine Airlangga University, Jl. Airlangga 60115, Surabaya, East Java, Indonesia.

E-mail: dwiyp@fpk.unair.ac.id

**Abstract.** Eel requires unsaturated fatty acids of linolenic acid for growth. Which can be supplied from earthworms. In this study, addition of earthworm in formulation feed aimsed to improve the fatty acid profile eel meat. This research used experimental method and randomized complete design method with five treatments. Each treatment was repeated four times. The use of earthworms in feeding treatment formulation was done for 21 days with different level i.e: 0 % (P0), 25 % (P1), 50 % (P2), 75 % (P3) and 100 % (P4). The result showed that the addition of eartworm significantly influenced the omega 3 contents (EPA & DHA) of eel meat.

## 1. Introduction

Eel (*Anguilla bicolor*) is a freshwater fish that has a large potential to be developed. According to Hameed *et al.* [1], eel contains 48.430 % saturated fatty acids, 50.639% unsaturated fatty acids. In addition, there is a 0.461 % EPA, 1.294 % DHA, 9.134 % linoleic acid and 0.472 % arachidonic acid. Eelcontains 1.337 mg/100 g DHA and 742mg/100g EPA. Consumers' demand for eels has increased due to its high nutrient content. Demand for eels in international markets has reached 300,000 tons/year. The market demand for eels is increasing because people consider the meat is savory and rich beneficial for the body [2]. Eels areknown as a fishery commodity that is rich in protein, fat, minerals, and vitamins compared to other fish species [3].

Polyunsaturated fatty acids functions as a nutrient in the body, such as EPA and DHA that give benefits to human health. EPA and DHA contained in fatty fish and they cannot be synthesized in the human body [4]. The ratio between omega-3 and omega-6 fatty acidsis a good indicator for comparing the relative nutritional value of different species of fish. The ratio of omega-3 and omega-6 fatty acids helps to prevent coronary heart disease by reducing the levels of plasma lipids and risk of cancers[5]. Omega-3 and omega-6 fatty acids are polyunsaturated are important components of cell membranes and are precursors to many other substances in the body such as those involved with regulating blood

IOP Conf. Series: Earth and Environmental Science **137** (2018) 012032 doi:10.1088/1755-1315/137/1/012032

pressure and inflammatory responses, thus theymust be obtained through food [6]. Therefore, this research was aimed to know effect of addition of earthworms (*Lumbricus rubellus*) in feed formula on fatty acid profile of eels.

## 2. Methodology

### *2.1 Preparation of earthworm (Lumbricus rubellus)*

Earthworms were obtained from Malang, East Java. They were mixed with feed formulation (commercial fishmeal) in doses of 0, 25, 50, 75, and 100 % and shaped like pasta. The feed then went through proximate analysis as shown in table 1.

**Table 1.** Nutritional feed ingredients.

| Feed Ingredients | Dry Ingredients | Crude Protein (%) | Crude Fat (%) | Crude Fiber (%) | Ash (%) | NFE (%) |
|---|---|---|---|---|---|---|
| Formulation Feed | 80.029 | 42.067 | 13.011 | 6.806 | 14.228 | 3.916 |
| Earthworms | 22.896 | 13.634 | 5.749 | 0.548 | 1.589 | 1.376 |

### *2.2 Experimental design*

Eels used were from fingerling stage weighing 20-25 g with the total of 100 g and they were supplied from Malang, East Java. Eelswere selected and then acclimatized for 30 min. Furthermore,eels wasadapted for a week. For 21 days, the eels were given feed containing 0 %, 25 %, 50 %, 75 %, and 100 % earthworm.

### *2.3 Fish composition*

The levels of fat, protein, fiber, and energy ingredients without nitrogen (BETN) in eel meat were analyzed based on AOAL methods [7].

### *2.4 Fatty acid derivatisation*

Eel meat was cut, chopped, weighed for 1 g, and put into test tubes. Sodium chloride (0.5 g) and 4 mL hexane were added and the mixture was vortexed for 2 min until it was clear. Clear hexane layer was taken and transferred into the next derivatisation tube and drained with stream of nitrogen. NaOH 2 % (2 mL) was added to methanol then close to temperate at 90 °C for 5 min. The result was left to cool before added with 2 mL methanol in BF3 further heated again for 30 min. Samples were then extracted with 3 mL of hexane to final stage. Extract was analyzed by GC-MS[8].

### *2.5 Gas chromatography (GC)*

Sampleswere analyzed using gas chromatogram Shimadzu GC-2014 with helium as a carrier gas and SGE forte BPX 70 column (film thickness of 30 m x 0.25 mmID x 0.25μm) (SGE Europe Ltd. Milton Keynes.UK) as the analytical column. The peaks were identified using standardmix of 38 external FAME (FAME Mix C4-C24. Supelco; Sigma – Aldrich). Initial column temperature was set at 50°C for 1 min. Temperature was raised at 2 °C/min until it reached 188 °Cand maintained for 10 min. next, the temperature was increased further to 240 °C and maintained for 4 minbefore it was returned to the initial temperature [9].

### *2.6 Statistical analysis*

The data were expressed as mean ± standard deviation. The data were analyzed using SPSS version 16.0 (SPSS Inc., Chicago, IL). Differences between means were analyzed by Analysis of Variance followed by Duncan's multiple comparison test. Significantdifferent was set at $p < 0.05$.

## 3. Results

### *3.1 Proximate analysis of eel meat*

ANOVA showed that there were no significant effect of different feed formulations (p>0.05) (Figure 1) on nutritional content of eel meat. The use of earthworms led to increasing levels of fat on eel meat with 1.612 % in P0, 1.242 % in P1, 1.256 % in P2, 1.690 % in P3, and 1.505 % in P4.Protein levels in eel meat were 17.073 % in P0, 16.957 % in P1, 16.599 % in P2, 17.413 % in P3, and17.846 % in P4. Meanwhile, ash content ranged from 0.083 % to 1.148 % and the energy content was within the range of 0.95 to 1.014 %.

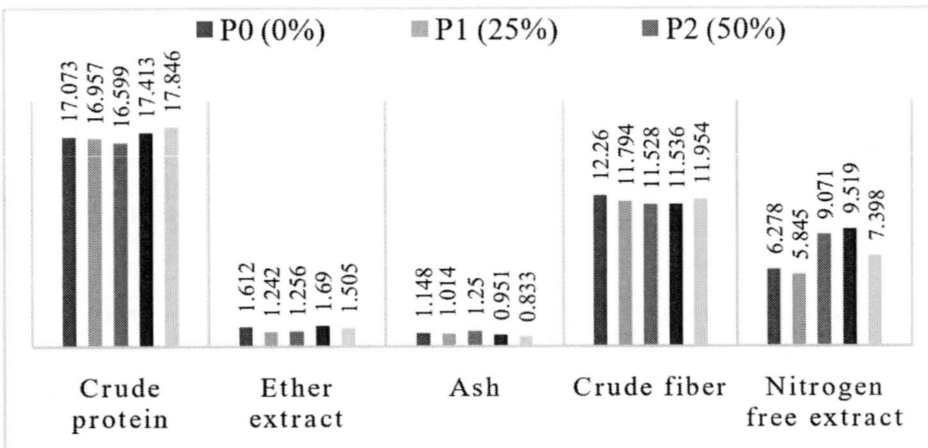


**Figure 1.** Proximatecomposition of eel meat as affected by different feed formula added with earthwormson the feed formulation against fish eel meat content.

### *3.2 Fatty acid profile*

There were no significant effect given by the different feed formula on saturated fatty acids, unsaturatedfatty acids, and omega 6 fatty acids of eel meat. Meanwhile, the treatments gave significant differences in EPA and DHA content of eel meat, where P4 had significantly higher EPA and DHA contents than P0 (table 1).

**Table 2.** Fatty acid profile of eel meat.

| FAMEs | Treatment (%) | | | | |
|---|---|---|---|---|---|
| | P0 | P1 | P2 | P3 | P4 |
| C12:0 Lauric acid | 0.347 | 1.003 | 0.624 | 0.397 | 0.830 |
| C14:0 Myristic acid | 5.291 | 5.417 | 5.507 | 4.942 | 5.465 |
| C13:0 Pentadecanoic acid | 0.665 | 0.696 | 0.629 | 0.701 | 0.739 |
| C16:0 Palmitic acid | 25.347 | 25.057 | 25.410 | 24.529 | 25.164 |
| C18:0 Stearic acid | 5.317 | 5.471 | 5.010 | 5.834 | 5.552 |
| ∑Saturated Fatty Acid (SFA) | 36.967 | 37.644 | 37.180 | 36.403 | 37.750 |
| C16:1Palmitoleic acid | 6.219 | 6.217 | 6.445 | 5.699 | 6.070 |
| C18:1 Oleic acid (ώ9) | 32.478 | 31.880 | 32.177 | 32.193 | 31.859 |
| C18:1 Elaidic acid | 3.894 | 4.179 | 3.978 | 3.898 | 4.099 |
| C20:1 cis 11 Eicosenoic acid | 2.268 | 2.648 | 2.521 | 3.403 | 2.638 |

IOP Conf. Series: Earth and Environmental Science **137** (2018) 012032 doi:10.1088/1755-1315/137/1/012032

| | | | | | |
|---|---|---|---|---|---|
| C22:1 Erucic acid | 0.293 | 0.256 | 0.372 | 0.458 | 0.273 |
| ∑Mono Unsaturated Fatty Acid (MUFA) | 45.152 | 45.180 | 45.493 | 45.651 | 44.939 |
| | | | | | |
| C18:2 Linoleic acid (LA) (ώ6) | 8.139 | 8.946 | 8.593 | 9.1635 | 8.661 |
| C20:4 Arachidonic acid (ώ6) | 1.432 | 1.481 | 1.314 | 1.488 | 1.392 |
| C20:5 Eicosapentaenoic acid (EPA) (ώ3) | $1.243^{a}$ | $1.014^{c}$ | $1.194^{ab}$ | $1.101^{abc}$ | $1.073^{bc}$ |
| C20: 3 Cis 8. 11. 14 Eicosatrienoic acid (hGLA) (ώ6) | 0.712 | 0.771 | 0.673 | 0.792 | 0.749 |
| C18:3 α Linolenic (ώ3) | 0.600 | 0.541 | 0.577 | 0.600 | 0.543 |
| C22:6 Docosahexaenoic (DHA) (ώ3) | $4.599^{a}$ | $3.257^{b}$ | $4.045^{ab}$ | $3.655^{ab}$ | $3.790^{ab}$ |
| ∑Poly Unsaturated Fatty Acid (PUFA) | 16.725 | 16.010 | 16.396 | 16.799 | 16.208 |
| ∑Unsaturated Fatty Acid (UFA) | 61.877 | 61.190 | 61.889 | 62.450 | 61.147 |
| SFA/UFA | 0.597 | 0.615 | 0.600 | 0.582 | 0.617 |
| UFA/SFA | 1.673 | 1.625 | 1.664 | 1.715 | 1.619 |
| ∑n3 | 6.442 | 4.812 | 5.816 | 5.356 | 5.406 |
| ∑n6 | 10.283 | 11.198 | 10.580 | 11.443 | 10.802 |
| n3/n6 | 0.626 | 0.429 | 0.549 | 0.468 | 0.500 |
| n6/n3 | 1.596 | 2.327 | 1.819 | 2.136 | 1.998 |
| EPA | $1.243^{a}$ | $1.014^{c}$ | $1.194^{ab}$ | $1.101^{abc}$ | $1.073^{bc}$ |
| DHA | $4.599^{a}$ | $3.257^{b}$ | $4.04^{5ab}$ | $3.655^{ab}$ | $3.790^{ab}$ |
| EPA/DHA | 0.270 | 0.311 | 0.295 | 0.301 | 0.283 |

P0 = commercial feed and earthworm (100%: 0%). P1 = commercial feed and earthworm (75%: 25%). P2: commercial feed and earthworm (50%: 50%). P3 = commercial feed and earthworms (25%: 75%). P4 = commercial feed and earthworm (0%: 100%)

## 4. Discussion

Figure 1 is in accordance with the results of Litzow *et al.* [10] who stated the fat content in fish feed should be about 15%.Fat content in eel meat is highly correlated with the content of essential fatty acids. Moreover, Kandemir and Polat [11] stated that the content of fatty acids in aquatic organisms can be influenced by the living condition, either wild in nature or in captivity. There was lack of linolenic acid found in feed formula although the fatty acid was found in the earthworms.

The fatty acid profile of eel meat as shown Table 2 is in accordance with Oku *et al.* [12] who reported fatty acid content of Japanese eel (*Anguilla japonica*) fresh meat consisted mostly of monounsaturated fatty acids (MUFA), while unsaturated fatty acids (PUFA) appeared in low amount. Variation of fatty acids in aquatic organisms can be influenced by seasons, geographical location, and environment salinity [13].

Different doses of earthworms in eel feed formula could increase EPA and DHA contents in eel meat. The content of the omega-3 fatty acids EPA and DHA was affected by the presence or absence of earthworms in the feed formulation. According to Robin *et al.* [14] stated that when feed isrich in omega-3 fatty, Then the fish meat composition would beinfluenced. This is in accordance with the results of Huang *et al.* [15] stating that fatty acids contained in fish meat is derived from the fatty acids consumed by the fish.

Omega 3 and omega 6 fatty acids arepolyunsaturated fatty acids (PUFA). Omega 6 in eel meat showed a higher percentage compared with omega 3 [1]. Extremely high Omega 6 can negatively affect the body. The number and ratio of omega 3 and omega 6 fatty acids are important to be considered in formulating fish feed. A good ratio of omega 3: omega 6 is 10:1, which means there should be higher omega 3 content compared with omega 6 [3]. The best ratio was found in treatment P1 (75 %: 25 %) that was 1: 1.8. The composition of fatty acids in feed formulation can affect the ratio of omega 3 and omega 6 [16].

Fatty acids n-3 and n-6 are required in fat biosynthesis, so that in the event of a shortage or excess of one of the fatty acids,it will inhibit the rate of biosynthesis of other fatty acids and eventually it will

IOP Conf. Series: Earth and Environmental Science **137** (2018) 012032 doi:10.1088/1755-1315/137/1/012032

affect the composition of fatty acids in fish. an imbalance ratio of omega 3 and omega 6 can lead to competition inutilizing enzymes in fat metabolism, which can affectgrowth. As a conclusion, the use of earthworms in eel feed did not significantly affect fatty acid profile in eel meat.

## 5. Conclutions

Additions of earthworm in eel feed formula increase EPA and DHA contents in eel meets. The best EPA and DHA ratio of commercial feed and earthworm ws oin treatment P1 (75 %: 25%).

## 6. References

[1]. Hameed A S, Hussain, Shabbir, Pasha and Song 2017 *Am. J. of Biochem. & Biotech.* **13** 15-26
[2]. Ozogul F and Ozogul Y 2006 *Food Chem.* **99** 574-578
[3]. Seo J S, Choi J H, Seo J H, Ahn T H, Chong W S, Kim S H, Cho H S and Ahn J C 2013 *Fish Aquat. Sci.* **16** 85-92
[4]. Kolanowski W and Laufenberg G 2006 *Eur Food Res Technol.* **222** 472-477
[5]. Simopoupus A. P 2002 *Biomed. Pharmacother.* **8** 36-79
[6]. Escott-Stump S and Mahan L K 2005 Krause, alimentos, nutricao and dietoterapia *Editoraroca* **9**
[7]. AOAC 2005 Official methods of analysis (18th edition) association of official analytical *Chemists International* maryland USA
[8]. Amer B, Caroline N, Hanne C B, Grith M, Kjeld H and Trine K D 2013 *Elsevier* **32** 199-203
[9]. Farhat J and Shakoor C 2011 *J. Food Chem.* **125** 991-996
[10]. Litzow M A, Bailey K M, Fredrick G and Prahl H R 2006 *Mar. Ecol. Prog. Ser* **315** 1-11
[11]. Kandemir S and Polat 2007 *Turkish J. Fish Aquatic Sci.* **7** 27-31
[12]. Oku T, Sugawara, Choundhury, Komatsu, Yamada and Ando 2009 *Food Chem.* **115** 436-440.
[13]. Ozyurt G O, Duysak, Akamca and Tureli 2006 *Food Chem.* **95** 382-385
[14]. Robin J H, Regost, Arzel and Kaulshik 2003 Fatty acid profile following a change in dietary fatty acid source model of fatty acid compossion with a dilution hypothesis **225** 10-13.
[15]. Huang W Y, Wu J T, Chiang Y R and Jane W N 2006 *Aquat. Toxicol.* **80** 38-45
[16]. Bae J Y, Han, Park and Bai 2004 *J. Aquacult.* **17** 275-281

IOP Conf. Series: Earth and Environmental Science **137** (2018) 012033 doi:10.1088/1755-1315/137/1/012033

# The effect of earthworms (*Lumbricus rubellus*) in feed formulation on growth and retention of eel (*Anguilla bicolor*)

**P C Jatmiko[1*], N AMadinah[1], Agustono[2] and T Nurhajati[3]**

[1] Aquaculture, Faculty of Fisheries and Marine , Airlangga University, Jl. Airlangga 60115, Surabaya, East Java, Indonesia.
[2] Management of Fish Health and Aquaculture Departement, Faculty of Fisheries and Marine, Airlangga University, Jl. Airlangga 60115, Surabaya, East Java, Indonesia.
[3] Livestock Department, Faculty of Veterinary Medicine, Airlangga University, Jl. Airlangga 60115, Surabaya, East Java, Indonesia.

E-mail: mikocahya@gmail.com

**Abstract.** Earthworms (*Lumbricus rubellus)* has high protein content. The addition of earthworms in formulation feed not only can increase the appetite of eel but also increase the nutritional content in feed. The purpose of this research was to know the potention of earthworms *L. rubellus*in feed formulation that can gives increase on the growth and retention. Research's method was using Complete Randomized Design (CRD) consisted of five treatments and four replication. Treatments in this research ware the different addition of earthworms *L. rubellus*in feed formulation which were 0 %, 25 %, 50 %, 75 % and 100 %. The result showed that there were significantly different on the growth and retention of eel during maintenance for 21 days. the best result was on the 100% of earthworms *L.rubellus*addition.

## 1. Introduction

Eel (*Anguilla bicolor*) has high economic and export commodity from the fisheries sector [1]. According to Afandi and Suhenda [2] eel needs feed with protein content more than 45%. The commercial feed mostly has protein content only 30-40 %, so to increase the protein content in feed, it can be done by adding high animal protein ingredients, such as trash fish, eggs, and earthworms (*Lumbricusrubellus*) [3].Earthworms contain 64-76 % of easily-digested protein that can be hydrolyed into amino acids that are useful for the body. Earthworms produce many enzymes and they have fishy smell that attracts its predators [4]. The use of proper attractant can stimulate fish to eat the feed quickly, thus giving additional nutrition, such a protein, and increasing the growth and feed efficiency [5]. Growth is closely related to protein and energy retention. Protein retention is the amount of protein that has been consumed and absorbed or utilized to build or repair the damaged body cells also for daily metabolism [6]. While energy retention is the amount of the saved energy in the form of tissues in the fish body that is divided by the amount of energy in the consumed feed [7]. Thus, this research was aimed to know the potential of

IOP Conf. Series: Earth and Environmental Science **137** (2018) 012033 doi:10.1088/1755-1315/137/1/012033

earthworms (*Lumbricus rubellus)* to be used in feed formulation to increase growth and retention value of eels.

## 2. Methodology

### *2.1 Preparation of earthworms (Lumbricus rubellus)*

Earthworms were obtained from Malang, East Java. It was then mixed with formulated feed (commercial feed and fish meal) as much as 0, 25, 50, 75, and 100 % in the form of paste. Feed was then analyzed for proximate analysis (table 1).

**Table 1.** Nutritional content of feed ingredients.

| Feed Ingridients | Dry Ingridients | Crude Protein (%) | Crude Fat (%) | Crude Fiber (%) | Ash (%) | NFE (%) |
|---|---|---|---|---|---|---|
| Formulation Feed | 80,029 | 42,067 | 13,011 | 6,806 | 14,228 | 3,916 |
| Earthworms | 22,896 | 13,634 | 5,749 | 0,548 | 1,589 | 1,376 |

### *2.1.1 Experimental design*

About 100 Eels that has been used were at fingerling state with average weight 20-25 gs each as much as 100 eels that came from Malang, East Java. Eels were selected and than acclimatized for about 30 minutes. After that eels were weighed and measured than divided and placed in the aquarium each aquarium contained five eels. During 21 days eels were feeded with treatments feedthat were 0%, 25%, 50%, 75% and 100% of earthworms. Eel's media constantly were controled and also dark or low light. Every week eels were taken to be weighed and measured for the growth rate, feed effiency and feed convertion ratio.

### *2.1.2 Fish meat compotitions*

Crude protein content, crude fat, crude fibber, ash, and nitrogen-free extract of eel's meat were proximately analyzed based on [8].

### *2.1.3 Growth Rate (GR)*

Measurement of growth rate used the formula from [7]:

$$GR = \frac{Wt - W_0}{t} \quad (1)$$

Description :
GR = Growth Rate (g/day)
$W_t$ =Final Weight (g)
$W_0$ = First Weight (g)
T = Time (day)

### *2.1.4 Feed Convertion Ratio* (FCR)

[8] stated that the feed conversion ratio could be calculated by using this formula:

$$FCR = \frac{F}{Wt + D - W_0} \quad (2)$$

Description :
FCR = Feed convertion ratio
F = Consumed Feed (g)
$W_t$ = Final Weight (g)
D = Weight of dead fish (g)
$W_0$ = First Weight (g)

*2.1.5 Feed Efficiency* (FE)
Feed efficiency was done by using thefollowing formula [10]:

$$EP = \frac{\mathrm{Wt} - \mathrm{W_0}}{\mathrm{F}} \text{ x } 100\% \quad (3)$$

Description :
FE = Feed Efficiency
F = Consumed Feed (g)
$W_t$ = Final Weight (g)
D = Weight of dead fish (g)
$W_0$ = First Weight (g)

*2.1.6 Protein Retention* (PR)
The measurement of protein retention used:

$$GR = \frac{(\mathrm{Wt} - \mathrm{W_0})\ \mathrm{g}}{\mathrm{Consumed\ \ feed'sprotein\ \ (g)}} \text{ x } 100\% \quad (4)$$

Description :
$W_t$ = Final Weight (g)
$W_0$ = First Weight (g)

*2.1.7 Energy Retentio* (ER)
The measurement of energy retention was as follows :

$$ER = \frac{(\mathrm{Et} - \mathrm{E_0})\mathrm{kkal}}{\mathrm{Energy\ \ total\ \ of\ given\ \ feed\ \ (kcal)}} \text{ x } 100\% \quad (5)$$

Description :
$E_t$ = Final Body Energy
$E_0$ = First Body Energy

IOP Conf. Series: Earth and Environmental Science **137** (2018) 012033 doi:10.1088/1755-1315/137/1/012033

*2.1.8 Statistical Analysis*

The experiment was conducted following complete randomized design (CRD) and data collected were analyzed by analysis of variance (ANOVA). If there were significant differences, then it was continued with Duncan's multiple range test with the level of confidence of 0.05 by using SPSS v.16.

## 3. Results And Discussion

Growth is the increase of length or weight in certain period. In this research, parameters that were used to measure growth were weight and length. After being treated for 21 d with five different treatments, Eels generally increased their weight and length (table 2). The highest weight increment was from treatment P4. However, it was not significantly different than P3. However, P4 was significantly different from P0, P1, and P2. The difference in growth rate was due to the amount of amino acid in given feed. According to Gusrina [11] stated that growth can be reached if feed contains optimal amino acid, while feed with lack of amino acid balance can decrease fish ability to grow., balance of amino acids can be obtained by using combination of animal and vegetable proteins. So, if the feed has good nutritional profile, it will increase the growth well [12].

ANOVA analysis showed that there was significant differences among treatments and the best result was from Treatment P4 with 100 % of earthworms, as can be seen in Table 2. the lowest growth was seen in Treatment P1 with 25 % of earthworms. According to Hariyadi *et al.* [13], earthworms (*Lumbricussrubellus)* has fishy smells that will stimulate the appetite of fish. Eels are fish that eat on the bottom of the pond and they are active and places with low light density or at night (nocturnal). Smell and touch sensor are used by eels to look for food [14]. One of the criteria of feeding for nocturnal fish is by using feed that has odor stimulation [15]. Carnivore fish has the ability to utilize protein better than carbohydrates and fiber. Therefore, fish can utilize the energy to growby digesting more protein than carbohydrates and fiber.

Feed conversion ratio (FCR) is a ratio of the amount of feed needed to obtain specific weight [16]. Higher value of feed conversion ratio shows that the feed does not effectively affect the growth of the fish. The best feed conversion ratio was gained from Treatment P4 (2.55) with 100 % of earthworms. Earthworms are easilydigested by fish[11]. However, in Treatment P0 with the 100 % of earthworms, the FCR was the lowest (7.35). This was related to the ability of fish in digesting the feed. on the feed in Treatment P0 was not 100 % consumed. Treatment P4 gave the best result of 2.55 that means 2.25 kg of feed was needed to give one kg of weight increase. Feed is not 100 % turned into meat, but some of it is used for energy, metabolism, etc. According to Lesmana [17], feed consumed by fish is used for production of meat, energy metabolism, and energy to move.

**Table 2.** Growth Rate (GR), Feed Convertion Ratio (FCR) and Feed Efficiency (FE).

| Treatments | Measured Parameters | | |
|---|---|---|---|
| | GR(gs/day) | FCR | FE(%) |
| P0 (0 % of earthworms) | $0{,}17^{bc} \pm 1{,}03$ | $7{,}35^{d} \pm 0{,}42$ | $13{,}5^{c} \pm 0{,}57$ |
| P1 (25 % of earthworms) | $0{,}12^{c} \pm 0{,}005$ | $5{,}77^{bc} \pm 0{,}46$ | $16{,}7^{bc} \pm 1{,}50$ |
| P2 (50 % of earthworms) | $0{,}17^{bc} \pm 0{,}02$ | $4{,}06^{c} \pm 0{,}51$ | $24{,}2^{c} \pm 3{,}59$ |
| P3 (75 % of earthworms) | $0{,}25^{ab} \pm 0{,}09$ | $3{,}48^{b} \pm 0{,}81$ | $29{,}5^{ab} \pm 7{,}50$ |
| P4 (100 % of earthworms) | $0{,}33^{a} \pm 0{,}10$ | $2{,}55^{a} \pm 0{,}88$ | $39{,}0^{a} \pm 11{,}16$ |

Description :different alphabet of superscript shows significantly difference (P < 0,05)

Feed efficiency is a comparison value between weight increase with consumed feed [12]. The higher the feed efficiency, the faster the growth is [18]. The highest value of feed efficiency came from

Treatment P4 with 100% of earthworms and it gave efficiency value as much as 39.0 %. Feed efficiency can be seen from some factors and one of them is feed conversion ratio [19]. Factors that affects feed efficiency value are types of nutritional source and the nutritional content in the feed [18].

Result from table 3 shows that there were no significantly difference (P>0,05) between P0 P1 P2 P3 and P4 treatments. Undifferent result can be state that each treatment with feed formulation and earthworms *(Lumbricus rubellus)* gave same effect on the protein retention. This happened because the difference of protein content of the given feed has no difference. Beside, generally protein content in the fish body is always kept to remain stable. This matter is suitable with Buwono (2000) who states that protein retention is an image of the amount of protein that has been given and to be asbrobed or utilized to build or repairs the damaged body cells also for daily metabolism. The lowest energy retention value was on P0 treatment with value of 6.6600 and the highest value was on P4 treatment with value of 23.6900. Increasing of energy retention can be happened because in earthworms *(Lumbricus rubellus)* there is leucine that one of essential amino acids that can be synthesis by body so it has to be added in feed. In this reseach was using earthworms. By proper feeding and proper dogase will increse the energy retention, because on fish the main energy source is protein.

**Table 3.** Protein and energy retention of eels (*A.bicolor*) after treatment.

| Treatments | Measured Parameters | |
|---|---|---|
| | Protein Retention | Energy Retention |
| P0 (0 % of earthworms) | 8.8150 $^{a}$ ± 4.02946 | 6.6600 $^{b}$± 4.2823 |
| P1(25 % of earthworms) | 14.6775 $^{a}$ ± 7.39997 | 7.0575 $^{b}$± 3.17454 |
| P2(50 % of earthworms) | 11.1925 $^{a}$ ± 5.30501 | 17.4700$^{ab}$± 14.33125 |
| P3(75 % of earthworms) | 21.1000 $^{a}$ ± 17.5551 | 23.0025$^{a}$± 11.75915 |
| P4(100 % of earthworms) | 27.8075 $^{a}$ ± 18.49406 | 23.6900$^{a}$± 6.11639 |

Description : same alphabet of superscript shows no significantly difference (P > 0,05), different alphabet of superscript shows significantly difference (P < 0,05).

Table 3 shows that there were no significant differences (P > 0.05) among treatments, due to the lack of difference in protein content of the given feed. Besides that. protein in the fish body is generally always kept stable. According to Buwono [6] stated that protein retention is the amount of protein given with the absorbed or utilized protein, which is used to build or repair the damaged body cells as well as for daily metabolism. The lowest energy retention value was from Treatment P0 (6.6600) and the highest value was from treatment P4 (23.6900). Increased energy retention was due to the existence of leucine, one of the essential amino acids, in earthworms. proper feeding will increse the energy retention because the main energy sourcefor fish is protein.

## 4. Conclusion

This study has demonstrated that earthworms (*Lumbricus rubellus*) can be potentially used in formulation of feed to improve the growth and retention of eels (*Anguilla bicolor*).

## 5. References

[1] Affandi R 1999 Environmental Optimization and Nutrition on Local Fowling Farming Anguilla bicolor LPPM IMB Bogor
[2] Affandi R and N Suhenda2003 Cultivation Technique of Sidak Fish (*Anguilla bicolor) Prosiding Sumberdaya Perikanan Sidat Tropik*
[3] Afrianto Eand E Liviawaty2005*Feed fish* Kanisius Yogyakarta pp 9-77

IOP Conf. Series: Earth and Environmental Science **137** (2018) 012033 doi:10.1088/1755-1315/137/1/012033

[4] AOAC 2005Official Methods of Analysis (18th edition) Association of Official Analytical,Chemists International, Maryland, USA
[5] Arief M, N Fitriani and S Subekti 2014 *J. Fis. Mar.* **6** 49-53
[6] Buwono I D 2000 *Essential Amino Acid Requirement in Fish Rations* (Yogyakarta: Kanisius) p 52
[7] Buwono I D 2010 *Essential Amino Acid Requirement in Fish Feed Rations* (Yogyakarta: Kanisius) pp 24-39
[8] BSNI 2006Artificial feed for eel fish (*Angilla* spp) on intensive cultivation
[9] Djarijah A 1995 *Natural Fish Feed* (Yogyakarta: Kanisius) pp 21-25
[10] Effendie M I 1997 *Fisheries Biology* (Yogyakarta: Yayasan Pustaka Nusatama) pp 39-73
[11] Gusrina 2008F ish cultivation Direktorat Pembinaan Sekolah Menengah Kejuruan Departemen Pendidikan Nasional pp 167-249
[12] Hariati A M 1989 Fish food Diktat Kuliah Universitas Brawijaya Malang
[13] Hariyadi B, Haryono and U Susilo 2005 Evaluation of Food Efficiency and Protein Efficiency in Grass Carp Fish (Ctenopharyngodon idella) feed with carbohydrate and Different Energy. Fakultas Biologi Unseod **4** 87-92
[14] Haryono 2008 Eel Eel: Potential and Aspects of Budidayanya Fauna Indonesia **8** 22-26
[15] Herawati, Vivi Endar 2005 Developing the Course Prog Course of Fish Feeding Management Fakultas Perikanan and Kelautan Universitas Diponegoro Semarang pp 4-16
[16] Herianti I 2005 Environmental Engineering To Spur The Development Of Ovarian Fish Sidat Oscanologi and Limnologi Indonesia. **3** 725-41
[17] Lesmana D S 2005*Water Quality For Fresh Water Fish*. (Jakarta: Penebar Swadaya). pp 7-25
[18] Mudjiman 2000 *Fish Food* (Jakarta: CV Simplex)
[19] Mudjiman A 2002 *Fish Food* (Jakarta: Penebar Swadaya) pp 88-89
[20] Palungkung, R. 2010. *Earthworms Livestock* (Jakarta: Business Swadaya)
[21] Polat A and G Beklevik 1999 The Importance Of Betaine and Some Attractive Substances as Fish Feed Additives *In*:Brufau, J. and A, Tacon (Eds) *Feed Manufacturing In The Mediterranean Region: Recent Advances in Research and Technology Zaragoza CIHEAM*, IAMZ, Spain. pp: 217-220
[22] Purwanto J 2007 *J. Aquac* **6** (in Indonesia)
[23] Sasongko A, J Purwanto, U Arie 2007 *Sida*t (*Anguilla bicolor*) (Jakarta; Penebar Swadaya) pp 5-74
[24] Suitha I M and A Suhaeri 2008 *Farming Sidat* (*Anguilla bicolor)* (Jakarta: PT Agromedia pustaka) pp 1-27
[25] Tacon, F W 1997The Nutrision And Feeding Of Farmed Fish *The Essensial Nutrients Food And Agriculture Organization* The United Nations, Brazil
[26] Waluyo J, B Sugiharto and NC Zaini 2007. *Basic Sci.* **8** 37-40 (in indonesia)

IOP Conf. Series: Earth and Environmental Science **137** (2018) 012034 doi:10.1088/1755-1315/137/1/012034

# Increasing β-carotene content of phytoplankton *Dunaliella salina* using different salinity media

**J Hermawan[1], E D Masithah[2], W Tjahjaningsih[2] and A A Abdillah[2]**

[1]Aquaculture undergraduated, Faculty of Fisheries and Marine, Airlangga University, Jl. Airlangga 60115, Surabaya, East Java, Indonesia.
[2]Departement of marine Faculty of Fisheries and Marine, Airlangga University, Jl. Airlangga 60115, Surabaya, East Java, Indonesia.

E-mail: annur.ahadi@fpk.unair.ac.id

**Abstract.** *Dunaliella salina* have got great attention in the nutritional, pharmaceutical and cosmetic companies because contain β-carotene. β-carotene functions as antioxidants and precursors of vitamin A and can treat tumors and cancer in humans. The content of β-carotene in *D. salina* can be increased by increasing salinity levels in the culture medium. The aim of this study was to determine whether increasing salinity may increas β-carotene content of phytoplankton *D. salina*. The research use data collection method with direct observation and then analyzed the result with descriptive method. The results showed that different salinity of media can influenced β-carotene content of *D. salina*. The highest β-carotene content of *D. salina* was at treatment B (30 ppt) which equal to 2.312 mg/L on 10$^{th}$ day. The production of β-carotene in *D. salina* can be increased was other environmental stress treatments in the form of stress-temperature, light and nutrients using.

## 1. Introduction

*Dunaliella salina* is one of the most potent natural feeds as used feed additive and feed supplement in aquaculture [1]. Based on research conducted by abd EL-Baky [2], *D. salina* accumulates a high amount of carotenoids (12.6 % of the dry weight) which comprising β-carotene (60.4 %), α-carotene (17.7 %), zeaxanthin (13.4 %), lutein (4.6 %) and cryptoxanthine (3.9 %). Salinity is one of environmental stressors that can affect the increase of β-carotene content of *D. salina*. Based on the above discussion, this study was expected to see whether the difference of salinity concentration can influence the increase of β-carotene production in *D. salina* , so that the best salinity concentration can be identified to obtain the maximum amount of β-carotene in *D. salina*.

## 2. Methodology

### *2.1. Research method*

The phytoplankton culture of *D. salina* was conducted in Education Laboratory of Faculty of Fishery and Marine, Airlangga University, Surabaya. Meanwhile, the measurement of β-carotene content was carried out at Pharmacy Testing, Airlangga University, Surabaya. The research data were collected through direct observation, which is an approach or technique to get primary data by observing research objects directly [3]. Data obtained from culture process and calculation of β-carotene content were then analyzed using descriptive method [4] .The results are presented in tables and figures.

*2.2. Materials*
The equipments used in this research were glass tube, aeration hose, aerator, reaction tube, reaction tube shelf, measuring cylinder, digital balance, dropper pipette, micro pipette, binocular microscope, haemocytometer, scissors, cuvet, handtally counter, spectrophotometer, culture shelf, refractometer, thermometer, pH pen, lux meter, autoclave, glass cover, vortex, centrifuge, dark plastic, styrofoam, 40 Watt lamp, and aluminum foil. The materials used in this research were *D. salina* seed and Walne medium (Brackish Water Cultivation Center (BBAP) Situbondo), clean seawater from Kenjeran Surabaya (30 ppt), aquades, petroleum ether (PE) alcohol, chlorine and Na-Thiosulfate.

*2.3. Research procedure*
*2.3.1. Equipment and material preparation*
The heat-resistant equipment was sterilized with autoclave at 121 °C for 15 minutes. Heat-resistant apparatus was also sterilized by immersion in a chlorine solution and neutralized with Na-Thiosulfate. Sterilization was also done on the culture medium and sea water. Before being sterilized, the salinity level of the sea water was measured using refractometer. Seawater that had been treated was further sterilized by giving a solution of chlorine and neutralized using a solution of Na-Thiosulfate [5].

*2.3.2. Culture media for Dunaliella salina*
The culture of *D. salina* in this study was incubated at 25-30°C with light intensity of 1200 lux using 40 Watt lamps with a dark period of 12 hours and light 12 hours [6].

*2.3.3. Seedling Dunaliella salina*
Planting of *D. salina* seeds was done after calculating stock density by taking samples of plankton from the stock medium and counting under a microscope with haemocytometer to the sample population calculation. Seed *D. salina* is fed to a medium with a density of 105 cells/ml [7].

*2.3.4. Harvest of Dunaliella salina*
The harvesting of phytoplankton *D. salina* was done partially on the 2$^{nd}$, 4$^{th}$, 6$^{th}$, 8$^{th}$, 10$^{th}$, 12$^{th}$ and 14$^{th}$ days [6]. The harvesting process of *D. salina* was done up to the stationary phase. Harvesting *D. salina* was carried out by putting 10 ml samples into the reaction tube. It was then stored in the refrigerator at 5°C-15°C [5]. After that, the β-carotene content in *D. salina* was calculated.

*2.3.5. Extraction of β-carotene from Dunaliella salina*
Extraction of β-carotene content was carried out by taking 1 ml of *D. salina* yield using pipette 8 ml of distilled water was added and homogenized using vortex. The mixture (2 ml) was added to the test tube contained 2 ml of 96 % alcohol and 10 ml of petroleum ether (PE). They were then homogenized with vortex and centrifuged for 5 min. After centrifugation, two layers were formed was the lower supernatant and petroleum ether layer (PE) at the top. The petroleum ether (PE) layer formed was separated and marked as layer I while the remainder was added again with 10 ml petroleum ether (PE). The mixture was homogenized again for 2 minutes using a vortex and then centrifuged for 5 minutes. The petroleum ether (PE) layer formed was separated again and marked as layer II. Layers I and II were homogenized 2 ml of each layer was then taken and read on spectrophotometers with wavelengths of 480 nm, 645 nm and 663 nm [8].

*2.3.6. β-carotene count on Dunaliella salina*
The calculation of β-carotene content in phytoplankton *D. salina* was determined by using spectrophotometer. The extraction results of the petroleum ether (PE) layer formed subsequently were incorporated into the cuvet and read using a spectrophotometer with blanks ie petroleum ether (PE) solution [9]. The calculation formula of β-carotene content in *D. salina* [10] is as follows:

$$\beta\text{-carotene (mg/L)} = \frac{(\text{Abs } 453 - \text{Abs } 665)}{3.91} \times 3.657 \times 3 \times X \quad (1)$$

IOP Conf. Series: Earth and Environmental Science **137** (2018) 012034 doi:10.1088/1755-1315/137/1/012034

Note :
Abs 453 : Absorbance at 453 nm
Abs 665 : Absorbance at 663 nm
3,657 : Calibration factor on β-carotene content analysis
*X* : Dilution factor for measuring absorbance in spectrophotometer

## 3. Result and discussion

The result of the calculation of β-carotene content was obtained by using the specimen. Samples were taken on the 2nd, 4th, 6th, 8th, 10th, 12th and 14th day.

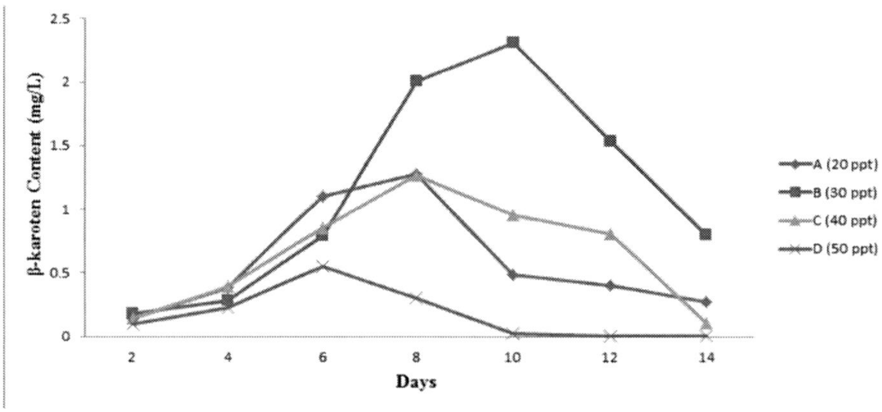


**Figure 1.** Average of β-carotene content of *D. salina* cultured with different salinity.

Based on the average graph of β-carotene content (figure 1), it can be seen that the highest β-carotene content was in treatment B (30 ppt) i.e: on the 10th day of 2.312 mg/L, while treatment A (20 ppt) and C treatment (40 ppt) occurred on day 8 of 1.275 mg/L and 1.2648 mg/L and treatment D (50 ppt) of the highest β-carotene content occurred on the 6th day of 0.5525 mg/L.

Salinity is one of environmental stressors that affects the increase of β-carotene accumulation in *D. Salina*. Higher salinity (hypertonic) may cause shrinkage of cells and lower salinity (hypotonic) may result in cell swelling [6]. According to Ramos *et al.* [11], *D. salina* has two ways to defend itself from the salinity stress, that is by producing glycerol in the short term and producing carotenoids in the long term.

The results of the research show that the highest β-carotene content in *D. salina* was found in treatment B (30 ppt), that was 2.312 mg/L on the 10th day, while the β-carotene content in normal treatment was A (20 ppt), having the highest amount of 1.275 mg/L obtained on day 8. This shows that salinity can increase the β-carotene content in *D. salina* . It was proven that the content of β-carotene in treatment B (30 ppt) is higher than that of treatment A (20 ppt) as control or normal condition.

According to Shariati and Hadi [12], the content of β-carotene in *D. salina* can be increased with environmental stress of one salinity; the higher the salinity given, the higher the β-carotene content. However, in the C treatment (40 ppt) and D treatment (50 ppt), there were noticeable decreases in the total β-carotene content of 1.2648 mg/L and 0.5525 mg/L obtained. It is suspected that the *D. salina* microalgae have tolerance limits to the salinity stress given, so that too high salinity stress conditions can affect the decrease of β-carotene content in *D. salina* due to plasmolysis. The growth of *D. salina* is calculated daily for 14 days of maintenance by means of a haemocytometer device.

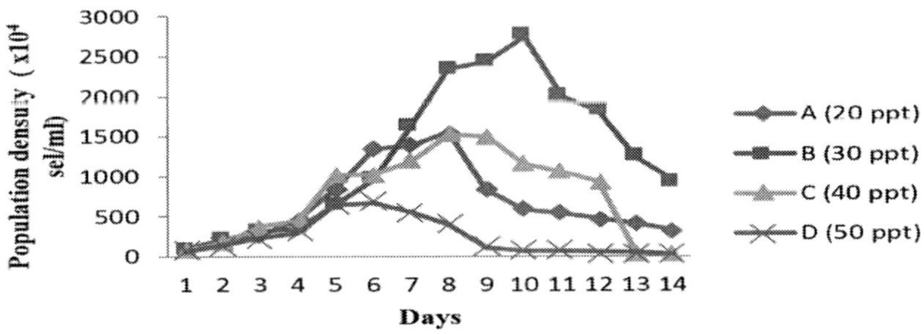


**Figure 2.** Average population density *D. Salina.*

All treatments had lag phase on day 1. According to Ayustama and Sari [13], in the phase of the lag, microalga cell was still in the adaptation stage as an effort to adjust it self to the changing conditions of the initial media environment to the new media. The exponential phase occured after the lag phase i.e.: on the 2nd day until the 7th day (for treatment A and C) and the 2nd day until day 9 (for treatment B) and the 2nd day until the 5th day (for treatment D), *Dunaliella salina* in this exponential phase grew rapidly.

The stationary phase occured at day 8 (treatment A and C), the $10^{th}$ (treatment B) and the 6th (treatment D). This phase occured because the nutrients in the media had been greatly reduced so that it was insufficient for growth and cell division [14]. The phase of culture death occured after the stationary phase at each treatment until the 14th day of maintenance. This phase is characterized by the color change of the culture water, the froth surfaces of the media culture and faded colors and clumps of algae cells that settle based on culture [13].

Based on the research, it was found that the highest density was obtained at treatment B (30 ppt) (figure 2), which was 6154.3 x 104 cells/ml on day 10, while treatment A (20 ppt) and C (40 ppt) had peak density on day to-8 with respective densities of 1555.85 x 104 cells/ml and 1523.1 x104 cells/ml and treatment D (50 ppt) which experienced peak density at day 6 of 672.5 x 104 cells/ml.

Figure 2 also showed that at 30 ppt salinity, the amount of *D. salina* cell density increased compared to other treatments (20 ppt, 40 ppt and 50 ppt) because at salinity of 30 ppt, this *D. salina* microalgae could still tolerate salinity stress given so that the growth tended to be normal as an adaptation system in the face of salinity stress *D. salina* forms β-carotene. Meanwhile, the amount of density of *D. salina* cells in salinity 40 ppt and 50 ppt tended to decrease because *D. salina* microalgae have plasmolysis or cell rupture due to too high osmotic pressure which cannot be tolerated. It affects cell death so that the production of β-carotene also decreased.

Water quality measurements were performed daily in the morning and afternoon during the maintenance period. Water quality parameters measured were temperature, pH, salinity, and dissolved oxygen (DO) (table 1).

**Table 1.** Water Quality Range during Maintenance Period of *D. Salina.*

| Parameter | Result | Satuan |
|---|---|---|
| Temperature | 28,2-30,8 | °C |
| pH | 7,5-8,6 | - |
| Salinity | 20-50 | ppt |
| *Dissolved Oxygen* (DO) | 6,3-6,9 | mg/L |

Based on the research results, the average temperature data ranged from 28.2 to 30.8 °C. The optimum temperature range can be regarded as being suitable for *D. salina* environment. was between 1 °C-30°C [15]. The mean pH value in the study ranged from 7.5 to 8.6. Based on these results, it can be assumed that the pH value is under normal conditions. According to Tafreshi and Shariati [16], *D. salina* has a high pH tolerance of 0-11 and the optimum pH for growth culture of *D. salina* microalgae is between 9-11. the mean DO values in the study ranged from 6.3-6.9 mg/L. The range of DO can be categorized suitable for the environment of *D. salina*. The optimum Dissolved Oxygen (DO) value for the growth of *D. salina* is between 6-8 mg/L [16].

## 4. Conslutions

Differences in salinity concentration can increase the content of β-carotene in *D. salina* . The best salinity to obtain the highest β-carotene content in *D. salina* was 30 ppt with the amount of β-carotene of 2.312 mg/L on day 10.

## 5. References

[1] Zainuri M H P Kusumaningrum and E Kusdiyantini 2006 Microbiological and Ecophysiological Characterization of Green Algae *Dunaliella* sp. for Improvement of Carotenoid Production. Faculty of Fisheries and Marine Sciences. Diponegoro University pp 1-1

[2] Abd El-Baky, H, F K El Baz and G S El-Baroty 2007 *J. Agric Environ.* **2** 792-800

[3] Djaelani A R 2013 *Veteran Semarang Sci Magazine Pawiyatan* pp1-11

[4] Dharminto D 2007 Research Method and sampel research http://eprints.undip.ac.id/ 17/12/2013 p 8

[5] Isnansetyo A dan Kurniastuty 1995 P*hytoplankton dan Zooplankton Culture Tecnique* (Yogyakarta: Kanisius) pp 40-83

[6] Pisal D S and S S Lele 2004 *Indian J. Biotechnol.* **4** 476-483

[7] Mamduh, A, E D Masithah and M A Alamsjah 2013 The Influence of *Azzola pinnata* Fertilizer on Chlorophyll Content In *Dunaliella salina* Thesis Faculty of Fisheries and Marine Affairs Airlangga University Surabaya p 73

[8] Nielsen, S 2009 *Food Analysis Fouth Edition.* (USA: Purdue University West Lafayette) 195 p.

[9] Dhanam, D S and K Dhandayuthapani 2013 *Int J Curr Microbiol Appl Sci.* **2** 37-43

[10] Kleinegris, D M M, M Janssen, W A Brandenburg, R H Wijffels 2011 *Journal Elsivier.* **48** 253-259

[11] Ramos, A A, J Polle, D Tran, J C Cushman, E Jin and J C Varela 2011 The Unicellular Green Alga Teod as a Model for Abiotic Stress Tolerance: Genetic Advances and Future Perspectives. Algae *The Korean Society of Phycol.* **26** 3-20

[12] Shariati, M and M R Hadi 2011 Microalgal Biotechnology and Bioenergy in *Dunaliella. In :* A. Carpi (Ed) Progress in Molecular and Environmental Bioengineering-From Analysis and Modeling to Technology Applications Intech Europa pp 483-506

[13] Ayustama, A L S and E A W Sari 2011 Microalgae Production Process in Banyan Pond Mini Photobioreactor for Biodiesel Fuel *Thesis* Faculty of Engineering. Diponegoro University 65 p

[14] Prihantini, N B, B Putri and R Yuniati 2005 *Makara J Sci.* **9** 1-6

[15] Ben-Amotz, A, J E W Polle and D V S Rao 2009 *The Alga Dunaliella : Biodiversity, Physiology, Genomics and Biotechnology* (America:Science Publisher) pp 357-493

[16] Tafreshi, A H and Shariati 2009 *J Appl Microbiol.* **107** 14-35.

**Acknowledgments**

The authors would like to thank the Faculty of Fisheries and Marine, University of Airlangga for funding the research.

IOP Conf. Series: Earth and Environmental Science **137** (2018) 012035 doi:10.1088/1755-1315/137/1/012035

# Abnormalities of hybrid grouper (*Epinephelus fuscoguttatus* x *Epinephelus lanceolatus*) in Situbondo

**J Triastuti[1], K T Pursetyo[1], A Monica[1], L Lutfiyah[2] and D S Budi[2]**

[1]Departement of Marine, Faculty of Fisheries and Marine Airlangga University, Jl. Airlangga 60115, Surabaya, East Java, Indonesia.
[2]Departement of Fish Health Management and Aquaculture, Faculty of Fisheries and Marine, PSDKU Banyuwangi of Airlangga University, Jl. Airlangga 60115, Surabaya, East Java, Indonesia.

E-mail :jt.astuti@gmail.com

**Abstract.**Grouper is one of consumption fish which is demanded excessively by local consumers and foreign consumers. Hybridization of grouper has been performed considerably that produce the good genetic quality of hybrid variants. One of grouper fish which has good genetic in its growth is kertang grouper fish. Nowadays many hatcheries performing hybridization between kertang grouper fish and tiger grouper fish, however observation of the hybrid abnormality has not been performed yet. Abnormality is able to increase since genetic causes, so that observation of abnormality occurrence in cantang hybrid grouper fish in Situbondo, JawaTimur, Indonesia in May – July in 3 times grading of juvenile stadia was performed. Results showed abnormalities were observed on mouth and operculum, branched of neural arch, fusion of neural arch, fusion of posterior truncus vertebrae, fusion of caudal vertebrae, fusion of anterior truncus vertebrae.

## 1. Introduction

Groupers are one of the main marine aquaculture commodities. Although they are relatively expensive,but there is a high demand for it. Hybridization carried out on various species of grouper has resulted in several types of hybrid groupers, which have more advantages than its original type. Cantang grouper is a type of hybrid grouper cultured by fish farmer.Itis the result of hybridization between tiger grouper and kertanggrouper The main advantage ofcantanggrouper is its fast growth with low FCR compared to tiger grouper orkertang grouper. The interest from fish farmers for grouper breeding is increasing, but the occurrence of abnormalities in the hybrid variety had never been observed. Abnormalities may increase due to genetic and environmental factors. Many factors still need to be assessed to suppress abnormalities, such as genetics, the environment, diseases, treatment, and maintenance conditions [1,2,3,4,5]. Therefore, it is necessary to conduct a study on type of abnormalitiesin hybrid grouper that is cultivated in hatcheries in Situbondo, East Java.

## 2. Methodology

This research was carried out in cantanggrouper hatchery located in Situbondo, East Java. The survey was conducted for 3 months, from June to August 2017. Materials used included KOH 35% and 40%,

IOP Conf. Series: Earth and Environmental Science **137** (2018) 012035 doi:10.1088/1755-1315/137/1/012035

glycerin,Alizarin Red S, alcian blue, 96 % alcohol, and fish feed. Observations were made including juvenile maintenance, grading at several juvenile ages, and seed harvesting. Juvenile maintenance was started 1 month after hatching (D30), which was the time for early grading of juvenile maintenance. Grading was conducted in 3 stages, when the juveniles aged 60 days; 67 days, and 74 days. Each stage of grading was carried out by taking sample as much as 30 % of the population to know the type of abnormality that had occurred. The type of disability occurred was used as additional information .The bone and cartilage staining was conducted using Alizarin Red S and alcian blue to provide a picture of abnormality that occurs in the vertebra and fins of cantang grouper .

## 3. Results and Discussion

### *3.1. Results*

Survey was conducted on juvenile maintenance and divided into 3 times of grading. The first grading was conducted when the juveniles entered the age of 60 days, which consisted of 3 size variations, the first category of size was 1.8-2 cm with population density of 150.000. The second size category was2.1-2.3 cm with a population density of 150.000. The third size category was 2.4-2.5 cm with a population density of 200.000. The second grading was conducted1 week after the first grading (Day 67), consisting of 3 variations of size: 2.4-2.6 cm; 2.7-2.8 cm; > 2.9 cm, with a population density of 100.000. The third grading was conducted 1 week after the second grading (Day 74) covering 3 variations of size: 2.6-2.8 cm; 2.9-3.1 cm, and 3.2-3.3 cm, with a population density of 100.000. Abnormalities of hybrid grouper are presented in table 1.

**Table 1**. Type of hybrid grouper abnormality (*Epinephelus fuscoguttatus* x *Epinephelus* sp.) cantang in Situbondo- East Java.

| Grading | Dense Stocking | Size of Juvenile (cm) | Type of Abnormality |
|---|---|---|---|
| I | 150.000 | 1.8-2 | Mouth abnormality.opeculum abnormality. |
| I | 150.000 | 2.1-2.3 | Mouth abnormality.opeculum abnormality. |
| I | 150.000 | 2.4-2.5 | Mouth abnormality.opeculum abnormality. neural arch fusion |
| II | 100.000 | 2.4-2.6 | Branches of neural arch. neural arch fusion |
| II | 100.000 | 2.7-2.8 | Branches of neural arch |
| II | 100.000 | >2.9 | Fusion of posterior truncus vertebrae. fusion of anterior truncus vertebrae |
| III | 100.000 | 2.6-2.8 | Fusion of posterior vertebrae. fusion of anterior truncus vertebrae |
| III | 100.000 | 2.9-3.1 | Fusion of caudal vertebrae. fusion of anterior truncus vertebrae |
| III | 100.000 | 3.2-3.3 | Fusion of posterior vertebrae |

IOP Conf. Series: Earth and Environmental Science **137** (2018) 012035 doi:10.1088/1755-1315/137/1/012035

### *3.2. Discussion*

The survey results showed that abnormality incidence consisted of several types of disabilities in several different sizes. Abnormality is an irreversible irregularity that is caused by changes in quality of morphology, etiology, and physiology [6]. On the other hand, abnormality comes from mutations and teratogenicity effects damage environmental factors (water temperature, oxygen solubility, salinity, and parasitic infections) [7]. Abnormalities of the chepal (mouth and operculum) occurred in smaller fish (1.8-2.5 cm)and vertebral defects occurred in larger fish (> 2.5 cm). The abnormality of oral performance might cause the fish to not respond to feeding properly, which may lead to death. Abnormalities in the mouth were usually found just after the fish grew into spindles. Allegedly, other than genetic factors, abnormality can also be caused by nutritional factors of the feed that can lead to malnutrition. , which triggered the occurrence of abnormalities in the mouth [8]. The defective fish had difficulty in getting the feed, so the fish growth was slow with low endurance that made them easily get attacked by disease [9]. Poor aquatic environment conditions will result in a decrease or abnormality of aquatic organisms that live in it [10].

Some researchers argue that marine fish seeds produced from hatcheries have many abnormalities caused by many factors such as nutritional deficiencies [11], including deficiency of vitamin C and vitamin D or phospholipids. Moreover, tryptophan may cause unnatural bone growth [12, 13]. The vertebral deformities in fish were in the form of lordosis, kyphosis, scoliosis, spinal fusion, and mandibular joint deformity. Vertebral abnormalities in fish are caused by chemical pollutants, nutritional deficiencies, disease and genetic infections [14]. The vertebrae abnormalities in *Hippoglossus hippoglossus* fish were in the form of branching in neural spine bone and haemal spine caused by environmental factors and malnutrition [15].

## 4.Conclusion

Abnormalities on hybrid grouper fish occurred in a hatchery in Situbondo, East Java.The form of abnormalities found were those on mouth and operculum, branched neural arch, fusion neural arch, fusion posterior truncus vertebrae, fusion of caudal vertebrae, and fusion of anterior truncus vertebrae.

## 5. Reference

[1] Barahona-Fernandes M.H 1982 *J. Fish. Biol.* **21** 239-249
[2] Taniguchi N K A zumiand and S U meda 1984 *Bull. Jpn. Soc. Sci. Fish.* **50** 787-792
[3] Divanach P, Papandroukalis, N Anastasiadis, P G Koumoundou-rosand and M Kentouri 1997 *Aquac.* **156** 145-155
[4] Koumoundouros G P D ivanach and M Kentouri 2001 *J. Mar.* **200** 285-304
[5] Ismi S 2014 *J. Mar. Sci. Tech.* **6** 109-119
[6] Cataudella S, Divanach P, Boglione C, Menu. B, Koumondouros G,Kentouri 1996 *Aqua.* **97** 169-180
[7] Jawad L A 2004 First Record of an Anomalous Mullet Fish (*Mugilcephalus*) from New Zealand *Museum of New Zealand Te Papa Tongarewa.* **15** 121-124
[8] Aslianti T 2010 *J. Sci. Tech. Mar.* **6** 26-33
[9] Kordi M G H 2004 *Preventation Pest and Disease of Fish* (Jakarta: PT Rineka Cipta and PT Bina Adikarsa) p 68
[10] Fitrani, M Burmansyah and Muslim 2013 *J. Swamp Aquac.* **1** 23-33
[11] Kanazawa A 1985 *Essential Fatty Acid and Lipid Requirement of Fish Nutrition and Feeding in Fish* (London: Academic Press) pp 281-298
[12] Weis J S and P Weis 1989 *Aqua. Sci.* **1** 45-73
[13] Akiyama T T Murai and K Mori 1989 *Bul.l Jpn. Soc. Sci. Fish.* **52** 1249-1254

IOP Conf. Series: Earth and Environmental Science **137** (2018) 012035 doi:10.1088/1755-1315/137/1/012035

[14] Yadegari M, Mehdi R, Mahsa A 2011 A Radiographical Study on Skeletal in Cultured Rainbow Trout (*Oncorhynchus mykiss*) in Iran. Faculty of Veterinary Medicine Islamic Azad (University Sharekord Branch Iran) pp 601-602
[15] Witten. P E Leah M L and SantoshP L 2004 *Aquac.* **241** 47–59

IOP Conf. Series: Earth and Environmental Science **137** (2018) 012036 doi:10.1088/1755-1315/137/1/012036

# Effect of probiotic culture water on growth, mortality, and feed conversion ratio of Vaname shrimp (*Litopenaeus vannamei* Boone)

**M Bachruddin, M Sholichah, S Istiqomah and A Supriyanto**

Department of Biology, Airlangga University, Jl. Airlangga 60115, Surabaya, East Java, Indonesia.

E-mail: bio_asp@yahoo.co.id

**Abstract.** This study was aimed to determine the effect of various dose of probiotics in the culture water to the growth and mortality of Vaname shrimp. This study consist of treatment control and treatment of various dose of probiotics. Control (0 mL/10 L water), P1 (1 mL/10 L water), P2 (2 mL/10 L water), P3 (3 mL/10 L water) and P4 (4 mL/10 L water) treatment, given to the Vaname shrimps with intervals once per week. This probiotic consist of *Lactobacillus plantarum*, *Lactobacillus fermentum*, *Bacillus subtilis*, *Bacillus licheniformis*, *Bacillus megaterium*, *Nitrobacter sp*., and *Nitrosomonas sp*. Dependent variables in this study are weight of shrimp, length of shrimp, mortality and feed conversion ratio. The results had different of various dose probiotics application in the water showed significance for each treatment on growth and mortality of Vaname shrimp. The best results were shown in treatment P2 (2 mL/10 water) with mean value of Vaname shrimp weight is 7.447 ± 1.193 g/shrimp, the length is 10,390 ± 0,469 cm/shrimp, mortality is 41%, and the value of FCR is 0.91.

## 1. Introduction

The Republic of Indonesia is nation set on an archipelago of islands and populated by some of the most frequent fish consumers in the world [1]. Aquaculture practices date back to at least the 15$^{th}$ century in Indonesia and are often typified by the Tambak, a traditional brackish-water pond system integrating shrimp, milkfish and other finfish [2].

Demand for shrimp is increasing steadily, while the world shrimp fishery is decreasing or remaining static; only aquaculture can meet this demand [3]. In Indonesia, while the shrimp aquaculture industry was being affected by white spot syndrome virus (WSSV), a bacterial pathogen that causes early mortality syndrome, *Vibrio,* appears with devastating consequences [4]. Thus affecting the process of shrimp farming. Such as in China, shrimp production was in decline from 200.000 tons in 1992, had lower to only 55.000 tons in 1994 [5]. This drastic decline is caused by several factors, such as infection of pathogenic bacteria and waste from feed remains that affect growth, mortality, and quality of culture water. Various pathogenic microorganisms live in aquatic environment such as *Vibrio* which commonly attack shrimp larvae at zoea, mysis, and the beginning of post-larvae stadium [6]. Thus it become an obstacle in supplying high number of shrimp hatchling for shrimp production.

In addition, the decline of shrimp production was also caused by contamination of organic materials in the aquaculture water. These materials originated from uneaten feed, dead plankton, fertilizer, and accumulated shrimp feces at the base of the pond [7]. Accumulation of organic matters at high level will raise the anaerobic decomposition rate, resulting in various toxic compounds, such as ammonium. Anaerobic condition at the base of the pond will also lower shrimp appetite and make it susceptible to disease. Level of ammonium can affect pH of the water; if ammonium level rises, pH will also elevate [8]. According to Raj and Raj [9], salinity is one of the environment factors that plays an important role on growth and survival of shrimp in facing rapid change of salinity and extreme condition of the environment. Feed consumption and efficient feed conversion are main components of growth and survival of Vaname shrimp that are affected by salinity and temperature [10].

According to Verschuere *et al.* [11], probiotic is living microbial agent that is able to give various advantages for its host by modifying microbe community or associating with the host, enhancing nutrition value and food utilization, improving host response to disease, and elevating quality of the environment. Thus, application of probiotic can be a solution to obtain optimal growth and feeding efficiency, lower production cost such as feeding cost, substitute use of antibiotic, and reduce environment burden due to waste accumulation in the water [12]. Several species of probiotic microbes can be used for breaking down organic material from feed remains, including *Bacillus subtilis*, *Bacillus licheniformis*, and *Bacillus megaterium*. In addition, nitrification bacteria such as *Nitrobacter sp.* and *Nitrosomonas sp.*, can also be added to reduce ammonium level. Similarly, lactic acid-producing bacteria, such as *Lactobacillus plantarum* and *Lactobacillus fermentum,* can also be utilized to improve shrimp digestive system and maintain water pH at optimal level.

Application of probiotic cannot be claimed as effective and efficient yet, due to lack of clarity in terms of dose and different application results. Generally, application in laboratory scale is much more controlled compared to field studies due to controlled environment of aquarium used in laboratory approach that is different from aquaculture ponds and eventually causes differences in probiotic dose, water quality, and aquaculture management.

This study was aimed to determine the effect of probiotic dose variation in aquaculture water on growth, mortality, and Feed Conversion Ratio (FCR) of Vaname shrimp (*Litopenaeus vannamei* Boone).

## 2. Methodology

### *2.1. Time and Place of Study*

This study was conducted in two locations: Microbiology Laboratory Airlangga University for probiotic preparation and fish ponds in Sidoarjo for fish culturing and data collection. This study was conducted for 9 months, starting from September 2015 to May 2016.

### *2.2. Materials*

Materials used in this study including Vaname shrimp hatchlings originated from Vaname shrimp aquaculture in Lamongan, Indonesia, water, microbial starter culture from Laboratory of Microbiology, Airlangga University that consisted of *Bacillus subtilis*, *Bacillus licheniformis*, *Bacillus megaterium*, *Nitrobacter sp.*, *Nitrosomonas sp.*, *Lactobacillus plantarum*, and *Lactobacillus fermentum*. Growth medium used for microbial culture was Nutrient Agar (NA) (Oxoid), 1% yeast extract (YE) (Oxoid), 1% glucose, sterilized distilled water, and molasses; while medium used for Total Plate Count (TPC) were produced from Nutrient Agar (NA) (Oxoid), 1 % glucose, and sterilized distilled water.

### *2.3. Design of Study*

This study was designed as experimental study using Completely Randomized Design. Probiotic was applied into aquaculture water every week with 5 varying doses; 1 mL/10 L (P1), 2 mL/10 L (P2), 3 mL/10 L (P3), 4 mL/10 L (P4), and a control treatment of 0 mL /10 L water (K). Each treatment consisted of 6 replications and for each replication, 5 shrimp samples were collected.

*2.3.1. Variables of study*

Three types of variables in this study could be elaborated as follows:

A. Independent variable : Dose of probiotic (mL/10 L water)

B. Dependent variable : Growth of shrimp including body length (cm), weight (g), and mortality of Vaname (percentage of dead shrimp).

C. Controlled variable : Variety of shrimp hatchling, shrimp age (days), $OD_{600}$=1 of microbes in the probiotic.

*2.4. Methods*

*2.4.1. Probiotic preparation*

A. Medium and isolates preparation. Sub culturing of microbial isolates was conducted on agar slant, consisted of seven bacteria species; *Bacillus subtilis*, *Bacillus licheniformis*, *Bacillus megaterium*, *Nitrobacter sp.*, *Nitrosomonas sp.*, *Lactobacillus plantarum,* and *Lactobacillus fermentum*. From each isolate, one *ose* of bacteria culture was inoculated on NA slant in sterile condition and then incubated for 24 hours.

B. Starter preparation. Microbes previously subcultured were taken two or three *oses* to be inoculated into 100 mL YE added with 1 % glucose. Inoculation was conducted in laminar flow cabinet. Culture was homogenized using shaker at 100 rpm for 5-6 h before it was incubated for 24 h at room temperature.

C. Measurement of microbe quantity. Each microbe culture was measured for its $OD_{600}$ = 1 using spectrophotometer before TPC analysis was performed. Pour plate was prepared in petri dish for the culture at 1 mL of microbe culture in ± 15 mL NA added with 1 % glucose. After colony grew in the dish, it was counted using colony counter and data was collected from dilution serial with results of around 30-300 colony.

D. Probiotic preparation. Liquid probiotic was prepared by mixing seven microbe starters in 100 mL YE added with 1 % glucose to obtain 700 mL of mixed starter culture composed of seven bacteria species. Probiotic consisted of 10 % starter and 90 % carrier. The carrier used for probiotic was composed of 3 % molasses and 87 % distilled water. Probiotic was prepared by mixing 700 mL of mixed starter culture with 189 mL molasses and 6111 mL distilled water.

*2.4.2. Preparation and culture of Vaname shrimp hatchling*

A. Preparation of Vaname shrimp growth medium. Five shrimp ponds were prepared from tanks with volume of 210 L each. An aerator connected to a hose was set in each pond to aerate the water.

B. Spreading of Vaname shrimp hatchling. Healthy hatchlings were selected randomly before spread into the pond in the morning. Hatchling used was PL-12 (aged 12 days) with initial weight of 0.016 g/shrimp and length of 0.7 cm. A basin filled with hatchlings was lowered slowly into the water and the hatchlings were let go into the water of the pond.

*2.4.3. Culturing of Vaname shrimp*

A. Feeding. Feed used was certified by Directorate General of Fishery and Aquaculture. It was given three times a day in the morning, noon, and afternoon.

B. Measurement of pH of water . Measurement of pH was performed at the start of the study and later on, once a week.

C. Measurement of temperature. Measurement of temperature was performed at the start of the study and later on, once a week.

D. Measurement of salinity. Salinity was measured the start of the study and later on, once a week after probiotic was applied.

IOP Conf. Series: Earth and Environmental Science **137** (2018) 012036 doi:10.1088/1755-1315/137/1/012036

*2.4.4. Vaname shrimp harvest*

Vaname shrimp was ready for harvest after 60 days. Growth of shrimp was measured, including body length and weight, and so was shrimp mortality. Shrimps were collected in net and then put into collection tank for data collection.

*2.4.5. Data collection*

A. Measurement of Vaname shrimp growth

Following [13], formula used for quantifying growth rate (GR) was as follows:

$$GR = Wt - Wo \tag{1}$$

In which:

GR = growth rate of weight (g)
Wt = Mean weight of shrimp at the end of caring period (g)
$W_0$ = Mean weight of shrimp at the start of caring period (g)

Growth of length was quantified using formula based on Effendi [14], as following:

$$P = Pt - Po \tag{2}$$

In which:

P = Absolute growth of shrimp (cm)
Pt = Length of shrimp at the end of culture period (cm)
$P_0$ = Length of shrimp at the beginning of culture period (cm)

B. Mortality of *Vaname* shrimp

Mortality was counted using hand counter and then entered into a formula based on [14] as follows:

$$M = \frac{A}{B} \text{ x } 100\% \tag{3}$$

In which:

M = Percentage of mortality
A = Number of dead shrimp during culture period
B = Number of shrimp spread into the pond

C. Feed Conversion Ratio

Feed Conversion Ratio (FCR) of shrimp was quantified using the following formula developed by Effendi [14]:

$$FCR = \frac{\text{Weight of feed given}}{\text{Growth of shrimp weight}} \tag{4}$$

*2.5. Data analysis*

Data of Vaname shrimp growth were compared and analyzed statistically using SPSS version 21 while mortality and FCR were analyzed descriptively.

## 3. Results and discussion

The effect of probiotic application at various doses towards Vaname shrimp growth was determined based on several parameters such as body weight, length, mortality, and FCR. Based on statistical test, application of probiotic to Vaname shrimp significantly affected its growth, indicated by significant increase in shrimp weight and length compared to control that was given no probiotic, as presented in table 1.

**Table 1.** The effect of various probiotic doses applied into aquaculture water towards growth (body weight and length) of Vaname shrimp (*Litopenaeus vannamei* Boone).

| No | Group | Mean weight (g) | Mean length (cm) |
|---|---|---|---|
| 1 | K (0 mL/10 L water) | $4.701 \pm 0.341^a$ | $8.450 \pm 0.688^a$ |
| 2 | P1 (1 mL/10 L water) | $5.467 \pm 0.305^a$ | $9.357 \pm 0.395^b$ |
| 3 | P2 (2 mL/10 L water) | $7.447 \pm 1.193^b$ | $10.390 \pm 0.469^c$ |
| 4 | P3 (3 mL/10 L water) | $7.204 \pm 0.498^b$ | $10.300 \pm 0.302^c$ |
| 5 | P4 (4 mL/10 L water) | $5.277 \pm 0.607^a$ | $8.914 \pm 0.322^{ab}$ |

*Difference of letters following number indicated statistical difference result of Duncan test at $\alpha = 0.05$

Data of Vaname shrimp mortality was collected by comparing the number of shrimp after 60 days of caring period with initial number of shrimp spread into the ponds. Mortality percentage was compared among groups treated with probiotic at various doses and untreated control. Mortality percentage was quantified using the formula based on [14]. The highest number of living shrimp after 60 days was found from P2 (2 mL/10 L), amounting of 59 shrimps with lowest mortality rate of 41 %; while the lowest number of living shrimp after 60 days was from P4 (4 mL/10 L), amounting of 50 shrimps with the highest mortality rate of 50 %. Effect of probiotic application at various doses on mortality is presented in table 2.

Feed Conversion Ratio (FCR) of shrimp was also quantified using formula based on [17]. FCR was the amount of feed given to the shrimp converted into its unit. it is measured from weight of feed given during caring period and the increase of shrimp weight from the start up to the end of caring period. Lowest FCR was found from P2 (2 mL/10 L) at 1 : 0.91, while highest FCR was K (0 mL/10 L) at 1 : 1.66 (table 2).

**Table 2.** Mortality percentage of Vaname shrimp (*Litopenaeus vannamei* Boone) and Food Conversion Ratio(FCR) after probiotic application at various doses for 60 days.

| No | Group | Starting shrimp number | Shrimp number after 60 days | Mortality (%) | Food Conversion Ratio(FCR) |
|---|---|---|---|---|---|
| 1 | K (0 mL/10 L water) | 100 | 51 | 49 % | 1.66 |
| 2 | P1 (1 mL/10 L water) | 100 | 56 | 44 % | 1.307 |
| 3 | P2 (2 mL/10 L water) | 100 | 59 | 41 % | 0.9 |
| 4 | P3 (3 mL/10 L water) | 100 | 57 | 43 % | 0.974 |
| 5 | P4 (4 mL/10 L water) | 100 | 50 | 50 % | 1.516 |

The varying percentage of mortality as presented in table 2 could be caused by environment impact, feed remains, and shrimp aggressiveness. Accumulated feed remains in the water medium could produce ammonium toxic for the shrimp. Stated that the growth of Vaname shrimp was affected by two factors; molting frequency (interval period) and growth. On every molting time, hard layer of carapace was released by the shrimp before it secreted new carapace. When the carapace was still soft, shrimp had higher chance to be preyed upon by other shrimp.

Feed conversion ratio was used in this study to determine the effectiveness of feeding, indicating level of feed remains and also profit of aquaculture. FCR was found to be varying from across treatment groups given different probiotic dose. FCR of P2 (2 mL/10 L) was found to be 1 : 0.91, meaning that for every 1 g of shrimp weight, 0.91 g feed is required (table 2).

Water quality during caring period had important role to support shrimp life and growth. Parameters of water quality were temperature, salinity, and pH or acidity level. Water quality in caring medium could affect growth level, mortality, molting frequency, and number of disadvantageous bacteria. As observed during the caring period of current study, the quality of aquaculture water used was still in the acceptable range for growth and mortality of Vaname shrimp with recorded salinity of 5 ppt while water temperature was at range of 28-32°C. Based on Haliman and Adijaya [13], fluctuation of temperature during caring period could be affected by the environment. during rainy season, temperature relatively normal, but rain water could affect water pH and salinity. Optimal temperature for Vaname shrimp caring was at a range of 26-31°C, while optimal pH was 7.5-8.1. Suitable water pH for intensive aquaculture of Vaname shrimp was at pH 7.4-8.9, with the most optimum pH of 8.0 [15].

Fluctuation of water pH during Vaname shrimp caring period was mainly influenced by the environment in the form of rain , which could be predicted to lower pH of culture water. Normal limit of acidity of rain was at pH 5.6, while pure water was produced at pH balances with global $CO_2$ concentration (330 ppm) in the atmosphere [16]. Activity of BAL microbe could lower pH periodically and also inhibit the growth of pathogenic bacteria [17].

## 4. Conclusions

- Treatment of P2 (2 mL/10 L) resulted in the highest value compared with other treatments in Vaname shrimp growth, which has an average weight of Vaname shrimp of 7.4 grams and a shrimp length of 10.40 cm.
- Treatment with P2 (2 mL/10 L) resulted in 59 live Vaname shrimp and a low percentage mortality rate of 41%.
- Treatment P2 (2 mL/10 L) had the lowest feed conversion ratio value with FCR value of 1: 0.91.

## 5. References

[1] Patrik J G H, Nhuong T, Chadag V M, Chin Y C, U-Primo R, Sharon S, Lara D M, Nur B P U, Stephen H, Michael J P 2017 *J. Cleaner Prod.* **17** 3
[2] Troell M 2009 *FAO Fish. Aquac. Tech. Paper* **529** 1-86
[3] Bondad-Reantaso M G, Subasinghe R P, Josupeit H, Cai J and Zhou X 2012 *J. Inver Pathol.* **110** 158-165
[4] Flagel T, Lightner D V, Lo C F and Owens L 2008 *Diseases in Asian Aquac.* **6** 355-378
[5] Sugita H Ahibuya K 1996 *Fish. Aquac.* **145** 195-203
[6] Lavilla-Pitogo CR, Baticados MCL, Cruz-Lacierda ER, De Ia Pena LD. 1990 *J. Aquac.* **91** 1-13
[7] Boyd C E 1990 *Water Quality in Ponds for Aqua Culture* (Auburn University : Alabama Agricultural Experiment Station) 482 p
[8] Raccota I, S E Palacios and A M Ibbara 2003 *J. Aquac.* **227** 107 – 130
[9] Raj P R and Raj P J S 1982 *Proc. Symp. Coastal Aquac.* **1** 236–243
[10] Venkataramaiah A, Laksmi G J and Gunter G 1972 *The Effect of Salinity, Temperature and Feeding Levels on the Food Conversion, Growth and Survival Rates of the Shrimp Penaeus aztecus* (Marine Technology Society : Food-Drugs from the Sea Proceedings) p 29–42
[11] Verschuere L, G Robaut, P Sorgeloos and W Verstraete 2000 *Microbiol. Mol. Bio.* **64** 655-671
[12] Iribarren D, P Dagá, M T Moreira, G Feijoo 2012 *Aquac. Int.* **20** 779-789
[13] Haliman R W dan D Adijaya S 2005 *Vaname Shrimp* (Jakarta: Penebar Swadaya)
[14] Effendi I 1997 *Fisheries Biology* (Yogyakarta: Yayasan Pustaka Nusantara)
[15] Wyban J A and Sweeney J N 1991 *Intensive Shrimp Production Technology* (USA: The Oceanic Institute Press)
[16] Seinfeld J H and Pandis S N 1998 *Atmospheric Chemistry and Physics from Air Pollution to Climate Change* (New York: John Wiley and Sons INC)
[17] Markas and De Vyust 2006 *Int.Dairy J.* **16** 1049-1057

**Acknowledgements**
This work was supported by the Research Program of Department of Biology, Airlangga University, Indonesia.

# The comparison of heavy metals (Pb and Cd) in the water and sediment during spring and neap tide tidal periods in Popoh Bay, Indonesia

**D Yona**[1,2], **R Febriana**[1] **and M Handayani**[1]

[1]Marine Science Department, Brawijaya University, Veteran street, Malang, East Java Indonesia
[2] Marine Resources Exploration and Management (MEXMA) Research Group, Brawijaya University, Veteran street, Malang, East Java Indonesia

E-mail: defri.yona@ub.ac.id

**Abstract.** This study attempted to investigate different concentration of lead (Pb) dan cadmium (Cd) in the water and sediment during spring and neap tidal periods in the Popoh Bay, Indonesia. Water and sediment samples were taken during spring and neap tides from eight sampling stations in the study area. The result shows higher concentration of Pb than the concentration of Cd in both spring and neap tides due to higher input of Pb from the oil pollution by boat and fisheries activities. Pb concentrations were doubled during neap tide in both water and sediments with the value of 0.51 and 0.28 ml/L in the water during neap and spring tide, respectively; and 0.27 ppm and 0.16 mg/kg in the sediment during neap and spring tide, respectively. On the other hand, Cd concentrations in the water were found in almost similar values between spring and neap tide (0.159 and 0.165 ml/L in spring tide and neap tide, respectively), but in the sediment, the concentration was a little higher during spring tide (0.09 and 0.05 mg/kg during spring and neap tide, respectively). This study shows that water movement during spring and neap tides has significant effect on the distribution of heavy metals.

## 1. Introduction

Heavy metals can be considered as pollutants to the environment due to its characteristics that are stable, persistent, non-biodegradable and toxic at certain levels. It enters the aquatic environment mostly through anthropogenic activities, either from domestic or industrial waste waters [1, 2, 3, 4]. Once in the water column, it can be transferred to sediments by the processes of adsorption onto suspended matter and sedimentation [5, 6]. The availability of metals in the water column that can be absorbed by soil depends on environmental conditions, such as pH [7, 8], redox potential [9] and organic content [10].

Water movement has been known to have roles in the circulation process of the aqueous environment such as influencing the distribution of nutrients, carbon and trace metals. Water movement, such as tidal exchange is important incontrolling heavy metal distribution, especially in the estuaries and coastal areas. Spring-neap variations in tidal flow dominate in narrow, shallow estuaries [6, 11, 12] could drive the mechanism of heavy metal transports by suppressing the stratification of water column. Moreover, tidal regimes could result in the variations of the redox potential in the

IOP Conf. Series: Earth and Environmental Science **137** (2018) 012037 doi:10.1088/1755-1315/137/1/012037

sediment that can lead to the differences in metal concentration, as [9] found that metal concentrations in the pore water were generally lower in sites with higher inundation frequency due to tidal effects.

Popoh Bay is a semi-enclosed waters connected to the Indian Ocean. It receives metal input derived from multiple sources. The western side of the bay is dedicated to aquaculture activities that might discharge heavy metals from the culture system and feed. The eastern side of Popoh Bay is highly influenced by fishery activities from the fishing port. Moreover, this area is also subject to contribute metal pollution from tourism activities from several beaches surrounding the bay that are popular among the local people. Water movement from tidal periods (spring-neap tides) may have a significant impact on the distribution of the metal concentrations in this semi-enclosed waters that is connected to the open ocean. Therefore, the aim of this study was to determine the concentrations of Pb and Cd in the seawater and sediments during the spring and neap tidal periods; and to compare the distribution patterns of Pb and Cd in the seawater and sediments between the spring and neap tidal periods.

## 2. Methodology

The study was conducted in Popoh Bay, Tulungagung, which is located in the southern part of Java Island (Latitude 8°26'84.85"S - 8°25'91.70"S and Longitude 111°77'36.18"E - 111°80'16.10"E). Samples were taken in May 2016 from eight sampling stations (figure 1) that could represent water quality parameters in the bay. Sampling was conducted in two periods, the first one was on May 11$^{th}$ (spring tide) and the second sampling was conducted one week after on May 18$^{th}$ (neap tide).

Heavy metal samples in the seawater were collected from the surface and placed in 250 ml bottles that were pre-washed with $HNO_3$ polyethylene. Seawater samples were then acidified by adding 10 ml $HNO_3$. Heavy metal samples in the sediment were collected using a grab sampler at 0-10 cm depth. About 1 kg of sediments were obtained and stored in tightly sealed plastic bags. The grab sampler was washed with seawater before subsequent sampling to avoid contamination from the previous sample. Both the seawater and sediment samples were preserved in iceboxesbefore further analysis in the laboratory. Water quality parameters measured were for temperature, salinity, pH, DO and current.

In the laboratory, the seawater samples were filtered through 0.45 μm filter paper and Pb and Cd concentrations were measured using flame atomic absorption spectrometry. Sediment samples were air dried at 105 °C for about 3 hours and ground using a mortar and pestle. About 1 g of sediment was placed in the erlenmeyer and then added with $HNO_3$ and $H_2SO_4$ and left to destruct for about 3 hours at 120°C. The samples were filtered and the concentrations of Pb and Cd were measured using flame atomic absorption spectrometry.

The Pearson`s correlation test was conducted to test the relationship between water quality parameters and heavy metal concentrations in both spring and neap tide tidal periods. Independence students' t-tests were analysed to compare heavy metal concentrations between spring and neap tidal periods in both seawater and sediments samples. Prior to conducting the test, the normality and homogeneity of data werechecked using the Shapiro-Wilk test. All statistical analyses were calculated using SPSS 16.0.

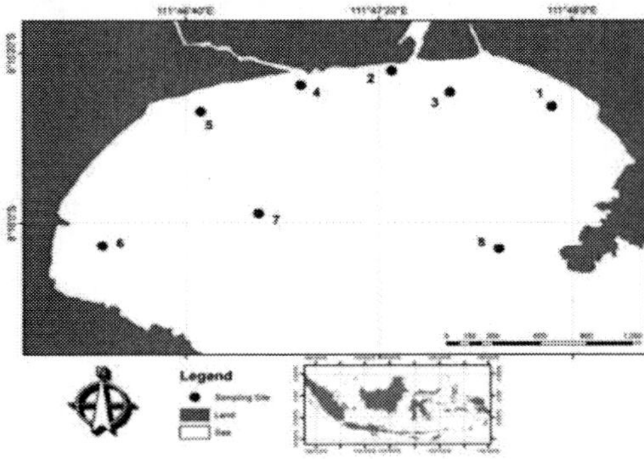


**Figure 1.** Study area and sampling sites in Popoh Bay, Tulungagung, Indonesia.

## 3. Result and discussion

The distribution of Pb concentrations in the seawater and sediments in Popoh Bay, Tulungagung are shown in fig. 2. The concentrations in the seawater were in the range of 0.40 – 0.57 ml/L and 0.16 – 0.43 ml/L during neap tide and spring tide, respectively. In the sediments, Pb concentration was in the range of 0.16 – 0.44 mg/kg during neap tide and in the range of 0.05 – 0.30 mg/kg during spring tide.Pb concentrations were found higher during neap tide for both the seawater and the sediments. This result is similar to the result of the study conducted by Teuchies et al. [9] that found lower metal concentrations in the site with higher inundation frequency compared to the one in the lower inundation frequency due to the mobility of metal which is lower under flooded conditions.

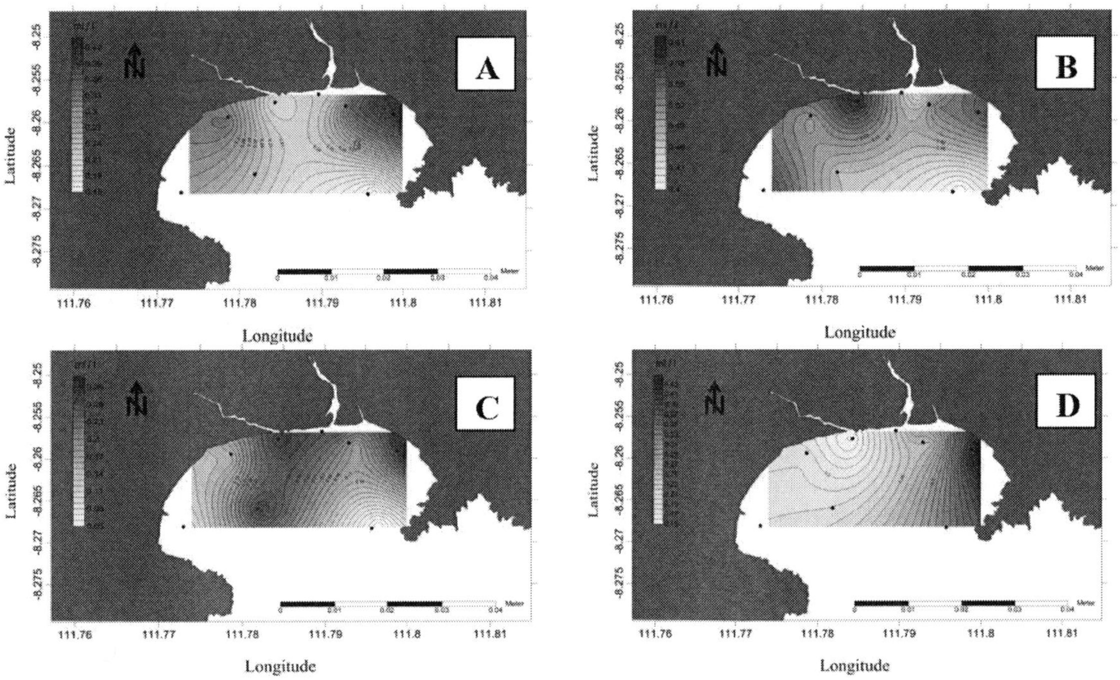


**Figure 2.** Distribution of Pb in the seawater during spring tide (A) and neap tide (B) and in the sediments during spring tide (C) and neap tide (D) in the Popoh Bay, Tulungagung.

Cd distribution patterns were found to be different between the spring and neap tidal periods for the seawater and sediments (fig. 3), but the concentrations were very much higher in the seawater than in the sediments. Cd concentrations in the seawater were found in the similar range between spring and neap tide (0.08 – 0.23 ml/L and 0.12 – 0.21 ml/L, respectively). However, higher concentrations were found in the sediments during spring tide (0.07 – 0.12 mg/kg) compared to during neap tide (0.02 – 0.08 mg/kg). Metals tend to accumulate in the sediment due to the lower mobility of metals under flooded conditions [9].

Spatial distribution of the metal concentrations were varied across the sampling areas between both tidal periods (fig. 2 and fig. 3). Higher Pb concentrations were observed in the seawater in the western and eastern sides of Popoh Bay during the spring and neap tidal periods. On the other hand, the Pb concentration in the sediments shows higher concentration only in the eastern side of Popoh Bay during neap tide.The spatial distribution patterns of Cd were observed and were rather similar to the spatial distribution of Pb. Higher concentrations were found in the western and eastern sides of the study areas. The significant contribution of heavy metals in these sites were mainly due to the heavy activities from aquaculture (western side) and fisheries (eastern side).

Results of the t-test showed that there were significant differences of Pb concentrations in the seawater and Cd concentrations in the sediments ($p < 0.05$) between spring tide and neap tide tidal periods (table 1). On the other hand, Pb concentrations in the sediments and Cd concentrations in the seawater did not show significant differences between both tidal periods. These differences might be the result of the variability in the water capacity [13] during the tidal periods that influences the concentration of heavy metals to the environment. This variability of water capacity may be related to the tidal current, and the result was supported with the correlation between the current and heavy metal concentrations especially Pb. According to table 2, a strong negative correlation can be observed between the Pb concentration in the seawater during spring tide with the current ($r = -0.790$, $p < 0.05$) and also between Pb concentrations in the sediments during neap tide with the temperature ($r = -0.916$, $p < 0.05$) and current ($r = -0.927$, $p < 0.05$).

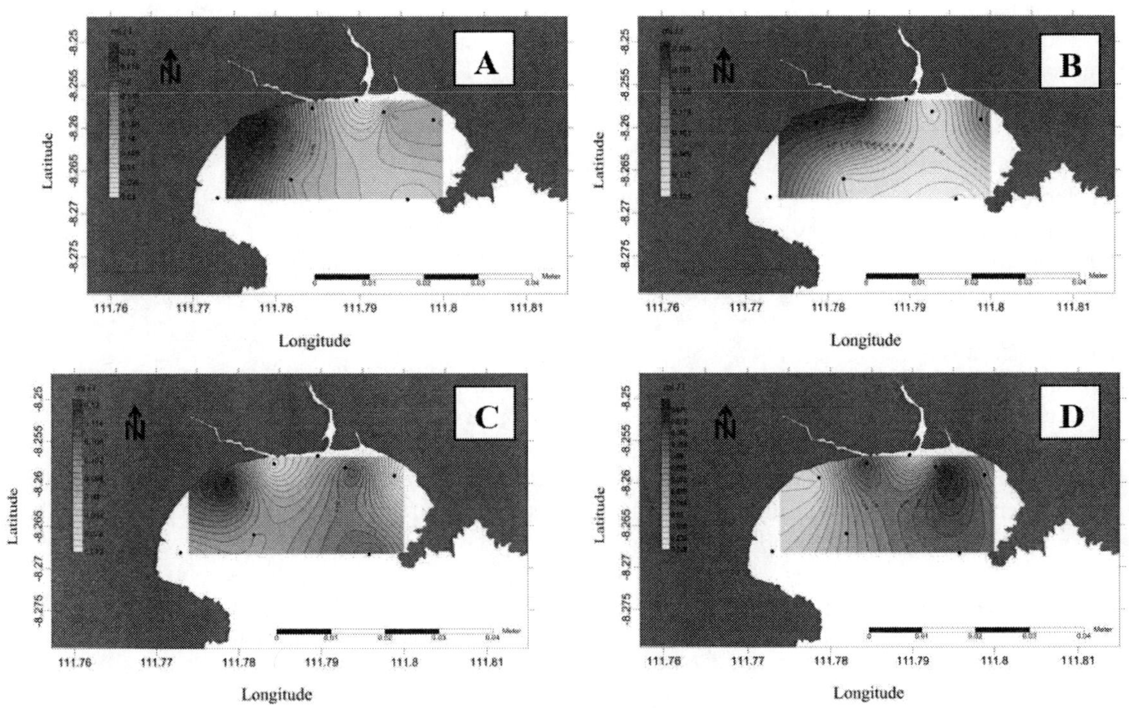


**Figure 3.** Distribution of Cd in the seawater during spring tide (A) and neap tide (B) and in the sediments during spring tide (C) and neap tide (D) in the Popoh Bay, Tulungagung.

**Table 1.** Results of t-test for tidal variability of heavy metal concentrations in the seawater and sediments

| | t value | P value |
|---|---|---|
| Pb in seawater | -5.741 | 0.000 |
| Pb in sediments | -1.329 | 0.211 |
| Cd in seawater | -0.286 | 0.779 |
| Cd in sediments | 4.114 | 0.002 |

Changes in the water flow during tidal periods have been understood to play a siginificant role in controlling the transport of materials into the estuary and coastal regions [11, 12, 14, 15]. Shaha et al.

[12] in their study revealed that spring tide can cause the well-mixed conditions and rapid exchange of materials due to larger tidal amplitude, while neap tide causes stratification of the water column due to the reduction in vertical mixing. Moreover, this phenomenon might also affect soil conditions and pore water, causing the release of metals to the overlying waters [9, 16]. The results of this study found a variability of heavy metal concentration in the seawater and sediment during the spring and neap tidal periods. Tidal periods caused significant differences in the distribution of Pb in the seawater and Cd in the sediments (table 1).

**Table 2.** Correlation between heavy metal concentrations and environmental factors during spring and neap tidal periods.

| | Temperature | Salinity | DO | pH | Current |
|---|---|---|---|---|---|
| Spring tide (n = 8) | | | | | |
| Pb in seawater | -0.590 | -0.695 | -0.373 | -0.147 | -0.790* |
| Pb in sediments | -0.637 | -0.670 | -0.685 | -0.518 | -0.298 |
| Cd in seawater | 0.318 | 0.364 | 0.122 | 0.079 | 0.023 |
| Cd in sediments | 0.460 | 0.579 | 0.382 | 0.795* | 0.395 |
| Neap tide (n = 8) | | | | | |
| Pb in seawater | 0.399 | -0.428 | -0.285 | -0.470 | 0.041 |
| Pb in sediments | -0.916* | 0.131 | -0.457 | 0.447 | -0.927* |
| Cd in seawater | 0.350 | -0.495 | -0.246 | -0.586 | -0.017 |
| Cd in sediments | 0.290 | -0.458 | -0.467 | -0.297 | 0.364 |

* correlation is significant at the 0.05 level (2-tailed).

## 4. Conclusion

This study shows that the concentrations of Pb and Cd in the seawater and sediments were varied between spring and neap tides with generally lower values of metals during spring tide. The variability was significantly due to the water movement which occurred during the tidal period,where the current and temperature have a rather significant effect especially on Pb concentration. Higher concentrations for both metals were observed in the western and eastern sides of the study area that wererelatedto heavy activities of aquaculture and fisheries.

## 5. References

[1] Ali M M, Ali M L, Islam M S, and Rahman M Z 2016 *Environ. Nanotechnol. Monit. Manag.* **5** 27–35
[2] Bi S, Yang Y, Xu C, Zhang Y, Zhang Y, and Zhang X 2017 *Mar. Pollut. Bull.* **121** 357–366
[3] Borges R C,Caldas V G,Filho F F L S, Ferreira M M, and Lapa C M F 2014 *Mar. Pollut. Bull.* **89** 75–84
[4] Ochieng E Z, Lalah J O, and Wandiga S O 2009 *Bull. Environ. Contam. Toxicol.* **83** 600–607
[5] FengC, GuoX, YinS, TianC, LiY, and ShenZ. 2017 *Chemosphere.* **185** 717–725
[6] Zhang A*et al.* 2017 *Reg. Stud. Mar. Sci.* **11**: 32–42
[7] KarbassiA R and Heidari M 2015 *Glob. J. Environ. Sci. Manag.* **1** 41–46
[8] Li H, Shi A, Li M, and Zhang X 2013 *J. Chem.* 1–11
[9] Teuchies J*et al.* 2008 *Environ. Pollut.* **155** 20–30
[10] Akan J C, AuduS I, Mohammed Z A, and Ogugbuaja V O 2013 *J. Environ. Prot.* **6** 618–628
[11] Gracia-Escobar M F, Millán-Núñez R, González-Silvera A, Santamaría-del-Ángel E, Camacho-Ibar V F, and Trees C C 2014 *Open J. Mar. Sci.* 480– 100
[12] Shaha D C, Cho Y K, Seo G H, Kim C S, and Jung K T 2010 *Hydrol. Earth Syst. Sci.* **14** 1465–1476
[13] Bastami K D*et al.* 2014 *Mar. Pollut. Bull.* **81** 262–267
[14] Reckhardt A *et al.* 2015 *Estuar. Coast. Shelf Sci.* **159** 1–14

[15] Riedel T, Lettmann K, Schnetger B, Beck M, and Brumsack H J 2011 *Geochim. Cosmochim. Acta.* **75** 134–147
[16] Atkinson C A, Jolley D F, and Simpson S L 2007 *Chemosphere.* **69** 1428–1437

**Acknowledgment**
This research did not receive any specific grant from funding agencies in the public, commercial, or not for profit section. However, we would like to thank Supriyadi for the help in creating the map of sampling location better.

IOP Conf. Series: Earth and Environmental Science **137** (2018) 012038 doi:10.1088/1755-1315/137/1/012038

# The implementation of vessel-sinking policy as an effort to protect indonesian fishery resources and territorial waters

**Nurdin[1,2], Ikaningtyas[1,2], Rika Kurniaty[1,2]**

[1]International Law Department-Faculty of Law, Brawijaya University, Jl.MT.Haryono 169 Malang
[2] Coastal and Marine Research Center, Brawijaya University Malang

E-mail: ninktyas@ub.ac.id

**Abstract.** This study aims to analysis the effectiveness of foreign ship sinking policies to eradicate illegal, unreported, and unregulated (IUU) fishing. There are many foreign fishing vessels were detained due to IUU fishing in Indonesia`s exclusive economic zone (EEZ) waters, particularly in the Natuna and Anambas region. In combating illegal fishing, the government of the Republic of Indonesia take concrete actions in protecting marine potentials by sinking foreign vessel policies. In the last three years more than 300 foreign ships are drowned by Indonesian government. This study revealed that regulations concerning the act of sinking the vessel have been in existence since 2009 but lack of socialization. The Indonesian government's policy regarding foreign-flagged vessel carrying out IUU fishing is regulated under Law Number 45 of 2009 on Fisheries, and internationally permitted with certain restrictions on conditions set forth in article 73 paragraph (3) of UNCLOS 1982. These policy is part of an effort to improve the deterrence effect of regional offenses that could harm and threaten the sovereignty of the state.

## 1. Introduction

As an archipelagic state, Indonesia benefits from its location which lies at the intersection of two Continents-Asia and Australia-and two oceans- Indian Ocean and Pacific Ocean. After the enactment of the United Nations Convention on the Law of the Sea in 1982 on 16 November, 1994, the area of Indonesia's territory has increased to 8,193,163 $km^2$, which consists of 2,027,087 $km^2$ of land and 6,166,163 $km^2$ of waters. The size of Indonesia's sea can be detailed into 0.3 million $km^2$ of territorial sea, 2.8 million $km^2$ of archipelagic waters, and 2.7 millions $km^2$ of EEZ [1].

The potential of Indonesian marine resources has four extremely strategic meanings: (1) as a provider of natural resources and source of livelihood; (2) as a unifier of the society; (3) as a defense medium; and (4) as a liaison medium. As we all know, two-thirds of trade traffic takes place at sea. Therefore, sea has a considerable benefit for the sustainability of the world's economy [2]. The abundance of natural resources and services of marine environment can be used for improving the prosperity of the nation, at least in 11 marine economic sectors: (1) marine fisheries; (2) marine cultivation; (3) fish processing industry; (4) marine biotechnology industry; (5) mining and energy; (6) marine tourism; (7) mangrove forest; (8) sea transportation; (9) small islands resources; (10) maritime industry and services; and (11) non-conventional natural resources. The overall potential economic value of all of the eleven marine sectors is estimated to reach USD500 billion (4,500 trilliums) each year (Indomiliter, 2013). Based on the data from the Food and Agricultural Organization (FAO), Indonesia's detriment caused by illegal fishing has reached 30 idr trillion each year, which is 25% of

IOP Conf. Series: Earth and Environmental Science **137** (2018) 012038 doi:10.1088/1755-1315/137/1/012038

the total value of Indonesia's fishery resources. It means that the detriment is 25% of 6.5 million tons, which equals to 1.6 million tons or 1.6 billion kilograms. The detriment is still lower than that estimated by the government, which reaches up to 300 idr trillion each year.

There are disastrous impacts of illegal fishing that not only harm the nation economically by causing loss of trillion Rupiahs, but also destroy the economy of fishermen. Illegal fishing may also trigger conflicts or tension in the diplomatic relation between nations, especially the bilateral relation between a country and its neighboring countries. This is due to the fact that sea serves as a boundary between two countries, and the boundary is determined based on bilateral and multilateral extraditions . It means that the power of a country is only valid within its territorial borderline [3]. Furthermore, conflicts in some cases of traditional fishing often happen at the frontier between Indonesia and Malaysia and between Indonesia and Australia. Therefore, there are some worries that illegal fishing will cause harm to the state's sovereignty and threaten the preservation of marine resources. It also raises concerns that there is an impression that Indonesia is unable to protect its marine resources. The impacts of illegal fishing can also be found in the context of environment and ecology. The government has made policies relating to fishing, the requirements and criteria for obtaining fishing permits, establishment of fishing ground, types of the fish to be captured, the amount of fish appropriate to the vessel qualification, the area of fishing activity, and the fishing gear.

From 2014 to 2017, more than 300 foreign ships have been drowned in Indonesia's EEZ, especially Natuna and Anambas waters. There are numerous modes used in fish theft, and the modes most frequently used were illegal fishing (catching fish without permission), violating the laws despite owning the permit, falsifying document, transshipment at sea, disabling the transmitter (especially the case with the ships that are required to use transmitter), and destructive fishing using chemicals, biological materials, explosives, equipment, tools, methods, or even construction that can endanger the sustainability of marine resources [4].

In eradicating illegal fishing cases in Indonesia, and as an effort to enforce law at sea, Indonesian government has enacted Law Number 31 of 2014 *juncto* Law Number 45 of 2009 on Fishery. According to this fishery law, the sanctions to be imposed against the person who conducts illegal fishing are quite severe (i.e. if a fishing vessel does not possess the permit or license to conduct activity or if the vessel with Indonesian flag violates the rules, the operator and the owner of the vessel are subject to imprisonment of 6 years or fine of 2 idr billion). Moreover, if the violation is conducted by a foreign-flagged vessel, the operator and the owner of the vessel are liable to imprisonment of 6 years or fine of 20 idr billion. The government of the Republic of Indonesia also takes a concrete action to protect its maritime potential. This action is taken by establishing Coordinating Ministry of Maritime Affairs. This ministry has asserted that Indonesia's sovereignty cannot be disturbed by taking a firm action in the form of sinking foreign vessels proven to have conducted illegal fishing in Indonesia's territorial waters. Based on this new policy, three vessels from Vietnam were detonated on 5 December, 2014 at Natuna Sea, Riau Archipelago. The three vessels with more than 2.1 tons of stolen fish and 33 crew members were caught in the northern part of Tarempa Island, Anambas, Riau Islands, on 1 November, 2014. Some other vessels will also be sunk as soon as possible. The vessels aforementioned include five vessels in Pontianak and three vessels in Berau. Exploding and sinking vessels are done to show Indonesia's firmness in protecting its territorial waters. Indonesia assured that it will take any necessary steps to protect its territory. The action of sinking vessels is expected to convey a message to other foreign fishermen not to violate the law in Indonesia's territory and rob the sea's wealth. It is also intended to give preventive effect and decrease violation of law in Indonesia's territorial waters. Based on the aforementioned facts, it is important to conduct a scientific study on the effect of the application of 'sink the foreign vessel' policy on illegal fishing in Indonesia's territorial waters.

## 2. Indonesia`s Vessel-Sinking Policy in Securing the Fishery Abundance

Illegal fishing activities in Indonesia's waters also have significant negative social impacts on fishermen and the workers of fish processing factories. For Indonesia and other countries in Southeast Asia, fishery and forestry sectors are the main sources of food sustainability. Massive and drastic

IOP Conf. Series: Earth and Environmental Science **137** (2018) 012038 doi:10.1088/1755-1315/137/1/012038

exploitation of the sea as a major effort to improve a country's economy and the welfare of its population often happens. Social impact increases with the vulnerability caused by conflict/dispute among traditional fishermen and trawl owners of different countries. Besides, with the abundant fishery resources, fish processing factories become very important. Catch reduction and illegal fishing activities will indirectly affect the survival of fish processing factory workers. Reduced fish supply disables some companies to operate, and many layoffs happen because there is no more supply of raw materials. The fish caught by foreign vessels or Indonesian fishing boats are usually directly delivered out of Indonesia through transhipment, which is inconsistent with the Regulation of the Minister of Marine Affairs and Fisheries Number 16 of 2006, which requires that fish catch should be landed and processed on land.

The rise of occurances of IUU fishing activities in Indonesia's sea has increasingly raised concerns. Based on the data released by the Ministry of Marine Affairs and Fisheries of the Republic of Indonesia, the loss of the state due to illegal fishing reaches 300 trillion rupiahs per year, which is 25% of the total potential of Indonesian fisheries. The loss affects the state and threatens the sustainability of marine and fishery resources. Some of the factors causing illegal fishing are the rise of global fish consumption level, dicreased fish resources in other countries, weakness of Indonesian fishery fleet in terms of technology, lack of monitoring by authorities at Indonesia's sea, and weak law enforcement at Indonesia's sea.

There are considerable impacts of illegal fishing practices that not only harm the state economically with the loss of trillions of rupiahs, but also destroys the fishermen's economy. Illegal fishing often triggers conflict or tension in the diplomatic relations between nations, especially bilateral relations between adjacent and neighboring countries. Negative impacts of illegal fishing can also be found in the environmental and ecological contexts. The government's policies which are related to fishing should comply with the rules and criteria of Fishing Permit, the establishment of fishing ground, the type of fish consumed, the catch capacity in accordance with the type of vessel and the total allowable catch, and the fishing equipment. The vastness of the sea area belonging to the jurisdiction of Indonesia and the fact that Indonesia's EEZ is widely exposed and adjacent to the open sea have attracted entry of foreign and local fishing boats to practice illegal fishing.

In 2014, 78 foreign fishing vessels were seized for conducting illegal fishing in Indonesian EEZ, especially in Natuna and Anambas waters. Illegal fishing makes the Anambas people, 90% of whom work as fishermen, to be eliminated. Many modes are applied in illegal fishing, but some modes/types of illegal activities are frequently used, namely unlicensed fishing, licensed fishing but violating the provisions of law, fraud/manipulation of documents, transshipment at sea, deactivation of transmitters (partially for ships required to install transmitters), and destructive fishing using chemicals, biological materials, explosives, tools and/or modes, and/or constructions that endanger the preservation of fish resources.

The impacts of illegal fishing crimes are detailed as follows.

1) Damage to the Sustainability of Fish in the Sea
Head of Research and Development Department of the Ministry of Marine Affairs and Fisheries, Zulficar Mochtar, explains that 85% of global fishery has experienced overexploitation. He said that if no anticipative action is taken, the fisheries in the territory of the Republic of Indonesia will also be threatened. Thousands of foreign ships have illegally exploited the fish in Indonesia by duplicating fishing licenses in Indonesia. The data recorded that 7,000–10,000 foreign ships have looted the fish in Indonesia every year.
2) Detriment to the state's economy
Indonesia's losses due to illegal fishing practices are considerable. The data from the Food and Agricultural Organization (FAO) noted that Indonesia loses 30 idr trillion per year due to illegal fishing. The number was assessed by the Minister of Marine Affairs and Fisheries, Susi Pudjiastuti, as quite small. According to her calculation, the state's losses due to illegal fishing can reach USD20 billion or 240 idr trillion per year.

IOP Conf. Series: Earth and Environmental Science **137** (2018) 012038 doi:10.1088/1755-1315/137/1/012038

3) Social Affairs
For Indonesia, IUU Fishing is a main concern because it happens every day in Indonesia's waters. In Southeast Asia, fishery sector is one of the main sources of food sustainability. Economic motives often serve as the reason for the massive exploration of fishery resources, which in turn, become a major cause for drastic reduction of fish stocks in Southeast Asia. This problem will adversely affect the survival of more than 100 million people. This has also led to disputes among local fishermen with trawler owners and between traditional fishermen of different countries. The decline in fish stocks in Indonesia's waters as a result of illegal fishing by trawler owners has also forced Indonesian traditional fishermen to engage in illegal fishing activities in Australia's waters, which has caused problems between the two countries. The direct impact is not only felt by the fishermen, but also the workers of factories, especially the fish processing factories.
4) Environmental Damage
In terms of environment, the use of large-scale fishing gear (trawl), which is not in accordance with the maritime requirements and conditions, has caused permanent damages and disturbed the marine ecosystem and biota. The depletion of fish resources in Arafuru waters is certain because for almost 3 years, there have been massive fishing activities which are arbitrary and exploitative in nature.
5) Illegal Fishing that violates Indonesian sovereignty
From the majority of illegal fishing cases occurring in Indonesia, the violation of Indonesian sovereignty is committed by all illegal fishing actors of the abovementioned countries. Foreign fishermen have crossed Indonesia's borders without permission, entered Indonesia's seas, and robbed Indonesia's fishery assets. Indonesia's law should be upheld in order to maintain Indonesia's sovereignty and prevent the border-crossing by countries that intends to rob the nation's assets. Such action is a form of criminal act occurring in Indonesia's waters.

The Minister Regulation Number KEP/50/MEN/2012 is the implementation of the Code of Conduct for Responsible Fisheries (CCRF) agreed in 1995 by the Food and Agriculture Organization (FAO) countries regarding the management and development of an orderly, responsible, and sustainable fishery, and as the implementation of international action to combat IUU Fishing as outlined in the International Plan of Action to Prevent, Deter, and Eliminate IUU Fishing (IPOAIUU Fishing) in 2001. Follow-up measures on IPOA-IUU Fishing should be taken by all countries, including Indonesia, by preparing action plans for the prevention and control of IUU Fishing at the national level.

Indonesian government's policy of explicitly sinking foreign vessels that commit illegal fishing in Indonesian territorial waters has received varied responses from many parties. This policy is supported by some as a form of assertiveness by the government of Indonesia in maintaining the sovereignty of the nation. For example, the Minister of Foreign Affairs, Retno Marsudi, said that this decisive action should be done as an effort of the state to conduct law enforcement without being bargained, and further stated that this action is a starting point for Indonesia to be respected by other nations [5]. It is also supported by the Minister of Marine Affairs and Fisheries, Susi Pudjiastuti, who stressed that the sovereignty of the state is a thing that must be maintained in any required ways [6]. However, some parties also doubt whether such decisive actions could be undertaken by the government of Indonesia which has been inclined to be gentle on border conflicts. It is understandable, given that the sunken ships so far were small vessels only.

Actually, the special action of sinking foreign vessels by Indonesian government is not a new issue. This can be seen from the regulation concerning that action, which has been listed since 2009 in the Law concerning Fisheries (Law of the Republic of Indonesia Number 45 of 2009 on Fisheries), Article 69 paragraph 4 which states;

"*... penyidik dan/atau pengawas perikanan dapat melakukan tindakan khusus berupa pembakaran dan/atau penenggelaman kapal perikanan yang berbendera asing berdasarkan bukti permulaan yang cukup.*" (... a fishery investigator and/or supervisor may

IOP Conf. Series: Earth and Environmental Science **137** (2018) 012038 doi:10.1088/1755-1315/137/1/012038

take a special action in the form of burning and/or sinking a foreign-flagged fishing vessel on the basis of sufficient preliminary evidence)

This regulation has been applied by Indonesian Navy. This was confirmed by the Head of Strategic Intelligence Agency for the period 2011–2013, Rear Admiral (Ret.) Soleman B. Ponto [7]. The actions taken by Indonesian ggovernment to sink illegal foreign vessels are part of the efforts to improve the deterrent or repellent effect of regional offenses that harm and threaten the sovereignty of the state. Although the regulation on sinking vessel action has been in place since 2009, the socialization of the regulation had been under-done by the government, both to the internal and external parties through Indonesian embassies in many countries, especially those bordering with Indonesia. That is what causes many foreign fishermen remain violating the territory of Indonesia. They do not feel threatened since they are unaware of the firm action against the committed violations. On the other hand, if the regulation is implemented without sufficient socialization, it is worried that it will disrupt Indonesia's diplomatic relations with countries whose fishermen commit violations.

Law Number 31 of 2004 which was amended with Law Number 45 of 2009 on Fisheries mentions that fish theft is classified as a criminal act. The penalty applies not only to the on-board operators, but also the owner of the ship and the owner of the company (Article 8). The foreign ship of fish thieves may also be burned and drowned (Article 69), even the owner is subjected to a fine of up to 20 idr billion (Article 93). The legal basis for the drowning of foreign fishing vessels is also clearly stated in Article 69 paragraph (4) of Law Number 45 of 2009 on Fisheries, which reads, "in the case of carrying out the supervisory functions, the investigator and the fishery supervisor may perform special actions in the form of burning and/or sinking foreign flagged fishing vessels on the basis of sufficient preliminary evidence".

Regarding sovereignty, even though Indonesia as a sovereign state is entitled to a total right to secure its national interests, in this case monitoring the territorial boundaries to secure existing marine resources, the implementation of such regulation will be difficult if Indonesia has a bilateral agreement with a particular country. For example, Indonesia has a Memorandum of Understanding (MoU) agreement with Malaysia on the treatment of fishermen, agreed in Bali, January 27, 2012. Regarding the violation of the territory by the fishermen, each country agreed on a precautionary measure with the details set forth in Article 3 Point B. This is contrary to Law Number 45 of 2009, On the one hand, the law requires strict action, but on the other hand, the rules of the MoU put forward more persuasive efforts.

The conflict in the regulation is one of the reasons why Indonesian government raised the issue of firm action against foreign fishermen into national and international scope. It is part of the Indonesian government's diplomacy strategy to reduce territorial violations by foreign fishermen. Indonesian government prefers not to take risks by changing the bilateral agreement because it will affect the relations between the two countries. Therefore, the current issue is directed more directly to the perpetrators of violations in order to provide psychological effects by spreading fear through the threat of burning and sinking ships. This is what causes deterrent effects later. Furthermore, the deterrent effect can be interpreted as an attempt to influence the judgment of others on its importance, which in its application, it is possible to use provocative strategies in accordance with the existing conditions [8]. This method will certainly be much safer than making the government of Indonesia to instantly sign a deal with the government of the concerned countries, which could adversely affect diplomatic relations between countries.

## 3. The Effectiveness of Vessel-Sinking Policy

The sinking of foreign vessels that conduct fish theft in the territory of Indonesia per se is not new in the era of President Joko Widodo. This is commonly practiced in the world as an effort to overcome illegal fishing. In February 2014, Indonesian fishermen's vessels that were caught entering Papua New Guinea's sea territory were also burned by the country's marine patrol. The head of the Indonesian Navy's Information Service, First Admiral Manahan Simorangkir, said that it was not the first time the

IOP Conf. Series: Earth and Environmental Science **137** (2018) 012038 doi:10.1088/1755-1315/137/1/012038

Navy had attempted to sink illegal vessels. The Navy has at least sunk foreign vessels that stole fish in the waters of the archipelago at least four times in 2003. Throughout 2003, the Navy had sunk four illegal foreign vessels with Philippine flag and 4 illegal foreign vessels with Thailand flag. Meanwhile, the Director General of Marine and Fishery Resources Control, Asep Burhanudin, also stated that the drowning of foreign vessels that conducted fish theft in Indonesian marine areas has often been done since the administration of President SBY. He stated that in the period 2007–2012, there were 38 foreign ships drowned. Asep detailed that since 2007 to 2012, there were 38 ships that have been drowned, with 1 ship being sunk in 2007, 32 ships in 2009, 3 ships in 2010, 1 ship in 2011, and 1 vessel in 2012. All of the vessels were captured by the Natuna Sea Fisheries Management Area. The majority of the captured and drowned vessels were from Vietnam.

Many strategic points within Indonesian archipelago can be used as illegal fishing spots, namely: North Sumatra Sea, Riau Islands, West Kalimantan, Maluku, North Sulawesi, and North Halmahera Papua. Other locations selected in Indonesia are: Natuna Sea, Karimata Strait, Banda Sea, Sulawesi Sea, Arafuru Sea, Maluku Sea, and Seram Sea. The location of the spots was chosen because, in addition to having abundant marine resources, they also lack supervision from the authorities and other surveillance equipment. Thus, foreign ships that conducted illegal fishing could freely take the sea wealth in the area. Illegal fishing perpetrators, viewed from the perspective of arrest operations, can be classified into two groups namely Indonesian ships and foreign ships or ex-foreign ships originating from various countries, which mostly are the neighboring countries. Foreign ships do have more modern technology t, larger capacity of catch, and wider net machine han those of Indonesian ships. As a coastal state, certainly Indonesia has the authority to protect its territorial waters, especially the natural resources contained there. The position of Indonesian waters in Indian Ocean has several significant points such as the Strait of Hormuz, the Strait of Malacca, the Strait of Lombok, the Sunda Strait, and the large extent of Indonesian waters, but it does not support adequate infrastructure and resources in every water region open to remote areas, and in the border areas of neighboring countries. Hence, frequently the perpetrators of illegal fishing can easily penetrate the sovereignty of Indonesia. Due to the privileges of international routes in the territorial waters of Indonesia and the aspects of sovereignty, which is the essential property of a state as its ultimate power, the state has a monopoly of power within its boundaries, in which a current nature of public and state organization cannot justify the individuals taking such disservice. Therefore, in an effort to overcome the violation of sovereignty or illegal fishing criminal acts, it should be done as a real action in the form of submerging the foreign ships.

In combating the problem of illegal fishing in Indonesia, and as a law enforcement effort at sea, Indonesian government has approved the Law number 31 Year 2004 *juncto* Law Number 45 of 2009 on Fisheries. Yet, it is better to understand that law enforcement on land is different from marine law enforcement at sea. Some opinions explain:

1. At sea, there are two conflicting interests, namely the interests of national and international law. While on land, only national interest is accomodated.
2. Law enforcement at sea is based on different legal regimes. Differently, differences in legal regime are not recognized on land.
3. At sea, the subject of law is a person (Indonesian citizen or foreigner) and the state shown in the flag of the ship, while on land, the subject of law is only person.
4. the provisions of national and international law grant law enforcement authorities the permit for warships against maritime security breaches

Regading law enforcement, including law enforcement against illegal fishing perpetrators, UNCLOS 1982 outlines marine regions in two categories, namely the seas under the sovereignty of a coastal state and the marine territory in which a country has the jurisdiction of a coastal state. The marine areas that are subject to the sovereignty of a coastal/archipelagic state cover inland waters and territorial sea or archipelagic waters and territorial waters. While the marine areas where a coastal/archipelagic state has sovereign rights[1] and jurisdiction are the EEZ and continental shelf

[1] Sovereignty is the highest authority that a country can perform to exercise its power over a territory and/or its

IOP Conf. Series: Earth and Environmental Science **137** (2018) 012038 doi:10.1088/1755-1315/137/1/012038

(article 77 of UNCLOS 1982). The EEZ region has a sui generis legal status (unique/different). The uniqueness lies in the existence of rights and obligations of coastal states and other countries over EEZ. Unlike in territorial sea, where coastal states have sovereignty, in EEZ, coastal states have only sovereign rights. Such sovereign rights are limited to the exploration and exploitation of marine resources, both biological and non-biological resources.

UNCLOS 1982 does not govern the IUU Fishing. The discourse on illegal fishing came together in the framework of IUU fishing practices during the CCAMLR (Commission for Conservation of Antarctic Marine Living Resources) forum that was held on 27 October to 7 November, 1997. IUU Fishing practice is conducted both in the region of the sea that is subject to sovereignty and in EEZ by the vessel of the concerned coastal state itself or by a foreign-flag ship. Although it does not regulate IUU Fishing, UNCLOS 1982 is still related to law enforcement at sea in general both in marine areas subject to the sovereignty and EEZ of a state.

If a violation of the coastal state legislation takes place in the territorial sea, inland waters or archipelagic waters of a country, then in accordance with the sovereignty granted by Article 2 of UNCLOS 1982, the coastal state may enact all of its laws and even its criminal law, against the vessel provided that the violation has an impact on the coastal state or disrupts the security of the coastal state as defined in Article 27 (1) of UNCLOS 1982. However, if the elements mentioned in Article 27 (1) UNCLOS 1982 are not met, the coastal state cannot apply its criminal jurisdiction to the vessel. The extent to which the authority of a coastal state is allowed to enforce its law for unlawful foreign ships in its territorial waters, inland waters, or archipelagic waters (meeting the provisions of article 27, paragraph 1) is the embodiment of territorial jurisdiction.

Article 27 (5) UNCLOS 1982 further refers to Chapter IX (Conservation and Protection of the Marine Environment) and Chapter V on EEZ. In the case of violations of coastal state's laws relating to exploration, exploitation, conservation, and management of fisheries resources, a coastal state is allowed to enforce its law. Coastal state law enforcement in EEZ is regulated in article 73 of UNCLOS 1982, which says:

1. The coastal State may, in the exercise of its sovereign rights to explore, exploit, conserve and manage the living resources in the EEZ, take such measures, including boarding, inspection, arrest and judicial proceedings, as may be necessary to ensure compliance with the laws and regulations adopted by it in conformity with this Convention.
2. Arrested vessels and their crews shall be promptly released upon the posting of reasonable bond or other security.
3. Coastal State penalties for violations of fisheries laws and regulations in the EEZ may not include imprisonment, in the absence of agreements to the contrary by the States concerned, or any other form of corporal punishment.
4. In cases of arrest or detention of foreign vessels the coastal State shall promptly notify the flag State, through appropriate channels, of the action taken and of any penalties subsequently imposed.

Under Article 73 of UNCLOS 1982, if a foreign ship fails to comply with the coastal state's fishery legislation in EEZ, the coastal state may board, inspect, arrest, and conduct litigation to the vessel, and notify the state of the vessel flag. However, the vessel and its arrested crew shall immediately be released with reasonable bonds provided to the coastal state. The imposed penalty shall not be in the

---

people. In the case of exercising its sovereignty, a country does not need to ask for 'permission' to other countries to exercise its power. This sovereignty, when associated with the geographical conditions of Indonesia, includes land, inland waters (internal waters), archipelagic waters, and territorial sea.

While Sovereign Right is the authority of a country to a certain region where its implementation must be subject to the rule of law applicable to the international community. What this means is, Sovereign Right of a country must be a consensus and receive approval from another country. Sovereign rights generally govern the use of natural and/or marine resources in certain areas which not covered by the territory of the State's sovereignty.

IOP Conf. Series: Earth and Environmental Science **137** (2018) 012038 doi:10.1088/1755-1315/137/1/012038

form of physical punishment, for instance, imprisonment. Thus, the law and the form of punishment for the ship and its crew in the sea area that is subject to sovereignty are different from that in EEZ. The authority of the coastal state against violations in EEZ is limited only to enforce the laws relating to fisheries. This difference is due to the fact that in EEZ, the coastal state only has sovereign rights. Therefore, it is limited to matters relating to sovereign rights owned by the coastal state or the archipelagic state. The provision of article 73 (2) requires the coastal State to immediately release the captured vessel and its crew after being given a proper security deposit or other form of security. The release procedure is immediately set forth in Article 292.

1. Where the authorities of a State Party have detained a vessel flying the flag of another State Party and it is alleged that the detaining State has not complied with the provisions of this Convention for the prompt release of the vessel or its crew upon the posting of a reasonable bond or other financial security, the question of release from detention may be submitted to any court or tribunal agreed upon by the parties or, failing such agreement within 10 days from the time of detention, to a court or tribunal accepted by the detaining State under article 287 or to the International Tribunal for the Law of the Sea, unless the parties otherwise agree.
2. The application for release may be made only by or on behalf of the flag State of the vessel.
3. The court or tribunal shall deal without delay with the application for release and shall deal only with the question of release, without prejudice to the merits of any case before the appropriate domestic forum against the vessel, its owner or its crew. The authorities of the detaining State remain competent to release the vessel or its crew at any time.
4. Upon the posting of the bond or other financial security determined by the court or tribunal, the authorities of the detaining State shall comply promptly with the decision of the court or tribunal concerning the release of the vessel or its crew.

The immediate release procedure of the detained ship and its crew is an innovation in international maritime law. However, while Article 292 Paragraph (1) requires that the security deposit or other financial security be "reasonable", UNCLOS 1982 does not provide details about the "feasibility" of the financial security.

In Fishery Law Number 31 of 2004, the sanctions given to the perpetrators of illegal fishing are quite heavy. An example is the obligation of each fishing vessel to have a fishing license, and for an Indonesian-flagged vessel violating such provisions, the operator and the owner of the vessel may be subject to a six year criminal penalty and a fine of Rp.2 billion. If the violation is done by a foreign-flagged vessel, the operator and the owner of the vessel are subjected to imprisonment of six years and a fine of Rp.20 billion. The judge must also have passed the decision within 30 days from the date of receipt of the submission of the case file from public prosecutor. The same period of time also applies to high court judges and the Supreme Court in deciding the request for Appeals and Cassation. It is regulated in Article 80 paragraph (1) of Law on Fisheries, which states: "*Dalam jangka waktu paling lama 30 (tiga puluh) hari terhitung sejak tanggal penerimaan pelimpahan perkara dari penuntut umum, hakim harus sudah menjatuhkan putusan*" (Within a maximum period of 30 (thirty) days from the date of receipt of the matter delegation from the public prosecutor, the judge shall have already rendered the verdict).

With this ad-hoc court, it is expected that the value of salvaged fish can increase while the damage of foreign vessels to be confiscated is reduced This can be donated to national fishermen. The existence of a fishery law is expected to bring positive results in law enforcement at sea. In combating illegal fishing, the government of the Republic of Indonesia has also taken concrete actions to protect its marine potential. The action referred to is establishing the Coordinating Ministry for Marine Affairs and Fisheries. This Ministry has shown that Indonesian sovereignty cannot be contested by taking decisive action to sink foreign fishing ships proven to have conducted illegal fishing in Indonesian waters.

IOP Conf. Series: Earth and Environmental Science **137** (2018) 012038 doi:10.1088/1755-1315/137/1/012038

In the implementation of supervision and guard of Indonesian maritime security, Fishery Supervisory Boat performs supervision and law enforcement in fishery field within the fishery management area of the Republic of Indonesia. The fishery supervisor performs its functions in accordance with Law Number 45 of 2009, especially Article 69. In carrying out the function, the fishery investigator and/or supervisor may take special actions in the form of burning and/or drowning the foreign-flagged fishing vessel based on "sufficient preliminary evidence" in accordance with the mandate of Law Number 45 of 2009. This "sufficient preliminary evidence" is a preliminary evidence to suspect a criminal offense in the field of fisheries by a foreign-flagged fishing vessel, for example a foreign-flagged fishing vessel which has no fishing license, and clearly captures and/or transports fish when it enters the area of the fisheries management of the Republic of Indonesia. "Sufficient preliminary evidence" in law enforcement efforts is only required in the preliminary examination stage or in searching and finding the truth. The word "sufficient preliminary evidence" is contained in Article 17 of the Criminal Procedure Code where it is stated that "sufficient initial evidence" forms the basis for arresting someone suspected of committing a crime".

Thus, "sufficient preliminary evidence" is a preliminary evidence for suspicion of a criminal offense. This provision indicates that arrest cannot be arbitrarily executed and must be addressed to those who actually commit a crime. Meanwhile, "sufficient preliminary evidence" according to the explanation of the Fisheries Law is an initial proof to suspect a fishery crime committed by a foreign vessel in which the foreign vessel clearly captures and or transports fish in the Territorial Waters and Fisheries of the Republic of Indonesia (TWFRI) without permission. Furthermore, in the explanation of the article, it is also stated that the action is only done if the investigator and/or fishery supervisor believe that the foreign vessels really have committed a fishery crime. Such a thing is very understandable because an investigator, in carrying out his duties, has a guideline or rule of law that must be obeyed. The guideline is in the form of applicable legislation to regulate what an investigator performs and how he performs the tasks in investigation

The vessel-sinking policy is aimed to demonstrate the firmness and dignity of Indonesian government in protecting its natural resources in its territory, and its sovereignty by applying a deterrent effect, and securing the sea from looting by foreigners, as well as to take a concrete action in an attempt to translate the vision of being a central maritime axis intensified by the government in the past year, especially with regard to full sovereignty at sea. The government of Indonesia has stated explicitly that it will combat illegal fishing in the fishery management territory of the Republic of Indonesia.

## 4. Conclusions

According to the analysis and discussion based on the theoretical above, the following two conclusions were drawn in accordance with the raised issues: The policy of sinking a foreign-flagged vessel in IUU fishing is part of an effort to increase the detterent or repellent effect of regional offenses that could harm and threaten the sovereignty of Indonesia. Although the regulations on vessel-sinking action have been in place since 2009, the socialization of the regulation has been under-done by the government both at home and abroad through Indonesian embassies in countries, especially those bordering Indonesia. Furthermore, the Indonesian government's policy of sinking a foreign-flagged vessel conducting IUU fishing is regulated in Law Number 45 of 2009 on Fisheries and is internationally allowed with certain restrictions under the provisions of Article 73 paragraph (3) of UNCLOS.

## 5. References

[1] Akhmad F 2007 *Fisheries and Marine Policy* (Jakarta: Gramedia)

[2] Charles T A 2001 *Suistainable fishery system, fish and aquatic resources series* (Oxford: Blackwell Science)

[3] Dikdik M S 2011 *International law of sea and its arrangements in indonesia* (Bandung: PT. Refika Aditama)

[4] Dewi R and C P F Luhulima 2008 *Problems of the indonesian sea border area in the south*

*china sea* (Jakarta: LIPI Press)
[5] Dinas Pembinaan Hukum Mabesal 2009 Authority of TNI AL Officers as Investigators Jakarta
[6] Heru P 2007 *International law of the sea* (Malang: Bayumedia Publishing )
[7] Ibrahim J 2006 *Theory and methodology of normative law research* (Bandung: PT. Remaja Rosdakarya)
[8] Kusumaatmadja M 1986 *International law of the sea* (Bandung: Bina Cipta)

IOP Conf. Series: Earth and Environmental Science **137** (2018) 012039 doi:10.1088/1755-1315/137/1/012039

# Analysis on traditional fishing grounds in Indonesia`s Natuna waters under International Law

**R Kurniaty[1,2], Ikaningtyas[1,2], P A Ruslijanto [,2]**

[1]International Law Department-Faculty of Law, Brawijaya University, Jl.MT.Haryono 169 Malang
[2] Coastal and Marine Research Center, Brawijaya University, Malang
E-mail: rika_kurniaty@ub.ac.id

**Abstract.** This paper examines the boundary tension between Indonesia and China regarding traditional fishing ground in Natuna. Indonesia`s Natuna island is claimed by the China government as its traditional fishing zone/ground. The inclusion of Natuna territory into China`s traditional fishing zone brings new problems to Indonesia, especially with the Chinese ships docked and entered Indonesia`s exclusive economic zone, as well as several cases of illegal fishing over the territorial waters of Indonesia. Claims on traditional fishing zones have the potential to threaten the sovereignty of the Indonesian territory. This study aims to analyze the claims of the traditional fishing rights of China over the waters of the Natuna Islands under international law, especially UNCLOS 1982. This study revealed that the china`s argument of traditional fishing ground in Natuna to the nine dash line map is a unilateral claim, there is no international legal norm that can be used as the legal basis. Indonesia and some ASEAN countries have Internationally validated bilateral agreement on the continental shelf (i.e. Indonesia-Vietnam and Indonesia-Malaysia) thus the inclusion of Natuna into China`s nine dash line map rejects the legal status of Indonesian water under UNCLOS 1982.

## 1. Introduction

Indonesia is one of the largest archipelago countries in the world consisting of thousands of islands. The Natuna Islands is a part of Indonesia, with its position protruding to the north and located in the South China Sea. Natuna has many natural resources (minerals and fish - reaching 1 million tons per year with total utilization of only about 36%), and has become an international maritime trajectory for ships coming from the Indian Ocean and heading towards industrialized nations around the Indian and Pacific Oceans. Furthermore, Natuna is one of the areas that is well known to generate income for Indonesia (oil and gas). Based on a study report by the Ministry of Energy and Mineral Resources (ESDM), Natuna's oil reserves amount to 308.30 million barrels. While the largest reserves of gas in Indonesia amounts to 54.78 trillion cubic feet. Oil and gas are the main source of revenue for Natuna.

However, Natuna with its strategic geographically position has been facing some problems such as the illegal fishing, and a boundary conflict with China [1]. The core problem of Natuna lies in the waterfront area around Natuna that has the potential to overlap the imaginary line of the Nine Dash Line [2]. The government of the People's Republic of China (PRC) have claimed Natuna in 2009 and entered the Natuna region into their territory map based on the nine imaginary lines (commonly called the Nine Dash Line). In this case the problem is not in the archipelago claim alone, but in the waters around Natuna as well, which affects the sovereignty of Indonesia's Exclusive Economic Zone (EEZ) area. The area of conflicting maritime claims could be as large as 94,000 square kilometers. China's traditional fishing grounds are enclosed by the Nine Dash Line and the 200 nautical mile (nm) EEZ

IOP Conf. Series: Earth and Environmental Science **137** (2018) 012039 doi:10.1088/1755-1315/137/1/012039

that Indonesia claims from Natuna in the South China Sea. Surrounding this region lies Indonesia's East Natuna gas field with 46 trillion cubic feet of proven assets, stressing the potential gamble in China's claims.

One of the incidents in Natuna waters occurred on June 18th 2016, when the Indonesian Navy seized a Chinese vessel that was operating within Indonesia's EEZ around Natuna. Similar incidents also occurred in March and May. On June 19th 2016, the Chinese government spokesperson described the incident occurring in Natuna waters as "waters that are Chinese fishermen's traditional fishing grounds." Moreover, she stated that it occurred "where China and Indonesia have overlapping claims for maritime rights and interests". China filed the Nine Dash Line claim based on on the principle of "historic waters" [3]. The claim is not just potentially threatening to Indonesian sovereignty over Natuna, but also the importance of Indonesia as an archipelago country, with the existence of conflict in Natuna between KM Kway Fey 10078 with KP Shark 11. Claims about traditional fishing grounds are potentially a form of Chinese denial of Natuna's legal status as part of Indonesia`s sovereign rights. Based on the provisions of Article 46 UNCLOS 1982: *"archipelagic State" means a State constituted wholly by one or more archipelagos and may include other islands; As for what is meant by The archipelago is a group of islands, interconnecting waters and other deep scientific characteristics such close ties thus forming a unity geographically, economically, and politically or historically indeed seen as such.*

It is debating the incident when the Chinese fishing boat KM Hwai Fey was arrested by the Indonesian law enforcement apparatus in March 19, 2016 and later came a large China Coast Guard ship which seemed to "*guard*" it; this raises a perception that there is intent from the Government of China to send fishermen to the region to show its presence. From the above description, this study examines the questions: what is meant by "*traditional fishing grounds*" claimed by PRC in Natuna waters? How is the legal position of the China`s claims in Natuna waters against Indonesian sovereignty according to UNCLOS 1982?

## 2. Methodology

The approach method used in this study was the normative juridical approach method [4, 5]. The juridical approach means within research, principles that are used for reviewing, viewing and analyzing the problem related to the application between the legal norms of "traditional fishing rights "as provided for in Article 51 UNCLOS 1982 with the norm the Chinese claim is "traditional fishing grounds". The normative approach means that through legal research sources, material is taken from legal material primers that are all related with implications of Chinese claims on traditional fishing grounds to the existence of Indonesian sovereignty as an archipelagic country. Specifications of this study are analytical descriptive, i.e. describes the rules of legislation in effect that is attributed to legal theories and practice implementation of the positive law concerning the above issues. Furthermore, describing the object that is the problem and analyzing the legal material obtained from the research.

## 3. Concept and Definition Concerning Traditional Fishing Ground

As an internal norm, the international law understanding of claims on traditional fishing grounds has not been found anywhere in any form, time or place until now, so various objections have emerged from various parties including countries in the region of the South China Sea waters, arguing that there is no juridical reference which underlies the claim. Indonesian marine law expert Prof. Hasyim Djalal stated that " *The traditional fishing grounds claimed by China in the Indonesia`s EEZ is incorrect, EEZ complies with provisions of international maritime law. There is no Chinese traditional fishing ground in the EEZ of Indonesia*". In another context, Hasjim Djalal distinguishes between traditional fishing rights with traditional rights to fish (traditional rights for catching fish) [6]. He explains that traditional rights to fish refers to the right of each country by tradition to catch fish in the open sea without regard, whether they have ever exercised that right or not. While traditional fishing rights is defined as the right to catch fish which arises because in practice they have fished in certain areas for generations and such activities have long been done [6, 7].

IOP Conf. Series: Earth and Environmental Science **137** (2018) 012039 doi:10.1088/1755-1315/137/1/012039

The basic reference that China has been doing activities in the waters is historical evidence that mentions that the Chinese people have long done activities and stayed in the Spartly and Paracel Islands (that used to be named as Xisha and Nansha) since the Tang and Song to Ming and Qing dynasties. At the end of World War II, under the Republic of China government, Chiang Kai Sek published a map with a U Shaped Line. This map was later known as the 11 Dash Line because it contains 11 broken lines which became the basis of claims by the government of China in the South China Sea dated 4 September 1958. China expressed her wishes related to South China Sea claims through the Territorial Sea Declaration. This declaration states ownership of South China Sea waters as part of Chinese territory. However, this declaration does not explain the legal claims on Chinese ownership of the South China Sea. The Chinese government affirmed their claim on the South China Sea in 2009 with controversial official maps which was followed by acts of reclamation on several islands in the Spratly Islands as well as Paracel Island [8].

Furthermore, the traditional fishing ground claim norm is a unilateral norm from the related claims of the Chinese Nine Dash Line which is against the sovereignty of a maritime feature (Spratly archipelago of Spratly and Paracel) in the waters of the South China Sea. Its assumption of emergence has juridical consequences to the rights of additional zones, EEZ and continental land. China`s claim to the territory of the cluster of the Spratly Islands as well as the Paracel Island will impact on the recognition of attribute of the accompanying law of such claims, as governed in the provisions of Article 121UNCLOS 1982 on the island regime that states: *1. An island is a naturally formed area of land, surrounded by water, which is above water at high tide.2. Except as provided for in paragraph 3, the territorial sea, the contiguous zone, the exclusive economic zone and the continental shelf of an island are determined in accordance with the provisions of this Convention applicable to other land territory.3. Rocks which cannot sustain human habitation or economic life of their own shall have no exclusive economic zone or continental shelf* [1, 7].

Moreover, the majority of countries in the world acknowledge and use the UN Convention on the Law of the Seas (UNCLOS) 1982 as a legal basis within set boundaries of its waters, while China uses the Nine Dash Line map which is their own. Many parties later interpret this Nine Dash Line as a form of China's claim to the South China Sea becomes overlapping claims against sovereign territory of other countries around its waters. Due to Chinese claims, Natuna waters are not just involved in fishery-related disputes, but more importantly as a violation of the maritime zone, sovereign rights, and sovereignty that cannot be tolerated. As a sovereign state that claims sea territory based UNCLOS 1982, Indonesia certainly cannot let foreign fishermen catch fish in the the area which belongs to it. For Indonesia, the area is not even just a claim but it is a right protected by international law, though it certainly still requires bilateral agreements with neighboring countries, especially Malaysia and Vietnam [9].

Disputes in this region cover two main issues, namely territorial and maritime sovereignty. Territorial sovereignty refers to ownership of the territory of the land / islands in the South China Sea area, while maritime sovereignty disputes relate to the allowable limit setting by Article 121 of the United Nations Convention on the Law of the Sea (UNCLOS III) 1982, namely territorial sea sovereignty and sovereign rights of EEZ, including the continental shelf and additional zones.

## 4. Setup and Implementation of Traditional Fishing Rights According to UNCLOS 1982

One of the legal regimes that is important from UNCLOS 1982 for Indonesia is the recognition of the archipelago state regime [3]. Under the provisions of Article 121 of UNCLOS 1982, countries with territorial sovereignty over the island have sovereign rights of 12 nautical miles of territory and 200 miles of EEZ around the island. Subject to the provisions of Articles 47 and 49 of UNCLOS 1982 if the state has sovereignty over the entire archipelago and becomes an archipelagic state, it shall have the right to draw a straight line between the outermost islands and shall have sovereignty over the natural resources on the seabed located at in the straight base line. Island countries draw their baseline by using the baseline method of the islands (archipelagic baseline). Consecution line drawdown when based in this way is the change of section status of the sea that was once free sea into the territorial sea

IOP Conf. Series: Earth and Environmental Science **137** (2018) 012039 doi:10.1088/1755-1315/137/1/012039

of an Archipelago state. Recognition of the island states are accompanied by various other settings such as providing guarantees of cross-rights of peace (right of innocent passage) and cross-rights through the archipelagic country lanes (the right of archipelagic sea-lanes passage) for foreign ships in the inland seas of the island country.

In addition, the island countries must also respect the traditional fishing rights of neighboring countries and existing agreements with other countries. Traditional fishing rights is a right which is given to the traditional fishermen to fish in a traditional manner in the waters of certain neighboring island countries under a bilateral agreement. Recognition of rights are accommodated in Chapter IV Article 51 paragraph (1) of UNCLOS 1982 stating that the archipelagic state must respect the agreements that exists with other states and must recognize the rights of traditional fishery and other activities This applies to a legitimate directly neighboring state in the archipelagic waters. Terms and provisions for the exercise of such rights and activities including the nature, scope and areas of of such activities apply, are above the request of one country concerned and should be arranged with bilateral agreements among them. Such rights should not be transferred or shared with a third state or a resident country [5]. As an archipelagic state, Indonesia has an obligation to respect traditional fishing activities that has been carried out for generations in the waters and have changed status into waters of the archipelago. The recognition of island countries shall implement these rights according to UNCLOS 1982.

Furthermore, Indonesia has signed various bilateral agreements with neighboring countries about the recognition of traditional fishing rights, such as Australia, Malaysia and Papua New Guinea. Indonesia had entered into bilateral agreements with Malaysia in 1982 called The 1982 Agreement on the Regime Laws of the State Archipelago and Malaysia's Rights in the Territorial Sea and Waters. The provisions formulated in this bilateral agreement is in line with the provisions in Article 47 paragraph (6) of UNCLOS which specifies that if a part of an island of archipelagic states lies between two parts of a neighboring state which are directly adjacent, the existing rights and other legitimate importance are implemented traditionally by that state at such waters, as well as all rights specified in the agreement between states will remain in effect and must be respected.

In addition to Malaysia, Indonesia has also made a basic agreement with Papua New Guinea (PNG) about border settings which recognizes traditional rights, including traditional fishing rights. If associated with the terms of Article 51 (1) UNCLOS 1982, the norm becomes the state's obligation to honor and recognize traditional fishing rights of the neighboring country that are directly side by side with the island state in a variety of International law instruments and practice. There is strong legitimacy to give protection to traditional fishermen. There are several international conventions which support the recognition of the rights of traditional fishermen. It can be concluded that in fact recognition of traditional fishing rights is not only an obligation for the island state. Traditional fishing rights are also attached to groups of fishermen of a particular State who have habitually fished over a long period in a certain area. A concrete example is in the intermediate agreement between Indonesia and Australia which provides protection of the traditional fishing rights of Indonesian fishermen in Australia [9]. Australia is not an island state, so granting rights is more based on the historical realities of some groups of traditional fishermen from generation to generation, such as searching for sea cucumbers and fishing in Australian waters, especially on Ashmore Island (Sand Island).

Through the provisions of Article 51 and Article 62 paragraph (3) UNCLOS 1982, the governments of Indonesia and Australia have accommodated the interests of the traditional fishermen in a Memorandum of Understanding, i.e. Memorandum of Understanding between the Government of Australia and the Government of The Republic Of Indonesia regarding the Operation of Indonesian Traditional Fishermen in the Area of the Australian Exclusive Fishing Zone and Continental Shelf in 1974 [9].

## 5. The Legal Position of Traditional Fishing Ground Claims by China in Natuna Waters

The Indonesian government reacts hard on the intervention of the Chinese authorities during the incident when the Marine and Fisheries Ministry (KKP) tried to catch the ship of the illegal fisherman from China around Natuna waters, Riau Islands. The Government of China reasoned that the territory is still in the area of traditional fishing grounds. China claims sovereignty on several islands and oceans around the South China Sea based on historical rights, factor of discovery, naming, and organization history of the government that has been going on for more than 2,000 years. Consequently, China's claim on the South China Sea garnered protests and became disputes with a number of ASEAN countries such as Vietnam, Philippines, Brunei, and Malaysia. Subject to the provisions of Article 121 UNCLOS 1982, the country that owns sovereign territory over the island has sovereign rights of 12 nautical miles and 200 miles of ZEE around the island. Subject to the provisions of Article 47 and 49 UNCLOS 1982, if the country has sovereignty over the whole islands and becomes an island country, then the country has the right to draw a baseline straight between the islands outer and will have sovereignty over the natural resources at the seabed lying within the straight baseline.

Indonesia was originally a Non Claimant State in the South China Sea dispute that finally reacted because of its impact on Traditional Fishing Ground claims. This is due to the Chinese statement that they have the right to conduct fishery activities in their traditional territory according to the Nine Dash Line Map which is against the sovereignty of maritime features (Spratly Islands and Paracel cluster) of the waters of the South China Sea with the assumption of the emergence of juridical consequences against the rights of its maritime zone which includes sovereignty and sovereign rights to all resources in it.

Regarding the boundaries with neighboring countries, Indonesia and Vietnam share a continental shelf boundary in the South China Sea. The two countries have agreed through an agreement between the Government of the Republic of Indonesia and Government of the Socialist Republic of Vietnam about the Determination of Continental Shelf Boundary, which was signed in Hanoi on June 26th 2003 and ratified with Law no. 18 of 2007 on the Sea Shelf Boundary Agreement. In addition, it is argued that China's claim to traditional fishing grounds is inconsistent with UNCLOS 1982. The convention's cardinal principle is that the land dominates the sea. Claims to maritime jurisdiction must be based on the ownership of land features. As China recognizes Indonesia's sovereignty over the Natuna Islands, any Chinese claim to maritime rights such as traditional fishing grounds would have to be based on an island in the South China Sea that could generate a 200 nautical miles of EEZ.

## 6. Conclusion

Based on research and results, the conclusion and suggestions are as follows: China`s claim on traditional fishing grounds is a unilateral claim of China fishery activities in the region of the South China sea. The basic argument is the Nine Dash Line map which up to now has no legal norm (international) that can be made as a legal basis. China`s claim on traditional fishing grounds (inherent in the Nine Dash Line map) has voided valid bilateral agreements between Indonesia-Vietnam and Indonesia-Malaysia as well as denying/violating the legal status of Indonesia as an Archipelago state under UNCLOS 1982.

## 7. References

[1] Dikdik M S 2011 *International Law of the sean and its arrangements in Indonesia* (Bandung: PT. Refika Aditama)

[2] Jinming L dan L Dexia 2003 The dotted line on the chinese map of the south china sea: a note ocean development and international law **34**

[3] Mohamed M 1995 *Ocean States: Archipelagic Regimes in the Law of the Sea* (Dordrecht: Martinus Nijhoff ) p 5

[4] Goldie L F E 2008 The Critical Date International and Comparative Law Quarterly **1204** 1963

[5] Rony H S 1998 *Legal research methodology and jurimetry* (Jakarta: Ghalia Indonesia)

[6] Djalal H and I Townsend-Gault 1999 *Preventive diplomacy: Managing Potential Conflicts in the South China Sea* (Washington D.C: United Stated Institute of Peace Press)

[7] Hasjim D 1995 *Indonesia and the law of the sea* (Jakarta: Centre for Strategic and International Studies) p 17

[8] Sichun W 2013 *Solving disputes for regional cooperation and development in the South China sea: a chinese perspective* (UK: Chandos Publishing) Oxford
[9] Songa, W W 2000 *Implementation of the agreement between indonesia and australia on traditional fisheries rights related to fishermen from East Nusa Tenggara* (Thesis: Padjajaran University)

IOP Conf. Series: Earth and Environmental Science **137** (2018) 012040 doi:10.1088/1755-1315/137/1/012040

# The effect of water immersion on decreasing copper (Cu) and granulocyte levels in *Crassostrea cucullata*

**D Arfiati, D P Arsanti, D R Suci, A Kurniawan, U Zakiyah and H F Kharismayanti**

Faculty of Fisheries and Marine Science, Brawijaya University, Malang, Indonesia

E-mail :d_arfiati@yahoo.com

**Abstract.** Oyster habitat in a body of water often contain excessive metal content. This research aimed to decrease of Cu level concentration and understand the change of granulocyte in oyster body after being treated for sea water for 12 hours.This research used experimental method with Completely Randomized Design consisting 4 treatments and 3 replications. Cu concentration was measured by using AAS method and the amount of hemocyte (granulocyte) was calculated by using haemocytometer. Cu concentration in oyster before treatment was 0.374 ppm. Immersion with the water discharge of 0.05 m/s for 12 hours was able to decrease Cu 46 % (0.158 ppm of Cu residue) with the lowest granulocyte (31.92 %).

## 1. Introduction

### *1.1. Background*

Coastal waters contain copper (Cu) of 0.063 ppm, and the level measured in *Crassostrea cucullata* is 0.374 ppm. According to the Regulation of the Minister of Environment of the Republic of Indonesia No. 51 2004 stated that, the Cu level in marine biota should be 0.008 ppm.Thus, the Cu concentration found in *Crassostrea cucullata*is under the standard quality. The amount of Cu in oysters can be reduced by immersing water. The immersion of *Crassostrea cucullata* in seawater sterilized with chlorinfor 3 days can significantly decrease the total number of hemocytes, thus the oyster's condition improves after the first 12 hours of immersion [1]. If the oyster's body contains a high level of heavy metal, the production of hemocytes will increase [2]. Oyster hemocytes consist of hyalinocytes, that recognize foreign materials, and granulocytes, that perform phagocytosis.

Copper or Cu in oysters is required for the formation of haemocyanin (oyster blood), and the permissible level is only 0.008 ppm. The absorption of Cu into the blood can occur in acidic conditions in the stomach [3]. An environment with toxic conditions will trigger the body of an oyster to adjust its metabolism by performing excretion or detoxification. The natural detoxification process begins with the binding of metal ions on the cell surface, since positive ions are bound to the reactive side to the negative charge of the extracellular polymer. The next stage is the transport of metal ions to the cytoplasm, and then the metal ions are accumulated by metallothioneins or MT proteins [4].

Various animal physiological devices such as rates of excretion and metal detoxification are typical of each species [5]. Not all organismsare able to excrete metals. Therefore, the rate of detoxification in animals becomes the only balancer of the uptake rates to prevent metal toxicity. When metal first enters the body, metal regulation immediately begins, and the metal ions are circulated throughout the body by the hemolymph fluid [6]. Toxic metals are excreted, while un-excreted metals will accumulate in the body. The accumulated metals will be regulated through metabolism or released through detoxification or excretion. All non-essential metals (including heavy metals) will be detoxified by the animal body by being strongly bonded to a particular site that is difficult to be removed.

By immersing clean water can dissolve the protein in the oyster body. The use of destilled water as a solvent can reduce protein levels because aqua is able to dissolve and modify the properties of biomolecules such as nucleic acids, proteins and carbohydrates through the formation of hydrogen bonds with polar functional groups [7]. According to Suaniti [8], the heavy metal concentrations are reduced byreleasing metal-protein complex bonds, so that the metal ions are out of the oyster stomach. Metal ions are naturally present in the animal body, and almost all of them bind to proteins.

## 2. Methodology

This research was conducted by using experimental method with Completely Random Design and consisted of four treatments of water flowing (3-hour, 6-hour, 9-hour, 12-hour flow) and 3 replications. Oysters were sterilized by water sterlizied with chlorine at a rate of 0.05 m/sec. Phytoplankton observations were conducted in the original habitat of oysters to determine the amount of feed administered in the treatment media, that was 9.070 cells/L. The height of the tidal water was also measured to determine the average water content of the treatment media, which was 30 cm in average. In addition, the oysters and water samples were taken to observe the levels of granulocyte and copper (Cu) prior to treatments.

The pailfor treatment was washed and chlorinated first, and it was given certain hole to adjust the water flowat the outlet. Furthermore, water filling in each treatment pailwas done at a height of 30 cm in accordance with the condition of the oysters' habitat. Each ± 20 liter basin was given an aerator and filled with four oysters. Water flowing was done for up to 12 hours based on the tides. Oyster sampling was done every 3 hours for measurement of granulocyte and Cu as well as water quality (temperature, pH, dissolved oxygen, and salinity).

## 3. Results and Discussion

### *3.1.Analysis of Cu and granulocyte in Crassostrea cucullata*

The levels of Cu and granulocyte decreased after the oysters wereimmersed by water for up to 12 hours. The granulocyte level decreased by 31.92 %, and the Cu level decreased to 46 % (figure 1). Treatments of 3-hour, 6-hour, 9-hour and 12-hour had significantly decreased the granulocyte and Cu levels in oysters. The best treatment was 12-hour because the granulocyte level decreased by 31 %, and the remaining Cu level was 46 %.

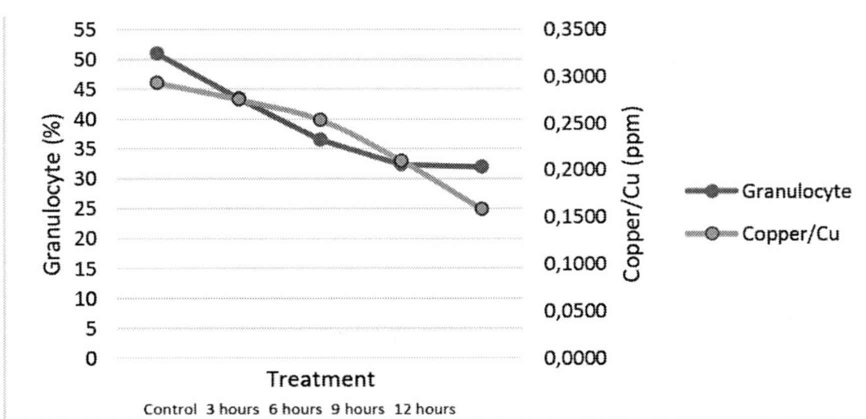


**Figure 1**. The Average of Cuand Granulocyte in *Crassostrea cucullata* after3 hours, 6 hours,9 hours and 12 hours Immersion.

### *3.2. Water quality parameters*

Water quality parameters such as temperature, pH, dissolved oxygen, and salinity at the sampling location and immersion treatment location were in the normal range (table 1). However, there was a difference of 5.8 °C between the sampling location and the treatment location .

**Table 1.** Water quality analysis result.

| Water Quality | Observation Result | | Standard |
|---|---|---|---|
| | Oyster Habitat | Laboratory | |
| Temperature (°C) | 30.5 | 24.7-26.3 | 28-32 [12] |
| pH | 8 | 8.3- 8.5 | 7.8-8.6 [13] |
| DO (mg/L) | 2.35 | 7.62 – 8.43 | >1 [14] |
| Salinity (ppt) | 33 | 32.5 – 35 | 27-33 [12] |

Water quality becomes one of the factors that supports oyster life.When the research was conducted, the water quality parameters were in the range appropriate for oyster life. The difference between the temperature of the sampling location and that of the treatment location was 5.8 °C, which was still in the range of the appropriate temperatures for oyster life [12]. Bivalves are able to live in waters at pH 7.8–8.6 [13].Oysters are still able to survive for 5 days in waters containing >1 mg/L of dissolved oxygen [14]. The optimal salinity range for bivalves in the waters is 27-33 ppt [12].

The average concentration of Cu in the oysters was 0.29297 ppm, and the number of oyster granulocytes (control) was 50.88 %. The levels of Cu and granulocyte decreased after the oysters were for 12 hours. The granulocyte level decreased 31.92 %, and the Cu level decreased 46 %. If the percentage of granulocytes is 31 %, it means the oyster is in good health or the environment has not been contaminated [9]. The granulocyte levels of bivalves in the uncontaminated waters ranged from 34 %, and the granulocyte levels of bivalves in polluted waters ranged from 52 %. Based on this research, the seoysters are safe to consume.

Heavy metals in animal body are absorbed by the blood, bound by blood proteins, then distributed throughout the body tissues [10]. Heavy metals can enter the oyster body in two ways: passive uptake (passive transport) and active uptake (active transport) because heavy metals and essential metals have similar physical-chemical properties [11].

The flowing using seawater over *Crassostrea cucullata* decreased the Cu and granulocyte levels. The 12-hour flowing was effective to reduce the amount of Cu in oysters to 46 %, and the number of granulocytes by 31 %.

## 4. References

[1]. Arfiati D, Nuriyani, Mufarika and Asiyah S N 2015 Effect of Deeping to Hemosit of Oyster from Waters Containing Hg, Cd and Pb *Proc. Int. Fish. Sym* (The Gurney Hotel, Penang, Malaysia 1st – 4th Desember 2015)
[2]. Ekawati AW, Nursyam H, Edi W and Marsoedi 2012 *J. Sci. Tech.* **14** 46-53 (In Indonesia)
[3]. Palar, H 2004 *Heavy metal pollution and toxicology (*Jakarta: Rineka Cipta) (In Indonesia)
[4]. Bernal-Hernandes Y Y, Medina-Diaz I M, Robledo-Marenco M L, Velazques-Fernandes J B, Giron-Perez M I, Ortega-Cervantes L, Maldonado-Vasquez W A andRojaz-Garcia A E 2010 *Ecotoxicology* **19** pp 819-825
[5]. Cardoso P G AI, Lillebo E, Pereira A, Duarte C and Pardal M A 2009 *Mar.Envi. Res.* **68** 1-22
[6]. Rainbow P S 2006 *Envir. Int.* **33** 576-582
[7]. Nuraini A and Sulistyorini L 2006 *J. Envi. Health* **2** (In Indonesia)
[8]. Suaniti N M 2008 *J. Envi. Sci.* **2** (In Indonesia)
[9]. Giamberini L, Michel Auffret and Jean-Claudie P 1996 *J. Mollusca Study.* **62** 367-379

[10]. Arifin 2005 Know the heavy metal http:smk3ae.wordpress.com.2009-02-02-mengenal-logam-berat-heavymetal. Diakses tanggal 10 Juni 2017
[11]. Ridhowati S 2013 *Know metal pollution* (Yogyakarta: Graha Ilmu) (In Indonesia)
[12]. KEPMENLH 2004 Regulation of the minister of environment No 51 of 2004 on the quality standard of sea water
[13]. Winanto T 2004 *Producing pearl oysters* (Jakarta: Penerbit Penebar Swadaya). pp 17-24 (In Indonesia)
[14]. Octavina C, Yulianda F and Krisanti M 2014 *Jurnal Depik.* **3** 108-117 (In Indonesia)

IOP Conf. Series: Earth and Environmental Science **137** (2018) 012041 doi:10.1088/1755-1315/137/1/012041

# Validation of potential fishing zone forecast using experimental fishing method in Tolo Bay, Central Sulawesi Province

**W E Rintaka and E Susilo**

Institute for Marine Research and Observation, Ministry of Marine Affair and Fisheries
Jl. Baru Perancak, Jembrana - Bali, Indonesia

E-mail: era09.bpol@gmail.com

**Abstract.** The national scale of Indonesian Potential Fishing Zone (PFZ) forecast system has been established since 2000. Recent times this system use Single Image Edge Detection algorithm to automatically identify thermal front from remote sensing images. Its generate two fishing ground/FG criteria: FG (high probability) and potential fishing ground/PFG (medium/low probability). To quantify the accuracy of this algorithm, an experimental fishing/EF was carried out in Tolo Bay, Central Sulawesi Province at September 2016 the late southeast monsoon period by using a pole and line fishing vessel. Four fishing activities (P1, P2, P3, and P4) were conducted during this study at a different location nearby the PFZ forecast position, two of them had good results. Based on distance measurement, these locations P1 and P4 were associated with PFZ forecast position. They were associated with PFG and FG criteria. The distance between EF to P1 and P4 were 9.7 and 6.69 nautical miles. The amount of catch for each location was 850 and 900 kg, respectively.The other locations P2 and P3 were also associated with PFG criteria, but there was no catch. We conclude that the number of the catch is influenced by the distance from PFZ forecast position and criteria.

## 1. Introduction

The information of fishing ground forecast in Indonesian waters is very important. The fishermen face limitation in determining fish migration areas and fishing grounds. Oceanographic variability is considered to be the dominant factor affecting the pelagic fish distribution and abundance in the ocean [1,2,3]. It also tends to be influenced by regional and global climatic conditions [4, 5]. In the last few decades, the availability of satellite images allows us to monitor marine environmental conditions in near real time for large areas, making them particularly suitable for determining the pelagic fishing ground [6,7,8,9].

The Institute for Marine Research and Observation (IMRO) has developed a prediction system of pelagic fishing ground in Indonesian waters. A national scale of Potential Fishing Zone (PFZ) forecast system called *Peta Prakiraan Daerah Penangkapan Ikan* (PPDPI) has been established in 2000. The IMRO PFZ forecast system has provided the prediction of pelagic fishing ground locations throughout Indonesian waters routinely. The information of potential fishing locations is generated based on oceanographic variables such as chlorophyll-a concentration, sea surface temperature and sea surface anomaly derived from satellite oceanographic images. Recently, the PFZ forecast system uses Single Image Edge Detection (SIED) algorithm to identify thermal front from sea surface temperature images automatically. This system will generate two fishing ground criteria, namely fishing ground/FG (high probability) and potential fishing ground/PFG (medium or low probability) [10, 11]. PFZ forecast system is expected to be helpful to determine the potential locations for fishing activities and to make fishing operations more effective and efficient.

In general, research and development activities are still underway to improve the availability and quality of IMRO PFZ forecast system based on three aspects, namely: 1) near real-time data retrieval processing; 2) algorithm development; and 3) accuracy assessment. The accuracy of PFZ information becomes an important aspect. Several studies have been carried out to determine the accuracy of this product based on fishing logbook and remote sensing data. After being compared with the fishing dataset, a significant accuracy value for Bali Strait area of 41.80 % was obtained. The highest value (69.57 %) occurred in October [12]. Furthermore, some radar images showed that the fishing vessel activities at Natuna Sea corresponded to PFZ information (40.97 %) with the highest value reaching 86.74 % in May [13]. Nevertheless, an accuracy assessment through direct fishing survey activities (experimental fishing) has not been done. The purpose of this study was to determine the variability of oceanographic conditions and trends and to validate PFZ forecast product requiring actual catch data at fishing grounds by collecting experimental fishing data at Tolo Bay, Central Sulawesi Province.

## 2. Methodology

### *2.1. Study area*

Tolo Bay is the water lying between the eastern and south-eastern peninsulas of Sulawesi (Celebes) Island in Indonesia. Being included in the Fisheries Management Area (FMA) 714, the oceanographic conditions of Tolo Bay are highly influenced by the Banda Sea since Tolo Bay lies open to and face directly to this area (figure 1). Tolo Bay has a huge amount of small and large pelagic fish production, approximately 68.456 tons per year, which is dominated by skipjack tuna and eastern little tuna [14].

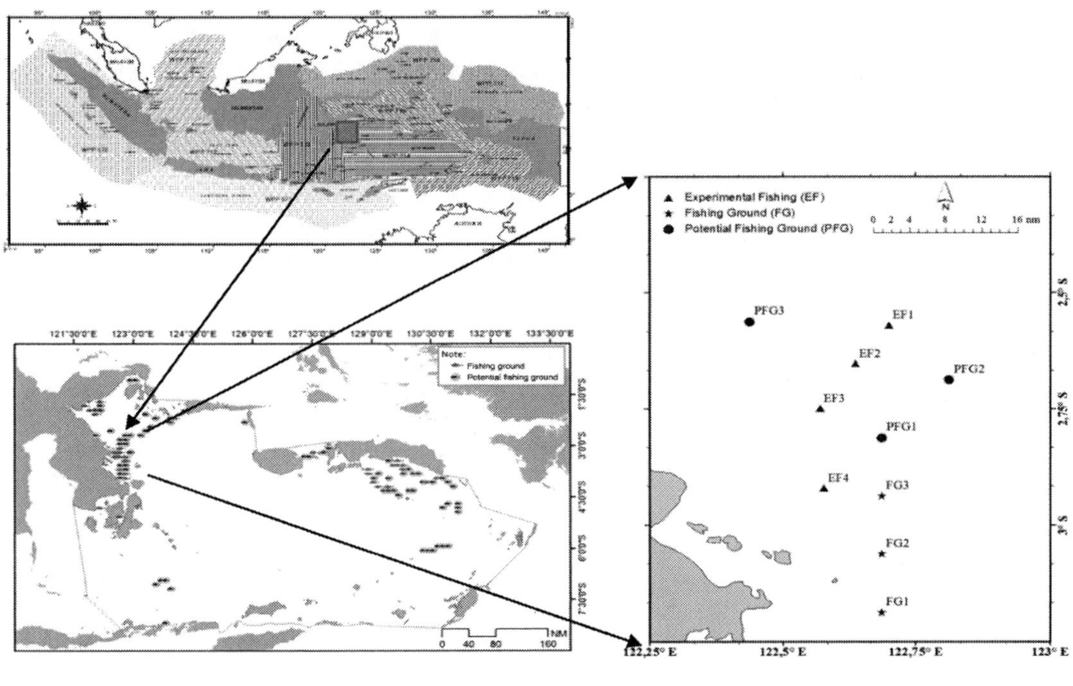


**Figure 1**. The study area (blue square). The location of experimental fishing (▲), PFZ forecast criteria (●: *PFG or* ★: *FG*).

### *2.2. Potential fishing zone data set*

The spatial distribution and trends of potential fishing zone were prepared based on the time series analysis of the IMRO PFZ forecast system in 2016, particularly in the Banda Sea. This forecast system provided information three times per week routinely (Monday, Wednesday and Friday). It generated two PFZ criteria (fishing ground/FG and potential fishing ground/PFG) based on the appearance of thermal front and favorable range of chlorophyll-a concentration. Both variables were derived from MODUS (Moderate Resolution Imaging Spectroradiometer)/Aqua or MODUS/Terra Level 3 images. The sea surface temperature images were used to identify thermal front by using Single Image Edge Detection (SIED) algorithm. Meanwhile, the favorable range of chlorophyll-a concentration was determined to be 0.2-0.5 $mg/m^3$ [10]. These satellite images were downloaded from the NASA's OceanColor web (http://oceancolor.gsfc.nasa.gov). The daily fishing ground locations were used for validation purpose. There were three pieces of FG information and three pieces of PFG information generated by the PFZ forecast system coinciding with the experimental fishing survey time (figure 1).

### *2.3. Validation method*

The experimental fishing activities were carried out for five days in the late southeast monsoon period (20-24 September 2016) in four different locations near the PFZ (figure 1). The fishing activities were conducted by using a pole and line vessel from Kendari Fisheries Port. The results of this survey include the information on the actual fishing activities, i.e. fishing location (longitude/latitude) and characteristics of fish catch (species, amount, and size). The accuracy assessment was done by comparing the PFZ information with the experimental fishing dataset based on the distance from PFZ and amount of catch. We also measure the oceanographic variables (temperature, salinity, chlorophyll-a) in the fishing location to know the relationship between oceanographic conditions, particularly the vertical profile of temperature and chlorophyll-a concentration, and the number of catches.

### *2.4. Oceanography variables*

Measurement oceanography variables (temperature, salinity, chlorophyll-a) to know environmental parameters in the fishing location. The oceanography variables in the fishing location to know relationship vertical profile of temperature and chlorophyll-a concentration, with the abundance variability and number of catch.

## 3. Results and Discussion

### *3.1. Relationship between potetial fishing zone and oceanographic conditions*

Time series analysis of the potential fishing zone generated by IMRO PFZ forecast system is important to know the occurrence of the potential fishing zone in experimental fishing areas and its relationship with the variability of the oceanographic conditions. The highest number of fishing grounds in 2016 occurred during the southeast monsoon period (June-July), which coincided with the peak of upwelling in the Banda Sea. The feature occurring peak upwelling shows low sea surface temperature (SST) and high chlorophyll-a concentration according to the satellite images. The time series of MODIS/Aqua or MODIS/Terra satellite images revealed that the SST and chlorophyll-a ranges during this period were 26.82-30.9°C and 0.13-1.63 $mg/m^3$, respectively. However, the number of fishing grounds decreased significantly during the rainy season (December-May). This is because of the fact that the satellite images in that period had high cloud cover so that the PFZ forecast system could not generate any PFZ information. Warmer SST and chlorophyll-a concentration fall during the northwest monsoon period trigger the decrease of number of fishing ground occurrences. The full temporal variability of SST, chlorophyll-a and number of fishing grounds in 2016 can be seen in figure 2.

The distribution and seasonal variation of fish involve the study of diverse factors such as the availability of food supply [14], as well as temperature, light, salinity and other chemical properties of water quality [15, 16]. SST will lead to changes in current pattern, upwelling and accumulation of plankton [17]. Water temperature and phytoplankton concentration usually influence the behavioral response of fish [18, 19]. During the northwest monsoon period, the average temperature of the Banda Sea was 4°C warmer than that in the southeast monsoon period [20]. Other research concludes that in February the wind blows from the northwest, which causes the waters in the Banda Sea, Java Sea and Flores Sea to be relatively warm [21].

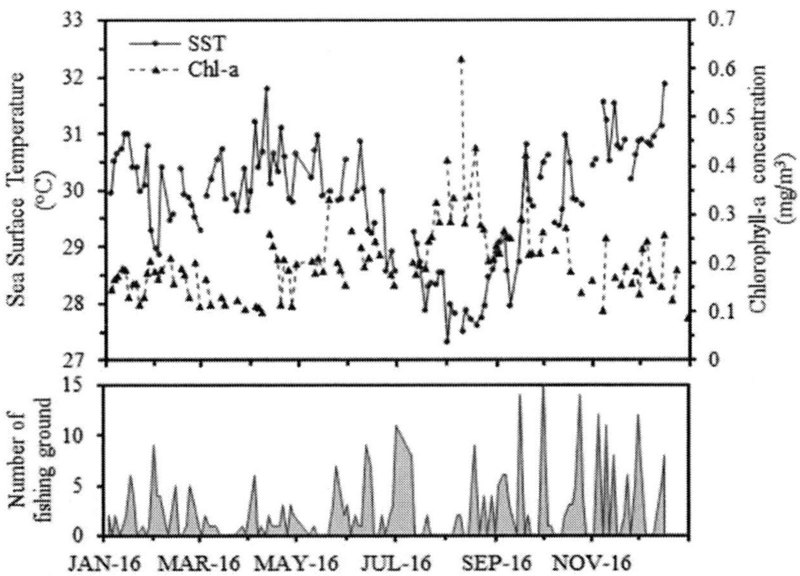


**Figure 2**. Temporal variability of sea surface temperature, chlorophyll-a concentration, and number of fishing ground in 2016.

The spatial and temporal analyses of SST according to satellite images show that low-temperature conditions occurred from mid-July to mid-September 2016, which ranged between 22-23 °C (Figure 3). This condition coincided with the period of the monsoon, which is characterized by the gusts of southeast monsoon winds that carry cold temperatures from the eastern part of the Banda Sea. Conversely, during the northwest monsoon period, the ocean temperature conditions become warmer in the range of 26-29°C. Being one of the waterways from the Pacific Ocean to the Indian Ocean through Arlindo system, the Banda Sea has characteristics and dynamics strongly influenced by the topography and configuration of the existing islands and the monsoon winds that occur in the region. The basic variation of the waters and the configuration of the archipelago can lead to mixing of water masses, changes in current direction and upwelling, which cause current deflection due to the threshold and the sea. Monsoon winds blowing in the Banda Sea, as in Indonesian waters in general, are the northwest monsoon in December-February and southeast monsoon in June-August [22]. Low SST and high chlorophyll-a concentration tend to be seen in near-shore areas. The high concentration of chlorophyll-a in these coastal waters as a result of upwelling occurs in coastal areas. High concentration of chlorophyll-a is also suspected to be caused by the high supply of nutrients derived from the mainland through river runoff. Conversely, they tend to be low in offshore area. The chlorophyll-a concentrations ranged between 0.65 mg/m$^3$ and 0.90 mg/m$^3$. In general, chlorophyll-a concentration has a negative correlation with SST, in that when the SST reaches a minimum value, chlorophyll-a concentration is relatively close to a maximum value. This can be seen clearly in December-January when the minimum surface-chlorophyll-a occurs and in July-August when the maximum occurs (figure 4). This result is close to the result of research [23], showing that the chlorophyll-a concentration of the Banda Sea reaches a maximum value (0.45 mg/m$^3$) in the the

IOP Conf. Series: Earth and Environmental Science **137** (2018) 012041 doi:10.1088/1755-1315/137/1/012041

eastern monsoon period and, on the contrary, reaches the minimum value (0.15 mg/m$^3$) in the western monsoon period.

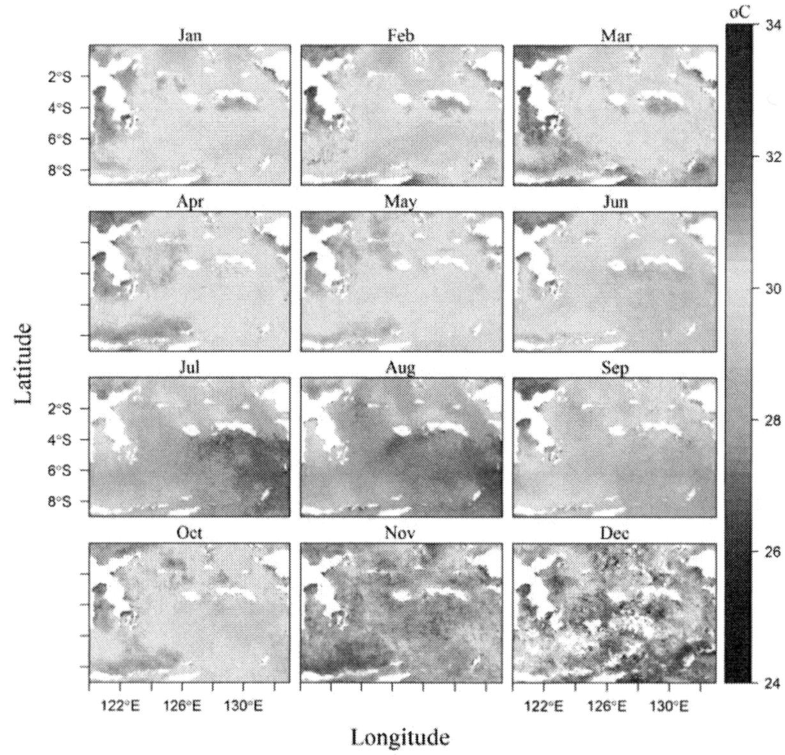


**Figure 3.** Spatial and temporal distribution of monthly sea surface temperature in 2016.

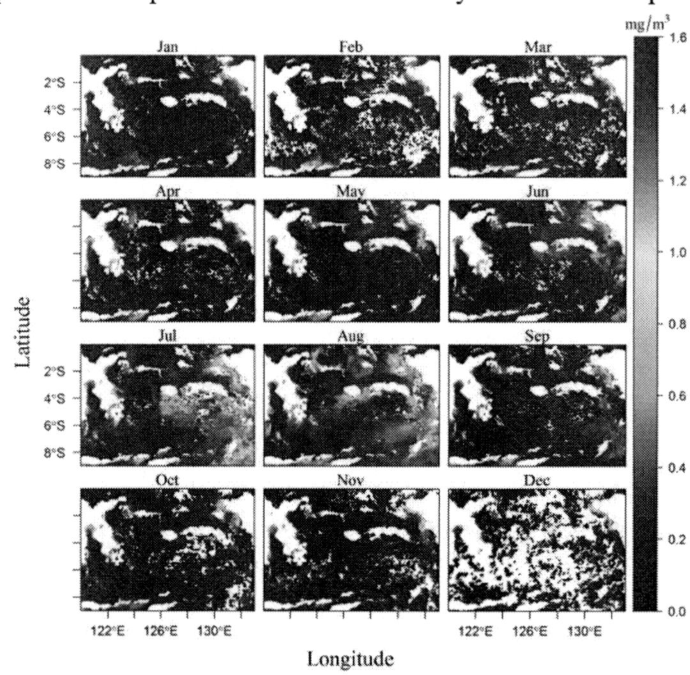


**Figure 4.** Spatial and temporal distribution of monthly chlorophyll-a concentration in 2016.

IOP Conf. Series: Earth and Environmental Science **137** (2018) 012041 doi:10.1088/1755-1315/137/1/012041

Upwelling causes nutrients enrichment on the ocean surface layer. High nutrients increase phytoplankton growth and primary productivity [5, 24]. In addition, the emergence of cold air mass increases the chances of the front temperature [25]. The front temperatures at the moment rise more clearly and strongly by the longer front line [11]. However, the use of image data from passive sensor recording, such as MODIS, is associated with cloud cover problem. This prevents the identification of front temperature and favorable chlorophyll-a concentration, which are the main parameters of PFZ forecast system, from being carried out. The horizontal variation in SST and chlorophyll-a concentration results in a change of current patterns indicative of upwelling process and accumulation of phytoplankton along the thermal front/boundaries. Upwelling is quite pronounced in the equatorial region when trade wind blows and surface water flows from east to west in the southeast monsoon period [17]. During the southeast monsoon period (July-September) strong winds from the Banda Sea blow to Tolo Bay until the southest coast of Sulawesi and cause upwelling in this area. Meanwhile, weak upwelling is observed during the northwest monsoon period (Februari-Maret).

The in situ SST from CTD measurenment during the southeast monsoon period ranges between 29.01°C and 29.51°C, which is relatively constant from the surface to a depth of 16-50 m. The temperature decreases with the increase of depth. Relative temperatures decrease drastically at a depth of 50 (28.32°C) to 100 m (19.76°C). The minimum temperature at a depth of 120-150 meters having been observed ranges between 19.46°C and 21.21°C. The surface salinity ranges between 33.61 psu and 33.75 psu. At a depth of 20 meters, it decreases between 34.17 psu and 34.21 psu. The salinity ranges between 34.08 psu and 34.28 psu at a depth of 50 to 100 m. Salinity increases with the increase in the depth and reaches the maximum at a depth of 120-150 meters. The surface chlorophyll-a concentration ranges between 0.04 mg/m$^3$ and 0.09 mg/m$^3$. It increases with the increase in depth until reaching a maximum value, and the concentration decreases until the maximum depth is reached. The maximum chlorophyll-a consentration is obtained at a depth of 45-50 meters, ranging between 0.66 mg/m$^3$ and 0.88 mg/m$^3$. The vertical distribution of temperature, salinity and chlorophyll-a is very important because it is closely related to the distribution of fish, especially pelagic fish such as skipjack tuna and yellowfin tuna. Skipjack tuna can be caught at a depth of 0-40 meters. Its spread in tropical waters is strongly influenced by the thermocline layer (figure 5).

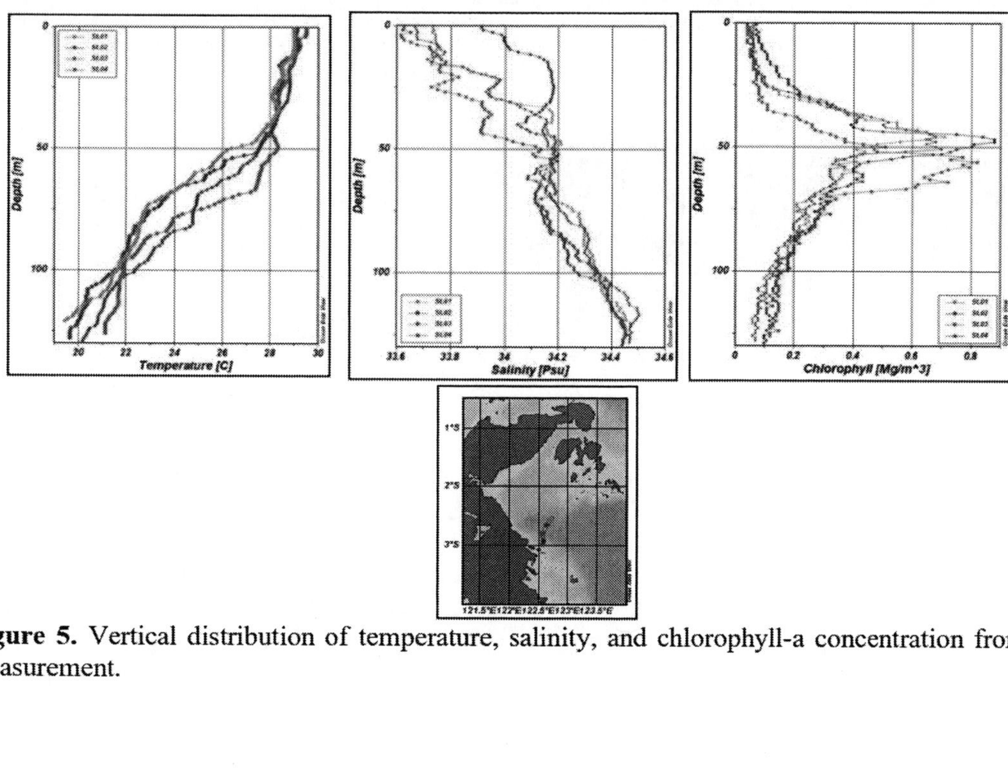


**Figure 5.** Vertical distribution of temperature, salinity, and chlorophyll-a concentration from in-situ measurement.

IOP Conf. Series: Earth and Environmental Science **137** (2018) 012041 doi:10.1088/1755-1315/137/1/012041

### *3.2. Validation of potential fishing zone forecast*

Validation of PFZ Forecast was conducted by using inter-cropping method between the location of experimental fishing (▲) and the PFZ location (★for fishing ground; ●for potential fishing ground). The location identification of the PFZ forecast system refers to the information of the existence of fronts temperature (*thermal fronts*) and water fertility. Thermal fronts are an oceanographic phenomenon characterized by the meeting of water masses that have different temperatures [26]. The identification of the thermal fronts in the PFZ forecast system is carried out by applying different temperature gradients in order to obtain the most accurate temperature gradient. The temperature gradient for detecting thermal fronts is 0.5 °C. This value was used as a proxy for the detection of two different water masses of temperature at the study site. Pelagic fish converge at the current boundaries that have horizontal temperature gradient (thermal front). Meanwhile, the value of water fertility indicated high abundance of phytoplankton at a limit of between 0.2 and 1 mg/m$^3$. The abundance of phytoplankton serves as a sign of potential fish presence in the prediction area. But at the time of the field survey the value of chlorophyll-a concentration in the northern part of Tolo Bay was detected by satellites to be < 0.2 mg/m$^3$. Cloud coverage was also seen at several points adjacent to the experimental fishing locations. This condition caused some prediction sites to be declared as potential fish areas because they did not meet the criteria of water fertility. By using this method on September 22-24, 2016, the PFZ forecast system provides information on three locations of fishing ground and three locations of potential fishing ground (figure 6).

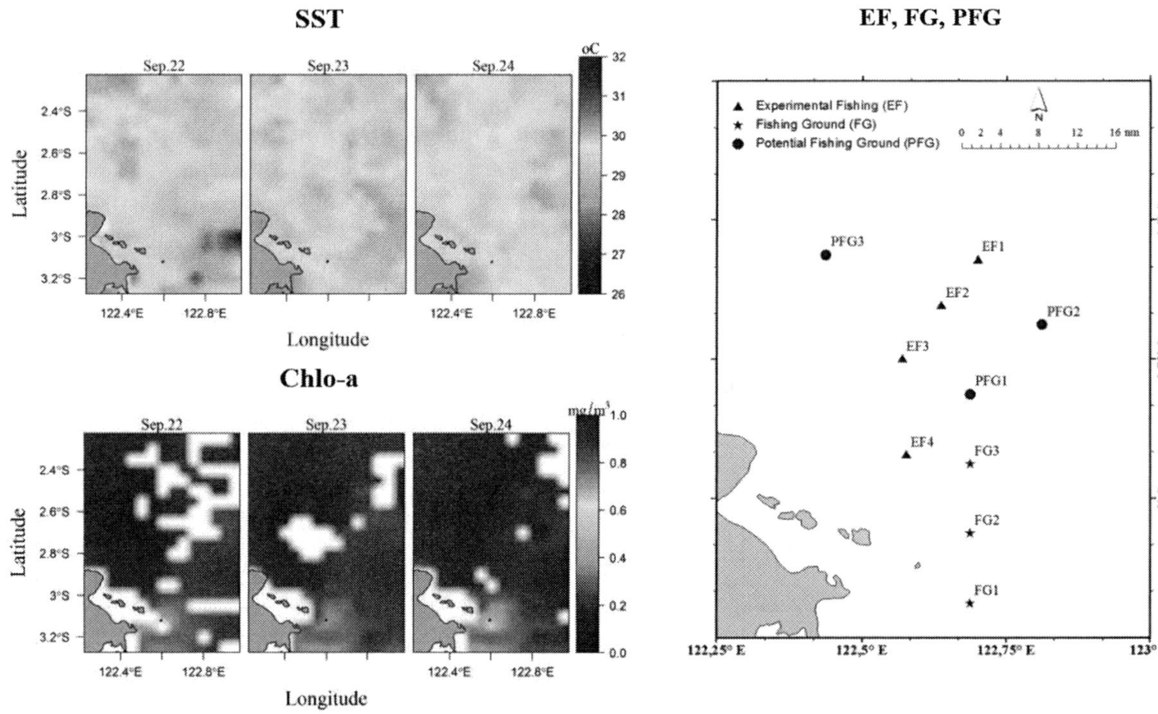


**Figure 6.** Sea surface temperature, chlorophyll-a concentration and the location of experimental fishing (▲), PFZ forecast criteria (●: *PFG or* ★: *FG*) on September 22-24 2016.

There were a total of four locations of experimental fishing (EF) sites. Based on the analysis of the nearest distance between experimental fishing (EF) location and fishing ground (PFG or FG) location, there were one location adjacent to the fishing ground (EF4) and three locations adjacent to the potential fishing ground (EF1, EF2, EF3). EF1 was 9.7 nautical miles (17.96 km) away from the

IOP Conf. Series: Earth and Environmental Science **137** (2018) 012041 doi:10.1088/1755-1315/137/1/012041

potential fishing ground (PFG2). The activity in EF1 yielded catch of 850 kg, which consisted of 680 kg of skipjack tuna (*Katsuwonus pelamis*) and 170 kg of yellowfin tuna (*Thunnus albacares*). The other experimental fishing locations (EF2 and EF3) were 10.05 nautical miles (18.52 km) and 8.01 nautical miles (14.83 km) away from the potential fishing ground (PFG1), respectively. The last experimental fishing location EF4 collected 900 kg of catch, which consisted of 540 kg of skipjack tuna (*Katsuwonus pelamis*) and 270 kg of yellowfin tuna (*Thunnus albacares*), 90 kg of mackerel tuna (*Euthynnus sp*) and dolphinfish (*Coryphaena hippurus*). EF4 location was the closest to FG. The distance between experimental fishing (EF) locations and fishing ground and catch is shown in Table 1. In the present study, the feedback analysis indicated that catch found to be high is a confluence of SST and chlorophyll features [12]. In this study, the validation of potential fishing zone (PFZ) forecast using experimental fishing in Tolo Bay, Central Sulawesi provides information about the shortest distance to the nearest PFZ location and the number of catches.

**Table 1.** The location conformance summary of experimental fishing and PFZ Forecast.

| Name | Locations | | PFZ Criteria | Distance (nautical mile) | No. of Catch (kg) | Chl-a ($mg/m^3$)[a] |
|---|---|---|---|---|---|---|
| | Latitude | Longitude | | | | |
| EF1 | -2.57016 | 122.70107 | PFG2 | 9.70 | 850 | 0.71 |
| EF2 | -2.65237 | 122.63751 | PFG1 | 10.05 | 0 | 0.70 |
| EF3 | -2.74917 | 122.56985 | PFG1 | 8.01 | 0 | 0.81 |
| EF4 | -2.92087 | 122.57716 | FG3 | 6.69 | 900 | 0.88 |

[a]Maximum of chlorophyll-a concentration at 40-50 m depth

Temperature, salinity and chlorophyll-a concentration affect fish behavior, fish distribution and number of catches at an EF location. Temperature and depth of water are the most important influence and contribute the most to pelagic fish. Most of the catches in the EF location were pelagic fish, dominated by skipjack tuna, yellowfin tuna and mackerel tuna. The optimum temperatures for yellowfin tuna and skipjack tuna were 20-28°C and 28-29°C, respectively. Skipjack tuna around the equator are commonly found from the surface layer, above the thermocline layer, to a depth of 40 meters. The optimum temperatures for skipjack tuna in the experimental fishing (EF) locations are found from the surface to a depth of 30-40 meters. The optimum temperatures at EF1, EF2 and EF4 locations are found from the surface to a depth of 30 meters, while at EF3 location, they are found at deeper depths (figure 5). The optimum salinity for yellowfin tuna and skipjack tuna was 18-38 $^o/_{oo}$ and 33-35 $^o/_{oo}$, respectively. The optimum salinity for yellowfin tuna and skipjack tuna at all experimental fishing locations is measured from the surface to a depth of 120 meters. The maximum chlorophyll-a concentration of 0.7-0.88 $mg/m^3$ at the experimental fishing site was found at a depth of 30-50 meters. The highest maximum chlorophyll-a concentration at EF4 location was found at a depth of 40 meters. The EF4 location which has the highest chlorophyll-a concentration has the highest number of fish catches as compared to the others fishing locations.

## 4. Conclusion

Environmental parameters such as SST, salinity, depth and chlorophyll-a concentration affect fish habitat. The validation of experimental fishing (EF) conducted synchronously with PFZ forecast indicates that the synergetic use of SST and chlorophyll information derived from the satellite was found to be very useful in the exploitation of fisheries resources. The result of the validation of potential fishing zone (PFZ) forecast using experimental fishing (EF) at Tolo Bay, Central Sulawesi Province showed that the number of fish catches is influenced by the distance from PFZ forecast position and PFZ forecast criteria. The closer the experimental fishing (EF) location to fishing ground (FG), the greater the likelihood of fish catch, and the greater the number of fish catches.

IOP Conf. Series: Earth and Environmental Science **137** (2018) 012041 doi:10.1088/1755-1315/137/1/012041

## 5. References


[1] Amri K, Suman A, Irianto HE, *et al* 2015 *Ind. Fish. Res.J.* **21** 75-90

[2] Hendiarti N, Suwarso, Aldrian E, *et al* 2005 *Oceanogr.* **18** 112-23

[3] Syamsuddin ML, Saitoh S-I, Hirawake T, *et al* 2013 *Fish. Bull.* **111** 175-88

[4] Susanto D and Marra J 2005 Effect of the 1997/98 *Oceanogr.* **18** 124-7

[5] Susanto RD, Gordon AL and Zheng Q 2001 *Geophys. Res. Lett.* **28** 1599-602

[6] Lumban-Gaol J, Leben RR, Vignudelli S, *et al* 2015 *Eur. J. Remote Sens.* **48** 465-77

[7] Santos AMP 2000 *Fish. Res.* **49** 1-20

[8] Setiawati MD, Sambah AB, Miura F, *et al* 2015 *Adv. in Space Res.* **55** 732-46

[9] Zainuddin M, Nelwan A, Farhum SA, *et al* 2013 *Int. J.of Geosci.* **4** 259-66

[10] Ardianto R, Setiawan A, Hidayat JJ, *et al* 2014 *IOP Conf Series: Earth and Environmental Sci.* **54**

[11] Jatisworo D and Murdimanto A 2013 *Proc. National Symp. on Geoinf. Sci. III* (Jogjakarta: PUSPICS Faculty of Geography Gadjah Mada University) p 440.

[12] Pertami D and Suniada KI 2014 *Biennial Conference of Pan Ocean Remote Sensing Conference (PORSEC 2014)*

[13] Hastuti AW, Suniada KI, Susilo E, *et al. Proc. National Sem. on Remote Sens.* p. 691-9

[14] Profile of Conservation Area of Central Sulawesi Province (Jakarta: Directorate of Area Conservation and Fish Type, Directorate General of Marine, Coastal and Small Islands, Ministry Marine Affair and Fisheries, Republic of Indonesia)

[15] Platt T, Fuentes-Yaco C and Frank KT 2003 *Nature* **423** 398

[16] Laevastu T and Larkins HA 1981 Marine Fisheries Ecosystem: Its Quantitative Evaluation and Management: Farnham, Surrey, England: Fishing News Books)

[17] Raja BTA 1974 *Proc. of the Indo-Pacific Fish. Council* **15** 241-52.

[18] Choudhury SB, Jena B, Rao MV, *et al* 2007 *Int. J. Remote Sens.* **28** (12) 2683-93

[19] Arnone RA 1987 *Remote Sens. of Environment* **23** 417-37

[20] Laurs RM, Fiedler PC and Montgomery DR 1984 *Oceanograph. Res. Papers* **31** 1085-99

[21] Gordon A and Ffield A 1994 *J. Geophy. Res.* **99** 18235-42

[22] Sprintall J and Liu WT 2005 *Oceanography* **18** 88-97

[23] Wyrtki K 1962 *Australian J.Mar.Ffreshwater Res.* **13** 217-25

[24] Upwelling in the Banda Sea and South Coast of Java as well as its relationship with ENSO and IOD Omni-Akuatika **12** 119-30

[25] Susanto RD, Thomas SM and Marra J 2006 *Geochemistry, Geophysics, Geosystems* **7**

[26] Jing Z, Qi Y, Du Y, *et al* 2015 *J. Geophy. Res.: Oceans* **120** 1993-2006



**Acknowledgements**

This study is part of research activities 'Validation of Potential Fishing Ground Areas' founded by the Ministry of Marine Affairs and Fisheries through Institute for Marine Research and Observation in 2016. The authors would like to thank Dr. I Nyoman Radiarta, M.Sc. for valuable advice and comment, and also thanks to surveying team members and PT Timur Laut Maritim and Northeast crew members for their contribution giving assistance during the experimental fishing survey.

IOP Conf. Series: Earth and Environmental Science **137** (2018) 012042 doi:10.1088/1755-1315/137/1/012042

# Characterization of elasticity and hydration of composite hydrogel based on collagen-iota carrageenan as a corneal tissue engineering

**M Rinawati[1], J Triastuti[2]* and K T Pursetyo[2]**

[1]Bachelor Programme of Aquaculture, Faculty of Fisheries and Marine, Universitas Airlangga, Surabaya 60115, Indonesia
[2]Department of Marine, Faculty of Fisheries and Marine, Universitas Airlangga, Surabaya 60115, Indonesia
Email: jt.astuti@gmail.com

**Abstract.** The cornea is a refractive element of the eye that serves to continue the stimulation of light into the eye it has a clear, transparent, elastic and relatively thick tissue. Factors caused corneal blindness, are dystrophy, keratoconus, corneal scaring. Hydrogels can be made from polysaccharide derivatives that have gelation properties such as iota carrageenan. Therefore, it is a need to develop composite hydrogel based collagen-iota carragenan as an engineeried corneal tissue with high elasticity and hydration properties. Collagen hydrogel has a maximum water content an has equlibrium up to 40 %, less than the human cornea, 81 % and under normal hydration conditions, the human cornea can transmit 87 % of visible light. In addition, the refractive index on the surface of the cornea with air is 1.375-1.380. Based on this study, it is necessary to conduct research on the development and composition of hydrogel composite collagen-iota carrageen hydrogen based on.The best result was K5 (5:5) treatment, which has the equilibrium water content of 87.07 % and viscosity of 10.7346 Pa.s.

## 1. Introduction

The cornea is a refractive element of the eye that serves to continue the stimulation of light into the eye; it has a clear, transparent, elastic and relatively thick tissue [1]. It is estimated that around 10 million people worldwide suffer from vision loss due to corneal damage [2]. Some factors causing corneal blindness are infection (45 %), trauma (25 %), heredity (5 %), and other factors such as nutritnal (13 %) of the total number of people suffering from blindness in the world. Factors that cause corneal blindness can lead to certain conditions on the cornea, including Fuch's dystrophy, keratoconus, keratophaty, dystrophies, and corneal scaring. This indicates that a corneal donor through transplantation (keratoplasty) is able to overcome the problem of blindness due to obstacles to refractive ability, clarity and corneal wholeness [3]. The donor cornea (graft) provides the cornea of a natural donor (allografts) and an artificial cornea (keratoprosthesis). Allografts have good biocompatibility, which can accelerate the proliferation of endothelial cells in the cornea. However, rejection by the recipient often occurs.

Based on medical approach, the ideal replacement cornea must be biodegradable and can be a tissue regeneration promoter. The application of synthesis materials such as glycopolymers produces corneal

IOP Conf. Series: Earth and Environmental Science **137** (2018) 012042 doi:10.1088/1755-1315/137/1/012042

membranes that cannot be degraded by the eye stromal tissue. Therefore, they are rarely used. The membrane is relatively unreactive, which will be rejected (immunologically rejected), <10 %. The development of replacement corneas in recent years has focused on such materials as soft polymeric hydrogels [4].

The natural ingredients of collagen and its composites have shown good biocompatibility and tissue regeneration in animals [2]. Collagen fibrillar, the composition of corneal stromal layer, is a type I collagen that can be extracted from tilapia species. Due to the increase of zoonotic infection cases in mammals, its use as a source of collagen is decreasing. Alternatively, fish are used as a potential source of low-potency collagen. In addition, fish collagen is easily digested and absorbed by the body with cow / pig collagen. This fibrillar structure can act as a good scaffold on corneal tissue techniques. The application of the collagen as a material for artificial cornea results in membranes with high mechanical strength (elasticity), which is low. So, it needs other materials. Hydrogels can be made from polysaccharide derivatives that have gelatin properties such as carotene iota [5]. The caroteneic ion will undergo a transition into a helix-shaped coil (helix) that carries it on gelation, resulting in a smooth elastic gel.

Patel *et al.* [6] showed that collagen-polymer hydrogel has a maximum water content equlibrium up to 40 %, while the human cornea has 81%. This characteristic is required in corneal tissue engineering for high mechanical and hydration strength. Based on this background, it is necessary to conduct research on the manufacture and composition of composite hydrogel based on collagen-iota karaginan (5: 0; 5: 1; 5: 2; 5: 3; 5: 4; 5: 5) so that the optimum composition characteristics approaching corneal characteristics can be identified.

## 2. Methodology

The research was conducted using fish collagen (pure), semi refined iota karaginan from *Eucheuma spinosum*, $Ca(OH)_2$, ethanol 96 %, HCl 0.1N, glutaraldehyde, Phospate Buffer Saline, glycerine and distilled water. This research employed experimental methods with completely randomized design, which consists of six treatments. Are K0 (5:0) as control, K1 (5:1), K2 (5:2), K3 (5:3), K4 (5:4) and K5 (5:5) with three repetitions for each treatment.

K0: 50 mg collagen: 0 mg iota carrageenan
K1: 50 mg collagen: 10 mg iota carrageenan
K2: 50 mg collagen: 20 mg iota carrageenan
K3: 50 mg collagen: 30 mg iota carrageenan
K4: 50 mg collagen: 40 mg iota carrageenan
K5: 50 mg collagen: 50 mg iota carrageenan

Hydrogel synthesis was performed with blinding technique. Ten percents (w/v) of collagen solution was prepared by dissolving 50 mg of granular collagen using 5 ml distilled water with a temperature of 50°C. It was then stirred at 600 rpm for 20 minutes. Furthermore, using the same step, 10% (w/v) iota carrageenan solution was prepared by dissolving iota carrageen into distilled water with a temperature of 50°C, which was then stirred at 500 rpm for 35 minutes. Iota carrageenan solution was then poured into the collagen solution with 200 rpm stirring for 10 min at 50°C. After that, a cross linker of 0.5 ml glutaraldehyde and 0.5 ml ethanol and 0.05ml 0.1 N HCl sequentially and 3 % of glycerin were added at each treatment to form viscous. Furthermore, it was printed in a 15x2 cm mica mold and left in room temperature for 15-30 minutes.

The main parameters of this study were equilibrium water content and viscosity of the hydrogel. Observation of equilibrium water content was conducted by cutting the gel into square with the size of 1x1 $cm^2$ as much as five pieces taken from each mold sampling point (top, bottom, middle) and weighed ($W_0$), then dipped into phosphate buffer saline (pH 7.4) solution. After being charged for 3 hours, the hydrogel was then removed and weighed wet weight (W). Then, the hydrogel was dried to a constant weight, and

IOP Conf. Series: Earth and Environmental Science **137** (2018) 012042 doi:10.1088/1755-1315/137/1/012042

the dry weight ($W_0$) was weighed. Equilibrium water content (Wt) was calculated using the following equation:

$$Wt = \frac{W - W_0}{W} x\, 100\, \% \qquad (1)$$

Viscosity test was aimed at measuring the elasticity value of hydrogel, which indicates the mechanical strength. The viscosity test was carried out at a temperature of 30°C. Hydrogel samples were cut in a circle form with a diameter of 2 cm and thickness of ± 1mm. The samples were thawed by heating and diluted using distilled water 1:4. The samples were then inserted into Oswald's viscometer and the time when hydrogel dropped from the first line to the second line was calculated. Viscosity number was calculated using the following Poiseville equation.

$$\eta = \frac{\pi P R^4 T}{8LV} \qquad (2)$$

Supported parameters are refractive index and light transmission. Refractive Index measurements were performed using a VEE GEE refractometer at a temperature of 30°C by dripping the gel on the prismic surface of the refractometer reading the refractive index scale. Transparency testing was performed by measuring the transmittance of the gel using a UV-Vis spectrophotometer at 600 nm wavelength. The hydrogel solution was taken ± 3 ml, put into the cuvette. Cuvette was inserted into UV-Vis Spectrophotometer. Then, the percentage of transmittance was calculated.

Data of sperm motility and motion duration that have been obtained were analyzed using ANOVA. The data were then analyzed using Duncan's Multiple Range Test at the resulting p-value lower than 0.05 ($p < 0.05$).

## 3. Results

Hydrogel characterization was conducted physically. The result of the composite hydrogel characterization is presented in table 1. It shows the average result of the composite hydrogel characterization from the first treatment (K0), K2, K3, K4, until K5. As the table indicates, the equilibrium water content is more varied than the literature. The viscosity and transparency were similar to those indicated by results in the literature. Meanwhile, the refractive index was smaller than the literature.

**Table 1.** Equilibrium water content of hydrogel at different treatments

| Parameter | Research result |
|---|---|
| Equilibrium water content | 0-87.07% |
| Viscosity | 1.4847-10.7346 Pa.s |
| Refractive index | 1.3473-1.3513 |
| Transparency | 17.67-97.80% |

The hydration of hydrogel was observed by the percentage of equilibrium water content, which was displayed in table 1. ANOVA test results indicated that the composition of hydrogel has a significant effect ($p < 0.05$) on the average equilibrium water content. Duncan's multiple range test indicates that the lowest average of equilibrium water content was present in K0 (5:0) treatment that significantly differs ($p < 0.05$) from K1 (5:1), K2 (5:2), K3 (5:3), K4 (5:4) and K5 (5:5) treatments. Meanwhile, the K1 (5:1), K2 (5:2), K3 (5:3), K4 (5:4) and K5 (5:5) are not significantly different.

**Table 2**. Equilibrium water content of hydrogel.

| Treatment | Eq. Water Content Average (%) ± SD |
|---|---|
| K0 (5:0) | $0^{a}$±0.00 |
| K1 (5:1) | $76.18^{b}$±12.08 |
| K2 (5:2) | $76.78^{b}$±10.11 |
| K3 (5:3) | $74.58^{b}$±14.35 |
| K4 (5:4) | $80.40^{b}$±8.51 |
| K5 (5:5) | $87.07^{b}$±7.28 |

Note: Different superscript in one column indicates significant difference ($p<0.05$)

K0 treatment is significantly different from the K1 (5:1), K2 (5:2), K3 (5:3), K4 (5:4) and K5 (5:5) treatments. K1 (5:1), K2 (5:2), K3 (5:3), K4 (5:4) and K5 (5:5) were not significantly different.

The mechanical strength of hydrogel was then observed by the viscosity of hydrogel using Oswald viscosimeter to measure the time of hydrogel flow. Then, the viscosity was calculated using Poiseuville equation. ANOVA test results indicate that the composition of hydrogel has a significant effect ($p < 0.05$) on the viscosity. Duncan's Multiple Range Test results show that the lowest average of viscosity is in K0 (5:0), and the highest viscosity is in K5 treatment that is significantly different ($p < 0.05$) from K1 (5:1), K2 (5:2), K3 (5:3), and K4 (5:4) treatments. However, both K2 and K3 treatments were not significantly different. table 3 below illustrates the viscosity of hydrogel.

**Table 3.** Viscosity of hydrogel.

| Treatment | Average of viscosity of hydrogel (cP) |
|---|---|
| K0 (5:0) | $1.4847^{a}$ |
| K1 (5:1) | $3.6433^{b}$ |
| K2 (5:2) | $5.0931^{c}$ |
| K3 (5:3) | $6.1111^{c}$ |
| K4 (5:4) | $8.1274^{d}$ |
| K5 (5:5) | $10.7346^{e}$ |

Note: Different superscript in one column indicates significant difference ($p< 0.05$)

The refractive index of hydrogel was observed using refractometer at a temperature of 29°C. The refractive index of hydrogel is shown in table 4. ANOVA test results indicate that the composition of hydrogel has no significant effect ($p < 0.05$) on the refractive index of hydrogel.

IOP Conf. Series: Earth and Environmental Science **137** (2018) 012042 doi:10.1088/1755-1315/137/1/012042

**Table 4.** Refractive Index of hydrogel.

| Treatment | Average of Refractive Index |
|---|---|
| K0 (5:0) | 1.3513[a] |
| K1 (5:1) | 1.3570[a] |
| K2 (5:2) | 1.3490[a] |
| K3 (5:3) | 1.3480[a] |
| K4 (5:4) | 1.3477[a] |
| K5 (5:5) | 1.3473[a] |

Note: Different superscript in one column indicates significant difference ($p<0.05$)

The transparency of hydrogel was observed in terms of its light transmission using UV-Vis spectrophotometer at 600 nm wavelength as shown in table 5.

**Table 5.** Light Transmission Index of hydrogel.

| Treatment | Light Transmission (%) |
|---|---|
| K0 (5:0) | 97.80[a] |
| K1 (5:1) | 75.00[b] |
| K2 (5:2) | 59.90[bc] |
| K3 (5:3) | 50.00[cd] |
| K4 (5:4) | 39.47[d] |
| K5 (5:5) | 17.67[e] |

Note: Different superscript in one column indicates significant difference ($p<0.05$)

ANOVA test results indicate that the composition of hydrogel has a significant effect ($p < 0.05$) on the light transmission. Duncan's Multiple Range Test results show that the highest light transmission is in K0 (5:0), and the lowest light transmission is in K5 treatment, that is significantly different ($p < 0.05$) from K1 (5:1), K2 (5:2), K3 (5:3), and K4 (5:4) treatments. However, both K1 and K2, as well as K2 and K3 treatments were not significantly different.

IOP Conf. Series: Earth and Environmental Science **137** (2018) 012042 doi:10.1088/1755-1315/137/1/012042

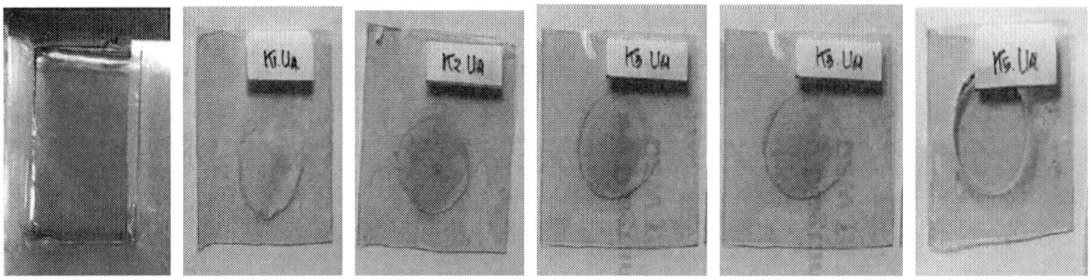

A B C D E F

**Figure 1.**Composite hydrogel Collagen-Iota Carrageenan Based each treatment, A: K0 (5:0), B: K1(5:1), C: K2(5:2), D: K3(5:3), E:K4(5:4), F:K5(5:5).

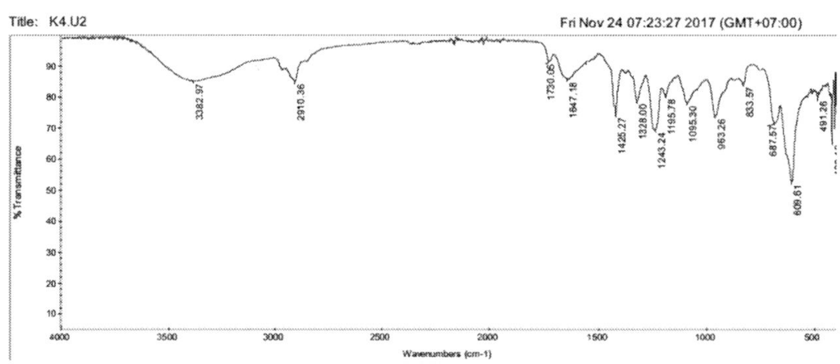


**Figure 2.** FTIR result of Composite hydrogel Collagen-Iota Carrageenan Based.

The FTIR test showed a high intensity peak at a wavelength of 687.57 $cm^{-1}$, an amine group. The others, the peak is at the wavelength of 1243.24 $cm^{-1}$, which can be interpreted as the vibrational range of the C-N group, 1425.27 $cm^{-1}$ for the carboxylate group and 2910.36 $cm^{-1}$ for the C-C-C-H range of medium intensity.

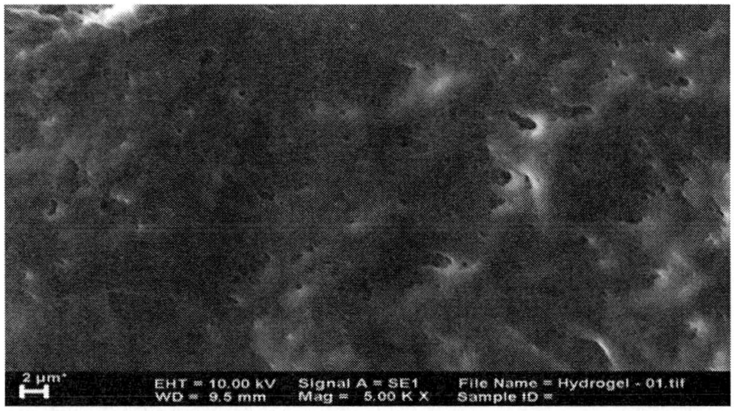


**Figure 3.** SEM image of Composite hydrogel Collagen-Iota Carrageenan Based.

The SEM test result at 5.000x magnification shows that the composite hydrogel morphology is solid and has an uneven surface. The absence of a film layer formed also signifies the linked (collagen) with iota carrageenan because of crosslinking. According to Gong *et al.* [7], polymer scaffolds should be able to provide space for appropriate cells to grow into new tissue on transplants. High porosity is required for growth and invasion of surrounding tissues. Furthermore, the connected pore tissue is essential for the growth, visualization and diffusion of nutrients in the cell.

## 4. Discussion

The main parameters observed in this study are equilibrium water content and viscosity, while the supporting parameters consisted of refractive index and transparency of hydrogel. Based on the results of variance analysis, it is known that the composition of collagen- iota carragenen gives a significant effect on the equilibrium water content of the hydrogel. Equilibrium water content indicates the hydration state of the cornea. According to Hayes *et al.* [8], the addition of iota carrageenan on the hydrogel affects the equilibrium water content of hydrogel. The test results show that the greater number of iota carrageenan added on the composite, the greater the value of equilibrium water content. This is because the iota carrageenan is a hydrophilic substance. Iota carrageenan is anionic. Some of the hydroxyl piranol groups are substituted by sulfate-half esters. These sulphate groups possess highly acidic and hydrophilic properties. So, water will rapidly bond into sulfate grups. K5 treatment has the highest equilibrium water content, up to 87.07 %. This value exceeds the water content limit of human cornea (78-80 %). According to Hayes *et al.* [8], an increasingly decreased hydration rate will reduce Bragg's average intermolecular spacing of collagen. If hydration gets higher, the intermolecular spacing of hydrogel will be higher, causing hydrogel to be more porous. In addition, Dumitriu [9] stated that the higher water content in hydrogel, the faster the degradation, resulting in lower biocompatibility. However, from the physical feature, hydrogel of K5 treatment has better physical properties. This is because, the more water content within hydrogel, the higher the swelling degree that hydrogel has. This means that hydrogels have a high water resistance capability [10].

Viscosity is the degree of consistency at a certain concentration and temperature. According to variance analysis, the addition of iota carrageenan on the composition gives a significant effect to the viscosity value of the composite hydrogel produced. The greater the addition of iota carrageenan to the composition of hydrogel, the higher its viscosity. This is in accordance with the statement of Nussinovitch [11], that the addition of gelling agents such as carrageenans can increase the viscosity or other physical properties of the gel, such as elasticity. The carrageenan ion may form a gel in low concentration due to the electrostatic interaction between the positive charge of collagen and the negative charge of the sulphate group in the carrageenance, which leads to an increase in the reactivity of collagen [12]. The more iota carrageenan concentration added, the more the negative charge of the sulphate group in the carrageenan induces a positive charge on collagen electrostatically, resulting in an increase in the reactivity of collagen which will lead to an increase of its viscosity. In addition, viscosity is also influenced by crosslinking activities. These activities can easily affect collagen fibers during intrafibrils crosslink process. When the collagen fibrils are connected to iota carrageenan groups, the composite structure becomes more rigid, making it difficult to separate. This makes the hydrogel viscosity higher. Consequently, the hydrogel elasticity becomes higher [13]. The higher the elasticity, the higher the mechanical strength of the hydrogel is.

Refractive index is related to total suspended solid in the sample. Based on the results of the analysis of variance, it is known that the addition of iota carrageenan composition did not give a significant effect on the refractive index value of the resulting composite hydrogel. There is a correlation between refractive index and corneal hydration, but no detailed explanation has been found on this subject. According to Patel *et al.* [6] theoretically, it is assumed that the natural optics and water distribution in the cornea in mammals are uniform. In the anterior and posterior portions of the stroma, there are differences in the characteristics of swelling, hydration and collagen fibril structure. The anterior part of the cornea is more likely to be low in hydration, and more resistant to water flow than the posterior part. Thus, refractive index measurements on the lower portion of hydration will result in higher refractive index values. Similarly, with regard to the refractive index hydrogel value, if the sampling is not homogeneous, then the refractive index becomes inaccurate. The test results show that the higher iota carrageenan added, the lower the refractive index. This is associated with hydrogel hydration, the more iota carrageenan addition, the higher the resulting hydration. This causes the refractive index measurement to become smaller, in accordance with the theories that have been proposed by Patel *et al.* [6]. In all treatments, the refractive index value is less than the refractive index value of the human cornea. So, no treatment is eligible for this parameter. However, all treatments have a refractive index value approaching the refractive index of human cornea.

Transparency is measured using the percentage of light transmission. Transmission is the amount of light passed on the speculator. Transmission of light is correlated with refractive index hydrogel. Organic materials do not absorb light. Therefore, more light damping is caused by the spreading of light. Due to differences in the refractive index between the nearest fibrils and the aqueous environment, a glimmer of light will be scattered from each fibril. However, due to the arrangement of random hydrogel fibrils, the result is bias in light interference in all directions [14]. In addition, according to Lambert-Beer's Law, the thickness of the sample will affect the visible light transmission. From the variance analysis, it is known that the addition of iota cargeenan composition gives a real effect to the value of the light transmission of hydrogel. The more iota carrageen composition, the smaller the light transmmission value of the hydrogel. This may be possible because the arrangement of collagen fibrils is randomized by the collagen-iota molecule, which results in uneven distribution of light, so that more light is absorbed in the sample than passed through the sample [14]. The decrease in the value of the light transmission may also be related to the presence of inorganic materials present in the composite hydrogel. Carrageenous ions are known to have sulfate groups [5], wherein this group is an inorganic group. The presence of this inorganic group causes the light to be absorbed into the inorganic material, so that the light that is passed will decrease, resulting in the smaller transmission [15].

## 5. Conclusion

It can be concluded that the higher composition of collagen: iota carrageenan leads to higher equilibrium water content, viscosity but lower refractive index and transparency. The best equilibrium water content and viscosity were found in K5 (5:5) treatments. The best refractive index and transparency were found in K0 treatment. This study suggests that K5 (5:5) can be performed as a point to create good hydrogel for corneal tissue engineering because of its equilibrium water content and viscosity. Accordingly, the development of hydrogel collagen-iota carrageenan based can use the composition of (5:5) with a blending method.

## 6. References

[1] Heckenlively J R, Yoser S L, Friedman L H and Oversier J J 1988 *Am. J. Ophtalmol* **105** 504-511.

[2] Chao D, Feng L, Joanne M H 2009 *Acta. Biomater.* **6** 187–194

[3] Saldanha M and Mendonca N 2013 *J. Dental Med. Sci.* **5** 18-20

[4] Griffith M, Hakim, M Shimmura S *et al* 2002 *Artificial human corneas scaffolds for transplantation and host regeneration dalam cornea* Philadelphia. Lippincocc Williams and Wilkins, Inc.

[5] Distantina S, Rochmadi R, Fahrurrozi M and Wiratni W 2013 *J. Eng.* **17** 57–66.

[6] Patel S, Marshall J, Fitzke F W 1995 *J. Refract. Surg.* **11** 100-105.

[7] Gong S, Dong J, Xue, S T, and Wang, J Y 2005 *Engineering in Medicine and Biology* **27** 4884-4887.

[8] Hayes S, Lewis, P and Meek K M 2015 *Acta. Biomater.* **40** 286-293

[9] Dumitriu, S 2001 *Polymeric biomaterials, revised and expanded* (New York: CRC Press, Inc.)

[10] Padhi J R, Nayak D Nanda A, Rauta P R, Ashe S, Nayak B 2016 Development of highly biocompatible gelatin & i-carrageenan based composite hydrogels: in depth physiochemical analysis for biomedical applications *Carbohydr Polym* Inpress.

[11] Nussinovitch A 2003 Water-Soluble polymer applications in foods Blackwell Science Ltd.

[12] Saha D and Bhattacharya S 2010 *J. Food Sci. Tech.* **47** 587–597.

[13] Evingur G A and Acay F G 2015 *The effect of crosslinking to elasticity of cornea with various composite solutions* Composite Structures **127** 395-398.

[14] Maurice D and Giardini A 1951 *British J. Ophthalmology* **37** 791-797.

[15] Benedek G B 1971 *Theory of transparence of the eye* Application Ophtalmology **10** 459-473

IOP Conf. Series: Earth and Environmental Science **137** (2018) 012043 doi:10.1088/1755-1315/137/1/012043

# The biomass, abundance, and distribution pattern of starfish *Asterias* sp. (Echinodermata: Asteroidea) in East Coast of Surabaya

**N N Dewi[1*], K T Pursetyo[2], L Aprilianitasari[1], M H Zakaria[3], M R Ramadhan[3], R A Triatmaja[3]**

[1]Department of Fish Health Management and Aquaculture, Faculty of Fisheries and Marine Universitas Airlangga, Campus C Unair 60115 Surabaya, Indonesia
[2]Department of Marine Faculty of Fisheries and Marine Universitas Airlangga, Campus C Unair 60115 Surabaya, Indonesia
[3]Aquaculture Study Program, Faculty of Fisheries and Marine Universitas Airlangga, Campus C Unair 60115 Surabaya, Indonesia

**Abstract.** This study aims to determine the biomass, density, and distribution patterns of *Asterias* sp.Samples were collected from three locations such as Wonokromo, Dadapan and Juanda, each divided into 3 zones. In each zone, samples were taken as many as 5 repetitions using swept area method. Temporarily, the highest biomass of starfish was 2.95 gr/m$^2$ in Dadapan Zone on January. Spatially, biomass of starfish was found in Dadapan Zone (3,35 gr/m$^2$). Similarly, the high density was also found in Dadapan Zone on January (9 ind/10 m$^2$). In general, the distributionpattern of starfish in East Coast Surabaya throughspatial and temporal showed that the pattern of starfish was grouping distribution (Id value> 1) for Dadapan and Juanda, and uniform for Wonokromo.Oceanographic condition, antropogenic activity, and water quality in East Cost of Surabaya become important things which is affected the biomass, densityand distribution pattern of starfish. The knowledge of starfish biomass and density is very important given that this biota has ecological value as a balancing ecosystem in the waters.

## 1. Introduction

The east coast of Surabaya is close to Sidoarjo Regency, is an industrial development area of the city of Surabaya. These industries can potentially cause pollution from industrial waste disposal through rivers and finally to the waters of the east coast of Surabaya. According to Pourvali *et al.* [1], echinoderms are one of the oldest and most important groups of marine organisms which play an important role in marine ecology. Asteroidea is a class of the Echinoderm and are commonly found in shallow waters with a depth of 10 meters. These animals are commonly found in tropical coral reefs, rocky shores, tidal pools, mud and sand, kelp forests, seagrass meadows and the deep-sea floor down to at least 6.000 m. Starfish are biota that are sensitive to marine pollution and considered to be a bioindicator for marine ecosystems [1,2]. *Asterias* sp. is one genus of the Asteroideaclass. This species is usually found by fishermen during their shellfish fishing activities in the east coast of Surabaya. According to Shah [3], *Asterias* are known by the name of starfish. There are about 150 species of *Asterias*, all of which have different geographic distributions. The habitat of the *Asterias* is exclusively marine, bottom dwelling or benthonic, inhabiting various types of sea floors, mainly in the littoral zones where they are cruising or concealed.

The knowledge of starfish biomass and abundanceis very important given that this biota has the ecological value of balancing the ecosystem in the waters. There is not much publication and information about the biomass, abundance and distribution patterns of starfish, especially in the east coast of Surabaya. International research on *Asterias* sp. has been conducted by others such as Anger *et al.* [4] and Dalgic *et al.* [5]concerning in-situ investigations on the echinoderm *Asterias rubens* as a predator of soft-bottom communities in the western Baltic Sea.In addition, other studies of *Asterias vulgaris* and *A. forbesi* in several heterogeneous environments were carried out by Menge [6]. Other studies on the abundance and distribution patterns of Asteroidea include Freeman *et al.* [7] regarding *Astropecten irregularis* in the coastal waters of North Wales and also research on *Oreaster reticulatus* in St. Croix, U. S. Virgin Islands by Scheibling [8].Research on *Asterias* sp. is still minimal in Indonesia. The research conducted so far has been more focused on Echinodermata in general as conducted by Aziz [2] and Supono and Arbi [9] in Pari Island. Therefore, this particular research is important as basic information on the ecology aspect of this biota and how environmental factors affect them. Here we report on the biomass, abundance, and pattern of spatial distribution of *Asterias sp.*populations, where the species has not been commercially harvested and its habitat remains largely impacted by human activities, but has a function in balancing the ecosystem.

## 2. Methodology

The study on starfish (*Asterias* sp.) wasconducted in the east coast of Surabaya between January and March 2017. Samples were collected from three zones namely Wonokromo, Dadapan, and Juanda where each location was divided into three sites (figure 1) as five repetitions. Samples were collected usingthe swept area method [10] by using a local fishing tool names *garit* for 1 minute at the speed of about 3 km/hour for each repetition. Samples were preserved in a cold state. The number and biomass of starfish from each repetition were measured.

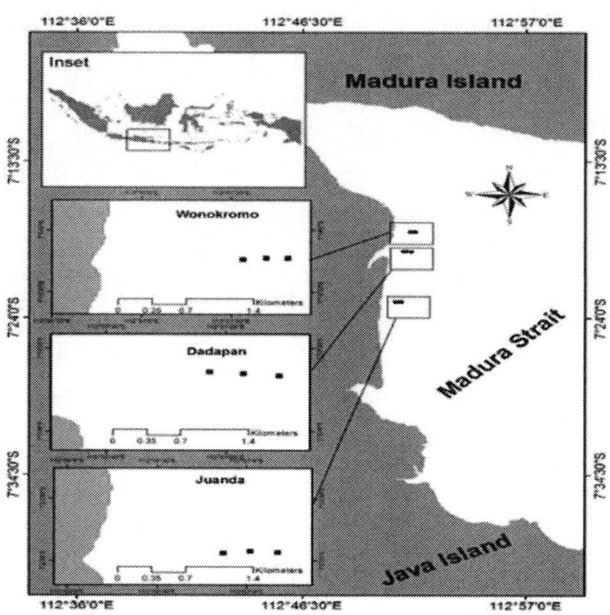


**Figure 1.** Map locations of the research.

IOP Conf. Series: Earth and Environmental Science **137** (2018) 012043 doi:10.1088/1755-1315/137/1/012043

## 3. Data Analysis

### *3.1. Biomass and abundance*

The density (individual / $m^2$) and biomass (kg / $m^2$) of starfish were calculated based on the swept area method, based on the relationship between the speed of the fraction, the duration of the pull, the length of the *garit* and the length of the opening of the *garit* mouth. Stock density was mathematically calculated by using the following formula [10]:

Track distance: $$D = V.t \quad (1)$$

Information:
D = track distance
V = The speed of the pull of a *garit* (km / h)
t = Length of withdrawal (hours)

Sweep area determination: $$a = D.\ h. \quad (2)$$

Information :
a = Sweep area ($m^2$)
h = Length of garit (m)
D = Distance of the covered sweep path

Density / biomass was calculated using the formula

$$Q = \frac{Cw/a}{ef} \quad (3)$$

Information :
Q = Density per sweep area (ind / $m^2$) or biomass (kg / $m^2$)
Cw = The catch per sweep area
a = Area of sweep ($m^2$)
ef = Factor of breakthrough

### *3.2. Distribution pattern*

To find out the distribution pattern of starfish, the method of Morisita distribution pattern [11] was used. The formula for its calculation is as follows:

$$Id = \frac{n \sum Xi^2 - N}{N\ (N-1)} \quad (4)$$

Information:
Id = Morisita distribution index
n = Number of sampling plots
N = Total number of individuals gained
Xi = Number of individuals in the i sampling sequence

The results from the Morisita index are grouped as follows:
Id <1: Distribution patterns are uniform
Id = 1: Individual distribution patterns are random
Id> 1: Individual distribution patterns are grouping

To test whether the distribution pattern is significantly different between the three zone, the Chisquarestatistical test [12] was subsequently caried out.

IOP Conf. Series: Earth and Environmental Science **137** (2018) 012043 doi:10.1088/1755-1315/137/1/012043

## 4. Results

The latest results show that the highest biomass of starfish was 2.95 tons/km$^2$ in the Dadapan Zone in January. Spatially, the highest biomass of starfish was found in the Dadapan Zone (figure 2). Similarly, the highest density was also found in the Dadapan Zone in January (9.3 ind/10 m$^2$) and the Dadapan zone generally hadthe highest density compared tothe other locations (figure 3). In general, the distribution pattern of starfish in the east coast of Surabaya throughspatial and temporal testing showed that the patterns of starfish was found to be as a clumped distributionfor Dadapan and Juanda (Id value > 1) and uniform distribution for Wonokromo (id value < 1) (table. 1).

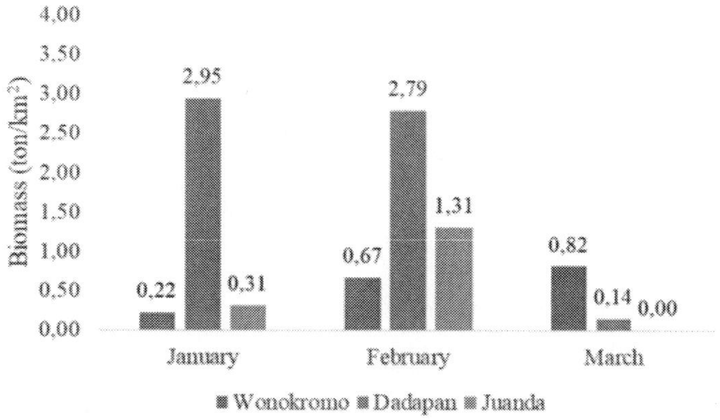


**Figure 2.** Biomass of *Asterias* sp. in East coast of Surabaya.

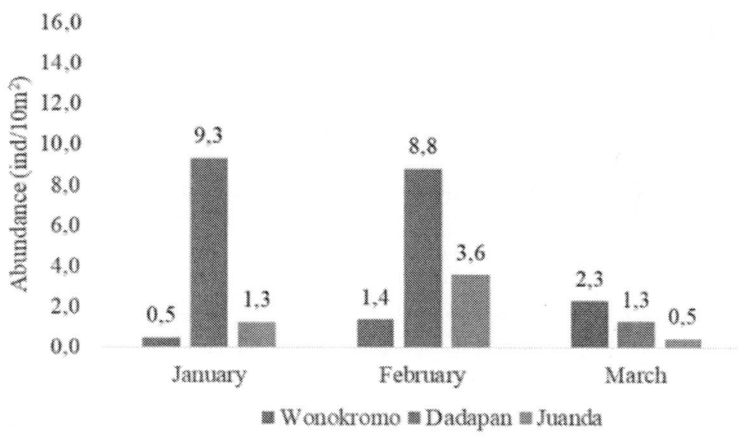


**Figure 3.** Biomass of *Asterias* sp. in East coast of Surabaya.

**Table 1.** Distribution pattern of *Asterias* sp. in East Coast of Surabaya.

| Zone | Id value | Distribution pattern |
|---|---|---|
| Wonokromo | 0.06 | Uniform distribution |
| Dadapan | 3.15 | Clumped distribution |
| Juanda | 9.13 | Clumped distribution |

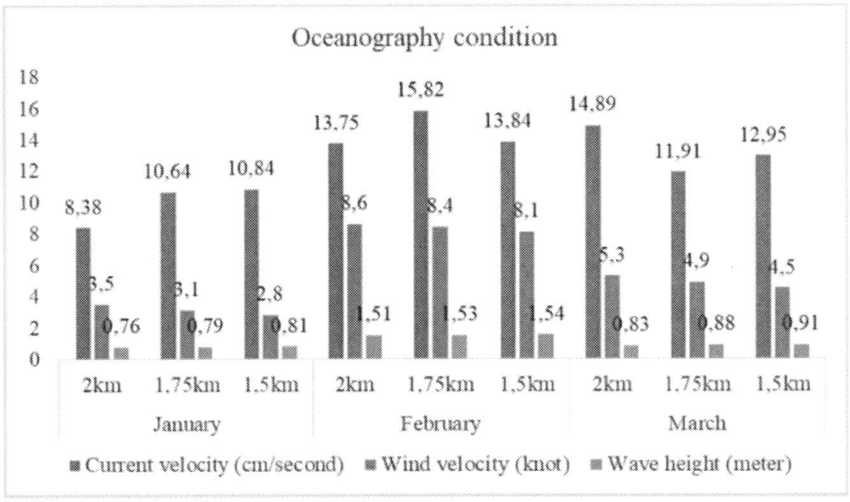


Source [13]

**Figure 4.** Oceanography condition of East Coast of Surabaya.

Environmental factors can affect the biomass, abundance, and distribution patterns of *Asterias* sp.such as oceanography conditions and water quality. Based on information from the Indonesian Agency for Meteorology, Climatology and Geophysics in Perak I [13], in general the value of the current velocity, wind velocity, and wave heights at the three zones observed in January was lower than February and March (figure 4). The water quality of the east coast of Surabaya is shown in table 2. Based on BOD value, all three zones have exceeded the quality standard of 20 mg/L [14]. Dadapan had the highest BOD value among the three zones. Based on the dissimilarity index, Wonokromo and Juanda had similar water quality characteristics (figure 5).

**Table 2.** Water Quality of East Coast of Surabaya.

| Parameters | Units | Wonokromo | Dadapan | Juanda |
|---|---|---|---|---|
| Temperature | °C | 28.67 ± 0.58 | 29.33 ± 1.53 | 30.00 ± 0 |
| Salinity | ppt | 31.67 ± 4.93 | 30.67 ± 1.15 | 30.00 ± 1.00 |
| pH | - | 8.13 ± 0.12 | 8.17 ± 0.15 | 6.77 ± 5.95 |
| Transparency | Cm | 56.67 ± 15.28 | 55.00 ± 21.79 | 96.67 ± 25.17 |
| DO | mg/l | 5.00 ± 0 | 4.50 ± 0.87 | 7.33 ± 2.31 |
| TSS | mg/l | 32.67 ± 16.92 | 11.83 ± 6.25 | 15.83 ± 9.8 |
| BOD | mg/l | 27.60 ± 15.65 | 39.26 ± 10.01 | 35.10 ± 14.26 |

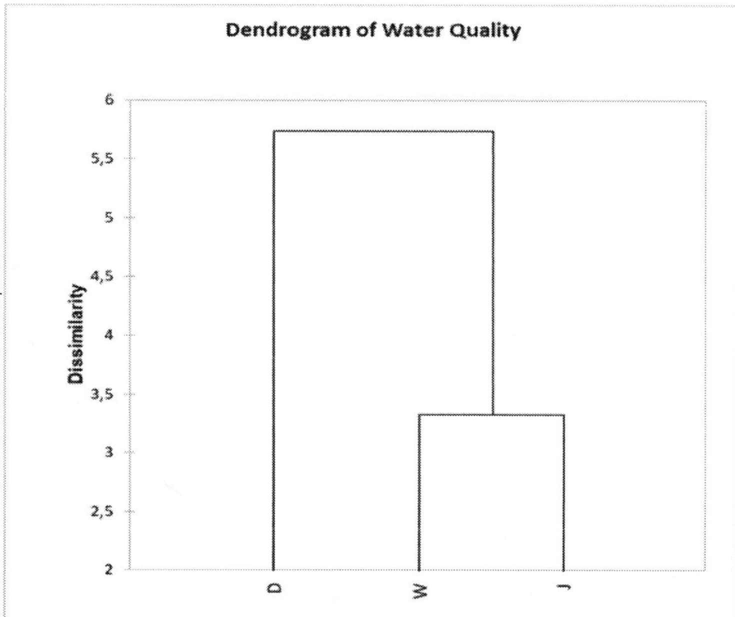


**Figure 5.** Dissimilarity index of water quality in East Coast of Surabaya.

## 5. Discussion

Information on biota distribution helps to preserve species and also provides basic information for ecological studies [15]. In general, the high biomass and abundance in January especially in Dadapan was due to the low levels of oceanographic conditions which include wind velocity, current velocity and wave height (figure 4). This indicated that there was no high dispersion of starfish in that zone. The Wonokromo and Juanda zones almost had the same biomass and density. This is in accordance with the water quality habitat conditions that possess a high degree of similarity (figure 5). While the Dadapan zone has a lower level of similarity. The high biomass and abundance of starfish in Dadapan was indicated by the presence of organic material which can be seen from the high BOD value. This is because Dadapan has more than one river flowing into its waters indicating high organic material entering it. This is in contrast to Wonokromo and Juanda which only has one river flowing into them. Thisindicates that Dadapan has a richer food source compared to the other zones. According to Helm *et al.* [16] the presence of aquatic animals is strongly influenced by abiotic factors (eg temperature, salinity, dissolved oxygen and substrate) as well as biotic factors (eg food, predation and competition). All or a combination of several environmental variables can affect the growth and distribution of species.

The distribution pattern of *Asterias* sp. for Dadapan and Juanda are grouping, while it tends to be uniform for Wonokromo. This is also related to the type of substrate and availability of food in the living locations as well their adaptability. The grouping of macrozoobentos in certain places is suspected to be due to the movement of the slow macrobenthos [17]. The grouping distribution pattern is the most common form of dispersal in the ecosystem. This is because individuals in the population tend to form groups of various sizes. The grouping distribution occurs as a result of a response to the habitat locally [18]. The pattern of distribution in groups is due to the strategy of groups of individuals in response to changes in weather and seasons, as well as changes in habitat and the reproduction process [19]. The distribution of a population occurs due to habitat selection, predation and competition against other species, as well as both physical and chemical environmental factors [20]. Predation or competition is characterized by the presence of other species and also influences the distribution of species as they compete for space, food and other factors [21].

IOP Conf. Series: Earth and Environmental Science **137** (2018) 012043 doi:10.1088/1755-1315/137/1/012043

## 6. Conclusion

This research shows that the highest biomass and abundance of starfish was found in January. Spatially, *Asterias* sp. are commonly found in Dadapan. The oceanographic conditions, anthropogenic activity and water quality ofthe east coast of Surabaya are important factorsthat affect the biomass, densityand distribution patterns of the starfish *Asterias* sp.

## 7. Reference


[1] Pourvalli N, Fatemeh EP, Arash S, and Hamid R 2014 *J. Biodivers. Envi. Sci.* **5** 14-20

[2] Aziz A 1981*Ocean. Ind.* **14** 41 – 90 (In Indonesia)

[3] Shah R 2018 Asterias (Starfish): History, Habitat and Development. http://www.biologydiscussion.com/invertebrate-zoology/starfish/asterias-starfish-history-habitat-and-development/27860

[4] Anger K, Rogal U, Schriever G, and Valentin C 1977 *Helgoliinder wiss. Meeresunters.* **29** 439-459

[5] Dalgic G, Ceylan Y, and Sahin C 2009 *Aquat. Invasions.* **4** 485-86

[6] Menge BA 1979 *Oecol.* **41** 245-72

[7] Freeman SM, Richardson CA, and Seed R *Estuar. Coast. Shelf Sci.***53** 39-49

[8] Scheibling RE 1980 Abundance, Spatial Distribution, and Size Structure of Populations of Oreaster reticulatus (Echinodermata: Asteroidea) on Sand Bottoms*Mar Biol.* **57** 107-119

[9] Supono, Arbi UY 2012 *J. of Trop Mar Sci and Tech.* **4** 14-20

[10] Sparre PandVenema SC 1999 *Introduction to tropical fish stock assessment* (Jakarta : Center for Fisheries Research and Development, Agricultural Research and Development Agency) p 436 (In Indonesia)

[11] Brower JE and Zar JH 1977 *Field and laboratory method for general ecology* (United States : Wm. C. Brown Compaby Publisher dubuque, iowa) p 28

[12] Steel RGD and Torrie JH 1991*Principles and procedures of statistics* (Jakarta: Gramedia Pustaka Utama) p 772 (In Indonesia)

[13] Indonesian Agency for Meteorology, Climatology and Geophysics Perak I. 2017. Data of current velocity, wind velocity, wave height in East Coast of Surabaya (In Indonesia)

[14] Indonesian state minister of environment decree no. 51 year 2004 (In Indonesia)

[15] O'Hara T and Poore GCB 2000 *J. Biogeogr.* **27** 21–35.

[16] Helm MM, Bourne N, and Lovatelli A 2004 *Hatchery culture of bivalves a practical manual* FAO Fisheries Technical Paper No 471 (Rome : Food And Agriculture Organization Of The United Nations) p 25

[17] Nybakken J W 1988 *Marine biology: an ecological approach* (Jakarta : Gramedia) p 459 (In Indonesia)

[18] Komala R 2012 Analysis of ecobiology as the basis of the management of blood clam resources (Anadara granosa) in the Lada Bay, Sunda strait (Bogor : Bogor Agricultural University) p 231(In Indonesia)

[19] Odum EP 1993 *Fundamental of ecology* (Yogyakarta : Gadjah Mada Univ. Press) p 697(In Indonesia)

[20] Krebs CJ 1985 *Ecology: the experimental analysis of distribution and abundance* 3th Edition(United States : Harper and Row Publishers, Inc) p 798

[21] Leimena HEP, Tati SS, Subadarand Adianto 2005 *J. Math. Sci.* **10** 75- 80 (In Indonesia)

IOP Conf. Series: Earth and Environmental Science **137** (2018) 012044 doi:10.1088/1755-1315/137/1/012044

# The exploration of trophic structure modeling using mass balance Ecopath model of Tangerang coastal waters

**N N Dewi[1], M Kamal[2], Y Wardiatno[2] and Rozi[1]**

[1]Department of Fish Health and Aquaculture, Airlangga University, Airlangga street 60115, Surabaya, East Java, Indonesia.
[2]Department of Aquatic Resources Management, Bogor Agricultural University, Lingkar Akademik Street, Campus IPB Bogor, Indonesia 16680

E-mail: ninanurmaliadewi@fpk.unair.ac.id

**Abstract.** Ecopath model approach was used to describe trophic interaction, energy flows and ecosystem condition of Tangerang coastal waters. This model consists of 42 ecological groups, of which 41 are living groups and one is a detritus group. Trophic levels of these groups vary between 1.0 (for primary producers and detritus) to 4.03 (for tetraodontidae). Groups with trophic levels 2≤TL<3 and 3≤TL<4 have a range of ecotropic efficiency from 0 to 0.9719 and 0 to 0.7520 respectively.The Mean transfer efficiency is 9.43% for phytoplankton and 3.39% for detritus. The Mixed trophic impact analysis indicates that phytoplankton havea positive impact on the majority of pelagic fish, while detritus has a positive impact on the majority of demersal fish. Leiognathidae havea negative impact on phytoplankton, zooplankton and several other groups.System omnivory index for this ecosystem is 0.151. System primary production/respiration (P/R) ratio of Tangerang coastal waters is 1.505. This coastal ecosystem is an immatureecosystem because it hasdegraded. Pedigree index for this model is 0.57. This model describes ecosystem condition affected by overfishing and antropogenic activities. Therefore,throughEcopath model we provide some suggestions about the ecosystem-based fisheries management.

## 1. Introduction

Coastal areas are ecosystems of waters that have high productivity. These ecosystems are the nursery ground for various fish biotas with economic value [1]. Nowadays, the rapidly growing human population and various human activities such as industrial waste and household garbage can increase the nutrient inputs in the waters. In addition, fishing activities and changes of global climate are also the aspects influencing fisheries system in an ecosystem, such as changing productivity, biodiversity and interaction withinthe ecosystem [2,3,4].

Tangerang Regency has a coastal area directly bordering the Java Sea influenced by anthropogenic activities, including massive exploitation through construction, exceeding the ecological carrying capacity of the coastal area. The growing number of population during the period of 2003 to 2010 had caused uneven population distribution, coastal erosion and sedimentation. Damage at the coastal area due to abrasion in Tangerang coastal waters has reached 51 %. This was caused by heavy coastal currents, weak water catchment, lack of mangrove where the potential of mangrove in Tangerang coastal waters has been declining in the last 10 years, and the remnants of illegal sea sand exploitation occurring 3-5 years ago. Waste from industries and households are also the causes of the damage of

IOP Conf. Series: Earth and Environmental Science **137** (2018) 012044 doi:10.1088/1755-1315/137/1/012044

natural resources at Tangerang coastal waters. There were around 100 industries in Tangerang Regency in 2010. Furthermore, fishing by using explosives, toxin and trawl has contributed to overfishing, proved by the shrinking number of fish catch landed at the fish landing site. This intensified exploitation is threatening the preservation of coastal resources [5].

Generally, management of fisheries is still conducted through single species or stock approach [2,6].However, this is considered inadequate. So, ecosystem-based managementis getting much attention in fisheries management and ecosystem health [6]. Management through the food chain approach becomes an attempt to comprehendthe ecosystem-based management. This is because each organism in an ecosystem is related to one another, forming trophic level [7]. Knowledge about food web interaction is an ecosystem-monitoring program, which is relevant to develop knowledge about the structure and functions of ecosystems [8,9].

In recent years, Ecopath with Esocim has been used as one of the tools for ecosystem-based management. Using this tool, the impact of fishing activities and climate changes can be identified. It can even evaluate the cumulative impacts of environmental and human activities in marine food webs [10].Ecopath software was first introduced by Polovina [11] to provide a simple method related to standing stock and energy flows within anecosystem. This software was subsequently developed, modified and widely applied to aquatic systems, both freshwater and seawater [12,13]. Ecopath with ecosim (EwE) is open source software consisting of three sub-models including Ecopath, Ecosim and Ecospace [12,14].Currently,Ecotracer has become the $4^{th}$sub-model in the Ecopath with Ecosim software. At this moment, this study is only focusedon the Ecopath model.

There has been no research employing tropicalanalysisconducted in Tangerang coastal waters. Studies on trophicanalysisusingEcopathmodeling approach in Indonesia, especiallyin coastal areas, have not yet been performed. This type of research was only done in Cirata Reservoir[15] and the Gulf of Ekas [16]. Therefore, this research is a new study related to coastal resource management using Ecopath model approach. The purpose of this research is to analyze the food web interaction of different functional groups and to describe the flow of energy in Tangerang coastal water using mass balance EwE version 6.

## 2. Methodology

### *2.1. Study Site*

The research location was theTangerang coastalwaters(Figure 1). Tangerang coastal waters is a coast into which various rivers, including Cipasilian, Cimandiri, Cimauk, Cileuleus, Cirarab, Cisadane, Tanjung Burung and Teluk Jakarta flow. This shallow coastal ecosystem is influenced by various anthropogenic activities and is exploited by fishing. In addition, it has a gently sloping condition with a height ranging from 1 to 10 meters above the sea level. Our study area covers an inshore water body of 305.92 $km^2$within an average water depth of 12.5 m.

**Figure 1.** Map of Tangerang coastal waters. The study area is in circle.

### *2.2. Overview of Ecopath Model*

Ecopath model is a static and descriptive model used to identify trophic interactions in aquatic systemsin a given period [12,14]. Although this software is progressing and is constantly updated, the principles have not changed[17]. This model represents an ecosystem in which there is an inter-connection between trophic groups based on biomass and is linked by mass transfers [11,12]. The Ecopath model refers to the principle of mass balance, which assumes that the production of any given preys is equal to the biomass consumed by the predators plus the biomass caught (eg. in fisheries) plus any exports from the system (Eq.1)

$$B_i\left(\frac{P}{B}\right)_i * EE_i = Y_i + \sum B_j * \left(\frac{Q}{B}\right)_j * DC_{ij} + EX_i \quad (1)$$

where Bi = biomass of prey group i; P/Bi = production/ biomass ratio of group i; EEi = ecotrophic efficiency; Yi = yield (=fishery catch); Bj = biomass of predator group j; Q/Bj = food consumption per unit biomass of j, and DCji = fraction of i in the diet of j. Input data was standardized, B as wet weight (ton $km^{-2}$) and P/B and Q/B rates per year for each component group. The ecotrophic efficiency (EE) has no unit.

The current version of Ecopath does not limit the number of ecological groups that should be in the model, but detritus group must exist. For each ecological group, B, P/Band Q/B parameters must be inputted, but EE parameter is difficult to determine. So, it is estimated from other parameters through the software. A full description of Ecopath can be seen inChristensen and Pauly [12], Christensen and Pauly [18] and Pauly *et al.* [13].

### *2.3. Construction Model*

#### *2.3.1. Ecological grouping*

Ecological groups were grouped based on species with similarity in size, population parameters, prey equality and predators [19]. Based on these assumptions, resources in Tangerang coastal waterswere grouped into 42 ecological groups (table 1).

IOP Conf. Series: Earth and Environmental Science **137** (2018) 012044 doi:10.1088/1755-1315/137/1/012044

Data on basic parameters, diet composition and other information wereobtained from primary data and literature study. If there were incomplete data, the complementary data were taken from an ecosystem having the same characteristics. In addition, Fishbase [20] was also used to supplement the inadequate data where possible. The data on annual landing obtained from Kronjo Fish Landing Siteare presented in table 1.

**Table 1.** Ecological groups and their annual landings in 2013.

| No | Group name | Annual landing (ton/km$^2$) |
|---|---|---|
| 1 | Ambassidae | - |
| 2 | Apogonidae | - |
| 3 | Ariidae | 0.008027 |
| 4 | Caesionidae | 0.002266 |
| 5 | Carangidae | 0.003063 |
| 6 | Clupeidae | 0.01161 |
| 7 | Cynoglossidae | - |
| 8 | Drepanidae | - |
| 9 | Engraulidae | 0.003236 |
| 10 | Gobiidae | - |
| 11 | Haemulidae | - |
| 12 | Lactariidae | - |
| 13 | Leiognathidae | 0.017446 |
| 14 | Lutjanidae | 0.00106 |
| 15 | Menidae | - |
| 16 | Mugilidae | 0.001799 |
| 17 | Mullidae | 0.008003 |
| 18 | Nemipteridae | 0.00598 |
| 19 | Platycephalidae | 0.004696 |
| 20 | Polynemidae | 0.000866 |
| 21 | Pristigasteridae | - |
| 22 | Psettodidae | 0.004321 |
| 23 | Scatophagidae | - |
| 24 | Sciaenidae | 0.005111 |
| 25 | Scombridae | 0.001628 |
| 26 | Serranidae | 0.002291 |
| 27 | Siganidae | - |
| 28 | Sillaginidae | - |
| 29 | Sphyraenidae | 0.001722 |
| 30 | Stromatidae | - |
| 31 | Synodontidae | - |
| 32 | Tetraodontidae | - |
| 33 | Terapontidae | - |
| 34 | Triachantidae | - |
| 35 | Trichiuridae | 0.001722 |

| | | |
|---|---|---|
| 36 | Cephalopoda | 0.002686 |
| 37 | Crabs | - |
| 38 | Shrimps | - |
| 39 | Benthos | - |
| 40 | Zooplankton | - |
| 41 | Phytoplankton | - |
| 42 | Detritus | - |

*2.3.2. Basic Parameters*

*2.3.2.1. Biomass*

Average biomass of fish per families (t/km$^2$) was estimated from swept area method [21]. Where B is biomass (ton/km$^2$), Cw is caught in weight per hour, a is swept area and $X_1$ is the fraction of the biomass in the effective path swept by the trawl (0.5).

$$B = \frac{\frac{Cw}{}}{X_1} \quad (2)$$

Phytoplankton biomass value was estimated from chlorophyll a value. The mean chlorophyll a value in this study was 12.01 mg m-3[22]. This value was converted into a square meter unit (12.01 mg m-2) by multiplying it with the euphotic depth (1 m). This value was subsequently converted to the wet weight of phytoplankton using conversion twice [23]. In the first conversion, chlorophyll a value was converted to carbon, where the ratio of carbon and chlorophyll a was 40:1. It was subsequently converted back into the wet weight where carbon was 10% of wet weight. So, the phytoplankton biomass of Tangerang Coastal waters is estimated to be 4.8052 ton/km$^2$.

Zooplankton biomass was calculated y looking at the biomass and abundance of zooplankton in Pari Island conducted by [24], which is assumed to have the same composition of zooplankton as inTangerang coastal waters. By comparison method, through the data abundance of zooplankton [25], the biomass of zooplankton in Tangerang coastal waters is0.0028 ton/km$^2$.

Detritus biomass was calculated as a function of primary production and euphotic depth based on the formula of [18]. Where D = detrital biomass (gC/m2); PP = primary production (in gCm-2 year-1); E = euphotic depth in meter (1.4 m)

$$\text{Log } B_D = 0.954 \log P_f + 0.863 \log E - 2.41 \quad (3)$$

*2.3.2.2. Production/biomass (P/B) ratio*

Fish production/biomass ratio (P/B) value is difficult to estimate directly. So, the value was assumed to be total mortality (Z) [13]. Total mortality value is the sum offishing mortality (F) and natural mortality (M). Natural mortality value was obtained from Fishbase [20], while fishing mortality was obtained from the equation B = Y / F[26], where B is biomass of the biota, Y is yield, and F is fishing mortality. In addition, the P/B values of several groupswere obtained from literature studies such as [27], [19], [28], [16],[29], [30], [31], [32]and[6].

P/B value of phytoplankton was predicted from primary production [33]. Primary production value in Tangerang coastal waters is 941.54 gr C/m$^2$/year. This value was subsequently converted to wet weight to be adjusted with phytoplankton biomass so that the primary production became 9415.43 gr/m$^2$/year. The average biomass of phytoplankton is 4.8052 gr/m$^2$.So, the P/B phytoplankton value is 1959 per year. P/B value of Cephalopods was obtained from Indriyati (2005). Shrimps, crabs, and benthos values were obtained from Mohamed *et al.* (2008).

*2.3.2.3. Relative food consumption (Q/B)*
Q/B value for fish was estimated from the empirical relationship below (Palomaros and Pauly 1999) using FishBase database (Froese and Pauly, 2014):

$$\text{Log}\ \frac{Q}{B} = 7.964 - 0.204\ \log W_{inf} - 1.965T + 0.083A + 0.532h + 0.398d \quad (4)$$

Winf is the asymptotic weight, A is the ratio of the square of the height of caudal fin and its surface area, indicating metabolic activity.

Values of h and d are variables indicating feeding category of fish species, namely herbivore (h = 1, d = 0), detritivore (h = 0, d = 1) or carnivore (h = 0, d = 0). T is the average annual habitat temperature for fish population [expressed as 1000/(Tc + 273.15)] where Tc is average annual sea surface temperature (31°C). The height measurement of caudal fin and its surface area was implemented in Macro Biology Laboratory, Department of Aquatic Resources Management, Bogor Agricultural University. Q/B values of some ecological groups were obtained from literature studies, while cephalopods and zooplankton values were obtained from Mohamed et al. [19]. Shrimps value was obtained from Ullah et al. [6] and benthos value was obtained from Indriyati [16].

*2.3.2.4. Diet composition of groups*
Information on diet composition becomes an important information to understand the dynamics of ecosystems, given that food networks connect different ecological groups[34]. Diet composition of an ecological group was estimated through stomach content analysis [33]. In this study,the data on diet composition wereobtained fromprimary data through stomach content analysis and some secondary data through literature studies such asMohamed et al. [19], Hajisamae [35], Zahid [36]and Fishbase [20].

*2.3.2.5. Ecotrophic efficiency (EE)*
Ecotrophic efficiency (EE) is the proportion of a group used in the system[37] or a fraction of production of an ecological group consumed by another ecological group or caught by fishing activities.EE value is difficult to calculate directly. So, itwas estimated by Ecopath software based on other parameters inputted. EE value ranges from 0-1[38].

*2.4. Balancing the model*
A model is said to be mass-balanced if the catches, consumption, biomass accumulation and export do not exceed the production of a group. This is in accordance with the Ecopath equation (Eq.1). A balanced model is characterized by an EE value not exceeding 1 [6]. EE value exceeding 1 means that demand for the group is very high [19]. Therefore, modification of diet composition and biomass (mean or maximum biomass) is conducted until a balanced model is achieved.

## 3. Result
*3.1. Model validation*
Ecotrophic efficiency (EE) value is a parameter used in the validation of the Ecopath model [38]. EE values in the Ecopath model in coastal Tangerang range from 0 to 0.9719. According to [6], EE value close to 0 means that the group is not consumed by another group in the system, whereas if the EE value is close to 1 then the group is highly consumed by predators (another group) or caught by fishing activities. Biota with tropical $2 \leq TL < 3$ and $3 \leq TL < 4$ levels have EE 0-0.9719 and 0-0.7520 respectively. Ecological groups that have high EE values include Leiognathidae Ariidae, Carangidae, Scombridae, Trichiuridae, cephalopods, shrimps and detritus (table 2).

IOP Conf. Series: Earth and Environmental Science **137** (2018) 012044 doi:10.1088/1755-1315/137/1/012044

**Table 2.** Basic estimates obtained after the Ecopath model was mass-balanced.

| No | Groups | Trophic level | B (ton/km²) | P/B (/year) | Q/B (/year) | EE |
|---|---|---|---|---|---|---|
| 1 | Ambassidae | 3.00 | 0.0177 | 2.88 | 57.90 | 0.0000 |
| 2 | Apogonidae | 2.76 | 0.0047 | 2.14 | 27.90 | 0.0000 |
| 3 | Ariidae | 2.98 | 0.0039 | 2.50 | 8.70 | 0.8233 |
| 4 | Caesionidae | 3.00 | 0.0195 | 0.88 | 7.50 | 0.1321 |
| 5 | Carangidae | 2.64 | 0.0172 | 3.47 | 14.97 | 0.9719 |
| 6 | Clupeidae | 2.11 | 0.0060 | 10.37 | 34.82 | 0.5945 |
| 7 | Cynoglossidae | 3.00 | 0.0067 | 1.20 | 11.00 | 0.0027 |
| 8 | Drepanidae | 3.00 | 0.0028 | 0.47 | 7.50 | 0.0000 |
| 9 | Engraulidae | 2.33 | 0.0437 | 5.77 | 56.90 | 0.5701 |
| 10 | Gobiidae | 3.00 | 0.0048 | 4.17 | 37.00 | 0.0000 |
| 11 | Haemulidae | 3.30 | 0.0056 | 1.32 | 6.40 | 0.0000 |
| 12 | Lactariidae | 3.47 | 0.0103 | 1.64 | 22.90 | 0.0110 |
| 13 | Leiognathidae | 2.50 | 0.4436 | 5.28 | 45.00 | 0.8951 |
| 14 | Lutjanidae | 3.00 | 0.0030 | 1.57 | 6.90 | 0.2251 |
| 15 | Menidae | 2.50 | 0.0043 | 1.00 | 11.40 | 0.0000 |
| 16 | Mugilidae | 2.50 | 0.0707 | 1.57 | 32.80 | 0.0162 |
| 17 | Mullidae | 3.00 | 0.0817 | 2.56 | 30.20 | 0.0405 |
| 18 | Nemipteridae | 2.94 | 0.1945 | 5.04 | 32.90 | 0.1143 |
| 19 | Platycephalidae | 3.18 | 0.0061 | 2.95 | 5.50 | 0.3199 |
| 20 | Polynemidae | 3.09 | 0.0070 | 0.68 | 4.40 | 0.1820 |
| 21 | Pristigasteridae | 3.00 | 0.0301 | 0.96 | 17.10 | 0.0000 |
| 22 | Psettodidae | 3.50 | 0.0972 | 3.94 | 9.10 | 0.0113 |
| 23 | Scatophagidae | 2.61 | 0.0029 | 2.15 | 22.20 | 0.0000 |
| 24 | Scianidae | 3.33 | 0.0146 | 8.36 | 12.33 | 0.6927 |
| 25 | Scombridae | 2.14 | 0.0248 | 6.24 | 20.47 | 0.6190 |
| 26 | Serranidae | 2.40 | 0.0014 | 3.80 | 9.00 | 0.4307 |
| 27 | Siginidae | 2.00 | 0.0017 | 2.10 | 29.70 | 0.0000 |
| 28 | Sillaginidae | 2.50 | 0.0013 | 2.10 | 32.60 | 0.0000 |
| 29 | Sphyraenidae | 3.81 | 0.0287 | 0.73 | 11.00 | 0.0822 |
| 30 | Stromateidae | 2.18 | 0.0378 | 1.27 | 6.00 | 0.0000 |
| 31 | Synodontidae | 3.36 | 0.0021 | 0.94 | 11.60 | 0.1580 |
| 32 | Tetraodontidae | 4.03 | 0.0354 | 0.91 | 49.90 | 0.0000 |
| 33 | Terapontidae | 3.19 | 0.0176 | 1.69 | 10.75 | 0.0000 |
| 34 | Triachantidae | 2.90 | 0.0040 | 0.42 | 8.40 | 0.0000 |
| 35 | Trichiuridae | 3.52 | 0.0166 | 2.66 | 8.70 | 0.7520 |
| 36 | Cephalopods | 3.01 | 0.0257 | 4.87 | 36.50 | 0.7679 |
| 37 | Crabs | 3.21 | 0.4933 | 6.42 | 8.50 | 0.4818 |
| 38 | Shrimps | 2.80 | 0.2733 | 6.68 | 19.20 | 0.8263 |
| 39 | Benthos | 2.00 | 405.5554 | 6.57 | 27.40 | 0.0108 |
| 40 | Zooplankton | 2.00 | 0.0028 | 119.70 | 300.00 | 0.8008 |

IOP Conf. Series: Earth and Environmental Science **137** (2018) 012044 doi:10.1088/1755-1315/137/1/012044

| | | | | | | |
|---|---|---|---|---|---|---|
| 41 | Phytoplankton | 1.00 | 4.8052 | 1959.00 | | 0.0015 |
| 42 | Detritus | 1.00 | 35.7398 | | | 0.7788 |

### *3.2. Trophic flows*
Values of tropical ecological groups in Tangerang coastal waters range from 1.0 for primary producers and detritus to 4.03 for the top predator, that is Tetraodontidae. The grey circle indicates the amount of biomass, in which benthos has the highestamong other groups. Other biotas that still have high biomass include Leiognathidae, crabs, and shrimps (figure 2).

### *3.3. Niche overlap*
Overlap index was summarized to prey overlap and predator overlap[39,40]. Prey overlap is an index comparing preys with similar predators, whereas predator overlap is the similarity of two predators in consuming the same prey. This index compares each pair of ecological groups and the value ranges from 0-1[41]. In Figure 3,the rectangular shape is an example of an ecological group that has a prey overlap value close to 1; the triangle shapehas a predator overlap value close to 1, and the circle shape is an example of a group having a prey and predator overlap valueclose to 1. Biotaswithprey overlap close to 1 arebetween Clupeidae (6) and Engraulidae (9), and Nemipteridae (6) and shrimp (38), whereas the groupswith predator overlap close to 1 arebetween Mullidae (17) and Synodontidae (31), and Sciaenidae (24) and Trichiuridae (35). The groups with prey and predator overlap close to 1have the same preys and predators, which are between Cynoglossidae (7) and Mullidae, and Engraulidae (9) and Scombridae (25).

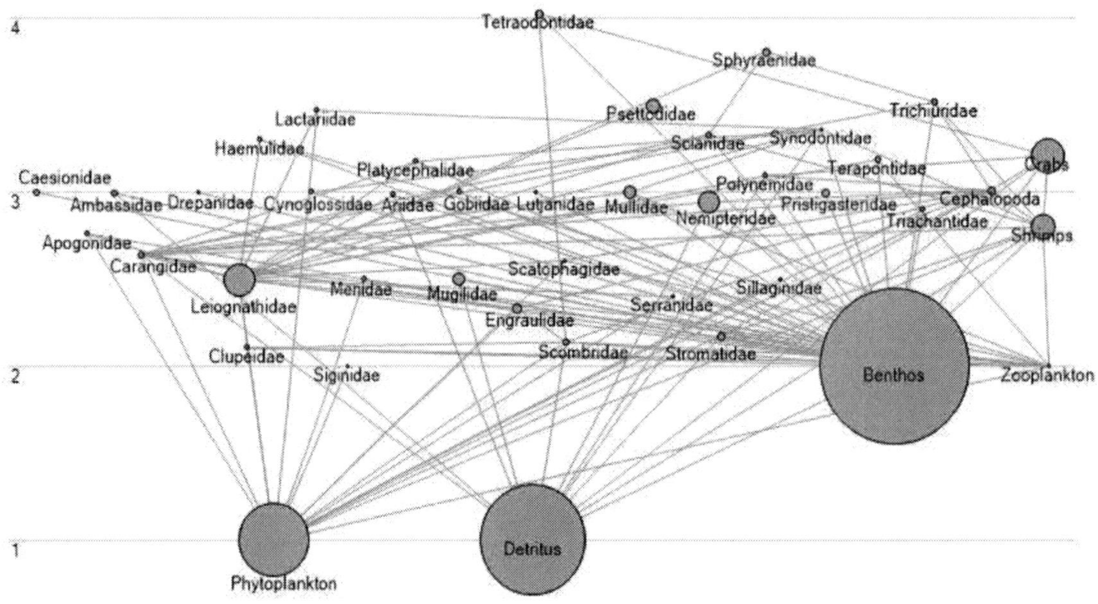


**Figure 2.** The flow diagram shows trophic flows in Tangerang coastal waters. The grey circles show biomass.

### *3.4. Transfer efficiency*
Transfer efficiency is a fraction expressing total flows from a trophic level to another trophic level [37]. Based on Ecopath software,the grouping of the trophic level isgenerally more than 4,commonly called Lindeman spine [42]. For the ecosystem of Tangerang coastal waters, it consists of 8 trophic levels, where Ecopath presents the flow of each trophic level (Figure 4). Based on Table 3, the transfer

rate efficiency for TL II is 0.3%. For TL III and IV, thetransfer efficiency are 11.6% and 12.4%, respectively. The average transfer efficiency is relatively low, 9.43% for phytoplankton and 3.39% for detritus, and the total transfer efficiency in Tangerang coastal waters is 3.42%.

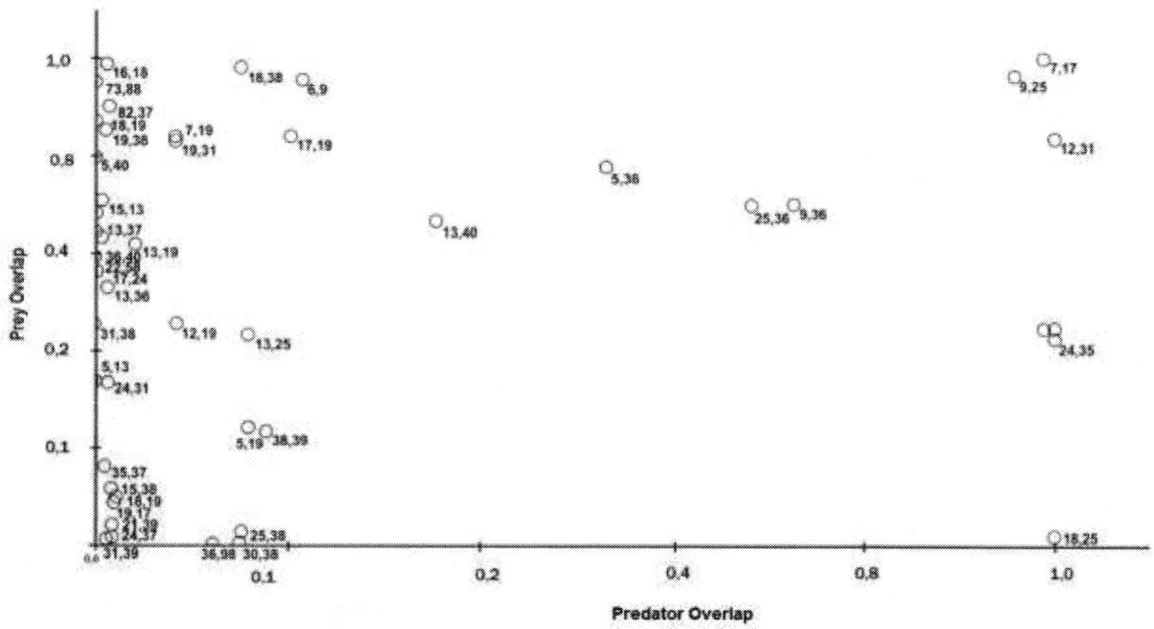


**Figure 3.** Niche overlap in Tangerang coastal waters.

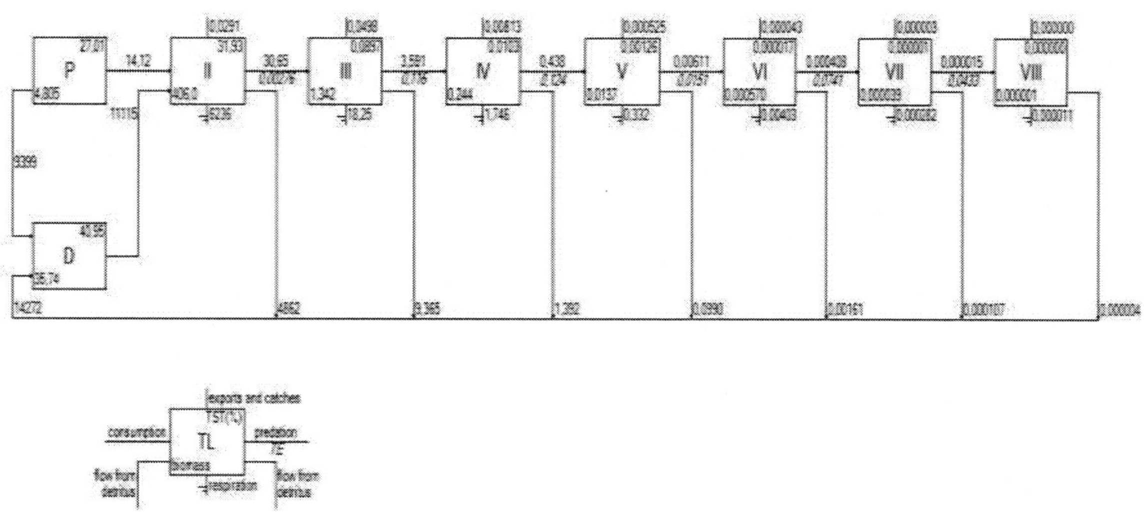


**Figure 4.**Trophic flows transmitted between discrete trophic levels in Tangerang coastal waters. TST: total system throughput; TL: trophic level; TE: transfer efficiency. Units: ton/km$^2$/year.

**Table 3.** Flow (ton/km$^2$/year) and transfer efficiency (%) for each trophic level of Tangerang coastal waters.

| *Transfer efficiency – (%) by trophic level* | | | | | | | | |
|---|---|---|---|---|---|---|---|---|
| Source\TL | I | II | III | IV | V | VI | VII | VIII |
| Producer | | 11.3 | 6.4 | 11.7 | 5.1 | | | |
| Detritus | | 0.3 | 11.9 | 12.5 | 1.4 | 7.5 | | |
| All flows | | 0.3 | 11.6 | 12.4 | 1.5 | 7.4 | 4.3 | 2.4 |

Proportion of total flow originating from detritus: 0.73
Transfer efficiency (calculated as geometric mean for TL II-IV)
From Primary Producer: 9.43%
From detritus: 3.39%
Total: 3.42%

### *3.5. Mixed Trophic Impact*

In Tangerang coastal waters, phytoplankton and detritus havea positive impact on most other groups (figure5). The impact of phytoplankton is the highest for some pelagic fish like Stromateidae, Sillaginidae, Siganidae, Serranidae, Scombridae, Scatophagidae, Menidae, Engraulidae, Clupeidae, Carangidaeand Apogonidae. Detritus hasa positive impact on some groups like shrimps, benthos, Nemipteridae, Mugilidae and Cynoglossidae. Leiognathidae have a negative impact on some groups and themselves, showing competition for the same food resources within and between the groups. Gears also negatively impact many groups. Trawl negatively impacts some of the economic groups such as Sphyraenidae, Serranidae, Psettodidae, Polynemidae, Platycephalidae, Mullidae, Mugilidae, Lutjanidae, Caesionidae and Ariidae. Therefore, increasing fishery activitiesin Tangerang coastal waters ecosystem would have a more influential negative impact on fish groups, particularly on the economic groups.

Through MTI analysis, the Ecopath model provides information related to key species. Ecological groupsincluded in key species are the groupsplaying important roles in the ecosystem. The key species are represented by plot point approaching or greater than 0[43](figure 6). The key species groups in Tangerang coastal waters are Leiognathidae, Sphyraenidae, Cephalopods and Phytoplankton.

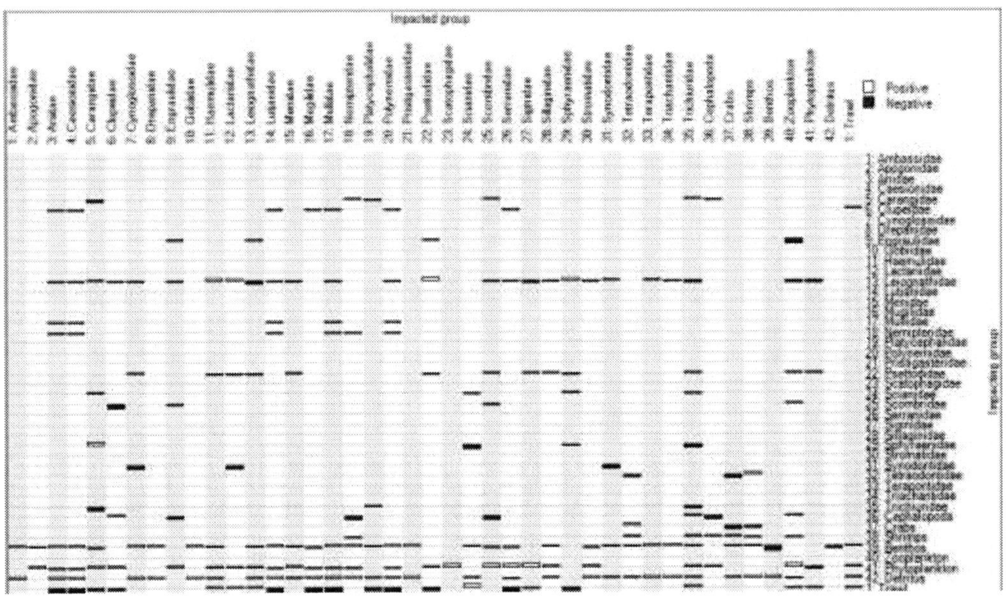


**Figure 5.** Mixed trophic impacts of the ecological groups of Tangerang coastal waters. Positive impacts are shown above the baseline in light columns, while negative impacts are shown below the baseline in dark columns.

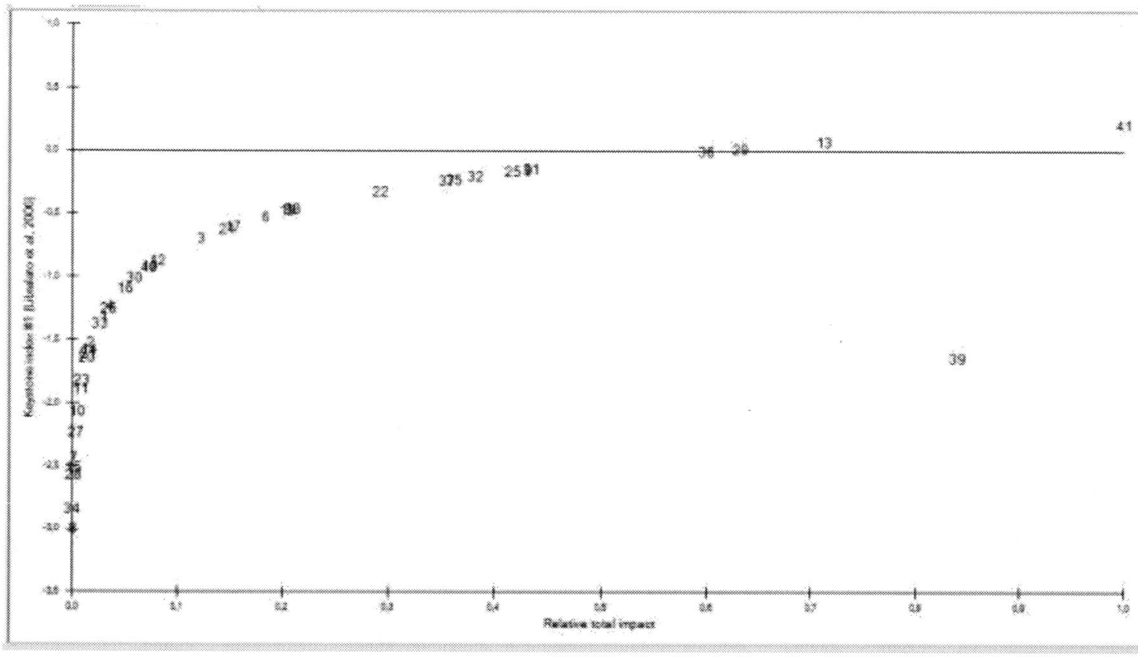

**Figure 6.** Relative total impacts and Keystone index values of the ecological groups of Tangerang coastal waters

### *3.6. Total System Statistics*

Total system statistic of Tangerang coastal waters is shown in Table 4. Total system statistic consists of total system throughput, which is the sum of all existing flows in the system. The flowsconsist of consumption, exports, respiration and flows into detritus. Total system throughput estimated for Tangerang coastal waters is34851 ton/km$^2$/year. The value of the mean trophic level of the catch for Tangerang coastal waters is 2.778 and its gross efficiency is 0.00001.

The maturity of the ecosystem is described by the value of primary production/respiration (P/R) ratio [44]. P/R ratio of Tangerang coastal waters is 1.505. The value of net system production (or yield) for this ecosystem is 3156.84 ton/km$^2$/year while the PP/B value for the present model is 22.825. The system biomass/throughput ratio is 0.012 ton/km$^2$/year. Finn's cycling index and Finn's mean path length are 24.85% and 3.702% respectively. Pedigree index for this model is 0.57. The current version of Ecopath model provides the Shannon diversity index, and the value of the diversity index in Tangerang coastal waters is 0.107.

**Table 4.** Total system statistic of Tangerang coastal waters (The language of the table below is genuinely from the software results).

| Parameter | Value | Units |
|---|---|---|
| Sum of all consumption | 11164.510 | t/km$^2$/year |
| Sum of all exports | 3157.473 | t/km$^2$/year |
| Sum of all respiratory flows | 6256.581 | t/km$^2$/year |
| Sum of all flows into detritus | 14272.390 | t/km$^2$/year |
| Total system throughput | 34850.960 | t/km$^2$/year |
| Sum of all production | 12088.440 | t/km$^2$/year |

IOP Conf. Series: Earth and Environmental Science **137** (2018) 012044 doi:10.1088/1755-1315/137/1/012044

| | | |
|---|---|---|
| Mean trophic level of the catch | 2.778 | |
| Gross efficiency (catch/net p.p.) | 0.00001 | |
| Calculated total net primary production | 9413.416 | t/km$^2$/year |
| Total primary production/total respiration | 1.505 | |
| Net system production | 3156.835 | t/km$^2$/year |
| Total primary production/total biomass | 22.825 | |
| Total biomass/total throughput | 0.012 | t/km$^2$/year |
| Total biomass (excluding detritus) | 412.426 | t/km$^2$/year |
| Total catch | 0.088 | t/km$^2$/year |
| Connectance Index | 0.072 | |
| System Omnivory Index | 0.15 | |
| Throughput cycled (including detritus) | 8659.56 | t/km$^2$/year |
| Ascendency | 53.2 | |
| Overhead | 46.8 | |
| Finn's cycling index | 24.85 | % of total throughput |
| Finn's mean path length | 3.702 | |
| Ecopath pedigree index | 0.57 | |
| Shannon diversity index | 0.107 | |

### 3.7. Model Comparison

The comparison of Ecopath model with other marine and coastal ecosystemsis as follows (table 5).

**Table 5.** Comparison of Ecopath model of Tangerang coastal waters with other ecosystems.

| Ecosystem | TST | TPP | TPP/TR | GE | NSP | CI | OI | FCI | FML | PI |
|---|---|---|---|---|---|---|---|---|---|---|
| Laguna Alvarado, Mexico [45] | 2683 | 1291 | 1.30 | >0.001 | 303.5 | 0.27 | 0.3 | - | - | - |
| Karnataka Arabian Sea [19] | 11522.00 | 4095 | 1.283 | 0.0016 | 904 | 0.382 | 0.299 | 6.03 | 2.81 | 0.52 1 |
| Northern Hangzhou Bay [46] | 19323 | 6785.80 | 2.56 | - | 4139.15 | 0.31 | 0.35 | 25 | 2.174 | 0.61 |
| The Bay of Bengal [6] | 2628.00 | 1017.363 | 1.351 | 0.0015 | 264.235 | 0.42 | 0.224 | 10 | 2.58 | - |
| Jimo Coastal Ecosystem [43] | 5112.73 | 1993.60 | 3.78 | - | 1466.05 | 0.40 | 0.36 | 4.92 | 2.57 | 0.48 |
| Tropical coastal estuarine ecosystem around Bight of Benin, Nigeria [2] | 34385.762 | - | 6.325 | - | 12598.790 | 0.327 | 0.288 | 1.7 | 2.298 | 0.36 |
| Red Sea [47] | 7335 | 2996 | 2.25 | 0.000085 | 1665.03 | 0.31 | 0.24 | - | - | - |
| Tangerang coastal waters(Present study) | 34850,960 | 9413,416 | 1.505 | 0.00001 | 3156,835 | 0.072 | 0.15 | 24.85 | 3.702 | 0.57 |

IOP Conf. Series: Earth and Environmental Science **137** (2018) 012044 doi:10.1088/1755-1315/137/1/012044

(TST:Total system throughput, TPP: Total primary production, GE: Gross efficiency, NSP: Net system production, CI:Connectance Index, OI: Omnivory Index, FCI:Finn's cycling index, FML:Finn's mean path length, PI: Pedigree index)

### *3.8. Discussion*

Leiognathidae havea high ecotrophic efficiency value. So, it must be predated by other groups because their abundance and biomass in Tangerang coastal waters are so high. Whereas, Ariidae, Carangidae, Scombridae, Trichiuridae, Cephalopods, and shrimps, in addition to being affected by predators, are also affected by fishing. According to Kepmen KP 45 [48], fish resources in coastal area around the Java Sea are overexploitated. EE value of the detritus describes the incoming stream that becomes the detritus and that comes out of detritus. The EE value of detritus in Tangerang coastal waters is less than 1, indicating that a stream that becomes detritus is higher than the stream coming out (Table 2).

Many coastal biota groups in Tangerang coastal watersare found in $2 \leq TL < 3$and $3 \leq TL < 4$ trophic level. Indriyati [16] stated that the more groups in the same trophic level mean that there is a competition in getting food sources. Biota having a trophic level of $\geq 4$ is only one group, which is Tetraodontidae (Figure 2). Biotas that are predators or carnivorous fish such as Scombridae, Carangidae and Trichiuridae have low trophic level values,meaning that the food network in Tangerang coastal waters has short food web. This indicates the low ecological efficiency of the ecosystem.

The low value of trophic levels is inseparable from anthropogenic activities and fishing activities in these ecosystems. High anthropogenic activities around Tangerang coastal waters cause a decline in water quality and affectbiota in the ecosystem. It is indicated by the size of biota that tends to be small sothat the biomass is also low. This contaminated condition can cause fish to experience environmental stress, which may cause the decrease in population. This becomes an indication of low fish biomass in Tangerang coastal waters. According to Zahid et al. [49], biological richness in an ecosystem reflects the health of the environment.

In addition to anthropogenic activities, other factors affecting trophic levels are fishing activities. Fishing activities can alter the abundance and spatial distribution of fish, especially piscivores or carnivores. This has an important impact on the interaction of species and trophic structures[50,51].Loss of carnivorous fish has an impact on the entire ecosystem [52]. According to Pauly [53], the decline in the number of carnivorous or predatory fish in water indicates the occurrence of fishing down the food web. Fishing down the food web is a condition where predatory fish are depleted and fishing activities target small fish that have low trophic levels.

Overlap value close to one means higher level of similarity in food and predators,which includes similarity in consuming prey (predator overlap) and the similarity of predators (prey overlap). The similarity is influenced by habitat similarities, such as between Clupeidae and Engraulidae groups and between Engraulidae and Scombridae. which are pelagic fish, Nemipteridae and shrimps, and between Mullidae and Synodontidae, which are demersal biota. Sciaenidae fish which are demersal fish have predator overlap with Trichiuridae, which are pelagic fish. This can happen because Tangerang coastal watersare shallow waters. So, it becomes an indication of both groups in having the same food niches (Figure 3).

A group that becomes a prey has a positive impact on its predator, while a predatory group hasa negative impact on its prey Based on MTI analysis (Figure 7), phytoplankton and detritus have positive impacts on most groups. Phytoplankton affect most groups that are phytoplankton predators such as pelagic fish, whereas detritus has a positive impact on biota groups dominated by demersal biota. Leiognathidae have negative effects on phytoplankton and zooplankton, because phytoplankton and zooplankton are part of their prey. Leiognathidae also negatively affect some biotas, even themselves, due to the existence of competition in getting food sources. Trawler fishing equipment negatively affects the biomass of several groups, especially economic biotas, due to the ongoing fishing activities that cause the stock of biota to decrease.

Based on Table 3, the transfer rate efficiency for TL II is 0.3%. This value is very lowcompared to the typical value for coastal areas, which is 15% [54], or ranging between 10-20% [55,56]. This result is similar to that of Ullah et al. [6] in Bay of Bengal which is 0.5% but different from Su [43]'s research in Jimo coastal ecosystem,which is 27.4%. The low transfer efficiency on TL II in Tangerang coastal waters is caused by the absence of predators and the high respiration flow of TL II as shown in Lindeman spine (Figure 3). The TL III and IV efficiency transfer is in the range of 10-20% for coastal zones [55,56]. These results are comparable toSu [43] )in Jimo coastal ecosystem but it is different from Bay of Bengal which has higher transition efficiencies for trophic III and IV than the standard range due to high exports of fishery yields and the presence of major dominating fish groups with higher EE value around trophic level III [6].

Based on total system statistics of the Ecopath model (Table 4), Tangerang coastal waters have a high total system throughput value. This is consistent with tropical coastal ecosystems in general which tend to have high turnover rates [19]. The mean trophic level of catch in Tangerang coastal waters is low,indicated by the lack of predatory species on Tangerang coastal waters and the domination by species that have low tropical levels. This can be provedby the low biomass of predatory fish in Tangerang coastal waters. In addition, low energy efficiency value makes the food chain of Tangerang coastal waters short. According toUllah et al. [6], mean trophic level of the catch describes the overall level of exploitation in the food web in an ecosystem and how it impacts predators and preys. The gross efficiency value in Tangerang coastal waters is only 0.00001. This is because, in general, the total catch in this ecosystem has a very small value, which is 0.088 ton/km$^2$/year. Predatory fish have already been overfished. So, their biomass tends to be low.

The Ecopath model also presents several parameters related to ecosystem maturity described byOdum [44]. The first parameter is primary production/respiration (P/R) ratio system. A mature ecosystem has P/R value equal to 1 which means that there is an ecosystem balance between production and respiration(44). Research results in Tangerang coastal waters show a value of>1 which means that Tangerang coastal waters are immature ecosystem. According to Fetahi and Mengistou [34],this could be caused by the effect of human activities including fishing and the declining water quality due to pollution, water diversion, etc.

Net system production (or yield) isreduction in total primary production and total respiration. Net system production will be large for systems that are still in the process of maturation and close to zero for mature systems [19]. The net system production value of Tangerang coastal waters is 3156.835-ton km$^{-2}$ year$^{-1}$ which indicates that the ecosystem is immature.

The ratio between primary production and biomass also reflects the maturity of the ecosystem. According to Ullah et al. [6], immature ecosystems tend to have greater production value than respiration. This causes the accumulation of biomass so that it impacts the PP/B value. The PP/B value for this model is 22.825. The next parameter that describes ecosystem maturity is the ratio of total biomass to total system throughput. According to Christensen [14], total biomass/total system throughput value is proportional to the maturity of the ecosystem, of which the value will be low in the ecosystem that is still undergoing the maturation process. For this model, the ratio of total biomass and total system throughput is 0.012 per year. This means that the total biomass existing in Tangerang coastal waters is smaller than the total system throughput.

Total biomass is the sum of all biomass includingdetritusis 412.426 ton km$^{-2}$. This result is greater than that in the Arabian Sea, which is 136 ton km$^{-2}$ [19], in the Bay of Bengal 69.241 ton km$^{-2}$(Ullah et al., 2012), and in Indian Vellar Estuary 250.397 ton km$^{-2}$[57]. This is due to the high biomass of benthos obtained. According to Meilana [58], the content of organic materials such as high total organic carbonin the coast of Tangerang becomes an indication of the high biomass of benthosin the ecosystem.

System omnivory index is the average omnivory index of all consumers. This index measures the complexity of eating interactions distributed among trophic levels[19]. System omnivory index value in this study is 0.151. Overall,the majority of biota groupsin Tangerang coastal watersare specialist in choosing food.

IOP Conf. Series: Earth and Environmental Science **137** (2018) 012044 doi:10.1088/1755-1315/137/1/012044

Finn's cycling index (FCI) is the proportion of flow in a system that is recycled compared with total system throughput while Finn's mean path length (FCL) describes the average number of ecological groups flowing from entry to exit from the system (Xu et al., 2011). The low FCI and FML in this model mean that the proportion of matter recycled through the detrital pathway is still low. This signifies an immature ecosystem.

Ascendency is the key index that characterizes the rate of development and maturity of the system [59], while overhead describes the existing reserve energy in the system [60]. The relative values of ascendancy and overhead in Tangerang coastal waters are 53.2% and 46%. This ascendency value is greater than that in the Bay of Bengal which is only 38.7%, but its overhead is smaller than BOB which reaches 61% [6]. Overall, these results indicate the level of maturity and stability of ecosystems. These results indicate that Tangerang coastal waters have energy reserves to be resistant or resilient to perturbations.

The output obtained from the Ecopath model in Tangerang coastal waters was compared with some other coastal and marine waters (Table 5). The total system throughput (TST) value of Tangerang coastal waters is the highestamong other ecosystems; the second highest is the tropical coastal estuarine ecosystem around the Bight of Benin, Nigeria. The gross efficiency value of coastal Tangerang is the lowest among other ecosystems. TPP/TR value describes the maturity of the ecosystem. The TPP/TR value of Tangerangcoastal waters is moderate compared to other ecosystems but exceeding 1 which means that it is an immature system. According to Su [43], this condition means that many nutrients are not used in the system. So, the ecosystem is said to be immature. This is related to the large number of nutrient inputs in Tangerang coastal waters derived from human activities as well as industry. Connectance index (CI) and omnivory index (OI) values show the internal state ofecosystem complexity. A mature ecosystem tends to have CI and OI values close to 1 [43]. The CI and OI of Tagerang coastal waters have moderate values compared to other ecosystems, but these values are still low. The low value of the omnivory index is due to the low diversity of feeding interactions within the ecosystem. Finn's cycling index (FCI) and Finn's mean path length (FML) also describe the maturity condition of an ecosystem. This is related to the proportion of nutrient materials or flow through the food chain. According to Christensen and Walters [61], high FCI value will be followed by high FML value. Tangerang coastal waters have the second highest FCI value after Northern Hangzhou Bay and its FML value is the highest among other ecosystems. This means that Tangerang coastal waters have higher and more energy as well as material flow compared to other ecosystems. The results of Ecopath model research in Tangerang coastal waters have a relatively high pedigree index value compared to other studies. Based on the data from Morissette et al. [62], 150 Ecopath research survey models in the world have pedigree index ranging0.16-0.68. The pedigree index value of Ecopath model of Tangerang coastal wateris in the range of these values. This means that the data sources in this study are quite accurate and reliable. So, basedon Christensen et al. (2000), this model is acceptable.

### *3.9. Management Recommendations*

This Ecopath model can be used to evaluate the impacts of fishing and environmental changes. It can be employed as a management strategy which has been proven to be a useful tool for the development of policies for tropical fisheries catchment and multispecies fisheries management (Ullah et al., 2012). The management of Tangerang coastal waters should pay attention to the aspects of capture fisheries and coastal biology. Tangerang coastal watershave generally collapsed in terms of fishery resources. Predatory fish have been decreasing in number, biomass and size. This is due to the high catching activities that cause the existing biota to be overfished. Based on the trophic model obtained, fishery resources that can still be utilized and optimized include Leiognathidae, Nemipteridae fish, crabs and swimming crabs, and shrimp,especially Leiognathidae fish which are very abundant in nature and their utilization is not optimal. Based on the information by fishermen, Leiognathidae are only used for duck feed. These fish have low economic value because the price is only around Rp. 2000.Actually, the fish can be processed into various processed foods that have high economic value.

In addition, Tangerang coastal watersare still classified as an ecosystem that is still atdevelopment stage. So, it is more vulnerable to disturbance, both by natural and anthropogenic activities. Thepollutants from various industries and garbage along the coast can impact on the existing biota. Biota will experience environmental stresses so that more energy is used to adapt to environmental conditions than for growth. Therefore, it is necessary to monitor and evaluate pollutants entering the waters. Further, related institutions need to regulate/manage industrial waste and garbage that have been causing pollution in the ecosystem.

## 4. Conclusion

This model is able to describe trophic interaction, energy flows and ecosystem conditions that have been affected by fishingand anthropogenic activities. This model has an important role related to the availability of ecological data on Tangerang coastal waters ecosystem. It also serves as a basis for new data which can subsequently be used to develop models, especially Ecosim, Ecospace, and Ecotracer models. The research on Tangerang coastal waters ecosystem is expected to contribute to the policy making by the stakeholders.

## 5. References

[1] Brando V 2004 *Ecol.Modell.* **172** 213-232
[2] Abdul WO and Adekoya EO 2016 *Environ.Biol.Fish.* **99** 909-923
[3] Chen Z, Xu S, and He P 2010 *Helgol. Mar. Res.* **65** 217-231
[4] Daskalov GM, Abdoli A, Akhundov M, Annachariyeva J, Isbekov K, Khodorevskaya R, Kim Y, Mammadli T, Morozov B, Muradov O, and Shahifar R2013 Ecosystem modelling in support of ecosystem-based management in the Caspian*The Handbook of Environmental Chemistry.* Springer Berlin Heidelberg p 1-36
[5] Dewi NN, Kamal MM, and Wardiatno Y 2014Spatial and temporal variation of fish community biomass in coastal waters of Tangerang Regency, Banten *J. Trop Mar Sci Technol.* **8** 39-55
[6] Ullah MH, Rashed-Un-Nabi M, and Al-Mamun MA 2012 Trophic model of the coastal ecosystem of the Bay of Bengal using mass balance Ecopath model*Ecol.Modell.* **225** 82-94
[7] Supongpan M, Poonsawat R, and Christensen V 2005 Trophic levels of multi-species in the Gulf of Thailand ECOST
[8] Harvey CJ, Williams GD, and Levin PS 2012 *Estuaries and Coast.* **35** 821-838
[9] Pasquaud S, Pillet M, David V,Sautour B, and Elie P 2010 *Estuar.Coast. Shelf. Sci.* **86** 237-246
[10] Coll M, Akoglu E, Arreguín-Sánchez F, Fulton EA, Gascuel D, Heymans JJ, Libralato S, Mackinson S, Palomera I, Piroddi C, Shannon LJ, Steenbeek J, Villasante S, and Christensen V 2015 *Fish Biol. Fish.* **25** 413-424
[11] Polovina 1984 *Coral Reefs.* **3** 1-11
[12] Christensen V and Pauly D 1992 *Ecol.Modell.* **61** 169-185
[13] Pauly D, Christensen Vand Walters CJ 2000 *J. Mar. Sci.* **57** 697-706
[14] Christensen V 1995 *Ecol.Modell.* **77** 3-32
[15] Kartamihardja ES 2007 Spectra of plankton biomass size and potential utilization for fish communities in Limnetic zone of Reservoir Ir. Hj. Juanda, West Java(Bogor : Bogor Agricultural University) p 137
[16] Indriyati E 2005 *Sci Mar.* **10** 85-89
[17] Geers TM, Pikitch EK and Frisk MG 2014 *Top. Stud. Oceanogr.* **129** 319-331
[18] Christensen V and Pauly D1993 Trophics Model of Aquatic Ecoystem (Manila: ICLARM Conference Proceedings) p 390
[19] Mohamed KS, Zacharia PU, Muthiah C, Abdurahiman KP and Nayak TH 2008Trophic model of the arabian sea ecosystem off karnataka and simulation of fishery yield (Kerala india: CMFRI Publication Central Marine Fisheries Research Institute) p 140
[20] Fröese R and Pauly D 2014 Fishbase 2014 [cited 2014]Available from: www.fishbase.org

[21] Sparre P andVenema SC 1999 *Introduction of Tropical Fish Stock Assessment* (Jakarta: E-manual Book *(Translate edition)* )
[22] Wulandari DY 2015 *Phytoplankton Community Structure and Fertility Level of Coastal Waters of Tangerang* (Bogor : Bogor Agricultural) p 31
[23] Jones JG 1979 A guide to methods for estimating microbial numbers and biomass in fresh waterFreshwater Biological Association, Scientific Publication No. 39
[24] Puspasari R 2012Trofodinamik phytoplankton - zooplankton as a determinant of the survival of fish larvae in Laguna Pulau Pari Kepulauan Seribu Bogor Agricultural University p219
[25] Ardhito 2015Spatial Distribution of Zooplankton in Coastal Waters of Tangerang RegencyBogor Agricultural University p 21
[26] Gulland JA 1971 *The Fish Resorces of the Ocean Surrey* (England: Fishing News (Books) Ltd fo FAO) p 255
[27] Mahmoud HH, El Haweet AAKand Dimech M 2013 *Egypt. J. Aquat. Res.***40** 443-450
[28] Dwiponggo A, Hariati T, Banon S, Palomaros ML and Pauly D 1986 Growth, Mortality, and Recruitment of Comercially Important Fishes and Penaid Shrimps in Indonesian Waters ICLARM Technical Report 17, Research Institute for Marine, Jakarta Indonesia and Fisheries for International Center for Living Aquatic Resources Management (Manila) p 91
[29] Haputhantri SSK, Villanueva MCS and Moreau J2008 *Estuar.Coast Shelf. Sci.***76** 304-318
[30] FAO Food and Agricultur Organization 2001The living Marine of the western central pasific (Rome) volume 5 Bony fishes part 3 Edited by Carpenter KE and Niem VH
[31] Ashraful HA 1998 Population dynamics of five commercially important marine fishes in north-eastern part of the Bay of Bengal University of Chittagong (Bangladesh)
[32] Moradinasab AA, Ehsan K and Mahin A 2014 *Oceano.* 5.
[33] Fetahi T 2005 Trophic analysis of Lake Awassa using mass-balance Ecopath model (Ethiopia: Addis Ababa University) p 99
[34] Fetahi T and Mengistou S 2007 *Ecol. Modell.* **201**: 398-408
[35] Hajisamae S 2009 *Estuar.Coast Shelf. Sci.* **82**:503-514
[36] Zahid A 2013Trophic ecology of the fish community as the basis for the management of fish resources in the estuary ecosystem Segara Menyan Subang, West Java Bogor : Bogor Agricultural University p 75
[37] Dutta S, Chakraborty K and Hazra S 2017 *Fish. Scien.* **83** 145-159
[38] Christensen V, Walters CJ and Pauly D 2000Ecopath with Ecosim: A User's Guide Fisheries Center, Univeristy of British Columbia, Vancouver and ICLARM, Malaysia
[39] Hurlbert SA 1978 *Ecology.* **59** 67-77
[40] Loman 1986 *Ecol.Modell.* **24** 231-243.
[41] Guénette S 2014 User's guide to the "Ecopath with Ecosim" model of the Bay of Bengal Large Marine Ecosystemp 23
[42] Ulanowicz RE 1995 The part-whole relation in ecosystems In: Patten, BC, Jørgensen, SE(Eds.), Complex Ecology. Prentice-Hall, Englewood Cliffs, New Jersey) p 549–560
[43] Su M 2016 *J. Ocean Univ. China.* **15** 1059-1066
[44] Odum EP 1969 *Scie.***104** 262-270
[45] Cruz-Escalona VH, Arreguín-Sánchez F and Zetina-Rejón M 2007 *Estuar.Coast Shelf. Sci.***72** 155-167
[46] Xu S, Chen Z, Li S, He P 2011 *Estuar and Coasts.* **34** 351-363
[47] Tesfamichael D 2016 *Fisheries.* **7** 111-134
[48] Kepmen KP 45 2011
[49] Zahid A, Simanjuntak CPH, Rahardjo MF and Sulistiono 2011 *J. Ichtiolog Ind.* **11** 77-85
[50] Garrison L 2000 *J. Mar. Sci.***57**:723-730
[51] Russ GR and Alcala AC 1989 *Mar. Ecol. Prog. Ser.* **56** 13-27
[52] Grigg RW 1994 *Mar. Ecol. Prog. Ser.***103** 25-34
[53] Pauly D 1998 *Scie.* **279** 860-863

IOP Conf. Series: Earth and Environmental Science **137** (2018) 012044 doi:10.1088/1755-1315/137/1/012044

[54] Ryther JH 1969 *Scie.***166** 72–76
[55] Odum EP 1971Fundamentals of Ecology, 3rd edition(Philadelphia : W.B. Saunders)
[56] Barnes RSF and Hughes RN 1988An Introduction to Marine Ecology, 2nd ed. Blackwell Scientific Publications, Oxford
[57] Murugan S, Joseph AP and Khan S A 2012 *Int. J. Pharma. Bio. Sci.* **3** 662-676
[58] Meilana L 2014 The meiobhentos abundance as bioindicator of tangerang coastal waters (Bogor : Bogor Agricultural Universityp 17
[59] Ulanowicz RE 1986 *Growth and Development: Ecosystem Phenomenology* (NewYork.Springer-Verlag)
[60] Ulanowicz RE and Norden JS 1990 *Int. J. Syst. Sci.* **21** 429–437.
[61] Christensen V and Walters CJ 2004 *Ecol. Modell.* **172** 109-139
[62] Morissette L, Hammill MO and Savenkoff C 2006 *J Mar Mammals Sci.* **22** 74-103

IOP Conf. Series: Earth and Environmental Science **137** (2018) 012045 doi:10.1088/1755-1315/137/1/012045

# Phytochemical compounds of *Enhalus acoroides* from Wanci Island (Wakatobi) and Talango Island (Madura) Indonesia

**C S U Dewi[1,2], R D Kasitowati[1] and J A Siagian[3]**

[1] Department of Marine Science, Fisheries and Marine Science Faculty, University of Brawijaya, Jl. Veteran, Malang, East Java, Indonesia 65145
[2] Coastal and Marine Research Centre, Institute for Research and Community Services, University of Brawijaya, Jl. Veteran, Malang, East Java, Indonesia 65145
[3] Reef Check Indonesia, Jl. Tukad Balian Gg. 43 No. 1A Renon, Denpasar Bali, Indonesia 80226

E-mail: satryacitra@ub.ac.id

**Abstract.** The existence of *Enhalus acoroides* certainly gives important influence to the ecosystem, both as a producer in the food web and as a living habitat. In the last decade, *Enhalus acoroides* was widely used as an object of marine bioprospection research. That research showed potential results as antibacterial, antifungal, even as antifouling. This is a good reason to know the content of phytochemical compounds in *Enhalus acoroides* from two different locations. The purpose of research purpose to determine (1) the crude extract produced by *Enhalus acoroides* from two different locations; and (2) the phytochemical compounds contained in the crude extract of *Enhalus acoroides* from two different locations. This study this research was to used samples collected from Wanci Island (Wakatobi), and Talango Island (Madura), Indonesia. The extraction process and phytochemical test were conducted at the Marine Science Laboratory, FPIK, University of Brawijaya, and lasted for two months, from June 2017 to September 2017. The extraction was done by three solvent, are methanol, ethyl acetate and chloroform. Furthermore, phytochemical test was performed qualitatively. The results provided that the yield produced by *Enhalus acoroides* from Wanci Island, Wakatobi is relatively lower than Talango Island, Madura. *Enhalus acoroides* is also renowned to contain phytochemical compounds of tannins and saponins.

## 1. Introduction

*Enhalus acoroides* is a type of *Enhalus acoroides* that is easy to find in Indonesia. This species can easily be recognized and identified because it has a different morphology. *Enhalus acoroides* has larger roots, stems and leaves compared to other *Enhalus acoroides* species found in Indonesia [1]. The type of *Enhalus acoroides* leaf is long, hard and stiff like a belt. It can be found in all major islands in Indonesia and even in small islands [2].

The presence of *Enhalus acoroides* certainly has an important influence on the ecosystem, both as a producer in food networks and as a living habitat of biota [3]. Moreover, *Enhalus acoroides* leaf is also renowned to help coral reefs reduce the waves, as well as catch solid sediment particles in the water column. In the last decade, *Enhalus acoroides* is also widely used as a marine bioprospective research object and given potential results as antibacterial, antifungal, even as an antifouling [4, 5, 6].

IOP Conf. Series: Earth and Environmental Science **137** (2018) 012045 doi:10.1088/1755-1315/137/1/012045

Previous research showed that environmental conditions would affect the content of bioactive compounds produced by the organism. This condition is a strong reason for the identification of basic information about phytochemical compounds in *Enhalus acoroides* from two different locations. The purpose of this study is to determine (1) the yield produced by *Enhalus acoroides* from two different locations; (2) a class of phytochemical compounds in the yield of *Enhalus acoroides* from two different locations.

## 2. Methodology

This research began with the collection of fresh *Enhalus acoroides* samples from Wanci Island, Wakatobi and Talango Island, Madura. The aamples of fresh *Enhalus acoroides* es obtained were then dried naturally under the sun and extracted at the Marine Science Laboratory, Fisheries and Marine Science Faculty, University of Brawijaya. The *Enhalus acoroides* extract obtained was then used as a test material, namely phytochemical test. There were three sections for processing; there were sampling fresh *Enhalus acoroides* , extraction and evaporation, and the phytochemical test. This research was conducted from September to November 2017.

The tools and materials used in this study were divided into three major sections (table 1), i.e. tools and materials for *Enhalus acoroides* sampling, extraction, and phytochemical test. This research was conducted in several stages, i.e. collection of *Enhalus acoroides* samples, extraction, phytochemical test. The samples of fresh *Enhalus acoroides* leaves were taken directly from their habitat, then cleaned from the biofouling. Then, the leaves of the *Enhalus acoroides* were dried naturally until the *Enhalus acoroides* leaves became constant. For instance, the dried *Enhalus acoroides* leaves were prepared three times as replicates.

**Table 1.** The tools and materials used during research.

| Stages | Tools | Materials |
|---|---|---|
| 1. Sampling *Enhalus acoroides* | Mask<br>Snorkel<br>Plastic<br>Dive knives | |
| 2. Extraction of bioactive materials | Glass<br>Evaporator<br>Sbath<br>Whatman<br>The scales | For instance of *Enhalus acoroides*<br>Methanol<br>Klorofom<br>Ethyl acetate |
| 3. Test Of Phytochemicals | | |
| a. Alkaloid | Test tubes<br>Pipette<br>Glass cup | Sulfuric acid 2N<br>Reactant meyer<br>Reagents dragendorf<br>Reactant wagner |
| c. Flavonoid | Test tubes<br>Pipette<br>Glass cup | Magnesium<br>Amyl alcohol<br>Alcohol |
| d. Saponin | Test tubes<br>Pipette | $HCl_2N$ |

| | | |
|---|---|---|
| e. Tannin | Glass cup<br>Test tubes<br>Pipette<br>Glass cup | Water, Solvent of $FeCl_3$ |

The extraction process was carried out by modification of some previous studies [4, 5], in which exemplified *Enhalus acoroides* samples were weighed 40 gram each, then soaked in non-polar solvents, chlorofomes (120 ml) with a ratio of 1: 3 (w/v) in a glass bottle. Then, it was macerated for 24 hours. Subsequently, the *Enhalus acoroides* sample solution was filtered using a filter paper. The exemplified *Enhalus acoroides* sample filtrate was then re-soaked with semi-polar solvent, ethyl acetate (120 ml) in a glass bottle, then macerated for 24 hours. Next, the *Enhalus acoroides* sample solution was filtered using filter paper. Exemplared *Enhalus acoroides* sample filtrate was re-soaked with polar solvent, methanol (120 ml) and macerated for 24 hours. After 24 hours, the *Enhalus acoroides* sample solution was filtered using filter paper. After that, *Enhalus acoroides* sampling solution of filtration result was evaporated using Rotary evaporator applied at a temperature of 50 $^0$C so that raw extract of *Enhalus acoroides* in the form of paste was obtained. The raw extract pasta of the *Enhalus acoroides* was weighed in order to know the percentage of rendement.

Phytochemical test is one of qualitative tests for identifying the bioactive compounds that are conceivable within an organism. There are four classes of compounds to be tested: Alkaloids, Flavonoids, Saponins, and Tanin [7, 8, 9]. The quality of the phytochemical compounds in *Enhalus acoroides* extract can be identified by observing the color changes that occur in the extract.

Alkaloid test was performed by adding 1.5 – 2 % HCl into three tubes. Tube 1 solution plus 0.5 ml of dilute acid solution as control, 2 tubes plus 2 - 3 drops of Dragendorff reagent, and tube 3 plus 2 - 3 drops of Mayer reagent. If orange precipitate was formed in tube 2 and yellowish brass precipitate was formed in tube 3, the alkaloids were present. The flavonoid test was performed by dissolving the *Enhalus acoroides* extract in 1 - 2 ml of 50 % hot methanol. Then, Mg metal and 4 - 5 drops of concentrated HCl were added. The test is positive if the solution became red or orange.

The saponin test was done by adding water (1 : 1) into the *Enhalus acoroides* extract, and shaking for 5 minutes. The foam formed which can last for 30 minutes indicates the presence of saponin compounds. The tannin compound was tested by dissolving the *Enhalus acoroides* extract in 1 - 2 ml of water and 2 drops of $FeCl_3$ solution; the presence of a blackish blue color indicated the presence of a tannin compound error and if the color was blackish, it showed a tannin catechol compound.

## 3. Result and Discussion

The *Enhalus acoroides* extracts produced in this study showed different values between *Enhalus acoroides* taken from Wanci Island (Wakatobi) and Talango Island (Madura). The crude extract of *Enhalus acoroides* collected from Talango Island (Madura) was higher yield than Wanci Island (Wakatobi). These can be seen from the percentage of the crude extract of two *Enhalus acoroides* samples as shown in table 2.

**Table 2**. *Enhalus acoroides* extract yield from two different locations.

| No | Example of *Enhalus acoroides* | Solvents | | |
|---|---|---|---|---|
| | | Metanhol | Ethylacetate | Chloroform |
| 1 | *Enhalus acoroides* in Wanci Island, Wakatobi | 1.125 | 0.25 | 0.55 |
| 2 | *Enhalus acoroides* in Talango Island, Madura | 1.725 | 0.125 | 0.8 |

IOP Conf. Series: Earth and Environmental Science **137** (2018) 012045 doi:10.1088/1755-1315/137/1/012045

The yield produced by *Enhalus acoroides* es through extraction with methanol solvent was higher when compared to ethyl acetate and chloroform. This could happen because methanol is polar solvent polar compounds in the body of the organism, can extract polar compound. The crude extract of chloroform and ethyl acetate solvent is less, since the ethyl acetate solvent is semi polar, and the chloroform is non polar. Meanwhile, the semi polar and non polar compound of *Enhalus acoroides* is less than that of the polar compound. This condition also has been reported in previous research.

The phytochemical qualitative test of the six extracts of *Enhalus acoroides* showed almost the same result, a positive response to the saponin and tannin test (table 3). A positive response to the saponin test was demonstrated by the appearance and persistence of the foam during the test, in chloroform solvent extraction. Furthermore, a positive response of tannin test was demonstrated by the ethyl acetate extract of *Enhalus acoroides* from Wanci Island (Wakatobi), as well as chloroform extract from two sites.

Chemically, saponins have a wide diversification of structures, and certain saponin compounds with their surfactant properties can cause lysis of the protozoan cell wall, thereby being used for protozoal defaunation [10]. Tannins are generally defined as polyphenol compounds that have a fairly high molecular weight (over 1000) and can form complexes with proteins. Tanin has a biological activity as an antioxidant, so that the tannin will have an effect on the antioxidant activity [9].

**Table 3**. *Enhalus acoroides* phytochemical compounds from two different locations.

| No | Phytochemical compounds | *Enhalus acoroides* in Wanci Island | | | *Enhalus acoroides* in Talango Island | | |
|---|---|---|---|---|---|---|---|
| | | Metanhol | Ethyl Acetate | Chlorofo rm | Metanhol | Ethyl Asetat | Chlooroform |
| 1 | Alkaloid | - | - | - | - | - | - |
| 2 | Flavonoid | - | - | - | - | - | - |
| 3 | Saponin | - | - | + | - | - | + |
| 4 | Tanin | - | + | + | - | - | + |

Note: (-): Nothing, (+): Any

Alkaloid and flavonoid test in this study showed a negative value, meaning that there were no alkaloid and flavonoid compounds in the *Enhalus acoroides* extract. This result differs from some previous studies, which reported that *Enhalus acoroides* extracts generally contain flavonoid group chemicals. This is due to the different environmental conditions of *Enhalus acoroides* sampling. Flavonoid compounds become one of interests in bioprospective research. This is because some pure compounds belonging to the flavonoid group have the potential to be used as antifouling. Flavonoid compounds, according to Jensen *et al.* [11], are also able to inhibit the growth of sticking microorganisms, fungi type Schizichytrium aggregatum.

## 4. Conclusions

*Enhalus acoroides* collected from two different sites have slightly different bioactive potentials. Crude extract of *Enhalus acoroides* from Wanci Island (Wakatobi) is relatively higher yield than that from Talango Island (Madura). Meanwhile, phytochemical tests of both extracts showed the same bioactive compounds, were tannins and saponins.

## 5. References

[1] Azkab M H 1999 *Oseana* **24** 1–16
[2] Tomascik T, Mah A J, Nontji A and Moosa M K 1997 *The Ecology of the Indonesian Seas* (Part II Periplus Edition (HK) Ltd) Chapter pp 13–23

[3] Waycott M, K McMahoon, J Mellors, A Calladine and D Kleine 2004 *A guide tropical Enhalus acoroides es of the Indo-West Pacific* (Australia: James Cook University)
[4] Dewi C S U, D Soedharma dan M Kawaroe 2012 *J. Tekn. Per. Kel.* **3** 23–28
[5] Dewi C S U, R D Kasitowati dan A Yamindago 2015 *Conf. on The 5th Annual Basic ScienceInternational (Bioprospecting of Enhalus acoroides from South Malang as marine natural product)* (Malang: Oral Presentation February 11th – 12th Hotel Atria)
[6] Riniatsih I dan W A Setyati 2009 *J. Ilmu Kelautan* **14** 138–141
[7] Indrayani, Lany, H Soetjipto dan L Sihasale 2006 Skrining Fitokimia dan Uji Toksisitas Ekstrak Daun Pecut Kuda (*Stachytarpheta jamaicensis* L. Vahl) Terhadap Larva Udang Artemia salina Leach. Berk (12) (Penel Hayati) :57– 61.
[8] Prameswari O M dan S B Widjanarko 2014 *J. Pangan dan Agroindustri* **2** 16–27
[9] Malangngia L P, M S Sangia dan Jessy J E Paendonga 2012 *J. Mipa Unsrat Online* **1** 5–10
[10] Thalib A 2004 *JITV* **9** 164-171
[11] Jensen P R, Jenlins K M, Porter D and Fenical W 1998 *AEM* **64** 1490–1496

## Acknowledgment

Marine Science Laboratory, Fisheries and Marine Science Faculty, Universitas Brawijaya, has provided excellent facilities and services during this research.

IOP Conf. Series: Earth and Environmental Science **137** (2018) 012046 doi:10.1088/1755-1315/137/1/012046

# Development of an aquaculture system using nanobubble technology for the optimation of dissolved oxygen in culture media for nile tilapia (*Oreochromis niloticus*)

**G Mahasri[1], A Saskia[2], P S Apandi[3], N N Dewi[1], Rozi[1] and N M Usuman[1]**

[1] Department of Fish Health Management and Aquaculture, Faculty of Fisheries and Marine, Universitas Airlangga, Campus C Mulyorejo 60115
[2]Nano Center Indonesia, Jl. Witanaharja, West Pamulang, Pamulang, South Tangerang City, Banten 15417 Indonesia
[3]Jenderal Soediman University, Jl. HR Boenyamin 708, Grendeng, Purwokerto Utara, Grendeng, Purwokerto Utara, Grendeng, Purwokerto Utara, Kabupaten Banyumas, Jawa Tengah 53122 Indonesia

E-mail: mahasritot@gmail.com

**Abstract.** The purpose of this research was to discover the process of enrichment of dissolved oxygen in fish cultivation media using nanobubble technology. This study was conducted with two treatments, namely a cultivation media without fish and a cultivation media containing 8 fish with an average body length of 24.5 cm. The results showed that the concentration of dissolved oxygen increased from 6.5 mg/L to 25 mg/L. The rate of increase in dissolved oxygen concentration for 30 minutes is 0.61 pp/minute. The rate of decrease in dissolved oxygen concentration in treatment 1 is 3.08 ppm/day and in treatment 2 is 0.23 ppm/minute. It was concluded that nanobubble is able to increase dissolved oxygen.

## 1. Introduction

Fishery cultivation is growing rapidly in Indonesia. Many new kinds of technology have emerged to increase the production of aquaculture. The increase in fishery production is achieved by developing cultivation technology from the traditional into intensive and even into super intensive technology. Super intensive technology is characterized by high stocking density. High level density in intensive cultivation will result in high production. However, intensive cultivation systems have a disadvantage that can harm cultivators, one of them is the increase in ammonia levels that can reduce the concentration of oxygen in the water. Dissolved oxygen is the most important factor for sustaining fish lives because it relates to respiration. Low level dissolved oxygen can degrade the quality of cultivation water, cause growth disturbance and decrease body resistance to infection [1].

A solution to this problem is the utilization of aquaculture technology called the nanobubble system as an activity or effort performed in the cultivation process to increase the productivity of cultivation yields through the changes and modifications of aquaculture systems. Using nanobubble technology, the water quality in the cultivation containers is maintained so that the survival and growth of fish are optimized [2].

Nanobubble is a technology to increase the concentration of dissolved oxygen in cultivation water. Nanobubble can also play a role in the processing of (organic and inorganic) waste in cultivation containers by supplying dissolved oxygen to the water to be used by microbes that

IOP Conf. Series: Earth and Environmental Science **137** (2018) 012046 doi:10.1088/1755-1315/137/1/012046

decompose organic waste. Nanobubble is a development of microbubble technology. In a study, microbubble technology was applied in the cultivation of *Crassostrea gigas* and pearl shells, *Pinctada martensii* [2].

In general, this study aims to determine the enrichment process of dissolved oxygen in fish cultivation media using nanobubble technology. The existence of nanobubble is expected to enrich the concentration of dissolved oxygen in the water.

## 2. Methodology

### 2.1 Time and Location

This research was conducted for 4 weeks from 1 February 2016 to 26 February 2016. The location of this research was at LIPI Innovation Center Laboratory, Jalan Raya Jakarta-Bogor KM 47, Cibinong, Indonesia.

### 2.2 Tools and Materials

The tools and materials used during this research include a 47x32x27 cm tub, aerator, measuring tape, DO meter, water pump, nanobubble generator, bucket, stationery, camera, eight tilapia with the size of 24.5 cm and fish feed. The figure of the nanobubble generator device is shown by table 1.

**Table 1.** Major specification of nanobubble generator device.

| Name | Specification | | | | | Figure |
|---|---|---|---|---|---|---|
| | Mixer | Size (mm) | Weight (kg) | Power (V) | Flow rate (L/h) | |
| Nanobubble generator device | Acry flat honeycomb equipped | 490×560 ×270 | 40 | 200 | 12000 |  |
| Nanobubble device (lab scale) | RNM030A-S RNM050A-S.H.L | 650×1325 ×600 | - | Main 200 | RNM050A-S 280L/h(1Mpa) RNM050A-L 600L/h(1Mpa) |  |

The working principle of the nanobubble device is that water and gas are supplied inside using the speed of the mixer velocity and high pressure of the gas toward the ramond which will mix and split water and gas particles into nano sizes, which are subsequently flowed out with a set discharge frequency (figure 1).

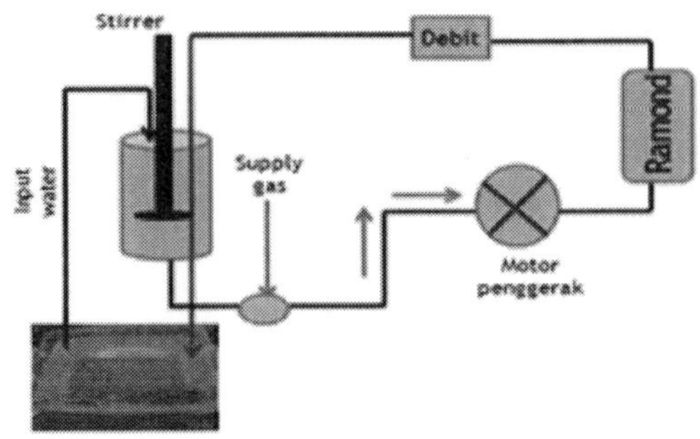


**Figure 1.** Working scheme of the nanobubble device.

*2.3 Treatment process of cultivation media*

The sample water was measured as 30 L in each treatment, then its concentration of dissolved oxygen was measured. Subsequently, the water in the treatment media were given nanobubbles with a discharge of 3 L/minute, gas flow rate 100 mL/min, driven gas pressure 0.17 Mpa, rotation speed 3050 rpm and was treated with an aerator to add oxygen. The operation time of each device was 30 minutes. After the nanobubble and aerator tools were turned off, the resulting concentration of dissolved oxygen was measured as well as the rate of decrease in oxygen in each treatment.

*2.4 Measurement of rate of decrease in oxygen concentration*

Eight tilapia (*Oreochromis niloticus*) with an average length of 24.5 cm was added into the water containing nanobubble oxygen with a previously adjusted oxygen concentration as the first treatment. While in the second treatment there was water containing nanobubble oxygen without fish in it. The oxygen concentration in both was measured periodically until the water was stable again at the normal water DO of 6.5 mg/L, then the time required for the dissolved oxygen in the water to return to normal was recorded.

## 3 Results and Discussion

*3.1 Increase in oxygen concentration using nanobubble*

Oxygen is one of the most important chemical elements as it is the main support of life for various organisms. Oxygen dissolved naturally in water is derived from the air diffusion and photosynthesis results of chlorophyllous organisms living in the water [3].

Dissolved oxygen is obtained through the diffusion process from the atmosphere to water and also the diffusion from the photosynthesis of algae or chlorophyllous plants in the water. Water quality is also influenced by how well the diffusion process from air to water was carried out [1].

Several species of fish can survive in water with an oxygen concentration of less than 4ppm but their appetite will start to decline. Therefore, it is necessary to increase the concentration of dissolved oxygen so that the quality of water cultivation is maintained, thus increasing the appetite and growth of fish.

The process of dissolved oxygen enrichment using nanobubble technology for a period of 30 minutes is able to increase the oxygen concentration from 6.5 ppm to 25 ppm. Rudiyanti and Ekasari [4] It is said that the a good level of dissolved oxygen for carp cultivation is between the range of 6.5 - 8.5 ppm. Increased oxygen concentrations can be seen in table 2.

IOP Conf. Series: Earth and Environmental Science **137** (2018) 012046 doi:10.1088/1755-1315/137/1/012046

**Table 2.** Rate of increase in dissolved oxygen.

| Time interval (minute) | Concentration of dissolved oxygen (ppm) |
|---|---|
| 0 | 6.5 |
| 5 | 9.9 |
| 10 | 13.2 |
| 15 | 16.4 |
| 20 | 19.4 |
| 25 | 22.5 |
| 30 | 25.0 |

*3.2 Decrease of oxygen concentration in cultivation media*

A decrease in oxygen concentration will result in the increase in ammonia concentration, which can be a dangerous threat to aquatic animals since a low oxygen concentration will increase the rate of respiration, decrease respiration efficiency, inhibit growth and cause mass death [5].

Dissolved oxygen is one of the indicators of water fertility. The level of dissolved oxygen decreases with the increase of organic waste in the water. This is because oxygen is also required by decomposing bacteria to decompose organic substances into inorganic substances [3]. Here are the results of measurements of the treatment in the media without fish (table 3). From the data above, it can be calculated that the rate of decrease in dissolved oxygen in the media without fish is 3.08 ppm/day.

**Table 3.** Rate of decrease in dissolved.

| Time interval (day) | Concentration of dissolved oxygen (ppm) |
|---|---|
| 1 | 25 |
| 2 | 14.8 |
| 3 | 11.4 |
| 4 | 9.3 |
| 5 | 7.9 |
| 6 | 6.6 |

In addition to being used for fish respiration, dissolved oxygen in water is also used by other aquatic microorganisms that live in the water, resulting in a high rate of decrease in dissolved oxygen in the water. The result of decrease in dissolved oxygen in the media treatment containing fish can be seen in table 3.

IOP Conf. Series: Earth and Environmental Science **137** (2018) 012046 doi:10.1088/1755-1315/137/1/012046

**Table 4.** Rate of decrease in dissolved oxygen in the media containing fish.

| Time interval (Minute) | Concentration of Dissolved Oxygen (ppm) |
|---|---|
| 0 | 25 |
| 4 | 20.9 |
| 8 | 19.6 |
| 12 | 18.3 |
| 16 | 17.1 |
| 20 | 16.1 |
| 24 | 15.1 |
| 28 | 14 |
| 32 | 13 |
| 36 | 11.9 |
| 40 | 10.9 |
| 44 | 10 |
| 48 | 9.3 |
| 52 | 8.6 |
| 56 | 7.9 |
| 60 | 7.2 |
| 64 | 6.7 |
| 68 | 6.1 |
| 72 | 5.6 |
| 76 | 5.2 |
| 80 | 4.9 |
| 84 | 4.5 |
| 88 | 4.1 |

Oxygen is required by fish in the catabolism process which produces energy for the fish to swim, breed, and grow. Therefore feed conversion and growth rates are highly determined by the availability of oxygen as well as other factors [1]. From the table above, it can be calculated that the rate of decrease in dissolved oxygen containing eight carp with an average body length of 24.5 cm in 30 L water is 0.23 ppm/minute (table 4).

The decrease in dissolved oxygen concentration will decline with the increasing density of fish stocking. Fish stocking density is the number of fish spread or nurtured in a certain area. Dense stocking is a very important factor affecting the success of fish cultivation.

Overly high fish density can degrade water quality, which in turn will inhibit the growth rate even though the food needs of the fish is met [6]. This is because there is competition for space and oxygen consumption in the water. Increased stocking density will be followed by an increased amount of feed, waste from body metabolism and oxygen consumption. It can also degrade water quality, causing stress to the fish so their growth decreases and they become susceptible to death [7]. Therefore, the high stocking of fish should be balanced with good environmental conditions in order to support the lives of fish and other aquatic organisms in order to produce a high production value.

## 4 Conclusion

From the results of treatment, it was discovered that the working principle of the nanobubble device is to combine gas and liquid and then break it down through a mixing process and high pressure so that the water and gas particles become nano-sized, enabling them to last longer in the water. The nanobubble generator can increase the dissolved oxygen in the cultivation media from the initial DO of 6.5 ppm to 25 ppm. During the nurturing period of eight carp broodstock with an average body length of 24.5 cm in 30 L of water, the nanobubble generator can be turned on for 30 minutes every 1 ½ hours. The higher the density level of the fish, the longer the nanobubble must be turned on.

## 6. References

[1] Rahman A and Khairoh.L W 2012 *J. Ekosains* **4** 1-10 (Text in Indonesia)
[2] Hakim L 2015 Aplikasi micro-bubble generator dalam pembentukan bioflocs pada budidaya udang vaname (*Litopenaeus vannamei*) synopsis of research proposal Faculty of Biology of UGM Yogyakarta (Text in Indonesia)
[3] Simanjuntak M 2007 *J. Ilmu kelautan* **12** 59 – 66 (Text in Indonesia)
[4] Rudiyanti S and Ekasari A D 2009 *J. Saintek Perikanan* **5** 49-54 (Text in Indonesia)
[5] Izzati M 2008 Changes in the concentration of dissolved oxygen and pH pond waters after the addition of seaweed *Sargassum plagyophyllum* Laboratory of Biology for Structure and Functions of Plants Department of Biology FMIPA UNDIP 60 – 69 (Text in Indonesia)
[6] Widiastuti 2009 The growth and survival rate of mas fish (*Cyprinus Carpio*) maintained in controlled containers with solid distribution (Central Sulawesi: Media Litbang) **2** 126-130 (Text in Indonesia)
[7] Diansari R R V R, Arini E, Elfitasari T 2013 *J of Aquaculture Management And Technology* **2** 37-4 (Text in Indonesia)

## Acknowledgement

The authors would like to express gratitude to those who have assisted the authors in conducting this research. This research is a part of practical work in the Aquaculture Study Program at the Faculty of Fisheries and Marine Sciences in Jenderal Soedirman University. This research was funded by research grants from the Nano Center Indonesia.

IOP Conf. Series: Earth and Environmental Science 137 (2018) 012047 doi:10.1088/1755-1315/137/1/012047

# Clustering and estimating fish fingerling abundance in a tidal river in close ploximity to a thermal power plant in Southern Thailand

**S Chesoh, A Lim and C Luangthuvapranit**

Faculty of Science and Technology, Prince of Songkla University, Pattani Campus, Pattani, 94000 Thailand.

E-mail: chesoh.s@hotmail.com

**Abstract.** This study aimed to cluster and to quantify the wild-caught fingerlings nearby thermal power plant. Samples were monthly collected by bongo nets from four upstream sites of the Na Thap tidal river in Thailand from 2008 to 2013. Each caught species was identified, counted and calculated density in term of individuals per 1,000 cubic meters. A total of 45 aquatic animal fingerlings was commonly trapped in the average density of 2,652 individuals per 1,000 cubic meters of water volume (1,235–4,570). The results of factor analysis revealed that factor 1 was represented by the largest group of freshwater fish species, factors 2 represented a medium-sized group of mesohaline species, factor 3 represented several brackish species and factor 4 was a few euryhaline species. All four factor reached maximum levels during May to October. Total average numbers of fish fingerling caught at the outflow showed greater than those of other sampling sites. The impact of heated pollution from power plant effluents did not clearly detected. Overall water quality according the Thailand Surface Water Quality Standards Coastal tidal periodic and seasonal runoff phenomena exhibit influentially factors. Continuous ecological monitoring is strongly recommended.

## 1. Introduction

Water temperature is a key environmental factor affecting aquatic organisms and ecosystems. Normally, aquatic animals are cold-blooded or poikilothermous; the organism whose internal temperature varies considerably depending on their environment in which they inhabit. Fish larvae are highly sensitive to water temperature change, particularly having a tolerance limit to survive. The upstream area of a tropical tidal river is home to fertile living organisms that have played an important linkage role between truly terrestrial and marine ecosystems [1]. Thermoelectric power plants are one of the main causes of thermal pollution; once steam has passed through a turbine to spin around for generating electricity, it must be cooled back to water state before it can be reused to produce more electricity and finally drained into natural waters.

These power plant cooling water systems have a range of adverse potential impacts on the environment including: the huge volumes of water used, fish killed due to impingement on water intake screen, entrainment and passage of organisms through the cooling system, the addition of biocides to the cooling water to control biofouling, and the discharge and dispersal of the elevated water temperature of the effluent [2, 3]. The 1,571.8 MW combined cycle power plant is located on the Na Thap tidal river bank in Songkhla province, southern Thailand. Here, water withdrawals for

IOP Conf. Series: Earth and Environmental Science **137** (2018) 012047 doi:10.1088/1755-1315/137/1/012047

thermoelectric power generation was designed to utilise approximately 38,880 cubic meters daily and then discharge warm water back into the natural river. Hence there is great concern about the ecological impacts of this power plant on aquatic organism biodiversity and fishery productivity that may affect the livelihoods of the local community [4]. How to cluster and summarize the huge number of data regarding species distribution and measured environmental variables has always been a challenge to the community of biologists. Factor analysis is a statistical method used to cluster species or environmental variables according to a common pattern of abundance and allows an objective outcome of species association and their distribution [5, 6].

Protecting the biodiversity of such ecosystems is the challenge faced by scientists. Therefore, this study aims to investigate the effects of the Chana power plant effluents on fish fingerling distribution surrounding the upstream waters of the Na Thap River, which connects to the Gulf of Thailand. Our findings can provide useful information for the authorities to reduce the adverse impact and to propose practical management guidelines for sustaining aquatic ecosystems and rehabilitating resources.

## 2. Methodology

### *2.1 Site description*

The Na Thap River, 26.5 km in length, is a tidal river in the Songkhla province of Southern Thailand. The river originates at the confluence of the Thailand–Malaysia border and empties into the Gulf of Thailand, supporting agriculture, fisheries, agro-industry as well as the surrounding communities. The river water body can be classified into three parts, comprising of freshwater in upstream areas, brackish water in the middle and near shore ecosystems in downstream areas. This study includes only the upstream ecosystem in which four sites of the river are covered (site 1: Ban Pa Ching located about 4 km upstream from the inflow station, site 2: Inflow, site 3: Outflow and site 4: Ban Kuan Hua Chang about 2.6 km downstream from the outflow) as shown in Fig 1. Site 2, 3 and 4 normally are affected by salinity due to seawater intrusion from the Gulf of Thailand.

Water samples were collected and water quality parameters were analyzed by applying the standard method in fieldwork and laboratory inspections [7]. Fish fingerling data were gathered from aggregated individual numbers of all collected aquatic organisms of each sampling site by using 200 sq. m x 1 m purse seine net. The aquatic organisms sample was classified by species according to Rainboth [8] and Taki [9] guidelines. Bongo nets were also used for the rechecking of fish larvae [10].

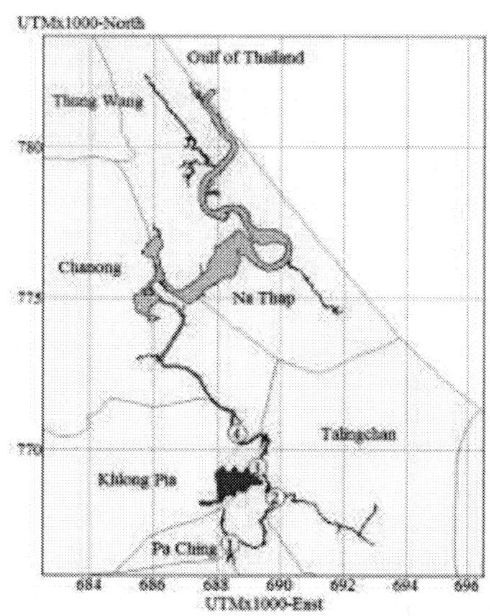


**Figure 1**. Map of The Na Thap tidal river and all four upstream sampling sites.

IOP Conf. Series: Earth and Environmental Science 137 (2018) 012047 doi:10.1088/1755-1315/137/1/012047

*2.2 Statistical analysis*

The studied population comprises of the monthly 45 fish fingerling species collected during January 2008–December 2013. For each species, there were 72 monthly observations for six years from the four different sites. Spearman's rank correlation was used to identify and test the strengths of relationships between the fish species [11]. Factor analysis based on maximum likelihood was used for allocating the fish species into a smaller number of interpretable groups [12]. The Promax rotation method was applied and then data were reduced as shown in the equation 1.

$$y_{ij} = \mu_j + \sum_{k=1}^{p} \lambda_j^{(k)} f_i^{(k)} \tag{1}$$

If $y_{ij}$ is number of fish fingerling in month $i$ species $j$, the factor model is formulated as (1) where $\mu_j$ is the average fish fingerling in species $j$, the $p$ column vectors $f_j^{(k)}$ in this model are called common factors and the $p$ row vectors $\lambda_j^{(k)}$ are called their loadings.

Factor models also provide the uniqueness value that is unique to the variable species which have high uniqueness and cannot be grouped together [13].

After factor analysis, factor scores were calculated using the average fish fingerling of each factor. A linear model was fitted to investigate the effect of month and location on the average number of fish fingerlings for each factor using an additive model as

$$f_{ij} = \mu + \alpha_i + \beta_j \tag{2}$$

Where $f_{ij}$ the average is number of fish fingerling for factor $k$ in calendar month and location (1, 2, 3 and 4), $\mu$ is an overall mean, $\alpha i$ is the coefficient for each month of the year and $\beta_j$ is the coefficient for the collected site [14]. Furthermore, multiple linear regression was also used to assess the relationship between fingerlings in each factor and water quality parameters. All statistical and graphical analysis were carried out using the R program [15].

## 3. Results and discussion

*3.1. General fish fingerling distribution*

During the six years study period of January 2008 to December 2013, a total of 45 aquatic animal fingerlings were commonly caught with the average density of 2,652 individuals per 1,000 cubic meters of water volume (1,235–4,570). However, 25 commonly dominant fingerling species, two Phyla (Chordata and Arthropoda), two Classes (all selected fish were members of Class Actinopterygii, and two freshwater prawns were under Class Malacostraca), and eight Orders and sixteen Families, were selected to calculate the prioritizing component, as shown in table 1.

Four factors (groups of species) were fitted and then a Promax rotation was cleated giving loadings as shown in table 2. This table shows that Factor 1 has fourteen freshwater fish species. Factor 2 includes six species of five vertebrates and one invertebrate (Lanchester's freshwater prawn), which is a group of mesohaline species that frequently inhabit in water salinity between 5 and 18 part per thousand (ppt). Factor 3 includes two species of low salinity brackish fish namely short-nose pony fish and spotted scat. Factor 4 incoporated a few euryhaline species including 2 vertebrates of small eyes silverside fish and soldier catfish, and one invertebrate species of giant freshwater prawn. Most prefer to inhabit their life in salinity ranging between 0.5–35 ppt.

The loading factors of all 25 species were clearly separated in one group. The last column represents the uniqueness of all species and only one species, namely the giant freshwater prawn, with a high uniqueness (0.814). Even though this species had high uniqueness, we included it in factor 4 because the species of this group has the loading factor (0.303) which is also high in factor 4.

The factor analysis indicated that four factors were sufficient and can explain 69.4% of the total variance of 25 upstream inhabiting species trapped in the river.

**Table 1.** Selected fingerling species found in the Na Thap River from January 2008-December 2013.

| No | Common name | Scientific name | Family | Order |
|---|---|---|---|---|
| 1 | Catopra | *Pristolepis fasciata* | Pristolepididae | Perciformes |
| 2 | Hard lipped barb | *Osteochilus vittatus* | Cyprinidae | Cypriniformes |
| 3 | Croaking gourami | *Trichopsis sp.* | Osphronemidae | Perciformes |
| 4 | Transverse bar barb | Hampala macrolepidota | Cyprinidae | Cypriniformes |
| 5 | Schwanenfeld's tinfoil barb | *Barbonymus schwanenfeldii* | Cyprinidae | Cypriniformes |
| 6 | Longfin mojarra | *Parambassis siamensis* | Ambassidae | Perciformes |
| 7 | Grey Feather back | *Notopterus notopterus* | Notopteridae | Osteoglossiforme |
| 8 | Gunther's walking catfish | *Clarias sp.* | Clariidae | Siluriformes |
| 9 | Three spot gourami | *Trichopodus sp.* | Osphronemidae | Perciformes |
| 10 | Common silver barb | *Barbonymus gonionotus* | Cyprinidae | Cypriniformes |
| 11 | Common climbing perch | *Anabas testudineus* | Anabantidae | Perciformes |
| 12 | Snake Skin Gourami | *Trichogaster pectoralis* | Osphronemidae | Perciformes |
| 13 | Snake head fish | *Channa striata* | Channidae | Perciformes |
| 14 | Minnow | *Rasbora tornieri* | Cyprinidae | Perciformes |
| 15 | Round-tail garfish | *Dermogenys sp.* | Hemiramphidae | Beloniformes |
| 16 | Sumatran Tiger Barb | *Barbodes tetrazona* | Cyprinidae | Cypriniformes |
| 17 | Green Puffer fish | *Tetraodon nigroviridis* | Tetraodontidae | Tetraodontiforme |
| 18 | Lanchester's freshwater prawı | *M. lanchesteri* | Palaemonidae | Decapoda |
| 19 | Dwarf goby | *Eugnathogobius sp.* | Gobiidae | Perciformes |
| 20 | Java tilapia | *O. mossambicus* | Cichlidae | Perciformes |
| 21 | Short-nose pony fish | *Leiognathus splendens* | Leiognathidae | Perciformes |
| 22 | Spotted scat | *Scatophagus argus* | Scatophagidae | Perciformes |
| 23 | Small eyes silverside | *Aplocheilus panchax* | Aplocheilidae | Cyprinodontiforr |
| 24 | Soldier Catfish | *Hemibagrus sp.* | Bagridae | Siluriformes |
| 25 | Giant freshwater prawn | *M. rosenbergii* | Palaemonidae | Decapoda |

The correlation matrix of fish fingerling numbers in each species is displayed as a bubble plot, ordered by the four factors. The correlation matrix of fish fingerlings between each species before the factor model (figure 2A; upper graph) and the correlation matrix of residuals reduced after the factor model to see how well the factor model reduces these correlations. There remains a substantial correlation between each factor (figure 2 B; lower graph).

After factor analysis, the data for each factor was explored against year elapsed data during the study period, as shown in the figure 3. This figure shows boxplots of average individual fish fingerling numbers in four factors. The Y axis represents the number of the fish (individual per 1,000 cubic meters of water volume) and the X axis represents the years during the study period. Each box represents the distribution of fish for six years from four different sites. The box size shows the interquartile range of data, the middle line in each box shows the median of the fish fingerling counted numbers. We can see that the amount of fingerlings in each factor was higher compared to those in the year 2013, particularly at site 3 (out flow).

IOP Conf. Series: Earth and Environmental Science 137 (2018) 012047 doi:10.1088/1755-1315/137/1/012047

**Table 2.** Loadings greater than 0.1 and uniqueness of different fish species obtained from factor analysis, where shaded values show species in the same factor.

| Species | Factor1 | Factor2 | Factor3 | Factor4 | uniqueness |
|---|---|---|---|---|---|
| Catopra | 1.012 | | 0.139 | | 0.109 |
| Hard lipped barb | 0.998 | | | | 0.085 |
| Croaking gourami | 0.991 | | | | 0.093 |
| Transverse bar barb | 0.983 | | | | 0.103 |
| Schwanenfeld's tinfoil barb | 0.954 | | | | 0.1 |
| Longfin mojarra | 0.951 | | | | 0.108 |
| Grey Feather back | 0.935 | 0.16 | | | 0.126 |
| Gunther's walking catfish | 0.933 | | | | 0.097 |
| Three spot gourami | 0.933 | | | | 0.107 |
| Common silver barb | 0.915 | -0.137 | | | 0.08 |
| Common climbing perch | 0.891 | | -0.1 | | 0.097 |
| Snake Skin Gourami | 0.882 | | -0.105 | | 0.122 |
| Snake head fish | 0.813 | | -0.195 | | 0.135 |
| Minnow | 0.519 | 0.289 | | | 0.175 |
| Round-tail garfish | 0.104 | 0.720 | | -0.33 | 0.504 |
| Sumatran Tiger Barb | 0.145 | 0.690 | 0.117 | -0.196 | 0.523 |
| Green Puffer fish | -0.138 | 0.573 | -0.169 | 0.413 | 0.393 |
| Lanchester's freshwater prawn | -0.215 | 0.553 | | 0.16 | 0.598 |
| Dwarf goby | -0.2 | 0.373 | 0.165 | 0.357 | 0.473 |
| Java tilapia | -0.187 | 0.351 | | | 0.799 |
| Short-nose pony fish | | | 0.897 | 0.102 | 0.172 |
| Spotted scat | -0.142 | | 0.741 | | 0.295 |
| Small eyes silverside | 0.228 | -0.207 | 0.117 | 0.732 | 0.484 |
| Soldier Catfish | | | | 0.595 | 0.664 |
| Giant freshwater prawn | 0.145 | 0.19 | | 0.303 | 0.814 |

### *3.2. Relationship between fingerlings in each factor and water quality parameters*

To examine the annual fluctuation of fish fingerling numbers, a time series distribution of average numbers of some dominant fingerling species for the four factors by site and year (2008–2013) in the upstream area of the Na Thap River was constructed figure 4. Then linear regression was used to investigate the association between each of the four factors and the determinants (sampling site, season and some water quality parameters).

Factor 1 reached the maximum peak in August and was significantly associated with season, sampling site and water transparency. Factor 2 was significantly associated with season, sampling site, and water turbidity and dissolved oxygen. Factor 3 was significantly associated with only season and sampling site. Factor 4 was significantly associated with season, sampling site and water turbidity. All three factor reached maximum levels during May to October. The total average number of fish fingerlings caught at the outflow showed to be greater than those of other sampling sites. A time series analysis of water temperature variations in the Na Thap River adjusted to month of year and sampling site from June 2005 to October 2015, shows a slightly declining trend of water temperature during the study period figure 5.

IOP Conf. Series: Earth and Environmental Science **137** (2018) 012047 doi:10.1088/1755-1315/137/1/012047

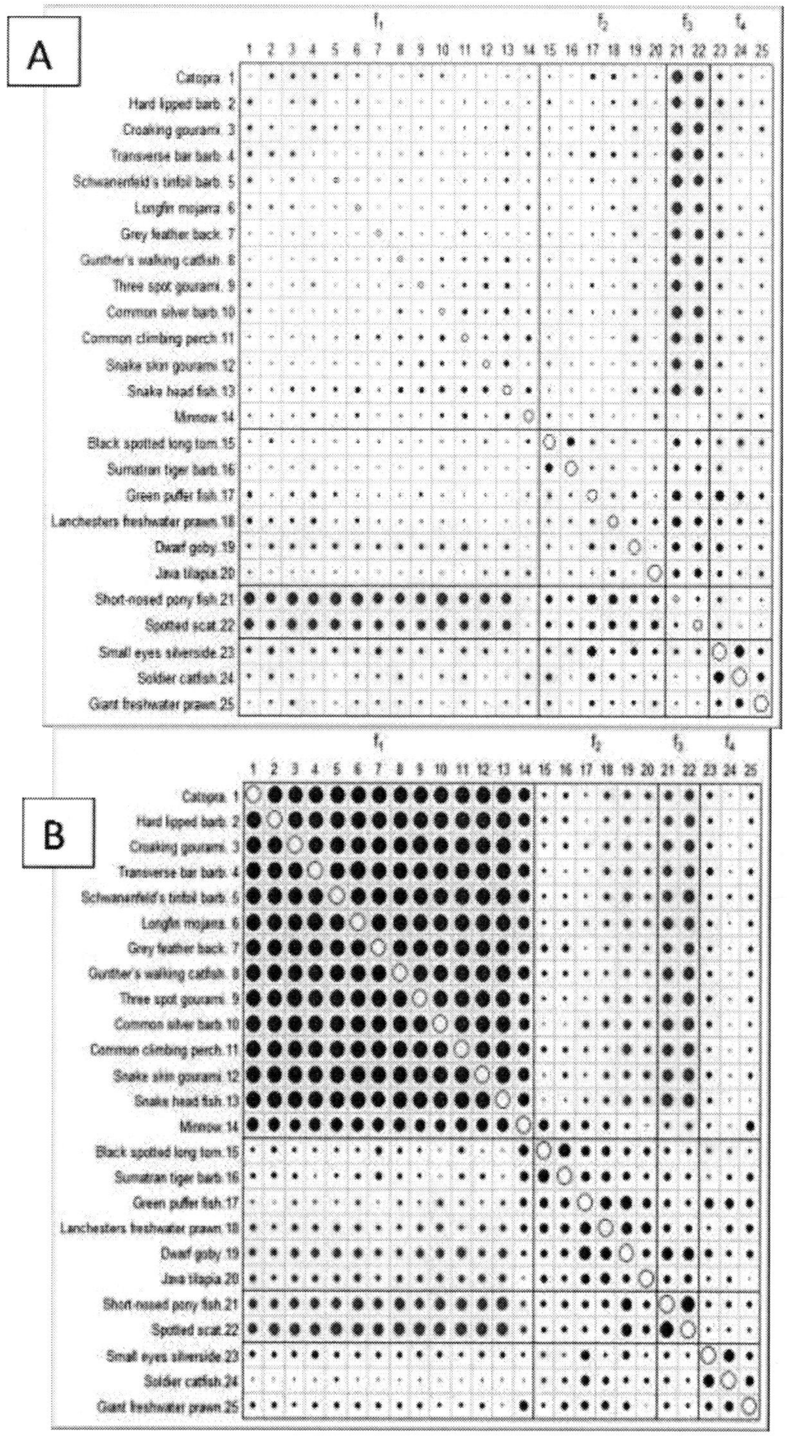


**Figure 2.** Bubble plots of correlations between fish fingerling in each species: (**A**) Before fitting the factor model; (**B**) After fitting the factor model. The size of each bubble indicates the magnitude of correlation for the matrix cell.

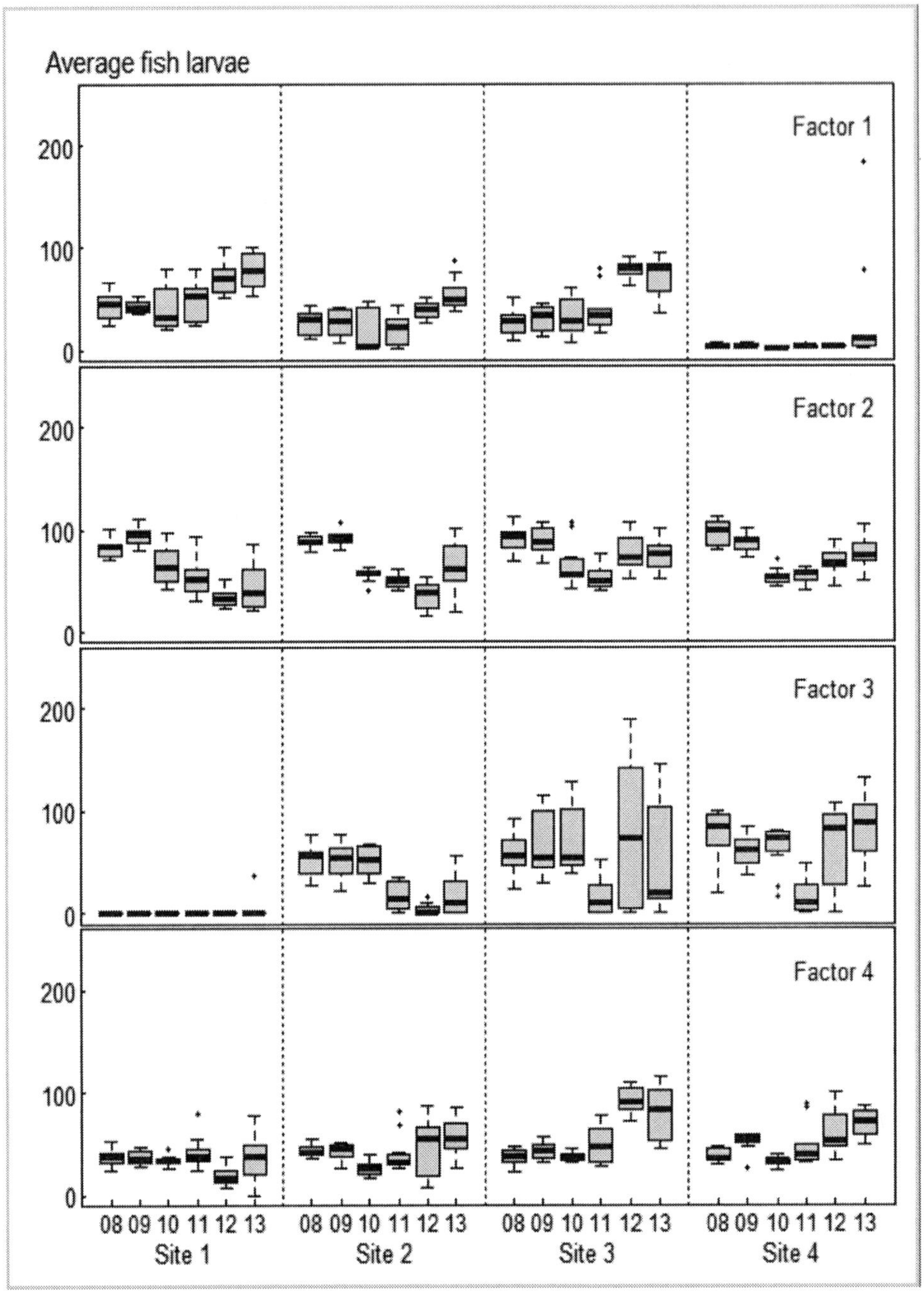


**Figure 3.** Box plots of average numbers of fish fingerling for four factors by site and year (2008–2013) for 25 species in the upstream Na Thap River. (Each box represents the distribution of amount of fish fingerling for six years from the four different sites. The box size shows the interquartile range of data, the middle line in each box shows the median of the fish fingerling in counted numbers, the vertical bars show the range of data.)

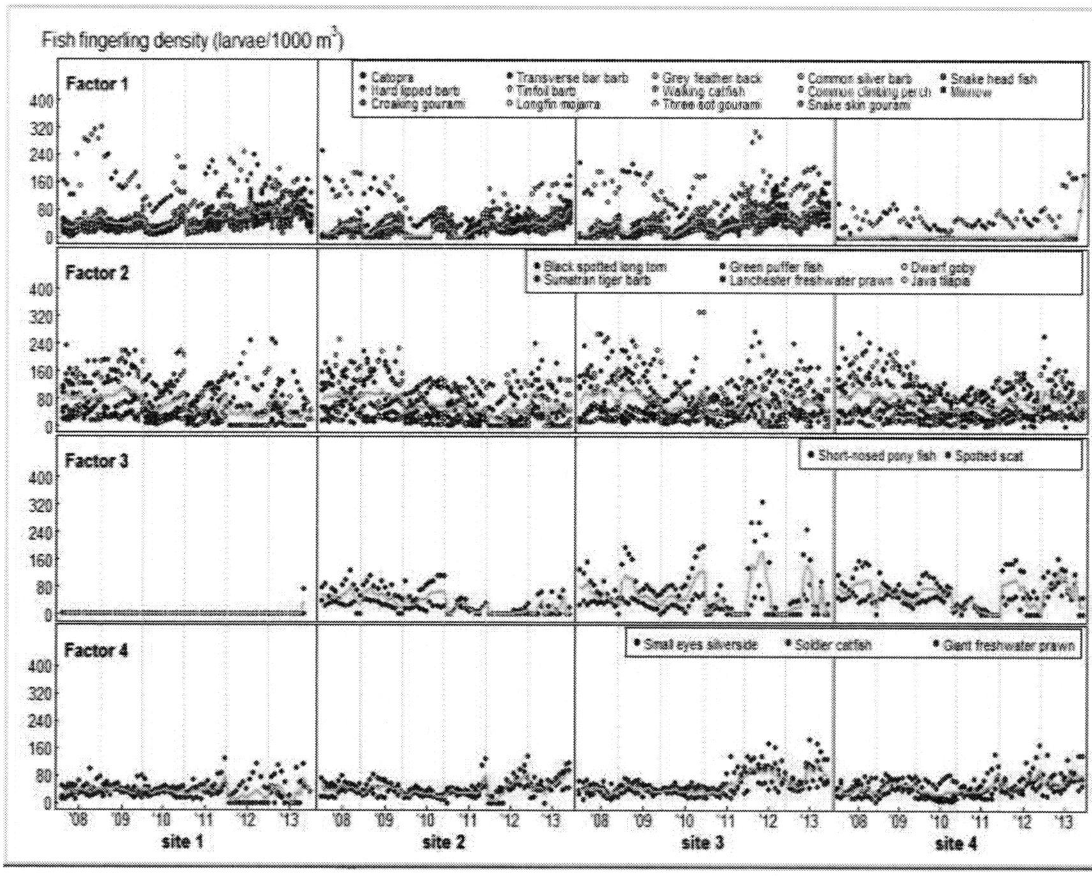


**Figure 4.** Time series distribution of average numbers of some dominant fish fingerling species for four factors by site and year (2008–2013) in upstream adjacent of the Na Thap River.

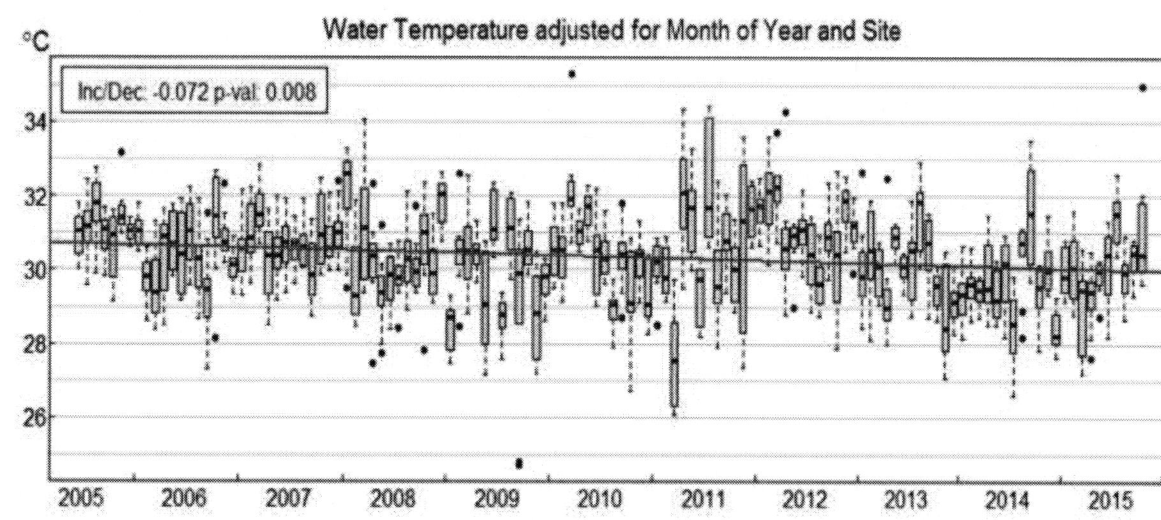


**Figure 5.** Time series analysis of overall water temperature (degree Celsius) variation in the Na Thap River adjusted for month of year and sampling site from June 2005 to October 2015.

IOP Conf. Series: Earth and Environmental Science **137** (2018) 012047 doi:10.1088/1755-1315/137/1/012047

## 4. Discussion

During the period of January 2008–December 2013, a total of 45 fish fingerling species were collected in four upstream areas of the Na Thap River where the Chana power plant is located. The average density was 2,652 individuals per 1,000 cubic meters of water volume (1,235–4,570). In this study, 25 commonly dominant fingerling species were selected as constructed factor analysis components comprising of two Phyla, two Classes (all selected fish were members of Class Actinopterygii, and two freshwater prawns were under Class Malacostraca), and eight Orders and sixteen Families. Four cluster of fingerling were obtained and prioritized.

Cluster 1 (freshwater fish) had fourteen freshwater fish species. All members of this cluster are catopra, hard lipped barb, croaking gourami, transverse bar barb, Schwanenfeld's tinfoil barb, longfin mojarra, grey feather back, Gunther's walking catfish, three spot gourami, common silver barb, common climbing perch, snake skin gourami, snake head fish and minnow. Moreover, the species in this group reached maximum peak in August each year and was significantly associated with season, water transparency and sampling site, where Site 1 (Ban Pa Ching is located about 4 km upstream from inflow station) demonstrates numbers that are statistically higher than those of average fingerling numbers. These results were consistent with freshwater fish in the upper Songkhla Lake Chesoh and lim [16] and Beamish *et al.* [17] confirms that Cyprinid fish prefer to inhabit in the transparent and slightly circulating waters of the wetland ecosystem.

Cluster 2 (mesohaline species) includes six species of five vertebrates and one invertebrate namely round-tail garfish, Sumatran Tiger Barb, green puffer fish, dwarf goby, Java tilapia and Lanchester's freshwater prawn. This component is a group of mesohaline species that frequently inhabit in water salinity between 5 and 18 part per thousand (ppt). Similarly, this species group are significantly associated with season. Moreover, the sampling site at the outflow was at maximum peak in the year 2009, thus relates to water turbidity and dissolved oxygen.

Factor 3 (low salinity brackish fish) includes two species of low salinity brackish fish namely short-nose pony fish and spotted scat. This species group was significantly associated with only season and sampling site, the highest amount of average fingerling was reached in the year 2012 at the outflow followed by Kuan Hua Chang station during the highest peak in 2013. The characteristics of species members in both Factor 2 and 3 were consistent with the report on fish assemblages across a salinity gradient in the Zeeschelde estuary in Belgium reported by Breine *et al.* [18] which stated that the mesohaline zone served as an important habitat for most estuarine species, diadromous species and marine migrants.

Factor 4 (euryhaline species) comprised of a few euryhaline species including 2 vertebrates of small eyes silverside fish and soldier catfish, and one invertebrate species of giant freshwater prawn. Most of them prefer to inhabit in salinity ranging between 0.5–35 ppt. This species group was significantly associated with season and water turbidity, and the sampling site reached the highest amount of average fingerling in the year 2012 at the outflow as well. These results were consistent with freshwater fish in the Kuan Kreng peat swamp forest in Nakhon Sri Thammarat province where catfish favored turbid water [19].

Fish fingerling populations in the upstream areas of the river fluctuate from year to year in response to seasonal effects and frequently reach maximum level during the rainy season of August or between May to October annually. The highest number of fish fingerlings occurred at the outflow throughout the study period.

The impact of heated pollution from power plant effluents was not clearly detected due to no statistically significant difference between water temperatures among the sampling sites. This result agreed with the report of Craddock and Donovan [20] which suggested that in order to minimize the power plant's impact on marine environment and biota, measures should be made regardless of the proposed location of the plant- whether it be offshore, coastal, or inland. In regards to onshore cooling at most plants, adequate planning must be implemented to protect the environment. Although overall water quality can be classified as medium clean according to Thailand Surface Water Quality Standards, it must be ordinarily treated before being used for consumption and agriculture [3]. We conclude that seasonal effects from climate,coastal tides and annual runoff variability interact in driving river biotic communities and transitional aquatic habitats. Therefore, continuous ecological monitoring is strongly recommended.

## 5. Conclusion

We concluded that seasonal effects from climate and coastal tide and annual runoff variability interact to drive river biotic communities and transitional aquatic habitats. Therefore, continuous ecological monitoring is strongly recommended to be implemented.

## 6. References

[1] Rahaman M B Biswas S, Rahaman K S, Ghosh K A, Sarder. L, Siraj S M S, and Islam S S 2014 *Ecol. Processes* **3** 18 (*Preprint*)
[2] Langford T E . 1991 Ecological effects of thermal discharges *J. Mar. Biol. Assoc. U.K.* **71** 915-916
[3] Chesoh S and Lim A 2015 Monitoring of thermal power plant operation on aquatic ecosystem in tropical estuarine river of Thailand 2008-2013 *E-Proc. of the 36th IAHR World Cong.* 28 June – 3 July 2015 The Hague the Netherlands.
[4] Chesoh S and Lim A 2014 *Asian Social Science* **10** 168-175
[5] Saheem N, Lim A and Chesoh S 2014 *J. Sci. Technol.* **36** 13–19.
[6] Saheem N, Chesoh S and Lim A 2014 *Kasetsart J. Nat. Sci.* **49** 188–199
[7] American Public Health Association (APHA) American Water Works Association (AWWA) and Water Environmental Federation (WEF) 2012 Standard Methods for the examination of Water and Wastewater (Washington: American Public Health Association) 2nd ed.
[8] Rainboth W J 1996 FAO Species Identification Field Guides for Fishery Purposes: Fishes of the Cambodian Mekong FAO Rome.
[9] Taki Y 1974 Fishes of the Lao Mekong Basin. U.S. Agency for International Development Mission to Laos Vientiane Laos 232 p
[10] Backiel T and Welcomme R L 1980 Guidelines for sampling fish in inland waters EIFAC 1980 *Technical Paper* No. 33
[11] Conti L, Grenouillet S, Lek and Scardi 2012 *Fish. Res.* 119–120: 1–12.
[12] Johnson R.A. and Wichern D W 2007 *Applied multivariate statistical analysis* (USA: Prentice Hall N J) 6th ed. 773 p
[13] Mardia K V, Kent J V and Bibby J M 1980 *Multivariate Analysis* (London UK: Academic Press Inc.) 521 p.
[14] Venables W N and Ripley B D, 2002 *Statistics and computing*: *Modern Applied Statistics with S* (New York: Springer-Science) 495 p
[15] R Development Core Team 2010 *R*: *A Language and Environment for Statistical Computing* (Austria: Vienna-R Foundation for Statistical Computing) 1706 p
[16] Chesoh S and A Lim 2008 *Sci. Asia* **34** 335-340
[17] Beamish H W, Sa-ardrit P and Tongnunui S 2006 *Environmental Biology of Fishes* **76** 237-253
[18] Breine J Maes J Ollevier F and M Stevens 2011 *Belg. J. Zool.* **141** 21-44.
[19] Chesoh S Rungruang S Sangkapaitoon S Kaewsrithong C and Kaewrit N 2010 Influence of water quality and plankton communities on fish species diversity and distribution in Kuan Kreng peat swamp forest of Nakhon Si Thammarat province of Thailand *Thai. Fisherie Gazette* **63**(6) 532-545
[20] Craddock R and Donovan S 1976. Impact of cooling waters on the aquatic resources of the Pacific Northwest *Marine Fisheries Review* 27-33.

IOP Conf. Series: Earth and Environmental Science **137** (2018) 012047 doi:10.1088/1755-1315/137/1/012047

**Acknowledgements**
The authors express their deepest gratitude and appreciation to Prof. Don McNeil for making it possible to complete this report. The authors would also like to thank the Electricity Generation Authority of Thailand (EGAT) for providing valuable information for the study and the Prince of Songkla Scholarship (RDO) and PSU Collaborative Research Fund PRMP ID 16287 (SAT ASEAN) for financial support.

IOP Conf. Series: Earth and Environmental Science **137** (2018) 012048 doi:10.1088/1755-1315/137/1/012048

# First records of bentfin devil ray (*Mobula thurstoni*) and the examination in physical factors of its habitat in the western waters of Morotai Island (North Moluccas)

**D A Mukharror[1,4], I T Baiti[2], S A Harahap[3], D J Prihadi[3], M Ichsan[1], N Pridina[1]**

[1]Shark Diving Indonesia Morotai, Kampung Yayasan, Daruba, Morotai
[2]Marine Science Student, FPIK, Universitas Padjadjaran, Jl. Raya Bandung-Sumedang KM 21, Bandung Indonesia
[3]Marine Science Department, FPIK, Universitas Padjdjaran, Jl. Raya Bandung-Sumedang KM 21, Bandung Iindonesia

E-mail : darmawan_ahmad_m@yahoo.com

**Abstract.** Bentfin devil ray (*Mobula thurstoni*) was recorded for the first time in Morotai waters on 3rd March 2017. In a conservation management context, it is important to clarify the population of *Mobula thurstoni* and their relations with their habitat. Thus, we examined the existence of *Mobula thurstoni* with the physical parameters: depth, temperature, visibility, current, weather, and tidal. We measured the existence of *Mobula thurstoni* with the Underwater Visual Census (UVC) combined with Diver Operated Video (DOV) census. The research from 3rd March to 14th July 2017 (50 dives) showed the Frequency of Occurence (FO) is 50% per single dive. The highest aggregation of 30 *Mobula thurstoni* was recorded at 14th May 2017 and the average sighting was 3.1 *Mobula thurstoni* per single dive. Among the examined parameters, it was found that strong factor affecting the sighting of *Mobula thurstoni* were at the depth of 30-35 m, temperature of 30°C, visibility of 16-20 m, low current (< 1 knot), sunny weather, and tidal category of B. Analysis of UVC and DOV results show that the research location was categorised as feeding location for the *Mobula thurstoni*.

## 1. Introduction

Mobula rays and manta rays are categorized in mobulidae family and they are distributed globally in both tropical and subtropical oceans. The mobulidae consists of two genera and 11 species [1] that are harmless and feed mainly on zooplankton, shrimps, crabs, and small fishes [2, 3]. Recently, the manta genera were redefined and they have at least two genus: the reef manta *Manta birostris* (Walbaum, 1792) and the giant manta *Manta alfredi* (Krefft, 1868). The mobula genus comprises of nine recognized species: the giant devil ray *Mobula mobular* (Bonnaterre, 1788), the Atlantic devil ray *Mobula hypostoma* (Bancroft, 1831), the spinetail devil ray *Mobula japanica* (Mueller and Henle, 1841), the shortfin devil ray *Mobula kuhlii* (Mueller and Henle, 1841), the Chilean devil ray *Mobula tarapacana* (Philippi, 1893), the bentfin devil ray *Mobula thurstoni* (Lloyd, 1908), the pygmy devil ray *Mobula eregoodootenkee* (Bleeker, 1959), the lesser Guinean devil ray *Mobula rochebrunei* (Vaillant, 1979), and the Munk's devil ray *Mobula munkiana* (Notarbartolo-di-Sciara, 1988).

*Mobula thurstoni* was known to occupy shallow neritic waters [4]. It has worldwide distribution with the species reported from Indonesia [5, 6], the Arabian sea [7], Peru [8], Oman [9], the Philippines [10], Chile [11], Mexico [12, 13], the Western Atlantic [4], the South China Sea [15], the Red Sea [16], the

IOP Conf. Series: Earth and Environmental Science **137** (2018) 012048 doi:10.1088/1755-1315/137/1/012048

Gabon, Congo, Senegal and Angola [17], the Sao Tome e Principe [18], the Saint Peter and Saint Paul Archipelago [19], the Australia, Mariana Islands, Ecuador, Maldives [20], the tropical Eastern Pacific [21], India [22], and Sri Lanka [23].

*Mobula thurstoni* is classified as near threatened (NT) by the International Union for Conservation of Nature and Natural Resources IUCN [4] with the exact population size is largely unknown. It is estimated that of 9% of all mobulids (409) that were taken as bycatch in drift gillnets fishing for skipjack tuna in the southern waters of Indonesia [5] is *Mobula thurstoni*. In the southern water of Indonesia, approximately 155 *Mobula thurstoni* have landed at Tanjung Luar, Lombok Island and 109 others at Cilacap, Central Java between 2001 and 2005 [24]. In Lamakera, East Nusa Tenggara, 200-300 *Mobula spp.* were caught annually as they were harpooned during annual traditional harvesting [6]. 6there was a decline for the *Mobula thurstoni* catch's number data in 2013-2014 shows the 75% decline in Tanjung Luar and 77% decline in Cilacap.

*Mobula thurstoni* is presumed to be long lived and slow growing according to Coutirier *et al.* [25] with the number of pup of one per birth and disc wide of 650-850 at birth [13]. The largest size of *Mobula thurstoni* was ever recorded was 1,870 mm disc width from Bohol Sea, the Philippines [26]. It has estimated that *Mobula thurstoni* has age of maturity when its disc width (DW) is 1,538 mm [5]. In all recorded specimens, female *Mobula thurstoni* were potentially having the larger size than of male [3, 26, 5].

Bentfin devil ray in the Bohol Sea, the Philippines, feed almost exclusively on the specific krill family of euphausiids: *Euphausia diomedeae* (89.66 %) with the trace of copepod, polychaete larvae, and plant seed as the remaining diet [26]. In the Eastern Pacific, the examination of the stomach of *Mobula thurstoni* indicated that this species is extremely specialized in its feeding habits, which comprise of euphausiids species of *Nyctiphanes simplex* (86.7 %), *Mysidium sp.* (11.1 %), and the remaining is copepod, megalopa larvae, hyperiids amphipods, fish eggs, *Nemastocels diff.*, and stomatopods larvae [3]. However, bentfin devil ray also has the capability to alter its diet depends on the availability of krill by feeding predominantly on Euphausiids spp. (*Nyctiphanes simplex*) in the summer and *Mysidium* spp. in the winter [27].

## 2. Methodology

Fifty research dives were conducted on one coral reef (research station) around the west waters of Morotai Island in March – June 2017. Selection of the station is based on previous accidental sighting of unknown Mobulid in 2013. The research station was located at N 02° 09' 18.15" E 128° 12' 24.43, which is shown on the Morotai Island's map (figure 1) below as Devil Ray Point.

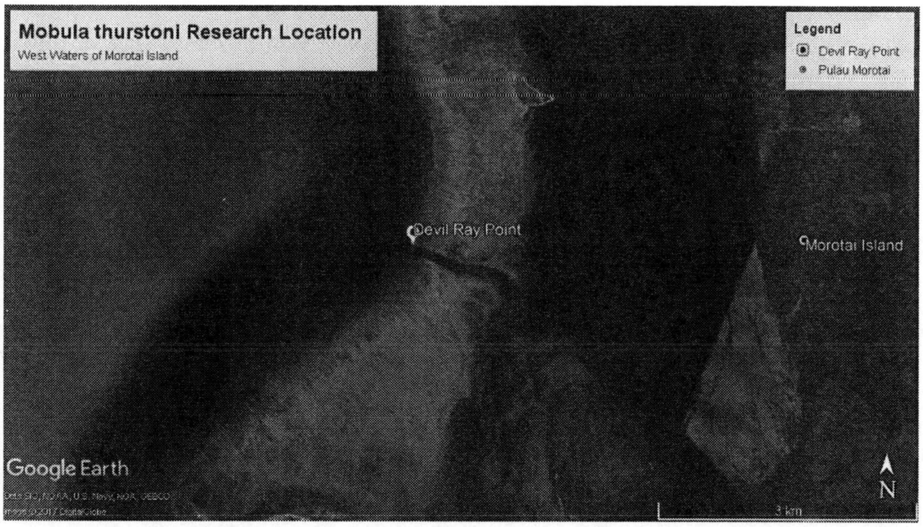


**Figure 1.** Research station in west waters of Morotai Island (source: Google Maps).

Materials used for this research were: SCUBA diving gear, underwater slate, transect meter, squeezable empty mineral water bottle, GPS locator, and underwater camera. For analysis purposes, the following tools were used: SPSS™, Microsoft Excel™ 2007, and Field Guide to the Identification of Mobulid Rays. Survey was conducted at the research station regularly between February to June 2017.

Examination of the abundance of *Mobula thurstoni* was performed using Underwater Visual Census (UVC) combined with Diver Operated Video (DOV) census. UVC has been used to estimate the densities of the reef fishes since 1950s [28]. There were two methods of UVC: transect count (an experienced diver swam at constant speed along the transect and noted the reef fish species observed on a slate) and point count (an experienced diver stayed at a vantage point and noted the surrounding reef fish species observed on a slate). UVC with transect and point counts have relatively similar level of significance ($p = 0.054$ and 0.048 for 10 ha count respectively), therefore both census are deemed reliable Watson *et al.* [29] to be selected for small survey area. A novel method, which is a combination of UVC and diver operated video (DOV), presents great potential to count the abundance of reef fishes in a research station [30].

The physical factors measured and recorded during the survey were: depth, temperature, visibility, current, weather, and tides. As an addition, visual observation was performed to categorize the behavior of the bentfin devil ray during research dives. Reef fish survey was performed in order to obtain the abundance of cleaner fish in the research location. Identified reef fishes were then classified into their cleaning habits based on classification on www.fishbase.org. The density of zooplankton in the research location was also examined to observe the specific species of zooplankton existed that was related to *Mobula thurstoni*.

## 3. Results

### *3.1 Mobula thurstoni existence*

The first devil ray documented in Morotai was on March 3rd, 2017, as shown below in figure 2. Based on the *Field Guide on the Identification of Mobulid Rays Indo-West Pacific* [31], *Mobula thurstoni* has the following distinctive features: 1) ventral mouth; 2) white-tipped dorsal fin; 3) lack of spine at the base of the tail; 4) long tail that equals with disc width in length when fully intact; 5) short necked appearance; 6) short cephalic fins, length from the tip of each fin to the corner of mouth is less than 16% of the total disc width; 7) white ventral marking does not extend above the eyes; 8) tail base that was dorsoventrally compressed closed to the dorsal fin; 9) pectoral fin anterior margin that has a distinctive

IOP Conf. Series: Earth and Environmental Science **137** (2018) 012048 doi:10.1088/1755-1315/137/1/012048

double curvature; 10) small sub-circular spiracle under the margin of the pectoral fin where it meets the body; 11) coloration: general deep blue-back on dorsal surface; white ventral surface anteriorly, while the distal half has a silver-brown sheen; 12)the top of the head has a dark band which stretches across the head behind the eyes.

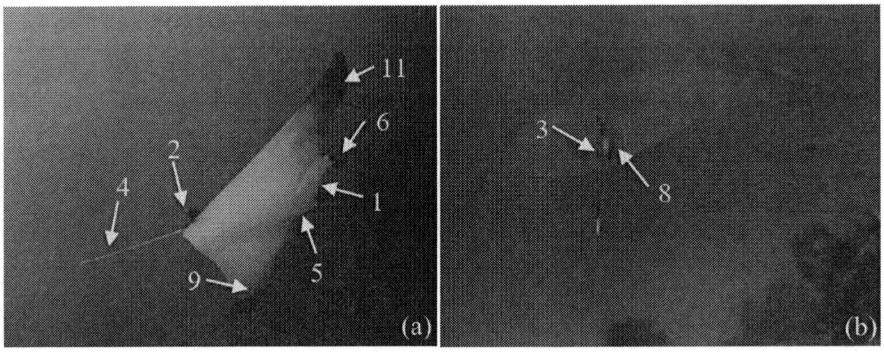


**Figure 2.** Appearance of Mobula thurstoni in research station (a) Ventral (b) dorsal.

*3.2 Mobula thurstoni abundance*

Research dives between March 3rd and July 14th, 2017 showed the average sighting of *Mobula thurstoni* of 3.1 per dive with the maximum number of individual sighted in one dive was 30 *Mobula thurstoni* on May 14th, 2017. The existence of *Mobula thurstoni* in every research dive in the research stations is shown in figure 3, while table 1 shows the abundant indices of *Mobula thurstoni*. Figure 4 shows 6 probabilities of physical parameters that affected the sighting of *Mobula Thurstoni* and the dominant sighting was at the depth of 30-35 m, temperature of 30 °C, visibility of 16-20 m, low current (< 1 knot), sunny weather, and tidal category of B.

**Table 1.** Abundant Indices of *Mobula thurstoni.*

| Abundance Indices | Sighting (dive) |
|---|---|
| Number of dives | 50 |
| Total sighting | 24 |
| Total abundance of 1 individu | 8 |
| Total abundance of 2-10 individu | 6 |
| Total abundance of > 10 individu | 8 |
| Mean sighting frequency (%) | 50 |
| Number individuals per sighting | 6.2 |
| Number individuals per dive | 3.1 |

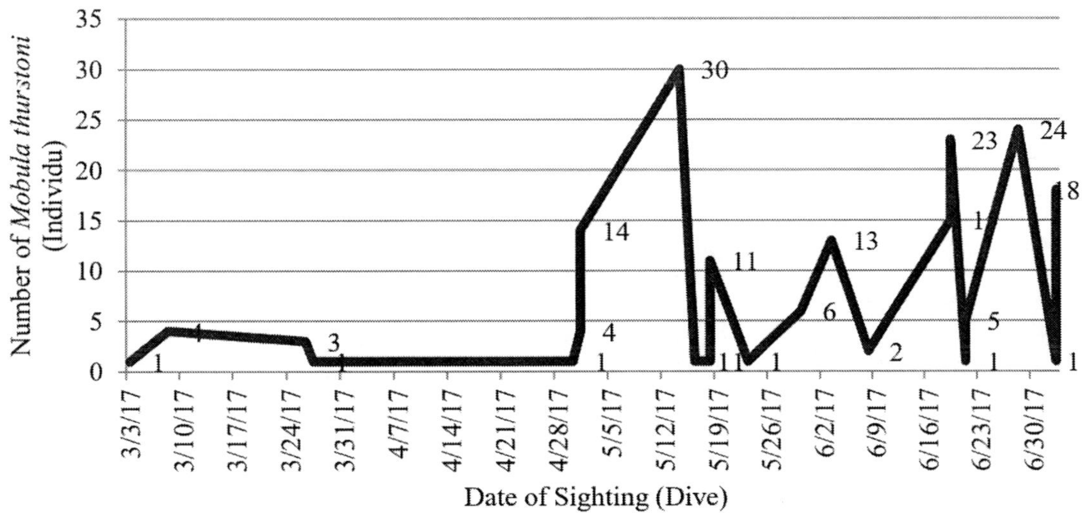


**Figure 3.** Existence of *Mobula thurstoni* in research stations.

We examined the existence of *Mobula thurstoni* with 6 physical parameters. Each parameters had 4 categories that are formed from the results. The categories of tidal in figure 4(f) shows that A is high to low tide, B is low tide, C is low tide, and D is low to high tide. Figure 4 shows probability of *Mobula thurstoni* sighting that was calculated from total sighting (24 dives) and each parameter's categories.

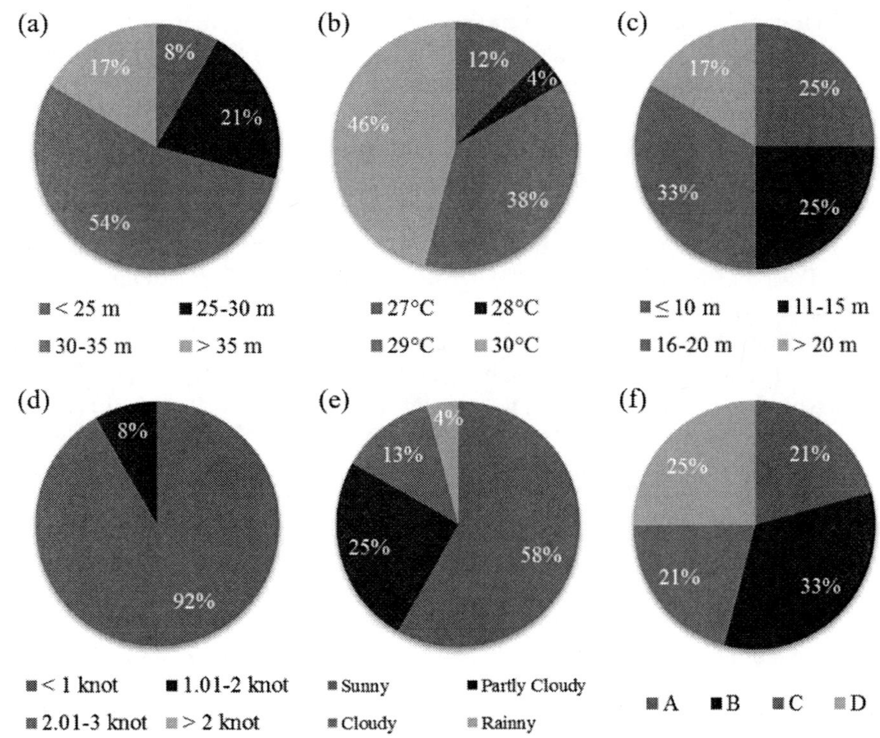


**Figure 4.** Probability of *Mobula thurstoni* sighting on various physical parameters (a) depth (b) temperature (c) visibility (d) current (e) weather (f) tidal.

IOP Conf. Series: Earth and Environmental Science **137** (2018) 012048 doi:10.1088/1755-1315/137/1/012048

*3.3 Behavior of Bentfin Devil Ray*
The research of *Mobula thurstoni* behavior in Morotai Island waters posed several states: feeding, swimming, and cleaning. Pictures captured on their behavior are shown in figure 5 below, representing all those three states mentioned above.

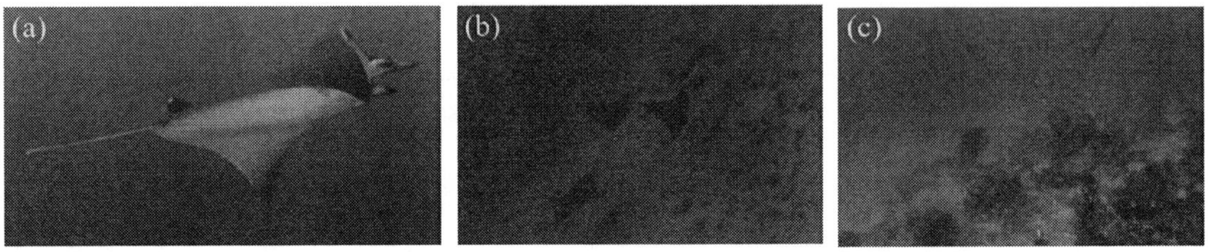


**Figure 5.** Behavior of *Mobula thurstoni* in research station (a) feeding (b) swimming (c) cleaning.

## 4. Discussion
*4.1 Mobula thurstoni existence and abundance*
Examination of the existence of mobula ray confirmed that the species of *Mobula thurstoni* existed in the Western Waters of Morotai Islands. even though there was no real specimen collected from the fishermen, the photographic physical identification resulted in 8 out of 12 of distinctive features of *Mobula thurstoni* that were confirmed, as shown in the figure 2. No size measurement was conducted during research survey due to the difficulty in getting near the *Mobula thurstoni* to obtain a minimum parallax measurement error as they always got away when they were approached.

Despite the number of samplings that were far lower (50 dives compared to 4.222 dives), the probability of sighting of *Mobula thurstoni* in Western Water of Morotai (50 %) was extremely higher than the sighting likelihood in Eastern Pacific that was only 0.5 % [32]. The mean number of individuals per sighting was 6.3 and sighting record of 30 individuals indicated that the research area has a potential for large school of *Mobula thurstoni* for recreational diving attraction.

The dominant physical factors that affected the sighting of *Mobula thurstoni* were at the depth of 30-35 m, temperature of 30 °C, visibility of 16-20 m, low current (< 1 knot), sunny weather, and tidal category of B.

Figure 6 shows R-square from 6 parameters: depth (0.14), temperature (0.75), visibility (0.1), current (0.67), weather (0.9), and tidal (0). R-square is always between 0-1. If R-square is close to 1, then the model fits your data. Temperature, current, and weather had R-square close to 1. Thus, the 3 parameters had correlation to Mobula sighting at the research station.

Temperature can significantly influence poikilotherm physiological process, and this is likely a key benefit to staying in warm shallow waters, despite the decreased opportunities for foraging [33]. As a filter feeder, current helps *Mobula thurstoni* to catch more prey. The research was done during the summer. it was predicted that in the summer, there was abundance of giant devil rays in the area occupied by the Pelagos Sanctuary, the Sardinian Sea, and the Central Tyrrhenian Sea. Seasonal movements may be related to possible energetic advantages deriving from warmer waters if this species thermoregulates like other mobulids Alexander [2] or to the localized availability of high density of prey Canese [35] e.g. mesopelagic and epipelagic fish in the Strait of Messina [36]. Research in other seasons is recommended to know the abundance of *Mobula thurstoni* in each season

By having these dominant parameters, a potential underwater tourism of mobula ray sighting can be developed in the future as an alternative to fisheries. To increase the quality of research, it is suggested to explain scientifically why each factor contributes to the sightings.

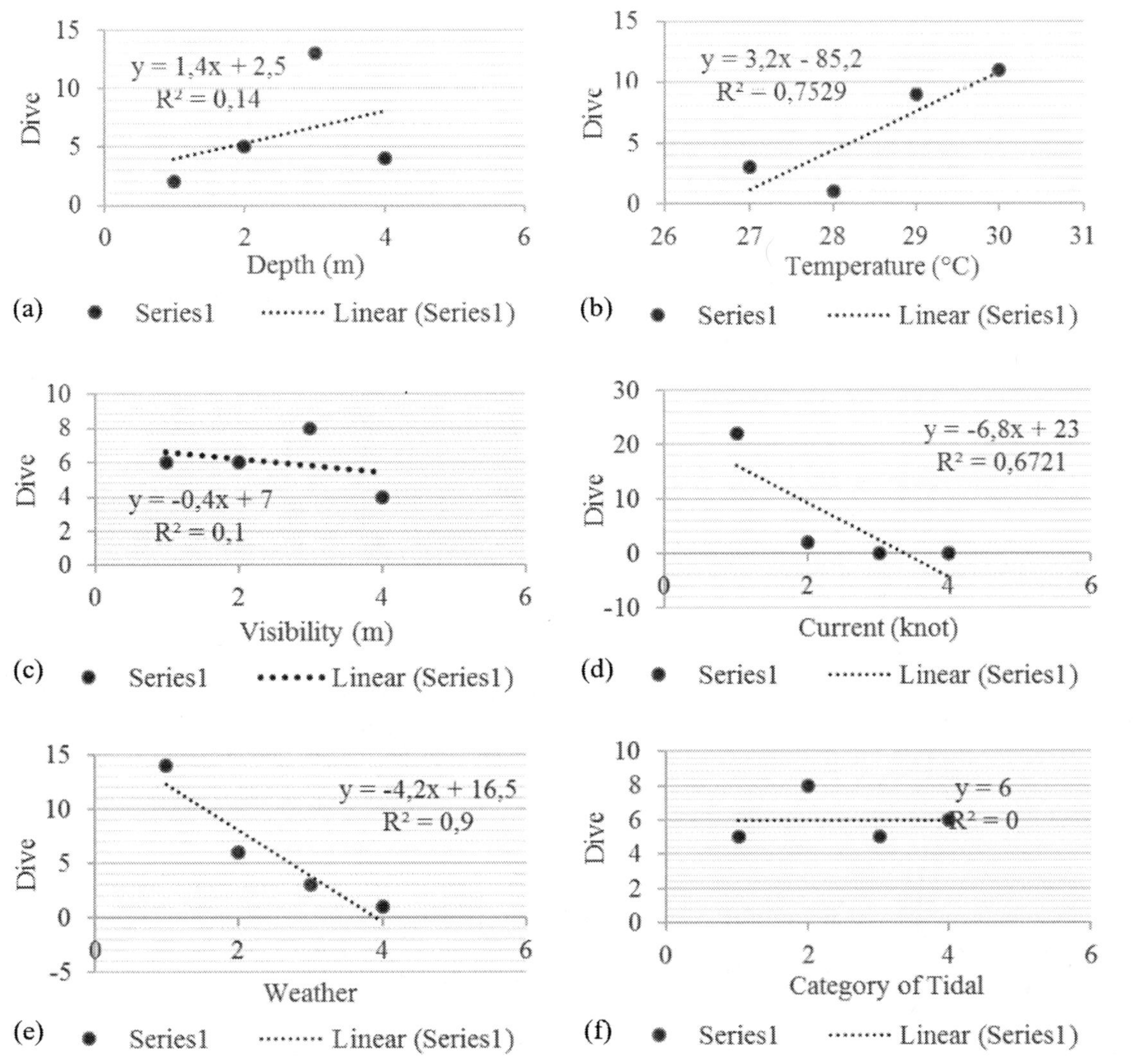


**Figure 6.** Correlation and regression of *Mobula thurstoni* sighting and its physical parameters (a) depth (b) temperature (c) visibility (d) current (e) weather (f) tidal.

### *4.2 Mobula thurstoni behavior*

One of the early indications of the sighting was that the research location was considered as a feeding station for *Mobula thurstoni*. Table 2 shows that the highest number of *Mobula thurstoni* sighting was during feeding (14 dives). during feeding, mobulids were generally observed to swim forward with an open mouth creating a passive water flow through the gill raker apparatus, a behavior referred to as ram filter feeding [37, 38, 39]. Understanding the foraging ecology of threatened species will aid in their conservation and management since feeding behavior often determines critical habitat use and spatial patterns that are important in preventing or mitigating targeted or incidental capture and other human impacts [40].

The cleaning behavior of *Mobula* spp. has never been documented prior to 2016 Murie *et al.* [25] and none of the research on the *Mobula thurstoni* cleaning has been conducted. Interaction between *Mobula kuhlii* and the blue streak cleaner wrasse, *Labroides dimidiatus*, were observed in Mozambique [25]. The difference between *Mobula kuhlii* cleaning behavior in Mozambique and the *Mobula thurstoni*

IOP Conf. Series: Earth and Environmental Science **137** (2018) 012048 doi:10.1088/1755-1315/137/1/012048

behavior in Morotai was the latter's cleaning activity can be conducted even with only single individual while the previous one cited that the *Mobula kuhlii* preferred to aggregate (more than one individual) while cleaning [41].

**Table 2.** Behavior of *Mobula thurstoni* in research station.

| Behavior | Sighting (dive) |
|---|---|
| Feeding | 14 |
| Swimming | 6 |
| Cleaning | 4 |

### *4.3. Mobula thurstoni and zooplankton density*
Morotai Island Waters lies in the tip of NECC (North Equator Counter Current) area and classified as very rich area of zooplankton [42]. Previous marine survey showed that the research location had zooplankton's density of 0.757 – 1.070 mg/m3 and categorized as the highest (approximately 600 individuals per $m^3$) in Morotai Waters [43]. *Mobula thurstoni* study in the Philippines revealed that the zooplankton species of Euphausiids was their dominant preys [26].

## 5. Conclusion
The existence of *Mobula thurstoni* (Lloyd 1908) was confirmed through the identification of the photographs and videos. Among the examined parameters, it was found that factors that strongly affecting the sighting of *Mobula thurstoni* were at depth (30-35 m), temperature (30 °C), visibility (16-20 m), low current (< 1 knot), sunny weather, and tidal category of B. Early indication of research site categorized as feeding station was supported by the fact the most sighting was observed during feeding behavior.

## 6. References

[1] Compagno L J V and Last P R 1999 Dasyatididae Stingrays In K.E. Carpenter and V. H. Niem (eds.) FAO species identification guide for fishery purposes p 1479-1505

[2] Bigelow H B and W C Schroeder 1953 *Fish. Bull.* **53** 1-577

[3] Nortarbartolo-di-Sciara, Giuseppe 1988 *Fish Bull* **86** 45-66

[4] Clark T B, Smith W D and Bizzarro J J 2006 Mobula thurstoni The IUCN Red List of Threatened Species 2006: e.T60200A12317473

[5] White W T et al 2006 *Fish. Res.* **82** 65-73

[6] Dewar Heidi 2002 Preliminary Report: manta harvest in lamakera. pfleger institute of environmental research in WWF Indonesia Project

[7] Nortarbartolo-di-Sciara G et al 2016 *Aquatic Conserv: Mar Fresw Ecosyst.*

[8] Cornejo R *et al.* 2015 *J. Biodiv. Data* **11** 1-7

[9] Al-Jufaili S *et al.* 2001 *JKAU Mar. Sci* .**21** 3-51

[10] Acebes J M V and Malcolm Tull 2016 *PLO Sone* **11** 1-27

[11] Bustamante C *et al.* 2012 *Mar. Biodiv. Rec.* **48** 1-4

[12] Eschmeyer W N, Herald E S and Hamann H 1983 *A field guide to the pacific coast of north america* (Boston: Houghton Mifflin Co.)

[13] Nortarbartolo-di-Sciara, Giuseppe 1987 *Zool. J .Linn. Soc.* **91** 1-91

[14] Gadig O B F *et al.* 2003 *J. Mar .Bio. Assoc. U.K.* **83** 869-870

[15] Randall J E, Kevin K P L 2000 *Raffles Bull Zool.* **8** 569-667 15

[16] Golani D, Sergey V. Bogorodsky 2010 *Zootaxa.* **2463** 1-135 16

[17] Weir C et al 2012 *Mar. Biodiv. Rec.* **5** 1-6

[18] Rodrigues N V, Fontes J, Bertoncini A A 2016 Ten new records of marine fishes for Sao Tome, West Africa *Act .Ichthy. Pisc.* **46** 123-129

[19] Mendonca S A *et al.* 2012 *Pan-A J. Aq. Sci.* **7** 21-26
[20] Lawson J M *et al.* 2017 *Peer. J.* (*Preprints* 1-41)
[21] Ward-Paige C, Brendal D and Boris W 2013 *PLOSone* **8** 1-9
[22] Mohanraj G *et al.* 2009 *Asian Fish. Sci.* **22** 607–615
[23] Fernando D, Guy S 2011 *A study of srilanka's manta and mobula fishery* Manta Trust Report
[24] Lewis S A *et al.* 2015 *Peer. J.* (*Preprints* 1-39)
[25] Couturier L I E *et al.* 2012 *J. Fish. Bio.* **80** 1075-1119
[26] Rohner C A *et al.* 2017 *R .Soc. Op. Sci.* **4** 1-14
[27] Sampson L *et al.* 2010 *J. Mar. Bio. Ass. U.K.* **90** 969-976
[28] Brock V E 1954 A *J. Wildl. .M'ment.* **18**:297-308
[29] Tessier A *et al.* 2013 *Aquatic. Biology* **18** 229-241
[30] Watson R A, T J Quin I I 1997 *Ecol . Mod.* **104** 103-112
[31] Stevens G 2011 *Field guide to the identification of mobulid rays (Mobulidae) Indo West Pacific.* Manta Trust.
[32] Ward-Paige C A *et al.* 2012 *J. Fish. Bio.* **80** 1844-1869
[33] Hochechka P W, George S 2002 *Bio-Chemical Adaptation: mechanism and process in physiological evolution* (New York: Oxford University Press.)
[34] Alexander R L 1996 *Zool. J. Linn. Soc.* **118** 151-164
[35] Canese S *et al* 2011 *Endang Spec. Res.* **14** 171-176
[36] Celona A 2004 *Annales* **14** 11-18 36
[37] Cortes E *et al.* 2008 An Overview of the Feeding Ecology and Physiology of Elasmobranch Fishes. In *Feeding and Digestive Functions of Fishes* (Cyrino E E P, Bureau D and Kapoor B G eds), pp 393-443 Edenbridge Science Publisher
[38] Sanderson S L, Wassersug R 1990 *Sci .Am.* **262** 96-101
[39] Sanderson S L, Wassersug R 1993 Convergent and alternative designs for vertebrate suspension feeding In *The Skull* (Hanken J. and Hall B. K. eds) **3** 37 – 112
[40] James M C *et al.* 2006 *Can. J. Zool* .**84** 754-765
[41] Murie Callum J G, Andrea D M 2016 *Peer. J.* (*Preprints* 1-23)
[43] Windyana N 2001 *J. Indo. Mar. Sci.* **2** 147-158
[43] Yang G *et al* 2017 2014 *Cont. Shelf. Res.* **135** 14-22

**Acknowledgements**
We thank the research team in Morotai Island for their support during data collection as well as to Marine Science Department of Universitas Padjadjaran for offering many suggestions and comments on the manuscript.

IOP Conf. Series: Earth and Environmental Science **137** (2018) 012049 doi:10.1088/1755-1315/137/1/012049

# The percentage of coral reef cover in Saonek Kecil Island, Raja Ampat, West Papua

**D A Wiguna[1*], E D Masithah[2] and A Manan[3]**

[1]Aquaculture Study Programme, Faculty of Fisheries and Marine Universitas Airlangga, Surabaya, Indonesia
[2]Department of Marine Affairs, Faculty of Fisheries and Marine Universitas Airlangga, Surabaya, Indonesia
[3]Departement of Fish Health Management and Aquaculture, Faculty of Fisheries and Marine, Universitas Airlangga, Surabaya, Indonesia

E-mail: dwiadi17@gmail.com

**Abstract.** Raja Ampat archipelago is located in the heart of the world's coral triangle which is the center of the richest tropical marine biodiversity in the world. The Saonek Kecil Island has a location close to the Waisai Harbour (±2 km of sea routes). The Island that has no inhabitants and has a location close to harbour activities potentially damage coral reefs. This research was conducted by Line Intercept Transect (LIT) method that calculate the length of each colony form of growth (life form) of coral reefs on the line transect which stretched along the 50 metres parallel to the coastline at each station to obtain the percentage cover data, diversity index, uniformity index, and dominance index. The results of research precentage cover of coral reeef in the waters of Small Saonek Island reach 68.80% – 79.30% by category according to the decision of the Minister of State for the Environment number 4 of 2001 about the damage the reefs criteria included in the category of good – very good. As for the value of diversity index (H') of 0.487 – 0.675 (medium-high), uniformity index (J) 0.437 – 0.606 (medium-high), and dominance index (C) 0.338 – 0.502 (medium-high).

## 1. Introduction

Indonesia has a marine area, including ZEEI, of 5.8 million $km^2$, which is three fourths of the total area of Indonesia. The area contains 13.466 islands and is surrounded by 95.200 km long coastline, the second longest after Canada [1]. The ocean has the greatest wealth and biodiversity in the world thats is one of them is the coral reef ecosystem [2]. The Raja Ampat Island located at the heart of the world's triangle coral (Coral Triangle) and is the center of the richest tropical marine biodiversity in the world. The archipelago is located in the westernmost point of main island in Papua, stretching over an area of approximately 4.6 million hectares. Raja Ampat has a high number of animal species with the discovery of 1.104 species of fish, 699 species of mollusks, and 537 species of coral animals [3].

According to Coremap [4], stated that coral reefs can be affected by natural conditions and human activities. threats can come from water pollution, waste, tourism activities, cultivation activities, and destructive fishing practices. Unlike the large Saonek Island that has a population, the condition of uninhabited Saonek Kecil island can be potentially vulnerable to excessive fish and coral reefs exploitation because of minimal supervision. Periodic observation of the structure of coral reef ecosystem communities that suffer from natural or human damage should be managed. Observation of

IOP Conf. Series: Earth and Environmental Science 137 (2018) 012049 doi:10.1088/1755-1315/137/1/012049

the percentage of hard corals and coral recruitment indicated the extent of its ability to recover naturally [5].

Saonek Kecil is located near the Wasai Port (± 2 km sea lane). The island is near to harbor activity that can potentially damage the coral reef ecosystem since coral reefs and benthic foraminifera always change as a response to the environmental changes, either naturally or anthropogenically [2]. The anthropogenic damages include sedimentation, eutrophication, waste pollution, and exploitation of fish and coral reefs in large numbers [6]. Thus, to support the efforts of coral reef ecosystem's management, a complete and accurate data are required. The data obtained can be a reference to the authorities to determine the direction of the management policy.

## 2. Methodology

The preparation was done by checking all research equipment. hand refractometer and water quality checker's sensor were washed using distilled water. The sampling location was in Saonek Kecil island due to the presence of coral reefs that spread along the coast of the Island. Furthermore Adi *et al.* [7], explained that determination of the research station was done purposefully (purposive sampling) based on the consideration that the location of the selected station represented the waters of the island as a whole. Survey of the research location was conducted by using motorized rubber boat around the waters of Saonek Island. The data collection was determined by 3 points of purposive sampling, where the points' coordination was then plotted with the help of GPS (Global Positioning System). Station I was at the north of the island, Station II was on the eastern part of the island, and Station III was on the western part of the island. No data was collected in the southern area of Saonek Kecil Island because of the coastal topographical condition of the area that was not an area of protected waters. measurement of the coral reef growth (life form) used line transects by mounting a 50 meter roll meter parallel to the coastline [8].

Data collection of percentage of coral cover used Line Intercept Transect (LIT) method. This method is often used to determine the percentage of coral cover [10]. The researcher snorkeled along the line transect that was stretched for 50 meters along the research stations while recording and collecting the data. The data collection was done in the morning at 09.00 WIT since it has the lowest tidal of ± 1 m depth. Percentage of cover is the percentage of area covered by coral growth. It is obtained by measuring the coral colony intercepted by the transect line. The length of coral intercept along the transect line was divided by the length of the line transect and multiplied by 100% to give the percentage value of the cover [10]. Water quality measurements were performed at each research station. Measurement of salinity used hand refractometer, current velocity was measured with steel ball, brightness was determined with sechi disk, while temperature, pH, and TDS (Total Dissolved Solids) were all assessed using water quality checker.

### *2.1 Place and Date of Research*

The study was conducted in April 2016. The data collection of the percentage of coral cover was conducted in the waters of Saonek Kecil Island, South Waigeo District, Raja Ampat Regency as part of the *Ekspedisi NKRI Koridor Papua Barat 2016.*

### *2.2. Tools and Materials*

The tools used were swimming aids for feet (fins), snorkeling equipment, underwater camera, steel ball, drop pipette, Global Positioning System (GPS), hand refractometer, buoy (life jacket), motorized boat, water quality checker, chest table, waterproof paper, roll meter, raffia strap, sechi disk, and bottle marker; while the material for the research was distilled water.

### *2.3. Data Analysis*

The data collected was quantitative, while the presentation was done descriptively. The coral cover percentage was obtained from measurement of coral life form (coral intercept) by using the formula [8, 10] :

IOP Conf. Series: Earth and Environmental Science **137** (2018) 012049 doi:10.1088/1755-1315/137/1/012049

$$L = \frac{Li}{N} \times 100\% \quad (1)$$

Notes:
L = Percentage of coral cover (%)
Li = Length of life form (colony intercept) of category i
N = Transect length (50 m)

The percentage of coral cover data was compared with the Decree of the State Minister of Environment No. 4 of 2001 on the criteria of coral reef damage with 0 - 24.9 % categorized as broken, 25-49.9 % as medium, 50 - 74.9 % as good, and 75 – 100 % as excellent. According to Legendre [11] calculations of diversity index (H′), uniformity (E), and dominance (C) in coral analysis used the following formula:

$$H' = -\sum_{i=1}^{q} pi \log pi \,; pi = ni/N \quad (2)$$

Note:
H′= Shannon diversity index
ni = Length of coral i species on the line transect
N = The length of total coral colonies on the line transect.

The uniformity index (E) is used to see balance of individuals within the coral reef community. The value is a comparison between the value of diversity with its maximum diversity that ranges from 0 to 1. Pielou in [16] stated the formula for uniformity (J) as the following:

$$J = H' / H_{max} = (-\sum_{i=1}^{q} pi \log pi)/\log q \quad (3)$$

Notes:
J = Evenness uniformity index
H′ = Shannon diversity index
pi = ni / N
q = Total number of coral species

The dominance index (C) is used to determine the extent to which a group of coral biota dominates other groups. The value of this index ranges from 0 to 1. If the index value is close to 0 then there is no dominant coral group, whereas if the index value is close to 1, then there is a dominant coral group. [11] states the formula of dominance index as the following:

$$C = \sum_{i=1}^{q} pi^2 = \sum_{i=1}^{q} (ni/N)^2. \quad (4)$$

Note:
C = Dominance Index
ni = The length of the i-coral coral species on the line transect
N = The length of the total coral colonies on the line transect
q = The total number of corals.

## 3. Results and Discussion

### *3.1. The Geographical condition of Saonek Kecil island*

Saonek Kecil Island is part of the South Waigeo District. The island has an area of ± 1.175 $m^2$ and is an uninhabited island. Access to the island takes ± 15 min from the navy post of TNI AL Waisai by motorboat. The island has a wooden dock, lush vegetation, and there is a church that is no longer used by the people. The northern boundary of the island lies at coordinates 0 ° 26'56,490 "North Latitude, 130 ° 48'3,645" East Longitude, and it is bordering with the Siwindores Bay and Waigeo Island. Northeast of the island is bordering with the Port of Waisai with a distance of ± 2 km of sea lane. East Boundary of the island lies at coordinates 0 ° 27'4,279 "North latitude, 130 ° 48'12.451" East longitude; while the southern boundary of the island is at coordinates 0 ° 27'15.325 "North latitude, 130 ° 48'6,349" East longitude, and the western boundary of the island is at coordinates 0 ° 27'5,051 "North Latitude, 130 ° 47'57,542" East Longitude. from its eastern to the western part, the island is surrounded by the Dampir Strait, however the southwestern part of the island is bordering with the island of Saonek Besar with a distance of ± 2.5 km of sea lane.

*3.2 The Percentage of coral coverage and index*

The samples from Station 1 (North) consisted of 7 types of coral growth form with total coral cover percentage of 70.42 % containing Soft Coral (SC) 30.40 %, Coral Submassive (CS) 8.90 %, Coral Massive (CM) 9.14 %, Coral Encrusting (CE) 14.38%, Acropora Submassive (ACS) 2.44%, Acropora Digitate (ACD) 2.78 %, and Acropora Branching (ACB) 2.38%. Another 29.58 % of the sample included Sand (S) 21.70 %, Rubble (R) 6.48 %, and Dead Coral Algae (DCA) 1.40 %. Samples from Station 2 (East) were 5 types of coral growth form with total coral cover percentage reached 79.30 % consisting of: Soft Coral (SC) 31.20 %, Coral Massive (CM) 22.90 %, Coral Entrusting (CE) 23.40 %, Acropora Submassive (ACS) 0.40 % and Acropora Branching (ACB) 1.40 %. Another 20.70 % included Sand (S) 5.60 % and Rubble (R) 15.10 %.

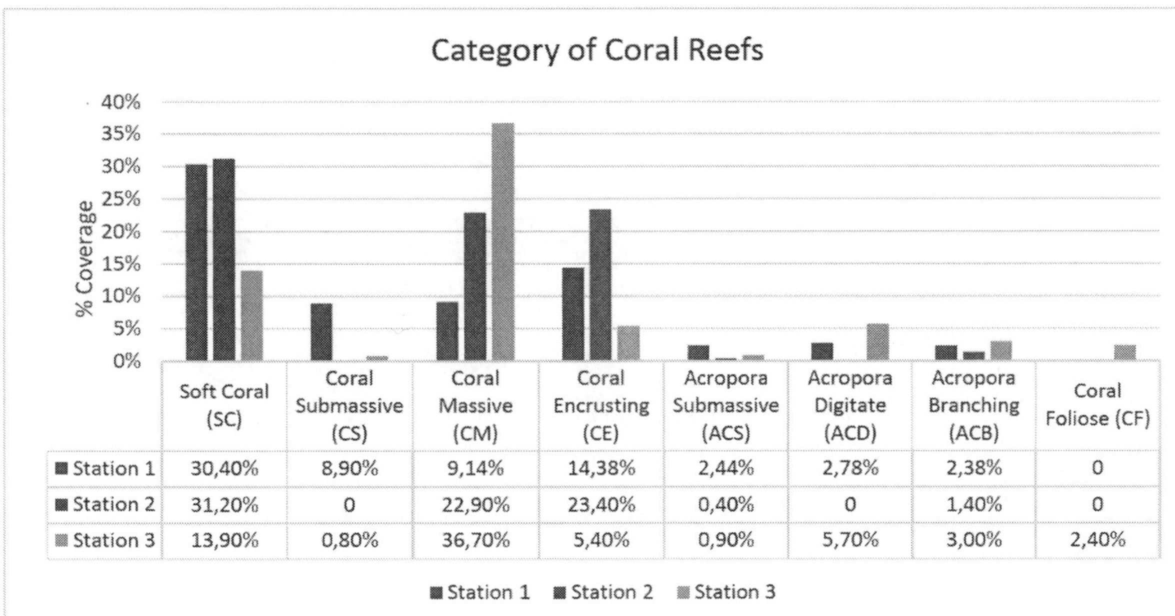


| | Soft Coral (SC) | Coral Submassive (CS) | Coral Massive (CM) | Coral Encrusting (CE) | Acropora Submassive (ACS) | Acropora Digitate (ACD) | Acropora Branching (ACB) | Coral Foliose (CF) |
|---|---|---|---|---|---|---|---|---|
| Station 1 | 30,40% | 8,90% | 9,14% | 14,38% | 2,44% | 2,78% | 2,38% | 0 |
| Station 2 | 31,20% | 0 | 22,90% | 23,40% | 0,40% | 0 | 1,40% | 0 |
| Station 3 | 13,90% | 0,80% | 36,70% | 5,40% | 0,90% | 5,70% | 3,00% | 2,40% |

**Figure 1.** Percentage diagram of coral reef category.

The samples from Station 3 (West) were 8 types of coral growth form with total coral cover percentage reaching 68.80 % with the details: Coral Foliose (CF) 2.40 %, Soft Coral (SC) 13.90 %, Coral Subclassive (CS) 0.80 %, Coral Massive (CM) 36.70 %, Coral Encrusting (CE) 5.40%, Acropora Submassive (ACS) 0.90 %, Acropora Digitate (ACD) 5.70 %, and Acropora Branching

(ACB) 3.00 %. While another 31.20 % was in the form of Rubble (R). Percentage diagram for coral cover category can be seen in figure 1, while for non-coral reef category is shown in figure 2.

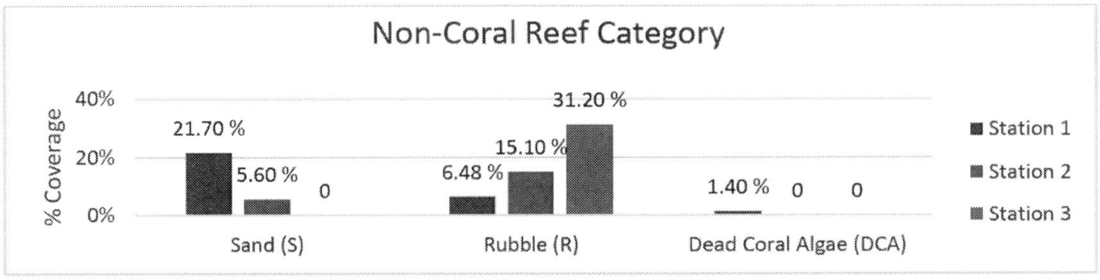


**Figure 2**. Percentage diagram of non-coral reef category

The data of coordinate's location of the research are shown in table 1. Diversity Index (H'), Uniformity Index (J), and Index Dominance (C) at each research station can be seen in table 2. water quality at each research station is displayed in table 3. Conditions at all research stations are shown in figure 3.

**Table 1.** Coordinates of research stations.

| Station 1 (North) | Station 2 (East) | Station 3 (West) |
|---|---|---|
| S 00° 26' 984" | S 00° 26' 956" | S 00° 27' 072" |
| E 130° 48' 023" | E 130° 48' 077" | E 130° 48' 007" |
| UTM 0036-5026 | UTM 0070-5038 | UTM 0033-5010 |

**Table 2.** Index value of research station.

| Index | Station 1 (North) | Station 2 (East) | Station 3 (West) |
|---|---|---|---|
| Diversity (H′) | 0,675 | 0,511 | 0,487 |
| Uniformity (J) | 0,606 | 0,459 | 0,437 |
| Dominance (C) | 0,502 | 0,323 | 0,338 |

**Table 3.** Water quality of saonek island station.

| Parameter | Station 1 (North) | Station 2 (East) | Station 3 (West) |
|---|---|---|---|
| Salinity (‰) | 30 | 30 | 31 |
| Current Velocity (m/s) | 0,033 southwest direction | 0,05 northwest direction | 0,033 southwest direction |
| Depth (m) | 0,5 | 1 | 1 |
| Temperature (°C) | 31,8 | 33,9 | 30,5 |
| pH | 8,05 | 8,41 | 8,13 |
| TDS (mg/l) | 0,8 | 0,9 | 0,9 |

### *3.3. Analysis of coral cover*

The result obtained by LIT (Line Intercept Transect) method was compared with the Decree of the State Minister of Environment No. 4 of 2001 on the standard criteria for coral reef damage. At Station 1 (North) the percentage of coral cover total reached 70.42 % and it was categorized as good. The dominant coral species was Soft Coral (SC) 30.40 %. The Acropora Branching (ACB) was the least with a percentage of 2.38 %. Abiotic rate of closure at this station was 28.18 %, which consisted of Rubble (R) 6.48 %, and Sand 21.70 %. it is better for a coral reef ecosystem if the percentage of live coral cover is greater than the percentage of abiotic cover [12]. The percentage of corals in this water was in good category because the location of the research station is leeward waters. Leeward is a reef area protected from the coming wind [4]. closed coral habitats rarely experience changes in temperature, salinity, and nutrient levels caused by strong current and wave movements with average annual temperature and salinity of 23-25 °C and 32-35 ‰, respectively. Growth of soft coral species blooms when the hard coral is in a critical or damaged condition. These corals grow well in low-brightness water conditions and high nutrient content [13]. According to the results, the dominant coral species where? was the Soft Coral (SC) of 31.20 %.

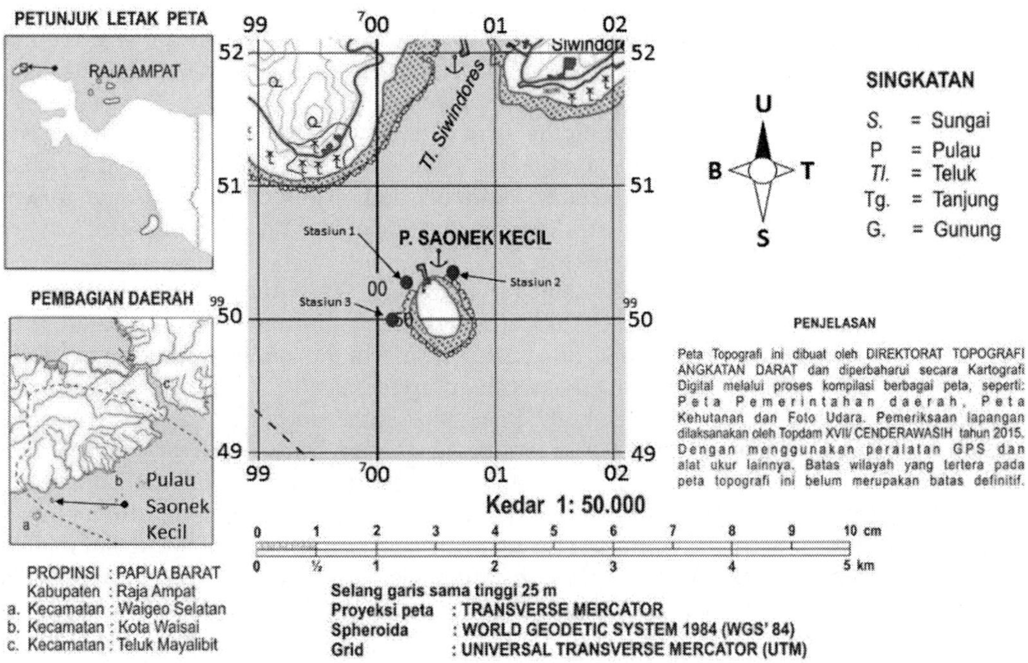


**Figure 3.** Research location.

Station 2 (East) had 79.30 % of total coral cover and it was included in the excellent category. The dominant coral species was Soft Coral (SC) at 31.20 %. Acropora Submassive (ACS) had the lowest percentage of 0.40%. Abiotic closure rate was seen from the amount of Rubble (R), which was 15.10 %, and Sand (S) (5.60 %). At this station, the percentage value of coral cover was the highest compared to other research stations, due to there was strong current at this station. According to Manuputty [14], dynamic (choppy) waters have benefits for coral growth, such as the availability of higher oxygen than in calm waters. When there is a wave, most of the oxygen is tied up and accumulated in the water column and leaching occurs easily on the surface of the reef. At this station, the dominant coral species was soft corals because during the lowest tide phase, the water level was not lower than the height of the coral reefs, thus coral reefs were not exposed to direct sun and air.

IOP Conf. Series: Earth and Environmental Science **137** (2018) 012049 doi:10.1088/1755-1315/137/1/012049

Moreover, soft coral is a coral group that could grow easily in a short time compared to rock corals. Water movement caused by The currents makes the soft corals to survive, adhere to hard substrates, and support high growth processes [14].

The percentage of coral cover in Station 3 (West) reached 68.80 % that made it included in the good category. The dominant coral type there was the Coral Massive (36.7 0%)and Coral Submassive had the lowest percentage of 0.80%. Coral damage rate of 31.20 % included Rubble and it was the highest if compared to the two previous research stations. The lowest percentage of coral reefs was found in this location since the waters of the western part of Saonek Island included reef zone flat that was adjacent to the waters of the Dampir Strait. Moreover Luthfi [15], stated that reef zone flat has uncertain environmental conditions due to the very low water volume at the lowest tide causing many hard corals to be exposed to the air. Salinity that rises sharply during the lowest tide limits the types of hard corals that can grow. The lack of currents and waves to supply nutrients is also another factor inhibiting the growth of hard corals.

The water salinity of Saonek Island was 30-31 ‰, which was in accordance with the salinity of Indonesian waters in general that ranges from 30-35 ‰ [16]. According to the Decree of the State Minister of Environment No. 51 of 2004 on sea water quality standards, salinity tolerance for marine life, especially coral reefs, is ± 5 ‰ of natural salinity value that is 33-34 ‰ at seasonal average. Therefore, having such salinity made it possible for coral reef to survive and grow at the research stations. Moreover Seveso *et al.* [17], stated that salinity changes can provide cellular damage to coral reefs caused by disruption of physiological regulatory process.

Current velocity at the study site including the calm current was 0.033-0.05 m/s. Saonek Island is part of the South Waigeo District, which is a protected water area [3]. Current velocity may change at any time depending on the weather and seasonal conditions. Current velocity affects coral growth both directly and indirectly since the strength of a current is associated with distribution of oxygen and nutrients and decreased rate of precipitation [18].

The Temperature at the research location ranged from 30.5-33.9 °C. According to the Decree of the State Minister of Environment No. 51 of 2004 about sea water quality standard, temperature tolerance for marine life, especially for coral reefs, is ± 2 °C from natural temperature that is 28-30 °C. Tolerance of coral reefs towards high temperature can go up to 36- 40 °C. At temperature lower than 18 ° C, coral reefs cannot develop [19]. The high temperature at the study site was caused by the depth of the research location, which was only 0.5 m so that penetration of the sunlight could increase the temperature to be higher than the normal temperature. Temperature affects the metabolic rate of an organism. A temperature rise of 10 ° C will cause the required oxygen consumption for an organism's metabolic activity to increase twice. Some corals can withstand temperature of 14 °C but their rate of calcification decreases. The effects of temperature changes on corals can lead to decreased feeding responses, reduced reproduction rate, mucus, and decreased photosynthesis and respiration [20].

The degree of acidity or pH at the study site did not have a significant difference, which was between 8.05- 8.41. According to the Decree of the State Minister of Environment No. 51 year 2004 about seawater quality standard, pH tolerance for marine life especially coral reefs was ± 0.2 from 7 - 8.5 unit pH. In general, the pH of sea water did not change much, because the carbon dioxide the water has a strong buffering capacity. Good pH values for coral reefs range from 6-9 [16].

Climate change and the various disasters that often occur these days are alarming. The impacts are varied, including coastal changes in some areas. Some areas are reduced due to the erosion and abrasion of the sea, but there are also areas that are increasing due to the high rate of sedimentation [21]. The TDS measurements resulted in 0.8-0.9 mg/L. According to the Decree of the State Minister of Environment No. 51 of 2004 on seawater quality standards, suspended solids tolerance for marine life especially coral reefs, was ± 10% from seasonal concentration of a maximum value of 20 mg/L. Effendi [22] stated that TDS is usually caused by inorganic materials in the form of ions that are commonly found in waters.

IOP Conf. Series: Earth and Environmental Science **137** (2018) 012049 doi:10.1088/1755-1315/137/1/012049

### *3.4. Coral reef index analysis*

The diversity index value of 0.487- 0.675 shows the diversity of coral reefs at the research locations, which fell into the medium-to-high category. The coral reefs had low ecological pressures to achieve ecosystem balance. It caused abundant types of coral found in the study sites. The results showed that the highest diversity index of 0.675 was obtained at Station 1 (North) with 70.42 % percentage of closure, while the highest percentage was in Station 2 (East). The result of the total coral cover was 79.30%.

The uniformity index value of 0.437- 0.606 belonged to the medium - high category. This value indicated that each form of the coral reef (life form) at the study site has relatively equal number and they did not vary much from one to another. The high uniformity index values in the study site indicated a balanced coral reef ecosystems; a relatively similar number indicated that the level of competition between coral species was low. The distribution of uniform species (life form) can have an inverse relationship with biological activity in an environment. The lower the value of uniformity, the higher the biological activity,, such as food production, life cycle, and inter-tropical energy flow. Seasonal weather changes and pollution can increase competition among species and decrease the value of uniformity [10].

The dominance index value of 0.323- 0.502 was included in the medium- high category. The numbers indicated that there was a group of dominant forms of coral growth (life form) at the study site. If the value of the dominance index (C) is near zero, it indicates that there is no dominant biota and it is usually followed by a high uniformity (E) value. Conversely, if the value of the dominance index (C) is close to one, then it illustrates that in the waters there is one dominant biota and it is usually followed by a low uniformity value [23].

Overall, the coral reefs on the island of Saonek Kecil are in good condition. The extent of damage caused by natural factors can be seen from the minimal coral bleaching that is due to the exposure of corals to the air and direct sunlight at the lowest tide phase. In addition to the absence of anthropogenic activity on the island of Saonek Kecil, it is also worth to consider that the stress on the coral reef ecosystem does not come from humans. Damage to the coral reefs caused by humans, such as destructive fishing with explosives, is no longer found because it is now strictly regulated. This condition is also supported by the local wisdom in the form of prohibition of fishing at a certain time (*sasi*). This customary law was even written on a dock at Mayalibit Bay District shown in figure 4.

**Figure 4.** Local rule written on a dock.

This regulation shows that the Raja Ampat's people participate in keeping the sea as renewable resources by giving the ecosystem time to recover before re-using it. Management of coastal areas like this is expected to continue in order to create sustainable natural resources to generate a positive impact on the environment as well as for the wider community. By analyzing the dynamics and the importance of the coastal areas, it is absolutely necessary that the management is carried out seriously using an integrated system, considering many interests may intersect with each other [24]. In addition, according to Pursetyo [25], the coastal ecosystems, such as mangroves, can provide jobs, whether by providing seed for the fishery industry or wood for firewood, paper materials, and construction materials with sufficiently high economic value. today, mangrove ecosystem is getting developed for recreational facilities or as a tourism place to increase state revenues. To improve the welfare of the people dependent on coral reef ecosystems, there is an adaptive management plan that includes strong fisheries regulations, consistent periodic monitoring, and integration of land and marine management practices [26]. The custom regulation indirectly becomes one of the supporting factors that can improve fishery yield and increase tourist attraction. This is in accordance with data of production volume of fishery products at Raja Ampat Regency in 2010-2014, which was 222.36-4,352.48 tons. the number of foreign and domestic tourists visiting the area also increased in 2011-2015 from 6,405 people to 14,190 people [27]. Coral reef-based tourism in the Coral Triangle has a positive impact on the economic sector, but it may have a negative potential for the social and environmental sectors, thus an approach is needed to evaluate the sustainability of the activity [28].

## 4. Conclusions

Based on the research it can be concluded that the percentage of coral cover of Saonek Kecil Island reached 68.80- 79.30% that was categorized as good- excellent according to the Minister of Environment Decree No. 4 in 2001 on the standard criteria of coral reef damage . the index value of diversity (H ') was 0.487- 0.675 (medium- high), uniformity index (J) was 0.437- 0.606 (medium-high), and dominance index was (C) 0.338- 0.502 (medium- high).

It is necessary to observe at different depth and to add more research stations to find out more accurate result of coral reef condition in Saonek Kecil island. Research on the percentage of coral cover on the Saonek Kecil island should be done periodically, at least every 5 years according to the Decree of the Minister of State for Environment no. 4 of 2001 on the standard criteria of coral reef damage, to be able to determine the condition of coral reefs regularly related to update the status of the territory.

## 5. References

[1]. Dahuri R 2014 *For you Indonesia* (Bogor: Roda Bahari) p 16.
[2]. Aulia K N, Hikmat K, Tatang S E and Suhartati M N 2012 *J. Sci. and Tech. Trop. Marine* **4** 335-345
[3]. Raja Ampat District Government 2006 *Atlas of coastal resources of raja ampat regency of west irian jaya province* (Indonesia: Conservation International Indonesia) 1
[4]. Coremap 2014 Technical guidance of basic habitat mapping of shallow marine waters (Jakarta: Research Center of Oceanography, Institute of Science) p 19.
[5]. Toruan L N L, Soedharma D and Dewi K T 2013 *J. Sci. and Tech. Trop. Marine* **5**
[6]. Sunarto 2006 *Biodiversity and degradation of coral reef ecosystem* (Semarang: Faculty of Fisheries and Marine Science, Diponegoro University) p 30
[7]. Adi A B, Mustofa and Ketjulan 2013 *J. Idn. Mina Marine* **1** 49-60
[8]. Fachrul M F 2008 *Sampling Method for Bioecology* (Jakarta: PT Bumi Aksara) p 136
[9]. Eleftheriou A 2013 *Methods for the study of marine benthos* vol 4 (United Kingdom:Wiley Blackwell) pp 155-156
[10]. Romimohtarto K and Juwana 2009 *Marine biology* (Jakarta: Djambatan) pp 404-406
[11]. Legendre L and Legendre 1983 *Numerial ecology* (New York: Elsevier) pp 239-245
[12]. Muqsit A D, Purnama and Ta'alidin 2016 *J. Enggano* **1** 75-87
[13]. Panggabean A S and Setiadji 2011 *J. Bawal* **3** 255-260
[14]. Manuputty A E W 2012 Coastal ecosystem of Pangkajene Island of South Sulawesi Province (Jakarta: Research Center of Oceanography, Institute of Science) p 5
[15]. Luthfi O M 2003 Spatial distribution of hard reef (Scleractina) in Pulau Panjang Jepara Island (Semarang: Faculty of Fisheries and Marine Science, Diponegoro University) p 97
[16]. Edward dan Z. Tarigan. 2003. *Monitoring of hydrological condition in Raha waters, Muna Island, Southeast Suawesi in relation to coral reef condition Makari Sci.* **8** vol 2 pp 73-82.

[17]. Seveso D, Montano, Strona, Orlandi, Galli and Vai 2013 Exploring the effect of salinity changes on the levels of Hsp60 in the tropical coral *Seriatopora caliendrum Marine Env. Research* **90** 96-103.
[18]. Yulius N, Novianti, Arifin, Salim, Ramdhan and Purbani 2015 *J. Sci. and Tech. of Trop. Marine* **7** 59-69
[19]. Kepel T L and Abrar 2013 *J. Segara* 1-84
[20]. Nugroho S C 2008 Level of survival and growth rate of soft coral *Sinularia dura* and *Lobophytum strictum* in Pramuka Island, Kepulauan Seribu Jakarta (Bogor: Faculty of Fisheries and Science, Bogor Agricultural Institute) p 76
[21]. Andriyono S and Mubarak A S 2011 *J. Fish. and Marine* **3** 139-143.
[22]. Effendi H 2003 *Water quality study for the management of water resources and environment* (Yogyakarta: Kanisius) pp 64-66
[23]. Nugraha M A, Purnama, Wilopo and Y Johan 2016 *J. Enggano* **1** 43-56
[24]. Andriyono S 2010 *J. Fish. and Marine* **2** 171-177
[25]. Pursetyo K T, Tjahjaningsih and Andriyono 2013 *J. Fish. and Marine* **5** 129-137
[26]. Eastwood E K, Clary and Melnick 2016 *Coral reef health and management on the verge of a tourism boom : a case study from Miches* (Dominican Republic: Ocean and Coastal Management) **138** pp 192-204
[27]. Central Bureau of Statistic of Raja Ampat Regency 2016 *Raja Ampat regency in 2016* (Raja Ampat ) pp 154-192
[28]. Huang Y and Coelho 2017 *Sustainability performance assesment focusing on coral reef protection by The Tourism Industry in The Coral Triangle Region Tourism Management* **59** pp 510-527.

IOP Conf. Series: Earth and Environmental Science **137** (2018) 012050 doi:10.1088/1755-1315/137/1/012050

# Marine tourism and the locations of protected turtles on Sukamade Beach, Meru Betiri National Park, East Java

**D J Prihadi[1], A Shofiyullah[1] and Y Dhahiyat[2]**

[1]Marine Science Program Study, Faculty of Fisheries and Marine Science, Indonesia
[2]Lecturer of Faculty of Fishery and Marine Science, University of Padjadjaran, Indonesia

E-mail: donny.juliandri.prihadi@unpad.ac.id

**Abstract.** The research was conducted in Sukamade Beach, Meru Betiri National Park, East Java. The purpose of this research was to identify marine tourism activity and to determine the differences in the characteristics of turtle-nesting beaches towards the number and species of turtles that came to the beach. Data collection conducted in August-September 2014. The method used in this research was a survey method at 7 reseach stations to collect primary data (biophysical characteristics) and secondary data. The Primary data was collected by monitoring turtles, width and slope of the beach, temperature, pH, moisture, sand texture, and beach vegetation conditions at each station. The results of the research shows that marine tourisms always involve tourists who attend to see turtle nesting, when turtles arrive at the beach, and turtles return to the sea, how large the turtles and how they lay eggs on the beach, and the release of little turtles (tukik). The number of turtles that landed from station 1 to station 7 is as many as 311 individuals of three species. The most dominant species of turtles that arrived at the beach is green turtle (*Chelonia mydas*), followed by olive ridley turtles (*Lepidochelys olivaceae*) and leatherbacks turtles (*Dermochelys coriacea*).

## 1. Introduction

Indonesia is a maritime country with approximately two-thirds of its territory in the form of water, thus possessing a high potential of marine resources. These potential marine resources should be organised and preserved as optimally as possible in order to improve the well-being of communities without interfering with environmental sustainability. One of the marine resources with potential is the existence of turtles in Indonesia. Indonesia's aquatic habitats and tracks are suitable for turtles migration. The abundance of biological resources and the very high biodiversity are the main attraction for the park. Six out of seven types of turtles that exist in the world have habitats in Indonesia, namely the dark turtle (*Lepidochelys olivacea*), leatherbacks (*Dermochelys coriacea*), green sea turtle (*Chelonia mydas*), hawksbills (*Eretmochelys imbricate*), flatback (*Natator depressus*) and loggerhead turtles (*Caretta caretta*). Whereas Kempi turtles (*Lepidochelys kempi*) are only found in the waters of Florida and the sea of Mexico [1].

Tourism development is one of the methods suitable in managing and utilizing the area, especially for areas with potential and attractions that can trigger the interests of domestic tourists to visit the region. The main potential of the region is the appearance of Sukamade (aggregation) turtles around the beach which serves as a main tourist attraction. The fertile waters with abundant of food resources make the waters in Sukamde region an area that is suitable for turtle migration and egg laying on the beach.

Turtles are reptiles that inhibit the sea and their presence have long been threatened both by natural or human activities alike. Internationally, sea turtles are categorized into "endangered" species by the International Union for Conservation of Nature and Natural Resources (IUCN), which means they are listed on the Red Data Book. Whereas in the Convention on International Trade in Endangered Species of Wild Fauna and Flora (CITES), all sea turtles are entered in Appendix I, which means that no sea turtles may be bought due to the number traded in nature and the fear of turtles being an endangered species [1]. Therefore, turtle conservation efforts are important to protect and save turtle populations.

Turtles have a high economic value. Almost all parts of the turtle can be utilized to support human life. Turtle shells can be manufactured into a variety of products like frames, glass eyes, bracelets, necklaces, household furniture and various ornaments. Their meats and eggs are also a source of animal protein for humans [1]. One single turtle serves a large number of benefits which also endangers their existence and preservation when it is done without control.

Turtle protection is carried out due to declining population in the wild. Data from various sources show that the turtle population in Indonesia has dropped drastically, especially during the last two decades. A number of observations on spawning locations show that population decline could reach 80% (average 70 %) compared to their population 15 years earlier [2]. No wonder that lately the turtle population is reported to have declined drastically in some areas as a result of hunting and taking too many of their eggs without giving them a chance to hatch naturally. The extinction of turtles in the region will increase if their egg-laying habitat suffer damage.

There are several places in Indonesia that have become turtle spawning grounds, which incorporates the South and North beaches of Indonesia. One of them is in the coast of the Meru Betiri National Park in Sukamade, East Java. The Sukamade UPKP (Turtle Conservation Management Unit) located on the beach is a special group that manages the conversion of the Sukamade coast into a turtle conservation area. There are 4 types of turtles in the Sukamade Coast region i.e. green turtles (*Chelonia mydas*), the Dark Turtle (*Lepidochelys olivaceae*), Hawksbills (*Eretmochelys imbricata*), Leatherbacks (*Dermochelys coriaceae*). Currently green turtles are the types found most often during the search.

Sukamade Beach is one of the places that is suitable for green turtles to lay eggs. The characteristics of the Sukamade coast is generally the same as the beaches of the south coast of Java namely waves and clear water. The beach spawning area tends to have distinctive characteristics that may affect how the turtles land, including season, availability of food in the ocean and biotic and abiotic components of the spawning beaches. The beach in Sukamade is the most productive location for turtle spawning in East Java. The difference in the number and type of turtles that land are caused by turtles which tend to choose their own spawning locations. The selection of spawning locations are influenced by biotic and abiotic component factors that can be viewed from the type of sand, slope of the beach, sand beach width, temperature, vegetation and the atmosphere around the coast.

Until now, information about the characteristics of the protected environment around the Sukamade Coast is still relatively scarce. This research was conducted at the Sukamade Coast protected area by describing the physical and biological factors for each location or observation station. Knowledge about the relationship between the environmental characteristics with the number of turtles that migrate is indispensable in order to serve as input for the management of turtle conservation and preservation. As for the purpose of this research is to know the difference and the characteristics of the turtle spawning beaches at several points and locations organised around the Sukamade Coast station, the spread of the number and types of turtles, as well as the turtles' tendency of choosing their preferred spawning locations and nesting locations.

This rich natural potential has been supported by suitable management and policies, a method of collaborative management has already been applied and carried out after the promulgation of the legislation regarding collaborative management systems namely the Regulation of the Minister of Forestry (Permenhut) Number P. 19/Menhut-II/2004. It refers to a collaboration in the management of the area and the preservation of the natural sanctuary. Collaborative management includes

IOP Conf. Series: Earth and Environmental Science **137** (2018) 012050 doi:10.1088/1755-1315/137/1/012050

stakeholders managing the national parks directly or indirectly which include the National Park Hall, local authorities, indigenous people, NGOs, TNI/POLRI, Scientific Institutions, mapping and training, as well as religious institutions [3].

## 2. Methodology

This research was conducted from May to October 2014. Data retrieval was performed on September 2014 at the Sukamade Coast National Park and the Meru Betiri National Park, Banyuwangi, East Java. The study area at TNMB is located in the province of East Java with the longitude of 113°58 ' 38 "– 113°58 ' 30" and 8°20 ' 48 "– 8°33 ' 48" S whereas administratively, Meru Betiri National Park is located in two districts namely the Jember and Banyuwangi districts.

The tools used in this study consist of GPS (Global Positioning System), meter rolls, soil testers, thermometers, refractometers, pH meters, DO meters, TMD software (Tide Driver Model), KUMMOD software, Google Earth software, and vegetation life forms. The materials are used to gather primary data which is then utilized to gather secondary data. The primary data include interviewing 10 tourists on site, composing physical or abiotic component parameters (width of the beach, slope of the beach, great composition and grains of sand, sand temperature, pH, and moisture), biotic or biological parameters (condition of vegetation and fauna), water quality parameters (pH, DO, salinity and temperature), and monitoring turtles for research. The secondary data is comprised of the amount and types of turtles that migrate from UPKP (Turtle Conservation Management Unit) at the Sukamade Resort, tidal data at the Sukamade beach during August-September 2014 which was obtained from the tidal model Global TMD (Tide Model Drivers), and the average precipitation per-day over the span of one month on the beaches of Sukamade during August-September, satellite data obtained from TRMM (Tropical Rain shower Measuring Mission) that can be correlated with the map from Google Earth software. Additionally, tourists and the manager of Sukamade Beach were interviewed to gain insight on tourist activities.

The method used was a survey method by observation, inspection, interview and direct measurement on the object of study in the field and its description. The research location is Sukamade Coast with stations in seven locations. Below is a picture of the overview map of the data retrieval station.

IOP Conf. Series: Earth and Environmental Science **137** (2018) 012050 doi:10.1088/1755-1315/137/1/012050

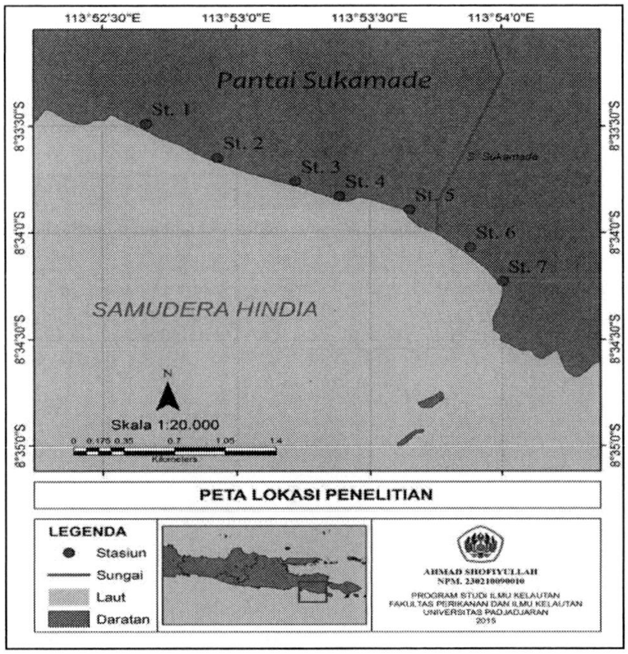


**Figure 1.** Data capture stations on the coast of Sukamade.

The methods of data analysis used are the descriptive comparative method which compares data observations of one station to the other, and discussing them through the use of literature study. The data obtained at the time of observation is tabulated and presented in the form of text, tables, charts and images. Further observations at each station are in accordance with the descriptive analysis compliance which is based on the spawning location of sea turtles. Interviews and observing tourist activities in Sukamade Beach was also performed.

## 3. Result and Discussion

The general condition of the Sukamade Coast was measured during the research. The measurements include water quality parameters, namely temperature, salinity, dissolved oxygen (DO), and pH. The measurement was performed in the morning and randomly spread across the sea waters and estuaries of Sukamade. The results are indicated on table 1 below.

**Table 1.** Water Quality Parameters.

| Parameter | River | | Sea | | The Raw Quality * |
|---|---|---|---|---|---|
| | Range | Average | Range | Averange | |
| Temperature (°C) | 25-26 | 25.33 | 24-26 | 25 | 24-32 |
| Salinity (ppt) | 2-4 | 3.33 | 32-34 | 33 | 29-34 |
| DO (mg/l) | 4.60-5.80 | 5.23 | 4.80-6,20 | 5.37 | > 4 mg/L |
| pH | 7.18-8.16 | 7.74 | 7.18-8.44 | 7.93 | 7-8.5 |

Note: The decision of the Minister of the environment the number 51 of the year 2004

The results of the water quality parameter measurements at Sukamade Beach during the research show that some parameters are within the ideal range for the life of green turtles. This is based on the

IOP Conf. Series: Earth and Environmental Science **137** (2018) 012050 doi:10.1088/1755-1315/137/1/012050

raw quality of physical-chemical parameters of marine life as described in the Minister of Environment Decree number 51 of 2004.

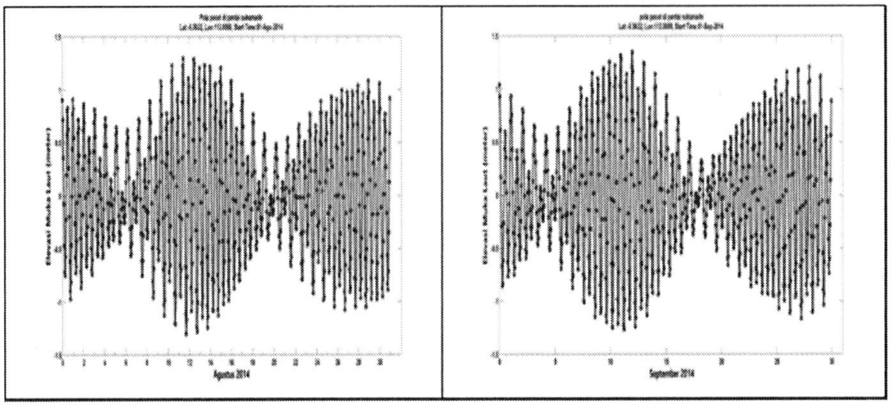

**Figure 2.** Elevation of the sea.

*3.1 The difference in the characteristics of each Station.*

**Table 2.** Description of the environmental characteristics of each station in Sukamade Coast with literature.

| Parameter | | Research results at each Station | | | | | | | Literature range |
|---|---|---|---|---|---|---|---|---|---|
| | | 1 | 2 | 3 | 4 | 5 | 6 | 7 | |
| Beach Width (m) | Min | 46.74 | 64.29 | 73.51 | 45.86 | 66.76 | 54.92 | 1090 | > 30-80 m from the furthest tide |
| | Max | 78.05 | 89.43 | 99.32 | 66.58 | 73.82 | 64.52 | 19.05 | |
| The slope of the Beach (α°) | | 8.08 | 6.42 | 7.40 | 7.16 | 15.36 | 9.57 | 11.89 | < 30° |
| The temperature and depth of the sand 30-50 cm (°C) | | 28.41 | 28.22 | 27.52 | 27.67 | 26.59 | 27.22 | 27.15 | 24°-33° C |
| Composition and large grains of sand | Total deposit of sand substrate (%) | 99.85 | 99.99 | 99.98 | 99.99 | 99.99 | 99.99 | 99.97 | Sand Fraktion ≥ 90 % & large-sized grains of sand 0,10-0,50 mm (fine-medium) |
| | Medium-fine sand (%) | 98.85 | 92.38 | 74.40 | 88.00 | 93.67 | 96.97 | 99.,16 | |
| Sand pH | | 6.20 | 6.80 | 6.60 | 6.60 | 6.60 | 6.80 | 6.20 | 6-9 |
| Moisture Beach (%) | | 2.00 | 1.00 | 2.00 | 3.00 | 3.00 | 2.50 | 3.50 | 2-12 % |
| Vegetation condition | | Dominant | Dominant | HT, IP | IP, CD | Dominant | - | Dominant | Dominated by vegetation: |

| | | | | | |
|---|---|---|---|---|---|
| | (HT), the rest of it (PT, SL, IP, CD, WD) | (PT, HT) & sisanya (WD | (IP, CD) | (IP), & the rest of it (TC, AP, CI, SC) | Pandan sea (*Pandanus tectorius*), katang-katang, (*Ipomea pescaprae*), waru sea (*Hibiscus tiliacus*) |

Note: Vegetation condition
HT (*Hibiscus tiliacus*), PT (*Pandanus tectorius*), IP (*Ipomea pescaprae*), CD (*Cynodon dactylon*), SL (*Spinifex littoralis*),WD (*Wedelia biflora*), TC (*Terminalia catappa*), AP (*Abrus precatorius*), CI (*Callophylum inophylum*), dan ST (*Scaevola taccada*).

Beach tourism is a type of tourism that prioritizes cultural activities and coastal communities such as recreation, water sports, scenery and climate enjoyment [4]. Shore excursions in Sukamade would be categorized as ecotourism involving special interests. Tourist satisfaction is obtained from the service provided by nature directly. The comfort level decreases when tourists are concerned for the sustainability of resources, especially when green turtles are concerned. The development of ecotourism should avoid direct exploitation of the green turtles at Sukamade Beach. Additionally, an alternative form of management is expected to benefit the community financially by preserving environmental sustainability.

The sustainability of the beach tourism area consists of 6 parameters, which are: the daily increase of the number of turtles, beach width, slope of the beach, depth, the basic materials of the beaches, water, closure of the land and surrounding beach [4]. There is also beautiful, natural panorama which include broad white sand, crystal clear water, dense and distinctive coastal vegetation, waves suitable for surfing, and increased activities of green turtles [4]. In this area travelers can enjoy the natural beauty of Sukamade Beach with both day and evening activities.

### *3.2 The number and types of turtles landed.*

The observation results of the number of turtle landings in each station on August-September 2014 in the coast of Sukamade show differences in the number of turtle landings at each station (table 3).

**Table 3.** The number of Turtle landings At each Station.

| Station | Turtle type | The Number Of Landings | Landing | | | |
|---|---|---|---|---|---|---|
| | | | Spawn | % | Memeti | % |
| 1 | Green | 95 | 37 | 38.95% | 58 | 61.05 % |
| 2 | Green | 66 | 33 | 50% | 33 | 50% |
| | Olive Ridley | 2 | 2 | 100% | 0 | 0% |
| 3 | Green | 50 | 27 | 54% | 23 | 46% |
| | Olive Ridley | 1 | 1 | 100% | 0 | 0% |
| | Leatherback | 1 | 1 | 100% | 0 | 0% |
| 4 | Green | 62 | 34 | 54.84% | 28 | 45.16% |
| 5 | Green | 23 | 10 | 43.48% | 13 | 56.52% |
| 6 | Green | 11 | 5 | 45.45% | 6 | 54.55% |
| 7 | Green | 0 | 0 | 0% | 0 | 0% |
| Total | | 311 | 150 | 48.23% | 161 | 51.77% |

In table 3, we can argue that there are many green turtle nesting and spawning areas in Sukamade Beach. Tourists always want to see how the turtles arrive at the beach, how big the green turtles are, how they lay eggs. Tourists also want to be involved in the release of turtle hatchlings (*tukik*). Such activities which include ecotourism may or may not impact the natural environment. The possibility of such impact is always there, particularly the negative impacts for turtles [4]. Fostering tourists towards nature for the good of society, requires a thinker who has discernment and wisdom to help people. NGOs and experts were able to protect the turtles from extinction by making society aware of their plight and ensured their survival in Sukamade Beach. Legal support is needed to support ecosystems, as well as social, economic, and cultural factors.

More tourists are willing to enhance the quality and integrity of the area so that the ecosystem habitats of the green turtles will be more authentic. Tourists are expected to be more aware of the importance of conservation efforts. Tourism contributes a great increase in the income for both the local area and its community when it is managed in an integrated and sustainable manner. The current paradigm of tourism in the beach and sea areas is mainly focused on economic profits, which is attracts tourists without regard to preserving the environment [5].

Tourist are always interested to see:

1. Turtle nesting,
2. When the turtles migrate to the beach,
3. When the turtles return to the sea,
4. The large size of the turtles
5. How the turtles lay eggs on the beach
6. The release of turtle hatchlings *(tukik)*.

Green turtles generally have the potential to attract tourists. Arrangements should be made to support the required area and to regulate tourists activities. Tourist activities are difficult to control and they often reduce the comfort of turtles [4]. This condition is threatening the existence of green turtles at Sukamade Beach. Turtle population decline in Sukamade Beach is caused by tourist visits, camping grounds, campfires, the usage of camera flash when taking photos, theft of turtle eggs, and fish nets.

Nature tours which focus on the turtles can only be enjoyed in intact ecosystems with a strong connection to their habitats [4]. The area has been determined as unsuitable for the development of ecotourism in the form of shore excursions due to fear of causing damage to the environment as well as the loss of species. Changes in ecosystems are difficult to be restored to its original state because their properties and recovery behaviours are not the same. Efforts to accelerate recovery time often require great manpower and funds. The design of turtle tours must be as exclusive tours. The total number of tourists must be restricted and shall not exceed the carrying capacity of the area [5]. For this reason, we will be conducting a carrying capacity study of the area in the next research.

An increasing number of tourists are attracted by the green turtles' natural, beautiful and unique habitats and spawning areas [6]. The uniqueness of the turtle is the main reason tourists come to Sukamade's beautiful beach. Being far away from home is also one of the considerations for tourists to come to Sukamade. Humans live in social communities with natural surroundings and whole layouts of natural environment, and are also one with intellectual power, and serves as a natural means of economic nature layout. The Turtle Spawning Manager has the task of making rules that should be followed by tourists who come to see the turtles spawn, land, return to the sea, how large the turtles are when they lay eggs and the release of turtle hatchlings *(tukik)*. The Manager is obligated to create rules of how many tourists are allowed to come and see the turtles while at the same time maintain the sustainability of Sukamade Beach so it will still be favored by the turtles to spawn.

IOP Conf. Series: Earth and Environmental Science **137** (2018) 012050 doi:10.1088/1755-1315/137/1/012050

## 4. Conclusion

Marine tourism was developed at Sukamade Beach for turtle viewing. Tourists come to see the turtles nesting, when they land, when they return to the sea, how large they are, how they lay eggs on the beach and also the process of releasing the turtle hatchlings (*tukik*). Green turtles are found in seven stations, while Olive Ridley turtles are found in two. Leatherback turtles are only found in one station. The favorite vegetation chosen by the turtles are HT (*Hibiscus tiliacus*), PT (*Pandanus tectorius*), IP (*Ipomea pescaprae*), CD (*Cynodon dactylon*), SL (*Spinifex littoralis*), WD (*Wedelia biflora*). For Leatherback turtles, they prefer HT (Hibiscus tiliacus) and IP (Ipomea perscaprae). The manager has the task of show the nesting sites, accompany tourists to the site and try to give interesting information about the turtles. This can help avoid the descruction of the turtle nesting sites.

## 5. References

[1] Dahuri R, Rais J, Ginting S P, Sitepu M J 2003 Integrated *Resource management of coastal and Ocean Areas* (Jakarta: PT Pradnya Paramita)

[2] Darsoprajitno H, Saewarno 2013 *Ecological tourism: Object Management Governance and tourist attraction* (Bandung: Publisher ANGKASA)

[3] Nugraha A 2004 The Life Cycle Of Sea Turtles. Workshop papers at WWF. The translation of the Turtle information Kt GBRMPA WWF Bali p 14-16

[4] Nugroho I 2011 *Ecotourism and sustainable development* (Pustaka Pelajar)

[5] Nuitja I N S 1992 *Biology and ecology of the preservation of sea turtles* (Bogor: IPB Perss) 128 p

[6] Nurbaeti N 2016 The management of beach tourism Berbesis Koservasi Turtle Green Turtle (*Chelonia mydas*) at Pangumbahan Sukabumi district of West Java Thesis IPB

IOP Conf. Series: Earth and Environmental Science **137** (2018) 012051 doi:10.1088/1755-1315/137/1/012051

# The prevalence and intensity of gastrointestinal endoparasite worms of cantang grouper (*Epinephelus fuscoguttatus - lanceolatus*) on floating net cages at Lamong Bay Surabaya, Indonesia

**L D Agustina[1], S Subekti[2] and Kismiyati[3]**

[1]Bachelor programme in Aquaculture, Faculty of Fisheries and Marine, Airlangga University, Jl. Airlangga 60115, Surabaya, East Java, Indonesia
[2]Departemen of Marine, Faculty of Fisheries and Marine, Airlangga University, Jl. Airlangga, 60115, Surabaya, East Java, Indonesia
[3]Departemen of Fish Health Management and Aquaculture, Faculty of Fisheries and Marine, Airlangga University, Jl. Airlangga 60115, Surabaya, East Java, Indonesia

E-mail : kismiyati@fpk.unair.ac.id

**Abstract.** Cantang groupers (*Epinephelus fuscoguttatus-lanceolatus*) is a hybridized grouper fish of Brackishculture Center, Situbondo. In Indonesia, currently information about the parasite infection in cantang groupers is still few. This study aims to determine the prevalence and intensity of endoparasite worms that infect the gastrointestinal of cantang groupers (*E. fuscoguttatus-lanceolatus*) on the floating net cages at Lamong Bay, Surabaya. The method used in this study is survey method and analyzed descriptively. The endoparasite worms found in the gastrointestinal of cantang groupers were *Anisakis physeteris* and *Neoechinorhynchus longnucleanus*. The highest prevalence is single infection of *Neoechinorhynchus longnucleanus* was 3 % (occasionally) with intensity of 1 individual/fish and the lowest prevalence was single infection of *Anisakis physeteris* is 1 % (occasionally) with intensity of 1 individual/fish.

## 1. Introduction

Groupers are one of leading Indonesian export commodities widely cultivated. The production levels of groupers from 2009 to 2013 in Indonesia were 8.791 tons, 10.398 tons, 10.580 tons, 11.950 tons and 18.864 tons respectively [1]. Cultivation of groupers in Surabaya is located in Lamong Bay uses a floating net cage system. Cantang grouper (*Epinephelus fuscoguttatus-lanceolatus*) is hybridized a grouper of female tiger groupers (*E. fuscoguttatus*) with male kertang groupers (*E. lanceolatus*) [2].

In 2014, mass mortality occurred in cantang groupers cultivated on floating net cages at Lamong Bay. The body of the dead fish has a lot of mucus and there were wounds on some parts of the body. In addition, the growth of the fish during cultivation became slower. The mortality in the cantang groupers was suspected to be related to infectious diseases caused by parasites. One of the effects of the spread of endoparasites in the fish is the presence of invertebrates around the floating net cages that act as intermediate hosts of some endoparasite species [3]. Therefore, carnivorous fish including groupers are more likely to be infected by endoparasite worms than herbivores and omnivorous fish [4].

IOP Conf. Series: Earth and Environmental Science **137** (2018) 012051 doi:10.1088/1755-1315/137/1/012051

The endoparasite worms found in the gastrointestine of several species of groupers are: *Prosorhynchus lafii* [5], *Prosorhynchus maternus* [6], *Callitetrarhynchus gracilis* [7], *Echinostoma* [8], *Camallanus carangis*, *Procamallanus variolae* [9], *Capillaria plectropomy* [10], *Neoechinorhynchus* sp. and *Serrasentis sagittifer* [11].

The results of sea water monitoring in Lamong Bay Surabaya, Indonesia in 2014 showed that the sea water quality is below the seawater quality standard for biota; it also showed the existence of pollution [12]. Decreased water quality can be a trigger of infectious diseasees in fish. The development of endoparasites in the fish body can be influenced by environmental factors, such as temperature and chemicals in waters [13]. Research on the prevalence and intensity of endoparasite worms in the gastrointestine of cantang groupers (*Epinephelus fuscoguttatus-lanceolatus*) is important to determine the prevalence and intensity of endoparasite worms infecting cantang groupers.

## 2. Methodology

Life specimens of *E. fuscoguttatus-lanceolatus* were taken from floating net cages at Lamong Bay Surabaya, Indonesia on March to April 2016. The samples used in this study were 100 groupers (three months old with a length of 10-20 cm). This research used survey method and the data were analyzed descriptively. The fish were examined for endoparasite under a dissecting microscope. The staining of endoparasite used Semichen-acetic Carmine method, which refers to Kuhlmann's modification [14]. The illustrations used a lucida camera. The key identifications of endoparasite worms are Amin et al. [15], Chen and Shih [16], Grabda [17], Hoffman [18], Kabata [19] and Pavlovskaya [20]. The prevalence of endoparasite worms infecting the gastrointestines of cantang groupers was calculated based on Bush et al. [21] and the intensity was examined according to Dogiel et al. The category of prevalence and intensity refers to Williams and Williams [23].

## 3. Results and discussion

From the identification of endoparasite on gastrointestines of cantang groupers (*Epinephelus fuscoguttatus-lanceolatus*) on floating net cages at Lamong Bay Surabaya, Indonesia, two species of endoparasite worms attached to the intestinal wall were found, which are third stadia larvae of *Anisakis physeteris* and adult worm of *Neoechinorhynchus longnucleanus*. The two species of endoparasite worms can be seen in figure 1 to figure 4.

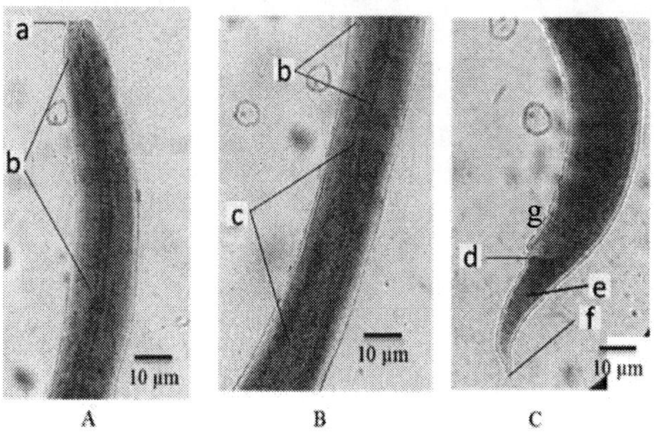


**Figure 1.** Third Stadia Larvae of *Anisakis physeteris*, scale bar = 10 μm. Coloring images based on binoculars microscope magnification 400x. Description; (A) Anterior part, (B) Mid body part, (C) Posterior part, (a) Booring tooth, (b) Esophagus, (c) Ventriculus, (d) Anal, (e) Appendix, (f) Tail, (g) Vulva.

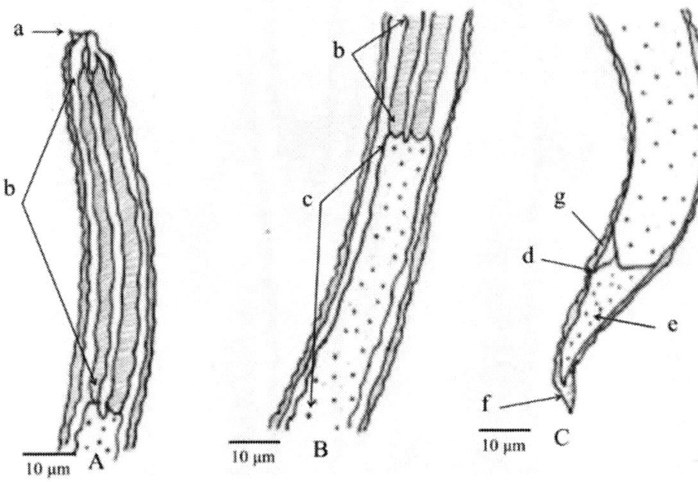


**Figure 2.** Third Stadia Larvae of *Anisakis physeteris*, scale bar = 10 µm. Images with binocular microscope equipped with camera lucida. Description; (A) Anterior part, (B) Mid body part, (C) Posterior part, (a) Booring tooth, (b) Esophagus, (c) Ventriculus, (d) Anal, (e) Appendix, (f) Tail, (g) Vulva.

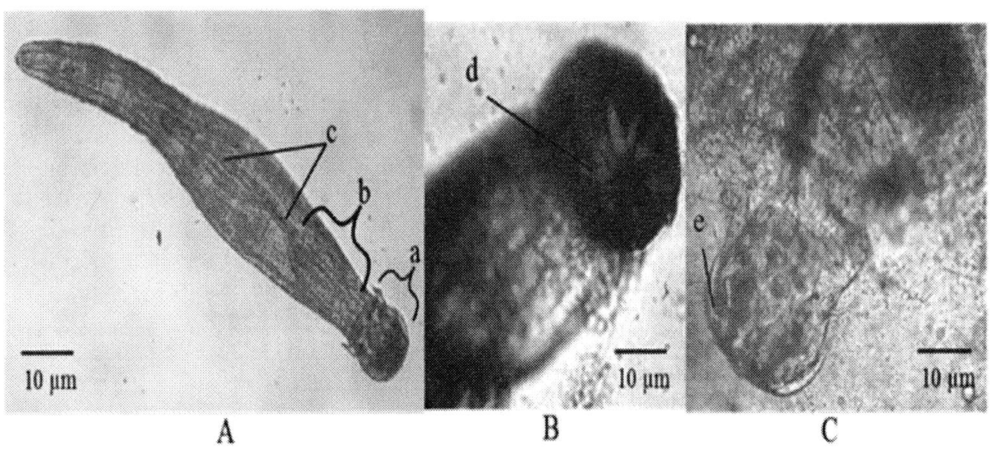


**Figure 3.** Adult worm of *Neoechinorhynchus longnucleanus*, scale bar = 10 µm. Native on a magnification binocular microscope magnification 100x (in figure A) and magnification 400x (in figures B and C). Description; (A) whole body, (B) Anterior end portion, (C) Proboscis, (a) Proboscis, (b) Proboscis receptacle, (c) Leminisci, (d) Hook the first line, (e) Hook.

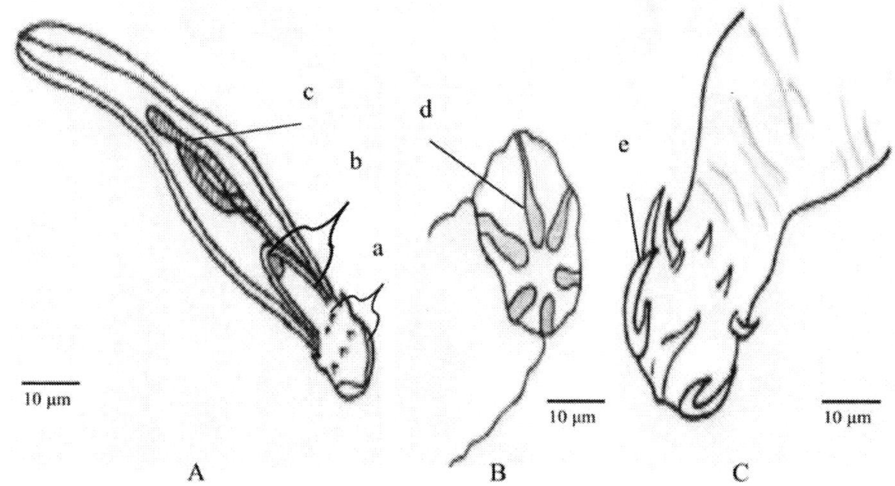


**Figure 4.** Adult worm of *Neoechinorhynchus longnucleanus*, scale bar = 10 µm. Images with binocular microscope equipped with camera lucida. Description; (A) whole body, (B) Anterior end portion, (C) Proboscis, (a) Proboscis, (b) Proboscis receptacle, (c) leminisci, (d) Hook the first line, (e) Hook.

*Anisakis physeteris* are included in Phylum Nemathelminthes, Class Nematoda, Order Ascaridida, Suborder Ascaridata, Family Anisakidae, and Genus *Anisakis* [12]. This worm has an elongated cylindrical body, with a total length of 4.113 mm and width of 0.13 mm. There were cuticle, booring tooth (larval tooth), esophageal with a length of 0.63 mm, ventriculus with a length of 0.13 mm, intestine, conical tail and on the tail, there was no mucron. This is in accordance with the statement of Anshary [24] and Hoffman [18] that the worm *Anisakis* sp. could be distinguished from other Anisakidae parasites by looking at the anterior end (booring tooth) and the ventricular shape that looks elongated and looks like white spots when observed with a binocular microscope. *Anisakis physeteris* worms were included in *Anisakis* type II which has a conical tail without mucron, according to [24, 16, 17]. *Anisakis physeteris* found was female larvae of third stadia larvae characterized by the presence of vulva in the posterior body. The life cycle of *Anisakis* starts from eggs which develop to second stadia larvae that infect the first intermediate host (small crustaceans), then ingested by the fish, developed into third stadia larvae in the fish body and adult in marine mammals. Nematode worms usually infect the gastrointestine of fish whereas larvae was found in almost every organ of fish [24, 25, 18]. The study of Chen and Shih [16] found third stadia larvae of *Anisakis physeteris* infecting the gastrointestine of *Scomber australasicus*.

*Neoechinorhynchus longnucleanus* belongs to the Phylum of Acanthocephala, Class of Eoachanthocephala, Order of Neoechinorhynchida, Family of Neoechinorhynchidae and Genus of *Neoechinorhynchus* [15]. This worm has a short round proboscis equipped with three row hooks; each line has six hooks. The hook on the first row is the largest compared to those on the second and third rows (fig 3B,C; 4B,C). According to Hoffman's ,the genus *Neoechinorhynchus* has a small, bowed or straight, short proboscis body equipped with three rows of hooks with each row of six hooks; the anterior hook is longer and greater than the others. *Neoechinorhynchus longnucleanus* has an anterior hook with a simple root shape, two leminisci of different lengths of 0.81 mm and 0.67 mm (fig 3A,C; 4A,C). This is in line with the statement of Amin et al. that *Neoechinorhynchus longnucleanus* has an anterior hook root with a simple shape and has two different lengths of leminisci and a large nuclei. The worm was an adult *Neoechinorhynchus longnucleanus*. The life cycle of the genus *Neoechinorhynchus* developed into adulthood in marine and freshwater fish, frogs and turtles; larvae developed in the crustacea. Adult worms of *Neoechinorhynchus* were attached to the small intestine and some of them in the peritoneal cavity [18,26]. Amin et al. found *Neoechinorhynchus longnucleanus* infecting the intestine of *Strongylura strongylura* and Ruckert et al. found it in the gastrointestine of tiger grouper (*E. fuscoguttatus*).

### *3.1. Prevalence of endoparasite worms*

Prevalence refers to the precentage of fish infected by endoparasites in the population. The prevalence of cantang groupers infected by endoparasite worms could be seen in table 1.

**Table 1.** Prevalence of cantang groupers infected by endoparasite worms on floating net cages at Lamong Bay Surabaya, Indonesia (100 samples).

| No | Species of Endoparasite | Infected fish (individual) | Prevalence (%) | Infection Category (Wiliams and Williams, 1996) | Predilection of endoparasites |
|---|---|---|---|---|---|
| 1 | *Anisakis physeteris* | 1 | 1 | *Occasionally* | Intestinal wall |
| 2 | *Neoechinorhynchus longnucleanus* | 3 | 3 | *Occasionally* | Intestinal wall |
| Total | | 4 | 4 | | |

The total prevalence of endoparasite worms infecting the gastrointestine of cantang groupers was 4%. The highest prevalence was a single infection of *Neoechinorhynchus longnucleanus* with a prevalence rate of 3 % and the lowest prevalence was a single infection of *Anisakis physeteris* with a prevalence rate of 1 %. According to Williams and Williams [23], the prevalence value was included in the occasional category. The prevalence of cantang groupers infected by *Anisakis physeteris* and *Neoechinorhynchus longnucleanus* was low compared with some studies that found some prevalences of *Anisakis* and *Neoechinorhynchus* worm such as third stadia larvae of *Anisakis physeteris* infecting *Strongylura strongylura* from Taiwanese waters with prevalence of 4 % [16], the prevalence of *Anisakis* sp. in gastrointestine of *Lutjanus malabaricus* on the auction of fish at Brondong Lamongan was 67 % for fish with the size of 21-24 cm and 80 % for fish with the size of 25-37 cm [27], while *N. longnucleanus* was found to infect the intestinal wall of cultured *E. fuscoguttatus* in Lampung Bay, which was 2.9 % [11], the prevalence of *N. agilis* in *Cheon labrosus* in Beymelek Lagoon Lake in Atalya, Turkey was 24% for male and 8.3 % for female [28] and the prevalence of *N. villaldoi* in *Austrolebias bellottii* from Punta Indio, Argentina was 80% [19].

Fish could act as a definitive host of *Neoechinorhynchus* [30, 18] and as an intermediate host of *Anisakis* [24, 31, 18]. The presence of *Anisakis physeteris* and *Neoechinorhynchus longnucleanus* infections on the intestinal wall of cantang groupers is due to the eating habit of the fish as predators, such as eating trash fish and small shrimp (which could act as intermediate hosts of the endoparasite worms). The above statement was confirmed by Heemstra and Randall [16] that *Epinephelus* were epibethic predators feeding on macro invertebrates (mainly crustaceans). It was also supported by Ruckert et al. who stated that fish could be attacked by food-borne diseases and the existence of invertebrates around the floating net cage as an intermediate host of some endoparasite worms. Therefore, these may affect the spread in fish. Fidyandini et al. also stated that low prevalence rates were due to parasite adaptation abilities in host body and host compatibility for parasitic survival and environmental quality.

### *3.2. Intensity of endoparasite worms*

Intensity refers to the endoparasites that infect each individual. The intensity of cantang groupers infected by *A. physeteris* and *N. longnucleanus* could be seen in table 2.

**Table 2.** Intensity of cantang groupers infected by *A. physeteris* and *N. longnucleanus* on floating net cages at Lamong Bay Surabaya, Indonesia.

| No. | Species of Endoparasite Worm | Number of Infected fish (%) | Number of Endoparasite Worm | Intensity (Individual/fish) |
|---|---|---|---|---|
| 1. | *A. physeteris* | 1 | 1 | 1 |
| 2. | *N. longnucleanus* | 3 | 3 | 1 |
| Total | | 4 | 4 | 1 |

The total intensity of endoparasite worms found in the gastrointestinal of cantang groupers was 1 individual/fish, that was regarded as mild infection [23]. The intensity of cantang groupers infected by *Anisakis physeteris* and *Neoechinorhynchus longnucleanus* was low compared with some research that found some intensity of *Anisakis* and *Neoechinorhynchus* worms such as; the intensity of *Anisakis* sp. in the gastrointestine of *Lutjanus malabaricus* on the auction of fish at Brondong Lamongan was 5 individual/fish (fish size 21-24 cm) and 18 individual/fish (fish size 25-37 cm) [27]. The intensity of *Anisakis* sp. in *Chanos chanos* on the ponds of Ketapang Village, Mauk, Tangerang District, Province of Banten was 4 individual/fish [20], while the intensity of *N. agilis* in *Cheon labrosus* in Beymelek Lagoon Lake in Atalya, Turkey was 2 individual/fish (male) and 4 individual/fish (female) [28] and the mean intensity of *N. villaldoi* in *Austrolebias bellottii* from Punta Indio, Argentina was 3.13 individual/fish [29].

The intensity value was also influenced by the endoparasite egg survival factor [35, 36]. This study found that endoparasite worms only infected the intestinal wall of cantang groupers. This is in line with the statement of Murata et al. that the small intestine and lumen of the small intestine provide nutrients. The structure and physiology of the intestine could be parasitic microhabitat affecting the presence of parasites. The seawater quality at Lamong Bay was not much different from the standard of seawater quality for biota (based on the Decree of the Minister of Environment No. 51 of 2004 on the Quality Standard of Sea Water). The pH of 7.85 is within the normal range (pH 7-8.5) and the temperature of 29.1$^0$C is in the normal range of sea water temperature (28-30$^0$C). Meanwhile, the salinity was 29.3 ppt, less than the normal of sea salinity of 33-34 ppt. The low salinity of this study is due to seasonal changes. Komarawidjaja explained that water quality degradation can be a driving force for the development of pathogens in fish.

## 4. Conclusions

The endoparasite worms infecting cantang groupers were *Anisakis physeteris* and *Neoechinorhynchus longnucleanus*. The total prevalence of endoparasite worms infecting the gastrointestine of cantang groupers was 4% with total intensity of 1 individual/fish. The predilection of both species of worms was on the intestinal wall of cantang groupers.

## 5. References

[1] Ministry of Marine and Fisheries 2013 Production volume of kerapu-seaweed-tilapia in 2009-2013. Directorate General of Aquaculture Ministry of Marine Affairs and Fisheries

[2] Brackishculture Center, Situbondo. 2012. Grouper fish: hybrid between female tiger grouper and male kertang grouper. http://bbapsitubondo.com (Accesed 11 January 2015)

[3] Ruckert S, Klimpel S, Al-Quraishy S, Mehlhorn H and Palm H W 2009 *J. Parasitol. Res.* **104** 523-32

[4] Sarjito dan Desrina 2005 Analyze the infection of the endoparasites worm in white snapper (Lates calcarifer Bloch) from Demak coastal waters Activity Report of Lecturers Research Result Faculty of Fisheries and Marine Science Diponegoro University Semarang

[5] Boot N J and Cribb T H 2009 *Syst. Parasitol.* **72** 57–69

[6] Bray R A and Justine J L 2006 *J. Folia Parasitol.* **53** 181–8

IOP Conf. Series: Earth and Environmental Science **137** (2018) 012051 doi:10.1088/1755-1315/137/1/012051

[7] Kleinertz S, Damriyasa I M, Hagen W, Theisen S and Palm H W 2014 *J. Helminth.* **88** 50 – 63
[8] Ulkhaq M F, Kismiyati dan Kusdarwati R 2012 *J. fish. Mar.* **4** 6–15
[9] Justine J L, Beveridge I, Boxshall G A, Bray R A, Moravec F, Trilles J P and Whittington I D 2010 *J. Folia Parasitol.* **57** 37-62
[10] Moravec F and Justine J L 2014 *Capillaria plectropomi* n. Sp. (Nematoda: Capillariidae), A new intestinal parasite of the leopard coral grouper *Plectropomus leopardus* (Serranidae) Off New Caledonia. http://parasite-journal.org (Accesed 09 July 2015)
[11] Ruckert S, Klimpel S and Palm H W 2010 *Aqua. Res.* **41** 58-69
[12] Susanto C 2015 Result of monitoring of seawater quality environmental information systems BLH Surabaya Http://kotasurabaya.silh.menlh.go.id (Accesed 16 May 2015)
[13] Hassan M 2008 Parasites of native and exotic freshwater fishes in the South-West of Western A ustralia Thesis Murdoch University. Perth, Western Australia
[14] Kuhlmann W F 2006 Preservation, staining, and mounting parasite speciment. http://www.facstaff.unca.com (Accesed 14 January 2015)
[15] Amin O M, HA N V and HA D N 2011 *J. Parasit.* **18** 21 – 34
[16] Chen H -Y and Shih H -H 2015 *Acta Tropica* **145** 61- 7
[17] Grabda J 1991 *Marine Fish Parasitology* (Polish Scientific Publishers, Warsawa) pp 142-155
[18] Hoffman G L 1999 *Parasites of North American Freshwater Fishes/Glenn L. Hoffman;* with a foreword by Ernest H. Williams, Jr. – 2nd ed. Cornell University Press. Ithaca and London
[19] Kabata Z 1985 *Parasites and diseases of fish cultured in the tropics* Taylor and Francis. London pp 167-170
[20] Pavlovskaya I E B, Gusev A V, Dubinina M N, Izyumova N A, Smirnova T S, Sokolovskaya I L, Shtein G A, Shul'man S S and Epsthein V M 1964 *Key to parasites of freshwater fish of the U.S.S.R.* Israel Program for Scientific Translations. Jerussalem.
[21] Bush A O, Lafferty K D, Lotz J M and Shostak A W 1997 *J. Parasitol.* **83** 75–83
[22] Dogiel V A, Petrushevski G K and Polyanski I 1970 *Parasitologi of fishes* T. F. H. Publisher. Hongkong
[23] Williams E H and Williams I B 1996 *Parasites of offshore big game fishes of Puerto Rico and The Western Atlantic.* Puerto Rico. Departement of Natural and Environtmental Resources.
[24] Anshary H 2011 *J. Fish. Sci.* **8** 70-7
[25] Audicana M T, Ansotegui I J, de Corres L F and Kennedy M W 2002 *Trends in Parasitol* **18** 20-4
[26] Jithedran, K.P. and S. Kannappan 2010 *J. Parasit. Dis.* **9** 11-3
[27] Muttaqin M Z and Abdulgani N 2013 *J. Sains Seni Pomits.* **2** 2337 – 3520
[28] Aydogdu A, Emre N and Emre Y 2015 *Turk. J. Zool.* **39** 43–51
[29] Montes M M, Barneche J, Garcia I, Preisz S and Martorelli S R 2017 *Check List* **13** 53 – 9
[30] Al-Sady R S 2009 *J. Pure. Appl. Sci.* **22** 61-6
[31] Anggraini F, Kismiyati dan Subekti S 2014 *J. Aqua. Fish Hlth.* **1** 1 – 10
[32] Heemstra P C and Randall J E 1993 **16** 11 – 5
[33] Fidyandini H P, Subekti S dan Kismiyati 2012 *J. Mar. Coast. Sci.* **1** 91 – 112
[34] Junardi, E., Mustahal and A. N. Putra 2014 *J. Fish. Mar.* **4** 251–57
[35] Hibur O S, Detha A I R, Almet J and Irmasuryani 2016 *J. Kajian Vet.* **4** 40 – 51
[36] Woolmark 2002 Sheep worm control; Online Book. Http://sydney.edu.au (Accesed 20 October 2017)
[37] Murata R, Suzuki J, Sudamasu K and Kai A 2009 *Parasitol. International* **60** 193–8
[38] Komarawidjaja W 2006 *J. Hidrosfir.* **1** 32 – 7

**Acknowledgments**
The author would like to thank Mr. Suharyanto, A.Pi, M.M, head of the KPIM Class I Surabaya II, for allowing the use of facilities at KPIM Class I Surabaya II and Retno Desi Tary for her help during the examination of the cantang groupers.

IOP Conf. Series: Earth and Environmental Science **137** (2018) 012052 doi:10.1088/1755-1315/137/1/012052

# Crab and shellfish occurrences in the newly-grown mangrove habitats in southern Thailand

**P Yeesin, S Bautip and S Chesoh**

Faculty of Science and Technology, Prince of Songkla University, Pattani, Thailand, 94000.

E-mail: Chesoh.s@hotmail.com

**Abstract.** Mangrove crabs and shellfish populating in Prince of Songkla University's new grown mangrove forest were investigated from January 2011 to December 2011 and then repeated annually. A total of 12 species under 6 families of crab and 11 species under 5 families of shellfish were recorded. The most abundant family of crab was Sesarmidae (64.18 %), followed by Ocypodidae, Varunidae, Macrophthalmidae, Portunidae and Dotillidae. *Episesarma mederi* (H. Milne Edwards, 1853) showed highest dominant species. In addition, the most dominant family of shellfish was Potamididae (13.79 %), followed by Melampidae, Assimineidae, Onchidiidae and Littorinidae. Sea snail (*Cerithidae quadrata*; Sowerby, 1866) presented the most dominant coastal mollusc species. Abundance and diversification crabs and mollusks show important component of food web of this type ecosystem. However, only trapped hold samples during low tide were collected but this preliminary finding enables reasonable specified regulation measures.

## 1. Introduction

Mangroves are recognized as a crucial link between land and sea. This coastal ecosystem is significant not only as a protection against shoreline erosion and disaster, but also as the home of rich biodiversity they represent that provides food, dwelling, and energy for animal and mankind [1]. Tropical mangroves are highly productive and greatly contributive to food chain and food web [2]. They provide habitats, nesting, place for breeding, and feeding grounds for countless species of fish, crustaceans, marine fauna and wild birds.

Pattani Bay is known as a unique ecosystem into which two main rivers, Pattani River and Yaring River, directly flow [1]. Rapid sedimentation process in the Pattani Bay area has resulted in the development of new plain mudflats naturally forming along the 2-mile coastline of Pattani River mouth which has yet to be occupied by any plant species. This huge sediment land has been claimed de facto as the territory of Prince of Songkla University of Pattani. In 1987, the university launched a project to plant mangroves along the coasts of Pattani bay. The formerly mudflat in which sediments deposited now is a 40-hectare habitat of planted mangroves. It has been well-managed as a mangrove study center and supported as a feeding and shelter protected area for many organisms. However, there has been scarcely any published report on the species composition and distribution of fauna or habitat

IOP Conf. Series: Earth and Environmental Science **137** (2018) 012052 doi:10.1088/1755-1315/137/1/012052

restoration in human-planted mangroves. Moreover, there has been no scientific information regarding species composition and distribution of the organisms living in this habitat, especially crabs and molluscs.

Therefore, this study aimed to investigate and identify the species composition and distribution of crabs and molluscs in human-planted mangrove forests.

## 2. Method

The study area was located in the northern part of Prince of Songkla University of Pattani in southern Thailand (6° 53′ 27.61″ N, 101° 14′ 21.29″ E–6° 53′ 21.01″ N, 101° 14′ 29.01″ E and 6° 52′ 47.87″ N, 101° 13′ 37.32″ E—6° 52′ 41.25″ N, 101° 13′ 39.69″ E), connected to Pattani Bay and South China Sea (Figure 1). Sampling was conducted in each of three sub-sampling sites by digging holes sized 1.0x1.0 $m^2$ at a depth of 50 cm for trapping crabs and molluscs in this area. The holes were left overnight, and the sample trapped in the holes was collected in the next morning. The specimens were identified according to Ng [3], Swennen *et al.* [4], Sanpanich *et al.* [5], Sottiyothin and Kulabtong [6], weighted, counted and stored in 10% formalin solution. Additional observation was carried out on the organisms living on the bottom surface outside the trap holes for a comparison purpose. All activities were investigated from January 2011 to December 2011 and then repeated annually until 2013.

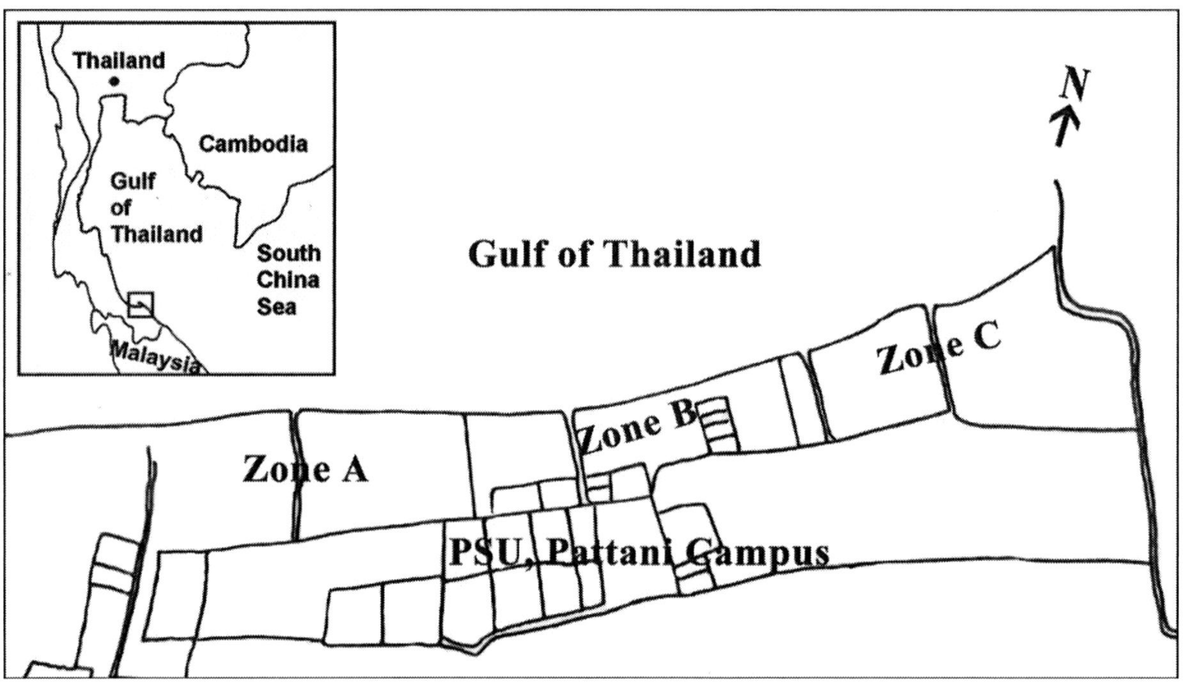


**Figure 1.** Location of the study area in the Campus of Prince of Songkla University, Pattani, Thailand.

## 3. Results

A total of 24 species and 11 families of crabs and molluscs were collected. Out of these, 73.7% were crabs and 26.3% were molluscs of various families. The crab class Malacostraca was the dominant group, and most of the molluscs were gastropods. Decapoda was the most dominant order, followed by Mesogastropoda, Acteophila, and Neotaenioglossa. In addition, Sesarmidae was the most dominant family of crab, followed by Potamididae, Melampidae, Ocypodidae, Varunidae, Macrophthalmidae, Assimineidae, Onchidiidae, Portunidae, Dotillidae, and Littorinidae (table 1).

IOP Conf. Series: Earth and Environmental Science **137** (2018) 012052 doi:10.1088/1755-1315/137/1/012052

**Table 1.** Relative contribution by family of crab and mollusks collected in mangrove forest of Prince of Songkla University, Pattani.

| Animal | Order | Family | Number | % |
|---|---|---|---|---|
| Crab | Decapoda | Sesarmidae | 405 | 64.18 |
| | | Macrophthalmidae | 13 | 2.06 |
| | | Ocypodidae | 20 | 3.17 |
| | | Portunidae | 6 | 0.95 |
| | | Dotillidae | 4 | 0.63 |
| | | Varunidae | 17 | 2.69 |
| Mollusks | Acteophila | Assimineidae | 9 | 1.43 |
| | Neotaenioglossa | Littorinidae | 3 | 0.48 |
| | Acteophila | Melampidae | 59 | 9.35 |
| | Onchidiacea | Onchidiidae | 8 | 1.27 |
| | Mesogastropoda | Potamididae | 87 | 13.79 |

The most numerous species of crabs was *Episesarma mederi* (average length of 1.48 cm), followed by *Parasesarma plicatum* (average length of 1.32 cm), *Perisesarma eumolpe* (1.58 cm), *Metaplax elegans* (1.25 cm) and *Macrophthalmus latreillei* (1.56 cm). The most numerous species of molluscs was *Cerithidae quadrata* (average length of 3.02 cm), followed by *Cassidula nucleus* (average length of 1.13 cm), *Cassidula* sp. (1.46 cm) and *Cassidula aurisfelis* (1.61 mm) (table 2).

Additionally, the observation of crabs and molluscs outside the sampling holes found that some species that were not trapped and recorded in the hole appeared in this habitat, including *Scylla serrata*, *Scylla tranquebaricus*, *Varuna yui*, *Uca* perplexa, *Uca urvillei*, *Paracleistostoma* sp., *Cerithidae cingulata*, *Ellobiumaurismidae* and *Assiminea brevicula*.

IOP Conf. Series: Earth and Environmental Science **137** (2018) 012052 doi:10.1088/1755-1315/137/1/012052

**Table 2.** Distribution and number of species of the shellfish in mangrove forest, Prince of Songkla University Pattani campus.

| Order /Family | Species name | N | Sex M | Sex F | Mean length (cm) | SD | Length range (cm) | Mean weight (g) | SD | Weight range (g) |
|---|---|---|---|---|---|---|---|---|---|---|
| **Decapoda** | | | | | | | | | | |
| Dotillidae | *Ilyoplax orientalis* (De Man, 1888) | 4 | 4 | 0 | 1.33 | 0.09 | 1.2-1.4 | 1.15 | 0.33 | 0.69-1.53 |
| Acrophthalmidae | *Macrophthalmus latreillei* (Desmarest, 1822) | 13 | 8 | 5 | 1.56 | 0.28 | 0.9-2.1 | 1.4 | 0.79 | 0.12-3.61 |
| Ocypodidae | *Uca annulipes* (H. Milne Edwards, 1837) | 11 | 6 | 5 | 1.76 | 0.33 | 1.3-2.5 | 2.22 | 0.78 | 0.97-3.48 |
| Ocypodidae | *Uca forcipata* (Adams & White, 1848) | 1 | 1 | 0 | | | 2.2 | | | 5.3 |
| Ocypodidae | *Uca paradussumieri* (Bott, 1973) | 8 | 7 | 1 | 1.53 | 0.25 | 1.2-1.8 | 1.64 | 0.74 | 0.87-2.6 |
| Portunidae | *Scylla olivacea* (Herbst, 1796) | 6 | 6 | 0 | 4.43 | 1.31 | 2.8-6.5 | 21.2 | 16.7 | 4.2-52 |
| Sesarmidae | *Episesarma mederi* (H. Milne Edwards, 1853) | 234 | 206 | 28 | 1.48 | 0.79 | 0.4-4 | 4.58 | 9.39 | 0.02-56 |
| Sesarmidae | *Episesarma versicolor* (Tweedie, 1940) | 8 | 5 | 3 | 2.78 | 0.48 | 1.9-3.6 | 18.6 | 9.16 | 10.93-40 |
| Sesarmidae | *Episesarma chengtongense* (Serene &Soh, 1967) | 2 | 2 | 0 | 3.2 | 0.12 | 3.1-3.3 | 30 | 3.46 | 27-33 |
| Sesarmidae | *Parasesarma plicatum* (Latreille, 1806) | 106 | 62 | 44 | 1.32 | 0.29 | 0.6-2 | 1.47 | 1.01 | 0.1-4.56 |
| Sesarmidae | *Perisesarma eumolpe* (De Man, 1895) | 55 | 40 | 15 | 1.58 | 0.23 | 1-2.1 | 2.8 | 1.64 | 0.74-10.57 |
| Varunidae | *Metaplax elegans* (De Man, 1888) | 16 | 10 | 6 | 1.25 | 0.31 | 0.6-2.2 | 1.12 | 1.23 | 0.2-5.61 |
| Varunidae | *Metaplax dentipes* (Heller, 1830) | 1 | 0 | 1 | | | 1 | | | 0.46 |
| **Acteophila** | | | | | | | | | | |
| Melampidae | *Cassidula aurisfelis* (Brugulere, 1789) | 12 | | | 1.61 | 0.24 | 1.2-2 | 0.58 | 0.27 | 0.1-1 |
| Melampidae | *Cassidula* sp. | 21 | | | 1.46 | 0.19 | 1.1-1.9 | 0.55 | 0.17 | 0.3-0.9 |
| Melampidae | *Cassidula nucleus* (Gelin, 1791) | 23 | | | 1.13 | 0.21 | 0.5-1.4 | 0.27 | 0.09 | 0.06-0.5 |
| Melampidae | *Cassidula mustelina* (Deshayes, 1830) | 3 | | | 2.07 | 0.05 | 2-2.1 | 1.31 | 0.08 | 1.22-1.39 |
| **Littorinimorpha** | | | | | | | | | | |
| Assimineidae | *Assiminea brevicula* (Pfeiffer, 1854) | 9 | | | 0.41 | 0.13 | 0.2-0.6 | 0.04 | 0.02 | 0.01-0.07 |
| **Mesogastropoda** | | | | | | | | | | |
| Potamididae | *Cerithidae* sp. | 3 | | | 2.67 | 0.88 | 2.1-3.8 | 2.41 | 0.7 | 1.85-3.3 |
| Potamididae | *Cerithidae quadrata* (Sowerby, 1866) | 77 | | | 3.02 | 0.44 | 2.2-3.8 | 1.83 | 0.59 | 0.9-3.82 |
| Potamididae | *Cerithium coralium* (Kiener, 1841) | 6 | | | 3.3 | 0.18 | 3-3.5 | 2.24 | 0.37 | 1.8-2.87 |
| Potamididae | *Telescopium telescopium* (Linnaeus, 1758) | 1 | | | | | 6.7 | | | 23 |

IOP Conf. Series: Earth and Environmental Science **137** (2018) 012052 doi:10.1088/1755-1315/137/1/012052

| | | | | | | | | |
|---|---|---|---|---|---|---|---|---|
| **Neotaenioglossa** | | | | | | | | |
| Littorinidae | *Littoraria melanostoma* (Gray, 1839) | 3 | 1.8 | 0.15 | 1.7-2 | 0.61 | 0.02 | 0.6-0.64 |
| **Onchidiacea** | | | | | | | | |
| Onchidiidae | *Onchidiidae* sp. | 8 | | | | 4.94 | 1.36 | 3.23-6.45 |

IOP Conf. Series: Earth and Environmental Science **137** (2018) 012052 doi:10.1088/1755-1315/137/1/012052

## 4. Discussion and Conclusion

The inner part of PSU's planted mangrove forest, once a bared mudflat, was converted into a productive home of mangrove trees as a feeding, living or refuge area of wild crabs and molluscs. At least 24 species from 11 families of crabs and molluscs were members of this human-modified rehabilitative forest.

Most of the crabs found in this habitat were ground-dwelling species. Sesarmidae, especially *Episesarma mederi* and *Parasesarma plicatum,* dominated the daily catch. They were important economic species collected by the villagers and fishermen living near the university. They were widely used to make native favourite menu called salted crab. These crabs are "semi-terrestrial" and tolerant to fluctuating salinity [7]. They can breathe by circulating the water in the mesh for oxygenation and then back to their gills for air exchange. Although they were more abundant in number, these crabs were rarely found because they are very sensitive to disturbance, and always escape into cracks or tree roots. They feed mainly on rotten mangrove leaves, seeds and branches [8], [9]. Furthermore, they are capable of consuming a large amount of mangrove litter and generating a significant relation of the energy flow within the mangrove forest. Sesarminae crabs usually produce air circulation around mangrove roots due to their burrow-doffing behaviour. Another economically important species is *Scylla olivacea.* This species also digs their burrow inside mangrove forests. However, most of the collected crabs were considered as non-economically important species but more appreciated for their ecological values. For molluscs, most of them were gastropods and slugs.

The presence of *Cerithidae quadrata* presented mollusc-dominated community. These snail are capable of climbing up a tree to feed on algae growing on the roots and stems of mangrove. They are consumed as food in rural areas, but not in Pattani region. They always show a relation with *Cassidula nucleus* and *Cassidula aurisfelis.* These small-sized snails prefer to climb up the mangrove trees and feed by grazing algae growing on the trees and on the ground.

In addition, the abundance and diversity of crabs and molluscs occurring inside the PSU's planted mangrove forest are an important part of the food web of this mudflat ecosystem. However, this study has a limitation, i.e., the sample was only trapped into the holes during low tide. Further study will need more complete data collection and specimen identification of this mangrove forest.

## 5. References

[1] Hajisamae S and Yeesin P 2014 *Raffles Bull. Zool.* **62** 107-114
[2] Bouillon S and Connolly RM 2009 Carbon exchange among tropical coastal ecosystems Springer 45-70
[3] Ng P K L 1998, FAO Species Identification Guide for Fishery Purpose vol 3, ed K.E. and Niem, V.H. 87
[4] Swennen C, Moolenbeek RG, Ruttanadakul N, Hobbelink H, Dekker H and Hajisamae S 2001 The Molluscs of the Southern Gulf of Thailand. Thai Studies in Biodiversity (The Biodiversity Research and Training Program, Bangkok) **4** 1-210
[5] Sanpanich K Wells F E and Chitramvong Y 2004 *Records of the Western Australian Museum* **22** 241-251
[6] Sottiyothin S and Kulabtong S 2011 *Veridian E-Journal SU.* **4** 911-916
[7] Naiyanetr P 1998 Check List Crustacean Fauna in Thailand. Office for Environmental Policy and Planning Bangkok 13-23
[8] Printrakoon C 2010 Biogeographic distribution of gastropod associated mangrove from Thailand, Tropical Natural History. 17th World Congress of Malacology, 18-24 July 2010. Royal Phuket City Hotel Phuket Thailand

IOP Conf. Series: Earth and Environmental Science **137** (2018) 012052 doi:10.1088/1755-1315/137/1/012052

[9] Printrakoon C 2011 *Burapha Sci. J.* **16** 114-124

**Acknowledgements**
We would like to thank Associate Professor Dr. Sukree Hajisamae for his assistance, valuable discussion and encouragement. Mr. Suriya Kosuthum for some sample collecting. Moreover, we also thank to all fishermen in Rusamilae Village who supported us catch crabs. Thanks to the Faculty of Science and technology for research facility support.

IOP Conf. Series: Earth and Environmental Science **137** (2018) 012053 doi:10.1088/1755-1315/137/1/012053

# Monogenean parasites on cantang grouper (*Epinephelus fuscoguttatus- lanceolatus*) wilture in floating net cage for mariculture center Lombok, West Nusa Tenggara, Indonesia

**N T B Dewi [1], I F Aryadi [1], A F T Arrizal [1], D R Mardika[1], P A Syahputra[1], S Subekti [2], Kismiyati [3] and P D W Sari [3]**

[1]Bachelor programme in Aquaculture, Faculty of Fisheries and Marine, Airlangga University, Jl. Airlangga 60115, Surabaya, East Java, Indonesia.

[2]Departement of Marine, Faculty of Fisheries and Marine, Airlangga University, Jl. Airlangga 60115, Surabaya, East Java, Indonesia.

[3]Departement of Fish Health Management and Aquaqulture, Faculty of Fiheries and Marine, Airlangga University, Jl. Airlangga 60115, Surabaya, East Java, Indonesia.

E-mail : kismiyati@fpk.unair.ac.id

**Abstract.** Cantang groupers (*Epinephelus fuscoguttatus-lanceolatus*) is a hybrid grouper species. This hybrid species has several advantages, one of the advantages is it has better resistance to disease. However, it does not guarantee that grouper fish can not be infected by parasite that leads to disease. This study aims to determine the prevalence and intensity of monogenean parasites that infest cantang groupers (*Epinephelus fuscoguttatus-lanceolatus*) from floating net-cages farm, at Mariculture Center Lombok, West Nusa Tenggara Indonesia. The study conducted on grouper with size ranging from 15 to 27.5 cm, body weight ranging from 55 to 233 g. The result showed that monogeneans was found in cantang groupers were *Neobenedenia girellae* (3.8-5.4 x 1.5-2.5 mm) that infested the fins, eyes, and fish skin and *Pseudorhabdosynochus seabassi* (0.52-0.68 x 0.12-0.25 mm) only found on gills lamella. The highest prevalence of monogeneans was found in mixed infestation of *P. seabassi* and *N. girellae* 50% in most of categories. The highest intensity of monogeneans was 10.09 individuals/fish in single infestation of *Pseudorhabdosynochus seabassi*.

## 1. Introduction

Grouper is one of the most important fishery commodities that have a very high export demand in Asia. The number of grouper fish production in Indonesia during 2015 was up to 8,972 tons and had increased to 15,089 tons in the third quarter of 2016 [1]. The total number of exported alive groupers during 2016 reached 4,006,852 tons [2].

Grouper have many species, one of them is Cantang Grouper (*Epinephelus fuscoguttatus-lanceolatus*). Cantang Grouper is a hybrid product of female tiger grouper *E. fuscoguttatus* and male kertang grouper *E. lanceolatus*. In addition, cantang groupers has another advantages, such as better resistance to disease, better tolerance to environmental changes, and can easily live in the narrow space [3].

Cantang groupers which is cultivated using floating net-cages system can not easily avoid ectoparasite infestation. Parasites in fish rarely cause death, but it can lead to several body injuries and open the access of other pathogens such as viruses and bacteria. Monogenean parasites infestation on cultivated activities can cause various losses such as economic, health and ecological losses [4]. Monogenean parasites commonly infest cantang groupers such as *Benedenia epinepheli, Diplectanum grouperi, Dactylogyrus, Neobenedenia girellae, Haliotrema epinepheli,* and *Pseudorhabdosynochus seabasi* [5]. Based on this background, it is necessary to conduct a research on the prevalence and intensity of Monogenean parasites in the cantang grouprs which is located at the Mariculture Center Lombok, West Nusa Tenggara.

## 2. Methodology

Life specimens of *E. fuscoguttatus-lanceolatus* were taken from floating net-cages farm at Mariculture Center Lombok, West Nusa Tenggara. These samples were taken from March to April 2017. The number of samples were taken from 10% of the population (400 fishes). Therefore, samples used in this study were 40 groupers. The fish samples were examined for ectoparasites under dissecting microscope. All the obtained parasites were fixed in 5% of alcohol glycerin, stained with acid carmine, dehydrated in a graded ethanol series, cleaned and finally mounted by Hung's I and Hung's II in accordance to Kuhlmann [6] modified methods. Illustrations were made with the aid of a camera lucida. The Monogeeneans infestation prevalence was calculated according to Bush [7] and intensity by Dogiel [8]

## 3. Results and Discussion

In the present study of the occurrence of monogenean parasite on cantang groupers (*Epinephelus fuscoguttatus-lanceolatus*) were monitored. Some of them are *Neobenedenia girellae* which infested skin, fins, eyes and *Pseudorhabdosynochus seabassi* which attached to the gill's lamella. Occasional haemorrhagic spots were noticed in the head region near to the eyes of the fish. Heavily infested fish showed excessive mucus secretions on the body surface. *Neobenedenia girellae* and *Pseudorhabdosynochus seabassi*, the two monogeneans species can be seen in figure 1.

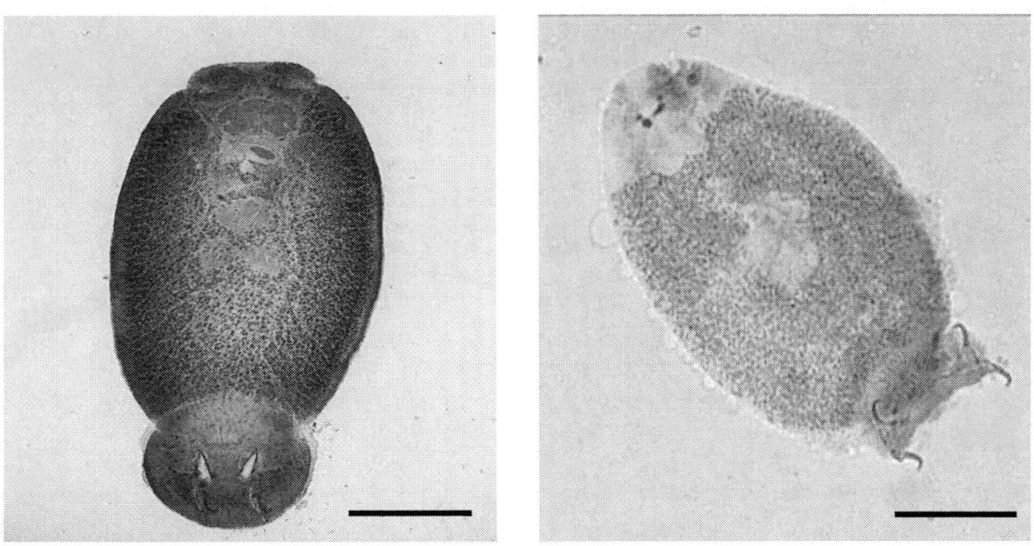

(A) (B)

**Figure 1**. (A) Adult worm of *Neobenedenia girellae,* scale bar = 100 μ m (B) *Pseudorhabdosynochus seabassi,* scale bar = 50 μm.

IOP Conf. Series: Earth and Environmental Science **137** (2018) 012053 doi:10.1088/1755-1315/137/1/012053

*3.1. Description of Neobenedenia girellae*

Based on 14 adult worm samples that had been examined, the parasite found in the fish was *Neobenenedia girellae* (Monogenean trematode, Monopisthocotylea order, Capsalidae family). *Neobenedenia girellae* had elliptical body and dorsoventral-shaped with total body length 3.8-5.4 mm and width was 1.5-2.5 mm. Anterior attachment organ (sucker) 0.28-0.37 mm long and 0.29-0.34 mm, another sucker on posterior was ophisthaptor 0.84-1.1 mm long and 0.78-1.42 mm wide. Haptor had oval shaped and aseptate with a pair of sclerites accessories, anterior hamuli and posterior hamuli in linier series. Seven pairs of hooklets arranged radially in haptor were not clear. Sclerites accesory hadscoop-shaped and distally-branched with blunt tips 0.15-0.21 mm long. Anterior hamuli 0.20-0.31 mm long, strong, sharp, posterior terminus recurved directed points. Posterior hamuli 0.08-0.11 mm long, straight and fine recurved point.

Two pairs of eyes between Anterior attachment organ and pharynx. Pharynx glandular had six lobes ranging from 0.36-0.42 mm length and 0.41-0.58 mm width. Intesninal crura dendritic medially and laterally, extending posteriorly to the end of proper body, terminating blindly. Two testes was in mid body, lobed, post ovarian, different size 0.28-0.46 mm long. Vas deferens ascending sinistral to ovary and vitelline reservoir. Male copulatory organ situated in posterior to pharynx, consisting of a penis, a large pear shaped muscular sac containing seminal vesicle and an accessory gland reservoir. Penis slightly tapering anteriorly. Ovary 0.18-0.38 mm long, 0.24-0.48 mm wide, medial. Vagina absent. Vitelline reservoir transversely wide anterior to ovary. Vitellaria follicular extending over its body, from behind anterior attachment organ to posterior end of the body.

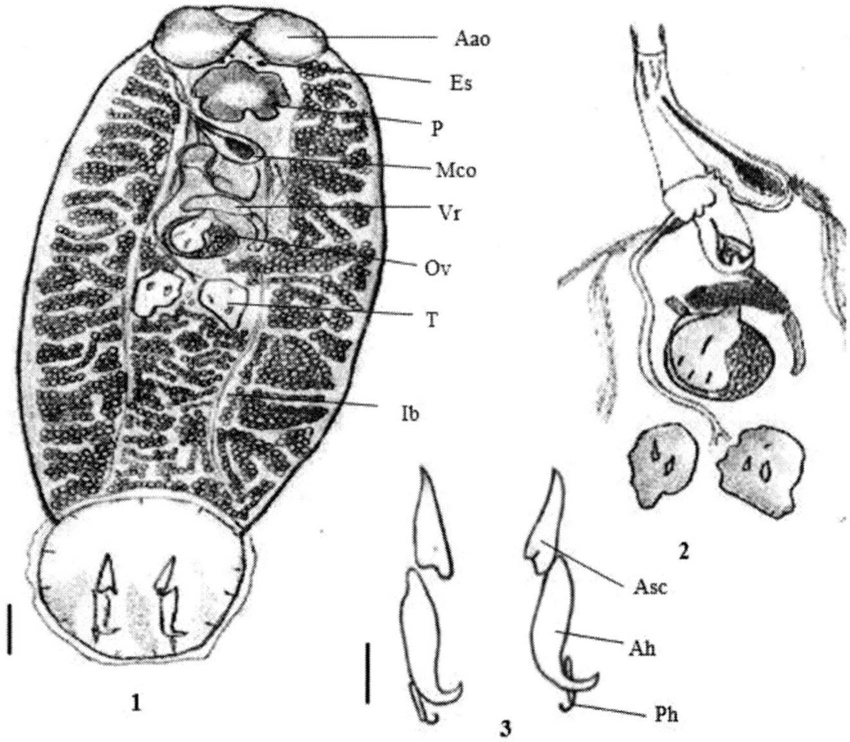


**Figure 2**. *Neobenedenia girellae*. Images taken with binocular microscope equipped with camera Lucida.

1) *Neobenedenia girellae* ventral view, scale bar 0= 100 μ m 2) Reproductive organs, scale bar = 100 μ m 3) Median sclerites, scale bar = 10 μ m, Description: (Aao) *Anterior attachment organ*, (Es) *Eye spot*, (P) *Pharynx*, (Ov) *Ovary*, (T) *Testis*, (Ib) *Intestinal branch*, (Mv) *Marginal valve*, (Vr) *Vitelline resevoir* (Asc) *Accessory sclerite*, (Ah) *Anterior hamuli*, (Ph) *Posterior hamuli*.

IOP Conf. Series: Earth and Environmental Science **137** (2018) 012053 doi:10.1088/1755-1315/137/1/012053

### *3.2 Description of Pseudorhabdosynochus seabassi*

*Pseudorhabdosynochus seabassi* was Monogenean Trematode, Dactylogyrydea order, Diplectanidae family. *Pseudorhabdosynochus seabassi* had an ovoid body, slender-shaped, with total length including haptor to ovary 0.52-0.68 mm and width 0.12-0.25 mm. Tegument had directed scales in the posterior half of the body. Two pairs of dissimilar eyes spot were in the anterior of the body. Mouth was in median-subterminal part of the body. This worm had a short esophagus which connects the posterior part of the body with pharynx, a simple intestinal caeca which blindly terminates on the posterior margin of vitelline field. The glandular-shaped pharynx had 0.05-0.071 mm for total length and 0.047-0.107 mm for total width. The subpherical testes was located in the posterior of ovary. Vas deferens was located in the left side of testes, extending along the anterior part of the body and widen in the seminal vesicle part which was located in the central part of the body. Vas deferens also connected the testes to the quadriloculate organ which its terminal section enlarged. The quadriloculate organ consists of four sclerotized chambers. In addition, the fourth chamber has shorter size. The ejaculatory duct had spiral muscular which bands the posterior region of male copulatory bulb organ with male accessory reservoir. The characteristic of each species of *Pseudorhabdosynochus* was different copulatory organs [9, 10].

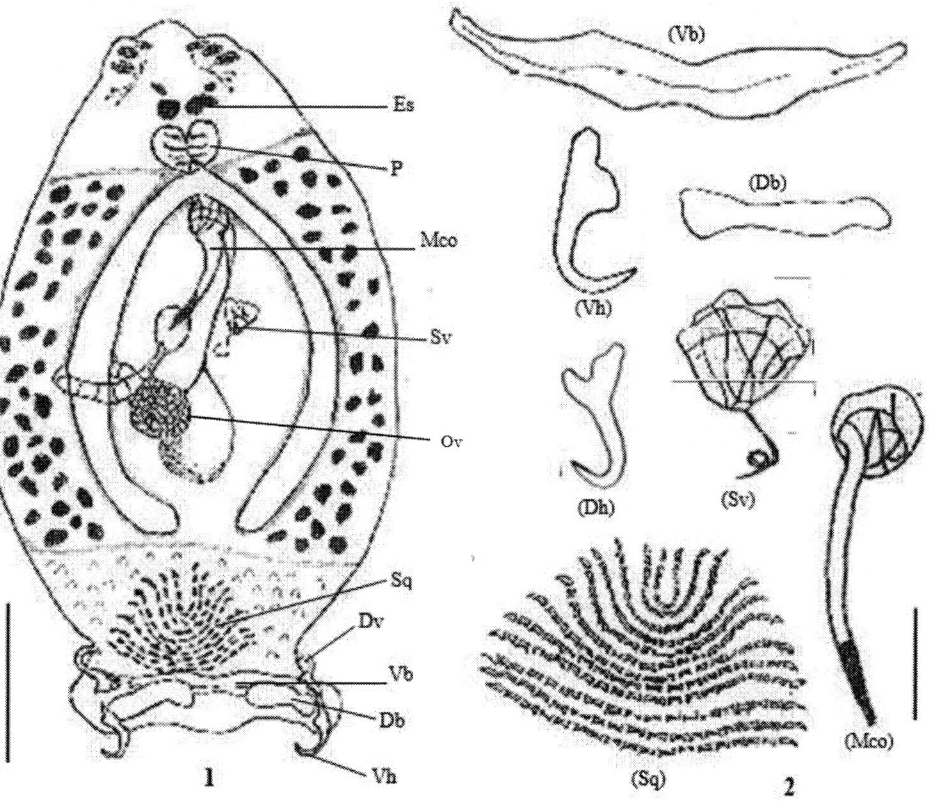


**Figure 3.** *Pseudorhabdosynochus seabassi*. The images were taken using binocular microscope equipped with camera Lucida.

(1) ventral view, scale bar= 100 µm, (2) Eye spot (Es), Male copulatory organ (Mco), pharinx (P), Sclerotised vagina (Sv), Squamodisc (Sq), Ventral hamuli (Vh). Bar Scale 100 µm. Male copulatory organ (Mco), Sclerotised vagina (Sv), Squamodisc (Sq), Ventral hamulus (Vh), Dorsal hamulus (Dh). Bar Scale 20 µm.

IOP Conf. Series: Earth and Environmental Science **137** (2018) 012053 doi:10.1088/1755-1315/137/1/012053

The equatorial ovary covered up the intestinal caecum on some parts. Vagina was divided into two parts. The first part of vagina was located close to ovary, partly sclerotized on the proximal part, and had a lotus flower shape. The second part was sclerotized on the distal part and fibriform-shaped. The vagina was located in sinistral part of the male copulatory organ with a complex sclerotized structure. The sclerotized vagina was 0.104-0.123 mm long and 0.124-0.161 mm wide, comprised sclerotized trumpet, primary canal, primary chamber, secondary canal and secondary chamber. Haptor was located in posterior part of the body, differentiated from the rest of the body, provided with two similar squamodiscs in ventral and dorsal part of the body. This haptor also had two pairs of lateral hamuli, three transverse bars and seven pairs of marginal hooklets but its shape were not clear. The Squamodiscs had a rounded shape and was made of rows of rodlets. Squamodisc had 9-12 concentric rows of rodlet and two inner most rows forming circles. The rodlet sometimes were equiped with visible spurs.

Bars were located between hamuli, strong and concaved on the central part. It also getting thinner in the edge of the body. Bars were straight, with flattened medial extremity and thick cylindrical lateral extremity. Ventral bar was 0.187-0.195 mm long, had flat shaped, massive, with slightly constricted median portion and blunt was extremities.The dorsal hamuli inner was 0.026-0.033 mm long and the outer was 0.04-0.045 mm long. The ventral hamuli inner length was 0.036-0.046 mm and outer length was 0.047-0.057 mm.

### *3.3 Prevalence, intensity andpredilection*

The parasite prevalence, intensity, amd predilection were recorded. There were three types of Monogenean parasites that infested each individual. *N. girellae* has found on fish body surface, eyes and fins. *P. seabassi* has found only in fish gills lamella. Prevalence and intensity of cantang grouper fish has infested by Monogeneans could be seen in table 1.

**Table 1.** Prevalence, Intensity and Predilection of Cantang Grouper Fish Infested by Monogeneans.

| No | Species of Ectoparasite | Infested fish (ind) | Prevalence (%) | Intensity (ind/fish) | Predilection of parasites |
|---|---|---|---|---|---|
| 1 | *Pseudorhabdosynochos Seabassi* | 11 | 27.5 | 10.09 | Gills lamela |
| 2 | *Neobenedenia girellae* | 4 | 10 | 2.75 | Eyes, fins and body surface |
| 3 | *P. seabassi x n. girellae* | 20 | 50 | 9.55 | |
| | Total | 35 | 87.5 | - | |

Grouper fish is a valuable and high-demanded fish in Southeast Asia. *Epinephelus coioiodes* and *Epinephelus bleekeri* are intensively cultured in sea cages in Vietnam. *E. coioiodes* is also a common species for grouper aquaculture in China. The hybrid grouper fish, the result of *E. fuscoguttatus* and *E. lanceolatus* crossbreeding, is the main commodity for grouper mariculture in Indonesia. Koesharyani, et al [11] reported that *Cromileptes altivelis* (humpback grouper) were infested by Capsalids monogeneans *(Benedenia epinepheli* and *Neobenedenia girellae)* in Indonesia. Wiyatno, et al [5] also reported *Neobenedenia sp.* mainly infested humpback grouper in sea cages in Situbondo, Indonesia. In Malaysia, *N. girellae* was also found in some species of *E. coicoides* (greasy groupers), *E. malabaricus*, and *E. bleekeri* [12]. *N. girellae* was first reported in California [13]. *N. girellae* (Capsalids) had been reported widely across the world, from subtropical to tropical areas [14]. This parasite demonstrated a very low host specificity [15]. The capsalids that infested on the body surface and fins can caused excessive and lost appetite. *N. girellae* was infested almost on head and eyes area. Hard infestation of *N. girellae* can caused opaque eyes and blindness [16].

Diplectanids are monogenean groups that mostly infest the gills of serranid fish and have a wide

IOP Conf. Series: Earth and Environmental Science **137** (2018) 012053 doi:10.1088/1755-1315/137/1/012053

world distribution [17]. Justine [18] mentioned that there were 12 monogeneans species on the highfin grouper *Epinephelus maculatus*, including 8 new species of the diplectanid *Pseudorhabdosynochus*. The latter study was recently described from Wu, et al [19] that *P. seabassi* have been found in *Lates calcarifer* that cultured in Guangdong Province, China. The report said that *Pseudorhabdosynochus* which infested grouper fish in Indonesia were rarely found, but according to the research from Ruckert, et al [20] *Pseudorhabdosynochus* sp. that infested *Epinephelus fuscoguttatus* from 25 (cultured grouper) and 30 (wild grouper) parasite species/taxa were identified in Lampung Bay, South Sumatra and the most abundant parasites are the monogeneans *Pseudorhabdosynochus epinepheli* and *Pseudorhabdosynochus lantauensis* for both, cultured and wild grouper. In addition, Wiyatno, et al and Fidyandini, et al [5, 9] stated that the gill worm *Pseudorabdosynochus* was also found infesting the Humpback grouper (*Cromileptes altivelis*) and Milkfish (*Chanos chanos*) at the Situbondo marine waters. Based on the results, cantang groupers that is cultivated in Mariculture Center Lombok has possibility infested by gill worm *Pseudorabdosynochus* because the groupers has imported from Situbondo area. *P. seabassi* caused haemorrahagic and abrasive lesions on the body. *Pseudorhabdosynochus* sp. infested all over the gills lamella, a heavy infestation from this parasite on gills lamella caused abrasive lesion and respiratory disorder which could caused death. All of fish samples has been infested by *P. seabassi were* showing clinical symptoms of excessive mucus on the body surface.

Cantang groupers (*Epinephelus fuscoguttatus-lanceolatus*) from 40 samples were examined, 35 samples were infested by *N. girellae* and *P. seabassi*. The total prevalence of monogeneans that infested cantang grouper fish was 87.5 %. Total intensity of monogeneans has found in cantang grouper was 8.94 individual/fish from 35 samples infested by monogeneans. The highest number of prevalence appeared in mixed infestation between *N. girellae* and *P. seabassi,* with 50 % prevalence. Therefore, the greatest number of intensity occurred in infestation of *P. seabassi* with 10.09 individual/fish. The mortality which was caused by infestation of monogeneans has rarely occurred in Cantang grouper. The intensity of monogenean parasites that infested grouper fish was classified as upper-middle category and severely did not show any haemorrhagic or abrasive lesions on body surface. The prevalence and intensity of the monogeneans could be associated with poor management of fish maintenance such as dirty sea cage condition and also uncontrolled water quality because it depends only on the season [5]. It also could be associated with the life cycle of Monogeneans that has a direct life cycle. Onchomiracidium has infective phase and swam freely to found the appropriate host. If parasite had found suitable host, it would stick and develop [21]. According to Kennedy [22], the important factor is the presence of host contact with the parasite. The parasite can survive when the host condition is appropriated.

## 4. Conclusion

The monogeneans which has been found infesting cantang gropuer were *Neobenedenia girellae* and *Pseudorhabdosynochos seabassi*. Total Prevalence of infestation ectoparasites was 87.5 %. The prevalence of *Neobenedenia girellae* was 27.5 % with intensity was 2.75 individual/fish. The prevalence of *Pseudorhabdosynochus seabassi* was 10% with intensity was 10.09 individual/fish. The prevalence of mix infestation between *N. girellae* and *P. seabassi* was 50 % and intensity was 9.55 individual/fish. Predilection of monogenean has infested the grouper (*Ephinephelus fuscoguttatus-lancolatus*) that were *N. girellae* on fish body surface, eyes, dorsal fin and pectoral fin. *P. seabassi* only on gills lamella.

## Acknowledgments

We thanks to all staff of the Mariculture Center Sekotong Lombok and Faculty of Fisheries and Marine, Airlangga University for use of facilities during the experiment.

## 5. References

[1] Directorate General of Aquaculture 2016 (Indonesia: Dirjen Perik. Bud. KKP) p 176

[2] Fish Quarantine and Inspection Agency 2017 BKIPMSTATISTIK *http://bkipm.kkp.go.id/bkipmnew/?r=stats/#_ops_volume/E/Ekor/m/0/2016/0/nm_umum* (accessed 14 April 2017)
[3] Brackishculture Center Situbondo 2012 *http://bbapsitubondo.com* (Accesed January 6, 2016)
[4] Sauyai K, Longdong S N J and Kolopita M E F *Bud. Per.* 2 **3** 76 – 8
[5] Wiyatno F H, Subekti S and Kusdarwati R 2012 *J. Ilmu Perik. Kel.* 4 **1**.
[6] Kuhlmann W F 2006 Preservation, staining, and mounting parasite spesiment p 8
[7] Bush A O, Lafferty K D, Lotz J M And Shostak A W 1997 *J. Parasitol* **83**.
[8] Dogiel V A G, Petrushevski G K and Polyanski I 1970 *Parasitology of Fishes* (Hongkong: T.F.H. Publiser) p 384
[9] Fidyandini H P, Subekti S and Kismiyati 2012 *J. Mari. Coast. Sci.* 1 **2**.
[10] Martinez V M V and Franco E F M 1998 *J. Folia Parasitologica* **45**.
[11] Koesharyani I, Zafran, Yuasa K and Hatai K 1999 *Fish Pathology* 34 **3**:165 – 6
[12] Seng L T and Colorni A 2002 Bruno D W and Lim L H S 2002 *Disease and Disorders of Finfish in Cage Culture* (U. K: CABI Publishing) p 354
[13] Hargis W H Jr 1955 A new species of Benedenia (Trematoda: Monogenea) from *Girella nigricans* the opaleye. *J. Parasitol* **41**:48-50
[14] Koesharyani I, Zafran, Yuasa K and Hatai K 1999 *Fish Pathology* 34 **3**.
[15] Ogawa K, Bondad-Reantaso M G, Fukudome M and Wakabayashi H 1995 *J. Parasitol* 81 **2** 223 – 7
[16] Ogawa K, Miyamoto J, Wang H, Lo C and Kou G 2006 *Fish Path* 41 **2**.
[17] Dang B T, Glen A B, Christoffer S and Bjorn B 2013. *Int. J. Aqua. Sci* 4 **2**.
[18] Justine J L 2007 Syst parasitol **66**.
[19] Wu X Y, Li A X, Zhu X Q and Xie M Q 2005 *Folia Parasitologica* **52**.
[20] Ruckert S, Klimpel S and Palm H W 2010 *Aqua. Res.* **4**.
[21] Kabata Z 1985 *Parasites and diseases of fish cultured in The Tropics* (London and Philadelphia: Tailor and Feancis Inc) p 381
[22] Kennedy C R 1975 Ecological Animal Parasitology (New York: Halsted Press) pp 65-73

IOP Conf. Series: Earth and Environmental Science **137** (2018) 012054 doi:10.1088/1755-1315/137/1/012054

# Preliminary design of a low-cost greenhouse for salt production in Indonesia

**A A Jaziri[1,3], Guntur[2], W Setiawan[4], A A Prihanto[1,3], A Kurniawan[2,3]**

[1]Department of Fishery Product Tecnology, Faculty of Fisheries and Marine Science, Universitas Brawijaya, Veteran Street, Malang 65145, East Java, Indonesia
[2]Faculty of Fisheries and Marine Science, Universitas Brawijaya, Malang 65145, Indonesia
[3]Coastal and Marine Research Center, Universitas Brawijaya, Malang 65145, Indonesia
[4]Kencana Tiara Gemilang, PT, Malang 65153, Indonesia

E-mail: andi_k@ub.ac.id

**Abstract.** Salt is an assential material of industry, not only in food industry point of view but also in various industries such as chemical, oil drilling, and animal feed industries, even less than half of salt needs used to household consumption. It is crucial to ensure salt production in Indonesia reaches the national target (3.7 million tons) due to relatively low technology and production level. Thus salt production technology is developed to facilitate farmers consisted of geomembrane and filtering-threaded technology. However, the use of those technologies in producing salt was proved less effective due to unpredictable weather conditions. Therefore, greenhouse technology is proposed to be used for salt production for several good reasons. This paper describes the preliminary design of a low-cost greenhouse designed as a pyramid model that uses bamboo, mono-layer and high density polyethylene plastics. The results confirmed that the yield of salt produced by greenhouse significantly incresed compared with prior technology and the NaCl content increased as well. The cost of greenhouse was IDR 5,688,000 and easy to assembly.

## 1. Introduction

Salt production technology is one of the important factors to produce salt in the world. There are several technologies have been applied extensively in many areas, particularly in Indonesia. Among the technologies, traditional technology is widely used as a simple salt production technology compared to other technologies due to low-cost, low-energy, and easy to use technology. However, traditional technology has many disadvantages such as low productivity, poor quality, and high efforts to produce salt. Prior studies reported that the productivity of salt produced by traditional technology can yield in the range of 60 tons/ha/year [1], the NaCl purity reaches less than 90 % [2], the colour of salt is brown to grey, and the price is generally inexpensive with the value of IDR 300-500 per kg [3].

Other technologies implemented in salt production are geomembrane liner and filtering-threaded technology. Geomembrane liner is a improved traditional technology by lining the salt pans with appropriate plastic membrane. The geomembrane liner commonly used is high density polyethylene (HDPE) [4]. It has a wide range of properties, for instances high anti-seepage coefficient, chemical stabability, anti-aging, resistant to high and low temperature, and high mechanical strengh. On the other hand, filtering-threaded technology is a modified geomembrane liner by passing water through a series of shallow channels to saturate the water into brine with an additional filtering membrane to

IOP Conf. Series: Earth and Environmental Science **137** (2018) 012054 doi:10.1088/1755-1315/137/1/012054

purify the water. Several prior studies have been explored that both the geomembrane liner and filtering-threaded technology can increase productivity, improve quality, dan shorten production time [5], but the filtering-threaded technology is more effective than traditional ones.

On the contrary, one of the most factors challenge salt production in Indonesia is changes in climate resulting in unreliable weather patterns. Those technologies of salt production described above cannot deal with unpredictable wether. Therefore, innovation of new salt production technology is needed to figure out unreliable weather patterns in producing salt. In this study, we focus on designing a new salt production technology by proposing low-cost greenhouse technology. According to Suseela and Rangaswami [6] greenhouse technology can maintains air temperatures and relative humidity. This paper aims to design a low-cost greenhouse that uses bamboo, mono-layer and high density polyethylene plastics. Through this a low-cost greenhouse may be breaktrhough in producing salt and facilitate farmers in several areas throughout Indonesia.

## 2. Material and Methods

### *2.1. Main materials required*

The following materials are required to construct a low-cost greenhouse for salt production:

- Bamboos, as a main framework to build a low-cost greenhouse. It is selected due to inexpensive, light weight, and non-corrosive material. In addition, it is widely used and readily available in many places, especially in Java, Indonesia. For building geenhouse, it is usually in medium size with diameter of 10-12 cm and the length of bamboos used in the range of 1 m, 7 m and 10 m.

**Figure 1.** Bamboo framework of a low-cost greenhouse.

- UV plastic, as a glazing material (plastic covering). Typically, the plastic covering the greenhouse is replaced many times before the framework fails due to corrosion and mechanical loading [7]. The specifications of UV plastic used in covering greenhouse are UV plastic of 14% in reducing intensity of sunlight, 200 micron thick, and semi transparent plastic. It is designed suitable for salt production greenhouse.

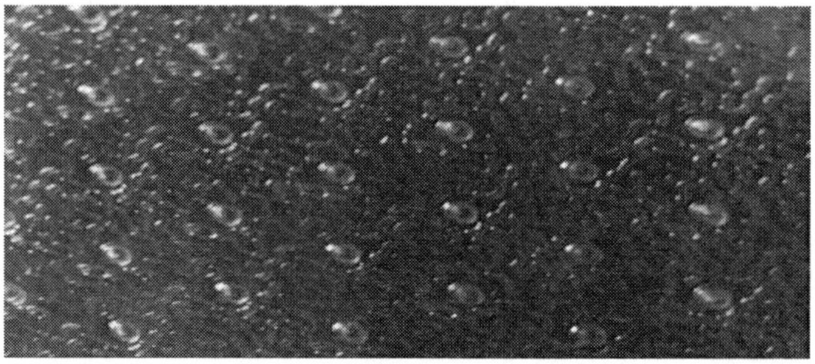

**Figure 2.** UV plastic with 14% reducing sunlight intensity.

- High density polyethylene (HDPE), as geomembrane liner in salt pan of greenhouse. It is generally fabricated by adding carbon black and antioxidant additives, which are extensively used to retard and inhibit the UV degradation and temperature degradation of geomembranes [8]. The specifications of HDPE can be seen in table 1.

**Table 1.** HDPE geomembrane properties.

| Properties | Test Method | GM 13 STANDARD 1,0 mm | EXTERNAL LABORATORY 1,0 mm |
|---|---|---|---|
| Laboratory | | | CTT Group * |
| Thickness | ASTM D5199-12 | | 1.16 mm |
| Density | ASTM 729-13 Method A | 0,94 g/cc | 0.941 g/cc |
| Carbon Black Content | ASTM D1603-14 | 2-3 % | 2,53% |
| Carbon Black Dispersion | ASTM D5596-03 (2009) | | 1 |
| Tensile Properties | ASTM D6693-04 (2015)e1 | | |
| MACHINE DIRECTION | | | |
| - Strength at Yield | | 15 kN/m | 17 kN/m |
| - Elongation at Yield | | 12% | 17% |
| - Strength at Break | | 27 kN/m | 42.5 kN/m |
| - Elongation at Break | | 700% | 772% |
| CROSS DIRECTION | | | |
| - Strength at Yield | | 15 kN/m | 17.1 kN/m |
| - Elongation at Yield | | 12% | 17% |
| - Strength at Break | | 27 kN/m | 41.6 kN/m |
| - Elongation at Break | | 700% | 786% |
| Tensile Strength | SNI ISO 37:2015 (IDT-2011) | | |
| Secant Modulus at 2% | ASTM D5323-92 (2011) | | |
| MACHINE DIRECTION | | | 308.4 kN/m |
| CROSS DIRECTION | | | 311 kN/m |
| Tear Resistance | ASTM D1004-13 | | |
| MACHINE DIRECTION | | 125 N | 160 N |
| CROSS DIRECTION | | 125 N | 153 N |
| Puncture Resistance | ASTM D4833/D4833M-07 (2013)e1 | 320 N | 320 N |
| Oxidative Induction Time | | | |
| Standard OIT | ASTM D3895-14 | 100 min | 151 min |
| Stress Crack Resistance | ASTM D5397-07(2012) (App) | 300 hours | >500 hours |

*) CTT Group Geosynthetics - Canada

### *2.2. Skecth of a low-cost greenhouse*

The sketch of a low-cost greenhouse by using AutoCAD mechanical desktop that is a computer-aided design application for 3D mechanical design, simulation, visualization, and documentation. The sketch designed as pyramid model, which is expected as a suitable model of greehouse for salt production located in seashore areas in Indonesia. It has length around 7.1 m with the hight of 3.5 m.

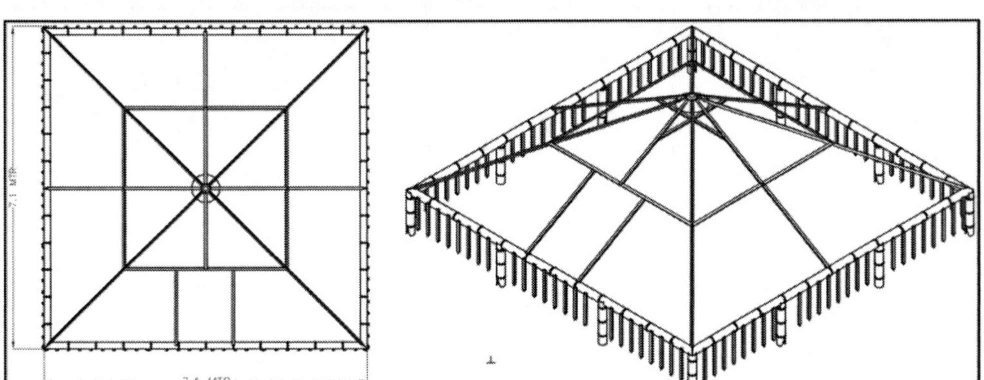


**Figure 3.** The sketch of a low-cost greenhouse for salt production.

## 3. Results and Discussion

### *3.1. Design of a low-cost greenhouse*

The greenhouse model is generally designed as the hoop and gable. These models are typically constructed of steel or galvanized steel as a framework of greenhouse, but the material costs are expensive and also the maintenance costs are relatively unaffordable. In addition, they are vulnerable to corrosion because steel or metal material tend to get rapid corrosive if it is placed in the seawater [9]. Another constuction of greenhouse is made by wooden frame, but it is heavy and high price, partticulary in East Java. Those materials are not suitable for developing greenhouse technology of salt production, in particular for salt farmers in East Java. Hence, a low-cost greenhouse proposed to facilitate farmers in producing salt with innovative technology, inexpensive, and easy to operate. The design of a low-cost greenhouse technology of salt production is displyed at figure 4.

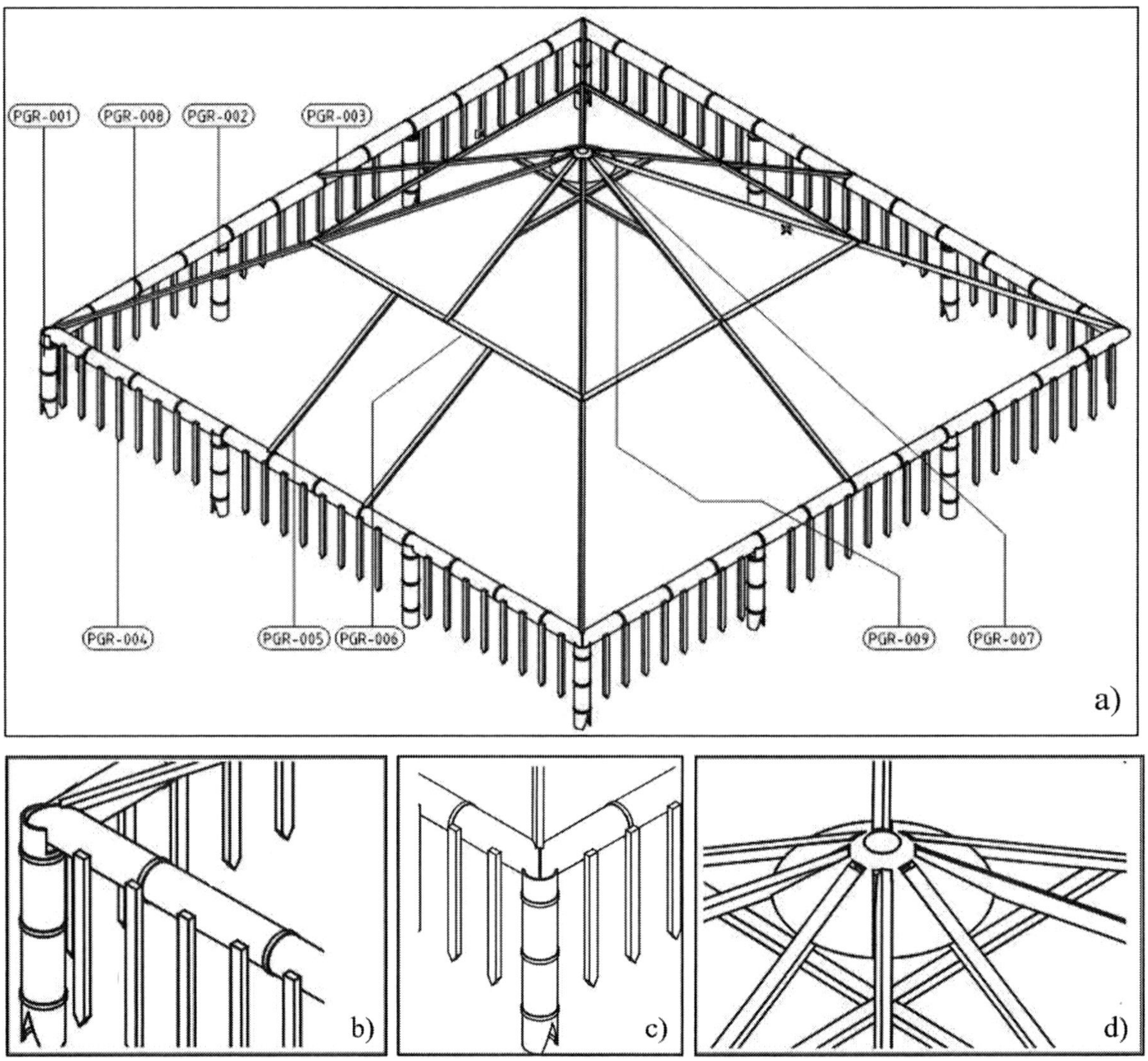


**Figure 4.** The skech of a low-cost greenhouse with pyramid model. a) the prespective view of pyramid greenhouse; b) the connecting in engle of pyramid greenhouse; c) the angle of pyramid greenhouse; d) the connecting roof of pyramid greenhouse.

The a low-cost greenhouse technology designed as a pyramid model, which was composed by several materials, such as bamboos, UV plastics and geomembrane liners. The pyramid model of a low-cost greenhouse was choosen due to suitable for sesshore areas, generally the salts pan located in the seashore, especially in East Java, Indonesia. The prespective view of the a low-cost grenhouse as shown in figure 3a that has connecting part in every angles and roof of the preliminary design by using bamboos as seen in figure 3b, 3c, 3d, respectively. Bamboos, as a primary construction of a low-cost greenhouse bacause they were low-price, light weight, non-corosive, and easy to obtain materials. The diameter size of bamboos in the range of 16 cm and the length of bamboos used in the range of 1 m, 7 m and 10 m, repectively. Furthermore, for glazing material of a low-cost greenhouse applied UV plastic with the specifications of UV plastic was 14% aiming to reduce intensity of sunlight, 200 micron thick, and semi transparent plastic. According to Ghany *et al.* [10] UV plastic used in glazing material for greenhouse is suitable to increase temperature and reduce sunlight intensity. For the liner in the a low-cost greenhouse performed by HDPE, which is generally manufactured by carbon black and antioxidant additives are widely used to inhibit the UV degradation and temperature degradation of geomembranes [4].

### *3.2. Field testing results*

After designing the greenhouse, we conducted the performance and feasibility testing of greenhouse technology in producing salt. The field testing performed in salt pan located in Lamongan, East Java, by investigating the salt productivity and quality. The data were described in table 2.

**Table 2.** Field testing results in producing salt.

| Parameters | Value |
|---|---|
| Average temperature | 45 °C |
| Productivity | 150 kg/ha |
| Sodium chloride/NaCl | 95 % |
| Colour | white |

The results of this study revealed that the productivity of salt was in the range of 150 kg/ha, which significantly increase compared with the traditional technology. According to Mustofa and Turjoyo [1] the salt productivity by using traditional technology around 60 tons/ha. In addition, the quality of salt produced by greenhouse improved with the content of sodium chloride about 90% and the colour of salt was white (purified). Arwiyah *et al.* [2] stated that the quality of salt produced by traditional technology displayed the NaCl content less than 90% and its colour is brown to gray.

### *3.3. Investment analysis*

Investment analysis is carried out to evaluate a total cost required to construct a low-cost greenhouse. As a low-cost greenhouse of salt production, we decided to use not only affordable materials, but also easy to assembly and operate for farmers. The invesment analysis is shown at table 3.

**Table 3.** Investment analysis of a lo-cost greenhouse technology.

| Materials | Lifetime | Total | Price per items (IDR) | Total (IDR) |
|---|---|---|---|---|
| Screws | 5 | 800 pcs | 350 | 280,000 |
| Bamboos | 5 | 20 pcs | 50,000 | 1,000,000 |
| labourage | - | 3 people | 200,000 | 600,000 |
| Geomembranes | 5 | 15 kg | 119,000 | 1,785,000 |
| UV plastics | 5 | 17 kg | 119,000 | 2,023,000 |
| | | | Total | 5,688,000 |

IOP Conf. Series: Earth and Environmental Science **137** (2018) 012054 doi:10.1088/1755-1315/137/1/012054

The investment analysis of a low-cost greenhouse of salt production might account for IDR 5,688,000 per units. It is more affordable that that of greenhouse constructed by metal materials. The predicted lifetime of geenhouse around 5 years, which is appropiate technology for salt production. The benefits of the low-cost greenhouse of salt production, as follows:

- Greenhouse costs: Less than IDR 5,688,000 for an affordable greenhouse with a minimum planting area of 64 square meter.
- Ease of assembly: Three individuals with minimal training should be able to assemble the greenhouse from the ground-up in a day.
- Lifetime: Minimal lifetime of five years for the greenhouse structure and five years for the glazing; only limited maintenance should be required by the salt farmer.
- High productivity: It can improve productivity compared with traditional technology.
- Purity of salt: Sodium chloride can reach more than 95%.

## 4. Conclusion

This study deduced that the preliminary design of a low-cost greenhouse for salt production designed as a pyramid model constructed by bamboos, UV plastic, and HDPE. It can increase the salt productivity and quality. In addition, it is an affordable, easy to assembly, and suitable for salt farmers.

## 5. References


[1] Mustofa and Turjoyo E 2015 *J. WIGA* **5** (In Indonesia)

[2] Arwiyah, Zainuri M, and Efendy M 2015 *J. Kel.* **8**

[3] Susanto H, Rokhati N, Gunawan and W Santoso 2015 Development of traditional salt production prosess for improving product quality dan quality in Jepara District, Central Java, Indonesia *Proceeding Environmental Science* **23**:175-178.

[4] Suhendra A 2016 *EJGE* **11**:4272-4280

[5] Bramawanto R, Sagala S L, Suheimi I R, and Prihatno 2015 *J. Segara* **11**:1-11(in Indonesia)
Sudarto 2011 *J. Triton* **7**:13-15

[6] Suseela P and Rangaswami M V 2012 *Karnataka J. Agric. Sci.,* **25** 100-103

[7] Tzouramani I, Mattas K, and Grafiadellis M 1995 *Economics of greenhouse construction decisions for cultivating tomatoes*, Acta Hort.

[8] Denis R, Tan D, and Cao D 2012 A literature review on lifetime prediction of thin hdpe geo-membranes in the exposed environment, *Proc. 5th Asian Regional Conference on Geosynthetics* (Thailand: Bangkok)

[9] Rahmanto W H, Gunawan, and Nuryanto R 2002 *J. Coast. Develop.* **5** 67-74

[10] Ghany A M A, Al-Helal I M, Alzahrani S M, Alsadon A A, Ali I M, and Elleithy R M 2012 *Sci. Wor. J. 2012*

IOP Conf. Series: Earth and Environmental Science **137** (2018) 012055 doi:10.1088/1755-1315/137/1/012055

# Proximate composition of several fish from Jatigede Reservoir in Sumedang district, West Java

**T Herawati, A Yustiati, A Nurhayati and R Mustikawati**

Faculty of Fisheries and Marine Science Padjadjaran University, Jalan Raya Bandung - Sumedang KM 21 Jatinangor Indonesia

E-mail: herawati.h19@gmail.com

**Abstract.** The aim of research is to determine the nutritional contents of fishes in the Jatigede Reservoir, the research uses survey method from November 2016 to October 2017. Sample collecting is done in the Wado district and Jatigede district. Samples are identified by its species and the nutritional contents are determined. The analyzed parameter includes water content, ash content, crude fiber content, protein content, fat content and carbohydrate content. Research shows that the nutritional contents of 13 species of fish vary. The water contents ranges between 67.50 % to 77.75 %, the highest percentage found in *Chana striata*. Ash contents ranges between 1.25 % to 4.87 %, the highest percentage found in *Rasbora argyrotaenia*. Crude fiber percentage cannot be found. Protein ranges between 14.43% to 21.93 %, the highest percentage found in *Chanos chanos*. Fat ranges 0.72 % to 6.86 %, the highest percentage found in *Hampala macrolepidota*. Carbohydrate ranges between 0.12 % to 2.64 %, the highest percentage found in *Osteochillus vittatus*.

## 1. Introduction

The fish found in the Jatigede Reservoir are derived from the indegenous fish of Cimanuk River, as well as fish deliberately stocked by the Ministry of Fisheries and Marine for specific purposes. Fish from the Cimanuk River consist of *Barbodes balleroides*, *Barbodes gonionotus*, *Osteochillus* sp, *Osteochillus vittatus*, *Rasbora argyrotaenia*, *Diplocheilichthys pleurotaenia* , *Tor duorensis*, *Cyprinus carpio*, *Hampala macrolepidotta*, *Mystacoleucus marginatus*, *Chana striata*, *Mystus gulio*, *Oreochromis niloticus*, *Tilapia mozambica*, *Trichogaster pectoralis*, *Liposarcus pardalis*, and *Mastacembelus erythrotaenia*. Fishes that were intentionally stocked are fishes that already inhabited the Cimanuk River and were restocked, such as *Osteochillus haseltii, Barbodes gonionotus, Barbodes balleroides, Tor duorensis, Cyprinus carpio,* and *Oreochromis niloticus.* Additionally, new fishes that previously did not inhabit the Cimanuk River were also stocked, such as *Barbodes orphoides, Pangasius hypopthalmus*, *Chanos chanos*, *Stenoparingodon iddella,* and *Helostoma teminkii.*

The number and species of fish stocked to Jatigede Reservoir in October 2015 to December 2015 numbered to 3,355.600 fish; in January 2016 to July 2016 5,240,000 fish; and again on 30th August 2016 to the amount of 2,105,000 fish. In the framework of World Water Day on April 27th 2017, 550,000 fish consisting of 1,000,000 *Pangasius hypopthalmus* fishes, 500,000 *Cyprinus carpio* fishes, 2,000,000 *Oreochromis niloticus* fishes, 350,000 *Helostoma teminkii* fishes, 200,000 *Barbodes* sp fishes, 650,000 *Osteochillus haseltii* fishes, and 250,000 *Chanos chanos* fishes were stocked.

The consumption level of fish in West Java in 2015 was only 27 kg per capita per year, meaning that it is still far from the ideal consumption of the Word Health Organization (WHO) standard [1]. The

IOP Conf. Series: Earth and Environmental Science **137** (2018) 012055 doi:10.1088/1755-1315/137/1/012055

Governor of West Java has utilized Jatigede Reservoir as a new center of fish production as a means to increase the fish consumption of West Java citizens. Capture fishery is the utilization system used at the Jatigede water reservoir.

Fish is one of the food commodities and is a source of protein and fat from the waters. The chemical composition of fish meat consists of water, ash, fat, carbohydrates, vitamins, and minerals, whose composition varies greatly between one fish species to the other. The differences in the chemical composition of fish meat for both the amount and the constituent components are due to the environmental and biological factors of the fish. Biological factors which are commonly called intrinsic factors are factors derived from the type of fish such as the species or class, age and sex of fish [2]. The species or class of fish is very influential on the difference in the composition of fish meat; each type of fish has a different kind of chemical composition in its meat. Age also affects the chemical component, the older the fish, the more fat content it will tend to have. Additionally, gender is closely related to gonad maturity; while an actively moving fish encourages it to find enough food to meet its energy needs. Habitats also affect the composition of meat; predatory fish will be different from plant-eating fish. Environmental factors that affect the nutritional composition of fish meat include the source of food. Based on the above description, research on the composition of nutrition in the fish inhabiting Jatigede Reservoir is necessary.

## 2. Methodology

Fish identification was carried out by using books by [3,4,5,6,7,8,9,10,11]. The types of fish taken as samples to analyze nutritional content were fish caught during the research. Samples were taken randomly from the catch of fishermen. The nutrient parameters analyzed were moisture content, ash, crude fiber, protein, fat, and carbohydrates. The analysis methods used: were the Oven Method for water and ash content, the Kjeldhal Method for protein content, and the Soxhlet Method for fat content. The work procedure of analysis refers to the guidelines of [12]. Carbohydrate content is determined by difference with the calculation = 100 % - (% water + % ash + % fat + % protein). Food standarization was referenced from the book by [13,14]. The composition of 100 gram material of fresh fish is 76 % water content, 17 % protein, and 4.5 % fat [13]. The composition of 100 gram material fresh fish, is 70 % to 85 % water content, 15 % to 25 % protein, 0.5 % to 7 % fat, 0.1 % to 1 % carbohydrate, and 0.1 % to 4 % ash [14]. The analysis was conducted at the Laboratory of Animal Feed Technology at the Faculty of Animal Husbandry of Padjadjaran University and at the laboratory of the Faculty of Agricultural Industrial Technology of Padjadjaran University.

## 3. Result and discussion

The results of the nutritional composition measurements consisting of water content, ash, protein, fat, and carbohydrates from 13 species of fish found in Jatigede Reservoir are as follows: The nutritional composition of Barbodes, *Barbodes balleroides*: 67.50 % water content, 4.36 % ash, 19.30 % protein, 6.54 % fat and 2.30 % carbohydrate. *Barbodes gonionotus*: 76.53 % water content, 3.63 % ash, 14.43 % protein, 5.37 % fat and 2.04 % carbohydrate. The research results of the Research Centre for Marine and Fisheries Product Processing and Biotechnology showed: 82 % water content, 1.5 % ash, 9.7 % protein, 5.1 % fat and 1.7 % carbohydrate. According to Muchtadi [2], 66% water content, 19 % protein, 13 % fat. The water content of *Barbodes gonionotus* fish that live in the Jatigede Reservoir is lower than the results from the Research Centre for Marine and Fisheries Product Processing and Biotechnology, but higher than the research result of [2]. The difference of water content in *Barbodes gonionotus* fish is caused by environmental factors associated with the osmoregulation process in the body of the fish. The levels of ash, protein, fat and carbohydrates of this particular fish found in Jatigede Reservoir was higher when compared to the research results of the Reserch Centre for Marine and Fisheries Product Processing and Biotechnology. The high levels of nutrients of Barbodes fish living in Jatigede Reservoir is caused by environmental factors, Cirata Reservoir is suspected to have food supplies in the form of detritus and other materials which is a source of quality feed and exists in huge numbers.

IOP Conf. Series: Earth and Environmental Science **137** (2018) 012055 doi:10.1088/1755-1315/137/1/012055

The composition of the fish nutrition of *Diplocheilichthys pleurotaenia* are as follows: 70.91 % water content, 4.73 % ash, 19.05 % protein, 2.89 % fat and 2.42 % carbohydrate. The nutritional composition of fish from the genus Osteochyllus, *Osteochyllus* sp is: 75.60 % water content, 3.24 % ash, 17.18 % protein, 3.59 % fat and 0.39 % carbohydrates. For *Osteochyllus hasseltii*: 70.35 % water content, 4.30 % ash, 18.63 % protein, 4.04 % fat and 2.64 % carbohydrate. In the case of Fish Hike: 73.30 % water content, 4.24 % ash, 17.67 % protein, 2.82 % fat and 1.97 % carbohydrate. The nutritional composition of the fish *Hampala macrolepidota*: 71.15 % water content, 2.21 % ash, 19.31 % protein, 6.86 % fat and 0.47 % carbohydrate. The nutritional composition of the *Rasbora argyrotaenia* fish: 70.80 % water content, 4.87 % ash, 19.88 % protein, 2.58 % fat and 1.87 % carbohydrate. Research results of the Research Centre for Marine and Fisheries Product Processing and Biotechnology measured the Rasbora fish protein level at 10 %, and fat at 3.2 %. Compared to these results, levels of Rasbora fish protein that inhabit Jatigede Reservoir are higher with a smaller fat content. The nutritional composition of *Mastacoleucus marginatus* was measured at: 75.93 % water content, 3.59 % ash, 15.64 % protein, 4.40 % fat and 0.74 % carbohydrate. When compared with the results from the Research Centre for Marine and Fisheries Product Processing and Biotechnology, the fish protein and fat content of *Mastacoleucus marginatus* found at Jatigede Reservoir is higher, because results from the Research Centre for Marine and Fisheries Product Processing and Biotechnology showed protein to be at 14.8 %, and fat 2.3 %.

*Chanos chanos* has the following nutritional composition: 74.22 % water content, 1.98 % ash, 21.93 % protein, 1.66 % fat and 0.21 % carbohydrate, while the research results of the Research Centre for Marine and Fisheries Product Processing and Biotechnology measured 75.85 % water content, 2.81 % ash, 21.49 % protein, 0.72 % fat and 0.11 % carbohydrate. Water content was found to be at 74.0 %, protein 20.0 %, fat 4.8 % [2]. When comparing research results of protein, fat and carbohydrate levels between the milkfish from Jatigede Reservoir, the Research Centre for Marine and Fisheries Product Processing and Biotechnology and Muchtadi R, the milkfish from Jatigede Reservoir measured higher. The nutritional composition of *Trichogaster pectoralis* from Jatigede Reservoir had a water content of 71.31 %, 3.73 % ash, 18.96 % protein, 4.18 % fat and 1.82 % carbohydrate. Protein content performed by the Agency for Biotechnology and Food Processing Marine and Fisheries was 8.79 %, indicating that the protein content of the Three Spot Gourami from Jatigede Reservoir is higher. The nutritional composition of *Oreochromis niloticus* from the Jatigede Reservoir is as follows: 76.81 % water content, 1.67 % ash, 18.75 % protein, 1.58 % fat and 1.19 % carbohydrate. The nutrition of *Chana striata* from the Jatigede Reservoir: 77.75 % water content, 2.85 % ash, 18.57 % protein, 0.72 % fat and 0.12 % carbohydrate. *Mystus gulio* from Jatigede reservoir: 75.95 % water content, 1.25 % ash, 16.90 % protein, 4.94 % fat and 0.96 % carbohydrate.

Water content is very influential on the quality of food, hence water content is often removed or reduced by means of evaporation or drying during food processing. Fish found in the Jatigede Reservoir with the lowest water content was obtained from *Barbodes balleroides* at 67.50 %, while the highest was obtained from *Chanos chanos* at 77.75 % (figure 1). The content of water in foodstuffs is an indication of food resistance against microbial attack, therefore the food to be processed should be dry so that resistance to microbial growth can be suppressed [13].

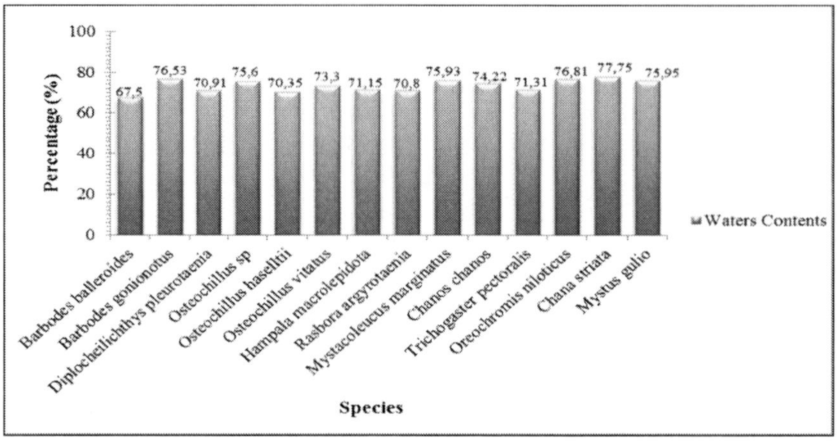


**Figure 1.** Water contents of fish in Jatigede reservoir.

*Mystus gulia* had the lowest ash level at 1.25 %, while the highest was *Rasbora argyrotaenia* at 4.87 % (figure 2). The high level of ash in *Rasbora argyrotaenia* comes from bones and scales, because the fish is rather small so every organ consisting of head, body, tail and scales were measured.

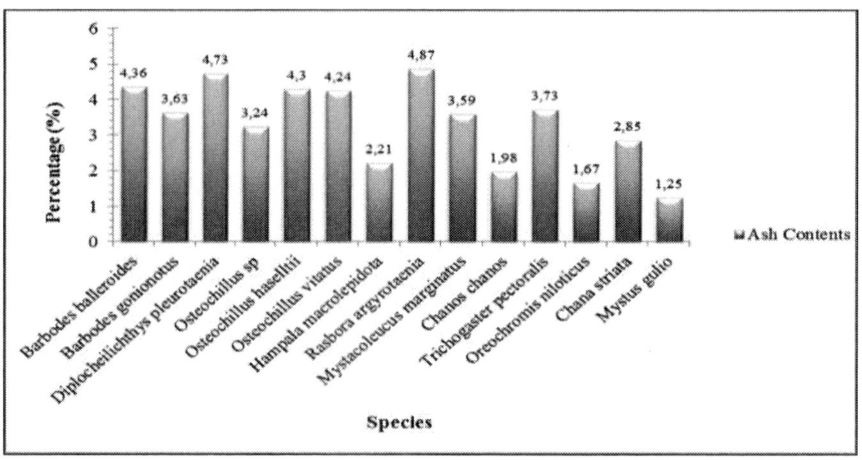


**Figure 2.** Ash contents of fish in Jatigede reservoir.

*Barbodes gonionotus* had the lowest protein level at 14.43 %, while the highest was found in Chanos chanos at 21.93 % (figure 3). Protein is a nutrient group that is very important in enzyme, antibody, and some hormone production.

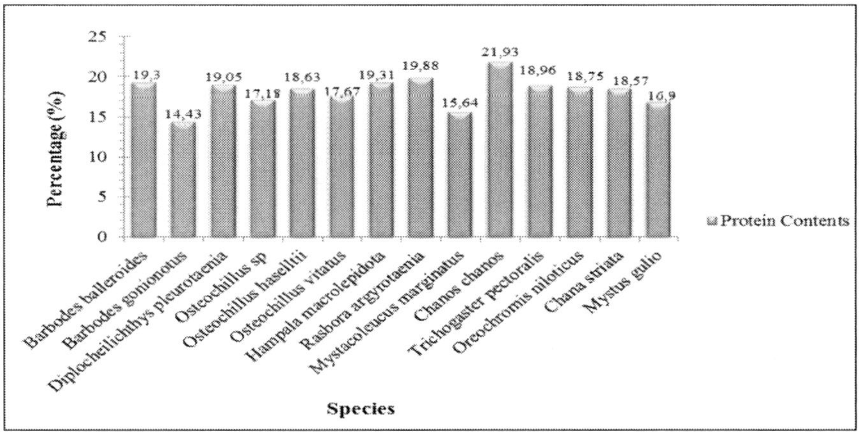


**Figure 3.** Protein contents of fish in Jatigede reservoir.

Hike fish had the lowest fat level at 0.72 %, while the highest was *Barbodes balleroides* at 6.52 % (figure 4). Each gram of fat consists of 2.25 times the calories from the total calorie made by one gram of protein of carbohydrate [13]. Fat acts as a supplier of some of the energy necessary to perform daily activities, and are also providers of fatty acids and a component of food taste and vitamins.

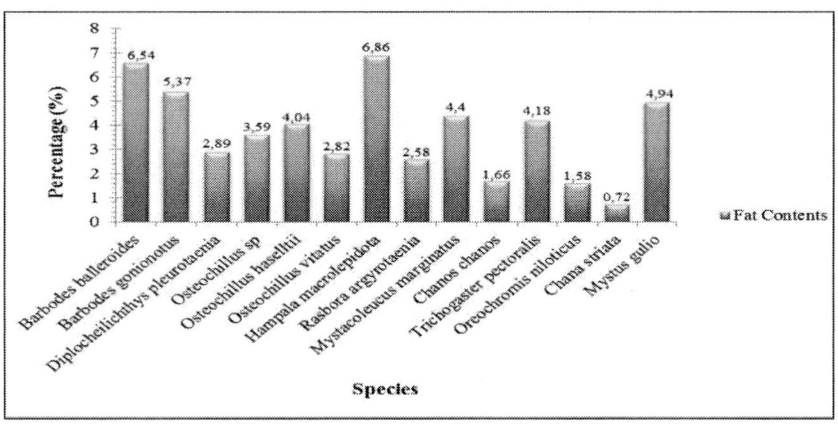


**Figure 4.** Fat contents of fish in Jatigede reservoir.

*Chana striata* had the lowest level of carbohydrate at 0.12 %, while the highest was measured in *Osteochyllus hasseltii* at 2.64 % (figure 5). The low level of carbohydrate in Chana is caused by their main food source which comes from fish or is in the carnivore fish group. The high level of carbohydrate in *Osteochyllus hasseltii* is because the fish is a herbivore.

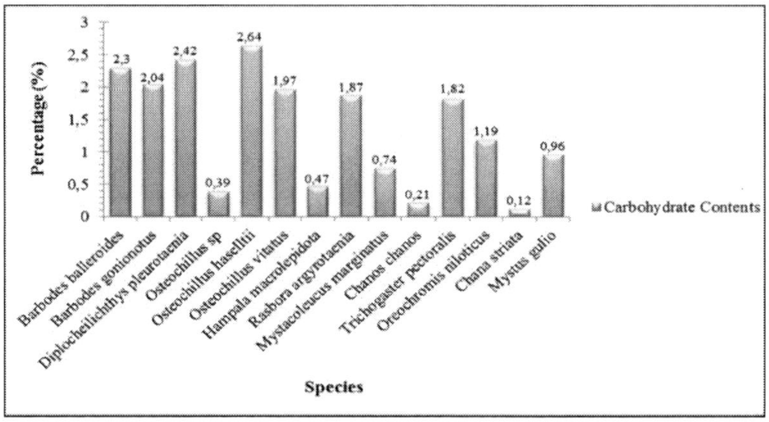


**Figure 5.** Carbohydrate contents of fish in Jatigede reservoir.

Silver barb, silver rasbora, spotted barb, milkfish, and three spoted gouramy that inhabit Jatigede Reservoir had a relatively higher protein level when compared to the same fish that were researched by the Research Centre for Marine and Fisheries Product Processing and Biotechnology, and [2], research results suggest that these are caused by environmental factors.

## 4. Conclusion

According to research results from 13 fish species that inhabit Jatigede, we can conclude that:

1. Fish that live in the Jatigede Reservoir contain good nutrients to be consumed as a food ingredient because it contains important substances for the body such as proteins, fats and carbohydrates. The levels of the nutrients are different for each species.
2. Silver Barb, Silver Rasbora, Spotted Barb, Milk Fish, and Three Spoted Gouramy that inhabit the Jatigede Reservoir had a relatively higher protein level when compared to the same fish that were researched by the Research Centre for Marine and Fisheries Product Processing and Biotechnology, and [2].

We suggest that further research needs to be done on complete nutrition content analysis including mineral and vitamin content.

## 5. References

[1] Kuswandi R 2015 Fish consumption rate : people can free take fish in Jatigede Reservoir. *Kompas* (In Indonesia)

[2] Muchtadi Tien R, Sugiono M and Ayustaningwarno S F 2016 *Food science* (Bandung : Alfabeta) (In Indonesia)

[3] Hadiaty R K and Siebert D J 1998 *Revue Franchaise d'aquariologie herpetologie Journal* **25** 1-4

[4] Karnasuta J 1993 *Kasetsart University Fish.Res. Bull.* **19** 1-105

[5] Kottelat M, Whitten A J, Kartikasari S N and Wirjoatmodjo S 1993 *Freshwater Fishes of Western Indonesia & Sulawesi* (Jakarta : Periplus Edition) (In Indonesia)

[6] Kottelat M and Whitten A J 1996 Freshwater Fishes of Western Indonesia & Sulawesi: Additions and Corrections

[7] Kottelat M 2001 Fishes of Laos *WHT Pub. (Pte). Ltd. Srilanka.*

[8] Kottelat M 2003 *Raffles Bull. Zoo. Supplement* **27** 1-663

[9] Nelson J S 2006 *Fishes of the word* fourth edition (New Jersey: John Willey & Sons, Inc. Hoboken)

[10] Roberts T R 1989 *Memoirs of the California Academy of Science* **14** 1-210

[11] Tan H H and M Kottelat 2009 *Ichtiological Exploration of Freshwaters* **20** 13-69

[12] Winarno F G 2002 *Food Chemistry and Nutrition* (Jakarta : Gramedia Pustaka Utama) (In Indonesia)
[13] Nurjanah and Abdullah 2010 *Intelligent In choosing Fish and Preparing the Processed* (Bogor: IPB Press) (In Indonesia)

**Acknowledgement**
Funding for this research came from the flagship research Universities, The Ministry of Research, Technology and Higher Education of Republic Indonesia. Number of MoU 855/UN6.3.1/PL/2017, date March 8, 2017

IOP Conf. Series: Earth and Environmental Science **137** (2018) 012056 doi:10.1088/1755-1315/137/1/012056

# The isolation and identification of endophytic bacteria from mangrove (*Sonneratia alba*) that produces gelatinase

**H Nursyam*, A A Prihanto, N I Warasari, M Saadah, R E Masrifa, N A Nabila, N Istiqfarin, and I J Siddiq**

Department of Fishery Product Technology, Faculty of Fisheries and Marine Science, Brawijaya University. Jl. Veteran Malang, 65145, East Java Indonesia

Email : happy_nsy@ub.ac.id

**Abstract**. Gelatinase is an enzyme that hydrolyze gelatin into gelatin hydrolyzate. The purpose of this study was to isolate and to identify endophytic bacteria from *Sonneratia alba* mangrove which able to produce gelatinase enzyme. *Sonneratia alba* mangroves was obtained from Bajul Mati Beach, Malang Regency. The samples in this study were, stems, and leaves. Pure cultured bacteria were investigated for its capability for producing gelatinase enzyme by using gelatin media. Best producer would further be analyzed its species using microbact system. Screening process resulted in 3 positive isolates, namely code isolate of R, B, and L. R which was isolate from root of *S. alba* was the best producer for gelatinase. Identification process with morphology and microbact system revelaed that A. SBM is a Gram-negative bacterium that has a basil cell shape, with a diameter colony of 2.19 mm. Based on the microbact system test carried out, the bacteria is *Pseudomonas aeruginosa*.

## 1. Introduction

The metalloproteinase gelatinase is an important enzyme not only for the chemical industry, but also for the food industry as a solvent, binder, stabilizer, water binder as well as an anticancer, antibiotic, and other medicinal products [1]. Hence, the exploration of new sources of this enzyme is important and is always attracting the interests of not only microbiologists, but food chemists and pharmacists as well. The gelatinase enzyme can be produced by animals, plants and microorganisms. Microorganisms such as *Micrococus, Pseudomonaaeruginusa, Seratia marcescens, Bacillus subtilis, Corynebacterium, Salmonella, Vibriocomma, Chromobacterium violaceum, Staphylococcus aureus* and *Proteus* are widely known as gelatinase producers [2]. Microorganisms, especially bacteria is a potential enzyme producer due to the fact that it can rapidly grow. Hence, the efficiency for mass production is very promising.

Mangroves are reported to have unique bioactives that have the potential to produce several enzymes for industry. Mangroves grow and develop in coastal areas and have a unique adaptation to environmental stress. Mangroves live in two different habitats, namely the terrestrial and ocean habitats at the same time. This harsh environment may trigger the production of the microorganism to produce unique bioactives such as enzymes [3].

This study investigates the probability of the bacterial endophytic mangrove in producing gelatinase enzyme. The isolate would further be identified its species by using morphology and the microbact system analysis.

IOP Conf. Series: Earth and Environmental Science **137** (2018) 012056 doi:10.1088/1755-1315/137/1/012056

## 2. Methodology

Materials for this study were the leaves, bark and root of the mangrove Sonneratia alba from Bajul Mati Beach, Malang Regency which is located in East Java, Indonesia. Other media and reagents were Luria Bertani agar (LB agar), and (0.15 grams yeast extract, 0.3 grams pepton, 0.3 grams NaCl, 0.45 gram agar) and medium for gelatinase screening (0.15 gram pepton, 0.09 gram beef extract, 3.6 gram gelatin) per 30 mL aquades that were pure grade.

### *2.1. Isolation of endophytic bacteria*

The isolation of endophytic bacteria was carried out according to the method of Rajendra et al [4] with modification. One gram of sample (leaves, bark and root) were grinded and diluted until 10-4. The sample was then cultured in LB agar and incubated for 48 hours at the temperature of 37 °C. Each colony was pure cultured by using the streak quadrant method. The pure isolates were kept at 4°C until further used.

### *2.2. Screening of gelatinase-producing bacteria*

The pure culture of bacteria was further analyzed to discover its capability for producing gelatinase by using a medium for gelatinase screening following the method of Ballan *et al.* and Della *et al.* [2, 5] with slight modification. The best producer of gelatinase would be identified by using morphology and the microbact analysis system.

### *2.3. Species identification*

Species identification was performed with morphology and biochemical characterization. The Gram stain was performed according to the standard method. The microbact sistem is a method for identifying microorganisms based on their biochemical characteristics and depends on the outcome of the oxidase and non-oxidase bacteria. If it is a positive oxidase, the microbact 24E system is the most appropriate; while microbact 12E should be used when the bacteria shows a negative oxidase. All protocols followed the manufacturer's instructions [6].

## 3. Result and discussion

Based on the results of the streak plate quadrant, the isolation steps obtained 3 bacteria isolates. The code of R, B, and L were the isolates from the root, bark and leaves of S. alba, respectively. The three isolates which were successfully isolated from the mangrove S. alba were further analyzed for its capability to produce gelatinase. The gelatinase test was performed on the three isolates and results indicated that all of the isolates produced gelatinase. The production of gelatinase was marked with the remaining liquid media on the gelatinase screening media after the incubation at 4 °C. The results showed that only two isolates were positive for producing gelatinase (figure 1), namely from the root and bark. It also showed that the isolate from the root (isolate R) was the best producer of gelatinase.

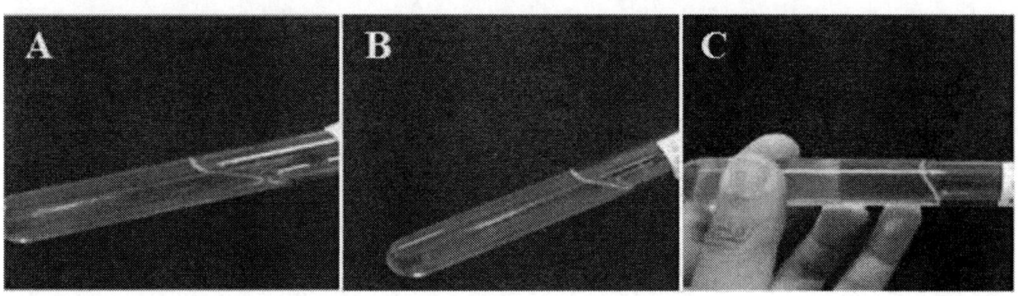


**Figure 1.** the result of screening of gelatinase-producing bacteria A isolate R, B isolate B, C isolate L.

Further analysis of the species was performed only on the best producer. Hence, isolate R was analyzed by its morphology and biochemical characteristics in order to determine the species. Gram positive and morphology analysis revealed that the isolate R is a gram negative bacteria (figure 2.) with a diameter of about 2.19 mm.

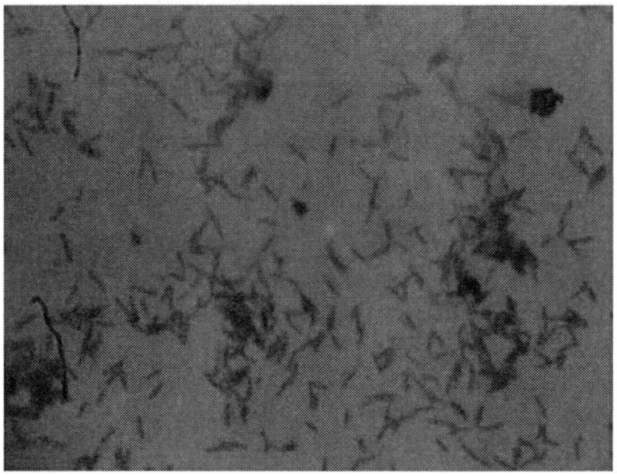

**Figure 2.** Gram identification of the isolate R.

The endophyric bacteria from the root of S. alba (isolate R) was an oxidase negative bacteria, therefore the 24E microbact system was used. An analysis of the oxidase, spores, nitrate, lysine, ornithin, H2S, glucose, mannitol, xylose, ONPG, indole, V-P, citric, TDA, gelatin, moalonat, inositol, rhammnosa, sucrose, lactose, arabinose, adonitol, raffinosa, salicin, arginine, catalase, coagulase, hemolysis, starch hydrolysis, and casein hydrolysis was conducted. The results of the microbact analysis are described in table 1.

**Table 1.** The results of the biochemical characteristic of isolate R based on Microbact analysis system.

| No. | Biochemical test | Result | Bergey's |
|---|---|---|---|
| 1 | Spore | + | + |
| 2 | Oxsidase | + | + |
| 3 | Motility | - | + |
| 4 | Nitrate | + | + |
| 5 | Lysine | - | Tidak diuji |
| 6 | Ornithine | - | Tidak diuji |
| 7 | $H_2S$ | - | Tidak diuji |
| 8 | Glukose | + | + |
| 9 | Manitol | + | + |
| 10 | Xylose | + | + |
| 11 | ONPG | + | Tidak diuji |
| 12 | Indole | - | Tidak diuji |
| 13 | Urease | - | Tidak diuji |
| 14 | V-P | + | Tidak diuji |
| 15 | Citric | - | + |
| 16 | TDA | - | Tidak diuji |
| 17 | Gelatine | - | + |
| 18 | Malonate | - | Tidak diuji |

| | | | |
|---|---|---|---|
| 19 | Inositol | - | Tidak diuji |
| 20 | Rhamnose | - | Tidak diuji |
| 21 | Sucrose | - | + |
| 22 | Lactose | - | - |
| 23 | Arabinose | + | + |
| 24 | Adonitol | - | Tidak diuji |
| 25 | Raffinose | - | Tidak diuji |
| 26 | Salicine | - | Tidak diuji |
| 27 | Catalase | + | + |
| 28 | Arginin | - | Tidak diuji |
| 29 | Coagulase | - | Tidak diuji |
| 30 | Beta Hemolysis | + | Tidak diuji |
| 31 | Novobiosin sensitivity | - | Tidak diuji |
| 32 | Starch hydrolysis | + | + |
| 33 | Casein hydrolysis | + | + |

Based on the analysis of the microbact system, the isolate R is Pseudomonas aeruginosa. It has the characteristic of positive in glucose, xylose, d malonic and rhamnosa fermentative. Contrastingly, it is negative in mannitol, arabinose, inositol, sucrose, lactosa, adonitol, rafinose fermentation, salicin and arginine. Simple sugar was used by bacteria as a source of carbon source. In this bacteria, it seemed that if the source of simple sugar is absent then it will utilize the complex sugar.

*Pseudomonas aeruginosa* is a gram negative bacteria, rod-shaped in form, and aerobic. It has flagella for their motility. This bacteria is capable of making their own pili type IV that serves as a process of the adhesin to bind on the host cell. *P. aeruginosa* is able to attach and colonize on the various types of cells [7]. This bacteria is a pathogen bacteria and causes acute ventilator-associated pneumonia and chronic lung infections in patients with cystic fibrosis (CF). It is also difficult to treat naturally because it has developed protective shield from neutrophils [8, 9].

## 4. Conclusion

Gelatinase-producing bacteria was successfully isolated from the mangrove of *S. alba.* The highest bacteria which produced gelatinase was *P. aeruginosa*. Although this bacteria is a potential gelatinase producer, it is unrecommended for further development due to safety issues.

## 5. References

[1] Hisano T, Abe S, Wakashiro M, Kimura A, Murata K 1989 *J. Fermen Bioeng* **68** 399-403
[2] Balan S S, R Nethaji, S Sankar, S Jayalakshmi 2012 *Asian Pac. J. Trop. Biomed.* **2**(3) 1811-1816
[3] Wang K W, Wang S W, Wu B, Wei J G 2014 *Mini Rev. Med. Chem.* **14** 370-91
[4] L Rajendran, R Samiyappan, T Raguchander and D Saravanakumar 2007 *J. Plant Interact.* **2** 1-10
[5] Dela Cruz T E E and J M O Torres 2016 *Gelatin hydrolysis test protocol* (American: Socienty for Microbiology) 10 p
[6] Oxoid 2005 *Microbact TM gram negatif identification system* London 231 p
[7] Radji M 2005 *The Role of Biotechnology and Microbial Endofit in the Development of Herbal Medicines of* Pharmaceutical Science Magazine 218 pp
[8] Klockgether J and Tümmlera B 2017 F1000 *Res.* 2017 **6** 1261
[9] Shielding C P 2009 *Microbiology* **155**:3474–3475

IOP Conf. Series: Earth and Environmental Science **137** (2018) 012057 doi:10.1088/1755-1315/137/1/012057

# Alternative bioenergy through the utilization of *Kappaphycus alvarezii* waste as a substitution of substrate for biogas products

**R Yulita[1], Agustono[1], D Y Pujiastuti[2] and M A Alamsjah[2]***

[1]Department of Fish Health Management and Aquaculture, Faculty of Fisheries and Marine, Universitas Airlangga, Jl. Airlangga Surabaya 60115, East Java
[2]Department of Marine, Faculty of Fisheries and Marine, Universitas Airlangga, Jl. Airlangga Surabaya 60115, Eat Java

E-mail: alamsjah@fpk.unair.ac.id

**Abstract.** Biogas is one of the renewable energy resources which are able to be developed by providing some sufficient renewable substances and manufactured from the fermentation process of organic substances metabolized by anaerobic bacteria. In this research, *Kappaphycus alvarezii* seaweed waste from carrageenan processing and contents of rumen were used. This research aims to comprehend the carrageenan processing waste of macroalga *K. alvarezii* can be used as alternative source generating biogas. The research method is P0 (100 % of the contents of rumen), P1 (75 % of the contents of rumen and 25 % of seaweed waste), P2 (50 % of the contents of rumen and 50 % of seaweed waste), and P3 (25 % of the contents of rumen and 75 % of seaweed waste), and P4 (100 % of seaweed waste). The result showed that according to the quality determination of biogas based on SNI (Indonesia National Standard) 8019:2014, the *K. alvarezii* seaweed waste from carrageenan processing can be utilized as the alternative source of manufacturing biogas and got the methane gas resulted from the comparison method is P2 (50 % of the contents of rumen and 50 % of seaweed waste), with value of 58.61 %.

## 1. Introduction

The need for fuel has increased along with population growth, regional development and energy crisis in the world. The scarcity of fuel caused by significant rise of world's oil prices has pushed the government to invite people to deal with energy. Furthermore, many studies have been conducted in order to find alternative energy sources as renewable resources [1]. An increase in the population's demand for energy, the depletion of world's oil reserves and the problem of fossil fuel emissions put pressure on every country to immediately produce and use renewable energy [2]. Indonesia has various natural resources which can be utilized as energy. Natural resources in the form of oil, gas, coal, geothermal heat, water and so forth. They can be used in various development activities directly or exported to earn state revenues. These materials are not renewable. The use of non-renewable resources causes crisis problems. One of the signs of energy crisis occurring lately is scarcity of oil fuels, such as kerosene, liquefied petroleum gas, gasoline and diesel fuel.

Scarcity happens because the need for fuels is very high and keeps on increasing every year. On the contrary, crude oil as the raw material for making fuel is limited and it takes time for the fuel to be produced [3]. Hence, alternative energy is necessary for preserving oil reserves at the present time.

IOP Conf. Series: Earth and Environmental Science **137** (2018) 012057 doi:10.1088/1755-1315/137/1/012057

Biogas is one of energy sources that can be developed by providing raw materials available in the nature in sufficient amount (renewable condition). Renewable energy sources are relatively easy to obtain, require low operating expanses, and do not cause any waste problem thus oil shortage can be overcome.

A biogas is a flammable gas derived from the process of fermentation of organic materials by anaerobic bacteria that do not need oxygen to survive and reproduce. Generally, all types of organic materials like rumen contents and animal's urine suitable for simple biogas system can be processed to produce biogas [4]. Biogas can be used to produce methane as a substitute for fuels, especially kerosene, and can be used for cooking. On a large scale, biogas can be used as the material for power plant. In addition, the process of biogas production will produce residue of cattle's rumen contents that can be used directly as organic fertilizers for crops or agricultural cultivation. More importantly, it can be used to reduce the dependence on non-renewable fuels [5].

One of the commodities in Indonesian waters that is potential to be developed is seaweed. Seaweed is widely used in and profitable for the industries of food and beverage, personal care, cosmetics, animal feed, steel, ceramic, paint, ink, mining, coal briquette and asphalt, paper and pulp, textile, fertilizer and medicine (pharmacy) as food fibers [6]. Seaweed can be processed into various food and will produce waste (unused), so what is available can be utilized. According to Alamsjah and Prayogo [7], seaweed waste is usually left accumulating at landfills, but it has the potential to be processed into biogas. An example of seaweed is *Kappaphycus,* that can be extracted into carrageenan. The waste from the processing of carrageenan is still underutilized. According to Saputra *et al.* [2], biogas may come from various organic matters such as cattle's rumen contents, feces, scrap paper and aquatic plants (e.g. hyacinth, filamentous algae), seaweed and seaweed waste. The C/N ratio in the making of biogas should be noted. The C/N ratios for *Kappaphycus* and carrageenan waste are 43.98/L and 55.01/L, respectively.

Cattle's rumen contents are the most suitable starter in observing the use of biogas because they have C/N ratios of between 11 and 30 [8]. Another important thing found in the substratum of rumen of a cow is bowel bacteria producing methane. The bacteria in the bowel of ruminants can break down organic compounds using fermentation, so that the waste processing in anaerobic animals can produce gases consisting of methane ($CH_4$) and carbon dioxide ($CO_2$). The substrate in cattle's rumen contents consist of bacteria that produce methane in the stomach of ruminants. The bacteria in the bowel of a ruminant help with fermentation process, thus the biogas production in digester's body can be done faster. In addition, fresh rumen contents are easier to process compared with long and/or dried rumen contents during the drying time [9].

## 2. Methodology

A research method was used to solve a problem, which may be done by data collection through observation, survey and experiments [10]. The method used in this research was descriptive method. Descriptive method is a planned attempt to express new facts, corroborate a new theory, or even deny the existing research results. The study was done by observing and comparing the amount and volume of methane and carbon dioxide and the C/N ratio resulting from each research treatment.

The materials used in this study were 10,000 g of carrageenan processing waste of *Kappaphycus*, 10,000 mL of fresh rumen contents and 4,000 mL of water. Into each treatment was added 200 mL of water. The following was used in this research: P0 = 1,000 mL of rumen contents + 0 g of carrageenan processing waste (100 % : 0 %); P1 = 750 mL of rumen contents + 250 g of carrageenan processing waste (75 % : 25 %); P2 = 500 mL of rumen contents + 500 g of carrageenan processing waste (50 % : 50%); P3 = 250 mL of rumen contents + 750 g of carrageenan processing waste (25 % : 75 %); and P4 = 0 mL of rumen contents + 1000 g of carrageenan processing waste (0 % : 100 %). The materials used in the preparation of anaerobic fermentation process were carrageenan seaweed waste (substrate), rumen contents (co-substrate) and water. The addition of water as one of the materials was aimed to meet the water levels required for biogas production.

IOP Conf. Series: Earth and Environmental Science **137** (2018) 012057 doi:10.1088/1755-1315/137/1/012057

The number of bioreactor or digester used in this research was 20. There were 5 treatments and they were repeated 4 times. The volume of a bioreactor was 1,500 mL, and the bioreactor was made from plastic, with an airtight cover, faucet and hose. The cover was connected to the faucet, that was connected to the hose. The hose was connected to top gas in the form of a plastic bag fastened with rubber band and cling warp. The substrate consisted of cattle's rumen, carrageenan waste and water. Cow's rumen and carrageenan waste were mixed at a ratio as specified. Every tube digester will contain filling materials (processed carrageenan, seaweed and rumen contents) and water at 80% of the fermenter volume [11].

The fermentation system in the biogas production in this research used a closed system (batch fermentation) for 21 days of fermentation. For every treatment, four tests were conducted. The biogas produced filled the space in the bioreactor, then moved toward the hose to be accommodated at the gas on the top. The gas to be accommodated was directly analyzed according to the length of time of fermentation. According to Anggraini *et al.* [12], the length of time of fermentation required to produce methane ($CH_4$) was 21 days. The biogas was analyzed by considering the levels of methane parameters, volume of gas and carbon dioxide. The biogas sample was extracted into a plastic bag, then the volume of the biogas was measured using a volume meter and the levels of methane ($CH_4$) and carbon dioxide ($CO_2$) was measured using gas chromatography. The analysis of C/N ratio was conducted by measuring the C-organic content and the total nitrogen of the sample.

The analysis of C-organic was conducted in this research using cremate method. Included in the procedure of C-organic analysis was the weighing of an empty cup (a) of each sample treatment at about 1 g. Then, the sample was put into an oven for 4 hours at a temperature of 105 °C. Then, the cup was cooled in desiccator for 15 minutes. The sample and the cup were weighed again after being removed from desiccator (c). In the next step, the cup that contained the sample was put in a furnace (600 °C for 4 hours), then the cup was incorporated into a desiccator after being removed and weighed for the final measurement (d). The calculation for this analysis used the equation below:

$$\text{\% of water content} = \frac{b-c}{b-a} \text{ x } 100\% \quad (1)$$

$$\text{\% of ash content} = \frac{d-a}{b-a} \text{ x } 100\% \quad (2)$$

$$\text{\% of organic substrate} = 100 - (\text{\% of water content - \% of ash content}) \quad (3)$$

$$\text{\% of C-organic} = \text{\% of organic substrate x } \frac{58}{100} \quad (4)$$

Note: $\frac{58}{100}$ is the conversion of organic substrate to carbon

The analysis of the total nitrogen in this research used Gunning method. This analysis was conducted in three stages, namely destructor, distillation and titration. The destructor stage was aimed to destroy material (sample) through the addition of $K_2SO_4$ and $CuSO_4$. One g sample was put into Kjeldahl squash and added with 10 mL of $H_2SO_4$, 5 g of $K_2SO_4$, 0.3 g of $CuSO_4$ and stones boiling. Next, the mixture was heated on an electric heater, firstly at low heat and then at high heat. Warming ended after the solution became colorless. As much as 100 mL of dd$H_2O$, 1 g of Zn and 50 mL of 45% NaOH were added to the Kjeldahl squash. Then, the Kjeldahl squash was mounted in the stage distillation. The mixture was distilled until the volume of distillation reached a value of 75 mL.

The titration was started by pouring 75 mL of distillate on titration squash, added with 50 mL of 0.1 N HCl and PP indicator. The titrant used was 0.1 N NaOH. Titration was carried out to figure out the volume of NaOH needed to neutralize sample. The levels of N (%) can be calculated using the equation below:

$$\text{N (\%)} = \frac{\textit{Volume of NaOH blank} - \textit{Volume of NaOH sample}}{\textit{massa x } 1000} \text{ x N NaOH x 14.008 x } 100\% \quad (5)$$

The C/N ratio could be calculated after obtaining the C-organic level and the total nitrogen value. The C/N ratio can be calculated using the equation below:

$$\text{Ratio of C/N} = \frac{C\text{-}organik}{N\text{-}total} \quad (6)$$

## 3. Results and discussion

This research yielded the methane levels, C/N ratio (before and after fermentation), pH and temperature value (before and after fermentation). The C/N ratio of rumen contents to seaweed waste was 41.04:36.5. The organic carbon of rumen contents was 28.15 and the total nitrogen was 0.68, while the organic carbon of seaweed waste was 26.80 and the total nitrogen was 0.73. The measurement of the biogas production in this research was conducted in terms of quality (levels of methane in biogas) and quantity (biogas volume). The methane levels and the biogas volume produced in each of the rumen contents and seaweed carrageenan waste are displayed in table 1.

**Table 1.** The average levels of methane (%) and biogas volume (mL) from the comparison of rumen contents and seaweed waste.

| Treatment | Methane Level (%) | Biogas Volume (mL) |
|---|---|---|
| P0 (100 %:0) | 40.70 ± 2.97 | 198 ± 183.85 |
| P1 (75 %:25 %) | 56.51 ± 5.67 | 409 ± 74.95 |
| P2 (50 %50 %) | 58.61 ± 2.56 | 1182.5 ± 109.60 |
| P3 (25 %:75 %) | 56.50 ± 7.07 | 1306 ± 113.14 |
| P4 (0:100 %) | 44.35 ± 2.47 | 746.5 ± 160.51 |

Note: P0 (100 % : 0 %) = 1000 mL of rumen contents + 0 g of carrageenan processing waste; P1 (75 % : 25%) = 750 mL of rumen contents + 250 g of carrageenan processing waste; P2 (50 % : 50 %) = 500 mL of rumen contents + 500 g of carrageenan processing waste; P3 (25 % : 75%) = 250 mL of rumen contents + 750 g of carrageenan processing waste; P4 (0 % : 100 %) = 0 mL of rumen contents + 1000 g of carrageenan processing waste.

Table 1 showed that there was a dynamic of methane levels for each ratio of rumen contents to seaweed waste in all treatments. Treatment P2 yielded the highest average levels of methane (58.61%), while the highest average value of biogas volume was yielded by treatment P3 (1306 mL). If the levels of methane and the biogas volume were related, the following equation was obtained:

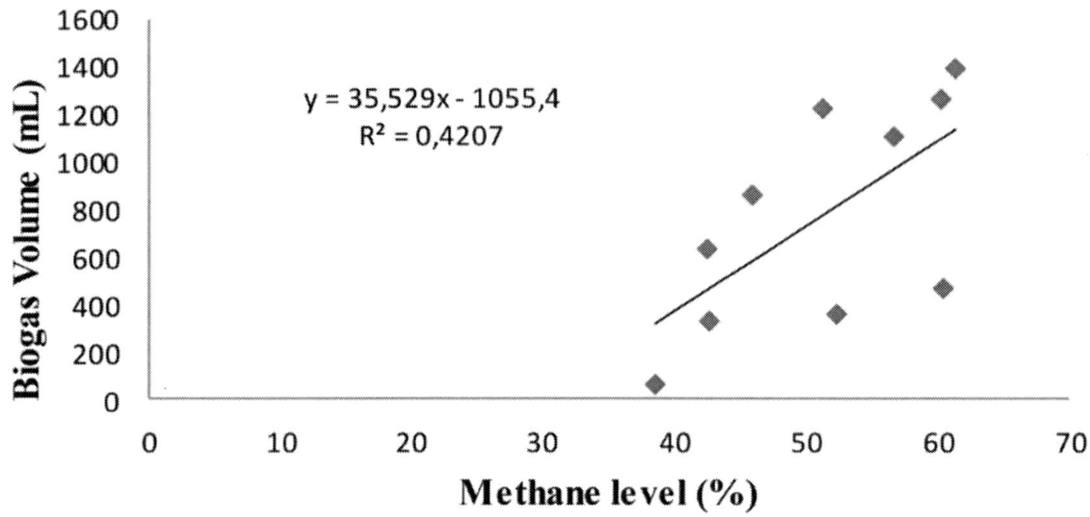


**Figure 1.** The relation between the methane levels and biogas volume in each ratio of rumen contents and seaweed.

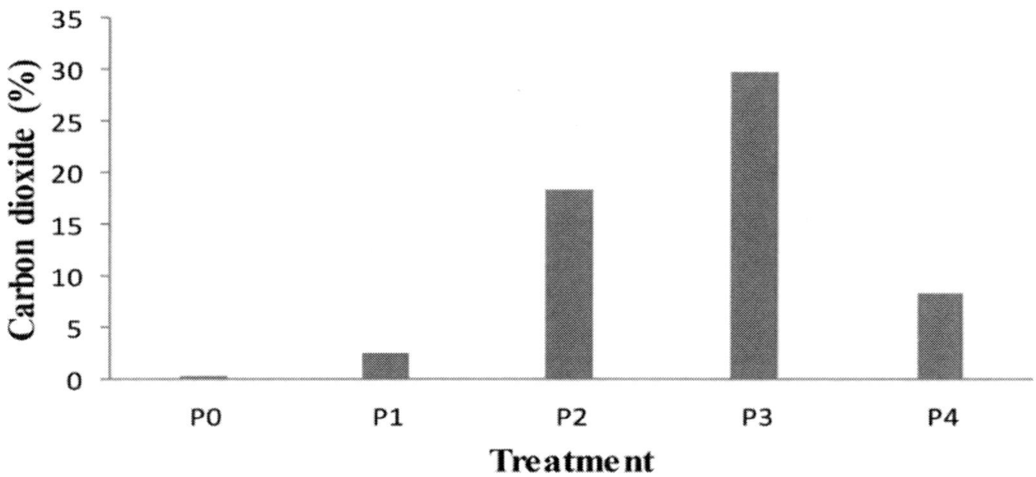


**Figure 2.** The average levels of carbon dioxide (%) in each ratio of rumen contents and seaweed waste.

Note: P0 (100 % : 0 %) = 1000 mL of rumen contents + 0 g of carrageenan processing waste; P1 (75 % : 25 %) = 750 mL of rumen contents + 250 g of carrageenan processing waste; P2 (50 % : 50 % ) = 500 mL of rumen contents + 500 g of carrageenan processing waste; P3 (25 % : 75 %) = 250 mL of rumen contents + 750 g of carrageenan processing waste; P4 (0 % : 100 %) = 0 mL of rumen contents + 1000 g of carrageenan processing waste.

The biogas production can also be linked with the C/N ratio. The C/N ratios in the beginning and the end of the treatments for 21 days of fermentation are shown in Figure 3.

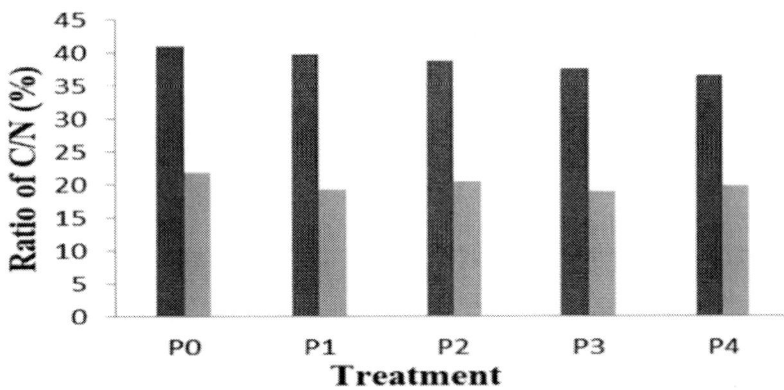


**Figure 3.** The C/N ratio in the beginning and the end of fermentation on every comparison of rumen contents and seaweed waste.

Note: P0 (100 % : 0 %) = 1000 mL of rumen contents + 0 g of carrageenan processing waste; P1 (75 % : 25 %) = 750 mL of rumen contents + 250 g of carrageenan processing waste; P2 (50 % : 50 %) = 500 mL of rumen contents + 500 g of carrageenan processing waste; P3 (25 % : 75 %) = 250 mL of rumen contents + 750 g of carrageenan processing waste; P4 (0 % : 100 %) = 0 mL of rumen contents + 1000 g of carrageenan processing waste; ▒ ratio of C/N in the beginning of fermentation; ■ ratio of C/N in the end of fermentation

The C/N ratio decreased with research observation. The C/N ratios at the beginning of fermentation for P0, P1, P2, P2, P3 and P4 were 41.04 %, 39.84 %, 38.69%, 37.58 %,and 36.5 %, respectively. Furthermore, C/N ratios at the end of fermentation for P0, P1, P2, P2, P3 and P4 were 21.85 %, 19.25 %, 20.4 %, 18.9 % and 19.7 %, respectively. In addition, the pH value and temperature before and after fermentation changed. The changes are presented in figure 4 and 5.

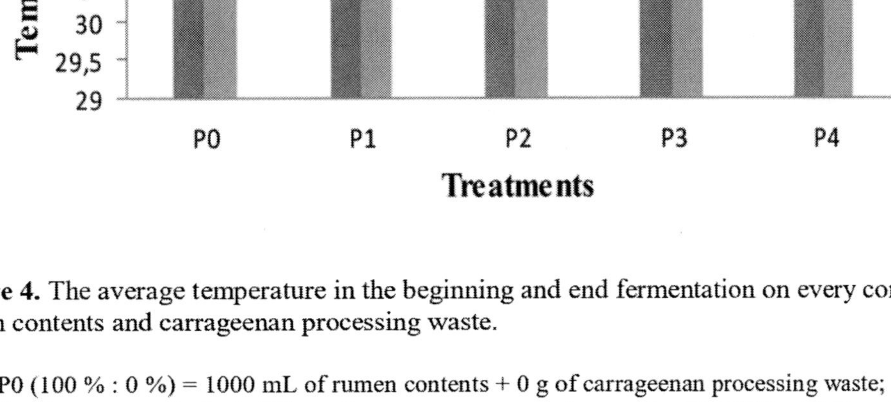


**Figure 4.** The average temperature in the beginning and end fermentation on every comparison of rumen contents and carrageenan processing waste.

Note: P0 (100 % : 0 %) = 1000 mL of rumen contents + 0 g of carrageenan processing waste; P1 (75% : 25%) = 750 mL of rumen contents + 250 g of carrageenan processing waste; P2 (50% : 50%) = 500 mL of rumen contents + 500 g of carrageenan processing waste; P3 (25 % : 75%) = 250 mL of rumen contents + 750 g of carrageenan processing waste; P4 (0% : 100%) = 0 mL of rumen contents + 1000 g of carrageenan processing waste; average temperature in the beginning of fermentation; average temperature in the end of fermentation.

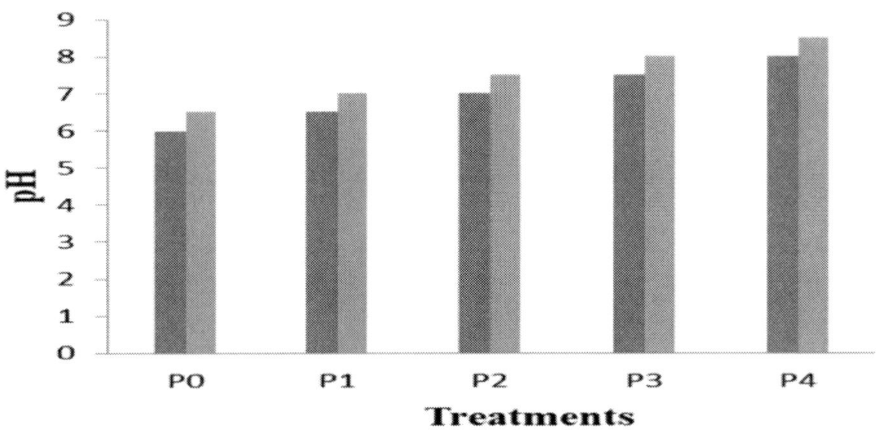


**Figure 5.** The average pH for the beginning and the end fermentation on every comparison rumen contents and carrageenan processing waste

Note: P0 (100 % : 0 %) = 1000 mL of rumen contents + 0 g of carrageenan processing waste; P1 (75 % : 25 %) = 750 mL of rumen contents + 250 g of carrageenan processing waste; P2 (50 % : 5 0 %) = 500 mL of rumen contents + 500 g of carrageenan processing waste; P3 (25 % : 75 %) = 250 mL of rumen contents + 750 g of carrageenan processing waste; P4 (0 % : 100 %) = 0 mL of rumen contents + 1000 g of carrageenan processing waste; average pH in the beginning of fermentation; average pH in the end of fermentation.

The pH and temperatures rose at the end of fermentation, whereas temperature in the beginning of fermentation ranged 30.38—31.63 °C, which increased to 31.38—32.75 °C at the end of fermentation. Meanwhile, the pH in the beginning of fermentation ranged at 6-8 and increased to 6.5-8.5 at the end of fermentation. In the anaerobic fermentation process to form biogas, bacteria have a very important role in reorganizing organic matter. All reactions during fermentation involved bacteria since the stage of hydrolysis, acidogenesis, acetogenesis until methanogenesis. The process of biogas formation must be in anaerobic condition, which means that the digester must be airtight and no air can come into it. The processing of organic compound contained in waste into methane and carbon dioxide does not require oxygen. If there is a leak in the digester during the fermentation, methane will not be formed. In general, the process of biogas making uses cow dung because cattle dung is the most suitable starter for the formation of biogas [9]. In this research, the starter used was rumen contents, while a substrate used were carrageenan processing waste. According to Triwisari [13], reports that the main compositions of carrageenan processing were cellulose (26.72 %) and carbon (46 %). Different organic contents can affect the formation of biogas and methane [14]. According to Anggraini *et al.* [12], the methane content in biogas produced on a reactor depends on the type of feeds, input composition, fermentation time and reactor capacity. The highest methane was produced in P2 (50 %:50 %) with 500 mL of rumen contents and 500 g of carrageenan processing waste. This may happen because the raw materials of carrageenan processing waste was not degraded by bacteria completely, while rumen contents as a source of methane which have organic components were degraded by the bacteria. Organic component is a source of nutrients for bacteria to produce biogas. According to Saputro [15], Mara and Ida [16], add that organic component affects bacteria productivity in producing biogas. The biogas volume produced was different from the levels of methane but was still influenced by the ratio of rumen contents to carrageenan processing waste. The measured biogas volume was influenced by the length of fermentation. In this research the duration of time required for fermentation was 21 days. The duration of time applied in this research was referred to the research of [12], which states that the length of fermentation time required to produce the highest amount of methane ($CH_4$) recorded is 21 days.

According to Mara and Ida [16], the main factor affecting the difference in the biogas volume produced is the physical properties of material caused by water content and acidity of media (pH levels). The ratio of the rumen contents to carrageenan processing waste and the addition of water will result in different properties in the blending condition of each composition . Furthermore accoording to Budiharjo [17] claims that the increase in the volume of biogas volume is an indication that the system has been running well and stable. According to Calzada *et al.* [18], Indarto [19], the composition and the volume of biogas depends on the different characteristics of the substrate, while the highest volume of biogas is not always offset by high levels of methane. According to Yulistiawati [20] says that the lowest C/N ratio in the substrate will produce biogas with characteristics such as low level of $CH_4$ and $H_2$ and high level of $CO_2$ and $N_2$. According to Indonesia National Standart [21] reports that biogas generally consists of $CH_4$ (40—70 %) and $CO_2$. According to Sooerawidjaya [22], biogas consists of $CH_4$ (50—80 %), $CO_2$ (20—50 %) and several other gases such as $H_2$, CO, $N_2$, $O_2$ and $H_2S$. The results of the test conducted on methane to find out the ratio of rumen contents to carrageenan processing waste in this treatment showed 58.61% of methane, thus this level of methane met the standard amount of biogas. Biogas production was also linked to the change in C/N ratio that showed a decline during fermentation (21 days). The decrease of C/N ratio shows that nutrition has been used well. The C/N concentration declined with fermentation time. The decline in carbon and nitrogen was the result of biodegradation of organic matter by bacteria originating either from rumen contents or carrageenan processing waste. It is used as a source of energy for the growth of bacteria and the formation of biogas. This phenomenon was reported by Yani and Darwis [23], who states that the decrease in the C/N ratio means that carbon and nitrogen are used as nutrients by microbes to grow and develop.

Furthermore, according to Siallegan [24], shows that the reduction in C/N ratio can cause the production of biogas to stop, which in this case, the C/N ratio cannot help bacteria to produce biogas. This means that this substance can be used as fertilizer. The C/N ratio allowed for an ingredient to make solid fertilizer ranges at 12-25 [25]. The process of fermentation also affects the temperature and pH [26, 27, 28].

According to Sofyan [29], the enzyme of bacteria will change the substrate into a product and yield heat. The release of the heat involved an exothermic reaction for organic matter decomposition to the process of biogas formation in the acidifying stage. Material decomposition will result in acids, carbon dioxide and heat. The process of organic matter decomposition takes place in anaerobic condition, thus the system will produce heat, and consequently, the temperature in the reactor increases and affects the microorganisms and microbes in the fermentation process. The other thing that changed during the fermentation process is pH. Anaerobic biodegradation to organic matter will be influenced by the environment in biodigester. Acid environment is appropriate for anaerobic biodegradation. The degradation by bacteria will be affected by pH. Furthermore, [17,30,31] conclude that pH will increase after the fermentation. Bacteria can live at pH of 6.5-8.5 (optimally at pH of 7.0-8.0). This is due to the fact that the acid production will convert into ammonia and alkali compound. Afterwards, pH will decline again and become stable at the end of fermentation. The role of pH is important because such compounds produce acids and alkalis, such as organic acids and ammonium ion for the degradation of organic compounds in digester. According to kresnawaty *et al.* [32] states that at the beginning of biogas formation, acid bacteria will be active and cause pH in the digester to decrease. Then, methanogen bacteria will use the acid as a substrate, and the pH increases. This is supported by Sejati [28], who explains that the acid formed in the acidification will be used by methanogen bacteria as a substrate for producing methane and others. Subsequently, it will increase pH in digester. This indicates that in the process of biogas production, pH is regulated naturally. The pH values impact on the activity of methanogen bacteria, thus the bacteria are very vulnerable to acid condition.

## 4. Conclusion

The results of the research show that according to the requirement of quality of biogas of SNI (Indonesia National Standard) 8019:2014, *K. alvarezii* waste from carrageenan processing can be utilized as an alternative source of manufactured biogas. The highest amount of methane gas resulted from the ratio method was that in P2 (50 % of the rumen contents and 50 % of seaweed waste), amounting approximately 58.61 %.

## 5. References

[1] Putri R D A, Tsani S T 2015 Effect of cow rumen temperature and concentration on biogas production from Vinasse *J. Bahan Alam Terbarukan.* **4** 1-5
[2] Saputra A, Wikem A E P, Riris A 2011 Biogas production from Caulerpa racemosa and Sargassum duplicatum as alternative energy resources *Maspari J.* **3** 30-35
[3] Wahyuni S 2013 Biogas alternative energy substitute fuel gas and electricity (Jakarta: Agro Media Pustaka)
[4] Silalahi E F, Melki, Heron S 2012 Characterization of Methane Gas Producing Bacteria in Gracilaria sp. *Maspari J.* **4** 83-91
[5] Fahri A 2013 Biogas production technology from livestock sewage Agricultural Technology Assessment Center (BPTP) Riau
[6] Alamsjah, M A 2010 *Res. J. Fish. Hydrobiol.* **5** 159-167
[7] Alamsjah M A, Prayogo 2014 *J. Nat. Sci. Res.* **4**
[8] Mohee R 2007 *J. Environ. Inform.* **9** 87-99
[9] Triyatno J 2007 The effect of water content comparison with cow sewage on biogas productivity in sealed digester Department of Machine Faculty of Engineering STTI Bontang
[10] Kusriningrum 2012 Experimental design Faculty of Veterinary Medicine Airlangga University Surabaya
[11] Windyasmara L, Ambar P, Lies M Y 2012 *Bul. Peternak.* **36** 40-47
[12] Anggraini D, Pertiwi M B, Bahrin D 2012 *Tek. Kim.* **18** 17–23
[13] Triwisari D A 2010 Polysaccharide fractionation from several types of seaweed Bogor Agricultural University
[14] Firdausiyah N 2015 Variety comparison of compost raw materials and cow manure as well as fermentation time in production with addition of hidrolytic bacterial consortium Thesis Department of Biology Faculty of Science and Technology Airlangga University
[15] Saputro R R 2004 Biogas production from livestock waste Diponegoro University Semarang
[16] Mara I M, Ida B A 2011 Biogas quality and quantity analysis of livestock dung Department of mechanical engineering Mataram Univerity **1**
[17] Budihardjo M A 2009 *J. Presipitasi Media Komun. dan Pengemb. Tek. Lingkungan* **6** 27-34
[18] Calzada J F, De P E, Yurrita A, De A M C, De M F, Rolz C, Menchu J F 1984 *Agric. Wastes.* **9** 217-230
[19] Indarto K E 2010 Biogas production of tapioca liquid waste industry through temperature improvement and urea addition in anaerobic repair Sebelas Maret University Surakarta
[20] Yulistiawati E 2008 Effect of temperature and c/n ratio on production of biogas made from organic garbage Bogor Agricultural University
[21] Indonesia National Standard 2014 High quality standards for biogas SNI 8019
[22] Soerawidjaja T H 2009 Biogas technology kemurgi bandung (ID): Institut Teknologi Bandung. pp1-30
[23] Yani M, A A Darwis 1990 Biogas technology dictation Center of biotechnology Bogor Agricultural University

IOP Conf. Series: Earth and Environmental Science **137** (2018) 012057 doi:10.1088/1755-1315/137/1/012057

[24] Siallegan N S R 2010 Effect of retention time and composition of raw materials on fermentation process of industrial liquid wastes on biogas production *Biogas* Thesis University of Sumatera Utara Medan
[25] Suriadikarta D A 2004 Technical guidance on quality test and effectiveness of inorganic alternative fertilizer Soil Research Institute: Ministry of Agriculture
[26] Darmanto A, Sudjito S, Denny W 2012 *J. Rekayasa Mesin* **3** 317-328
[27] Nurdimansyah R, Amaliyah R I U, Ahmad Q 2015 Analysis of substrate level influence on anaerobic digester scale laboratory for methane production Physics Engineering Program Faculty of Electrical Engineering Telkom University Bandung
[28] Sejati S P 2015 Circulation system lindi on anaerobic digester for biogas production from vegetables waste Faculty of Agricultural Technology Bogor Agricultural University
[29] Sofyan F 2015 Anaerobic biodegradation in macroalga Eucheuma sp. Marine Science and Technology Bogor Agricultural University
[30] Haryati T 2006 *Wartazoa.* **16** 160-169
[31] Seadi T A, Rutz D, Prassl H, Kottner M, Finsterwalder T, Volk S, Janssen R 2008 Biogas University of Southern Denmark Esbjerg: Denmark
[32] Kresnawaty I, Susanti I, Siswanto, Panji T 2008 *Menara Perkeb.* **76** 23-35

IOP Conf. Series: Earth and Environmental Science **137** (2018) 012058 doi:10.1088/1755-1315/137/1/012058

# Methallothionein expression on the gills and stomach of Chinese pond mussels exposed to lead (Pb)

**H Kartikaningsih, A M Suryanto and D Arfiati**

Faculty of Fisheries and Marine Science, Brawijaya University, Malang.

E-mail: hartatikartikan@ub.ac.id

**Abstract.** In freshwaters area, Pb originates from rocks (naturally), industries, and pesticides. The ability of Chinese pond mussels as biofilters to absorb heavy metal (Pb) was demonstrated in water circulation system using ten 8 cm mussels. $PbNO_3$ (0, 10, 20, and 30 ppm) was administered into water containing mussels. Carp culture was done for 30 days, and Pb accumulation in carps was measured every week (week 0, 1, 2, and 3). The results showed that the highest Pb ion accumulation was found in the gills of mussels. The examination using hematoxylin-eosin showed that tissues were damage due to haemorrhage, cell ruptures, and cell deaths. The results of the measurement of metallothionein (MT) showed that MT molecular weight was 12.84 kDa.

## 1. Introduction

Biofilter is usually used in closed cultivation system (recirculation). The recirculation system is used mainly to reduce water usage. In addition to aquatic plants, biofilters common to aquatic environment are organisms that live at the bottom of waters, e.g. gastropods and bivalves. Mussels have the potential to be biofilters because they fareilter feeder [1]. An adult's capacity of food filtering is very high up to 2.5 litres/hour [2]. Mussels are also resistant to high levels of pollution and able to filter out water rapidly, giving them the potential to be biofilters [3].

Organisms adapt to the heavy metal pollution in the environment through a defence system, including exclusion, compartmentalization, and complex formation and synthesis of binding proteins such as metallothionein (MT) and phytochelatins (PC) [4]. Methallothionein is a protein possessed by all organisms to bind both essential and non-essential metals [5]. Plants, however, also have phytochelatins in addition to metallothionein.

The expression of metal-binding proteins (MT) in each individual in response to the presence of heavy metal pollution varies. MT is a non-enzymatic protein with a low molecular weight (10 kDa [6], 3-15 kDa [5], 6-7 kDa [7], 14.3 kDa [4]), which has high cysteine content and no aromatic amino acids, and is unstable by heat. The thiol group of MT (-SH) is a cysteine residue which allows MT to bind heavy metals. MT was reported to be present in vertebrates, including many species of fish [8], and aquatic invertebrates, particularly molluscs [4, 6]. MT is a protein that has sulphur content and binds metal ions. The sulfur content reaches 20 % of the total molecular weight of MT. In mammals, MT is characterized by a molecular weight of 6-7 kDa, 60-68 amino acid residue content, and binds total 7 equiv. bivalent metal ions [9].

Chinese pond mussels (*Anadonta woodiana*) live on the bottom of a pond or lake. Chinese pond mussels can "devour" pollutants, including heavy metal Pb that is suspended in the water. Because they are more resistant to pollutants than fish, Chinese pond mussels are used as 'the cleaner' of the environment [3]. As a filter feeder, an *Anadonta woodiana* gets food by filtering the water entering its body. The volume of water that can be filtered by an adult Chinese pond mussels is 2.5 litres per hour. The food that comes with the water is removed, squeezed, and digested with the help of cilia that

IOP Conf. Series: Earth and Environmental Science **137** (2018) 012058 doi:10.1088/1755-1315/137/1/012058

move 2—20 times per second. Foods that enter can be zooplankton, phytoplankton, bacteria, flagellata, protozoa, detritus, algae or various other substances suspended in the water where the mussels live [10].

The purpose of this study was to determine the ability of Chinese pond mussels as biofilters based on Pb concentration in the stomachs and gills of Chinese pond mussels, the metallothionein expression as well as its molecular weight.

## 2. Methodology

Chinese pond mussels (*Anodonta woodiana*) (8 cm in length) and carps (*Cyprinus carpio*) fingerlings (5 cm) were obtained from the Fish Seed Bureau, Punten, Batu, East Java. Chinese pond mussels were placed in a 20-litre tub. Mussels and carps were exposed to $PbNO_3$ for 96 hours [11] at doses of 0 %, 12.5 %, 25 %, and 37.5 %. The $PbNO_3$ doses used in the chronic exposure of this study were 0, 10, 20, and 30 ppm. Plastic tubs measuring 40 cm in height and 100 cm in diameter were filled with ten 8 cm Chinese pond mussels. The water in this tub was channelled into another fibre-sized tub that was filled with ten 5 cm carps *(Cyprinus carpio*). The water in the tub containing the fish was channelled back into the tub containing gravestone, resulting in closed system circulation. During rearing, the fish were fed with Hi-Pro-Fit 781-1 (CP Prima). On the other hand, the Chinese pond mussels were fed with phytoplankton and pellet (Feng Li Code FL0 flour, PT Matahari Sakti). The rearing tubs were aerated to ensure sufficient oxygen. On day $8^{th}$, the Chinese pond mussels were inserted into a recirculation-model tank and were left for a day without being fed (acclimatization process).

The Pb levels in the gills and stomachs of Chinese pond mussels were observed 96 hours after the exposure, and the Pb levels in carps were observed in weeks 0, 1, 2 and 3. The accumulation of Pb ions in the gills and stomach was observed using SEM/EDX. In this stage, qualitative measurement was performed with SDS PAGE and western blotting to determine the molecular weight of expressed MT in kilodalton (kDa). Observations on gill tissue damage and gastric gravity were also carried out with hematoxylin-eosin staining.

The expression of MT in the gills and stomachs of Chinese pond mussels was observed with the Olympus SN 3K19322 microscope. The quantitative measurement of MT was performed according to Ana and Garcia-Vasquez [12] with the standard curve of glutathione (GSH), assuming that 1 mole of MT contains 20 moles of cysteine.

MT molecular weight measurements were performed with SDS-PAGE according to Murthy *et al.* [4] method with standard BM from Thermo Scientific. The microanatomy condition of Taiwanese gills exposed to Pb was identified using SEM method, and a test with Energy Dispersive X-ray (SEM/EDX) was conducted to determine the levels of Pb accumulation in the target tissues.

## 3. Result and discussion

One day after the exposure, there was an increase in the mucus excretion around the Chinese pond mussels. High levels of $PbNO_3$ in the water resulted in the inhibition of the activity of enzyme and cell membrane due to the bonding of toxic materials with enzymes and cell membranes. This metabolism may cause stress which eventually resulted to caused excessive mucus excretion. This is a self-defence instinct of aquatic organisms to adapt to poor environmental quality. According to Palar [13], the effect of lead toxicity (Pb) primarily depends on the length of contact time, which caused intoxication of aquatic organisms as their response to heavy metals by increasing the mucus fraction of the gills, causing the respiratory and metabolic processes to be disrupted. Binelli *et.al* [14] described that heavy metals enter the body along with water or food. At first, the heavy metals will accumulate in the gills, then in the digestive tract, and eventually in the kidneys. Subsequently, the Chinese pond mussels will die if the heavy metals entering has exceeded the tolerable threshold. Pb is known to cause oxidative effects of the enzyme d-aminolevulinic acid dehydratase in molluscs. It is especially disruptive in the synthesis pathway.

The gills and stomachs of the Chinese pond mussels were collected 4 days (96 hours) after the beginning of the exposure. This was done by referring to the study conducted by Geffard *et al.* [11], indicating that the induction of metallothionein in *Mytilus galloprovincialis* larvae exposed to metal gives a noticeable effect at 96 hours after exposure. By that time, the metal will experience dilution and strong correlation with the increase in metal on the cytosolic fraction.

### *3.1. Molecular Weight (MW) MT*

Metallothionein (MT) is a non-enzymatic protein with a molecular weight of 6—7 kDa [9], 10 kDa [6], 3—15 kDa [5] or 14.3 kDa [4]. MT has high cysteine content and no aromatic amino acids and is unstable to heat. The thiol group of MT (-SH) is a cysteine residue which allows MT to bind heavy metals. The first MT was found on horseshoe cortex. In addition, MT has been reported to be found in vertebrates, including many species of fish [8], and invertebrates [9], especially molluscs [6].

The estimated molecular weight of MT is presented in tables 1 and 2, with an estimated MT molecular weight of Chinese mussel gills of 12.18 kDa. MT is a protein rich in cysteine, low molecular weight, and forms complexes with heavy metals such as Cd, Co and Zn which are essential metals for growth. MT also acts as in detoxification process for toxic metals such as Cd and Hg [5]. The results of Vergani *et al.* [16] showed that MT in mussel and fish had MW of 12 kDa, but MT on mussel was more reactive which was structurally different which was suspected due to life style differences in both organisms. Jenny *et al.* [17] mentioned that MT is a metal-binding protein with MW 6-7kDa with cysteine repetition (Cys-X-Cys or Cys-Xn-Cys) that plays a role in homeostatic metal, detoxification and stress response.

**Table 1.** Estimation of MT molecular weights with standard MW marker using Western Blotting method

| Band No. | BM | Range (cm) | Rf (x) | Log BM (y) | Equation |
|---|---|---|---|---|---|
| 1 | 40 | 0.8 | 0.1495 | 1.60 | Y = -1.1461x + |
| 2 | 25 | 1.75 | 0.3271 | 1.40 | 1.7848 |
| 3 | 15 | 2.9 | 0.5421 | 1.18 | R=0.9438 |
| 4 | 10 | 4.25 | 0.7944 | 1 | |
| 5 | 4.6 | 4.7 | 0.8785 | 0.66 | |

**Table 2.** Molecular weight estimation after the exposure of PbNO3

| Well | Rf | Log MW | BM |
|---|---|---|---|
| 1 | 0.6075 | 1.0886 | 12.26 |
| 2 | 0.6075 | 1.0886 | 12.26 |
| 3 | 0.6075 | 1.0886 | 12.26 |
| 4 | 0.6168 | 1.0776 | 11.96 |
| MW estimation of MT | | | 12.18 |

### *3.2. Metallothionein in the stomachs and gills of Chinese pond mussels*

MT induction is a "specific biomarker" in response to a heavily polluted environment. Increased MT synthesis relates to metal binding capability and increased resistance to metallic toxicity [15]. The expression of MT is a metal accumulation [18]. The results of measurements of MT values in each treatment on the gills and stomachs of Chinese pond mussels are presented in table 3.

**Table 3.** Data of MT content in gills and stomach.

| Content of $PbNO_3$ (ppm) | Gill | | | | Stomach | | | |
|---|---|---|---|---|---|---|---|---|
| | Repetition | | | | Repetition | | | |
| | 1 | 2 | 3 | Average | 1 | 2 | 3 | Average |
| 0 | 2600 | 2600 | 2600 | 2600 | 9250 | 9250 | 9250 | 9250 |
| 10 | 14150 | 5700 | 5600 | 8483 | 18000 | 14750 | 10100 | 14283 |
| 20 | 27350 | 26450 | 9050 | 20950 | 17900 | 17150 | 18150 | 17733 |
| 30 | 31350 | 32800 | 29000 | 31050 | 11600 | 11850 | 11000 | 11483 |

Metallothionein is a protein class that has a low molecular weight, high cysteine levels, and ability to bind IB and IIB metals [6]. MT has a protein and polypeptide content which is affected by metal exposure. MT primarily functions in metabolic processes and detoxification of essential and non-essential metals. The results of previous studies explain that MT has a role in the regulation and control of intracellular availability of Cu, Zn, and Cd as well as other metals that have been studied. MT is able to control the removal of Cu and Zn with the help of appropriate receptor molecules such as metalloenzymes. Thus, metal setting is very specific. Because of MT's selectivity and role in the process of metabolism of metals in organism bodies, MT is believed by researchers to be a protein that plays a role in protecting organisms from metal toxicity. This protein binds to toxic metals such as Cd, where MT acts as a Cd receptor [8]. Many studies have been conducted on MT in relation to metal pollution, allowing this protein to be used as a biological marker for metal pollution in the water [19]. MT level in the stomach decreased in the exposure of Pb at 30 ppm with an average value of 11.483 ng/mL. It was suspected as a metabolic disorder because the Chinese pond mussels had been poisoned as the dose of $PbNO_3$ became untolerable. The poisoning in the organism was caused by the decrease in the immune system. According to Dafre *et al.* [20], poisoning or decrease in the immune system in bivalves results from the oxidative reaction of the enzyme d-aminolevulinic acid dehydratase due to the high levels of heavy metals accumulated in bivalves.

### *3.3. Pb levels in the gills and stomachs of Chinese pond mussels*

The results showed that the Pb content was directly proportional to the doses of $PbNO_3$ administered. The lead content in the gills tended to be higher at higher doses of $PbNO_3$ exposure. According to [4], lead (Pb) is a heavy metal that will be continuously accumulated in organisms because it has properties that tend to be difficult to regulate by living things. The lead content in the stomach tends to be higher at higher exposure doses as well. This is due to the fact that the absorption of heavy metals also takes place along the digestive tract. The higher the level of Pb directly given, the higher the heavy metal content mixed with the food and accumulated in the digestive tract. According to [21], the absorption of heavy metals by organisms generally takes place in the gills and digestive tract. Therefore, Pb tends to accumulate in the gills and stomach. According to Indonesian National Standard [22], the safe limit of concumption of heavy metal Pb contained in bivalves is 1.5 ppm. The Pb levels in the stomach and gills are presented in figure 1.

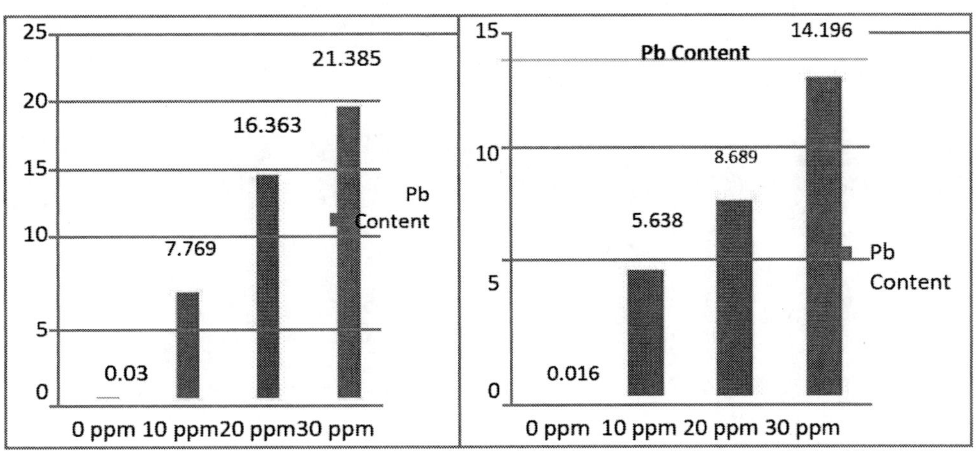


(a) b

**Figure 1**. Pb content in the gills (a) and stomachs (b) of Chinese pond mussels

### *3.4. Pb content in carp*

Carps are frequently used in toxicity tests because they have properties that are sensitive to environmental changes. The most sensitive, and the first, organ in contact with the environment is the gill. Carps are among the fish species that have low tolerance to water quality as well as to the toxic substances contained. The Pb content of carp fish gill decreased in week 3 and week 4 as the heavy metals were reduced. According to Liu *et al.* [23], Chinese pond mussels (*Anodonta woodiana*) are able to accumulate good heavy metals in freshwater because they serve as bio-accumulators based on high heavy metal content.

*3.5. Chinese Pond Mussels Histology*

*Anodonta woodiana* breathes with two gills and the mantle section. These gills are sheets of structures (lamellae) that contain a lot of gill stems. Meanwhile, between the body and the mantle is a mantle cavity. This cavity is the air exit path [24]. Therefore, the gills of *Anodonta woodiana* are very sensitive to the effects of heavy metal toxicity. The normal gill histology of Chinese pond mussels is presented in figure 3a. In normal gills, the gill lamellae arrangement is still well-constructed. At a $PbNO_3$ exposure dose of 10 ppm, the Chinese pond mussels suffered from edema and hyperplasia. Edema is the swelling of cells or accumulation of fluid in the body [25]. At a $PbNO_3$ exposure dose of 20 ppm, the gills showed pathology of hyperplasia, edema, atrophy, and lysis. In abnormal growth, epithelial cells slowly erode the epidermal layer of gill lamellae and push out the lamellae, supported by a liquid dam on the basal membrane, causing erosion of the basement membrane and lysis, or outbreak, of the lamellae. Severe damages occurred in the gills of the Chinese pond mussels at a dose of 30 ppm as the gills could no longer repair cell damages that occurred. Cells that are necrotic will escape their supporting tissues (basement membranes), and nearby tissues become susceptible to irritants [26].

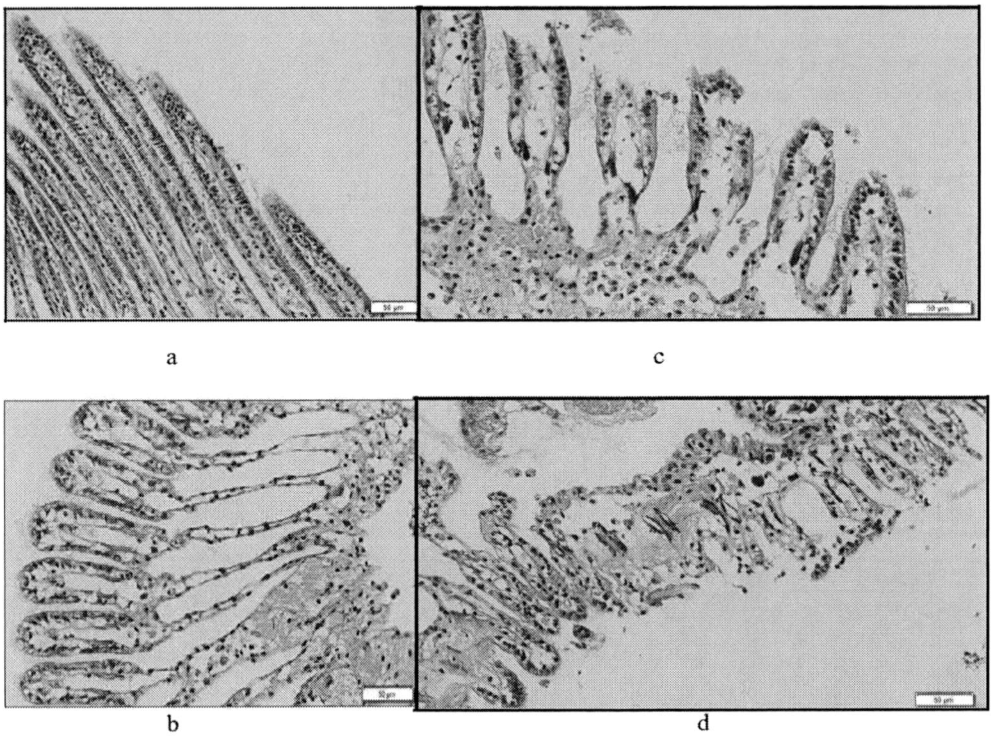

a c

b d

Note: (a) 0 ppm of $PbNO_3$, (b) 10 ppm of $PbNO_3$, (c) 20 ppm of $PbNO_3$, (d) 30 ppm of $PbNO_3$

**Figure 2.** Chinese pond mussel gill histology after $PbNO_3$ exposure.

*3.6. Stomach Histology of Chinese Pond Mussels*

Food enters the Chinese pond mussel mouth and continues its way down the esophagus, which is shaped like a dorsoventral flat sponge and seems as if it is the front of the stomach, then into the stomach. The protective outside of the gastric wall is called digestive diverticulum, which is a thin but strong tissue. The thickness of the wall is low to high, ranging between 44.4 μ and 93.3 μ.

Before being exposed to Pb, the outer portion the Chinese pond mussels is oval, and the inside is squiggly. Silia line up neatly over the surface of the stomach wall.

Digestive diverticula that tend to be round or oval spread out on the outside of the abdomen. According to the results of this observation, the stomachs of the Chinese pond mussels were seen to be in a good condition and had not been damaged yet. At a $PbNO_3$ exposure dose of 10 ppm, the stomachs of the Chinese pond mussels had edema and hyperplasia in the digestive diverticula. At an exposure dose of above 10 ppm, thickening occurred on the stomach wall, which was accompanied by fluid cavity located at the bottom of cilia. The irregular formation of cilia looked like a pile of cilia leading to the inside of the stomach. Slowly, the pile of cilia merged into one. The histology of the stomachs of the Chinese pond mussels is shown in figure 3.

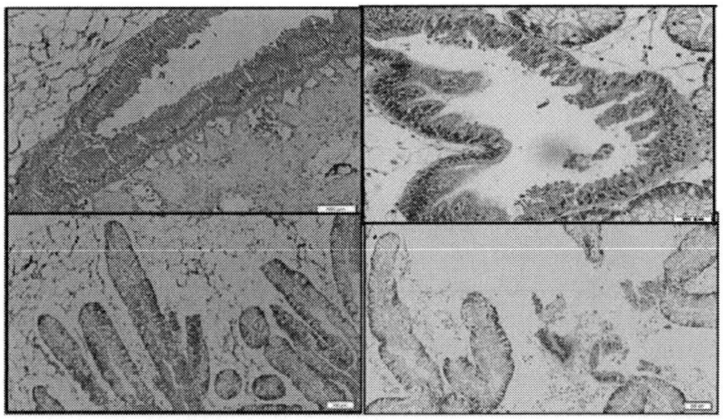

**Figure 3.** The histology of the stomachs of the Chinese pond mussels

### *3.7. Observation of Pb content with SEM/EDX*

The Pb in the gills of the Chinese pond mussels observed using an electron microscope is shown in figure 5. The exposure of 30 % $PbNO_3$ provided more Pb binding effects in the gills of the Chinese pond mussels (table 4). Similar results were obtained by Bebiano and Serafim [27], who induced *Mytilus galloprovicialis* with Cd at 100 ug/L. The concentration of metal in the organ due to heavy metal induction (table 1, table 4) shows the role of MT as a detoxifier. Similarly, Ghedira [28] also showed that metal concentration and MT induction strengthen the role of MT as in the metal homeostatis and detoxification.

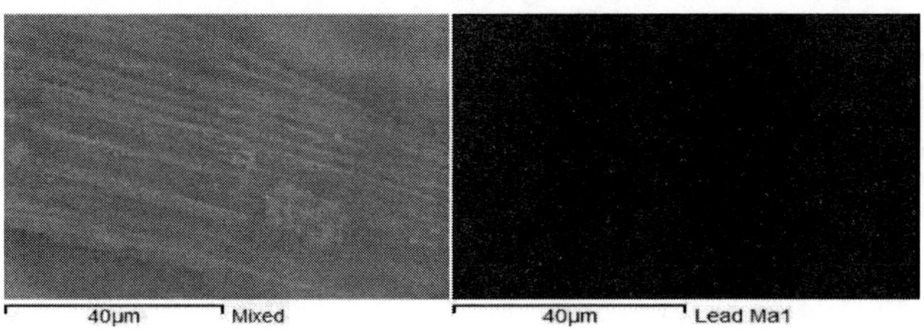


**Figure 4.** The Pb in the gill of a Chinese pond mussel observed using an electron microscope being marked with red spots.

IOP Conf. Series: Earth and Environmental Science **137** (2018) 012058 doi:10.1088/1755-1315/137/1/012058

**Table 4.** The Accumulation of Pb in the gills of Chinese pond mussels observed using SEM/EDX.

| Treatment | Weight (%) | Atomic (%) |
|---|---|---|
| $PbNO_3$ 0 % | 0.179 | 0.001 |
| $PbNO_3$ 10 % | 0.281 | 0.029 |
| $PbNO_3$ 20 % | 0.718 | 0.072 |
| $PbNO_3$ 30 % | 0.683 | 0.072 |

## 4. Conclusion

Chinese pond mussels can be used as biofilters for carp cultivation in the closed system. The highest Pb accumulation in gills of carps was detected in the first week of treatment, and the accumulation continuously decreased over 30 days. Accumulations of Pb ions in the gills and stomach were detected at a $PbNO_3$ exposure dose of 30 ppm. The MT molecular weight of Chinese pond mussels was 12.14 kDa.

## 5. Reference


[1]. Hafiz M 2009 Mussel (*Pilsbryoconcha exilis*) characterization in Situ Gede (Bogor : Bogor Agriculture Institute)

[2]. Hasim 2008 Mussels as heavy metal biofilter http://www.kompas.com/kompas-cetak.html

[3]. Dave A 2009 Mussels as heavy metal biofilter http://kimia9.blogspot.com/2009/01/kerang-sebgai-biofilter-logam-berat.html

[4]. Murthy S, Bali G and Sarangi K 2011 *Afr. J. Biotechol.* **10**(71): 15966-15972.

[5]. Couillard Y, Campbell P G C, Pellerin J M and Auclair J C 1995 *J. Fish. Aquac. Sci* **52**: 703-715.

[6]. Couillard Y, Peter G C C and Tessier A 1993 *Limnology Ocenogaphy* **38** 299-313.

[7]. Ivanina A N, Cherkasov A S and Sokolova I M 2007 *J Experiment Biology* **211**: 577-586.

[8]. Roesijadi G, Rezvankhah S, Perez M A, Mitelberg A, Torruellas K and Veld V P 2009 *Mar.Envir. Res. J.* **67** 25–30.

[9]. Fowler B A, Hildebrend C E, Webb Y M 1987 *Experientia Supply* **52** 19-22

[10]. Mamun A A and Khan Md.A 2011 *World Appl.Sci.* **12** 580-585.

[11]. Geffard A, Olivier G, Edouard H and Amiard J C 2002 *Mar.Ecology Progress Series* **233** 131-142

[12]. Ana R L and Garcia-Vazquez E 2006 *Biochemical. Moecularl. Biology. Education* **34** 360–363

[13]. Palar, H 2008 *Pollution and toxicology of heavy metals* (Jakarta : Rineka Cipta)

[14]. Binelli A, Riva C and Provini A 2010 *Chemospere* **79** 518-528

[15]. Lemoine S, Bigot Y, Sellos D, Cosson R P and Laurier M 2000 *Mar.Biotech.* **2** 195-203

[16]. Vergani L, Grattarola M, Borghi C, Dondero F and Viarengo A 2005 *FEBS Journal* **272** 6014-6023

[17]. Jenny M J, Wan G W, Ringwood A M., Boltzegar D A and Chapman R W 2000 *Gene* **379**:156-165

[18]. Wen-Xing W and Rainbow P S 2005 *Ecotoxicology and Environmental safety* **61** 145-159

[19]. Krizkova S, Adam V, Petrlova J, Zitka O, Stejskal K, Zehnalek J, Sures B, Trnkova L, Beklova M and Kizek R 2007 *Electroanalysis* **19** 331-338

[20]. Dafre A L, Medeiros I D, Muller I C, Ventura E C and Bainy A C D 2004 *Chem. Biol. Interact* **149** 97-105

[21]. Suaniti N M 2007 The effect of EDTA in the determination of lead and copper content in green shells (*Mytilus viridis*) **2**

[22]. Indonesian National Standard (SNI) 2009 Determination of heavy metal content of lead (Pb) and Cadmium (Cd) on fishery products **87** 14-20

[23]. Liu J, Gu B, Bian J, Hu S, Cheng , Ke, Q and Yan H 2008 *Eur. Food Res. Tech.* **227** 919-924

[24]. Imaningtyas *et al* 2008 Invertebrae resource guide 2 www.modulonline.com

IOP Conf. Series: Earth and Environmental Science **137** (2018) 012058 doi:10.1088/1755-1315/137/1/012058

[25]. Laksman T 2003 *Medical dictionary* (Jakarta : Djambatan)
[26]. Widayati W E 2011 Sugar cane-spesific bacteria can increase sugar content and weight *Sugar rod. Article* 12-33
[27]. Bebianno M J and M A Serafim 1996 *Sci. Total Envi* **214** 123-131
[28]. Ghedira. J, J Jebali, Z Bouraoli., M Banni., H Guerbej., H Bousetta. 2010. *Fish. Phys. Biochem.* **36** 101-107

IOP Conf. Series: Earth and Environmental Science **137** (2018) 012059 doi:10.1088/1755-1315/137/1/012059

# The effect of amino acid lysine and methionine addition on feed toward the growth and retention on mud crab (*Scylla serrata*)

**Y R Alissianto[1], Z A Sandriani[1], B S Rahardja[2], Agustono[2] and Rozi[2]***

[1]Bachelor Programme of Aquaculture, Faculty of Fisheries and Marine, Universitas Airlangga, Surabaya 60115, Indonesia
[2]Department of Fish Health Management, Faculty of Fisheries and Marine, Universitas Airlangga, Surabaya 60115, Indonesia

E-mail: rozi@fpk.unair.ac.id

**Abstract.** High market demand of mud crab (*Scylla serrata*) encourages farmers to increase the production of mud crab. However, mud crab can not synthesize essential amino acids, so it is necessary to supply essential amino acids such as lysine and methionine in the diet. This study aims to determine the effect of lysine and methionine on feeds to increase growth and retention of mud crabs (*Scylla serrata*). In this study the amount of lysine amino acid and methionine added to the trash fish diet were: P0 (0: 0%); P1 (0.75: 0.75%); P2 (1: 1%); P3 (1.25: 1.25%); P4 (1.5: 1.5%) with the ratio of lysine and methionine 1: 1. The parameters observed in this study were Survival Rate (SR), Specific Growth Rate (SGR), Feed Conversion Ratio (FCR), Efficiency Feed (EF), protein retention and energy retention. The results of the 35-day maintenance study showed significant differences ($P < 0.05$) against Specific Growth Rate (SGR), Feed Conversion Ratio (FCR), Efficiency Feed (EF), protein retention and no significant effect ($P > 0.05$) on energy retention and Survival Rate (SR) on mud crab. The best results in this study were found in P4 treatment with addition of lysine amino acids and methionine (1.5: 1.5%).

## 1. Introduction

Mud crab (*Scylla serrata*) is an important economic fishery commodity. The demand of the global mud crab market is increasing every year. The demand for mud crab increases because of its high nutritional content so that it is demanded by many people [1]. The high market demand encourages farmers to increase the production of mud crab (*Scylla serrata*). Mud crabs (*Scylla serrata*) have the potential to be cultivated because of their advantages, including rapid growth, larger size than other species, disease resistance, and high selling prices [2].

Feed is a factor that must be met to achieve maximum production in the cultivation of mud crabs. The use of trash fish feed is more advantageous because the price is relatively inexpensive, easy to obtain, and suitable with the eating habit of mud crab (*Scylla serrata*) [3]. Essential amino acids are amino acids derived from additional food ingredients. Because they cannot be produced by the body, they need to be obtained in order to be able to meet the daily needs of amino acids [4].

Some of the essential amino acids needed by mud crabs (*Scylla serrata*) are lysine and methionine [5, 6]. Lysine is an amino acid that makes proteins accelerate growth [7]. Methionine is an essential amino acid containing sulfur, a substance necessary for natural antioxidants [8]. The presence of lysine and methionine amino acids in the diet can improve the utilization balance of other amino acids in

IOP Conf. Series: Earth and Environmental Science **137** (2018) 012059 doi:10.1088/1755-1315/137/1/012059

order to increase growth [9]. Research on the addition of lysine and methionine amino acids to determine the growth, energy retention and protein retention in mud crab has not been done. So, further research is required.

## 2. Methodology

### *2.1. Sample preparation*

The samples were male mud crabs (*Scylla serrata*) with the size of 80-90 gram suitable for enlargement [9]. Male mud crabs (*Scylla serrata*) can grow faster than the female ones, because the feed consumed by male crabs is only used for growth process and not for egg maturation process [10].

### *2.2. Feed feeding*

Sampling was carried out once every 7 days to determine the amount of feed given during maintenance. The amount of feed required for mud crab was 5% of its body weight in order to grow optimally [11]. Feed given in the form of trash fish that had been mixed with lysine and methionine amino acids with a dose of P0 (0: 0 %); P1 (0.75: 0.75 %); P2 (1: 1 %); P3 (1.25: 1.25 %); P4 (1.5: 1.5 %).

### *2.3. Feed making*

In the process of making the feed, the composition of each material needed was calculated and weighed in accordance with the needs of the feed. Then, the type of yellow fish was cut and smoothed using chopper. Further, lysine and methionine were added to the amino acids, further binded in the form of tapioca starch with a predetermined dose into the trash fish feed. It should be noted that the mixing of feed ingredients should be done gradually, starting from the smallest to largest materials needed. Trash fish feeds that had been added with lysine and methionine amino acids and tapioca starch were aerated about 5-10 minutes before being given to mud crabs (*Scylla serrata*) [12].

### *2.4. Proximate analysis*

The mud crab samples were tested in terms of their protein content using Kjeldahl method. The principle of kjeldahl method includes 3 stages of destruction, distillation, and titration. At the stage of destruction, the sample was taken as much as 1 gram and then put into kjeldahl flask. HgO 40gr, $K_2SO_4$ 1.9 mg and $H_2SO_4$ 2 ml were added. Then, the gourd was heated in the acid chamber until the solution became clear. The result of the destruction was diluted with 10-20 ml of distilled water. The distillation was carried out by washing and rinsing the destruction sample with distilled water on the distillation apparatus, then 8-10 ml of sodium thiosulfate was added. The result was then diluted to about 50 ml. Prepare a 125 ml containing 5 ml of boric acid and 2-4 drops of indicator of (mixture of 2 parts methyl red 0.2 % in alcohol and 1 part methylene blue 0.2 % in alcohol). Titrations were performed by dripping HCl 0.02 N from the burette on the sample until the solution color turned pink [13].

### 2.5. *Protein and energy retention*

Protein and energy retention can be calculated based on the following formula of [14]:

Protein retention:

$$\frac{Final\ body\ protein\ (gram) - initial\ body\ protein\ (gram)}{Total\ protein\ consumption\ (gram)} \ x\ 100\ \% \quad (1)$$

Energy retention:

$$\frac{Final\ energy\ (kkal) - Initial\ energy\ (kkal)}{Total\ energy\ consumption\ (kkal)} \ x\ 100\ \% \quad (2)$$

IOP Conf. Series: Earth and Environmental Science **137** (2018) 012059 doi:10.1088/1755-1315/137/1/012059

2.6. *Statistical analysis*
Statistical analysis was conducted using Analysis of Variance (ANOVA) to determine the effect of the treatment given. If there were significant results, then the calculation was continued with Duncan Multiple Range Test.

## 3. Results and Discussion

The results showed a significantly different on each treatment and the highest growth yield was in P4 treatment with the addition of lysine and methionine of 1.5: 1.5%. The lowest crab growth was found in P0 treatment without the addition of lysine and methionine in feed. Meanwhile, for SR, it is not significantly different in each treatment (table 1).

**Table 1.** The average growth data.

| Treatment | SR ± SD | GR± SD | SGR ± SD | FCR ± SD | EF± SD |
|---|---|---|---|---|---|
| P0 | 1 ± 0.000 | $0.47^{a} \pm 0.041$ | $0.53^{a} \pm 0.04$ | $2.80^{b} \pm 0.321$ | $0.36^{a} \pm 0.037$ |
| P1 | 1 ± 0.000 | $0.52^{ab} \pm 0.034$ | $0.59^{ab} \pm 0.03$ | $2.92^{b} \pm 0.210$ | $0.34^{a} \pm 0.024$ |
| P2 | 1 ± 0.000 | $0.55^{bc} \pm 0.040$ | $0.61^{bc} \pm 0.04$ | $2.78^{b} \pm 0.205$ | $0.36^{a} \pm 0.025$ |
| P3 | 1 ± 0.000 | $0.56^{bc} \pm 0.015$ | $0.62^{bc} \pm 0.14$ | $2.72^{b} \pm 0.152$ | $0.36^{a} \pm 0.021$ |
| P4 | 1 ± 0.000 | $0.59^{c} \pm 0.055$ | $0.65^{c} \pm 0.06$ | $2.04^{a} \pm 0.193$ | $0.49^{b} \pm 0.046$ |

Note: The results ($p < 0.05$) were significantly different, and the results ($p > 0.05$) were not significantly different. P0: Lysine dose and methionine (0.0 %) P1: Lysine dose and methionine (0.75:0.75 %); P2: Lysine dose and methionine (1:1 %); P3: Lysine and methionine dosage (1.25:1.25 %); P4: Lysine dose and methionine (1.5:1.5 %). Survival Rate (SR), Growth Rate (GR), Specific Growth Rate (SGR), Feed Conversion Ratio (FCR), Efficiency Feeds (EF).

The results of protein retention showed significantly different results on each treatment with different dose. Protein retention increased in P1 (0.75: 0.75 %), P2 (1: 1 %), P3 (1.25: 1.25 %), and significantly in P4 treatment with lysine and methionine doses of 1.5: 1.5%. The results of energy retention showed no significant difference on each treatment with different dose. The highest energy retention at P4 treatment was 148.1375 kcal/kg with the addition of lysine amino acid and methionine in feed at 1.5 %: 1.5 % methionine (table 2).

**Table 2.** The average retention data.

| Treatment | Protein retention (%) ± SD | Energy retention (Kkal/kg) ± SD |
|---|---|---|
| P0 | $20.280^{a} \pm 7.652$ | 143.775 ± 27.601 |
| P1 | $36.485^{ab} \pm 17.181$ | 118.612 ± 6.761 |
| P2 | $46.517^{b} \pm 12.719$ | 132.585 ± 6.452 |
| P3 | $50.040^{b} \pm 11.419$ | 125.022 ± 13.010 |
| P4 | $84.137^{c} \pm 21.841$ | 148.137 ± 19.839 |

Note: The results ($p < 0.05$) were significantly different, and the results ($p > 0.05$) were not significantly different. P0: Lysine and methionine dosage (0.0 %) P1: Lysine and methionine dose (0.75:0.75 %); P2: Lysine dose and methionine (1:1 %); P3: Lysine and methionine dosage (1.25:1.25 %); P4: Lysine dose and methionine (1.5:1.5 %).

The Increase of GR and SGR occured because the food consumed can be digested and absorbed well by the body. This is indicated by the increased weight of mud crab (*Scylla serrata*) at the end of treatment [15]. The highest increase of GR and SGR were in P4 treatment with lysine and methionine dose of 1.5: 1.5 %. The presence of lysine and methionine in the diet can improve the balance of other amino acids so that the process of absorption of nutrients into the body become faster, leading to the increase of growth [9].

The lowest FCR and the highest EF were identified in P4 treatment with lysine and methionine dose of 1.5: 1.5 %. Good quality feed has low FCR and highest EF [3]. Undigested or undesirable feed is a factor that causes a high Feed Convertion Ratio (FCR) [16]. Factors influencing the value of EFF are feed type, feed quality, and feeding technique [17]. Lysine and methionine are simple compounds. When entering the stomach, they will be rapidly metabolized by digestive enzymes and absorbed by the body quickly [18]. Addition of lysine and methionine may increase the utilization of feed consumed, resulting in a low feed conversion ratio [19].

The level of SR was not significantly different in each treatment because lysine and methionine absorbed by the body only affected the weight gain of mud crab (*Scylla Serrata*) [20, 21]. Increased protein content of mud crabs is due to lysine and methionine, an important amino acid in protein synthesis (transcription process, which translates the sequence of nitrogenous bases in DNA to form RNA) because the methionine code is the same as the initial code for one set of RNAs. The protein content in the body of its existence is influenced by the retention of feed and the pile of proteins that are positively correlated with the protein content of the feed [22]. Retention of protein has increased from the control treatment. This is because trash fish added with lysine and methionine have fulfilled the nutritional needs of feed as well as absorption of essential amino acids [23]. The study of Palavesam *et al.* [24], in pearl spotted fish also showed an increase in protein metabolism after feeding was given L-Lysine. For example, fish given 25 % L-Lysine had higher consumption, assimilation and protein production compared to the control treatment.

The decrease of energy in mud crabs caused by excessive protein supply without the addition of energy resulted in the increase of energy for the deamination process of lysine and methionine amino acids so that it will reduce the portion of energy for vital growth [25, 26]. Energy retention is not significantly different from control treatment. This is because the size of the crab is too small. So, most of the energy absorption from lysine and methionine is used for growth [23]. Crustaceans use energy savings in feed for growth, metabolism, maintenance needs, ammonia excretion, feces and molting [27].

## 4. Conclusion

this study has demonstrated that amino acid lysine and methionine can potentially be used in addition on feed toward the growth and retention on mud crab (*Scylla serrata*).

## 5. References

[1] David H O 2009 *West Indian Ocean J.* **8** 127-137.
[2] Viswanathan C and S M Raffi 2015 *Saudi J. Biol. Sci.* **47** 99-105
[3] Suwirya K, A D Nyoman and M Marzuqi 2009 *Indonesian Aqua. J.* **2** [Text in Indonesia].
[4] Mohanty B 2014 *J. Amino Acids* **2**
[5] Hardy R W and Barrowa F T 2002 *Diet Formulation and Manufacture* Fish Nutrition **3** 505-600
[6] Gatlin D M, F T Barrows and P Dabrowski 2007 *Aquc. Res.* **38** 551-557
[7] Giri I N A, A S Sentika, K Suwirya and M Marzuqi 2009 *J. Aquac. Res. Dev.* **4** 357-366
[8] Latiful M I and K Yahya 2016 *Intern. J. Fish. Aqua. Studies* **4** 270-278
[9] Veronica D P 2004 *Isr. J. Aquac.* **56** 111-123.
[10] Islam M D and H Kurokura 2013 *J. Fish.* **1** 7-13
[11] Trino A T, O M Millamea, and C P Keenan 2001 *Asian Fish. Sci.* **14** 191-200
[12] Dani N P, B Agung, dan L Shanti 2005 *Jurnal Biosmart* **7** 83-90 [Text of Indonesia]
[13] AOAC 1995 *Official methods of analytical of the association of chemists* ($16^{th}$ edition) AOAC Arlington VA
[14] Thung P H and S Y Shiau 1991 *Aquaculture* **92** 343-350
[15] Jamabo N A, R I Fubara and H E Dieny 2015 *Int. J. Fish. Aquat. Sci.* **3** 353-356

IOP Conf. Series: Earth and Environmental Science **137** (2018) 012059 doi:10.1088/1755-1315/137/1/012059

[16] Sutarmat T 2006 *preliminary study determination of red snapper in KJA with different stock sizes* Prosiding Konferensi Akuakultur Indonesia p 193-197 [Text in Indonesia]
[17] Boyd C 2005 *Feed efficiency indicators for responsible aquaculture* Global Aquaculture Advocate.pp 73
[18] Hamid S N I N, M F Abdullah, and Z Zakaria 2016 Formulation of fish feed with optimum protein-bound lysine for african catfish (*Clarias Gariepinus*) Fingerlings *Rocedia Engineering* **1**(3) 361-369.
[19] Abdel A, F Yan, S Cerrate, and C Coto 2010 *Int. J. Poult. Sci.* **9** 21-27
[20] Baker H D 2007 *J. Nutrition*
[21] Saha M R, P K Roy, and U Ahmed 1996 *Bangladesh J. Fish.* **1** 47-52.
[22] Phumee P, Hashim R, Paiko M A, Chien A C S *Aquat. Res.* **40** 456-463.
[23] Yamamoto T, T Sugita, and H Furuita 2005 *Aquaculture* **246** 379-391.
[24] Palavesam A, S Benna and G Immanuel 2008 *Turkish J. Fish. Aquat. Sci.* **8** 133-139.
[25] NRC (National Research Council) 1993 *Nutrient requirement of fish* (Washington DC : National Academy of Press)
[26] Hernandez M D, Egea M A, Rueda F M, Aguado F, Martinez F J, Garcia B 2001 *Growth and Nutrient Utilization Aquaculture* **195** 321-329.
[27] Bhavan P S, S A Ruby, Poongodi, C Seenivasan and S Radhakrishnan 2010 *J. Ecobiotechnology* 2:5 (09-19). 10 p.

IOP Conf. Series: Earth and Environmental Science **137** (2018) 012060 doi:10.1088/1755-1315/137/1/012060

# The antagonistic activity of lactic acid bacteria isolated from *peda*, an Indonesian traditional fermented fish

**T F Putra[1], H Suprapto[2], W Tjahjaningsih[3], and H Pramono[3 4]**

[1] Study Program of Aquaculture, Faculty of Fisheries and Marine, Airlangga University, Surabaya 60115, Indonesia
[2] Departement of Fish Health Management and Aquaculture, Faculty of Fisheries and Marine, Airlangga University, Surabaya 60115, Indonesia
[3] Department of Marine, Faculty of Fisheries and Marine, Airlangga University, Surabaya 60115, Indonesia

E-mail: heru.pramono@fpk.unair.ac.id

**Abstract.** *Peda* is an Indonesian traditional fermented whole fish prepared by addition of salt prior to fermentation and drying process. Salt used to control the growth of the lactic acid bacteria for the fermentation process. The objectives of this study were isolating and characterize the potential lactic acid bacteria (LAB) from *peda* as culture starter candidate, particularly its activity against pathogenic bacteria. A total of five samples from five regions of East Java Province was collected and subjected to LAB isolation. Fifty-seven of 108 colonies that show clear zone in de Man, Rogosa and Sharpe (MRS) agar supplemented with 0.5% $CaCO_3$ were identified as LAB. Twenty-seven of the LAB isolates were exhibit inhibition against *Staphylococcus aureus* ATCC 6538 and *Pseudomonas aeruginosa* ATCC 27853. Isolate *Aerococcus* NJ-20 was exhibited strong inhibition against *S. aureus* ATCC 6538 (7.6 ± 1.35 mm inhibition zone) but was not produce bacteriocin. This finding suggests that the isolate *Aerococcus* NJ-20 can be applied as biopreservative culture starter on *peda* production. Further analysis on technological properties of isolates will be needed prior to application.

## 1. Introduction

*Peda* is an Indonesian traditional fermented fish product that contains 20-30 % (w/w) salt [1]. *Peda* is often produced from mackerel (*Rastrelliger* sp.) [2] with the addition of salt and put through a drying process to prolong the shelf life of the fish. Salt in fermented fish inhibits the growth of spoilage and pathogenic bacteria, as well as maintaining lactic acid bacteria for the fermentation process [3].

Lactic acid bacteria (LAB) is Generally Recognized as Safe (GRAS) by WHO and plays an important role in the process of fermentation of food such as inhibiting spoilage bacteria and producing the flavour, aroma, and texture of fermented food [4,5]. LAB may be isolated from fermented fish such as *plasom*, *bekasam*, fish sauce, and *peda* [6,7,8,9].

The antagonistic activity of LAB is due to antimicrobial compounds produced by the bacteria such as lactic acid, hydrogen peroxide ($H_2O_2$), diacetyl, reuterin, and bacteriocin [10]. Lactic acid can disrupt the outer membrane of bacteria that causes lysis, whereas bacteriocins are antimicrobial proteins or peptides synthesized by ribosomes [11] that initiate pore formation on the bacterial membrane. Bacteriocin has bactericidal or bacteriostatic properties against spoilage and pathogens bacteria and is used as a biopreservative in food products [12,13].

The antagonistic activity of LAB was isolated from various fermented fish such as *bekasam* [14]. The different regions of production and methods performed during the production of *peda* may lead to differences in LAB isolation as well as their antagonistic activity against foodborne pathogens. The objective of this study was to isolate and analyze the antagonistic activity of LAB from *peda* as a culture starter of *peda* production.

## 2. Methodology

### *2.1 Samples collection*

Sample collection was conducted by the method of Cho [15]. Mackerel *Peda* (*Rastrelliger* sp.) samples were collected from five regions of East Java (Surabaya, Sidoarjo, Gresik, Kediri, and Nganjuk). All samples were wrapped in sterile plastic bags, stored in an insulated cool box (4°C) and transported to the laboratory within 5-8 hours.

### *2.2 Isolation of lactic acid bacteria (LAB)*

The isolation of lactic acid bacteria was done according to Angmo et al [16] and Pranomo et al [17]. A total of 10 g sample of mackerel *peda* was homogenized with 90 ml of normal saline (0.85 % w/v) and diluted. One mL of the diluted sample was spread on de Man, Rogosa and Sharpe (MRS, Merck, Germany) Agar supplemented with 0.5 % $CaCO_3$, and then incubated at 30°C for 48 hours. Individual colonies were picked and streaked on MRS supplemented with $CaCO_3$ and incubated at 30°C for 24 hours. Purified cultures were tested for catalase and Gram staining. All catalase negative and Gram-positive strains were stored in MRS broth with the addition of sterile glycerol (1:1 v/v) at -20°C.

### *2.3 Antagonistic activity of LAB*

The isolates of LAB were cultured on de Man, Rogosa and Sharpe (MRS, Merck, Germany) broth and incubated at 30°C for 24 hours twice prior to the antagonistic test. The indicator bacteria (*Staphylococcus aureus* ATCC 6538 and *Pseudomonas aeruginosa* ATCC 27853), which represented Gram-positive and Gram-negative bacteria respectively, were cultured on Trypticase Soya Broth (TSB, Oxoid, UK) and incubated at 30°C for 24 hours. Each indicator bacteria ($10^7$ CFU/ml) was swabbed on a Mueler Hinton Agar (MHA, Oxoid, UK) medium using a sterile cotton-swab. LAB culture was applied at 20 µl into a sterilized paper disk. The paper disk was then placed on the surface of the MHA medium that had been swabbed with indicator bacteria and incubated at 30°C for 24 hours. LAB isolates that formed the inhibited zone ($\geq$ 6 mm) were selected for the bacteriocin production test [18].

### *2.4 Bacteriocin-producing test*

The LAB isolates which inhibited indicator bacteria were cultured on de Man, Rogosa and Sharpe (MRS, Merck, Germany) broth medium and incubated at 30°C for 24 hours. The culture was then centrifuged (Hettich Rotanta 420 R, Germany) at 4600 rpm at 4°C for 10 minutes, then the pH adjusted to 7.0 with 1 N NaOH. The neutralized-cell free supernatant is heated at 80°C for 10 minutes to eliminate hydrogen peroxide and kill the cells. The neutralized-cell free supernatant was filtered with 0.22 µm millipore filter. The antimicrobial activity of the supernatant was confirmed by the Agar Well Diffusion Assay (Pringsulaka et al. 2012) against the indicator bacteria *S. aureus* ATCC 6538. MHA (Merck, Germany) soft agar (0.75 % agar w/v) which was inoculated with 1 ml of indicator bacteria $10^7$ CFU/ml. A 50 µL cell free neutralized supernatant was loaded to each well and incubated at 30 °C for 24 hours. The isolate that exhibited antimicrobial activity in the AWD test was then characterized by the nature of its activity by adding proteinase K. The neutralized-cell free supernatant that lost activity by the addition of proteinase K indicates bacteriocin production [30]. Briefly, the proteinase K of 200 µg (w/v) was dissolved in a 1 ml phosphate buffer solution and then added to the neutralized-cell free supernatant. The neutralized-cell free supernatant was incubated at 30 °C for 2 hours. The activity of the proteinase K enzyme was stopped by heating it at the temperature of 100 °C for 30 minutes. The neutralized-cell free supernatant was tested against *S. aureus* ATCC 6538 with Agar Well Diffusion Assay.

*2.5 Identifitication of LAB*
Identification was performed to find the genus of LAB that showed high antagonistic activity. The characterization included performing tests on gas production (fermentation type), pH effect, temperature influence, and salinity and comparing them to the manual identification of LAB.

*2.6 Data analysis*
All measurements of antagonistic activity were performed in triplicate. The data was expressed in the mean and standard deviation of the three experiments.

## 3. Results and Discussion

*3.1 Isolation of LAB*
A total of 108 colonies that exhibited a clear zone on de Man, Rogosa and Sharpe agar supplemented with $CaCO_3$ 0.5% was purified in MRS. Among 108 isolates, only 57 isolates (52.7 %) were included as lactic acid bacteria due to it being Gram-positive, rods and cocci-shaped, and negative catalase. LAB shows a clear zone on MRS supplemented with $CaCO_3$ due to the production of organic acid (such as lactic acid, propionic acid, and/or acetic acid) [5] that dilutes the $CaCO_3$ in the medium, then the organic acids can react with $CaCO_3$ forming Ca-lactate and forms a clear zone [6,19]. LAB is a Gram-positive, non-spore forming, rod or cocci-shaped, catalase negative, and non-motile bacteria. In addition, from the data of Gram staining (*data not shown*), 12 and 45 isolates out of 57 isolates LAB were rod and cocci-shaped respectively.

The isolating rate of lactic acid bacteria was correlated with various factors such as the location, method, and source of LAB. Angmo et al [16] was an isolated LAB from a fermented beverage of Ladakh and found 46 LAB isolates from dried cottage cheese and beverage. Pringsulaka et al [20] isolated 152 LAB from 93 samples of Thai fermented meat and fish product. Furthermore, 160 LAB strains were isolated from the digestive tracts of marine fish and evaluated of potential use as probiotics by Buntin et al [21]. Additionally, there are 51 strains of LAB isolated from the intestinal tract of Kuruma shrimp (*Marsupenaeus japonicus*) and the selected strain for probiotic [22].

The total bacterial count on the samples is well known to be correlated with the source of the raw materials. In the case of fermented fish, the main source of LAB is the gastrointestinal tract of the fish. Ringø [23] reported a minor number of LAB from the Arctic charr, *Salvelinus alpines*, approximately 10 % from the gastrointestinal microflora of *Arctic charr, Salvelinus alpines*. In contrary, Buntin at al [21] reported the high number of LAB from warm-water or tropical fish, approximately 72.5 % of the total isolated LAB. In this study, the low number of LAB (52.7 %) may be due to the high concentration of salt employed to produce the *peda* fermented fish. The addition of salt during *peda* fermentation reduces the water activity of its food matrix. This condition inhibits the growth of Gram-negative bacteria and promotes the growth of Gram-positive bacteria such as LAB. Therefore LAB plays a central role in fish fermentation along with the addition of salt [24]. Salt is used to control the growth of desirable bacteria such as lactic acid bacteria [25].

*3.2 Antagonistic activity of isolated LAB*
Among 57 isolates, 27 isolates exhibited inhibition activity (Table 1) and only one isolate exhibited strong activity against *S. aureus* ATCC 6538, namely NJ-20 with the inhibition zone of 7.6 ± 1.35 mm. The inhibition of pathogenic bacteria by LAB is due to the acidification of the medium. Organic acids are substances produced by LAB that inhibit Gram-negative bacteria such as *Yersinia enterocolitica*, *Salmonella* sp., *Escherichia coli, Pseudomonas* sp. [16,10,6]. The inhibition of pathogenic bacteria may be due to more than one mechanism, which includes suppression of growth, production of organic acids (such as lactic acid), and active compounds (bacteriocins) [24,26].

IOP Conf. Series: Earth and Environmental Science **137** (2018) 012060 doi:10.1088/1755-1315/137/1/012060

**Table 1**. Antagonistic acticity of isolated lactic acid bacteria from Peda against *S. aureus* ATCC 6538 and *P. aeruginosa* ATCC 27853.

| Isolates | Inhibition Zone (mm)[a] | |
|---|---|---|
| | *S. aureus* ATCC 6538 [a] | *P. aeruginosa* ATCC 27853 [a] |
| NJ-15 | 2.9 ± 0.46[a] | 0 ± 0 |
| NJ-17 | 2.2 ± 1.46 | 0 ± 0 |
| NJ-18 | 2.6 ± 0.75 | 0 ± 0 |
| NJ-20 | 7.6 ± 1.35 | 0 ± 0 |
| NJ-21 | 4.7 ± 1.38 | 1.1 ± 0 |
| KD-1 | 2.3 ± 1.25 | 0 ± 0 |
| KD-3 | 2.3 ± 1.35 | 0 ± 0 |
| KD-6 | 1.3 ± 0.40 | 0.6 ± 0.55 |
| GR-2 | 0 ± 0 | 4.5 ± 0.51 |
| GR-3 | 0 ± 0 | 1.8 ± 0 |
| GR-8 | 0 ± 0 | 3.5 ± 0 |
| GR-10 | 0.9 ± 0 | 0 ± 0 |
| GR-11 | 0 ± 0 | 1.5 ± 0 |
| GR-13 | 0 ± 0 | 2.3 ± 0.2 |
| GR-14 | 0.9 ± 0 | 0.7 ± 0.75 |
| SD-18 | 0 ± 0 | 0.9 ± 0 |
| SD-19 | 0 ± 0 | 2.2 ± 0 |
| SD-20 | 0 ± 0 | 0.8 ± 0 |
| SD-21 | 0 ± 0 | 1.1 ± 0 |
| SD-22 | 0 ± 0 | 1.2 ± 0 |
| SD-24 | 0 ± 0 | 2.7 ± 0 |
| SD-25 | 0 ± 0 | 0.7 ± 0 |
| SD-28 | 3.1 ± 0.36 | 0 ± 0 |
| SB-1 | 1.7 ± 0 | 0 ± 0 |
| SB-4 | 2.8 ± 0.83 | 0 ± 0 |
| SB-19 | 0 ± 0 | 1.7 ± 0 |
| SB-20 | 0 ± 0 | 3.4 ± 0.87 |

[a]Mean ± SD

The LAB NJ-20 isolate was then selected for the detection of bacteriocin producing activity against *S. aureus* ATCC 6538 due to its inhibitory zone of ≥ 6 mm [18]. Gram-negative is susceptible to organic acids because the cytoplasmic membrane of Gram-negative bacteria is damaged by organic acids [27]. In this study, the strong activity of LAB NJ-20 isolate against Gram-positive bacteria (*S. aureus* ATCC 6538) indicates the potential for bacteriocin production.

IOP Conf. Series: Earth and Environmental Science **137** (2018) 012060 doi:10.1088/1755-1315/137/1/012060

*3.3 Bacteriocin-producing test of isolate LAB NJ-20*

The bacteriocin-producing test employed an Agar Well Diffusion Assay (AWDA) of LAB NJ-20 isolate which was obtained from *peda*. However, the neutralized-cell free supernatant of the NJ-20 did not exhibit a clear zone around the well against *S. aureus* ATCC 6538. This finding indicates that the neutralized-cell free supernatant did not contain an antimicrobial substance such as bacteriocin. Bacteriocin is an antimicrobial peptide or protein that is heat-stable, resistant to pH, sensitive to protease and inactivated by the digestive enzyme [12].

The incubation temperature and pH play an important role in the production of bacteriocin, because an optimum temperature and pH are required for bacterial growth so as to produce bacteriocin. The optimal range of temperature and pH to produce bacteriocins depend on the strains of the producer bacteria [4,28]. The optimal temperature and pH stimulate LAB to produce bacteriocin at the beginning of the exponential phase. According to Ponce *et al.* [29] bacteria produces bacteriocin exponentially in the initial phase and is inactivated at the stationary phase. The phase of lactic acid bacteria in producing bacteriocin varies depending on the strains of the producer bacteria [29]. The long incubation time resulted in the bacteriocin being degraded by a bacterial secreted enzyme [30].

However, LAB is known to produce a wide range of antimicrobial compounds (organic acids, bacteriocin, diacetyl and hydrogen peroxide) [10,24,26] that can still be applied as biopreservation. LAB had been granted Generally Recognized As Safe (GRAS) status by the US Food and Drug Administration [24]. Meat-borne LAB generally inhibits other LAB species as well as Gram-positive bacteria such as *Staphylococcus aureus*, *Listeria monocytogenes*, *Clostridium perfringens*, *Clostridium botulinum*, *Enterococcus faecalis*, and *Bacillus* spp. due to the formation of pores in the cytoplasmic membrane of sensitive strains [31].

**Table 2**. Identfication of NJ-20 isolate.

| Identification | Isolate LAB NJ-20 | *Aerococcus* |
|---|---|---|
| Shape | Cocci | Cocci |
| Shape tetrad | + | + |
| Gram-positive | + | + |
| Catalase | - | - |
| $CO_2$ from glucose | - | - |
| Growth at 10 °C | + | + |
| Growth at 45 °C | - | - |
| Growth in NaCl 6.5 % | + | + |
| Growth in NaCl 18 % | - | - |
| Growth at pH 4.4 | - | - |
| Growth at pH 9.6 | + | + |

*3.4 Identification of LAB NJ-20*

The genus identification based on morphology and physiology tests indicate that NJ-20 belong to the *Aerococcus* genus. The results of identification of NJ 20 isolates are shown in table 2. These results were different with the previous study on the identification of LAB from *peda* which found Lactic acid bacteria isolated from *peda* consist of *Lactobacillus plantarum*, *L. curvatus*, *L. murinus*, *Streptococcus thermophilus*. The difference may be due to the different sampling locations as well as the raw materials used. *Aerococcus* is included in the *Lactobacillales* order which is closely related to the *Streptococcus*, *Lactococcus*, *Carnobacterium*, *Vagococcus* and *Enterococcus* species. The *Aerococcus* species lives in a wide range of environments including air, vegetation, meat-curing brines, soil, and marine sources [32].

IOP Conf. Series: Earth and Environmental Science **137** (2018) 012060 doi:10.1088/1755-1315/137/1/012060

## 4. Conclusion

The antagonistic activity of lactic acid bacteria isolated from *peda*, a traditional fermented fish from Indonesia, suggested that 27 isolates exhibited an antagonistic effect against *S. aureus* ATCC 6538 and/ or *P. aeruginosa* ATCC 27853. Strong inhibition to *S. aureus* ATCC 6538 was performed by the LAB NJ-20 isolate. The inhibitory activity of *Aerococcus* NJ-20 was not due to the production of bacteriocin. This finding suggests that Aerococcus NJ-20 may be applied as a biopreservation ingredient for fishery products as well as a culture starter of *peda* fermentation. Nevertheless, further studies are needed to evaluate the enzyme activity, safety, growth of isolates on the fermentation process and biogenic amine production prior to applications.

## 5. References


[1] Tanasupawat S and Visesssanguan W 2014 *Seafood processing: technology, quality and safety* 1st edn.(United Kingdom: John Willey and Sons) pp 177-207

[2] Huda, N 2012 *Handbook of animal-based fermented food and beverage technology Indonesian fermented fish product* 2nd, ed Y H Hui (New York: CRC Press) p 717-737

[3] Albarracin W, Sanchez IC, Grau R, and Barat J M 2011 *Int. J. Food Sci and Technol* **46** 1329–36

[4] Akkoc N, Ghamat A, and Akcelik M 2011 *Int J. of Dairy Tech* **64** 3

[5] O'Bryan C A, Crandall P G, Ricke S C, and Ndahetuye J B 2015 *Handbook of natural antimicrobials for food safety and quality: Lactic acid bacteria (LAB) as antimicrobials in food products: types and mechanisms of action* ed T M Taylor (United Kingdom: Woodhead Publishing,) pp 117-129

[6] Hwanhlem N, Buradaleng S, Wattanachant S, Benjakul S, Tani A, and Maneerat S 2011 *Food Control* **22** 401-7

[7] Rahayu ES 2003 *Agritech* **23** 75-84

[8] Sukmarini L, Mustopa A Z, Normawati M, and Muzdalifah I 2014 *J. Biosci* **21** 144-150

[9] Udomsil N, Rodtong S, Tanasupawat S, and Yongsawatdigul J 2010 *Int J. Food Microbio* **141** 86-194.

[10] Cizeikiene D, Juodeikiene G, Paskevicius A, and Bartkiene E 2013 *Food Control* **31** 539-5

[11] Sonomoto K, Nagao J I, and Nishie N (2012) *Biocontrol Sci.* **17**(1) : 1-16

[12] Hwanhlem N, Chobert J M, and H-Kittikun A H 2014 *Food Control* **41** 202-211

[13] H-Kittikun AH, Biscola V, El-Ghaish S, Jaffres E, Dousset X, Pillot G, Haertle T, Chobert JM, Hwanhlem N 2015 *Food Control* **54** 126-134

[14] Desniar, Rusmana I, Suwanto A, and Mubarik NR 2012 *J. Akuatika* **3** 135-145

[15] Cho GS and Do HK 2006 *Ocean Sci J.* **41** 9-113

[16] Angmo K, Kumari A, Monika, Savitri, Bhalla TC 2016 *Food Biosci.* **13** 26-31

[17] Pramono H, Suciati P, and Andriyono S 2015 *Mar. Sci.* **20** 33-37 (In Indonesia)

[18] Heredia-Castro PY, Romero JIM, Mendoza AH, Félix EA, Córdova AFG, and Cordoba BV 2015 *J. Dairy Sci* **98** 1-9

[19] Nuryadi M M, Istiqomah T, Faizah R, Ubaidillah S, Mahmudi Z, and Sutoyo 2013 *J. UNEJ* **1** 1-11 (In Indonesia)

[20] Pringsulaka O, Thongngam N, Suwannasai N, Atthakor W, Pothivejkul K, Rangsiruji A 2012 *Food Control* **23** 547-551

[21] Buntin N, Chanthachum S, and Hongpattarakere T 2008 *J. Sci Technol* **30** 141-8

[22] Maeda M, Shibata A, Biswas G, Korenaga H, Kono T, Itami T, and Sakai M 2014 *Mar. Biotechnol.* **16** 181–192

[23] Ringø E 1993 *Aquac. Fish. Man.* **24** 133-5

[24] Françoise L 2011 *Food Microb.* **27** 698-709

[25] Yang E, Fan L, Jiang Y, Doucette C, and Fillmore S 2012 *EMB express* **2** 48

[26] Sofyan A, Aswari AN, Purwoko T, and Damayanti T 2013 *Vet. Med.* **36** 216-223 (In Indonesia)

IOP Conf. Series: Earth and Environmental Science **137** (2018) 012060 doi:10.1088/1755-1315/137/1/012060

[27] Schirru S, Favaro L, Mangia NP, Basaglia M, Casella S, Comunian R, Fancello F, Franco BDGdeM, Oliveira RPdeS, Todorov SD 2013 *Microbiol.*
[28] Mandal V, Sen SK, and Mandal NC 2008 *J. Biochem & Biophy* **45** 106-110
[29] Ponce AG, Moreira MR, del Valle CE, and Roura SI 2008 *LWT* **41** 432-441
[30] Pal A, Ramana KV, and Bawa AS 2010 *J. Food Sci Technol* **47** 258-265
[31] Ammor MS and Nayo B 2007 *Meat Sci.* **76** 138–146
[32] Zhang H And Cai Y 2014 *Lactic acid bacteria : fundamental and practice* (London: Springer)

IOP Conf. Series: Earth and Environmental Science **137** (2018) 012061 doi:10.1088/1755-1315/137/1/012061

# Biochemical and physicochemical analysis of fish protein isolate recovered from red snapper (*Lutjanus* sp.) by-product using isoelectric solubilization/precipitation method

**H Pramono, D Y Pujiastuti, and A M Sahidu**

Department of Marine, Faculty of Fisheries and Marine, Universitas Airlangga, Surabaya, Indonesia

E-mail: heru.pramono@fpk.unair.ac.id

**Abstract.** The effect of acid- and alkali-process on biochemical and physicochemical characteristics of fish protein isolate from red snapper (*Lutjanus* sp) by-product was evaluated. Protein recovered by alkali process (16.79%) was higher compared to acid process (13.75%). Reduction of lipid content and total volatile basic nitrogen (TVB-N) exhibited in both treatments indicated both process improved fish protein isolate recovered from red snapper by-product. In addition, the increasing of water holding capacity and oil binding capacity were observed. However, high peroxide value of fish protein isolate was showed in both treatment. This finding indicated that acid and alkali process can be used as a useful method to recover proteins from red snapper by-product. Alkali process gave a protein isolate with better overall quality compared to acid process.

## 1. Introduction

Consumer demand forfunctional food has beenpredicted to increase to a value of USD100 million in 2013 and has showed a positive trend from year to year [1]. Seafood-based functional food isdeveloped from various sources such as krill, squid, and seaweed. Red snapper (*Lutjanus* sp.) is important for fillet production in Indonesia for its high quality and the demand for the fillet. During the filleting process, a large amount of low-value by-product is produced, and it is used as fish mealfor animal feed. Head by-product fromfish processing is about 20 % of the fish weight [2], while the total by-product consisting of head, frame flesh and scale comprises 60 % of the fish weight [3]. Due to the high value offish head, many restaurants in Indonesia use red snapper head as traditional food. Meanwhile, the frame, skin, and scale are discharged or used as fish meal, plant fertilizer or human value-added foods.

Fish frame, scale, and skin contain valuable proteins such as myofibrillar proteins, stromata,sarcoplasmic proteins, and others that can be recovered forhuman consumption [4]. There have been some reports about the development of functional proteins and peptone using hydrolysis method from red snapper by-product forsilage or fish feed [5, 6]. Currently, protein recovery with pH-shift method or isoelectric solubilization/presipitationhas beenstudied. Protein recovery of bigeye snapper (*Priacanthus* spp.) with pH-shift method was evaluated to have a yield of 30% and acceptable characteristics [4].

Isoelectric solubilization/precipitation (ISP) processing is selective, pH-induced water solubility and removal of lipid, bone, scales and skin [7]. This method was proposed by Hultin and Kelleher [8]. The basic concept of ISP is about protein side-chains that have different electrostatic charges depending on the pH condition. Atcertain pH levels, fish protein will solubilize (on), and atisoelectric point(pI), fish protein will precipitate (off) [9]. The pH for protein solubilizationislow (acid) at around 2-3.5 or high (alkali) at around 11-13 [10].

Fish protein isolate produced with ISP can be applied as a raw material in the making of Frankfurter-type fish sausages [11], functional fish protein gel [12], or fish ball [13]. The yield of the fish protein isolate rangesat 50–87 % [4]. Therefore, ISP process is an effective way for recovering proteins from the by-product of fish processing industry. The aim of this study was to isolate and characterize the fish protein isolate recovered from red snapper (*Lutjanus* sp.) filleting industry by-product.

## 2. Methodology

### *2.1.Materials*

The fish by-product was purchased from a filleting factory located in Sidoarjo, East Java, Indonesia. The sample was wrapped in apolyethylene bag under frozen condition (-18°C) and transported to the laboratory. The sample was directly stored in the laboratory freezer (-18°C) prior to further process. The hydrochloric acid and natrium hydroxide were of p.a.quality.

### *2.2.Fish by-product preparation*

The frozen fish by-product sample was thawed at 4°C overnight prior to processing. The skin, bone, flesh, and frame of thethawed sample were then separated from each other. The frame and flesh were homogenized with laboratory food processor (Bosch, Germany) three times for 3 minutes. The homogenized sample was stored in a polyethylene bag and frozen until further process.

### *2.3.Protein recovery with isoelectric solubilization/precipitation (ISP)*

The thawed red snapper by-product (RSB) homogenate was processed to recover protein inacid and alkaline process. The solubilization was conducted at pH 2.5 and 11.5 for acid process and alkaline process,respectively. The recovery process followed Gehring *et al.* [3] method with a modification in thecentrifugation speed. A total of 100 g of RSB was mixed with 900 mL of distilled water (4°C) and homogenised for 2x30 seconds using Food Processor (Bosch, Germany). The pH of the homogenate was then adjusted to either 2.5 or 11.5 using 1N HCl or 1N NaOH with constant stirring (400 rpm). The pH was monitored with calibrated pH meter (Labortechnik, Germany). The homogenate, the pH of which has been adjusted, was centrifuged at 6.000 rpm and a temperature of 4°C for 25 minutes. The supernatant was collected by filtering usingthree layers of cotton sheet. The pH of the filtrate was then adjusted to 5.5 using 1N NaOH or 1N HCl. Second centrifugation was performed to separate the pellet (recovered protein/fish protein isolate). The recovered protein was then collected in apolyethylene bag and weighed.

### *2.4. Proximate analysis*

The proximate analysis of fish protein isolate was performed following AOAC (2002). The moisture, protein, fat, and ash content were evaluated. The total nitrogen of the fish protein isolate was determined with Kjeldahl method. The total crude protein was measured by multiplying thetotal nitrogen of the sample by 6.26.

### *2.5. Total volatile basic nitrogen (TVB-N)*

The TVBN of the fish protein isolate was measured with Conway microdilution method. Briefly, 2 g of sample was added to 8 mL of 4 % (w/v) TCA and homogenized at 10,000 rpm for 2 minutes. The homogenate was centrifuged at 3. 000 rpm for 15 minutes at room temperature. The supernatant as sample (1 mL) was placed in the outer ring of Conway apparatus. The inner ring consisted of 1 % boric acid solution. The reaction was initiated with an addition of 1 mL of $K_2CO_3$ to the sample at the outer ring. The Conway unit was closed and incubated at room temperaturefor 16 hours. The solution at the inner ring was then titrated with 0.02 N HCl until the green colour turned pink. The TVB-N value was expressed as mg of nitrogen released/100 gram of sample.

*2.6.Peroxide value*
The peroxide value was measured using Panpinat and Chaijan [4] method. A total of 2 g of sample was treated with 25 mL of organic solvent (with ratio of chloroform to acetic acid of 2:3). The mixture was shaken and added with 1 mL of saturated KI solution. The mixture was stored in the darkness for 5 minutes, then 75 mL of distilled water was added. A total of 0.5 mL of starch solution (1 % w/v) was added as an indicator. The PV was measured by titrating the iodine liberated from KI with standardized 0.01 N sodium thiosulfate solution. The PV was expressed as milliequivalents of free iodine per 100 g of lipid.

*2.7.Water holding capacity*
The water holding capacity was measured usingmethod by Panpinat and Chaijan [4]. A total of 100 mg of fish protein isolate was homogenized with 10 mL of distiled water and centrifuged at 10.000 rpm for 30 minutes, then decanted. The difference between the initial weight and the final weight was measured as the water holding capacity of the fish protein isolate.

*2.8.Oil binding capacity*
The oil binding capacity was determined by adding 10 mL of coconut oil into100 mg of fish protein isolate, and then the mixture was homogenized. The homogenate was then centrifuged at 2.500 rpm for 30 minutes. The free oil was decanted, and the weight difference was stated as oil binding capacity [14] with modification.

*2.9.Amino acid profile*
The amino acid profile was measured with a HPLC method [15].

*2.10. Statistical analysis*
All treatments except the amino acid profile weretriplicated, and the differences between treatments were analyzed with analysis of variance (ANOVA), while the comparison of means was made by conducting Duncan's multiple-range test for stating the significant difference.

## 3. Results and Discussion

*3.1.Protein recoveryof red snapper by-product*
The protein recovery consisting of acid and alkaline process had different yields based on the protein content. The yield of fish protein made by alkali process was higher (77.30 %) than that of fish protein made by acidprocess (63.28 %). These results were in line with Chen and Jaczynski[10], who reported protein isolates recovered from trout processing by-product. High yields of alkaline process of salmon and pollock head of 50 % and 87 % were also reported by Bechtel *et al.*[16]. However, theseresultswere contrary to Panpina and Chaijan [4], who reported that the recovery protein from head by-product of big eye snapper hadhigher yield by acid process than byalkaline process. These results indicated that different raw materials had different suitability at different pH levels of the processing.

*3.2.Proximate analysis*
The protein, lipid, and moisture contents of fish protein isolate areexhibited in table 1.

**Table 1**. Proximate analysisof red snapper by-product (RBP) and fish protein isolate (FPI).

| Compounds | RBP | Acid-made FPI | Alkali-made FPI |
|---|---|---|---|
| Protein | 21.73±0.83$^{b}$ | 13.75±1.56$^{a}$ | 16.79±1.70$^{ab}$ |
| Lipid | 3.74±0.32$^{b}$ | 0.66±0.12$^{a}$ | 1.18±0.07$^{a}$ |
| Moisture | 70.72±1.42$^{a}$ | 83.63±1.17$^{b}$ | 75.35±0.73$^{ab}$ |

Note: values are given as mean±standard deviation from triplicate measurement. Different letter in the same row indicate significant differences (P<0.05)

Based on the proximate analysis, somereduction of protein occurredin the acid and alkaline process. This result was contrary to Chomnawang and Yongsawatdigul [17], who reported that 85–90% of protein content in fish protein isolate was produced from tilapia by-product. The reduction of protein might bedue to the denaturation of muscle proteins induced by pH shift, which caused the aggregation of protein and led to separation of bone, skin, and debris during centrifugation [4]. Another reason might be due to the use of centrifugation speedthat was lower in this study than in the previous study.

Some significant reduction of lipid content showed in acid and alkaline process compared to the raw material. Lipids are susceptible to oxidation, which leads to rancidity or fishy odour [10].Thus, the removal of lipid is an importantcharacteristic. Both treatments (acid and alkali) were effective in lipid removal. The effectivenessof alkaline process in lowering lipid content might bedue to the saponification reaction of alkali and lipid, whichproducedsoap that precipitated during centrifugation. This result was in accordance with [17], who reported that alkaline process was more effective than acidprocessinlipid removal.

**Table 2**. Total volatile basic nitrogen (TVB-N) and Peroxide value of red snapper by-product and fish protein isolates.

| Parameter | RBP | Acid-made FPI | Alkali-made FPI |
|---|---|---|---|
| TVB-N | 28.06±6.08$^{b}$ | 10.52±0.01$^{a}$ | 10.51±0.01 $^{a}$ |
| Peroxide value | 0.01±0.04$^{a}$ | 0.06±0.06$^{b}$ | 0.04±0.01 $^{ab}$ |

Note: values are given as mean±standard deviation from triplicate measurement. Different letter in the same row indicate significant differences (P<0.05). TVBN value (mg N/100 g) and PV (mEq/1000 g)

### *3.3. Total volatilebasic-nitrogen (TVB-N)*
The total volatile basic nitrogen of fish protein isolate and raw material is exhibited in table 2. There wasno significant difference between acid process and alkaline process (P <0.05) in terms of TVB-N value. Both processes improved the stability of cellular structure and preventedthe spoilage of fish protein isolate produced. TVB-N value indicates the measurement of dimethylamine (DMA), trimethylamine (TMA), ammonia, and another compound, which indicates seafood spoilage due to the degradation of muscle proteins. The decreaseinTVB-N might bedue to the washing mechanism of the centrifugation and precipitation process when the structural protein at pH 5.5 precipitated, while other compounds (hydrophilic compounds such as DMA and TMA) leached out into the supernatant [4].

### *3.4. Peroxide value*
Peroxide value depicts the formation of unstable compound (peroxide) due to lipid oxidation process. Fish muscle contains high amount of unsaturated fatty acids, particularly n-3 fatty acid that is susceptible to oxidation [4]. The peroxide value of this study (table 2) exhibited that the fish protein wassusceptibletooxidation. Both acid process and alkaline process resulted inhigh peroxide values. Rapid lipid oxidation was reported in thefish protein isolaterecovered by acid process [9]. However, this study showedthat there wasno significant differenceinlipid oxidation between acid and alkaline process.

IOP Conf. Series: Earth and Environmental Science **137** (2018) 012061 doi:10.1088/1755-1315/137/1/012061

Thesefindingswere in accordance with Panpipat and Chaijan [4],who reported that both processes(acid and alkaline process) enhanced the oxidative instability of lipid in bigeye snapper head by-product.

**Table 3**. Water holding capacity (WHC) and oil binding capacity (OBC) of red snapper by-product and fish protein isolate produced by acid- and alkali-process.

| Parameter | RBP | Acid-made FPI | Alkali-made FPI |
|---|---|---|---|
| WHC (mL/g) | 0.13±0.02$^{a}$ | 0.50±0.26$^{b}$ | 0.56±0.18$^{b}$ |
| OBC (mL/g) | 0.06±0.04$^{a}$ | 0.17±0.06$^{b}$ | 0.15±0.05$^{b}$ |

Note: values are given as mean±standard deviation from triplicate measurement. Different letter in the same row indicate significant differences (P<0.05).

### *3.5. Water holding capacity*

Water holding capacity indicates the ability of the food product to store water during processing. Sarcoplasmic protein has low water holding capacity, which does not allow water for a well-developed gel matrix formation [18]. The WHC of fish protein isolate produced by acid and alkaline process in this study was significantly higher than that of the raw material ($P < 0.05$), butthere was no difference between acid and alkaline process. These results were in accordance with Foh *et al.* [19], who reported that the fish protein isolate produced from tilapia by acid process and alkaline process was around 2.63 and 2.51 mL/g, respectively. This finding indicated that both methods reduced the sarcoplasmic proteins and isolated the muscle/myofibrilar proteins. The isoelectric solubilisation/precipitation method (ISP) employing pH shift during processing was believed to affect the water holding capacity of the fish protein isolate produced [19].

### *3.6. Oil binding capacity*

The oil binding capacity of fish protein isolate from red snapper by-product produced byacid and alkaline process was significantly higher than that of the raw material used (table 3). The acidprocess showed a higher score (0.17 mL/g), but the score was not significantly different from that ofalkaline process (0.15 mL/g). This finding was in accordance with Foh *et al.* [19], who reported that the fish protein isolate of tilapia produced by acid and alkaline process had high oil binding capacity (3.38 mL/g). High oil absorption of fish protein is an important characteristic of formulation of food system such as cake, sausage, mayonnaise, butter, and saladdressing [19].

### *3.7. Amino acid profile*

The amino acid profile of the fish protein recovered withISP was affected by the pH treatment. Chen, Tou, and Jaczynski [20] reported that extreme pH during ISP process improvedthe essential amino acid profile of the FPI isolated from whole antarctickrill (*Euphausiasuperba*). Glutamic acid and lysine were higher than other amino acids. Meanwhile, threonine was the lowestamino acid in acidprocess, and histidine was the lowest in alkaline process. Thesefindingswerein accordance with [20], who reported that glutamic acid was the highestamino acid contained in krill protein isolate produced with isoelectric solubilisation/precipitation method.

**Table 4.** Amino acid profile of fish protein isolate produced with acid-and alkali-process.

| Amino acid | Acid process | Alkali process |
|---|---|---|
| Aspartic acid | 8.05 | 8.46 |
| Glutamic acid | 14.11 | 12.91 |
| Serine | 3.07 | 3.28 |
| Glysine | 5.24 | 5.63 |
| Histidine | 5.35 | 1.17 |
| Arginine | 5.29 | 5.59 |
| Threonine | 0.98 | 2.48 |
| Tyrosine | 3.03 | 3.15 |
| Valine | 5.08 | 4.37 |
| Methionin | 2.85 | 3.62 |
| Isoleusine | 4.57 | 4.84 |
| Phenilalanine | 3.94 | 4.06 |
| Lysine | 12.12 | 9.51 |
| Leusine | 7.16 | 6.87 |
| Alanine | 5.20 | 5.72 |

## 4. Conclusion

Isoelectric solubilisation/precipitation with acid and alkali process can be an alternative process to recover proteins from red snapper by-product. This study demonstrates that alkali process gave a protein isolate with better overall quality compared to acid process.

## 5. References

[1] Smith J and Charter E 2010 Microencapsulation in functional food product development Functional Food Product Development (Sanguansri, L, Augustin, M, ured)
[2] Gildberg A R, 2004 *J. Aquatic Food Product Tech.* **13** 3-11
[3] Gehring C K, Gigliotti J C, Moritz J S, Tou, J C and Jaczynski, J 2011 *Food chemistry* **124** 422-431
[4] Panpipat, W and Chaijan, M, 2016 *Turk. J. Fish. Aquat. Sci.* **16** 041-050
[5] Himaya S W A and Kim S K 2014 *Functional proteins and peptides from fish skin in seafood processing by-products* (New York: Springer) p 197-205
[6] Hernández A, García B G, Jordán, M J and Hernández M D 2014 *Anim. Feed Sci. Technol.* **195** 112-119
[7] Gehring C K, Gigliotti J C, Moritz J S, Tou J C and Jaczynski J 2011 *Food chemistry* **124** 422-431
[8] Hultin H O and Kelleher S D Advanced Protein Technologies Inc 2002 Protein composition and process for isolating a protein composition from a muscle source US Patent 6,451,975
[9] Undeland I, Hultin H O and Richards M P 2003 *J. Agric. Food Chem.* **51** 3111-3119
[10] Chen Y C and Jaczynski J 2007 *J. Agric. Food. Chem.* **55** 9079-9088
[11] Pires C, Batista I, Fradinho P and Costa S 2009 *J. Aquati. Food Product Tech.* **18** 170-190
[12] Tahergorabi R, Beamer S K, Matak K E and Jaczynski J 2012 *LWT-Food Sci. Tech.* **48** 89-95
[13] Shaviklo G R, Arason S, Thorkelsson G, Sveinsdottir K and Martinsdottir E 2010 *J. Sens. Stud.* **25** 316-331
[14] Jemil I, Jridi M, Nasri R, Ktari N, Salem R B S B, Mehiri M, Hajji M and Nasri M 2014 *Process Biochemistry* **49** 963-972
[15] White J A, Hart R J and Fry J C 1986 *J. Anal. Methods Chem.* **8** 170-177

[16] Bechtel P J, Sathivel S and Oliveira A C 2006 Production and characterization of alkali extracted protein powders from pollock and salmon heads *In Joint Trans Atlantic Fisherie Tech. Conf.* (Canada: TAFT Quebec City) pp 10-29
[17] Chomnawang C and Yongsawatdigul J 2013 Protein recovery of tilapia frame by-products by pH-shift method *J. Aquat. Food Product Tech.* **22** 112-120
[18] Sikorski Z 2012 Seafood proteins Springer Science and Business Media
[19] Foh M B K, Wenshui X, Amadou I and Jiang Q 2012 *Food Bioproc. Tech.* **5** 2192-2200
[20] Chen Y C, Tou J C and Jaczynski J 2009 *J. Food Sci.* **74**

IOP Conf. Series: Earth and Environmental Science **137** (2018) 012062 doi:10.1088/1755-1315/137/1/012062

# Biofilm as a bioindicator of Cr VI pollution in the Lotic Ecosystems

**A Kurniawan[1,2,*] Sukandar[2], C Satriya[1,2] and Guntur[2]**

[1] Coastal and Marine Research Center, University of Brawijaya, Indonesia
[2] Faculty of Fisheries and Marine Science, University of Brawijaya, Indonesia

E-mail: andi_k@ub.ac.id

**Abstract.** Biofilm is ubiquitous in aquatic ecosystems such as river. Biofilm have been reported to have high sorption capacities that promote the accumulation of nutrient ions inside biofilm matrix. The ion that can be accumulated inside the biofilm is not only nutrient ions but also other ions such as heavy metal ions. The pollution of heavy metal ions emerge as one of the biggest aquatic ecosystem problems. Thus, the effort to monitor the heavy metal pollution in the aquatic ecosystem in the aquatic ecosystems is needed. The difficulty to monitor the water pollution particularly in the lotic ecosystems is mainly related to the water flow. Therefore, the utilization of indicator of pollution in such ecosystem is fundamentally important. The present study investigated the accumulation of Cr VI inside biofilm matrices in the river ecosystems in order to develop biofilm as a bioindicator for pollution in the lotic ecosystems. The result indicates that biofilm can accumulate Cr VI from the surrounding water and reserve the ion. According to the result of this study, biofilm is a promising bioindicator to monitor the Cr VI pollution in the lotic ecosystems.

## 1. Introduction

Biofilm is found in most natural aquatic environments [1,2] and has been found to constitute the predominant mode of growth of microbe in streams and lakes in all parts of the world and to be both viable and metabolically active [3,4,5]. Various definitions of the term biofilm have been proposed. Generally, biofilm can be considered as an aggregate of microorganisms imbedded in a matrix composed of microbially produced extracellular polymer substances (EPS) and is attached to a surface [6].

The nutrient ion concentration in the interstitial water of biofilms formed on reed and stone surfaces were hundreds to thousands of times higher than those in lake waters regardless of the biofilm type and sampling site [7]. These particular nutrient-rich environments inside of these biofilm matrices are established from a very early stage of biofilm formation [5]. The adsorption of nutrient ions from the surroundings to the biofilm seems to form this nutrient-rich environment [8,9].

The ions that can be attracted and accumulated into biofilm are not only nutrient ions but also others such as heavy metal ions. Heavy metal pollution emerges as the one of the most serious problems in aquatic ecosystems [10,11]. One heavy metal that has become a pollutant is Cr VI [12]. Many efforts have been carried out to monitor water pollution. In the case of lotic ecosystems, the main problem of monitoring is related to water flow. Hence, the utilization of biological substances that can attract and retain Cr VI may be used as a bioindicator for the contamination. The present study investigated the accumulation of Cr VI inside biofilm matrix in order to develop biofilm as a bioindicator of pollution in lotic ecosystems. The result indicates that biofilm can accumulate Cr VI from the surrounding water and

IOP Conf. Series: Earth and Environmental Science **137** (2018) 012062 doi:10.1088/1755-1315/137/1/012062

reserve the ion. According to the results of this study, biofilm is a promising bioindicator to monitor Cr VI pollution in the lotic ecosystems.

## 2. Materials and Method

### *2.1. Sample preparation*

Samples were collected from 3 sites at Badek River in Malang, Indonesia. Stones were taken from a depth of about 20 cm and brought back to the laboratory in a plastic container filled with surrounding river water; the container was maintained at 4°C. The biofilm on the surfaces of the stones were removed using a toothbrush and suspended in sterilized distilled water. The filtrates and biofilm pellets were prepared by centrifuging (8.000 × g at 4°C for 10 min) the biofilm suspensions. The pellets of the biofilms (approximately 1 wet-g) were taken and then dehydrated for 3 days until the weight was stable to give a dry weight. This data was used to convert the dry weight per the wet weight of biofilm polymer.

### *2.2. Electrophoretic mobility*

The pellets of biofilm were washed three times as follows. The pellets were resuspended in 40 mL of 10 mM NaCl aqueous solution. The suspension was centrifuged (8.000 × *g* at 4°C for 10 min), and the supernatant was discarded. The obtained biofilm pellets (ca. 0.03 g) were suspended in 1 mL of 10 mM NaCl aqueous solution. The suspension was mixed vigorously with a vortex for 5 min, then sonicated (2510J-MT, Yamato Scientific, Tokyo, Japan; 42 kHz, 125 W) for 10 min, followed by a vortex for 10 s. The detailed method of electrophoretic mobility measurement is described by Kurniawan and Fukuda [13].

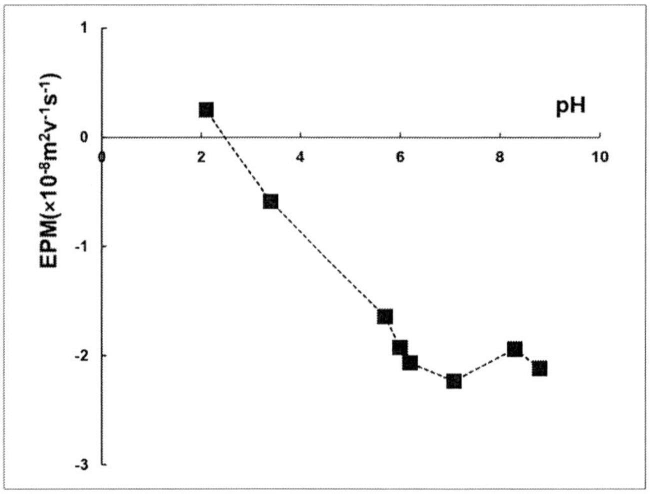


**Figure 1**. The electrophoretic mobility of biofilm measured in various pH

### *2.3. Cr VI concentration*

The sample used to measure the concentration of Cr VI inside the biofilm was a biofilm suspension that was prepared by brushing the biofilm from the surface of the stones in sterilized distilled water. The concentration of Cr VI in the surrounding river water was measured from the surrounding river water of the stones (about 15 cm). The concentration of Cr VI was measured using the Atomic absorption spectroscopy method.

## 3. Result and Discussion

### *3.1. Characterization of biofilm polymers*

The accumulation of various ions into biofilm matrix is promoted by the ability of biofilm to attract the ions [14,15,16]. This ability is mainly promoted by the electrostatic interaction between the biofilm

IOP Conf. Series: Earth and Environmental Science **137** (2018) 012062 doi:10.1088/1755-1315/137/1/012062

polymers and the ions. In order to analyse the electrostatic characterization of biofilm polymers, the electrophoretic mobility measurement was conducted in the present study.

Electrophoretic mobility (EPM) of biofilm polymers were measured in various pH (pH 2 until pH 9) (figure. 1). The biofilm showed a net negative charge at around neutral pH, which is typically the pH of aquatic ecosystem. This indicates that biofilm carries negative charge sites in a natural aquatic ecosystems. The EPM of biofilm moves to the positive charge when the pH is down to the acid condition, which suggests that the decrease of negative charge is due to the protonation of functional group carrying negative charge. The great change of the EPM value around pH 4 indicates the presence of a functional group with pKa around pH 4 such as carboxylic groups [17]. The EPM of biofilm shows a positive value around pH 2 indicating the presence of positively charged sites such as amino groups [18]. The result of EPM measurement indicates that biofilm carries both negative and positive charge sites in aquatic ecosystems.

The presence of positively and negatively charged sites becomes a key factor in promoting the ability of biofilm to attract cationic and anionic substances from the surrounding water including heavy metals such as Cr VI. Hence, biofilm should accumulate Cr VI if the ion exists in the river water. Thus, to analyse the presence of Cr VI in the aquatic ecosystem, the measurement of the concentration of Cr VI inside the biofilm matrix may be used as an indicator.

### *3.2. Cr VI accumulation*

The concentrations of Cr VI inside the biofilm and in the surrounding river water were measured (table 1). The concentrations of Cr VI inside the biofilm ranges from 1.8 μmol/wet-g to 2.3 μmol/wet-g. These concentrations were much higher than those in surrounding river water (0.03 μmol/wet-g until 0.4 μmol/wet-g). Interestingly, although the concentration of Cr VI in the river water became lower due to the input of rain water or the cessation of pollutant input, the concentration of the ions inside the biofilm was still much higher. The attraction and retention of the Cr VI in the biofilm is mostly due to the electrostatic interaction [14] and ion exchange mechanism [8]. The Cr VI can be retained on the charge sites of biofilm polymers and in the regions between the biofilm polymers [8].

**Table 1**. The Cr VI accumulation inside biofilm and surrounding river water.

| Sampel | Cr VI concenteration (ppm) | | |
|---|---|---|---|
| | Sinte 1 | Site 2 | Site 3 |
| Biofil | 2.3±0.3 | 1.9±0.6 | 1.9±0.3 |
| Water | 0.032±0.002 | 0.036±0.005 | 0.042±0.007 |

### *3.3. Biofilm as an bioindicator of water pollution*

Water pollution promotes various problems not only to aquatic organisms but also to humans [12,17]. Many technologies have been suggested to solve these problems [18]. Moreover, to monitor the pollution and hinder more negative effects from it, the monitoring of pollution is critically important. For the case of water pollution in aquatic ecosystems, the pollution monitoring of lotic ecosystems such as rivers is much more difficult to monitor than lentic ecosystems such as lakes. The difficulty is mainly related to the presence of water flow. Hence, the contamination will be not detected if the measurement uses only the river water as a sample and is conducted during an unsuitable time such as after rainfall or when pollution sources such as factory plants do not operate temporarily. Therefore, the utilization of bioindicators such as aquatic organisms to monitor the pollution in lentic ecosystems is fundamentally required.

The utilization of aquatic organisms as a bioindicator of water pollution is believed to be an effective alternative to other technologies and is considered environmentally safe [19]. In particular case, the biofilm that can attract and retain pollutants such as Cr VI may serve as an alternative. This current study indicates that biofilm can attract and retain Cr VI in the lotic ecosystems. This toxic substance will be accumulated continuously if the contamination of Cr VI to the river is not stopped. Even if the discharge of Cr VI is stopped temporarily to the river or there is dilution due to the input of rain water, the Cr VI will still exist in the biofilm. By monitoring the concentration of Cr VI inside the biofilm formed in the river, the level of pollution of Cr VI in the river can be monitored. Moreover, the biofilm is directly connected to the food web of the aquatic ecosystems. If the contaminated biofilm is consumed by higher trophic level organisms such as fish, the Cr VI may be accumulated in the fish. Hence, if the fish is consumed by a human, the Cr VI may also accumulated in the human. The ability to attract and retain heavy metal ions such as Cr VI makes the biofilm a promising biosorbent and bioindicator for water pollution in lotic ecosystems [20, 21].

## 4. Conclusion

Biofilm carries electric charge sites which can attract and retain heavy metal ions such as Cr VI. These mechanisms cause the concentration of Cr VI inside the biofilm to be much higher than the surrounding river water. The results of the present study indicates that biofilm may accumulate Cr VI from surrounding river water and reserve the ions. Cr VI will be accumulated continuously if the input of Cr VI to the river is not stopped. Even if the discharge of Cr VI is stopped temporarily to the river or there is dilution due to the input of rain water, the Cr VI will still exist in the biofilm. Hence, even if the river water in one site changes due to water flow that makes it difficult to detect the Cr VI in the site, the history of Cr VI pollution can still be tracked through the measurement of Cr VI in the biofilm. According to the results of this study, biofilm is a promising bioindicator to monitor the Cr VI pollution in lotic ecosystems such as rivers.

## 5. References

[1] Costerton J W, Lewandowski Z, Caldwell D E, Korber D R, Lappin S H M 1995 *Ann.u Rev. Microbiol.* **49** 711–745
[2] Kurniawan A, Yamamoto T, Tsuchiya Y, Morisaki H 2012 *Microbes. Environ.* **27** 399–406
[3] Lappin-Scott and Costerton JW 1995 *Microbial biofilms* (Cambridge: Cambridge University Press)
[4] Hall-Stoodley L, Costerton J W, Stoodley P 2004 *Nat. Rev. Microbiol.* **2** 95–108
[5] Hiraki A, Tsuchiya Y, Fukuda Y, Yamamoto T, Kurniawan A, Morisaki H 2009 *Microbes. Environ.* **24** 265–272
[6] Lewandowski Z, Beyenal H 2007 *Fundamental of biofilm research* (New York: CRC press)
[7] Tsuchiya Y, Hiraki A, Kiriyama C, Arakawa T, Kusakabe R, Morisaki H 2011 *Microbes. Environ.* **26** 113–119
[8] Kurniawan A, Tsuchiya Y, Eda S, Morisaki H 2015 *Colloids. Surf. B. Biointerfaces.* **136** 22–26.
[9] Watnick P, Kolter R 2000 *J. Bacteriol.* **182** 2675–2679
[10] Quintelas C, Rocha Z, Silva B, Fonseca B, Figueiredo H, Tavares T 2009 *Chem. Eng. J.* **152** 110–115
[11] Okabe S, Oshiki M, Kamagata Y 2010 *Microbes Environ.* **25** 230–240
[12] Sukandar, Kurniawan A 2017 Biosorption of Cr VI using Rice Straw Waste *Ponte* **73** 185–190.
[13] Kurniawan A, Fukuda Y 2016 Electric charge characteristics of biofilms formed on various surfaces *J.Pure App. Chem. Res.* **5** 95–100
[14] Kurniawan A, Yamamoto T, Tsuchiya Y, Morisaki H 2012 *Microbes. Environ.* **27** 399–406
[15] Kurniawan A, Fukuda Y 2015 *Microbiol. Indones.* **9** 106–112
[16] Gadd, G.M 2009 *J. Chem. Technol. Biotechnol.* **84** 13–28
[17] Freifelder D 1985 *Principles of physical chemistry with application to the biological sciences* (Boston: Jones and Bartlett Publisher) 2nd ed.
[18] Walton H F, Rocklin R D 1990 *Ion exchange in analytical chemistry* (Florida: CRC Press)
[19] Volesky B 2007 *Water Res.* **41** 4017–4029.
[20] Kurniawan A, Yamamoto T 2013 Biofilm polymer for biosorption of pollutant ions *Procedia Environ. Sci.* **17** 179–187

IOP Conf. Series: Earth and Environmental Science **137** (2018) 012062 doi:10.1088/1755-1315/137/1/012062

[21] Kjelleberg S, Givskov M 2007 The biofilm mode of life. In: Kjelleberg S, Givskov M (eds) the biofilm mode of life: mechanisms and adaptations, original, horizon bioscience, norfolk UK, p 5–21

IOP Conf. Series: Earth and Environmental Science **137** (2018) 012063 doi:10.1088/1755-1315/137/1/012063

# The potential of mangrove *Avicennia marina* and *A. Alba* from Nguling district, Pasuruan,East Java as an antioxidant

**F Iranawati[1,2], F Muhammad[1], H Fajri[1], R D Kasitowati[2] and S Arifin[3]**

[1]Marine Science Department, Fisheries and Marine Science Faculty, Brawijaya Unversity, Veteran Street, Malang 65145, Indonesia
[2]Marine Resources Exploration and Management (MEXMA) Research Group, Brawijaya University, Veteran Street, Malang 65145, Indonesia
[3]Academy of Pharmacy, Yayasan Putra Indonesia Malang

E-mail: fe_iranawati@ub.ac.id

**Abstract.** Free radicals are highly reactive molecules due to unpaired electron in their outer orbital. Excess of free radicals inside human body as consequences of environmental exposure such cigarette smoke may lead to degenerative diseases such as diabetic, cancer etc. This negative effect can be limited by the utilization of natural antioxidant substances, especially produced from plant. *Avicennia alba* dan *A. marina* are mangrove species that widely distributed in Indonesia and are expected potential as antioxidant. The objective of this study is to evaluated Avicennia alba dan A. marina potency as antioxidant performed with DPPD (1,1-diphenyl-β-picryl hydrazyl) method. Leaf and bark of *Avicennia alba* dan *A. marina* were collected from Nguling District, Pasuruan, East Java. Results shows that based on 50% inhibition Concentration (IC50), Avicennia alba leaf were categorized had a very high antioxidant potential (IC50 14,85 ppm) whereas the bark were categorized had a weak antioxidant potential IC50 167,17 ppm). For *A. marina*, the leaf were categorized had a moderate antioxidant ($IC_{50}$ 123,23 ppm) whereas the bark were categorized had a weak antioxidant potential ($IC_{50}$ 198,15 ppm).

## 1. Introduction

Free radical molecules can originate from environment pollutants such as cigarettes, vehicle smoke and radiation exposure. These molecules affect susceptible compounds such as proteins and lipids and lead to degenerative diseases such as atherosclerosis, heart disease, premature aging and cancer [1]. This could happen due to limited antioxidants in the body, therefore additional antioxidants are needed. Antioxidants are substances that can prevent, delay and eliminate oxidative damage that occur in molecular targets such as DNA, protein and lipid [2]. These substances have a function to regulate the concentration of free radicals so that the level of damage in the body can be minimized. The action mechanism of antioxidants is by breaking free the radical chain, detoxifying and activating oxidative enzymes such as glutathione peroxidase, catalase, and superoxide dismutase [3]. Antioxidants can be found in several forms such as minerals, phytochemicals and vitamins [4]. By source, antioxidants can be derived from natural and synthetic materials. Natural ingredients are often tested for antioxidant content such as plants and fruits.

The mangrove is a potential plant that may contain antioxidant compounds. In some Asian countries such as India and Malaysia, mangrove plants are widely used as traditional medicine for diabetes,

IOP Conf. Series: Earth and Environmental Science **137** (2018) 012063 doi:10.1088/1755-1315/137/1/012063

cancer, asthma, cancer, injury and even AIDS. One mangrove species that is used as a drug is *Avicennia* spp. which locals often call as Api api [5,6]. This species is found widely in Indonesia and locals have consumed its fruit as a snack and as medicine to treat diarrhea. Plants that have been used as traditional medicine are potential sources of natural antioxidant substances and can be developed as modern medicines [6]. This study aims to assess the potency of *Avicennia* spp. as an antioxidant.

## 2. Methodology

Samples of the bark and leaf of *A. alba* and *A. marina* were collected from the location of Nguling Mangrove Ecotourism, Pasuruan District, East Java. Samples were then water dried and grinded to powder. Further laboratory work was performed at the Exploration Laboratory of Fisheries and Marine Faculty of Brawijaya University.

### *2.1 Extraction*

The sample powder was immersed in a methanol solvent (1: 3) for 24 hours. After 24 hours the filtrate was separated (using *Whatman* paper no. 42) and the residual was immersed and filtered again for another three times. Then the whole filtrate was evaporated at a temperature of 42°C for 1 hour to separate the solvent and the sample extract.

### *2.2 Antioxidant test*

The antioxidant activity was tested by the DPPH (1.1-diphenyl-2-picrylhydrazyl) method. The DPPH method was used because the substance is capable of capturing free radicals causing paired electrons which then cause a color repellent proportional to the number of electrons captured. DPPH provides strong absorption at 517 m wavelength with a dark violet color [7]. The concentration of DPPH solution used for the antioxidant testing was 0.5 mM [8], whereas the four different concentrations of the sample extracts (leaf and bark) used for the testing were 31.25 ppm, 62.5 ppm, 125 ppm, and 250 ppm. Each sample concentration was added with 1 ml of 0.5 mM DPPH and then incubated in a dark room at room temperature for 30 minutes. The positive controls used were vitamin C or ascorbic acid with concentrations of 2 ppm, 4 ppm, 6 ppm, and 8 ppm. According to [8], the classification of antioxidants based on IC50 (50% Inhibitor Concentration) can be seen in Table 1, as follows:

**Table 1.** Classification of Antioxidant Categories.

| Concentration | Classification |
|---|---|
| <50 ppm | Very Strong |
| 50 – 100 ppm | Strong |
| 100 – 150 ppm | Moderate |
| 151 – 200 ppm | Weak |
| >200 ppm | Very weak |

The color change (qualitative test) that occurred when the sample was mixed with DPPH (from light purple to yellow) indicated the presence of antioxidant activity in the sample solution [9]. The quantitative test can be evaluated based on percent inhibition value. Percent inhibition is the value of free radical inhibition while IC50 is the concentration of samples needed to reduce 50 % of DPPH. Percent inhibition is calculated using the following formula:

$$\% \text{ inhibition} = \frac{(\text{blanko absorbance} - \text{sample absorbance})\ 100\ \%}{\text{sample absorbance}} \quad (1)$$

The concentration values of the samples as well as vitamin C and the percent of their inhibition were plotted on the x-axis and y-axis in the linear regression equation. The linear regression equation was then used to calculate IC50.

## 3. Result and Discussion

Antioxidants are substances or molecules that can prevent or restore free radicals by cutting off the chain reaction of free radicals. The color change occurs when the sample extract gives its hydrogen atom to DPPH thus it becomes stable.

The result of the antioxidant activity test of the leaf and bark of *A. alba* and *A. marine* showed the color change to yellow, indicating that the leaf and bark extract of *Avicennia spp.* can reduce DPPH radicals. The results of the qualitative antioxidant activity test on *A. alba* can be seen in Figure 1, while the results of *A. marina* can be seen in Figure 2.

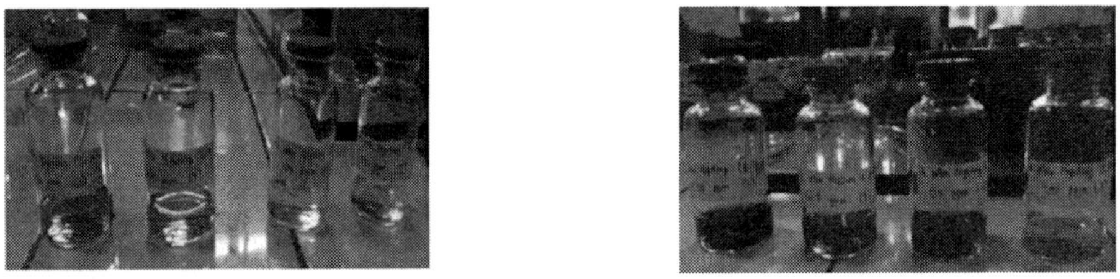

**Figure 1.** Antioxidant activity of *A. alba* leaf (left) and bark (right).

**Figure 2.** Antioxidant activity of *A. marina* leaf (left) and bark (right).

The subsequent percentage of the inhibition value and IC50 value *of Avicennia spp.* samples and positive controls were calculated as presented in Table 2 and 3.

**Table 2.** Results of antioxidant activity test on *A. alba.*

| Sample | Concentration (ppm) | Absorbance | % Inhibition | IC50 (ppm) |
|---|---|---|---|---|
| Leaf | 31.25 | 0.49 | 40.67 | 14.85 |
| | 62.5 | 0.27 | 66.88 | |
| | 125 | 0.1 | 87.39 | |
| | 250 | 0.04 | 95.3 | |
| Bark | 31.25 | 1 | -21.77 | 167.17 |
| | 62.5 | 0.86 | -4.02 | |
| | 125 | 0.49 | 40.74 | |
| | 250 | 0.11 | 86.3 | |

**Table 3.** Results of antioxidant activity test on *A. marina.*

| Sample | Concentration (ppm) | Absorbance | % Inhibition | IC50 (ppm) |
|---|---|---|---|---|
| Leaf | 31.25 | 0.76 | 7.05 | 123.23 |
| | 62.5 | 0.58 | 28.94 | |
| | 125 | 0.27 | 66.93 | |
| | 250 | 0.10 | 88.41 | |
| Bark | 31.25 | 0.81 | 1.09 | 198.15 |
| | 62.5 | 0.79 | 4.26 | |
| | 125 | 0.73 | 11.03 | |
| | 250 | 0.22 | 73.61 | |

The results of antioxidant activity presented in Tables 2 and 3 show that the different mangrove species and in different organ had different potential of antioxidant. Based on IC50 value, *A. marina* leaf and bark had a lower potential of antioxidant than *A. alba* leaf. In addition, *A. alba* leaf had a higher antioxidant potential than its bark. *A. marina leaf* can be categorized as a moderate antioxidant whereas the bark can be categorized as a weak antioxidant.

Weak antioxidant activity was also found for *A. alba* bark whereas a very strong antioxidant activity was found for the leaf. Higher antioxidant activity in the leaf than the bark of *A. marina* was also found in [10] and [11] whereas the opposite result was found in [12]. A very strong antioxidant potential of *A. alba* leaf may be the result of the presence of vitamin C in the leaf [13]. The different results of this study compared to previous studies is likely due to different sample collection locations that may result in different bioactive compounds found in the plants. Every plant or organ of plant has a different metabolic seconder, with different polarity and even a different level of concentration [11].

## 4. Conclusion

*Avicennia* spp. has a potential as an antioxidant with higher potency found in the leaf than the bark. *A. alba* had a higher antioxidant potency than *A. marina.* The highest potency of antioxidant was found in the *A. alba* leaf.

## 5. References

[1] Kusumowati I T D, Sudjono T A, Suhendi A, Da'I M, Wirawati R 2012 *J. Farm. Ind. PHARMACON.* **13** 1-5

[2] Purwaningsih S 2012 Antioxidant Activity and Chemical Composition of Red Eyed Conch *J. Mar Sci.* **17** 39-48

[3] Harliansyah 2005 *Paksi J. Ind. Stud. Assoc. Malay.* **8**:92-97

[4] Nurjanah N, Izzati L, Abdullah A 2011 *Ind. J. Mar Sci.* **16** 119-124

[5] Banerjee D, Chakrabarti S, Hazra A K, Banerjee S, Ray J, Mukherjee B 2008 *Afr. J. Biotechnol.* **7** 805-810

[6] Haq I, Hossain A B M, Khandaker M M , Merican A F , Faruq G, Boyce A N, Azirun M S, others 2014 *Int. J. Agric. Biol.* **16** 707–714

[7] Umayah E and Amrun M 2007*) J Pharm Sci.* **8** 83-90

[8] Pramesti R 2013 *Bul. Osean Mar.* **2** 7-15

[9] Romansyah Y 2011 Content of Bioactive Compounds Antioxidant Coral Soft Sarcophyton sp. Natural and Transplantation in Pramuka Kepulauan Seribu Island Skripsi IPB Bogor

[10] Handayani S 2013 Flavonoid Leaves of Bark and Leaf-Fire Tree *(Avicennia marina (Forks.) Vierh.)* As an Active Compound Antioxidant Skripsi IPB Bogor

[11] Moteriya P, Ashish D, Sumitra C 2015*J. Coast. Mar. Life Med.* **3** 713-717.

[12] Suwendiyanti R 2016 Effectiveness of Root, Trunk, Stem, Leaf, and Fraction Extract *Avicennia marina* as antioxidant Skripsi FPIK Universitas Padjajaran Bandung

[13] Munandar A, Mustopa A Z, Tarman K, Nurhayati T 2014 Antibacterial Activity of Kapang Xylaria psidii KT30 Protein Against *Eschercia coli* and *Bacillus subtilis* *J. Technol.Ind. food.* **25** 146-151

**Acknowledgment**
The author would like to give gratitude for the support provided by the Faculty of Fisheries and Marine Science of Brawijaya University.

IOP Conf. Series: Earth and Environmental Science **137** (2018) 012064 doi:10.1088/1755-1315/137/1/012064

# The reducibility of heLa cell viability by *Sargassum polycystum* extracts

**M Firdaus*, D Setijawati, I Islam, H Nursyam, H Kartikaningsih, H S Yufidasari, A A Prihanto, R Nurdiani and A A Jaziri**

Department of Fishery Product Technology, Faculty of Fisheries and Marine Science, Brawijaya University, Malang – 65145, Indonesia

E-mail: muhamadfir@ub.ac.id

**Abstract**. Cervical cancer is the second largest cause of death-related cancer in women. The efficacy of cancer drugs is still low. Bioactive of brown seaweed has been studied by *in vitro* and *in vivo* as anticancer. The aim of this study was to evaluate the cytotoxicity of *Sargassum polycystum* extracts on HeLa cell, to recognize bioactive on extract and estimate the interaction between the bioactive and target protein. *S. polycystum* was found from Talango Island waters and HeLa cell was obtained from Indonesian Science Institute. Sample was extracted by ethanol, ethyl acetate and hexane, concentrated and finally, extracts were assayed on HeLa cell. The viability of this cell was quantified on ELISA-Reader. The bioactive compounds of the extract were elucidated by GC-MS. The interaction between bioactive and target protein was evaluated by using *in silico* method. The result showed that the lowest viability of HeLa cell on n-hexane extracts treatment. The n-hexane extract of this seaweed contained benzenepropanoic acid. This compound reduced HeLa cell viability by reducing of thrombin concentration. In conclusion, the benzene propanoic acid of *S. polycystum* was the cytotoxic agent and it is potential agent for anti-cervical cancer.

## 1. Introduction

Cervical cancer is the second highest cause of death in women after heart disease. More than three hundred thousand women died from this cancer each year and mostly occurs by the persistent infection of human papillomavirus [1]. In the future, this disease is predicted to be the main causative of death in women if early detection and curative measures on this disease is not done. Nowadays, the efficacy of cancer drugs distributed and consumed by patients is still low and also has side and resistant effects. The World Health Organisation (WHO) predicts that more than 80 % of the world population relies on plant-based medicine for healthcare. These substances are safer and more natural [2].

Brown seaweed is a natural marine resource that contains many secondary metabolites or bioactives and also shows capability as an anticancer ingredient. Fucoidan is the bioactive of this seaweed that is known to have anti-proliferation, antitumor and anticancer properties by inducing apoptosis and inhibiting invasion, metastasis, and angiogenesis of cancer cells. This substance has been known to be able to inhibit the growth of several cancers [3, 4, 5]. There are several active compounds of this seaweed that is also capable in inhibiting and destructing cancer cells. These substances kill cancer cells by activating apoptosis, inhibiting the binding between the oncoprotein, tumor suppressor protein, and function of HPV-16-harboring cervical cancer cells [6].

IOP Conf. Series: Earth and Environmental Science **137** (2018) 012064 doi:10.1088/1755-1315/137/1/012064

*Sargassum* sp. has been grouped as brown seaweed and this alga is also found in Indonesia waters. One species is *S.npolycystum* that is obtained and grows abundantly in the waters of Talango Island, District of Sumenep, East Java. Brown seaweed has been known for their anti-cancer potency, however, the exploration and usage of *S. polycystum* as sources of bioactive compounds is very limited and its inquiry as an anti-cervix cancer ingredient has not yet been done. Therefore, the purpose of this study was to obtain the cytotoxic activity of brown seaweed (*S. polycystum*) extract on HeLa cells, find the identity of their bioactives and their interaction with the target protein.

## 2. Material and methods

### *2.1. Material*

*S. polycystum* was obtained from the waters of Talango Island, Sumenep District, East Java in September 2013. HeLa cells were obtained from the Research Center of Biotechnology, Indonesian Science Institute, Cibinong. Hexane, ethyl acetate, and ethanol were obtained from Merck. 3-(4,5-dimetilthiazolil-2)-2,5-diphenyl tetrazolium bromide (MTT) was purchased from Sigma-Aldrich.

### *2.2. Methods*

*S. polycystum* was washed and dried under the sun until dry. The dried sample was crushed and then sieved to get powder matter. This material was macerated by n-hexane, ethyl acetate and ethanol (1:3; b:v) at 4°C overnight. Afterwards it was filtered by Whatman paper to obtain a supernatant. This was concentrated by a rotary evaporator and was collected in a tight bottle and saved until used for assaying.

*2.2.1. Cell Viability.* This assay used 3-(4,5-dimethylthiazol-2)-2,5-diphenyltetrazolium bromide (MTT) to observe the viability of HeLa cells [7]. A hundred μL of RPMI 1640 was poured and mixed with HeLa cells in 96 microwells to obtain 3 x $10^4$ cell/mL of HeLa cell suspension. Each well was poured with 100 μL of HeLa cells. One well filled with HeLa cell was stated as the negative control. Six wells were mixed with 15 μL of doxorubicin as the positive control and the others were added with 125 μL of ethanol, ethyl acetate and hexane extract of *S. polycystum*, respectively. The viability of cells was observed by the ELISA reader. Absorbance value related positively to the color formation of formazan that shows metabolism activity of the cell.

*2.2.2. Gas Chromatography Determination.*

This analysis used GC-17A Agilent 5890 series II Plus and GC-MS Agilent Technologies 5975C inert MSD with Triple-Axis Detector and the column HP 5-MS. One μL of sample was injected in the column port and carried by Helium with a flow rate 0.7 mL each minute. The injector temperature of GC was 295°C and the oven temperature was programmed starting from 70°C for 2 minutes and was increased by 10°C each minute until 295°C for 28 minutes.The identity of its mass spectra was confirmed by the Wiley Library.

*2.2.3. In-silico.* HITPICK was used to predict the protein target of *S. polycystum* bioactive and STITCH was exercised to determine the interaction between the *S. polycystum* bioactive and protein target.

### *2.3. Statistical analysis*

Data of HeLa cell viability were analyzed by one way methods. The significant level was set at $P< 0.05$.

## 3. Results and Discussion

### *3.1. HeLa cell Viability*

Figure 1 illustrates the percentage of HeLa cell viability exposed by the control and n-hexane, ethyl acetate and ethanol of *S. Polycystum* extract. The percentage of HeLa cell viability of n-hexane extract was the lowest among other treatments.

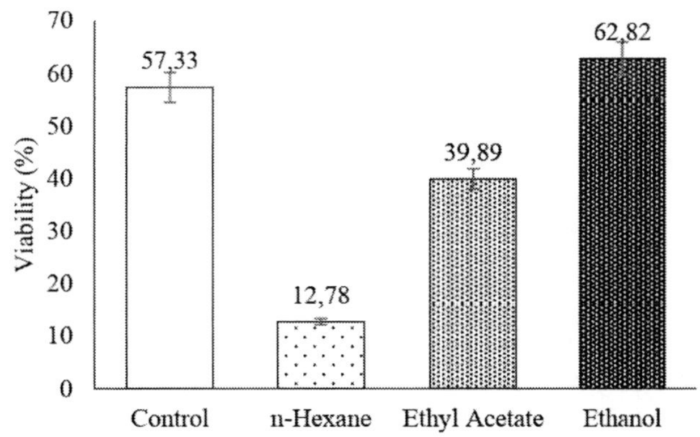


**Figure 1**. Percentage of HeLa cell viability was treated positive control and n-hexane, ethyl acetate and ethanol extract of *S. polycystum*.

The viability of cell is a method of measurement on the life or death of the cell. Cytotoxicity of the cell is generally indicated by decreasing cell proliferation, nucleic acid or protein synthesis and cell viability. The alteration on membrane permeability or disturbance on several metabolic pathways is the causality of cell death that is observed by cell viability. Therefore, cell viability can used as the cytotoxicity indicator of a substance [8]. The life cells are able to produce succinic dehydrogenase in their mitochondria. This enzyme will react with the MTT substance to form a formazan crystal that will turn to the color violet. The darkness of the color is indicated and positively correlated with cell life [9].

Figure 1 showed that the HeLa cell viability that was exposed to the n-hexane *S. polycystum* extract exhibited the lowest measurement. It means that the n-hexane extract of *S. polycystum* has the capability to kill HeLa cells. The methanol extract of brown algae (*Dictyotaciliolata*) extract showed the strongest capability to kill HeLa cells [10]. Meanwhile, the dichloromethane extract of brown alga (*Cystoseiraabies-marina*) acted as a HeLa cell anti-proliferation substance [11]. Although this experiment shows different results from the previous studies, it indicates that brown seaweed is a natural marine resource that contains many bioactives that can be developed and used to act as anticancer drugs [12]. The ability of brown alga extracts to eliminate HeLa cells is caused by their being able to change the morphology and biochemical of HeLa cell. These phenomena are characterized by apoptosis that is marked by loss of cell viability, chromatin condensation, phosphatidylserine externalization and phase accumulation of sub-G1 cycle [10].

### *3.2. Bioactive Identity*

Figure 2 shows the spectra of *S. polycystum* n-hexane extract. Based on the Wiley Library, it exhibited that the identity of *S. polycystum* n-hexane extract is grouped in oncarboxylic and fatty acids, i.e.: hexadecanoic acid, benzenepropanoic acid, and arachidonic acid.

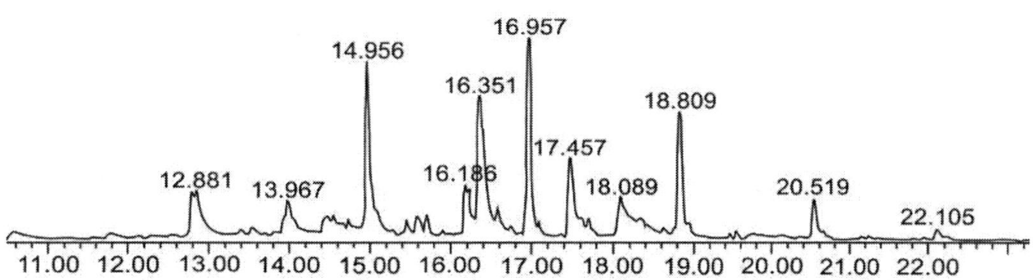


**Figure 2.**Spectra of *S. polycystum* n-hexane extract by gas chromatography mass spectra.

IOP Conf. Series: Earth and Environmental Science **137** (2018) 012064 doi:10.1088/1755-1315/137/1/012064

Figure 2 shows that bioactive compounds that were contained in the n-hexane extract of *S. polycystum* are grouped in the derivation of fatty and carboxylic acids. Many studies have reported that fatty acids were anticancer agents in a variety of human cancer cell lines derived from colonic, pancreatic, prostate, and breast cancers. The fatty acids compound is capable as an anticancer agent by attenuation growth and inducing apoptosis [13]. The mechanisms underlying the anti-tumor effects of omega3-FA are complex. The incorporation of omega3-FA in biological membranes alters the profile of lipid mediators generated during inflammatory reactions. Furthermore, omega3-FA acts as ligands of nuclear peroxisome proliferator-activated receptors that attenuate transcription of NF-kappaB-dependent genes. Thereby, the cyclooxygenase-2/prostaglandin E(2)-dependent production of pro-angiogenic vascular endothelial growth factor and levels of anti-apoptotic Bcl-2 and Bcl-X(L) are decreased. Eicosanoid-independent pro-apoptotic pathways include enhanced lipid peroxidation, modulation of mitochondrial calcium homeostasis and enhanced production of reactive oxygen species as well as activation of p53 [13].

### *3.3. Target ProteinPrediction and Interaction*

Based on HITPICK and STITCH software analysis, it is suggested that benzene propanoic acid is capable of reducing the viability of HeLa cells by the reduction of thrombin concentration. This protein is known as an anticoagulant, however if its concentration is high in the blood, it will induce tumor growth, angiogenesis, and metastasis. The depressing of thrombin concentration will inhibit several mechanisms of cancer cell viability, such as (1) tumor cell adhesion to platelet cells, endothelial cells, and sub-endothelial matrix protein, (2) enhancement of tumor cell growth, and (3) increasing of tumor cell seeding and spontaneous metastasis [14].

## 4. Conclusion

The benzene propanoic acid of the n-hexane extract of *S.polycystum* was the cytotoxic agent on HeLa cells by reducing the thrombin concentration mechanism.

## 5. References


[1] Anonymous 2017 *Human Pappilomavirus and related diseases report* (Barcelona: Institut Catala d'Oncologia)

[2] Anonymous 2002 WHO *Traditional medicine strategy* 2002-2005 Geneva

[3] Marudhupandi T, Kumar T T A, Lakshmanasenthil S, Suja G, Vinothkumar T 2015 *Int. J. Biol. Macromol.* **72** 919-923

[4] Han Y, Lee J H, Lee S H 2015 *Biomolecular Therapeutic* **23** 225–232.

[5] Senthilkumar K, Manivasagana P, Venkatesan J, Kim S K 2013 *Int. J. Biol. Macromol.* **60** 366–374

[6] Tariq 2015 *J. Ethnopharmacol.* **164** 96–119

[7] Stockert J C, Blázquez-Castroa A, Canete M, Horobinb RW, Villanueva Á 2012 *Acta Histochemica* **114** 785–796

[8] Riss T L, Moravec R A, Niles A L 2011 *Methods Molecular Biology* **740** 103-14

[9] Wang P, Henning S M, Heber D 2010 Limitations of MTT and MTS-Based assays for measurement of antiproliferative activity of green tea polyphenols PLoS ONE **5**(4) e10202

[10] Gomes D L, Telles C B, Costa M S, Almeida-Lima J, Costa L S, Keesen T S, Rocha H A 2015 *Molecules*.13 **20** 6573-91

[11] Barreto M C, Mendonça E, Gouveia V, Anjos C, Medeiros J S, Seca A M L, Neto A I 2012 *Life and Marine Sciences* **29** 53-58

[12] Moussavou G, Kwak D H, Obiang-Obonou B W, Maranguy C A O, Dinzouna-Boutamba S D, Lee D H, Pissibanganga O G M, Ko K, Seo J I, Choo Y K. 2014 *Marine Drugs* **12** 4898-4911

[13] Wendel M, Heller A R 2009 *Anticancer Agents Medical Chemistry* **9** 457-70

[14] Anonymous 2017 The cancer pathways Kanehisa Laboratories Kyoto Encyclopaedia of Gene and Genomes

IOP Conf. Series: Earth and Environmental Science **137** (2018) 012065 doi:10.1088/1755-1315/137/1/012065

# Blue carbon content of mangrove vegetation in Subang district

**I Nurruhwati, S D Purwita, Sunarto and Zahidah**

Fisheries and Marine Science Faculty, Padjadjaran University, Jl.Bandung Sumedang km 21, Hegarmanah, Jatinangor, Sumedang, West Java 45363

Email: purwitashin@gmail.com

**Abstract.** The purpose of this research was to know the carbon content of mangrove parts such as leave, stems and roots and to know its ability to absorb carbondioxide ($CO_2$). The research was conducted in 27th April until 16th May 2017 in Blanakan Village, Langensari Village and Jayamukti Village. The samples are dried at Pilotplane Laboratory Faculty of Industrial Engineering Padjadjaran University. The method in this research is explorative survey method. The results showed that there were two dominant mangroves species in three research stations, they are *Avicennia marina* and *Rhizophora mucronata*. Index of Important value of each mangrove type on the three stations in the medium criterion with a range of values is 106,86 %- 193,13 %.The highest carbon content was found in *Rhizophora mucronata* at station 1 (93,43 %) which was equivalent with 342,87 % absorption of $CO_2$ which was The lowest carbon content was in *Avicennia marina* at station 1 (67,49 %) which was equivalent with 247,70 % absorption of $CO_2$.

## 1. Introduction

Global warming is one important phenomenon that has become a problem not only in Indonesia but has also evolved into an issue discussed at the international level. Global warming affects climate change which can cause various disasters such as the rise of sea levels, prolonged drought, flooding duringthe rainy season, failure of reproduction of various animals living in the wild, declining biodiversity and so on [1].

Global warming followed by climate change is caused by excessive energy use which can increase Green House Gas (GHG) emissions especially $CO_2$ (carbon dioxide). The increase in population causesthe $CO_2$ concentration in the atmosphere to rise due to the increase of the anthropogenic activity of humans. Undernormal conditions, $CO_2$ will be utilized by autotrophic organisms located in terrestrial and aquatic ecosystems for photosynthesis which then results in the process of food and oxygen gas ($O_2$)formation. If there is a decrease in the number of plants acting as $CO_2$ absorbers, the presence of $CO_2$ in the atmosphere becomes more abundant and will cause global warming [2].

The high $CO_2$ gas in the atmosphere is one cause of climate change that can be avoided because the earth can act as a natural carbon storage, among others through terrestrial forests, photosynthesis processes, as well as stored in biomass and sediments known as green carbon. Green carbon is now being applied in the GHG emission reduction mechanism) known as the Reducing Emissions from Deforestation and Degradation Plus by referring to the method developed by the 2006 Intergovermental Panel Climate Change Guide Line (IPCC GL), however it does not take into account

IOP Conf. Series: Earth and Environmental Science **137** (2018) 012065 doi:10.1088/1755-1315/137/1/012065

the existence of coastlines and seas. Seventy percent of the earth's surfaceare oceans that can absorb, bind and store carbon through photosynthesis mechanisms with a very large carbon pool that is potentially larger than green carbon [2].

Lowland or coastal areas are particularly vulnerable to the effects of climate change, this is because coastal areas are directly affected by rising sea levels caused by rising earth temperatures. These coastal areas include seagrass, mangrove and coral reef ecosystems that are also vulnerable to the impacts of climate change and are feared to disrupt and threaten its sustainability and the living biota inside [1].

Coastal ecosystems can transfer and store carbon in vegetation and sediments with much greater capability and reach thousands of years longer than land forests which can hold hundreds of years. The carbon stored for thousands of years is called blue carbon [3]. Mangrove ecosystems with seagrass beds and coral reefs and brackish swamps are important ecosystems that act as carbon absorbers and binders. The entire ecosystem can bind 235-450 million tons of carbon per year which is equivalent to almost half of carbon emissions through transportation worldwide [4].

Subang Regency is one of the districts in West Java with a considerably large mangrove area of 3033 ha (Forest service of west java 2013). The mangrove areas of Subang Regency are located in four districts namely the Blanakan, Pamanukan, Legonkulon and Pusakanagara Sub-Districts. Subang Regency has a relatively highpotential of mangrove resources, however, its carbon stockstoring and its $CO_2$absorption ability is yet uncertain. Therefore, research is necessary to identify the potential of blue carbon content.

## 2. Methodology

The research was conducted in Subang Regency with research stations in three villages (Blanakan, Jayamukti and Langensari). The research was conducted between April 27th to May 16th 2017. The analysis of carbon content was carried out at the Pilot plant Laboratory Faculty of Agricultural Industrial Technology, Padjadjaran University.

The methods used in this research were the field survey and explorative methods. An explorative method is a method that examines and reveals something from the field as a finding used to construct models and draw conclusions [5].

The measurement of biomass in vegetation can provide information on overall nutrients and carbon stocks. Mangrove biomass is a calculation of production and density based on the diameter, height and weight measurement of the tree species [6].

Biomass or dry weight is calculated using the following formula.

$$\text{Biomass} = \frac{\text{Dry Weight Samples}}{\text{Weight Wet Samples}} \text{ x } 100\ \% \quad (1)$$

### *2.1. Carbon Stock*

Calculations of Carbon biomass can use [7] and the International Panel on Climate Change (IPCC, 2003).

$$\text{Carbon Stock} = \text{Dry Weight of the plant (Biomass) x C } \% \quad (2)$$

Information :

1. Dry weight obtained from biomass sampling
2. C% based on IPCC stipulations in 2006 of 0.5

### *2.2. Carbondioxide Absorption*

Calculation of carbondioxide absorption ($CO_2$) is done by using the formula [9]:

$$CO_2\ (\%) = \frac{Mr\ CO_2}{Ar\ C} \text{x Carbon content} \quad (3)$$

Information :
Mr $CO_2$ = Relative Molecular Mass Relative Value of Carbon Dioxide 44
Ar C = Relative Carbon Atomic Value of 12

## 3. Result and Disscussion

### *3.1. Biomass and Carbon Content of Mangrove Vegetation*

The measurement of biomass and carbon content was done on tree-level mangrove plants with the tree samples of *Avicennia marina* and *Rhizophora mucronata,* which is the dominant type at the Blanakan, Langensari and Jayamukti research stations. The trees were determined based on Diameter at Breast at Highest (DBH) or highest chest diameter in the observation plot. Leaves, stems, roots and sediments were taken from each tree as to determine biomass and carbon content.

An autotrophic organism processes photosynthesis by absorbing CO2 from the atmosphere and converting it into organic carbon and storing it in its body's biomass such as in leaves, stems, roots, tubers, fruits and other parts. Biomass is a plant dry weight of 50 % consisting of carbon content (IPCC 2003).

The carbon content of the*Rhizophora mucronata* mangrove species has a higher value than the *Avicennia marina* type. The total carbon content of *Avicennia marina* was 67.5 % while it was 93.42 % in *Rhizophora mucronata* (table 1).

**Table 1**. Biomass and Carbon Content in Station 1.

| Mangrove Types | Sample | Gross Weight (g) | Dry Weight (g) | Biomass (%) | Carbon Content (%) |
|---|---|---|---|---|---|
| *Avicennia marina* | Leaf | 263 | 71 | 27,00 | 13,50 |
| | Stem | 540 | 284 | 52,59 | 26,30 |
| | Root | 213 | 118 | 55,40 | 27,70 |
| | Total | 530 | 473 | 134,99 | 67,5 |
| *Rhizophora mucronata* | Leaf | 288 | 100 | 34,72 | 17,36 |
| | Stem | 270 | 191 | 70,74 | 35,37 |
| | Root | 317 | 258 | 81,39 | 40,69 |
| | Total | 875 | 549 | 186,85 | 93,42 |

Results from Station 1 showed thatthe highest carbon content wasfound in the *Rhizophora mucronata* root at 40.69 % which is equal to 129 grams, while the *Avicennia marina* root has a value of 27.70 % which equals to 59 grams. The carbon content of the *Rhizophora mucronata* stem was 35.37 % which is equal to 95.5 grams, while the value of *Avicennia marina* is 26.30 % which equals to 142 grams. The carbon content of *Rhizophora mucronata* leaves was 17.36 % (50 grams), while *Avicennia marina* leaves measured 13.5 % (35.5 grams).

Comparisons of carbon content at Station 1 based on sample sections ofboth types indicated that the root has the highest carbon content while the leaves had the lowest. The high carbon content value of the root is caused by environmental conditions which are more extreme at the roots levelcompared tothe carbon content of the above-ground biomass. Additionally, the value of the carbon content of the roots did not vary much with the stem because the characteristics of its constituent substances are almost the same.

The low carbon content of the leaves did not depend on the age of the stand. The differences inleaf biomass in each mangrove species can be due to species characteristics, leaf shape and light availability [10].

The carbon content in *Rhizophora mucronata* compared to*Avicennia marina* was different from Mandari *et al.* [11] in the Bandar Bakau Dumai area, Riau Province. By using allometric method, the measurement of carbon content in the *Rhizophora mucronata* type was 0.34 ton C/ha, while the carbon content of *Avicennia marina* type was 0.93 ton C/ha. This occurs because of the differences in the density [11] *Avicennia marina* has a much higher density and has larger average tree diameter of 8.51 cm while the type *Rhizophora mucronata* has an average tree diameter of 6.54 cm.

At Station 2, the value of carbon content of the *Rhizophora mucronata* mangrove is greater than that of *Avicennia marina* which is 87.98 % and 76.68 % respectively (table 2).

**Table 2**. Biomass and carbon content in station 2.

| Mangrove Type | Sample | Gross Weight (g) | Dry Weight (g) | Biomass (%) | Carbon Content (%) |
|---|---|---|---|---|---|
| *Avicennia marina* | Leaf | 267 | 85 | 31,84 | 15,92 |
| | Stem | 300 | 191 | 63,67 | 31,83 |
| | Root | 242 | 140 | 57,85 | 28,93 |
| | Total | 809 | 416 | 153,36 | 76,68 |
| *Rhizophora mucronata* | Leaf | 409 | 131 | 32,03 | 16,01 |
| | Stem | 343 | 250 | 72,89 | 36,44 |
| | Root | 456 | 324 | 71,05 | 35,53 |
| | Total | 1.208 | 705 | 175,97 | 87,98 |

The highest carbon content was found in the stem of *Rhizophora mucronata* at 36.44 % which is equivalent to 125 grams, while the *Avicennia marina* stem has a carbon content value of 31.83% which is equal to 95.5 grams. The value of carbon content at the root of *Avicennia marina* was 28.93% (70 grams), while results of the*Rhizophora mucronata* root measuredto 35.53 % (162 grams). The lowest carbon content was found in the leaf, with the value of *Avicennia marina* at 15.92 % (42.5 grams) and *Rhizophora mucronata* at 16.01 % (65.5 grams). Comparisons of carbon content at Station 3 obtained from sample sections of both types indicated that the rod had the highest carbon content while the leaf portion had the lowest.

Results from Stations 1 and 2 showed that *Rhizophora mucronata* leaves were measured to have higher a carbon content than *Avicennia marina* leaves. This is due to the characteristics of *Rhizophora mucronata* leaves that are wider and larger than *Avicennia marina* leaves. Higher levels of carbon content were also present in the stem of *Rhizophora mucronata* at the two research stations. The method used in the research was a drying method, hence the carbon content of the stem was not affected by the diameter of the tree. This allows the carbon content value ofthe *Rhizophora mucronata* rod to measure higher although the the *Avicennia marina* has a greater tree diameter. The value of carbon content of the *Rhizophora mucronata* root was greater than that of *Avicennia marina* at Stations 1 and 2, this may be due to the root characteristic of *Avicennia marina* which has a pencil-like vertical root, while the *Rhizophora mucronata*has above ground root that has a larger structure and is harder than the *Avicennia marina* root.

The *Rhizophora mucronata* at Stations 1 and 2 had a higher carbon content value compared to the *Avicennia marina*. The result was in accordance with the research conducted by Rahmawati *et al.* [13] in the coastal area of Muara Gembong of Bekasi Regency, which used the method of allometric equation and measured the carbon content value of *Rhizophora mucronata* mangrove at 17.60 ton C / ha while it is 5.27 ton C/ha for the *Avicennia marina* type.

### *3.2. Carbondioxide Absorption*
The amount of carbon content in a plant can illustrate how much the plant can absorb carbon dioxide ($CO_2$). The greater the carbon content in the plant, the greater the ability to absorb carbon dioxide ($CO_2$).

**Table 3**. Carbondioxide absorption.

| Types of Mangrove | Station 1 | | Station 2 | |
|---|---|---|---|---|
| | Carbon Content | $CO_2$ | Carbon Content | $CO_2$ |
| *Avicennia marina* | 67,49 % | 247,70% | 76,68% | 281,40% |
| *Rhizophora mucronata* | 93,43% | 342,87% | 87,98% | 322,90% |

The highest amount of $CO_2$ absorption measured was at 342.87 % in *Rhizophora mucronata* (Station 1), while the lowest was in *Avicennia marina* at Station 1 at 247.70 %. The ability of mangroves to absorb $CO_2$ depends on the carbon content in the trees.

The ability of *Rhizophora mucronata* to absorb $CO_2$ is higher than that of *Avicennia marina*. This was caused by the average value of the carbon content of the *Rhizophora mucronata* which is also higher. Density values and biomass content also influenced the mangrove's ability to absorb $CO_2$. This is in accordance with research by Heriyanto*et. al.* [9] which states that the *Rhizophora mucronata* can absorb carbon dioxide at 398.60 tons $CO_2$/ha in Alas Purwo National Park. *Rhizophora* sp. has a very important and potential role in carbon sequestration when compared to other types.

The content of the carbon store in vegetation depends on the amount of biomass contained in the tree. Plants absorb carbon from the air and converts it through photosynthetic processes where theresults are used for vertical and horizontal growth [12].

## 4. Conclusions
The highest carbon content was found in *Rhizophora mucronata* in Station 1 with the amount of 93.43% which is equivalent to 342.87 % $CO_2$ absorption. Meanwhile, the lowest carbon content was measured in *Avicennia marina* at Station 1 with theamount of 67.49 %, equivalent to 247.70 % absorption of $CO_2$. At Station 2the carbon content of *Rhizophora mucronata* was found to be at 87.98 % and equivalent to 322,90 % absorption of $CO_2$ and also in *Avecinna arina* at 76.86 % which is equivalent to 281.4 % absorption of $CO_2$.

## 5. References
[1] Riani E 2012 *Climate change and aquatic biota life (impact of bioaccumulation of hazardous and toxic & reproductive materials)* (Bogor: IPB Press)
[2] Rustam A, Adi N S, Ati R N A, Kepel T L, Daulat A, Kususmaningtyas M A, Suryono D D, Sudirman N, Heriati A, Mangindaan P and Salim H L 2015 Blue carbon: blue island indonesia initiative program derawan island - berau east kalimantan Research and Development Center for Marine and Coastal Resources Agency for Research and Development of Marine and Fisheries North Jakarta
[3] Lawrence A 2012 Blue carbon a new breakthrough to reduce climate change impacts through conservation and conservation of coastal ecosystems in coral triangle area World Wild Foundation (WWF)
[4] Ghufran H K 2012 *Mangrove ecosystem: potency, function and management* (Jakarta: Rineka Cipta)
[5] Cahyaningrum S T, A Hartoko and Suryanti 2014 *Diponegoro J. Maquares* **3** 34-42
[6] Hairiah K, S M Sitompul, M van Noordwijk and C Palm 2001 Methods for sampling carbon stocks above and below ground International Center for Research in Agroforestry Bogor p 31

[7] Fourqurean J, B Johnson, J B Kauffman, H Kennedy, C Lovelock, J P Megonigal, A F Rahman, N Saintilan and M. Simard 2014 *Coastal blue carbon, methods for assessing carbon stock and emissions factors in mangroves, tidal salt marshes and seagrass meadows conservation international, intergovernmental oceanographic commission of UNESCO* (USA: Arlington) p184
[8] Intergovermental Panel on Climate Change 2003 *Good practice guidance for land use, land-use change and forestry* (Japan: institute for global enviromental strategies*(IGES)*) p 590
[9] Heriyanto, N M and E Subiandono 2012 *J. For. Nat Conserv.* **9** 023-032
[10] Khan, M N I, R Suwa and A Hagihara 2007 *Wetlands Ecol Manage.* **15** 141-153
[11] Mandari, D Z, H Gunawan and M N Isda 2016 *J. Riau Biol.* **1** 17-23
[12] Ati, R N A, A Rustam, T L Kepel, N Sudirman, M Astrid, A Daulat, P Mangindaan, H L Salim and A A Hutahaean 2014 *Segara J.* **10** 119-127
[13] Rachmawati, D, Setyobudiandi, I, Hilmi, E 2014 *Omni-Akuatik.* **13** 85 -91

IOP Conf. Series: Earth and Environmental Science **137** (2018) 012066 doi:10.1088/1755-1315/137/1/012066

# Polyculture Engineering technology of *larasati* red tilapia (*Oreochromis niloticus*) and white shrimp (*Litopenaeus vannamei*) based for protease enzyme

**I Samidjan and D Rachmawati**

Aquaculture Departement, Faculty of Fisheries and Marine Science, Diponegoro University, Jl.Prof.H.Soedarto SH, Tembalang, Semarang, 50275,Indonesia.

E-mail : istiyanto_samidjan @ yahoo.com

**Abstract.** The objective is polyculture technology of red tilapia larasati fish and white shrimp with different combinations density. The material is saline red tilapia *larasati* 3.29 ± 0.018 g and white shrimp with initial weight 1.39 ± 0.025 g. Seeds are density of red tilapia larasati larvae 5 and 10 larvae / $m^2$ fish. And white shrimp 5 larvae / $m^2$ and 10 larvae / $m^2$. An artificial feed used enzyme dose of 2.25 g / kg. The experimental using complete randomized design 4 treatments and 3 replications that is given seeds 5 larvae / $m^2$ larvae red tilapia larasati and given 5 larvae / $m^2$ white shrimp (A), 5 larvae / $m^2$ red tilapia ) and 10 $m^2$ / $m^2$ of white shrimp (B), 10 $m^2$ larvae and 5 $m^2$ white shrimp (C), 10 $m^2$ larvae and 10 $m^2$ white shrimp (D) ). The data were growth of absolute weight, survival rate, FCR, and water quality data (temperature, salinity, pH, $O_2$, $NO_2$, $NH_3$). Data were analyzed of variance (F test).The results showed significantly effect ($P < 0.01$) on the growth. The highest absolute growth in D treatment were red tilapia larasati (185.75 ± 0.50g) and white shrimp (25.25 ± 0.95 g).

## 1. Introduction

Red tilapia and white shrimp are important protein sources worldwide, especially in Indonesia. In Indonesia the majority of fish consumption is caught in the ocean, with amounts around 90 million tons in recent years [1,2,15,45]). Meanwhile production of inland fisheries in 2003 in Central Java reached 339,000 tons (Marine and Fisheries Agency of Central Java Province, 2004) [37]. Fishery production was dominated by marine fisheries for about 236,240 tons (approximately 74 % of total fishery production) with a value of IDR 0.77 trillion. In 2003, aquaculture and fisheries businesses in public waters of Central Java, both in terms of production and value, were higher than the previous year. The production of aquaculture and fisheries businesses in public waters reached 75 to 88 thousand tons and 14.33 thousand tons with production value reached IDR 0.88 billion and IDR 91.90 billion, respectively. currently, the fish farmers on the northern coast of Central Java, particularly in Semarang, still use the conventional monoculture fisheries system, thus, their production value is low [3,4,5,6].

The problems that often arise in fisheries include high mortality (60-90 %) of *larasati* red tilapia and white shrimp in fisheries that is caused by lack of nutrient intake and poor water quality [5,7].

*larasati* Red tilapia and white shrimp are commodities that have important economic value and they can be cultured together, which bring benefits of increased growth of red tilapia in 3-5 months to reach the weight of 200 -300 g of and the white shrimp to reach the size of 25-30. Hassan *et al.* [11] reported that the most important polyculture [20, 21, 22, 23]. Hassan *et al.*[11] stated that one solution of monoculture fish farming problems is to replace it by polyculture fish and shrimp simultaneous farming where there is no competition so both species grow simultaneously. Bardach *et al.* [5] Bautista [7] described polyculture fisheries in pond supplied with manure is a widespread practice in many countries. explained this system usually use a combination of planktopagic silver carp, bighead carp, macrophytophagic grass carp, benthphagic common carp, and omnivorous tilapia sp. To take advantage of the various feeding pond riches [9, 10, 11].

Other advantages of combined farming of *larasati* red tilapia and white shrimp are their synergism, and they can be cultivated in ponds together using seaweed as biofilter system that is able to absorb organic material to improve water quality. In aquaculture environment, red tilapia and white shrimp with seaweed biofilter system can quickly adapt, can be fed with various types of food from artificial feed, clashing, plankton to disposal in the form of organic materials, have a fast growth, are easy to multiply, and relatively resistant against disease. The studies suggested that to improve polyculture by using enriched artificial feed with the protease papain could enhance feed utilization and digestibility in most polyculture of *larasati* red tilapia and white shrimp.

The importance of papain in artificial feed to improve the quality of red larvae and vannamei prawns in ponds is affected by abrasion or the highest tides of seawater. Papain is enzyme found naturally in unripe papayas with is usually used as meat tenderizer, which helps to break down the tough band between fibers in muscle tissues. Papain is extracted from carica papaya with an activity of approximately 100.000 mU/mg. it is used to catalyze proteolysis and speed up the digestive process in fish [45] [57] [58[59] [60] [61].

The study was mainly aimed to investigate the role of polyculture technology of *larasati* red tilapia and white shrimp with different ratio of *larasati* red tilapia and white shrimp on their growth and survival rates.

## 2. Methodology

### *2.1. Experimental fish*

The test feed used in this study were made in the form of pellet with 35 % protein content and energy of 300.36 kcal (1 kcal = 4 186.8 J) [22]. The diet contained fish meal, soybean meal, corn meal, rice bran, wheat flour, fish oil, corn oil, vitamin mix, and mineral mix modification [28].

The material in this study were seed of marine *larasati* red tilapia (average initial weight 3.29 ± 0.018 g) and *Litopenaeus vannamei* white shrimp (average initial weight 1.25 ± 0.025 g). The artificial feed contained 35 % protein and enriched with a protease (100 mg / kg of feed) with the amount of feed given at 3 % biomass/day based the method modified from Mo et al (2016) [45], who reported the use of the enzyme papain at a dose of 170 mg / kg of feed that produced the best growth in goldlined seabream (*Rhapdosargus sarba*), mud grouper (*Ephinephelus Sp* Bleeker), and *Trachinotus blochii*. Feed ingredients and proximate analysis can be seen in Table 1.

### *2.2. Diet preparation*

Ingredients and proximate composition of experimental diets are presented in Table.1.

**Table 1.** Proximate analysis of compositiont of Feed Composers Materials

| Ingredient (g per 100 g diet) | Analysis of materials | | | | | |
|---|---|---|---|---|---|---|
| | Moisture | Ash | Crude lipid | Coarse Fiber | Crude Protein | BETN |
| Fish meal | 10,89 | 22,75 | 7,98 | 9,25 | 45,40 | 3,73 |
| Soybean meal | 11,06 | 5,65 | 9,23 | 5,46 | 38,71 | 29,88 |
| Corn meal | 13,71 | 1,77 | 2,03 | 0,01 | 9,38 | 73,09 |

| | | | | | | |
|---|---|---|---|---|---|---|
| Rice meal | 12,43 | 9,25 | 10,97 | 18,94 | 13,62 | 34,79 |
| Dextrin | 10,60 | 0,20 | 0,59 | 0,00 | 0,10 | 88,51 |

Source: Laboratory of Animal Nutrition and Feed Science, Faculty of Animal Husbandry, Diponegoro University 2017

**Table 2.** Composition of experimental diets

| Ingredient (g per 100 g diet) | (g per 100 g diet) |
|---|---|
| Enzim protease | 2.25 |
| Fish meal | 33 |
| Soybean meal | 33 |
| Corn meal | 8.25 |
| Rice meal | 8.1 |
| Dextrin | 10.2 |
| Fish oil | 1.5 |
| Corn oil | 1.5 |
| Vitamine& mineral | 1.1 |
| CMC | 1.1 |
| Total | 100 |
| Results of Proximate Analyses | |
| Protein (%) | 35.01 |
| Fat (%) | 11.50 |
| BETN (%) | 33.72 |
| Energy (kcal) | 300.36 |
| Ratio E/P(cal · g–1) | 8.57 |

Procedure for making artificial feed for the experiment: an appropriate dose of papain was dissolved first into warm water (45 °C), and then mixed evenly with artificial feed. The mixture was stored in airtight container for 24 h [16] [17[18] [24]. The artificial feed was prepared by mixing the least amount of the ingredients first; then the more voluminous ingredients were gradually added and mixed. The corn oil and fish oil were added last to the mixture. After that, granules with diameter of 1-2 mm were formed from the mixture. Lastly, the artificial feed was dried in the oven at 40 °C [28].

Material (g) Composition protease papain enzyme (g/kg feed) 0.08 Fish meal 33 Soybean 33 Corn meal 8.25 Rice meal 8.1, Dextrin 10.2, Corn oil 1.5 Fish oil 1.5, Vitamin& mineral 1.1 ,CMC 1.1 Total 100 Energy (kkal) 300.02 Ratio E/P 8.7.

The material used in this research was *larasati* marine tilapia (3.29 ± 0.018 g) and Vannamei shrimp with initial weight of 1.39 ± 0.025 g. Seeds were stocked with the density of red larvae at 5 and 10 tail/$m^2$ fish And vannamei shrimp at 5 tail/$m^2$ and 10 tail/$m^2$. An artificial feed with protein content of 35% enriched with protease at 22.5g / kg of feed was given at a dose of 3 % per biomass per day. The dose of papain used in this study were modified from the study by Mo et al (2016) [45]. The results suggested that the protease papain enzyme as much as 100 mg kg–1 diet was the optimum level for the growth of red tilapia *larasati* and white shrimp with the average weight of 3.29 ± 0.018 g and Vannamei shrimp with initial weight 1.39 ± 0.025 g · fingerlings-1. The experimental was conducted in the brackish water pond using Randomized Complete Design (RCD) with four treatments and three replications namely, that is given seeds 5 tails / $m^2$ *larasati* red tilapia fish and given 5 tails / $m^2$ white shrimp (A), 5 tails / $m^2$ red tilapia Larvae) and 10 $m^2$ / $m^2$ of white shrimp (B), 10 $m^2$ larvae and 5 $m^2$ white shrimp (C), 10 $m^2$ larvae and 10 $m^2$ white shrimp (D) ). The data obtained were data of growth of absolute weight, life, FCR, and water quality data (temperature, salinity, pH, $O_2$, $NO_2$, $NH_3$). Data were analyzed by various analysis (F test) and descriptive. The study was conducted in media maintenance of polyculture technology of ± 1200 $m^2$, with each research plot area of 100 $m^2$,

### *2.3. Absolute Growth*

The absolute growth formula is calculated by the formula Steffens (1989)[13][53][55], with the following formula:

W (Absolute growth weight) = Weight of end test animal research – Weight of animal initial test of

$$FCR : \frac{\text{The amount of feed consumed}}{(\text{Final weight} + \text{Total weight of dead fish}) - \text{Initial weight})} \times 100\,\%$$

$$SR\ (\text{Survival}): \frac{(\text{Final count})}{\text{Initial count}} \times 100\%$$

### *2.4. Statistical Analysis*

The experimental design used Completely Randomized Design (CRD) four treatments and three replications that is given seeds 5 larvae / $m^2$ red tilapia *larasati* fish and given 5 larvae / $m^2$ white shrimp (A), 5 larvae / $m^2$ red tilapia *larasati* Larvae) and 10 $m^2$ / $m^2$ of vannamei shrimp (B), 10 $m^2$ larvae and 5 $m^2$ vannamei shrimp (C), 10 m2 larvae and 10 $m^2$ vannamei shrimp (D). The data obtained were weight growth, survival rate, FCR, and water quality data (temperature, salinity, pH, $O_2$, $NO_2$, and $NH_3$). Data were analyzed by analysis of variance (F test) and descriptive analysis. The study was conducted in ponds (± 1200 $m^2$) with polyculture technology , with each research plot area sized 100 $m^2$.

The effect of the treatments was tested using an analysis of variance (ANOVA) [56. Before analysis, the normality, additivity, and homogeneity of the data were first tested. If the analysis of variance was significant ($p<0.05$) or highly significant ($p<0.01$), Tukey test was conducted to find out the difference among treatments [56]. To determine optimal dose of papain, analysis was conducted using Minitab 16. Water quality data were descriptively analyzed.

## 3. Results and Discussion

The results showed that there was a highly significant effect artificial feeding enriched with papain (P <0.01) on growth, food conversion ratio (FCR), survival rate, and feed conversion ratio (FCR).

### *3.1. The absolute growth of Larasati Red Tilapia .*

The results showed that treatment D (10 $m^2$ larvae of *larasati* red tilapia and 10 $m^2$ / $m^2$ white shrimp) gave the highest weight growth in *Larasati* Red Tilapia with 185.75±0.50 g and in vannamei shrimp with 19.97 ± 1.75 g (Table.3). This result showed that Treatment D gave better weight growth due to higher protein content in the feed that affected the absolute weight of *larasati* red tilapia . Protein is useful for growth and it highly affected the growth of red tilapia and white shrimp protein quality energy content of feed and balance can increase growth rate d the nutrient and feeding rate [17]. Protein deficiency could decrease the weight of red tilapia and white shrimp due to muscle loss in the muscles and vital organs [2]. Similar with previous researches, the current study has shown that artificial feeding with protein content of 35 % enriched with papain enzyme was able to increase the weight growth of *larasati* tilapia red from 179.5 g to 185.25 g [13][51][55].

The use of papain at 2.25 g/kg feed to enrich the artificial feed could increase the absolute growth of tilapia [13] [51] [52] [55]. Addition of exogenous protease such as papain could mitigate the adverse effect of plant protein. In our previous experiment, common carp *Cyprinus carpio* (freshwater fish) fed with papain-digested soybean residues mixed with beef liver showed significantly higher weight gain (12.43%) after 93 days, (Mo et al., 2016). In addition, Mo et al. (20016)[45] reported using papain at a dose of 17 g/kg pellets in artificial feed that could increase relative weight gain (%)of gold lined sea bream from 157.1±20.3 % to 320.9±67 % with the final weight between 150.8 ± 20.8 to 189.3±29.6 g. Mo et al (2016)[45] also reported an increase in final weight of brown spotted grouper from

IOP Conf. Series: Earth and Environmental Science **137** (2018) 012066 doi:10.1088/1755-1315/137/1/012066

100.0±27.4 g to 163.0± 47.8 g and relative weight gain from 305.6 % to 599.4±49.8 by using similar dose of papain. But application of technology polyculture red Tilapia *larasati* and white shrimp different growth of absolute was better good that the fish fed with digested soybean residues mixed with effective digestion was useful to maintain body condition, hence, contributes also in the growth. Papain is needed by fish even in small amounts because the body cannot synthesize it; thus, fish should get extra papain from artificial feeds [45]. Therefore, complete and balanced nutrient feed formula is necessary for *larasati* red tilapia and white shrimp fish. Moreover, polyculture system is one of the important systems that can increase growth and survival rate of red tilapia and white shrimp [6][49][50][51][52][53][54][55].

**Table 3.** Absolute growth based on weight (g), survival rate, food conversion ratio on a variety of treatments and replications on the polyculture of red tilapia Larasati and white shrimp.

| | Treatment | | | |
|---|---|---|---|---|
| | A (5N+5V) | B (5N+10V) | C (10N+5V) | D (10N+10V |
| 1. Absolute growth of red tilapia of larasati (g) | $174.52{\pm}2.45^{b}$ | $176.19{\pm}4.46^{ab}$ | $178.97{\pm}5.87^{ab}$ | $185.75{\pm}0.50^{a}$ |
| 2. Absolute growth of white shrimp (g) | $21.72{\pm}0.54^{b}$ | $21.90{\pm}0.53^{b}$ | $23.29{\pm}0.94^{ab}$ | $25.25{\pm}0.95^{a}$ |
| 3. Survival rate of red tilapia of larasati (%) | $75.48{\pm}0.47^{c}$ | $88.55{\pm}1.24^{b}$ | $91.49{\pm}1.66^{b}$ | $97.25{\pm}0.33^{a}$ |
| 4. Survival rate of white shrimp (%) | $75.99{\pm}0.55^{c}$ | $89.58{\pm}1.23^{b}$ | $91.88{\pm}1.72^{b}$ | $95.75{\pm}0.52^{a}$ |
| 5. FCR of red tilapia larasati and white shrimp. | $3.16{\pm}0.11^{a}$ | $2.32{\pm}0.07^{b}$ | $1.95{\pm}0.05^{c}$ | $1.22{\pm}0.13^{d}$ |

Note: N = *larasati* red tilapia , V= white shrimp
Different superscript letters within the same column indicate significant differences between samples at the level of ($P < 0.01$)

### *3.2. Absolute growth of White Shrimp*

The highest rate of absolute weight growth was shown by Treatment D (10 $m^2$ larvae of *larasati* red tilapia and 10 $m^2$ / $m^2$ white shrimp) in ponds culture in Kandang Panjang, Pekalongan. Effect of the difference in the density of *Larasati* red tilapia of and white shrimp on the absolute growth of white shrimp (g) in polyculture system was highly significant ($P < 0.01$) ( Table 3).

The highest value was from Treatment D (10 individuals per $m^2$ for red tilapia of *Larasati* and 10 individuals per $m^2$ for white shrimp) with the value of 25.25±0.95. g. .*larasati* red tilapia could use the feed well, and the addition of protease in the artificial feed helped to break down protein into amino acids that were easy to be breed by *larasati* red tilapia properly According to [23][24][25][26][27], fish growth is influenced by two main factors: internal factors related to the fish itself including genetic and physiological characteristics, and external factors related to the environment, such as chemical composition of water, temperature, metabolic rate, oxygen availability, and feed.

Growth occurs when there is excess energy after it is used for body maintenance, basal metabolism, and activities. Growth is supported by feeding that is tailored to the nutritional needs of the fish and it should have a high digestibility value. In the current study, artificial feed was given with a protein content of 35 % to 40 %. polyculture system that simultaneously maintaining white shrimp and milkfish (30 individuals per $m^2$ of white shrimp and 30 individuals per $m^2$ of milkfish) was able to provide the highest absolute weight growth, because the amount of feed and the pond's density were appropriate. Indicators of The growth was change in length and weight in a given period. Individual growth is due to the addition of tissue mitotic cell division that causes changes in size [29][30][31][32][33][34]. Marine and fisheries agency of central java province (2004) [35][36[37][38][39][40] reported that growth factors that affect the feed ratio and the weight of the fish are external and internal factors. External factors include water and environmental conditions, while the internal factors are the species, sex, genetic, and physiological status of fish. Indicator of physical growth includes number or size of cells that build up the body tissue and morphologically-visible growth of body. Growth occurs when the energy needed for metabolism and tissue maintenance fulfill the needs of the fish [23][24][41], or when the amount of consumed feed was greater than the amount needed for body maintenance, and when feed is used as an energy source for the fish [42][43][44][45][46].the highest value was obtained from polyculture technology with 30 individuals of white shrimp and 30 individuals of milkfish per $m^2$ in a polyculture system.

### *3.3. Survival rate of Larasati Red Tilapia .*

The results showed that the highest survival rate in *larasati* red tilapia was obtained from Treatment D (10 individuals per $m^2$ for red tilapia of *Larasati* and 10 individuals per $m^2$ for white shrimp) with 97.25±1.15 % (Table.3). analysis of variance showed a highly significant($P < 0.01$) survival of *larasati* red tilapia . This is likely because of the innovation in polyculture system that used recirculation and biofilter system from seaweed. Other factors also contributed such as administration of artificial feed containing 35 % of protein enriched with papain at a dose of 2.25 g/kg feed to improve survival rate and growth rate, good environmental media, and proper number of stocking density, where all of them helped to increase survival rate, i.e. 97.25±1.15 . Result of the current study was better by that of [51], which showed polyculture with white shrimp per $m^2$ and 30 individuals of milkfish per $m^2$ that had survival rate of 75±0.13 % . this shows the Effect of the difference in density of white shrimp *L. van*namei and milkfish on the survival rate of milkfish in polyculture system .

Survival is the ratio between the number of fish or individuals living at the end of maintenance period with the number of fish or individuals living at the start. According to [3][13][45][56][63], fish survival is influenced by biotic and abiotic factors. Biotic factors include parasites, density, population, competitors, while abiotic factors include physical and chemical properties of aquatic environment. [2] [25] [56] explained that good growth of fish will increase the production volume that depends on the rate of growth and survival of the cultivated fish.

### *3.4. Survival of White Shrimp*

The results showed that the highest survival rate in white shrimp reared in polyculture system was obtained from Treatment D (10 individuals per $m^2$ of *Larasati* red tilapia of and 10 individuals per $m^2$ of white shrimp) with the value of 95.76±1.02 %(Table.3).

Table 3 shows the effect of density difference between *larasati* red tilapia and white shrimp on the survival rate of white shrimp in polyculture system. analysis of variance showed a significant influence of density ($P < 0.01$) on survival of white shrimp. The survival rate was relatively high at Treatment D (95.76±1.02$^a$ %). This is more likely due to several factors: e application of biofilter with a high protein content that was enriched by 35 % with papain at a dose of 2.25 g/kg feed, which increased growth and survival rates. Artificial feed that was added with papain also contributed to proper tissue maintenance. the stocking density of *larasati* red tilapia and white shrimp was relatively sufficient, thus, contributing to high survival rate (95.76±1.02$^a$ %), better than the result of [51] who reported 95 ± 0.33 % of survival of white shrimp and milkfish in polyculture system . Result of the current study is consistent with previous researches by [11][14][20][21][22][45][63] who reported good water quality is essential in fish cultivation. Water quality affects survival of white shrimp and milkfish, which led to better growth rate [30][31][32][42][56]. Furthermore, maintaining water quality using biofilter and application of seaweed could increase the survival rate of white shrimp and

IOP Conf. Series: Earth and Environmental Science **137** (2018) 012066 doi:10.1088/1755-1315/137/1/012066

milkfish polyculture up to 80 to 90 % [18][19][20][21][22][45][63]. Eventually it could improve production and economic returns of white shrimp and milkfish reared in polyculture system higher than those in the crop rotation polyculture system. Furthermore, previous research on polyculture of tilapia shrimp showed an improvement on water quality in white shrimp and milkfish cultivation ponds by reducing diseases and chemicals that could increase survival rate.

### *3.5. Food conversion Ratio (FCR) of Larasati Red Tilapia and White Shrimp.*

feed conversion ratio was a very important parameter to examine whether the feed was able to increase the growth of *larasati* red tilapia and white shrimp. feed conversion values are also useful to see how far the feed was converted further into meat of *larasati* red tilapia or white shrimp. better FCR could be obtained from treatment D feed conversion values are also useful to see how far the feed was converted further into meat of red tilapia *larasati* and white shrimp (Table.3). In the current study, the FCR value was of 1.22 ± 0.13 was better compared to the research of [51] who reported FCR of 1.25 ± 0.02, since *larasati* tilapia fish and vannamei shrimp can utilize feed well. Amal et al (2008) [3][11][47][60] also reported Food Conversion Ratio (FCR) of white shrimp and milkfish reared in polyculture system. it can be concluded that the feed formulated was suitable for polyculture system; thus, it increases growth rate better. Moreover, lower FCR value means more efficient feed administration [15] [25] [39] [40].

### *3.6. Water quality*

The water quality was suitable for *larasati* red tilapia and vannamei shrimp. The existence of application technology of red tilapia culture cultivation as a tourism object of maritime education in Pekalongan city.

Table 3 shows that all parameters , i.e. the dissolved oxygen content (4,75 mg L–1 to – 6,75 mg L–1), temperature (26.5 – 29,5 °C), salinity (20.5 L–1 to 29,5 ng L–1), pH (7.5 to 8.5), and ammonia (0.01 mg L–1 to 0.14 mg L–1 ), were still in the range of viable and capable for the life of *larasati* red tilapia and white shrimp, which were reared in polyculture system [3][8][15][25][39][40][52]. Several other researchers reported the application of closed polyculuture system of tilapia shrimp combined with tagelus could increase productivity up to 28 %, which was higher than polyculture system for shrimp and tagelus.

**Table 4.** Water quality maintenance media polyculture system red tilapia larasati and white shrimp used biofilter system

| Water quality variable | Range | Worthiness (literature) |
|---|---|---|
| Oksigen terlarut (mg/l) | 4,75 – 6,75 | >4 mg/l [a,b] |
| Temperature ($^0$C) | 26.5 – 29,5 | 25,5 – 35 $^0$C [c,d,f] |
| Salinity (ppt) | 20.5 – 29,5 | 15 – 30 [c,d] |
| pH | 7.5 – 8,5 | 7,5 – 8,7 [c,d,e] |
| Ammonia (mg/l) | 0.01– 0,14 | <1 mg/l [c,d.e] |

Note: [40][a], [39][b], [52][c],[3][d], [8][e], [25][f].

Based on Table 4, it is shown that biofilter helped to produce decent water quality for polyculture system maintenance and the source should be environmentally friendly, because it uses seaweed as biofilter placed in inlet and outlet at 7.25 - 5.85 mg / L. The temperature of 25.5 mm was found to increase the survival rate of fish (93.73 ± 0.39c%) and vannamei shrimp (96.71 ± 0.85c%). The water quality shows the feasibility of the system for maintenance of milkfish and vannamei shrimp [15][25][39][40][52].

IOP Conf. Series: Earth and Environmental Science **137** (2018) 012066 doi:10.1088/1755-1315/137/1/012066

## 4. Conclusion

The results showed that the density difference of *larasati* red tilapia and white shrimp gave a highly significant effect (P <0.01) on growth, survival rate, and feed conversion ratio (FCR) of *larasati* red tilapia and white shrimp. The highest absolute weight growth was from treatment D (10 $m^2$ larvae and 10 $m^2$ / $m^2$ white shrimp) with 185.75 ± 0.50g *larasati* red tilapia () and 25.25 ± 0.95 g of white shrimp. The survival rate of *larasati* red tilapia was 97.25 ±0.33%, white shrimp 95.75 ± 0.52%, FCR 1.22 ± 0.13

The quality of the water was still feasible for the life of *larasati* red tilapia and . white shrimp The existence of application technology of red tilapia *larasati* culture cultivation.

## 5. Reference

[1] Akegbejo and Y Samsons 1999 J Appl Trop Agric **49** (1): 37–41

[2] Akiyama D M, W G Dominy, and A L Lawrence 1991 Penaid Shrimp Nutrition for The Commercial Feed Industry. In Akiyama, D M and R K H Tan (eds.). [Proceedings of the Aquaculture Feed Processing and Nutrition Workshop, Thailand and Indonesia Singapore 19– 25 September 1991 American Soybean Association Singapore] 80–89.

[3] Amal S H, S H Sayed and E M Ibrahim 2008 Effect of Stoking Rates and Supplementary Feed on The Growth Performance of Blue Tilapia (Oreochromis aureus) and Grass Carp (Ctenopharyngodon idella) Reared in Earthern Pond. [In Elghobashy, H., K].

[4] Fitzsimmons and A S. Diab (eds) 2008. Proceedings of 8th International Symposium on Tilapia in [Aquaculture] 12–24 October 2008 949–964.

[5] Aslam A, G S Hossain, M R Biswas, S K Barman and K Anisulhuq 2009 Int J Sustain Crop Prod **4**(4): 23–27

[6] Badan Pengkajian dan Penerapan Teknologi 2007 Tiger Shrimp Cultivation Traditionally. [Research Report. Jakarta: BPPT].

[7] Bautista M N 1986 J.Aquaculture **3**(3–4): 229–242

[8] Boyd H E, Burgess, Pronek and Walls 1982 Water Quality in Warm Water Fish Pond [Auburn: Auburn University, Aquaculture Experiment Station].

[9] Davis J 2011 Polyculture Opportunities in The Mid-hills of Nepal for Resource Poor Farmers. Ecological Aquaculture Studies & Reviews. [Kingston: University of Rhode Island].

[10] De Silva S S and F Y Anderson 1995 Fish Nutrition in Aquaculture [New York: Chapman and Hall].

[11] Dirisu S O, B Muinat and D M Yakubu 2007 J Animal Research International **4**(3): 737–740

[12] Edwards) in Koilsagar Reservoir of Mahabubnagar District (TS), India International Journal of Fisheries and Aquatic Studies **2**(4): 147–152.

[13] Effendie M I 1979 Methods of Fisheries Biology. [Bogor: Yayasan Dewi Sri]. [Bahasa Indonesia].

[14] Endrawati, H., S. Istiyanto, and A. Indarjo 2001 Application and Cultivation Technology Community Business Group Polyculture Nila Gift and Tiger Prawn in Ponds in An Effort to Empower Coastal Communities] Journal Info. **4**(1): 6– 18

[15] FAO (Food and Agriculture Organization) 2016 FishStat: a tool for Fishery statistics analysis version 2.12.2 http://.fao.org/fishery/statistics/software/fishstat/en(accessed 14.07.16.

[16] Fox J M, Addison LL, Anthony J S, Allen D, Denis RM, Elizabeth C D,and Tzachi MS 2006

[17] Furnichi, M 1988 Dietary Requirement in Fish Nutrition in Marriculture. [Japan: Japan International Coorporation Agency].

[18] Halver J E 1980 Fish Nutrition. [New York: Academic Press Inc].

[19] Halver J E and T Lovell 1989 Nutrition and Feeding of Fish. New York: [Van Nostrand Reinhold].

[20] Hassan A S, S H Sayed and 2008 Effect of stocking rate and supplementary feed on the growth performance of blue tilapia (Oreochromis aureus) and grass carp (Ctenopharigodon idella) reared in earthern pods.[International symposium on tilapia in aquaculture] P 949-964

[21] HASSAN M S, SOLTAN M A, AGOUZ H M, and BADR A M 2013 Egyptian Journal of Aquatic Research **39**(3):205–213.
[22] Hassan S, Altaff K, Satyanarayana T 2009 Pakistan Journal of Nutrition **8**(4):341–344
[23] Herpher B 1988 Nutrition on pond fisheries Cambridge, USA: Cambridge University Press. 388p.
[24] Hepher B 1988 Nutrition of Pond Fishes, Formerly of Fish and Aquaculture Research Station. [Cambridge: Cambridge University Press].
[25] Hepher B and Y Praginin 1981 Comercial Fish Farming. New York: [John Wiley Sons].
[26] He H L and R Liv 1992 J.Aquacultur 103: 177–185.
[27] Huet M 1971 Fish Culture, Breeding and Cultivation of Fish. [London: Fishing New (Books)] Ltd.
[28] [NRC] National Research Council 1993 Nutrient requirements of fish. [Washington DC: National Academy Press]. 124p.
[29] Nurjana M 2007 Proceedings of the National Seminar of, Aquaculture Society (MAI) Indonesia. Surabaya, Indonesia. 5–7 June 2007].
[30] Samidjan I 2001 Effect of Various Combinations of Natural Feed (Tetraselmis chui, Chlorella sp, Brachionus plicatilis Muller, Nauplius Artemia salina Leach) on Growth and Survival Rate Mangrove Crab (Scylla paramamosain)]. [Paper Presented at the National Seminar of Crustacean 2001. Institut Pertanian Bogor, Bogor. 20 – 21 July 2001].
[31] Samidjan I 2001 Enlargement of Mude Crabs (Scylla paramamosain) in Pond with Feed Trash in Different Dose]. [Paper Presented at the National Seminar of Crustacean 2001. Institut Pertanian Bogor, Bogor. 20–21 July 2001].
[32] Samidjan I 2008 Engineering of Monoculture Technology for Superintensive System on Mudcrab (Scylla paramamosain) Using Different Feeds on The Growth and Survival Rate. In Hartoko, A. (ed.). [Proceedings of International Conference, Geomatic, Fisheries and Marine Science for a Better Future and Prosperity Marine Geomatic Centre (MGC). Semarang, Indonesia. 21– 22 October 2008].
[33] Samidjan I, E Arini, and D Rachmawati 2012 Applicaton of Technology and Science in (IbM) Business Group Polyculture of Shrimp, Fish and Seaweed (Gracyllaria sp.) Based on The Biological Filter Mangkang Wetan Village, District Tugu, City Semarang]. [Research Report. Semarang: Universitas Diponegoro].
[34] Samidjan I 2009Use of Various Types of Probiotic Bacteria (Bacillus, Alcaligenes, Flavobacterium, and Lactobacillus) as The Commercial Feed in Crumble from Vannamei]. [Research report] Semarang: Universitas Diponegoro].
[35] Jaspe J C, C M A Caipang, and B J G Elle 2011 J. ABAH Bioflux. 3(2): 96–104.
[36] Kanazawa A 1985 Nutrition of Penaeid and Shrimp. In: Taki Y, J H Primavera and J A. Liobrera (eds). [Proceedings of the First International Conference on Culture of Penaeid / Shrimp. Aquaculture Dept. SEAFDEC. Iloilo, Philipphines]. 4–7 December 1984. 123–130.
[37] Kementerian Kelautan dan Perikanan 2004 Production of Shrimp and Milkfish in Fisheries Central Java, Indonesia]. [Jakarta: KKP].
[38] Kurmaly K 1995 Shrimp Nutrition and Disease: Role of Vitamins and Astaxanthin. [Bangkok: Roche Aquaculture Centre].
[39] Laxmappa B, S M Khrisna 2015 Polyculture of The Freshwater Prawn Macrobrachium malcolmsonii in (H.M. 98 Istiyanto Samidjan & Diana Rachmawati / Jurnal Teknologi (Sciences & Engineering) **78**:4–2 ;91–98
[40] Li D S and S L Dong 2000 J Chinese Journal of Oceanology and Limnology 8: 61–66.
[41] Makwinja R and F Kapute 2015 Net Journal of Agricultural Science **3**(3): 62–67.
[42] Mangampa M and Burhanuddin 2014 Jurnal Saintek Perikannan **10** (1): 30–36
[43] Marine and Fisheries Agency of Central Java Province 2004. Basic Data Production Potential and Fisheries Central Java [in The Figures. Semarang: Marine and Fisheries Agency].
[44] Miroslav C, T Dejana L Dragana Đ and Vesna 2011 J. Tehnologija Mesa **52**(1): 106–121.
[45] Mo W, WY, R S S.Lau, A C Kwok, M H Wong 2016 J. Fish growth and water quality

environmental pollution :1-6.
[46] Murachman, N Hanani, Soemarno, S Muhammad 2010 Journal of Sustainable Development and Nature **1**(1): 2087–3522
[47] Nikolova L 2013 J. Bulgarian Journal of Agricultural Science **19**(6): 1391–1395
[48] Phytase supplementation in aquaculture diets improves fish, shrimp growth performance. Retrieved from http://www.ag.auburn.edu/~davisda/publications/ publication_files/ca30_gaa-fox-feb06.pdf on 10 Jun 2016.
[49] Porchas M M, L. R M, Cordova M A P, Cornejo and J A L Elias 2010 J Reviews in Aquaculture 2: 73–85.
[50] Reksono B H, Hamdani and Yuniarti 2012 Journal of Fisheries and Marine **3**(3): 41– 49
[51] Samidjan I and D Rachmawati 2016 Jurnal Teknologi (Sciences & Engineering) 78:4–2 ;91–98.
[52] Solomon J R and M N Ezigbo 2010 Polyculture of Heteroclarias / Tilapia Under Different Feeding Regimes New York Science Journal **3**(10): 42–57
[53] Steffens 1989 Principles of Nutrition England: Ellis Horwood Limited
[54] Stickney R R 1979 Principle of Warm Water Aquaculture New York: John Weley and Sons Inc
[55] Stell R G D, Torrie J H and Dickey D A 1996 Principles and procedures of statistics, [3rd ed. New York: McGraw Hill International Book Company, Inc] 672p.
[56] Suyono, I Samidjan, D Rachmawati, and T Yasman 2010 Application Science and Technology in (IbM) Groups of Fish Farming Milkfish and Seaweed (Gracylaria sp.) in The Village of Muara Church, West Tegal Tegal]. In Suyono, N. Isdarmawan, and N. Zuhri (eds.). Proceeding of National Seminar on Development Strategy for Environmentally-Based Fisheries and Marine. [Pancasakti University, Tegal, Indonesia. 9 December 2011. 123– 46].
[57] Tacon 1987 Nutrition and Farmed Fish and Shrimp Training Manual Brazil: The Essential Nutrients Food and Agricultural Organization of the United Nations
[58] Watanabe 1988 Fish Nutrition and Marineculture. [Tokyo: Department of Aquatic Biosciences].
[59] Xie B, W Jiang and H Yang 2011 J. Bulgarian Journal of Agricultural Science **17**(6): 851–858
[60] Yang Y and K Fitzsimmons 2002 Tilapia Shrimp Polyculture in Thailand Research Report. Thailand: Asian Institute of Technology
[61] Yasin M 2013 Scientific Journal Edition March Agriba 1: 86–99
[62] Yuvaraj D ,R Karthik and R Muthezhilan 2015 Asian Journal of Crop Science **7**(3):219-232.
[63] Rachmawati D and I Samidjan 2017 Philippine J. of Science **146** (3):237-245

**Acknowledgements**
This work was supported by grants from Marine and Fisheries of Government. The authors are indebted to the Mr.T.Miftahuddin heads of POKDAN fish farmers Muara Rejeki in Pekalongan, who has lent his polyculture in pond culture to research.

# The physico-chemical properties of pangas catfish *(Pangasius pangasius)* skin gelatin

**K A Pradarameswari[1], K Zaelani[1], E Waluyo[1] R Nurdiani[1,2]**

1 Faculty of Fisheries and Marine Sciences, Brawijaya University, Malang, Indonesia
2 BIOSEAFOOD Research Group (Aquatic Biotech Exploration Sustainable), Brawijaya University, Malang, Indonesia

E-mail: rahmi nurdiani@ub.ac.id

**Abstract.** Gelatin can be used as emulsifier and stabilizer in food products. Until now, the most widely used raw materials for the production of gelatin industry are cow bone, cow skin and pig skin. Fish gelatin has been highlighted as a better alternative to replace mammals gelatin based on ethical and religious perspective. Fish gelatin was extracted from Pangas catfish skin to determine its physico-chemical properties. Different temperatures (45 °C, 50 °C, 55 °C) were employed during gelatin extraction. Higher temperature increased the yield and fat contents of Pangas catfish skin gelatin. In contrary, higher water, protein, ash contents were observed during lower temperature. Temperature significantly ($p < 0.05$) affected the gel strength, viscosity, melting point, and gelling point of fish skin gelatin. Based on the FTIR spectrum catfish skin gelatin functional groups can be identified as N-H, O-H, C = H, C-O and C-H.

## 1. Introduction

Gelatin is a hydrocolloid product obtained by hydrolyzing collagen proteins found in the skin, bone, and connective tissue. Sobral [1] explained that gelatin is a denaturated protein derived from collagen and is an important function of biopolymers with a very wide application in various industrial fields. Its functional properties depend on the processing conditions and raw materials. The quality of gelatin depends on its chemical physics, rheological properties and manufacturing methods. Gelatin has been applied in food as gel-forming agents, thickeners, emulsifiers, and is widely used in the pharmaceuticals, healthcare, cosmetics and photography industries because of its unique functioning [2].

Until recently, the most widely used raw materials for the production of gelatin are cattle bones, cattle skin and pig skin. The utilization of gelatin from mammals is still experiencing many obstacles such as the beliefs held by consumers, such as Hindus who are prohibited from consuming products derived from cows. Some people are also worried about consuming cattle byproducts due to mad cow disease, foot and mouth disease, and Bovine Spongiform Encephalopathy (BSE). In addition, ingredients derived from pigs cannot be consumed by Muslims [3].

Strong competition occurs between producers for the procurement of pig skin or other mammal sources, which has created an increased demand and raised costs. Researchers have not only continued to look for an alternative to gelatin but also to discover new sources of gelatin. In the last decade, there has been an increased interest in the market for gelatin derived from fish and poultry. The meat, skin,

IOP Conf. Series: Earth and Environmental Science **137** (2018) 012067 doi:10.1088/1755-1315/137/1/012067

and bone of poultry and fish are expected to produce gelatin in the near future, but commercial production is currently limited due to low yields [4].

With this in mind, fish gelatin has been highlighted as a better alternative to mammalian gelatin from an ethical and religious perspective. However, the quality is at lower melting point, so the faster solubility in the mouth and the absence of residual 'supple' feelings have affected commercial applications. Therefore, the production of fish gelatin is still in its infancy, contributing to only about 1% of the world's annual gelatin production [5]. 78% of total fish catch has been utilized as human food, leaving about 21% for non-food purposes [6]. Processing leads to the generation of a large biomass (50-75% of total catch ) of fish waste (eg, skin, bone, and fins), which are generally discarded [7, 8].

Freshwater fish are seldom used as a source of raw material for gelatin extraction. Several studies, however, have ensured that certain freshwater fish have a large amount of by-product after the removal of edible parts, and high gelatin yields can be expected from Tilapia [9] Nile perch [10], striped catfish *(Pangasius sutchi fowler)*.

In a study conducted by Ratnasari *et al.* [11] on gelatin extraction and the characterization of different freshwater fish as alternative sources of gelatin, the gelatin obtained from catfish *(Pangasius pangasius)* had the highest rheological properties compared to the others. It has the following properties: gel strength of 273.58 g, 36.5 cP viscosity, melting point at 32 °C, gelling temperature at 12 °C, melting temperature at 29 °C and a total amino acid content of 754.47 mg / g.

## 2. Methodology

### 2.1. Method of research

The method used was the experimental method. The analysis of data used in this research was through Completely Randomized Design (RAL) with three replications. There are three extraction temperatures employed i.e 45 °C, 50 °C, and 55 °C. The dependent variables are the proximate analysis of gelatin product was also conducted, yield, viscosity, strength gel, melting point, gel point and pH.

### 2.2. Procedure of research

The procedure of gelatin production from catfish skin includes the process of obtaining fresh skin and storing it in ice as described by Sae-Law and Soottawat [12], with some modifications. To remove the non-collagen protein, the skin was immersed in 0.05 M NaOH with a 1: 6 (w / v) skin / solution ratio and stirred for 2 hours at room temperature (± 26-28$^{0}$C), alkaline solution was replaced every hour. The immersion process of the NaOH solution resulted in swelling that can remove unwanted material (fat and non-collagen proteins). The alkaline skin is then washed with running water until neutral pH was reached. Furthermore, the skin was immersed in 0.05 M acetic acid with a 1 : 6 (w / v) skin / solution ratio for 1 hour with gentle stirring until the collagen swelled in a skin matrix. The treated skin was thoroughly stirred with acid as described previously. After swelling, the swollen skin is immersed in distilled water with a 1 : 6 (w / v) skin / water ratio in a waterbath shaker with a temperature of 45 °C, 50 °C and 55 °C to extract the gelatin from the skin. The extract was filtered and dried using an oven at 55 °C for ± 48 hours. The dried gelatin sheets are then put through a grinder.

## 3. Result and discussion

The results of the proximate test and physico-chemical properties of catfish skin gelatin with the extraction temperatures of 45 °C, 50 °C and 55 °C and compared to commercial gelatin, SNI standard for gelatin and GMIA standard for gelatin is presented in table 1.

IOP Conf. Series: Earth and Environmental Science **137** (2018) 012067 doi:10.1088/1755-1315/137/1/012067

**Table 1.** Proximate Test Results and Physics-Chemical Gelatin Fish Skin Patin.

| Proximate and Physicochemical Parameters | Temperature of extraction | | | Commercial gelatine * | Gelatine Standart ** | Gelatine *** |
|---|---|---|---|---|---|---|
| | $45^0C$ | $50^0C$ | $55^0C$ | | | |
| Rendemen (%) | 18.65±1.75 | 20.48±1.59 | 24.65±2.84 | - | - | - |
| Water content (%) | 11.20±1.18 | 9.48±0.90 | 9.62±0.92 | 12.937 | Max. 16 | - |
| Ash content (%) | 1.12±0.09 | 1.09±0.09 | 0.83±0.06 | 1.633 | Max. 3.25 | 0.3 |
| Protein content (%) | 89.61±7.85 | 89.55±7.6 | 88.11±5.48 | 87.707 | - | - |
| Fat content (%) | 1.47±0.12 | 2.35±0.22 | 3.72±0.31 | 0.23 | - | - |
| Gel strength (Bloom) | 79.59±1.07 | 84.49±1.13 | 105.30±1.08 | 318.300 | - | 50 – 300 |
| Viscocity (cP) | 7±0.56 | 5±0.33 | 4±0.35 | 9.75 | - | 1.5 – 7.5 |
| pH | 4.43±0.38 | 4.55±0.34 | 4.47±0.33 | 4.00 | - | 3.8 – 5.5 |
| Melting point (°C) | 28±1.22 | 28±1.22 | 30±1.22 | 29.60 | - | - |
| Gelling point (°C) | 11±0.71 | 12±0.71 | 12±1 | 19.50 | - | - |

### *3.1. Yield*
The results showed that the higher extraction temperature used, the higher yield obtained. The increase of temperature causes higher yields by increasing the breakdown of collagen fiber structure so that it binds with water to form more gelatin. This is in accordance with the results of the study by Mayangsari *et al.* [13]. The optimum temperature for extraction were 50 °C, 60 °C and 70 °C. If temperature is too low it will result in low yields, while temperatures that are too high will result in low gelatin quality. Increased long ripening (extraction time) or heating in water increases the solubility of collagen so that the yield of the gelatin will increase [14].

### *3.2. Water content*
From the results obtained, there is a tendency that when a higher extraction temperature was used, the value water content was lower. This was due to higher extraction temperatures that result in weak water holding capacity in gelatin and will cause the water to evaporate at the time of drying. According to Astawan and Aviana [15], the decrease in moisture content is due to an increasingly open collagen structure with a weak stroke, resulting in gelatin with a weak structure, and hence its water holding capacity is also weaker. The weak water holding capacity of the gelatin will make the water volatile at the time of drying.

### *3.3. Ash content*
During the extraction process at high temperature, there was a decrease in ash content due to the decomposition of mineral elements. According to Wijaya [16], the low ash content is caused by the release of mineral components such as carbon, calcium, phosphate from the collagen at the time of washing after immersion, as well as the elements dissolving upon extraction by heating so that there is a decrease in ash content .

### *3.4. Protein content*
From the research, the tendency seen is that the higher the temperature of extraction, the lower the value of the resulting protein. Gelatin can be obtained by the heat denaturation of collagen. The heating of the collagen will gradually cause the structure to break down and chains to be separated. The molecular weight, shape, and conformation of the collagen solution are sensitive to temperature

IOP Conf. Series: Earth and Environmental Science **137** (2018) 012067 doi:10.1088/1755-1315/137/1/012067

changes that can destroy its macromolecules. Furthermore, according to Lehninger [14], the protein will be damaged and denatured by not only heat, but by the influence of pH, which is a change in the main structure of peptide chains in proteins. If the denatured protein of the polypeptide chain bond arrangement is disrupted, the open protein molecule becomes randomized and subsequently coagulated, resulting in lower amounts of collagen being extracted.

### *3.5. Fat content*
Fat content is influenced by the use of different temperatures during the extraction process at the temperatures of 45°C, 50°C and55 °C for 6 hours, so there are differences in the erosion of fat on the catfish skin. Another factor that affects the fat content of catfish skin gelatin is the process of immersing NaOH for 2 hours. According to Tazwir and Rosmawaty [17], caustic soda which is called NaOH (Sodium hydroxide) in chemistry is a kind of strong metal base. In the medical world, it is known as a substance that dissolves fat tissue.

### *3.6. Gel strength*
Higher temperatures will lead to more amino acid chain breaks. This is due to the continued hydrolysis of collagen that has become gelatin and causes a short-chain of amino acid so that the gel strength is low. Short-chain amino acids cause lower interactions with water molecules and is unable to form into gel [18].

### *3.7. Viscocity*
The higher the extraction temperature, the more open the structure of the amino acid chain will be and will cause the chain to become shorter and viscosity to decrease. The higher extraction temperature causes the triple helical structure of the collagen to transform into a single chain structure. Changes in the collagen chain structure leads to a decrease in the molecular weight of gelatin. Purnomo [19] explains that the high temperature will break the bonds between the joined molecules and form smaller units that will facilitate the fluid to flow.

### *3.8. pH*
The different extraction temperatures in the production of catfish skin gelatin had no effect on the pH value. Basically the pH is not affected by heat or high temperature but will be affected by the decomposed medium at high temperatures that produce acids or bases. It is this acid or base that will affect the pH. According to Amiruldin [20], during immersion (curing), the skin collagen fibers undergo a swelling process and result in the decreased internal cohesion of the skin fibers. When swelling occurs, the amino acid bonding structure of the collagen molecule is exposed and the curing material is trapped within the bonding structure and is insoluble during the neutralization process.

### *3.9. Melting point*
The higher the extraction temperature, the lower the melting point value of the catfish skin. It is suspected that high temperatures will cause the hydrolysis of proteins to break the amino acid chain so that it will weaken the ability to bind protein and water. According to Astawan and Aviana [15], another factor that affects the melting point is the initial state of gel formation. When the gel forms quickly, the resulting gel is less stable and melts faster. In addition, the gelatin that experiences drying at higher temperatures generally indicates a higher melting point as well.

### *3.10. Gelling point*
The gelatin gel point of the research is thought to be caused by the increase in protein content along with the increase of the extraction temperature of the treatment. The protein content of the gelatin determines the amount of amino acid content of hydroxyproline in it. According to Amiruldin [20] who had conducted a study on the amino acid of gelatin derived from from tuna bones, the gel point was influenced by the amount of amino acid hydroxiprolin, the gel point would be lower if the amount

of amino acid hydroxiprolin was also low and the low hydroxiprolin would cause the hydrogen bond in the gelatin to be weaker .

### *3.11. Analysis of FT-IR*

The result of the catfish skin gelatin spectra is presented in figure 1.

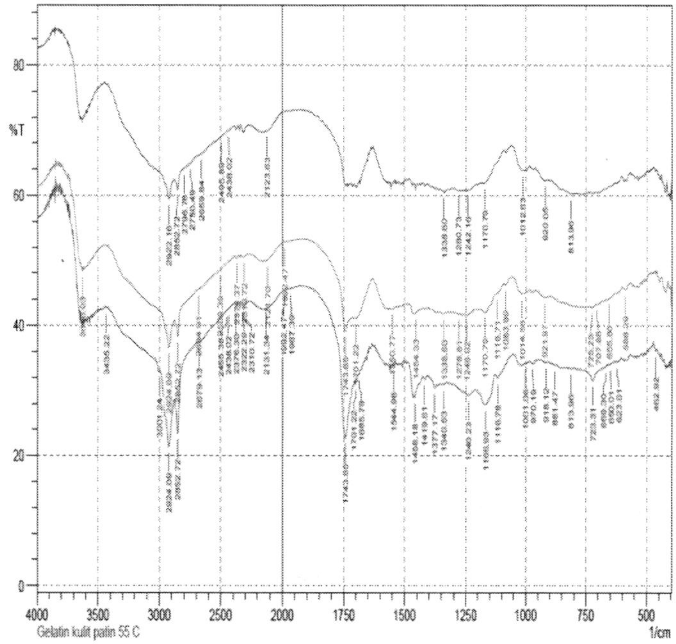


**Figure 1.** Spectrum FTIR of Gelatin Skin Catfish

Overall the FTIR spectra curve for patin skin gelatin extracted at the temperatures of 45 °C, 50 °C, and 55 °C has an intensity from amide A to larger amide II. The peaks in amide III are almost invisible. This is in accordance with the theory that collagen has been successfully denatured into gelatin. Based on FTIR analysis, patin skin gelatin extracted at a temperature of 55 °C is the most prominent uptake of a typical gelatin-functional group having a complete absorption area.

**Table 2.** Peak Position and functional spectrum FTIR of patin fish skin gelatin.

| Absorbansi area | Absorpstion peak ($cm^{-1}$) | | | | Notes |
|---|---|---|---|---|---|
| | $45^0$ | $50^0$ | $55^0$ | *References* | |
| Amida A | 2922,16<br>2852,72 | 3630,03<br>2924,09<br>2852,72 | 3435,22<br>3001,24<br>2924,09<br>2852,72 | 3650-3580<br>3600-2300 | Free NH<br>NH *streching* of an amide group associated with a hydrogen bond , OH group<br>$CH_2$ *stretching* symmetrical/ C-H alkana/ aldehide<br>$CH_2$ *stretching* asymmetrical /C-H alkana/ aldehide |
| Amida I | - | 1743,65<br>1701,22<br>1550,77<br>1454,33 | 1743,65<br>1701,22<br>1685,79 | 1700-1600 | Stretching C=O ; OH group pair with $COO^-$ |
| Amida II | - | - | 1544,98<br>1458,18<br>1419,61<br>1377,17 | 1560-1335 | NH bending pair with CN stretching<br>*$CH_2$ bending* |
| Amida III | 1242,16<br>1170,79<br>1012,63<br>920,05<br>813,96 | 1278,81<br>1246,0 | 1240,23 | 1240-670 | NH *bending*<br>C – O |

## 4. Conclusion

Gelatin is a denaturated of protein derived from collagen and is an important function of biopolymers witha very wide application in various fields of industry. Based on this research, the best treatment was found at the yield of 24.65 % at 55 °C; the yield of water content was 9.4 % at the temperature of 50 °C. The result of ash content was at 0.83 % at the temperature of 55 °C; protein content was 89.61 % at 55 °C; while fat content was at 1.47 % at 45 °C. Meanwhile, the gel yield strength was of 105.30 blooms at the temperature of 45 °C; the viscosity was 7 cP at 45 °C; the yield of the melting point was 28 °C at 45 °C; gel point result was 11 °C at 45 °C; and the FTIR results produced NH, CH, OH, and CO at 55 °C.

## 5. References

[1] Sobral P J A and Habitante A M Q B 2001 **15** 377–382

[2] Sompie M, A D Mirah, and CH M Karisoh 2015 Effect of temperature difference of extraction on gelatin characteristics of chicken leg skin *Pros Sem Nas Masy Biodiv Indon* **1**(4) 792-795 (in Indonesia)

[3] Wahyuni M and Peranginangin R 2007 Improving the competitiveness of fishery processing industry through utilization of non-economical waste fish into gelatin *Artikel DKP* (in Indonesia)

[4] Schrieber R and H Gareis 2007 Gelatine handbook,*Wiley-VCH GmbH & Co, Weinhem.*

[5] Arnesen J A and Gildberg A 2007 *Bioresource Technology* **98** 53–57

IOP Conf. Series: Earth and Environmental Science **137** (2018) 012067 doi:10.1088/1755-1315/137/1/012067

[6] Vannuccini S 2004 Overview of fish production, utilization, consumption and trade: based on 2002 data *FAO of the United Nations*

[7] Koli J M, Subrata B, Binay B N, Surendra B P, Ashif U P 2012 *Food and Bioproducts processing* **90** 555–562.

[8] Nurdiani R, Dissanayake M, Street, W A , Vasiljevic T, Donkor O N and Singh T K 2015 *Int. J. of Food Sci. and Tech.* **50** 2329–2339

[9] Grossman S and Bergman M 1991 Process for the production of gelatin from fish skins *European Paten Aplication*

[10] Muyonga J H, C G B Cole and K G Duodu 2004 *J. Food Hydrocolloids* **18**:581-592

[11] Ratnasari I, Sudarminto S Y, Nusyam H, Simon B W 2014 *Food and Public Health* **4** 140-150

[12] Sae-law T and Soottawat B 2015 *Food Bioscience* 59-68

[13] Mayangsari E, Fatimah N, Rizky A R 2013 Preparation of gelatin extract from bone fish mackerel (*Scomberomorus commerson*) trough hydrolysis process with alkaline solution *Fakultas farmasi dan Sains* Univesritas Muhammadiyah Prof. Dr. Hamka (in Indonesia)

[14] Lehninger A L 1982 *Basic of chemistry* Translation by Maggy Thenawijaya (Jakarta: Erlangga) 1st ed. (in Indonesia)

[15] Astawan M and Aviana T 2003 *Jurnal TIP* **14** 7-12 (in Indonesia)

[16] Wijaya I M 1998 The effect of protein concentration and pH on the bloom strength of gelatin. Gitayana *Majalah I. T. Pertanian* IPB Bogor (in Indonesia)

[17] Tazwir D L A and Rosmawaty P 2007 *J. Postharvest Biotechnol. Mar. Fish.* **2** (in Indonesia)

[18] Hafidz 2011 Halal gelatin making of bone milkfish (*Chanos chanos Forskal*) (as an alternative preparation of gelatin halal) *Laporan Penelitian, LEMLIT UIN Malang* (in Indonesia)

[19] Ockerman H W and Hansen C L 2000 *Animal by product processing and utilization* (USA: CRC Press)

[20] Amiruldin M 2007 Gelatin making and analysis of gelatin characteristics of tuna fishbone (*Thunnus albacares) FTP IPB* (in Indonesia)

IOP Conf. Series: Earth and Environmental Science **137** (2018) 012068 doi:10.1088/1755-1315/137/1/012068

# Evaluation of the proximate quality of the combination of Tuna *(Thunnus albacares)* and white oyster mushroom *(Pleurotus ostreatus)* nuggets

**H S Yufidasari, A A Prihanto, R Nurdiani and A A Jaziri**

Department of Fishery Product Technology, Faculty of Fisheries and Marine Science, Brawijaya University, Jl. Veteran Malang 65145, Indonesia

E-mail: heftisalis@gmail.com

**Abstract**. Nugget is a processed meat product which has great market demand but need variations to increase its nutritional content. Tuna is rich in omega-3 protein, vitamins, and minerals. White oyster mushrooms have high nutritional content which are about 23-33% protein, 36-68 % carbohydrates and 12-22 % amino acids. The purpose of this research is to evaluate the chemical quality of Tuna nugget (*Thunnus albacores*) with combination of white oyster mushroom (*Pleurotus ostreatus*). Complete Randomized Design (RAL) with parameters of Tuna and white oyster mushroom formulation, TJ1 (70 % Tuna: 30 % white oyster mushroom), TJ2 (50 % Tuna: 50 % white oyster mushroom), TJ3 (30 % Tuna: 70 % white oyster mushroom), and Control or K Treatment (100 % Tuna) is used. Results of Tuna nuggets with white oyster mushroom combination showed the highest value of water content in TJ3 50.14 %, protein K 19.6 %, fat TJ3 22.98 %, ash K 3.99 % and 2.47 % crude fiber. From these results, there is a need for further research on fat, ash and coarse fiber content that is used in the manufacture of fish nuggets combined with oyster mushrooms because it failed to meet Indonesian National Standard (SNI).

## 1. Introduction

Fish nugget is an alternative for fish processing with the aim to extend the durability of fish, it also increases the added value in fish marketing. There are many processed fish products.Nuggets isa type of product which is processed by crushing meat-based materials, steamed and then fried. Nuggets can be stored in the form of frozen food [1]. Tuna *(Tunnus albacores*) is a marine fish that has a high protein content at approximately 22 gr / 100 gr of meat. Tuna is also rich in omega 3, vitamins and minerals [2].

The processing of meat into nuggets needsother ingredients. White oyster mushroom (*Pleurotus ostreatus*), can be used as a filler in the manufacturing of the nuggets. The white oyster mushroom is a type of wood mushroom that can be consumed or falls into the category of edible mushroom. White oyster mushrooms contain high protein, fat, fiber, amino acids, and minerals. Based on dry weight, it contains 23-33 % protein, 36-68 % carbohydrates, 12- 22 % amino acids and 3.3-4.7 % fats, which is lower than beef [3].

Chicken nuggets that were substituted up to 50 % with oyster mushrooms have met the standards of protein and water content [4]. Other results explained that the best treatment of white oyster mushroom nuggets was obtained from 100 gr white oyster mushrooms, 25 gr fish, and 15 gr fluor. The making of nuggets with Tuna formulations (*Tunnus albacores*) and different white oyster mushrooms are new innovations in the manufacture of foodstuffs. The formulation of Tuna and white oyster mushrooms is expected to increase the quality of nuggets produced, including the texture, taste, aroma and nutrition of the nugget. The high nutrient content and texture of the fish meat is dense and soft in texture, while the texture of white oyster mushrooms is good with a similar texture to chicken meat. This is the main reason in the selection of these materials.

## 2. Methodology

This research used the experimental method with Complete Randomized Design (CRD), one factor, 4 treatments and 3 replications.The formulation of the Tuna and white oyster mushroom were: TJ1 (70 % Tuna: 30 % of white oyster mushroom), TJ2 (50 % Tuna: 50 % of white oyster mushroom), TJ3 (30 % Tuna: 70 % of white oyster mushroom), and Control or K Treatment (100 % Tuna).

### *2.1. Procedure of research*

The process of making oyster mushroom nuggetsbegan with the washing of the white oyster mushrooms to clean them and then steaming them with bay leaves to remove any unpleasant flavors, then mashed. The Tuna meat was washed and milled. Two pieces of white bread were soaked in milk, then stirred until smooth. One tablespoon of margarine was melted to sauté 25 grams of onion and 2 cloves of garlic that werefinely blended.

Minced Tuna meat, mashed white oyster mushrooms, sautéed onion, ½ teaspoon pepper, ½ teaspoon salt, 20 g carrot, and one egg were mixed until homogeneous. The dough was poured into a pan that was smeared with margarine and steamed for 15-20 minutes. After cooling down, the dough was cut to 5 cm x 3 cm x 1 cm in size and then was proceeded with the coating process. The cut dough was covered with whiteegg and breadcrumbs and then fried.

### *2.2. Parameters of test*

The test parameters used in this study were proximate quality (water content, protein, fat, ash and crude fiber tests) following the standard method.

## 3. Result and Discussion

This study aims to determine the proximate quality of the Tunanugget formulation with the white oyster mushroom substitution by testing its moisture, protein, fat, ash and fiber. The main raw materials in this study were Tuna *(Tunnus albacores)* and white oyster mushroom *(Pleurotus ostreatus)* which can be seen in figure 1.

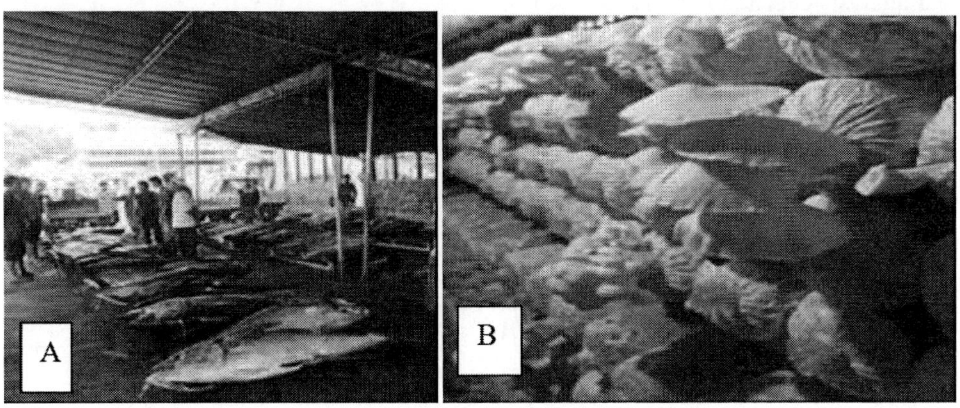


**Figure 1.** Tuna (A) and white oyster mushroom (B)

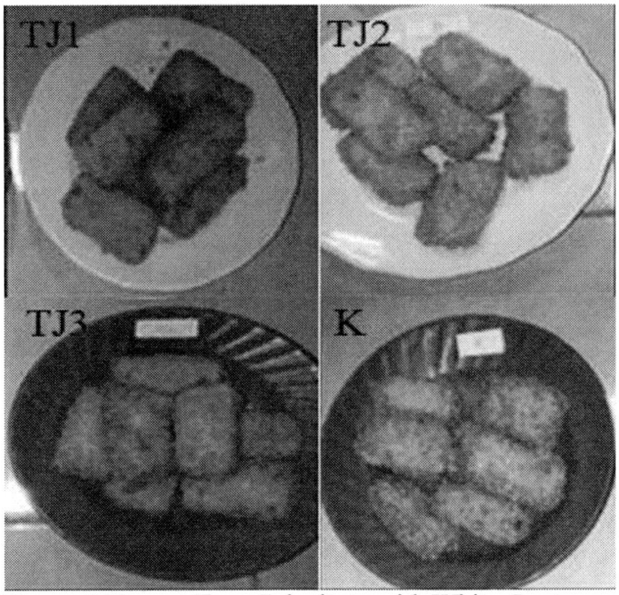


**Figure 2.**Nugget Tuna Substitute with White Oyster Mushroom (TJ1 = Tuna 70 %, Oyster Mushroom 30 %; TJ2= 50 % Tuna, 50 % Oyster Mushroom; TJ3 = 30 % Tuna, 70 % Oyster Mushroom and K = Control, 100 % Tuna fish).

Figure 2 showed that there is no difference in the physical appearance of the four treatments, because the coating and the frying processeswere the same. The coating of the nugget product is to attract the attention of the consumer [1], which describes that the sensory nugget characteristic depends on the coating technique.

### *3.1. Water content*

Water in food acts as a solvent of some components in addition to acting as a reagent, while water forms can be found as free water and bound water. Free water can easily disappear such as in the case of evaporation or drying, whereas bound water is difficult to be liberated in such a manner. The result of water analysis depicted in table 1.

**Table 1.** Results of test water content of Tuna nugget with white oyster mushroom substitutions.

| Treatments | Repetition (%) | | | Total(%) | Average(%) |
|---|---|---|---|---|---|
| | 1 | 2 | 3 | | |
| TJ1 | 51.29 | 48.19 | 49.74 | 149.22 | 49.740* |
| TJ2 | 49.54 | 50.05 | 49.79 | 149.38 | 49.793 |
| TJ3 | 49.95 | 50.33 | 50.14 | 150.42 | 50.140** |
| K | 46.19 | 53.16 | 49.68 | 149.03 | 49.677 |
| | | | | 598.05 | |

*Note: * lowest water content, ** the highest water content*

From the water content test, it was found that the highest water content was in the TJ3 treatment (30 % Tuna meat and 70 % g of white oyster mushroom) at 50.14 %. The lowest water content was showedin the control treatment (100 % Tuna meat) at 49.68 %.

### *3.2. Protein content*

The result of protein content can be seen in table 2.

**Table 2.** Results of protein content of Tuna nugget with white oyster mushroom substitution.

| Treatments | Repetition (%) | | | Total (%) | Average (%) |
|---|---|---|---|---|---|
| | 1 | 2 | 3 | | |
| TJ1 | 17.45 | 18.46 | 17.96 | 53.87 | 17.96** |
| TJ2 | 16.4 | 17.27 | 16.84 | 50.51 | 16.84 |
| TJ3 | 15.44 | 16.2 | 15.82 | 47.46 | 15.82* |
| K | 19.12 | 20.08 | 19.6 | 58.8 | 19.6 |
| | | | | 210.64 | |

*Note: * lowest protein content, ** the highest protein content*

From the test of protein content, it was found that the highest protein content was in the K treatment (100 % Tuna meat and 0 % white oyster mushroom) at 19.6 %. The lowest protein content was found in the TJ3 treatment (30 % Tuna meat and 70 % white oyster mushroom) at 15.82 %.

### *3.3. Fat content*

The fat content test results can be seen in table 3.

**Table 3.** Results of fat content of Tuna nugget with white oyster mushroom substitution.

| Treatment | Repetition (%) | | | Total (%) | Average (%) |
|---|---|---|---|---|---|
| | 1 | 2 | 3 | | |
| TJ1 | 21.45 | 23.26 | 22.36 | 67.07 | 22.36* |
| TJ2 | 22.04 | 23.75 | 22.9 | 68.69 | 22.90 |
| TJ3 | 22.18 | 23.77 | 22.98 | 68.93 | 22.98** |
| K | 20.64 | 22.18 | 21.41 | 64.23 | 21.41 |
| | | | | 268.92 | |

*Note: *lowest fat content, **highest fat content*

From the fat content test, we found that the highest fat content was in the TJ3 treatment (30 % Tuna meat and 70 % white oyster mushroom) at 22.98 %. The lowest fat content was found in the K treatment (100 % Tuna meat and 0 % white oyster mushroom) at 21.41 %.

IOP Conf. Series: Earth and Environmental Science **137** (2018) 012068 doi:10.1088/1755-1315/137/1/012068

### *3.4. Ash content*

The ash content test results can be seen in table 4.

**Table 4.** Results of ash content of Tuna nugget with white oyster mushroom substitution.

| Treatment | Repetition (%) | | | Total (%) | Average (%) |
|---|---|---|---|---|---|
| | 1 | 2 | 3 | | |
| TJ1 | 3.52 | 3.72 | 3.62 | 10.86 | 3.62** |
| TJ2 | 3.37 | 3.54 | 3.46 | 10.37 | 3.46 |
| TJ3 | 3.1 | 3.27 | 3.19 | 9.56 | 3.19* |
| K | 3.9 | 4.09 | 3.99 | 11.98 | 3.99 |
| | | | | 42.77 | |

*Note: *lowest ash content, **highest ash content*

The ash content test found that the highest ash content was in the K treatment (100 % Tuna meat and 0 % white oyster mushroom) at 3.99 %. The lowest fat content was found in the TJ3 treatment (30 % Tuna meat and 70 % white oyster mushroom) at 3.19 %.

### *3.5. Crude fiber content*

The crude fiber content test results can be seen in table 5.

**Table 5.** Results of crude fiber content of Tuna nugget with white oyster mushroom substitution.

| Treatment | Repetition (%) | | | Total (%) | Average (%) |
|---|---|---|---|---|---|
| | 1 | 2 | 3 | | |
| TJ1 | 0.9 | 0.95 | 0.93 | 2.78 | 0.93 |
| TJ2 | 0.53 | 0.56 | 0.55 | 1.64 | 0.55 |
| TJ3 | 0.83 | 0.87 | 0.85 | 2.55 | 0.85 |
| K | 2.41 | 2.53 | 2.47 | 7.41 | 2.47 |
| | | | | 14.38 | |

From the crude fiber content test we found that the highest crude fiber content was in the K treatment (100 % Tuna meat and 0 % white oyster mushroom) at 2.47 %. The lowest protein content was found in the TJ2 treatment (50 % Tuna meat and 50 % white oyster mushroom) at 0.55 %.

### *3.6. Discussion*

The results of the variation analysis on the nugget water content showed thattheTuna nugget with white oyster mushroom combinationslightly affecting ($P < 0.05$) to all treatments. However, the water content value when compared with the standard in SNI 7758: 2013 still shows a decent range of production, where the SNI standard of maximum water content for nuggets is 60 %. The value of water content will increase along with the addition of the oyster mushroom substitution. Oyster mushrooms have a higher amount of water content in a fresh state which is 90.97 % [5].

The results of the variation analysis of the nugget protein content showed that the Tuna with white oyster mushroom combinationnugget gave a very real effect ($P > 0.01$) to all treatments. The percentage of Tuna use in the nuggets manufacturedwas decreased until 30 % in line with the increase rate of the mushroom mixture to 70 %. This can affect the value of protein content in the nuggets because Tunais the ingredient that is thesource of protein, while the added oyster mushroom has a lower protein content than chicken meat protein, thus reducing the protein content in the nuggets.

IOP Conf. Series: Earth and Environmental Science **137** (2018) 012068 doi:10.1088/1755-1315/137/1/012068

Tuna protein content is 22-24 % per 100 grams of meat [3]. Oyster mushrooms have lower amounts in fresh state at 2.67 % per 100 grams [5]. This is in line with the results of the study [4], which stated that the results of the chicken nugget protein content showed decreased levels whensubstituted with white oyster mushrooms; the higher the amount of oyster mushrooms in the chicken nuggets,the lower the protein content. However, the value of protein content when compared with the standard level of protein of nuggets in SNI 7758: 2013 still shows a decent range of production, where the SNI standard level of protein for nuggets is at least 12 %.

The results of the variation analysis of fat content showed that Tuna with combined white oyster mushroom nuggets did not give a significant difference of influence ($P < 0,05$). The addition of oyster mushroom in the nugget will decrease the fat content of the Tuna until 30 % in line with the increase of mushroom mixture to 70 %. Thus, the fat content in the fish meat and oyster mushrooms will affect the fat content of the fish - mushroom nugget. The fat content of Tuna meat is 2.7 % per 100 grams [2], in fresh state the oyster mushroom have a lower amount of fat content at 0.33 grams per 100 grams [5]; hence the higher the percentage of mushrooms mixed into thedough, the lower the fat content of the chicken nuggets will be. Moreover, the fat content results compared to the SNI 7758: 2013 has still not shown a decent range of production, where the SNI standard of protein for nuggets is maximum 15 %. The high fat content of the nugget is caused by the process of frying it using the deep frying method where the material is immersed in cooking oil, which will increase fat content.

The results of the variation analysis on ash content of the nuggets showed that the Tuna with white oyster mushroomnugget combination gave a very real effect ($P > 0.01$) to all treatments. The value of ash content will decrease as the oyster mushroom composition is substituted. This is suitable with the statement of Well yalina *et.al* [6] that the use of more Tuna meat will increase the ash content in nuggets, because meat contains minerals such as phosphorus and iron;additionally,it is also affected by the result of the burning residue of added ingredients in the making of the nuggets including spices [7]. Compared to the standard ash content of nuggets in SNI 7758: 2013, it still does not show a proper production range, where the SNI standard of maximum ash contentfor nuggets is 2.5 %.

The results of the variation analysis on crude fiber content of the nuggets show that the Tuna nugget with the white oyster mushroom combination gives a very real effect ($P > 0.01$) to all treatments. The value of crude fiber content will decrease as the oyster mushroom composition is substituted. When compared to the study Warsino *et.al* [8], the crude fiber content of fresh oyster mushrooms was 3.5 % in 100 grams. Meanwhile, according to Suharjo [9], the fiber in white oyster mushrooms is 1.56 %, but its ability to bind water to white oyster mushroom is not as good as chicken meat so that its substitution on chicken meat is limited.

## 4. Conclusion

The results of the quality evaluation of Tuna nuggets substituted with white oyster mushroom showed that the highest water content was found in the TJ3 treatment (30 % Tuna meat and 70 % white oyster mushroom) which measured at 50.14 %, while the lowest was found in the control treatment (100 % Tuna meat) at 49.68 %. The protein content test results showed that the highest level was found in the treatment of K (100 % Tuna) at 19.6 % and the lowest in the TJ3 treatment (30 % Tuna meat and 70 % white oyster mushroom) at 15.82 %. The highest level of fat content was found in the TJ3 treatment (30 % Tuna meat and 70 % white oyster mushroom) at 22.98 %,while the lowest was found in the K treatment (100 % Tuna) at 21.41 %. The highest ash content result was obtained in the treatment of K (100 % Tuna) measured at 3.99 % and the lowest was in the TJ3 treatment (30 % Tuna meat and 70 % white oyster mushroom) at 3.19 %. From the test of crude fiber content, the highest results were measuredin the K treatment (100 % Tuna) which was 2.47 %. The lowest crude fiber content was in the TJ2 treatment (50 % Tuna meat and 50 % white oyster mushroom) at 0.55 %. Therefore, the best quality nugget formulation of Tuna and white oyster mushroom was found in the TJ2 treatment (50 % Tuna: 50 % white oyster mushroom).

IOP Conf. Series: Earth and Environmental Science **137** (2018) 012068 doi:10.1088/1755-1315/137/1/012068

## 5. References

[1]. Chen S D, Chen, Chao and Lin 2009 *J.Food Eng.* **95** 59–364.
[2]. Departement of Health, education and walfare 1972 Food composition for use in east Asia. USA.
[3]. Suriawiria U 2002 *Oyster mushroom cultivation* (Jogjakarta:Yayasan kanisius) (In Indonesia)
[4]. Laksono MA 2012 *Anim. Agri. J.***1** 692
[5]. Muchtadi T R 1990 *Preservative technology of earl mushrooms* (*Pleuratus ostreatus*) (Bogor : Bogor Agricultural Institute) (In Indonesia)
[6]. Wellyalina F, Azima and Aisman 2013 *J. Food Tech. Appl.* **1** (In Indonesia)
[7]. Afrianti L H 2013 *Food preservation technology* (Jakarta : Penerbit Alfabeta) p 260 (In Indonesia)
[8]. Warisno and Dahana 2010 *Oyster, sowing mushrooms, reap rupiah* (Jakarta : Gramedia) (In Indonesia)
[9]. Suharjo E 2008 *Mushroom cultivation with cardboard media* (Jakarta : Agro Media Pustaka ) (In Indonesia)

IOP Conf. Series: Earth and Environmental Science **137** (2018) 012069 doi:10.1088/1755-1315/137/1/012069

# Stability of prebiotic, laminaran oligosaccharide under food processing conditions

**A Chamidah**

Department of Fishery Product Technology, Faculty of Fisheries and Marine Science, University of Brawijaya, Jl. Veteran, Malang, 65145, East Java, Indonesia

E-mail: achamidah@yahoo.co.id

**Abstract.** Prebiotic stability tests on laminaran oligosaccharide under food processing conditions were urgently performed to determine the ability of prebiotics deal with processing. Laminaran, oligosaccharide is produced from enzymatic hydrolysis. To further apply this prebiotic, it is necessary to test its performance on food processing. Single prebiotic or in combination with probiotic can improve human digestive health. The effectiveness evaluation of prebiotic should be taken into account in regards its chemical and functional stabilities. This study aims to investigate the stability of laminaran, oligosaccharide under food processing condition.

## 1. Introduction

Several species of pathogenic bacteria have been found to cause acute gastroenteritis, ulcerative colitis, colon cancer, and pseudomembranous colitis [1], while certain beneficial genera such as bifidobacteria and lactobacilli are known to provide protection against infection [2]. Prebiotic is non digested carbohydrate which promote the growth of good bacteria in the intestine [3].

Non-digested carbohydrate and from the algae has an important health supporting effect. They are bowel function, preventing colon cancer, promoting the growth of certain beneficial microorganisms, lowering serum cholesterol, increasing mineral absorption, edema reducing agent, and antihyperglycemic [4,5]. Generally, commercial prebiotics are produced by terrestrial plant oligosaccharides. Commercial prebiotics often cause consumer anxiety due to the excessive amount of released gas during rapid fermentation in the digestive tract [6]. The majority of prebiotics producing gas are commercial prebiotics fermented by microorganisms [7].

Brown algae are sources of several bioactive polysaccharides such as alginate, FCP, and laminar [4, 8, 9, 10]. Seaweed or algae, according to Chamidah *et al.* [11], show the effects of prebiotics as antibacterial. Laminaran is a component of algae polysaccharide that may have prebiotic activities. stating that laminaran polysaccharides could not be digested by digestive enzymes. About 80% of laminar are fermented by human fecal bacteria after 24 hours [12,13,14, 15].

Prebiotics have been added to yogurt and other fermented dairy products, as well as a variety of other foods such as biscuits, candies, cereals, dairy products, beverages, baby food formulas and weaning foods [16,17]. However, data on prebiotic stability during food processing in complex matrices are very limited [18, 19].

Food processing affects prebiotics, either in quantity or conformation. Food processing and cooking operations change the concentration of prebiotics in food. Prebiotic stability during food processing is an important requirement as their biological activity may depend on their structural integrity. For example, hydrolyzed or degraded prebiotics caused by food processing may no longer be active [16, 20]. Therefore, it is important to know the effects of treatment and treatment on prebiotic carbohydrates to maintain optimum prebiotic concentration in processed foods as an effort to maintain intestinal health. The objective of this study was to investigate the stability of the laminaran under food processing condition.

## 2. Methodology

### *2.1. Material*

Prebiotics used in this study were crude laminaran by adopting a method described by Huebner *et al.* [21] and Moore [22] with slight modifications. The chemicals used include 96% ethanol, $H_2SO_4$, HCl (technical) and aquadest. Reagents used for chemical analysis were anhydrous glucose, Nelson reagent, aluminum hydroxide slurry, and glucose test/ Glucose Oxidation Method, acetone and chloroform, ethanol, $CaCl_2$, HCl, $Na_2CO_3$, and polyvinylpyrrolidone $(C_6H_9NO)n$ (Merck, with grade pro analysis, pa). Whatman no 40 paper, laminaran standard from Laminariadigitata and laminariaceae enzyme from Trichodermasp (Sigma-Aldrich), endo enzyme (1,3)- ß-D glucanase from *Trichoderma viridae* (Megazyme) were also used.

### *2.2. Preparation of laminaran oligosaccharide*

Crude Laminar was enzymatically hydrolyzed using endo (1,3) -β -D-glucanase enzyme from *Trichodermavir*ide. This hydrolysis procedure referred to Bohm *et al.* [23] and Crritenden and Playne [24]. Hydrolysis was carried out at 45°C. This study was conducted using a simple randomized design with one level of hydrolysis 0, 5.10, 15, 20, 25, and 30 mins. As shown in Figure 1. after obtaining the optimum time for the laminar oligosaccharide manufacture, a process for obtaining short chain laminar oligosaccharides by combining Sudarmo *et al.* [25] and Alderkamp *et al.* [26] with slight modifications was conducted (figure 1).

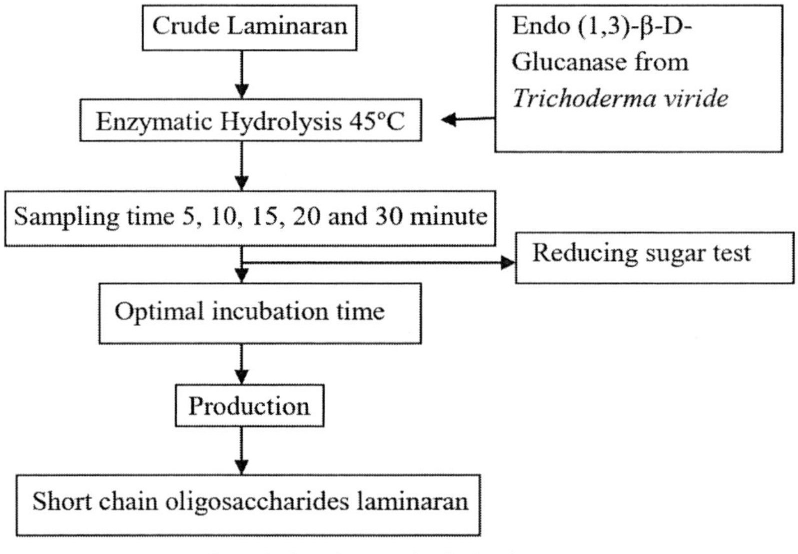


**Figure 1**. Process of crude laminaran hydrolysis.

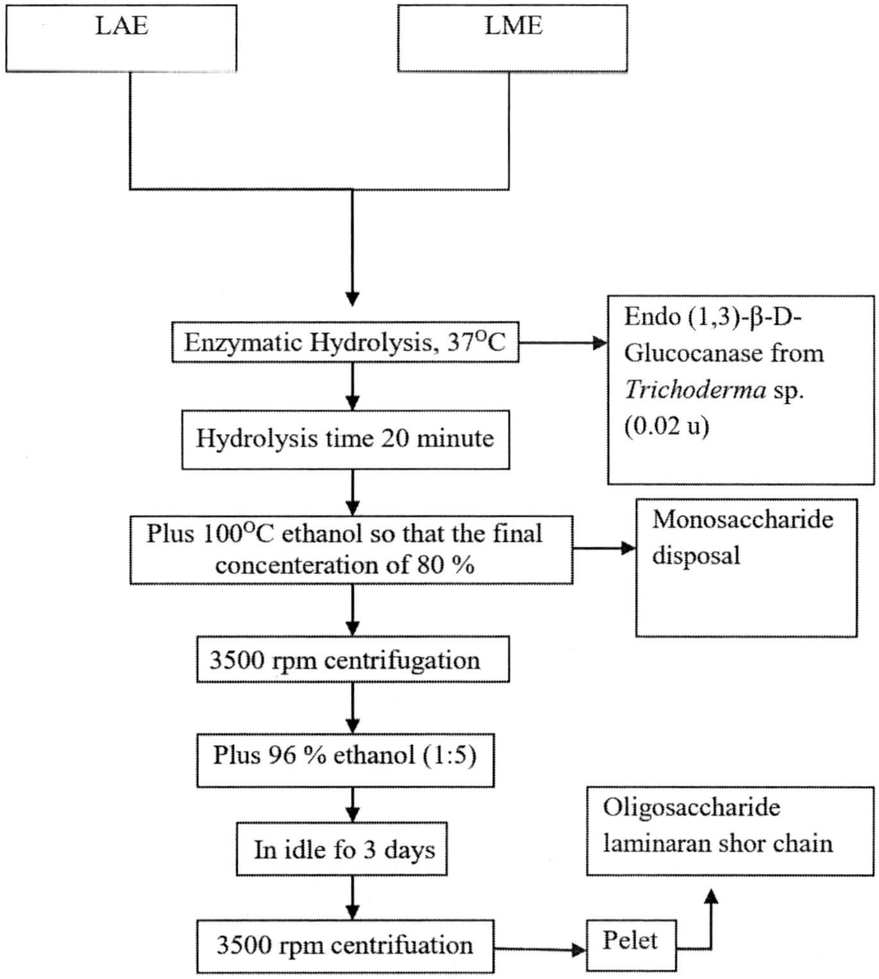


**Figure. 2**. Production Oligosaccharides Laminaran.

*2.3. Data Analysis*

The obtained quantitative data were tested by using analysis of variance (ANOVA) and then continued by Duncan's test ($\alpha$= 0.05) using SPSS 16.0 for Windows application program and Statistic Version 6 program.

*2.4. Stability test procedure of laminaran oligosaccharide in food processing conditions*

Stability test procedure on processing conditions was displayed on figure 3-4.

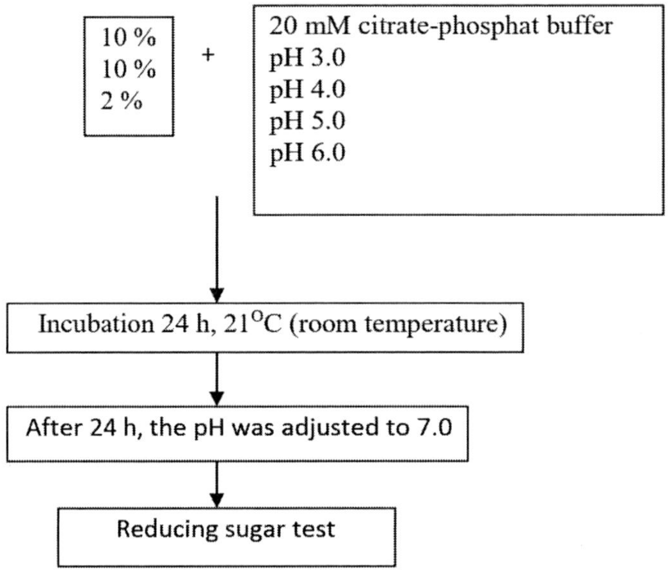


**Figure 3.** Oligosacharide stability at low pH.

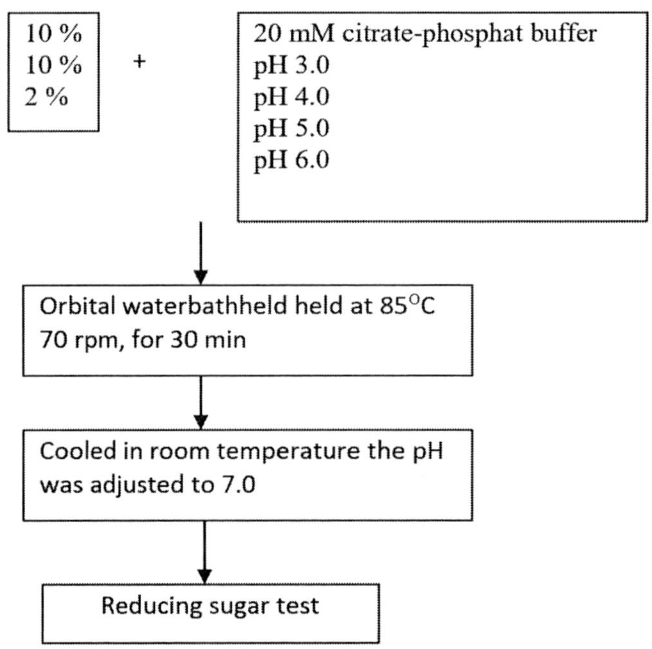


**Figure 4.** Heat at low pH oligosakarida laminaran stability.

## 3. Results and Discussion

After hydrolysis for 20 minutes with Endo-(1.3)-ß-D-Glucanase enzyme from *Trichoderma* sp. and monosaccharide was discharged, both OLAE and OLME were obtained (data not shown).

*3.1. Low pH effect*

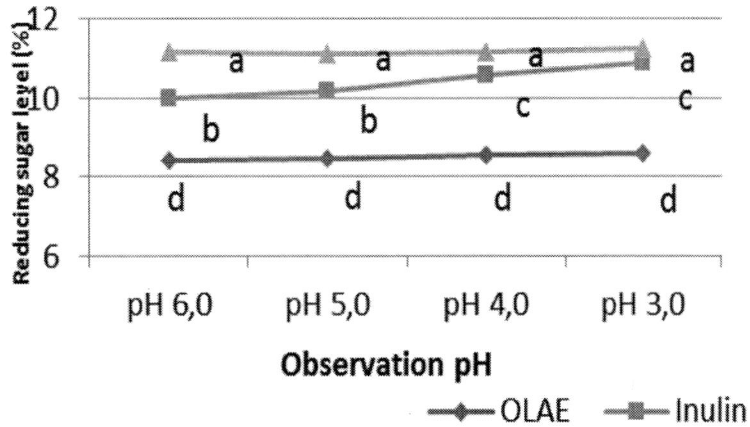


**Figure 5.** Reduced sugar levels of laminaran oligosaccharide at low pH effects.

Food material must be stable during food processing. Low pH stability is important for the prebiotic since they will experienced low pH in the digestive tract. Prebiotics may undergo chemical changes, but their functional activity may remain or even increase after processing [16]. The results for the stability test was shown in figure 6.

Chemical degradation was not detected in the test. Laminaran showed its stability even at the pH of 3. Reduced sugar content between OLAE and OLME is very significant (p <0.05), OLAE <OLME. This is because OLME contains larger carbohydrate components (laminaran) that are more easily degraded by acid, thus producing higher reducing sugars . The results of this study indicated similar trend with the result of [16,27,28] which used four commercial prebiotic samples. Thus, both OLAE and OLME are stable at low pH treatments.

*3.2. Heat effect and low pH*

During food processing, food material will contact with the heat. Hence, such condition needs to be tested on prebiotic candidates. The test is feasible to be applied to determine the biological stability in food material when exposed to heat. The result of heat and low pH treatment could be seen in Figure 6.

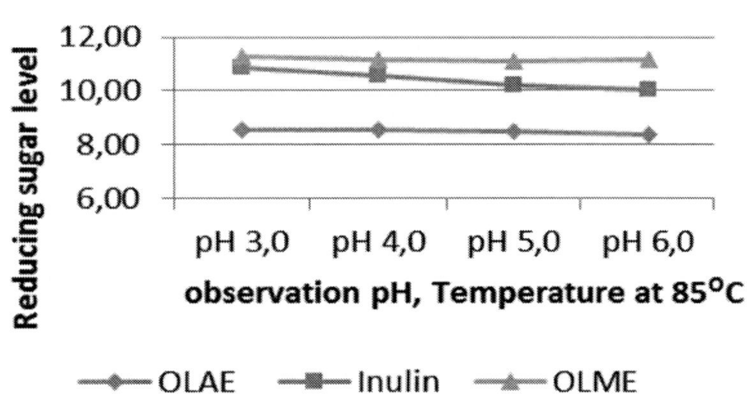

**Figure 6**. Reduced sugar content of laminaran oligosaccharide at 85 °C and low pH.

The leminaran still stable under heat and low pH. When compared with that of commercial inulin prebiotics, the reduction value of OLME laminaran oligosaccharide is higher. The decrease of sugar in this inulin prebiotic is due to biological matrial degradation when inulin is exposed to acidic and moderate heat (70 °C) [29]. Crittenden and Playne [30] showed that heated inulin at 195 °C for 30 mins resulted in perfect degradation of the fructant chain and new products with lower molecular weight were obtained. Reduced sugar level of OLAE and OLME is relatively unchanged during the food processing treatment. This means that the laminaran oligosaccharides are relatively stable against heat and low pH.

### *3.3. Browning calculation*
Calculation of its relative browning percentage in several laminaran types was shown in table 1.

**Table 1.** Relative percentage of browning laminaran and inulin oligosaccharides.

| Substrat | PersenRelatif Browning | | | |
|---|---|---|---|---|
| | 0 hour | 1 hour | 2 hour | 3 hour |
| OLME /Glc 10% | 340 | 369 | 373 | 380 |
| OLAE /Glc 10% | 24 | 26 | 26 | 29 |
| INULIN/Glc 2% | 44 | 42 | 36 | 37 |

It could be inferred from table 1 that the degradation during browning process is very high for OLME substrates compared to OLAE ones, moreover when compared to inulin, which is a tested as a commercial prebiotic.

## 4. Conclusion
Laminaran in the form of both OLAE and OLME are stable against the effect of low pH treatment, where the performance of OLME is better than OLAE. Laminaran in the form of OLAE and OLME is stable to heat and low pH but not resistant to Maillard reactions.

## 5. References
[1] Steer T, Carpenter H, Tuohy K and Gibson G R 2000 *Nutr. Res. ReV* **13** 229-54.
[2] Collins M D and Gibson G R 1999 *Am. J. Clin. Nutr.* **69** 1052S- 7S.
[3] Gibson GR, Probert HM, van Loo J, Rastall RA, Roberfoid MB 2004 *Nutr. Res. Rev.* **17** 259–75.
[4] Chamidah, A. , Prihanto A A , Hardoko, H. 2017 *Asian J. Pharm. Clin. Res.* **10** 311-313.
[5] Firdaus M , Nurdiani, R. , Prihanto, AA 2015 Marine Algae Extracts: *Processes, Products, and Applications* 2-2 381-394
[6] Anonymous, 2004 ( Publication number : WO2004002240 A3)
[7] Rycroft C E, Jones M R, Gibson G R and Rastall R A 2001 *J. Appl. Microbiol.* **91** 878–887.
[8] Burtin P 2003 *Electrom. J. Environ. Agric. Food Chem.* **2** 24–28.
[9] Deville C, Gharbi M, Dandrifosse G and Peulen O 2007 *J. Sci..Food. Agric.* **87** 1717-1725.
[10] MacArtain P, Gill C I R, Brooks M, Campbell R and Rowland I R 2007 *Nutr. Rev.* **65** 12
[11] Chamidah A, Hardoko, Prihanto A A 2017 A I P Conference Proceedings 1844,020011
[12] O'Sullivan L, Murphy B, McLoughlin P, Duggan P, Lawlor P G, Hughes H and Gardiner G E 2010 *Mar. Drugs* . **8** 2038–2064.
[13] Rioux L E, Turgeon S L and Beaulieu M 2007 *Carbohydr Polym.* **69** 530–537.
[14] Michel C, Barry J, Lahaye M, Bonnet C and Mabeau S 1993 *Proc. Nutr. Soc.* **52** 159A.
[15] Lahaye M and Kaeffer B 1997 *Sciences des aliments.* **17** 563-584

[16] Huebner J, Wehling R L and Hutkins R W 2008 *Int. Dairy J.* **18** 287-293.
[17] Duar R M, Ang P T, Hoffan M, Wehling R and Hutkins R W 2015 (University of Nebraska – Lincoln)
[18] Johnson C R, Thavarajah D, Thavarajah P, Payne S, Moore J and Ohm JB 2015 *J. Food Compost. Anal.* ***38*** 106–111.
[19] Nakakuki T 2002 *Pure Appl. Chem.* **74** 1245-1251
[20] Yvin J C, LeVasseur F and Hud'Homme F 1999 (US Patent :59 80 916).
[21] Mohsen A M S, Mohamed S F, Ali F M and El-Sayed O H 2007 *J. Appl. Sci. Res.* **3** 1178-1185.
[22] Nobe R, Sakakibara Y, Fukuda N, Yoshida N, Ogawa K and Suiko M 2003 *Bioschi. Biotechnol Biochem.* **67** 1349-1357.
[23] Sova V V, Shirokova N I, Kusaykin M I, Scobun A S, Elyakova L A and Zvyagintseva T N 2003 *Biochemistry* (Moscow), **68** 529-533.
[24] AOAC 1990 ( USA : Arlington)
[25] Ebihara K, Nakai Y and Kishida T 2006 *J. Food Sci.* **71** S163-S168
[26] Wang X and Gibson G R 1993 *J. Appl. Bacteriol.* **75** 373–380.
[27] Huebner J, Wehling R L and Hutkins RW 2007 *Int. Dairy. J.* **17** 770–775.
[28] Moore K E 2011 Biological Analysis of prebiotics in various processed food matrices (Thesis: Faculty of The Graduate College at the University of Nebraska)
[29] Bohm A, Kleessen B and Henle T 2006 *Europ. Food Res. Tech.* **222** 737–740.
[30] Crittenden R and M J Playne 2006 *Gastrointestinal Microbiology* 285-314.

IOP Conf. Series: Earth and Environmental Science **137** (2018) 012070 doi:10.1088/1755-1315/137/1/012070

# Isolation and Identification of cellulolytic bacteria from mangrove sediment in Bangka Island

**A Kurniawan[1,2], A A Prihanto[3], S P Sari[4], D Febriyanti[1], A Kurniawan[5], A B Sambah[6] and E Asriani[7]**

[1] Departement of Aquaculture, Bangka Belitung University, Balunijuk, Merawang, Bangka City, Bangka Belitung Archipelago, Indonesia
[2] Doctoral Student, Graduate School of Fisheries Science, Faculty of Fisheries and Marine Science, Brawijaya University, Jl. Veteran, Malang 65145, East Java, Indonesia
[3] Departement of Fishery Product Technology, Faculty of Fisheries and Marine Science, Brawijaya University, Jl. Veteran, Malang 65145, East Java, Indonesia
[4] Departement of Marine Science, Bangka Belitung University, Balunijuk, Merawang, Bangka City, Bangka Belitung Archipelago, Indonesia
[5] Departement of Aquatic Resource Management, Faculty of Fisheries and Marine Science, Brawijaya University, Jl. Veteran, Malang 65145, East Java, Indonesia
[6] Departement of Utilization of Fisheries Reources, Faculty of Fisheries and Marine Science, Brawijaya University, Jl. Veteran, Malang 65145, East Java, Indonesia
[7] Departement of Agrotechnology, Bangka Belitung University, Balunijuk, Merawang, Bangka City, Bangka Belitung Archipelago, Indonesia

E-mail: ardian_turen@yahoo.co.id

**Abstract.** Cellulolytic bacteria is bacteria which hydrolyze cellulose to reducing sugars. This research aims to obtain cellulolytic bacteria from the sediment of mangroves in Bangka island. Reasearch was conducted from March to August 2017. Sampling was conducted at Sungailiat, and Tukak Sadai, South of Bangka. Bacteria was isolated using 1% Carboxymetyl Cellulosa (CMC). The isolation resulted in four isolates from Sungailiat and nine isolates from Tukak Sadai. Total five isolates, namely *Bacillus pumilus*, *Pseudomonas* sp., *Bacillus amyloliquefacien*, *Bacillus alvei*, *Bacillus coagulant* were identified. The best isolates that produced cellulose was *Pseudomonas aeruginosa*.

## 1. Introduction

Estuaries ecosystem is a reservoir of microorganism which usually produces enormous and unique metabolites. Mangrove is an ecosystem found in tropical and sub-tropical regions. This ecosystem provides ecological benefits such as protection against tsunami danger, and oxygen supplier. Mangroves can store large quantities of organic carbon and sediments which mainly derive from litter falls and underground roots of mangrove plants [1]. Microbes in mangroves play an important role in the organic carbon cycle. Cellulolytic bacteria are bacteria which capable of degrading organic cellulose to simple sugar derivatives [2].

Cellulolytic bacteria obtained from nature are expected to be useful for degrading cellulose for both industrial and environmental conservation aplications. Cellulolytic bacteria are abundant in mangrove ecosystem. They easily be found in mangrove forests [2, 3, 4]. The bacteria are usually *Pseudomonas flurescent var cellulose, Cellulomonas fimi, Bacillus subtilis, Clostridium thermocellum, Acetobacter xylinum* [1].

Bangka Island is an area with mangrove-covering area for about 48.090 hectares. Nowadays, there is no exploration was recorded in this specific area. Contrastingly, study by Prihanto and Wakayama suggested that the ocean and coastal region is mainly unexplored source for enzymes [5]. In this area is also the location of tin mining. It activities surely affecting the physical and chemical characteristic of the mangrove ecosystem. Hence, by exploring this area, the unique enzyme will be achieved.

The objectives of this study was to investigate the potency of the mangrove sediment in Bangka Island as a source of cellulolytic bacteria. Further detail of the isolates would be identified using molecular method.

## 2. Materials and methods

### *2.1. Sampling sites*

Mangrove sediment were collected from Sungailiat and Tukak Sadai, South Bangka district (figure 1.). Sample taking was done on the depth of 10-20 cm. The samples were packed and brought to the Laboratory using ice box by maintaining its temperature at 4°C.

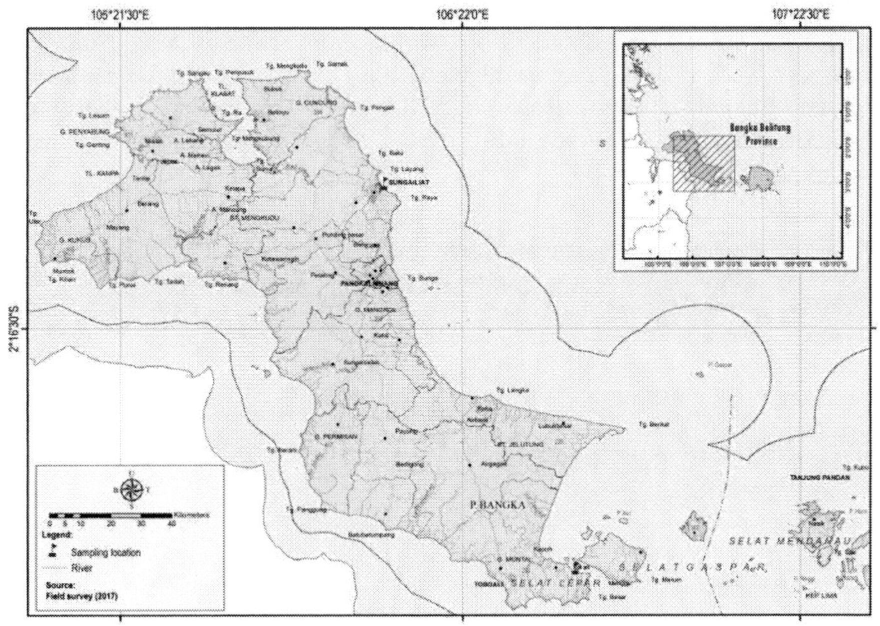


**Figure 1**. Sampling location in Sungailiat and Tukak Sadai.

### *2.2. Culture and screening of cellulolytic bacteria*

Prior to culture, Mangrove sediments weighed and proper diluted. The diluted samples were grown onto 1% CMC (Carboxy Methyl Cellulose) medium with a pour plate method. One loop of bacteria is scratched on the medium by forming a line with approximately one cm. The culture was incubated for 72 hours at 30 ° C. The Lugol solution (2g of potassium iodine and 1g iodine in 300 ml of aquadest) was flooded to cover the entire medium and is allowed to stand for a minute. The isolates

IOP Conf. Series: Earth and Environmental Science **137** (2018) 012070 doi:10.1088/1755-1315/137/1/012070

formed clear zone accounted as cellulose-degrading bacteria. Further identifications were performed by using two different methods namely, Microbact and 16s rDNA method.

### 2.3. Biochemical characterization of bacteria

Biochemical characterization was performed using Microbact system following company manual procedures. In order to determine the bacterial identity, their biochemical characters would be compared to Bergey's Manual of Determinative Bacteriology.

### 2.4. Identification of bacteria molecular 16S rRNA

The protocol for the the identification of best producer of cellulose, the methods by Prihanto *et al.*, (2016) was used with slight modification. DNA isolation was performed using the procedure of the DNA Isolation Kit (Wizard of Genomic DNA Purification Kit from Promega). PCR composition with a total volume of 20 μl / tube consists of 6 μl ddH2O, 10 μl PCR kit GoTaq® Green Master Mix (10 x taq polymerase buffer, dNTP, MgCl2, primer, Taq DNA polymerase, ddH2O), 1 μl forward primer, primary reverse 1 μl and 2 μl isolated DNA samples.
Primers for this analysis were 20F (52 - GTAATCGTCGGCCAGTA GAGTTTGATCCTGGCTC-32 ) and 1510R (52 - CAGGAA ACAGCTATGACCGGCTACC TTGTTACGACT-32 ). The sequence was BLAST (*Basic Local Alignment Search Tool*) analyzed. Phylogenetic tree was constructed using online phylogenetic analysis Phylogeny.fr.

## 3. Result and discussion

Thirteen isolates were successfully isolated from two locations. Nine bacterial isolates (TS1, TS2, TS3, TS4, TS5, TS6, TS7, TS8, TS9) were collected from Tukak Sadai and four isolates (SL1, SL2, SL3, SL4) from the Sungailiat mangrove samples. All thirteen isolates were further screened for its capability to hydrolyze CMC. The screening analysis suggested that only five isolates were cellulose positive bacteria (table 1).

**Table 1.** Screening of cellulose producing bacteria.

| Isolates Code | Screening result | Diameter (mm) |
|---|---|---|
| TS1 | + | 3.4 |
| TS2 | + | 4.6 |
| TS3 | - | - |
| TS4 | - | - |
| TS5 | - | - |
| TS6 | - | - |
| TS7 | + | 4.2 |
| TS8 | - | - |
| TS9 | - | - |
| SL1 | - | - |
| SL2 | + | 3.6 |
| SL3 | + | 4.1 |
| SL4 | - | - |

From the result it can be concluded that five isolates were cellulose positive. Three isolates from the location of Tukak Sadai (TS1, TS2, TS7) and two isolates from Sungailiat (SL2, SL3). All positive isolates were further analyzed using microbact analysis system. This analysis based on the biochemical characters of the isolate. The result for the analysis was provided in table 2.

**Table 2.** Results of biochemical characterization and test of bacterial isolates.

| Parameter | Result | | | | |
|---|---|---|---|---|---|
| | TS1 | TS2 | TS7 | SL2 | SL3 |
| Gram | + | + | + | + | + |
| Cell form | rod-shaped | rod-shaped | rod-shaped | rod-shaped | rod-shaped |
| Motility | - | + | - | - | - |
| Oxsidase | - | + | - | - | - |
| Catalase | + | - | - | - | + |
| Citrate | - | + | - | - | - |
| TSIA | As/As,G-H2S- | Alk/As,G-H2S- | As/As,G-H2S- | As/As,G-H2S- | As/As,G-H2S- |
| VP | + | - | + | + | - |
| Spore | + | - | + | + | - |
| Nitrate | + | - | - | + | - |
| Lysin | + | + | - | + | + |
| Ornithin | - | - | - | - | - |
| $H_2S$ | - | - | - | - | - |
| Glucosa | + | + | - | + | + |
| Mannitol | + | - | - | + | - |
| Xylose | - | - | - | - | - |
| ONPG | - | + | + | - | + |
| Indole | - | - | - | - | - |
| Urease | - | - | - | - | - |
| V-P | - | - | - | - | - |
| TDA | - | - | - | - | - |
| Gelatin | - | - | - | - | - |
| Malonat | - | - | - | - | - |
| Inositol | - | - | + | - | - |
| Rhamnose | - | - | - | - | - |
| Sucrosa | + | - | - | + | - |
| Lactosa | - | - | - | - | - |
| Arabinose | + | + | - | + | + |
| Adonitol | - | - | - | - | - |
| Raffinose | - | - | - | - | - |
| Salicin | - | - | - | - | - |
| Arginin | - | - | - | - | - |
| Catalase | + | - | + | + | - |
| Coagulase | - | - | - | - | - |
| Hemolysis | β | α | β | β | α |
| Novobiosin sensitivity | - | - | - | - | - |
| Starch hydrolysis | - | - | - | - | - |

| | | | | | |
|---|---|---|---|---|---|
| Casein hydrolysis | + | - | + | + | - |

The microbact analysis revelaed that the five bacteria for TS1, TS2, TS7, SL2 and SL3 were predicted as *Bacillus pumilus*, *Pseudomonas* sp., *Bacillus amyloliquefacien*, *Bacillus alvei* and *Bacillus coagulant*. The best cellulose producer was *Pseudomonas* sp. Hence, only this isolate were further comfirmed by using 16s rDNA analysis. The result of the identification revealed that Pseudomonas sp. was new strain of *Pseudomonas aeruginosa.* It has 91% similar identity of *Pseudomonas aeruginosa strain ZJHG29.* Hence, we named this isolate as *Pseudomonas aeruginosa* UBB Phylogenetic tree of this isolate was depicted in figure 2.

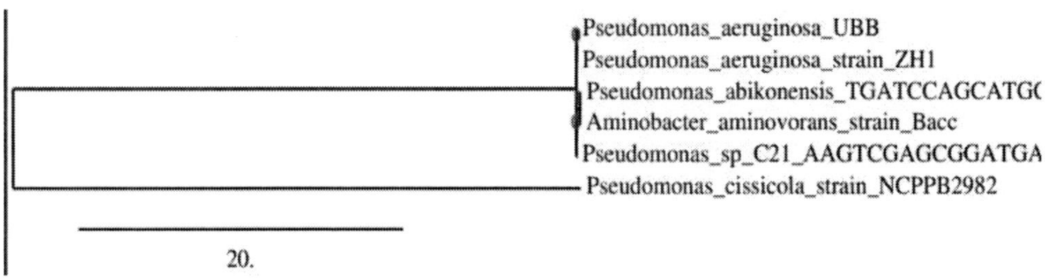


**Figure 2.** Phylogenetic tree of *Pseudomonas aeruginosa* UBB.

Bacillus bacteria in mangroves is important to degrade organic matter into nutrients. This is supported by the statement which explains that Bacillus bacteria are a phosphate decomposer. Phosphate (P) decomposing bacteria play a role in the re-provision of phosphate compounds in the seagrass ecosystem by decomposing the litter. The bacteria will enzymatically decompose the cellulose material through the active role of proteolytic, cellulolytic and chitinolytic enzymes [9]. The Bacillaceae (60 %) family is the predominant family in mangrove rhizospheres in the southern part of the Gulf of Kachchh, Gujarat and followed by the Rhodobacteraceae family constituting of 11.43 %; Micrococcaceae constitutes 8.57%; and the Nocardiaceae and Pseudoalteromonadaceae families constitute 5.71 % [10].

Cellulolytic bacteria have an important role in the nutrients providing for mangrove. Bacteria are an important source of cellulase with a variety of industrial and biotechnological applications. Non-haemolytic bacterial strains tolerant of various environmental pollutants (heavy metals and organic solvents) are identified as *Bacillus* sp [11].

The Bacillus genus were dominant of cellulolytic bacteria found in the mangroves ecosystem of Bangka Island in the biochemical characterization. They are *Bacillus pumilus*, *Bacillus amyloliquefacien, Bacillus alvei,* and *Bacillus coagulant*. Bacillus sp effectively reduced the crude fiber of palm oil cake from 17.74 % to 5.8 % through a fermentation process [12, 13]. *Bacillus pumilus* which was isolated from the mangrove ecosystem was identified with the potential production highest in chitinase [14]. *Bacillus pumilus* bacteria can be utilized in rumen cow fermentation to enrich the culture media of zooplankton [15].

The results of 16s rDNA analysis revealed that the bacteria is *Pseudomonas aeruginosa* UBB. The bacteria was closely related to the *Pseudomonas aeruginosa* strain ZH1. *Pseudomonas aeruginosa* may be present in contaminated mangrove sediments as in the opinion [12], that the *Pseudomonas aeruginosa* strain KVD-HM52 was isolated from oil contaminated mangrove sediments which showed significant antifungal activity against Fusarium oxysporum wilt disease in tomato plants using the purified RL biosurfactant .

## 4. Conclusion

Cellulolytic bacteria can be found in Sungailiat and Tukak Sadai island. Five bacteria, namely *Bacillus pumilus, Pseudomonas* sp., *Bacillus amyloliquefacient, Bacillus alvei, Bacillus coagulant.* Molecular identification revealed that the best producer for cellulose is *Pseudomonas aeruginosa* UBB..

## 5. References

[1] Rigonato J, Kent AD, Gumiere T, Branco L H Z, Andreote F D, Fiore M F 2017 *Braz. J. Microbiol.* 1–10
[2] Chantarasiri A. Aquatic 2015 *Egypt J. Aquat. Res.* **41** 57–64.
[3] Nurrochman F 2015 *Exploration of cellulolytic bacteria from land mangrove Forest Baros Kretek, Bantul* (Yogyakarta: Publication manuscript) Faculty of Teacher Training and Education Muhammadiyah Surakarta University
[4] Maulani S H 2014 *Isolation and identification of cellulolitic bacteria from land of estuary river mangrove Gunung Anyar* (Surabaya: ADLN Perpustakaan Universitas Airlangga)
[5] Prihanto A A, Wakayama M 2016 *Adv. Food Nutr. Res.* **79** 1–25.
[6] Prihanto A A, Jaziri A A, Perwira I Y *Biosci. Biotech. Res. Asia* **13** 1409-1413
[7] Ventorino V, Amore A, Faraco V, Blaiotta G, Pepe O 2010 *J. Biotechnol.***150** 181–181 [8] Koeck D E, Pechtl A, Zverlov V V, Schwarz W H 2014 *Curr. Opin. Biotechnol.* **29** 171–83.
[9] Yahya Y, Nursyam H, Yenny R, Soemarno S 2014 *Ilmu kelautan* **19** 35–42 [10] Sakhia N, Prajapati S, Shetty V, Bhatt S, Bhadalkar A. 2016 *J. Mar. Sci.* **6** 23–31 [11] Thomas L, Ram H, Singh VP. 2017 *Braz J Microbiol* 1–14.
[12] Sastrawibawa S, Safitri R. 2012 Crude Fiber In Feed Material From Agricultural Waste **2** 100–5.
[13] Pamungkas W and Ikhsan K. 2007 *Proc. of the Aquaculture Technology Innovation Forum* 2010 769–74
[14] Rishad K.S. Sharrel Rebello, P.S. Shabanamol , M.S. Jisha. 2017 *Pestic. Biochem. Physiol.* **137** 36-41
[15] Masithah E D, Ariesma N, Cahyoko Y 2011 *Jurnal Kelautan* **4**
[16] Deepika K V, P Ramu Sridharb, P V Bramhacharia 2015 *Biocatal Agric Biotechnol* **4** 608-615

## Acknowledgement

This research was part of a university research cooperation in 2017 between Bangka Belitung University and Brawijaya University which was supported by the Directorate General of Research and Technology Research and Development at the Ministry of Research, Technology and Higher Education. This publication was conducted during the LPDP doctoral scholarship. The researchers would like to thank the students of aquaculture at Bangka Belitung University and fishery technology students at Brawijaya University who assisted with the research.

IOP Conf. Series: Earth and Environmental Science **137** (2018) 012071 doi:10.1088/1755-1315/137/1/012071

# The characterization of edible coating from tilapia surimi as a biodegradable packaging

**E Saputra, A Alamsjah and A A Abdillah**

Marine Department, Fisheries and Marine Faculty, Airlangga University.

E-mail: ekasaputra@fpk.unair.ac.id

**Abstract.** One of the problems that often arise in the fisheries sector is maintaining the quality. In the room temperature, the fish more quickly enter the phase of rigor mortis and lasted shorter. The retention of fresh fish can be extended by adding antibacterial compounds in the form of synthetic chemicals or natural ingredients. One of the safe natural ingredients used to extend the freshness of the fish is the edible coating. Edible coatings may be composed of hydrocolloid, lipids and composites. In the food industry surimi can be used as an ingredient to make edible packaging or better known in the form of edible film and protein-based edible coating. Edible film and potential coatings are used as packaging materials as they may affect food quality, food safety, and shelf life. Protein-based edible film have superior inhibitory and mechanical properties compared to polysaccharide-based ones. This is because protein contains 20 different amino acids and has most special characteristics that produce functional characteristics when compared with polysaccharides used as an ingredient in edible film and coating making most homopolymers.

## 1. Introduction

One of the problems that often arises in the fisheries sector is maintaining quality. The quality of fish can be maintained if the fish is handled carefully, clean, stored in a room with refrigeration, and quick. At room temperature, the fish enters the phase of rigor mortis more rapidly and does not last for long. If the rigor phase can no longer be maintained, then decay by enzyme and bacterial activity will take place more quickly. The activity of enzyme and bacteria causes change very rapidly so that the fish enters the phase of post rigor. This phase indicates that the quality of fish is low and unfit for consumption in terms of maintaining the quality of fish and its relation to shelf life. The longer the shelf life of a fish, the quality of the fish will be better maintained.

The retention of fish freshness can be extended by adding antibacterial compounds in the form of synthetic chemicals or natural ingredients. These compounds can diffuse into the surrounding environment and inhibit or stop bacterial growth. Antibiotic substances are grouped into antibiotics that are effective against several types of bacteria (narrow spectrum) and antibiotics that are effective against many types of bacteria (broad spectrum). Synthetic antibiotics such as tetracyclines have been banned for health reasons, hence no more effective antibiotic agents are used in handling fish catches. The use of natural ingredients can be a solution that is not harmful to health.

One of the safe natural ingredients used to extend the freshness of fish is the edible coating. Edible coatings may be composed of hydrocolloid based (proteins, polysaccharides), lipids (fatty acids, acyl glycerol, waxes or waxes) and composites (mixed hydrocolloids and lipids) [1]. Surimi in the food industry can be used as an ingredient to make edible packaging or is better known in the form of edible film and protein-based edible coating. Edible film and other potential coatings are used as packaging materials as they may improve food quality, food safety, and product shelf life.

In addition to acting as inhibitors of mass diffusion (moisture, gas, volatile), edible films and coatings also serve as food and additive carriers including flavors, antioxidants, vitamins and dyes, as well as to improve food handling [2]. Edible films and protein-based coatings have superior inhibitory and mechanical properties compared to polysaccharide-based ones. This advantage is due to the fact that proteins contain 20 different types of amino acids and have special characteristics that produce functional characteristics, which is more varied when compared with polysaccharides used in the preparation of edible films and coatings, which are mostly homopolymers.

## 2. Methodology

### *2.1. Time and place of reasearch*

This research was held on August 2016 in the Laboratory of Aquaculture, Faculty of Fisheries and Marine, Airlangga University.

### 2.2. *Tools and ingredients*

The tools used in this research were tilapia fish fillets obtained from several markets in Surabaya city, water, salt, ice, cryoprotectant (sugar), filter paper, packing materials such as cling film, styrofoam and chemicals used for physical and chemical analysis. Other tools used include tools for manufacturing surimi such as knives, scales, plastic trays, grinders, gauze, cutting boards, cool boxes, refrigerators for storage, containers for coating, and equipment for proximate analysis, amino acids, TVB, pH meter, aw meter, TPC, and viscometer.

### 2.3. *Working procedures*

This research was conducted in two stages: preliminary research and the main research. The preliminary research included the testing of raw materials used as the basic material for edible coating. Main studies include the manufacture of surimi from fish waste filet and various surimi washing (1, 2, 3) of the edible coating. The frozen surimi is thawed for 20 minutes. The resulting edible coating of surimi was then analyzed for its viscosity.

### *2.4. Test procedures*

*a. Potential hydrogen value*

pH measurement was done using a digital pHmeter. Prior to use, the pH meter was rinsed with distilled water and dried with a tissue. It was further calibrated by the use of the buffer solutions pH 4 and pH 7. The tool was dipped in the buffer and was allowed a moment to steady.

*b. Total volatile base value (TVB)*

The Total Volatile Base Test is one of the methods of measurement to determine the freshness of fish \based on the evaporation of the basic compounds. TVBN analysis is conducted by weighing a sample of 100 grams and adding 300 mL of 7 % TCA and then mashed. The solution was filtered with filter paper to obtain a clear filtrate. To make the distillation, the distillate was accommodated with 15 mL of HCl 0.01 M. A few drops of phenol red indicator was added in the distillate and then titrated with NaOH 0.01 M until it turned pink.

*c. Total plate count value (TPC)*

The microbiological test was done by calculating the number of microbes in the sample by diluting as necessary and was done in duplicate. A mixture of 1 mL was taken and placed into a tube containing 9 mL of sterile 0.85 % saline solution in order to obtain dilution 10-2. A similar procedure was then performed for the dilution of 10-3 and so on up to the 10-5 dilution. To sterile, the sample was placed into a sterile petri dish and allowed to clot. 0.1 mL of the diluted sample is pipetted on the agar

IOP Conf. Series: Earth and Environmental Science **137** (2018) 012071 doi:10.1088/1755-1315/137/1/012071

surface. The sample was leveled on the surface of agar using a sterile glass rod and incubated at 10 °C for 5 days.

d. *Water content*
The cleansed porcelain cup is then dried in the oven for 1 hour at 105$^0$C, then cooled in a desiccator for 30 minutes and weighed (A gram). The 2 grams of smoothed sample were weighed in a cup (B gram) and then dried in an oven at 105$^0$C for 6 hours. Next, the sample was chilled with the desiccator for 20 minutes then weighed several times until the fixed weight was obtained (C gram) [3].

$$\text{Water content} = \frac{(B-C)}{(B-A)} \text{ x } 100\% \tag{1}$$

### 2.5. *Data analysis*
This research used Complete Randomized Design (CRD) that consists of four treatments and three replications. The treatment in this research is the edible coating on various surimi concentrations (2, 4, 6, and 8 %). Data analysis in this research utilized Analysis of Variance (ANOVA) and continued with the Duncan's Multiple Range Test to determine which treatment gives a different effect compared to the other treatments [4].

## 3. Results and Discussions
The pH values of the washing effect at one, two and three washings are 7.01, 6.54 and 6.45 respectively. The value is still considered in a good gel-producing category, although the pH value decreases with the amount of washing. The degree of acidity (pH) of the surimi at one-timewashing was still close to the normal pH state. According to Babji and Kee [5], the high pH value at one-time washing is caused by the loss of acid residues (lactic acid as an anaerobic glycolysis process) in muscle protein due to the washing effect [5]. It is also asserted that pH value is very important in relation to gel formation, where the gel formation process will be difficult if the pH value is below 6 [6]. The average pH value of tilapia surimi can be seen in table 1.

**Table 1.** The average value of pH surimi of tilapia.

| Washing | pH value |
|---|---|
| 1 | 7.01 |
| 2 | 6.54 |
| 3 | 6.45 |

The pH value affects the strength of the gel (ashi). The gel strength will be high if the pH of the meat ranges from 6.0 to 7.0 because the myosin protein is easily soluble in the pH range. The gel strength will be lower outside the pH range, either in a more alkaline state (pH> 7) or in more acidic state (pH <6) [7].

The Total Volatile Base Test is one of the methods of measurement to determine the freshness of fish based on the evaporation of the basic compounds. The higher the value of TVB shows the decline in meat quality. The TVB value of the surimi-coated tilapia fillet during storage ranged from 9.03 up to 24.44 mg N/100 g sample. The TVB value of the tilapia fillet increased with the length of storage time. The increased TVB value of fish during storage occured due to the degradation of protein or derivatives, which produces a number of volatile bases such as ammonia, histamine, hydrogen sulfide, and the foul-smelling trimethylamine. The average value of TVB for the fillet of tilapia can be seen in table 2.

**Table 2.** The average value of TVB fillet of tilapia with surimi coating solution for room temperature storage.

| Storage time (hours) | Coating surimi |
|---|---|
| 0 | 9.03 |
| 6 | 12.56 |
| 12 | 20.54 |
| 18 | 24.44 |

The TVB value for the fillet of tilapia obtained during these observations was still categorized as fit for consumption as it is below the standard value of TVB, which is 30 mg N/100 g sample. This refers to the standard freshness of the fish based on the value of TVB. An increase in the value of TVB during storage was due to the degradation of the protein, resulting in a number of volatile bases such as ammonia, histamine, and trimethylamine. Surimi has a positively charged polikation able to bind to proteins, one of which is an enzyme. Surimi which binds to the enzyme is able to minimize the action of the enzyme. The binding of the enzyme by surimi only takes place on the surface of the fillet coated with surimi.

The number of bacteria that grew on the tilapia fillet sample ranged from $1.34 \times 10^4$ to $7.55 \times 10^6$ colonies/g sample. The average results of the analysis of microbes on the tilapia fillet with chitosan coating during storage at room temperature are presented in table 3.

**Table 3.** The average value of TPC fillet of tilapia with chitosan coating solution for room temperature storage.

| Storage time (hours) | Coating surimi |
|---|---|
| 0 | $1.34 \times 10^4$ |
| 6 | $2.54 \times 10^4$ |
| 12 | $3.76 \times 10^5$ |
| 18 | $7.55 \times 10^6$ |

Surimi has an antibacterial ability in inhibiting the growth of microbes. A thin layer (edible coating) of surimi covering the entire surface of the fish will inhibit the entry of O2 and water through the surface of the fish's body and can hinder the evolvement of microbes [8].

The water content of surimi after the washing effect is presented in table 4. Based on the surimi test, the water content ranged from 82.42 % to 89.42 %. The average value of the water content of tilapia surimi can be seen in table 4.

**Table 4.** The average value of water content of tilapia surimi.

| Washing | pH value |
|---|---|
| 1 | 89.42 |
| 2 | 88.45 |
| 3 | 84.37 |

IOP Conf. Series: Earth and Environmental Science **137** (2018) 012071 doi:10.1088/1755-1315/137/1/012071

## 4. Conclusion

From the research conducted, results show that washing and storage influenced the surimi and kamaboko produced, where there was a decrease of pH value at each increased washing. More research needs to be done on the storage of whole fish at chilling temperatures according to the K-value method to determine more specific levels of fish freshness. Moreover, additional analysis need to be carried out for further research including proximate analysis, amino acid content, free fatty acids, and the activity of katepsin enzymes. It is also advised to undertake research to test the activity of surimi against the whole fish at chilling temperature storage with the K-value method to see more specific levels of the freshness fish. Other additional analyses are needed for future studies including studies on free fatty acids, and enzyme activity katepsin.

## 5. References

[1]. Gennadios, A dan L W Curtis 1990 *J. Of Food Tech.*

[2]. Krochta J M, Baldwin, Carrriedo 1994 *Edible coating and films to improve food quality Technomic Pub. Ca. Inc.* Lancaster.

[3]. [AOAC] Association of Official Analytical Chemist 1995 *Official Methods of Analysis.* Washington DC.

[4]. Kusriningrum R S 2012 *Experiment design* University of Airlangga Surabaya p 43-63 [Text in Indonesia]

[5]. Babji A S, Kee G S 1994 *Asean Food Journal* **9** 63-68.

[6]. Suzuki T 1981 *Fish and krill protein in processing technology* (London: Applied Science. Publishing. Ltd.)

[7]. Shimizu Y, Toyohara H, Lanier T C 1992 *Surimi production from fatty and dark Flesh sp. Dalam Lanier TC dan Lee CM (ed) Surimi Technology* (New York: Marcel Dekker, Inc.)

[8]. El Ghaouth A, Grenier J A, Benhamoun N, Asselin A, Belenger 1994 *Phytopathology* **84**: 3.

IOP Conf. Series: Earth and Environmental Science **137** (2018) 012072 doi:10.1088/1755-1315/137/1/012072

# Effect of maggot (*Hermetia illucens*) flour in commercial feed on protein retention, energy retention, protein content, and fat content in tilapia (*Oreochromis niloticus*)

**D R Kurniawan[1], M Arief[2], Agustono[2] and M Lamid[2,3]***

[1]Aquaculture, Faculty of Fisheries and Marine, Airlangga University, Surabaya 60115
[2]Department of Fish Health Management and Aquaculture, Faculty of Fisheries and Marine, Airlangga University, Surabaya 60115
[3]Department of Animal Husbandary, Faculty of Veterinary Medicine, Airlangga Univeristy, Surabaya 60115

E-mail:mirnylamid@yahoo.com

**Abstract.** Nile tilapia is a fish with a nature of breeding throughout the year and has a high adaptation. Feed is one of the most important factors that will affect the growth and the sustainability of fish. Maggot is one of the alternative feeds which has function as source of protein.The aim of this research is to ascertain the influence of maggot (*Hermetia illucens*) flour substitution on commercial feed which affect the retention of energy, retention of protein, and also rough fat and protein of Nile tilapia flesh. The treatment was to substitute the commercial feed with different doses of maggot flour. In this study, the amount of maggot flour substituted on commercial feed is as follows: treatment P0 (0 %), P1 (12 %), P2 (14 %), P3 (16 %) and P4 (18 %). The parameter observed in this study is an effect on to protein retention, energy retention, and also rough fat and protein in Nile tilapia flesh. The result showed the highest protein content was gained from the treatment P4 with 17.267 %. The lowest protein content was 16.344 % obtained in the treatment P0. There was an increasing protein content in Nile tilapia along with the increase of the maggot flour addition in each treatment.

## 1. Introduction

One of the freshwater fish commodities favored by consumers considerably is Nile tilapia (*Oreochromis niloticus*). Nile tilapia (*Oreochromis niloticus*) is a fish with a nature of breeding throughout the year and has a high adaptation [1]. The production of Nile tilapia fishery has enhanced significantly from year to year, from 567.078 tons in 2011 to 695.063 tons in 2012 and 909.016 tons in 2013 [2]. Nile tilapia is a freshwater fish that contains high protein and has a preeminence of breeding rapidly. Nutritional content of Nile tilapia are 16-24 % of protein, 0.2-2.2 % of fat, carbohydrate, minerals, and vitamins [3]. According to Venugopal [4], fish is a good source of nutrients to fulfill nutritional needs of humans since fish contains protein, vitamins, minerals, calcium, taurine, and it is a source of good fats such as omega 3 (EPA and DHA) and omega 6.

IOP Conf. Series: Earth and Environmental Science **137** (2018) 012072 doi:10.1088/1755-1315/137/1/012072

Feed is one of the most important factors that will affect growth and sustainability of fish life [1]. Maggot is one of the alternative feed which has a function as a source of protein, maggot contains about 45.01 % protein [5]. Maggot has amino acids, 44.26 % protein, 29.65 % fat, and 0.7 % linoleic acid and 2.24 % linolenic acid as the source of EPA and DHA [6]. Maggot can be given in fresh form to the fish and also in a pellet form. It can be processed as maggot flour before it is formed as a maggot pellet. The maggot which is made into a pellet must be steamed with a certain time where the time of maggot steaming process could affect the nutritional content of the maggot pellet [5].

To improve the quality of the fish meat, one thing to do is the application of feed with a balanced nutrition. Protein is one of the important nutrients for fish and it needs to be fulfilled to achieve optimal growth. On the other side, according to Diana [7], protein inside the body is the main source of energy other than carbohydrates and fats, it also functions as building substance, and as regulatory substance. Proteins regulate metabolic processes in the form of enzymes and hormones, as well as the defense mechanism of the body against various microbes and other toxic coming from outside the body, along with retaining cells and body tissue.

The inclusion of maggot flour in commercial feed is expected to be an alternative feed for Nile tilapia farmers to yield efficient feed cost, and high retention of energy, retention of protein, protein content, and fat content of Nile tilapia flesh.

## 2. Methodology

### *2.1. Maggot flour making process*

Maggot (*Hermetia illucens*), which was cultivated for 17 days, was obtained from Puspa Agro Sidoarjo. The maggot flour making process was as follows: (1) 17 days old maggots were harvested, then it was cleaned to remove dirt, (2) clean maggots were boiled for 5 minutes, and subsequently squeezed using muslin cloth to reduce water and fat content, (3) afterward, maggots were steamed for 20 minutes at 80 °C, (4) after steaming, the maggots were put into the oven at 50v°C for 2-3 days, and (5) dried maggots were mashed with a blender and sieved to obtain maggot flour which was ready to use as feed ingredients.

### *2.2. Feed making preparation*

The first step of feed making process is to calculate the composition of each feed ingredient in every treatment. The Second step was mixing the ingredients gradually starting from the lightest weight to the heaviest weight into a tub and stirred until it mixed well before added with water. The third step was, after the feed ingredients were evenly mixed, they were put into the pellet maker machine. After the pellet feed had been shaped with the diameter of 1-2 mm, they were either chilled or sun-dried. Lastly, the pellet feed was ready to be given to Nile tilapia.

### *2.3. Data analysis*

Data was analyzed using Analysis of Variance with Completely Randomized Design. When there were any differences, then Duncan's Multiple Range Test was used with a degree of confidence 0.05 to know the differences among treatments [8].

## 3. Results and Discussion

### *3.1. Protein retention*

The results showed that protein retention of Nile tilapia (*Oreochromis niloticus*) flesh ranged between 58.217 % - 91.505 %. The average data of protein retention of Nile tilapia fish flesh can be seen in table 1.

IOP Conf. Series: Earth and Environmental Science **137** (2018) 012072 doi:10.1088/1755-1315/137/1/012072

**Table 1**. The average of nile tilapia fish flesh protein retention.

| Treatments | Protein Retention ± SD | Transformation($\sqrt{y + 0.50}$)±SD |
|---|---|---|
| P0 (0 %) | $89.322^{b}$± 13.439 | $9.456^{b}$± 0.727 |
| P1 (12 %) | $73.840^{ab}$ ± 14.978 | $8.587^{ab}$± 0.897 |
| P2 (14 %) | $91.505^{b}$± 22.274 | $9.538^{b}$± 1.165 |
| P3 (16 %) | $61.322^{a}$± 11.295 | $7.837^{a}$± 0.722 |
| P4 (18 %) | $58.217^{a}$± 14.210 | $7.621^{a}$± 0.920 |

Note: Different superscripts in the same column are indicated (p <0.05)differences

The highest protein retention (91.505 %) was obtained from the Treatment P2 with 14 % of maggot flour substitution. The lowest protein retention was 58.217 % that was found in Treatment P4 with 18 % of maggot flour substitution. Treatments P1 and P3 showed an enhancement in protein retention of Nile tilapia fish flesh, yet the protein retention in P0 (89.322 %) was still greater than P1 and P3. The treatment P2 showed the highest protein retention (91.505 %) compared to other treatments (P0, P1, P3, and P4). results of the Analysis of Variance (ANOVA) showed a significant difference (P<0.05) between each treatment.

Protein retention indirectly describes the amount of feed protein consumed and used to build the body's protein tissue (beneficial for growth) [9]. Based on the results of statistical analysis, it indicated that maggot substitution on commercial feed significantly affected (P<0.05) protein retention in Nile tilapia. This occurred because the proteins in the feed with high biological values will lead the protein in the body to have good biological value as well [10].

The highest protein retention rate was found in treatment P2 with 14 % maggot flour that contained 9.538% feed protein. This exemplifies that feeding Nile tilapia fish with 29.8677 % protein in resulted in 9.538% protein retention.

### *3.2. Energy retention*

Energy retention in Nile tilapia fish flesh (*Oreochromis niloticus*) ranged between 26.312 kcal/kg - 47.672kcal/kg. The average data of energy retention can be seen in table 2.

**Table 2**. The average energy retention of Nile tilapia fish flesh.

| Treatments | Energy Retention ± SD | Transformation($\sqrt{y + 0.50}$)±SD |
|---|---|---|
| P0 (0 %) | $47.672^{c}$± 14.926 | $6.876^{c}$± 1.084 |
| P1 (12 %) | $30.112^{ab}$± 4.274 | $5.522^{ab}$± 0.397 |
| P2 (14 %) | $42.485^{bc}$± 10.692 | $6.517^{bc}$± 0.826 |
| P3 (16 %) | $27.217^{a}$± 4.567 | $5.251^{a}$± 0.433 |
| P4 (18 %) | $26.312^{a}$± 6.163 | $5.152^{a}$± 0.590 |

Note: Different superscripts in the same column are indicated (p<0.05) differences

The highest energy retention was produced from the control treatment (P0), which was 47.672 %. The lowest energy retention was 26.312 % found in treatment P4 with 18% maggot flour substitution. Treatments of P1 and P3 showed an enhancement of energy retention, yet the energy retention in P2 (42.485 %) was still higher than P1 and P3. Treatment P0 showed the highest energy retention by 47.672% compared to other treatments. Analysis of Variance (ANOVA) showed significant difference ($P<0.05$) among treatments.

The use of energy in Nile tilapia is influenced by the amount of feed consumed. Energy is obtained from reshuffling the chemical ligament through the oxidation reaction process towards feed components, specifically on proteins, fats, and carbohydrates to produce simpler compounds (amino acids, fatty acids, and glucose), so that it can be absorbed by the body to be either used or saved. Energy retention is a depiction of the amount of energy stored in tissues in the fish's body divided by the amount of energy in the feed consumed [11].

Average energy retention ranged from 26.312 kcal/kg–47.672 kkal/kg, with the highest energy retention found in control treatment (P0). Analysis of Variance (ANOVA) showed significant differences ($p<0.05$) among treatments. This signifies that the energy content in feed comes from proteins, causing it to be used to synthesize proteins in the body instead of as energy storage.

### *3.3. Crude fat*

Crude fat of Nile tilapia fish flesh ranged from 3.572% - 4.779%. Analysis of Variance (ANOVA) showed a significant difference ($P<0.05$) among treatments. Average data of fat content can be seen in table 3.

**Table 3.** The average of crude fat content of Nile tilapia (*Oreochromis niloticus*) flesh.

| Treatments | Rough Fat Content ±SD | Transformation $\sqrt{y+0.5}$±SD |
|---|---|---|
| P0(0 %) | $3.574^a \pm 0.288$ | $2.017^a \pm 0.071$ |
| P1(12 %) | $3.572^a \pm 0.305$ | $2.016^a \pm 0.075$ |
| P2(14 %) | $3.898^a \pm 0.301$ | $2.096^a \pm 0.071$ |
| P3(16 %) | $4.772^b \pm 0.137$ | $2.295^b \pm 0.029$ |
| P4(18 %) | $4.779^b \pm 0.321$ | $2.296^b \pm 0.069$ |

Note: Different superscripts in the same column are indicated ($p<0.05$) differences

The highest crude fat content was obtained from treatment P4 with 4.779 % with 18 % maggot flour substitution. The lowest crude fat content was 3.572 % that was found in treatment P2 with 12% maggot flour substitution. The Duncan test results showed that substitution of maggot flour in commercial feed had increased crude fat content in Nile tilapia, but the substitution of 14 % maggot (P1) had decreased the crude fat content which was not significantly different with P0 and P2. it can be seen that the substitution of maggot flour in commercial feed can increase the crude fat content of Nile tilapia fish flesh between 3.572 % - 4.779 %. Nile Tilapia is a low fat type of fish with 2.54 % fat content [12]. The increase in crude fat of Nile tilapia was caused by high content of fat in maggot flour that was 13.945 %. The commercial feed ration substituted with maggot flour yield a crude fat content of 11.242 % - 11.709 %. A proper growth will be generated when the fat content in the fish ration is between 4 – 8 %, but the content of 10 – 11 % of crude fat in the feed is still tolerable by the fish in general [13]. The decrease of the crude fat content in treatment P1 was caused by the low temperatures during the maintenance. Fat content decreases at low temperatures and increases at high temperatures. Water temperature greatly affects the

IOP Conf. Series: Earth and Environmental Science **137** (2018) 012072 doi:10.1088/1755-1315/137/1/012072

activity of the digestive tract of Nile tilapia [12]. Nile tilapia is a type of thermopilc animal whose feeding ability will rise according to the increasing of the water temperature [14].

### *3.4. Protein*

Protein content of Nile tilapia fish ranged between 16.344 % - 17.267 %. Analysis of Variance (ANOVA) showed a significant difference of ($P<0.05$) among treatments. Protein content of tilapia fish can be seen in table 4.

**Table 4**. The Average protein content of nile tilapia fish flesh.

| Treatments | Protein Content ± SD | Transformation $\sqrt{y+0.5}$ ± SD |
|---|---|---|
| P0(0 %) | $16.344^{a} \pm 0.252$ | $4.104^{a} \pm 0.030$ |
| P1(12 %) | $16.876^{a} \pm 0.518$ | $4.168^{a} \pm 0.061$ |
| P2(14 %) | $17.061^{ab} \pm 0.270$ | $4.190^{ab} \pm 0.032$ |
| P3(16 %) | $17.094^{b} \pm 0.647$ | $4.194^{b} \pm 0.077$ |
| P4(18 %) | $17.267^{b} \pm 0.093$ | $4.215^{b} \pm 0.011$ |

Note: Different superscripts in the same column are indicated ($p<0.05$) differences

The highest protein content was gained from the treatment P4 with 17.267 %. The lowest protein content was 16.344% obtained in the treatment P0. There was an increasing protein content in Nile tilapia along with the increase of the maggot flour addition in each treatment. maggot flour substitution in commercial feed can increase the protein content of Nile tilapia fish with the range between 16.344 % - 17.267 %.

Nile tilapia is the type of fish which has a low protein (16 %) [3]. The increase of protein content in Nile tilapia found in this study was caused by the high content of protein in the maggot flour (29.8 %). The commercial feed that was substituted with maggot flour gained 29.22 % - 29.29 % of protein content. Protein in the Nile tilapia flesh indirectly represented the amount of protein in the feed consumed and used to build tissues in the body and for growth [9]. Fish proteins is influenced by the feeding quality and their energy rate. the right balance of protein energy with the right amount of feeding will produce the best growth, feed conversion, and feeding efficiency. The amount of protein in the feed will affect the feed consumption leading to the level of protein absorption in the fish flesh [15].

## 4. Conclusion

The current study has demonstrated that the addition of maggot flour can improve protein retention, energy retention, protein content, and fat content in Nile tilapia. The treatment P4 with the addition of maggot flour 18% has effect on increasing the fat and protein contents of 4.779% and 17.267%, respectively.

## 5. References

[1] Anggriani I and Ankiq T 2012 *J. Per. Kel.* **3**
[2] Directorate General of Aquaculture-Ministry of Marine Affairs and Fisheries 2014 Aquaculture Production Supports National Economy and Increases Welfare
[3] Widaryati R 2016 *J. Ilm. Hewani Trop* **5**
[4] Venugopal 2010 *J. Apl. Tek. Pang.* **1**

IOP Conf. Series: Earth and Environmental Science **137** (2018) 012072 doi:10.1088/1755-1315/137/1/012072

[5] Haryati 2010 *Effect of substitution of fish starch with maggot flour on retention and efficiency of nutrition utilization in the milkfish body (Chanos chanos)* (Makasar : Universitas Hasanudin)
[6] Fahmi M R, Saurin, H, and I Wayan S 2009 Maggot potential as one of protein source of animal feed loka research on freshwater ornamental fish national *Seminar on Technology Support To Improve Animal Food Products In Order To Fulfill People's Nutrition*
[7] Diana F M 2010 *J. Kes. Masy.* **4**
[8] Kusriningrum R S 2012 *Experiment Design* (Surabaya : Airlangga University Press)
[9] Halver J E 1988 *Fish Nutrition* (London Academis Press, INC) p 798
[10] Johnson E G, W O Watanabe and S C Ellis 2002 *North American J. Aquacult.* **64** 47-54
[11] Haryati E, Saadedan A, Pranata 2011 *The effect of substitution rate of fish flour with maggot flour on the retention and efficiency of nutrient utilization in milkfish body* (Makassar : Universitas Hassanudin) p 31
[12] Isa M, Rinidar, Zalia T, Harris A, Sugito dan Herrialfian 2015 *J. Med. Vet.* **9** 60-63
[13] Benyamin B 2002 *The effect of using kaliandra leaf flour in an artificial feed against growth of freshwater pomfret seeds (Collosuma macropomum)* (Bandung : Unversitas Padjajaran)
[14] Turker C S 2004 *Biology and culture of channel catfish* (Elsevier) **34** p 686
[15] Fran S and Akbar J 2013 *Effect of Protein Level Difference and Feed Protein Ratio on Growth of Trichogaster pectoralis* (Banjarmasin : Universitas Lambung Mangkurat)

IOP Conf. Series: Earth and Environmental Science **137** (2018) 012073 doi:10.1088/1755-1315/137/1/012073

# Carrageenan :the difference between PNG and KCL gel precipitation method as *Lactobacillus acidophilus* encapsulation material

**D Setijawati, H Nursyam and H Salis**

Department Of Fish Processing Technology, Fisheries And Marine Sciences Faculty, Brawijaya University, Jl Veteran Malang, Indonesia

E-mail : dwisetyawati@ub.ac.id/Setijawatis@gmail.com

**Abstract.**The study on the effects of using of materials and methods in the preparation of the microcapsules *Lactobacillus acidophilus* towards the viability has been done. The research method used is experimental laboratory design. Variable research was kind of material (A) as the first factor with sub factor (A1 =*Eucheuma cottonii*) (A2 =*Eucheuma spinosum*) (A3 =mixture of *Eucheuma cottonii* and *Eucheuma spinosum* 1:1 ratio), while the second factor is a method of extraction to produce caragenan (B) with sub factor (B1 = Philipine Natural Grade modification) (B2 = KCl gel Press Precipitation). Analysis of different influences uses Analysis Of Varians followed by Fisher's test. Analysis of data uses Mini tab 16. The results shows that the kind of extraction factors and methods gave significantly different effects on the viability of *Lactobacillus acidophilus*. The highest mean of Viablity obtained in the treatment of materials with a mixture of *Eucheuma cottonii* and *Eucheuma spinosum* and used KCl Gel Press method is equal to 7.14 log (CFU / mL). It is ssuggested using of kappa-iota carrageenanmixture asencapsulation material with KCl Gel Press method on *Lactobacillus acidophilus* microencapsulation process because it treatment gavethe highest average of *Lactobacillus acidophilus* viability.

## 1.Introduction

Microencapsulation is one of the means commonly used to protect probiotic bacteria from the pressure of external conditions such as temperatures and pH. *Lactobacillus acidophilus* is a probiotic bacterium that gives health benefits. Microencapsulation process is influenced by core material, shell material and method. One of the groups for encapsulation materials originating from hydrocolloid is carrageenan [1]. Carrageenan is an extraction of red hydrocolloid such as *Eucheuma* sp., *Chondrus* sp., *Gigartina* sp. and Fuscellaran. *Eucheuma* sp. commonly found in Indonesia is *Eucheuma cottonii* and *Eucheuma spinosum*. *Eucheuma cottonii* produces kappa carrageenan, while *Eucheuma spinosum* produces iota carrageenan in different extraction condition. Extraction condition difference uses chemical substances and the methods. Several extraction methods commonly used in producing carrageenan are the Philippine Natural Grade (PNG) and KCl gel Press methods producing Semi Refine Carrageenan (SRC), whereas alcohol precipitation method produced carrageenan in the form of Refine carrageenan (RC). The differences in the use of methods in the production of carrageenan result in different qualities in terms of gel strength and

IOP Conf. Series: Earth and Environmental Science **137** (2018) 012073 doi:10.1088/1755-1315/137/1/012073

viscosity. Gel strength and viscosity can affect the formation of gel beads and stability in the production of microencapsulation. Moreover, both types of materials have different properties. *Eucheuma spinosum* produces iota carrageenan with soft gel properties and elastic Meanwhile, the gel properties of kappa carrageenan are strong, sturdy, but easy syneresis. The mixture of Kappa carrageenan and *Locust Bean Gum* will provide cohesive elastic gel texture. These properties are silmilar to the mixture of kappa and iota carrageenan [2].

Carrageenan used as the encapsulation material using emulsification method produces microcapsules in a semi-solid form [3, 4]. Probiotics encapsulated by carrageenan and xanthan gum are similar to alginate, having the ability to protect the viability of probiotics compared to no encapsulation on a condition of pH and a high concentration of bile salts [5, 6]. In another research, K-carrageenan material was used to encapsulate probiotic through emulsification methods [3,7] and the application of the product chedar cheese [7]. However, probiotic encapsulation using biopolymer materials such as calcium alginate, k-carrageenan, and gellan gum directly is still applicable in a laboratory scale [8]. The k-carrageenan microencapsulation technique using emulsification method is easier in determining the scale-up for larger productions [9]. The use of probiotic of *Lactobacillus bulgaricus* using K-carrageenan concentration of 3 % and soybean oil as the continuous phase does not need special treatments and can be applied for mass production used 2 % k-carrageenan in the process of microencapsulation *Bifidobacterium longum* with vegetable oil as the continuous phase, given no special treatment, and applied to set yogurt [10]. Meanwhile, the microencapsulation of bacterium *Bifidobacterium longum* uses k-carrageenan as the material with concentration of 2 %, vegetable oil as a continuous phase, given no special treatment, while further applications were on stirred yogurt [4]. Moreover, research on the use of encapsulation materials such as *Eucheuma cottonii* and *Eucheuma spinosum* as well as the mixture of *Eucheuma cottonii* and *Eucheuma spinosum*in the forms of kappa, iota and kappa-iota carrageenan mixture from the methods of different extraction towards the viability of *Lactobacillus acidophilus* has not been done yet.

## 2. Methodology

### *2.1. Research materials*

*Eucheuma cottonii, Eucheuma spinosum*, water, KOH, $Ca(OH)_2$, KCL, plastics, rubber, distilled water, filter cloth), cultureof *Lactobacillus acidophilus*, MRS Agar, MRS broth, saline, alcohol, NaFis, KCl3M, NaCl, distilled water, corn oil, tween 80. Autoclave, water bath, incubator, vacuumdryer, homogenizer, centrifuge, microbiology testing tools

### *2.2. Stages of Research are an experimental laboratory study*

The stages of the research are as follows:

1. The making of carrageenan using PNG method with *Eucheuma cottonii* and *Eucheuma spinosum* as materials was done through the following steps: seaweed *Eucheuma cottonii* and *Eucheuma spinosum* were weighed, and then washed, and extracted using an alkaline solution KOH 6 % for *Eucheuma cottonii* and $Ca(OH)_2$ for *Eucheuma spinosum.* After 2 hours, then, they were neutralized to pH 6-8, after that, cut, dried, and pulverized to 80 mesh.
2. The making of carrageenan using KCl Gel Press Precipitation method with *Eucheuma cottonii* and *Eucheuma spinosum* as materials wasconducted through the following steps: seaweed *Eucheuma cottonii* and *Eucheuma spinosum* were weighed, washed and put in the extraction solution using KOH 6 % for *Eucheuma cottonii* and $Ca(OH)_2$ for *Eucheuma spinosum* at a temperature of 85-90 °C for 30 minutes. After that, the materials were pulverized using a blender. Later, hot water was added with 1: 10 composition (comparison of seaweed and hot water). HCl was then added to neutralize the pH. Besides, 1.5 % KCl was also added into the filtrate. After 2 hours of extraction, the paste was filtered using a filter cloth. Then, it was frozen, dried and finally made into flour.

3. The making of microcapsules of *Lactobacillus acidophilus* was done using the emulsification method which Adhikari [4] has been modified Setijawati [11] through the preparatory steps of the carrageenan materials, kappa, iota, a mixture of kappa-iota (1: 1) at a concentration of 3 % (w / v). Then, the materials were heated at a temperature of 95 °C for 5-6 minutes and thoroughly mixed until homogeneous.The culture of *Lactobacillus acidophilus* were prepared at $10^9$ CFU / mL of 10 ml. After that, the temperature of the carrageenan in the sol condition was lowered to 42 °C. After reaching thetemperature, 10 ml culture of *Lactobacillus acidophilus*was mixed with sol carrageenan. The second step was to prepare 50 mL vegetable oil asemulsifier, heated on a hotplate stirrer to a temperature of 40 °C for 2-3 minutes. Then, the mixture of cells and carrageenan was added to the oil, stirred for 10 minutes using a speed of 250 rpm, later added a solution of KCl 3,9 M. After the oil phase was removed from the mixture, the microcapsules were harvested centrifugally. Furthermore, the microcapsules were washed using a solution of KCl 3.9M twice. The microcapsules thenwere stored at a temperature of 4 °C in semi-solid form. After that, the viability of *Lactobacillus acidophilus* microcapsules was analyzed.

*2.3. Variables of Treatment*
A1 =*Eucheuma cottonii*; B1 = the making process of carrageenan using PNG method
A2 =*Eucheuma spinosum*; B2 = the making process of carrageenan using Gel KCl Press
A3 = mixture of *Eucheuma cottonii* and *Eucheuma spinosum*

*2.4. Data Analysis*
To measure the quality of carrageenan, some tests were conducted,such as the test of gel strength, viscosity, water content, FTIR, gelling point, and melting point. Meanwhile, the viability of *Lactobacillus acidophilus* was tested using the pour method (pour plate) with MRSA media. The effect of treatment used ANOVA factorial, and the difference test used Fisher's test. Moreover, the analysis of data used Minitab software 16.

## 3. Result and Discussion

*3.1. The materials characteristics of functional groups kappa, iota caragenan using FTIR*
One of the ways to identify the quality of carrageenan is through differences in the characteristics of functional groups. The characteristics of functional groups of carrageenan *Eucheuma Cottonii* (*Kappaphyzuzalvarezii*) producing kappa caragenan and *Eucheuma spinosum* producing iota caragenan using FTIR can be seen in figure 1 and figure 2.

IOP Conf. Series: Earth and Environmental Science **137** (2018) 012073 doi:10.1088/1755-1315/137/1/012073

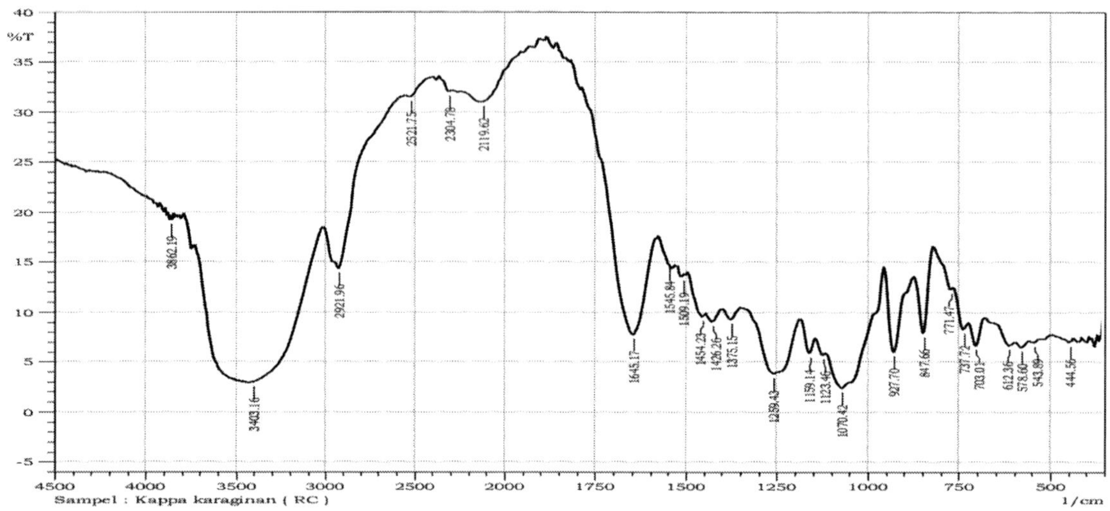


**Figure 1.** The characteristics of functional groups of Carrageenan species *Eucheuma cottonii* using FTIR.

Figure 1 shows that the functional groups results of the analysis respectively sulfate ester functional groups at the wave number of 1243 $cm^{-1}$, glycosidic bond at 977-1000.47 $cm^{-1}$, anhydrogalaktosa at 960 $cm^{-1}$, galactose sulfate at 848 $cm^{-1}$, and galactose 2 sulfate at 800-805 $cm^{-1}$. The absorption intensity indicated by sulfate ester glycosidic bond and galactose is very strong. In the fingerprint region, the absorption is sharp and broad at a wavenumber of 1210-1260 $cm^{-1}$. It is also showed by other functional groups.

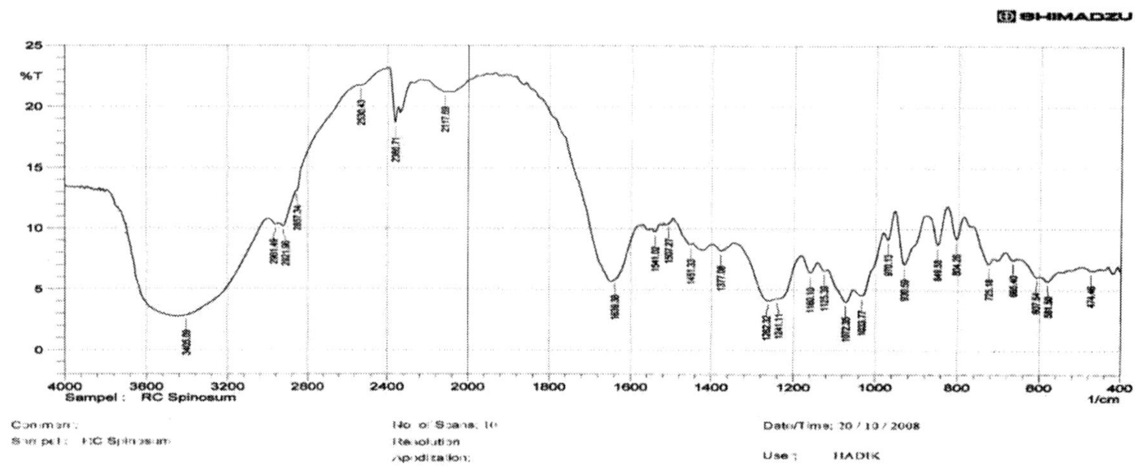


**Figure 2.** The characteristics the functional groups of carrageenan species *Eucheuma spinosum* producing iota caragenan using FTIR.

IOP Conf. Series: Earth and Environmental Science **137** (2018) 012073 doi:10.1088/1755-1315/137/1/012073

Figure 2 shows that the functional groups results of the analysis respectively sulfate ester functional groups at the wave number of 1180-1241.11 $cm^{-1}$ glycosidic bond at 1033.77 - 1072.23 $cm^{-1}$ anhydrogalaktosa at 930 9 -970.13 $cm^{-1}$ galactose sulfate at 846.56 $cm^{-1}$ and galactose 2 sulfate at 804.25 - 849.56 $cm^{-1}$. The absorption intensity indicated by sulfate ester glycosidic bond and galactose is very strong. This is supported by the results of testing IR on functional groups and testing on *Eucheuma cottonii* producing kappa carrageenan and *Eucheuma spinosum* producing iota caragenan standards using Sigma product. Sulfate ester functional group is at the wave number of 1210-1260 $cm^{-1}$ glycosidic bond at 1010-1080 $cm^{-1}$ an hydrogalaktosa at 928-933 $cm^{-1}$ galactose sulfate at 840-850 $cm^{-1}$ and galactose 2 sulfate at 800-805 $cm^{-1}$. The absorption intensity indicated by sulfate ester glycosidic bond and galactose is very strong. In the fingerprint region, the absorption is sharp and broad at a wavenumber of 1210-1260 $cm^{-1}$.

Similarly, it is showed by other functional groups. The functional groups using standard caragenan using Sigma products are shown in table 2.

**Table 1.** The wave number of carrageenan polysaccharide functional group in spectroscopy FTIR 8108 Shimadzu.

| | Wavelength of functional group ($cm^{-1}$) | | | | | | |
|---|---|---|---|---|---|---|---|
| Sample | Ester sulfate | Glycoside bond | 3,6 anhydride D-galactan | D-galactan -4 SO4 | D-galactan -2 SO4 | D-gal-6S | 3,6-anhydro galactan-2SO4 |
| Kappa Carrageenan-(standard)* | 1261.8 | 1068.7 | 929.8 | 844.9 | - | - | 802.5 |
| iota Carrageenan (standard)* | 1260 | 1072.6 | 931.7 | 848.8 | - | - | 804.4 |
| Kappa caragenan ** | 1258 | 1070 | 937.7 | 847.7 | - | - | - |
| Iota caragenan** | 1241.1-1262.32 | 1010-1080 | 930.59 | 930.59 | - | - | 804.26 |

: *[12]**Research Result

One of the ways to demonstrate the differencesof carrageenan and its type is by infrared spectroscopy, which will show caragenan functional groups and the intensity of absorption at certain wavenumbers.The absorption intensity of glycosidic bonds and ester sulfatesis very strong in all three types of carrageenan. The absorption intensity shown by 3.6-anhidrogalactose is very strong on carrageenan kappa type, while carrageenan iota type is very weak, and the type of carrageenan lambda is completely none. In addition, the importance of determining these types relates to the use of carrageenan in various industries. Carrageenan with different types provides different physical properties, especially viscosity and gel strength, which in turn affect the applications in food, beverages and other industries [13].

IOP Conf. Series: Earth and Environmental Science **137** (2018) 012073 doi:10.1088/1755-1315/137/1/012073

*3.2. The physicochemical quality of encapsulation material*

The results of the analysis of the physicochemical quality identification in carrageenan-making process using PNG and Gel KCl Press methods of seaweed species *Eucheuma cottonii* and *Eucheuma spinosum* can be seen in Table 2.

**Table 2.** The physicochemical quality of kappa, iota, kappa-iota carrageenan with different methods of extraction.

| quality | Eucheuma cottonii | | Eucheuma spinosum | | Mixture of Eucheuma cottonii and Eucheuma spinosum (1:1) | |
|---|---|---|---|---|---|---|
| | PNG | KCL Gel Press | PNG | KCl Gel Press | PNG | KCl Gel Press |
| Water content (%) | 10.64 | 9.61 | 11.72 | 11.04 | 10.67 | 10.54 |
| Ash (%) | 1.27 | 1.37 | 1.33 | 1.6 | 1.4 | 1.2 |
| Sulfate (%) | 24.2 | 21.71 | 28 | 25.77 | 24.5 | 24.2 |
| Gel strength (N) | 1.06 | 1.18 | 0.6 | 0.9 | 0.94 | 0.98 |
| gelling point ($^{o}$C) | 25.2 | 30.21 | 23.8 | 25.5 | 25.5 | 32.0 |
| melting point ($^{o}$C) | 80.2 | 76.35 | 80.3 | 80 | 76 | 79 |
| Viscosity (mPas) | 225 | 245 | 180 | 210 | 224 | 189 |

*3.3. The different effect of the treatments to the encapsulation materials and methods of extraction on the viability of Lactobacillus acidophilus*

The result of analysis of variance shows that the treatment of materials, methods of extraction and their interactions significantly give different effects on the viability ($P \leq 0.05$) with R-Sq = 98.23% R-Sq (adj) = 97.49%.

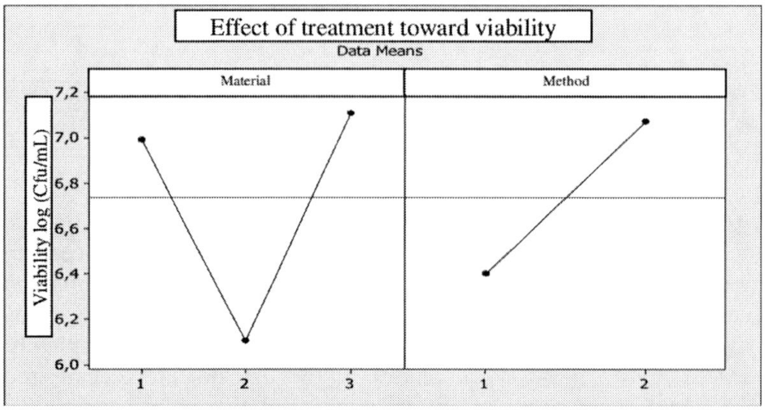


**Figure 3.** The effects of the treatments to the materials and methods of extraction toward the *Lactobacillus acidophilus* viability.

IOP Conf. Series: Earth and Environmental Science **137** (2018) 012073 doi:10.1088/1755-1315/137/1/012073

Figure 3 shows that the treatments of encapsulation materials with *Eucheuma cottonii* producing kappacaragenan and *Eucheuma spinosum* producing iotacaragenan are different but the treatments of encapsulation materials with the mixture of *Eucheuma cottonii* and *Eucheuma spinosum* producing kappa-iotacaragenan are similar. The treatments using the mixture of *Eucheuma cottonii* and *Eucheuma spinosum*evenly give *L.acidophilus* higher viability than *Eucheuma cottonii* and *Eucheuma spinosum*. Meanwile, PNG and Gel KCl Press precipitation process are different on the viability of *L.acidophilus*.The treatments using KCl gel Press precipitation process evenly give *L.acidophilus* higher viability than the PNG processing method. The effect of treatments of the interaction between the materials and methods of extraction provides significantly different effects on the average viability, as can be seen in figure 4.

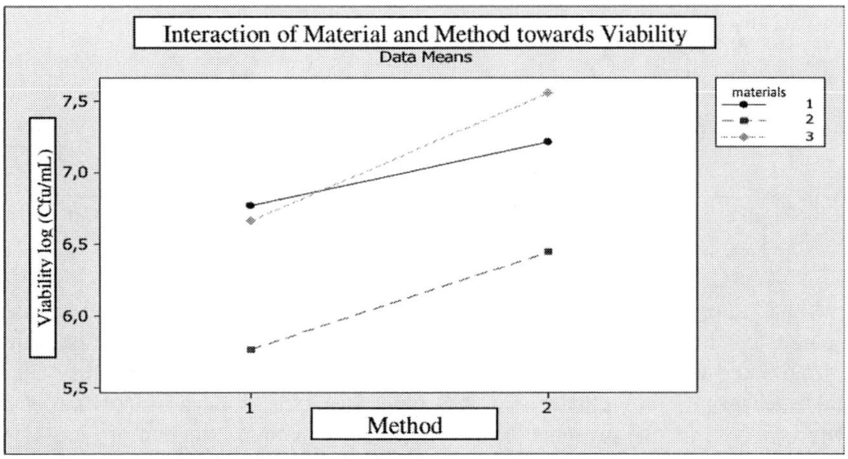


**Figure 4.** The effect of the treatments, interaction of materials, and method of extraction on the *L.acidophilus* viability.

Figure 3 and 4 show that the encapsulation material of kappa caragenan-iotacaragenan using KCl Gel Press precipitation process gives the highest viability. This is due to the mixture of kappa-iota using KCl Gel Press precipitation, a mixture of materials having elastic, cohesive and not syneresis properties. Therefore, it can affect the formation of microcapsules matrix to improve the viability. *Eucheuma cottonii,* on average, has a higher gel strength, a lower sulphate content, and a higher gelling point than *Eucheuma spinosum*.The species of seaweed and carrageenan methods of extraction process differences are factors that affect the quality observed from the strength of the gel, and the sulfate content of the gelling point or melting point. They are alsorelated to the use of temperature and type of cations as extraction materials. KCl Gel Press precipitation process method produces a higher gel strength, a lower sulphate content, and a higher gelling point than PNG process of extraction method. This is related to the content of 3.6-AG and sulphate content. PNG extraction process uses the extraction process at temperatures of 80°C to 82°C, whereas KCL Gel Press extraction process uses the extraction process at temperatures of 85°C to 90°C. Therefore, the closer the optimum process with the temperature, the more perfect the termination of sulfate ester group into cluster 3.6 -AG will be. The existence of group 3.6 - AG can affect the strength of the gel. The gel strength of Carrageenan is noted as *breaking force* defined as the maximum load required for breaking the polymer matrix in the burdened areas. Meanwhile, carrageenangel formationis a process which involves the deposition ofionic bonds between metal cation with negative charge from sulfate ester.

If the sulfate ester group is a lot more than the sulfate, the sulfate will bindwith water. A higher sulfuric content in carrageenan causing three-dimensional structures formed absorbs more water. Thus, it reduces the gelation strength. Ester sulfate is "hydrophilic" while 3.6 AG is "hydrophobic"Besides, the species *Eucheuma cottonii* and *Eucheuma spinosum* have differences on the formation of double helix and single helix structure. So, it affects the formation of aggregation. The aggregation formation of kappa carrageenan leads to the formation of the matrix layer thicker than iota carrageenan. It Is also supported by Chandrasekaran [14] that describes double helix structure and single helix structure on kappa and iota carrageenan through X-ray diffraction, as shown in figure 5.

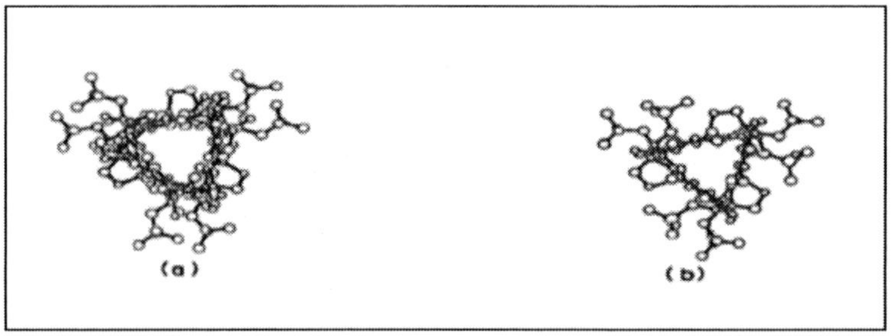


**Figure 5.** Double helix structure structure in (a) Kappa carrageenan and (b) Iota carrageenan using X-ray Diffraction [14].

Figure 5 shows that kappa carrageenan molecule projection shows a disorganized and denser molecular structure compared to the structure of iota carrageenan molecules. Kappa carrageenan has molecular bonds that are fragile and easily broken [14]. The presence of $K^+$ that is needed by kappa caragenan in breaking sulfate ester group into cluster 3.6 AG generates gel properties that are strong, rigid, and syneresis. On the other hand, the presence of $Ca^{2+}$ needed by iota carrageenan produces gel properties that are weak, elastic and not syneresis. So, the mixture of *Eucheuma cottonii* and *Eucheuma spinosum* producing kappa-iota carageenan have such gel properties as strong, elastic and no syneresis. In line with the description from Phillip [2] and Chandrasekaratan [14], it is presumably that the differences on the gel properties can affect the viability of *Lactobacillus acidophilus*, because the structure of microcapsules matrix formation and stability of matrix are also affected by the gel properties. The structure of microcapsules matrix formation and stability of matrix will protect the *L.acidophilus* from external conditions so that the viability is improved. It is supported by Setijawati [11] that the mixture of kappa caragenan and iota caragenan from *Eucheuma cottonii* and Eucheuma spinosum gave the highest viability. Species and concentration factors gave different influences toward *Lactobacillus acidophilus* viability through simulation of*GI Tract* pH condition.The best result was mix kappa-iota *carrageenan* (75:25), 6 % concentration with the highest viability average 6.1097 cfu/ml (log).

## 4. Conclusion

The encapsulation materials with process extraction methods give different effects on the viability of *L. acidophilus*. The highest average *L. acidophilus* viability was obtained from the treatment of the mixture of *Eucehuma cottonii* and *Eucheuma spinosum* (1:1) producing kappa-iota caragenan with KCl Gel Press precipitation process method of 7.14 log (CFU /mL).

## 5. References

[1]. Vashista 2002 *Microencapsulation* (San Antonio : Southwest Research Institute TM)
[2]. Phillip G O and William P A 2001 *Handbook of hydrocolloids* (Boston New York Washington DC : Woodhead Publishing, CRC Press Boca Rotan)
[3]. Adhikari K, Mustapha A, Grun and Fernando L 2000 *J. Dairy Sci.* 1946-1951
[4]. Adhikari 2003 *J. Food Sci.***68** 275-280
[5]. Ding W K and Shah N P 2009 *J. Food Sci.***74** 100-107.
[6]. Begley M, Gahan CGM and Hill C 2005 The interaction between bacteria and bile. *FEMS Microbiol. Rev.* **29** 625–651
[7]. Dinakar P, and Mistry VV 1994 *J Dairy Sci.* **81** 2804-2816.
[8]. Deloyres and Lacroix 2005 *Appl. Microbiol. Biotech.* **60** 168 – 173
[9]. Shah NP and RavulaRR 2000 *J. Dairy Technol.* **55** 139-144.
[10]. AudetPC, Paquin and Lacroix C 1988 *App.Microbiol. Biotech. J.* **29** 11-18.
[11]. Setijawati D 2014 *J. Basic. Appl. Sci. Res.* **4** 261-268.
[12]. Rachmanian R1996 *Introduction of types of indonesian seaweed* (Jakarta : Center for Oceanography LIPI) (In Indonesia)
[13]. Satari R 1996 *Characteristics of Polysaccharide Carrageenan from Eucheuma sp. and Hypnea sp.* (Jakarta : Oceanology Research Center LIPI) (In Indonesia)
[14]. Chandrasekaratan R J S 2002 *Carbohydrate Res.* **337** 525-535

**Acknowledgment**
Our deepest gratitude is presented to Brawijaya University and the Faculty of Fisheries and Marine Sciences that have provided us the fund for this research through the Research Fund BOPTN 2015.

IOP Conf. Series: Earth and Environmental Science **137** (2018) 012074 doi:10.1088/1755-1315/137/1/012074

# Antibacterial activity of red algae *(Gracilaria verrucosa)* extract against *Escherichia coli* and *Salmonella typhimurium*

**S Dayuti**

Faculty of Fisheries and Marine Science, University of Brawijaya, Malang. Indonesia

E-mail: sdayuti@gmail.co.id

**Abstract.** Red alga was widely used in several fields, including food, feed, phamacy and industrial point of view. The chemical analysis showed that red alga contained terpenoid, acetogenic, and aromatic compounds, which have a wide range of biological activities, such as anti-micobial, anti-inflammatory and anti-viral. The objectives of this research was to evaluate the effect of extraction solvent and time on antibacterial activity of red alga (*Gracilaria verrucosa*), and to explore the bioactive compound contained within *Gracilaria verrucosa*. The method in this study used descriptive reseach. These findings revealed that the highest inhibition activity among all extracts was obtained with the ratio of methanol:aquades (75:25) and extraction time around 72 hours against *Escherichia coli* and *Salmonella typhimurium*. The bioactive compounds of *Gracilaria verrucosa* tested by phytochemical analysisi consisted of flavonoid, alkaloid, and saponin. Those secondary metabolites may be approximated as antibactial substances.

## 1. Introduction

Currently, the use of seaweeds have been extensively studied in many fields, including food and pharmacy. they are also used as a new material for cosmetics, herbal medicines and fertilizers [1]. In the animal indutry poin of view, seaweeds can improve nutritional value of animal feed. One of the most potential sources of seaweeds is red algae. It has a brilliant color due to the pigment phycoerythrin and phycocyanine. This algae can live at greater depths than brown and green algae because it absorbs blue ligh.

Red algae is considered as a source of bioactive compounds as it is able to produce secondary metabolites [2] characterized by a wide spectrum of biological activities, such as antiviral, antibacterial and antifungal [3]. Previous studies have suggested that red algae has a great variety of secondary metabolites than brown and green algae. Hence, red algae provides an alternative approach to the use of the synthetic antimicrobial agents. However, the potent antimicrobial effect of red algae resides in the efficiency of the extraction method [4], the red algal species, and the solvents being used [5].

According to Shekhar [6], red algae (*Gracilaria edulis*) had antibacterial activity extracted by ethanol against *Escherichia coli, Enterobacter aerogenes*, *Pseudomonas aeruginosa, Staphylococcus aureus* and *Streptococcus faecalis*. On the other hand, the research of *Gracilaria verrucosa* as a antibacteria substance and had bioactive compounds especially isolated from Ponjuk, Talango, Madura (East Java) are poorly documented yet. Therefore, in this study we focus on evaluating evaluate the effect of extraction solvent and time on antibacterial activity of *Gracilaria verrucosa*, and to explore the bioactive compounds contained within *Gracilaria verrucosa*.

## 2. Material and Methods

### *2.1. Read algae collection and preparation*

The red algae used as raw material in this research was *Gracilaria verrucosa*, which was obtained from Ponjuk Village, Talango Island, Madura. *Gracilaria verrucosa* was washed by seawater to clean impurities material. *Gracilaria verrucosa* was put into a polyethylene bag. Then, it was transfered into cool box. *Gracilaria verrucosa* was dried, then cleaved into small pieces using a grinding machine. The extraction was conducted by maceration technique in which *Gracilaria verrucosa* powder have been dried was subsequently soaked (maceration). The maceration technique was used to extract the bioactive compounds within *Gracilaria verrucosa*.

In this study used descriptive method, which can be described as a statement of affairs as they are at present with the researcher having no control over variable. We applied 60 g of each treatment macerated by solvent with different extraction time in the range of 24, 48, and 72 hours, respectively. The fisrt treatment was conducted with solvents ratio (ethanol with different water content or ethanol : destilled water) were 75 % : 25 %; 50 % : 50 %; and 25 % : 75 %. The second treatment used methanol extraction in the rasio of 75 % : 25 %; 50 % : 50 %; and 25 % : 75 % destilled water.

### *2.2. Phytochemical analysis*

The phytochemical analysis used Nurdiani *et al.* [7]. It was conducted to investigate a wide range of active compound contained in the fraction. Alkaloids, flavonoids, saponins, and terpenoids were analyzed. The data analysis was described elsewhere.

### *2.3. Antibacterial activity test*

Antibacterial activity was carried out using agar diffusion method. Pure cultures of *E. coli* and *S. typhimurium* were grown in liquid media nutrient broth and incubated at 35°C for 3 h, then the suspension was cultured using TSA media. The fraction was dissolved in DMSO 10 %, then 10 μL of the fraction was impregnated into a paper disc and was put on TSA media consisted of bacteria. It was kept for 24 h at temperature of 30°C. Zone of inhibition was measured.

## 3. Results and Discussion

### *3.1. Extraction yields of Gracilaria verrucosa*

The yields of the crude extraction of *Gracilaria verrucosa* using distilled water, ethanol, and methanol were presented in table 1.

**Table 1.** Extraction yield of *Gracilaria verrucosa*.

| Treatment | | Weight of sample (g) | Average of weight extract (g) | Yield (%) ± SD |
|---|---|---|---|---|
| Solvent | Extraction time | | | |
| Ethanol | 24 | 60 | 3.37 | 5.61±0.55 |
| | 48 | 60 | 4.33 | 7.22±0.80 |
| | 72 | 60 | 4.81 | 8.01±1.05 |
| Methanol | 24 | 60 | 4.86 | 8.09±1.50 |
| | 48 | 60 | 5.54 | 9.23±0.88 |
| | 72 | 60 | 5.55 | 9.26±1.13 |

The extraction yield of *Gracilaria verrucosa* treated by the different concentration of solvents and extraction time revealed that highest yield of *Gracilaria verrucosa* extraction was 9.26 in the rasio of 75 : 25 (methanol : destilled water) for 72 h of extraction time. According to Wijaya *et al.* [8], this solvent mixture will result in a maximum amount of yield. Mustafa and Turner [9] added that methanol solvent can result in the hingest yield. Siregar *et al.* [10] stated that the longer extraction time between solvent and sample can get more extract.

*3.2. Qualitative phytochemical analysis*

The qualitative phytochemical analysis of *Gracilaria verrucosa* extract was conducted to find out the types of secondary metabolites existing in the extract. The secondary metabolites assesed were alkaloids, flavonoids, saponins, and terpenoids. The results of secondary metabolite screening from *Gracilaria verrucosa* can be seen in table 2.

**Table 2.** Qualitative phytochemical analysisi of *Gracilaria verrucosa* extract.

| Secondary Metabolite | Ethanol : Destilled water[a] 75:25 | | | Methanol : Destilled water[a] 75:25 | | |
|---|---|---|---|---|---|---|
| Time (h) | 24 | 48 | 72 | 24 | 48 | 72 |
| Flavonoid | - | - | - | - | + | + |
| Alkaloid | + | + | + | + | + | + |
| Saponin | - | - | + | - | - | + |
| Terpenoid | - | - | - | - | - | - |

[a] *(+) = Contains test compound; and (-) = Does not contain test compounds*

The results of the phytochemical analysis showed that the highest number of secondary metabolite compounds was in the rasio of methanol-distilled water around 75 : 25 with extraction time of 72 h. The metabolite compounds detected in the phytochemical analysis were flavonoid, alkaloid, and saponin. It means that the highest yield of metabolite compounds is suggested by using 75 % of methanol as a extraction solvent. Phytochemicals can be extracted with an appropriate solvent [11]. The degree of polarity affects the components of the extracted phytochemicals. Nafisyah *et al.* [12] stated that the secondary metabolite compounds, such as flavonoids, alkaloids, and saponins have capability in dissolving in the polar solvents.

*3.3. Antibacterial activity of Gracilaria verrucosa against E. coli and S. typimurium*

The results of the antibacterial activity of *Gracilaria verrucosa* extract against *E. coli* and *S. typhimurium* can be seen in figure 1 and figure 2.

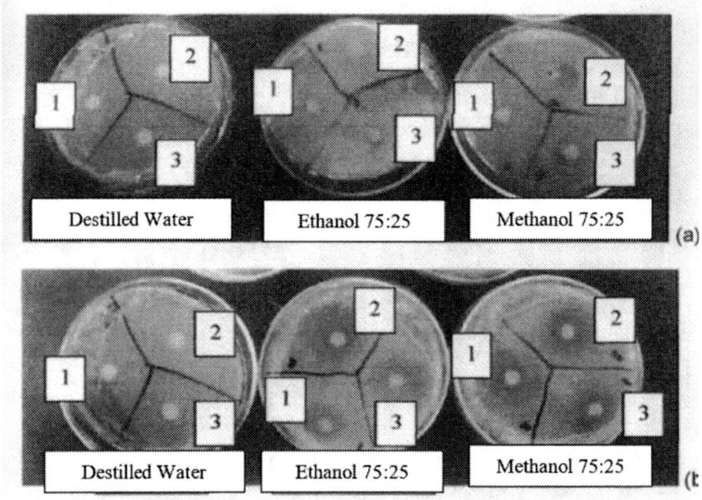


**Figure 1.** The results of *Gracilaria verrucosa* extract against *E. coli* and *S. Typhimurium*. The code a and b indicated *E. coli* inhibition and *S. typhimurium* inhibition, respectively. The number '1, 2, 3' showed the extraction time for 24 h, 48 h, and 72 h, respectively.

IOP Conf. Series: Earth and Environmental Science **137** (2018) 012074 doi:10.1088/1755-1315/137/1/012074

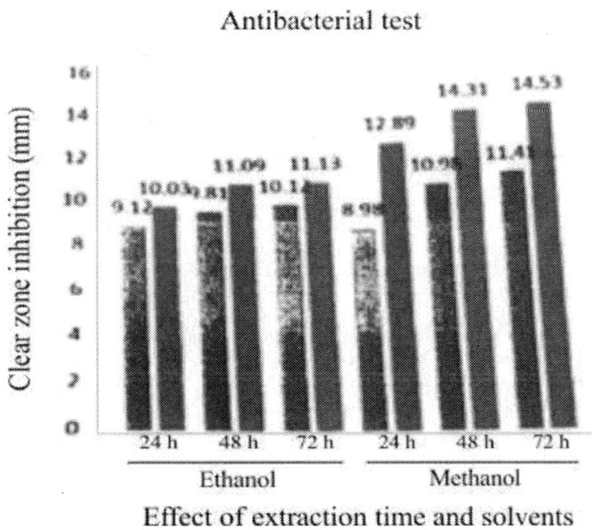


**Figure 2.** The Inhibition zone of *Gracilaria verrucosa* extract.

The figure 2 above showed that the activity of antibacterial compounds extracted from *Gracilaria verrucosa* had positive effect to inhibit pathogenic bacteria such as *E. coli* and *S. typhimurium*. The antibacterial inhibition formed clear zones around the disc test through the paper treated by *Gracilaria verrucosa* extract. However, the inhibition activity of antibacterial compounds has different inhibitioin zone.

The lowest inhibition zone of antibacterial activity against *E. coli* and *S. typhimurium* was displyed in the methanol extraction with the range of time for 24 h and the concentration of methanol around 25 %. On the contrary, the highest inhibition zone was performed in the concentration of methanol 75 % for 72 h of extraction time. The lowest antibacterial activity was due to the presence of only one secondary metabolite compound (alkaloid), and the highest antibacterial activity was caused by the activity of three phytochemicals, they are alkoloid, saponin dan steroid. The extraction of natural sources like red algae is suggested to use more methanol concentration as an agent of extracting secondary metabolites compounds that will obtain higher antibacterial activity [13]. Secondary metabolites extracted by using methanol will produce better inhibitory power than that of produced by chloroform [14].

## 4. Conclusion

The conclusions of this research are the qualitative physicochemical screening test showed that only appropiate secondary metabolites could be extracted by ethanol and methanol solvents. the highest inhibition activity among all extracts was obtained with the ratio of methanol:aquades (75:25) and extraction time around 72 hours against *Escherichia coli* and *Salmonella typhimurium*.

## 5. References

[1]. Benjama O and Masinyom P 2012 *J. Sci. Technol.* **34** 223-230
[2]. Barbosa M, Valentao P, and Andrade P B 2014 *Mar. Drugs* **12** 4933-4972
[3]. Suleria H A R, Osborne S, Masci P, and Gobe G. 2015 *Mar. Drugs.* **13** 6336–6351
[4]. Ksouri R, Falleh H, Megdiche W, Trabelsi N, Hamdi B, Chaieb K, Bakhrouf A, Magné C, and Abdelly C. 2009 *Food Chem. Toxicol.* **47** 2083–2091.
[5]. Aboshora W, Lianfu Z, Dahir M, Qingran M, Qingrui S, Jing L, Al-Haj N Q M, and Ammar A. 2014 *Trop. J. Phar. Resear.* **13** 2057-2063
[6]. Shekhar S 2012 Asi. J. Bio. Sci. **21** 219-222
[7]. Nurdiani R, Furdaus M, and Prihanto A A 2012 *J. Ba. Sci. Technol.* **1** 2-27

IOP Conf. Series: Earth and Environmental Science **137** (2018) 012074 doi:10.1088/1755-1315/137/1/012074

[8]. Wijaya D P, PaendongJ E and Abidjulu J 2014 *J. MIPA Unsrat Online***3**(1):11-15
[9]. Mustafa A and Turner C 2010 *Analyctica Chimica Acta* 8-18
[10]. Siregar A F, Sabdono AandPringgenais D 2012 *J. Mar. Res.* **1** 152–160
[11]. Yudiati E,Sedjati S, Sunarsih and Agustian R 2011 *Mar.Sci.* **16** 187-192
[12]. Nafisyah A L, Tjahjaningsih W, Kusdarwati RandAbdillah A A 2015 *J. Fish. Res* **7** 87-93
[13]. Pushparaj A, RaubbinR S andBalasankar T 2014 *Pharm. Tech.* **6** 1-5
[14]. Pushparaj A, RaubbinR S andBalasankar T 2014 *Pharm. Tech.* **6** 1-5

IOP Conf. Series: Earth and Environmental Science **137** (2018) 012075 doi:10.1088/1755-1315/137/1/012075

# Potential of mangrove *Avicennia rumphiana* extract as an antioxidant agent using multilevel extraction

**L Sulmartiwi[1], D Y Pujiastuti[1]*, W Tjahjaningsih[1] and Jariyah[2]**

[1]Department of Marine, Faculty of Fisheries and Marine, Airlangga University, Jl. Airlangga Surabaya 60115, East Java, Indonesia
[2] Pembangunan Nasional University "Veteran" Surabaya

E-mail: dwiyp@fpk.unair.ac.id

**Abstract.** *Avicennia rumphiana* is one of abundant mangrove found in Indonesia. Multilevel extraction methods were simultaneously conducted to screen the antioxidant activity from mangrove. The leaves, fruits and barks were consequently extracted using n-hexane, ethyl acetate and ethanol. The presence of phenolic, flavonoids and tannins compounds were characterized by quantitative and qualitative phytochemical assay as well as the antioxidant activity was examined using DPPH-free radical scavenging assay. The phytochemical test revealed that all of the extracts showed positive result. The fruits extract exhibited the highest phenolic, flavonoid and tannin (23.86 mg/g, 13.77 mg/g and 74.63 mg/g), respectively. The extracts were further confirmed for antioxidant using $IC_{50}$ value and revealed that ethyl acetate extract has antioxidant activity better than n-hexane and ethyl acetate extract. Furthermore, this study indicated that mangrove *Avicennia rumphiana* could be subsequently explored for other biological activities due to their potential secondary metabolites.

## 1. Introduction

Natural plants are commonly used as primary medicine due to their pharmacological properties. The inquiry of the efficacy of plant-based drugs has been paid great attention because of their milder, safer, and has more availability compared to chemical medicine [1]. *Avicennia rumphiana*, also known as black mangrove, is the best mangrove species in stabilizing the soil habitat. It spreads in the first zone or often called as pioneer zone, which is directly faced to the sea. pioneer zone is a habitat for mangrove that has strong resistance to withstand wave blows, and they are able to assist in the process of sediment accumulation.

Free radicals are atoms or molecules that are unstable and highly reactive because they contain one or more unpaired electrons in their outermost orbital. Therefore, free radicals easily react with surrounding molecules to obtain electron pairs to achieve atom or molecular stability. excess Free radicals are known to cause degenerative diseases [2]. Free radical detection through the addition of antioxidant compounds is associated with the ability of antioxidants to donate hydrogen protons. Antioxidant compounds are able to counteract free radicals by inhibiting reaction rates at the propagation and initiation stage. Ordinarily, many synthetic antioxidants, such as butylated hydroxyltoluene (BHT), tert-butylhydroquinone (TBHQ), and butylated hydroxyanisol (BHA), have been applied to reduce oxidative damage in the human body. However, the use of synthetic antioxidants in humans might cause several side effects, such as carcinogenesis and liver damage [3].

IOP Conf. Series: Earth and Environmental Science **137** (2018) 012075 doi:10.1088/1755-1315/137/1/012075

Antioxidants are compounds that can inhibit or retard oxidation of lipids and other biomolecules. They prohibit the start of an oxidizing chain reaction or quench the propagation [4]. Natural antioxidants are found in plants on their stems, leaves, roots, and fruits [5]. Bioactive compounds that act as antioxidants can be obtained through the extraction process. Multi-stage extraction is used to obtain component with higher purity than a single-stage extraction. The DPPH method is one of the most frequently-used methods to assess the ability of compounds to act as free radical scavengers or hydrogen donors and to evaluate the antioxidant capacity of samples [6]. No biological activities have been reported for mangrove *Avicennia rumphiana* leaves, barks, and fruits extract. This study was aimed to screen the antioxidant activity of *Avicennia rumphiana* using DPPH free radical scavenging assay as a rapid screening tool for crude extracts. The sample was extracted using multilevel extraction to produce specific compounds based on solvents differences. Furthermore, the phytochemical contents were also determined quantitatively and qualitatively.

## 2. Methodology

### *2.1. Materials and chemical reagents*

Mangrove *Avicennia rumphiana* was collected from Wonorejo, Surabaya, Indonesia. Ethanol, ethyl acetate, n-hexane, hydrochloric acid (HCl), magnesium (Mg), ferric chloride ($FeCl_3$), sulfuric acid ($H_2SO_4$), ascorbic acid, and 1.1-diphenyl-2-picrylhydrazyl (DPPH) were purchased from Sigma Chemical Co. (St. Louis, MO, USA) and all other chemicals used in this study were of reagent grade.

### *2.2. Sample preparation*

In order to prove the species of the mangrove, the fresh and young leaves, barks, and fruits were identified at Purwodadi Botanical Conservation Center, Indonesian Institute of Science (LIPI), Indonesia. The samples were dried for two weeks and then powdered using mechanical grinder. The powder (200 g) was then extracted using n-hexane, ethyl acetate, and ethanol, consecutively. The extraction using n-hexane was conducted for 24 h using ratio of 1:5 (w/v) and then the solution was filtered. The filtrate was collected as n-hexane extract. The residue was extracted further using ethyl acetate and was left for 24 h with ratio of 1:5 (w/v). The filtrate was collected as ethyl acetate extract. Subsequently, the residue was extracted using ethanol with ratio of 1:5 (w/v) for 24 h and the filtrate was collected as ethanol extract. All of the extraction was conducted using orbital shaker at 180 rpm. Extracts were concentrated under rotary evaporator and the crude extract kept -4 °C for subsequent phytochemical analysis. The yield percentage was measured from the dry extract powder.

### *2.3. Phytochemical assay*

In order to screen the presence of secondary metabolites content from each extracts, the analysis of phytochemicals was conducted according to the previous study by Harborne with slight modification. The crude extract (0.5 mg) was dissolved in 5 mL distilled water then heated. Two mL of the extract then was added with 0.2 mL 5 % $FeCl_3$. The presence of blue or violet color indicated the phenolic content. Next, 2 mL extract was added with 0.4 mL amyl alcohol containing 37 % HCl and 96 % ethanol, 0.5 mg magnesium, and 70 % alcohol. The presence of red or orange color indicated the flavonoid content. Subsequently, in order to identify the tannin content, 2 drops of 1 % $FeCl_3$ and $H_2SO_4$ was added in 2 mL extract. The presence of brown precipitate indicated the tannin content.

### *2.4. Antioxidant assay*

The antioxidant activity of crude extract was measured by the free radical scavenging 1.1-diphenyl-2-picrylhydrazyl according to previous study by Sun *et.al.* with slight modification. The mixture solution was composed of DPPH and crude extract as the inhibitor. The ascorbic acid was used as positive control. All of the solution was incubated at 37 °C for 30 min. The solution was detected using a UV spectrophotometer at 517 nm. The antioxidant activity (ppm) was calculated according to the following equation:

$$\text{Antioxidant activity (\%)} = (\Delta A \text{ control-} \ \Delta\Delta \text{ inhibitor})/\ (\Delta A \text{ control})$$

IOP Conf. Series: Earth and Environmental Science **137** (2018) 012075 doi:10.1088/1755-1315/137/1/012075

where ΔΔ inhibitor and ΔA control are absorption of sample with and without inhibitor, respectively. The $IC_{50}$ value was defined as the concentration of inhibitor which was able to inhibit 50% of DPPH activity.

*2.5. Statistical analysis*
The experiments were run in triplicate and data were expressed as the mean ± standard deviation. The data was analyzed using SPSS version 16.0 (SPSS, Inc., Chicago, IL). Differences among mean values were analyzed by Analysis of Variance (ANOVA) followed by Duncan's multiple comparison test if the resulting p-value was lower than 0.05.

## 3. Results and discussion

*3.1. The yield of crude extract*
The yield of leaves, bark, and fruits extract of mangrove *Avicennia rumphiana* was obtained from percentage of weight of extract divided by the original weight of *Avicennia rumphiana* used. The resulting yield value was different for each extraction solvent. The efficacy of solvent extraction is affected by many factors such as the type of solvent, solvent concentration, time, temperature, pH, number of steps, liquid-to-solid ratio and particle size of the plant material [9]. In our study, we used the multilevel extraction method with n-hexane, ethyl acetate, and ethanol. The purpose of using n-hexane as the first solvent is due to the coated cell wall of phospholipid. Dissolution of phospholipid causes opening of the cell to make it easier for the compounds stored in cells to be extracted using the next solvent. The residue from the extraction process using n-hexane was dissolved in a semi-polar solvent (ethyl acetate). Ethyl acetate is capable of attracting hydrophilic and lipophilic compounds [10]. extract from the ethyl acetate extraction was then dissolved iin ethanol . The yield of n-hexane, ethyl acetate, and ethanol extracts from mangrove *Avicennia rumphiana* can be seen in table 1. According to Jacoeb *et.al.*, extraction of *Avicennia marina* leaves using methanol solvent resulted in 9.61% yield, while yield of ethyl acetate and n-hexane extraction were 1.28% and 0.62%, respectively [11].

**Table 1**. The yield of crude extract from mangrove *Avicennia rumphiana*.

| Sample | Weight (g) | Weight of extract (g) | | | Yield (%) | | |
|---|---|---|---|---|---|---|---|
| | | Bark | Fruit | Leave | Bark | Fruit | Leave |
| n-hexane | 200 | 0.78 | 6.31 | 6.28 | 0.39 | 3.15 | 3.14 |
| Ethyl acetate | 200 | 5.71 | 7.33 | 2.76 | 2.85 | 3.66 | 1.38 |
| Ethanol | 200 | 6.73 | 10.55 | 12.48 | 3.36 | 5.27 | 6.24 |

*3.2. Screening of phytochemical contents from Avicennia rumphiana*
In order to comprehensively screen the phytochemical contents of *Avicennia rumphiana*, various solvent was used to extract the diverse compounds. Typically, the leave, bark, and fruit of mangrove *Avicennia rumphiana* contain most of the phytochemicals like flavonoid, phenols,and tannin (table 2). Test solution turned greenish when added with $FeCl_3$ 5%. It showed that there were phenol compounds in the extract. When ethyl acetate and ethanol extracts were added with amyl alcohol and Mg, orange color appeared in amyl alcohol extraction. It shows there were flavonoid compounds in the extract, but in the n-hexane test the resulting color appeared vague. Results showed that there were flavonoid compounds in n-hexane extract although in small amount. Samples added with 1% $FeCl_3$ and $H_2SO_4$ solution showed brownish green color and brown precipitate, which indicated that there were tannin in the extract. The color changes in phytochemical test can be seen in figure 1. phytochemicals, plant-derived non-nutritive compounds, are the results of the plant synthesis process and they play an important role in various functions in the human body [12]. According to the result above, all of the extracts contained phenols, flavonoids, and tannins. It showed that phenolic

IOP Conf. Series: Earth and Environmental Science **137** (2018) 012075 doi:10.1088/1755-1315/137/1/012075

compounds can be dissolved using non-polar, semi-polar, and polar solvents. Flavonoids are intermediates that act as lipophilic and hydrophilic antioxidants [13]. The red color produced in flavonoid test is due to the formation of the flavilium salt [14]. While tannin compounds have OH groups that affects polarity, so by having tannins, the extract was able to be dissolved in polar compounds.

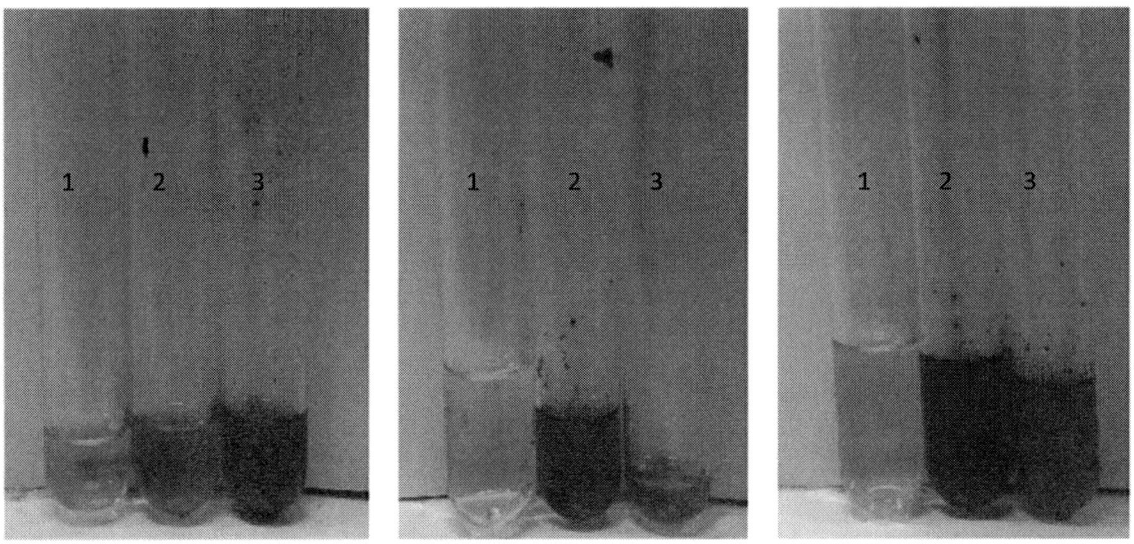


**Figure 1**. Phytocemical test solution of *Avicennia rumphiana* (A) Result of phenolic assay (1) n-hexane extract (2) ethyl acetate extract (3) ethanol extract. (B). Result of flavonoid assay (1) n-hexane extract (2) ethyl acetate extract (3) ethanol extract. (C) Result of tannin assay (1) n-hexane extract (2) ethyl acetate extract (3) ethanol extract.

**Table 2**. Quantitative analysis of phytochemicals in *Avicennia rumphiana* crude extract.

| No. | Sample (crude extract) | Weight of sample (g) | Phytochemical | | |
|---|---|---|---|---|---|
| | | | Total of phenol (mg ekv. Gallic acid/gram extract) | Total of flavonoid (mg ekv. Quersetin/gram extract) | Total of tanin (mg ekv. Tannic acid/gram extract) |
| 1 | Ethanol extract of bark | 0.1 | 0.9072 | 4.8151 | 4.1008 |
| 2 | Ethyl acetate extract of bark | 0.1 | 1.2413 | 6.1796 | 5.7699 |
| 3 | n-hexane extracte of bark | 0.05 | 0.9654 | 2.0258 | 2.9869 |
| 4 | Ethanol extract of fruit | 0.1 | 9.9428 | 13.1323 | 11.0055 |
| 5 | Ethyl acetate extract of fruit | 0.02 | 23.8643 | 13.7796 | 74.6339 |
| 6 | n-hexane extracte of fruit | 0.1 | 0.2881 | 1.4075 | 5.0033 |
| 7 | Ethanol extract of leave | 0.1 | 0.8761 | 2.3043 | 2.3536 |
| 8 | Ethyl acetate extract of leave | 0.1 | 1.4360 | 13.8000 | 21.2940 |

| | | | | | |
|---|---|---|---|---|---|
| 9 | n-hexane extracte of leave | 0.1 | 0.2556 | 5.0839 | 1.5142 |

**3.3.** *Antioxidant activity*

In order to determine whether the crude extract of different solvents had antioxidant activity, the $IC_{50}$ value was calculated. From different solvent extraction, ethyl acetate exhibited lower $IC_{50}$ compared with others (table 3). DPPH radical is scavenged by antioxidants through donation protons forming reduced DPPH [15].

**Table 3**. $IC_{50}$ value of *Avicennia rumphiana* extracts.

| Sample | | $IC_{50}$ value (ppm) | Strength of antioxidant activity |
|---|---|---|---|
| Ascorbic acid | | 9.06 ± 0.08 | Very strong |
| Leave | n-hexane | 1122.99 ± 1.30 | Very weak |
| | Ethyl acetate | 251.98 ± 1.59 | Very weak |
| | Ethanol | 492.22 ± 1.03 | Very weak |
| Bark | n-hexane | 1717.10 ± 1. 26 | Very weak |
| | Ethyl acetate | 515.77 ± 1.95 | Very weak |
| | Ethanol | 1088.95 ± 2.08 | Very weak |
| Fruit | n-hexane | 1140 ± 2.32 | Very weak |
| | Ethyl acetate | 611 ± 1.03 | Very weak |
| | Ethanol | 884 ± 1.67 | Very weak |

The results showed the low antioxidant activity of leaves extract from the mangrove *Avicennia rumphiana*. N-hexane, ethyl acetate, and ethanol extracts had different antioxidant activity that depended on to the type of antioxidants contained in each extract. Phenolic compounds commonly found in plant and flavonoids are a group of phenolic compounds that have antioxidant activity [16]. In our study, even though there were qualitative proofs of the existence of phenolic compounds in leaves of *Avicennia rumphiana*, the antioxidant activity was relatively low, which might be due to its low quantity. On the other hand, the n-hexane extract had a high yield, but the antioxidant activity value was low. Most of the compounds extracted by n-hexane were potentially non-antioxidant compounds. N-hexane is capable of extracting non-polar compounds such as fats and oils. *Avicennia rumphiana* leaves have high salt content, which is a mineral with polar property, thus they are able to dissolve in a polar solvent such as ethanol.

On the other hand, the high inhibition activity of ethyl acetate extract from the bark was due to the nature of ethyl acetate that is a polar solvent which can extract polar or nonpolar antioxidant active compounds even though the yield of ethyl acetate extraction was lower than in ethanol. Hardiningtyas *et.al.* stated that the highest antioxidant activity was present in ethyl acetate extract of *Avicennia marina* mangrove ethyl acetate is a semi polar solvent, thus it can extract both lipophilic and hydrophilic antioxidants. Antioxidants are categorized based on their solubility into lipophilic and hydrophilic antioxidants [18]. Hydrophilic and lipophilic antioxidants work synergistically to create a strong ability to inhibit the attack of free radicals. Tannin was also positively detected in n-hexane, ethyl acetate, and ethanol extracts from the bark. Tannins consist of a complex mixture of polyphenols compounds and they have many hydroxyl (OH) groups which are polar so that it can optimally be extracted with polar solvents. Tannins are very effective as electron donors as well as chelating agent on metals because of their hydroxyl group and conjugated double bonds, which allow the occurrence of electron delocalization [19]. It can be seen in table 2 that tannins are the most abundant compound found in *Avicennia rumphiana* bark.Nuraini reported that tannins are produced through polymerization of flavonoids and they are usually rich in wood plants that have the ability in the tannery leather. The presence of secondary metabolite compounds in *Avicennia rumphiana* bark is influenced by several factors such as genetics, age of the plant, climate, habitat, pests, and disease.

## 4. Conclusion
The current study hasdemonstrated that multilevel extraction (n-hexane, ethyl acetate and ethanol) can potentially be used to screen phytochemicals and antioxidant activities of *Avicennia rumphiana* mangrove. The crude extract of of the mangrove contained phenolics, flavonoids, and tannins. The ethyl acetate extract had better antioxidant activity compared to n-hexane and ethyl acetate extracts in leaves, bark and fruits of *Avicennia rumphiana*, with $IC_{50}$ values of 152.42 ± 1.30 ppm, 182.32 ± 2.03 ppm, and 232.43 ± 1.43 ppm, respectively.

## 5. References
[1] Bhattacharjee I, Chatterjee, S K, Chandra G 2010 *Asian Pac. J. Trop. Med.* **3** 547–551
[2] Selawa W, Revolta M, Runtuwene J, Citraningtyas G 2013 *J. Ilm. Farm* UNSRAT **2** 2302–2493
[3] Ko S C, Kim D, Jeon Y J 2012 *Food Chem. Toxicol* **50** 2294–2302
[4] Wollinger A, Perrin E, Chahboun J, Jeannot V, Touraud D, Kunz W 2016 *Comptes. Rendus. Chim.* **19** 754–765
[5] Nuraini A D 2007 Extraction on component of antibacteria and antioxidant of lotus seed Nymphaea pubescens Willd Bogor Agricultural Institute pp 1-94
[6] Hidayat M A, Fitri A, Kuswandi B 2016 *Acta Pharm. Sin.* **B 7** 395–400
[7] Harborne J B 1998 Phytochemical methods Journal of Chemical Information and Modeling
[8] Suh S S, Hwang J, Park M, Park H S, Lee T K 2014 *Asian Pac. J. Trop. Med.* **7** 531–535
[9] Tan M C, Tan C P, Ho C W 2013 *Int. Food Res. J.* **20** 3117–3123
[10] Hardiana R, Rudiyansyah, Zaharah T A 2012 *Jkk.* **1** 8–13
[11] Jacoeb A M, Purwaningsih S, Rinto 2011 *J. Proces. Fish. Prod.* **14**: 143–152
[12] Imaga N A, Gbenle G O, Okochi V I, Adenekan S, Duro-Emmanuel T, Oyeniyi B, Dokai P N, Oyenuga M, Otumara A, Ekeh F C 2010 *Sci. Res. Essays* **5**: 2201–2205
[13] Middleton E J, Kandaswami C, Theoharides T C 2000 *Pharmacol. Rev.* **52**: 673–751
[14] Marlinda M, Sangi M S, Wuntu A D 2012 *J. MIPA* Unsrat Online **1** 24–28.
[15] Salusu H D, Ariani F, Obeth E, Rayment M, Budiarso E, Kusuma I W, Arung E T 2017 *AGRIVITA J. Agric. Sci.* **39** 214–218
[16] Pourmorad F, Hosseinimehr S J, Shahabimajd N 2006 *African J. Biotechnol.* **5** 1142–1145
[17] Hardiningtyas S D, Purwaningsih S, Handharyani E, Agatis J, Hewan F K 2014 *J. Proces. Fish. Prod.* **17** 80–91
[18] Yang J, Kim J S, Sa Y J, Kim M O, Jeong H J, Yu C Y, Kim M J 2011 *African J. Biotechnol.* **10** 9438 – 9444
[19] Utari S P S D 2016 Potential lindur *Bruguiera gymnorrhiza* from mangrove as antioxidant and inhibitor α-glucosidase (Bogor: Thesis Bogor Agricultural Institute) p 45

IOP Conf. Series: Earth and Environmental Science **137** (2018) 012076 doi:10.1088/1755-1315/137/1/012076

# Antimicrobial resistance prevalence of *Aeromonas hydrophila* isolates from motile *Aeromonas septicemia* disease

**R Kusdarwati [1], Rozi [1], D Dinda N [1] and I Nurjanah[2]**

[1]Department of Fish Health Management and Aquaculture Faculty of Fisheries and Marine,Airlangga University, Jl. Airlangga 60115, Surabaya, East Java, Indonesia.

[2] Aquaculture, Faculty of Fisheries and Marine, Airlangga University, Jl. Airlangga 60115, Surabaya, East Java, Indonesia.

E-mail: rahayukusdar@gmail.com

**Abstract.** Fish suffer, from bacteria, fungi, virus and parasites or by physical ailments. Gurami (*Osphronemus gouramy*), nila (*Oreochromis niloticus*), carp (*Cyprinus carpio*), catfish (*Clarias* sp.) were the most reported infections caused by *Aeromonas* are bacterial hemorrhagic septicemia or Motile *Aeromonas* Septicemia (MAS). Antibiotics are drugs of natural or synthetic origin that have the capacity to kill or to inhibit the growth of micro-organisms included MAS. However, the use of antibiotics in the long term can cause negative impacts, among others, feared the occurrence of bacterial resistance in certain antibiotics. The results showed five of isolates were sensitive to antibiotics of chloramphenicol, gentamycin, oxytetracycline, cefradoxil and nalidixic acid but resistant to vancomycin colistin sulphate, rifampisin, cephalosporin and novobiocin.

## 1. Introduction

Motile *Aeromonas septicemia* (MAS) disease can cause death between 80-100% of dumbo catfish seeds in a period of approximately one week [1]. MAS disease has also been reported to have infected freshwater fish in the Banyumas region. The Livestock and Fishery Service Office of Banyumas Area (2005) reported that there were at least 72,000 freshwater fish that were infected with *Aeromonas hydrophyla* in 2003, namely 52,100 gurami and 19,900 catfish; while in 2004 the number amounted to 43,000 freshwater fish, which included 29,900 gurami and 13,100 dumbo catfish.

MAS disease control in cultivation is usually carried out by using antibiotics such as chloramphenicol, novobiocin, gentamicin, oxytetracycline, vancomycin, nalixidic acid, colistin sulphate and several other types of antibiotics. However, the use of antibiotics in the long term can cause negative impacts, among them is the fear of the occurrence of bacterial resistance to certain antibiotics.

The main cause of antibiotic resistance is its inappropriate dosage of use [2]. There have been case reports of bacterial resistance to these chemicals, namely from *Aeromonas caviae, Aeromonas sobria, Aeromonas bestiarum* and *Aeromonas hydrophyla* which are resistant to tetracycline [3]. The resistance that occurs will cause the bacteria to be resistant to the drug given so that the use of the drug becomes ineffective. Therefore, it is necessary to perform resistance tests of *Aeromonas hydrophyla* isolates against some types of antibiotics to find out which isolates are resistant.

IOP Conf. Series: Earth and Environmental Science **137** (2018) 012076 doi:10.1088/1755-1315/137/1/012076

## 2. Methodology

### *2.1. Bacteria isolate*

Three isolates of *Aeromonas hydrophyla* bacteria were obtained from Sukabumi, West Java; Jepara, Central Java; and Surabaya, East Java.

### *2.2. Antibiotics*

The antibiotics used were 30 (mcg) vancomycin, 30 (mcg) chloramphenicol, 5 (mcg) novobiocin, 10 (mcg) gentamicin, 30 (mcg) oxytetracycline, 10 (mcg) colisin sulphate, 5 (mcg) rifampicin, 5 (mcg) cefixime, 2 (mcg) cefadroxil and 30 (mcg) nalidixic acid.

### *2.3. Media*

The media used in this research was Tryp ticase Soya Agar (TSA), Mueller Hinton Agar (MHA) and NaCl Physiological.

### *2.4. Research design*

This study was conducted in July-August 2017 as an experimental study by observing, measuring and comparing the inhibit zone formed in the resistance test and comparing it with each antibiotic standard [4]. Three isolates of *Aeromonas hydrophyla* collected from Surabaya, Sukabumi and Jepara were used in this study. Each isolate was tested using seven antibiotics namely chloramphenicol, gentamycin, oxytetracycline and nalidixic acid, but resistant to vancomycin, colistin sulphate and novobiocin

### *2.5. Reidentification of bacteria*

The identification of the *Aeromonas hydrophila* bacterial isolate was done by the analytical profile index (API) method by using KIT API.

### *2.6. Bacterial culture*

After the reidentification, the bacterial isolates were cultured in Trypticase Soya Agar (TSA) media to obtain sufficient quantity of bacterial stock. The amount was then determined for the resistance test.

### *2.7. Production of bacterial suspension*

Bacteria turbidity was likened to Mc Farland's cloudy number 1 with the number of bacteria at $3x10^8$ / mL [5]. A reaction tube containing physiological NaCl with a volume of 9 ml was prepared. An amount of 1 ml of bacteria was taken from the bacteria turbidity and then fed into a reaction tube containing physiological NaCl with a volume of 9 ml.

### *2.8. Resistance test*

The resistance test was performed by using diffusion discs or diffusion test discs. Petri dishes contained MHA media were first prepared. An *Aeromonas hydrophyla* suspension of 100 μl was taken using a micropipette and dispersed on the entire surface of the agar plate and flattened using a drigalski to obtain even growth. After 10-15 minutes, the disc paper was placed on the agar medium by using sterile tweezers. The bacteria-planted medium and paper discs were incubated at 30°C for 18-24 hours [6]. The media was observed and the clear zone that formed around the disc paper was measured using the sliding term. The measurement of the inhibit zone diameter was done by reducing the diameter of the obstacle area by the diameter of the disc paper (6 mm) [7].

### *2.9. Data analysis*

After the data was obtained, namely the clear zone diameter seen around the disk paper that has been overgrown with *Aeromonas hydrophyla* bacteria, the results were then compared to the standard

IOP Conf. Series: Earth and Environmental Science **137** (2018) 012076 doi:10.1088/1755-1315/137/1/012076

inhibition zone of each antibiotic according to CLSI (Clinical Laboratory Standards Institute) and descriptively analyzed.

## 3. Results and discussions

### *3.1. Reidentification of Bacteria*

The results of the identification of *Aeromonas hydrophyla* isolates from Sukabumi, Surabaya and Jepara are presented in table 1.

**Table 1.** Results identification of isolates *Aeromonas hydrophyla* from Sukabumi(K), Surabaya (S) and Jepara (J).

| Physiological character | Isolate | | | *Aeromonas hydrophyla* |
|---|---|---|---|---|
| | **S** | **K** | **J** | |
| Gram | - | - | - | - |
| Motility | + | + | + | + |
| Citrate utilization | - | - | - | V |
| $H_2S$ | - | - | - | - |
| Oxidase | + | + | + | + |
| Indol production | + | + | + | + |

The result of examination using KIT API shows that the three isolates identified are *Aeromonas hydrophila* with a 90.16 % isolation for the Sukabumi isolate, 90.05 % for the Surabaya isolate and 88.31 % for the Jepara isolate.

### *3.2. Resistance test*

The results of the resistance tests of each isolate on several antibiotics are presented in table 2.

**Table 2.** Resistance test result of *Aeromonas hydrophyla* isolate from Sukabumi, Jepara and Surabaya against some antibiotics.

| Antibiotic (mcg concentraion) | Isolate(Code) | | |
|---|---|---|---|
| | Sukabumi (K) | Jepara (J) | Surabaya (S) |
| Vancomycin (30) | - (R) | - (R) | - (R) |
| Chloramphenicol (30) | 29,8 (S) | 25,8 (S) | 27,6 (S) |
| Novobiocin (5) | 8,1 (R) | 5,8 (R) | 5 (R) |
| Gentamicin (10) | 17,2 (S) | 15,1 (S) | 14,8 (S) |
| Oxytetracycline (30) | 24,5 (S) | 20,2 (S) | 24,6 (S) |
| Colistin Sulphate (10) | 9,4 (R) | 8 (R) | 7,3 (R) |
| Nalidixic Acid (30) | 31,4 (S) | 28,1 (S) | 28,9 (S) |
| Rifampicin (5) | - (R) | - (R) | - (R) |
| Cefixime (5) | - (R) | - (R) | - (R) |
| Cefadroxil (2) | 1,775 (S) | 1,7 (S) | 1,95(S) |

Resistance test results of *Aeromonas hydrophyla* against some antibiotics were marked by the formation of clear zones around the antibiotic dish. Each isolate showed resistance to five antibiotics (prevalence 50 %), with the same type of antibiotic.

Doses of the 10 antibiotics used were standard doses of antibiotics or in general can be considered as effective doses of antibiotics, such as vancomycin 30 mcg, 30 mcg chloramphenicol, novobiocin 5 mcg, 10 mcg gentamicin, oxytetracycline 30 mcg, 10 mcg colistin sulphate, 5mcg rifampicin, cefixime

IOP Conf. Series: Earth and Environmental Science **137** (2018) 012076 doi:10.1088/1755-1315/137/1/012076

5mcg, cefadroxil 2mcg and nalidixic acid 30 mcg [4]. When administered correctly, they greatly affect the ability of antibiotics in inhibiting the growth of microorganisms, otherwise it may affect the resistance of these antibiotics [8].

Sensitive results indicate that antibiotics are capable of inhibiting microbes in recommended antibiotic concentrations that can be used for the treatment of microorganism infections. While resistant results indicate that isolates of microorganisms cannot be inhibited by antibiotics at normal doses and are no longer used in treatment [4]. Novobiocin is an antibiotic that has a working mechanism by inhibiting DNA synthesis and tricholic acid in bacterial cell membranes [9]. Test results show a drag zone was seen with a clear color around the antibiotic dish of novobiocin of 8.1 mm in the Sukabumi isolate (K), 5.8 mm in the Jepara isolate (J) and 5 mm in the isolate of Surabaya (S). Based on data from Clinical Laboratory Standards Institute (CLSI) [4], novobiocin antibiotics are categorized as resistant (R) if the inhibit zone formed is less than or equal to (≤) 17 mm. This means that novobiocin antibiotics are resistant to *Aeromonas hydrophyla* bacteria. The cause of resistance to this type of antibiotic is the suspected structural changes of the bacteria that result in these antibiotics not working optimally in carrying out its action and causing bacteria to remain resistant.

Vancomycin is an antibiotic of the polypetides group that has a working mechanism that inhibits the synthesis of bacterial cell walls. Recently increased use of vancomycin has reduced the sensitivity of this antibiotic [7]. Based on the results of the tests, the three isolates of *Aeromonas hydrophyla* namely Sukabumi (K), Jepara (J) and Surabaya (S) do not show any inhibition zone around the vancomicin antibiotics dish (0 mm). This suggests that vancomicin antibiotics are resistant to *Aeromonas hydrophyla* bacteria.

The mechanism of resistance and reduced vancomycin sensitivity is thought to be associated with changes and rearrangement of bacterial cell walls. In addition, the production of excess Penicillin Binding Protein-2 (PBP-2) is also considered an important factor for the expression of resistance to vancomycin. Resistance to vancomycin is mediated by a van A gene specific to glycopeptides. The presence of Van A results in a change in the target terminal of D-alanil-D-alanil or D-alanil-D-serine, which causes its bond with vancomycin to be poor as the critical point for the hydrogen bond is lost. This causes vancomycin to be unable to be bound, resulting in decreased sensitivity [11].

Colistin can inhibit the permeability of Gram negative cell wall bacteria. The cytoplasm of all living cells is limited by the cytoplasmic membrane acting as a barrier to selective permeability, performing an active transport function so that it can control the arrangement of cells. When the integrity of the cytoplasmic function is impaired so that the permeability of the cell wall changes or even becomes damaged, important components, such as proteins, nucleic acids, nucleotides and cells gradually die [11].

The colistin sulphate antibiotic also showed similar activity with vancomycin and novobiocin and is also resistant to all isolated *Aeromonas hydrophyla* isolates. Based on the data obtained, the inhibition zone formed around the antibiotic dish of colistin sulphate was 9.4 mm in the Sukabumi isolate (K), 8 mm in the Jepara isolate (J) and 7.3 mm in the isolate of Surabaya (S). Based on these results, it can be categorized as resistant (R) because it has a zone of <10 mm [4]. Colistin becomes resistant thrugh the modification of the outer membrane structure of LPS, the bacteria is negatively charged so that the LPS phosphate component becomes neutral and weakens the binding of colistin with LPS wall.

The emergence of resistance to an antibiotic can occur through several mechanisms: the bacteria synthesizes an enzyme inactivator or antibiotic destroyer; bacteria change their permeability to drugs; bacteria develop a change of target structure for drugs; or bacteria to develop metabolic path changes directly inhibited by drugs [12].

## 4. Conclusions

Based on the results of the study, it can be concluded that the isolates of *Aeromonas hydrophila* from Sukabumi, Surabaya and Jepara showed three resistant properties of vancomycin, novobiocin, rifampicin, cephalosporin and colistin sulphate (prevalence 50 %), while it showed sensitive results against 5 other antibiotics.

IOP Conf. Series: Earth and Environmental Science **137** (2018) 012076 doi:10.1088/1755-1315/137/1/012076

## 5. References

[1] Mulia, D S 2007 *JPP* **7** 43-52
[2] Krisnaningsih M W, Asmaradan M W 2005 *J. Sains Vet.* 1
[3] Nurjanah S, S B Prayitno and Sarjito 2014 *J. Man. Cul. Tech.* **3** 308-316
[4] Clinical Laboratory Standards Institute (CLSI) 2012 *Performance Standards for Antimicrobial Disk Susceptibility Tests 22th ed.CLSI 2012 document M100- S22* (USA : Clinical Laboratory Standards Institute Wayne,PA) **32**
[5] Whitman, K A and N G MacNair 2004 *Finfish and Shellfish Bacteriology Manual:Techniques and Procedures* (State Avenue : Iowa State Press)
[6] Behrman, Kligman and Arvin 2000 The science of child health Nelson Vol. 2 (Jakarta : Penerbit Buku Kedokteran EGC) 904 p
[7] Hermawan, A, W Hana dan T Wiwiek 2007 *The influence of betel leaf extract against the growth of Staphylococcus aureus and Escherichia coli* (Surabaya : University of Airlangga)
[8] Chambers, H F 2006 *Beta-Laktam Antibiotics and Other Inhibitors of Cell Wall Synthesis In :Katzung, Bertram G, et al. Basic and Clinical Pharmacology 10th* edition (New York : McGraw Hills) 754-773 p
[9] Jawetz, E 1986 *Mycrobiology for health professions XVI ed translated by dr. Bonang,G.* (Jakarta : EGC Press) pp 336-384
[10] Hiramatsu, K 2001 *Lancet Infect. Dis.* **1** 147-55
[11] Lecrere V, R Marti, M Bechet, P Fickers and P Jacques 2006 *Arch. Microbiol.* 186 475-483
[12] Jawetz E 1997 *Principle of Antimicrobial Drug Action Basic and Clinical Pharmacology Third edition* (Appleton and Lange Norwalk)

IOP Conf. Series: Earth and Environmental Science **137** (2018) 012077 doi:10.1088/1755-1315/137/1/012077

# *Bacillus subtilis* UBTn7, a potential producer of L - Methioninase isolated from mangrove, *Rhizophora mucronata*

**A A Prihanto**

Department of Fishery Product Technology, Faculty of Fisheries and Marine Science, Brawijaya University, Jl. Veteran Malang 65145, East Java, Indonesia.

**Abstract.** L-methioninase is an enzyme that degrades sulfur-containing amino acids to α-keto acids, ammonia, and thiols. L-methioninase could be found in plants, bacteria, and fungi. The aims of this study was to obtain L-methioninase-producing endophytic bacteria isolated from mangrove *Rhizophora mucronata.* The mangrove was collected from Jenu Beach, Tuban, East Java, Indonesia. The samples were roots, stems, and leaves of *Rhizophora mucronata.* Endophytic bacteria were pure isolated using LB agar medium. Each bacteria were screened its capability to produce L-methioninase using selective media namely modified Czapek Dox agar. The best producer of enzyme was further identified with morphological and biochemical analysis. The result showed that three bacteria produced L-methioninase. Based on the result of morphological and biochemical analysis, the best producer was *Bacillus subtilis* UBTn7.

## 1. Introduction

L-methioninase is an enzyme that degrades sulfur-containing amino acids to α-keto acids, ammonia, and thiols. It belongs to the family of lyases enzyme, a class of carbon-sulfur lyases [1]. This enzyme is found in many microorganisms and has a specific substrate that varies depending on the type of microorganism. In addition, this enzyme also has different physiological characteristics. For example, the type of bacteria *P. gingivalis* acts as a determinant of its pathogenicity, whereas in some groups of anaerobic bacteria, this enzyme plays a role in the metabolism of ATP [2].

The potential application of this enzyme in the health sector is as a cancer inhibiting agent, especially in acute lympoblastic leukemia. The basis of this application is the fact that cancer cells often have unique characteristics and generally require higher nutrients. Furthermore, the restriction on nutritional components that become precursors or limit the development of these cells will be able to inhibit the growth of cancer cells.

*Rhizophora mucronata* is an unexploited source for endophytic microorganism. Endophytci bacteria plays an important role in the ecosystem. Some of their capabilities include nitrogen fixation, sulfate reduction, etc. Endophytic bacteria is a source for new and unique metabolites [3, 4]. The expectation in the exploration of *Rhizophora mucronata* is to gain bacteria that produces unique metabolites, such as high salt stability due to its extreme habitat.

Jenu Beach, Tuban is fully covered by mangrove plants. This natural resource has largely been free from exploitation and has a large diversity of mangrove. However, there is no exploration of this diversity to the best of our knowledge. Moreover, several studies have indicated that the coastal area is a myriad source for mining bioactives such enzymes and pharmaceutical compounds [5]. In this study, the mangrove *Rhizopora mucronata* was explored as a source of endhopytic bacteria which produces L-methioninase.

IOP Conf. Series: Earth and Environmental Science **137** (2018) 012077 doi:10.1088/1755-1315/137/1/012077

We also report the preliminary exploration of L-methioninase producing endophytic bacteria isolated from the mangrove, *Rhyzophora mucronata.*

## 2. Methodology

The equipment used in this study were glass beakers, erlenmeyer, measuring glasses, autoclaves, incubators, pH-meters, digital scales, petri dishes, micro pipettes, stirrers, refrigerator, freezer, mortar and pestle, and thermocyclers. The chemical compounds for this research were DNA extraction kit, bromothymol blue (BTB), aquades, yeast exstract, pepton, NaCl, agar, D- glucose, L-methionine,$Na_2HPO_4$, $KH_2HPO_4$, $MgSO_4.7H_2O$, $CaCl_2$, NaOH, HCl. All chemical compounds were analytical grade.

### *2.1. Sample preparation*

*Rhizophora mucronata* were obtained from Jenu Beach, Tuban, which is located in East Java, Indonesia. The samples used were the parts of the mangrove such as the root and leaves. The isolation of the mangrove bacteria was according to the method of Prihanto *et al.* [6] with modification.

### *2.2. Culture of the endophytic bacteria*

All samples were washed with tap water to remove soil and dust. The samples were then surface sterilized using ethanol 75 % (1 min), 0.5 % sodiumhydrochloride (3 min), and ethanol 75 % (30 s). Lastly, they were rinsed with sterile aquadest. The samples were then grinded and diluted. One mL of aliquot was plated onto LB agar (LB; 10 g/L pepton,5 g/L Yeast extract, 10 g/L NaCl, 15 g/L agar) and the plates were incubated for 24-48 hours with the temperature of 37°C. Prior to the collection of pure culture, the purity of each colony was checked by using the steak quadran method.

### *2.3. Screening of L-methioninase producing bacteria*

The screening method followed Gullati *et al.* [7] while the medium for the screening of L-methioninase bacteria was Modified Czapek Dox (MCD). The composition was 6 gr/L $Na_2HPO_4$, 2 gr/L, $KH_2PO_4$, 0.5 gr/LNaCl, 20 gr/L L-methionine, 2 gr/L glucose, 0.2gr/L $MgSO_4$, 0.005 gr/L $CaCl_2$, 20 gr/L agar and 0.007 %.BTB. pH for the medium was set at 5.5 - 6. All tested bacteria were plated onto MCD and incubated at 37°C for 24-48 hours.

### *2.4. Bacterial identification*

Bacterial identification was carried out using the Microbact kit analysis. This test is highly dependent on the results of the preceding oxidase test; if the result of the oxidation is positive then the test is done using the microbact 24E, and if it is negative the microbact 12E will be used. The microbact 24E or 12E was performed based on the manufacture protocol. In addition, the bacteria was also identified its gram type with the standard gram analysis procedure.

## 3. Result and discusion

### *3.1. Isolation of endophytic bacteria*

L-methioninase enzyme-derived bacteria were isolated from the *Rhizophora mucronata* mangrove plant from the Tuban Regency. It was diluted until it reached the appropriate number for the isolation of pure colony. The culturable endophytic bacteria from each part of the mangrove is shown in Table 1. The colonies were derived from the roots, stems and leaves of *Rhizopora mucronata* and it appeared that each part of the sample had culturable bacteria.

IOP Conf. Series: Earth and Environmental Science **137** (2018) 012077 doi:10.1088/1755-1315/137/1/012077

**Table 1.** Culturable endophytic bacteria isolated from *R. mucronata.*

| Sampel | Code | Cultured colony |
|---|---|---|
| Root | R1 | 1 |
| Leaves | L1 | 1 |
| Stem bark | S1, S2 | 2 |

Four colonies were obtained from the mangrove. Each isolate was then subjected to the screening process to test its capability to produce L-methioninase. The result of the L-methioninase screening showed that only one of the bacterial isolates produced L-methioninase (figure 1.). This is marked by the formation of a blue color surrounding the tested colony on the screening media. The formation of the blue color is due to the production of ammonia that will subsequently react with BTB which indicates the pH change in the media.

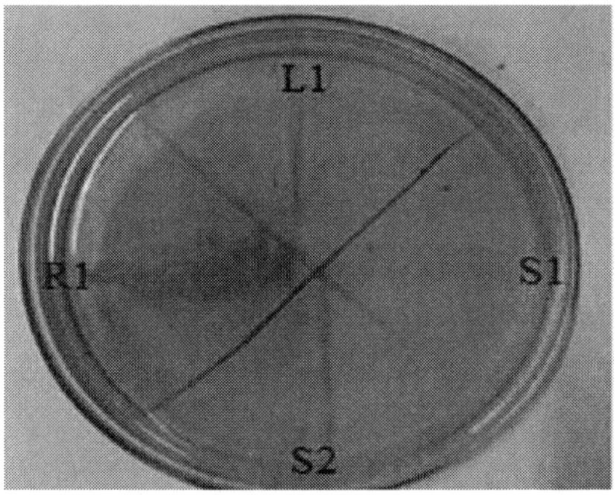


**Figure 1.** Positive L-methioninase producing bacteria. ($L_1$ = strain 1 from leaf, $S_1$ = strain 1 from stem, $S_2$ = strain 2 from stem, $L_2$ = strain 2 from leaf).

A larger the blue zone surrounding the isolates shows a greater activity of bacteria in producing L-methioninase. According to Suganya et al (8), L-methioninase generates ammonia from its catalisis. Selective media which was added with BTB (Bromothymol Blue) had a yellow color and turned to blue due to the change in the pH. The best isolate is the code samples of R1 which is the result of isolates from the root of the mangrove. It was further identified by using the Microbact system.

### *3.2. Bacterial Identification*

Gram staining revealed that the isolates bacteria is a gram positive bacteria (figure 2). Further analysis using microbact system was conducted using microbact 24E.

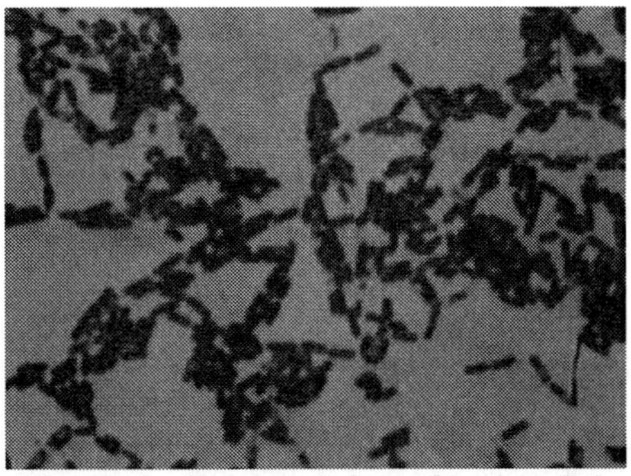

**Figure 2.** Gram staining of isolate R1.

The microbact system test is a test used to identify microorganisms which is highly dependent on the results of the preceding oxidase test. If the result of the oxidation is positive then the test is done using microbact 24E; while if the result is negative the microbact 12E will be used. This system is a standard micro-substrate system designed to simulate conventional biochemical substrates used to identify negative gram bacilli. The first stage of this test is placing the pure isolate in a centrifuge at 3000 rpm for 15 minutes. From the results of the centrifuge the supernatant and pellet was obtained but only part of the pellet was used. The resulting pellets were then given 5 ml of NaFis and then fed on Microbact 24E or 12E depending on the oxidation yield of 0.1 ml and incubated for 18-24 hours at 37°C. Evaluation of the results can be seen in the microbact wells, whether the results are positive or negative can be seen by comparing it with the color chart and the results are written on the Patient Record form. The octal numbers are derived from the sum of positive reactions alone from each group. The bacteria species can be identified from this total number by entering data into computer software [9].

The Microbact analysis revealed that the isolates R1 is *Bacillus subtilis* (table 2). Since this is a new strain from the coastal of Jenu, we have further named this bacillus as *Bacillus subtilis* UBTn7

**Table 2.** The result of microbact test.

| No. | Biochmeical test | Research | Bergey's | Reference |
|---|---|---|---|---|
| 1 | Spore | + | + | + |
| 2 | Oxidase | + | + | + |
| 3 | Motility | - | + | + |
| 4 | Nitrate | + | + | - |
| 5 | Lysin | - | Not tested | - |
| 6 | Ornithin | - | Not tested | - |
| 7 | $H_2S$ | - | Not tested | - |
| 8 | Glucosa | + | + | + |
| 9 | Mannitol | + | + | + |
| 10 | Xylose | + | + | + |
| 11 | ONPG | + | Not tested | + |
| 12 | Indole | - | Not tested | - |
| 13 | Urease | - | Not tested | - |
| 14 | V-P | + | Not tested | + |
| 15 | citric | - | + | - |
| 16 | TDA | - | Not tested | - |
| 17 | Gelatin | - | + | - |
| 18 | Malonat | - | Not tested | - |
| 19 | Inositol | - | Not tested | - |
| 20 | Rhamnose | - | Not tested | - |

IOP Conf. Series: Earth and Environmental Science **137** (2018) 012077 doi:10.1088/1755-1315/137/1/012077

| | | | | |
|---|---|---|---|---|
| 21 | Sucrose | - | + | - |
| 22 | Lactose | - | - | - |
| 23 | Arabinose | + | + | + |
| 24 | Adonitol | - | Not tested | - |
| 25 | Raffinose | - | Not tested | - |
| 26 | Salicin | - | Not tested | - |
| 27 | Catalase | + | + | + |
| 28 | Arginin | - | Not tested | - |
| 29 | Coagulase | - | Not tested | - |
| 30 | Beta Hemolisis | + | Not tested | + |
| 32 | Starch hydrolysis | + | + | + |
| 33 | Casein hydrolysis | + | + | + |

The bacteria is a motilbacteria with bacilform with a colony diameter of 3.11 mm, has a oxidase reaction and is catalase positive. The isolate were creamy and globular colonies with uneven edges to the colony, flat, and having a wet consistency. *Bacillus subtilis* is a gram-positive bacteria, with the form of the stems forming a chain; there are several species that are aerobic obligate and also anaerobic facultative; and have endospores as a structure to be able to survive harsh environmental conditions [10]. *Bacillus subtilis* is a non-pathogenic and non-tocix bacteria that is gram-positive and has characteristics that is much used in the production of food-grade enzymes. This bacteria is a saprophytic microorganism that can be found everywhere, usually in water, land, air, and residue of a decaying plant.

## 4. Conclusion

L-methioninase-producing endofit bacteria is able to be isolated from the *Rhizopora mucronata* mangrove derived from Jenu Beach, Tuban regency, East Java. The best isolate in producing l-methioninase is R1 which is known as *Bacillus subtilis* UBTn7.

## 5. References

[1] Sato D, Nozaki T 2009 *IUBMB Life* **61** 1019–1028
[2] Nozaki T, Ali V, Tokoro M 2005 *Advances in Parasitology* **60** 1–99
[3] Christina A, Christapher V, Bhore S J 2013 *Pharmacogn Rev.* **7** 11–16.
[4] Brader B, Compant S, Mitter B, Trognitz F, Sessitsch A 2014 *Curr. Opin. Biotechnol.* **27** 30–37
[5] Prihanto A A,Wakayama M 2016 *Adv. Food Nutrition Res.* **79** 1-25
[6] Prihanto A A, Firdaus M, Nurdiani R 2011 *J. Food. Sci. Eng.* **1** 386-389
[7] Gulati R, Saxena RK, Gupta R 1997 *Lett. Appl. Microbiol.* **24** 23-6.
[8] Suganya L, Govindan K, Prabha P, Murugan M 2017 *Biocata.l Agric. Biotechnol.* **12** 104–115
[9] Murtiningsih 1997 Evaluation of the serobact$^{TM}$ and microbact$^{TM}$ system for the detection and identification of *Listeria* spp. (Australia : Department of Food Science and Technology, The university of New South Wales)
[10] Korsten L dan Cook N 1996 *S Afri Avoc Grow Assosi Yb.* **19** 54-58.

IOP Conf. Series: Earth and Environmental Science **137** (2018) 012078 doi:10.1088/1755-1315/137/1/012078

# The effect of various concentration of tilapia (*Oreochromis* sp.) surimi for edible coating on the shelf-life of *Pangasius* sp. fillets

**M A P Purnama[1], Agustono[1] and A M Sahidu[1]**

[1]Faculty of Fisheries and Marine, Airlangga University, Jl. Airlangga 60115, Surabaya, East Java, Indonesia.

E-mail: adriana_monica16@yahoo.co.id

**Abstract.** *Pangasius* sp. fillets prone to deterioration the quality that will affect the appearance and the shelf life of fillets. The effort to extend the shelf life of fish fillet that is by using an edible coating. Surimi can be used as a protein-based edible coating because they have superior inhibitory and mechanical properties compared to the polysaccharides based material. Surimi can be made from freshwater tilapia (*Oreochromis* sp.) fish. The experimental design used was Completely Randomized Design (CRD) with five treatments of surimi (0 gr, 2 gr, 4 gr, 6 gr, 8 gr) with four replications. The results showed that *Pangasius* sp. fillets with an edible coating 8 gr surimi have the highest value in the organoleptic test. The pH testing on *Pangasius* sp. fillets with edible coating 2 gr, 4 gr, 6 gr, and 8 gr surimi from the $0^{th}$ hour to $18^{th}$ hour have increased but slower than *Pangasius* sp. fillets without edible coating surimi. The best value of Total Plate Count (TPC) test is in edible coating 6 gr and 8 gr surimi as it is in accordance with SNI 2696:2013 at room temperature storage until the $18^{th}$ hours.

## 1. Introduction

*Pangasius* sp. is a freshwater fish commodity with high potentials. This fish is much in demand by the community and has a wide export market [1]. Indonesian Fishery Statistics shows that the number of the national production increased from 229,267 tons in 2011 to 347,000 tons in 2012 and 410,883 tons in 2013 [2]. These data indicate that has an important role in national fishery industry.

There has been effort to develop of products in the form of fillets similar to what is known in the global market as a dory fillets [3]. is not scaly, has less spinal bones and reddish-white flesh, and is easy to peel, so it is relatively easy to make good fillet products from it [4].

The problems that often arise in the development of fish fillets are that they quickly decline in quality. fillets are prone to damage and deterioration that will affect the appearance and shelf life of the fillets. This is due to the characteristics of that contain high protein and water and the pH of its meat is close to neutral [5]. According to Killoincceker *et al.* [6], edible coating can be used to extend the shelf life and maintain product quality. Edible coating is formed directly on packaging materials and food products, which serve as a barrier to moisture, lipid, and gas and it can improve texture of the food products [7]. Edible coating is important for food products that are easily damaged, such as seafood, because it can inhibit the growth of microorganisms on the surface of the products [8].

IOP Conf. Series: Earth and Environmental Science **137** (2018) 012078 doi:10.1088/1755-1315/137/1/012078

Surimi is a hydrocolloid that can be used as an ingredient to make protein-based edible coating [9]. Protein-based films and coatings have superior inhibitory and mechanical properties compared to the homopolymer polysaccharides-based materials since proteins contain 20 different types of amino acids that produce more varied functional characteristics, such as the high molecular bonding [10]. Surimi can be made from freshwater fish such as from fillets of tilapia (*Oreochromis* sp.) since the flesh is white and the side of the body is thick so it is suitable to be made into fillets (boneless cutlets). According to Yoon *et al.* [11], tilapia has been used to produce surimi due to its good ability to form gel. There have not been many studies performed on surimi-based edible film or coating and its application, including edible film from Alaskan Pollack surimi [9] and from red snapper fillets [12]. Most edible films or edible coatings that have been produced are from marine fish, thus this study was aimed to use surimi from fresh water tilapia fillet. Therefore, the objective of this study was to determine the effect of various concentration of tilapia surimi tilapia for edible coating on the shelf life of fillets

## 2. Methodology

### *2.1. Materials and chemicals*

The tools used to produce surimi were knife, cutting board, muslin cloth, stirrer, food processor, kitchen scale, analytical balance, plastic tray, cooler box and equipments used for testing total volatile base (Conway Cup), pH (pH meter), and total plate count (beaker glass, measuring cup, hotplate, magnetic stirrer, tissue paper, gloves, mask, analytical balance, test tubes, test tube shelf, spoon, petri dish, Bunsen burner, inoculation needle, autoclave, spatula, pH meter, oven, incubator, heat-resistant plastic, paper label, knife, lighter, thermometer, vortex, Drigalski spatula, cotton, and clamp). For making edible coating from tilapia surimi, several equipments were used: hot plate, measuring cup, magnetic stirrer, pipette, analytical balance, pH meter, coating container, and glass bottle. The ingredients used in this study were tilapia, water, ice, cryoprotectant (sodium tripolyphosphate), NaOH 1 M, saline, distilled water, , 70% alcohol, 95% alcohol, and plate count agar (PCA).

### *2.2. Surimi manufacturing from tilapia fillets*

Total volatile base (TVB) analysis and pH measurement were performed before surimi production to know the freshness level of tilapia that was used. The manufacture of surimi from tilapia fillets used methods modified from [13]. Tilapia Fillet (skinless) was grinded until smooth and crushed using a food processor to produce meat without bones, spines, and dirt. After that, the fish meat was washed twice using cold water (±5 °C) and 0.3 % (w/w) salt solution. First immersion with cold water (water: meat = 3:1) was performed for 10 min to clean dirt that was still attached to the meat as well as to dissolve sarcoplasmic protein. The second immersion was with 0.3 % (w/w) salt solution (salt: meat = 3: 1) for 10 min and then it was filtered with a muslin cloth. After the process, cryoprotectant (0.3% sodium tripolyphosphate) was added and mixed using a food processor. Cryoprotectant was added to prevent protein denaturation in frozen surimi. Surimi produced was stored in polyethylene plastic at -20°C.

### *2.3. The production of edible coating surimi solution*

The production of edible coating was modified from [14]. First Frozen tilapia surimi of was thawed for 20 min. After that, it was weighed as much as 0 g, 2 g, 4 g, 6 g, and 8 g. The surimi was then dissolved in 1 M NaOH until the pH reached 11 and heated with 150 mL distilled water on a hot plate. The solution was homogenized and heated to a temperature of ± 55 °C. Next, it was filtered using muslin cloth and a filtrate of surimi solution was obtained.

### *2.4. Application of edible coating surimi tilapia fillets.*

First, fillet (7x3 cm) was soaked in 50 mL edible coating solution obtained in 2.3 mL for 30 s on each side, then it was drained and left to dry for 30 min. After that, fillets were stored at room temperature for 18 hand observed every 6 h [15].

## 3. Characterization of edible coating surimi

### *3.1. Total plate count*

Microbial testing was done by determining the total plate count (TPC) (Nugraheni, 2013).The maximum limit of microbial contamination in fish fillets according to SNI 2696: 2013 is $5x10^5$ colony/g. TPC was performed by creating a multilevel dilutions. One g of fillet was grinded gin nine mL of sterile saline with a ratio of 1: 9 ; thus, a $10^{-1}$ dilution was obtained. next, 1 mL of $10^{-1}$ dilution was taken using a sterile volumetric pipette and inserted into a reaction tube containing 9 mL of sterile saline and homogenized to obtain a $10^{-2}$ dilution. Further dilutions were prepared ($10^{-1}$, $10^{-2}$, $10^{-3}$, $10^{-4,}$ and $10^{-5}$) [16].

Sterile PCA media that had been cooled to 50 °C was inserted into 15 mL petri dishes. While pouring the medium, the lid must not be opened too wide to avoid contamination. The Petri dishes were then moved on the table in a circular motion mimicking the number eight, and after that it was left to solidify (Fardiaz, 1993). TPC calculation was done by spread plate method by taking 1 mL of the diluted sample with a sterile volumetric pipette, and then it was transferred to the surface of the media and flattened it using sterile Drigalski spatula. The petri dish was incubated at 37 °C for 24 h n reversed position [16].

### *3.2. pH measurement*

Measurement of pH of fillet meat was performed by dissolving 3 g in 50 mL sterile distilled water. Sample was homogenized and the then the pH was measured using a pH meter. Determination of pH is done after pH meter is calibrated first. After that, the electrodes are rinsed with distilled water and dried. The electrode is dipped in the sample solution, and the pH measurements can be set. The electrode is left immersed for a while until a stable reading is obtained, then the sample pH can be recorded.

*3.3. Organoleptic test.* Organoleptic testing is a way of testing using the human senses as a primary tool in assessing the quality of fish and fresh fishery products based on SNI 2696: 2013. There were 30 untrained panelists from Fisheries and Marine Faculty of Airlangga University. The trait of organoleptic testing is subjective because it only relies on the sense and sensibility of panelists [17].

## 4. Results and Discussion

Total Volatile Base (TVB) of the raw material of tilapia fillets was 20.07 mgN/100g with pH of 6.97. The value of TVB within the range of 20-30 mgN/100g indicates that the raw material is still fresh and in the edible boundaries [18]. Tilapia (*Oreochromis* sp.) fillets showed a neutral pH that would help to create high gel strength because the myosin is easily soluble [19].

Treatments with edible coating application were found to provide higher organoleptic value than control treatment without edible coating within similar storage time. This shows that the application of surimi as edible coating could maintain the quality of fillets. Organoleptic tests performed included appearance, odor, and texture. Appearance is an important organoleptic parameter because it is the sensory trait that will be seen first by the consumers [20]. The value for appearance given to all samples (control, 2 g, 4 g, and 6 g edible coating) was 7 in a hedonic scale. Meanwhile, the value given to treatment with 8 g surimi was 9 (table 1). Treatment with 8 g surimi could form good edible coating. Edible coating from surimi can cover the surface of the fillet perfectly and when applied, it can produce clear, transparent, shiny, and bright fish fillets [12].

**Table 1.** Organoleptic Average Value Result of Early (*Pangasius* sp.) Fillets.

| Organoleptic Test | Edible coating 0 gr surimi | Edible coating 2 gr surimi | Edible coating 4 gr surimi | Edible coating 6 gr surimi | Edible coating 8 gr surimi |
|---|---|---|---|---|---|
| Appearance | 7 | 7 | 7 | 7 | 9 |
| Odor | 9 | 9 | 9 | 9 | 9 |
| Texture | 7 | 9 | 9 | 9 | 9 |

At the end of the organoleptic test, the appearance value given to control treatment was 5 as it began to change its color and became dull. Meanwhile, the value given to other treatments with edible coating was 7 (table 2). This is probably due to the non-enzymatic browning reaction or Maillard reaction that may occur during storage. Maillard reaction describes the brownish reaction between reducing sugars and amino acid groups. Surimi, in the preparation for edible coating, was first dissolved with NaOH to reach pH 11. The pH value has a significant effect on Maillard reaction. The browning level increases with increasing pH [21]. The Maillard reaction has an optimum pH above 7. This alkaline condition induces the degradation of myofibrillar protein, leading to the availability of free amino groups for the browning reaction [22].

**Table 2.** Organoleptic Average Value Result of Final (*Pangasius* sp.) Fillets.

| Organoleptic Test | Edible coating 0 gr surimi | Edible coating 2 gr surimi | Edible coating 4 gr surimi | Edible coating 6 gr surimi | Edible coating 8 gr surimi |
|---|---|---|---|---|---|
| Appearance | 5 | 7 | 7 | 7 | 7 |
| Odor | 3 | 5 | 5 | 5 | 5 |
| Texture | 3 | 5 | 5 | 5 | 5 |

A sign of quality deterioration in fishery products is the emergence of off odor due to the production of free volatile compounds from proteins by microbial decomposition [20]. At the beginning of the organoleptic test, all samples were given the score 9 for their odor (table 1). Edible coating surimi has a neutral aroma, so when applied to fillet, it does not cause odors that deviate from the natural aroma of fillet [12].

At the end of organoleptic test, control sample received the score 3 for odor (acidic, little smell of ammonia, rancid). Meanwhile, other treatments were given the score of 5 (musty, slightly rancid) (table 2). Coating layer proved to be able to slow the emergence of off odor caused by the microbial decomposition of protein. Lipid oxidation in fish meat can cause unwanted rancid odors and carbonyl compounds resulting from the lipid oxidation can damage amino acid causing the formation of brown pigment and striking rancid odor [23]. However, if the organoleptic value is adjusted to the SNI 2696:2013 standard fillet was scored 7, the fillet coated with edible coating surimi cannot be accepted by consumers until the 18 h hat room temperature storage.

Good quality Fish fillets have a solid, compact, and elastic texture. When it has declined in quality, the texture becomes soft and crumbling [20]. This study indicated that the score for texture of control sample at the beginning of organoleptic testing was 7 (solid, less compact, less elastic). The score given to other treatments were 9 (solid, compact, and elastic) (table 1). At the end of organoleptic testing, control sample was scored 3 (soft, inelastic, and watery). Meanwhile, the score for other treatments was 5 (somewhat soft, less elastic, and slightly watery) (table 2). The organoleptic value of fillet tended to decrease during 18 h of h the texture of the fish to be not compact, soft, and tender, due to the autolysis process that causes changes in fish meat [24].

Degree of acidity (pH) is one indicator that is measured to determine the level of the fishery products freshness chemically. Generally, every microorganism has the optimum pH range for its growth that is usually between 6.5 and 7.5 [21]. fillet had fluctuating pH values at 0, 6, 12, and 18 h storage. Control sample increased its pH in 0 to 12 h. Meanwhile, the pH of other treatments was also increased (table 3), but in a slower rate because the fillets were protected by surimi edible coating with stable gel, so the process of protein degradation became slow due to the strong bond in gel such as by disulfide bonds that was involved in the stability of the film matrix from surimi [25].

**Table 3.** Average Result of pH (*Pangasius* sp.) Fillets.

| Observation time | pH Treatment | | | | | Average±SD |
|---|---|---|---|---|---|---|
| | A | B | C | D | E | |
| 0 Hour | $6.54^{a}$ | $6.77^{ab}$ | $6.93^{b}$ | $6.95^{b}$ | $6.91^{b}$ | 6.82±0,171 |
| 6 Hours | $6.89^{a}$ | $7.29^{b}$ | $7.35^{b}$ | $6.83^{ab}$ | $7.15^{ab}$ | 7.10±0,233 |
| 12 Hours | $6.98^{a}$ | $7.05^{a}$ | $7.14^{a}$ | $7.09^{a}$ | $7.10^{a}$ | 7.07±0,060 |
| 18 Hours | $6.91^{a}$ | $6.85^{a}$ | $7.00^{a}$ | $7.01^{a}$ | $6.88^{a}$ | 6.93±0,07I |

Note : A). Control without edible coating, B). edible coating 2 gr surimi, C). edible coating 4 gr surimi, D). edible coating 6 gr surimi, E). edible coating 8 gr surimi. The notation shown with different superscript letters in the same column shows the comparison between treatments having significant differences ($P < 0.05$).

However, pH of treatments with edible coating (2 g, 4 g, 6 g, and 8 g surimi) decreased after h18 h storage because protein-based edible coatings are a high barrier to oxygen and carbon dioxide. According to Aryanta [26], pH reduction in *Pangasius* sp. is caused by lactic acid accumulation that affects the increasing growth of lactic acid bacteria (BAL) due to reduced oxygen content. Moreover, protein-based edible coatings may inhibit bacterial activity, resulting in inhibition of production of non-protein nitrogen content that can lead to accumulation of bases [27].

The pH value of control treatment as well as treatments with edible coating increased with increasing storage time due to protein degradation and production of their derivatives by microorganisms that produce volatile bases, such as trimethylamine, dimethylamine, and ammonia. In addition, the increasing pH was caused by the alkaline condition of surimi filtrate that reached pH 11 due to the addition of NaOH. NaOH is added to surimi solution to dissolve the myofibrils protein. Microorganisms are the main cause of food damage. The growth of microbes in food will cause damage and quality deterioration. Total plate count (TPC) is one of the microbiological parameters to determine the level of deterioration. The TPC value of fillet decreased in Treatment D (edible coating with 6 g surimi) and E (edible coating with 8 g surimi) at 18 h of room temperature storage **(Table 4)**. If adjusted to SNI 2696: 2013, treatment E (edible coating 8 g surimi) and D (edible coating 6 g surimi) had been able to inhibit the growth of microorganisms in lower than the maximum limit of $5x10^5$ colonies/g up to 18 h of storage. Thus the fillets were still edible until h18 h of storage at room temperature. The increased surimi concentration added in the edible coating was able to increase the amino acid concentration and the bond in the film structure. According to Junianto [28], the bond in the structure of a film or coating is influenced by the molecular weight and the length of the amino acid chain. Greater length of amino acid chain will make the molecular weight larger, so the layer of film or coating that is generated will be tighter.

The results showed that surimi edible coating had stable gel strength so that when it was applied to the *Pangasius* sp. fillet, it could stick and cover the fillet surface properly due to the myofibril fish protein that has the ability to form a stable three-dimensional gel network [11]. The coating of *Pangasius* sp. fillets by surimi edible coating may affect the oxygen content present in the product. Edible coating surimi can inhibit the products from being in contact with oxygen, so that oxidation process gets inhibited. Moreover, protein-based edible coating is rich in amino acids that have good antioxidant properties such as cysteine and aromatic amino acids [29].

IOP Conf. Series: Earth and Environmental Science **137** (2018) 012078 doi:10.1088/1755-1315/137/1/012078

**Table 4.** Average result of Total Plate Count (TPC) (*Pangasius* sp.) Fillets.

| Observation time | Total Plate Count (TPC) Treatment | | | | |
|---|---|---|---|---|---|
| | A | B | C | D | E |
| 0 Hour | <30 colony | <30 colony | <30 colony | <30 colony | <30 colony |
| 6 Hours | $7x10^{6\ c}$ | $1.75x10^{5\ b}$ | $2.21x10^{4\ a}$ | $5.5x10^{4\ a}$ | $5.72x10^{4\ b}$ |
| 12 Hours | $8.3x10^{6\ c}$ | $7.57x10^{5\ b}$ | $7.65x10^{5\ b}$ | $7.23x10^{5\ b}$ | $6.45x10^{5\ a}$ |
| 18 Hours | $8.25x10^{6\ d}$ | $7.60x10^{5\ bc}$ | $7.03x10^{5\ ab}$ | $2.85x10^{5\ c}$ | $1.9x10^{5\ a}$ |

Note : A). Control without edible coating, B). edible coating 2 gr surimi, C). edible coating 4 gr surimi, D). edible coating 6 gr surimi, E). edible coating 8 gr surimi. The notation shown with different superscript letters in the same column shows the comparison between treatments having significant differences ($P < 0.05$).

## 5. Conclusions

Edible coating from tilapia surimi can slow down the shelf life of *Pangasius* sp. fillets at room temperature storage. The higher the surimi concentration, the higher the organoleptic value of the fillet and the lower the Total Plate Count value, which was in accordance with SNI 2696:2013. The highest scores of organoleptic testing for appearance, odor, and texture were obtained by sample of fillets with 8 g of surimi edible coatings. Control treatment increased its pH continuously from 0 h to 12 h. Meanwhile, the pH of treatments with edible coating was fluctuating but it tended to increase and approach the alkaline condition slower. The Total Plate Count of samples with 6 and 8 g of surimi edible coating after 18 h of storage at room temperature was in accordance with SNI 2696: 2013 with the value $2.85x10^5$ and $1.9x10^5$, respectively. Further research on utilization of surimi wastewater as the base material for protein-based edible coating could be of interest.

## 6. References

[1] Hutagalung S P 2009 Explore the potential of Indonesian fishery products. *Food Review.* **4** 10-13
[2] Ministry of Marine and Fisheries Affairs 2014 Marine and Fisheries ion figure 2014 (ID): Ministry of Marine and Fisheries Affairs
[3] Ikasari D, Dwi S 2014 Squallen Bulletin of marine and fisheries postharvest and biotechnology **9** 63-74
[4] Susanto H, Amri K 2002 *Pangasius sp. cultivation* (Jakarta: Penebar Swadaya)
[5] Kurniasih M 2013 The effect of adding the essential oil of red ginger rhizome (Zingiber officinale var Rubrum) to fillet flour on inhibition of microbiological and oxidative damage presented in cold temperature Thesis Surakarta (ID): Faculty of Agriculture, Universitas Sebelas Maret
[6] Kilincceker O, Dodan I S, Kucukoner E 2009 *Food Sci. Tech.* **42** 868-873
[7] Casariego A, B W S Souza, A A Vicente, J A Teixeira, L Cruz, R Diaz 2007 *Proc. Innov*
[8] Cagri A, Zeynep U, Elliot T R 2004 *J. Food Protect.* **67** 833-848
[9] Shiku Y, Hamaguchi P Y, Benjakul S, Visessanguan W, Tanaka M 2004 *J. Food Chem.* **86**:93-499
[10] Iwata K, Khizaki S, Handa A, Tanaka M 2003 *J. Food Sci.* **86** 493-499
[11] Yoon W B, Gunasekaran S, Park J W 2004 *Appl. Rheol.* **4** 133-139
[12] Rostini I 2011 Development of edible coating on boiled shrimp based surimi waste red fish fillet (Lutjanus sp.) Thesis Faculty of Fisheries and Marine Sciences Bogor Agricultural University
[13] Suzuki T 1981 *Fish and Krill Protein Processing Technology* (London: Applied Science Publisher Limited)

IOP Conf. Series: Earth and Environmental Science **137** (2018) 012078 doi:10.1088/1755-1315/137/1/012078

[14] Tanaka M, Visessanguan W, Benjakul S, Hamaguchi P Y, Shiku Y 2003 *J. Food Sci.* **86** 493-499
[15] Suptijah P 2006 Descriptive characteristics and application of chitin-chitosan Bogor (ID): Faculty of Fisheries and Marine Sciences Bogor Agricultural University
[16] Florensia D, Utami 2012 *J. Life Sci.* **I** 113-118
[17] Adawyah, R 2011 *Fish Processing and Preservation Earth* (Jakarta: Bumi Aksara)
[18] Farber L 1965 *Freshness Test* (New York: Academic Press)
[19] Shimizu Y, Toyohara H, Lanier T C 1992 *Surimi production from fatty and dark fleshed fish species* (New York : Marcel Dekker)
[20] Herliany N E 2012 Kappa caraginan application of seaweed Kappaphycus alvarezi as edible coating on peeled shrimp Thesis Bogor agricultural university
[21] Winarno F G 2008 *Food chemistry and nutrition* (Jakarta: PT. Gramedia Pustaka Utama)
[22] Prodpran T and Benjakul S 2005 *J. Sci. Technol.* **27** 563–574
[23] Rosari M I, Widodo F M, Tri W A 2014 *J. Proces Technol fish prod.* **3** 34-43
[24] Waryani S W, Rika S, Farida H 2014 *J. Chem Technol.* **3** 51-57
[25] Artharn A, Benjakul S, Prodpran T, Tanaka M 2007 *Food Chem.* **83** 867-874
[26] Aryanta I W R 2007 The role of lactic acid bacteria in the food processing industry Udayana University Denpasar
[27] Saputra E, W Tjahjaningsih, Patmawati 2017 Effect of coating chitosan inhibition to quality of fish fillet Oreochromis niloticus at room temperature storage Faculty of Fisheries and Marine Airlangga University
[28] Junianto 2003 *Fish capture technique* (Jakarta Penebar Swadaya) p5-13
[29] Le Tien C, Vachon C, Mateescu M A, Lacroix M 2001 Milk protein coatings prevent oxidative browning of apples and potatoes *J. Food Sci.* **66**: 512–516

IOP Conf. Series: Earth and Environmental Science **137** (2018) 012079 doi:10.1088/1755-1315/137/1/012079

# The effect of sea-water and fresh-water soaking on the quality of *Eucheuma* sp. syrup and pudding

**H Novianty and S M C Herandarudewi**

UPT LPKSDMO Pari Island LIPI, Jalan Raden Saleh 43, Cikini, Jakarta 10330

Email: hildanovianty2012@gmail.com

**Abstract.** *Eucheuma alvarezii* is one of marine commodity with great opportunities to be developed in Indonesia. This seaweed can be used as an additional material in cosmetic and pharmaceutical products or for syrup and pudding. Post-harvest technique conducted by the seaweed farmers will affects the quality of dried and processed products. The purpose of this study was to observe the effect of post harvest technique on the quality of dried seaweed and hedonic test (favorable test) of processed product (syrup and pudding). This study was conducted using descriptive method. The study compared dried, syrup, and puddings from two differents post-harvest technique, between salt and fresh-water draining products. The results showed that fresh-water draining technique obtained better quality results organoleptic test. The results showed that panelist prefered the fresh-water drained products of syrup and pudding. The hedonic scores were much higher for the fresh-water drained products in all three catagories of color, taste, and smell.

## 1. Introduction

*Eucheuma alvarezii* is a seaweed species with a high economic value. This species of seaweed is an hydrocolloid type which produces carrageenan that has functional properties as a gel-forming, stabilizer, suspension, and emulsion texture builder [1]. Due to its function, this species of seaweed is used as an additional ingredient in the food, cosmetics and pharmaceuticals industries. In addition, it also contains vitamins and minerals needed by the human body. *Eucheuma* seaweed has moisture content (11-13.90 %); crude protein (2.69-9.20 %); fat (0.09-0.37 %); carbohydrates (4.94-13.38 %); crude fiber (0.95-2.50 %); ash (4.79-17.09 %); calcium (22.39-69.25 ppm); vitamin B1 (0.019-0.21 mg/100 g); vitamin B2 (2,26-8,45 mg/100 g); and vitamin C (12.00-43.00 mg/100 g) [2]. Seaweed nutrients contain carbohydrates (sugar or vegetable gum), protein, fat and ash. Most carbohydrates are cellulose and hemicellulose that cannot be digested entirely by the enzymes in the body so it can part of a less calorie – diet [3]. Therefore, this seaweed has a high added value because of its diverse functional substances that also has nutritional levels needed by humans. In addition, Indonesia has a huge potential of *Eucheuma alverizii* cultivation, because Indonesian waters possess characteristics that makes it suitable for the optimum growth of *Eucheuma alverizii*.

The program from the Ministry of Marine Affairs and Fisheries to increase the production of seaweed cultivation of Indonesia must be balanced with the processing of cultivated seaweed. A guaranteed quality of seaweed cultivation will affect the processed seaweed. Hence, the post-harvest techniques (after cultivation) of seaweed cultivation have an important role in ensuring the quality of seaweed. The existence of good post-harvest management will ensure the quality of seaweed products that will ultimately encourage the sustainability of the seaweed processing industry [4]. Studies of seaweed post-harvest techniques have been carried out by Itung and Marthen [5] and Poncomulyo [2], by doing the same technique which was washing the seaweed with sea water and drying it under the sun. Meanwhile, the drying technique can be done by drying on the ground surface, drying with the method of para-drying and drying by the hanging method [4]. The most suitable post-harvest technique will determine the quality of the finished product [3]. The initial handling of seaweed (post-harvest) should be well carried out, because good quality raw materials will produce good products as well. This study will compare the effects of sea-water and fresh-water soaking to the hedonic test of syrup and pudding as seaweed processed products.

## 2. Methodology

The study was conducted in May 2016 using a descriptive study. This research began with the planting of seaweed for 45 days with the long-line method and then post-harvest techniques were done in two ways: sea-water and fresh-water soaking. The seaweed from both techniques was made into syrup and pudding. An organoleptic test was carried out on the syrup and pudding by 15 panelists. The water content test was done by PT. Saraswanti Indo Genetech, Bogor. The study was done by the following scheme of post-harvest proccess (figure 1).

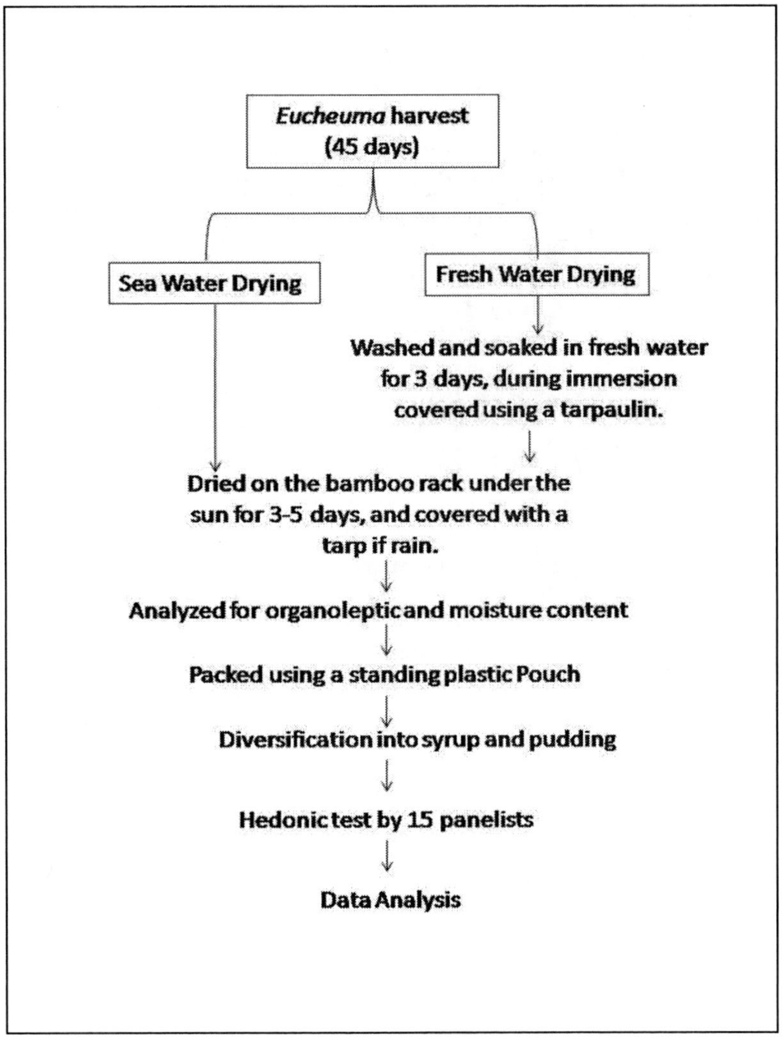


**Figure 1**. Seaweed post-harvest proccess

IOP Conf. Series: Earth and Environmental Science **137** (2018) 012079 doi:10.1088/1755-1315/137/1/012079

## 3. Result and Discussion

Economically, sea-water drying is less costly but fresh water drying was recorded to have resulted in better quality (figure. 2 and table 1). Fresh-water drying results in a white, clear, brighter appearance with less grains of salt attached to the dry thallus. On the other hand, the sea-water drying product has a darker appearance (brownish) with a considerable amount of dirt and sand granules attached to its thallus. Thus, physically, the post-harvest technique of the fresh-water drying system is better. The results of the moisture level test of the dried seaweed on both post-harvest treatment techniques showed (table 1) that the post-harvesting technique of fresh-water drying had a smaller moisture content of 14.36 % compared to sea-water drying which has a moisture content of 22.76 %. This is because in the sea-water post-harvest drying technique, the salt grains will remain attached to the thallus of the dried seaweed so that it will raise the weight of its water content. However, its water content still conforms to the dry seaweed SNI No.2690.1: 2009 [6].

**Figure 2.** Physical differences between sea water drying and fresh water drying seaweed.

**Table 1.** Qualitative data of *Eucheuma.*

| No | Sample | Moisture content (%) | Organoleptic | |
|---|---|---|---|---|
| | | | Appearance | Odor |
| 1 | Fresh water drying | 14,36 | White, clear, less salt | Seaweed specific ( not strong) |
| 2 | Sea water drying | 22,76 | Brown, lots of grains of sand and salt | Seaweed specific (strong) |
| SNI (2009)* : Dried seaweed | *Eucheuma* | Maks. 32 | - | Seaweed specific |

**in* [3]

IOP Conf. Series: Earth and Environmental Science **137** (2018) 012079 doi:10.1088/1755-1315/137/1/012079

The different treatments on seaweed post-harvest techniques were done to influence the quality of the processed products after harvest. As we can see from tables 2 and 3, table 2 shows the results of the hedonic test on the seaweed syrup beverage, while table 3 shows the results of the hedonic test on the seaweed pudding. The hedonic test is a test of preferences, in which the panelist (person who is tested) is asked for his/her personal response to preferences or dislikes using the hedonic scale [7]. The panelists who participated in this study were semi-trained panelists with a total of 15 people. A semi-trained panelist is a panelist who gets an explanation of hedonic testing and sensory traits with the number of panelists measuring from 15 to 25 people [7].

Tables 2 and 3 show that the seaweed syrup and seaweed pudding made from the fresh-water post-harvest technique had a greater hedonic value for all test categories (color, flavor and odor) compared to the products made using the sea-water drying technique. This indicates the panelists' preference for seaweed syrup and pudding products made using the fresh-water drying technique was preferred over the other. The specific odor of seaweed was stronger in the beverage and pudding product made by sea-water drying. This was not favored by the panelists. Table 1 also shows that dried seaweed made by the sea-water drying process resulted in a stronger specific seaweed odor compared to the other.

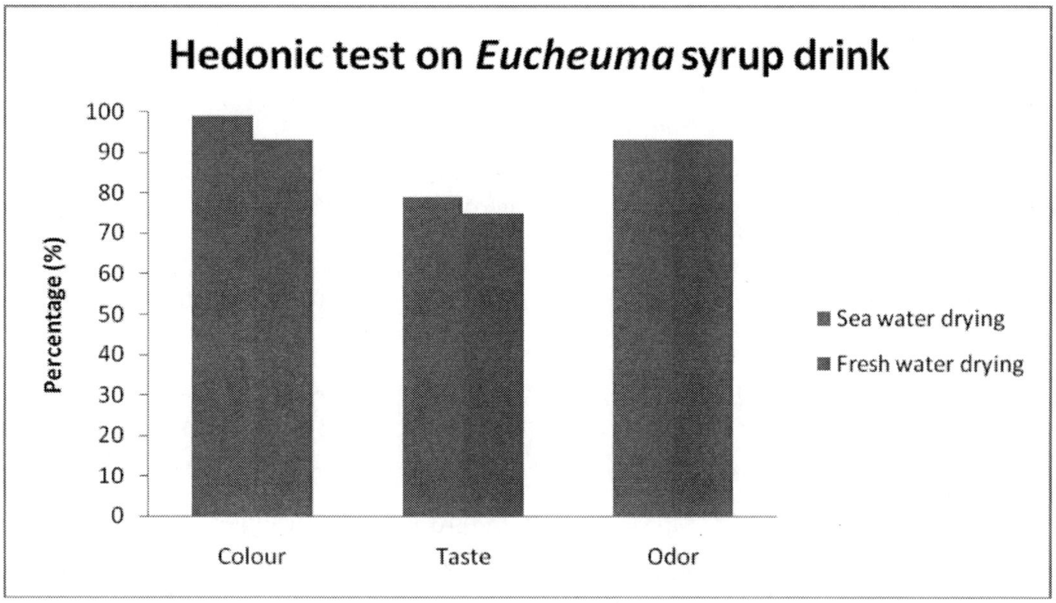


**Figure 3.** Hedonic test on *Eucheuma* syrup drink

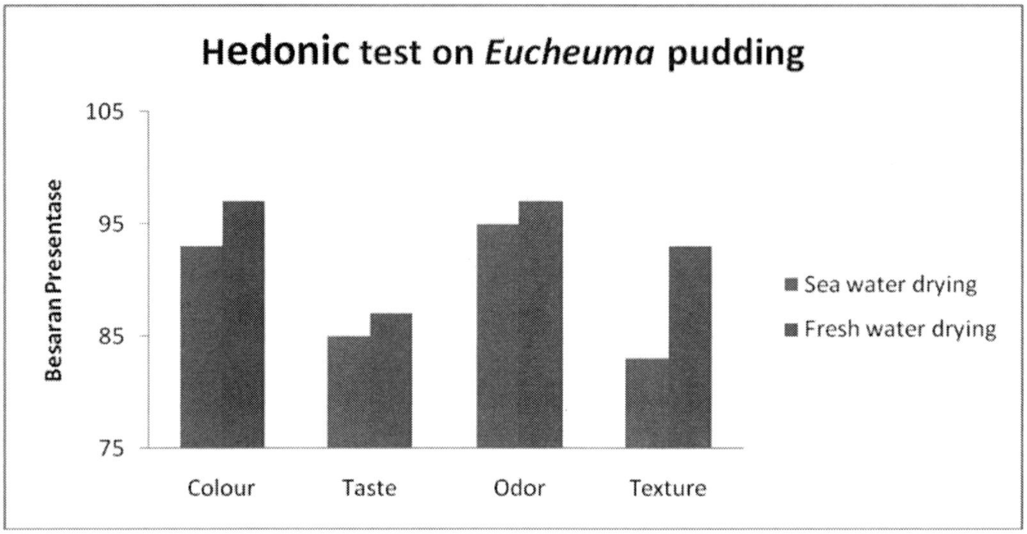


**Figure 4**. Hedonic test on *Eucheuma* pudding

## 4. Conclusion

There are differences in seaweed post-harvest techniques, namely fresh water drying and seawater drying. Freshwater drying is the leaching and soaking of harvested seaweed with fresh water; while in sea-water drying there is no washing and soaking treatment using fresh water. Both of these post harvesting techniques have influenced the results of the hedonic or preference test on the seaweed products of syrup and pudding, where the products made from fresh-water drying was more preferred.

## 5. References

[1] Susanto A B, Rini P and Alva W 2009 Seaweed Processing Technology in Indonesia *Biotechnol. and Seaweed Industri Proc. ISBN : 978-979-17585-3-6* 1-20 (Text in Indonesia)

[2] Poncomulyo T, Herti M, and Lusi K 2006 *Cultivation and Seaweed Processing* (Jakarta: Agro Media Pustaka) p 1-64 (Text in Indonesia)

[3] Wibowo S, Rosmawaty P, Muhammad D and Arif R H 2014 *ATC Proc. Techniques of Eucheuma* (Jakarta: Penebar Swadaya) 1-74 (Text in Indonesia)

[4] Direktorat Jendral Budidaya Perikanan (DJPB) 2016 Instructions to manage post harvest *seaweed htt://www.djpb.kkp.go.id/arsip/c265/PETUNJUK-PRAKTIS-MENGELOLA-PASCA-PANEN-RUMPUT-LAUT/?category id=13* Retrieved August 4 2016 (Text in Indonesia)

[5] Itung M and Marthen. Post Harvest Processing of *Eucheuma* and *Gracilaria* for export purpose *Marine Chimica. Acta.* **4** 5-8 ISSN 1411-2132 (Text in Indonesia)

[6] Peranginangin R, Ellya S and Muhammad D 2013 *Produce carrageen from seaweed* (Jakarta: Penebar Swadaya) p 1-74 (Text in Indonesian)

[7] Soekarto S T 1985 *Organoleptic assessment (for food and agricultural products)* (Jakarta: Penerbit Bhratara Karya Aksara) 1-122 (Text in Indonesia)

IOP Conf. Series: Earth and Environmental Science **137** (2018) 012080 doi:10.1088/1755-1315/137/1/012080

# The effects of using shell filters in the process of depuration for the survival of *Anadara* sp.

**K T Pursetyo, L Sulmartiwi, M A Alamsjah, W Tjahjaningsih, A S Rosmarini and M Nikmah**

Marine Department, Faculty of Fisheries and Marine, University of Airlangga, East Java, Surabaya, Indonesia 60115

E-mail: kustiawan@fpk.unair.ac.id

**Abstract**. *Anadara* sp. is one of the shellfish that has a source of animal protein is high and has a high economic value. However, to obtain a safe source of food, the products must meet the standard by the government, one of which is the limitations in heavy metals in the shells. In the standard of sanitation of shellfish is required to do the depuration process to remove the contaminants be it bacteria or heavy metals. In this study, randomized design with 5 treatments was used: P0 (control / without filter), P1 (25 % filter with shells), P2 (50 % filter with shells), P3 (75 % filter with shells), P4 (100 % filter with shells). Each treatment was replicated 4 times. The results showed that filtering of shell in depuration process could cause the highest shell death for 24 hours occurred in P4 of 24.39 % and the highest death during 48 hours also happened at the treatment of P4 which was equal to 61.71 %. During the research, water quality measurement was measured at 29-30 °C, pH 7.2-7.7, dissolved oxygen (DO) 4-4.4 mg/L and salinity 28-30 ppt.

## 1. Introduction

Cockles are potentially contaminated by heavy metals due to their filter feeder nature, so this biota is often used as a test animal to monitor the degree of the accumulation of heavy metals in marine organisms [1]. One type of shellfish that is abundantly found in Indonesian waters is the blood cockle (*Anadara* sp.). The blood cockle is a type of shellfish that has a high economic value and in general is a source of seafood in Southeast Asia and some Pacific regions [2]. Blood cockles have several uses, one of which is processed as food, however its shells can cause considerable waste. The discarded shells can act as a source of calcium. There has not been much utilization of blood cockle shells, except as raw materials for making local souvenirs.

Cockle shells contain 66.70% calcium carbonate [3], the high content of calcium carbonate can make use of them to be used as a water purifier. The calcium carbonate in shellfish can clean water, and can even reduce iron, manganese and other metals [4].

This method of depuration is principally a purification step of a biota under a controlled condition. The shellfish that are contaminated by heavy metals should be cleaned or depurated, the purpose of this depuration process is to reduce the risk of bacterial contaminants and some heavy metals harmful to human health [5].

Toxic depuration in shellfish is one of the efforts to avoid the impact of lead accumulation in shellfish that has the potential to harms humans who consume it [6]. According to the Decree of the Minister of Marine Affairs and Fisheries Number: KEP.17 / MEN / 2004, the depuration process is a cleaning process that uses a purifier with a recirculation system. This research uses a depuration method by using powdered shells as the filter. The shell contains 66.70 % calcium carbonate so as to absorb the heavy metal lead [3]. The high content of calcium carbonate can make the shellfish shells be used as a water purifier. The calcium carbonate in shellfish can clean water, and can even reduce iron, manganese and other metals [4].

## 2. Methodology

This research was conducted on 30th August to 30th September 2016 at the Faculty of Fisheries and Marine of the University of Airlangga, Surabaya. The testing of heavy metal content was conducted by using an Atomic Absorption Spectrophotometer (AAS).

The tools used during research were five tubs (95 cm x 70 cm x 75 cm), 20 plastic racks (42 cm x 33 cm x 14 cm), five plastic tubes with a height of 45 cm and a diameter of 26 cm, five water pumps, five flow meters, five UV meters, sponge filtration, pipes, hoses, scooters, measuring cylinders, digital scales and water quality gauges (thermometers, refractometers, pH pens, DO test kits, nitrate and nitrite test kits).

The materials used in the study where blood cockles (*A. antiquata*) originating from Sedati, Sidoarjo. The blood cockles used were 7-8 grams in weight and 10 kg of the shellfish were placed in each tub. Four shelves were placed in each tub along with sea water with a salinity of 30 ppt.

### *2.1. Work procedures*

The tubs were cleaned by using liquid soap followed by chlorine and then rinsed with clean water. The washed equipment was soaked with chlorine solution at 150 mg / l for 12-24 hours and rinsed with fresh water until the chlorine odor was gone [7]. The dried tubs were filled with 400 liters of seawater in each tub. After filling in the water, the installation of aeration, UV, water pumps, flowmeters and filters on each tub was carried out. The outer shells that were used as a filter were firstly washed and then pounded into a powder to a size of 0.5-2 cm. Next, the shell powder was placed in each tube as a filter. The amount of shell powder used was 15.3 kg for the 100 % filter, 11.5kg for the 75 % filter, 7.7k g for the 50 % filter and 3.8kg for the 25 % filter.

The method used in this research was the experimental method to find out the effects of the shell filter in the depuration process in decreasing the cadmium content of the shellfish. The experimental design was performed using Completely Randomized Design with a single source of diversity ie percentage of shell powder. This study used five treatments with four replications in each treatment. The treatments in this study are as follows:

P0: Without Filter Shells
P1: Using Filter Shells 25 %
P2: Using Filter Shells 50 %
P3: Using Filter Shells 75 %
P4: Using Filter Shells 100 %

## 3. Results and Discussion

Shellfish are biota that are potentially contaminated by heavy metals due to their filter feeder nature, so it is often used as a test animal to monitor the degree of accumulation of heavy metals in marine organisms [1]. The heavy metal cadmium will turn into poison for aquatic life when dissolved in the waters at a certain concentration. Heavy metals that enter the body of animals such as shells are generally unremoved from the shell's body [1]. The accumulation of heavy metals in the body of aquatic animals occurs because the rate of the collection of heavy metals by aquatic organisms is faster than the release process[8].

Based on the results of the research, the blood cockle *Anadara* sp. samples contained an average Cd content of 0.3598 ppm and Pb content of 2.144 ppm prior to being decontaminated. According to the Decree of the Minister of Marine Affairs and Fisheries Number: KEP.17 / MEN / 2004 regarding sanitary sanitation systems in Indonesia, there is a standard of the quality of live shellfish and its

IOP Conf. Series: Earth and Environmental Science **137** (2018) 012080 doi:10.1088/1755-1315/137/1/012080

processed products which are consumed directly. The threshold limits of heavy metals in processed products for consumption must meet the requirements of Cadmium (Cd) at a maximum of 1.0 ppm and lead (Pb) at a maximum of 1.5 ppm.

The process of depuration was conducted by using shell powder of different percentages as filters which was carried out for 48 hours [9]. Seawater was circulated continuously throughout the depuration process for about 48 hours. The percentage of death or mortality is calculated every 24 hours for 48 hours. The value of the mortality gap percentage is the number of dead shellfish divided by the number of shellfish at the beginning, as in the formula:

$$Mo = \frac{Mt}{N0} \times 100\% \qquad (1)$$

Information:
Mo: mortality
Mt: number of dead
N0: population

**Table 1.** Mortality of blood cockles.

| Treatment | Mortality ± SD 24 hours | Mortality ± SD 48 hours |
|---|---|---|
| P0 (Control) | 1.04 %c ± 0.6 | 3.12 % c ± 1.89 |
| P1 | 6.33 %b ± 3.82 | 30.61 % b ± 5.16 |
| P2 | 6.65 %b ± 5.35 | 31.33 % b ± 12.44 |
| P3 | 10.98 %b ±1,01 | 58.01 % a ± 0.84 |
| P4 | 17.71 %a ± 5,04 | 58.65 % a ± 3.06 |

Description: Different superscript on the same column shows a noticeable difference ($p < 0.05$).

Based on the analysis of variance (ANOVA) and Duncan's test, the results showed that the highest mortality found in the 48 hour treatment of P4 (58.65%) was not significantly different to the 48 hour treatment of P3 (58.01%). While the lowest mortality was found in the 24 hour treatment of P0 (1.04%).

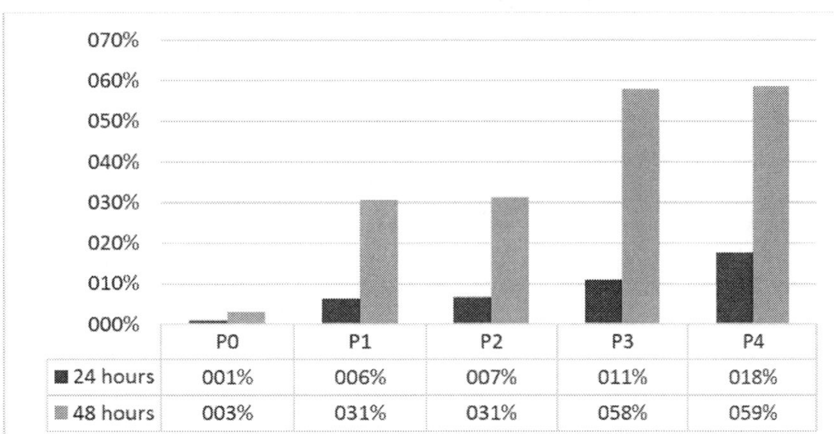


**Figures 1.** Mortality graphics.

High mortality was caused by the muddier water in the maintenance media so that the respiration process of the shellfish was disrupted and evidenced by the increasing Total Suspended Solid (TSS) content. The water quality measured during research is presented in table 2.

IOP Conf. Series: Earth and Environmental Science **137** (2018) 012080 doi:10.1088/1755-1315/137/1/012080

**Table 2.** Water quality.

| | Before | | | |
|---|---|---|---|---|
| | TDS (mg/L) | TSS (mg/L) | BOD (mg/L) | COD (mg/L) |
| P1 | 26280 | 12 | 185.73 | 3241.07 |
| P2 | 26280 | 12 | 185.73 | 3241.07 |
| P3 | 26280 | 12 | 185.73 | 3241.07 |
| P4 | 26280 | 12 | 185.73 | 3241.07 |
| P5 | 26280 | 12 | 185.73 | 3241.07 |
| | After | | | |
| | TDS (mg/L) | TSS (mg/L) | BOD (mg/L) | COD (mg/L) |
| P1 | 27150 | 13.5 | 121.945 | 1410.715 |
| P2 | 27300 | 19 | 118.87 | 1525.51 |
| P3 | 26830 | 39.5 | 140.64 | 1761.48 |
| P4 | 25370 | 142.335 | 158.925 | 1908.165 |
| P5 | 27150 | 62 | 171.285 | 2207.905 |

The above table shows that in the treatment of P4 the amount of Total Suspended Solid (TSS) at the 48th hour was 142.335 mg / L, TSS waters can reduce the value of waters and affect the lives of the organisms [10].The value of TSS 81-400 mg/L indicates that the suspended solid content was unsuitable for fishery[11]. TSS will affect the clarity of water, which in turn will affect the penetration of light and ultimately will affect primary productivity. High TSS can block the entry of sunlight into the water, which will interfere with the process of photosynthesis and cause the decrease of dissolved oxygen released by plants into the water [10].

## 4. Conclusion

The use of blood cockle shells as a filter had an effect on the number of deaths of the shellfish during the depuration process. The highest death occurred in P4 where 100% of the shells were used as a filter.

## 5. Reference

[1] Yennie Y, Murtini JT 2016 *J Mar. Fish. Sci. Ind* **2** 27–32 (In Indonesia)
[2] Shigeru S2009 ICMSS09 1-9
[3] Ulysses M, Lawrence RM, Valeriano B and Shigeru S 2009 *Comparative PSP toxin Accumulation in Bivalves, Paphia undulata and Perna viridis in Sorsogon Bay, Philippines* (France: Nantes) p 56
[4] Siregar S M 2009 Utilization of shells and epoxy resins on the characteristics of polymer concrete (MS thesis) p 89
[5] Simaremare S R 2013 *Healthy Work Envi.* **3** (In Indonesia)
[6] Gabr HR, Gab-Alla A-F 2008 *Oceanologia* **50** 83–93
[7] Chen CY, Chou HN 2001 *Toxicon* **39** 1029–34.
[8] Isnansetyo A Kurniastuty 1995 *Culture technique phytoplankton zooplankton : natural feed for the seeding of marine organisms* (Yogyakarta : Kanisius) p 67–71 (In Indonesia)
[9] Darmono 2006 Environment and pollution: relation to toxicology of metal compounds (Jakarta : University of Indonesia) p 67
[10] Murdinah M 2009 *Bull. Mar. Fish. Postharvest Biotechno.* **4**(2) 61–71
[11] Sumawidjaja K 1973 Limnology (IPB: Bogor) p 88 (In Indonesia)

IOP Conf. Series: Earth and Environmental Science **137** (2018) 012081 doi:10.1088/1755-1315/137/1/012081

# Analysis of consumer behavior in decision making of purchasing ornamental freshwater fish (case of study at ornamental freshwater fish market at Peta Street, Bandung)

**I Gumilar, A Rizal, Sriati and R Setiawan Putra**

Faculty of Fisheries and Marine Science, Padjadjaran University, Jl. Raya Bandung 45363, Bandung, West Java, Indonesia.

E-mail: iwang@unpad.ac.id

**Abstract.** This research aim was to analyzed process of decision making of purchasing ornamental freshwater fish at Peta Street, Bandung City and Analyzed what factors are driving consumers to buy freshwater fish Peta Street. The method used in this research is case study with rating scale and rank spearman analysis. The sampling technique is the accidental random sampling method consist of 30 respondents. The consumer's decision making process consist of five stages, namely the recognition of needs, information searching, alternative evaluation, process of purchasing, and the evaluation of results. The results showed that at the stage of recognition of needs the motivation of purchasing freshwater fish because respondents are very fond of ornamental freshwater fish, at the stage of information search, the information sources are from the print media and friends or neighborhood. At the stage of alternative evaluation, the reason consumers buy ornamental freshwater fish because the quality of good products. The stage of purchasing decision process consumers bought 1-5 fish with frequency of purchase 1 time per month. The evaluation of results of post-purchasing consumers feel very satisfied with the fish products and the price is very affordable. To observe the factors that influence purchasing motivation of consumers, spearman rank test is the method. The results showed that the quality and price of the product are the factors that most influence the purchase decision of ornamental freshwater fish with the range of student- t value 3,968 and 2,107.

## 1. Introduction

Ornamental fish is one of fishery commodities that have become a potential trading commodity in or outside the country. The production of freshwater ornamental fish in West Java in 2015 was 571.975.130 fish [1]. According to the world export statistics in 2010, the cost of ornamental fish reached above 350 million dollars with Singapore, Malaysia, and Thailand as the main exporters [2].

The trading activities of marine ornamental fish have become one of the main actors in fishery agribusiness, not only because the climate and natural resources of Indonesia could support it, but also because this business has proven to be able to survive in a long period of crisis. One of the large-scale trades of ornamental fish is Raiser that has the role to build the bridge for the cultivators of ornamental fish to sell domestic or international products. Raiser is located at Jalan Raya Bogor KM. 47, Cibinong, Bogor District. One of the small-scale trades of ornamental fish is Pasar Ikan Hias Jalan Peta (Ornamental Fish Market at Peta Street) that is located in Bandung. Even though in a small-scale,

IOP Conf. Series: Earth and Environmental Science **137** (2018) 012081 doi:10.1088/1755-1315/137/1/012081

the market could build the bridge between the producers and the consumers of ornamental fish in Bandung.

Ornamental Fish Market at Peta Street, Bandung, is a place for the freshwater or marine ornamental fish enthusiasts. Even though Ornamental Fist Market at Peta Street Bandung is located far from the coast, it has become the center of freshwater or marine ornamental fish trade. The price offered by the tradesmen there varies, depending on the quality of the fish itself. There are some social interactions in the buying and selling activities and the diverse behavior of the consumers in making decision of buying the ornamental fish. The information regarding the consumers' behavior could help the producers to set good marketing strategy.

This research was aimed to analyze the ornamental fish consumers' behavior at Peta Street, Bandung, and analyze the supporting factors of the consumers to buy freshwater ornamental fish at Peta Street.

## 2. Methodology

### *2.1. Place and time of research*

This research was conducted at Ornamental Fish Market at Peta Street, Bandung between March and April 2017.

### *2.2. Research method*

The research method used in this research was a case study method with the consumers that shopped in Ornamental Fish Market at Peta Street as the subject of the research.

### *2.3. Sampling Method*

The technique that was used to collect sample was accidental random sampling. The criteria of selecting the respondents were those who bought ornamental fish and were willing to do the interview. According to Sugiyono [3], the good sampling number in a research is between 30 to 500.

### *2.4. Data analysis method*

The data analysis method used in this research was quantitative descriptive analysis. This analysis was produced from the questionnaire and interview, which were tabulated and then analyzed. The feasibility test for the questionnaire was done by validity and reliability tests. The quantitative analysis was done by ranking the factors affecting the purchase decision making. The quantitative analysis methods used were the Ranking Scale and Spearman Rank Test.

#### *2.4.1. Validity and reliability tests*

A test is considered to have high validity when the tool runs the measurement functions appropriately or gives the correct measurement results. According to Kotler and Armstrong [4] stated that validity is closely related to the appropriateness of the measuring tool of the assessed concept so it could measure what should be measured appropriately. A test is considered reliable if it always produces the exact same result if it is tested to the same group in different time or different occasion.

#### *2.4.2. Rating scale*

##### *2.4.2.1. Likert scale*

According to Sugiyono [3], likert scale is used to measure somebody's or group's attitude, opinion, and perception towards social phenomenon.

##### *2.4.2.2. Guttman scale*

According to Sugiyono [3], stated that the measurement scale with Guttman type can get firm answer, which are: yes-no; have-have not; and positive-negative.

#### *2.4.3. Spearman test rank*

According to Karimah *et al.* [5], correlation test aims to test the relationship between two variables that do not show functional relationship (relation does not imply causation). Correlation test does not differentiate the variables (there is no dependent or independent variables) and the coherency of the relationship is stated in a coefficient form.

The hypothesis used in this research was:

$H_o$ : There is no relation between $Y$ (motivation) and $X$ (income, price, quality, sales service, family)

$H_1$ : There is a relationship between $Y$ (motivation) and $X$ (income, price, quality, sales service, family)

## 3. Result and discussion

### *3.1. Respondents characteristics*

There were 30 respondents consisting of 26 males and 4 females. The respondents' age was dominated by the age group of 40-50 years old who were mostly entrepreneurs. Sixteen of the respondents had bachelor degree , 4 respondents had family, 14 respondents had income above Rp. 5.400.000, and 24 respondents stayed in Bandung.

### *3.2. The process of ornamental fish purchasing decision*

#### *3.2.1. The introduction to the needs*

In this stage, the respondents' levels of fondness to ornamental fish were enthusiasm (56.7 % or as many as 17 people). The level of hobby for ornamental fish was as high as 70 % or as many as 21 people. The higher motivation for the respondents to buy ornamental fish was for amusement (56.7 % or as many as 17 people). And 70 % of the respondents stated that they purchased ornamental fish as an alternative to having mammals as pets.

#### *3.2.2. Information retrieval*

According to 16 respondents (53.3 %), the characteristics of healthy ornamental fish could be seen from its eyes and bright skin and the beauty of ornamental fish could be seen from its body colors. There were 12 respondents (40 %) who got the information of ornamental fish from printed/electronic media and 53.3 % or (16 respondents) got the information from their relatives or people around them.

#### *3.2.3. Alternative Evaluation*

The respondents that bought Koi fish were as much as 46.7 % or 14 people. The respondents that bought ornamental fish because of the good quality of the product were as many as 16 people (53.5%). Thirteen respondents (43.3 %) faced the unavailability of the product decided to cancel the purchase. The other alternative to buy ornamental fish in Bandung is small fish shops that were chosen by 13 respondents.

#### *3.2.4. Purchasing*

The respondents that bought 1-5 ornamental fish was as many as 20 people, while those having the frequency of purchasing that was once a month was 14 people. The respondents who bought Koi were as many as 15 people.

#### *3.2.5. The result evaluation after purchasing*

In this stage, 14 respondents stated that they would put the ornamental fish that they bought at the aquarium. According to 24 respondents, the level of cleanliness of the market could be categorized as sufficient. The respondents who thought the level of comfort was sufficient were 22 people. Those who believed that the market was satisfying were 16 people. There were 21 respondents that answered that the level of affordability of ornamental fish in this market was proper. The respondents who stated that they would buy smaller amount of fish if the price was increased were 18 people. The percentage

IOP Conf. Series: Earth and Environmental Science **137** (2018) 012081 doi:10.1088/1755-1315/137/1/012081

increase in price that was tolerated by the respondents was 30%, as stated by 14 people. There were 26 people who claimed that they would still buy the fish if they get an increase in their income.

*3.3. Validity test and questionairre reliability*

*3.3.1 The introduction to the needs*

the Cronbach's Alpha score was 0.788; this score was higher than 0.700 so it was statistically reliable. From the validity test on 4 questions that represented the dimension of recognition of needs, it had smaller correlation from 0.300 so it was statistically valid and could be used for a research.

**Table 1.** Validity and reliability of the dimension of recognition of needs.

| Statement | Correlation | Criteria | Info. | Alpha Cronbach | Info. |
|---|---|---|---|---|---|
| Statement 1 | 0.490 | > 0.300 | Valid | | |
| Statement 2 | 0.404 | > 0.300 | Valid | 0.778 | Reliable |
| Statement 3 | 0.692 | > 0.300 | Valid | | |
| Statement 4 | 0.780 | > 0.300 | Valid | | |

*3.3.2. Information retrieval*

The score of Cronbach's Alpha was 0.726, which is higher than 0.700, so it was statistically reliable. From the validity test score, 4 questions representing dimensional of information retrieval had score smaller than 0.300 so it was statistically valid and could be used for the research.

**Table 2.** Validity and reliability of dimensional information retrieval.

| Statement | Correlation | Criteria | Info. | Alpha Cronbach | Info. |
|---|---|---|---|---|---|
| Statement 1 | 0.548 | > 0.300 | Valid | | |
| Statement 2 | 0.570 | > 0.300 | Valid | 0.726 | Reliable |
| Statement 3 | 0.529 | > 0.300 | Valid | | |
| Statement 4 | 0.428 | > 0.300 | Valid | | |

*3.3.3. Alternative evaluation*

In the stage of alternative evaluation, the score of Cronbach's Alpha was 0.701, which is higher than 0.700, so it was statistically reliable. From the validity test score of the four questions that represented the alternative evaluation dimension, it had score smaller than 0.300 so it was statistically valid and could be used for the research.

**Table 3.** Validity and reliability of alternative evaluation dimension.

| Statement | Correlation | Criteria | Info. | Alpha Cronbach | Info. |
|---|---|---|---|---|---|
| Statement 1 | 0.521 | > 0.300 | Valid | | |
| Statement 2 | 0.461 | > 0.300 | Valid | 0.701 | Reliable |
| Statement 3 | 0.519 | > 0.300 | Valid | | |
| Statement 4 | 0.446 | > 0.300 | Valid | | |

*3.3.4. Purchasing*

In the stage of purchasing, the score of Cronbach's Alpha was 0,758 which was higher than 0.700 so it was statistically reliable. And from the validity test result from three questions representing purchasing dimension, it had a score smaller than 0,300 so it was statistically valid and could be used for the research.

IOP Conf. Series: Earth and Environmental Science **137** (2018) 012081 doi:10.1088/1755-1315/137/1/012081

**Table 4.** Validity and reliability of purchasing dimension.

| Statement | Correlation | Criteria | Info. | Alpha Cronbach | Info. |
|---|---|---|---|---|---|
| Statement 1 | 0.518 | > 0.300 | Valid | | |
| Statement 2 | 0.666 | > 0.300 | Valid | 0.758 | Reliable |
| Statement 3 | 0.683 | > 0.300 | Valid | | |

*3.3.5. The result evaluation after purchasing*

The score of Cronbach's Alpha was 0,819, which was higher than 0.700 so it was statistically reliable. And from the result of validity test, from eight questions that represented the dimension of evaluation after purchasing, it had score smaller than 0.300 so it was statistically valid and could be used for the research.

**Table 5.** validity and reliability of evaluation after purchasing.

| Statement | Correlation | Criteria | Info. | Alpha Cronbach | Info. |
|---|---|---|---|---|---|
| Statement 1 | 0.560 | > 0.300 | Valid | | |
| Statement 2 | 0.608 | > 0.300 | Valid | | |
| Statement 3 | 0.640 | > 0.300 | Valid | | |
| Statement 4 | 0.484 | > 0.300 | Valid | 0.819 | Reliable |
| Statement 5 | 0.555 | > 0.300 | Valid | | |
| Statement 6 | 0.528 | > 0.300 | Valid | | |
| Statement 7 | 0.541 | > 0.300 | Valid | | |
| Statement 8 | 0.392 | > 0.300 | Valid | | |

*3.3.6. Motivation*

In motivation stage, the score of Cronbach's Alpha was 0.781, which was higher than 0.700, so it was statistically reliable. And from the validity test result from five questions that represented motivation, it had score smaller than 0.300 so it was statistically valid and could be used for research.

**Table 6.** Validity and reliability of Motivation dimension.

| Statement | Correlation | Criteria | Info. | Alpha Cronbach | Info. |
|---|---|---|---|---|---|
| Statement 1 | 0.629 | > 0.300 | Valid | | |
| Statement 2 | 0.452 | > 0.300 | Valid | | |
| Statement 3 | 0.655 | > 0.300 | Valid | 0.781 | Reliable |
| Statement 4 | 0.583 | > 0.300 | Valid | | |
| Statement 5 | 0.515 | > 0.300 | Valid | | |

### *3.4. The relation between income, price, quality, service, and family effect with motivation in ornamental fish purchasing decision*

**Table 7.** The result of rank spearman from the relation between several variables with oriental fish purchasing motivation in ornamental fish market in peta street.

| Statement | Correlation | Criteria | Info. | Alpha Cronbach | Info. | Statement |
|---|---|---|---|---|---|---|
| 1 | The relation between motivation and the product quality | 0,600 | 0,36 | 3,968 | | Significant |
| 2 | The relation between motivation and price | 0,370 | 0,137 | 2,107 | | Significant |
| 3 | The relation between motivation and income | 0,292 | 0,085 | 1,616 | 2,048 | Not Significant |
| 4 | The relation between motivation and family | 0,127 | 0,016 | 0,678 | | Not Significant |
| 5 | The relation between motivation and service quality | 0,096 | 0,009 | 0,510 | | Not Significant |

The relation between ornamental fish purchasing motivation with the product quality had Spearman Ranks score of 0.600 and it got $t_{count}$ of 3.968. This score was higher than $t_{table}$ score which was 2,048, so $H_1$ was accepted since $t_{count}>t_{table}$. It means that there was a correlation between product qualities with ornamental fish purchasing motivation.

The Rank Spearman score between ornamental fish purchasing motivation with the product was 2.107. Where $H_1$ was considered significant since $t_{count}>t_{table}$, where $2.107 > 2.048$ which means that there as a correlation between price and the motivation in purchasing the product.

The factors that influence consumer loyalty are satisfaction, habits, commitment, product preferences, and transfer costs. Consumer loyalty is defined as a measure of customer loyalty in using a product or service over a period of time in situations where many choices of products or services can meet their needs [6].

Another factor that affects the ornamental fish purchasing motivation was income. The score of Rank Spearman of the correlation between income and ornamental fish purchasing motivation was 1.616. $H_0$ was not significant since $t_{count}<t_{table}$ where $1.616 < 2.048$ which means that the relation between purchasing motivation and the consumers' income was insignificant.

The score of Rank Spearman between the influence of family with the ornamental fish purchase motivation was 0.678. $H_0$ was not significant since $t_{count}<t_{table}$, where the score of $t_{count}$ was 0.678 and the score of $t_{table}$ was 2.048.

Another factor that affected the purchase decision was the sales service. The score of Rank Spearman for the relation between sales service and the ornamental fish purchase motivation was

IOP Conf. Series: Earth and Environmental Science **137** (2018) 012081 doi:10.1088/1755-1315/137/1/012081

0.510. $H_0$ was not significant since $t_{count} < t_{table}$, where the score of $t_{count}$ was 0.510 and $t_{table}$ was2.048, which means that the relation between purchase motivation and the sales service was insignificant.

## 4. Conclusion

The reasons why consumers bought ornamental fish in Peta Street ware because of the good quality of the product, the cleanliness of the market, and its affordable price. The respondents would still buy ornamental fish at Peta Street even though there was an increase in price, which they could tolerate as high as 30%. The alternative to buy ornamental fish was in small fish shop. The amount of fish purchased was around 1 – 5 fish. The consumers would cancel the purchase if the product was not available. The most purchased fish was Koi.

Product quality and price was the most significant influential factor with $t_{count}$ were 3.968 and 2.107, higher than $t_{table}$, which was 2.048; meanwhile, income, family, and sales service quality gave insignificant influence with $t_{count}$ of 1.616, 0.678, and 0.510, respectively, which was smaller than $t_{table}$ of 2.048.

## 5. References

[1] Lutfiadi R dan Fedinan 2013 *J. Agribus Reg Development.* **4**
[2] Rizal, Achmad 2013 *J. Aquatic.* **2**
[3] Sugiyono 2010 *Qualitative Quantitative Research Methods* (Bandung: Alfabeta)
[4] Kotler, P. dan Armstrong, G 2008. *Prinsip-Prinsip Pemasaran, Edisi 12.* (Jakarta : Erlangga)
[5] Karimah, Annisa, Iwang, Hasan 2012. *J. Fish Mar.* **3** 145-156
[6] Solomon 2004. *Consumer Behaviour* (New Jersey: Pearson Education, Inc.)

IOP Conf. Series: Earth and Environmental Science **137** (2018) 012082 doi:10.1088/1755-1315/137/1/012082

# Pattern variation of fish fingerling abundance in the Na Thap Tidal river of Southern Thailand: 2005-2015

**T Donroman**[1], **S Chesoh**[2], **A Lim**[3]

[1,3] Department of Mathematics and Computer Science
[2] Department of Science, Faculty of Science and Technology, Prince of Songkla University, Pattani Campus, Pattani, 94000 Thailand.

E-mail: chesoh.s@hotmail.com

**Abstract.** This study aimed to investigate the variation patterns of fish fingerling abundance based on month, year and sampling site. Monthly collecting data set of the Na Thap tidal river of southern Thailand, were obtained from June 2005 to October 2015. The square root transformation was employed for maintaining the fingerling data normality. Factor analysis was applied for clustering number of fingerling species and multiple linear regression was used to examine the association between fingerling density and year, month and site. Results from factor analysis classified fingerling into 3 factors based on saline preference; saline water, freshwater and ubiquitous species. The results showed a statistically high significant relation between fingerling density, month, year and site. Abundance of saline water and ubiquitous fingerling density showed similar pattern. Downstream site presented highest fingerling density whereas almost of freshwater fingerling occurred in upstream. This finding confirmed that factor analysis and the general linear regression method can be used as an effective tool for predicting and monitoring wild fingerling density in order to sustain fish stock management.

## 1. Introduction

Globally, people consume fish and other aquatic animals as their main sources of animal proteins. These aquatic animals are not only for consumption purposes, but in the ecosystem, they also serve as renewable resources. Food and Agriculture Organization predicted that the world fisheries production will reach 181 million tons by 2022, 161 million tons of which will be consumed by humans [1].

Fish recruitment, including fish birth, survival and catching rates, is an important predictor for the quality of fish population in rivers and other natural water bodies [2]. Abundance and species composition of fish fingerlings are very important as they increase productivity in aquatic ecosystem. Also, the amount of fish fingerling variation can be useful for forecasting fish stock in the future. In Thailand, the average per capita consumption of fish was 30.9 kg annually during the period 2000—2003, which was higher than the world's average fish consumption. This reflects that there is high demand of fish in Thailand [3]. Several published papers were found to have concerned the prediction of fish catch in Thailand's waters.

Several studies used linear regression [4], multiple regression [2], and multivariate analysis and ARIMA model [5] for predicting fingerlings. Linear regression and multiple regression are commonly used to find out the association between independent and outcome variables, while multivariate

IOP Conf. Series: Earth and Environmental Science **137** (2018) 012082 doi:10.1088/1755-1315/137/1/012082

analysis and autoregressive integrated moving average (ARIMA) model are regullary employed to describe data effectively with decreased weight for averaging of past observation. Additionally, ARIMA model is used for constructing the historic autocorrelation of the data and divining these into the future [6]. However, there is a rare study conducted on the prediction of wild fingerling pattern using three ecosystems in tidal river. Thus, this study aimed to investigate the pattern variation of spatial and temporal trends of fingerling abundance in order to develop an appropriate statistical model used as a key tool for carrying out an aquatic animal monitoring and assessment.

## 2. Methodology

The Na Thap tidal river, 26.5 km in length, was used as the sample area for this study (figure 1). This tidal river has its source from the range of mountains adjacent to the Thailand-Malaysia border and travels through Na Thawi and Chana Districts of Songkhla Province to form a large estuary before draining into the Gulf of Thailand. This river also serves as a habitat of productive aquatic fauna and flora [7]. The three aquatic ecosystems include saline water, brackish water, and freshwater. The upstream of freshwater ecosystem, covering an area of 7.5 km, is used for agriculture and peat swamp forest, whereas the downstream is used for household settlement, shrimp farm and fishing port. In this study, fingerlings were collected monthly by surrounding 10 sampling sites with nets along the Na Thap River from June 2005 to October 2015. The total number of fingerling species classified in taxa per 1.000 cubic meter of water volume was counted and recorded. In short, 58 species of fish fingerlings were obtained from 1.220 observations at 10 stations. The data were rechecked and cleaned for accuracy by field scientists. Any error was corrected.

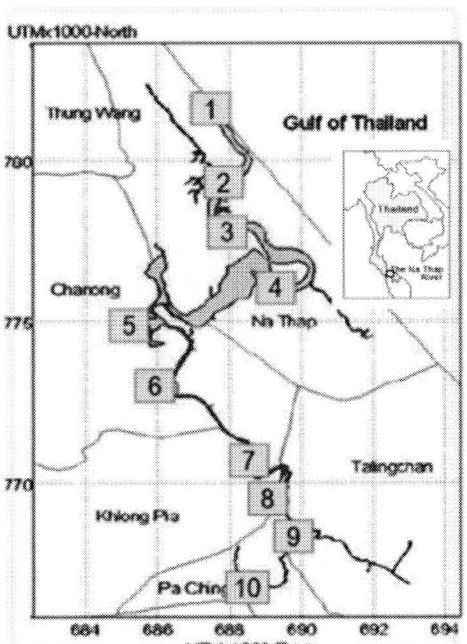


**Figure 1**. Map of The Na Thap tidal river and all 10 sampling sites.

### 2.1. Statistical analysis

In this study, fingerling density was employed as the dependent variable. Month, year and site were considered as independent variables. Square root transformation was used for satisfied normality distribution (before using factor analysis). Pearson's correlation was applied to identify and test the strengths of relationships between fingerlings in each species. Factor analysis was deployed to group

IOP Conf. Series: Earth and Environmental Science **137** (2018) 012082 doi:10.1088/1755-1315/137/1/012082

58 fingerlings into 3 interpretable factors based on maximum likelihood. The Promax rotation method was applied. Species with high correlation were grouped into the same factor based upon their unique value. The factor analysis provided loading score for individuals and selected the highest score for each species. After using factor analysis, factor scores were calculated from the sum of each species of fingerling in each month, year and site. The data were reduced as shown in the equation 1. If $y_{ij}$ is the fingerling density in month $i$ and species $j$ , the factor model is formulated as shown below

$$y_{ij} = \sum_{k=1}^{p} l_j^{(k)} f_i^{(k)} \qquad (1)$$

Where $p$ column vectors $f_i^{(k)}$ in this model are common factors and the $p$ vectors $l^{(k)}$ are their loadings. For the model assessment, r-square value, normal Q-Q plot, and fitted value against studentized residuals plot were evaluated. The linear model investigated the relationship between sum of species of density fingerling species in each factor by month, year and site

$$f_{ijk} = m + a_i + b_j + l_k + e \qquad (2)$$

In this model, $f_{ijk}$ is the sum of fingerling density in each factor, $m$ is a constant encapsulating the overall mean, $a_i$ is the coefficient for month of the calendar $i$ (January = 1, February = 2, ..., December = 12), $b_j$ is the coefficient for year $j$ (2005, 2006, …., 2015), $l_k$ is the coefficient for the data collection site $k$ (1,2,….,10) and $e$ is the error from the model. All of the statistical analyses and graphs were performed using R program.

## 3. Results and discussion

### *3.1 Preliminary analysis*

Figure 2 shows Q-Q plots of square root of fingerling weight in populated samples. The graph shows three different colors of circles for three different species categories. The brown color represents saline water fingerling species, blue color represents freshwater fingerling species, and yellow color represents ubiquitous fingerling species. The size of the circle denotes the average of fingerling occurrence in each species and q-q plot (black color, diagonal line passing through each circle) shows normality distribution. More than a half of species showing in figure 2 presented normal distribution after square root transformation was applied. Blue bar indicates the prevalence of fingerling in each species based on its common name. Overall density of fingerling can be described in table 1. The highest number of recorded prevalence was found in the ubiquitous fingerling (1.212), followed by saline water (1.025), and freshwater fingerling (516). The normality of each factor was checked using 3histogram and displayed in figure 4 (a). This figure shows that the observations of each cluster was right-skewed. Therefore, square root transformation for each factor was applied for normality and shown in figure 3 (b).

### *3.2 Results from model*

Multiple linear regression model was applied for each factor in order to examine the association between abundance and environmental factors. Model assessment was done by creating studentized residuals and fitted values versus studentized residuals plots. The plots for saline water, freshwater and ubiquitous fingerling density are displayed in figure 4, 5, 6 left (a) and right (b), respectively. The studentized residual plot for saline water and freshwater fingerling density illustrated in figure 4 (a) and figure 5 (a) performed better than the studentized residual plot for ubiquitous fingerling density shown in figure 6 (a). The r-squared value for saline water fingerling density (88.3 %) and freshwater

IOP Conf. Series: Earth and Environmental Science **137** (2018) 012082 doi:10.1088/1755-1315/137/1/012082

fingerling density (86.4 %) are nearly close to and higher than ubiquitous fingerling density (51.5 %). However, the fitted values versus studentized residuals plots for the three ecosystems shown in figure 4 (b), 5 (b) and 6 (b) followed normal distribution without any pattern.

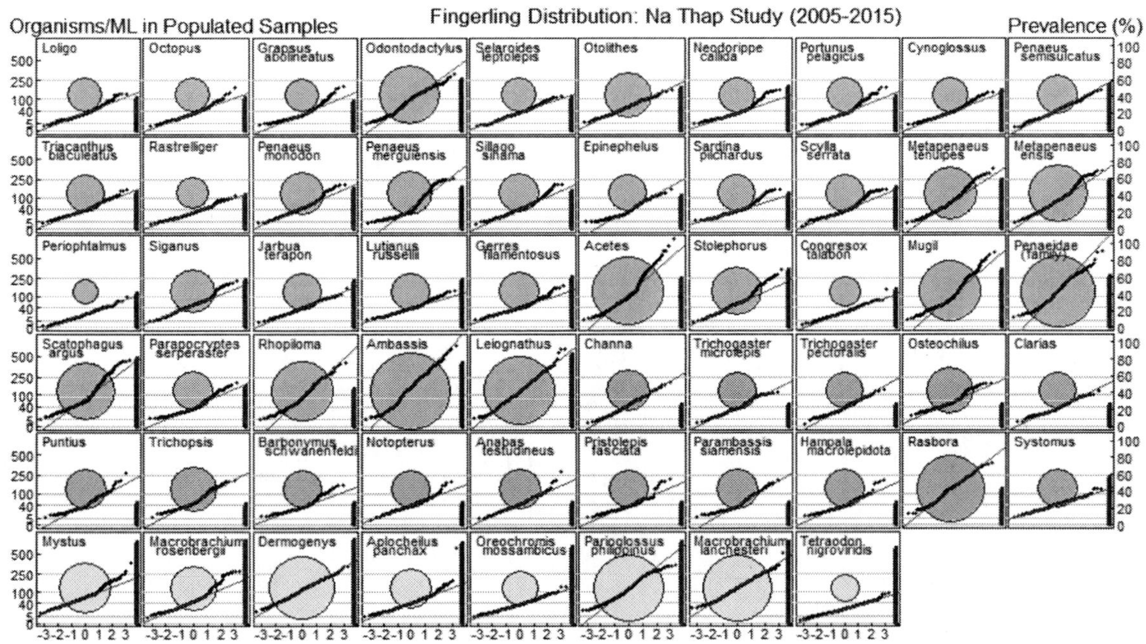


**Figure 2**. The abundance and distribution of fish fingerling density during study period. Blue bar denoted the prevalence of fingerling in each species.

**Table 1.** Overall density of fingerling (Organisms/mL) in each cluster (omitting zero).

| Factor | Transformation | Min | Median | Mean | Max | SD | Number |
|---|---|---|---|---|---|---|---|
| Saline water | Before | 1.0 | 1138.7 | 1295.7 | 5695.2 | 1183.4 | 1,025 |
| | After | 1.0 | 33.7 | 31.8 | 75.42 | 16.9 | |
| Fresh water | Before | 12.1 | 474.6 | 514.1 | 2283 | 432.6 | 516 |
| | After | 3.5 | 21.8 | 20.3 | 47.8 | 10.1 | |
| Ubiquitous | Before | 18.9 | 193.7 | 215.7 | 866.8 | 101.8 | 1,212 |
| | After | 4.3 | 13.9 | 14.32 | 29.4 | 3.2 | |

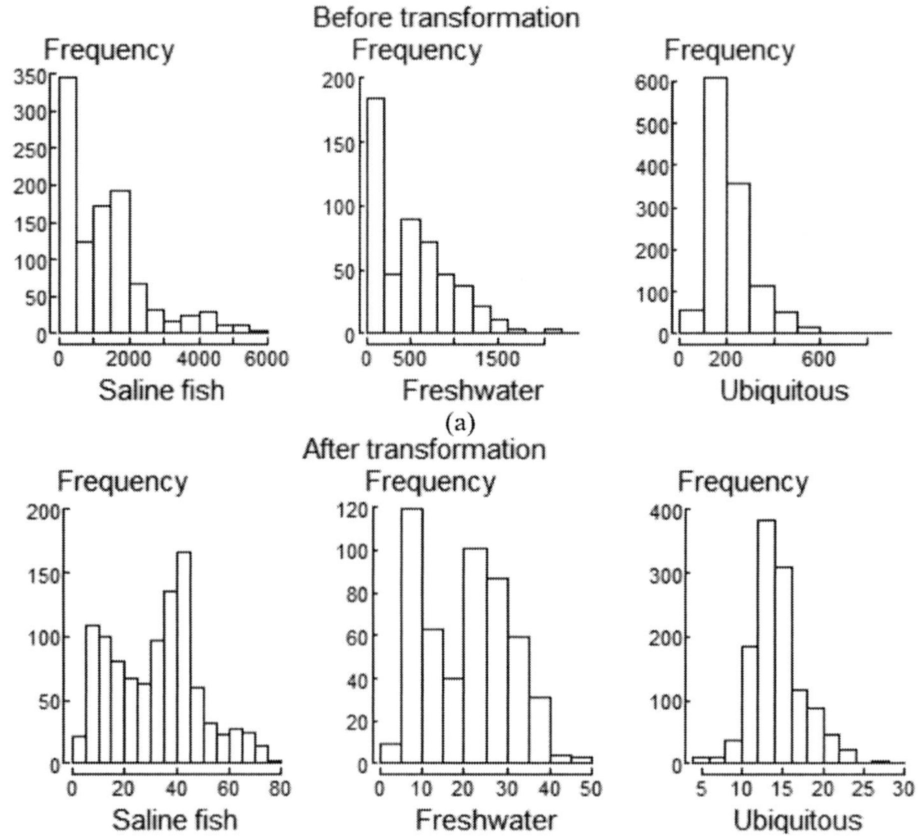


(b)

**Figure 3**. Histograms of density of fingerling (organism/mL) in each factor before and after transformation.

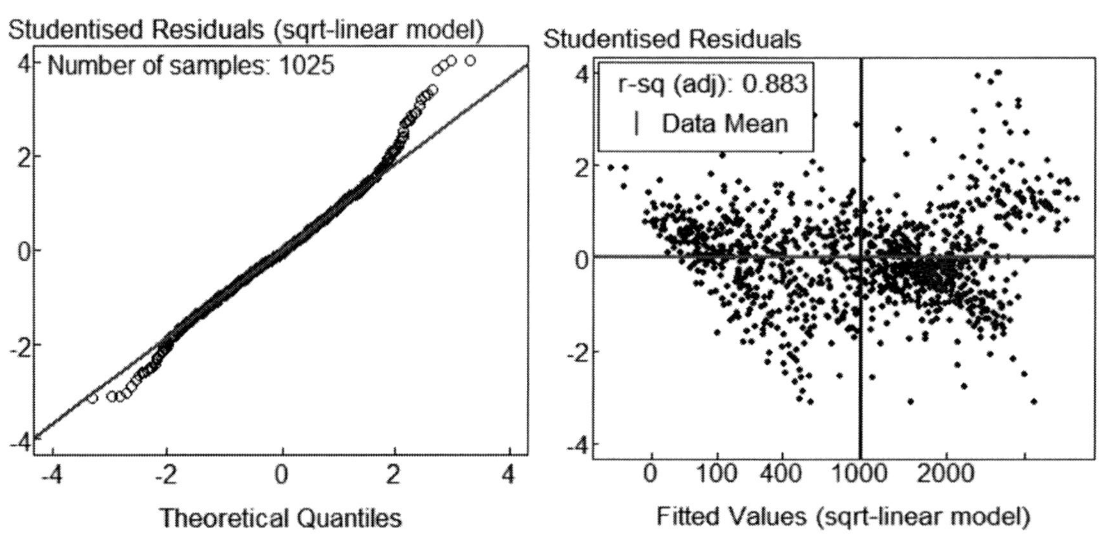


**Figure 4**. Studentized residuals plots from the linear model (a) and plots of the Studentized residuals versus fitted value from the model (b) (for saline water fingerling density).

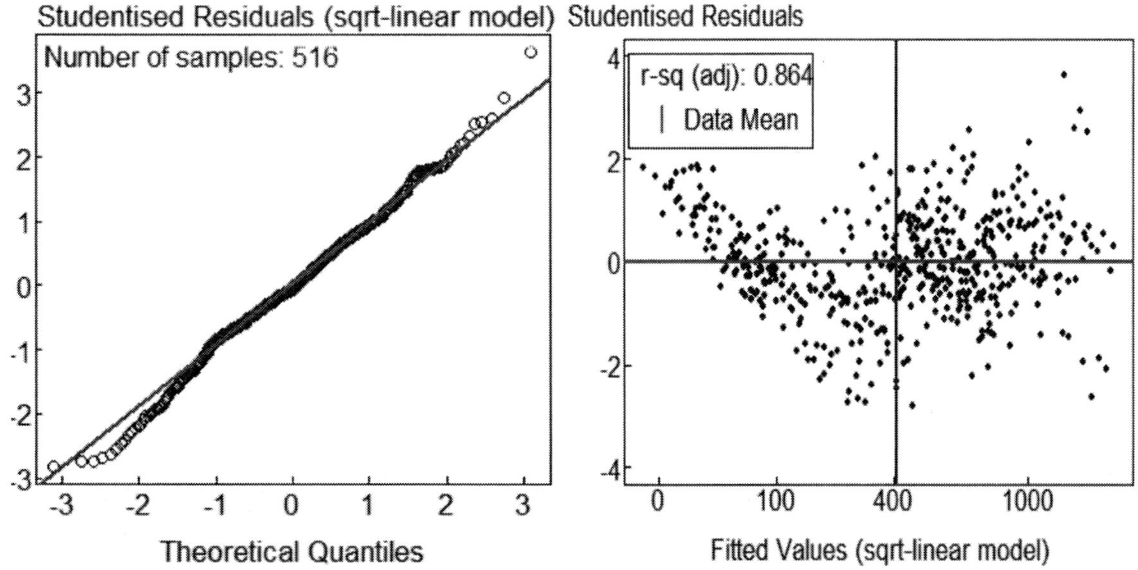


**Figure 5**. Studentized residuals plots from the linear model (a) and plots of the Studentized residuals versus fitted value from the model (b) (for freshwater fingerling density).

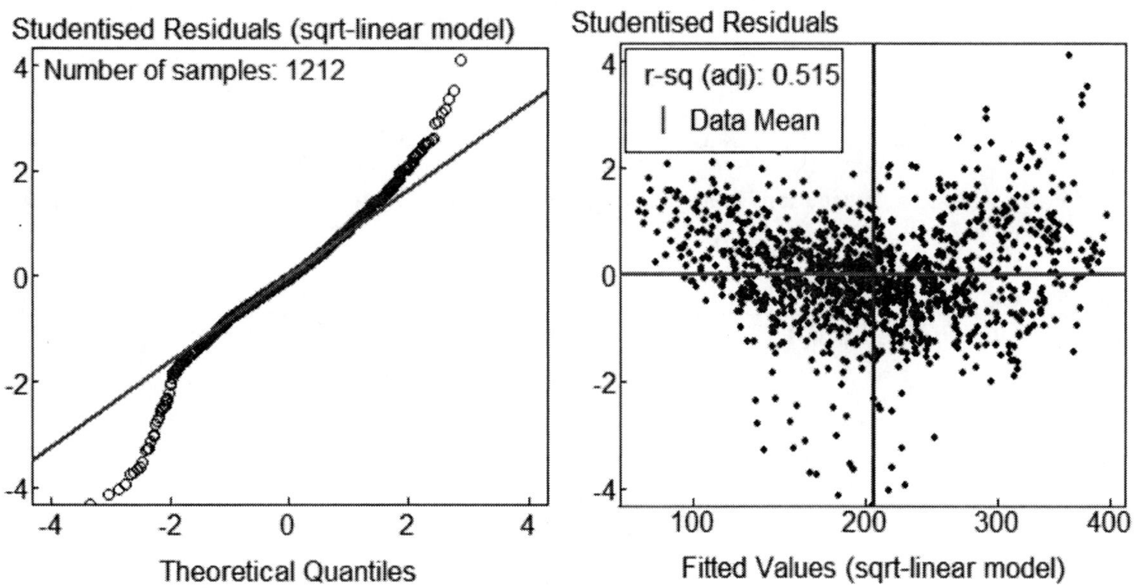


**Figure 6**. Studentized residuals plots from the linear model (a) and plots of the Studentized residuals versus fitted value from the model (b) (for ubiquitous fingerling density).

The results from the model using sum contrast method for the three factors was illustrated in figure 7, 8, and 9. The three figures showed the pattern distributions of fingerlings in month, year, and site. The red horizontal line is the overall mean of fingerlings. The first capital letter of the word of each month was used for identifying each month in order: J is January, F is February, M is March, etc. The abundance of fingerlings increased from January to April and gradually decreased from May to December. The overall abundance showed a constant pattern of fingerling density from 2005 to 2009 and a sharp decrease from 2010 to 2011. There was a gradual increase in density of fingerlings after 2011 to 2015. The highest abundance of fingerlings was noticed in site 1 (downstream), which decreased to the upstream.

Figure 8 shows the pattern distribution of freshwater fingerling density from the model. The fingerling abundance was constant from January to Feburary and slowly decreased from March to April. The lowest density of fingerlings was noticed in April, which steadily increased until December. It was a decreasing trend from 2005 to 2010 in general. However, its increased with the increase in year after 2011. Higher abudance of freshwater fingerlings was noticed in site 8 and 10 (upstream) compared to the abudance of other sites.

For ubiquitous fingerling density (figure 9), wave pattern was noticed for month. The abundance of fingerlings was less than the overall mean from 2005 to 2011. In contrary, it was greater than overall mean after 2011 till 2015. Site 1, 3 and 4 presented the highest abundance, and then the abundance sharply decreased until site 10.

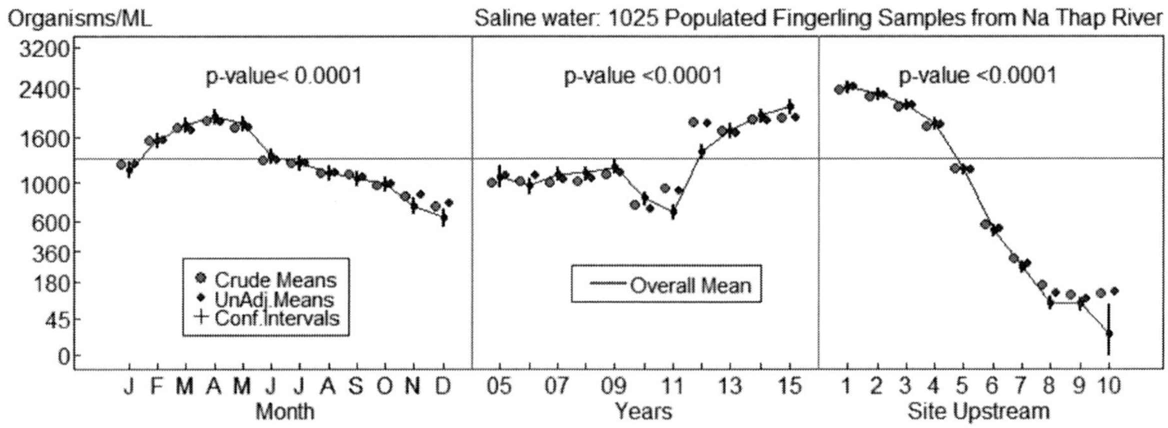


**Figure 7.** Pattern distributions of saline water fingerling density in month, year and sampling site.

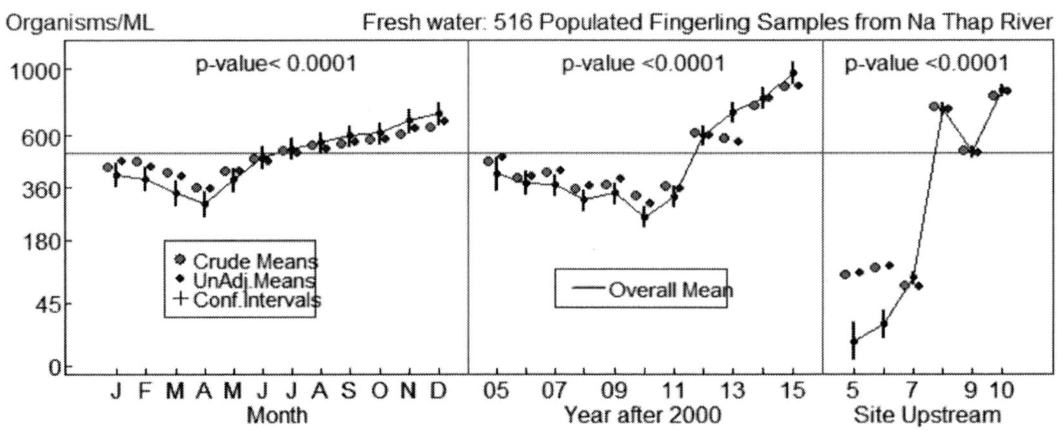


**Figure 8.** Pattern distributions of freshwater fingerling density in month, year and sampling site.

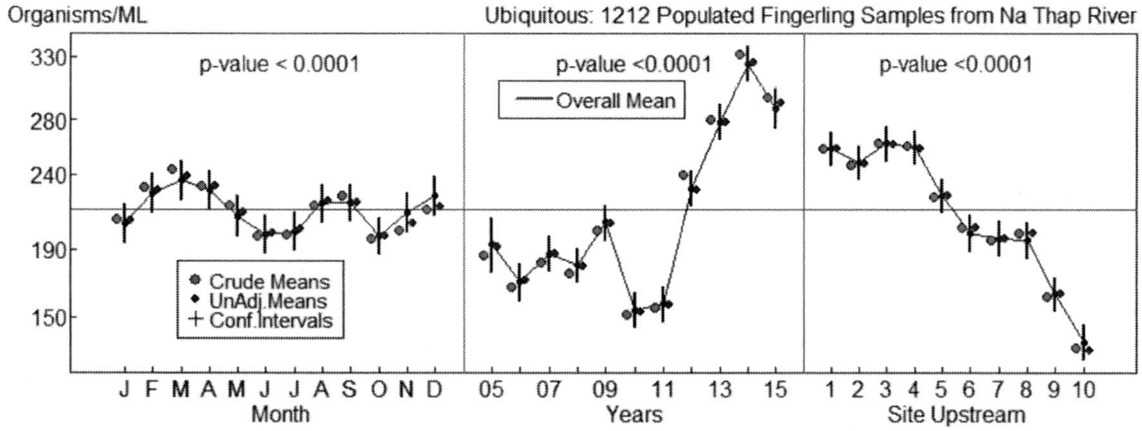


**Figure 9**. Pattern distributions of ubiquitous fingerling density in month, year and sampling site.

## 4. Discussion

Factor analysis is a powerful method generally used in studies such as socio-studies [8] and an effective tool for ecological assessment [9]. Therefore, factor analysis can reduce the dimensionality of those outcomes for removing duplicated information from a set of correlated variables. In this study, we applied this method with fingerling density. After using factor analysis, we could reduce 58 species of fingerlings into 3 intepretable factors based on salinity preference including saline water, freshwater, and ubiquitous fish fingerlings. These results were also consistent with their habitat charaterestics in the ecosystem.

Our model showed that saline and freshwater fingerlings had a good fit, whereas ubiquitous fingerlings showed a poor fit. Increasing trend in hot season and decreasing trend in rainy season occurred in saline fingerling density. Those occurances were opposite to the presenting trend of freshwater fingerling density. The saline water fingerling density was the highest in March and April. This evidence occurred because of the tropical monsoon effect. Our findings were consistent with previous report of the prediction of standing crop using lagged fingerling density of freshwater fish [9] and fish catch in Songkhla [10]. Moreover, the weight and density of freshwater fingerling species of standing crop were found at maximum levels during heavy rainy seasons, especially in November to December [11]. The ubiquitous fingerling density increases and decreases all the time. This finding may be the result of different habitat preferences of various ubiquitous fingerling species that exhibit fluctuative movement between the upstream and the downstream, and some of which migrate into and out of floodplain. In addition, the seasonal effects on the high percentage of estuarine fish species were found during dry season and the effects on the percentage of freshwater fish species were found during wet seasons in tropical lagoon in South-West Nigeria [12].

All of those three ecosystems of the Na Thap tidal river show a systematic pattern of three fish fingerling clusters based on their habitat preference throughout the year. The overall fish fingerling density (2005—2011) was less than the average, while from 2012 to 2015, the fingerling density increased. This is consistent with the data of weir construction in upstream and river mouth dredging to re-open the sand-choked mouth of the Na Thap River from 2009 to 2011. Saline water and ubiquitous fingerlings showed the same pattern of abundance, in which the highest fingerling density occured from the downstream to middle zone of the river where saline water to brankish water are found. On the other hand, high freshwater fingerling density can be found in the upstream or freshwater ecosystem. Although the presented model can fit well, some environmental factors, particularly water quality paramerters, catchability and ecosystem functional association to fingerling behaviour, are not taken into account. This presents the limitation of our study.

## 5. Conclusion
In conclusion, the abundance of fingerlings is associated with the month, year and sampling site and varied by distinct habitat preferences. This finding confirmed that factor analysis and general linear regression method can be used as an effective tool for predicting and monitoring wild fish fingerlings in order to sustain fish population management.

## 6. References
[1] FAO 2012 *The state of world fisheries and aquaculture* (Italy : Rome) 200 p
[2] Hilborn R and Walters C J 1992 *Qualtitative fisheries stock assessment* (USA : New York) 570 p
[3] FAO 2008 Food and agriculture organization of the united nations regional office for Asia and the Pacific Bangkok Thailand pp 8-12
[4] Fausch K D, Kawkes C L and Parsons M G 1988 *Pacific norhwest research station Oregon* USA 52 p
[5] Preciado I, Punzon A, Gallego J L and Villa Y 2006 *Boletin del Instituto Espanal de Oceanografia* **22** 83
[6] Venugopalan R and Srinath M 1998 *Indian J. Fish.* **45** 227
[7] Chesoh S, Luangthuvapranit C, Ruangchuay R and Lim A 2010 *Impact monitoring of thermal power plant operation on aquatic ecological environment in estuarine river of southern Thailand* (Thailand : Conf. Khon Kaen) 4-6
[8] Kaewdaisri S 2001 *Relationship between learning stule with demographic factors and socio-economic factors of high school students in Songkhla Province* (Thailand : Prince of Songkla University) pp 42-48
[9] Saheem N, Lim A and Chesoh S 2014 *Songklanakarin J. Sci. Technol.* **36** 13
[10] Chesoh S and Lim A 2008 *Sci. Asia* **34** 335
[11] Angsupanich S and Rakkheaw S 1997 *Netherlands J. Aquatic Ecology* **30** 297
[12] Olukolajo S O and Oluwaseun K A 2008 *J. Fish and Aquatic Sci* **3** 75

**Acknowledgments**
We are grateful to Prof. Don McNeil for his helpful guidance. Thanks to the Electricity Generation Authority of Thailand (EGAT) for providing raw data and this topic was financially supported by a PSU Collaborative Research Fund; PRPM ID 16287.

IOP Conf. Series: Earth and Environmental Science **137** (2018) 012083 doi:10.1088/1755-1315/137/1/012083

# The influence of waterweeds in the removal of phosphor in content aquatic environments

**L N Salamah[1,2], and A Kurniawan[2]**

[1] Graduated School of Life Science, Ritsumeikan University, 1-1-1 Noji-higashi, Kusatsu, Shiga, Japan 525-8577
[2] Coastal Marine and Research Centre, Fisheries and Marine Science Faculty, University of Brawijaya, Jl. Veteran, Malang, East Java, Indonesia 65145

E-mail: lutfinimatus@gmail.com

**Abstract.** Over the last few decades phosphorus removal methods have increasingly been designed. One of the methods is using waterweed for uptake the phosphorus (P) from water and sediment. However, lack information of suitable waterweed for P removal. The goals of this research are to investigate the effect of waterweeds in the P removal and to find the suitable waterweeds for P removal. Five types of waterweeds (*Phragmites australis, Egeria densa,, Bacupa myriophylloides, Eichornia crassipes,* and *Limnobium laevigatum*) were collected and cultivated in pots. Water and sediment were collected from natural small pond, and sampling were perfomed at 0, 7, 14, 21 and 28 days. The result showed that P of water and sediment were reduced during the experiment. *E. densa* accumulated highest TP (505 mg/kg). It was suggested that *E. densa* is suitable for phosphor removal.

## 1. Introduction

In nature, phosphorus is mainly sourced from the natural cycle (rain water) and fertilizers. Phosphorus presents in the soil and leaches to the water before accumulating in the sediment of aquatic environments. Therefore, the amount of phosphorus is higher in the sediment than in the water. Phosphorus tends to be the limiting nutrient for the aquatic environment, but too high an amount of phosphorus in the water and sediment can be harmful to the aquatic environment. The high amount of phosphorus may cause an over supply of nutrients, causing eutrophication [1]. This condition decreases the oxygen demand and is followed by the increase of chemical oxygen demand in the water.

In terms of phosphorus removal, several waterweeds are known to have the ability to remove phosphorus from sediment by absorbing it for their growth. Waterweeds are aquatic plants that dominate wetlands, shallow lakes, and streams. They grow in or near water. Waterweeds can be classified as emergent, submerged or floating, based on their predominant growth forms Brix [2] Submerged waterweeds are usually rooted or attached to the bottom. Their stems and leaves are located below the water surface (some plants may produce a few small floating or aerial leaves found at the water's surface). Submerged species include *Hydrilla, Egeria, Vallisneria* and curly leaf pondweed. Emergent waterweeds grow in shallow waters, and most have leaves above the water surface. Representative species of emergent plants include *Scirpus* spp, *Sagittaria* spp, *Sparganium* spp, *Typha* spp, *Phragmites australis* and *Lythrum salicaria.* Floating waterweeds are free floating on the water surface; waterlily

IOP Conf. Series: Earth and Environmental Science **137** (2018) 012083 doi:10.1088/1755-1315/137/1/012083

and spatterdock are floating-leaved species, whereas water hyacinth and water lettuce are free-floating species.

Waterweeds play an important role in healthy ecosystems as primary producers of oxygen through photosynthesis, providing substrates for algae and shelters for fish and many invertebrates. As a result, they assist in nutrient recycling to and from sediments and aid in stabilizing river and stream banks. Waterweeds, therefore, have the potential to be used to clean the static water owing to their metabolic functions. The metabolic function of waterweeds consists of nutrient uptake and $O_2$ release from the roots into the rhizosphere [3]. The ability and capacity of waterweeds in the nutrient removal from water bodies are currently of concern among researchers since the last two decades [4, 5]. Xu *et al.* [6] found that duckweed (*Spirodela oligorrhiza*) was able to remove of 83.7 % TN and 89.4 % of TP from swine lagoon water during a period of eight weeks.

Waterweeds improve the nutrient condition. Previous studies have attributed plant uptake of phosphorus that is responsible for 12-73 % of total phosphorus removal [7]. However, there is lack of information on the most suitable waterweed to remove phosphorus from sediment. Therefore, it is necessary to study the features of waterweeds in order to remove the phosphorus from the water system effectively. This study aimed to find out which type of waterweed had the highest ability to remove phosphorus from the sediment and water of aquatic environments.

## 2. Methodology

Sediment and water samples were collected from Kitanoshin pond located in Kusatsu city, Shiga, Japan (N: 34°98", E: 135°96") in July 2016. Five types of waterweeds were also collected from the natural water area. The waterweeds of *Phragmites australis, Egeria densa,, Bacupa myriophylloides, Eichornia crassipes* ,and *Limnobium laevigatum* were cultivated in pots containing 2.5kg sediment and 4L water as the treatments. Each treatment was performed in duplicate. Water and sediment sampling were perfomed at 0, 7, 14, 21 and 28 days to analyze Total Phosphor (TP).

TP of water was measured according to De Lucas [8]. The water sample was digested using the sulfuric acid digestion method. One mL of phenolphthalein was added into 50 mL of water sample in the erlenmeyer flask. After 5M sulfuric acid solution was added, the sample was mixed until the color disappeared. After the solution color turned to clear, 0.4 g of ammonium persulfate was added and heated at 90°C for 40 minutes. The solution was then cooled down at room temperature for 1 hour. One mL of phenolphthalein was added into the solution, then titrated using 1 N sodium hydroxide. After the solution was volumed up to 50 mL with distilled water, TP in 50 mL of each water sample was analyzed by using the ascorbic-acid method. A 50 mL of digested sample was placed into the erlenmeyer flask, 8 mL of combined reagent was added and mixed into the solution. The mixture was kept at room temperature for 30 minutes. The absorbance was measured by using an atomic absorption spectrophotometer (U-1900 Spectrophotometer, Hitachi, Japan) at 880 nm.

TP in sediment and waterweeds were measured using the molybdenum blue method. 100 μL of kjeldahl extract was mixed with 100 μL of molybdate solution and L-ascorbic acid solution (5:1) and kept at 30°C for 30 minutes for color development. The absorbance was measured using an atomic absorption spectrophotometer at 710 nm.

## 3. Results and discussion

Figure 1 shows TP concentration of water over time. TP was decreased in all treatments. At the end of experiment, the lowest TP concentration was found in *E. densa* (0.19 mg/L). Analysis of the sediment showed that TP was decreased in all vegetated treatments (Figure 2), the lowest concentration of TP in the sediment was found in *E. densa* treatment on 28 days of experiment (2.5 mg/kg). Waterweed treatments presented lower TP at the end of experiment, suggesting that waterweeds are directly involved in TP removal from the water and sediment. Results in this study are consistent with those reported by De Lucas *et al.* [9] who decribed that vegetated treatments removed TP more efficiently than unvegetated treatments.

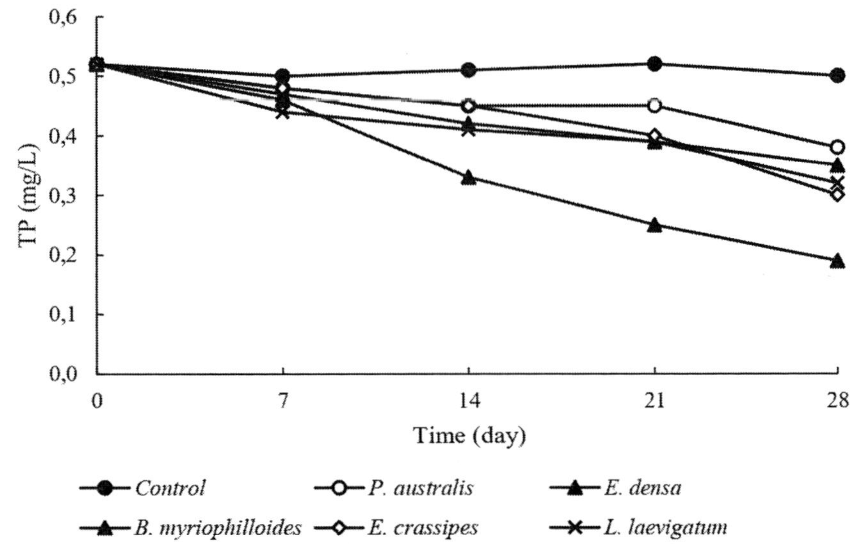


**Figure 1.** TP concentration of water during the experiment.

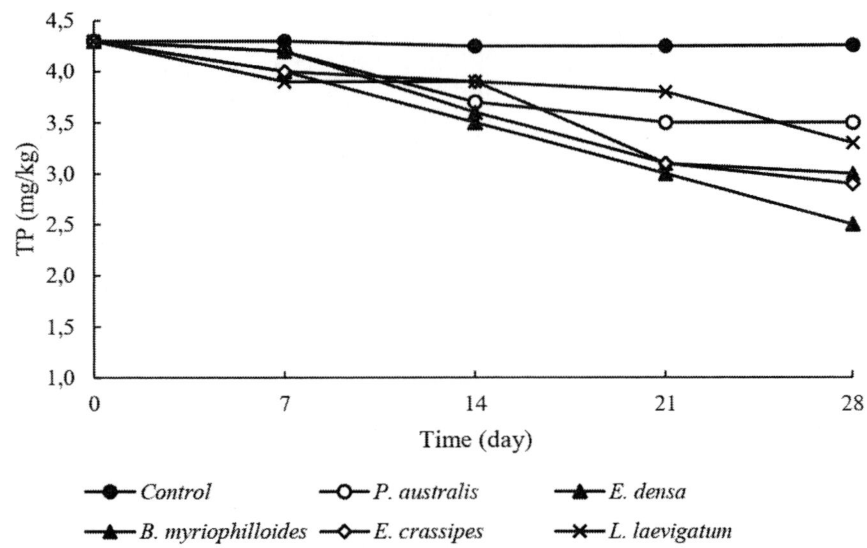


**Figure 2.** TP concentration of sediment during the experiment.

TP removal from water (figure 3) by vegetated treatments were greater than the control, this suggests that waterweeds are able to remove TP from water at up to 50 %. TP removal from water by *P. australis, E. densa, B. myriophylloides, E. crassipes* and *L. laevigatum* were 27.0, 52.0, 32.7, 28.9, and 38.4 %, respectively.

A previous study DE Lucas [9] reported that 68.1 % of phosphorus was successfully removed using several waterweeds. The highest TP removal was found in the *E. densa* treatment (52 %). Figure 4 shows percentage of TP removal from the sediment. TP removal from sediment in vetetated treatments were higher than the control. TP removal from sediment by *P. australis, E. densa, B. myriophylloides, E. crassipes* and *L. laevigatum* were 18.6, 42.0, 30.2, 32.6, and 23.3 %, respectively.

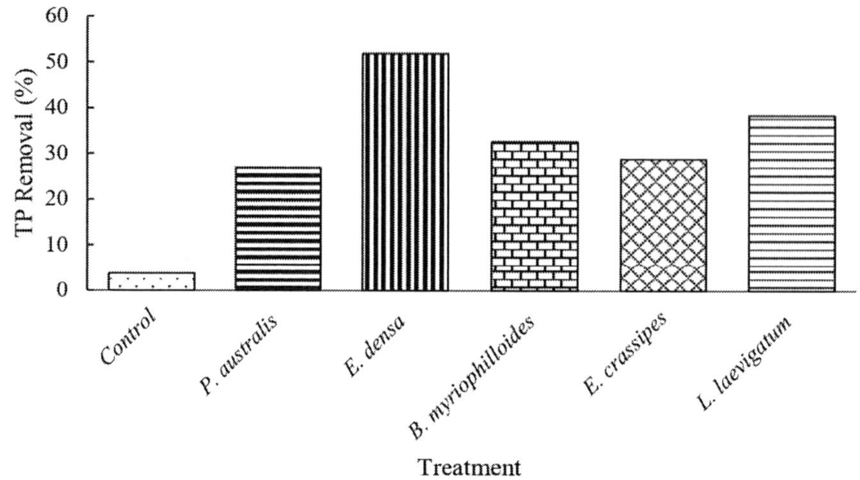


**Figure 3.** TP removal from water during the experiment.

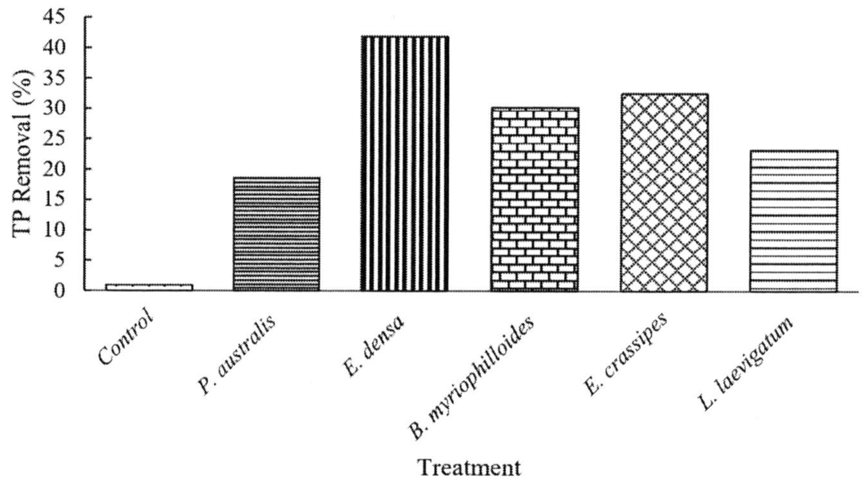


**Figure 4.** TP removal from sediment during the experiment.

Waterweeds were able to accumulate phosphorus into their tissue from the aquatic environment (water and sediment). TP accumulation in *P. australis, E. densa, B. myriophylloides, E. crassipes* and *L. laevigatum* were 445, 505, 465, 474, and 461 mg/kg, respectively. Among the waterweeds, the highest accumulation was found in the *E. densa* treatment. This result correlated with high TP removal by *E. densa* from water and sediment. Nutrient uptake of waterweeds were influenced by many factors such as water conditions, nutrient availability, plant species, and growth stage [10].

Waterweeds take up nutrients from water, sediment, or both [11]. Floating waterweeds generally obtain nutrients from water columns through the roots [12]. Submerged waterweeds generally acquire nutrients from both water and sediment [13], therefore, submerged waterweeds have important implications in the nutrient cycling within fresh water systems [14]. Floating-leaved waterweeds petioles hold the leaves floating on the water surface and are generally rooted in the sediment [12,13].

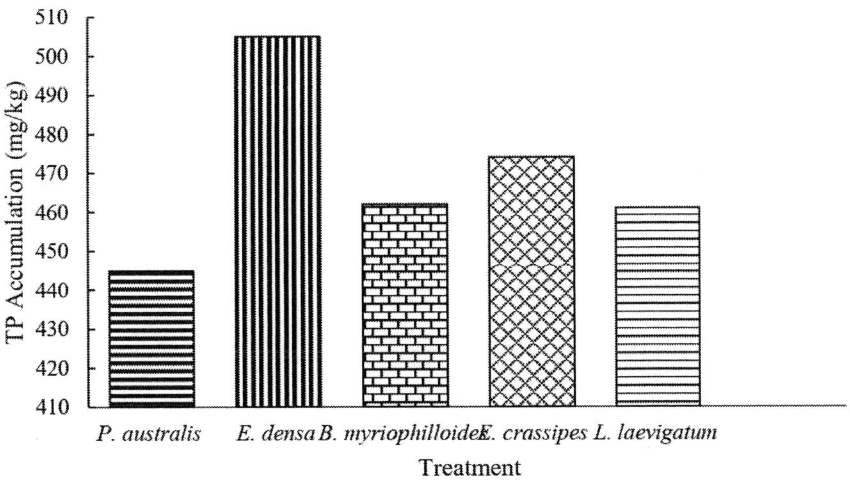


**Figure 5.** TP accumulation in the waterweeds during the experiment.

The five waterweeds could effectively be used in reducing TP concentration in the water environment. The TP concentration of sediment in the waterweed bed is often reduced, since waterweeds use sediment as their source of phosphorus and potassium [15]. The root of waterweeds are able to uptake TP and relocate from the root to the shoot [16].

As waterweed biomass wilts and decays, the P in biomass is released back into the water system [17]. Harvesting waterweed biomass is an effective and obligatory method to avoid the release of nutrient from waterweeds [6]. Apart from promoting continuous nutrient removal by plant uptake, the harvesting of waterweeds effectively prevents the release of nutrient from waterweed tissues into the water during decomposition by waterweeds senescence, wilting and decay in their decline phase and ultimately reduces the harmful effects to the treatment systems. The majority of the storage of P is temporary because plant tissues release nutrients to the water treatment systems when they die and decompose. If waterweeds are not harvested, most of the nutrients that have been incorporated into the plant tissue may be returned to the water during the decomposition processes.

## 4. Conclusion

The important contribution of waterweed uptake is the highest TP removal from aquatic environment and the lowest TP content in the aquatic environment. Present results point large TP removal from water and sediment of systems planted with *E. densa*. Therefore, *E. densa* is suitable for phytoremediation practice. TP removal from aquatic environments can be enhanced by utilizing wetland systems planted with *E.densa*.

## 5. References

[1] Mijani R, Herliwati, AA Prihanto 2017 *AACL Bioflux.* **10** 1091-1097.
[2] Champagne P 2008 Natural processes chapter 7 wetlands, A book report by the American society of civil engineers Canada.
[3] Brix H 1997 *Water Sci. Technol* **35**:11–17.
[4] Gersberg R B, Elkins C, Goldman 1983 *Water Res.* **17**:1009–1014.
[5] Mann R H, Bavor 1993 *Water Sci. Technol.* **27**:107–113.
[6] Xu G, Shen 2011 *Bioresour. Technol.* **102** :848–853.
[7] Menon R and M M Holland 2013 *Water air soil pollution* **224**:1602.
[8] American Public Health Association (APHA) 1999 Standard methods for the examination of water and waste water. 18th Edition, American Public Health Association, Washington DC.
[9] De Lucas A, Rodriguez L, Villasenor J, Fernandez F J 2007 . *Biochem. Eng. J.* **34**:51-61.

IOP Conf. Series: Earth and Environmental Science **137** (2018) 012083 doi:10.1088/1755-1315/137/1/012083

[10] Longhi D M Bartoli and P Viaroli 2008 *Bioresource Technology* 102:3471-3479.
[11] Shu L, Schneide P, Jegatheesan V, Johnson J 2006 *Bioresour. Technol.* **97**2211-2216.
[12] Jaffer Y, Clark T A, Pearce P, Parsons S A 2002 *Water Res* **36**:1834-1842.
[13] Kataki S, West H, Clarke M, Baruah D C 2016 *Resour Conservation Recycl* **107**:142-156.
[14] Xien L H, Hong-ying G Ke, S Ying-xue 2010 *Bioresour. Technol.* **101**:5494-5500.
[15] Mitsch W J, and Gosselink J G 2000. Wetland, Wiley, New York.
[16] Nowak B A P, Winter F 2013 *Fuel Process. Technol.* **105**:195-201.
[17] Pei Y S, Wang J, Wang ZY, Yang Z F 2010 *Proc Environ Sci* **2**: 1988-1996.

**Acknowledgement**
The authors are grateful to Monbukagakusho, The Ministry of Education, Culture, Sports, Science and Technology in Japan (MEXT), for financial support through the Super Global Program University (SGU) program.

# Porosity structure of green polybag of medium density fiberboard from seaweed waste

**M A Alamsjah[1]*, S Subekti[1], M Lamid[1], D Y Pujiastuti[1], H Kurnia[2] and R R Rifadi[2]**

[1]Department of Marine, Faculty of Fisheries and Marine, Universitas Airlangga, Surabaya 60115, Indonesia
[2]Department of Fish Health Management and Aquaculture, Faculty of Fisheries and Marine, Universitas Airlangga, Surabaya 60115, Indonesia

E-mail: alamsjah@fpk.unair.ac.id

**Abstract.** The last decade shown that the needs Medium Density Fibreboard (MDF) rapidly growing in Asia Pacific and Europe up to more 15 % per year. MDF made up of fibers lignoselulosa which combined with synthetic resin or tied other suitable but high temperatures and pressure. Technology engineering for green polybag of MDF from seaweed waste of *Kappaphycus alvarezii* and *Gracilaria verrucosa* is an alternative effort for ecosystem stability and technological innovations that is environmentally friendly. Structure porosity from the shape of green polybag shows that performance seaweed waste of *K. alvarezii* is better than seaweed waste of *G. verrucosa*. The circulation of water happened more optimal in green polybag formed from MDF of seaweed waste of *K. alvarezii* with size porosity 3.976 μm, while size porosity of seaweed waste of *G. verrucosa* measurable 4.794 μm. Structure of green polybag of MDF from seaweed waste showed that C components greater 50 % to *K. alvarezii* while C components less than 50 % to *G. verrucosa*. This resulted in the ties to structure of MDF stronger found in green polybag derived from seaweed waste of *K. alvarezii* than *G. verrucosa*.

## 1. Introduction

Polybags have been used for planting seeds until today. The use of polybag has some disadvantages, including the fact that the roots of plants grow in circle in the plastic bag and the plastic used as polybag material is not easily degradable by the environment or microorganisms living in the soil, causing the increasing in the accumulation of plastic waste [1]. Plant growth process often creates the problem of drought due to the use of water for watering plants. Drought occurs in the growth phase and causes considerable decline in the crop yields [2]. In the last decade, the needs for Medium Density Fibreboard (MDF) grew rapidly in Asia Pacific and Europe by more than 15 % per year. MDF is made from lignocellulosic fibers combined with synthetic resin under high temperatures and pressures. A major component of MDF is lignocellulose, that can be obtained from wood, hay, herbage, farming waste, forest, or industrial waste (wood, paper and other fibrous materials). Lignocellulose contains three major components, namely cellulose, hemicellulose, and lignin. The waste of *Kappaphycus alvarezii* and *Gracilaria verrucosa*, from which carrageenan is extracted, and agar that has enough lignocellulose content are the materials needed for making MDF. Ilknur and Cirik [3] stated that seaweed is not only

IOP Conf. Series: Earth and Environmental Science **137** (2018) 012084 doi:10.1088/1755-1315/137/1/012084

used as food. Its utilization as a pharmaceutical ingredient and industrial raw material also needs to be explored further. Indonesia's dried seaweed production of 800.000 tons/year since 2010 has contributed to 50 % of the world's production, and 85 % of that amount has been exported. By the importers, the seaweed is processed into food industrial materials as well as health and cosmetics products. Indonesia has 34 seaweed processing industries in which seaweed is processed into gelatin, alginate and carrageenan. However, the use of seaweed waste as a useful product has not been the center of attention. The waste generated is usually only left accumulating at landfills. Although it is harmless, the waste dump may cause pollution problems, especially if the landfills are no longer able to accommodate waste [4]. The engineering of green polybag of MDF made from the waste of *K. alvarezii* and *G. verrucosa* is an effort to maintain ecosystem stability and is an environmentally friendly technological innovation.

Based on the problem explained above, it is necessary to conduct research related to the waste of *K. alvarezii* and *G. verrucosa* as a raw material of MDF as a substitute for polybags with optimum and efficient water volume for the optimization of seaweed waste and reduction of the use of plastic to avoid environmental damages because of the use of polybags. The results of this research suggested that this prototype was able to reduce water consumption by exploring the structure porosity of the shape of green polybag.

## 2. Method

### 2.1. Materials

The tools used included: a cylindrical iron plate with an upper diameter of 7 cm, a bottom diameter of 6.5 cm and a height of 6 cm, functioning as a tool for the first printing; a cylindrical iron plate with an upper diameter of 11 cm, a bottom diameter of 9.5 cm and a height of 8 cm, functioning as a tool for printing both polybags; a lamp; an oven; a fitting; a stirrer; a basin; an analytical scale; and a ruler. The raw materials used were the waste of *G. verrucosa* obtained from farmers in Jabon, Sidoarjo and the waste of *K. alvarezii* obtained from seaweed farmers in Sumenep, Madura. The wood powder was obtained from the remnants of wooden furniture around Tempurejo, and the adhesive material was obtained from a hardware strore in Surabaya.

### 2.2. Sample preparation

Preparation included the collection of raw materials such as seaweed waste, wood powder and adhesive. The range of adhesive used in the dry process was 8—11% of the board dry weight. This experiment consisted of eight treatments, namely: G0: structure porosity from the shape of green polybag showing the performance waste of *G. verrucosa* (0 day); G7: structure porosity from the shape of green polybag showing the performance of waste of *G. verrucosa* (7 days); G14: structure porosity from the shape of green polybag showing the performance of waste of *G. verrucosa* (14 days); G21: structure porosity from the shape of green polybag showing the performance of waste of *G. verrucosa* (21 days); K0: structure porosity from the shape of green polybag showing that performance of waste of *K. alvarezii* (0 day); K7: structure porosity from the shape of green polybag showing the performance of waste of *K. alvarezii* (7 days); K14: structure porosity from the shape of green polybag showing the performance of waste of *K. alvarezii* (14 days); and K21: structure porosity from the shape of green polybag showing the performance of waste of *K. alvarezii* (21 days). The water volume was determined by the first particle board based on the pot to be filled with soil media for growing plants, and the pot based on the second particle board held water during the research. The water requirement of plants was determined based on the value of water content in the state of the field capacity [5].

Before doing the forging phase, the seaweed waste was cleaned. Then, it was dried naturally for ±3 days and by using an oven to decrease the water content. The next stage was grinding the seaweed to

IOP Conf. Series: Earth and Environmental Science **137** (2018) 012084 doi:10.1088/1755-1315/137/1/012084

obtain seaweed powder, which was filtered using a 40 mesh sieve. The other materials prepared were 100 mesh wood powder for mixture and synthetic adhesive additives.

### 2.3. *Medium density fiberboard (MDF)*

The making of MDF was done by a dry process using a hot press. After the raw material was mixed with adhesive, the mixture was pressed using a hot press at a temperature of 170 ºC and a pressure of 45 Pa for 25 minutes [6]. The first particle board was cylindrical with an upper diameter of 7 cm, a height of 6.5 cm and a bottom diameter of 6 cm. The second particle board pot had an upper diameter of 11 cm, a height of 8 cm and a bottom diameter of 9.5 cm. Then, re-conditioning was conducted for two days to get a high-quality particle board.

### 2.4. *Research parameters*

In this research, the main parameter observed was physical test of the structure porosity of MDF. The structure porosity of MDF from the shape of green polybag showing the performance of seaweed waste was tested using a scanning electron microscope in the study of identification of structure porosity and mineral contents. The other parameters measured during the research were temperature, soil pH and soil moisture that were to be used for testing the results of the research.

### 2.5. *Statistical analysis*

The experimental research and the data analysis were conducted using ANOVA (Analysis of Variance), and the design used was Completely Random Design, with the aim to determine the manufacturing of MDF with seaweed waste as the main material, which could replace polybags. If the results obtained were significant, they would be tested further using Duncan's test [7].

## 3. Results and Discussion

The structure porosity from the shape of green polybag showing the performance of waste of K. alvarezii was better than that of the waste of G. verrucosa. Within 21 days, the water circulation was more optimal in green polybag formed from MDF made of the waste of K. alvarezii with size porosity of 3.976 μm, while the size porosity of the MDF made of the waste of G. verrucosa was measured 4.794 μm. The structure of green polybag of MDF from seaweed waste showed that the carbon content in K. alvarezii was greater than 50%, but less than 50% in G. verrucosa. On the other hand, the structure of green polybag of MDF from seaweed waste showed that the oxygen content was greater than 50 % in G. verrucosa, but less than 50% in K. alvarezii. This caused the bonds in the structure of MDF stronger in green polybag derived from seaweed waste of K. alvarezii than G. verrucosa. Meanwhile, the structure porosity of MDF which stayed open allowed the greater availability of oxygen and increased the productivity of the growing media.

IOP Conf. Series: Earth and Environmental Science **137** (2018) 012084 doi:10.1088/1755-1315/137/1/012084

**Treatment | Structure Porosity | Mineral Content**

**$G_0$**

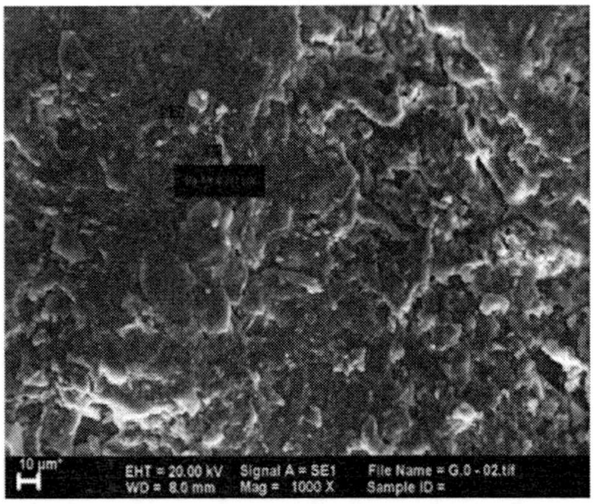


| El | AN | Series | unn. C [wt. %] | norm. C [wt. %] | Atom. C [at. %] | Error [%] |
|---|---|---|---|---|---|---|
| O | 8 | K-series | 49.32 | 49.32 | 45.97 | 30.8 |
| C | 6 | K-series | 39.57 | 39.57 | 49.13 | 13.2 |
| K | 19 | K-series | 2.77 | 2.77 | 1.06 | 0.1 |
| Ca | 20 | K-series | 2.28 | 2.28 | 0.85 | 0.1 |
| S | 16 | K-series | 1.75 | 1.75 | 0.81 | 0.1 |
| Cl | 17 | K-series | 1.47 | 1.47 | 0.62 | 0.1 |
| Si | 14 | K-series | 1.02 | 1.02 | 0.54 | 0.1 |
| Na | 11 | K-series | 0.66 | 0.66 | 0.43 | 0.1 |
| Al | 13 | K-series | 0.59 | 0.59 | 0.33 | 0.1 |
| Ti | 22 | K-series | 0.29 | 0.29 | 0.09 | 0.0 |
| Mg | 12 | K-series | 0.29 | 0.29 | 0.18 | 0.0 |
| | | Total: | 100.00 | 100.00 | 100.00 | |

**$G_7$**

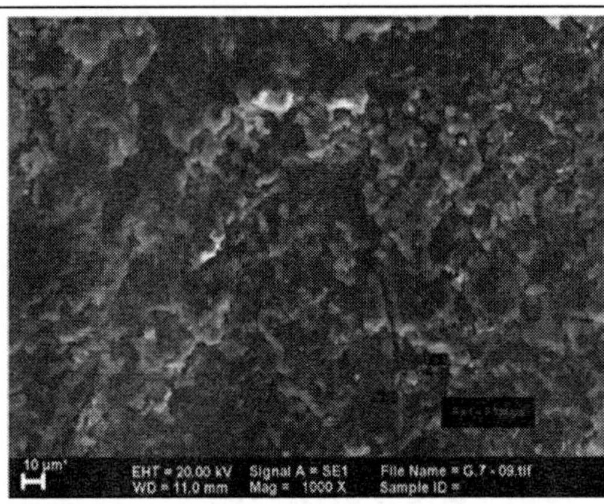


| El | AN | Series | unn. C [wt. %] | norm. C [wt. %] | Atom. C [at. %] | Error [%] |
|---|---|---|---|---|---|---|
| O | 8 | K-series | 51.53 | 51.53 | 47.83 | 16.4 |
| C | 6 | K-series | 38.71 | 38.71 | 47.86 | 13.0 |
| K | 19 | K-series | 3.71 | 3.71 | 1.41 | 0.1 |
| S | 16 | K-series | 2.61 | 2.61 | 1.21 | 0.1 |
| Cl | 17 | K-series | 1.30 | 1.30 | 0.55 | 0.1 |
| Na | 11 | K-series | 0.68 | 0.68 | 0.44 | 0.1 |
| Ca | 20 | K-series | 0.63 | 0.63 | 0.23 | 0.0 |
| Mg | 12 | K-series | 0.40 | 0.40 | 0.25 | 0.1 |
| Si | 14 | K-series | 0.28 | 0.28 | 0.15 | 0.0 |
| Al | 13 | K-series | 0.14 | 0.14 | 0.08 | 0.0 |
| | | Total: | 100.00 | 100.00 | 100.00 | |

**$G_{14}$**

| El | AN | Series | unn. C [wt. %] | norm. C [wt. %] | Atom. C [at. %] | Error [%] |
|---|---|---|---|---|---|---|
| O | 8 | K-series | 55.05 | 55.05 | 63.36 | 17.9 |
| C | 6 | K-series | 39.99 | 39.99 | 34.56 | 12.8 |
| S | 16 | K-series | 1.65 | 1.65 | 0.71 | 0.1 |
| Cl | 17 | K-series | 1.30 | 1.30 | 0.51 | 0.1 |
| K | 19 | K-series | 0.81 | 0.81 | 0.28 | 0.1 |
| Si | 14 | K-series | 0.51 | 0.51 | 0.25 | 0.0 |
| Ca | 20 | K-series | 0.28 | 0.28 | 0.10 | 0.0 |
| Na | 11 | K-series | 0.19 | 0.19 | 0.11 | 0.0 |
| Al | 13 | K-series | 0.14 | 0.14 | 0.07 | 0.0 |
| Mg | 12 | K-series | 0.08 | 0.08 | 0.04 | 0.0 |
| | | Total: | 100.00 | 100.00 | 100.00 | |

**$G_{21}$**

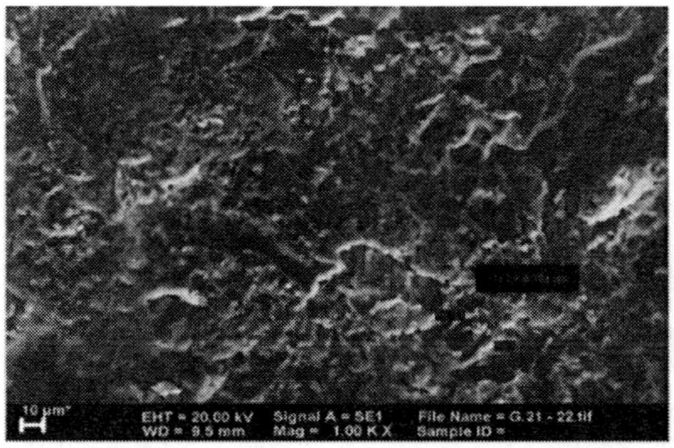

| El | AN | Series | unn. C [wt. %] | norm. C [wt. %] | Atom. C [at. %] | Error [%] |
|---|---|---|---|---|---|---|
| O | 8 | K-series | 52.19 | 52.19 | 50.25 | 16.7 |
| C | 6 | K-series | 33.45 | 33.45 | 42.91 | 11.4 |
| K | 19 | K-series | 5.13 | 5.13 | 2.02 | 0.2 |
| Cl | 17 | K-series | 3.13 | 3.13 | 1.36 | 0.1 |
| S | 16 | K-series | 2.61 | 2.61 | 1.25 | 0.1 |
| Na | 11 | K-series | 1.71 | 1.71 | 1.15 | 0.1 |
| Al | 13 | K-series | 1.22 | 1.22 | 0.69 | 0.1 |
| Mg | 12 | K-series | 0.57 | 0.57 | 0.36 | 0.1 |
| | | Total: | 100.00 | 100.00 | 100.00 | |

**Figure 1.** Structure porosity and mineral content from the shape of green polybag shows that performance seaweed waste of *Gracilaria verrucosa* (G0: treatment on 0 day; G7: treatment on 7 days; G14: treatment on 14 days; G21: treatment on 21 days).

| Treatment | Structure Porosity | Mineral Content |
|---|---|---|

**K0**

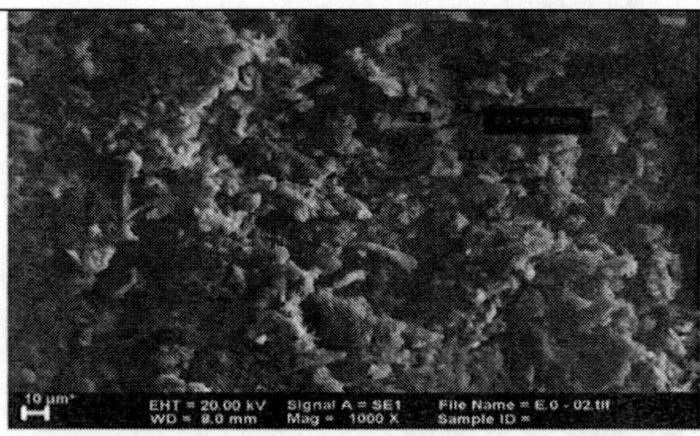

| El | AN | Series | Unn [wt.%] | C norm. [wt.%] | C Atom [at. %] | C Error [%] |
|---|---|---|---|---|---|---|
| C | 6 | K-series | 53.32 | 53.32 | 62.65 | 17.4 |
| O | 8 | K-series | 38.18 | 38.18 | 33.68 | 12.4 |
| Cl | 17 | K-series | 3.53 | 3.53 | 1.40 | 0.1 |
| K | 19 | K-series | 1.99 | 1.99 | 0.72 | 0.1 |
| S | 16 | K-series | 1.60 | 1.60 | 0.70 | 0.1 |
| Na | 11 | K-series | 1.18 | 1.18 | 0.72 | 0.1 |
| Mg | 12 | K-series | 0.21 | 0.21 | 0,12 | 0.0 |
| | | Total | 100.00 | 100.00 | 100.00 | |

**K7**

| El | AN | Series | Unn [wt.%] | C norm. [wt.%] | C Atom [at. %] | C Error [%] |
|---|---|---|---|---|---|---|
| C | 6 | K-series | 65.48 | 65.48 | 72.58 | 20.7 |
| O | 8 | K-series | 31.54 | 31.54 | 26.25 | 10.1 |
| Cl | 17 | K-series | 1.46 | 1.46 | 0.55 | 0.1 |
| K | 19 | K-series | 1.40 | 1.40 | 0.58 | 0.1 |
| S | 16 | K-series | 0.11 | 0.11 | 0.04 | 0.0 |
| Na | 11 | K-series | 0.01 | 0.01 | 0.00 | 0.0 |
| Mg | 12 | K-series | 0.00 | 0.00 | 0.00 | 0.0 |
| | | Total | 100.00 | 100.00 | 100.00 | |

**K14**

| El | AN | Series | Unn [wt.%] | C norm. [wt.%] | C Atom [at. %] | C Error [%] |
|---|---|---|---|---|---|---|
| C | 6 | K-series | 58.24 | 58.24 | 66.29 | 18.8 |
| O | 8 | K-series | 37.35 | 37.35 | 31.91 | 11.9 |
| Cl | 17 | K-series | 1.94 | 1.94 | 0.75 | 0.1 |
| K | 19 | K-series | 1.78 | 1.78 | 0.76 | 0.1 |
| S | 16 | K-series | 0.47 | 0.47 | 0.17 | 0.0 |
| Na | 11 | K-series | 0.20 | 0.20 | 0.12 | 0.0 |
| Mg | 12 | K-series | 0.01 | 0.01 | 0.01 | 0.0 |
| | | Total | 100.00 | 100.00 | 100.00 | |

**K21**

| El | AN | Series | Unn [wt.%] | C norm. [wt.%] | C Atom [at. %] | C Error [%] |
|---|---|---|---|---|---|---|
| C | 6 | K-series | 59.84 | 59.84 | 67.37 | 19.3 |
| O | 8 | K-series | 37.24 | 37.24 | 31.47 | 11.9 |
| Cl | 17 | K-series | 1.30 | 1.30 | 0.55 | 0.1 |
| K | 19 | K-series | 1.06 | 1.06 | 0.40 | 0.1 |
| S | 16 | K-series | 0.51 | 0.51 | 0.18 | 0.0 |
| Na | 11 | K-series | 0.05 | 0.05 | 0.03 | 0.0 |
| Mg | 12 | K-series | 0.00 | 0.00 | 0.00 | 0.0 |
| | | Total | 100.00 | 100.00 | 100.00 | |

**Figure 2.** Structure porosity and mineral content from the shape of green polybag shows that performance seaweed waste of *Kappaphycus alvarezii* (K0: treatment on 0 day; K7: treatment on 7 days; K14: treatment on 14 days; K21: treatment on 21 days.

**Table 1.** Data of soil quality.

| Parameter | Range |
|---|---|
| Temperature (°C) | 27 – 29 |
| Humidity (pF) | 3.0 – 6.2 |
| pH | 6.5 – 7 |

The physical data of the soil quality showed that the temperature, humidity and pH of the soil were normal for optimum growth of plants. Based on the results of this research, the waste of *K. alvarezii* and G. *verrucosa* can be used as raw materials of green polybag of MDF.

## 4. Conclusion

We demonstrated that *K. alvarezii* and G. *verrucosa* can be used as raw materials of green polybag of MDF. The porosity sizes of *K. alvarezii* and *G. verrucosa* were 3.976 µm and 4.794 µm, respectively. The carbon content was greater than 50% in *K. alvarezii,* and less than 50% in *G. verrucosa*. Therefore, the

IOP Conf. Series: Earth and Environmental Science **137** (2018) 012084 doi:10.1088/1755-1315/137/1/012084

structure of MDF is stronger in green polybag made from the waste of *K. alvarezii* than that in green polybag made from *G. verrucosa*.

## 5. References

[1] Budi S W, Sukendro A and Karlinasari L 2012 *J. Agron Indonesia* **40** 239-245 (Text in Indonesia)
[2] Suryanti S, Indradewa D, Sudira P and Widada J 2015 *J. Agritech.* **35** 114-120 (Text in Indonesia)
[3] Ilknur A and Cirik S 2004 *Pak. J. Biol. Sci.* **7** 2022-2023
[4] Alamsjah M A 2010 *Res. J. Fish. Hydrobiol.* **5** 159-167
[5] Marsha N D, Aini N and Sumarni T 2014 *Influence of frequency and volume of water supply on Crotalaria mucronata* (Malang: Desv. growth Faculty of agriculture, Brawijaya University) p 675
[6] Alamsjah M A Sulmartiwi L, Pursetyo K T, Amin M N G, Wardani K A and Arifianto M D 2017 *J. Indian Acad. Wood Sci.* **14** 32-45
[7] Kusriningrum R S 2012 *Experimental design* (Airlangga: University Surabaya) pp 31-35 (Text in Indonesia)

IOP Conf. Series: Earth and Environmental Science **137** (2018) 012085 doi:10.1088/1755-1315/137/1/012085

# Exploration of indigenous bacteria in an intensive aquaculture system of African catfish (*Clarias* sp.) in Banyuwangi, Indonesia

**Prayogo[1]*, B S Rahardja[1], A N Asshanti[1], N N Dewi[1] and M B Santanumurti[1]**

[1] Faculty of Fisheries and Marine, Airlangga University, 60115 Jl. Sutorejo, Campus C, Surabaya, Indonesia

E-mail: prayogo_unair@yahoo.co.id

**Abstract**. Intensive African catfish culture in tarpaulin pond was popular in Banyuwangi, Indonesia since the government supported the fisheries sector. Unfortunately, the failure of African catfish culture still occurred since the waste from fish metabolite process and feed residue decreased the water quality. Bacteria in the water could be the solution to increase the success rate of aquaculture by improving the water quality. This study purpose was to obtained indigenous bacteria in intensive aquaculture system of African catfish to improve water quality. This study successfully isolated bacteria contained with amylase, protease and lipase characteristic. Isolated bacteria in this study were identified as *Pseudomonas pseudomallei* (97.81%), *Bacillus subtilis* (95.81%) and *Pseudomonas stutzeri* (61.21%).

## 1. Introduction

In 2010, the Government of Banyuwangi Region in East Java launched a program which established 10,000 fish ponds [1]. The program was aimed to increase not only the incomes of the people, but also their nutrition since the fish contain high protein[2]. One of the fish species cultured in Banyuwangi is African catfish (*Clarias* sp). African catfish have a high economic value, are easy to maintain, and grow fast [3]. There are many systems used for African catfish aquaculture in Banyuwangi, including intensive aquaculture system in tarpaulin pond. Intensive aquaculture system in tarpaulin pond is low-cost and easy to maintain [4].

Unfortunately, the system is still facing challenges, especially water management issues. The issues are related to aquaculture waste including metabolite waste and feed residues, which often leads to failure in the African catfish culture due to decreased water quality [5]. Bacteria have an important role in overcoming aquaculture waste.

Bacteria act as biological control agents that can improve water quality through the degradation of organic materials specifically [6]. The process of degrading the organic materials requires an important role of proteolytic, lipolytic bacteria, and amylolytic bacteria [7, 8, 9].

Proteolytic bacteria is bacteria producing proteinase [7]. Proteolytic bacteria use proteinase to break down proteins into peptides and amino acids, and turn them into energy[9]. Lipolytic bacteria produce lipase [10]. Lipase is an enzyme for catalyzing the hydrolysis reaction of triacylglycerol into fatty acids and glycerol [8]. Amylolytic bacteria are bacteria producing amylase. Amylase can degrade amylum into simpler polymers or monosaccharide sugars, wherein the next monosaccharide will be

IOP Conf. Series: Earth and Environmental Science **137** (2018) 012085 doi:10.1088/1755-1315/137/1/012085

changed into energy [7]. Three kinds of bacteria are essential for managing water quality in African catfish culture.

This study aims to obtain the indigenous bacteria for degrading organic materials to improve water quality in the intensive aquaculture system of African catfish in the tarpaulin pond. This study also determines the effectiveness of the indigenous bacteria in degrading organic materials to overcome the waste in the intensive aquaculture system.

## 2. Methodology

### *2.1 Material*

The study was conducted from April until October 2016. The bacteria samples were collected from the intensive aquaculture system of African catfish in Muncar, Banyuwangi, Indonesia. The water containing the bacteria was put into bottles aseptically and taken using ice box to the microbiology laboratory of Science and Technology Faculty, Airlangga University. Each bottle contained 200 ml of water.

### *2.2 Method*

#### *2.2.1 Isolation*

Each sample was homogenized and removed 100 ml of each into collection tubes. The homogenized tubes were then diluted until $10^{-4}$. The 0.01 ml of diluted suspension was inoculated to the Bushnell Haas (BH) media using a poured plate method (10 samples were labeled as A, B, C, D, E, F, G, H, I and J). This study used three kinds of BH media. BH with 1% of fat (olive oil) was used to isolate fat-degrading bacteria. BH media with 1% of amylum was used to isolate the amylum-degrading bacteria. BH media with 2% of skim milk was used to isolate the protein-degrading bacteria. All media were incubated at 37° C for 24-48 hours.

#### *2.2.2 Hydrolysis index test*

Inoculated bacteria in media resulted in the growth of bacterial cells appearing as multiple colonies. Colonies of degrading bacteria could be shown by the formation of clear zones around the growing colonies. The potential of hydrolysis of each isolate could be calculated based on the ratio of the diameter of the clear zone to the diameter of the growing bacterial colony. Its value is called the hydrolysis index [11].

### *2.3 Identification*

Isolated indigenous bacteria were identified based on biochemical characterization (lipolytic, amylolytic and proteolytic test). Identification was performed by using Microbact™GNB 12A/B/E, 24E Identification Kits (Thermoscientific). The characterization results were referenced using Bergey's Manual of Determinative Bacteriology and International Journal of Systematic Bacteriology (IJSB)/International Journal of Systematic and Evolutionary Microbiology (IJSEM). The determination of species level was also done based on the characterization of the special decomposition properties of each isolate, referring to the latest literature and journals.

### *2.4 Water quality test*

The test was conducted on water samples from catfish pond added with degrading bacteria. Water quality parameters observed were residual protein, fat, carbohydrate, BOD(Biological Oxygen Demand) and COD(Chemical Oxygen Demand) levels. Water parameters were checked at day 0, day 3, day 6 and day 9. 5 mg/ml of bacteria used were *Pseudomonas pseudomallei* 97.81 % (A), *Bacillus subtilis* 95.81 % (B), *Pseudomonas stutzeri* 61.21% (C), *Pseudomonas pseudomallei* 97.81 % + *Pseudomonas stutzeri* 97.81 % + *Pseudomonas stutzeri* 61.21% (D). The use of bacteria was based on hydrolytic test results and identification.

IOP Conf. Series: Earth and Environmental Science **137** (2018) 012085 doi:10.1088/1755-1315/137/1/012085

## 3. Results and discussion

### *3.1 Hydrolysis test*

**Table 1**. Hydrolysis test from isolated bacteria.

| Test | Isolate | | | | | | | | | |
|---|---|---|---|---|---|---|---|---|---|---|
| | A | B | C | D | E | F | G | H | I | J |
| Proteolytic | 2 | 2.6 | - | 1.666 | 1.333 | 1.333 | 0.25 | 0.25 | - | 2.1 |
| Amylolitic | 0.25 | - | - | - | - | 0.142 | 0.334 | - | 1.5 | - |
| Lipolytic | - | 1 | 1.325 | 1 | 1 | 0.818 | 3 | - | 0.5 | - |

The table 1 shows that isolate B was the highest value of proteolytic with 2.6. Isolate G was the highest value of lipolytic with 3. Isolate I was the highest of amylolytic value with 1.5.

### *3.2 Identification of isolated bacteria*

Isolated bacteria chosen as potential waste-degrading aquaculture waste had a huge value of hydrolysis index. Isolated bacteria were B, G and I identified as *Pseudomonas pseudomallei* (97.81%), *Bacillus subtilis* (95.81%) and *Pseudomonas stutzeri* (61.21%).

### *3.3. Water quality test*

#### *3.3.1 Protein level*

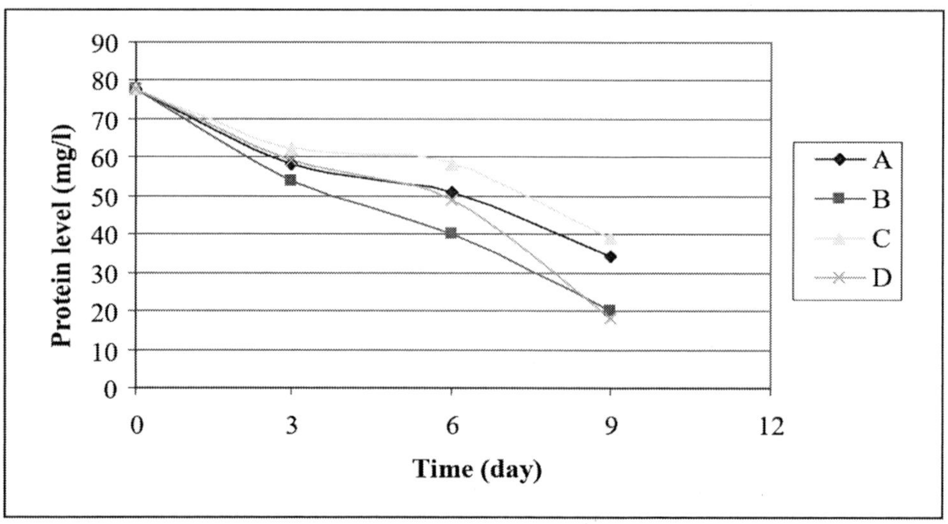


**Figure 1.** Protein level in the water.

Each day, the protein level decreased gradually. At day 3, the highest average protein level was the C treatment of 62 mg/l while the lowest was the B treatment of 54 mg/l. At day 6, the highest protein content was the C treatment of 58 mg/l while the lowest was the B treatment of 40 mg/l. At day 9, the highest protein level was in the C treatment of 39 mg / l, and the smallest was the D treatment of 18 mg/l. The results of the protein content of each bacterial inoculant are shown in figure 1.

*3.3.2 Fat level*

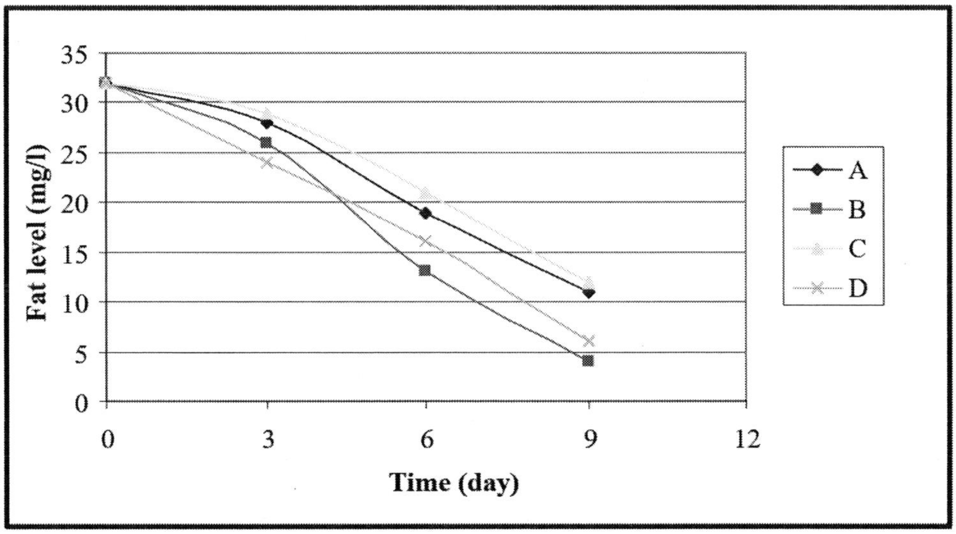


**Figure 2.** Fat level in the water.

This shows that the fat level decreased daily. At day 3, the highest fat level was the C treatment of 29 mg/l while the lowest was the D treatment of 24 mg/l. At day 6, the highest fat level was the C treatment of 21 mg/l while the lowest was the B treatment of 13 mg/l. At day 9, the highest fat level was the C treatment of 12 mg/l while the lowest was the B treatment of 4 mg/l. The data of fat level of each bacterial inoculant are shown in figure 2.

*3.3.3 Carbohydrate level*

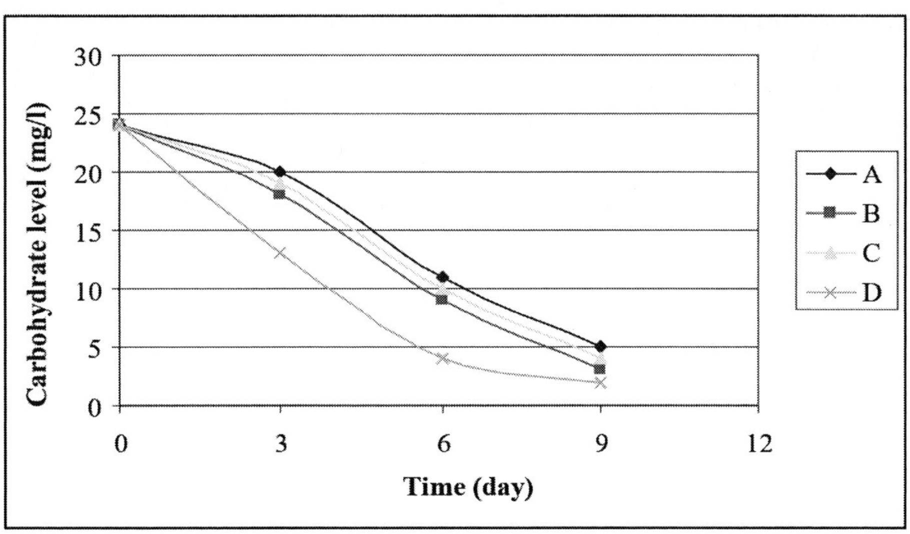


**Figure 3.** Carbohydrate level in the water.

IOP Conf. Series: Earth and Environmental Science **137** (2018) 012085 doi:10.1088/1755-1315/137/1/012085

At day 3, the highest carbohydrate level was the A treatment of 20 mg/l while the lowest was the D treatment of 13 mg/l. At day 6, the highest carbohydrate level was the A treatment of 11 mg/l while the lowest was the D treatment of 4 mg/l. At day 9, the highest carbohydrate level was the A treatment of 5 mg/l while the lowest was the D treatment of 2 mg/l. The data of carbohydrate level of each bacterial inoculant are shown in figure 3.

*3.3.4 BOD level*

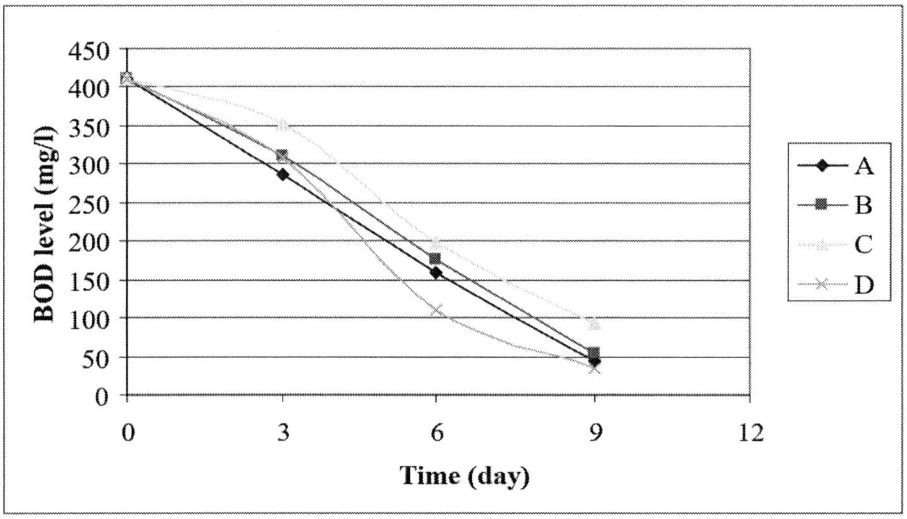


**Figure 4**. BOD concentration in the water.

The result showed that the BOD level at day 0 had the highest value in each bacterial treatment of 410 mg/l. The level of BOD decreased daily. At day 3, the highest BOD level was the C treatment of 353 mg/l while the lowest was the A treatment of 287 mg/l. At day 6, the highest BOD level was the C treatment of 197 mg/l while the lowest was the D treatment of 111 mg/l. At day 9, the highest BOD level was C treatment of 92 g/l while the lowest was the D treatment of 35 mg/l. The data of BOD level of each bacterial inoculant are shown in figure 4.

IOP Conf. Series: Earth and Environmental Science **137** (2018) 012085 doi:10.1088/1755-1315/137/1/012085

*3.3.5 COD level*

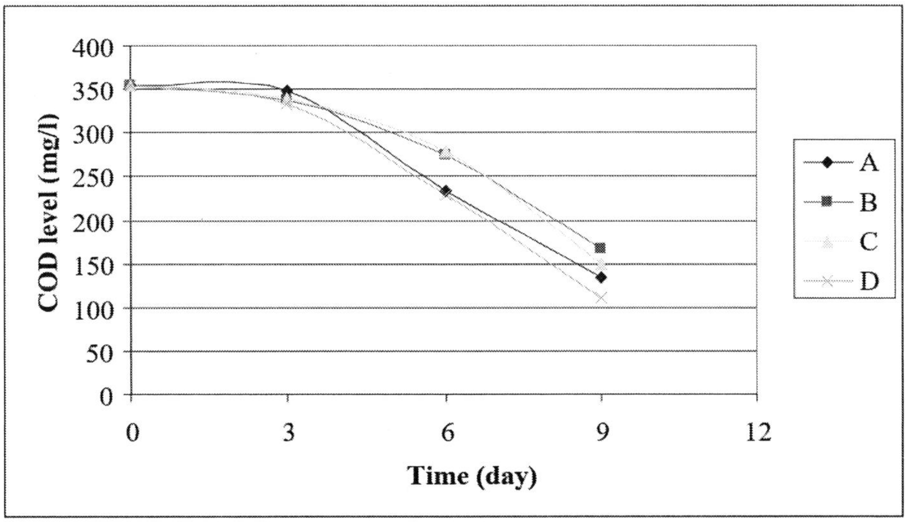


**Figure 5.** COD level in the water

The study shows that the COD level decreased daily. At day 3, the highest COD level was the A treatment of 349 mg/l while the lowest was D treatment of 334 mg/l. At day 6, the highest COD level was the C treatment of 280 mg/l while the lowest was the D treatment of 229 mg/l. At day 9, the highest COD level was the B treatment of 167 mg/l while the lowest was the D treatment of 110 mg / l. The data of COD level of each bacterial inoculant are shown in figure 5.

## 4. Discussion

This study identified the bacteria containing amylase, proteinase and lipase enzymes (table 1). The bacteria live in the water and act as biological control agents that can improve water quality [6]. The results of the isolation of organic degradation bacteria taken from intensive aquaculture system in African catfish (*Clarias* sp) in the tarpaulin pond showed 3 potential bacteria: *Pseudomonas pseudomallei*, *Bassilus subtilis,* and *Pseudomonas stutzeri*.

*Pseudomonas* and *Bassilus* are proteolytic bacteria [7]. Proteolytic bacteria are bacteria that produce extracellular proteinase enzymes, which are protein-soluble enzymes produced in cells and then released out of cells[12]. The aquaculture waste present in the water will be absorbed by bacterial cells and decomposed into amino acids by intracellular protease for protein synthesis or ultimately deaminated through intermediary metabolic pathways [13]. The protease was an extracellular enzyme released by microbes, which serves to hydrolyze peptide bonds in proteins to produce simpler peptides or also produce amino acids by peptidase [14]. Reduced levels of protein at day 3 to day 9 show the role of *Pseudomonas* and *Bassilus* as proteolytic bacteria to degrade the protein content in the waters (figure 1).

*Pseudomonas* and *Bassilus* are not only identified as proteolytic bacteria but also as lipolytic bacteria [15, 10]. Lipolytic bacteria are lipase-producing bacteria [10]. The ability of the bacteria in degrading fat is related to the bacteria's ability to produce lipase enzymes. The lipase enzyme serves to catalyze triglycerides into diglycerides and fatty acids [15]. Reduced levels of fat on day 3 to day 9 show that *Pseudomonas* and *Bacillus* as lipolytic bacteria can break down the protein content in the waters (figure 2).

*Pseudomonas* and *Bassilus* have other roles as amylolytic bacteria [13, 16]. Amylolytic bacteria are enzyme-producing amylase bacteria, enzymes that break down amylum into simpler polymers or

IOP Conf. Series: Earth and Environmental Science **137** (2018) 012085 doi:10.1088/1755-1315/137/1/012085

monosaccharide sugars, wherein the next monosaccharide will be degraded once more into energy[7]. Reduced carbohydrate levels at day 3 to day 9 show that *Pseudomonas* and *Bassilus* as amylolytic bacteria can degrade carbohydrate levels in the waters (figure 3).

Bacteria can perform a bioremediation process in the water. Bioremediation is a process that utilizes microorganisms, including bacteria, to transform harmful substances into non-toxic products[17]. Bioremediation is done by bacteria to activate metabolism system [11]. Metabolism is a process of chemical transformation and the formation of energy for bacterial life and proliferate. Bacterial metabolism is divided into two, aerobic and anaerobic. The basis of the reaction in aerobic metabolism[11]. Aerobic metabolism requires $0_2$ to produce energy. Reduced levels of BOD and COD at day 3 to day 9 show that *Pseudomonas* and *Bacillus* could reduce BOD and COD level in waters (figure 4 and figure 5).

Study of the water quality test of in vitro showed that the treatment using inoculant of 3 species of bacteria was more effective. This is indicated by the decreasing protein, fat, carbohydrate, BOD and COD levels occurring in the D treatment (figure 1, figure 3, figure 4 and figure 5). Bacterial concentrations are faster in degrading activity than only one species. The addition of combining inoculant bacteria stimulates the process of decomposition of organic materials into minerals more efficiently because it takes a shorter time [18].

## 5. Conclusion

In conclusion, we found successfully isolated bacteria contained with amylase, protease and lipase characteristic from intensive aquaculture system of African catfish in Muncar, Banyuwangi, Indonesia. Founded bacteria in this study were identified as *Pseudomonas pseudomallei* (97.81 %), *Bacillussubtilis* (95.81 %) and *Pseudomonas stutzeri* (61.21 %). Water quality test of protein, fat, carbohydrate, BOD and COD showed decrease level of each aspect by adding isolated bacteria. The combination bacteria showed faster degradating activity than individual bacteria.

## 6. References


[1] Akbar K, Abdul J A G and Riyanto 2014 *J. Public Adm.* **2** 591 (Text in Indonesia)

[2] Adriani R D, Jumirah and Ernawati 2015 *J. Univ. Sumatra Utara* **1** 1-10 (Text in Indonesia)

[3] Fauziah A F, Agustina T and Hariyati Y 2017 *Soc. Econom. J. Fish.***9** 20-32 (Text in Indonesia)

[4] Rosalina D 2014 *Maspari J.* **6** 20-24 (Text in Indonesia)

[5] Miller G E and George S L 1984 *Aquac. Eng.* **3** 39-57

[6] Irianto A 2003 *Aquaculture probiotics* (Yogyakarta: Gadjah Mada University Press) p 125 (Text in Indonesia)

[7] Fardiaz S 1993 *Analysis of Food Microbiology* (Jakarta: Raja Grafindo Persada) (Text in Indonesia)

[8] Gupta R, Pooja R, Namita G and Sapna B 2003 *J Appl. Biotechnol. Biochem.* **37** 63-71

[9] Volk W A 1992 *Basic Microbiology Seventh edition* (New York: Harper Collins Publishers) p 524 – 534

[10] Saxena R K, Ghost P K, Rani G, Davidson W S, Sapna B and Ruchi G 2004 *Curr. Sci.* **77** 101-105

[11] Suarsini E 2006 *Bioremediation of household liquid waste using consortium indigenous bacteria that potentially reduces pollutant* (Malang: Universitas Negeri Malang) p 211 (Text in Indonesia)

[12] Winarno F G 1993 *Food, nutrition, technology and consumers* (Jakarta: PT. Gramedia Pusat Utama) (Text in Indonesia)

[13] Schelegel 1993 *General Microbiology* (Cambridge: CU Press)

[14] (Warinangin, 1996 in Roosdiana et al., 2003)

[15] Haba E, Bresco O, Ferrer C, Marques A, Busquets M and Manresa A 2000 *Enzyme Microb Technol* **26** 40-44

[16] Hastuti U S, Nugraheni F S A and Al Asna P M 2017 Isolation and identification of amylolytic bacterial species derived from mangrove soils in Margomulyo, Balikpapan, East Kalimantan Conf (Malang: Universitas Muhammadiyah Malang) p 267-271 (Text in Indonesia)
[17] Dash H R, Neelam M, Jaya C, Supriya K and Surajit D 2013 *Appl. Microbiol. Biotechnol.* **97** 561-571
[18] Madigan M T 2003 *Photosynth Res.* **76** 157-171 (Text in Indonesia)

**Acknowledgement**
The study was funded by RKAT (Rencana Kerja dan Anggaran Tahunan) of the Fisheries and Marine Faculty, Airlangga University. The authors would like to acknowledgement Faculty of Fisheries and Marine Science for financial and support.

IOP Conf. Series: Earth and Environmental Science **137** (2018) 012086 doi:10.1088/1755-1315/137/1/012086

# Analysis of water quality on several waters affected by contamination in West Sumbawa Regency

**N N Dewi, W H Satyantini, A M Sahidu, L A Sari, A T Mukti**

Department of Fish Health Management and Aquaculture Faculty of Fisheries and Marine Universitas Airlangga, Campus C Mulyorejo, 60115 Indonesia

E-mail : ninanurmaliadewi@fpk.unair.ac.id

**Abstract.** This study reports the result of water quality in several waters in West Sumbawa Regency. The load of waste input from anthropogenic activity becomes an indication of the decrease of water quality in West Sumbawa Regency Waters. The existence of illegal mining activities around the water has the potential to cause water pollution. Sample of water were collected on April 2017 in four location such as Sejorong 1, Sejorong 2, Tongo, and Taliwang. Sample were analyzed as insitu and exsitu parameters. The result of this research showed that Sejorong 2 have the highest value of pollution index but generally four site on West Sumbawa Regency Waters were categorized lightly contaminated. Concentration of heavy metal cadmium at four locations exceed the water quality standard for fisheries and drinking water. However, the trophic classification using TSI and TRIX of all location was oligothropic water.

## 1. Introduction

Nowadays, the quality of surface water is an important concern. Waters such as rivers become water bodies that receive pollution input. Surface water is the area which most widely receives natural processes such as soil erosion, precipitation rates, and weathering processes, as well as anthropogenic activities such as industry. Rivers are land water resources essential for domestic activities, so control and reliable information is needed for effective management [1].

Batu Hijau area of West Sumbawa is an area that has a good biological and non- biological potential. The rivers located in the Batu Hijau area have an important economic value for the community be it for daily life, irrigation, or fishing for consumption [2].

Nowadays, pollution has become increasingly problematic in our daily life [3]. Along with the high potential of natural resources in the region of West Sumbawa, many industries have also been established there. The number of companies in the region have numbered to 625 industries covering textile, paper, chemical, petroleum, coal, plastics, metal goods, machinery and equipment, food, tobacco, and other processing industries [4]. The presence of acid mine overflow from the gold mining industry in West Sumbawa has the potential to decrease water quality. According to Onodera [5] the existence of various industrial activities causes various contaminants that can enter the waters. Phiri [6] said that these contaminants can interfere and harm the life of aquatic biota.

Based on information from people who live around the river, dead biota can often be found which include fish, shrimp, and crabs. The biota are suspected to be poisoned due to the contaminants that flow into the river [7,8]. This is an important concern that needs to be studied, especially regarding the description of water quality, quality status, and determination of trophic clasification in West Sumbawa Regency waters due to the problem of pollution. This is one of the monitoring efforts to

IOP Conf. Series: Earth and Environmental Science **137** (2018) 012086 doi:10.1088/1755-1315/137/1/012086

prevent the occurrence of water quality degradation in these waters which is an important habitat of various biota with great ecological and economical value for food consumption. The aim of this study is to analyze the water quality status and trophic classification of the West Sumbawa Regency waters.

## 2. Methodology

### *2.1. Monitoring site*

The study was conducted in April 2017 in several water sites in the West Sumbawa regency (figure 1). The research locations consist of 4 stations, namely Sejorong 1, Sejorong 2, Tongo, and Taliwang. Sejorong 1 is a river, Sejorong 2 is the estuary of the river, while Tongo and Taliwang are small lakes and tend to resemble stagnant waters. These four locations are indicated to be affected by anthropogenic activities, especially gold mining.

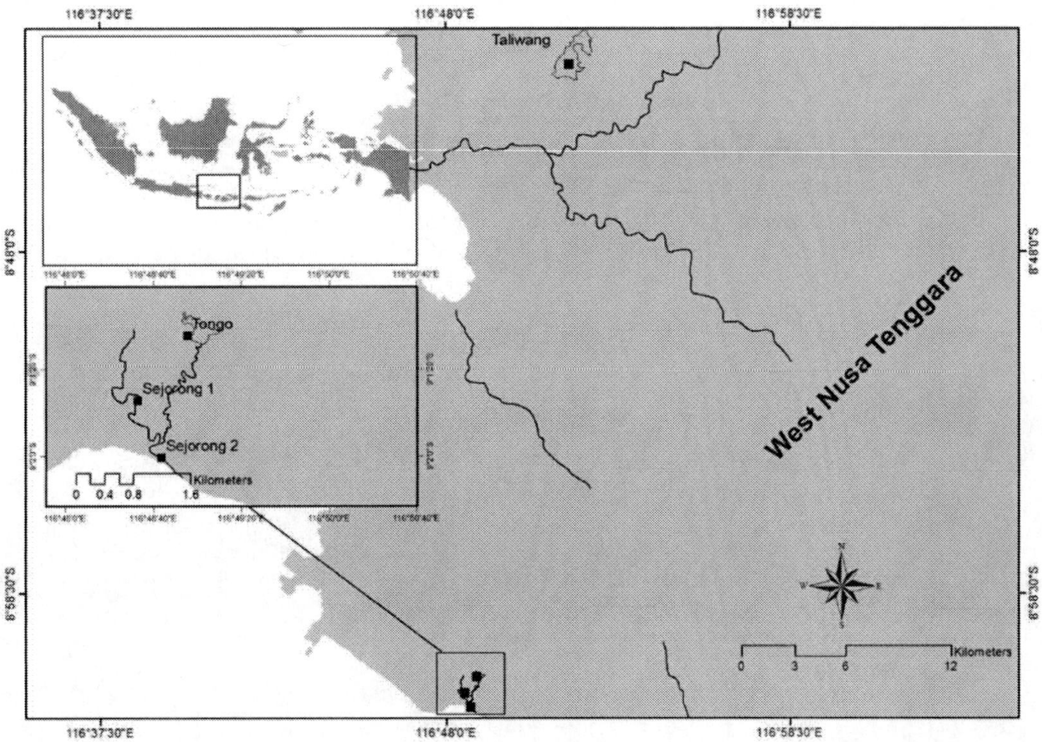


**Figure 1**. Map showing the water quality monitoring sites on several waters in West Sumbawa Regency waters.

IOP Conf. Series: Earth and Environmental Science **137** (2018) 012086 doi:10.1088/1755-1315/137/1/012086

The condition of West Sumbawa waters is shown by figure 2

Sejorong 1
Water bodies are narrower, tend to be shallow, greenish water color

Sejorong 2
More rapid currents, the most shallow point, the color is brown water, close to the mouth of the river

Tongo
brownish green water color, the existence of fishing activity

Taliwang
Conditions tend to be stagnant water, overgrown with aquatic plants (water hyacinth), presence of fishing activities by boat, and fishing, water color brownish green

**Figure 2.** Conditon of sampling locations.

### *2.2. Monitored parameters and analytical methods*

Water quality parameters measured consist of physical, chemical and biological parameters. The water samples taken were surface water samples. Water samples were measured in situ and some were prepared and analyzed exsitu at the Institute for Research and Standarization of Surabaya and Nutrition Laboratorium, Faculty of Public Health at Airlangga University (table 1).

**Table 1**. Water quality parameters, units and analytical methods as measured during April 2017 on West Sumbawa Regency waters.

| Parameters | Units | Method | Handling | Notes |
|---|---|---|---|---|
| Physical Parameters | | | | |
| Temperature | °C | Termometer (Hg) / (APHA, AWWA, WEF 2005) | | *Insitu* |
| Transparency | meter | *Secchi disk*/Visual (APHA, AWWA, WEF 2005) | | *Insitu* |
| Turbidity | NTU | HACH 2100 AN Turbiditimeter Handbook | | *Exsitu* |
| TDS | mg/L | SNI 06-6989.27-2005 | | *Exsitu* |
| TSS | mg/L | SNI 06-6989.3-2004 | | *Exsitu* |
| Chemical Parameters | | | | |
| Salinity | ‰ | Refractometer (APHA, AWWA, WEF, 2005) | | *Insitu* |
| pH | | pH meter | | *Insitu* |
| Alkalinity | mg/L | Standard Methods 20 th edition 1998 | Cooling | *Exsitu* |
| Dissolved oxygen | mg/L | DO meter | | *Insitu* |
| BOD | mg/L | SNI 06-6989.72 : 2009 | Cooling | *Exsitu* |
| Ammonia | mg/L | SNI 06-6989.30-2005 | $H_2SO_4$, pH <2 | *Exsitu* |
| Nitrate | mg/L | SNI 06-2480-1991 | $H_2SO_4$, pH <2 | *Exsitu* |
| Nitrite | mg/L | SNI-06-6989.9-2004 | $H_2SO_4$, pH <2 | *Exsitu* |
| Orthophosphate | mg/L | Standard Methods 20th edition 1998 | Cooling | *Exsitu* |
| Phosphate total | mg/L | Standard Methods 20th edition 1998 | Cooling | *Exsitu* |
| Lead (Pb) | mg/L | SNI- 06-6989.46-2005 | $HNO_3$, pH < 2 | *Exsitu* |
| Cadmium (Cd) | mg/L | SNI- 06-6989.38-2005 | $HNO_3$, pH < 2 | *Exsitu* |
| Copper (Cu) | mg/l | SNI- 6989.66-2009 | $HNO_3$, pH < 2 | *Exsitu* |
| Biological Parameters | | | | |
| Chlorofil a | mg/L | Aseton extract | Nathio-sulfat, dark condition | *Exsitu* |
| Phytoplankton | sel/m$^3$ | Microscope | lugol | *Exsitu* |

### *2.3. Data analysis*

#### *2.3.1 Pollution index*

All parameters of water quality were compared to river water quality standards using the quality standard of Indonesia [9]. The results obtained were then discussed descriptively for all observation stations. The quality status of the environmental quality of Sungai Tongoloka was calculated by the using pollution index. The pollution index was used to determine the level of pollution relative to the allowed water quality parameters [10]. The pollution index has an advantage in determining pollution levels at a point at one observation. The pollution index value results are then evaluated to determine the status of the water quality (table 2).

IOP Conf. Series: Earth and Environmental Science **137** (2018) 012086 doi:10.1088/1755-1315/137/1/012086

$$\mathrm{IP} : \sqrt{\frac{\left(\frac{C_i}{L_{ij}}\right)_M^2 + \left(\frac{C_i}{L_{ij}}\right)_R^2}{2}} \tag{1}$$

IP : Pollution index
Ci : Concentration of water quality parameters (i)
Lij : Standard of water quality parameters
(Ci/Lij)M : Maximum value Ci/Lij
(Ci/Lij)R : Mean value Ci/Lij

**Table 2.** Evaluation of pollution index value [10].

| Value | Water Quality Status |
|---|---|
| 0 < ≤ 1.0 | Good |
| 1.0 < ≤ 5.0 | Lightly contaminated |
| 5.0< IP ≤ 10 | Poluted moderately |
| IP > 10 | Heavily polluted |

*2.3.2 Trophic Classification using Trophic State Index (TSI)*
Trophic classification of the water was calculated based on several parameters that affect productivity in accordance with the calculation Trophic state index (TSI) [11]. TSI is based on three parameters namely total phosphate concentration (TSI-P), chlorophyll-a concentration (TSI-Chl-a) and Secchi disk depth value (TSI-SD). From these three parameters were obtained the value of TSI which is the average result of the value of TSI-P, TSI Chl-a and TSI-SD. Trophic classification based on Carlson Trophic state index (TSI) calculation is as follows:

$$\text{TSI-SD} = 10\left(6 - \frac{\ln SD}{\ln 2}\right) \tag{2}$$

$$\text{TSI-Chl-a} = 10\left(6 - \frac{2{,}04 - 0{,}68 \ln Chl\text{-}a}{\ln 2}\right) \tag{3}$$

$$\text{TSI-TP} = 10\left(6 - \frac{\ln \frac{48}{TP}}{\ln 2}\right) \tag{4}$$

$$\text{Average of TSI} = \left(\frac{\text{TSI SD} + \text{TSI Chl-a} + \text{TSI TP}}{3}\right) \tag{5}$$

SD = Secchi *disk* (m)
Chl-a = Chlorofil-a ($mg/m^3$)
TP = Total phosphate ($mg/m^3$)

Here is the category of trophic classification based on TSI Carlson (table 3).

**Table 3.** Category of trophic classification based on TSI Carlson.

| TSI | Fertility Status |
|---|---|
| <30-40 | Oligotrof |
| 40-50 | Mesotrof |
| 50-70 | Eutrof |
| 70-100+ | Hipereutrof |

IOP Conf. Series: Earth and Environmental Science **137** (2018) 012086 doi:10.1088/1755-1315/137/1/012086

*2.3.3 Trophic Classification using TRIX*

The parameters chosen in the TRIX index determination were dissolved inorganic nitrogen (DIN), dissolved inorganic phosphate (DIP / $PO_4$), chlorophyll-a (Chl-a), and saturation level (% $O_2$). On the determination of the TRIX index using the formula Vollenwieder [12] shown as follows.

$$TRIX = (\text{Log}\,[\text{Chl-a} \times \%\, sO_2 \times \text{DIN} \times \text{DIP}] + 1.5)/1.2 \quad (6)$$

% $O_2$ (saturation level) = (DO/DOi) x 100 % , DOi (saturatio oxygen) = 14.62 – 0,37(T°C) + 0,0045(T°C)2 - 0,097(Salinity) + 0,002(T°C)(Salinity) + 0,0003 (Salinity)2, (Weiss 1970), DIN*(Dissolved Inorganic Nitrogen)* = $NH_3 + NO_2 + NO_3$ , DIP*(Dissolved Inorganic Phospate)* = PO4

## 3. Results and discussion

*3.1. Physical Parameters*

The physical parameters are shown on figure 3. According to Goldman [13], temperatures are parameters that play a role in controlling aquatic ecosystems that may affect other physical, chemical, and biological parameters. This is related to the metabolic system of aquatic biota. Temperatures on West Sumbawa Regency waters range from 27.5-30.5°C.

Based on field observations, Sejorong 2 is the most shallow station among all of the stations (20 cm), while Taliwang has the deepest depth (100 cm). The color of the waters describe the content of substances that exist in the waters. The color of the waters at Sejorong 1 station is greenish and odorless, while in Sejorong 2 is brown but also odorless. The Tongo station tends to be brownish-green and odorless, Taliwang station is brownish-green and smells fishy.

Transparency is visually determined and measured using a Secchi disk [13]. Transparancy is one of the parameters that can be used to estimate the total turbidity of the waters due to organic or inorganic waste [11, 12]. Therefore, this technique can be used to predict the primary productivity of waters associated with the status of trophic classification. Sejorong 1 has the highest transparency value among other stations (72 cm) while Sejorong 2 has a low transparency value (7.3 cm) and Taliwang (8 cm).

The value of TDS for Sejorong 1, Tongo, and Taliwang has not exceeded the water quality standard based on [9] Class 1 regarding drinking water quality standard and Class 3 on the quality standard for fishery activities (1000 mg/L), while Sejorong 2 has exceeded the standard quality (1086 mg / L).

The TSS value describes the total of suspended solids in the water. TSS values range from 2-33 mg / L. Sejorong 2 has the highest TSS value of 33 mg/L. However, for the four stations the TSS value is still below the quality standard for fishing activities (400 mg / L).

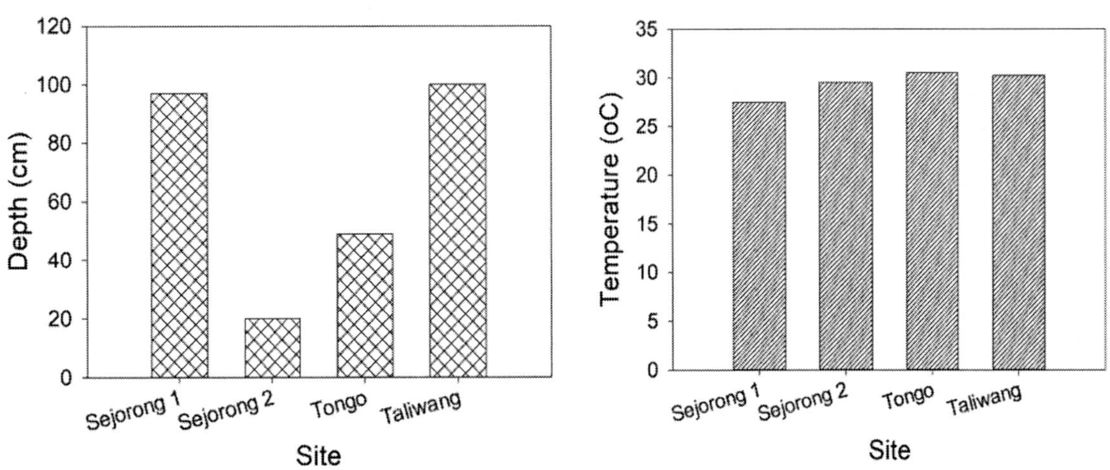

IOP Conf. Series: Earth and Environmental Science **137** (2018) 012086 doi:10.1088/1755-1315/137/1/012086

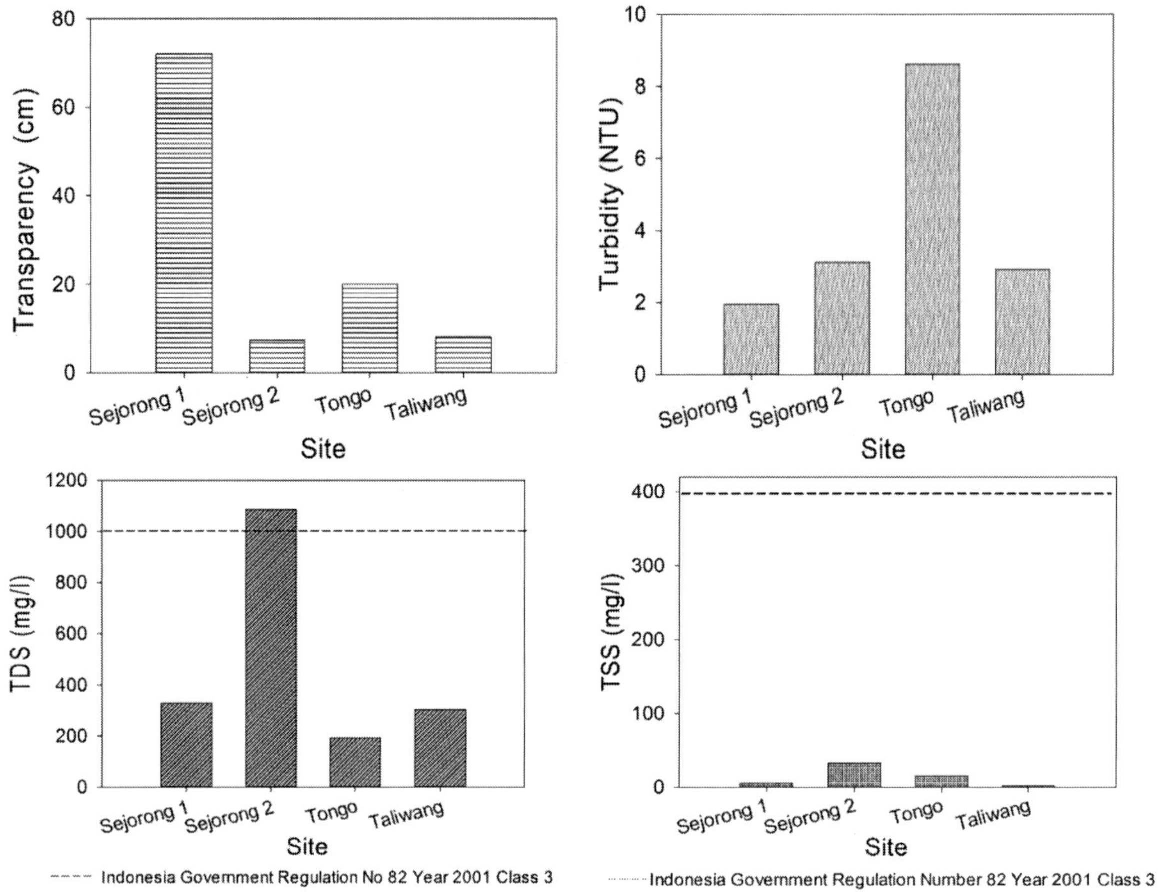


**Figure 3.** Physical parameter on West Sumbawa Regency waters.

*3.2 Chemical parameters*

The chemical parameters of water quality are shown by the figure 4,5 dan 6. The degree of acidity is influenced by the process of photosynthesis and respiration. The pH values at the study sites ranged from 6.9-7.95. Overall pH value is still in the range of drinking water quality standards and fishery activities. Oxygen values range from 4.51-6.1 mg / L. The oxygen content value at the four stations meets the quality standard for fisheries activities (minimum 3 mg / L).

IOP Conf. Series: Earth and Environmental Science **137** (2018) 012086 doi:10.1088/1755-1315/137/1/012086

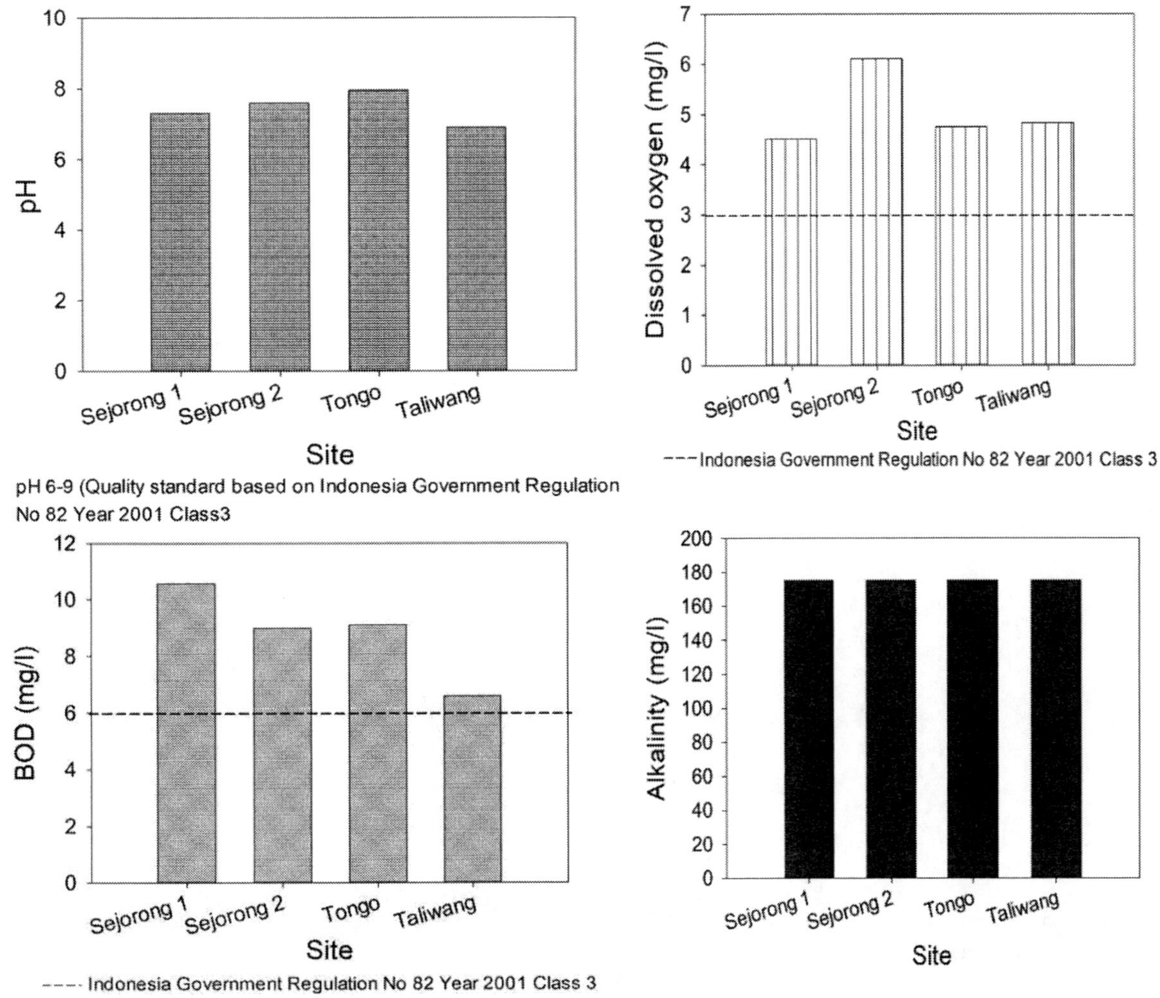


**Figure 4.** Chemical Parameter pH, DO, BOD, and Akalinity on West Sumbawa Regency waters.

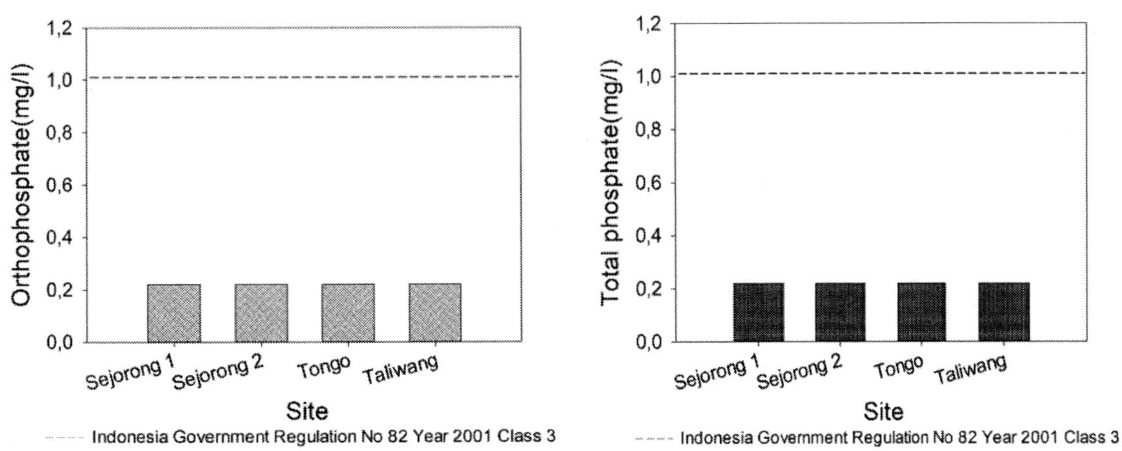

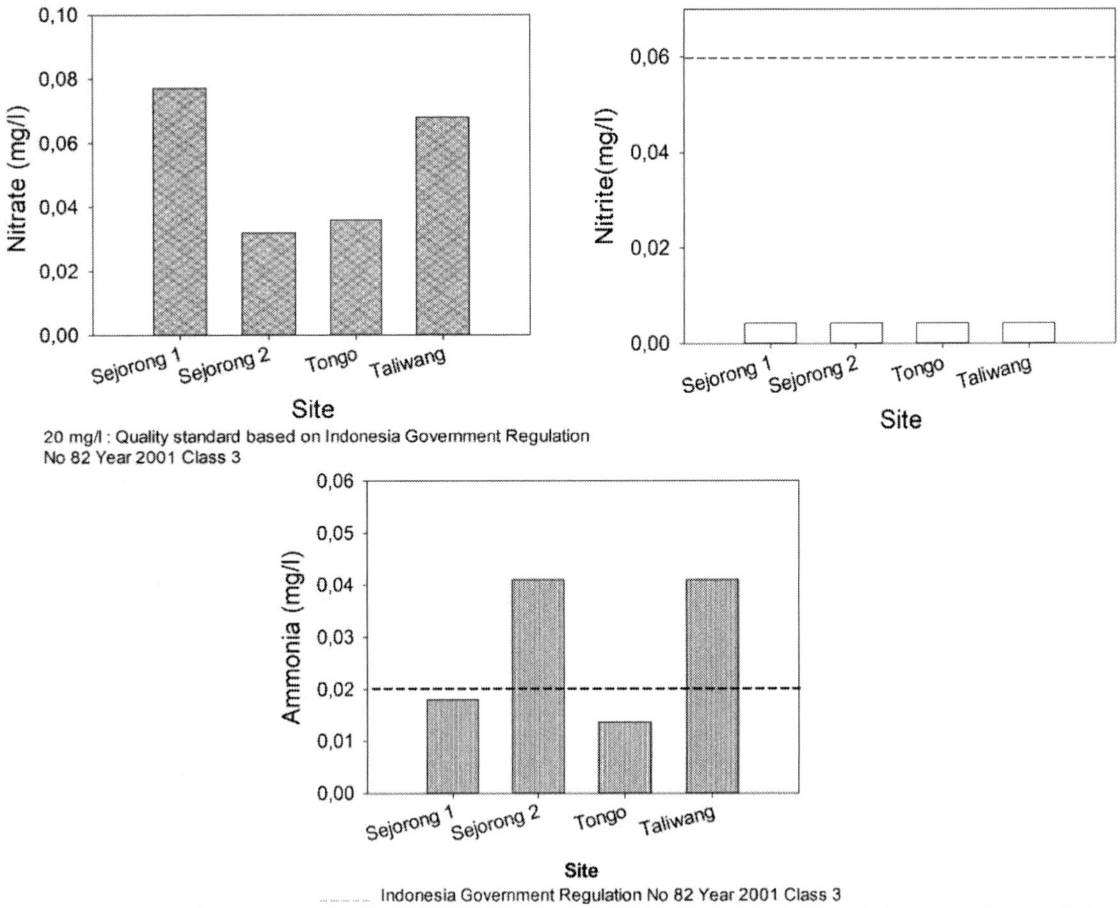


**Figure 5.** Chemical parameters orthoposphate, total phosphate, nitrat, nitrit, amonia on West Sumbawa Regency waters.

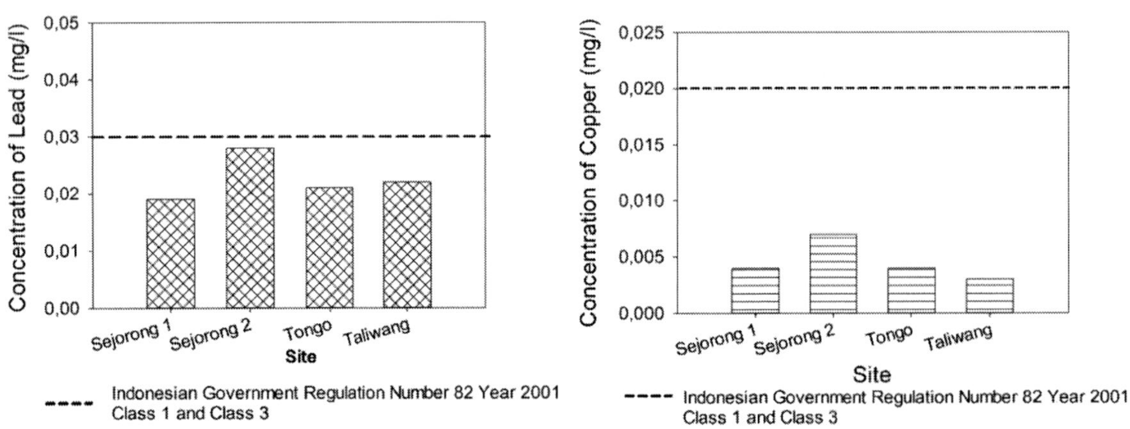

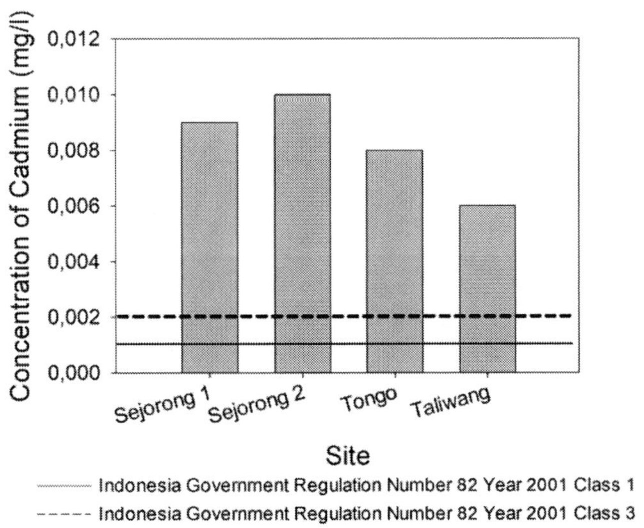


**Figure 6**. Concentration of Heavy metal on West Sumbawa Regency waters.

The value of alkalinity in all four stations had the same tendency value which was 175.05 mg $CaCO_3$ /L. BOD values obtained from the results ranged from 6.6 to 10.57 mg / L. The BOD value for all stations exceeded the good quality standard for fishery which is 6 mg/L. Phosphorus is one source of nutrients in the water. The total value of phosphate and orthophosphate in all four stations is <0.22 mg / L. The total value of phosphate and orthophosphate is still below the standard for fisheries (1 mg / L) and the quality standard of drinking water (0.2 mg / L).

Nitrate-nitrogen ($NO_3$-N) is the main form of nitrogen in natural waters and is a major nutrient for plant and algae growth, and is a limiting factor other than phosphorus [13]. The value of nitrate in the West Sumbawa Regency waters ranged between 0.032-0.037 mg/L. The highest nitrate value was in Sejorong 1 and the lowest was in Sejorong 2. The value was still below the drinking water quality standard (10 mg / L) and the fishery quality standard (20 mg / L). Nitrite is an unstable nitrogen parameter. The presence of high nitrite potentially becomes dangerous in the waters. The nitrite value for the four stations is <0.0042. The value still had not exceeded the fishery quality standard of 0.06 mg / L. Ammonia is a part of nitrogen which can be toxic to aquatic biota if its presence is in excess in the water. Levels of ammonia in the study sites ranged from 0.0137 to 0.041 mg / L. Sejorong 2 and Taliwang had a value above the fishery quality standard.

The heavy metal content of lead in the West Sumbawa Regency waters ranged from 0.014-0.028 mg / L with Sejorong 2 having the highest levels that almost exceeded the quality standard. Copper content ranged from 0.003-0.007 mg/L. In addition to lead concentration (Pb), Sejorong 2 also had the highest copper (Cu) heavy metal concentration among the stations. Although the value is still below the quality standard, it is necessary to monitor the levels in the water so the concentration does not increase. For cadmium (Cd), concentrations at all four stations exceed the standards for drinking water (class 1) and fishery (Class 3). The cadmium concentration values ranged from 0.006 to 0.01 mg / L. The Sejorong 2 station also had the highest cadmium metal concentration compared to the other stations.

*3.3 Biological parameters*

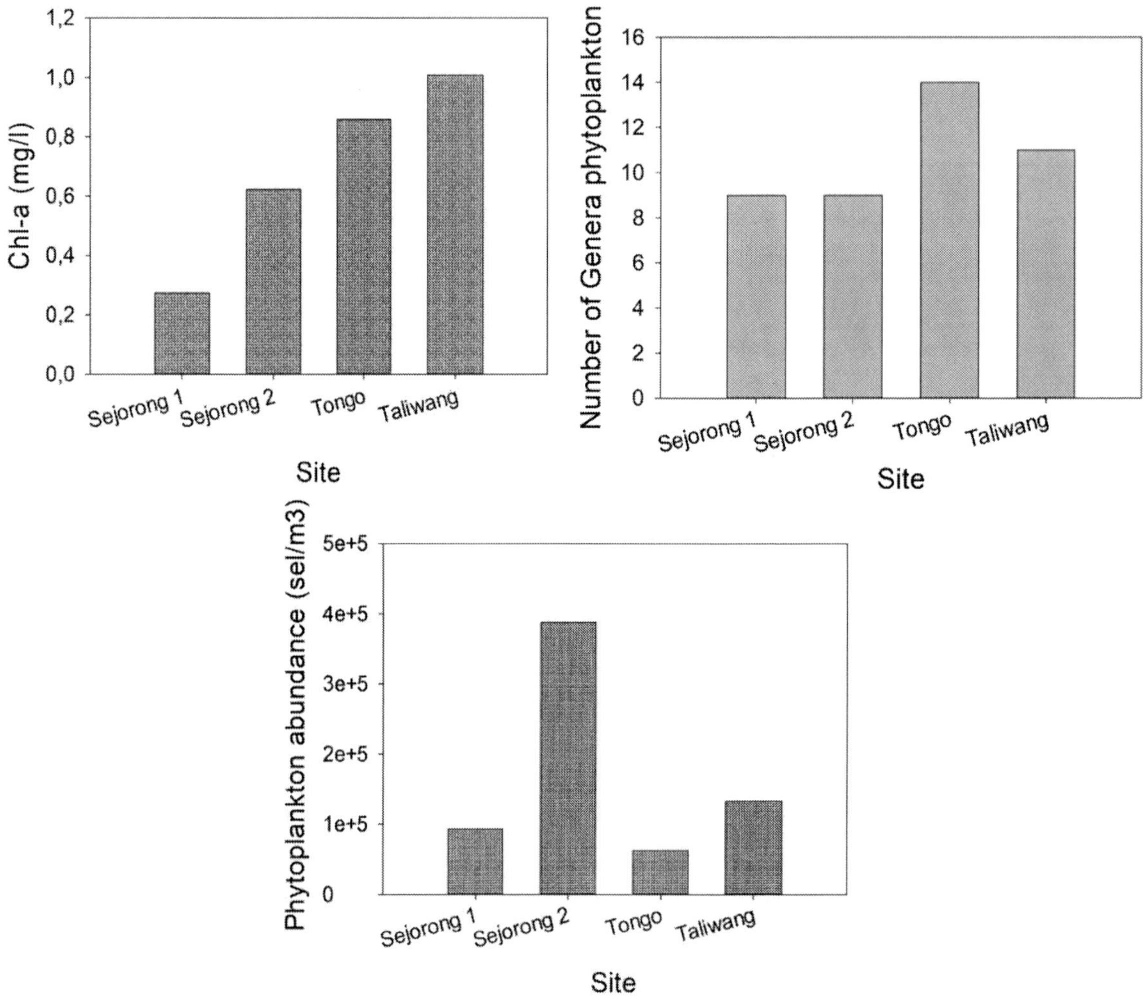


**Figure 7.** Biological parameters Chlorofil a, number of genera and phytoplankton abundance on West Sumbawa Regency waters.

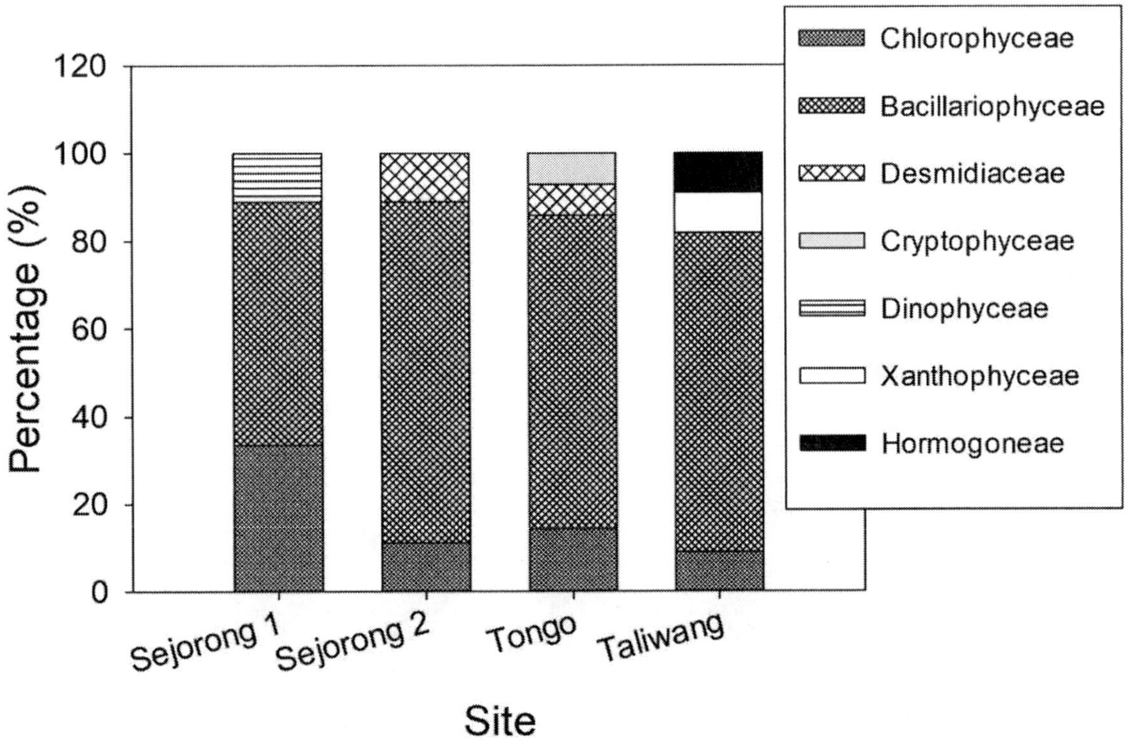


**Figure 8.** Composition of plankton on West Sumbawa Regency waters.

Chlorophyll a is a biological parameter that can predict the productivity of the water. Through chlorophyll, the biomass of phytoplankton in a waters can be known. The highest measurement of chlorophyll a was found in Taliwang Station at 1.008 mg / L and the smallest in Sejorong 1 with a value of 0.275 mg /L.

The number of plankton genera in Tongo was highest among the stations, which are 14 genera (figure 7). The highest abundance of plankton was Sejorong 2 (figure 7) due to the high abundance of *Synedra ulna* at the site. In general, the four stations were dominated by Bacillariophyceae with a range of 56-78%. The second dominant plankton was Chlorophyceae with a range of 11-33 %, while the other class was only about 7-11 % (figure 8).

Based on analysis of the pollution index, the four locations in West Sumbawa Regency waters was lightly contaminated (figure 9). However the trophic classification based on TSI and TRIX found all of locations were categorized as oligotrophic water (table 4).

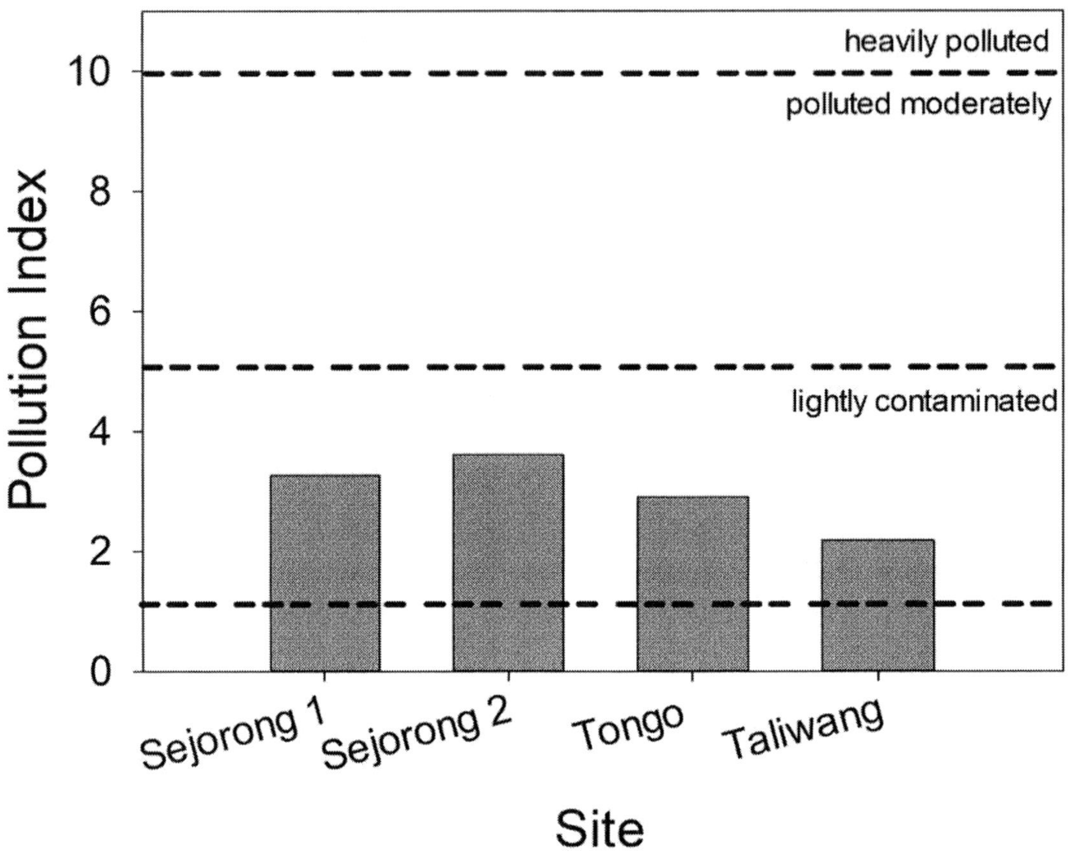


**Figure 9.** Water pollution index on West Sumbawa Regency waters.

**Table 4.** Trophic classification on West Sumbawa Regency waters.

| Site | TSI Carlson | TRIX | Trophic Classification |
|---|---|---|---|
| Sejorong 1 | 3.81 | 1.11 | Oligotrophic |
| Sejorong 2 | 16.67 | 1.51 | Oligotrophic |
| Tongo | 12.55 | 1.37 | Oligotrophic |
| Taliwang | 17.32 | 1.32 | Oligotrophic |

The existence of pollution was due to waste from the mining industry, illegal mining, and the excess mine acid from some time ago which has affected the aquatic environment as seen from the water quality parameters. The most heavily heavy metal waste was industrial waste, due to heavy metal elements being used as raw materials in industries, fungicide catalysts and additives. Industrial waste containing heavy metals will be carried by rivers or air to the sea [14]. In general, pollution waste derives from anthropogenic activities [15].

The physical parameters especially temperatures in this area are still within the range of aquatic biota thresholds, especially freshwater. Relating to the color parameters, Sejorong 1 is still greenish. One cause is the dominant plankton of Chlorophyceae. The brown color in Sejorong 2 is due to the station being closely located to the estuary, resulting in a lot of organic matter accumulating in the area. This is proven with the lowest t and highest TDS and TSS, with a turbidity that is also quite high.

IOP Conf. Series: Earth and Environmental Science **137** (2018) 012086 doi:10.1088/1755-1315/137/1/012086

Taliwang also has brown water color, but with a fishy smell. This is because the station is filled with water hyacinth that is likely to die and rot in the waters, causing unpleasant odors.

Overall transparency values in Sejorong 2, Tongo, and Taliwang tend to be small due to the high organic material at the site. This is in line with the high value of turbidity in the three locations. The pH value at Taliwang station is the lowest among the other locations. This is indicated from the high levels of $CO_2$ due to high respiration in these waters. However, the alkalinity value is quite high in all four locations, so the pH range is neither too low nor too high. Oxygen levels are still within optimum limits for fishery.

The content of nitrate is still below the quality standard for fishery, so in terms of nitrate it still can support fishery activities. However, the high ammonia in Sejorong 2 and Taliwang indicated that the decomposition of organic material was high in those locations. The rest of the organic matter were either from faeces or dead biota and can also increase ammonia levels in the water so that the value exceeds the quality standard.

The parameters of concern due to pollution caused by mining are heavy metals. The concentration of lead (Pb) and copper (Cu) at all locations was still below the quality standard, while cadmium (Cd) exceeded the quality standard. This needs to be monitored to determine whether the levels are increasing or decreasing.

The effect of the heavy metal Pb interferes with the enzyme oxidase, as a result it will inhibit the cell metabolic system, one of which inhibits the Hb system in the bone marrow [16]. In addition to the metal Pb, Cd metal also has the potential to poison aquatic biota. According to Tarigan [17] other than harmful to organisms, Pb metal is also harmful to humans. This is because humans consume aquatic biota such as fish, etc. Shrimp organisms will experience death within 24 - 504 hours from the exposure to metal Cd in concentrations of 0.005-0.15 ppm. This is evidenced by the conditions in the field where biota is very rarely found due to the effects of mine acid exposure and illegal mining that has killed off the biota.

The biological parameters observed were chlorophyll a and plankton. The value of chlorophyll a in Tongo and Taliwang was high. This indicates that Tongo and Taliwang are very productive waters compared to Sejorong 1 and Sejorong 2. In addition, the number of phytoplankton genera found in Tongo and Taliwang were higher than that of Sejorong 1 and Sejorong 2. However Sejorong 2 has the highest abundance due to the dominance of the species *Synedra ulna*.

Based on the pollution index analysis, the quality status of water quality in the four locations are classified as lightly contaminated. This is due to the parameters that have exceeded the standard. Based on the analysis of TSI and TRIX fertility rates, all stations in the West Sumbawa Regency waters are oligotrophic. This means that the waters are still clear waters and have little nutrient content.

## 4. References


[1] Singh K P, Malik K and Sinha S 2005 *Anal. Chim. Acta.* **538** 355-74.

[2] Nurdin R A 2004 Growth aspect, condition factors, and feeding habit (Sicyopterus cyanocephalus) on some rivers around the PT Newmont gold mine. Nusa Tenggara, Sumbawa Undergraduate Thesis (Bogor : Bogor Agricultural University) p 66 (In Indonesia)

[3] Ziyaadini M, Yousefiyanpour Z, Ghasemzadeh J, Zahedi MM 2017 *Iran J Fish Sci* **16** 23-34.

[4] Central Bureau of Statistics West Sumbawa Regency 2015 Number of companies, labor and production value by industry classification in west sumbawa regency, 2015 and number of companies by classification in west sumbawa regency

[5] Onodera S, Saito M, and Sawano M 2008 *Sci. Total Environ.* **404** 401–10.

[6] Phiri O, Mumba P, Moyo B H Z, and Kadewa W 2005 *Int. J. Environ. Sci. Technol.* **2** 237–44.

[7] Sumbawakini.com 2017 PT AMNT admits that the acid mine green rock drainage water overflows into the river : http://www.sumbawakini.com/2017/02/pt-amnt-akui-dam-air-asam-tambang-batu.html#ixzz4d9KZzyvM.

IOP Conf. Series: Earth and Environmental Science **137** (2018) 012086 doi:10.1088/1755-1315/137/1/012086

[8] Suarantb.com 2017 The mine acid flood from PT AMT is suspected to be polluted by river water http://www.suarantb.com.
[9] Indonesia Government Regulation No 82 Year 2001 Class 1 and Class 3 2001 Water quality management and water pollution control
[10] Minister Declaration of Environment Number 115 Year 2003 Guidelines for determination of water quality status
[11] Carlson's R E 1977 A trophic state index for lakes *Limnol and Oceanog* **22** 361-369.
[12] Vollenwieder R A, Giovanardi F, Montanari G, and Rinaldi A 1998 *Environmentrics* **9** 329-57.
[13] Goldman C R and Horne 1983 *Limnology* (Tokyo : Mc Graw-Hill International Book Company) p 464
[14] Kasari A F 2016 Status of contamination by heavy metal Pb, Hg, Cd, and Ag in water and sediment in estuary of donan river, East Segara Anakan Undergraduate Thesis (Bogor : Bogor Agricultural University) p 30 (In Indonesia)
[15] Fukue M C N, Sato Y, and Fujikawa T 2007 *Environ. Pollut.* **149** 70–8.
[16] Darmono 1995 *Metal in biological systems of living* (Jakarta: UI Press) p 140
[17] Tarigan Z, Edward, and Rozak A 2003 *Makara J. Sci.* **7** 119-27.

IOP Conf. Series: Earth and Environmental Science **137** (2018) 012087 doi:10.1088/1755-1315/137/1/012087

# Mercury Test on macroalgae from Burung and Tikus Island, Jakarta

**H Novianty[1], S M C Herandarudewi[1] and Suratno[2]**

[1]Technical Implementation Unit for Oceanography Centre in Pari Island-Indonesian Institute of Sciences, Jalan Raden Saleh 43, Cikini, Jakarta Pusat 10330, Indonesia
[2]Research Centre for Oceanography-Indonesian Institute of Sciencies, Jalan Pasir Putih 1, Ancol Timur, Jakarta Utara, Indonesia

E-mail: hildanovianty2012@gmail.com

**Abstarct.** Environmental pollution, caused by the introduction of hazardous substances such as heavy metals into coastal waters, affects not only the condition of the waters but also the source of food that will be contaminated by hazardous metals, one of them is mercury (Hg). Mercury is toxic metal which could cause damage to the human body in certain threshold amounts. The aim of this study was to determin the content of mercury in several species of algae from Burung and Tikus Island, Jakarta. This study was using a descriptive method. The samples were collected from Burung and Tikus Island by simple rundown sampling. Mercury level was measured by NIC3000 mercury analyzer tool. The results showed that none of the mercury levels have passed 0.5 mg/kg ( the safety standart level of mercury by SNI (Indonesian National Standard)7387 in 2019) mangrove. From tikus Island had lower total mercury than the ones from Burung Island. Burung Island is located near Pari Island which is a residential area, where pollution is more likely to occur.

## 1. Introduction

Indonesia is an archipelagic country which has 81,000 km long coastline, where many Indonesian people lived in coastal areas as coastal communities. The presence of pollution in the coastal areas, is very dangerous, not only to habitats but also humans who always have food intake daily from the sea. Miller [1] defines pollution as adding something to the water, air and soil and foods that endanger human health, endurance human activities or other living organisms. Whereas based on the Law No.23 1997 Anonymous [2] pollution is the entry or inclusion of living things, substances, energy, and or other components into the environment by human activities that cause mal function environment.

Coastal pollution can come from domestic waste products, oil spills from ships, and tailings. The source of the contamination contains heavy metals. Soemirat in Setiawan *et al.* [3] was stated that the toxicity of heavy metals in organism depending on aspects such as species, how to enter the body, frequency and duration of exposure, toxicity concentration and the susceptibility of various species to be toxicants. Mercury was high toxicity that can accumulate in the aquatic environment and became the complex compounds link together with organic and an organic materials [4].

The existence of heavy metals (mercury) in seawater can be derived from the activities of humans then enter through the river, can also come from the atmosphere that falls into the sea or come from volcanic activity [5]. The pollution activity involves exposure of heavy metals, including mercury

IOP Conf. Series: Earth and Environmental Science **137** (2018) 012087 doi:10.1088/1755-1315/137/1/012087

Setiawan *et al.* [6]. If mercury is accumulated in a food that consumed by humans it could be harmful, since mercury compounds can be stored permanently in the body, and enzyme inhibition or cell damage could occur permanently [4]. It is also stated by Setiawan *et al.* [3], that the bioaccumulation of heavy metals in aquatic organisms as a result of pollution activity is an important criterion of the impacts, particularly on food-exposed humans [3], or other sources of food originating from the sea, like sea algae. Marine algae are one of the leading commodities of fisheries that have great opportunities as a new food source, considering the various types. Therefore, a study on algae exposed to heavy metals (mercury) needs to be done.

## 2. Methodology

This study was conducted in January 2016. This study was using the descriptive method. The samples in this study were algae that found in the waters of the Burung and Tikus Island. Sampling of algae was done by a simple rundown sampling. The stages of data collection were preparation, collecting of secondary data, implementation and sample analysis phase. Mercury sample analysis was done using NIC3000 mercury analyzer tool.

### *2.1. Tools and materials*

The tools used in this study were mercury analyzer NIC3000, analytical scale, cup, 250 ml beaker glass, 250 and 100 ml measuring flask. Tools for sampling were snorkel, plastic, and labeling tools. The material used in making the standard curve mercury test were L-cysteine solution, concentrated $HNO_3$ and deionized water.

### *2.2. Stages of Testing*

Samples were collected into a transparent plastic and stored in the freezer. The preparation of samples, the making of standard mercury curve and the mercury test were then condueted. The standard mercury curve was made before mercury testing was performed. The use of L-cysteine was to stabilize the standard Hg solution prepared.

## 3. Result and Discussion

Table 1 showed a list of the algae obtained in the waters of Burung and Tikus Island, there are 7 species of algae. These algae were belonged to the groups of Phaeophyta (brown algae) were containing carotene pigment, xanthofil, fucoxanthin [7], were *Dictyota, Hormophysa, Padina*. While other species were found to be classified as Chlorophyta (green algae) were containing a and b pigment of chlorophyll, carotene, xantofil, lutein [8], were *Caulerpa* and *Halimeda macroloba*. The latter was from Rhodophyta (red algae) group were containing phycoerythrin pigment, fukoxantin, fukosianin [7], were *Galaxaura fasciculata* and *Hypnea*.

Algae found on Burung and Tikus Island were algae that have low economic value. But it has an opportunity as a new source of food from the sea because the functional properties of Phaeophyta, Chlorophyta and Rhodophyta. Rhodophyta have the content of karegenofit and agarofit, while Phaeophyta has alginofit content and Caulerpa (Chlorophyta) may serves as food. Karegenofit, agarofit and alginofit have functional properties as thickener, gelling, emulsifiers, stabilizers, suspending, flavor softener, crystallization prevention [8], which could be used as additives to the cosmetic, pharmaceutical and food industries. table 1 shows the total mercury content of algae:

**Table 1.** Result of mercury test in algae found in the Burung and Tikus Island.

| Names of Algae | Average of Hg (µg/kg) | |
|---|---|---|
| | Tikus Island | Burung Island |
| *Dictyota* sp | 31.3 | 40.16 |
| *Hormophysa* sp | 18.83 | 46.07 |
| *Padina* sp | 20.79 | 35.94 |
| *Caulerpa* sp | 33.33 | 64.81 |
| *Halimeda macroloba* | 6.64 | 42.2 |
| *Galaxaura fasciculata* | 23.98 | 64.32 |
| *Hypnea* sp | 29.12 | 40.1 |

All of marine algae species was found had a total mercury of 6 to 64 µg / kg. The quantity of mercury was still below the threshold specified in SNI 7387 [9]. The maximum limit of heavy metal contamination in food for mercury at about 0.5 to 1.0 mg / kg. This proves that the algae were safe to serve as a new food source from the sea.

Based on data in table 2, the algae found on the Tikus Island were still feasible or safe for consumption and algae found on Burung islands were unfeasible or safe for consumption partially.

All of the algae were found in the site of Burung Island had high mercury content compared to algae found in the Tikus Island. Burung Island had coordinates location of S.05.86654 and E.106.60470, while Tikus Island was S.06.00389 and E.106.77889. Burung Island is closer to Pari Island (S.05.8575 and E. 106.6181) (figure 1) than to Tikus Island. Pari Island is the type of island which have residential population, so domestic waste from Pari Island could spread to the Burung Island, so the accumulation of mercury total found in Burung Island was bigger than Tikus Island. Although for the both had mercury accumulation were still below the threshold of the Indonesian National Standard for the maximum limit of heavy metal (mercury 0,5 mg/kg) [9].

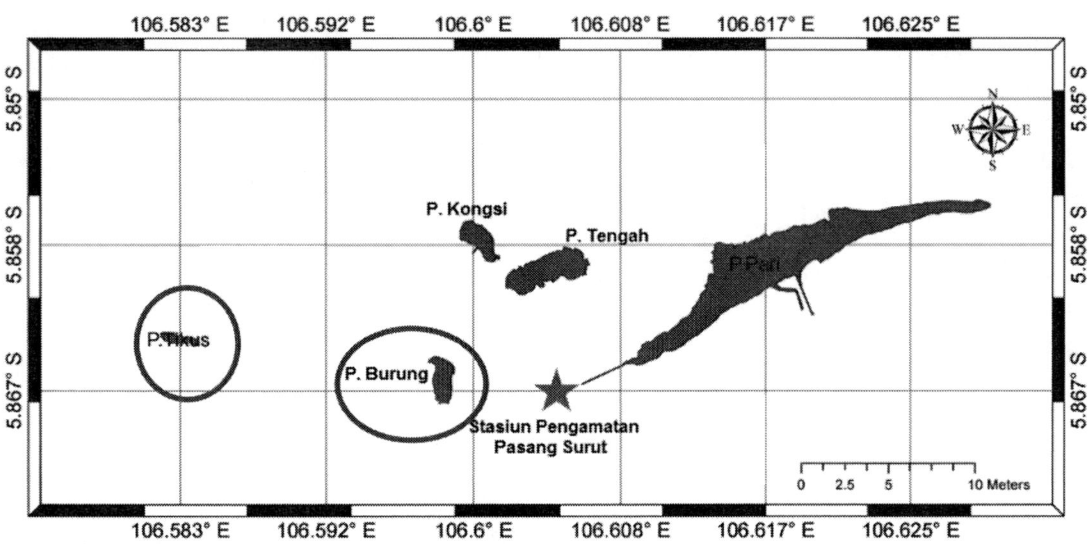


**Figure 1.** Map of Pari Island
Source: Corvianawatie *et al.* [10]

IOP Conf. Series: Earth and Environmental Science **137** (2018) 012087 doi:10.1088/1755-1315/137/1/012087

**Table 2.** The feasibility of algae consumption.

| Names of Algae | Mean Total Hg (μg/kg) | | Provisional Toreable Weekly Intake Hg (mg/kg) | Level of consumption feasibility | |
|---|---|---|---|---|---|
| | In The Tikus Island | In The Burung Island | | In The Tikus Island | In The Burung Island |
| *Dictyota* | 31.30 | 40.16 | 0,005 (SNI 7387, 2009) [9] | not feasible | not feasible |
| *Hormophysa* | 18.83 | 46.07 | | not feasible | not feasible |
| *Padina* | 20.79 | 35.94 | | not feasible | not feasible |
| *Caulerpa* | 33.33 | 64.81 | | not feasible | not feasible |
| *Halimeda macroloba* | 6.64 | 42.20 | | not feasible | not feasible |
| *Galaxaura fasciculata* | 23.98 | 64.32 | | not feasible | not feasible |
| *Hypnea* | 29.12 | 40.10 | | not feasible | not feasible |
| Names of Algae | Mean Total Hg (μg/kg) | | Hg (μg/kg) | Level of consumption feasibility | |
| | In The Tikus Island | In The Burung Island | | In The Tikus Island | In The Burung Island |
| *Dictyota* | 31.30 | 40.16 | 300 (WHO, 2003 in Dwiyitno *et al* 2008 [11]; WHO, 2004 in Simbolon *et al* 2010 [12]) | feasible | feasible |
| *Hormophysa* | 18.83 | 46.07 | | feasible | not feasible |
| *Padina* | 20.79 | 35.94 | | feasible | feasible |
| *Caulerpa* | 33.33 | 64.81 | | feasible | not feasible |
| *Halimeda macroloba* | 6.64 | 42.20 | | feasible | not feasible |
| *Galaxaura fasciculata* | 23.98 | 64.32 | | feasible | not feasible |
| *Hypnea* | 29.12 | 40.10 | | feasible | feasible |

## 4. Conclusion

Based on mercury total test all, of the algae found in the Burung and Tikus Island (Dictyota, Hormophysa, Padina, Caulerpa, Halimeda macroloba, Galaxaura fasciculata and Hypnea) have been contaminated by heavy metals mercury but have not exceeded the permissible limit according to SNI (Indonesian National Standard) 7387. It is necessary to develop the study of the impact of heavy metals on other marine biota or other sources food from the sea for safety food.

IOP Conf. Series: Earth and Environmental Science **137** (2018) 012087 doi:10.1088/1755-1315/137/1/012087

## 5. References

[1] Miller T G 2004 Enviromental Science : Working with The Earth 10th edition International Student Edition Thomson Learning Inc

[2] Anonim 2000 *Environmental Law and Environmental Impact Analysis* (Jakarta: Penerbit CV Tamita Utama) (Text in Indonesia)

[3] Setiawan A A, Ita E and Suheryanto 2013 Total Mercury Content On Different Types Of Cat Fish in Musi River – Palembang City (Lampung : *University of Lampung Science and Technology National Seminar V)* p 741-750 (Text in Indonesia)

[4] Setiyono A and Annisa D 2012 Consumption of Fish and Agricultural Products on Hg Levels in Blood (Text in Indonesia) *J. Health Soc.* **7** 110-116 (In Indonesia)

[5] Nugraha W A 2009 Heavy Metal Content In Water And Sediments In Socah and Kwanyar – Waters (Text in Indonesia) *J. Mar.* **2** 158-164 (In Indonesia)

[7] Susanto A B, Rini P and Alva W 2009 Seaweed Processing Technology in Indonesia (Text in Indonesia) *Proc. Biotechnology and Seaweed Industry ISBN : 978-979-17585-3-6* p 1-20

[8] Ahda A, Agus S, Imam B S, Ilham B, Iskandar I, I Made S, Restu Y, Setiawan, N Kurnia, Edward D, Sulistijo, Achmad Z, Jamal B, Irzal E, and Nico R 2005 *Indonesian Seaweed Profile (Text in Indonesia)* Department Marine and Fishery

[9] Anonim 2009 Contamination Maximum Limit of Heavy Metal In Food - Indonesian National Standard Number 7387 (Text in Indonesia) National Standard Organization

[10] Corvianawatie C, Sam W and Arvita R 2016 Sea Water Quality Assessment In Supporting Development of Marine Biodiversity Potential *UPT LPKSDMO LIPI p 77*(Text in Indonesia)

[11] Dwiyitno, Nugroho A and Ninoek I 2008 *Journal of Postharvest and Biotechnology of Marine and Fisheries* **3** 147-155 (Text in Indonesia)

[12] Simbolon D, Silvanus M S and Sri Y W 2010 *J. Mar. Sci.* ISSN 0853-7291 p 126-134 (Text in Indonesia)

IOP Conf. Series: Earth and Environmental Science **137** (2018) 012088 doi:10.1088/1755-1315/137/1/012088

# The effects of using shell filters in the process of depuration for the survival of *Anadara* sp.

**K T Pursetyo, L Sulmartiwi, M A Alamsjah, W Tjahjaningsih, A S Rosmarini and M Nikmah**

Marine Department, Faculty of Fisheries and Marine, University of Airlangga, East Java, Surabaya, Indonesia 60115

E-mail: kustiawan@fpk.unair.ac.id

**Abstract.** *Anadara* sp. is one of the shellfish that has a source of animal protein is high and has a high economic value. However, to obtain a safe source of food, the products must meet the standard by the government, one of which is the limitations in heavy metals in the shells. In the standard of sanitation of shellfish is required to do the depuration process to remove the contaminants be it bacteria or heavy metals. In this study, randomized design with 5 treatments was used: P0 (control / without filter), P1 (25 % filter with shells), P2 (50 % filter with shells), P3 (75 % filter with shells), P4 (100 % filter with shells). Each treatment was replicated 4 times. The results showed that filtering of shell in depuration process could cause the highest shell death for 24 hours occurred in P4 of 24.39 % and the highest death during 48 hours also happened at the treatment of P4 which was equal to 61.71 %. During the research, water quality measurement was measured at 29-30 °C, pH 7.2-7.7, dissolved oxygen (DO) 4-4.4 mg/L and salinity 28-30 ppt.

## 1. Introduction

Cockles are potentially contaminated by heavy metals due to their filter feeder nature, so this biota is often used as a test animal to monitor the degree of the accumulation of heavy metals in marine organisms [1]. One type of shellfish that is abundantly found in Indonesian waters is the blood cockle (*Anadara* sp.). The blood cockle is a type of shellfish that has a high economic value and in general is a source of seafood in Southeast Asia and some Pacific regions [2]. Blood cockles have several uses, one of which is processed as food, however its shells can cause considerable waste. The discarded shells can act as a source of calcium. There has not been much utilization of blood cockle shells, except as raw materials for making local souvenirs.

Cockle shells contain 66.70% calcium carbonate [3], the high content of calcium carbonate can make use of them to be used as a water purifier. The calcium carbonate in shellfish can clean water, and can even reduce iron, manganese and other metals [4].

This method of depuration is principally a purification step of a biota under a controlled condition. The shellfish that are contaminated by heavy metals should be cleaned or depurated, the purpose of this depuration process is to reduce the risk of bacterial contaminants and some heavy metals harmful to human health [5].

Toxic depuration in shellfish is one of the efforts to avoid the impact of lead accumulation in shellfish that has the potential to harms humans who consume it [6]. According to the Decree of the Minister of Marine Affairs and Fisheries Number: KEP.17 / MEN / 2004, the depuration process is a

IOP Conf. Series: Earth and Environmental Science **137** (2018) 012088 doi:10.1088/1755-1315/137/1/012088

cleaning process that uses a purifier with a recirculation system. This research uses a depuration method by using powdered shells as the filter. The shell contains 66.70 % calcium carbonate so as to absorb the heavy metal lead [3]. The high content of calcium carbonate can make the shellfish shells be used as a water purifier. The calcium carbonate in shellfish can clean water, and can even reduce iron, manganese and other metals [4].

## 2. Methodology

This research was conducted on 30th August to 30th September 2016 at the Faculty of Fisheries and Marine of the University of Airlangga, Surabaya. The testing of heavy metal content was conducted by using an Atomic Absorption Spectrophotometer (AAS).

The tools used during research were five tubs (95 cm x 70 cm x 75 cm), 20 plastic racks (42 cm x 33 cm x 14 cm), five plastic tubes with a height of 45 cm and a diameter of 26 cm, five water pumps, five flow meters, five UV meters, sponge filtration, pipes, hoses, scooters, measuring cylinders, digital scales and water quality gauges (thermometers, refractometers, pH pens, DO test kits, nitrate and nitrite test kits).

The materials used in the study where blood cockles (*A. antiquata*) originating from Sedati, Sidoarjo. The blood cockles used were 7-8 grams in weight and 10 kg of the shellfish were placed in each tub. Four shelves were placed in each tub along with sea water with a salinity of 30 ppt.

### *2.1. Work procedures*

The tubs were cleaned by using liquid soap followed by chlorine and then rinsed with clean water. The washed equipment was soaked with chlorine solution at 150 mg / l for 12-24 hours and rinsed with fresh water until the chlorine odor was gone [7]. The dried tubs were filled with 400 liters of seawater in each tub. After filling in the water, the installation of aeration, UV, water pumps, flowmeters and filters on each tub was carried out. The outer shells that were used as a filter were firstly washed and then pounded into a powder to a size of 0.5-2 cm. Next, the shell powder was placed in each tube as a filter. The amount of shell powder used was 15.3 kg for the 100 % filter, 11.5kg for the 75 % filter, 7.7k g for the 50 % filter and 3.8kg for the 25 % filter.

The method used in this research was the experimental method to find out the effects of the shell filter in the depuration process in decreasing the cadmium content of the shellfish. The experimental design was performed using Completely Randomized Design with a single source of diversity ie percentage of shell powder. This study used five treatments with four replications in each treatment. The treatments in this study are as follows:

P0: Without Filter Shells
P1: Using Filter Shells 25 %
P2: Using Filter Shells 50 %
P3: Using Filter Shells 75 %
P4: Using Filter Shells 100 %

## 3. Results and Discussion

Shellfish are biota that are potentially contaminated by heavy metals due to their filter feeder nature, so it is often used as a test animal to monitor the degree of accumulation of heavy metals in marine organisms [1]. The heavy metal cadmium will turn into poison for aquatic life when dissolved in the waters at a certain concentration. Heavy metals that enter the body of animals such as shells are generally unremoved from the shell's body [1]. The accumulation of heavy metals in the body of aquatic animals occurs because the rate of the collection of heavy metals by aquatic organisms is faster than the release process[8].

Based on the results of the research, the blood cockle *Anadara* sp. samples contained an average Cd content of 0.3598 ppm and Pb content of 2.144 ppm prior to being decontaminated. According to the Decree of the Minister of Marine Affairs and Fisheries Number: KEP.17 / MEN / 2004 regarding sanitary sanitation systems in Indonesia, there is a standard of the quality of live shellfish and its

IOP Conf. Series: Earth and Environmental Science **137** (2018) 012088 doi:10.1088/1755-1315/137/1/012088

processed products which are consumed directly. The threshold limits of heavy metals in processed products for consumption must meet the requirements of Cadmium (Cd) at a maximum of 1.0 ppm and lead (Pb) at a maximum of 1.5 ppm.

The process of depuration was conducted by using shell powder of different percentages as filters which was carried out for 48 hours [9]. Seawater was circulated continuously throughout the depuration process for about 48 hours. The percentage of death or mortality is calculated every 24 hours for 48 hours. The value of the mortality gap percentage is the number of dead shellfish divided by the number of shellfish at the beginning, as in the formula:

$$Mo = \frac{Mt}{N0} x\ 100\ \% \quad (1)$$

Information:
Mo: mortality
Mt: number of dead
N0: population

**Table 1.** Mortality of blood cockles.

| Treatment | Mortality ± SD 24 hours | Mortality ± SD 48 hours |
|---|---|---|
| P0 (Control) | 1.04 %c ± 0.6 | 3.12 % c ± 1.89 |
| P1 | 6.33 %b ± 3.82 | 30.61 % b ± 5.16 |
| P2 | 6.65 %b ± 5.35 | 31.33 % b ± 12.44 |
| P3 | 10.98 %b ±1,01 | 58.01 % a ± 0.84 |
| P4 | 17.71 %a ± 5,04 | 58.65 % a ± 3.06 |

Description: Different superscript on the same column shows a noticeable difference ($p < 0.05$).

Based on the analysis of variance (ANOVA) and Duncan's test, the results showed that the highest mortality found in the 48 hour treatment of P4 (58.65%) was not significantly different to the 48 hour treatment of P3 (58.01%). While the lowest mortality was found in the 24 hour treatment of P0 (1.04%).

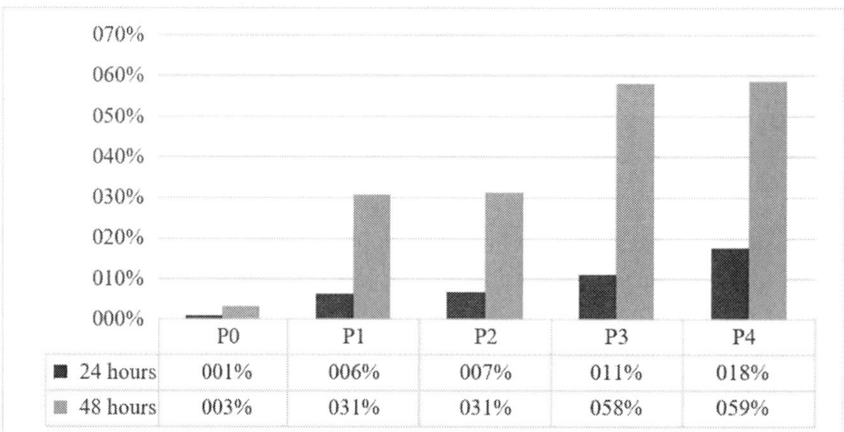


**Figure 1.** Mortality graphics.

High mortality was caused by the muddier water in the maintenance media so that the respiration process of the shellfish was disrupted and evidenced by the increasing Total Suspended Solid (TSS) content. The water quality measured during research is presented in table 2.

**Table 2.** Water quality.

| | Before | | | |
|---|---|---|---|---|
| | TDS (mg/L) | TSS (mg/L) | BOD (mg/L) | COD (mg/L) |
| P1 | 26280 | 12 | 185.73 | 3241.07 |
| P2 | 26280 | 12 | 185.73 | 3241.07 |
| P3 | 26280 | 12 | 185.73 | 3241.07 |
| P4 | 26280 | 12 | 185.73 | 3241.07 |
| P5 | 26280 | 12 | 185.73 | 3241.07 |
| | After | | | |
| | TDS (mg/L) | TSS (mg/L) | BOD (mg/L) | COD (mg/L) |
| P1 | 27150 | 13.5 | 121.945 | 1410.715 |
| P2 | 27300 | 19 | 118.87 | 1525.51 |
| P3 | 26830 | 39.5 | 140.64 | 1761.48 |
| P4 | 25370 | 142.335 | 158.925 | 1908.165 |
| P5 | 27150 | 62 | 171.285 | 2207.905 |

The above table shows that in the treatment of P4 the amount of Total Suspended Solid (TSS) at the 48th hour was 142.335 mg / L, TSS waters can reduce the value of waters and affect the lives of the organisms [10].The value of TSS 81-400 mg/L indicates that the suspended solid content was unsuitable for fishery[11]. TSS will affect the clarity of water, which in turn will affect the penetration of light and ultimately will affect primary productivity. High TSS can block the entry of sunlight into the water, which will interfere with the process of photosynthesis and cause the decrease of dissolved oxygen released by plants into the water [10].

## 4. Conclusion

The use of blood cockle shells as a filter had an effect on the number of deaths of the shellfish during the depuration process. The highest death occurred in P4 where 100% of the shells were used as a filter.

## 5. Reference

[1] Yennie Y, Murtini JT 2016 *J Mar. Fish. Sci. Ind* **2** 27–32 (In Indonesia)
[2] Shigeru S2009 ICMSS09 1-9
[3] Ulysses M, Lawrence RM, Valeriano B and Shigeru S 2009 *Comparative PSP toxin Accumulation in Bivalves, Paphia undulata and Perna viridis in Sorsogon Bay, Philippines* (France: Nantes) p 56
[4] Siregar S M 2009 Utilization of shells and epoxy resins on the characteristics of polymer concrete (MS thesis) p 89
[5] Simaremare S R 2013 *Healthy Work Envi.* **3** (In Indonesia)
[6] Gabr HR, Gab-Alla A-F 2008 *Oceanologia* **50** 83–93
[7] Chen CY, Chou HN 2001 *Toxicon* **39** 1029–34.
[8] Isnansetyo A Kurniastuty 1995 *Culture technique phytoplankton zooplankton : natural feed for the seeding of marine organisms* (Yogyakarta : Kanisius) p 67–71 (In Indonesia)
[9] Darmono 2006 Environment and pollution: relation to toxicology of metal compounds (Jakarta : University of Indonesia) p 67
[10] Murdinah M 2009 *Bull. Mar. Fish. Postharvest Biotechno.* **4**(2) 61–71
[11] Sumawidjaja K 1973 Limnology (IPB: Bogor) p 88 (In Indonesia)

IOP Conf. Series: Earth and Environmental Science **137** (2018) 012089 doi:10.1088/1755-1315/137/1/012089

# Studyof copper (Cu) contents in blood cockles (*Anadara* sp.) at Surabaya coastal waters

**S Alfionita, K T Pursetyo and A M Sahidu**

Marine Department, Fisheries and Marine Faculty, Airlangga University, East Java, Surabaya 60115

E-mail : kustiawan@fpk.unair.ac.id

**Abstract**. Heavy metal pollution categorized as harmful effect on the environment. This study aims to determine the content and the rate of accumulation of heavy metals copper (Cu) in the mussel of cockles (*Anadara* sp.) This study uses observation method with descriptive analysis. The test results Cu in sea water was <0.0223 mg/L. The results have exceeded the water quality standard according to state environmental ministerial decree number 51 2004 on marine water quality standards for marine biota maximum limit of Cu in sea water is 0.008 mg/L. Copper concentration in the sediment of 6.16 mg/kg at the Wonokromo station and at the Dadapan station is 4.03 mg/kg. According to the US-EPA in 2004 on the instructions of sediment pollution classification 49.98 mg Cu/kg. The results of shells fur Cu on Wonokromo was 1.833 mg/kg and Dadapan was 1.352 mg/kg. According to the MOH Decree No.0375/B/SK/1989 about Quality Standard Biota Consumption of Metals to the maximum limit of the existing Cu concentrations in marine biota was 20 ppm or 20 mg / kg so that the content of Cu in shellfish feathers still below the limit of existing quality standards.

## 1. Introduction

Surabaya coastal waters are potential to accumulate anthropogenic burden. Anthropogenic burden is a pollutant that enters the water due to human activities in the form of domestic activities (household), urban activities (cities), or industrial activities [1]. Anthropogenic burden is worsened by the misuse of the river as the waste disposal site. Thus, the pollutant can be distributed up to the estuary, even to the ocean.

The waste that enters the Pamurbaya waters includes solid, gas, and liquid waste. One of the types of wastes that is accumulated highly and pollutes the waters is solid waste including heavy metal pollution. Heavy metal pollution is categorized as pollution that causes dangerous effects to the environment and living organisms. Heavy metals are non-degradable. Heavy metal can be accumulated in the environment like in water column and sediments, and eventually can be absorbed into the marine biota [1]. Heavy metals that pollute the waters are Mercury (Hg), Lead (Pb), Zinc (Zn), Cadmium (Cd) and Copper (Cu). According to the Government Regulation No. 82 Year 2001 class I-IV on Water Quality Management and Pollution Control, the maximum level of heavy metals in the waters are Hg 0.001-0.005 mg/L, Pb $\leq$ 0.1 mg/L, Zn $\leq$ 5 mg/L, Cd 0.01 mg/L, and Cu 0.02 – 1 mg/L.

Cu is a heavy metal with essential elements that are needed by plants and animals as the composer of plastocyanin that functions in the transportation of electrons in photosynthesis process [2]. Cu levels that is over 0.5 mg/L would cause $LC_{50}$ (50 % death concentration) to sea and fresh water's

IOP Conf. Series: Earth and Environmental Science **137** (2018) 012089 doi:10.1088/1755-1315/137/1/012089

invertebrates [3]. Copper excess in human being may cause diarrhea, nausea, vomiting, anemia, stomach disorder, and kidney failure.

Sea water invertebrates that can be the parameters for the presence of Cu in the waters is, but not limited to, blood cockles. Blood cockles are animals that have a feeding habit by preying on particles that exists in the form of microorganisms or detritus. Blood cockles has a feeding pattern as filter feeder, which filters any type of foods around them, thus they can accumulate microorganisms and other foreign materials, including heavy metals that are absorbed and stored in their digestion system without poisoning the shells themselves [4]. Considering the high demands over blood cockles by the society that is in line with the increase environmental pollution; thus, it is necessary to conduct a research on the copper (Cu) heavy metal content in blood cockles in the east coast waters of Surabaya. The information will be beneficial for water pollution control and food safety.

## 2. Methodology

### *2.1. Research time and place*

This research was performed in February 2017 at the east coast waters of Surabaya. The blood cockles (*Anadara* sp.) were taken from two stations. The first one was located in Wonokromo, and the second one was in Dadapan. Analysis of copper in sea water, sediment, and blood cockles was done at laboratory of *Badan Riset dan Standarisasi Industri* (BARISTAND), Surabaya.

### *2.2. Tools and materials*

The research tools that used were boat, basket, dredges tool, water sampler, cool box, 2 kg plastic bags, refractometer, thermometer, Secchi Disk, pH pen, labeling paper, permanent marker, and GPS. The tools to analyze copper in water, sediment, and blood cockles were analytical balance (0.0001 g precision), 50 mL beaker glass, PVC, 50 mL measuring flask, knife, fume hood, micro pipette, and Atomic Absorption Spectrophotometry (AAS).

The materials used to support this research were the sea water sample, blood cockles, and sediments, concentrated $HNO_3$, solvent, and ice block.

### *2.3. Research procedure*

#### *2.3.1. Determination of sampling point*

Determination of The geographical coordinates of the sampling stations was done with Global Positioning System (GPS).

#### *2.3.2. Blood cockles sampling*

Sampling for blood cockles was done in two stations namely Wonokoromo and Dadapan. The sampling process in Wonokromo and Dadapan stations were separated into three points in each station. The blood cockles sampling were repeated twice for every points. Blood cockles sampling was done with dredges, a catching tool that is dragged with a fisherman boat with a speed of 2.7 to 4.6 m/min. Later, the catch is placed in baskets and plastic bags that are labeled according to the sampling station, and finally placed in a cool box. Later, the labeled blood cockles were handled by separating the shell and the meat. The meat was weighed every 50 g and placed in labeled plastic bags before it was brought to the laboratory to be tested for its copper content.

#### *2.3.3. Sea waters sampling*

Sampling of sea waters was done at both stations. Sample of sea water was taken with *water sampler* tool at a depth of 30 cm under sea level for as much as ± 100 mL and it was placed in a bottle. $HNO_3$ was added into the bottle and the bottle was placed in a cool box with ice blocks in it to ensure that there were no biological and chemical changes. Later, the cool box was transported to the laboratory to be examined.

### *2.3.4. Sampling of sediment*

Sampling of sediment was done at every station. The amount of sediment taken was 200 g. The sediment that was taken was placed in plastic bags, and given *air-dry* treatment so that there won't be any biological and chemical changes that occurs [5]. Later, the cool box is closed shut and is transported to the laboratory to be examined.

### *2.3.5. Copper analysis in blood cockles, sediment, and sea water.*

Copper analysis in water, sediment, and blood cockles sample was performed with Atomic Absorption Spectrophotometry (AAS) based on the Lambert-Beert law principle, which dictates that the amount of light absorbed is directly proportional with the level of the substance [6].

## 3. Research Parameters

The research parameters measured included main parameters and supporting parameters. The main parameters were copper in the water, sediment, and blood cockles from Wonokromo and Dadapan research stations in Pamurbaya Waters. The supporting parameters were water quality including temperature, dissolved oxygen(DO), salinity, water clarity, and pH. Data collected were analyzed with descriptive analysis and reported in the form of graphs, tables, and pictures.

## 4. Results and Discussion

### *4.1. General picture of east coast waters of Surabaya*

The general picture of East Coast Waters of Surabaya was reviewed in two stations, namely Wonokromo river station and Dadapan river station, as given in table 1.

**Table 1.** Result of water quality measurement in Wonokromo and Dadapan station.

| Sampling station | Temperature (°C) | pH | DO (mg/L) | Salinity (ppm) | Clearness (cm) | Suspended Solids (TSS) (mg/L) |
|---|---|---|---|---|---|---|
| Wonokromo Station | 29 | 8.2 | 5.5 | 35 | 40 | 37 |
| Dadapan Station | 28 | 8.3 | 4.4 | 40 | 60 | 7.5 |

Table 1 shows the results of water quality measurement including temperature, pH, DO, salinity, water clarity, and suspended solids (TSS). the temperature of East Coast Waters of Surabaya ranged from 28 to 29 °C. Degree of acidity or pH on the East Coast Waters of Surabaya ranged from 8.2 to 8.3.Dissolved Oxygen or DO in the East Coast Waters of Surabaya ranged from 4.4 to 5.5 mg/L. Salinity of the East Coast Waters of Surabaya ranged from 35 to 45 %. The water clarity of the East Coast Waters of Surabaya was 40-60 and the suspended solid or TSS in the East Coast Waters of Surabaya ranged between 7.5 and 37 mg/L.

With dissolved oxygen level of 1.0 - 5.0 mg/L, fish can survive, but their growth will be affected. While DO > 5.0 mg/L is preferred by almost every aquatic organisms [2]. In Dadapan River, it was found that the dissolved oxygen level was 4.4 mg/L. this value shows that the growth of aquatic organisms in the water were disturbed, while the Wonokromo River showed a DO level of 5.5 mg/L, which showed that the aquatic organisms living there were still normal. The water pH of Wonokromo River was 8.2 while water pH of Dadapan River was 8.3. According to Effendi [1] most aquatic biota are sensitive to a change of pH between 7-8.5. The pH levels in both rivers were still within the normal range for aquatic organism.

Salinity at the East Coast Waters of Surabaya ranged between 35 to 40 ppm. Effendi (2000) stated the sea salinity level ranged between 30-40 ppm. Water clarity of the East Coast Waters of Surabaya was 40-60 cm. Clarity is water transparency that is measured visually. It is affected by suspended solid or murk. Suspended solid level in Wonokromo River was 37 mg/L while in Dadapan River the TSS

IOP Conf. Series: Earth and Environmental Science **137** (2018) 012089 doi:10.1088/1755-1315/137/1/012089

value was 7.5 mg/L. Loyd (1982) stated that TSS value between l-25 mg/L does not affect fisheries while TSS value of 25-80 mg/L can mildly affect fisheries. From TSS point of view, the quality of the East Coast Waters of Surabaya was in a good condition.

*4.2.* Copper in the Sea Waters

Copper content of the sea water of East Coast Waters of Surabaya was analyzed from two research stations: Wonokromo River and Dadapan River and the results are shown in table 2.

**Table 2.** Copper (Cu) heavy metal substances in sea water.

| Testing Parameter | Unit | Level | |
|---|---|---|---|
| | | Wonokromo | Dadapan |
| Copper (Cu) | mg/L | <0.0223 | <0.0223 |

Level of copper in Wonokromo River and Dadapan River was <0.0223 mg/L. According to the Ministry of Environment Regulation Number 51 year 2004 on Sea Water Quality Standard, for the sea water biota, the maximum Cu is 0.008 mg/L. The concentration of Cu in East Coast Waters of Surabaya exceeded the quality standard. source of copper pollution in the East Coast Waters of Surabaya, allegedly, came from industrial waste, household waste, and shipwrecks. Naturally, copper can come from erosion, rock erosion, or from the atmosphere that was brought down by rain. human activities, industrial activities, mining, or shipyard industry, along with activities in docks, can increase the level of copper in the water [7].

*4.3.* Level of copper in sediment

Copper level found in the sediment of East Coast Waters of Surabaya is shown in table 3.

**Table 3.** Copper heavy metal substances in sediment.

| Testing Parameter | Unit | Level | |
|---|---|---|---|
| | | Wonokromo | Dadapan |
| Copper (Cu) | mg/kg | 6.16 | 4.03 |

The concentration of Cu in the sediment was 6.16 mg/kg in Wonokromo River and 4.03 mg/kg in Dadapan River. Classification of sediment pollution standard guide, the limit of Cu is 49.98 mg/kg, thus the Cu level in Wonokromo River and Dadapan River was still below the existing quality standard. Therefore, the sediment of the East Coast Waters of Surabaya was considered livable as aquatic organism habitat, including for blood cockles.

*4.4.* Cu level in blood cockles (*Anadara* sp.)

Copper level in the meat of blood cockles was taken from three points at the East Coast Waters of Surabaya: 2000 m, 1750 m, and 1500 m, with the measurement performed twice as shown in table 4.

**Table 4.** Cu heavy metal substances in the meat of blood cockles in each point with two repetitions.

| Station | Cu Levels in Shell's Meat (mg/kg) | | | | | |
|---|---|---|---|---|---|---|
| | point 1500m | | point 1750m | | point 2000m | |
| | 1st Repetition | 2nd Repetition | 1st Repetition | 2nd Repetition | 1st Repetition | 2nd Repetition |
| Wonokromo River Station | 1.44 | 2.05 | 1.79 | 2.64 | 1.92 | 1.16 |
| Dadapan River Station | 1.32 | 1.69 | 1.31 | 1.44 | 1.17 | 1.18 |

The copper level in blood cockles meat collected in Wonokromo River was 1.833 mg/kg while the blood cockles found in Dadapan River had copper level of 1.352 mg/kg. According to the Health Department of the Republic of Indonesia Regulation No.0375/B/SK/1989 on Standard for Metal in Consumable Biota, the maximum level of Cu in sea biota is 20 ppm or 20 mg/kg.

The concentration of heavy metal found in the meat from blood cockles taken from East Coast Waters of Surabaya did not exceed the existing standard. Results of the measurement of Cu in each point can be seen in table 6 and figure 8 showing that there were differences in concentration of metal found in each station and point. Copper concentration in Wonokromo River was larger than the one in Dadapan River. In Wonokromo River, the highest concentration was found at 1750 m, while in Dadapan River, the highest concentration was found at 1500 m. Concentration of copper decreases as the depth of the water increases [8]. Gulivey (year) stated that the depth sampling location, the lower the copper concentration.

## 5. Conclusion

Concentration of copper in the meat of blood cockles found in Wonokromo River and Dadapan River was 1.833 mg/L and 1.352 mg/L respectively. This level did not exceed the standard set by the Health Department of The Republic of Indonesia Regulation No.0375/B/SK/1989. However, the level of Cu found in the East Coast Waters of Surabaya showed a number that exceeded the quality standard, thus, efforts should be conducted to decrease the spread of heavy metal pollution in the East Coast Waters of Surabaya.

## 6. References

[1] Effendi H 2003 Water quality study for resource management and aquatic environments (Yogyakarta: Kanisius) p 23 (In Indonesia)
[2] Boyd C E 1979 Water quality in warmwater fish ponds (Alabama: Univ of Alabama Pr.) p 122 - 124
[3] Moore J W 2012 Inorganic contaminants of surface water: research and monitoring priorities (Berlin: Springer Science & Business Media) p 66
[4] Nurjanah Z Kustiyariyah 2005 *Bull. Fish. Tech.* **8** 15–24 (In Indonesia)
[5] Zhang H, Zhang Z, Li J, Cai S 2007 *Int. J. Electrochem. Sci.* **2** 788–96
[6] Supriyanto C Susana TS Validation of F-AAS methods for obtaining quality assurance on analysis of elements of Cd, Cu, Cr, Pb, and Ni in liquid waste test samples *Proc. PPIPDIPTN* pp 121-125 (In Indonesia)
[7] Widowati W Sastiono and Jusuf R 2008 *Toxic effects of metal: prevention and control of pollution* (Yogyakarta:Andi) pp 2–206. (In Indonesia)
[8] Guliyev JA Hadjiyeva SR Ghorbanpour S Aminbekov AF Guseynli AG 2014 *Int. J. Basic Sci. Appl. Res* .**3** 606–10

IOP Conf. Series: Earth and Environmental Science **137** (2018) 012090 doi:10.1088/1755-1315/137/1/012090

# The effect of sea-water and fresh-water soaking on the hedonic test of *Eucheuma* sp. syrup and pudding

**H Novianty and S M C Herandarudewi**

UPT.LPKSDMO PulauPari LIPI
JalanRaden Saleh 43, Cikini, Jakarta Pusat 10330

Email :hildanovianty2012@gmail.com

**Abstract.** Seaweed is a non-fishery marine commodity that has great opportunities to be developed in Indonesia. One of the seaweed with a high economic value is *Eucheuma alvarezii*. This seaweed can be used as an additional material in cosmetic and pharmaceutical products or directly used for syrup and pudding. Post-harvest technique conducted by the seaweed farmers will affects the quality of dried and processed products. The purpose of this study is to see the effect of post harvest technique on the quality of dried seaweed and hedonic test (favorable test) of processed product (syrup and pudding). This study was conducted using descriptive method. The study compared dried, syrup, and puddings from two differents post-harvest technique, between salt and fresh-water draining products. The results showed that fresh-water draining technique obtained better quality results organoleptic test. Supported by hedonic test, that showed more panelists were prefered the fresh-water drained products of syrup and pudding. The preference were much higher for the fresh-water drained products in all three catagories of color, taste, and smell.

## 1. Introduction

Seaweed is a non-marine fishery commodity with great opportunity to be developed. *Eucheumaalvarezii* is a seaweed species with a high economic value. This species of seaweed is an ahydrocolloid type which produces carrageenan that has functional properties as a gel-forming, stabilizer, suspension, and emulsion texture builder [1]. Due to its function, this species of seaweed is used as an additional ingredient in the food, cosmetics and pharmaceuticals industries. In addition, it also contains vitamins and minerals needed by the human body. *Eucheuma* seaweed has moisture content (11-13.90 %); crude protein (2.69-9.20 %); fat (0.09-0.37 %); carbohydrates (4.94-13.38 %); crude fiber (0.95-2.50 %); ash (4.79-17.09 %); calcium (22.39-69.25 ppm); vitamin B1 (0.019-0.21 mg/100 g); vitamin B2 (2,26-8,45 mg/100 g); and vitamin C (12.00-43.00 mg/100 g) [2]. Seaweed nutrients contain carbohydrates (sugar or vegetable gum), protein, little fat and ash which are mostly carbohydrates (cellulose and hemicellulose) that cannot be digested entirely by the enzymes in the body so it can part of a less calorie – diet [3]. Therefore, this species of seaweed has a high added value because of its diverse functional substances that also has nutritional levels needed by humans. In addition, Indonesia has a huge potential of *Eucheumaalverizii* cultivation, because Indonesian waters possess characteristics that makes it suitable for the optimum growth of *Eucheumaalverizii*.

The existence of a program from the Ministry of Marine Affairs and Fisheries to increase the production of seaweed cultivation of Indonesia must be balanced with the processing of cultivated seaweed. A guaranteed quality of seaweed cultivation will affect the processed seaweed.

Hence, the post-harvest techniques (after cultivation) of seaweed cultivation have an important role in ensuring the quality of seaweed. The existence of good post-harvest management will ensure the quality of seaweed products that will ultimately encourage the sustainability of the seaweed processing industry [4]. Studies of seaweed post-harvest techniques have been carried out by Itung and Marthen [5] and Poncomulyo [2], by doing the same technique which was washing the seaweed with sea water and drying it under the sun. Meanwhile, the drying technique can be done by drying on the ground surface, drying with the method of para-drying and drying by the hanging method [4]. The most suitable post-harvest technique will determine the quality of the finished product [3]. The initial handling of seaweed (post-harvest) should be well carried out, because good quality raw materials will produce good products as well. This study will compare the effects of sea-water and fresh-water soaking to the hedonic test of syrup and pudding as seaweed processed products.

## 2. Methodology

The study was conducted in May 2016 and it is a descriptive study. This research began with the planting of seaweed for 45 days with the long-line method and then post-harvest techniques were done in two ways: sea-water and fresh-water soaking. The seaweed from both techniques were made into syrup and pudding. An organoleptic test was carried out on the syrup and pudding by 15 panelists. The water content test was done by PT. Saraswanti Indo Genetech, Bogor. The study was done by the following scheme of post-harvest proccess (figure 1).

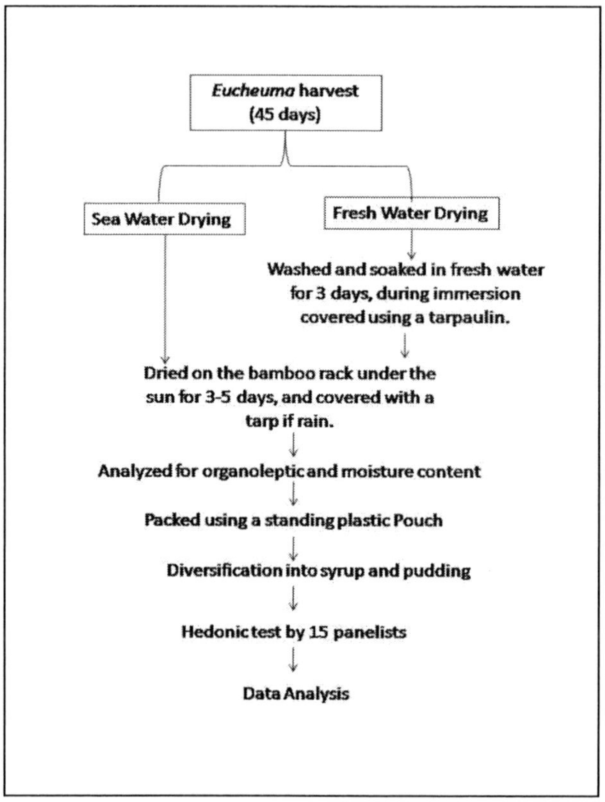


**Figure 1**. Seaweed post-harvest proccess

## 3. Result and Discussion

Economically, sea-water drying is less costly but fresh water drying was recorded to have resulted in better quality (figure. 2 and table 1). Fresh-water drying results in a white, clear, brighter appearance with less grains of salt attached to the dry thallus. On the other hand, the sea-water drying product has a darker appearance (brownish) with a considerable amount of dirt and sand granules attached to its thallus. Thus, physically, the post-harvest technique of the fresh-water drying system is better. The results of the moisture level test of the dried seaweed on both post-harvest treatment techniques showed (table 1) that the post-harvesting technique of fresh-water drying had a smaller moisture content of 14.36 % compared to sea-water drying which has a moisture content of 22.76 %. This is because in the sea-water post-harvest drying technique, the salt grains will remain attached to the thallus of the dried seaweed so that it will raise the weight of its water content. However, its water content still conforms to the dry seaweed SNI No.2690.1: 2009 [6].

**Figure 2.** Physical differences between sea water drying and fresh water drying seaweed.

**Table 1.**Qualitative data of *Eucheuma*.

| No | Sample | Moisture content (%) | Organoleptic | |
|---|---|---|---|---|
| | | | Appearance | Odor |
| 1 | Fresh water drying | 14,36 | White, clear, less salt | Seaweed specific ( not strong) |
| 2 | Sea water drying | 22,76 | Brown, lots of grains of sand and salt | Seaweed specific (strong) |
| SNI (2009)* : Dried seaweed | *Eucheuma* | Maks. 32 | - | Seaweed specific |

**in* [3]

The different treatments on seaweed post-harvest techniques were done to influence the quality of the processed products after harvest. As we can see from tables 2 and 3, table 2 shows the results of the hedonic test on the seaweed syrup beverage, while table 3 shows the results of the hedonic test on the seaweed pudding. The hedonic test is a test of preferences, in which the panelist (person who is tested) is asked for his/her personal response to preferences or dislikes using the hedonic scale [7]. The panelists who participated in this study were semi-trained panelists with a total of 15 people. A semi-trained panelist is a panelist who gets an explanation of hedonic testing and sensory traits with the number of panelists measuring from 15 to 25 people [7].

Tables 2 and 3 show that the seaweed syrup and seaweed pudding made from the fresh-water post-harvest technique had a greater hedonic value for all test categories (color, flavor and odor) compared to the products made using the sea-water drying technique. This indicates the panelists' preference for seaweed syrup and pudding products made using the fresh-water drying technique was preferred over the other. The specific odor of seaweed was stronger in the beverage and pudding product made by sea-water drying. This was not favored by the panelists. Table 1 also shows that dried seaweed made by the sea-water drying process resulted in a stronger specific seaweed odor compared to the other.

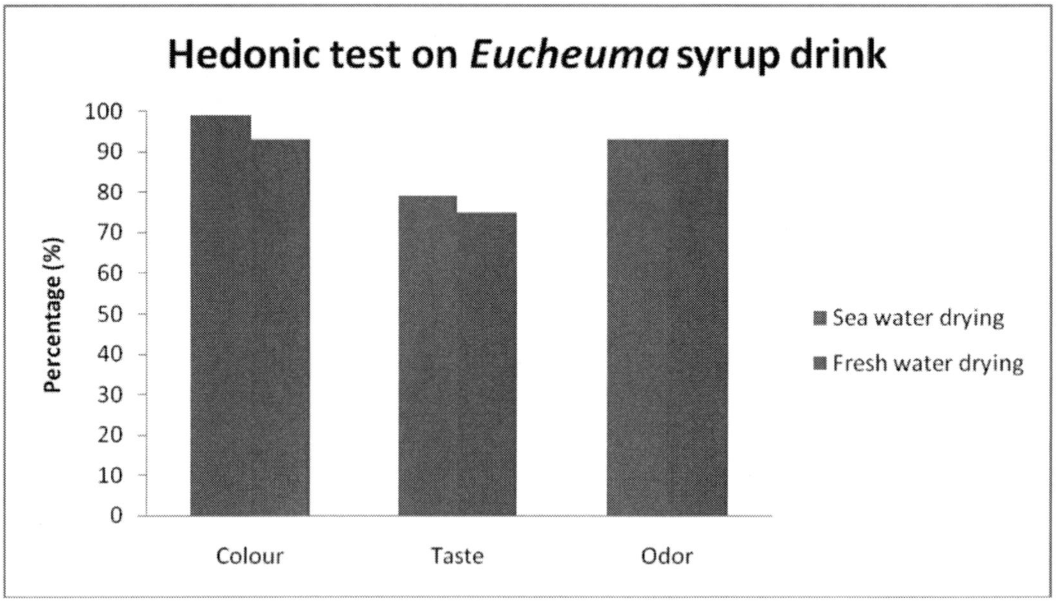


**Figure 3.** Hedonic test on *Eucheuma* syrup drink

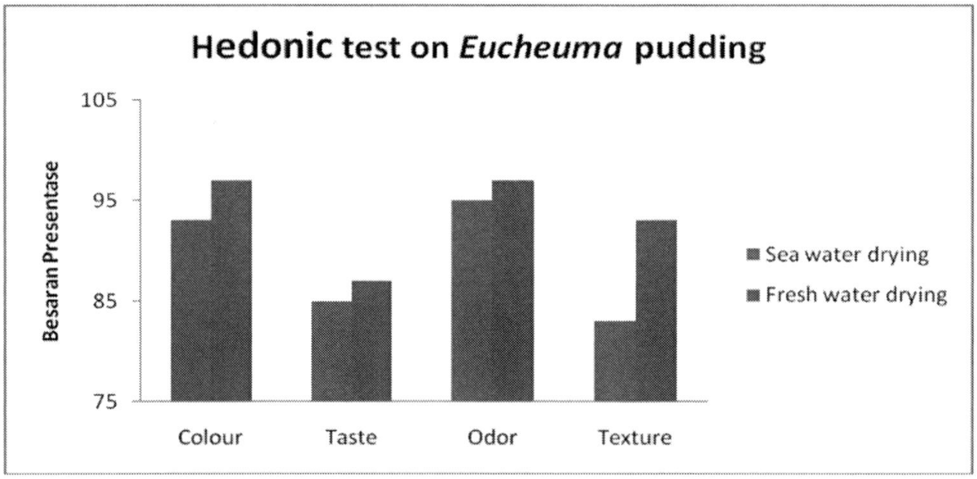


**Figure 4.** Hedonic test on *Eucheuma* pudding

IOP Conf. Series: Earth and Environmental Science **137** (2018) 012090 doi:10.1088/1755-1315/137/1/012090

## 4. Conclusion

There are differences in seaweed post-harvest techniques, namely fresh water drying and seawater drying. Freshwater drying is the leaching and soaking of harvested seaweed with fresh water; while in sea-water drying there is no washing and soaking treatment using fresh water. Both of these post harvesting techniques have influenced the results of the hedonic or preference test on the seaweed products of syrup and pudding, where the products made from fresh-water drying was more preferred.

## 5. References

[1] Susanto A B, Rini P and Alva W 2009 Seaweed Processing Technology in Indonesia *Biotechnol. and Seaweed Industri Proc. ISBN : 978-979-17585-3-6* 1-20 (Text in Indonesia)

[2] Poncomulyo T, Herti M, and Lusi K 2006 *Cultivation and Seaweed Processing* (Jakarta: Agro Media Pustaka) p 1-64 (Text in Indonesia)

[3] Wibowo S, Rosmawaty P, Muhammad D and Arif R H 2014 *ATC Proc. Techniques of Eucheuma* (Jakarta: Penebar Swadaya) 1-74 (Text in Indonesia)

[4] Direktorat Jendral Budidaya Perikanan (DJPB) 2016 Instructions to manage post harvest *seaweed htt://www.djpb.kkp.go.id/arsip/c265/PETUNJUK-PRAKTIS-MENGELOLA-PASCA-PANEN-RUMPUT-LAUT/?category id=13* Retrieved August 4 2016 (Text in Indonesia)

[5] Itung M and Marthen. Post Harvest Processing of *Eucheuma* and *Gracilaria* for export purpose *Marine Chimica. Acta.* **4** 5-8 ISSN 1411-2132 (Text in Indonesia)

[6] Peranginangin R, Ellya S and Muhammad D 2013 *Produce carrageen from seaweed* (Jakarta: Penebar Swadaya) p 1-74 (Text in Indonesia)

[7] Soekarto S T 1985 *Organoleptic assessment (for food and agricultural products)* (Jakarta: Penerbit Bhratara Karya Aksara) 1-122 (Text in Indonesia)

IOP Conf. Series: Earth and Environmental Science **137** (2018) 012091 doi:10.1088/1755-1315/137/1/012091

# Prediction of Supratidal Zones as Turtle Nesting Sites using Remote Sensing and Geographic Information System, A Case study in Pacitan, Southern Java Sea

**A Darmawan[1*], D K Saputra[2], D G R Wiadnya[3], and A M Gusmida[4]**

[1]Department of Aquatic Resource Management, Faculty of Fisheries and Marine Science, University of Brawijaya, Jl Veteran 65145 Malang, Indonesia
[2]Department of Marine Science, Faculty of Fisheries and Marine Science, University of Brawijaya,Malang, Indonesia
[3]Department of Fisheries, Faculty of Fisheries and Marine Science, University of Brawijaya, Malang, Indonesia
[4]Development Planning Agency at Sub-National Level, Pacitan Regency, East Java, Indonesia

E-mail: ariefdarma@ub.ac.id

**Abstract.** Turtles, the most threatened coastal-marine fauna, are protected through both national, and global regulations. However, many of their nesting sites have been degraded in the past years. Completing natal homing, adult females emerged at night to lay-down eggs in the upper intertidal and supra-tidal zone of sandy beach from where they hatched. This study explained coastal topology of beaches usually used for nesting sites, covering 117 km coastline at Pacitan Regency, Southern Java Sea. The shift in beach morphology through times was figured out based on Landsat 8 and Sentinel 2a satellite imagery and remote sensing (GIS methods). This was combined with in-situ data on current coastline features, slope, and tide variations. Results showed a typical sandy beach, called Taman Ria Beach, a long time identified as nesting site for *Lepidochelys olivacea*, locally named as Penyu Lekang. Also, there was approximatelly 3.49 ha of supratidal area predicted in Taman Ria Beach according this study

## 1. Introduction

A supratidal or supra-littoral zone can be defined as a zone near the shore but above the level of high tide [1]. In general, supratidal zone has been referred to as spray zone immediately above the highest high tide level [2]. It is also known as the splash zone in a coastal area. Furthermore, the zone sometimes splashed but not inundated or submerged by seawater. In other words, supra-littoral is a transitional region from land to sea floor above the spring high tide line [3]. As a result, supratidal zone is only covered by seawater during when very high tides or tsunamis or possibly large storm waves splash over the coastal zone.

The aim of the study was to predict the locations of supratidal zones with remote sensing and GIS approaches. Supratidal zone is well-known as an important area for completing natal homing of female sea turtles and their eggs hatching. Adult female turtles emerged at night to lay eggs in the upper intertidal and supra-tidal zone of a sandy beach from where they hatched. After several days, if the temperature and the humidity of the sand are suitable, those eggs that were left under sand will hatch as baby sea turtle or "hatchling". On the other hand, although sea turtles were protected by both national and global regulations, the degradation of many nesting sites in the past years has been disturbing the sea turtle life cycle. In the end, by predicting supratidal zones and measuring coastal physical characteristics, including coastline dynamics, the study could contribute to sea turtle conservation activity and coastal zone protection and serve as a baseline for further analyses related to coastal zone management in general.

## 2. Methodology

Morphological features such as coastline shape and type of coastal area in Pacitan directly facing to the Southern Java Sea were monitored from 2015 to 2017 via Landsat 8 OLI and Sentinel 2A remote sensing satellite images. Multi temporal Landsat 8 OLI images was aquired on 19 May 2014, 23 June 2015, 28 August 2016 and 28 June 2017. Meanwhile, Sentinel 2A images were aquired on 28 June 2017 and 25 November 2017. This study applied remote sensing techniques to obtain information through the analysis of images acquired by those satellite sensors [4]. Hence, during the process of detection and analysis, the researchers were not in contact with the investigated object or area.

According to McIntry [5], a shore is determined by the geology and topography of the coastline and adjacent land, the physical processes operating on them, the actions of rivers and waves created, the supply, and the distribution of sedimentary materials. One of the results is the presence of sand that indicates exposure to wave actions. As a consequences, the morphological type of a coastal zone serves as an important characteristic that determines whether a coastal zone is suitable for turtle nesting or not. The coastal type classification by [6] was applied. As a result, the sandy beaches surrounding Pacitan area were selected. The indication that sand beaches are much flatter than shingle beaches [5] was also used as a consideration during this process. Meanwhile, cliff and mudflat were removed because they did not meet the requirements for turtle nesting. In the end, the identification of sand beaches was carried out to predict the locations of supratidal zones, because the existing sand zones are more possible to be turtle nesting sites.

Meanwhile, the identification of sand beaches applied different methods by previous researchers [7]. In this study, multi-temporal images were used following the examples of Indian researchers [8], but IRS was substituted with Landsat 8 OLI for the mapping of physical features such as the characteristics of coastal morphology and the dynamics of coastline shapes, and for the identification of possible coastal zones for turtle nesting. Meanwhile, Sentinel 2A for June 2017 was used to detail and predict supratidal zones after the previous process with Landsat 8 OLI. Visual image interpretation methods were used on those satellite images to map physical features that could lead to the prediction of supratidal zones. Image enhancement was performed to increase the visual quality of the satellite images. Image enhancement was conducted through False Color Composite (FCC) and panchromatic sharpening (*pan-sharpen*). For Landsat 8 OLI, FCC RGB 567 was pan-sharpened with band 8 panchromatic. Band 8 of Landsat OLI that had better spatial resolution (15 meters) could enhance bands 5, 6, and 7 that only had 30 meter spatial resolution. Moreover, the supratidal zone prediction used Sentinel 2A images that were processed into reflectance images. Then, Band 8 and 3 of the Sentinel 2A images were selected to produce new images called Normalized Difference Sand Index (NSDI) images. In this study, NSDI algorithm [9] was applied with a modification to enhance sand features from the images. It means that Band 3 (green) of Sentinel 2A was used instead of Band 4 (red). The original NSDI algorithm [9] was formulated by rationing SWIR Band and Red Band as below:

$$NSDI = \frac{SWIR2 - R}{SWIR2 + R}$$

Where:
SWIR2 = Short Wave Infra Red Band
R = Red Band

Furthermore, NSDI images were combined with the elevation and average sea level data. In the end, the supratidal zones were predicted according to the average sea level data. We assumed that the supra-tidal zones would be up to the average high tide or average sea level during spring tide. The area would only be submerged if very high tide happens there.

## 3. Results and discussion

In the first step, the Landsat 8 OLI images acquired on 23 June 2015, 28 August 2016 and 28 June 2017 were used to delinieate the coastline in the study area and to identify the locations of marine-deposition coasts to be the locations of sandy beaches where turtle would land. As a result (figure 1), there were 4 (four) locations identified as marine-deposition coasts likely to meet turtle nesting requirements, namely: (1) Teleng Ria beach, (2) Soge beach, (3) Taman Ria beach, and (4) Kunir beach. Meanwhile, from 2015 to 2017, the coastlines of Teleng Ria beach and Taman Ria beach were very dynamic, especially near the river mouth. It is believed that the sand areas near river mouth will be less visited by turtles due to rapid shift. For further investigation, Taman Ria beach was selected as the area of interest because this location had a turtle conservation area managed by the local community in cooperation with the Marine Affairs and Fisheries Office at sub-district level.

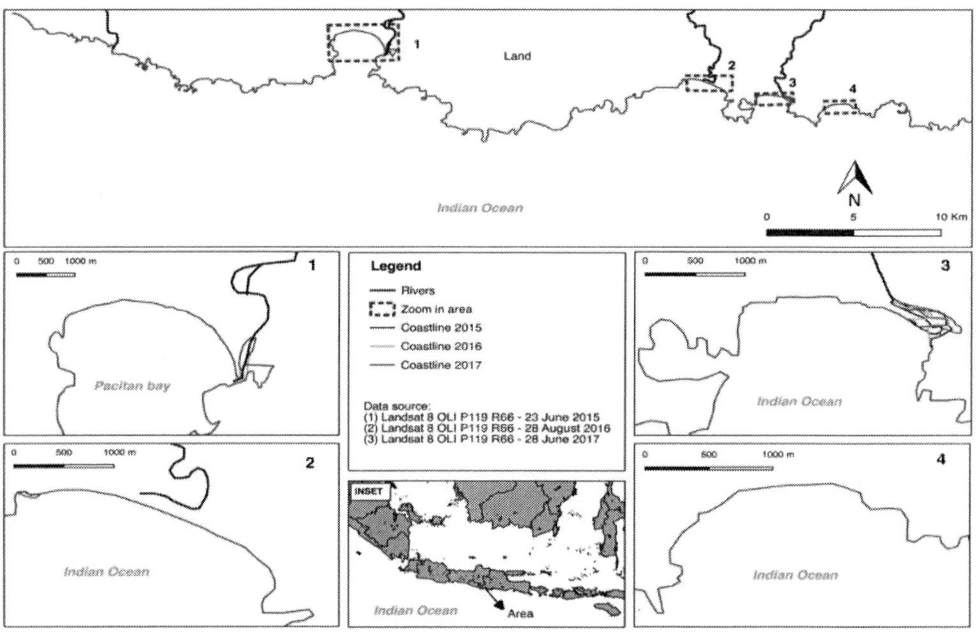


**Figure 1.** Map of coastline dynamics and identified marine-deposition coast

The NSDI images resulted from Sentinel 2A processing at Taman Ria sandy beach are as shown in figure 2 and figure 3. in figure 2 and figure 3, the NSDI images were set with pseudo color to make it easier to distingish sand from other features such as water, vegetation and built area. Furthermore, by observing the pixel value of the NSDI image, it was found that the sand had a value of between 0.0264 and 0.100. Meanwhile, vegetation and built area had NSDI values of more than of 0.1. Lastly, water body, including breaking wave, deep sea water and shallow water from the river were identified with NSDI to have negative values.

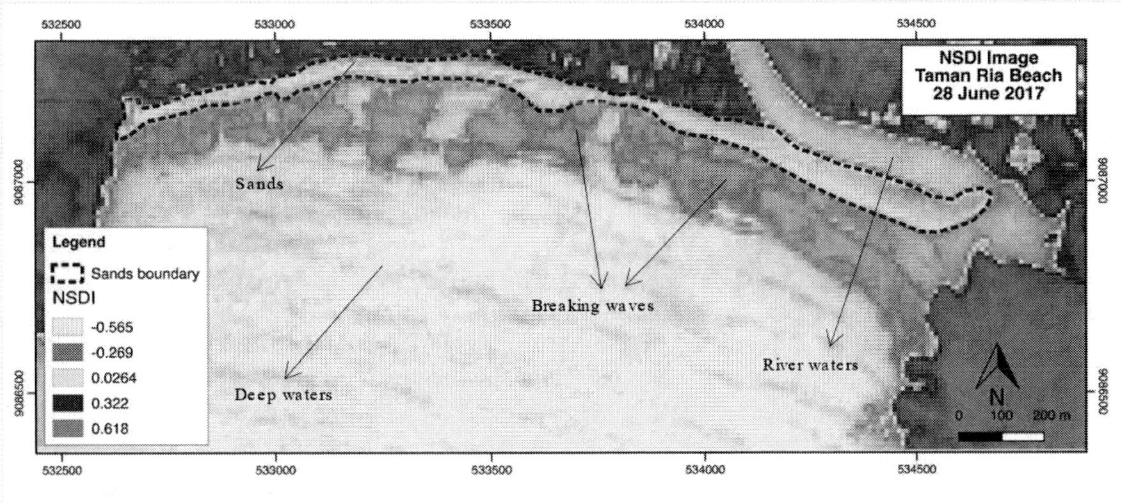


**Figure 2.** NSDI image of Taman Ria beach using Band 8 and Band 3 of Sentinel 2A that acquired 28 June 2017

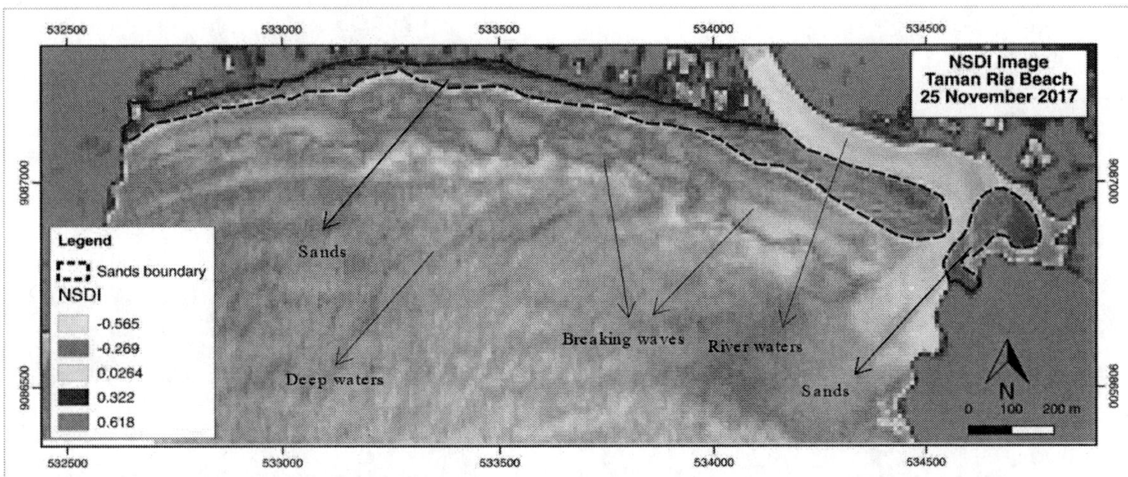


**Figure 3.** NSDI image of Taman Ria beach using Band 8 and Band 3 of Sentinel 2A acquired on 25 November 2017

The NSDI images (figure 4) were combined with the SRTM data. Then, 3D analysis was implemented to represent the *in situ* landscape. This process resulted in the 3D perspective of Taman Ria beach landscape as seen in figure 4.a.1 to figure 4.b.2. As initial condition, figure 4.a.1 and figure 4.b.1 clearly represent the situation at Taman Ria beach when it has normal sea level based on the images acquired on 28 June and 25 November 2017. In this condition, sand spread widely from the western to the eastern part of Taman Ria beach, including the sand spit as marine-deposition feature located near the river mouth. Meanwhile, figure 4.b.1 and figure 4.b.2 serve as a simulation for predicting the situation if the sea level rises approximately 50 cm from normal condition. Figure 4.b1 and figure 4.b.2 clearly show that the sand spit feature near the river mouth disappear or is submerged by the seawater. It indicates that the sand spit or sand bar area near the river mouth is not part of supratidal zone because it has already been submerged by the water with only a 50 cm sea level rise. As the result of the simulation, it is believed that female turtles will not land there for laying their eggs due to the instability of the sand area. Meanwhile, according to that simulation, the wide area next to the sand spit feature was

IOP Conf. Series: Earth and Environmental Science **137** (2018) 012091 doi:10.1088/1755-1315/137/1/012091

surprisingly not submerged by the water. However, further investigation was still required to provide evidence for it to be predicted as a supratidal zone.

As the investigation proceeded, 3D analysis was continued by using ±100 cm and ±200 cm sea level rises in sequence. The results of this investigation are presented on figure 5.a.1 to figure 5.b.2. Those figures showed that the area located at the western part of the sand spit near the river mouth was still stable and was not submerged by the seawater. These results clearly indicated that the sandy area was close to be a supratidal zone. Furthermore, to prove the prediction of the supra-tidal zone at Taman Ria beach, we conducted an interview with local people who worked in the turtle conservation since its establishment by the District Government in 2013. The results of the interview were consistent with the results obtained from the 3D analysis. Adult female turtles regularly complete natal homing at night to lay their eggs in that area as predicted by the analysis. Figure 6 shows the predicted supratidal zone according to this study according to the Sentinel 2a images acquired on 28 June 2017 and 25 November 2017 and SRTM data.

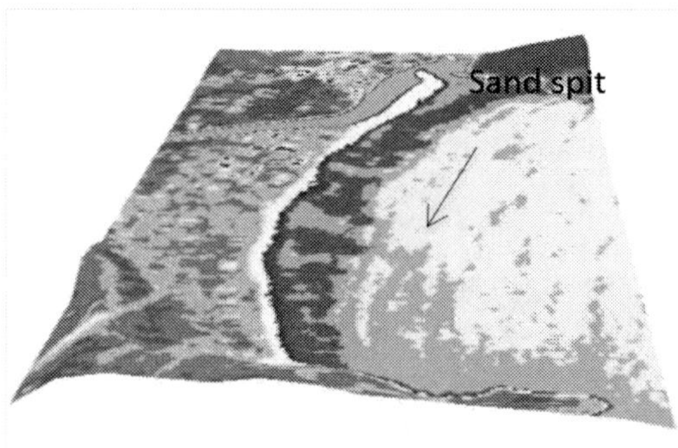


(a.1)

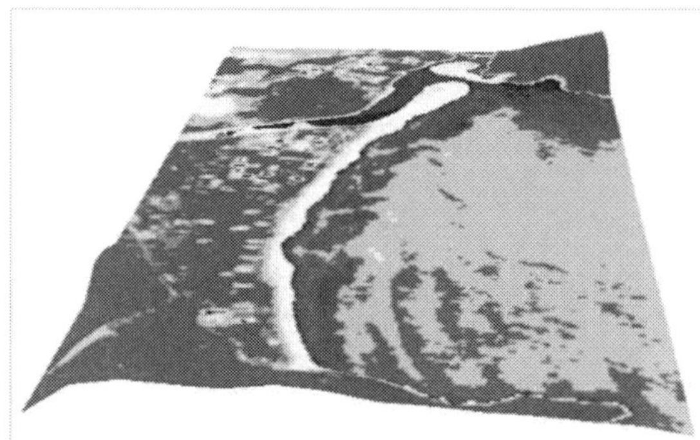

(b.1)

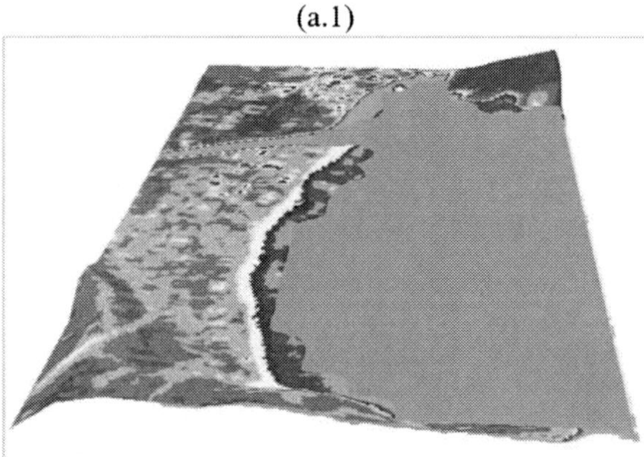

(a.2)

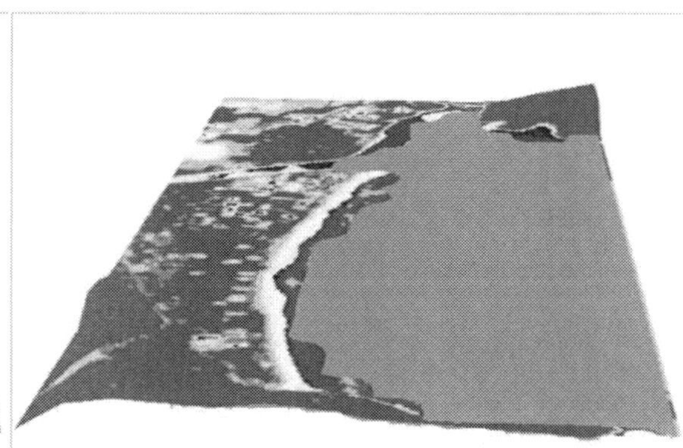

(b.2)

**Figure 4.** Simulation (a.1) Normal sea level condition on 28 June 2017, (a.2) Normal sea level condition 25 November 2017, (b1) sea level rise ±50 cm for 28 June 2017 and (b.2) sea level rise ±50 cm for 25 November 2017.

IOP Conf. Series: Earth and Environmental Science **137** (2018) 012091 doi:10.1088/1755-1315/137/1/012091

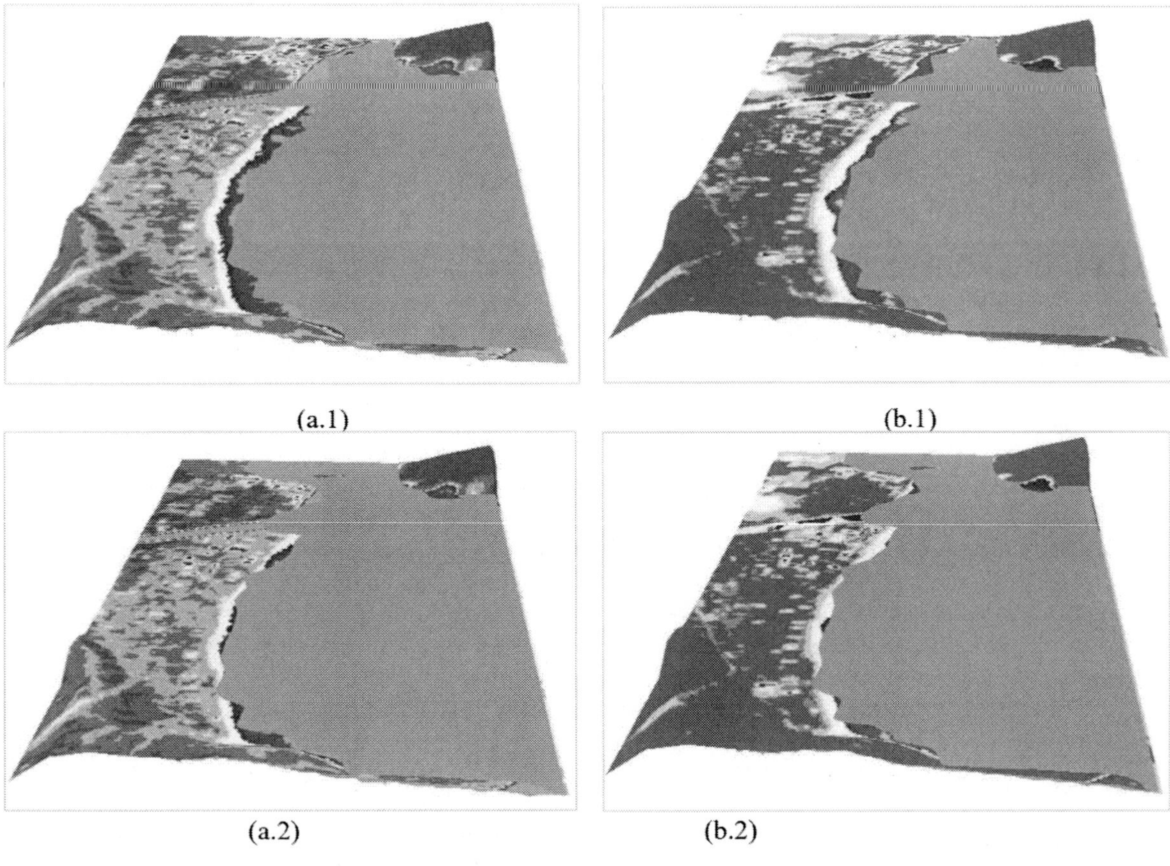


**Figure 5**. Simulation (a.1) sea level rise ±100 cm from normal conditon for 28 June 2017 and (b.1) for 25 November 2017, then (a.2) sea level rise ± 200 cm from normal condition for 28 June 2017 and (b.2) for 25 November 2017 as simulation of an extreme sea level rise condition or very high tides.

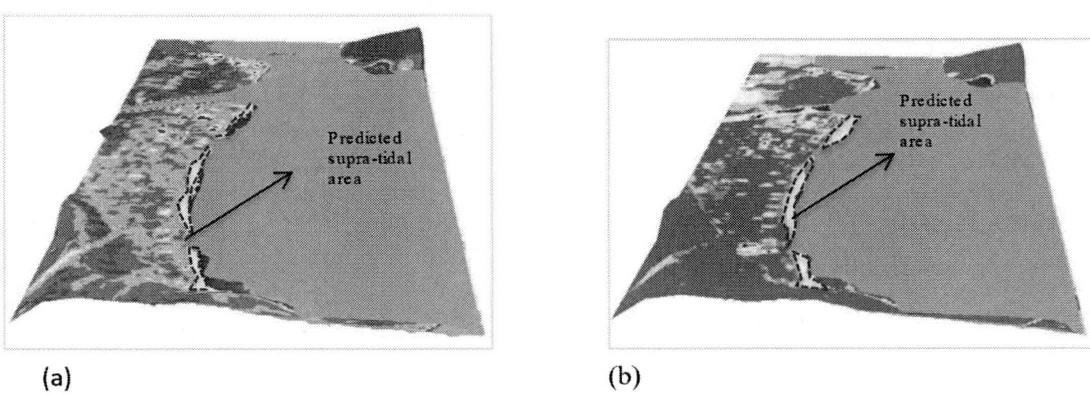


**Figure 6**. Predicted supratidal area at Taman Ria Beach, Pacitan according to Sentinel 2a acquired on (a) 28 June 2017, (b) 25 November 2017 and SRTM. The prediction is based on simulation of an extreme sea level rise events.

The final predicted supratidal zone according to Sentinel 2a images acquired on 28 June 2017 and 25 November 2017 is represented by figure 6. It was resulted from the overlay between the delineation in figure 6.a and figure 6.b and then subset by the existing sand area resulted by the NSDI image in figure 3 previously. The red polygon in figure 6 shows the predicted supratidal zone of approximatelly 34,986 $m^2$ or about 3.49 Ha. Furthermore, it was also the most stable part of Taman Ria sandy beach where adult female turtles likely lay their eggs during natal homing figure 6. Predicted supratidal zone at Taman Ria Beach according to this research

A study evidence shows that sea turtles lack of parental care behaviour that making nest site becomes particularly important for the survival of their eggs [10]. It seems like adult female turtles select the locations for nesting according to their instinct in a unique way. Moreover, elevation is another important factor in the determination of a place where female turtles will lay their eggs. For example, hawksbill turtles tend to place their eggs at an elevation of approximatelly 1.2 m, while green turtles tend to place their eggs at an elevation of 1—3 m [11]. Furthermore, a smoother beach slopes may be favourable for adult female turtles for them to approach the nesting site in a quicker and easier way [12]. Temperature [13], moisture and sand particle size are important factors considered by adult female turtles. McGehee [14] and Rubio [15] found that higher hatchling success is achieved when the moisture is about 25% and lower hatchling success is achieved when the moisture is below or above 25%.

GPS locations of individual nesting sites were not available due to limitation in equipment and data records from the turtle conservation community working at Taman Ria Beach. As a consequence, this research was unable to verify the predicted supratidal zone against the real nest location. Ideally, the prediction of supratidal zone using methods applied in this research should be completed *in situ* to obtain the data of the place where turles exactly put their eggs, supported by detailed beach profiles and detailed tide data. The researchers eventually agreed that the methods and results of this research can still be improved.

## 4. Conclusion

Supratidal zone is a dynamic area. Type of coast, height/elevation, and tide or sea level affect the boundary of this zone. As a result, the boundary of a supratidal zone could vary from time to time according to those variables. Furthermore, the most stable part of a supratidal zone of a sandy beach can be expected to be the main location for adult females to lay their eggs at night. However, due to the limitation of data of the site where turtle eggs are located in Taman Ria beach, tide data or sea level data, and satellite imagery spatial resolution, this study could not provide detailed prediction of supratidal zones. An improvement of those dataset is required for further studies.

## 5. Acknowledgement

All authors would like to appreciate BPPM Faculty of Fisheries and Marine Science and LPPM of Universitas Brawijaya, which financially supported this study. Also, we wish to thank the field team that consisted of dedicated students M. Naufal Irfansyah, Bayu Muhammad, Rizal Mustofa, Ari Bimi Suhendra, Farizko Fernando, and Nadzir Badriawan who have conducted the field work and gathered the necessary data.

## 6. References

[1] Schwartz M L 2005 Encyclopedia of Coastal Science Springer: Netherland
[2] Nybakken J W 1993 Marine Biology: An Ecological Approach 3rd edition New York: Harper Collins
[3] Trujillo, Alan P, Thurman, Harold V 2011 Essentials of oceanography 10th edition Prentice Hall: Boston
[4] Lillesand T M, Kiefer R W, Chipman J W 2004 Remote Sensing and Image Interpretation, Fifth Edition Wiley: New York
[5] McIntryre A D 1977 The Coastline. In Barnes, R S K (ed) London: John Wiley & Sons Ltd
[6] Shepard F P 1973 Seacoast classification, in Submarine Geology. Harper Row: New York

[7] Dunkin, Lauren, Reif, Molly., Altman, Safra., Swannack, Todd 2016 A Spatially Explicit, Multi Criteria Decision Support Model for Loggerhead Sea Turtle Nesting Habitat Suitability: A Remote Sensing-*Based Approach. Remote Sens* 8:573

[8] Prusty G, S Dash, and Singh M P 2007 Spatio-temporal Analysis of Multi-date IRS Imageries for Turtle Habitat Dynamics Characterization at Gahirmatha Coast, India *Int. J. Remote Sens.* 871–83

[9] Fadhil A 2013 Sand Dunes Monitoring Using Remote Sensing And GIS Techniques for Some Sites in Iraq *Int. Soc. Optical Eng*

[10] Kamel S J and Mrosovsky N 2005 Repeatability of nesting preferences in the hawksbill sea turtle, *Eretmochelys imbricata*, and their fitness consequences *Animal Behaviour* **70**:819–828

[11] Horrocks J A and Scott N M 1991 Nest site location and nest success in the hawksbill turtle *(Eretmochelys imbricata*) in Barbados, West Indies *Mar. Ecol. Prog. Ser.* **69**:1–8

[12] Garmestani A S, Percival H F, Rice K G, and Portier K M 1997 Sea turtle nesting in the Ten Thousand Islands of Florida *Fla. Coop. Fish and Wildl. Res. Unit USGS-Biological Resources Division Tech. Rep.* 56-96

[13] Wibbels T 2002 Chritical Approaches to Sex Determination in Sea Turtles *In* Lutz, Peter L., Musick, John A., Wyneken, Jeanette (eds) The Biology of Sea Turtles **II**:104-126 CRC Press: Boca Raton

[14] McGehee,M A 1990 Effects of moisture on eggs and hatchlings of loggerhead sea turtles *(Caretta caretta) Herpetologica* **46**:251–258

[15] Rubio M 2009 Nesting beach characteristics of endangered sea turtles in Las Perlas Archipelago, Panama *Dissertation Report* Center for Marine Biodiversity and Biotechnology School of Life Sciences Heriot-Watt University: Edinburg

IOP Conf. Series: Earth and Environmental Science **137** (2018) 012092 doi:10.1088/1755-1315/137/1/012092

# The host preference and impact of *Argulus japonicus* ectoparasite on cyprinids in Central Java, Indonesia

**Kismiyati, P D Wulansari and N N Dewi**

Department of Fish Health Management and Aquaculture Faculty of Fisheries and Marine, Airlangga University, Jl. Airlangga 60115, Surabaya, East Java, Indonesia.

E-mail: kismiyati@fpk.unair.ac.id

**Abstract.** The most widely cultivated freshwater fish are from Familia Cyprinidae, among others goldfish (*Carassius auratus*), koi (*Cyprinus carpio*) and comet goldfish (*Carassius auratus auratus*). One of the constraints of freshwater fish cultivation is ectoparasite infestation *Argulus japonicus*. Financial losses have been experienced by some farmers, caused by these ectoparasitic infestaions. This study was aimed to determine the impact of ectoparasite *Argulus japonicus* infestation on host (freshwater ornamental fish from Familia Cyprinidae), in order to find a preventive solution to treatment on the host. The results showed that prevalence of infested fish by *Argulus japonicus* were 57 % goldfish, 31 % comet fish and 65 % koi. Changes of histopathology on host were congestion, baoning degeneration, epithelium erosion and inflammatory cell infiltration. The image of infected leukocytes infested by *Argulus japonicus* were 8.5 % of lymphocytes, 4.7 % neurophils, 3.9 % monocytes, 1.45 % eosinophils and 0,17% basophils.

## 1. Introduction

One area of the *minapolitan* area development in Central Java that is based on aquaculture is Magelang regency. Here, the cultivation of freshwater ornamental fish being optimized are fish from the Cyprinidae family, such as koi fish, goldfish and comet fish. Disease is the main problem in fish farming, one of which is the *Argulus* ecoparasite, which is a major ectoparasite in fish from the Cyprinidae family with predilection of the gills, fins and skin (body surface) [1,2,3]. *Argulus* sucks blood and injects anti-coagulants that cause the immune response of the host to be disturbed [4]. These ectoparasites can also cause stunted growth and even death, resulting in the decreased production (weight) and quality of the ornamental fish that ultimately leads to economic losses for the cultivators.

The purpose of this study was to determine the prevalence, degree of infestation, histopathological changes and the description of leukocytes of goldfish, comet fish and koi fish due to *Argulus* infestation.

## 2. Methodology

The method used in this research is survey method, with the research taking place at the Center of Fish Culture in the Magelang regency of Central Java. Fish samples were taken from the Mungkid and Muntilan sub-districts from six cultivation ponds consisting of two goldfish farming ponds, two koi fish ponds and two fish comet fish ponds. 50 fish samples were taken from each pond, so that each type of fish amounted to 100 samples. After the prevalence is calculated, the degree of infestation is

IOP Conf. Series: Earth and Environmental Science **137** (2018) 012092 doi:10.1088/1755-1315/137/1/012092

then drawn up by blood to calculate the leukocyte image and the final stage of histopathologic preparation of the infected organs.

## 3. Results And Discussions

### *3.1. Prevalence, degree of infestation and leukocyte features*

The results covering prevalence are presented in table 1. We can see that more than 80% are classified as always there [5], with the highest degree of infestation in koi fish and the highest percentage of lymphocytes that is more than 88 %.

**Table 1.** Prevalence, degree of infestation and overview of leucocytes of Cyprinidae fish.

| Infested *Argulus japonicus* Fish Species | Prevalence | Infestation Level | Lymphocyte | Neutrop hil | Monocyte | Eosinophils | Basophil |
|---|---|---|---|---|---|---|---|
| Koi | 65 | medium | 89.5 | 4.9 | 3.98 | 1.45 | 0.17 |
| Goldfish | 57 | heavy | 88.3 | 5.95 | 3.9 | 1.70 | 0.15 |
| Fish comet | 31 | medium | 88.9 | 5.1 | 3.6 | 1.56 | 0.14 |

The highest prevalence was found in koi fish hosts, followed by goldfish and comet fish. Among the three species of ornamental fish, koi fish is proved to be the most susceptible to environmental changes and to *Argulus japonicus* ectoparasite, although the degree of infestation is moderate. In goldfish, the prevalence is lower but the degree of infestation is severe so that the impact is similar or even more severe. The comet fish was the most resistant to changes in aquatic environments and had the lowest prevalence value with a medium degree of infestation.

### *3.2. Pathological changes*

The pathological changes seen were: congestion, ballooning degeneration, epithelial erosion, and inflammatory cell infiltration. An *Argulus japonicas* infestation causes a pathological change of congestion, indicated by the presence of thickened blood vessels with a darker red color due to the accumulation of blood cells and forming special patterns such as circles. In the change of ballooning degeneration, the cell looks enlarged and there is an empty space inside just like balloon cells. Erosion of the epithelium is indicated by the erosion of soft tissue on the pectoral fins [6].

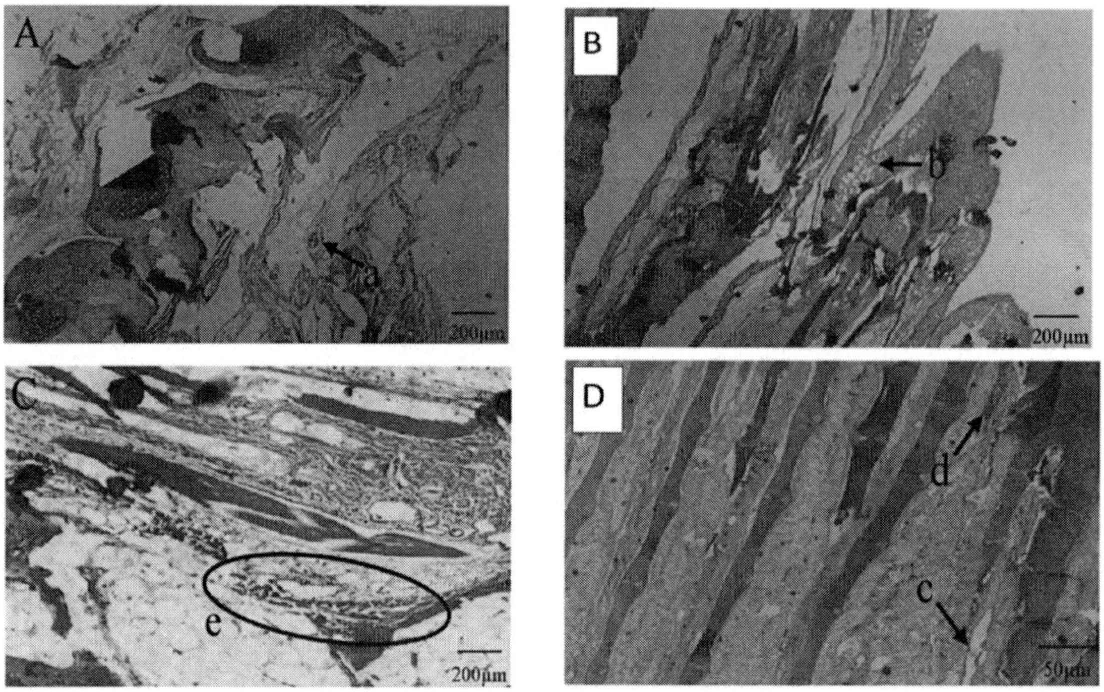


**Figure 1.** The histopathological features of koi fish infested by *Argulus (*A. pectoral final organs in mild infestations (100 x magnification); B. caudal fin organs on moderate infestation (magnification 40x); C. skin organ on mild infestations (100 x magnification); a. congestion; b. balloning degeneration; c. erosion of the epithelium).

## 4. Conclusion

As a conclusion, the highest prevalence was found in koi fish, this fish is most favored by *Argulus japonicas.* The most vulnerable host is the goldfish, with an average degree of infestation. The highest percentage of leukocytes is lymphocytes (more than 80 %).The dominant histopathological change due to the infestation *of Argulus* japonius is congestion.

## 5. References

[1]. Kabata Z 1984 *Parasite disease of fish culture in the tropics* (London : Taylor and Francis) 263 p
[2]. Kismiyati 2009 Ectoparasitic infestation (*Argulus japonicus*) on mascokoki fish (*Carassius auratus*) and efforts of fish control with fish of Sumatra (Surabaya: Airlangga University)
[3]. Prasetya N S, Subekti and Kismiyati 2013 *JIPK* **5** 113-116
[4]. Yildiz K and Kumantas A 2002 *Israel J. Vet. Med.* **57** 118-120
[5]. Williams E H and Williams L B 1996 Parasites of Offshore Big Gami Fishes of Puerio Rico and The Western Atlantic Puerio Rico Department of Natural and Environmental Resources. (San Juan : University of Puerio Rico) 383
[6]. Al-Darwesh A A, Al-Shabbani M A A and Faris B H 2014 *Global J. Bio-Sci Biotechnol.* **3** 384-387

IOP Conf. Series: Earth and Environmental Science **137** (2018) 012093 doi:10.1088/1755-1315/137/1/012093

# The current situation and environmental conditions of green mussel farming in the gulf of Thailand

**T Keawtawee[1], P Songsangjinda[2], Y Sangnoi[1] and A Uppabullung[1]**

[1]Department of Aquatic Science, Faculty of Natural resources, Prince of Songkla University, Songkhla, Thailand .
[2] Coastal Aquaculture Research and Development Division, Department of Fisheries, Kaset-Klang,Chatuchak, Bangkok, Thailand.

E-mail: teeyaporn.k@psu.ac.th

**Abstract.** This study investigated the current situation and environmental conditions of green mussel *Perna viridis* farming in Surat Thani, Phetchaburi and Samut Prakan provinces where were located in the gulf of Thailand. Samples were collected during April, August and November in 2014. Based on the results of this study, there were three major types of mussel culture patterns in Thailand: bouchot culture, longline culture (rope culture), and raft culture. Mussel farmers encountered problems on decreasing of larvae mussel in the water column, water quality degradation, high production cost and farmer's lack of negotiation power to control the price. The water quality of the mussel farms, such as dissolved oxygen (DO), pH, salinity, electrical conductivity (EC), fecal coliforms bacteria and the total coliforms bacteria met the standard for aquaculture. However, the nutrient, such as ammonium ($NH_4$-N), nitri-nitrogen ($NO_2$-N), nitrate-nitrogen ($NO_3$-N) and soluble reactive phosphorus ($PO_4$-P) were higher than the value of marine aquaculture standard. The results of the contamination of *Salmonella* spp. bacteria were not found in the mussel's tissue, however *Escherichia* coli was found at very low level.

## 1. Introduction

Green mussel is widely distributed in the coastal waters of the Asia-Pacific region [1]. It is one of the most valuable mariculture organisms produced with the largest quantity and profitable products for the country [2,3] . The culture of green mussel has been carried out in Thailand for more than 60 years. Green mussels aquaculture farms are located at intertidal areas and they are usually close to estuaries. Therefore, the chance of exposure to many contaminants from land-based activities through the riverine system as well as sea-based sources is high [4]. The aim of this study is to investigate the current situation and environmental condition in the mussel farming in the gulf of Thailand.

## 2. Methodology

The thirteen green mussel farms were from three provinces: Surat Thani, Phetchaburi and Samut Prakan provinces, located in the gulf of Thailand. Research samples were collected in April, August and November in 2014. Water quality, including dissolved oxygen (DO), pH, salinity and electrical conductivity (EC) was measured in the field by YSI multi-probe meter model Horiba U-52G. Water samples were collected using water sampler and put into plastic bottles and an ice compartment before being transported to the laboratory for further analysis. In the laboratory, the water samples were

IOP Conf. Series: Earth and Environmental Science **137** (2018) 012093 doi:10.1088/1755-1315/137/1/012093

filtered with GF/C filter paper for total ammonia (TAN), nitrite nitrogen ($NO_2$-N), nitrate nitrogen ($NO_3$-N) and soluble reactive phosphorus ($PO_4$-P) analysis. The green mussels were washed externally with a brush and distilled water to minimize contamination of tissue. Then, they were immediately put into a plastic bag and an ice compartment before transportation to the laboratory for further analysis. In the laboratory, the samples were stored frozen until the analysis was undertaken. Before dissection, the mussel samples were thawed at a room temperature (28 ± 1°C) with the posterior margin facing downwards in order to allow excess water to drain away. About 25-30 mussels from each sampling site were selected and dissected by removing the byssus and the shell before analyses. Contamination of the total coliforms and fecal coliforms bacteria in sea water and *Escherichia coli* and *Salmonella* spp. bacteria in mussel's tissue was investigated by using most probable number method (MPN Method).

## 3. Results and Discussions

There are three major types of green mussel culture patterns in Thailand: bouchot or bamboo poles culture (figure 1a-b), longline culture or rope culture (figure 1c-d) and raft culture (figure 1e). Thai mussel farmers encountered problems in decreasing larvae mussel in the water column, problems of water quality degradation, high production cost and farmers' lack of negotiation power to control the price.

**Figure 1.** Types of green mussel culture patterns in Thailand. Bouchot or bamboo poles culture (figure 1a-b), longline culture or rope culture (figure 1c-d) and raft culture (figure 1e).

The results of water quality in green mussel farms including dissolved oxygen (DO), pH, salinity and electrical conductivity (EC) were 6.30-9.80 mg/L, 6.79-8.88, 20.0-31.5 ppt and 34.2-50.7 ms/cm, respectively. The range of TAN concentration in April, August and November were 0.68-1.61, 1.32-13.88 and 1.80-6.25 mgN/L, respectively (table 1). The highest average of $NO_2$-N and $NO_3$-N concentrations were in August (1.16 mgN/L) and in April (1.50 mgN/L) at Samut Prakan province. The results showed that almost all nutrient concentrations were higher than the value of marine aquaculture standard in Thailand (table 1). The total coliforms bacteria concentrations in sea water were lower than 100 MPN/100 mL in all stations except in August at Phetchaburi and Samut prakan, which were 1600 and higher than 16000 MPN/mL, respectively. The concentration of fecal coliforms bacteria in the sea water samples were lower than the value of marine aquaculture standard in Thailand (70 CFU/mL), in all stations (table 2). *Escherichia coli* and *Salmonella* spp. bacteria were not found in the mussel's tissue and the level in the mussel's tissue from all stations was ver low (table 2). These results showed that green mussel were safe from fecal coliforms, *Salmonella* spp. and *E.coli* bacteria.

**Table 1.** Water qualities in sea water of green mussel farms in Surat Thani, Phetchaburi and Samut Prakan during April, August and November, 2014.

| Water qualities | | Surat Thani | Phetchaburi | Samut Prakan |
|---|---|---|---|---|
| Total Ammonia | April | 0.81-1.01 | 0.79-1.43 | 0.68-1.61 |
| (mgN/L) | August | 1.32-1.52 | 2.48-4.76 | 4.69-13.88 |
| STD: 0.1 mgN/L | November | 1.80-2.20 | 3.23-4.37 | 4.25-6.25 |
| NO2.-N | April | 0.11-0.13 | 0.09-0.15 | 0.08-0.11 |
| (mgN/L) | August | 0.02-0.04 | <0.002-0.004 | 0.49-1.16 |
| STD: 0.05 mgN/lL | November | <0.002-0.004 | <0.002-0.002 | <0.002-0.036 |
| NO3-N | April | 0.53-0.83 | 0.39-0.83 | 0.26-1.50 |
| (mgN/L) | August | 0.08-1.03 | 0.73-1.45 | <0.02-0.65 |
| STD: 0.06 mgN/L | November | <0.20 | <0.5 | <0.20-0.23 |
| PO4-P | April | <0.05 | <0.05-0.07 | 0.07-0.08 |
| (mgP/L) | August | <0.03 | 0.05-0.12 | <0.03-0.15 |
| STD: 0.045 mgN/L | November | 0.09-0.11 | <0.05-0.07 | <0.05-0.13 |
| TSS | April | 15.45-30.85 | 27.70-111.80 | 20.55-64.30 |
| (mg/L) | August | 24.80-104.85 | 38.75-79.63 | 16.15-138.00 |
| | November | 15.80-37.33 | 18.25-35.80 | 17.25-22.47 |

STD: Standard value of marine aquaculture standard in Thailand.

**Table 2.** Bacterial contamination in sea water and green mussel's tissue in Surat Thani, Phetchaburi and Samut Prakan during April, August and November, 2014.

| Bacterial contamination | | Surat Thani | Phetchaburi | Samut Prakan |
|---|---|---|---|---|
| Total coliforms bacteria in sea water (MPN/100ml) | April | <1.1 | <1.1 | <1.1 |
| | August | <1.8 | <1.8-1600 | <1.8->16000 |
| STD: 100 MPN/100 ml | November | 17-70 | <1.8-23 | <1.8-20 |
| Fecal coliforms bacteria in sea | April | <1.8 | <1.8-6.8 | <1.8-4.0 |
| water (CFU/ml) | August | ND | <1.8-23 | <1.8 |
| STD: 70 CFU/ml | November | 4.5-13 | <1.8-4.5 | <1.8 |
| *Samonella* spp. in mussel's tissue | April | ND | ND | ND |

| | | | | |
|---|---|---|---|---|
| (MPN/25g) | August | ND | ND | ND |
| STD: Not detectable | November | ND | ND | ND |
| *E.coli* in mussel's tissue | April | <3.0 | <3.0-23.0 | <3.0-23.0 |
| (MPN/g) | August | <1.8 | <1.8-2.0 | <1.8 |
| STD: 100MPN/g dry weight | November | 2.0-4.5 | <1.8-1.8 | <1.8 |

ND: Not detectable, STD: Standard value of marine aquaculture standard in Thailand.

In addition, the heavy metals contamination in the green mussel farms should be investigated as the green mussel's aquaculture farms are located in intertidal areas and they are usually close to estuaries. Therefore, the chance of exposure to many contaminants and heavy metals in sea water and mussel's soft tissue is high [5,6,7].

## 4. Conclusions

The results indicate that the concentration of the nutrients in sea water was high. However, green mussel farming in the gulf of Thailand is safe from fecal coliforms, *Salmonella* spp. and *E.coli* bacteria.

## 5. References

[1] Tanabe S 2000. *Poll. Bull.* **40** 651.
[2] Saraya A 1982 *Bivalve culture in Asia and the Pacific country reports: Thailand in Proc* (Singapore : Asia Pasific workshop)
[3] Mccoy, EW and Chongpeepain T 1988 Bivalve mollusc culture research in Thailand (Philippines : ICLARM Technical Reports 19 International Center for living Aquatic Resources Management Manila)
[4] Yap C K, Ismail A and Tan S G 2004 *Food Chemistry* **84** 569–575
[5] Goldberg E D, Bowen V T, Farrington J W, Harvey G, Marti J H, Parker P L, Risebrough R W, Robertson W, Schneider W and Gamble E 1978 *Envi. Conserv.* **5** 101–125
[6] Gorell JM, Johnson CC, Rybicki BA, Peterson EL, Kortsha GX, Brown GG and Richardson RJ 1997 *Neurology* **48** 650–658
[7] Grienke U, Silke J and Tasdemir D 2014 *Food Chem.* **142** 48–60

## Acknowledgments

This study was funded by the Office of Standard Development, National Bureau of Agricultural Commodity and Food Standards (ACFS). The writer would like to thank the staff of the Department of Fisheries and the owner of the green mussel Farms for their valuable help during the fieldwork.

IOP Conf. Series: Earth and Environmental Science **137** (2018) 012094 doi:10.1088/1755-1315/137/1/012094

# The Prevalence of *Cryptocaryon irritans* in wild marine ornamental fish from Vietnam

**K V Van and D T Nhinh**

Department of Aquatic Environment and Fish Pathology, Faculty of Fisheries, Vietnam National University of Agriculture. Hanoi, Vietnam

E-mail: kvvan@vnua.edu.vn

**Abstract.** Wild-caught ornamental marine fish from NhaTrang (Khanh Hoa) and Ha Long (Quang Ninh) in Vietnam were examined over the three sampling batches corresponding to the spring, summer and autumn times for the prevalence of *Cryptocaryon irritans*. Out of a total of 211 fish (15 species), 143 (67.7 %) were found to be infected with the mean intensity of 7.67 parasites per field of view (x4 magnification). The prevalence of *C.irritans* in fish caught during the spring (91.0 3% on average) was significantly higher than that of the fish caught during the summer (39.29 %). A wide variation in the prevalence of the parasite was shown among the fish species. The highest prevalence and intensity of the infection occurred in *Plataxteira, Diodon holocanthus, Paracanthurus hepatus*at 100 % of infection and density of 12 parasites/field of view (x4 magnification) while and the lowest prevalence of *C. irritans* appeared on *Rhinecanthus aculeatus, Zancluscornutus, and Zebrasoma veliferum* with less than 50% of fish infected. Clinical signs of fish infected of *C.irritans* showed such as tiny white spots on skin, gills, and fins; ragged fins, changes in skin colour, cloudy eyes and increase mucus production.

## 1. Introduction

Ornamental fishes are receiving an increased attention due to the growth of local and global demand and the consequent bloom of small and medium commercial aquariums and marine museums for leisure activities in Vietnam. A vast number of aquatic species have been successfully adapted to live in the aquarium. Marine ornamental fishes aremore colorful and shapely. These fish groups are, therefore, more attractive to aquarists. However, ornamental fishes kept in aquariums are susceptible to numerous diseases particularly parasitic diseases. *Cryptocaryon irritans* infections in fish are a significant disease problem for marine aquarists and commercial marine culture worldwide.

*Cryptocaryon irritans*, a ciliated protozoan parasite, is one of the most devastating parasites of marine fish cultured in temperate and tropical areas.This parasite is an obligate ectoparasite that causes Cryptocaryonosis, also known as the white spot disease in marine fish [1]. Although *C. irritans* is commonly found in tropical, subtropical and warm temperate waters at low infection intensity [2], infection by this parasite has emerged as a major problem in confined environments, such as in marine culture and aquariums due to the buildup of the parasite and high population density of fish in these systems [3].

*Cryptocaryon irritans* has a direct life cycle [4]. This means infection can spread within a group of

IOP Conf. Series: Earth and Environmental Science **137** (2018) 012094 doi:10.1088/1755-1315/137/1/012094

fish without the need for another animal host for development. However, the life cycle is complex. It includes 4 stages that develop on and off the fish, which is briefly described by Yanong [5] as follows: The trophont is the "feeding stage," during which the parasite is found embedded within the tissues of the fish. After the trophont leaves the fish, it becomes a protomont before encysting and transforming into a to mont, or "reproductive stage." The tomont develops and divides into numerous to mites, which eventually leave the cyst as the ronts, the free-swimming infective stages.Theronts actively seek fish hosts for a new life cycle.

The length of the entire life cycle varies, depending upon a number of factors, including the strain of *Cryptocaryon,* temperature, and fish host [4, 6-9]. Even for a specific strain of the parasite and fish host, the life cycle may vary by weeks or months [4]. An average life cycle appears to be 1 to 2 weeks. However, life cycle durations may range from 6 days to 11 weeks, primarily because of the unpredictability of tomont development [4, 9, 10].

The most commonly observed stage is the trophont or "feeding" stage. The trophont is found on the fish, usually underneath the outer skin layers. Spherical or pear-shaped, with cilia all over its body, the trophont will normally be seen "rolling" or rotating slowly under the epithelium. Because the trophont is embedded within the skin, it is relatively protected from any potential treatments [5].Trophonts can range in size from about 48 x 27 μmto 452 x 360 μm. The trophont feeds on the body fluids and cells of the fish for about 3–7 days before leaving the host. Trophonts will also actively leave fish that have died, but are not immediately infective. They require additional time to develop from protomonts to tomonts, just as they would if leaving a live host [5].

Although numerous studies have been conducted on various aspects of *C.irritans,* information on infections related to this parasite on marine ornamental fish in Vietnam is very limited. The report aims to present the current results and updated information on the infection of *C.irritans* on wild-caught marine ornamental fish in Vietnam. The study was conducted at Vinpearlland Aquarium Times City,which is one of the largest commercial marine aquaria in Vietnam.

## 2. Methodology

The study was conducted from 2/2016 to 12/2016 through three sampling batches corresponding to the times of the spring, summer and autumn of the year. In total 211 fish of 15 marine ornamental fish species captured from Ha Long and Nha Trang were examined (table 1). The fish were transported by closed system oxygen packed polythene bags [11]. After transportation, the fish were held in separated recovery tanks to examine the infection of *Cryptocaryon irritans* according to the method described by Dogiel (1929) cited by Ha Ky and Te [12]. During this acclimation period, infected fish were also visually observed in terms ofexternal signs and behaviour changes. To inspect the presence of the parasites, mucus samplesfrom gills and skin of the fish were taken and then smeared onto clean microscope slides to observe under the microscope with magnification powers of x4, x10, x40 and x100.

Based on the collected data, the prevalence and intensity of infection by seasons and fish specieswere calculated as follows:

Prevalence of infection = number of infected fish/total number of fish inspected x 100 %

Infection intensity = total number of parasites on fields(4x10)/total number of fields with the presence of parasite.

The data were collected and analysed on Excel 2010.

**Table 1**. Summary information on fish species, sample batches, sample size and sources of fish used in the study. The first, second and the third sample batches were in the spring, summer and autumn respectively. The fish used in the study were captured from coastal areas of Nha Trang (Khanh Hoa) and Ha Long (Quang Ninh).

| No | English name | Scientific name | Sampling batch (number of sampled fish) | | | Sample size | Sampling sites |
|---|---|---|---|---|---|---|---|
| | | | 1 | 2 | 3 | | |
| 1 | Picasso Triggerfish | *Rhinecanthus aculeatus* | 5 | 5 | 5 | 15 | |
| 2 | Spotted Parrotfish | *Cetoscarus ocellatus* | 5 | 4 | 5 | 14 | |
| 3 | Philippine Butterflyfish | *Chaetodon adiergastos* | 6 | 6 | 6 | 18 | NhaTrang-KhanhHoa |
| 4 | Blue Surgeonfish | *Paracanthurus hepatus* | 4 | 0 | 4 | 8 | |
| 5 | Tomato Clownfish | *Amphiprion frenatus* | 5 | 5 | 6 | 16 | |
| 6 | Long-spine porcupinefish | *Diodon holocanthus* | 3 | 0 | 4 | 7 | |
| 7 | Teira batfish | *Plataxteira* | 6 | 5 | 5 | 16 | |
| 8 | Domino damsel | *Dascyllus trimaculatus* | 6 | 5 | 6 | 17 | |
| 9 | Lionfish | *Pteroisvolitans* | 4 | 0 | 3 | 7 | |
| 10 | Oriental Sweetlips | *Plectorhinchus vittatus* | 5 | 5 | 5 | 15 | |
| 11 | Golden Rabbitfish | *Siganus guttatus* | 6 | 6 | 6 | 18 | QuangNinh |
| 12 | Sailfin tang | *Zebrasoma veliferum* | 4 | 3 | 4 | 11 | NhaTrang |
| 13 | Moorish Idol | *Zanclus cornutus* | 5 | 4 | 5 | 14 | KhanhHoa |
| 14 | Royal angelfish | *Pygoplites diacanthus* | 4 | 0 | 3 | 7 | |
| 15 | Common Clownfish | *Amphiprion ocellaris* | 10 | 8 | 10 | 28 | |
| | **Total** | | **78** | **56** | **77** | **211** | |

**3. Results and discussion**

In the study, a total of 211 marine ornamental fish of 15 species were sampled at 2 – 3 sampling batches corresponding to the spring, summer and autumn time. Three to ten fish individuals from each fish species were sampled for each sampling batch. Most of the fish used in the study were captured from NhaTrang – KhanhHoa (almost 90 %); other small numbers (10 %) were sourced from Ha Long – QuangNinh (table 1).

The prevalence of *Cryptocaryon irritans* infections ranged between 33.3 and 100 % for the 15 fish species examined with the overall prevalence of 66.8 % (table 2).Previousresearch by Diggles and Lester [7] has reported the similar prevalence (66.7%) of *C. irritans* infection on marine wild caught fish in Australia. However, a much lower prevalence (22.44 %) of freshwater "Ich" infecting on juvenile grass carp has been found by Van and Tho *et al.* [11].

**Table 2.**Summary data from 211 fish of 15 species sampled at three sampling batches from NhaTrang and Ha Long to examine for the infection of *C. irritans*. The data of fish length and mean prevalence are present as mean ± SE.

| STT | Fish species | No of Fish sampled | Mean Fish length (cm) | No of Fish infected *C. irritans* | Prevalence (%) |
|---|---|---|---|---|---|
| 1 | *Rhinecanthus aculeatus* | 15 | 15.2 ± 0.63 | 5 | 33.33 |
| 2 | *Cetoscarus ocellatus* | 14 | 13.1 ± 0.54 | 8 | 57.14 |
| 3 | *Chaetodon adiergastos* | 18 | 7.9 ± 0.34 | 15 | 83.33 |
| 4 | *Paracanthurus hepatus* | 8 | 8.4 ± 0.49 | 6 | 75.0 |
| 5 | *Amphiprion frenatus* | 16 | 8.9 ± 1.03 | 12 | 75.0 |
| 6 | *Diodon holocanthus* | 7 | 13.5 ± 2.57 | 6 | 85.71 |
| 7 | *Plataxteira* | 16 | 12.3 ± 1.35 | 13 | 81.25 |
| 8 | *Dascyllus trimaculatus* | 17 | 8.4 ± 0.68 | 10 | 58.82 |
| 9 | *Pterois volitans* | 7 | 11.2 ± 1.14 | 7 | 100 |
| 10 | *Plectorhinchus vittatus* | 15 | 12.0 ± 0.59 | 12 | 80.0 |
| 11 | *Siganus guttatus* | 18 | 7.8 ± 0.35 | 15 | 83.33 |
| 12 | *Zebrasoma veliferum* | 11 | 8.8 ± 0.53 | 5 | 45.45 |
| 13 | *Zanclus cornutus* | 14 | 7.8 ± 0.42 | 6 | 42.86 |
| 14 | *Pygoplites diacanthus* | 7 | 9.8 ± 0.63 | 5 | 71.43 |
| 15 | *Amphiprion ocellaris* | 28 | 5.8 ± 0.23 | 18 | 64.29 |
| Total/mean | | 211 | | 143 | 67.77±7.23 |

Wide variationsof the prevalence of *C. irritans* infections in fish were shown among the sampling batches for each fish species (Table 3). Fish caught in the spring infected by *C.irritans*are at the highest prevalence of 91.03% on average with various fish species infected at 100% of prevalence such as *C. adiergastos*, *P. hepatus*, *A. frenatus*, *P. volitans*. The lower prevalence of *C. irritans* infections was observed in the fish captured during the autumn (64.94 %), while the lowest value occurredon the fish captured during the summer (39.29 %). According to Yoshinaga [14], who conducted a study on the impacts of temperature and dissolved oxygen levels on the development of *C. irritans*, temperatures for the optimal growth of most strains of *Crypto caryon* are between 23 and 30 °C [10, 14]. The development of the parasite was significantly damaged at the temperature above 31$^0$C and no parasite was recovered from any fish reared at 34$^0$C. These could be the reasonsrelated to the variation of the higher prevalence of parasite infections on fish captured in the spring compared to the summer. A similar trend of growth was also reported on *I. multifiis*, a freshwater parasite by Van According to Duijn [15], and on *Centrocestus formosanus,* a trematode parasite infectingcommon carp [11].

There was a significant difference in the prevalence of *C. irritans* infection among the fish species from the same sampling batch. The extremely high prevalences of infection on the fish groups of lionfish and long-spine porcupinefish were observed at 85.56 – 100% while the other groups of fish such as *Rhinecanthus aculeatus, Zebrasoma veliferum and Zanclus cornutus* were infected with*C. irritans*at less than 50 % of prevalence. The variation of the prevalence of infections among species could be due to the difference in susceptibility of the fish to the parasites [4,6,16,]. The same results were also reported by Diggles and Lester [7] when these authors examined the infection of this parasite on wild-caught fish from Australia. Differences in the susceptibility of wild hosts may be due to unequal exposure of hosts to infective stages because of differences in host behaviour or ecological niche [7].

Regarding infection intensity, the present results show the mean intensity of *C. irritans* on ornamental fish, which is 7.64 parasites per field of view (4x10 magnification) (table 3).The highest mean intensity was at 2.20 ± 2.27 parasites per field of view on *P. teira*and the lowest intensity at 5.67

IOP Conf. Series: Earth and Environmental Science **137** (2018) 012094 doi:10.1088/1755-1315/137/1/012094

± 1.09 parasites per field of view on *D. trimaculatus*. Aprevious study done by Diggles and Lester [7] reported a much lower intensity of *C.irritans* at 12.9 parasites per infected fish on wild fish. Another study conducted by Van et al. Accprding to Van *et al.* [13], on infection of *Ichthyophthirius multifiliis*on juvenile of grass carp informed an infection intensity of6.68 parasites per field of view.

**Table 3**. Prevalence of *C.irritans* infection by seasons and fish species and mean intensity of infection calculated for each species. The symbol "-" means no sample. The data on mean intensity and prevalence are presented as mean ± SE. The column with different letters indicates a significant difference.

| No | Fish species | Prevalence (%) | | | Mean Intensity |
|---|---|---|---|---|---|
| | | Spring | Summer | Winter | |
| 1 | *R. aculeatus* | 60 | *0* | 40 | 6.80 ± 2.39 |
| 2 | *C. ocellatus* | 80 | *25* | 60 | 7.71 ± 1.92 |
| 3 | *C. adiergastos* | 100 | *66.7* | 83.3 | 8.90 ± 2.30 |
| 4 | *P. hepatus* | 100 | - | 50 | 10.86 ± 2.03 |
| 5 | *A. frenatus* | 100 | *40* | 83.3 | 6.13 ± 1.51 |
| 6 | *D.holocanthus* | 100 | - | 75 | 10.22 ± 2.29 |
| 7 | *P. teira* | 100 | *60* | 80 | 12.20 ± 2.27 |
| 8 | *D. trimaculatus* | 83.3 | *40* | 50 | 5.67 ± 1.09 |
| 9 | *P. volitans* | 100 | - | 100 | 7.10 ± 1.03 |
| 10 | *P. vittatus* | 100 | *60* | 80 | 8.72 ± 0.34 |
| 11 | *S. guttatus* | 100 | *66.7* | 83.3 | 8.22 ± 1.29 |
| 12 | *Z. veliferum* | 75 | *0* | 50 | 6.20 ± 1.27 |
| 13 | *Z. cornutus* | 60 | *25* | 40 | 7.67 ± 1.09 |
| 14 | *P. diacanthus* | 100 | - | 33.3 | 6.10 ± 1.03 |
| 15 | *A. ocellaris* | 100 | *25* | 60 | 6.78 ± 1.34 |
| Mean | | $91.0^{a}$ ± 8.08 | $39.3^{c}$ ± 12.08 | $64.9^{b}$ ± 10.72 | 7.64 ± 1.34 |

The infected fish with a mean intensity higher than 10 parasites per field of view showed clear white spots on the skin, gills and eyes (figure 1). At the advanced stages of infection, these white spots were visible to naked eyes. Other typical signs were pale fills, ragged fins, changes in skin colour, cloudy eyes and increased mucus production. The behaviours of the infected fish also changed including hanging at the surface or at the sites of incoming water, acting lethargic, scratching (flashing) or swimming abnormally, and rapidly breathing due to the presence of the parasites on gills leading to damages of gill tissue. At this stage, sporadic to mass mortalityof infected fish occurred. Similar typical signs of infected fish were described in various studies on infection of *C.irritans* [4,17].

Under the microscopes, the parasites appeared in a round, spherical or oval shapes, measuring from 180 to 700 μm, depending upon the age of infection and stage of maturity (figure 2). Throughout the body of the infected fish, the parasite emerged within the epidermis of skin and also in the epithelia of gills. Invasions of subepidermal tissues have previously been attributed to advanced stages of the disease when underlying tissues are exposed by erosion and peeling of the epithelium.

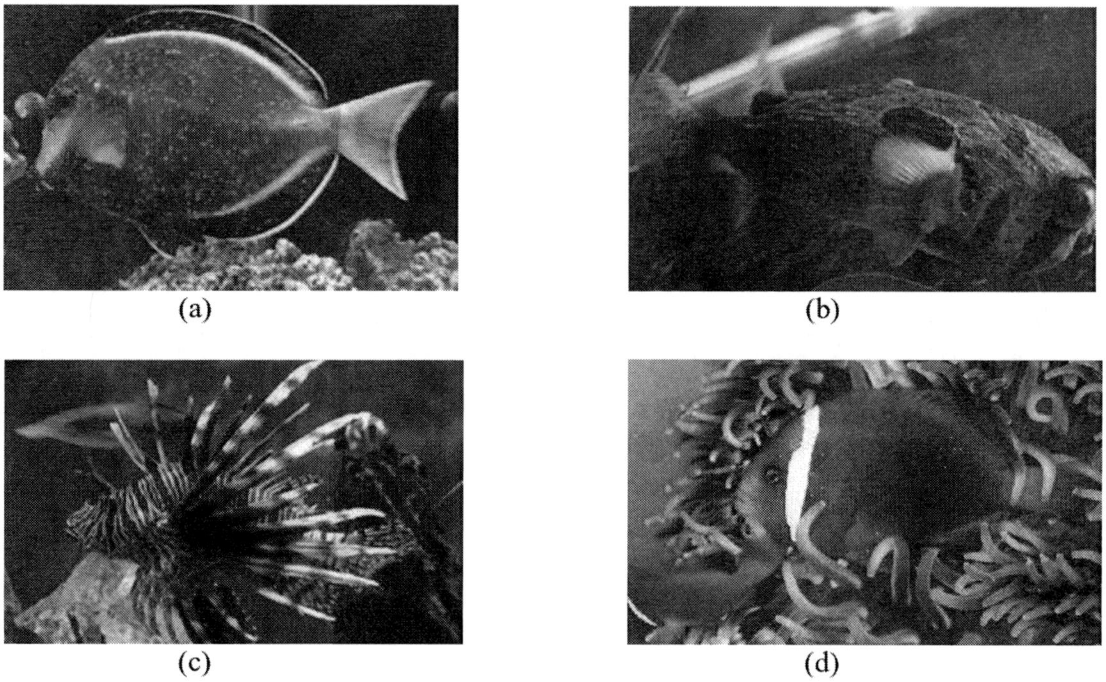


**Figure 1**. White spots appreared on skin and fins of infected fish (a) *P. hepatus; (b) D. holocanthus; (c) P. volitans (d) A. frenatus*

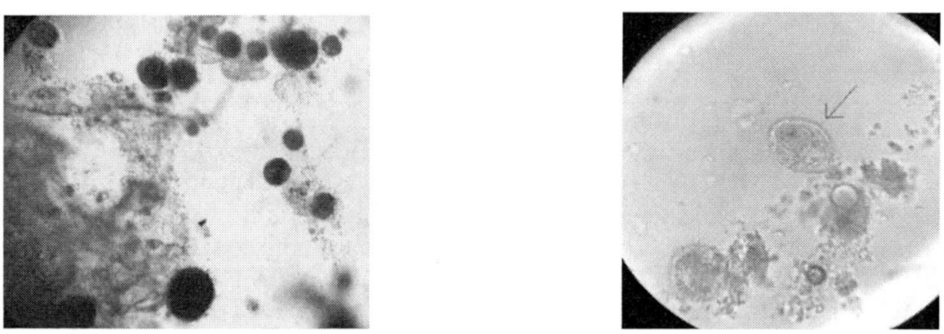

**Figure 2**. *Cryptocaryon irritans* identified by examining the fresh skin and gill tissues of the infected fish under the microscope (x4; x10). The parasites appeared in a round to oval shape and greyish structures due to various life stages of the parasite.

## 4. Conclusion

It is clear that *Cryptocaryon irritans* is common all yearround in ornamental marine fish sampled in Vietnam. However, the prevalence and intensity of the infection vary widely with the fish host and captured seasons. Some species are very susceptible to the parasite infection (up to 100 % infected) while others are infected at a much lower rate (less than 50 % infected). A seasonal variation in the prevalence and intensity of *C. irritans* infectionwas also clearly presented. The highest prevalence of infection occurred during the spring and the lower values were shown in the summer. These findings provide important information which can be employed by aquarists in keeping and treating this parasitefor ornamental fish in a confined condition.

IOP Conf. Series: Earth and Environmental Science **137** (2018) 012094 doi:10.1088/1755-1315/137/1/012094

**5. Referrences**

[1] Colorni A and Burgess P 1977 *Aquar. Sci. Conserv.* **1** 217-238
[2] Wrighta, AG and Colornib, A 2002 *Eur. Jof. Protistology.* **37** 375-378
[3] Yoshinaga T, Dickerson HW 1994 *J. Aquat. Anim. Health.* **6** 197-201
[4] Colorni A and Burgess P 1997 *Aquar. Sci. Conserv.* **1** 217-238
[5] Yanong R P E 2009 *Cryptocaryon irritans* infections (marine white spot disease) in fish. In: UF/IFAS Extension FA p 164
[6] Colorni A 1985 *Dis Aquatic Organism.* **1** 19–22
[7] Diggles B K and Lester R J G 1996 *Dis Aquatic Organism.* **25** 159–167
[8] Diggles B K and Lester R J G 1996 *Jof Parasitol.* **82** 45–51
[9] Yambot A V, Song Y L and Sung H H 2003 *Dis Aquatic Organisms* **54** 147–156
[10] Dickerson H W 2006 *Int. Cambridge MA* **1** 116–153
[11] Van K V and Tho N V 2013 Van K V and Tho N V 2013 *J Veterinary Sci Tech* **20** 74-81
[12] Ha Ky and Te B Q 2007 Vietnam Freshwater Fish. *J. Sci Technol.*212-213
[13] Vạn K V, Tuyen N V, Hoai T D and An N T 2015 *J.Sc. Devel.* **13** 200-205
[14] Yoshinaga T 2001 *Fish. Pathol.* **36** 231–235
[15] Van Duijn C Jr 1967 *Diseases of Fishesed 2* (London: Illiffe Books)
[16] Wilkie D W and Gordin H 1969 *California Fish and Game* **55** 227–236
[17] Noga E J 1996 Fish disease: diagnosis and treatment Mosby-Year Book Inc St Louis MO 367

IOP Conf. Series: Earth and Environmental Science **137** (2018) 012095 doi:10.1088/1755-1315/137/1/012095

# Inventory of the tropical coral reef fishes in Wondama Bay regency, West Papua, Indonesia

**K D P Madiyani[1], J Triastuti[2] and K T Pursetyo[2]**

[1] Aquaculture Study Programe, Faculty of Fisheries and Marine Airlangga University, Jl. Airlangga 60115, Surabaya, East Java, Indonesia.
[2]Department of Marine Affairs, Faculty of Fisheries and Marine Airlangga University, Jl. Airlangga 60115, Surabaya, East Java, Indonesia.

E-mail: jt.astuti@gmail.com

**Abstract.** Teluk Wondama Regency is one of the areas in West Papua that is part of the Cendrawasih Bay National Park due to its potential of coral fishes. Until now, there is no thorough information that gives the potential diversities of coral fishes species can be found in the waters, and one of the attempts to find out is to carry out an inventory. Inventories were conducted from March to May 2016 in Rariei and Auri (Roon District), Numamuran (Duairi District), Nusrowi (Rumberpon District) and Waprak (Roswar District). The results of these inventories found 28 species of coral fishes in Rariei consisting of 8 genus (*Balistapus, Synodus, Pentapodus, Thalassoma, Zebrasoma, Amblyglyphidodon, Chaetodon* and *Labroides*) in Rariei, 11 genus (*Abudefduf, Chaetodon, Amphiprion, Cheilinus, Dascyllus, Chromis, Parastromateus, Priacanthus, Epinephelus, Nemipterus* and *Lutjanus*) in Namamuran, 11 genus in Nusrowi (*Halichoeres, Pomacentrus, Dischistodus, Heniochus, Amblyglyphidodon, Arothron, Parupeneus, Lutjanus, Plectropomus, Epinephelus* and *Acanthurus*), 4 genus in the waters of Waprak (*Parupeneus, Pomacentrus, Chaetodon* and *Amblyglyphidodon*) and 10 genus in Auri (*Lutjanus, Pomacentrus, Amphiprion, Chaetodon, Chromis, Dascyllus, Zanclus, Halichoeres, Siganus* and*Amblyglyphidodon)*

## 1. Introduction

Most of the regions in Indonesia are an ocean which has potential of marine resources with the diversity of biological and other resources that need to be developed. The coastal area, which is a transitional region for terrestrial and marine ecosystem, has enormous potential for natural resources, especially in the three main ecosystems which are the mangrove ecosystem, coral reef, and seagrass. Those three ecosystems have a mutual supportive role for the entirety of each ecosystem [1]. According to Rani *et al.* [2] stated that reef fish is one of animals which are associated with coral reef; the existence is striking and can be found in coral reef environment.

One of the regions that hold the potential of reef fish is Teluk Wondama which is a part of the Teluk Cendrawasih National Park (TNTC). According to Marasabessy [3], reef fish can be consumed by the society. The fish can also be used as ornamental fishes that are traded in and outside the country [4]. Similarity, in Teluk Wondama in Rariei waters, Numamuran waters, Nusrowi waters, and Waprak waters, the majority of the fish is consumed. The location of Auri waters is far from the society, therefore, there is only a little intervention from human activities.

IOP Conf. Series: Earth and Environmental Science **137** (2018) 012095 doi:10.1088/1755-1315/137/1/012095

Currently, there is no scientific information that can show the potential diversity of reef fish species in the marine waters, at Teluk Wondama regency. One of the efforts to find out about the potential of reef fish in those areas by conducting an inventory of reef fish species, thus it can be developed further as a source of optimal and sustainable fisheries and tourism.

## 2. Experimental

### *2.1. Place and date of the research*

The research was conducted in Teluk Wondama regency, in the state expedition activity of the *Negara Kesatuan Republik Indonesia (NKRI) Koridor Papua Barat 2016* which was held by *Komando Pasukan Khusus (Kopassus), TNI*, and Coordinating Ministry of Human Development and Culture in March - May 2016.

### *2.2. Tools and materials*

The tools that used in the research were snorkeling equipment, underwater camera, Global Positioning System (GPS), buoy and stationery. The objects of the research were various types of reef fish found in the marine waters of Teluk Wondama.

### *2.3. Procedures of the research*

#### *2.3.1. Preparation stage*

This stage included getting the license to conduct the research from the this stage under permitted Marine Affairs and Fisheries office at Teluk Wondama as well as collecting information and secondary data about the general condition of research location at Maritime and Fisheries Department of Teluk Wondama regency and World Wildlife Fund (WWF) of Teluk Wondama regency.

#### *2.3.2. Location and sample*

The location of the data collection was in the marine waters of Teluk Wondama regency. To determine the research location, this research used purposive sampling method, which was by observing and considering the conditions of the location of the research [5]. The data collection was adjusted to the travel time and route of the survey vessel. There were five stations selected as the locations of the research. Location of station I was in the marine waters, south of Rariei Island, at Roon district, with the sample coordinate point of S: 02°21'07.5" E: 134°30'55.2". Station II was in the marine waters, south of Numamuran Island, Teluk Duairi district, with the sample coordinate point of S: 01°48'50.3'' E: 134°08'33.6". The location of station III was in the marine waters, south of Nusrowi Island, Rumberpon district, with the sample coordinate point of S: 01°48'50.3'' E: 134°08'33.6". Location of station IV was in marine waters, west of Kampung Waprak Coast, Roswar district, with the sample coordinate point of S: 02°06'25.9"E: 134°19'50.1". Location of station V was in the marine waters, south of Auri Island, Roon district with the sample coordinate point of S: 02°00'23.4" E: 134°41'50.1". All these information can be seen in Figure 1.

IOP Conf. Series: Earth and Environmental Science **137** (2018) 012095 doi:10.1088/1755-1315/137/1/012095

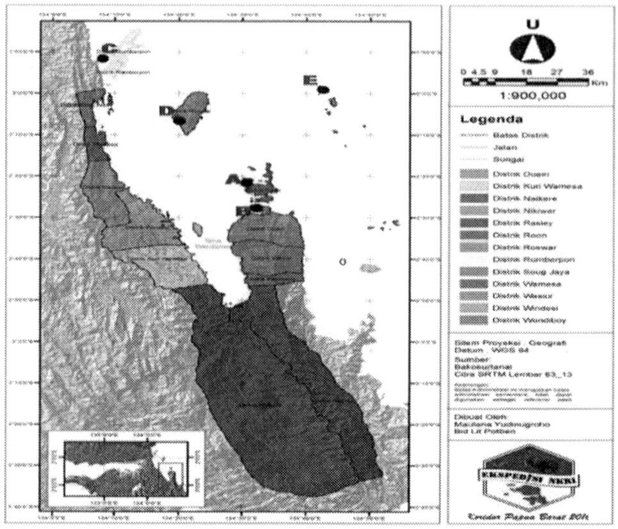


**Figure 1.** Research locations.

### *2.3.3. Data collection*

The data collection used *fish stationary plot survey* as the method of the research. According to Hill and Wilkinson [6] stated that this method is used to estimate the various types of fish. The basic of this method is using an imaginary tube by visually observing and listing all types of fish in the tube. In determining the diameter of the imaginary tube, this research used the length of the vessel. The measured diameter of the tube was ± 8 m following the length of the vessel. The depth itself is limited from the sea level to the bottom of the sea. The area inside the imaginary tube is the area to collect the data of reef fish. The water quality data is obtained from secondary data available at the Office of Marine and Fisheries of Teluk Wondama regency and WWF regency of Teluk Wondama including temperature, pH, salinity, Dissolved Oxygen (DO) and sea current.

### *2.3.4. Implementation of the research*

The research was done by snorkeling in imaginary tube around the vessel and observing the reef fish visually without touching it. During the observation, the collection of types of fish was done by taking several pictures at 1-4 m depth. [7] stated that various species can be found in more than one pictures. Those were necessary to show the difference in colors, markings, and physical features of the same species. This information is also very important when the pictures of the fish obtained have not been identified. The pictures of reef fish that have been obtained then grouped by the types based on their morphological characteristics. Reef fish is identified by referring to books of *Reef Fish Identification Tropical Pacific* [7], *Tropical Reef-Fishes of the Pacific Indonesia and Adjacent Waters* [8], *Fishes of the World Fourth Edition* [9] and *Field Guide to Saltwater Fish* [10].

## 3. Results and Discussion

### *3.1. Results*

### *3.1.1. The Types of Fish*

The reef fish obtained during the research in each station were identified based on their colors, shapes and spots on the body by reffering to the reef fish identification books (up to genus or species). Various types of reef fish that have been identified can be seen in Figure 2. The results of the types of reef fish found in five stations are shown in Table 1.

IOP Conf. Series: Earth and Environmental Science **137** (2018) 012095 doi:10.1088/1755-1315/137/1/012095

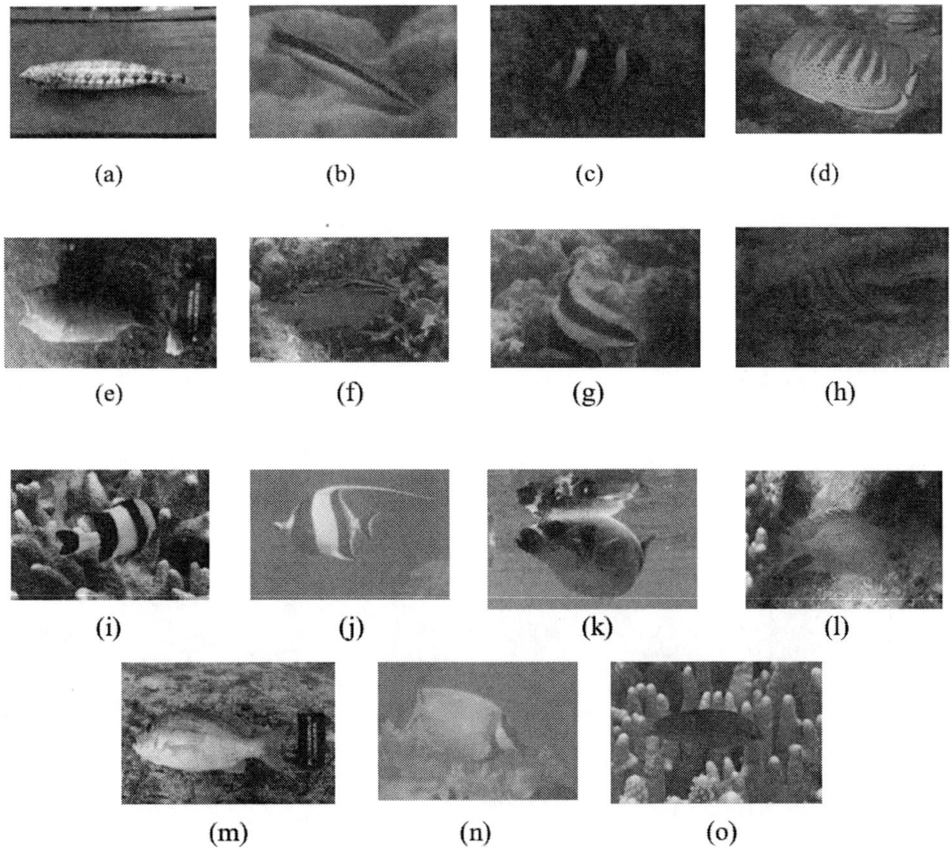


**Figure 2**. The Identified types of reef fish

Description: (a) *Synodus* sp., (b) *Labroides dimidiatus*, (c) *Amphiprion clarkii*, (d) *Chaetodon punctatofasciatus*, (e) *Lutjanus semicinctus*, (f) *Pomacentrus simsiang*, (g) *Heniochus chrysostomus*, (h) *Chaetodon octofasciatus*, (i) *Dascyllus melanurus*, (j) *Zanclus cornutus*, (k) *Arothron nigropunctatus*, (l) *Amblyglyphidodon curacao*, (m) *Lutjanus vita*, (n) *Siganus vulpinis* and (o) *Halichoeres leucurus*.

### *3.1.2. The measurement of the Water Quality*

The water quality data was obtained from secondary data available at the Office of Marine and Fisheries at Teluk Wondama regency and World Wide Fund (WWF) of Teluk Wondama regency. The data of marine waters parameter of Teluk Wondama Regency in each research station are shown in table 2.

**Table 1.** Genus and Species of reef fishes found in each station.

| Waters of Rariei Island | | Waters of the Numamuran Island | | Waters of the Nusrowi Island | | Waters of Kampung Waprak | | Waters of Auri Island | |
|---|---|---|---|---|---|---|---|---|---|
| Genus / Species | ∑ | Genus / Species | ∑ | Genus / Species | ∑ | Genus / Species | ∑ | Genus / Species | ∑ |
| Balistapus | | Abudefduf | | Halichoeres | | Parupeneus | | Lutjanus | |
| *B.undulatus* | | *A. vaigiensis* | | *H. leucurus* | | *P. multifasciatus* | | *L. kasmira* | |
| | 1 | | 1 | | 1 | | 1 | | 1 |
| Synodus | | Chaetodon | | Pomacentrus | | Pomacentrus | | Pomacentrus | |
| | 1 | | 3 | | 1 | | 1 | | 1 |

| | | | | | | | | | |
|---|---|---|---|---|---|---|---|---|---|
| *Synodus*sp. | | *C. baronessa* | | *P. simsiang* | | *Pomacentrus*sp. | | *Pomacentrus sp.* | |
| | | *C. falcula* | | | | | | | |
| | | *C.punctatofasciatus* | | | | | | | |
| Pentapodus | | Amphiprion | | Dischistodus | | Chaetodon | | Amphiprion | |
| *P.trivittatus* | 1 | *A. clarkii* | 1 | *D. perspicillatus* | 1 | *C. octofasciatus* | 1 | *A. ocellaris* | 1 |
| Thalassoma | | Cheilinus | | Heniochus | | Amblyglyphidodon | | Chaetodon | |
| *T.lunare* | 1 | *C. fasciatus* | 1 | *H. chrysostomus* | 1 | *A. curacao* | 1 | *C. lineoatus* | 1 |
| Zebrasoma | | Dascyllus | | Amblyglyphidodon | | | | Chromis | |
| *Z. scopas* | 1 | *D. fasciatus* | 1 | *A . curacao* | 1 | | | *C. viridis* | 1 |
| Amblyglyphidodon | | Chromis | | Arothron | | | | Dascyllus | |
| *A. curacao* | | *C. xanthura* | | *A. nigropunctatus* | | | | *D. fasciatus* | |
| | 1 | *C. viridis* | 2 | | 1 | | | | 1 |
| Chaetodon | | Parastromateus | | Parupeneus | | | | Zanclus | |
| *C.baronessa* | 1 | *Parastromateus*sp. | 1 | *P. bifasciatus* | 1 | | | *Z. cornutus* | 1 |
| Labroides | | Priacanthus | | Lutjanus | | | | Halichoeres | |
| *L. dimidiatus* | 1 | *Priachantus* sp. | 1 | *L. Vita* | 1 | | | *H. leucurus* | 1 |
| | | Epinephelus | | Plectropomus | | | | Siganus | |
| | | *E. areolatus* | | *P. oligacanthus* | | | | *S. vulpinis* | |
| | | | 1 | *P. leopardus* | 2 | | | | 1 |
| | | Nemipterus | | Epinephelus | | | | Amblyglyphidodon | |
| | | *Nemipterus*sp. | 1 | *Epinephelus*sp. | 1 | | | *A. curacao* | 1 |
| | | Lutjanus | | Acanthurus | | | | | |
| | | *L. semicinctus* | 1 | *A. xantopterus* | 1 | | | | |
| Total | 8 | | 14 | | 12 | | 4 | | 10 |

**Table 2.** Waters Quality.

| Parameter | Waters of Rariei Island | Waters of the Numamuran Island | Waters of the Nusrowi Island | Waters of Kampung Waprak | Waters of Auri Island |
|---|---|---|---|---|---|
| Temperature (°C) | 29,5 | 29,5 | 29 | 28,5 | 29 |
| pH | 7 | 7 | 6,5 | 7 | 7 |
| Salinity (ppm) | 29 | 31 | 31 | 31 | 31 |
| DO (ppm) | 7,1 | 8,1 | 8,5 | 8,5 | 9 |
| Current sea (m/det) and current direction (°) | 0,08 (155°) | 0,08 (163°) | 0,06 (295°) | 0,06 (170°) | 0,09 (125°) |

### *3.2. Discussion*

The location of the research was divided into five stations. Those stations were in the shallow coral reef area because it had 1-4 m depth. According to Kartikasari *et al.* [11], shallow coral reef area with 0-4 m depth and characterized by small wave movements is an area that protects and produces high abundance of fish and coral species due to optimal light penetration for coral growth. The fish in the research location had different body shapes and colors. There were various characteristics of fish in each of the five stations, such as bright pattern of body color. According to Connaughey BH and Zottoli R [12], reef fish has a bright color because it adapts to coral reef used as disguises to predator and the color pattern is a form of morphology adaptation to trick its predators (camouflage).

Station I (waters of Rariei Island) has similarity with station II (waters of Numamuran Island). Those areas are strain and closed waters. The closed waters area is an area where the water conditions are not easily affected by the open waters [13]. Both of the waters were directly protected from environmental factors due to the presence of the land. These conditions caused the environment of both areas not to have direct influence from the open sea. Thus, the sea currents those area was quite calm with the speed of 0.08 m/s. According to Nontji [14] stated that current is a flowing motion of water mass with the direction and current strength in the surface layer is greatly determined by wind or caused by long wavy motion. Rariei and Numamuran waters have similar temperature equation of 29.5 °C and a water pH of 7. Those values are the normal limit for reef fish life. According to Pandiangan [15] all types of fish have a low tolerance toward temperature changes. A good temperature range for fish life is between 25-32 °C. According to Souhoka and Patty [16] also stated that the degree of acidity (pH) in marine waters ranged from 6,91-8,18 and this variation of value is still safe for marine life.

Salinity in Rariei waters was 29 ppm and in Numamuran waters it was 31 ppm. However, those values were still tolerable for marine life such as reef fish. According to Nontji in [16] marine organisms have 27-40ppm of salinity tolerance. The lower salinity of Rairei waters is possibly caused by the entry of freshwater that came from the river on the island and Ron Island which was facing with the sample location. Those similar with Wahyuni [17] which stated that naturally high or low salinity is influenced by various factors such as turbulence mixing, evaporation, or the presence of river flow.

IOP Conf. Series: Earth and Environmental Science **137** (2018) 012095 doi:10.1088/1755-1315/137/1/012095

The dissolved oxygen (DO) in Rariei waters was 7.1 ppm and in Numamuran waters it was 8.1 ppm. These DO values could still support reef fish life. According to Nontji [16] stated that dissolved oxygen above 5 ppm is fairly supportive for reef fish life. The DO value in Rariei waters was lower because of household wastewater coming from residents across the location (Roon Island). That household wastewater is a cause of reduced oxygen-producing organisms. According to Salmin [18], the primary source of oxygen in waters comes from a diffusion process of free air and photosynthetic products of organisms living in the marine waters. According to Simanjuntak [19] also stated that one of the factors causing the decrease of dissolved oxygen level is usually the change of water quality mixes with waste containing organic carbon from the land and the waters.

Even though both waters have similar waters condition and geographical area however, they have different types of fish. The genus of Epinephelus and Lutjanus could not be found in Rariei waters because of the intensive fishing activity of those genus. Rariei Island is a sandy coast characterized by sand and rocks that spread out on the coast. Different conditions are shown on the coast of Numamuran Island where mangrove forests are the source of food and shelter. Therefore, the types of fish in the waters of Numamuran Island are more various compared to the Rairei waters. According to [20], mangrove water is an ideal place for any types of fish as their habitation place, food source, and juvenile growth. Related with food, mangrove forest provides organic food for fish which is formed from fallen leaves and various types of invertebrate animals, such as crabs and insects. The mangrove forest is also a place for spawning, permanent habitat, and breeding ground for the fish. Therefore, mangrove forest plays an important role as it provides habitation and protects the fish from predators.

Nusrowi Island is sandy type of beach which is characterized by sand coast and it has no mangrove and its common vegetations are coconut and ketapang. According to Effendi [21] stated that non mangrove vegetation found in coastal area with sand-dominated substrate are for example ketapang, pine, and coconut. Nusrowi waters has clean coral reefs and waters condition due to the location that is far from the society. According to Burke *et al.* [22] stated that the condition of coral reef is strongly influenced by human activities. Excessive human activity can threaten the condition of coral reefs and marine life.

The location of the research is in the south of the island and open sea is influenced from the north of Teluk Cendrawasih. These conditions cause the location of the research not to be affected by the current of the open sea. Thus, it has low sea currents of 0.06 m/s. According to Ruswahyuni [23], the strength of currents and waves in closed waters will decrease when it reaches to the land because the waters are blocked by the mainland or island in front of it. Water temperature in Nusrowi waters was 29 °C. This temperature is generally the average water temperature in the marine waters. According to Nybakken [24], the normal temperature value on sea surface is ranged from 20-30 °C. The pH of the water was 7 that was normal and suitable for the life of sea organism. According to Hutchinson [25] stated that good pH in the waters is 7. The value of DO of 8.5 ppm is considered good for fish life. According to Assyakur dan Wiyanto [26] the higher amount of DO (dissolved oxygen) the better the water quality.

Nusrowi Island has good environmental condition which can potentially become a location for fish cultivation with floating net cage system of *Keramba Jaring Apung Unit Penampungan Ikan Hidup (KJA UPIH)*, especially for grouper fish and lobster. KJA activity can contribute to high number of species of fish in the waters due to the remains of the feed given to the cultivation and consumed by other fish surrounding the area. According to Ondang dan Tumanduk [27], explained in the research that the presence of fish species around KJA showed that the remains of food from KJA carried by sea current were able to attract reef fish [28] also stated that various fish around KJA is related to food hunting, although the food varies. The remaining food that have not been eaten by fish cultivated inside the KJA, will over time come out of it. Then, it will be attract other fish around the cage to get closer and foraging around the *keramba*. In addition, the nutrients from the remaining food in KJA will cause an increase in water fertility. Thus, it will affect the high quantity of phytoplankton in the waters. Phytoplankton is an oxygen producer as well as a source of food for fish.

IOP Conf. Series: Earth and Environmental Science **137** (2018) 012095 doi:10.1088/1755-1315/137/1/012095

Station IV (waters of Roswar Island) is part of the inhabited Waprak Village. The coast is a muddy type with a stretch of coral reef far from the coast. Thus, mangrove vegetation and terrestrial are the main vegetation in this area. It is also supported by current with low speed of 0.06 m/s. According to Nybakken [24], mud sediments in marine waters can only form on the base which has low waves movement or deeper location. Thus, it is not influenced by the waves. Location of the research near the mangrove vegetation caused a lower water temperature of 28.5 °C. According to Hakidah *et al.* [29], the greater percentage of mangrove vegetation closure, the lower the temperature in the water. The presence of mangrove vegetation is helpful in reducing the absorption of light, so the temperature on the surface of the water is not too high.

The location of the research that is near the muddy coast and very close to the village resulting in a lower pH value of the water (6.5), which is good for the growth of marine organisms. According to mardi [30] stated that aquatic organisms generally live, grow and thrive at a pH between 6.5 and 8.5. According to Mahilda [31] stated that changes in pH value can be caused by household waste. This pH value is also influenced by the mangrove vegetation that grows around the coast. According to Yani [32] suggests that the mangrove usually grows in the water pH range between 6 to 8.5. The salinity in the research location was 31 ppm. Acoording to Patty [33] stated that salinity in Indonesian waters generally ranges from 30-35 ppm, the picture of salinity waters illustrates that the magnitude of salinity fluctuations is influenced by several factors, such as water circulation, evaporation, precipitation, and the presence of river flow. The DO (Dissolved Oxygen) water of research location is 8.5 ppm and still supports the life of marine organisms. According to Assyakur dan Wiyanto [26], the general condition of dissolved oxygen in coral waters that can still be tolerated by the marine organism growth ranges from 8.3 to 8.7 ppm. Coral reef in Waprak waters have aquatic colors that tend to get brown because it is affected by the coastal mud sediments. According to Saputra *et al.* [34] stated that mud can cover coral polyps caused by waves and currents which carry mud suspensions.

The types of fish found in the Waprak village waters are very few only 4 genus. It is because the coral reefs is located near the village, where many human activities take place such as fishing, household waste, and access to marine transportation. Waprak waters is also close to coast which has mud substrate. According to Supriharyono [35] stated that coral reefs cannot survive because of sediment that covers its feeding structure. The sediment can also cause the lack of sunlight required for photosynthesis. Thus, causing death to coral which also affects the decrease of reef fish species. Sedimentation and high turbidity levels in the waters will suppress the growth of coral reefs. Coral and reef fish require clear water to get enough sunlight. When the supply of sunlight decreases, coral growth becomes disturbed. Consequently, many marine life such as reef fish that utilize coral reef are decreasing in number [36].

Station V (Auri Island waters, Roon district) is the outermost island located between the clusters of islands in Teluk Cendrawasih. Therefore, the velocity of sea currents is relatively high compared to the other four waters with a speed of 0.09 m/s. Accoording to Ruswahyuni [23] stated that open water is strongly influenced by the waves. The waters facing the open sea without any obstacles, either islands or land. According to Hutomo [37] also added that sea currents from the open water plays an important role in supplying oxygen for coral growth. In addition, the pattern of ocean currents is also very important to carry nutrients, sediment materials and fish larvae.

The coast of Auri Island is a sandy coast and it has a cluster of coral reefs and non-mangrove vegetation such as pine and coconut. According to Efendi [21], non-mangrove vegetation is generally found in coastal areas with sand-dominated substrate. Good water conditions (temperature, pH, salinity, and DO) contribute to the growth of coral reefs. The water condition is minimally affected by human activities because Auri Island and the surrounding islands are unhabitated. According to Darmianti [14] , coral reefs are generally more developed in large undulating areas that can provide oxygen supply for corals. The-waves also provide new plankton for coral colonies that are also the source of food for reef fish. According to Sutono [38] in his research also stated that the fertility of the waters is supported by open water conditions. Thus, the circulation of running water and the coral reef ecosystem can be protected from human activity. Therefore, the coral reef ecosystem could maintain sustainability. The waters of Auri Island have a fairly high number of genus (10 genus). However, it was still lower in number than the Numamuran and Nusrowi waters. The large number of reef fish genus are supported by good water quality that allows good growth of the coral reefs.

Based on the results in each location, it can be seen that more types of reef fish were found in Numamuran Island because it is an area protected from the direct influence of the open sea. There were mangrove ecosystems that support the life of reef fish as spawning areas, feeding areas, and habitation. The less number of fish species were found in the waters of Waprak village. According to Nybakken [24], small waves of shrapnel on the sheltered sea side (leeward), could create a diminished algae bund development and dominant coral development. On the contrary, large wave fractures on the open sea side (windward) create the development of algae borders and minimal coral development. Algae bunds play a role in resisting wave velocity and resulting in calmer water conditions. The waters of Numamuran Island, Nusrowi Island, and Auri Island had more reef fish species compared to Rariei waters and Waprak village waters. This was due to the location of Numamuran, Nusrowi, and Auri waters that are located in coral reefs and away from human activities.

The diversity of reef fish in the waters of Numamuran Island, Nusrowi, and Auri is an illustration of the potential of coral reefs that are preserved due to the minimal intervention of human activities, protected waters, and sufficient supply of food. The Pomacentridae family is the largest fish found in each of the stations, but with different genus. According to Zulfianti [39] also stated that Pomacentridae family fish is the most dominant group on coral reef ecosystem.

## 4. Conclusions and Suggestions

Based on the research of the inventory of reef fish species in marine waters of Teluk Wondama regency, it can be concluded that there were 48 species which includes 28 genus of the fish. They were *Acanthurus, Zebrasoma, Balistapus, Parastromateus, Chaetodon, Heniochus, Cheilinus, Thalassoma, Halichoeres, Labroides, Lutjanus, Parupeneus, Nemipterus, Pentapodus, Abudefduf, Amblyglyphidodon, Amphiprion, Chromis, Dascyllus, Dischistodus, Pomacentrus, Priachantus, Epinephelus, Plectropomus, Siganus, Synodus, Arothron* and *Zanclus*.

It is necessary to conduct further research at different locations in order to complete the data collection on reef fish to provide scientific information on reef fish species found in Teluk Wondama regency. This research results are expected to be spread to the government and Teluk Wondama society to help planning efforts to maintain the marine ecosystem in those area.

## 5. References

[1] Olii A H, Muhlis and Sayuti 2015 *J. Fish Mar.* **20** 49-55 (in Indonesia)
[2] Rani C, Burhanuddin A I and Atjo A A 2010 *J Fish Mar.* 1-14
[3] Marasabessy M D 2010 *J. Oceanol Limnol.* **36** 63-84 (in Indonesia)
[4] Utomo S P R and Supriharyono C 2013 *J.l of Maquares.* **2** 81-90 (in Indonesia)
[5] Sugiyono 2013 *Quantitative Research Methods, Qualitative and R & D* (Bandung:Alfabeta) pp 218 - 219
[6] Hill J and Wilkinson C 2004 Method for Ecological Monitoring of Coral Reefs *Australian Institute of Marine Science.* pp 86-87
[7] Allen G, Steene R, Human P, *et al* 2003 *Reef Fish Identification Tropical Pacific* (Singapore: Star Standard Industries Pte Ltd) pp 1-484
[8] Kuiter R H 1992 *Tropical Reef-Fishes of the Western Pacific Indonesia and Adjacent Waters* (Jakarta: PT Gramedia Pustaka Utama) p 314
[9] Nelson J S 2006 *Fishes of the World Fourth Edition* (USA: John Wiley and Sons) p 601
[10] Schultz K 2004 *Field Guide to Saltwater Fish* (USA: John Willey and Sons) pp 1-285
[11] Kartikasari SN, Marshall A J and Beehler B M 2007 Series of Indonesian Ecology: Ecology of Papua *(https://books.google.co.id. 29 Mei 2017)* pp 335-346
[12] Mc Connaughey BH and Zottoli R 1983 *Introduction to Marine Biology* (London: The C.V Mosby Company) pp 1-410

[13] Dinas Manajemen Lingkungan Pantai Tertutup 2011 Introduction Guide Tital Pollutant Load Control System (TPLCS) *(http://http://www.env.go.jp/18 april 2017)* p 5
[14] Darmianti 2013 Hydrodynamics of Bau - Bau Coastal Waters and Wave Transformation over Natural Coral Reefs *Skripsi Universitas Hasanuddin.* p 89
[15] Pandiangan SL 2009 Study of Coral Reef Diversity in the Western Waters of Rubiah Island Nanggroe Aceh Darussalam (*Skripsi. Unversitas Sumatera Utara. Sumatera Utara*) p 76
[16] Souhoka J and Patty S I 2013 *J. Sci Platax.* **1** 138-147 (in Indonesia)
[17] Wahyuni P 2015 Concentration of Chlorine Waste (Cl2) in Sea Waters Bantam Profit from Electricity Cooling Process Activity Skripsi Institut Petanian Bogor. p 21
[18] Salmin 2005 *Jurnal Oceana.* **30** 21-26
[19] Simanjuntak M 2007 *J. Sci Mar.* **12** 59-66 (in Indonesia)
[20] Noor YR, Khazali M and Suryadiputra I N N 2012 Guide *to Introduction of Mangrove in Indonesia* (Bogor: PHKA/WI-IP) p 13
[21] Efendi E 2012 Coastal Vegetation (http://staff.unila.ac.id/.15 april 2017) p 5
[22] Burke L, E Seligand M Spalding 2012 *Threatened Coral Reefs in Southeast Asia* (Amerika Serikat:World Resources Institute) p 44
[23] Ruswahyuni 2010 Jepara *J. Sci Fish Mar.* **2** 11-20 (in Indonesia)
[24] Nybakken J W 1992 *Marine Biology: An Ecological Approach* (Jakarta: Gramedia Pustaka Utama) p 443
[25] Hutchinson W, Jeffrey M, O'Sullivan D, *et al.* 2004 Recirculating Aquaculture System Minimum Standards for Design, Construction and Management (South Australian Research and Development Institute http://www.fastonline.org/ 15 Juli 2017) p 70
[26] Assyakur A R and Wiyanto 2016 *J. Mar.* **9** 85-92 (in Indonesia)
[27] Ondang H M and Tumanduk N M 2014 Target Fish Associations around Artificial Reefs and Floating Net Cages in the Lembeh Strait of Bitung City (https://www.slideshare.net/ 15 Juli 2017) p 13
[28] Zainudin, Efizon D and Windarti 2015 Stomach Content Analysis of Pristolepis Grooti from the Koto Panjang Dam and Bencah Kelubi Swamp (Artikel. http://jom.unri.ac.id. 12 juni 2017) p 9
[29] Halidah, Qiptiyah M and Anwar C 2007 *Bogor Info for.* **4** 409-417 (in Indonesia)
[30] Mardi 2014 Linkage of Mangrove Vegetation Structure with Acidity and Organic Material of Total Sediment in Mampie Wildlife Reserve Area in Wonomulyo Sub-district Polewali Mandar Regency Skripsi Universitas Hasanuddin Makassar. p 71
[31] Mahida U N 1984 *Water Pollution and Industrial Waste Utilization* (Jakarta: Rajawali) p 543
[32] Yani E 2006 Structure and Ability to Grow Back Mangrove Forest Cikiperan Cilacap (http://journal.bio.unsoed.ac.id/ 17 Juli 2017) p 6
[33] Patty S 2013 *J Sci Platax.* **1** 148-157 (in Indonesia)
[34] Saputra AS, Yuda I P and Zahida F 2008. *Fish Mar.* **12** (in Indonesia)
[35] Supriharyono M S 2000 *Management of Coral Reef Ecosystems* (Jakarta: Djambatan)
[36] Siburian R 2016 *Conservation of Mangrove and Community Welfare* (http://books.google.co.id. 18 april 2017) p 114
[37] Hutomo M 1996 *Proc.* p 24
[38] Sutono D 2016 *Isr J Aquac.* **7** 169-176
[39] Zulfianti 2014 Distribution and Diversity of Coral Fish (Family of Pomacentridae) For Reference Plans of Marine Protected Areas (DPL) on Bonetambung Island Makassar Skripsi. Universitas Hasanuddin. Makassar. P 77

**Acknowledgements**
The present study was supported by expedition activity of the *Negara Kesatuan Republik Indonesia (NKRI) Koridor Papua Barat 2016, Komando Pasukan Khusus (Kopassus), TNI* and Coordinating Ministry of Human Development.

IOP Conf. Series: Earth and Environmental Science **137** (2018) 012096 doi:10.1088/1755-1315/137/1/012096

# A welfare study into capture fisheries in cirata reservoir: a bio-economic model

**Z Anna and P Hindayani**

Department of Fisheries Socio-Economics, Faculty of Fisheries and Marine Science, Universitas Padjadjaran. Jl. Raya Jatinangor Km 21, Jatinangor-Sumedang, Sumedang 40600, West Java, Indonesia. Tel. +62-22-87701519. Fax. +62-22-87701518.

E-mail:suzyanna18@gmail.com

**Abstract.** Capture fishery in inland such as reservoirs can be a source of food security and even the economy and public welfare of the surrounding community. This research was conducted on Cirata reservoir fishery in West Java, to see how far reservoir capture fishery can contribute economically in the form of resource rents. The method used is the bioeconomic model Copes, which can analyze the demand and supply functions to calculate the optimization of stakeholders' welfare in various management regimes. The results showed that the management of capture fishery using Maximum Economic Yield regime (MEY) gave the most efficient result, where fewer inputs would produce maximum profit. In the MEY management, the producer surplus obtained is IDR 2,610.203.099, - per quarter and IDR 273.885.400,- of consumer surplus per quarter. Furthermore, researches showed that sustainable management regime policy MEY result in the government rent/surplus ofIDR 217.891,345, - per quarter with the average price of fish per kg being IDR 13.929. In open access fishery, it was shown that the producer surplus becomesIDR 0. Thus the implementation of the MEY-based instrument policy becomes a necessity for Cirata reservoir capture fishery.

## 1. Introduction

Public waters such as reservoirs have potential for the development of fish resourcesthat can contribute to food security and even the economy and public welfare of localcommunity [1,2,3,4,5,6,7,8]. Fishery activities in reservoirsbecomeincreasingly important in the secondary use of reservoirs, particularly in developing countries, as reservoir fisheries tend to provide a relatively affordable source of animal protein forrural populations [1]. Inland capture fisheries, including reservoir fisheries, generally provide ecosystem services (i.e., fish for food, livelihoods and recreation) for the people; therefore, they have an economic value [9]. Hence, reservoir fisheries play an important role in the economy of a country and the livelihoods of the people living around the reservoirs [8]. Included in the potential of fishery business in reservoirs is capture fishery [10]. Unfortunately, the potential of capture fisheries is pretty much neglected because it is considered as less potentially economic than the fishery businesses having been widely developed in reservoirs, especially floating cage culture fishery.

IOP Conf. Series: Earth and Environmental Science **137** (2018) 012096 doi:10.1088/1755-1315/137/1/012096

Cirata Reservoir is one of the publicwaters in West Java that has thepotential to be used for fisheriesin addition to its use in the hydroelectric power plant (PLTA) unit of Java and Bali. The main fishing activityin this reservoiris floating-cage aquaculture fishery. In addition, there are other fishing activities generally carried out by the local community. Just like any other general problem in developing countries, despite their high potential, capture fiheriesin reservoirs receive less managerialattention [11]. This is also the case in the management of capturefisheriesin CirataReservoir. The lack of management in the reservoir capturefisheriescan be seen from various things, one of which istheir open access, which means that all membersof the society are entitled to use themwithout any restrictionsor regulations. This has ledto uncontrolled exploitation of fish resources, resulting in a drastic reduction in the production of capture fisheries in Cirata Reservoir during the period 2011-2015 [12]. The decrease of capture fishery production in Cirata was allegedly correlated with the decrease of water quality in the reservoir. It hasalso been revealed by[13], who found that there was abnormal growth of goldfish (slowing down)—whichwas unfavorable for the purpose of cultivation-due to the deterioration of the quality of the reservoir water. [14] stated that water condition greatly determines the abundance and spread of organisms in it, but every organism has different environmental needs and preferences for life, which are associated with its environmental characteristics. Thus, the decrease of water quality will cease the diversity of native or endemic fish in Cirata Reservoir, inhibit fish growth, cause diseases and pests for fish, and cause competition for the decreased amount of oxygenin Cirata Reservoir, which can lead to delayed growth of fish. The capture fisheries in Cirata has been inthe overfishing state [12]. This problem can be an indication of uneconomical capture fishery activities in the reservoir as well as low performance in and contribution tothe improvement ofthe welfare of the community, thus it does not receive proper managerialattention. This condition is like an endless loopof problems. In order to direct the attention of policy makers tosustainable fishery management, information on the role of capture fisheries in this case is important forthe economic welfare of beneficiaries as well as the government as a representative of public ownership of fish resources in the reservoir. Such economic information is expected to be one of the important drives for the management to be awareof the purpose of increasing the economic welfare of the stakeholders who utilize fish resources in the reservoir.

This paper describes the performance of capture fishery in Cirata Reservoir by analyzing the welfare of the stakeholders (producers, consumers, and the government) from the surplus they obtained in the actual condition in capture fisheries, i.e. The status quo of no management or business as usual as well as sustainable and optimal regime management with Copes model. In the economics of fisheries management, the analysis of management strategies to maximize resources is a central issue, and the importance of consumer and producer surplus (CS and PS, respectively) as illustrated in [15] is also central in thetotal economic surplus. According to Green in [10], using a surplus approach to measure the benefits of natural resources is an appropriate measure because resource utilization is judged on the best altenative use. Furthermore, in this study, Copes model was also used to analyze optimal fish prices for the welfare of fishermen, consumers, and the government.

## 2. Methodology

The research was conducted in Cirata Reservoir that is located in 3 regencies of West Java, namely: Cianjur, Bandung Barat, and Purwakarta. The data collected consisted of two types, namely primary data and secondary data, both qualitative and quantitative. The primary data were taken using purposive sampling technique. The primary data were obtained from interviews with300 fishermen who used gill nets, fishing lines, fishing nets, and fish pots, and from direct field observation. The data included thedata relatingto capture cost per unit fishing effort, catch per unit fishing effort, length and number of fishing effort units, and fish commodity price.

IOP Conf. Series: Earth and Environmental Science **137** (2018) 012096 doi:10.1088/1755-1315/137/1/012096

The secondary data werethe data relatingto production and time series capture effort. The data wereobtained from several related institutions such as the Office of Marine Affairs and Fisheries of West Java Province, Livestock and Fishery Service of Cianjur Regency, Technical Implementation Unitof Cirata Reservoir, Office of Marine Affairs and Fisheries of Cianjur Regency, Regional Technical Implementation Unit (UPTD) of Cianjur Fishery, and other related institutions.

The first stage before the analysis of capture fishery welfare was the standardization of fishing gear. The standardization of fishing gear was done by considering the variety of fishing gear that was operated in the research area i.e., gill net, fish pots, cast net, fishing line, chart, and other fishing gear.The standardization technique used followed the one developed by King [16], with the following formula:

$$E_{jt} = j_{jt} D_{jt} \tag{1}$$

where

$$j_{jt} = \frac{u_{jt}}{u_{st}} \tag{2}$$

$E_{jt}$ = standardized effort of fishing gear *j* during time period*t*

$D_{jt}$ = fishing days of fishing gear *j* during time period*t*

$j_{jt}$ = fishing power of fishing gear *j* during time period*t*

$u_{jt}$ = catch per unit effort (CPUE) of fishing gear *j* during time period*t*

$u_{st}$ = catch per unit effort (CPUE) of fishing gear being the basis of stardardization

In the analysis of Cirata capture fishery, after the fishing gear was standardized, ananalysis of biological estimation parameters usingSchaefer model wasperformed. The estimation of biological parameters r, K, and q was conducted to find out the maximum sustainable production function (MSY) value with the following formula:

$$h_{MSY} = qKE - \left(\frac{q^2K}{r}\right)E^2 \tag{3}$$

The equation above (3) wassimplified into:

$$h = \alpha E - \beta E^2 \rightarrow \alpha = qK,\ \beta = \frac{q^2K}{r} \tag{4}$$

The parameter values of sustainable production functions for coefficients α and β were estimated in aregression analysis between catch per unit efforts (*h/E*) at various levels of fishing effort.

The catch effort to achieve maximum sustainable production ($E_{MSY}$) was obtained by deriving the above catch effort (E) equation to obtain the following equation:

$$\mathrm{h_{MSY}} = \alpha.\mathrm{E_{MSY}} + \beta.\mathrm{E_{MSY}}^2 \tag{5}$$

$$\frac{\partial h}{\partial E} = \alpha - 2\beta\mathrm{E} = 0 \tag{6}$$

$$E_{MSY} = \left(\frac{\alpha}{2\beta}\right)\ or\ E_{MSY} = \left(\frac{r}{2q}\right) \tag{7}$$

The sustainable catch on MSY or $h_{msy}$ level was:

$$\mathrm{h_{MSY}} = \left(\frac{\alpha^2}{4\beta}\right)\ \text{or}\ \mathrm{h_{MSY}} = \left(\frac{\mathrm{rK}}{4}\right) \tag{8}$$

IOP Conf. Series: Earth and Environmental Science **137** (2018) 012096 doi:10.1088/1755-1315/137/1/012096

Afterobtainingthe two values on this MSY level, the biomass level on MSY level was obtained to be:

$$x_{msy} = \frac{h_{msy}}{qE_{msy}} = \frac{(rK/4)}{q(r/2q)} = \frac{K}{2} \quad (9)$$

Where
$h_{MSY}$ = Maximum sustainable yield (MSY)
$q$ = Catch coefficient
$K$ = Carrying capacity
$r$ = Intrinsic growth rate
$E$ = Catch effort level
$E_{MSY}$ = Catch effort level to reach MSY

After obtaining the sustainable production function (MSY), the optimization analysis of the fishery resource utilization was carried out using the Copes bioeconomic model to determine the bioeconomic balance of the fishery based on supply and demand [17]. In the microeconomic theory employed for the welfare analysis using the supply curve and demand curve, the actual supply curve is part of the marginal cost curve above the price received by the producer.

The demand curve of capture fishery products is strongly influenced by dependent variables such as price and income. The occurring prices are influenced by fish landing sites and other ship variables. In the static model study with changed prices used to determine the optimum level of fishery resource utilization, the price can be written as:

$$P = m - nh \quad (10)$$

The values of the demand function of the capture fisheriesin Cirata for coefficients m and n wereestimated usingregression analysis between fish price and total fish production in Cirata. According to Fauzi[17], to determine afishery supply curve, we must create a long-run marginal cost curve (LRMC) and long-term cost curve (LRAC). The mathematical equations for the LRAC and LRMC curves can be derived fromequation [17]:

$$TC = c\mathrm{E} \quad (11)$$

Information:
TC = Total catch cost
c = Average catch cost
$E$ = Catch effort level

By replacing*effort* (E) with catch (h),the formulabecomes:

$$E_{1.2} = \frac{-\alpha \pm \sqrt{\alpha^2 - 4\beta h}}{-2\beta} \quad (12)$$

With$\alpha^2 > 4\beta h$

The MEY point in the Copes model can be obtained from the intersection ofthe demand curve and the single ownership supply curve represented by the long-run marginal cost curve (LRMC). The solution wasa graphical approach using the software Maple 16 so that $h_{MEY}$could beobtained. To obtain a capture attempt from MEY, then $h_{MEY}$isas follows:

$$E_{MEY1.2} = \frac{-\alpha \pm \sqrt{\alpha^2 - 4\beta h_{MEY}}}{-2\beta} \quad (13)$$

IOP Conf. Series: Earth and Environmental Science **137** (2018) 012096 doi:10.1088/1755-1315/137/1/012096

The open access fishery point in the Copes model can be obtained from the intersection ofthe demand curve and the open access supply curve represented by the long-run average cost curve (LRAC). The catch inopen access fishery was obtained from the curve, and the catch effort (EoA) of open access is as follows:

$$E_{OA1.2} = \frac{-\alpha \pm \sqrt{\alpha^2 - 4\beta h_{OA}}}{-2\beta} \quad (14)$$

By replacing a positive E value, the total cost can be written as the followingfunction of output [17]:

$$LRAC = \frac{\delta TC}{\delta h} = -\frac{c}{2h\beta}(\alpha + \sqrt{\alpha^2 - 4\beta h}) \quad (15)$$

The long-run marginal cost curve can be derived from the long-run average total cost curve on h [17]:

$$LRMC = \frac{\delta AC}{\delta h} = \frac{c}{\sqrt{\alpha^2 - 4\beta h}} \quad (16)$$

The optimum profit in the utilization of capture fishery resources achieved in MEY condition is:

$$\pi_{MEY} = (m - n.h_{MEY})h_{MEY} - c.E_{MEY} \quad (17)$$

The optimum profit in the utilization of capture fishery resources achieved in the open access condition is zero ($\pi = 0$), as shown below:

$$\pi_{OA} = (m - n.h_{OA})h_{OA} - c.E_{OA} = 0 \quad (18)$$

Where:
AC = Average capture cost
MC = Marginal capture cost
c = Total cost per capture effort
TC = Total capture cost
MEY = Maximum Economic Yield
OA = Open Acces

The price variable of fish (p) is determined based on the average real price per year during the period 1995—2016 with the base price 2007 = 100. Furthermore, the long-run average cost curve and marginal cost curve in the consumer surplus and producer surplus can be obtained with the equations below:

$$CS = \int_0^{Q_0} P(Q)dQ - P_0Q_0 \quad (19)$$

$$PS(Q_0) = P_0Q_0 - \int_0^{Q_0} S(Q)dQ \quad (20)$$

Where:
CS = Consumer Surplus
PS = Producer Surplus
S(Q) = Supply function
P(Q) = Demand function
$P_0$ = Real price
$Q_0$ = Catch/output

IOP Conf. Series: Earth and Environmental Science **137** (2018) 012096 doi:10.1088/1755-1315/137/1/012096

## 3. Results and Discussion

### *3.1.Maximum sustainable yield (MSY)*

Fishery resources MSY conditions are the thresholds of the utilization of fish resources allowable without disturbing or destructing the sustainability to grow back. The utilization of fish resources in Cirata Reservoir inMSY condition and actual conditions can be seen in the figure below:

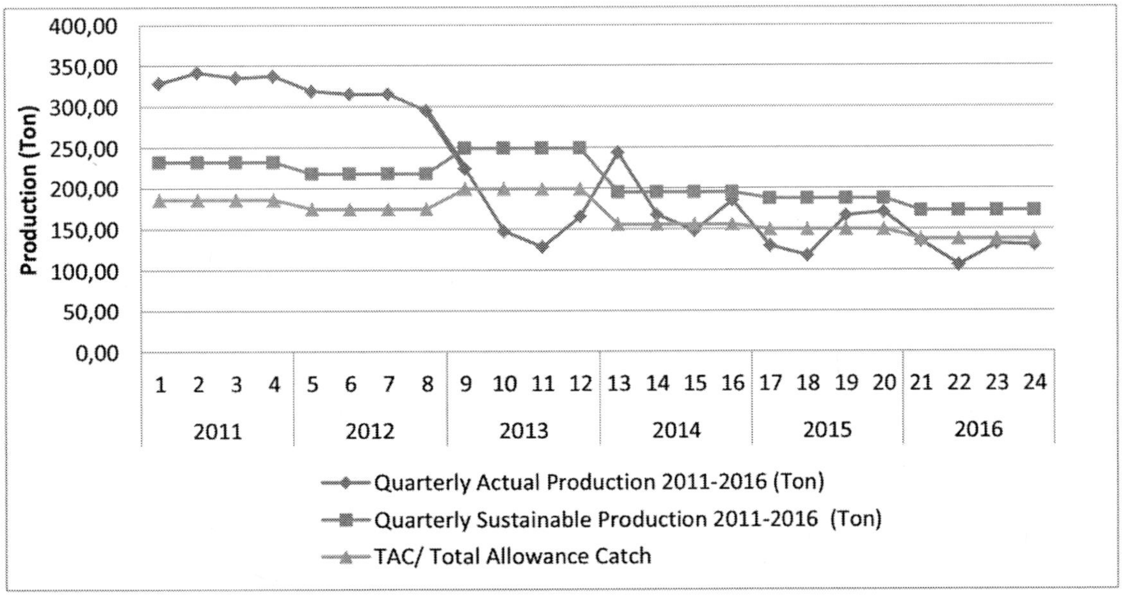


**Figure 1.**The comparison of actual production with sustainable production value.

As shown in the figure above, in the first quarter of 2011 through the last quarter of 2012 and in the first quarter of 2014, the catch exceeded the sustainable production value (MSY). The output limit was based on the Total Allowable Catch (TAC), where 80 % of the production value was sustainable (MSY). Referring to figure 1,the utilization of fish in Cirata Reservoir reached almost the average of the actual catch of 2011 and exceeded the allowable catch amount (TAC).Thus, it couldbe categorized as overfishing.

On the other hand, the actual catch value that continuedto declines annually as well as the low sustainable catch value basically describe the dynamics of fish resources in Cirata Reservoir which is a function of the exploitation itself and the condition of the aquatic habitat. The aquatic environment of Cirata Reservoir had long faced a problem, especially since its water had been used for massive fish farming with massive use of feed, which caused significant pollution. Thus, the environmental carrying capacity for capture fisheries dropped, and the sustainable catch dynamics lowered. The implication of the policy on this problem is the question of how the environmental carrying capacity can be improved to allow for the increase of stock capacity by improving the aquatic environment using physical, biological and other techniques. The reduction of the number of floating nets for the purpose of adjustment to the carrying capacity of the environment becomes a necessity for the development of capture fisheries, that can contribute to the economy of the local community as well as a to the food security.

IOP Conf. Series: Earth and Environmental Science **137** (2018) 012096 doi:10.1088/1755-1315/137/1/012096

*3.2.The optimization analysis of copes bioeconomic model for fishery management regime in Cirata Reservoir*

The optimization of copes bioeconomic model suggests that this model has the advantage in that it is based on output or production rather than on input;it enables the use of elastic demand curves;it enablesthe analysis of economic surplus, producer surplus, consumer surplus, and government rent; and it allows the analysis of imperfect economic structures such as monopoly, monopsony, and public ownership (government) [15,17]. The net social benefit wasevaluated using producer surplus and consumer surplus measures. According to Green in [17], using a surplus approach to measure the benefits of natural resources is appropriate because resource utilization is judged on the best alternative use. Thus, Copes bioeconomic model obtained the welfare effects of the perpetrators of the utilization of fisheries (producers) and consumers. The following isthe model curve:

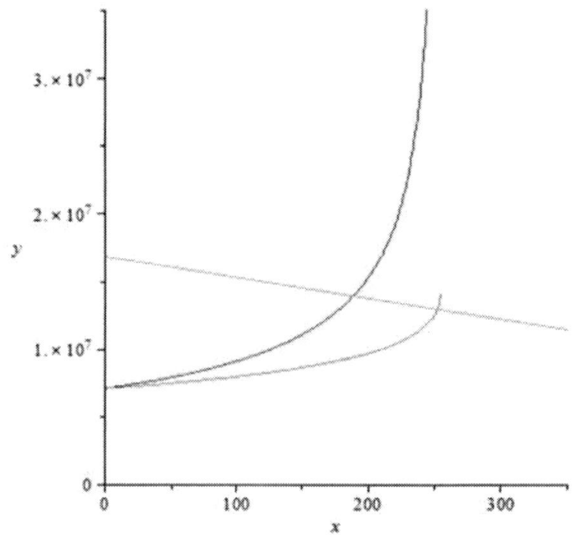


**Figure 2.** Copes model optimization on fish resources in Cirata reservoir(Y=Price (Rp), X= Catch (Ton)).

**Table 1.** Optimal value for every management rezime in copes model.

| Variable | Management Rezime | | |
|---|---|---|---|
| | MSY | MEY | OA |
| Production (h) (Ton) | 254.6150929 | 188.0716294 | 252.03 |
| Effort (E) (Trip) | 39201.70791 | 19160.88192 | 35252.9708 |
| Profit (pi) Rupiah | - | 847.280.297 | 0 |

Based on the table, the results obtained in the Copes model for each management regime wasnot significantlydifferent from the dynamics obtained in the GS model, the Gompertz model where the management of the MEY regime has the highest efficiency and more conservative biomass management.

IOP Conf. Series: Earth and Environmental Science **137** (2018) 012096 doi:10.1088/1755-1315/137/1/012096

*3.3.Producer surplus, consumer surplus, and government surplus on capture fishery resources in Cirata Reservoir*

The main principle in the Copes model is determining the bioeconomic equilibrium of fishery based on demand and supply. Surplus is derived from the demand and supply equations for the production of capture fishery resources in Cirata Reservoir. The producer surplus in capture fisheriesis the utility advantage gained as a result of market price on the sellers' willingness to sell their goods, resulting in the number of profitable manufacturers selling at market priceshigher than the prices of those willing to sell. The producers in questionwere fishermen/fish entrepreneurs. Meanwhile, consumer surplus refers to the utility advantage obtained by consumers because of the fact that market prices werelower than the consumer purchasing power per unit of goods. If the purchasing power of consumers per unit of goods is above the market prices or if the actual market prices are under the purchasing power of consumers, it means that the consumer gets utility profit. The assumption in this study was that the market was competitive, which means that the price was influenced by the market mechanism of demand and supply, thus the producers and consumers in the market could not affect the price and could only act as the recipients of the price.

*3.3.1.Consumer surplus and producer surplus in open access fisheries management regime (Uncontrolled)*

Open access or unregulated fishing means that the fishermen or fish farmers would compete to catch fish in Cirata Reservoir. Hence, more fish would be captured, whichcaused a decrease of fish prices due to the abuandance of fish available. In the open access curve, the consumer surplus is greater than the producer surplus because in the open access fishery, every business that utilizes fishery resources will try to catch as many fish as possible.

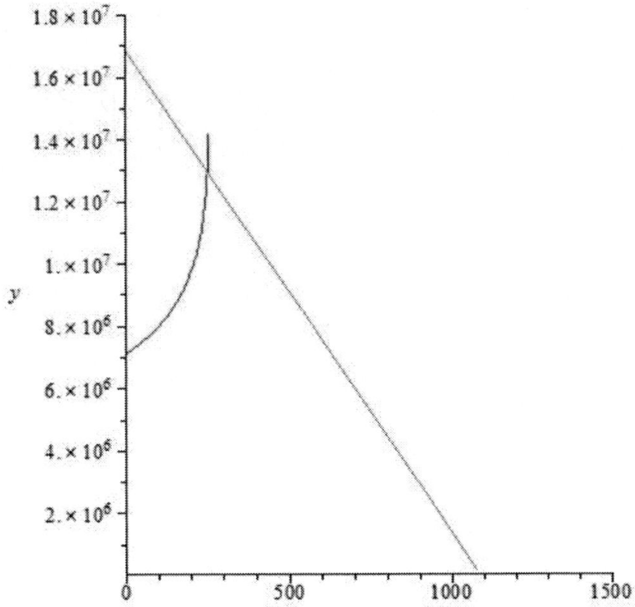


**Figure 3.** Consumer surplus and producer surplus on open access fishery(Y=Price (Rp), X= Catch (Ton)).

IOP Conf. Series: Earth and Environmental Science **137** (2018) 012096 doi:10.1088/1755-1315/137/1/012096

Based on the figure above, the producer surplus wasbelow the price of the equilibrium point and above the supply function, while the consumer surplus was in the area under the demand curve and above the equilibrium price. By plotting between the demand curve and open access supply curve,the value of producer surplus in open access fisheries was found to beIDR0, while the value of consumer surplus was found to be IDR 491.776.746. Hence, the producer surplus value in theopen access fisheries in Cirata Reservoir was not profitable at all (TC = TR), and the producers could even incur loss.

By determining the equilibrium point between the demand curve and the open access fishery supply curve in Cirata Reservoir, where the requested amount wasequal to the quantity supplied, an open access point for the catch $h_{OA}$was 252 Ton and the $P_{OA}$ = IDR 12.938.600. Thus, the increase of average fish price per kg applied was IDR12.938 per kg. In other words,the producers and the government didnot generate profit at all. Meanwhile, the interview results showed that the average price at which fish were sold by fishermen or fish farmers wasIDR12:000-13.000 per kg. Thus, the actual condition was still in open access fishery condition.

*3.3.2. Consumer surplus and government surplus in controlled fisheries management regime (MEY)*

The application of MEY regulation in the capture fisheries in Cirata would cause the consumer surplus to decrease because the price would no longer be low. According to the law of demand, if the price rises, the demand decreases; meanwhile, producer surplus wouldincrease because whenthe supply increases, the price also increases. Thus, in the long run, the controlled fisheries in Cirata Reservoir will benefit both consumers and producers because both consumers and producers will have sustainable capture fishery resources and obtain optimum economic benefits. The producer surplus resulting from the plotting between the demand curve and the marginal cost offering curve (MC) was IDR2.610.203.099 per quarter, which means that the average surplus of every fleet or fishing gear was about IDR914.651 per quarter while the consumer surplus as the region is under the demand curve and above the equilibrium price. The consumer surplus from theplotting between the demand curve and the marginal cost offer curve (MC) was IDR273.885.400. By determining the equilibrium point between the demand curve and the controlled supply curve in Cirata Reservoir,where the demanded amount was equal to the supplied quantity, the balance of open access conditions for the catch $h_{MEY}$ = 188.07 tonsand the price $P_{MEY}$ = IDR13.929.000.

IOP Conf. Series: Earth and Environmental Science **137** (2018) 012096 doi:10.1088/1755-1315/137/1/012096

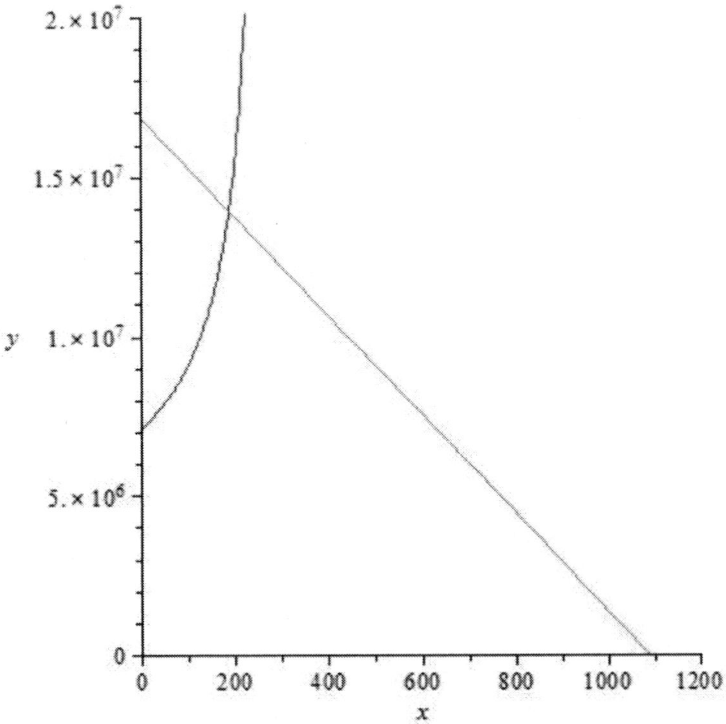


**Figure 4.** Consumer surplus and producer surplus on controlled fishery(Y=Price (Rp), X= Catch(Ton)).

As presented in the figure above, the results of the computation using the split in the Maple 16 program showed that the catch in the controlled fishery condition ($h_{MEY}$) was188.07 tons per year. This means that if the catch is as much as 188 tons per year, an optimal advantage will be gained. Furthermore, if the catch of fish in Cirata Reservoir is below 188.07 tons per year, it gets less proficiency. And if the catch is above 188 tons per year, reduced profit to the point of break-even or no profit at all will be experienced in the open access condition i.e.,252 tons per year.

*3.3.3.Government surplus*

Government surplus may be generated from tax or levy imposed to the fishermen in Cirata Reservoir, both on input and output. In the actual condition, the average price of fish in Cirata Reservoir was around IDR12.000-13.000 per kg, which means that the fishery management had yet to produce any government surplus. Furthermore, the government surplus could be obtained when the fish resources were utilized in the MEY condition. Thus, the application increased the average price of fish to IDR13,929 per kg (price under optimum/ MEY condition).This would result in a government surplus of IDR 217.891.345 per quarter.

## 4. Conclusion

The Welfare Analysis using Copes bioeconomic model yielded a producer surplus value in open access fishery (Open Access) of IDR 0 per quarter and a consumer surplus value of IDR 491.776.746 per quarter. The balance point of open access condition for the catch $h_{OA}$ = 252 tons and the price $P_{OA}$ = IDR

12.938.600 per ton. This means, fishermen who sold fish at IDR 12.930 per kg wouldexperiencea break-even or would not generate any profit at all. The producer surplus in regulated fishery management (MEY) was IDR 2.610.203.099 per quarter, while the consumer surplus was IDR 273.885.400 per quarter. The equilibrium point of MEY regime for the catch $h_{MEY}$ = 188.07 tons and the price $P_{MEY}$ = IDR 13.929.000 per ton.

Thus, the application of controlled fishery management regimes (MEY) increasedthe average price of fish to IDR 13.929 per kg (price under optimum/MEY condition), whichwouldgivethe government a surplus of IDR 217.891.345 per quarter. In the MEY management, theproducer surplus was IDR 2.610.203.099 per quarter, which means that the producer surplus per fleet or per fishing gear wasabout IDR 914.651 per quarter, and the consumer surplus was IDR 273.885.400 per quarter. Furthermore, research shows that sustainable management regime policy through MEY implementation will result in government rent/surplus of IDR 217.891.345 per quarter, with anaverage fish price of IDR 13.929 per kg. In the fisheries under an open access condition, the producer surplus is IDR0. Thus the implementation of the MEY instrument-based policy becomes an appropriate management of capture fishery in Cirata Reservoir.

## 5. References

[1]. Abery N W, Sukadi F, Budhiman A A, Kartamihardja E S, Koeshendrajana S, and Silva de S S 2005 *J. Fish. Man. and Eco.* **12** 315–330

[2]. Béné C and Russell A J M 2007 Diagnostic study of the volta basin fisheries part 2 - livelihoods and poverty analysis, current trends and projections (Cairo Egypt: WorldFish Center Regional Offices for Africa and West Asia) p 66

[3]. De Silva S S and Amarasinghe US 2009 Status of reservoir fisheries in five Asian countries (Bangkok : NACA Monograph No. 2. Network of Aquaculture Centres in Asia-Pacific) p 116

[4]. Paul P and Basak N K 2015 *Indian J. of Applied Res.* **5** 56-58

[5]. Paul P and Sandipan C 2016 *Inter.J. Fish.Aqua. Sci.* **6** 59-76

[6]. Renwick M E 2001 Valuing water in irrigated agriculture and reservoir fisheries: A multiple use irrigation system in Srilanka (Colombo: Research report 51 International Water Management Institute)

[7]. Sugunan V, Welcomme R, Béné C, Brummett R, Beveridge M 2007Inland fisheries, aquaculture and water productivity. Water for food, water for life: comprehensive assessment for agricultural water (London UK: Earthscan ed. Molden D., editor.) pp 459–484

[8]. Teame T, Natarajan P, ZelealemTesfay 2016 *Int. J. Fauna Bio.Stu.* **3** 105-113

[9]. Grantham R W and Rudd MA 2015. *Fish. Manag.and Eco.* **22** 458–471

[10]. Welcomme R L, Cowx I G, Coates D, Béné C, Funge-Smith S, Halls A and Lorenzen K 2010 *J. Phil. Trans. of the Royal Society B: Bio. Scie.s* **365** 2881–2896

[11]. Smith Laurence E D, Khoa S. Nguyen and Lorezen K 2005 *J. Water Pol.* **7** 359-383

[12]. Anna Z 2003 Dinamic embedded model economic interaction fisheries-pollution (Bogor : Bogor Agricultural Institute Disertation) (In Indonesia)

[13]. Komarawidjaja W, Sukimin, Sustrisno and Arman E 2011 *J. Env Tech. P3TL-BPPT.* **6** 268-273

[14]. Anwar N 2008 Physical and chemistry characteristics of aquatic and its relation to the distribution and abundance of fish larvae in the bay of palabuhan ratu (Bogor : Bogor Agricultural Institute Thesis) (In Indonesia)

[15]. Copes P 1972 *J. Manch. Sch. Soc. Econ. Study.* **40** 145-163

[16]. King M 1995 *Fisheries biology, assessment and management* (Great Britain: Fishing News Book)

[17]. Fauzi A 2010 *Fisheries economy* (Jakarta: Gramedia Pustaka Utama)(In Indonesia)

IOP Conf. Series: Earth and Environmental Science **137** (2018) 012097 doi:10.1088/1755-1315/137/1/012097

# The importance of aquaculture community group (ACG) in social media (Facebook) towards the aquaculture knowledge and financial improvement of small scale fish farmers (SSFF) in rural areas of Central Java

**T Elfitasari’ , R A Nugroho and A P Nugroho**

Aquaculture Department, Faculty of Fisheries and Marine Sciences, Diponegoro University, Indonesia

Email: titaelfitasari@yahoo.com

**Abstract.** Internet is now widely used by people all over the world, including small scale fisheries communities such as fish farmers. Many applications are being created including social media Facebook which are used by small scale fish farmers (SSFF) for its ease and convenience. The objective of this research is to identify the impact of aquaculture community group (ACG) in social media Facebook towards the improvement of aquaculture knowledge and financial condition of small scale fish farmers in Central Java. This research used quantitative approach where questionnaires were distributed into two groups: SSFF who are member of ACG in social media Facebook and who are not. Sampling technique used random sampling, used 60 samples of SSFF in Central Java. Data obtained were tested using the test statistic Independent t-test using SPSS v.20. Result showed a significant effect of group who are member of ACG in social media Facebook and those who are not, towards the aquaculture knowledge (t count -7.424 and sig 0.000) and financial improvement (t -3.775 and sig 0.000). The results of the average value of the SSFF who are ACG member in Facebook are also higher than farmers who are not.

## 1. Introduction

In this modern era, internet has reached all levels of society including small scale fish farmers (SSFF). Nowadays, the utilisation of internet is much easier. Moreover, the price of internet package is getting more affordable and there are more varieties of technological hardware to access the internet (computers, smart phones, and tablets) are available in markets.

There are more than 38 million users of internet in Indonesia, with 75 % of them are active on social media [1]. One of the well known and most used social media in Indonesia is Facebook. Indonesia is one of the largest users of Facebook in the world with user reaching up to almost 18 million people [2]. From all Facebook users in Indonesia, 26 % worked as entrepreneurs, where 68 % are entrepreneurs in the area of trade, services and agro industry including fisheries [3]. Facebook are chosen by small scale fish farmers for its easy use, wide network, and availability of groups to give opportunities to interact with other users with similarities in aquaculture commodity or location. Each year, Facebook aquaculture communities are getting larger and the number of members increases and distributes widely in Indonesia. The fact that there is no social limitation in social media has made

IOP Conf. Series: Earth and Environmental Science **137** (2018) 012097 doi:10.1088/1755-1315/137/1/012097

SSFF to become more open to share information and issues. Fish farmers can also widen their network since member of the fisheries community on Facebook came from different areas in Indonesia.

Currently one of the obstacles in aquaculture business is the slow distribution of information from the government to fish farmers, which results in lack of information to increase aquaculture production. Moreover, information on potential market cannot be easily accessed by small scale fish farmers. This has resulted in inability of most fish farmers to sell directly to market and it forces them to have to sell to middle men with lower selling price.

Social Media like Facebook offers speed, accuracy, validity, and reliability, and it has the solution for one the grounding issues faced by small scale fish farmers that is limitation in information and market. An SSFF utilizes community group to interact with other SSFF who have similar interest. It helps them to share information and knowledge on new technology, price and market. To date, there have been no formal studies on the potential and effect of aquaculture community group on Facebook towards the enhancement of SSFF quality in Indonesia.

## 2. Methodology

This research used quantitative approach in order to observe the impact of fisheries communities in social media Facebook towards the enhancement of knowledge and financial condition of SSFF.
This research used the following four variables:

1. Independent variable: SSFF that were Aquaculture Community Group members on Facebook
2. Independent variable: SSFF that were non ACG members on Facebook
3. Dependent variable: improvement on aquaculture knowledge and information among SSFF
4. Dependent variable: improvement on financial condition of SSFF

Population for this research were active SSFF member of fisheries community on Facebook and SSFF that do not utilize Facebook. This research limited the population to only SSFF that were located in Central Java, however the type of fish they farmed, their educational level, land width, and aquaculture system used were not limited, therefore it was assumed that the sample could represent the aquaculture population in Central Java.

The sampling technique used was purposive random sampling based on certain considerations. The Number of samples taken 15 % of the population according to Taro Yamame's theory where if the number of a population was 200 and the maximum error was set at 5 % (0.05) therefore, there should be 30 samples taken from each group. The questionnaire consisted of 40 questions, with 20 questions were related to aquaculture knowledge and 20 questions were related to economical aspects. The questions indicators in the questionnaire are as shown in table 1.

**Table 1.** Questions indicator.

| Measurement indicator | Percentage | # of Questions |
|---|---|---|
| Aspect on aquaculture knowledge | | |
| 1. Aquaculture cultivation technique indicator | 40 % | 8 |
| a. Aquaculture production preparation | | 2 |
| b. Feeding method | | 1 |
| c. Water quality management | | 2 |
| d. Disease and parasite management | | 1 |
| e. Production control | | 1 |
| f. Harvest | | 1 |
| 2. Aquaculture problem solving indicator | 20 % | 4 |
| a. Fish larvae issues | | 1 |
| b. Fish disease issues | | 1 |
| c. Water quality issues | | 1 |
| d. Feed issues | | 1 |
| 3. Aquaculture latest information indicator | 20 % | 4 |
| a. Aquaculture commodity | | 1 |
| b. System and technology | | 1 |
| c. Current issues on aquaculture | | 1 |
| d. Selling price on aquaculture commodity | | 1 |
| 4. Aquaculture information resources indicator | 20 % | 4 |
| a. When information are needed | | 1 |
| b. How often information are needed | | 1 |
| c. Information needed are important | | 1 |
| d. Easy to find information | | 1 |
| TOTAL | 100 % | 20 Questions |
| Financial Aspects | | |
| 1. Welfare indicator | 30 % | 6 |
| a. Income | | 3 |
| b. Cost for food | | 1 |
| c. Household purchasing | | 1 |
| d. Food security | | 1 |
| 2. Marketing indicator | 40 % | 8 |
| a. Advertising | | 2 |
| b. Marketing promotion | | 3 |
| c. Direct selling | | 3 |
| 3. Social network indicator | 30 % | 6 |
| a. Welfare | | 2 |
| b. Marketing | | 3 |
| c. Social network | | 3 |
| TOTAL | 100 % | 20 Questions |

The percentages of questionnaire were given according to priorities and the importance of indicators in each aspect. Validity of the questionnaire was then analysed using the Pearson Product Moment correlation, while reliability was analysed using split-half technique from Spearman-Brown.

IOP Conf. Series: Earth and Environmental Science **137** (2018) 012097 doi:10.1088/1755-1315/137/1/012097

During questionnaire distribution for the non-Facebook users group and data collection for the Facebook users, the background of respondent were taken into account such as education level and the width of ponds used for aquaculture business, in order to minimise differences between the two groups. Thus, it was expected that the results of the comparative study were purely based on the impact of utilising fisheries community network on Facebook.

Data collected were analysed using (1) descriptive analysis to analyse raw data and followed by normality and homogeneity tests, (2) inferential analysis to analyse sample data and the results were used for population and (3) independent t-test, to compare and find significant differences between the two sample groups in order to develop final conclusions.

## 3. Result and Discussion

### *3.1.Result*

Demographic condition of the respondents included age groups, educational level, and types of commodity. The results are shown in table 2-4.

**Table 2.** Distribution of respondent's age.

| Age (year) | ACG-FB member | Non member |
|---|---|---|
| 21 – 30 | 20 % | 0 % |
| 31 – 40 | 30 % | 33 % |
| 41 – 50 | 40 % | 30 % |
| 51 – 60 | 10 % | 33 % |
| >60 | 0 % | 3 % |

Table 2 shows that fish farmers, who were ACG members on Facebook, had a younger age group (21-30 years old), while fish farmers who were not ACG members on Facebook came from an older age group (31 years old and above).

Education level is one of the indicators that influence a small scale fish farmer in accepting, accessing, and understanding information on aquaculture production. Level of education can also influence the ability to access information technology. The education levels of respondents for both groups are shown in table 3.

**Table 3.** Distribution of educational level.

| Education | ACG-FB member | Non member |
|---|---|---|
| Primary School | 0 % | 10 % |
| Junior High School | 13 % | 30 % |
| Senior High School | 67 % | 50 % |
| University Graduate | 20 % | 10 % |

Table 3 shows that fish farmers, who were ACG-FB members, had a slightly higher educational level compared to those who were not members. Fish farmers who are utilizing ACG in FB for business are at least junior high school graduate with most population of senior high school graduates (67%).

The commodities cultured by respondent are various. The data shows the species of fish that are mostly cultured in Central Java as shown in table 4.

**Table 4.** Distribution of aquaculture commodity.

| Type of commodity | ACG-FB member | Non member |
|---|---|---|
| Catfish | 53 % | 43 % |
| Tilapia | 27 % | 27 % |
| Milkfish | 7 % | 10 % |
| Gouramy | 10 % | 20 % |
| Others (ornamental, etc) | 3 % | 0 % |

Data shown in table 4 suggested that there were similar pattern for respondents who were ACG members on Facebook with those who were not members. It indicated that catfish remained the top most favoured type of fish to be cultured in Central Java, followed by tilapia.

The knowledge aspects of small scale fish farmer comprises of several indicators: culture technique, problem solving, obtained information, and sources of information. Based on the data collected, scores of each indicator are shown in figure 1.

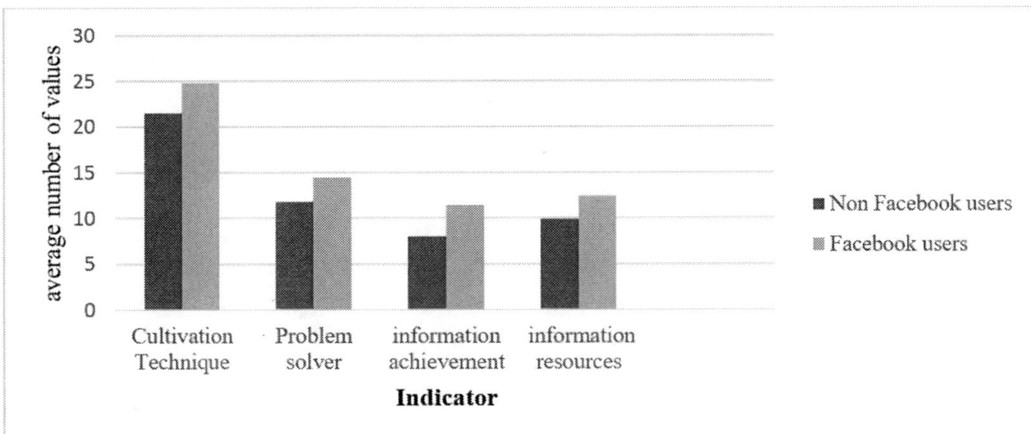


**Figure 1.** Histogram of knowledge aspect indicators.

The histogram showed that the scores for knowledge aspect for SSFF that were members of ACG on Facebook. All four indicators were found higher than the score for SSFF who were not members. Furthermore, the result of t-test shown in table 6 indicated there was significant difference between SSFF who were ACG members on Facebook and non-members for their aquaculture business in the knowledge aspect.

**Table 5.** Result of t-test for knowledge aspect of SSFF.

| | | | Result | |
|---|---|---|---|---|
| | | | *Equal variances assumed* | *Equal variances not assumed* |
| *Levene's Test for Equality of Variances* | F | | 15.803 | |
| | Sig. | | .000 | |
| *t-test for Equality of Means* | T | | -7.424 | -7.424 |
| | Df | | 58 | 42.321 |
| | *Sig. (2-tailed)* | | .000 | .000 |
| | *Mean Difference* | | -11.33333 | -11.33333 |
| | *Std. Error Difference* | | 1.52652 | 1.52652 |
| | *95% Confidence Interval of the Difference* | *Lower* | -14.38900 | -14.41329 |
| | | *Upper* | -8.27767 | -8.25338 |

Result of t-test showed that the significance value was 0.000, pointing to the significant difference between SSFF that were ACG members on Facebook with those who were not members in terms of enhancement of aquaculture knowledge.

**Table 6.** Average value of knowledge aspect of SSFF.

| Knowledge Aspects | N | *Mean* | *Std. Deviation* | *Std. Error Mean* |
|---|---|---|---|---|
| Non member | 30 | 49.3333 | 7.49866 | 1.36906 |
| ACG members | 30 | 60.6667 | 3.69840 | .67523 |

Results showed that ACG members had a higher mean (60.6667) compared to SSFF that were not members of ACG (49.3333). This suggested that SSFF who utilized Facebook had higher aquaculture knowledge than SSFF that did not utilize Facebook.

The economic aspect for SSFF consists of 3 indicators: welfare, market, and social network. Results of data analysis are shown in figure 2.

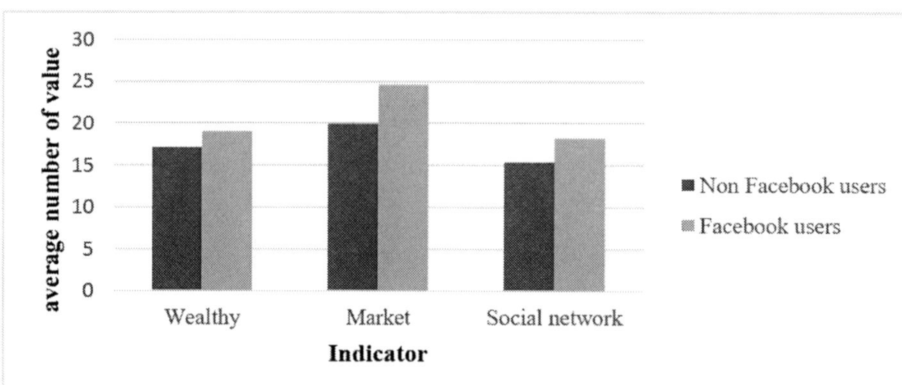


**Figure 2.** Histogram of economic aspects indicators in SSFF.

IOP Conf. Series: Earth and Environmental Science **137** (2018) 012097 doi:10.1088/1755-1315/137/1/012097

The economic aspect of SSFF, which is shown in figure 2, indicate that SSFF who were ACG members were found to have a higher score compared to SSFF that were not ACG members in all three indicators (welfare, marketing, and social network). Further analysis using t-test was performed to observe whether the two respondent groups had significant difference (table 7).

**Table 7.** T-test of economic aspects of SSFF.

| | | | Result | |
|---|---|---|---|---|
| | | | *Equal variances assumed* | *Equal variances not assumed* |
| *Levene's Test for Equality of Variances* | F | | 7.696 | |
| | Sig. | | .007 | |
| *t-test for Equality of Means* | T | | -3.775 | -3.775 |
| | Df | | 58 | 44.420 |
| | *Sig. (2-tailed)* | | .000 | .000 |
| | *Mean Difference* | | -9.43333 | -9.43333 |
| | *Std. Error Difference* | | 2.49859 | 2.49859 |
| | *95% Confidence Interval of the Difference* | *Lower* | -14.43480 | -14.46756 |
| | | *Upper* | -4.43186 | -4.39910 |

Analysis using t-test suggested that there was significant difference between the two SSFF groups (significance value of 0.000ACG members) in terms of economic aspect.

**Table 8.** Average value of economic aspect of SSFF.

| Economic aspect | N | *Mean* | *Std. Deviation* | *Std. Error Mean* |
|---|---|---|---|---|
| Non member | 30 | 52.4000 | 12.05905 | 2.20167 |
| ACG-FB member | 30 | 61.8333 | 6.47053 | 1.18135 |

The average value shown in table 8 suggests that SSFF who were ACG members on Facebook had a higher value (61.8333) compared to SSFF who were not ACG members (52.4000). This indicates that there were significant differences between both groups in terms of economic condition. Therefore, the results confirmed that Facebook gave significant impact towards the economic aspect and it could help to improve welfare, marketing, and social network of SSFF.

### *3.2. Discussion*

#### *3.2.1 Knowledge of SSFF.*

The knowledge level of SSFF who utilized Facebook by becoming ACG members (hereafter referred as Group 1) was higher compared to SSFF who did not utilize Facebook (hereafter referred as Group 2). This is due to the possibility that Group 1 obtain more information regarding better aquaculture techniques available from other the aquaculture community members on Facebook. On the contrary, Group 2 had limitation in obtaining new information on sustainable aquaculture system and other related issues. This indicates that there is an important role of information technology to increase the knowledge of SSFF through fast access to information. The development of Facebook community groups allows its members to share information, sell products and communicate with each other in order to increase social network and enhance the quality of business [1]. Similar opinion was emphasized by rural communities that confirmed that the internet and social media had widen and expand their knowledge by providing latest information that could increase their skills [4].

IOP Conf. Series: Earth and Environmental Science **137** (2018) 012097 doi:10.1088/1755-1315/137/1/012097

Group 1 also seemed to have more problem solving skills compared to those in Group 2. This was shown by the score of SSFF where Group 1 had higher score (14.5) than Group 2 (11.8). This phenomenon may be due to the fact that SSFF who belonged in Group 1 had more opportunities to interact and share issues and discuss how to solve problems with other members of the aquaculture community, who might have experience similar problems or who had expertise in dealing with such issues. On the other hand, SSFF in Group 2 did not experience this kind of exposure, thus they did not have access to share and discuss their problems with wider community and consequently, it limited their ability to solve aquaculture related issues. This was in line with the research from [4] who asserted that the internet and Facebook had help to ease communication for rural community to connect with the world beyond their area and increase their skills and knowledge in technology, culture, agriculture, etc.

Based on the results collected through questionnaire, it was found that SSFF in Group 1 were more "up to date" and they followed the latest information regarding technology and other aquaculture related information compared to Group 2. One of the examples was from one of the questions where SSFF were asked regarding latest information. SSFF in Group 1 obtained higher score than those in Group 2. This means that SSFF who were active on Facebook aquaculture community group were more aware on issues and development related to aquaculture while SSFF who did not follow latest information remain uninformed. This shows that community groups on Facebook gave positive impact towards rural society, such as better understanding of a particular issue [5]. Many Facebook users like to write an updated status that makes other Facebook users obtain information on that related topic. Therefore, virtually, people who use social network are connected with each other for social searching of information.

SSFF in Group 1 had a better and reliable source of information. Social media had wider scope and advanced technology that makes it possible for its users to access information accurately and instantly when needed. This has made SSFF in Group 1 were faster in obtaining information on aquaculture related issues. Social network is one of the social media that fulfils the various needs of its users, including communicating with other users with similar interest or background and discuss issues that are of interest for them [5]. The easy access to information is the aspect that was not experienced by SSFF in Group 2 considering that their only intensive interaction and source of information were limited to extension services and other SSFF peers located closely within their area.

*3.2.2.Financial condition of SSFF.*

Group 1 showed a significant difference in the level of welfare compared to Group 2. This may be due to the possibility that SSFF in Group 1 who were active members of ACG on Facebook had a wider social network and were more advanced in marketing. The scores for marketing indicator and social network were higher in Group 1 (marketing score of 24.6 and social network score of 18.6) compared to Group 2 (marketing score of 19.9 and social network score of 15.4). Better marketing skills and wider social network plays a role in determining the level of welfare of SSFF. This was supported by previous research that confirmed welfare is associated with economy, where one of the focuses is marketing [6]. Similar research also found that the key to enhance the economy is to increase income, whereas in marketing, the key is to employ a good promotional media and a wide social network [7].

SSFF in Group 1 also resulted in a higher value in the marketing indicator compared to SSFF Group 2. The higher value in marketing was influenced by the advertisement and promotion that were carried out by SSFF through ACG on Facebook. They uploaded photos of their product online through Facebook and through this marketing strategy they are able to increase selling by getting more new customers compared to the opposite group. Advertising through social media are statistically proven to influence the decision of customer to buy product, also the usage of commercial advertisement through creative media is important to influence customers to buy product [7]. Similar opinion was asserted [8] that social media based promotion is very effective to increase sales of product and services.

IOP Conf. Series: Earth and Environmental Science **137** (2018) 012097 doi:10.1088/1755-1315/137/1/012097

Result also showed that Group 1 scored higher (value score 18.2) in the social network indicator compared to Group 2 (value score 14.4). This means that SSFF who were ACG members on Facebook had a wider social network and good relationship with SSFF from other areas in Indonesia. Wider network and good communication and relation help ACG members to easily obtain information regarding latest aquaculture technique, problem solving, issues, as well as market development and opportunities. This has made this group of SSFF to be more advanced compared to SSFF who were not ACG members on Facebook. Social network on the internet is used widely by companies to attract customers [9]. Social media are used for its incredible connectivity between users and communities or groups that have been developed. It has also been emphasized that developing business network in social media is a sustainable marketing strategy. This was proven from the result of previous research that used two community groups on Facebook, where after one year; there was substantial growth in number of selling followed by an increase in the number of group members [1].
ACG on Facebook proved to have a significant effect towards the overall quality of SSFF. Social media is an efficient, effective and fast way to communicate and it covers a wide range of society regardless of their background [10]. ACG on Facebook is a communication media that provides information for SSFF regarding aquaculture techniques, problem solving methods, and current information on aquaculture. The aquaculture knowledge indicator was higher since Facebook assisted SSFF to communicate, in low cost, more easily and openly for everyone. Thus it had made the aquaculture society to develop a good and profitable discussion forum. The principle of openness in social media had made SSFF feel free to choose communities in which they wanted to join according to their interest and needs. The ease in communication has been successfully utilized to improve financial condition. It is effective to attract new consumers; therefore it has been used widely as means of promotion and advertising with economical value [11]. Based on the above discussion, it is fair to say that social media have now become one of the most important tools for SSFF to encounter business challenges and competitions, especially in improving the aquaculture knowledge and financial condition.

## 4. Conclusion

Based on the result of this research, it can be concluded that the SSFF in Central Java who were ACG members on Facebook had a significant improvement in aquaculture knowledge if compared to SSFF that were not members of ACG on Facebook. The score of SSFF knowledge for those who did not utilize Facebook was only 49.33, while those who utilized Facebook had a higher mean value of 60.66.

The SSFF who utilized Facebook also showed a significant improvement of their financial condition. A higher mean value score (61.83) was identified in SSFF who utilized Facebook, compared to SSFF who did not utilize Facebook (52.4). This research confirms that ACG on Facebook had a high level of importance in improving quality of knowledge and financial condition of SSFF in Central Java.

## 5. References

[1] Nugroho and Kastaman 2014 Effect of social media Facebook towards online business selling *Pros. Sem. Nat. Sci. Tech. App. SNAST)* 161-168 (In Indonesia)

[2] Enterprise J 2010 *Facebook goes to school* Yogyakarta: Gregorius Agung Founder 181. (In Indonesia)

[3] Puskakom UI 2015 Indonesian internet user profile Profil 2014 *Perpustakaan Nasional Republik Indonesia: Katalog Dalam Terbitan (KDT)* 56. (In Indonesia)

[4] Ekasari P and Dharmawan 2012 *J. Sos. Pedesaan.* **6** 57-71 (In Indonesia)

[5] Oktaviani K M 2014 Usage of social network facebook in Malaysian village society *Faculty of Human Ecology* (Bogor : Bogor Agricultural Institute) pp 57 (In Indonesia)

IOP Conf. Series: Earth and Environmental Science **137** (2018) 012097 doi:10.1088/1755-1315/137/1/012097

[6] Mudiarta K G 2011 Persective and socio economic role in society economic development *Forum RecearcherAgro Economy* **29** 55-66 (In Indonesia)
[7] Baransano I B, S Iswono and Handaini 2015 Impact of advertising media Facebook towards decision to buy organic eggs (case study advertising through Facebook at CV Laras Inti Agrisatwa Jember) *E-Sospol* **1** 1-14 (In Indonesia)
[8] Ekasari N 2014 *J. Pen. Universitas Jambi Seri Humaniora* **16** 81-102 (In Indonesia)
[9] Andriyanto R D and Haryanto 2010 *J. Man. Tech.* **9** 20-35 (In Indonesia)
[10] Prisgunanto I 2015 *J. Academic Communication Opinion Public* **19** 101-112. (In Indonesia)
[11] Riyantoro B and Harmoni 2013 Effectivity of advertisement through social network as a strategy to market Maicih spicy crackers *Proc.g PESAT* 256-263 (In Indonesia)

IOP Conf. Series: Earth and Environmental Science **137** (2018) 012098 doi:10.1088/1755-1315/137/1/012098

# The economic and social benefits of an aquaponic system for the integrated production of fish and water plants

**A Rizal, Y Dhahiyat, Zahidah, Y Andriani, A A Handaka[1] and A Sahidin**

[1]Faculty of Fisheries and Marine Science, Universitas Padjadjaran. Jl Raya Bandung-Sumedang Km 21, Jatinangor, Sumedang 45363, West Java, Indonesia. Tel./Fax.: +62-22-87701519 Indonesia

E-mail: achmad.rizal@unpad.ac.id

**Abstract.** Aquaponics is an evolving closed-system food production technology that integrates recirculating aquaculture with hydroponics. In this paper we give a brief literature overview of the benefit aspects of aquaponics by discussing its social, environmental, and economic impacts in different potential settings. The technology might be applied to commercial or community based urban food production, industrial scale production in rural areas, small scale farming in developing countries or as systems for education and decoration inside buildings. We concluded that due to the different potential applications and settings for installing the technology, benefit impacts need to be considered separately and that due the complexity, communities, urban and rural infrastructure and policy settings, further research and data acquisition is needed to be able to assess all benefit aspects.

## 1. Introduction

The lack of arable land area and degradation with water scarcity are some of the the current problems of agricultural production, especially in the most under developed areas with a scarcity of resources. This should make us re-evaluate the way in which food is produced. The consumer demand for fish has been increasing, but ocean fish catches continue to decline. Aquaculture, the cultivation of freshwater and marine plants and animals, is one of the fastest growing segments of Indonesian agriculture. The increase of farm-raised fish has lead to increased concern regarding discharges from those facilities.

Therefore, the treatment of fishery effluents needs to be considered when planning aquacultural production systems. However, it may be important to treat the nutrients in aquaculture effluents because the total nutrient mass loading can contribute significantly to environmental degradation, depending on the quality of receiving water. According to the current discourse, this involves an increase in productivity and resource use efficiency, solutions for small holder farmers, as well as a reduction in food waste [1].

Land-based recycle aquaculture facilities release dissolved nitrogen and phosphorus to the water environment, which contributes to the undesirable growth of macro and micro algae in the receiving waters. In land-based fish culture, water quality can be controlled by a high rate of either water exchange, which is costly; or water treatment and subsequent recirculation, which comes at a price. To offset treatment costs, the integration of aquaculture and plants offers an ideal solution to reduce nutrient discharge levels, increase profitability, and convert the excretion of fish culture into beneficial

IOP Conf. Series: Earth and Environmental Science **137** (2018) 012098 doi:10.1088/1755-1315/137/1/012098

products. Aquaculture as a business requires a stable run of the cultivation system, maintaining all environmental factors under control. Aquaponics is a relatively new concept to modern food production methods and can provide answers to many of the above-mentioned problems.

Aquaponics, the combined culture of fish and plants in recirculating systems, has become increasingly popular. Aquaponic systems offer several benefits. Dissolved waste nutrients are recovered by the plants, reducing discharge to the environment and extending water use (i.e., by removing dissolved nutrients through plant uptake, the water exchange rate can be reduced). Minimizing water exchange reduces the costs of operating aquaponic systems in arid climates and heated greenhouses where water or heated water is a significant expense. Having a secondary plant crop that receives most of its required nutrients at no cost improves a system's profit potential. The daily application of fish feed provides a steady supply of nutrients to plants and thereby eliminates the need to discharge and replace depleted nutrient solutions or adjust nutrient solutions as in the case of hydroponics.

In this article we would like to illustrate this complexity by using the example of aquaponics as a rapidly emerging technology that integrates recirculating aquaculture with hydroponics (production of plants in nutrient solution, without soil) [2, 3, 4, 5, 6], having its origins back in the 1970's [7, 8]. Aquaponic food production is highly efficient, because it re-uses the nutrients contained in fish feed and fish feces to grow the crop plants in an ecological cycle [9].

Essential technical components of aquaponic systems are the fish tanks and plant growth beds, while dedicated biofilters and settlers are optional and depend on the configuration of the system. The microbial community is central for the catabolism of the organic matter contained in the feces and feed residues and for the conversion of the fish-generated ammonia to nitrate [10, 11]. Fully contained and climate-controlled aquaponic systems potentially operate under water conserving and contaminant-free conditions.

At its highest level, aquaponics is a technology-intensive, capital-intensive and knowledge-intensive method of food production that is discerned based on definitions such as horizontal *vs.* vertical, and open *vs.* recirculating [12]. Systems are characterized according to the way plants are supplied with nutrient solutions in the hydroponic systems, e.g., floating polystyrene foam sheets (floating raft), nutrient film technique (NFT), or media filled growth beds arranged horizontally or vertically, while fish are kept in standard recirculating aquaculture conditions. Aquaponic technology is considered to be ecologically friendly: it uses nonrenewable resources with a very high efficacy as indicated by near zero-waste discharge [12].

In addition to its value as a food production system, smaller aquaponic units can be great assets as teaching tools for a wide range of subjects [13], demonstrating ecological cycles and may serve as decorative elements in homes or public places. Moreover, the principle of combining fish and crop production can be implemented from a low-tech level to a high-tech state-of-art system [14].

Although the basic arrangement of an aquaponic system is apparently simple, involving only three kinds of living organisms: fish, beneficial bacteria, and plants, the interrelations between the three are highly complex and interdependent [15]. In addition, the system inherently contains a toxic component: ammonia excreted by the fish [11]. The somewhat contrasting requirements of fish, plants, and bacteria makes it difficult to achieve maximum yield potentials [10]. More research is needed to manage the cycling of nutrients (especially nitrogen and phosphorus), and pH levels so that aquaponics can be economically viable, which will also affect its overall sustainability (e.g. the ASTAF-PRO approach by [10]).

The food produced by aquaponics are fish and plants: the healthiest human diet according to current nutritional science [12]. Recent publications on the sustainability of aquaponics give a broad perspective on the technology, and conclude that these systems can be sustainably managed only with a thorough knowledge of the fish, bacteria, and plant components on both an individual and systems level [6, 10, 12, 15, 16, 17, 18,]. These authors indicate the problems with waste from the nitrogen cycle (e.g., the toxicity of ammonium) and the advantages of higher yields and reduced water use, and suggest avenues for future research, such as the integration of nutrient flows (availability of key macro

IOP Conf. Series: Earth and Environmental Science **137** (2018) 012098 doi:10.1088/1755-1315/137/1/012098

and micronutrients), the need for technological advancements and fish feed alternatives. Unfortunately, other important benefit issues are often neglected such as resource scarcity, climate change and social aspects. We argue that the lack of reliable data and documented practice is the main knowledge gap to assess the benefit of aquaponics.

Benefit assessment of a new technology is a complex and data-intensive exercise, because in addition to the material and energy considerations, various environmental, societal and social factors have to be taken into account [19, 20, 21, 22]. As a result of the lack of data, most publications look at partial aspects of benefit and do not consider all "three pillars" – ecological, economic and social. Due to the numerous interdependencies within the technology and various application settings, the societal and social aspects are difficult to quantify [12]. As for other technologies, benefit assessment is typically a mixture of potential outcomes rather than a pure black or white answer, i.e., the use of the technology under different developmental, human, and climatic conditions will lead to different benefit scenarios.

The benefit analysis of complex systems includes manifold approaches focused on different levels (technology, enterprise, business model types, value chains, target groups, etc.) and different, sometimes multilayered sets of indicators and factors [23, 24]. For example, in terms of agricultural technologies, the comprehensive review of [24] lists 18 economic, 51 environmental, 21 social, and 14 technical indicators. Since aquaponics is still developing rapidly, we lack clear definitions, classifications and demarcations towards similar technologies. With regard to social aspects, there is ongoing discussion about how to qualitatively and quantitatively conceptualize and measure all aspects in indicators. The available impact assessment methodologies do not address the full range of specific activities and impacts of food systems and technologies, as discussed in a review for aquaculture [25].

Assessments are mainly ex-post, while in the case of aquaponic technology development ex-ante methods are needed. In the following paragraphs we briefly discuss the challenges of benefit assessments in three fields of application: urban agriculture, developing world aquaponics and industrial scale aquaponics. Based on a literature review we will consider questions regarding the economic, environmental and social benefist, depending on the potential setting where aquaponics is implemented.

## 2. Social Benefit

Aquaponics has already been used extensively in the education in natural sciences at the primary and secondary school levels and also in vocational training [26, 27]. However, little has been done to assess the social aspects (health, wellbeing, learning of education and demonstration projects [26]. There are still problems regarding technical and school settings that need to be overcome before claiming that aquaponic units facilitate education in benefit [28]. Another social aspect with good potential is community cohesion. However, the setup of such systems will be different to those for commercial urban or industrial production, so the benefit assessment would be different. There is probably a trade-off between technology and knowledge input (high-tech vs. low tech) on one side, and the potential for social impact on the other.

The greatest increase in worldwide human population will occur in urban areas. Food security and infrastructure will become a central issue and aquaponics may prove to be a solution. Even today, many urban areas around the world face the challenge of a food supply infrastructure (e.g. so called "food deserts") [29]. Aquaponics implemented either as professional urban agriculture or as community farming could help alleviate these food deserts. However, in urban settings, aquaponics can fulfill other functions besides food production. For example, it may serve as an educational tool in schools [26], interior greening (providing better climate in public buildings and homes), and as a unit in social institutions.

Aquaponics has the potential to be an integral part of the "blue and green" infrastructure of cities. It can be integrated into the local water cycle (using treated grey water and rainwater instead of freshwater), local energy flows (for example, the "watergy" concept [30]), and local biomass cycles

IOP Conf. Series: Earth and Environmental Science **137** (2018) 012098 doi:10.1088/1755-1315/137/1/012098

(re-use of nutrients).

Aquaponic operations installed in urban areas can meet the demands of consumers and thus will attain premium prices, which in turn will allow fast return on investment [31]. We have observed that several aquaponic businesses integrate the value chain vertically e.g. adding services (such as catering, selling of equipment, system planning services), because production itself is not yet economically viable when compared to specialized horticulture or aquaculture. The development of short value chains, e.g. selling directly to consumers, restaurants or supermarkets, can also be a viable option. Approaches to food production in urban areas on a commercial scale are only the beginning; hence there is a lack of benefit information for decision makers. In the long run, there are many visions for urban areas in temperate zones that include building-based food production [32]. However, these scenarios rely on increased technological and capital intensity Hase [33] that has to be assessed in the light of the development of food prices and income. It is still unclear, however, how sustainable cities will be developed based on existing infrastructure, and how that will affect the benefit of aquaponics. To solve water problems in cities around the world, aquaponics can be incorporated into building concepts to enlarge the local water cycle Viljoen *et al.* [34] or integrated into the matrix of the city [35]. An example of integrating aquaponics into cities as a part of the blue-green structures is the Roof Water Farm concept [36]. However, we still lack quantitative data for a comprehensive benefit assessment of aquaponics in urban environments.

Aquaponics can be used to improve the livelihoods of households and communities. Fish is an important source of protein in low and medium income countries while vegetables improve nutrition [37]. Aquaponics can help to increase food security and food sovereignty [38]. However, the costs of modern aquaponic systems might exclude the poor from its potential benefits: The dependency on electricity and water might limit its use in unplanned urban sprawl and rural areas where nutrition deficits in terms of food variety and protein are most predominant [39]. However, under favorable climatic conditions (tropics and subtropics), aquaponic systems may be very simple, consisting of un-insulated outdoor units (low-tech). [39] states that very few inputs are needed for a basic unit (e.g. fingerlings and seeds). Yet these inputs are often locally limiting factors to food security. Depending on the specific conditions, aquaponics can provide a sustainable food source in low and medium income countries, especially where climate conditions are favorable.

## 3. Economic benefit

Today crop production and fish farming occupy vast regions of the Earth's surface and have a strong negative impact on the environment by inducing soil erosion; polluting the soil and groundwater by pesticides, fertilizers, and animal waste; the production of greenhouse gases; and in many other ways [40]. A combination of plant production and fish farming in closed aquaponic systems results in a significant reduction on the environmental impact. Aquaponic systems can be operated almost waste-free; therefore they have no measurable effects on the soil if no new area is consumed for installing aquaponics. Even the relatively small amount of waste produced (in the form of sludge) can be easily composted and converted to beneficial products.

The viability of industrial scale aquaponics depends on achieving efficient and high yield systems. Fish feed is the biggest cost factor in intense aquaculture [41]. Both the environmental and economic benefits could be improved significantly by either formulating alternative fish feeds, and/or by reducing the fish meal and fish oil in the feeds [36, 42]. Also, the contamination of feed with mycotoxins, which can originate from feed-borne ingredients or from bad storage conditions, is often overlooked, which is dangerous since they can cause many health problems to fish, reduce yield and economic benefits [43, 44].

The cost of labor and energy are the main critical factors in industrial greenhouse vegetable production. Aquaponics is a labor-intensive technology: the operation and maintenance of such systems generates employment and income, but also high labor costs, as the monitoring has to be performed daily, including weekends. The claims of nutrient and water efficient food production depends upon the extent of recycling/recirculation of the nutrients and water in the system. The water

IOP Conf. Series: Earth and Environmental Science **137** (2018) 012098 doi:10.1088/1755-1315/137/1/012098

saving aspect, however, is expected to be most advantageous in areas with water scarcity [39]. In Indonesia, where water is more abundant, the discourse of benefit and alienation between consumers and producers as a result of highly specialized value chains opens economic potential for direct marketing aquaponic farms.

A basic requirement for an economically viable system is the acceptance of the products by consumers. Yet, as these fish and vegetables need to compete with conventionally grown products, the acceptance of the products by consumers remains to be studied.

## 4. Conclusions

Aquaponics, due to its integrative character and multiple application scenarios from high-tech to low-tech, is an atypical and complex food production technology. The complexity of the systems and their application in different settings potentially affects the delivery of all aspects of benefit: economic, environmental and social.

Our literature review demonstrates that due to the lack of data on the operation of commercial aquaponic systems in different environmental (climatic, social, and technological) conditions, a comprehensive benefit assessment is difficult. In addition, as of yet there are no reliable empirical data available on energy use, accidents, repairs, and social change pertaining to the technology. Prototypes used in research and development can only provide certain types of data, so more cooperation is needed with the few industrial operations to characterize appropriate and scalable indicators.

The challenge ahead is the simultaneous development of methodological approaches for technology-specific ex-ante and ex-post benefit assessments, while at the same time, the technology needs to spread in order to fully achieve the benefit potentials promised by the advancement of the technology. A co-development of technology, business models, and benefit data generation could contribute 1) to achieve the multiple potentials of the technology, and 2) to develop sustainable food systems from production to consumption. Benefit assessments could then enable policy makers, entrepreneurs and the general public to differentiate between food production systems with limited negative benefit externalities.

Furthermore, it would indicate processes within the system that have the highest environmental impact and thereby allow to effectively improve the environmental performance of the product under consideration. Yet, for the economic and social benefit aspects we see the need for conceptionalisation, empirical validation and operationalization and more data in order to inform the development of aquaponic technology with regard to delivering its potentials to contribute to sustainable food production.

## 5. References


[1] Rockstörm, Johan, Jeffrey D. Sachs, Marcus C. Ohman, and Guido Schmidt-Traub 2013 Sustainable development and planetary boundaries

[2] Rakocy J E, Donald S, Bailey, R. Charlie S and Eric S Thoman 2004 Update on tilapia and vegetable production in the uvi aquaponic system) in: new dimensions on farmed tilapia: *Proc. of the 6th International Symposium on Tilapia in Aquaculture* p 12 – 16

[3] McMurtry M R, P V Nelson, D C Sanders and L Hodges 1990 *J. Applied Agric. Res.* **5** 280–84

[4] Lennard W A and Brian V L 2006 *Aquat. Int.* **14** 539–50

[5] Pilinszky K, Andras B, Gabor G and Tamas K 2015 *Aquaculture* **435** 275–76

[6] Palm H W, M Nievel and Ulrich K 2015 Significant Factors Affecting the Economic Sustainability of Closed Aquaponic Systems. Part III: Plant Units *AACL Bioflux* **8** 89–106

[7] Naegel L C A 1977 Combined production of fish and plants in recirculating water *Aquaculture* **10** 17–24

[8] Sneed K, K Allen and J E Ellis 1975 *Aquaculture and the Fish Farmer* **2** 18–20

[9] Love D C, Jillian P F, Ximin Li, Elizabeth S H, Laura G, Ken S and Richard E T 2015 *Aquaculture* **435** 67–74

IOP Conf. Series: Earth and Environmental Science **137** (2018) 012098 doi:10.1088/1755-1315/137/1/012098

[10] Kloas W, Roman G, Daniela B, Johannes G, Henrik M, Uwe S, Georg S, Johanna S, Martin T, Bernd W, Sven W, Andrea Z and Bernhard R 2015 *Aquac. Environ. Interact.* **7** 179–92
[11] Bittsánszky A, Katalin P, Gábor G and Tamas K 2015 *Plant Sci.* **231** 184–90
[12] Sommerville C, Moti C, Edoardo P, Austin S, and Alessandro L 2014 *Small-Scale aquaponic food production - integrated fish and plant farming FAO* (Italy: Fisheries and Aquaculture Technical Paper Rome) no. 589
[13] Junge R, Sandra W and Urs H 2014 Aquaponics in classrooms as a tool to promote systems thinking *In Proc. of the Conf. VIVUS – on Agric., Envi., Horticulture and Floristics, Food Production and Proc. and Nutrition* **1** 234–44
[14] TrangN T D and Hans B 2014 *Aquat. Res.* **45** 460– 69
[15] Tyson R V, Danielle D T and Eric H S 2011 *HortTechnology.* **21** 6–13.
[16] Palm H W, Karl B and Ulrich K 2014 *AACL Bioflux* **7** 162–75
[17] Goddek S, Boris D, Utra M, Kristin V R, Haissam J and Ragnheidur T 2015 *Sustainability* **7** 4199–4224
[18] Loomis D K, Peter B O, Christopher R K and Shona K P 2014 Developing integrated ecosystem indices Ecological indicators, tools to support ecosystem based management of south florida's coastal resources **44** 57–62
[19] Carr E R, Philip M W, Sara C Y, Mary C T, Natalie K J and Justin R 2007 *International journal of sustainable development & world ecology.* **14** 543–55
[20] Jerneck A and Lennart O 2014 *International Journal of Agricultural Sustainability* **12** 1–22
[21] Klerkx L, Barbara van M and Cees L 2012 Evolution of systems approaches to agricultural innovation: concepts, analysis and interventions In Farming Systems Research into the 21st Century: The New Dynamic, edited by Ika Darnhofer, David Gibbon, and Benoît Dedieu, 457–83. Springer Netherlands
[22] Dunmade I 2014 *Int. J. Eng. Technol.* **3**
[23] Kriesemer S K and Detlef V 2012 *Analytical framework for the assessment of agricultural technologies stuttgart* (Germany: Food Security Center (FSC), University of Hohenheim)
[24] Samuel-Fitwi B, Ven W, Jan P S and C S 2012 *J. Clean. Prod.* **32** 183–92
[25] Junge
[26] Graber A, Nadine A and Ranka J 2014 The Multifunctional aquaponic system at zhaw used as research and training lab *In Conference VIVUS: Transmission of Innovations*
[27] Hart E R, James B W and Andy J D 2013 Implementation of aquaponics in education: an assessment of challenges and solutions Science Education International **24** 460–80
[28] Beaulac J, Kristjansson E and Cummins S 2009 *Prev. Chronic. Dis.* **6** A105
[29] Vadiee A and Viktoria M 2012 Energy management in horticultural applications through the closed greenhouse concept, state of the art *Renewable and Sustainable Energy Reviews* **16** 5087–5100
[30] Edwards P 2015 Aquaculture environment interactions: past, present and likely future trends *Aquaculture Research for the Next 40 Years of Sustainable Global Aquaculture* **447** 2–14
[31] Caplow T 2009 Building integrated agriculture: philosophy and practice (In Urban Futures 2030. Urban Development and Urban Lifestyles of the Future) *Publication Series on Ecology Heinrich Böll Stiftung*
[32] Kiss G, Hermen J, Veronica L C, Lydia O 2015 The 2050 city *Int. Conf. on Sustainable Design, Engineering and Construction Proc. Engineering* **118** 326 – 355
[33] Haase D 2015 Reflections about blue ecosystem services in cities *Sustainability Of Water Quality and Ecology, Modelling Ecosystem Services: Current Approaches, Challenges and Perspectives* **5** 77–83
[34] Viljoen A, Katrin B and Joe H 2005 Continuous productive urban landscapes: designing urban agriculture for sustainable cities (Architectural Press)

[35] Million A, Grit B, Anja S and Wolf R 2014 Roof water farm. participatory and multifunctional infrastructures for urban neighborhoods *In Finding Places for Productive Cities. Leeuwarden, the Netherlands.*
[36] Tacon A G J and Marc M 2013 *Reviews in Fisheries Science* **21** 22–38
[37] Ericksen P J 2008 *Global Environmental Change* **18** 234–45
[38] Little D C and Stuart B 2015 Urban aquaculture for resilient food systems (In Cities and Agriculture Developing Resilient Urban Food Systems ed. by Henk Zeeuw and Pay Drechsel. Earthscan Food and Agriculture, Taylor & Francis)
[39] Al-Hafedh Y S, Aftab A, and Mohamed S B 2008 *J. World. Aquac. Soc.* **39** 510–20
[40] Goudie A S and Heather V 2013 The earth transformed: an introduction to human impacts on the environment (John Wiley & Sons)
[41] FAO 2007 The state of world fisheries and aqua-culture 2006 FAO Fisheries and Aquaculture Department. FAO UN. Rome, 2007. ISBN 978-92-5 105568-7
[42] World Bank 2013 *Fish to 2030: prospects fo fisheries and aquaculture, agriculture and enviromental services discussion paper 3* (Washington DC: World Bank Group)
[43] Pietcsh C, Susanne K, Hana V, Sven D, Carsten S, Patricia B H and Ranka J 2015 *Toxins* **7** 3465-3480
[44] Pietsch C, Susanne K, Patricia B-H , Hana V and Sven D 2013 *Toxins* **5** 184-192

**Acknowledgements**
Thanks to Universitas Padjadjaran, Indonesia which funded the study of aquaponics, through Academic Leadership Grant program (ALG).

IOP Conf. Series: Earth and Environmental Science **137** (2018) 012099 doi:10.1088/1755-1315/137/1/012099

# The concept of community poverty reduction in coastal area of Surabaya based on sustainable livelihood approach

**A M Gai, I Soewarni and M M Sir**

Urban and Regional Planning Department, Faculty of Civil Engineering and Planning, National Institute of Technology (ITN) Malang Indonesia

E-mail: ardy_06pl@yahoo.co.id

**Abstract.** Multidimensional poverty becomes a trademark of fisherman community including the community in Surabaya. The fishermen in Surabaya belong to a society with quite apprehensive welfare in all aspects covering economy, social, and environment. Therefore, this research aims to organize poverty reduction concept in coastal area of Surabaya based on sustainable livelihood which assesses poverty through 5 (five) livelihood assets i.e. human asset, natural asset, social asset, physical asset, and financial asset. This research is a qualitative research using rationalistic approach with explorative, descriptive, and perspective nature. Primary data collected using Participatory Poverty Assessment (PPA) and secondary data collected through agency and literature survey. Purposive sampling was employed in getting the sample. Then, the data were analyzed using content analysis, statistics descriptive analysis, and delphi analysis. The results show that sustainable livelihood level in coastal area of Surabaya indicates the human asset is 65% at the SLA level and the lowest is social asset which is 20%, and financial asset is the most affecting factors of poverty in coastal area of Surabaya since the expense for fuel cannot be compared to the fish catched. Community empowerment is the concept proposed to overcome the poverty problems in coastal area of Surabaya.

## 1. Introduction

A never-ending discussion about poverty is essentially inseparable from environmental management issues since poverty and environment are two crucial phenomena that should be discussed as one unit. The reality of social, economic, and ecological problems in coastal area and the poverty in fishermen community occurred in Surabaya as well. The fishery profile described that most fishermen in Surabaya belong to low-level welfare category. Besides that, coastal area of Surabaya has limited access to education, which is reflected on the high number of fishermen who only graduated from elementary school and even no education background at all (around 45 %) [1], moreover, their partners were only elementary school graduates as well [2].

The low quality of human resources in the coastal area of Surabaya lead to low productivity, in which the income was 29.30% below the average income in Surabaya, with the living expenses over IDR 1,200,000. In addition, the fishermen's houses were 34.80% semi-permanent with 49 % of the houses having incomplete assets due to the high cost of usage and maintenance, for instance for boat and motorcycle fuel, electricity bill, and cell phone credits [2].

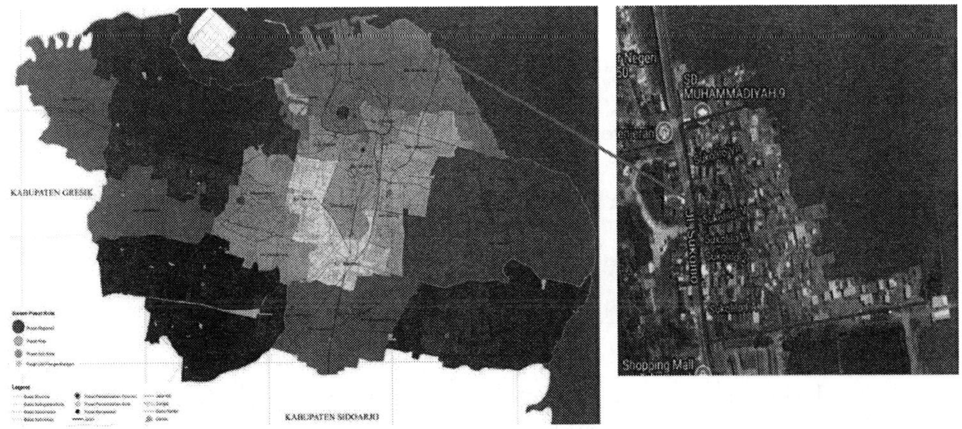


**Figure 1**. Coastal area of Sukolilo, Bulak District, Surabaya.

Realities of poverty, vulnerability, and environment degradation require comprehensive concept in overcoming the coastal area condition of Surabaya. The concept that needs to be developed should accommodate the needs of the coastal society in every aspect of their life and support welfare improvement and sustainable livelihood.

Sustainable Livelihood Approach is a method to improve the understanding of poor households. Unlike other methods, SLA is a multidimensional, integrated and rational approach to reduce poverty [3]. It attempts to fulfil the degree of social, economic, and ecological fulfilment fairly and balanced by combining activities and utilization of assets existing in the system of life [4]. This research was aimed to know the sustainable livelihood level and factors affecting poverty, and to develop a concept to overcome those problems in the coastal area of Surabaya.

## 2. Literature Review

Sustainable livelihood is a way of thinking about goals, scopes, and priorities of human development. Meanwhile, livelihood itself covers the ability, assets (including social resources), and activities needed as a means of survival. livelihood will turn into a sustainable livelihood when the community can overcome a destructive problem, recover from damages, sustain the condition, and even improve the ability and own assets in the present and future, by not relying on natural resources only [5, 6].

Livelihood concept is important in understanding coping strategies since it is a part of--or even considered the same as--livelihood strategies. Livelihood concept tries to define critical factors affecting vulnerability or the power to survive for individual or even families. This idea involves people's assets in activities that set proper standard of living to fulfil other goals, such as risk reduction, and factors that facilitate or inhibit people from getting easy access to assets and activities [7]. A livelihood covers income (cash and materials), social institutions, gender relation, and ownership rights needed to support and guarantee a life [8].

Development approach of sustainable livelihood is contemporary and it tries to correct the modern development approach that is well-known as unfriendly to the environment. Sustainable approach tries to reach the degree of fulfilment that covers social, economic, and ecological aspects fairly and balanced. Meanwhile, social welfare can be achieved through combination of activities and assets utilization [8].

There are five assets important for developing a system of a community reflecting conservatism and populism, which represents sustainable livelihood approach. Those assets are human asset, natural

IOP Conf. Series: Earth and Environmental Science **137** (2018) 012099 doi:10.1088/1755-1315/137/1/012099

asset, social asset, physical asset, and financial asset, all with equal status [9]. Furthermore, local asset consists of five different assets, namely human asset (including abilities, knowledge, and manpower), natural asset (i.e. environmental services), economic and financial assets (e.g. cash, credit, and saving), social asset (such as networking, social relation, and affiliation), and physical asset (for example, infrastructure). An assessment of people's access to the mentioned assets is a first step to diagnose a livelihood [10]. The following is the assets and their indicators table 1.

**Table 1**. Assets of sustainable livelihoods approach.

| Assets | |
|---|---|
| Natural asset | Natural asset from useful resources for a livelihood (for instance, soil, water, wild animal, biological diversity, and environmental resources). |
| Social asset | Network, group membership, trustful relationship, wide access to social institutions; based on the people who are interested in pursuing livelihood. |
| Human asset | Ability, knowledge, manpower, and good health to pursue the strategy of having different livelihood options. |
| Physical asset | Basic infrastructure (transportation, residence, water, energy, and communication), and tools production that can be used in the related livelihood. |
| Financial asset | The financial resources which are available to people (whether savings, supplies of credit, regular remittances or pensions), and also providing them with different livelihood options. |

*Source: Bennett (2010), adapted from Scoones, 1998; in Carney, 1998, p. 7*

Sustainable livelihood approach clearly shows that different stages of analysis are needed to understand the importance of micro and macro connections of livelihood. Furthermore, the understanding can be applied to develop a sustainable livelihood of rural areas, either in individual stage or family stage [11].

## 3. Methodology

To analyze the level of sustainable livelihood in the coastal area of Surabaya, descriptive statistics was used. Descriptive statistics is used for data analysis by describing or delineating collected data without any means for generalizing the conclusion [12]. Meanwhile, to analyze the factors affecting poverty condition in the research area, content analysis was used. Content analysis summarizes the quantitative analysis of messages that depend on scientific method (including objectivity attention, intersubjectivity, a priori design, reliability, validity, generalizability, imitation, and hypothesis testing), without being restricted only for countable variable or in a context of when the messages are made or given [13]. Content analysis technique is used to analyze categories that consist of data reduction by using coded and thematic organisation tools [14]. The steps involved are: determination of code, coded data classification, and data prediction [15].

To determine the poverty concept of the research area based on sustainable livelihood approach, Delphi analysis was used. Delphi analysis technique uses a systematic procedure to conclude a consensus of opinions from experts. This technique is based on data collection and knowledge filtering from experts through questionnaire, with opinions, as the feedback. This technique depends on subjective decision of a group that consists of professional experts in the related field [16].

## 4. Result and Discussion

### *4.1. Analysis of Sustainable Livelihood Level*

The result shows that in the fisherman village of Surabaya's coastal area, there were imbalances on the livelihood asset achievement, or in other words, the household's access to the assets was very low (figure 2).

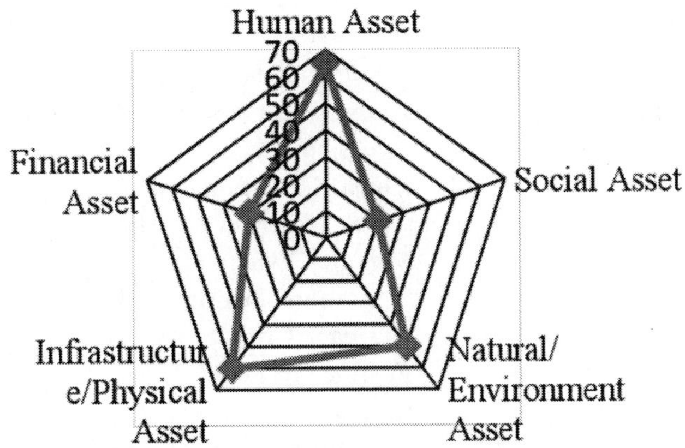


**Figure 2**. Level of Sustainable Livelihood in the fisherman village of Sukolilo, Surabaya.

Fishermen in Sukolilo, in fact, had not been able to reach the level of sustainable livelihood. Most of them could only optimize each asset one at a time. Figure 2 shows that: (1) human asset reached 65 %, (2) social asset reached only 20 %, (3) natural/environment asset reached 50 %, (4) infrastructure/physical asset reached 60 %, and (5) financial asset reached 30 %. Based on the analysis, human asset reached the highest level in sustainable livelihood approach, and the lowest level was achieved by social asset. Nevertheless, this condition needs a serious concern.

### *4.2. Analysis of factors affecting poverty based on Sustainable Livelihood Approach*

#### *4.2.1. Structured interview to the fishermen community in Sukolilo on factors affecting poverty*

Based on structured interview, the researcher obtained information on factors affecting poverty of fishermen community in the coastal area of Surabaya. The structure of the interview was organized based on literature study of related variables. The interview questions asked to obtain characteristics of the community were:

IOP Conf. Series: Earth and Environmental Science **137** (2018) 012099 doi:10.1088/1755-1315/137/1/012099

**Table 2**. Characteristic aspect in the structured interview.

| No | Asset | Indicator |
|---|---|---|
| 1 | Human | Frequency of Patients with Severe Disease<br>Education Level<br>Skills<br>Work Ability |
| 2 | Social | Membership of Organization<br>Level of trust among residents and within organization<br>Networking of society to increase efficiency |
| 3 | Natural / Environmental | Environmental Health<br>Water Production<br>Land Ownership<br>Fisheries Production |
| 4 | Physical | Amount and type of seafood production<br>Accessibility<br>House Condition<br>Clean Water Availability<br>Vehicle Ownership<br>Sanitation Condition<br>Drainage Condition |
| 5 | Financial | Income per Day<br>Expenses<br>Amount and type of saving<br>Debt |

*4.2.2. Content Analysis on Identifying the Factors*

Content analysis process was adapted from Martadwiprani [15] with following steps:

1. Finding the code

Based on the interview transcripts, several codes were made that showed congruency between the collected data and the research variables. Coding process was adjusted with the theme and indicators of research since the content analysis was employed to identify the factors affecting poverty in the coastal area of Surabaya. The code used can be seen in figure 3.

IOP Conf. Series: Earth and Environmental Science **137** (2018) 012099 doi:10.1088/1755-1315/137/1/012099

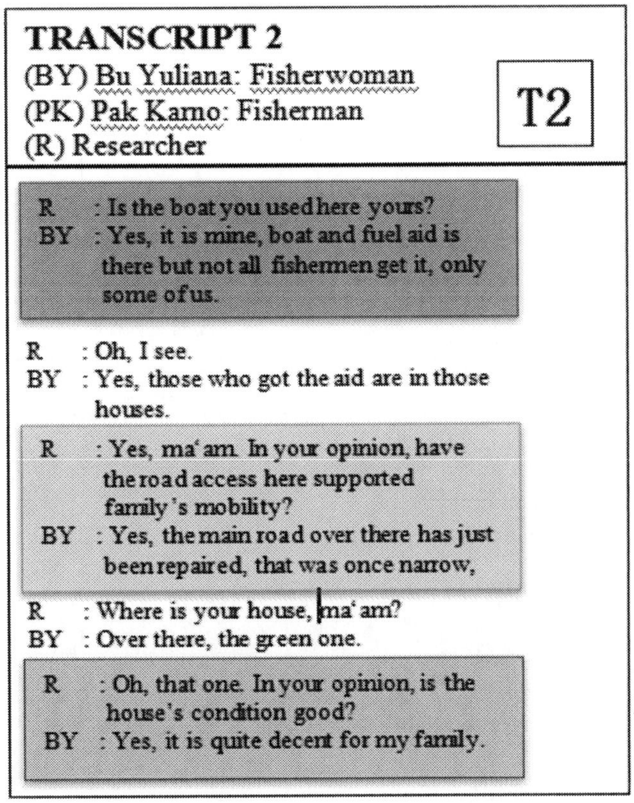

**TRANSCRIPT 2**
(BY) Bu Yuliana: Fisherwoman
(PK) Pak Karno: Fisherman
(R) Researcher

T2

R : Is the boat you used here yours?
BY : Yes, it is mine, boat and fuel aid is there but not all fishermen get it, only some of us.

R : Oh, I see.
BY : Yes, those who got the aid are in those houses.

R : Yes, ma'am. In your opinion, have the road access here supported family's mobility?
BY : Yes, the main road over there has just been repaired, that was once narrow,

R : Where is your house, ma'am?
BY : Over there, the green one.

R : Oh, that one. In your opinion, is the house's condition good?
BY : Yes, it is quite decent for my family.

**Figure 3**. Quotation of codes in interview transcript.

2. Data Classification based on Codes

Several new variables were found in the field data collection process, one of those was the government's involvement in providing physical aid in the form of boats and fuel for all fishermen, but yet, distribution of the aid was uneven. Those variables served as input and supporting data that indicate factors affecting poverty. The indicator codes used in the interview transcript can be seen in table 4.

**Table 3**. The Coding of indicator and interview transcript.

| Variable 1 | Human | Source Text | Remarks |
|---|---|---|---|
| Indicator | Frequency of patients with severe disease | T1.2 , T2.1, T3.3, T4.1, T5.1 | Confirmed |
| | Education Levels | T1.3 , T2.3, T3.4, T4.2, T5.2 | Confirmed |
| | Skills | T1.4 , T2.3, T3.5, T4.3, T5.3 | Confirmed |
| | Work ability | T1.5 , T2.4, T3.6, T4.4, T5.4 | Confirmed |
| **Variable 2** | **Social** | **Source Text** | **Remarks** |
| Indicator | Followed organization | T1.6 , T2.5, T3.7, T4.5, T5.5 | Confirmed |
| | Trust level between community and organization | T1.7 , T2.6, T3.8, T4.6, T5.6 | Confirmed |

IOP Conf. Series: Earth and Environmental Science **137** (2018) 012099 doi:10.1088/1755-1315/137/1/012099

| | | | |
|---|---|---|---|
| | Community bonding networks improving community efficiency | T1.8 , T2.7, T3.9, T4.7, T5.7 | Confirmed |
| Variable 3 | Natural/Environment | Source Text | Remarks |
| Indicator | Environmental health | T1.9 , T2.8, T3.10, T4.8, T5.8 | Confirmed |
| | Water production | T1.10 , T2.9, T3.11, T4.9, T5.9 | Confirmed |
| | Land ownership | T1.11 , T2.10, T3.12, T4.10, T5.10 | Confirmed |
| | Production and fisheries products | T1.12 , T2.11, T3.13, T4.11, T5.11 | Confirmed |
| Variable 4 | Physical | Source Text | Remarks |
| Indicator | Number and types of marine production equipment | T1.13 , T2.12, T3.14, T4.12, T5.12 | Confirmed |
| | Government involvement | T1.14 , T2.13, T3.15 | New |
| | Accessibility | T1.15 , T2.14, T3.16, T4.13, T5.13 | Confirmed |
| | House condition | T1.16 , T2.15, T3.17, T4.14, T5.14 | Confirmed |
| | Clean water availability | T1.17 , T2.16, T3.18, T4.15, T5.15 | Confirmed |
| | Vehicle owned | T1.18, T2.17, T3.19, T4.16, T5.16 | Confirmed |
| | Sanitation condition | T1.19 , T2.18, T3.20, T4.17, T5.17 | Confirmed |
| | Drainage condition | T1.20 , T2.19, T3.21, T4.18, T5.18 | Confirmed |
| Variable 5 | Financial | Source Text | Remarks |
| Indicator | Daily income | T1.21 , T2.20, T3.22, T4.19, T5.29 | Confirmed |
| | Expenses | T1.22 , T2.21, T3.23, T4.20, T5.20 | Confirmed |
| | Numbers and types of savings owned | T1.23 , T2.22, T3.24, T4.21, T5.21 | Confirmed |
| | Loans | T1.24 , T23.1, T3.25, T4.22, T5.22 | Confirmed |

3. Data Prediction

The results of identification process indicated that factors affecting poverty in coastal area of Surabaya were as follows:

a. Financial Asset
   - Expenses for boat's fuel were not comparable to the catches delivered to collectors.

- Many expenses must be spent on household needs such as daily meals, children's school needs, and medication, so that there was no saving done by the fishermen household in coastal area of Surabaya.

b. Human Asset
- The fishermen's low level of education, especially those who went out to the sea who were mostly only Elementary and Junior High graduates.

c. Social Asset
- The low awareness and knowledge of the importance of intensive and comprehensive community involvement in improving community network that might lead to improvement of community efficiency.
- The low level of community involvement in the existing fishermen community organization.

d. Natural or Environmental Asset
- Prominent fishery products in the Fisherman Village were shrimp, sea cucumber (3 kg/day), mackerel (4 kg/day), tuna (6 kg/day), and bamboo clam (2 kg/day). Appropriate technology for fish drying, boat manuals, fumigation and burning, cultivation of milkfish and tilapia also still exists to improve the products. The stipulation and marketing were done through collectors and they were distributed to the customers directly. Environmental sanitation (e.g drainage or river) had been improved, average family size was three people, average income per fishing period was around IDR 100,000 to 160,000. Other types of business existed in the village were grocery kiosks, food stalls, food vendors, shell craftsmen, and *ojek* (motorcycle drivers). Fish processing methods performed were salting, fumigation, cracker making, and shrimp paste making. The fishermen in Sukolilo sometimes did not go to fish for two days. Local Fishermen formed a community called *Maju Makmur* where the fishermen can borrow money for their business. Another existing social activity was *pengajian*

e. Physical Asset
- Until then, fishermen in Sukolilo still used traditional fishing gear such as *waring* or black nets and wooden boats under 5 GT. The fishermen's catch in Sukolilo before the construction of a new bridge at the UPTD tourism area *Taman Hiburan Pantai Kenjeran* Surabaya was abundant. The fishermen who also worked as guide for fishing tours could take passengers. The bridge construction in the UPTD tourism area *Taman Hiburan Pantai Kenjeran* Surabaya had made most of people in Sukolilo experienced profit loss since their catch was less and they could not go to the beach anymore. The regulation given by the management of UPTD tourism area *Taman Hiburan Pantai Kenjeran* Surabaya to the surrounding community only covered a selling location in the tourism area. Meanwhile, the fishing tour guides were not regulated. The catch from fishermen in Sukolilo was sold to the collectors since the fish market in Bulak was already closed. Considering the minimum catch and profit, assistance from the Agency of Marine and Fisheries including fishing gear and boat machine could help the fishermen in Sukolilo.

### 4.3. *The Analysis of Poverty Reduction Strategy Concept in Surabaya Coastal Area based on Sustainable Livelihood Approach*

From the results of several stages (iteration) of Delphi analysis, factors agreed by all respondents were found. These factors would be recommended as the Concept of Poverty Reduction Strategy in the Surabaya Coastal Area based on Sustainable Livelihood Approach.

IOP Conf. Series: Earth and Environmental Science **137** (2018) 012099 doi:10.1088/1755-1315/137/1/012099

Systematically, based on the research variables, the results of merging some concepts of fishermen tourism area development as a means of overcoming poverty would be presented as follows:

a. Physical Asset
   - Utilization of local resources by optimizing local fish processing kiosks.
   - Regular infrastructure maintenance (e.g. roads) and existing facilities to support village development.
   - Improvement of fishermen village through the development of infrastructure and facilities for coastal village or fisherman village.

b. Financial Asset
   - Strengthening local or professional institutions in managing the opportunity for society's savings and business loans with or without village improvement program.
   - Establishing cooperation with various parties to support the fishermen village development in the form of capital investment, promotion, and technology.

c. Human Asset
   - Providing appropriate incentive for society to be more interested in participating intensively in programs and routine activities.
   - Improving society's skills in managing the catch and in fish processing by providing trainings that suit the needs of the fishermen, either formal or informal, in order to formulate policy for fishermen community development.
   - Improvement of human resources and quality of education in fishermen communities by utilizing other communities with higher education and more broad-minded members.
   - Empowerment of coastal women in handicraft production and marketing management of the products (development of processed shrimp, sea cucumber, mackerel, tuna, and *lorjuk*/bamboo clam).

d. Social Asset
   - Improving trust between citizens in terms of familial ties in the location.
   - Providing easy access to information for community members; (2) engaging the society in deciding the development planning related to their needs, (3) developing solidarity; (4) enabling the mobilization of community resources; (5) enabling mutual attainment; and (6) establishing community togetherness and community behaviour.

e. Natural or Environmental Asset
   - Counselling to the fishermen community in terms of condition and weather that can affect fishing activity and should be considered by every fisherman.
   - Providing knowledge for the fishermen community regarding location or where the fish are located or gathered at a particular time and season.

## 5. Conclusion

The fishermen community in coastal area of Sukolilo, Surabaya, had not been able to achieve sustainable livelihood. Most of them were not able to optimally access every asset of sustainable livelihood. From the asset pentagon chart, it can be seen that: (1) human asset on SLA was 65%, (2) social asset on SLA was 20%, (3) natural/environmental asset on SLA was 50%, (4) infrastructure asset on SLA was 60%, and (5) financial asset on SLA was 30%.

The Concept of Poverty Reduction Strategy in Coastal Area of Surabaya based on Sustainable Livelihood Approach was an empowerment concept involving society and stakeholder by performing utilization of local resources, strengthening local professional institutions, establishing cooperation to

IOP Conf. Series: Earth and Environmental Science **137** (2018) 012099 doi:10.1088/1755-1315/137/1/012099

support fishermen village development, improving the quality of human resources and education, as well as empowering coastal women in handicraft production and products' marketing management.

## 6. References


[1] The Agency of Urban Planning and development (BAPPEKO) Surabaya 2011 Strategic Planning of Coastal Area in Surabaya

[2] Harlianingtyas I, Kusrini, Dwi E, Susilaningrum, Destri 2013 *J. Sci. Arts. Pomits.* **2**(1) 2337-3520 (*Preprint* 2301-928)

[3] Kamaruddin, Roslina 2014 *J. Soc. Econom. Res.* **1** 108-117

[4] Saragih S, Lassa J, Ramli A 2007 The Framework of Sustainable Livelihood

[5] Maas L T 2015 The Effect of social capital on governance and sustainable livelihood of coastal city community Medan 2015 *Glob. Conf. on Business and Social Science* 2nd GCBSS-2015 Bali Indonesia.

[6] Babuloa B 2008 *Agric. Syst.* **98** 147–155

[7] Allison, Edward H 2001 *Marine Policy* **25** 377–388

[8] Ellis, F 2000 *Rural Livelihoods and Diversity in Developing Countries* (Oxford: Oxford University Press)

[9] Allison, Edward H and Horemans, Benoit 2006 *Marine Policy* **30** 757–766

[10] Morse S and N Monaniara 2013 *Sustainable livelihood approach* A Critique of Theory and Practice (Dordrecht: Springer Verlag)

[11] Lisocka B 2015 Sustainable rural development or (sustainable) rural livelihoods strategies for the 21st century in peripheral regions University of Warsaw Poland

[12] Indriantoro N, Bambang S 2009 Business research methodology for accounting and management (Yogyakarta : BPFE Yogyakarta) 1st ed.

[13] Kaur M 2012 Content analysis of journal literature published from UK and USA. University of Nebraska – Lincoln DigitalCommons@University of Nebraska - Lincoln

[14] Marques C A N 2016 Comparing pss design models based on content analysis *Proc. CIRP* **47** 144 – 149

[15] Martadwiprani H dan Dian R 2014 Content Analysis in Identifying the Characteristics of Economic Society in Brondong Coast, Lamongan. *J. Eng.* POMITS **2**(2) ISSN: 2337-3539 (*Preprint* 2301-9271)

[16] Yan H 2016 The lectotype optimization study on submerged floating tunnel based Delphi method *Proc. Eng.* **166** 118 – 126.

IOP Conf. Series: Earth and Environmental Science **137** (2018) 012100 doi:10.1088/1755-1315/137/1/012100

# Promoting innovative business in the fishery sector in West Java, Indonesia

**A Nurhayati [1], I Aisah[2] and A K Supriatna[2]**

[1] Faculty of Fisheries and Marine Science, Padjadjaran University Indonesia
[2] Faculty of Mathematics and Natural Sciences, Padjadjaran University Indonesia

E-mail: atikah.nurhayati@unpad.ac.id

**Abstract.** West Java represents an important fisheries production center in Indonesia, owing to the abundant capture and aquaculture resources. However, the intrinsic characteristics of fish products such as perishable, voluminous, and seasonal currently prevent fisheries from having brought significant economic contribution to the province. In line with it, this research was aimed to analyze and identify leverage factors that will lead to fishery-based innovative business in West Java. Data used in this research were primary and secondary ones, which were collected through surveys involving 30 respondents representing fish processors and the same number representing consumers. A *Focus Group Discussion* (FGD) was also carried out to verify the collected data. Analytical tools adopted in this research were fishery *triangle product model.* Based on the analyses, it was found factors influencing the success of a fishery innovative business in West Java, Indonesia were consecutively: the existence of derivative products, product processing innovativeness, product price competitiveness, market place, and promotion. Based on the fishery *trianggle product model, it was found that fish onboard handling*, post-harvest *handling*, and *procesing was in the* development stage and therefore these production nodes need a particularly high attention.

## 1. Introduction

West Java is one of the provinces in Indonesia which have potential fisheries and marine resources. West Java province has varied tourism potential. It has tourism potential in the areas around its south and north coasts. Marine tourism can be developed on Java, for both of the beauty of its panoramic beaches and the various kinds of biota it has.

Some marine tourist destinations such as Pangandaran Beach and Pelabuhan Ratu have been well-known and crowded by domestic and foreign tourists. Pangandaran is the most popular tourist destination among domestic tourists as it has fantastic attractions. Thus, there is a possibility that Pangandaran Beach becomes a leading tourist destination in West Java. The same goes to Pelabuhan Ratu. West Java has the potential to develop fishery products, in the form of both processed and fresh fish, as well as culinary attraction. The trend of fish consumption in 2011–2015 showed an increase by 6.27%. The increase in fish consumption over the last 5 years is the result of various campaign efforts and activities themed the fond of fish for the community. The average fish consumption was 36.12 kg/cap/year.

Fisheries are all activities related to the management and utilization of fish resources and the environment, which range from pre-production, production, processing, marketing, to implementation in a fishery business system (Law 31/2004 Chapter l Article 1, paragraph 1). Fishery production must

IOP Conf. Series: Earth and Environmental Science 137 (2018) 012100 doi:10.1088/1755-1315/137/1/012100

have a market orientation value. Market orientation is a key factor in the effort to increase the value obtained from fishery commodities [1, 2, 3]. Marketing consists of social and managerial process, in which individuals or groups obtain what they need and want by creating, offering, and exchanging products and value to others [4].

Fishery marketing is defined as the performance of all business activities involved in the flow of production, which include capture fisheries and aquaculture in both fresh and processed forms from the point of production until reaching the hands of end consumers [5]. The definition of fishery marketing management also recognizes the interdependence between fishermen or cultivators and merchants or marketing wholesalers. The fishery production process does not stop at the fisherman or cultivator level; fishery marketing activities complement semi-fishery production process, which is not subsystem or traditional. While it is true that there will be no fish supply without the fishermen and cultivators, the relationship between fishermen, farmers, and fish marketing companies, both competitive and complementary, is still deemed necessary [5].

The orientation of fishery products marketing is related to innovation, product development, and competitiveness. The development of the fishing industry still faces some limitations in the design and creation of sustainable innovation, which can not occur naturally. However, West Java has the advantage of fishery resources that require further handling in that market-oriented innovation for its regional economic development. Market-oriented innovation strategy is a major factor in the increase of industrial capacity [6, 7, 8], particularly in the fishery sector.

In order to improve the competitiveness of fishery products, it is necessary to analyze various aspects influencing it in relation to the driving and inhibiting factors of the fishery business which disrupt the production, price, distribution, and promotion of fishery products in West Java. Mapping the innovation of fishery products by adopting market needs is key in the strategy to enhance the competitiveness of fishery products. The innovation of fishery products that meet the market demand is important for the fishery industry in West Java for them to be more acceptable and competitive.

The challenges of the development of the fishery industry are related to the upstream and downstream sectors, which range from production to marketing. The fishery industry in West Java may get benefits by increasing its supply to satisfy the needs in the fish product processing chain. The competitive market forces [9]. The VRIO model mentions that sustainable competitive advantage depends on four major factors: 1) value, which is positive customer evaluation of the product offered; 2) rareness, which is positive customer evaluation of the scarcity of marketed products compared to other products that satisfy customers; 3) immitation, which is the extent to which the product offered can be protected from imitation by competitors; and 4) organization, which is the ability of a reliable marketing organization. The value and rareness of the product offered may include customer ratings of all 4P in the marketing concept, product quality, and strategic location.

Market-oriented innovations can be defined as "ideas, products or practices that are new to adopters and provide new potential in the marketplace" [7]. Market-oriented innovations can drive performance and create more value for customers. Market-oriented innovation is acceptable to customers and able to meet consumer tastes [11]. The synergy between the industry and the government and even community groups in some cases encourages innovation and new partnerships to improve the quality standards of fishery products [1].

The innovation system is a network of institutions in the public and private sectors whose interactions initiate, import, modify and diffuse new technologies [12]. Meanwhile, according to Nelson [13], innovation system refers to a group of actors who, together, play an important role in influencing innovative performance (innovative performance). Arnold *et al.* [14] used the term "national research and innovation system", which means all actors and activities in the economy necessary for industrial and commercial innovations and that can lead to economic development. Innovation system creates new knowledge and guides the direction of the provider's search process and technology users [15]. Thus, innovation system actually includes the basis of science and technology, including educational, research, development, and engineering activities.

IOP Conf. Series: Earth and Environmental Science **137** (2018) 012100 doi:10.1088/1755-1315/137/1/012100

The effectiveness of a company's marketing is strongly influenced by the marketing communication system. The more effective the communication is, the more proportional it is to the marketing. Marketing communication can be carried out in several ways of communication called marketing promotion mix, which consists of five main communications [16].
Promotion as: promotion as communication that informs custemers of the existence of products, and persuades them that they have satisfying capabilities[17, 18]. Iinterpret the promotion as; promotion is an exercise in information, persuasion and conversely, a person who is persuaded is also being informed [19]. Promotion as: "coordination of all efforts initiated by the seller of goods and services or introducing an idea [20] .

## 2. Methods

This research was descriptive research. It was conducted to describe the aspects of marine tourism development. It was conducted by employing fishery triangle product model with multiple linear regresion. Multiple regression offers a way to address these issues. If we put all of the variables we have into one analysis, we can assess the impact of one factor when another factor is taken into account.

This research used survey method, by which information was obtained from 30 respondents that were taken by snowball sampling technique. Focus Group Discussion was also held to obtain opinions from the research objects. Based on the types of data and analysis used, the design of this study was a combination of quantitative and qualitative research.

## 3. Results and Discussion

West Java has coastal areas, especially in its northern part, including Pelabuhan Ratu in Sukabumi and Pangandaran. Coastal areas have a strategic meaning because they are transitional. The wealth of these resources attracts various parties to utilize the resources and various agencies to regulate the utilization of one of them as marine and culinary tourism. Many domestic and foreign tourists have enjoyed the beaty of Palabuhan Ratu and Pangandaran Beach, and many of them buy typical souvenirs of the area.

The restaurants outside hotels are classified into several types according to the target and market segments, main purposes, locations, and types of foods and beverages offered. The types are as follows: (1) automat restaurant, which uses automatic machines with which anyone can take their desired food by entering a number of coins according to the stated price of the meal; (2) delicatessent, which is a specialty restaurant that sells specialty foods like seafood; (3) bistro, which is a small restaurant [21] usually located in a crowded shopping center where the crowd pass through the place; (4) canteen, which is a restaurant that provides snacks. At first this restaurant was under a military camp. Now the term canteen is often used as a restaurant for employees in companies or office buildings; (5) cafe, which is an informal restaurant like hotel coffee shop, but it stands alone outside the hotel. It prioritizes the sale of fast food and provides musical accompaniment like a hotel lounge; (6) rail road catering, which is a restaurant inside a train. Guests can dine at the restaurant or tell their orders to the waiter, and the orders are served with a stroller or tray; (7) cafetaria, which is a self-service restaurant, where guests take their own preferred dishes; (8) steak house, which is a restaurant specializing in selling roasted food; (9) coffee pot, which is an informal little restaurant selling at prices affordable by economic group, usually in the form of a stall, and is placed on the side of the road; (10) drive inn, which is in a car, and the food is served in cars. It only sells practical foods such as sandwiches, ice cream, and so on; (11) rathkller, which is a typical German restaurant usually located at the bottom of a staircase or in a basement; (12) common, which is a restaurant that serves food for many people in one long table; its pride is in one institution; and (13) specialies restaurant, which is a specialty restaurant that serves typical dishes from a certain region or country. According to Government Regulation of the Republic of Indonesia No. 16 of 1997 dated June 18, 1997, franchise is a form of cooperation, in which the franchisor grants the franchisee the right to use its intellectual property right, such as the name, trademark of its products and services, and operating system. In return, the franchisee pays a certain amount of money for the intellectual property held by the bidder for a certain period of time.

IOP Conf. Series: Earth and Environmental Science **137** (2018) 012100 doi:10.1088/1755-1315/137/1/012100

Fishery triangle product model is an institutional approach that is carried out for handling competitive fishery products, involving producers—fishermen and fishery product processors, distributors—intermediary traders, and the government, before they reach the end-users (figure 1). In consequence, there is a need for consolidation and integration between fishery business branches, both horizontally and vertically, in order for them to be more rational, effective, and productive.

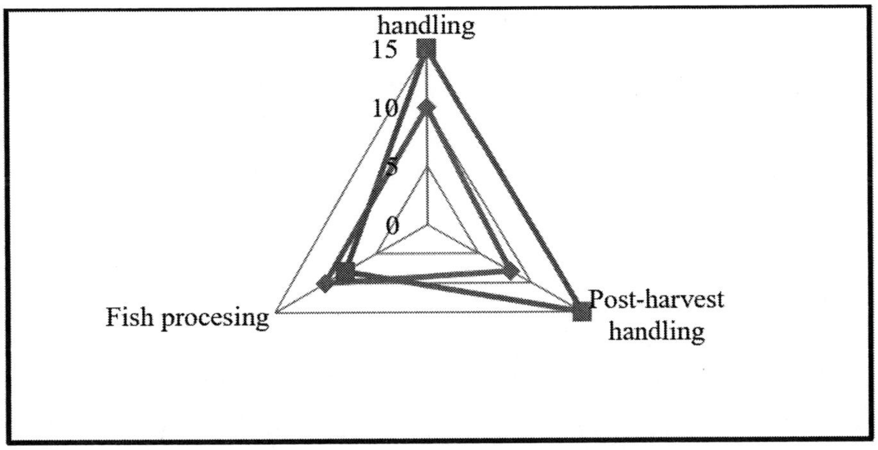


**Figure 1.** Fishery triangle product model

However, in the network economy, e-commerce application in the field of fish trade is inevitable. Electric commerse has provided a good opportunity for the business in fishery development to increase sale volume and profit by using the fisheries triangle product model. The promotion of marketing platform of fishery products is targeted to direct and indirect customers and food processors. Direct customers are individual customers, who purchase lesser, and the products purchased are generally a wide variety of rough, finishing class fish. Indirect customers include all kinds of hotels, enterprises, institutions, supermarkets, and chain stores. For these customers, the platform is mainly used to obtain information. Meanwhile, food processors buy main raw materials through online auction of the fish landed. They purchase the whole seafood and related production equipment. During the transaction of the landed fish, customers must register as ordinary members first, then submit some necessary personal information.

IOP Conf. Series: Earth and Environmental Science **137** (2018) 012100 doi:10.1088/1755-1315/137/1/012100

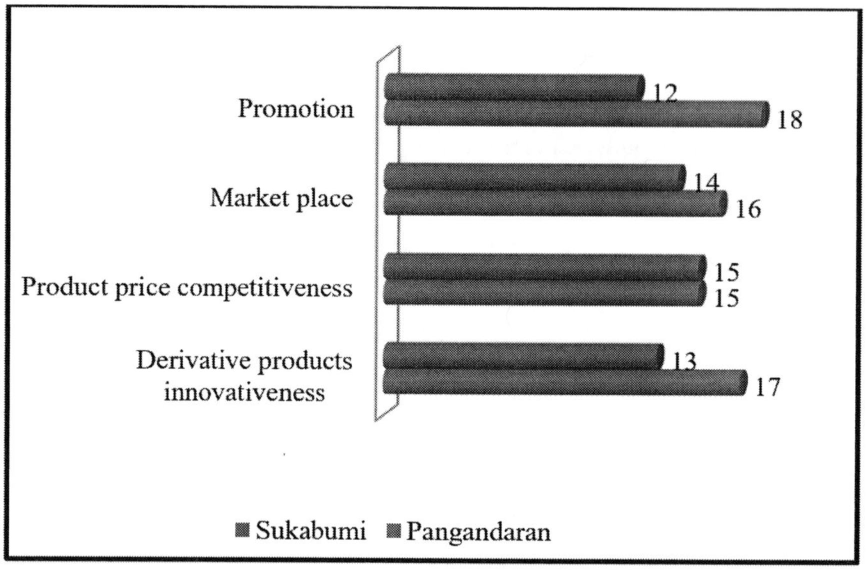


**Figure 2.** Promoting innovative business in fisheries

According to figure 2, innovative business in fisheries is related to promotion, marketplace, competitiveness of product prices, and derivative product innovations. Fishery derivative product processing is aimed to create added value for a product to give it a high economic value [5]. Farming system development is a way to increase the income of farmers and their families. bio-physical, socio-cultural, institutional, and economic conditions dynamic. The high potential of capture fisheries in Pangandaran and Sukabumi gives an opportunity for fish processing development. With this fish processing development, the people or fishermen do not only sell raw fish. The development of processed fish business in the micro, small, and medium enterprise (SME) sector has a characteristic of not-too-large capital, and its scale is limited. Micro, small, and medium enterprises (SMEs) in Pangandaran have penetrated the market of various regions and even abroad. However, a lot of SMEs in Pangandaran do not have any license, patent, or halal label. Some SME products such as processed fishery products developed by the Business Development Center (BDC) of Pangandaran have entered the international market. Patent and halal lebel are very important for ready-to-market food and beverage products.

The processing fisheries industry can encourage many supporting industries, including auxiliary material industry, ice factory, packaging material industry, and fish processing and packaging machine industry. The problem having been faced by fish processors to date is the frequent scarcity of raw material supply. The scarcity of fish used as raw materials of processed products can occur due to the following reasons: 1) During the season with no fish, fishermen prefer exporting the fish to selling them to the domestic processors; 2) the fish stock in the sea decreases; or 3) there is no harmonious relationship between fishermen and processors.

The prices set from fishermen to end-consumers are very volatile, which depends on the demand and fish supply or availability. Certain fish commodities are sought by many exporters. The four main factors determining the competitiveness of a commodity are factor conditions, demand conditions, related industries, and related supporting industries. In addition, structural conditions, competition, and industry strategy also determines the competitiveness of a commodity.

Marketing is an integrated communication arrangement process that aims to provide information about goods or services to satisfy human needs and wants. Marketing begins with the fulfillment of human needs, which eventually grows into human desire. The process of fulfilling human needs and wants is the concept of marketing. The pattern is built on the mission of parenting from the big to the small (large capital, large human resources, great technology, etc.), and from the strong to the weak, but in the same needs position and still on the basis of mutual benefit, mutual need, and strength. The core pattern of plasma, is the pattern of relationships between partner groups and partner companies, in which the partner group acts as the core plasma. A marketing channel is a series of interdependent organizations engaged in a process that makes a product or service ready to be consumed by its customers. The marketing channels in the capture fishery of Sukabumi Regency are divided into marketing channels of fresh commodities and marketing channels of processed fish products. Fresh fish commodity marketing channels include two channels as follows.

Fisherman → Fish Auction Place → Fish Basket → Market → Wholesaler → Customers

Fisherman → Fish Auction Place → Fish Basket → Customers

**Figure 3.** Fresh fish commodity marketing channels

A booth is a place where people sell by typical region. Souvenir outlets in this case perform the exchange function because of the activities of buying and selling processed fish. Souvenir outlets also do some sort of standardization in determining which goods to be sold. Therefore, the souvenir outlets perform the function of the facility. In performing its physical function, the souvenir store stores processed goods. The following innovations should be developed to accompany the fishery industrialization program: 1) innovation of environmentally friendly fishing gear technology; 2) technological innovation and superior seed biotechnology; 3) innovation of cultivation technology of both natural and artificial feed; 4) biotechnological technology innovation of the cleaning of water and the environment; 5) biotechnological innovation of vaccine and fish medicine; 6) innovation of cultivated fish processing technology, including technology of fish waste utilization; 7) biotechnological innovation of active and nutrasetikal compounds from marine materials; 8) innovation system or industrial model according to resources, be it fish resources, facilities and infrastructure or human resources; 9) innovation of industrial policy system in accordance with the physical, social, economic, political, and cultural conditions of Indonesia.

The Bussines Development Center of Pangandaran emphasizes materials on SME management trainings. It is expected that the dynamic development of the tourism sector can encourage SMEs to develop more interesting and challenging creative economy. The Bussines Development Center of Pangandaran can bring a positive value for the economic development of small businesses in Pangandaran, especially through the creation of new opportunities for Business. The 2017 Sukabumi Expo in principle is an event for all society levels of Sukabumi District, thus it engages representatives from several elements of society as well as the local government of Sukabumi Regency

## 4. Conclusions

Based on the analysis above, it was found that the factors influencing the success of a fishery innovative business in West Java, Indonesia are the existence of derivative products, product processing innovations, product price competitiveness, marketplace, and promotion. Based on the fishery triangle product model, it was found that fish onboard handling, post-harvest handling, and procesing are in the development stage;therefore, these production nodes need a particularly high attention.

## 5. Reference

[1] FAO and OECD 2011 *Price volatility in food and agricultural markets* (Policy Responses)

[2] Grunert K G, Trondsen T, Campos E M and Young J A 2010 Market orientation in the mental models of decision makers: Two cross-border value chains *International Marketing Review* **27**(1) 7-27
[3] Tondsen T 2012 Value chains, business conventions, and market adaptation: A comparative analysis of Norwegian and Icelandic fish exports *The Canadian Geographer* **56** 459-473
[4] Kotler P 1980 *Management marketing* (Yogyakarta: Fakultas Ekonomi Universitas Gadjah Mada)
[5] Nurhayati A 2016 Derivatif analysis of economic and social aspect of added value minapadi (paddy-fish integrative farming) a case study in the village of sagaracipta ciparay sub district Bandung West Java Province Indonesia *2nd Int. Symp. on Aquatic Products Proc. and Health* (Bandung: Aquatic Procedia) **7** 12–18
[6] Porter M E 2008 The five competitive forces that shape strategy *The Five Competitive Forces That Shape Strategy* (USA: Harvard Business Review) **86** p 78-93
[7] Rogers E T 2003 *Diffusion of innovations* (New York: Free Press) 5th ed.
[8] Afuah A 2009 *Strategic Innovation: new game strategies for competitive advantage* (New York: Routledge)
[9] Barney J B 2002 *Gaining and sustaining competitive advantage* (New Jersey: Prentice-Hall) 2nd ed.
[10] Grunert K G, Trondsen T ,Campos E M and Young J A 2010 Market orientation in the mental models of decision makers: Two cross-border value chains *International Marketing Review* **27**(1) pp 7-27.
[11] Lindkvist K B and Sanchez J L 2008 Conventions and innovation: a comparison of two localized natural resource-based industries *Regional Studies* **42** pp 343–354
[12] Freeman 1987 http://id.wikipedia.org/wiki/Sistem_inovasi, diakses 28 Nov. 2011
[13] Nelson R 1993 *National innovation systems* (New York: Oxford University Press) ed.
[14] Arnold E, Stefan K, and Barend van der M 2001 A Singular Council: evaluation of the research council of norway *Technopolis*
[15] Liu X, White S 2001 Comparing innovation systems: a framework and application to China's transitional context
[16] Kotler 2015 *Marketing an Introducing Prentice Hall twelfth edition* (England: Pearson Education Inc. Kotler)
[17] Alma B 2007 *Management marketing and marketing sercvice* (Bandung :Alfabeta) (Text in Indonesia)
[18] Enis B M 1974 Marketing principle *Goodyear Publ. Coy. Inc.* California (Text in Indonesia)
[19] William J S 2004 *Marketing principles* (Jakarta: Erlangga) 7th ed. chapter 1 (Text in Indonesia)
[20] Michael L R 1982 *Advertising and communication management* (New Delhi : Doubleday) (Text in Indonesia)
[21] Torsina M. 2000 *Successful Restaurant Business* (Jakarta : PT. Bhuana Ilmu Populer) (Text in Indonesia)

**Acknowledgement**
The authors thank Padjadjaran University for the funding of this work through the scheme of Academic Leadership Grant (ALG) with contract number 855/UN6.3.1/PL/2017.

IOP Conf. Series: Earth and Environmental Science **137** (2018) 012101 doi:10.1088/1755-1315/137/1/012101

# Analyzing the development of Indonesia shrimp industry

**L A Wati** [1]

[1]Fisheries and Marine Science Faculty, Brawijaya University, Veteran street, Malang, East Java, Indonesia.

E-mail: linaasmara@ub.ac.id

**Abstract.** This research aimed to analyze the development of shrimp industry in Indonesia. Porter's Diamond Theory was used for analysis. The Porter's Diamond theory is one of framework for industry analysis and business strategy development. The Porter's Diamond theory has five forces that determine the competitive intensity in an industry, namely (1) the threat of substitute products, (2) the threat of competition, (3) the threat of new entrants, (4) bargaining power of suppliers, and (5) bargaining power of consumers. The development of Indonesian shrimp industry pretty good, explained by Porter Diamond Theory analysis. Analysis of Porter Diamond Theory through four main components namely factor conditions; demand condition; related and supporting industries; and firm strategy, structure and rivalry coupled with a two-component supporting (regulatory the government and the factor of chance). Based on the result of this research show that two-component supporting (regulatory the government and the factor of chance) have positive. Related and supporting industries have negative, firm and structure strategy have negative, rivalry has positive, factor condition have positive (except science and technology resources).

## 1. Introduction

As an archipelago country, Indonesia has a great potential in the fisheries industry, especially it has a fairly broad sea area of 7.9 million $km^2$ and a long coastline of 80.791 $km^2$, with extensive aquaculture and fish ponds spread in some parts of the country. Indonesia has a vast 81 percent water area of the total area of Indonesia, so it is not impossible that Indonesia could rule business of fisheries world wide. Since 1987, Indonesia has become one of the most important suppliers of shrimp in the world [1]. Fresh and frozen shrimp are the main export commodity of marine and fisheries sectors. The contribution of shrimp exports to national income is relatively large, when compared with the biggest export of agricultural commodities such as coffee, tea, spices, tobacco, and cocoa beans. Indonesia is one of the world's largest shrimp exporting countries, especially to the market in Japan, the United States, and the European Union. However, the position of Indonesia in recent years has been declining and outnumbered by Thailand, China, and Vietnam. This is mainly due to the decline in the competitiveness of Indonesian shrimp in the international market, either because of the efficiency of the business, the ability to penetrate the market, as well as the quality and sanitation issues. Therefore, integrated and synergistic measures are needed to improve competitiveness and business efficiency so that the performance of the national shrimp industry can quickly bounce back. Nevertheless, it is estimated that shrimp will still be excellent fishery product in the next decade and it has a very bright market prospects. In the era of globalization, every country cannot avoid the multinational trade in order to survive. Shrimp market is on the rise despites the difficult global economic situation [2]. In

IOP Conf. Series: Earth and Environmental Science **137** (2018) 012101 doi:10.1088/1755-1315/137/1/012101

Japan, import of raw frozen shrimp fell in January-July 2011, but increased by 8% for processed shrimp.

In Indonesia, the importance of small shrimp farmers cannot be ignored although their production contribution is less significant. Constraints for shrimp production in Indonesia include low marketing efficiency, limited shipping port, and the lack of optimal government policies related to efforts to improve productivity and quality. In the first 5 months in 2011, Thai shrimp exports earned 552 million USD, a drop of 2.23 % from the same period in the previous year. Export to the EU in that period was recorded at 79 million USD, a decrease of 16 percent due to crisis in the Europe in 2011, Thai shrimp exports were as high as 260 million US dollars, rising 7.9 percent from 2010. Europe has been the third largest shrimp market for Thailand after USA and Japan, while Thai frozen shrimp in 2011 was valued at 1.7 billion dollars [3].

Local constraints for Indonesian shrimp industry are low productivity and quality. Other constraints for Indonesia in global shrimp market is low marketing efficiency, limited shipping port, and lack of optimal government policies related to efforts to improve productivity and quality. Thus, the problem discussed in this research was how the development of Indonesian shrimp industry is.

## 2. Analysis methodology

### *2.1. Analysis of porter's diamond theory*

The pure competition model implies that risk-adjusted rates of return should be constant across firms and industries. Michael Porter provided a framework that models an industry as being influenced by five forces. A strategic business manager seeking to develop an edge over rival firms can use this model to better understand the industry context in which the firm operates.

**Figure 1.** Porter's five forces analysis.

Porter's five forces analysis is a framework for industry analysis and business strategy development developed by Michael Porter (figure 1). There are five forces that determine the competitive intensity in an industry, namely (1) the threat of substitute products, (2) the threat of competition, (3) the threat of new entrants, (4) bargaining power of suppliers, and (5) bargaining power of consumers. To determine the factors affecting competitiveness in the liberalized market, Porter's Diamond Theory analysis was used in this study along with the use of qualitative method to complete the second objective of this research. Porter's diamond theory has four main factors, which are factor condition, demand condition, firm strategy, structure, and rivalry, related and supporting industries.

IOP Conf. Series: Earth and Environmental Science **137** (2018) 012101 doi:10.1088/1755-1315/137/1/012101

## 3. Results and discussion

Development of shrimp industry in Indonesia cannot be separated from the influence and development of the international shrimp market. The position of shrimp export from Indonesia in international market is influenced by competitiveness with other exporters and the condition of other exporters, which are considered as competitors. The development of Indonesia's shrimp industry is influenced by many factors, with competitiveness as the major factor. Than to analyze the development of Indonesian shrimp industry can use approach Porter's Diamond Theory. Analysis of Porter Diamond Theory through four main components namely factor conditions; demand condition; related and supporting industries; and firm strategy, structure and rivalry coupled with a two-component supporting (regulatory the government and the factor of chance).

### *3.1 Factor condition*

a. Natural resources

Indonesia, as a tropical archipelago, make up ground water is fertile and contains abundant natural wealth results. In terms of fisheries, Indonesia is supported by a large sea area that covers 5.8 million $km^2$ with 0.8 million $km^2$ of territorial waters, 2.3 million $km^2$ of archipelagic waters / sea, and 2.7 million $km^2$ of the Exclusive Economic Zone (EEZ). Fish and shrimp are among the fisheries resources in Indonesia . Shrimp from Indonesia is in demand from domestic and international markets.

b. Human resources

It is known that the employment in shrimp aquaculture reached 494.302 people in 2010. It was projected that in 2013 or 2014, there would be 624,025 people work in shrimp Industry due to the success of shrimp culture revitalization program [6].

Due to the growth of employment in the aquaculture industry, there is an increasing number of shrimp exporting firms that reached more than 200 companies. There are many stakeholders involved in shrimp industry, namely small and large enterprises, shrimp farmers, and government agencies.

c. Science and technology resources

Lack of advanced science and technology in shrimp cultivation and processing has caused the shrimp from Indonesia to have low quality; thus, their price is low and it gets less competitive. Unlike in Indonesia, Thailand, as one of the world's largest shrimp exporter, has implemented food traceability system in the shrimp processing industry since 2006, which traces the trade record based on Radio Frequency Identification (RFID) technology .

d. Capital resources

Capital resource is an important factor in the development of shrimp aquaculture. For all sectors of fisheries, the Ministry of Marine Affairs and Fisheries Resources has formed a partnership with the National Bank to overcome the problem of capital resources.

e. Infrastructure resources

Shrimp industry is one of the fishery sectors that have been equipped with infrastructure facilities. For example, one of the leading shrimp exporters in Indonesia, PT Citra Dipasena Darmaja (PT Aruna Wijaya Sakti) that is now owned by PT Central Protein, has been built since the early 1990s and not it is the largest and highly successful aquaculture company in Indonesia.

### *3.2. Demand condition*

*a.* Domestic demand

Production of shrimp in Indonesia for domestic consumption is not large, only 5%, while as many as 95% of the production is exported. This trend is due to the preference of Indonesian towards red meat. Domestic Demand of Fresh Shrimp is shown in Figure 2.

*b.* Export demand

Demand for Indonesia's shrimp export can be measured by the volume and value of the export that tends to increase. According to WTO [7], the economic value of the shrimp worldwide increases by 11.5 % per year in average. There was also an increase in the world's consumption of shrimp from

IOP Conf. Series: Earth and Environmental Science **137** (2018) 012101 doi:10.1088/1755-1315/137/1/012101

1,537,967 tons in 2004 to 1,670,925 tons in 2007. Table 2 shows that the average volume and value of Indonesia's shrimp export increased in 2002-2011. It reached 124,765 tons in 2002 and 158,062 tons in 2011, with the value of 836,563 USD in 2002 to 1,309,674 USD in 2011.

### *3.3. Related and supporting industries*

*a.* Related industries

Shrimp export-related industries include seed supplier, shrimp supplier, and shrimp feed supplier. Shrimp seed industry is still underdeveloped in Indonesia due to the difficulty of obtaining high-quality *benur* (shrimp seed). Local breeding centers are not able to grow high quality shrimp sees, thus causing them to have fewer buyers. There is a new shrimp breeding brood stock development hall in Situbondo, East Java. However, there are no research centers specializing in shrimp breeding yet. One of the obstacles faced by shrimp breeding centers are the facilities provided including technology to select parent shrimp with good quality, Considering the production system in Indonesia, efforts to meet the quality standard have been the major challenge for Indonesia's shrimp production in the near future. It was reported that there was 380,000 hectares of brackish-water shrimp ponds in Indonesia with 80% of it was cultivated traditionally. Production of Shrimp using this system is vulnerable to virus infestation; thus, the use of antibiotics, such as chloramphenicol, is common.

*b.* Supporting industries.

Processed shrimp has not given significant role in Indonesia's shrimp exports. This is due to the limited ability of shrimp processing industry to create new value-added innovation to increase competitiveness. The existing processed products are limited only to domestic market and cannot be exported. Indonesia's shrimp exports currently consist of frozen shrimp (80 %)and non-frozen shrimp in the form of as fresh shrimp and packaged shrimp (20 %).

### *3.4. Firm strategy, structure, and rivalry*

The competitiveness of Indonesia's shrimp in the international market is relatively good and Indonesia is currently one of the largest shrimp producers in the world next to China and Thailand. There are 182 shrimp exporting companies in Indonesia, however, the Southeast Asia domestic market is 50% dominated by the Chakroen Phokphand Group. Unfortunately, Indonesia has not developed any specific strategies to increase the competitiveness of its shrimp industry. This situation may lead to a decline in market and market value of Indonesia's shrimp export as compared to other competitors.

### *3.5. Government regulations (role of the government)*

In developing shrimp industry in Indonesia, the government already has prepared some strategies, such as by creating Indonesia High Commission on Shrimp (Komisi Udang Indonesia (KUI)). KUI was formed by the Minister of Marine Affairs and Fisheries (Number.KEP.42/MEN/2004) of Indonesian Shrimp Commission. KUI's duties include establishing special programs such as Health Management of Shrimp and the Environment and Effluent and Solid Waste Management. Shrimp Health Management program is aimed to prevent environment-based diseases. The program describes the steps that must be taken to prevent diseases. Meanwhile, Effluents and Solid Waste Management program aims to improve the quality of wastewater ponds that has been contaminated by waste. Post-harvest management program aims to guarantee quality assurance and food safety.

### *3.6. Opportunities*

There is a promising opportunities for Indonesia's shrimp in international market. Indonesia has vast ocean, great fisheries resources, and it is not one of the biggest shrimp exporters in the world, next to China and Thailand. Japan is one of the main market targets for shrimp export from Indonesia due to the local preference towards seafood. Based on the analysis of Porter's Diamond, there are strengths and weaknesses of Indonesia's shrimp industry. Shrimp production in Indonesia is likely to increase each year.

IOP Conf. Series: Earth and Environmental Science **137** (2018) 012101 doi:10.1088/1755-1315/137/1/012101

The shrimp industry in Indonesia has created a large number of jobs and it is predicted that it will keep increasing. However, the use of advanced science and technology in shrimp farming is still very minimum and majority of the farming activities are still performed traditionally.

The government of Indonesia has shown great support for the shrimp industry by cooperating with financial institutions to provide loans for those involved in the industry. Similarly, the development of infrastructure needed to support the shrimp industry has also been carried out. Currently, the production of shrimp in Indonesia has covered for domestic demand. Concurrently, international demand is also increasing and it has attracted shrimp exporters from Indonesia and Indonesia's shrimp export value has been increasing. In 2007, Indonesia was second to Thailand in terms of frozen shrimp export.

Unfortunately, the local shrimp market is still largely dominated by the Charoen Phokphand Group. Moreover, the local shrimp farmers still have not found a working strategy to increase their competitiveness in the world market. There are also few other hindering factors for the development of the local shrimp industry, such as the lack of supporting industries and lack of supply of high quality shrimp seeds. Those factors affect the Indonesia's shrimp export.

Indonesia has a great opportunity to increase the competitiveness of the shrimp industry in the world market. Therefore the government needs to take more proactive approach and set regulations that can help in improving the quality of shrimp from Indonesia.

Figure 2 shows the strength and weaknesses of Indonesia's shrimp industry after analysis with Porter's Diamond Theory. Positive sign represents a strong point, while a negative sign shows a weak point.

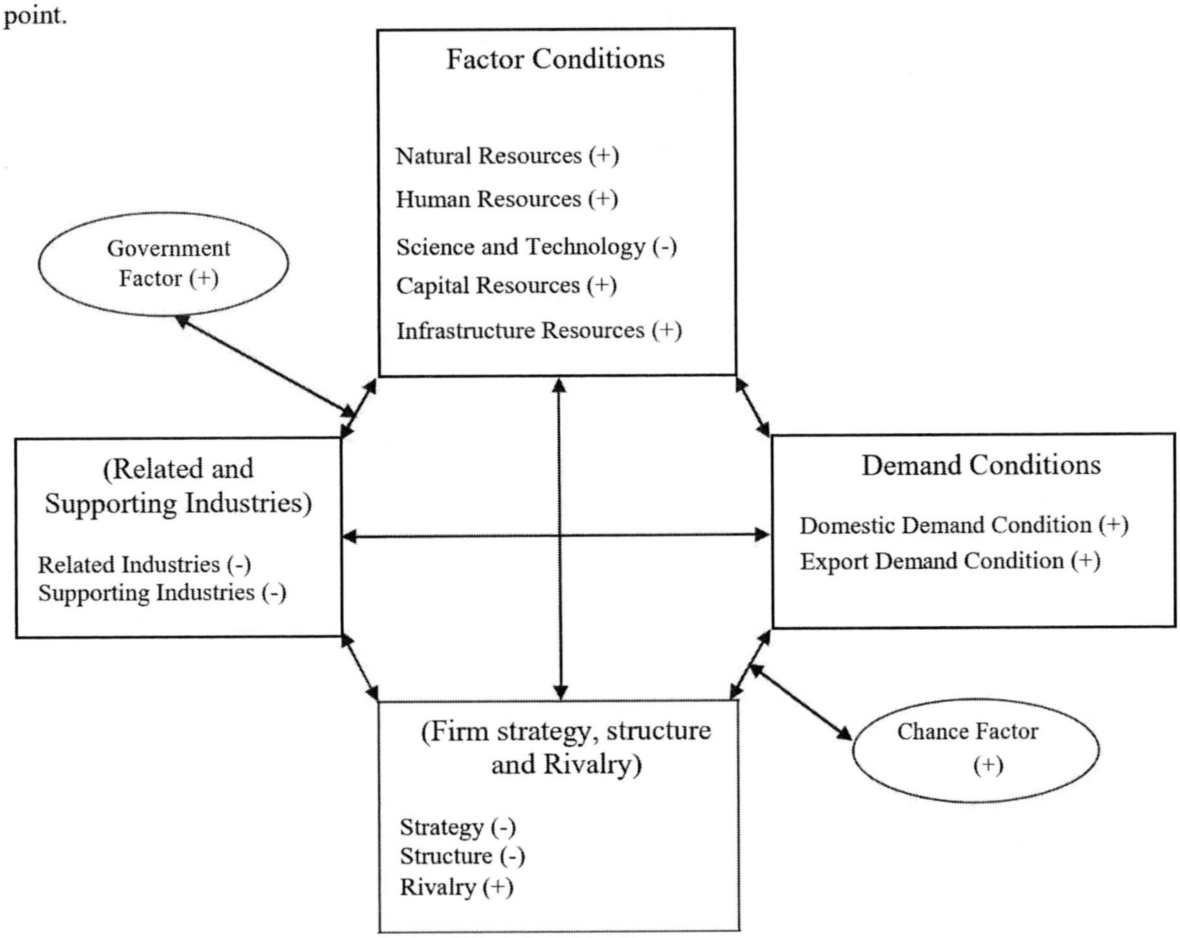


**Figure 2**. The results of analysis using Porter's diamond theory.

IOP Conf. Series: Earth and Environmental Science **137** (2018) 012101 doi:10.1088/1755-1315/137/1/012101

## 4. Conclusion

As analyzed using Porter Diamond Theory, the Indonesia's shrimp industry can be categorized as 'good'. There were four main components used in the analysis, namely factor condition; demand condition, related and supporting industries, and firm strategy, structure and rivalry, and supported with a two supporting components, which are the government regulations and opportunities. The supporting components were rated as positive, while related and supporting industries were given negative value, similar to firm and structure strategy that also had negative point. Meanwhile, rivalry was considered positive and so was factor condition, except for science and technology resources.

It is necessary for fishery industry to adopt the marketing strategy, structure industry, related support industries and supporting industries from the major shrimp exporter in the world (based on the result of Porter's Diamond analysis and econometrics analysis).

## 5. References

[1] Directorate of Agricultural and Forestry Product Export 2012 Indonesian Agriculture for world. Marketing Department of Indonesia Jakarta

[2] Globefish 2011 The Shrimp Market is on the Rise Despite the Difficult Global Economics Situation Market Reports of Shrimp www.fao.org/in-action/globefish/market-reports/resource-detail/en/c/338087/

[3] Ministry of Commerce 2014 Proogram for Export Improvement of Indonesia shrimp Ministry of Commerce Jakarta

[4] Ministry of Marine Affairs and Fisheries Republic of Indonesia 2017 Development of Fisheries Republic of Indonesia 2016 Ministry of Marine Affairs and Fisheries Republic of Indonesia Jakarta

[5] Ngamsaithong 2012 Thai Shrimp Exports to 6 Main Market to be Expanded The Thai Financial Post Publish on July 6th, 2012 Thailand

[6] Statistics Indonesia 2015 *Statistical Yearbook of Indonesia 2015* (Jakarta: Central Bureau of Statistics)

[7] WTO 2012 World Trade Organization. Liberalization Market

IOP Conf. Series: Earth and Environmental Science **137** (2018) 012102 doi:10.1088/1755-1315/137/1/012102

# The use of *matic* (automatic transmission of motorcycle) as catamaran boat engine to support economic activities in Ujung Pangkah, Gresik

**Sunardi[1], Sukandar[1], B Setionohadi[1], E Sulkhani[1], A B Sambah[1] and S Pamungkas[2]**

[1]Faculty of Fisheries and Marine Science, Brawijaya University, Malang, Indonesia
[2]Centre for Fisheries Development, Semarang, Indonesia

E-mail: sunardi@ub.ac.id

**Absract.** River is the only access to enter the pond area in Ujung Pangkah District, Gresik. The area is difficult to access due to the siltation of the river with only 50 cm depth. In addition, rent for traditional boat is expensive. This research developed a design of a small boat efficient enough to transport fish farmer from and to the pond area. Engine from motorcycle was used as it was easy to operate and economical use of fuel. The fundamental change for this design was a small double-hulled low-lane vessel. The motor matic engine was adjusted to a propeller to get maximum boost. The results showed that a ship with 3 meters length, 1.2 meters width and 1 m height, 2 persons in charge and speed of 7 knots, could use 150cc automatic motor engine by lowering engine rotation (RPM) from 7000 RPM to 4500 RPM using a Gear Box and 4-leaf B-Serries blades with a diameter of 14 cm.

## 1. Introduction

Gresik is located in the northern coast of East Java Province which has a large pond area of around 16,767.95 hectares. The main fishery production is milkfish and vaname shrimp. The main access to the pond area is from a nearby river with boat as the only mode of transportation. Traditional boats that exist today are only able to operate when the waters are in tidal conditions. When the water depth subsides (until only 50 cm), the boats cannot operate. This inhibits the daily activities of the farmers of the pond areas. This research aims to convert a *matic* (automatic transmission motorcyle) engine into boat propulsion for a small catamaran boat that can load two people and operate in the shallow pond area.

## 2. Methodology

The method used was the Parametric Design [1], which includes all design parameters to obtain the optimal ship design and function in accordance with the desired ship design objectives, namely a small catamaran boat with a *matic* engine.

## 3. Result and Discussion

Shallow river waters have considerable influence on the performance of a vessel, the vessel will have greater obstacles and will face more difficult maneuver controls than when in deep water [2]. The

IOP Conf. Series: Earth and Environmental Science **137** (2018) 012102 doi:10.1088/1755-1315/137/1/012102

magnitude of catamaran resistance in shallow waters is greater than in deep waters or free waters due to the influence of sinkage and ship bank interaction effect.

### *3.1. Catamaran motorboat desain*

Boats with double hulls have smaller ship resistance compared to single hull (monohull) boats [3]. Design parameters that must be fulfilled in this boat design were: having the ability to load two people; equipped with storage box supplies; has enough space to set up the motorcycle engines; crew position is similar as riding a motorcycle; and easy boat operation, as easy as people riding a motorcycle on the highway. The main size of the boat is 3 meters in length, 1.3 meters wide with a height of 0.95 meters. The general design of the boat are as follows:

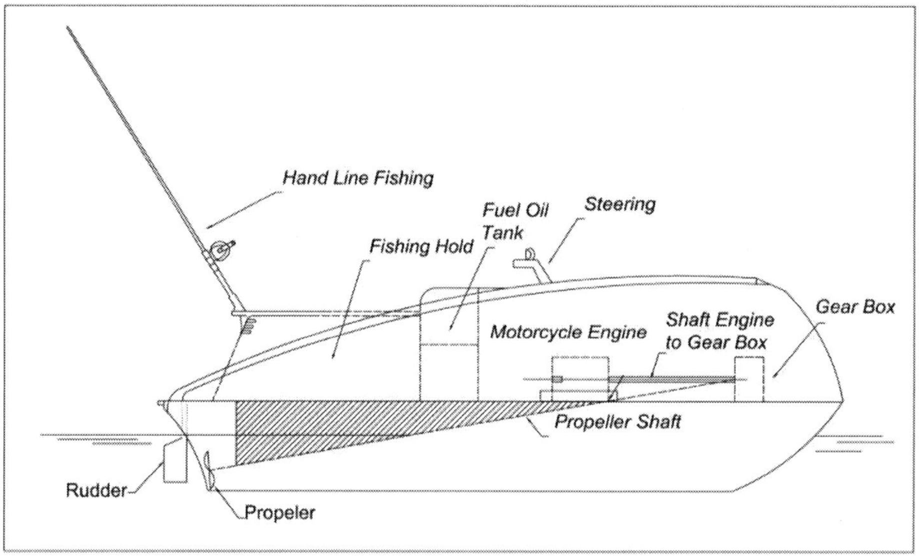


**Figure 1.** *Catamaran Motorboat* design.

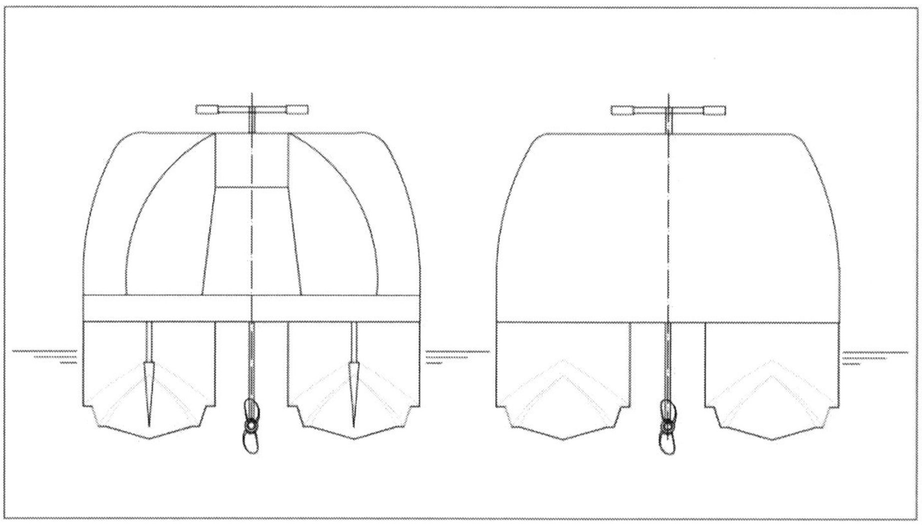

**Figure 2**. *Catamaran Motorboat* desain from back and front view.

IOP Conf. Series: Earth and Environmental Science **137** (2018) 012102 doi:10.1088/1755-1315/137/1/012102

The boat was designed to operate in limited waters (not seas) and for no more than one day of operation. The fuel provided does not exceed the capacity of the motorcycle tank placed between the seats and the steering wheel.

The hydrostatic calculations of ships with 0.25 meter boat load are:

| | |
|---|---|
| Displacement | : 0.43 ton |
| Volume (displaced) | : 0.42 $m^3$ |
| Draft Amidships | : 0.25 m |
| Immersed depth | : 0.25 m |
| WL Length | : 2.918 m |
| Beam max extents on WL | : 1.299 m |
| Wetted Area | : 4.263 $m^2$ |
| Max sect. area | : 0.188 $m^2$ |
| Waterpl. Area | : 2.205 $m^2$ |
| Prismatic coeff. (Cp) | : 0.769 |
| Block coeff. (Cb) | : 0.444 |

Based on these hydrostatic calculations, for a boat load of 0.25 meters, the boat's displacement is 0.43 tons. Thus, if the weight of the boat and machinery system is 0.23 tons, then it can be loaded with 0.20 tons which is enough to load two people with fuel and supplies.

### *3.2. Catamaran Motorboat resistance calculation*

According to Lammeren and Vossers [4], ship resistance is influenced by the ship's hull, speed and reynold number. Resistance components consist of friction resistance, wave making resistance and air resistance. The calculation of boat resistance by using the numerical method is by modeling the vessel then calculated by approaching the calculation of resistance with the slender body method. The results of the calculation of resistance are as follows:

**Table 1.** The relationship between speed and ship resistance.

| Velocity (knots) | Resistance (kilo Newton) | Power (HP) |
|---|---|---|
| 3 | 0.1 | 0.532 |
| 4 | 0.2 | 1.006 |
| 5 | 0.2 | 1.352 |
| 6 | 0.4 | 3.031 |
| 7 | 0.5 | 4.923 |
| 8 | 0.6 | 6.287 |
| 9 | 0.6 | 7.148 |
| 10 | 0.6 | 7.84 |

By using a 50 % propeller efficiency with a vessel speed of 8 knots, the boat resistance was 0.6 kilo Newton or 6.3 HP. Considering that the model operates in calm waters, while the vessel will be operated in free waters, it is necessary to add 15 % to the vessel's resistance to 7.245 HP. This is needed because the driving engine works only on average 85 % of the Maximum Continuous Rating (0.85 MCR), then there is an additional 15 % resistance to 8.33 HP. Additionally, due to the influence of the gear box and friction on the propulsion system which is around 15 %, the total boat resistance at 8 Knot velocity is 9.58 HP.

IOP Conf. Series: Earth and Environmental Science **137** (2018) 012102 doi:10.1088/1755-1315/137/1/012102

### *3.3. Selection of motor drive*

The selection of the driving engine must take into account reductions to the main engine power was caused by several factors including the efficiency of the hull and propulsion efficiency [5]. The power requirement of the engine to drive the ship at 8 knots is 9.58 HP, so the recalculation for the selected engine specifications are as follows:

1. Brand = Honda
   Type = Vario 150 CC
   Power = 12.6 HP
2. Brand = Yamaha
   Type = Mio 115 CC
   Power = 8.35 HP

Thus, to achieve a minimum speed of 8 knots, the engine used was a motorcyle engine with 12.6 HP power. While an engine with 8.35 HP power will produce a boat speed of less than 8 knots. More precisely, by using a 12.6 HP power engine, the catamaran motorboat will be able to run at a speed of 10 Knots. While using the 8.35 HP engine power, the boat will be able to run at the speed of 7 Knots [6]. Once the machine built, it was placed in the boat (figure 3.).

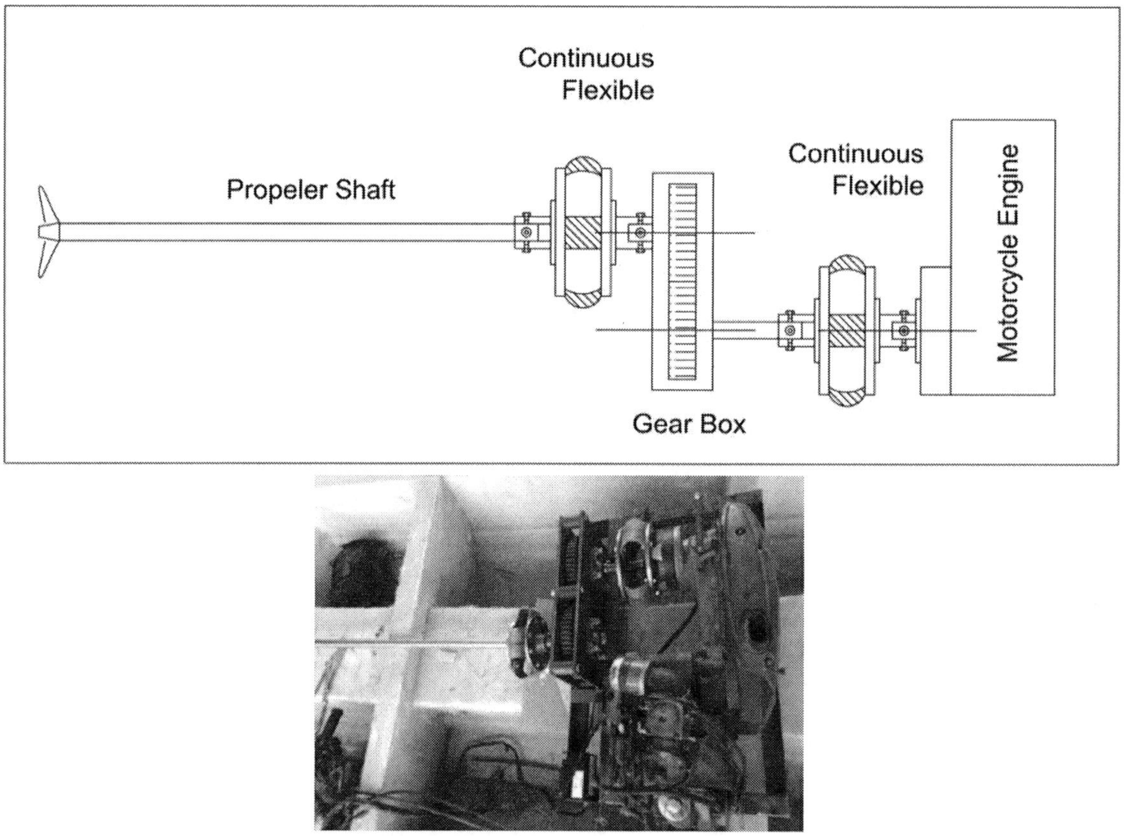


**Figure 3**. Setting the motor engine as catamaran motor boat driving force.

IOP Conf. Series: Earth and Environmental Science **137** (2018) 012102 doi:10.1088/1755-1315/137/1/012102

### *3.4. Catamaran Motorboat production*

Overall, the production process can be done on a traditional boat with fairly simple technology. Fiber material was chosen because of its ease of production. The main material needed was 9 mm and 25 mm plywood for mold formation. Laminating materials include resin, mat layer and roving, talc powder, catalyst and pigment dye. The main equipment needed were brushes, cutting machines, and drills. Completely built catamaran motorboat are shown in figure 4 – 7.

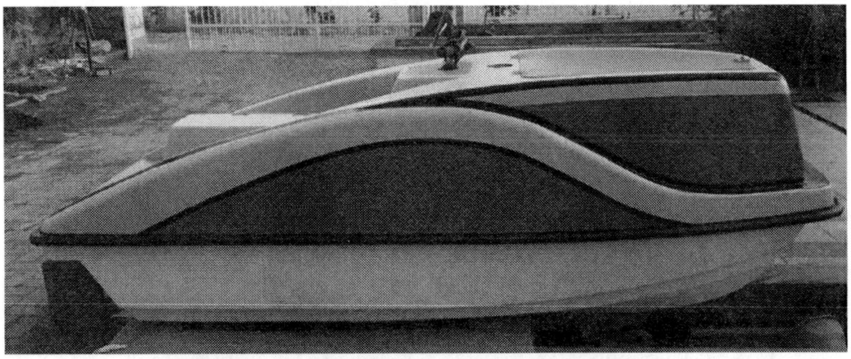

**Figure 4.** *Catamaran Motorboat* (side view).

**Figure 5.** *Catamaran Motorboat* (back view).

**Figure 6.** Aperture to engine room.

IOP Conf. Series: Earth and Environmental Science **137** (2018) 012102 doi:10.1088/1755-1315/137/1/012102

**Figure 7.** *Catamaran Motorboat* (front view).

### *3.5. Sea trial Catamaran Motorboat*

Trials were carried out to ensure the catamaran motorboat can operate properly (figure 8.). The sea trial results are as follows:

1. The speed of the boat only reaches 4 knots, less than the planning design that reached 8 knots.
2. When maneuvering the boat for the O test, it is capable to rotate with a diameter of 8 meters.

**Figure 8.** Trial in pond area

## 4. Conclusion

The catamaran motor boat with matic motorcycle engine speed only 4 knots as result. The development of design and engine sistems will be continue to get better performance and speed.

## 5. Reference

[1] Watson and David G M 2002 *Gulf Professional Pub.* **1**

[2] Lee, Seung-Hee, Young-Gill Lee, and Sang-Hyun Kim 2007 *Ocean Eng.* **34** 2061-2073

[3] Sunardi, Sunardi, Efani A, and Luthfi O M 2016 *Res. J. Life Sci.* **2** 199-204.

[4] Lammeren W P A and Vossers G 1962 *Resistance, Propulsion and Steering of Ships*

[5] Molland, Anthony F, Stephen R, Turnock, Dominic A and Hudson 2011 Ship resistance and propulsion: practical estimation of propulsive power (UK : Cambridge university press)

[6] Ali A, Maimun A and Ahmed Y M 2014 *J. Tech.* **66** 163-167.

**Acknowledgement**

Ministry of Research and Higher Education and Brawijaya University for financial support through Applied Research Project of Universities in 2017.

IOP Conf. Series: Earth and Environmental Science **137** (2018) 012103 doi:10.1088/1755-1315/137/1/012103

# Household economic modelsof gill net fishermen at Madura strait

**M Primyastanto**

Faculty of Fisheries and Marine Science, Brawijaya University, Veteran Street, Malang 65145, Indonesia

E-mail:mimitp@ub.ac.id; mimitprimyastanto@gmail.com

**Abstract.** The purposes of this research was to analyze household economic models of gill net fishermen at Madura strait. 30 families of gillnet fishermenwere used for purposive sampling. Data analysis used descriptive qualitative and quantitative (regression analysis). Quantitative descriptive analysis was used to analyze research and compare to factors that affecting household economic models of gill net fishermen family. Research results showed tha thousehold economic models of gill net fishermen at Madura strait was production value level or fishermen revenue at sea was strongly influenced byp roduction asset production, education level, fuel, and work flow. Work flow rate of fishermen families affected by asset production, non fisheries workflow and number of male workforce. Non fishing income level was strongly influenced by non-fishery business assets, number of family members owned andnon-fishing work flow. Spending levels of gill net fishermen at Madura strait was affected by fishing income, non-fishing income, fishermen wife education and fishermen family members.

## 1. Introduction

*And those, who, when they spend, are neither extravagant nor sparingly, but hold a medium (way) between those (extremes), Al-Furqon: 67.* Indonesiaisa maritime country with territorial waters area three times theentire land area. Therefore, Indonesia has high potential for fisheries.The development potential for Indonesian marine business todayisdirected atthe exploitationof marine and seabed resources as well asthe utilization of the function of the national territorial waters, including Exclusive Economic Zone, in a harmonious and balanced way with respect to marine capacity and sustainability to improve people'swelfare and expand business opportunities and field works[1]. Indonesiaisa maritime country with the longest beach in the world with a coastline of more than 81.000 km. Out of67.439 villages in Indonesia, 9.261 villages arecategorized as coastal villages and most of their people arepoor. Coastal villages are potential pocket of structural poverty. The difficulties to overcome poverty in coastal villages force their residents to beara burden without knowing when it will be over [2].

The resource potential in Indonesianwaters isquite large, especially withthe announcement of Exclusive Economic Zone (EEZ) withinup to 200 nautical miles limit. The areas that can be exploited and utilized become wider. Based on the evaluation results andthe available information,the overall potential of marine fishery resources is4.5 million tons per annum for Indonesianwaters and 2.1 million tons per annum forExclusive Economic Zone.The total potential estimated includes 3.5 million tons of pelagic fish per year, 2.5 million tons of demersal fish per year, 208 thousand tons of tunas per

IOP Conf. Series: Earth and Environmental Science **137** (2018) 012103 doi:10.1088/1755-1315/137/1/012103

year, 69 thousand tons of shrimp per year, 275 thousand tons of skipjack tunas(*Katsuwonus pelamis*) per year and 48 thousand tons of reef fish per year[3].

The phenomena of overfishing and povertyarecomplex and multidimensional. Low livelihood statusisoften used as a measure of poverty, but it essentially isonly the first link of a number of factors of poverty. In the perspective of politcal economy, povertyisunderstood as a product of economic forces, relations of production and power relations, all of whichlead to poverty process. Poverty and socio-economic pressures faced by fishermen families arealso rooted in interrelated complex factors. Natural factors are associated with the fluctuations in natural structure of fishing season and village economic resources. The factors associated with non natural limitationarefishing technology, poor marketing network as well as negative impact of fisheries modernization policies that has been underwayin thelast quarter century [4].

The characteristics of fishing communities differ from those offarming communitiesdue to the differences in the existingresources. Farming (agricultural) communities work oncontrolledland resources to produce a kind of commodity with predictable results. With these characteristics,production locationscan be settled. The business mobility is relatively low and the business risk factorsarerelatively small [5]. In dry season,the income gained from fishing activities is minimal, and oftentimesthere is no catch at all. This law catch-season lasts for about eightmonths, andthe savings from the yield obtained duringthe fishing season will not sufficeto overcomethe difficulties in fulfilling thedaily life needs of fishermen's households. In low-income periods, usually thewivesand children of fishermen struggle to earn money by doing allw orks to generatean income.

At times fishermen donot go to the sea.Theydo any kind of work on land to earn income to ensurehousehold survival. However,the extent to which theemployment opportunities can be obtained by the members of fishermen'shouseholdsisaffected bythe structural characteristics of local village's economic resources[6]. The structural characteristics of coastal village'seconomic resources depend entirely on marine fisheries production. The employment opportunities there arevery limited. Other business sectors generally relyon the procurement of raw materials fromthe sea. They will cease to operate. Fishing business isan exploitative business, whichisat risk ofnot getting any yieldbecause it relies entirely on the nature. According to Reardon *et al.* [7], in areas with poor agroclimate conditions, risky agriculture and no insurance market, non-farm activities allow these households to cope with severe downturns in the agricultural productivity.

Theprimary problemin this studyisrelated to the household economics of fishermen, which is stated as follows:How is the household economic model of gill net fishermen at Madura strait?

## 2. Methodology

### 2.1. Research method

The research conducted wasqualitative descriptive research under natural setting, and the data collected werequalitative data. Thequalitative methods used were based on phenomenological philosophy that promotes the appreciation toward and tries to understand and interpret the meaning of human behavior in certain situations. Qualitative descriptive analysis was used to look intoand to knowthe characteristics in this study. The problem of this research was poverty alleviationand food security for small-scale fishermen at Madura Strait. The analysis unit of this study werethe individuals, families, groups and community members who became small-scale gillnet fishermen [8].

In this research,the sample was taken from the population, and questionnaires were employed asthe main data collection tool. The survey results can be used to make certain predictions about social phenomena and to conduct an evaluation of the study results [9].

### 2.2. Data collection method

This study usedqualitative and quantitative data. Qualitative data were used to provide additional explanation of a phenomenon. The data collection techniques were interviews, observation, documentation and questionnaires [10].

### *2.3. Sources and data types*

This study collected two types of data, namely primary data and secondary data. Primary data were obtained from primary resources directly as a result of data collection [11]. The primary data were derivedfrom the parties inthe researchlocation. The primary data were collected using questionnaire technique from fishermen conducting small fishing, especially gillnet fishing. Structured questionnaires containing alist of questions were prepared in order to obtain more effective and accurate data in accordance with the research objectives.

Secondary data wer ecollected by data-collecting agencies,and they werepublished to users [12]. These data were derivedfrom various literaturesuch as textbooks, research journals, research reports, data from relevant agencies and so forth. Secondary data arealso called processed data.

### *2.4. Sampling techniques*

It was stated that sampleispart of pulation that becomes real object [13]. Population is the total number of analysis units with presumed traits [14]. The population of this study consisted of small-scale fishermen, particularly fishhook skippers and gill net fishermen.

In accordance withthe problems, objectives and focus of thisstudy, the researchers interviewed several informants who were considered to have considerable knowledge of or to be directly involved in the business activities ofsmall fishing households, and knowthe problems offishermenat Madura Strait.The sample taken was comprised of two groups, namely fishhook skippers and gillnet fishermen. Several sample units were taken from each group. The respondents were determined usingpurposive sampling technique, which means thattheelement within the sample were taken purposively if they were representative [15].

Researchers took 30 small-scale gillnet fishermen families using purposive sampling technique. The sampling criteria were as follows: (1) the catchingtechnology used was simple, with small boats less than 30 gross tonnage(GT) in size; (2) the fishermen usedmachinesof less than 12 PK; (3) the fishermen's venture capital was limited; (4) the catching membergenerally were the fishermen'srelatives, neighbors and/or close friends; (5) the fishermen's economic orientation was to fulfil everyday needs; and (5) fishermen needs.

### *2.5. Data analysis techniques*

The data were analyzed qualitatively and quantitatively (regression analysis). Quantitative descriptive analysis technique was used to analyzethe research environment, workflow behavior of fishermen families and fishermen'sproduction expenses. Meanwhile,qualitative descriptive analysistechnique was used to analyze and compare factors affecting fishermen families'economy. It includedvariables family income, total revenue, family consumption and fishermen's workflow time[16].

### *2.6. Regression analysis*

In general, regression analysisisused to analyzethe relationship between two or more variables. There are variables influenced by the other variables.The variables affected are called dependent variables, whilethe affecting variables arecalled independent variables [17].

Multiple regression analysisis a dependency technique. To use this technique, we need tobe able to divide variables into dependent and independent variables. Regression analysisisa statistical tool used when dependent and independent variables are metric variables. However, in certain circumstances, there arenon-metric variables (dummy, ordinal or nominal variable) that can also be used [18].

Multiple linear regression analysis wasused to analyzethe relationship between factors that affectedfood security in fishermen households. These factors includedthevariablesincome, total revenue , consumption and workflow time.

This study useddependent variablessailingproduction income (SPI) or (Q), non-fishing income (NFI), workflow time at sea (SWF) and total staple consumption (TSC).

IOP Conf. Series: Earth and Environmental Science **137** (2018) 012103 doi:10.1088/1755-1315/137/1/012103

So that a regression model isused to identify therelationship between one dependent variable and several independent variables. This model iscalled multiple regression model [19]. The results of multiple linear regression analysis used in thisresearch are as follows.

a. Sailing Production Income (SPI)

$$Y = a + b_1x_1 + b_2x_2 + b_3x_3 + \ldots\ldots + b_kx_k + e \quad (1)$$
$$SPI\ Q = a\ + b_1SAV\ + b_2OFS\ + b_3SWF\ + b_4 EDC + e^u$$

Description: Q (SPI): Sailing production income (kg/yr). SAV: Ship asset and fishing toolvalue(IDR). OFS:Operationfuel amount for sailing (L/yr). SWF: Sailing workflow (HOK/yr) EDC: Education (yr)

b. Sailing workflow (SWF)

$$Y = a + b_1x_1 + b_2x_2 + b_3x_3 + \ldots\ldots + b_kx_k + e^u \quad (2)$$
$$SWF\ = a + b_1SAV\ + b_2FFM\ + b_3 NFW\ + e^u$$

Description: SWF: Sailing workflow (HOK/yr). SAV: Ship asset and fishing tool value (IDR) FFM : Fisherman'familysize(People).NFW: Non-fishery workflow (HOK/yr)

c. Non-Revenue Fisheries (NRF)

$$Y = a + b_1x_1 + b_2x_2 + b_3x_3 + \ldots\ldots + b_kx_k + e^u \quad (3)$$
$$NRF\ = a + b_1NFB + b_2NFW + b_3FFM + e^u$$

Description: NRF: Non-revenue fisheries (IDR/yr). NFW: Non-fishery workflow (HOK/yr) FFM: Fisherman'sfamily size(People).NFB: Non-fisherybusiness asset (IDR)

d. Fisherman's Household Goods Consumption (HGC)

$$Y = a + b_1x_1 + b_2x_2 + b_3x_3 + \ldots\ldots + b_kx_k + e^u \quad (4)$$
$$HGC\ Scene = a + b_1ROF + b_2NFI + b_3\ FFM + b_4 FWE\ + e^u$$

Description: HGC: Fisherman'sHousehold Goods Consumption (IDR/yr). ROF: Revenue of fishery activity (IDR/yr). NFI: Non-fisheryincome (IDR/yr).FFM: Fisherman'sfamily size(People). FWE: Fisherman's wife'seducation (Years)

## 3.Results

### *3.1. Regression analysis of sailing production income*

The factors that may have affected the sailing production income (SPI) were production asset/IDR(SAV),amount of Fuel/L (OFS),sailing workflow/HOK (SWF) and Education /yr (EDC).The results of the calculation to determinethe relationship between independent variable and the dependent variable sailing production income (SPI)areas follows:

a. Regression coefficients

The results of the regression analysisareas follows:

$$Y = a + b_1x_1 + b_2x_2 + b_3x_3 + \ldots\ldots + b_kx_k + e \quad (5)$$
$$SPI\ = a + b_1SAV\ + b_2OFS + b_3\ SWF\ + b_4\ EDC + e^u$$
$$SPI = -\ 50\ x10^7 + 6.34\ SAV + 0{,}03\ OFS - 0.04\ SWF + 1.60\ EDC + e^u$$

**Table 1**. Results of the Regression Analysis of Factors Affecting a Fisherman Fishermen Household Sailing ProductionIncome (SPI).

| Variables | Coef. | t | Sig. |
|---|---|---|---|
| Constant (a) | -5,00E+08 | -1.71 | 0.09 * |
| (SAV) | 6.34 | 2.07 | 0.05 ** |
| (OFS) | 0.03 | 3.36 | 0.00 *** |
| (SWF) | -0.04 | -1.06 | 0.30ns |
| (EDC) | 1.60 | 0.77 | 0.45 ns |
| Adjusted $R^2$ | 0.60 | | |
| F | 8. 82 | | 0.00*** |
| N | 30 | | |

The constant 50E107 denotes that when a fisherman is not effected by production assets, fuel amount, sailing work flow and education,he will incur a loss of IDR 50.000.000. A regression coefficient of 6.34 denotes that an increase in production assets by 1 unit will increase fish production revenue by 6.34. A regression coefficient of 0.03 denoted that any addition of fuel by 1 liter will increase fish production revenue by 0.03. A regression coefficient of -0.04 denotes that any addition of sailing work flow by 1 HOK will reduce sailing revenue by 0.04. This was because one works 8 hours at mostin1 day. When working time increases, the revenue will decrease. Furthermore, overfishing in sea fisheries hinders optimum. A regression coefficient of 1.60 denoted that an increase in production assets by 1 unit will increase production revenue by 1.60 at sea.Thus,it is necessaryto improve ship technology and fishing tool to increase a fisherman's productivity [20].

3.1 *Multiple regression analysis of sailing work flow time*

The factors affecting the sailing work flow (SWF) were asset production/IDR (SAV), fisherman'sfamilies size (FFM) and non-sailing work flow/HOK (NFW).The results of the calculation to determinethe relationship between independent variable and the dependent variable non-fishery incomeareas follows.

Regression coefficientsBased on the data analysis, the regression equationis:

$$Y = a + b_1x_1 + b_2x_2 + b_3x_3 + \ldots\ldots + b_kx_k + e \quad (6)$$

**Table 2.**Results of regression analysis of factors affecting fishermen's Household pouring out Sailing Work Flow (SWF).

| Variables | Coef. | t | Sig. |
|---|---|---|---|
| Constant (a) | 378 | 3.20 | 0.00*** |
| SAV | - 1.50 | -1.19 | 0.24ns |
| FFM | -17.27 | -0.43 | 0.67ns |
| NFW | -0.19 | -0.54 | 0.59ns |
| Adjusted $R^2$ | 0.08 | | |
| F | 0.71 | | |
| N | 30 | | |

IOP Conf. Series: Earth and Environmental Science **137** (2018) 012103 doi:10.1088/1755-1315/137/1/012103

A constant of 378 denotes that if a fisherman is not affected by production assets, total workforce and non-fishery work flow, he will have a sailing work flow of 378 HOK. A regression coefficient of -1.50 denotesthat an increase in production assets will reduce work flow by1.50 day labourer at sea fishing. This negative result was due tothe fact that the addition of production assets will increase a fisherman's productivity as technological advances will lower physical labor. Weather/season factor is unpredicatable, and there is an indication that the sea has been overfished. In this case, local wisdom like *andun* should be developed to look for other untapped fishery resources [21]. Theregression coefficient of -17.27 denotes that an increase in work force will lower sailing work flow by17.27 HOK. The negative result shows that the waters at Madura Straithave already been overfished. It corresponds to theresults obtained from fihermen families , because they worked as artisans, workshop worker and construction workers [22]. A regression coefficient of -0.19 denotesthat every additional sailing workflow for off fishing. This is because the work was convertedfrom fishery to non-fishery to get maximum results in a period of 8 hours. Non-sailing workflow per seisvery small. Therefore, theincome is mainly earned fromsailing. Lower workflow time for fisheries will allow for time to seek alternative income.This is especially the case for fishermen'swives, where they voluntarily work in fish processing as local wisdom in fishing communities [23].

### *3.2. Regression analysis of Non-Revenue Fisheries(NRF)*

The factors presumed to have affected non-revenue fisheries (NRF) were the variables non-fishery business asset/IDR (NFB),non-fishery work flow/HOK (NFW) andfisherman's family size (FFM).The results of the calculationto determine the relationship between independent variable and the dependent variable non-fishery income areas follows:
Regression coefficients:
Based on thedata analysis, the regression equation is:

$$Y = a + b_1x_1 + b_2x_2 + b_3x_3 + \ldots\ldots + b_kx_k + e^u \tag{7}$$

Non Revenue Fisheries (NRF) :

$$NRF = -0.07 + 0.01\ NFB + 0.04\ NFW + 0.02\ FFM + e^u \tag{8}$$

**Table 3.**Results of regression analysis of factors affecting household economic of Non Revenue Fisheries (NRF).

| Variables | Coef. | t | Sig. |
|---|---|---|---|
| Constant | -0.07 | -0.03 | 0.98 |
| (NFB) | 0.01 | 2.74 | 0.01*** |
| (NFW) | 0.04 | 2.76 | 0.01*** |
| (FFM) | 0.02 | 0.32 | 0.75ns |
| Adjusted $R^2$ | 0.45 | | |
| F | 6.80 | | 0.02** |
| N | 30 | | |

A constants of -0.07 denotes that if a fisherman is not affected by non-fishery business assets, non-fishery workflow and number of family members,he will lose non-fishery income at0.07. A regression coefficient of 0.01denotes that an increase in non-fishery by 1 unit of business assets will increase non-fishery income by0.01. A regression coefficient of 0.04 denotes that each addition of non-fishery

IOP Conf. Series: Earth and Environmental Science **137** (2018) 012103 doi:10.1088/1755-1315/137/1/012103

work flow by1 HOK will increase non-fishery income by0.04. A regression coefficient of 0.02 denotes that each addition of1 family member will increase non-fishery income by0.02.

*3.3. Multiple regression analysis of fisherman's Household Goods Consumption (HGC)*
Fisherman's Household Goods Consumption (HGC)

$$Y = a + b_1x_1 + b_2x_2 + b_3x_3 + ....... + b_kx_k + e^u \quad (9)$$
$$HGC = a + b_1ROF + b_2NFI + b_3 FFM + b_4 FWE + e^u$$

Description: HGC: Fisherman's Household Goods Consumption (IDR/yr). ROF: Revenue of fishery (IDR/yr). NFI: Non-fisheryincome (IDR/yr) FFM: Fisherman'sfamily size(People). FWE: Fisherman's wife's education (yr)

The factors presumed to haveaffectedthe Fisherman's Household Goods Consumption (HGC) were: Revenue of fisheries (ROF),fishermen family members (FFM) and fisherman's wife education (FWE).The results of the calculation to determine therelationship between independent variable and thedependent variable Fishermen's Household Goods Consumption (HGC) areas follows:
Regression coefficients:
Based on thedata analysis, the regression equation is:

$$Y = a + b_1x_1 + b_2x_2 + b_3x_3 + ....... + b_kx_k + e \quad (10)$$
$$HGC = 13.82 - 0.01ROF - 0.16NFI - 0.51 FFM + 0.21FWE + e^u$$

**Table 4.** Results of the regression analysis of factors affecting the analysis of Household Goods Consumption fishermen (HGC).

| Variables | Coef. | t | Sig. |
|---|---|---|---|
| Constant | 13.82 | 3.98 | 0.00*** |
| ROF | - 0.01 | - 3.13 | 0.00*** |
| NFI | - 0.16 | - 6.03 | 0.00*** |
| FFM | - 0.51 | - 8.40 | 0.00*** |
| FWE | 0.21 | 0.50 | 0.60ns |
| Adjusted $R^2$ | 0.83 | | |
| F | 8.90 | | 0.00*** |
| N | 30 | | |

A constant of 13.82 denotes thatif a fisherman is not affected by food and consumption, number of his family membersand hiswife education,hehasa total staple consumption of13.82 per year. A regression coefficient of ROFof -0.01 denotes that revenue increase will reduce thetotal staple consumption by0.01 per year. This is related to the desire to increase business investment in order to increase revenue alternatives.This also effects the fishery revenue when themarine resources have already been overfished [24]. A regression coefficient of NFI of -0.16denotesthat an addition of non-fisheryincome will lowerthe total staple consumption by0.16. Non-fisheryincome islow because very few fishermen work in the non-fishery fields. This means that the fishermen still relyon sailing in come [25]. A regression coefficient of FFM of-0.51 denotes thatthat an increase inthe numberof family members will lowerthe total staple consumption by0.51. This is because the increase in thenumber of family members linearly will increasethe amount of staple and non-staple consumption [26]. A regression coefficient of 0.21 denotesthat an increasein a fisherman's wife's education(FWE) will increase total staple consumption by 0.21.

IOP Conf. Series: Earth and Environmental Science **137** (2018) 012103 doi:10.1088/1755-1315/137/1/012103

The Household Economic Model of Madura Strait Fishermen was as follows: Theproduction value or fishermen's revenue at sea was strongly influenced by production asset, fuel, work flow and education level. The work flow rate of fishermen's families was affected by production asset, number of male workforce and non-fishery work flow. Non-fishing income level was strongly influenced by non-fishery business assets, non-fisheryworkflow and number of family members. The spending levels fishermen at Madura Strait were affected by fishing income, non-fisheryincome (NFI), family size (FFM) and fisherman's wife's education (FWE).

Thesuggestions of this study are as follows: (1) The examination of fisherman's household economic model shows that fishermen should meetbasic needs such as food, clothing, shelter, health and education. (2) Local production infrastructure and facilities (production asset) should be provided for Madura Strait fishermen communities to obtaincheap, quality infrastructure and facilities. (3) The institutional role of Madura Strait fisherman communities as a collective action platform should be enhanced to achieve individual goals. (3) There should be a promotion ofproductive economic activities in areas with characteristics corresponding to local resources and have a clear markets, which should be carried out continuously with special attention paid to the resource capacity and implemented in the local communities, andto the appropriate use of advanced technology according to the assessment and research. (4) Economic structure of Madura Strait fishermen communities should be realized based on the economic activity in coastal and marine areas as a form of natural sea resources utilization.

## 4. References

[1] Primyastanto M, Soemarno, Efani A and Muhammad S 2013 *Inter. Inst. Sci. Tech. Edu. J.* 313-21.

[2] PrimyastantoM ,Pudji P and Yahya, 2014 *Inter. Rev. Soc. Sci.***2**

[3] Zain, Jonny, Syaifuddin and danK Rohmatin2009 *Fish. Mar. J.*121-127

[4] Primyastanto, M,Soemarno,EfaniA and M Sahri 2013 *J. Innov. Soc.Sci.& Human Res.***1** 43-51

[5] Rachman, S, Purwanti, P and Primyastanto, M *ECSOFiM Journal.* **1**

[6] Primyastanto, M 2012*Policy human resource management on fishing overfishing* (Malang : Brawjaya University) (In Indonesia)

[7] Reardon, T, Berdegue´ , Julio, Barret, Christopher B, and Stamoulis, Kostas 2006*Household income diversification in to rural non farm activities. In Steven Hagg blade, Peter Hazell, and Thomas Reardon (Eds), Transforming the rural non farm economy* (Baltimore : John Hopkins University Press)

[8] Usman, Husaini dan Akbar, P S 2006 Social research methods(Jakarta: PT.Bumi Aksara)

[9] Arikunto, S 2006 *Research Procedures A Practice Approach,5$^{th}$ edition* (Yogyakarta : RinekaCipta)(In Indonesia)

[10] Khoiriyah, A 2005 *Alternative capital in efforts empowerment and poverty reduction traditional fishermen (study of traditional fishermen capital condition in Weru village, paciran sub-district, Lamongan regency, East Java )*(Malang : *Essay* Faculty of Fisheries University of Brawijaya) (In Indonesia)

[11] Kartini, 1990 *Introduction to social research methodology mandar maju*(Bandung) (In Indonesia)

[12] Nazir, M 2005*Method of research* (Jakarta : PT Ghalia Indonesia) (In Indonesia)

[13] Koentjoroningrat 1991*Community research methods*(Jakarta : Gramedia Pustaka Utama)(In Indonesia)

[14] Singarimbun, M Effendi,S 1987*Survey research methods*(Jakarta : Penerbit LP3ES)(In Indonesia)

[15] Marzuki, 2005 *Research Methodology* (Yogyakarta: Faculty of Economy University of Moeslim Indonesia)(In Indonesia)

[16] Primyastanto, M,Soemarno,EfaniA and M Sahri 2013*Inter. J. Innov. Soc.Sci. Human Res.***1** 43-51

IOP Conf. Series: Earth and Environmental Science **137** (2018) 012103 doi:10.1088/1755-1315/137/1/012103

[17] Ghozali, Imam 2009*Econometrics (theory concepts and applications with spss 17 )*(Semarang: Diponegoro University)(In Indonesia)
[18] Sulaiman 2010*Alternative Model of Laotian Legal-Based Fisheries Management in Aceh's Historical District, Towards a Sustainability of a People-Oriented Welfare Environment*(Semarang: UNDIP)(In Indonesia)
[19] Nachrowi,N D Hardius and Usman 2002 *Use of econometric techniques: a popular and practical approach equipped with data analysis and processing techniques using the spss program packages* (Jakarta : Radja Grafindo Persada) (In Indonesia)
[20] Primyastanto M 2011 *Sustainable fish resource management (sdi): a bioeconomic approach book* (Malang : Brawijaya University) (In Indonesia)
[21] Primyastanto M 2016 *Asian J Microbiol.Biotech. Env. Sci.* **18** 297-303
[22] Primyastanto M, Agus T, Efani A Setiawan, Riniwati H and Ismadi 2013 *Study model of fishermen household economy in madura strait (*Malang : BrawijayaUniversity) (In Indonesia)
[23] Primyastanto M 2011 *Local Wisdom for the Management of Fish Resources*(Malang : BrawijayaUniversity) (In Indonesia)
[24] Primyastanto M 2011 *More capture fisheries policy* (Malang : Brawijaya University) (In Indonesia)
[25] Primyastanto M, Soemarno and Anthon E 2014 *Inter.J. Civil & Env. Eng.* **14**
[26] Primyastanto M2015 *Inter.J.Oceans and Oceano.* **9** 97-104

**Acknowledgements**

The authors are grateful to Prof. Dr. Ir. M. Bisri, MS, Prof. Dr. Ir. Happy Nursyam, MS and anonymous referee for constructive comments. The authors would like to thank all those who have contributed, either directly or indirectly, in writing this article until finished. And individuals who provided help during the research (e.g., providing language help, writing assistance or proof reading the article, etc.).

**Institute of Physics**
Dirac House, Temple Back
Bristol BS1 6BE UK

ISSN: 1755-1307
ISBN 978-1-5108-6152-7

9 781510 861527

CURRAN ASSOCIATES INC.
proceedings.com